MODERN ELECTRONICS
basics • devices • applications

MODERN ELECTRONICS
basics • devices • applications

David Bruce

Reston Publishing Company, Inc.
A Prentice-Hall Company
Reston, Virginia

Library of Congress Cataloging in Publication Data

Bruce, David,
  Modern electronics.

  Includes index.
  1. Electronics. I. Title.
TK7816.B785 1984   621.381   84-2014
ISBN 0-8359-4546-4

© Copyright 1984 by
Reston Publishing Company, Inc.
A Prentice-Hall Company
Reston, VA 22090

Text prepared and book designed by
  Robert Scharff and Associates, Ltd.

10 9 8 7 6 5 4 3 2 1

Printed in the United States of America.

CONTENTS

Acknowledgments

We would like to extend our thanks to Prentice-Hall, Inc., Englewood Cliffs, NJ, for permission to use the following materials: J. J. De France, *Electrical Fundamentals*, 2d ed., © 1983; Frederick F. Driscoll, *Analysis of Electric Circuits*, © 1973; Norval R. Ekeland, *Basic Electronics for Engineering Technology*, © 1981; Joel Goldberg, Ph.D., *Fundamentals of Electricity*, © 1981; Heathkit/Zenith Educational Systems, *Electronic Circuits* A Step-By-Step Introduction, Training in Computers & Electronics Series, © 1983; Herbert W. Jackson, *Introduction to Electric Circuits*, 5th ed., © 1981; John D. Lenk, *Handbook of Practical Electronic Circuits*, © 1982; Daniel L. Metzger, *Electronic Circuit Behavior*, 2d ed., © 1983; Gary M. Miller, *Modern Electricity/Electronics*, © 1981; Charles M. Thomson, *Fundamentals of Electronics*, © 1979.

Material was also obtained from the following books published and copyrighted by Reston Publishing Company, Inc., Reston, VA: William D. Cooper and Henry B. Weisbecker, *Solid-State Devices and Integrated Circuits*, © 1982; Mannie Horowitz, *Practical Design with Solid State Devices*, © 1979; Vester Robinson, *Manual of Solid State Circuit Design and Troubleshooting*, © 1977; Robert J. Traister, *DC Power Supplies: Application and Theory*, © 1979; Ben Zeines, *Transistor Circuit Analysis and Application*, © 1976.

Finally, special thanks for their contributions to this book go to Anthony Ferrari and Carl Sawejko.

PREFACE

Modern Electronics: basics • devices • applications is written for the student with no prior knowledge of electronics. While there are relatively few fundamental laws and mathematical relationships that make up the foundation for all studies in the field of electronics, it is not enough that these laws be simply committed to memory. They must be totally comprehended and their interrelationships completely understood before a true understanding of electronic circuits and devices can be realized.

The book begins with the fundamental principles of electricity and progresses through the basic circuit laws to DC and AC theory. Semiconductor theory is introduced. Diodes, transistors, and control devices are presented along with their applications in power supplies, amplifiers, and oscillators. Material on transducers is also included. And because of their historical significance and continued, although limited, use, vacuum tubes are added as a supplementary chapter.

This book is designed as a "working text." Whenever possible, space has been left in the margins for additional note-taking, instructor's points, or cross-referencing to lab work. *Modern Electronics: basics • devices • applications* sets a strong foundation for anyone entering the field of electronics and will be a valuable reference book throughout the reader's career.

CHAPTER 1

Principles of Electricity

Electricity is everywhere. Our nervous system uses it, nature displays it to us as lightning, we light our homes with it, and our society, as well as a major portion of the world, is dependent upon its continuous presence. Although electricity has been a part of our daily lives for many years, it has been only recently that we can explain what electricity really is. That is, we knew the principles of electricity—how to produce it, how to transport it, how to use it—but not what it was. We now know that electricity, the kind we use, is produced by small atomic particles known as electrons. It is the movement of these particles that produces the effects of heat in the toaster and light in the lamp.

The pressure that forces these atomic particles to move, the effects they produce when they encounter opposition, and how these forces are controlled are some of the basic *principles of electricity*.

THE NATURE OF ELECTRICITY

Accepted atomic theory states that all matter is electrical in structure. Any object you care to name, even your own body, is largely composed of a combination of positive and negative particles of electricity. Electric current will pass through your body, over a wire, or along a stream of impure water. It can be established, in some substances more readily than others, that all matter is composed of electric particles despite some basic differences in materials. The science of electricity then must begin with a study of the structure of matter.

Matter

Matter can be defined as anything that has mass (weight) and occupies space. It exists in three forms or states: gas, liquid, and solid. Matter in the gaseous state will conform perfectly to the shape of its container. It possesses neither a fixed shape nor a fixed volume. Some common examples of gases are: the atmosphere which we breathe, the carbon dioxide which we exhale, and water vapor.

A liquid differs from a gas in that it has a fixed volume. It is similar to a gas in the respect that it has no fixed shape and will conform to the shape of its container. Examples of common liquids are water, petroleum, and mercury.

In the solid state, matter has a fixed shape and volume. Common solids are iron, quartz, and carbon.

One of the fundamental properties of matter is its ability to change state. A change of state is most conveniently observed in the substance known as water. We are all familiar with the fact that water, which is normally a liquid, is easily converted to a solid (ice) or to a gas (steam). A moment's consideration of the three states of water indicates that some external influence must be involved in producing a change of state. Extending our reasoning along this line, the first thing that comes to mind is that steam is very hot and ice is very cold. From this one may correctly assume that heat is one of the factors involved in a change of state. Water at a temperature of 0° Fahrenheit (-18° Celsius) is solid ice. If the temperature is raised to 32°F (0°C), the solid ice becomes the liquid form of water. If the temperature is raised still higher to 212°F (100°C), the liquid vaporizes into the gas known as steam.

Composition of Matter

Matter is made up of small particles called *elements* and *compounds*. Elements are basic substances that cannot be decomposed into simpler substances. There are, at the time of this writing, 106 known elements with the possibility of the discovery of many more. They range from the abundant elements such as silicon, carbon, and oxygen to the rare elements such as lanthanum, samarium, and tuletium, which are extremely difficult to process. During World War II many elements were synthesized (man-made). The names of man-made elements are interesting because in many cases they indicate their origin by their names. Elements such as americium, californium, and berkelium are examples of elements of this type.

To make the discussion of elements and the subsequent material more meaningful, a table of elements called the Periodic Table is provided in Appendix C. Notice that this table is separated into vertical and horizontal columns that form blocks into which are placed symbols that represent the 106 different elements. The vertical columns are called groups, and the horizontal rows are called periods. The full significance of this table will be explained later, but a cursory description of the table is justified at this time. The symbol used for iron is Fe. It is located in period four, group eight B. The symbol used to represent copper (Cu) is also in period four, but it is in group one B. Notice that the elements in the B groups are all heavy metals, the elements in groups one and two A are light metals, and the elements in groups three to seven A are nonmetals. The elements in group eight A are called the inert gases. They are called inert because they will not combine chemically with other substances. The elements boron, silicon, germanium, arsenic, antimony, tellurium, and polonium are called metalloids because under certain conditions they can possess the properties of metals and nonmetals. The two remaining columns contain the lanthanide series, which are the rare earth elements, and the actinide series, part of which include the man-made elements 95 through 106.

Although many substances are composed of a single element, a far greater number of substances are composed of a combination of different elements. When two or more different elements are chemically combined, they form compounds. A common example of a compound would be a substance such as water, which is composed of the element hydrogen and the element oxygen. The process whereby the elements are chemically combined to form a compound is called *synthesis*. During the synthesis of water, one part of the oxygen element is combined with two parts of the hydrogen element. A compound once formed by synthesis can also be examined and broken down into its elements.

When various elements or compounds are mechanically combined without the occurrence of a chemical change, the result is called a *mixture*. The component elements or compounds of a mixture do not lose their chemical or physical properties. Though the mixture may acquire an appearance that differs from any of its parts, each ingredient that blends into forming the mixture will retain its identity. Thus, it is possible to easily separate a mixture into its individual parts.

Basic Particles of Matter

An *atom* is defined as the smallest particle into which an element can be divided and still retain the chemical properties of that element.

The atom is the smallest part of an element that enters into a chemical change, but it does so in the form of a charged particle. These charged particles are called *ions,* and they are of two types—*positive* (+) and *negative* (–). A positive ion may be defined as an atom that has become positively charged. A negative ion may be defined as an atom that has become negatively charged. One of the properties of charged ions, as shown in Fig. 1-1, is that ions of the same charge tend to repel one another, whereas ions of unlike charge will attract one another. (The term charge has been used loosely. At present, charge will be taken to mean a quantity of electricity which can be one of two kinds, positive or negative.)

The combination of two or more atoms to form the smallest part of a compound comprises a structure known as a *molecule.* For example, when the compound water is formed, two atoms of hydrogen and one atom of oxygen combine to form a molecule of water (Fig. 1-2). A single molecule is very small and is not visible to the naked eye; therefore, a few drops of water may contain as many as a million molecules. A single molecule is the smallest particle into which the compound can be broken down and still be the same substance. Once the last molecule of

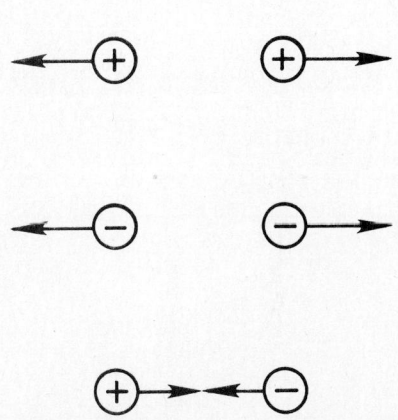

Fig. 1-1: Like charges repel and unlike charges attract each other.

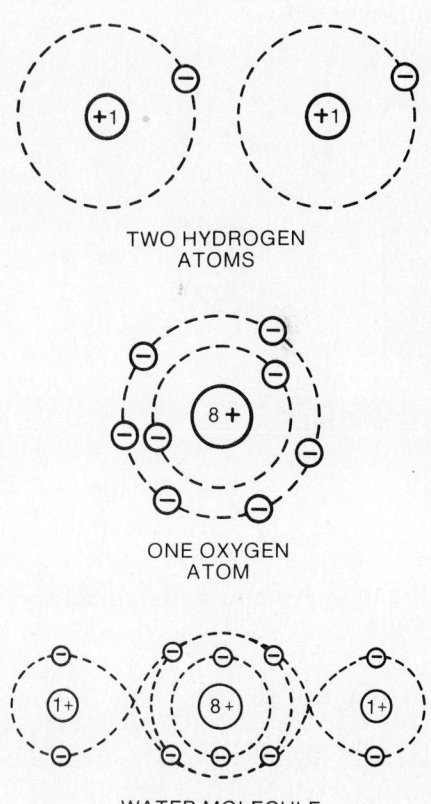

TWO HYDROGEN ATOMS

ONE OXYGEN ATOM

WATER MOLECULE

Fig. 1-2: How a molecule of water is formed.

3

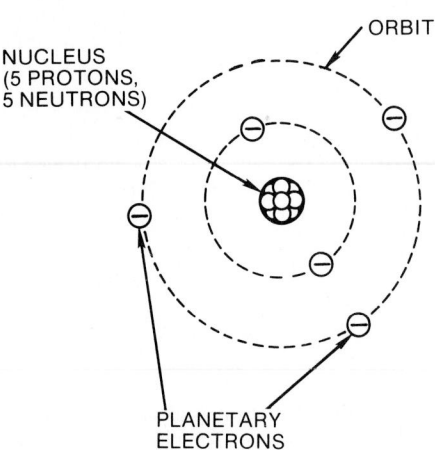

NUCLEUS
(5 PROTONS,
5 NEUTRONS)

ORBIT

PLANETARY
ELECTRONS

Fig. 1-3: Orbital path within a boron atom.

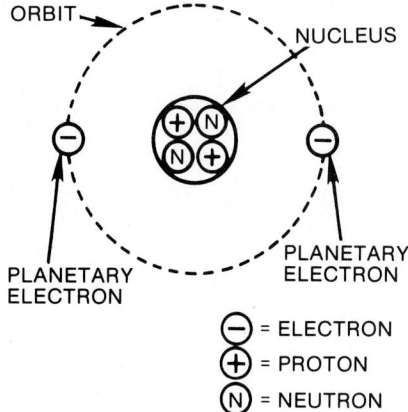

ORBIT

NUCLEUS

PLANETARY
ELECTRON

PLANETARY
ELECTRON

⊖ = ELECTRON
⊕ = PROTON
Ⓝ = NEUTRON

Fig. 1-4: Atomic model of helium atom.

a compound is divided into atoms, the substance no longer exists.

All atoms of all elements are similar in structure because their components are alike. Although nobody has ever seen an atom, scientists have been able to discover that all atoms basically consist of *electrons, protons,* and *neutrons.*

STRUCTURE OF THE ATOM

The atom is believed to consist of a group of positive and neutral particles (protons and neutrons), called the *nucleus,* surrounded by one or more negative orbital electrons. Figure 1-3 shows the arrangement of these particles for an atom of the element boron. This concept of the atom can be likened to our solar system in which the sun is the massive central body and the planets revolve in orbits at discrete distances from the sun. The nucleus commands a position in the atom similar to the position held by the sun in the solar system. The electrons orbit the nucleus of the atom much as the planets orbit the sun. In both the solar system and the atom, practically all the matter in the system is contained within the central body.

The Atomic Nucleus. Excluding short-lived subatomic particles such as mesons, neutrinos, and positrons, which are of little importance to electronics, the nucleus of an atom is made up of heavy particles called *protons* and *neutrons,* as we mentioned previously. An atom of one element differs from an atom of any other in the number of protons contained in the nucleus. The number of protons in the nucleus is called the atomic number of the element and varies from 1 for the element hydrogen to 92 for the element uranium. (The number is even higher for the new man-made elements mentioned in Appendix C.) The atomic number of helium, whose structure is shown in Fig. 1-4, is 2.

The negative charge of the planetary electron is equal and opposite to the positive charge of the proton. Since the nucleus contains all the protons, it carries a total positive charge that is equivalent to the number of protons present. Inasmuch as a normal atom is neutral—that is, it has no external electrical charge—the positive charges of the nucleus are exactly neutralized by the negative charges of the planetary electrons orbiting about it.

It follows that the neutral atom has as many planetary electrons as there are protons in its nucleus. Consequently, the number of electrons orbiting around the nucleus varies from 1 for hydrogen to 92 for uranium (and higher, for the man-made elements). Note that in the helium atom shown in Fig. 1-4, there are two protons and two planetary electrons.

All atoms, except those of ordinary hydrogen, contain one or more neutrons in the nucleus. While the helium atom contains two neutrons, the uranium atom may contain 146 neutrons. Although the opposite electrical charges carried by an electron and a proton are equal in magnitude, the mass, or weight, of the proton is about 1,840 times as great as that of the electron. The mass of the neutron is about equal to that of the proton. It is apparent that practically the entire mass, or weight, of the atom is contained in its nucleus.

The combined number of protons and neutrons in the nucleus of an atom determines its weight, or atomic mass. The atomic mass varies from 1 for ordinary hydrogen, which is the lightest of the elements (one proton and no neutrons in its nucleus), to 238 for uranium, which, until recently, was the heaviest element (92 protons and 146 neutrons in its nucleus). The new and heavier man-made elements have even greater atomic mass numbers (see Appendix C).

All of the atoms of the same type contain the same number of protons in their nuclei, and the atoms of one type differ from those of all other types in the number of protons so contained. For example, each hydrogen atom contains one proton in its nucleus and each uranium atom has 92 protons. But the atoms of the same type may differ in atomic mass, owing to different numbers of neutrons in their nuclei.

For example, three different forms of hydrogen atoms have been found. All of these atoms have the same atomic number of 1 (one proton in the nucleus). However, one of these atoms (the form most commonly found in nature) has a mass of number 1—that is, one proton and no neutron in its nucleus. A second form has an atomic mass of 2—one proton and one neutron in the nucleus. A third form has an atomic mass of 3—one proton and two neutrons. Except for the differences in mass, all three forms of atoms have identical properties.

These different forms of the same atom are called *isotopes.* Most atoms are known to have two or more isotopes. It is interesting to note that scientists have been able to produce artificial isotopes by bombarding the nuclei of atoms with neutrons. In this way, for example, an atom of uranium with an atomic mass of 238 (92 protons and 146 neutrons in its nucleus) sometimes captures a neutron, raising the atomic mass to 239 (92 protons, 147 neutrons).

The Planetary Electrons. Up to this point in the chapter, the focus has been on the nucleus of the atom. Of greater importance in the field of electricity and its practical applications are the planetary electrons that exist outside the nucleus. These will vary in number from 1 for hydrogen to 92 for uranium, when the atoms are in their normal condition.

These electrons do not revolve around the nucleus in a disorderly fashion. As already mentioned, they follow concentric paths, or *orbits,* about the nucleus in a manner somewhat similar to the orbiting of the planets around the sun of our solar system. But there is one great difference between the solar system and the atomic structure. The planets are held in their orbits around the sun by a combination of their motions and their relatively weak gravitational attraction to the sun. If a planet were to lose some of its motion by slowing down, it would simply take a new orbit closer to the sun. The planetary electrons in the atom, with their negative charges, are strongly attracted to the positively charged nucleus. Yet they do not fall into the nucleus. Obviously, the behavior of the charged particles in the atom is different from what would normally be expected.

This situation puzzled scientists until Niels Bohr, a Danish scientist, came up with his theory. Since the electron has a definite mass, the orbiting electron possesses a certain amount of energy. The greater the radius of the orbit (that is, the farther the electron is from the nucleus), the greater must be the velocity and, hence, the energy of the electron. Thus, the orbits correspond to the energy levels of the electrons that occupy them. Electrons with lower energy levels occupy orbits closer to the nucleus. Those with higher energy levels occupy orbits farther away from the nucleus. Should an electron absorb additional energy from some external source, it would jump to an orbit farther from the nucleus. Should it lose some of its energy, it would fall to an orbit closer to the nucleus.

Bohr stated that planetary electrons could only lose or gain energy in specific amounts if they were to change orbits. These "releases" of energy are called *quanta*. When an electron loses a quantum of energy, it jumps to the next inner orbit; when it gains a quantum of energy, it jumps to the next outer orbit. Since it may occupy only an orbit that corresponds to its energy level, only certain orbits are permissible. These permissible orbits are bunched in groups, called *shells*, arranged in concentric layers about the nucleus, somewhat as the layers of an onion. There is a certain maximum number of electrons that each shell may contain. The number of shells, up to seven, depends upon the atomic number of the atom—that is, the number of electrons in orbit around its nucleus. The regions between shells are forbidden zones for the electrons.

ORBITAL SHELLS AND VALENCE ELECTRONS

The structure of an atom is best explained by a detailed analysis of the simplest of all atoms, that of the element hydrogen. The hydrogen atom (Fig. 1-5) is composed of a nucleus containing one proton and a single planetary electron. As the electron revolves around the nucleus, it is held in this orbit by two counteracting forces. One of these forces is called *centrifugal force,* which is the force that tends to cause the electron to fly outward as it travels in its circular orbit. This is the same force that causes a car to roll off a highway when rounding a curve at too high a speed. The second force acting on the electron is *centripetal force.* This force tends to pull the electron in towards the nucleus and is provided by the mutual attraction between the positive nucleus and negative electron. At some given radius, the two forces will exactly balance each other, providing a stable path for the electron.

Energy Levels. If an external force is applied to an atom, one or more of the outermost electrons may be removed. This is possible because the outer electrons are not attracted as strongly to the positive nucleus as are the inner electrons. When atoms

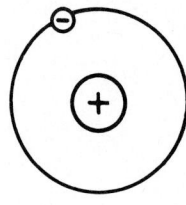

Fig. 1-5: Atomic model of hydrogen atom.

combine to form an elemental substance, the outer electrons of one atom will interact with the outer electrons of neighboring atoms to form bonds between the atoms. These atomic bonds constitute the binding force which holds all matter together. When bonding occurs in some substances, each atom retains its full complement of electrons. In other substances, one or more outer electrons will be gained or lost as a result of bonding. As indicated by the above statements, the electron configuration of the atom is of great importance. The chemical and electrical properties of a material are almost wholly dependent upon the electron arrangement within its atoms.

The orbit shown in Fig. 1-5 is the smallest possible orbit the hydrogen electron can have. In order for the electron to remain in this orbit, it must neither gain nor lose energy. Electrons can gain energy from light or heat sources or through collisions with other particles. For example, a light source consists of tiny energy releases called *photons*. Each photon contains a definite amount of energy, depending on the color (wavelength) of the light it represents. If a photon of sufficient energy collides with an electron, the struck electron will absorb the photon's energy. The electron will now have a greater than normal amount of energy and will jump to a new orbit farther away from the nucleus (Fig. 1-6). It must be emphasized, however, that the electron cannot jump to just any orbit. The electron will remain in its lowest orbit until a sufficient amount of energy is available, at which time the electron will accept the energy and jump to one of a series of permissible orbits. An electron cannot exist in the space between permissible orbits or energy levels. This indicates that the electron will not accept a photon of energy unless it contains enough energy to elevate the electron to one of the allowed energy levels.

Once the electron has been elevated to an energy level higher than the lowest possible energy level, the atom is said to be in an excited state. The electron will not remain in this excited condition for more than a fraction of a second before it will radiate the excess energy and return to a lower energy orbit.

Shells and Subshells

Although hydrogen has the simplest of all atoms, the basic principles just developed apply equally well to the atoms of more complex elements. Actually, the difference between the atoms, insofar as their chemical activity and stability are concerned, is dependent upon the number and position of the particles included within the atom. Atoms range from the simplest, the hydrogen atom containing one proton and one electron, to the very complex atomic structures such as silver, containing 47 protons and 47 electrons. How then are these electrons positioned within the atom? In general, the electrons reside in groups of orbits called *shells,* as we mentioned previously. These shells are elliptically shaped and are assumed to be located at fixed intervals. Thus, the shells are arranged in steps that correspond to fixed energy levels. The shells, and the

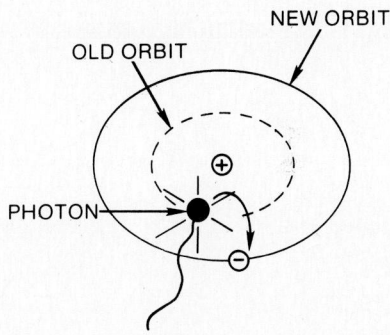

Fig. 1-6: Excitation by a photon.

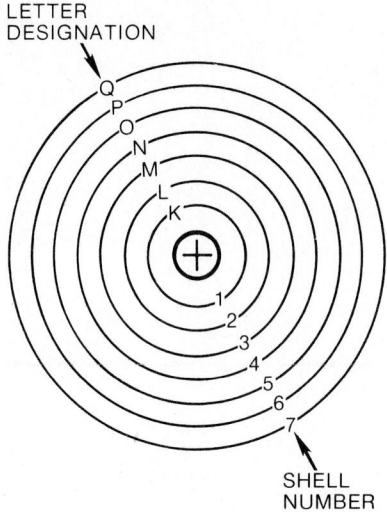

Fig. 1-7: Shell designations.

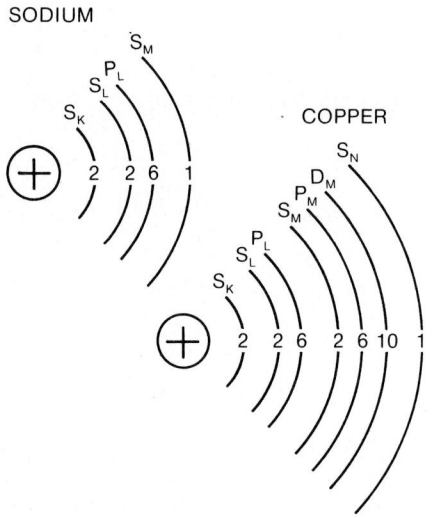

Fig. 1-8: Copper and sodium atoms.

number of electrons required to fill them, may be predicted by the employment of *Pauli's Exclusion Principle*. Simply stated, this principle specifies that each shell may contain no more than $2n^2$ electrons, where n corresponds to the shell number starting with the one closest to the nucleus. By this principle the second shell, for example, would contain $2(2)^2$ or 8 electrons when full. In addition to being numbered, the shells are also given letter designations as pictured in Fig. 1-7.

Starting with the shell closest to the nucleus and progressing outward, the shells are labeled K, L, M, N, O, P, and Q respectively. The shells are considered to be full or complete when they contain the following quantities of electrons: two in the K shell, eight in the L shell, 18 in the M shell, and 32 in the N shell. (The formula $2n^2$ can be used to determine the number of electrons only in the four shells closest to the nucleus of an atom. The O shell has 18 electrons maximum, the P shell has 12, and the Q shell has 2.) Each of these shells is a major shell and can be divided into subshells of which there are four labeled s, p, d, and f. A subshell exists at a given energy level; that is, at a given distance from the nucleus. There may be one or more subshells at a specified distance from the nucleus with the electron(s) of each moving in a different direction.

Like the major shells, the subshells are also limited as to the number of electrons which they can contain. Thus, the s subshell is complete when it contains two electrons, the p subshell when it contains six, the d subshell when it contains 10, and the last subshell when it contains 14 electrons.

Inasmuch as the K shell can contain no more than two electrons, it must have only one subshell, the s subshell. The M shell is composed of three subshells: s, p, and d. If the electrons in the s, p, and d subshells are added, their total is found to be 18, the exact number required to fill the M shell. This relationship exists between the shells and subshells up to and including the N shell. Notice the difference between the electron configurations for copper and sodium illustrated in Fig. 1-8.

Atomic Weight

There are a wide variety of atoms, a different type of atom comprising each of the 106 known elements. Each atom is similar in that all atoms consist of protons and electrons, as we have mentioned. However, atoms of different elements contain varying numbers of basic particles, thus causing a difference in weight. A classification, based on the atomic weight and atomic number of the atoms, has been devised to differentiate between different atoms.

Although atoms are far too small to be weighed, a system has been set up whereby the weight of one atom is given with reference to a universally accepted standard. This system of weights, called the atomic weight of the elements, uses the element oxygen as a reference. The atomic weight of oxygen is assigned a numerical value of 16, and the atomic weights of other elements are determined by comparison with oxygen. No element will have an atomic weight less than 1. The lightest element known, hydrogen, has an atomic weight equal to 1.008. The complete listing of atomic weights of the atoms of different elements is found in the periodic table shown in Appendix C.

Atomic Valence

The number of electrons in the outermost shell determines the *valence* of the atom (Table 1-1). For this reason, the outer shell of an atom is called the *valence shell;* and the electrons contained in this shell are called the valence electrons. The valence of an atom determines its ability to gain or lose an electron, which, in turn, determines the chemical and electrical properties of the atom. An atom that is lacking only one or two electrons from its outer shell will easily gain electrons to complete its shell, but a large amount of energy is required to free any of its electrons. An atom having a relatively small number of electrons in its outer shell in comparison to the number of electrons required to fill the shell will easily lose the valence electrons.

Table 1-1: Random Elements and Their Atomic Values

Element	Symbol	Atomic Number	Atomic Weight	Atomic Valence
Aluminum	Al	13	26.9815	+3
Carbon	C	6	12.01115	±4
Chlorine	Cl	17	35.453	−1
Copper	Cu	29	63.54	+1
Fluorine	F	9	18.9984	−1
Gold	Au	79	196.967	+1
Germanium	Ge	32	72.59	±4
Helium	He	2	4.0026	0
Hydrogen	H	1	1.00797	±1
Iron	Fe	26	55.847	+2
Lead	Pb	82	207.19	+2
Mercury	Hg	80	200.59	+1
Neon	Ne	10	20.183	0
Nickel	Ni	28	58.71	+2
Nitrogen	N	7	14.0067	−3
Oxygen	O	8	15.999	−2
Selenium	Se	34	76.96	±4
Silicon	Si	14	28.086	±4
Silver	Ag	47	107.868	+1
Sodium	Na	11	22.99	+1
Tin	Sn	50	118.69	+2
Uranium	U	92	238.03	+2, +3, +4, +5, +6

The valence shell always refers to the outermost shell, whether it be a major shell or a subshell. The copper (Cu) and silver (Ag) atoms each have one electron in the outermost shell. Even though the atomic weights and atomic numbers of copper and silver are quite different, the atoms are similar in that they both contain one valence electron.

In atoms that have more than one ring or shell of electrons, such as copper and silver, some electrons are farther away from the nucleus (which contains protons) than others. There is less attraction between the nucleus and those electrons that are farthest away from it than between the nucleus and those that are closer. These electrons that are farthest away (in the outermost orbit or shell) from the nucleus permit us to have electron flow. They are called "loosely bound" electrons. This does not mean that there are some electrons that are wholly detached from any nucleus. If there were, the material would be continually charged because it would always have an excess of electrons over the number required to make the atoms neutral. That is, the material has some electrons that are so loosely bound to the nucleus that they can easily be torn loose and made to drift.

A *negative valence* indicates an atom's tendency to gain electrons to complete its outermost orbital shell. The chlorine atom shown in Fig. 1-9A is an example of an atom having a valence of -1. This means that one electron is needed to complete the outer shell of the chlorine atom. Atoms having a valence of -2 require two electrons, -3 require three electrons, and so forth. These gained electrons cause the atom to become negatively charged, since the additional negative charge of the electrons is not offset by an increase in positive charge.

An atom with a *positive valence,* on the other hand, has a tendency to give up an electron when it bonds. One example of this is the sodium atom shown in Fig. 1-9B. It has a valence of +1, which means it has a single electron in its outer orbital shell which it will readily give up to another atom. An atom with a valence of +2 can donate a pair of electrons and so on. Donating electrons in this manner leaves an excess of protons in the atom's nucleus and creates a net positive charge.

There are some atoms which exhibit both negative and positive valence. They show no preference for gaining or losing an electron and, depending upon conditions, will do either. Hydrogen is an example of such an element. It has both a -1 and +1 valence.

Ionization

Ionization occurs when a valence electron is freed from an atom or when an atom acquires an extra electron (Fig. 1-9C).

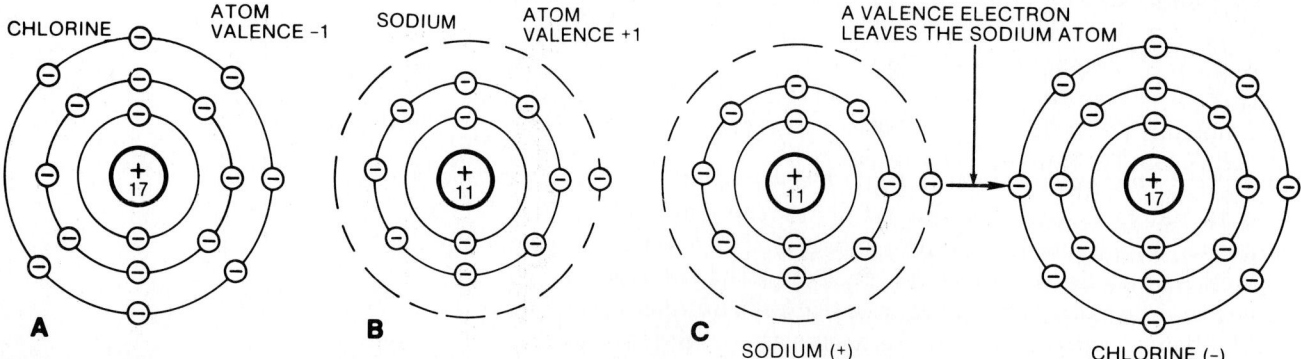

Fig. 1-9: Creation of negative (chlorine) ions and positive (sodium) ions.

For ionization to take place, there must be a transfer of energy which results in a change in the internal energy of the atom. An atom having more than its normal count of electrons is called a *negative ion*. The atom that gives up some of its normal electrons is left with fewer negative charges than positive charges and is called a *positive ion*. Thus, ionization is the process by which an atom loses or gains electrons.

The probability of an atom ionizing positively or negatively is based primarily on the number of electrons present in the valence shell. For example, an atom having seven electrons in the valence shell would not likely lose any electrons, but would more likely pick up loosely bound electrons to fill out its valence shell. An atom having only one electron in its valence shell may easily give up this electron to attain a more stable condition.

Driving electrons out of the shell of an atom requires raising the internal energy of the atom. This energy may be obtained through bombardment by photons (light energy) and phonons (vibrational energy) or by subjecting the atom to electric fields. The amount of energy required to free electrons from an individual atom is called the *ionization potential*.

The ionization potential necessary to free an electron from an inner shell is much greater than that required to free an electron from an outer shell. Also, more energy is required to remove an electron from a complete shell than an unfilled shell.

CONDUCTORS, INSULATORS, AND SEMICONDUCTORS

In the study of electronics, the association of matter and electricity is of paramount importance. Since every electronic device is constructed of parts made from ordinary matter, the effects of electricity on matter must be well understood. As a means of accomplishing this, all the elements of which matter is made may be placed into one of three categories: conductors, insulators, and semiconductors.

Conductors. Substances that permit the free motion of a large number of electrons are called conductors. Copper wire, for instance, is considered a good conductor because it has many loosely bound electrons.

As noted in Table 1-1, each copper atom has a +1 valence. In fact, conductors can be classified as materials having valence shells of 1, 2, or 3 electrons (Fig. 1-10A). Electrical energy is transferred through conductors by the migration of these loosely bound electrons from atom to atom within the conductor (Fig. 1-11). Each electron moves only a short distance to the neighboring atom, where it replaces other atoms and forces them out of their orbits. This chain reaction continues until the movement is completed through the entire length of the conductor. Conductive properties are directly proportional to the number of electrons which move through the material under force. A good conductor is said to have low opposition or low resistance to the electron movement or flow, which is commonly referred to as *electric current*.

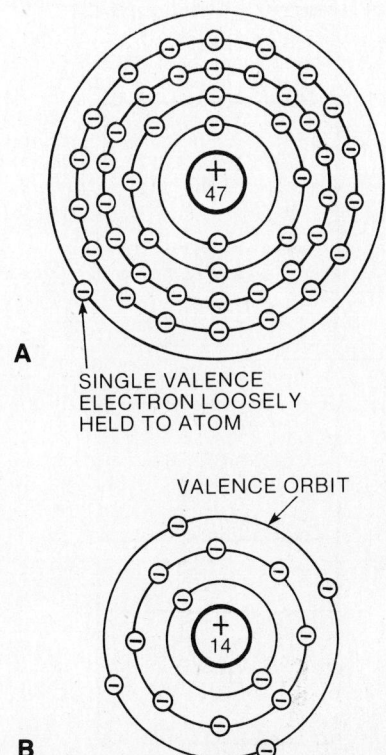

A
SINGLE VALENCE
ELECTRON LOOSELY
HELD TO ATOM

VALENCE ORBIT

B

Fig. 1-10: The atomic structure of (A) a conductor (silver) and (B) a semiconductor (silicon).

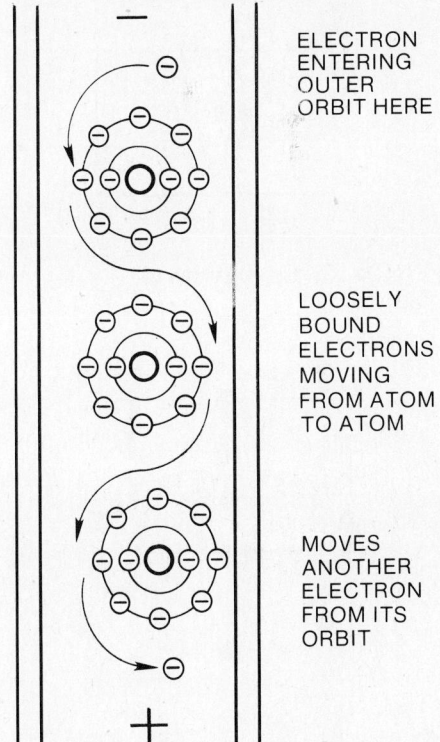

ELECTRON
ENTERING
OUTER
ORBIT HERE

LOOSELY
BOUND
ELECTRONS
MOVING
FROM ATOM
TO ATOM

MOVES
ANOTHER
ELECTRON
FROM ITS
ORBIT

Fig. 1-11: How electrons move through a conductor.

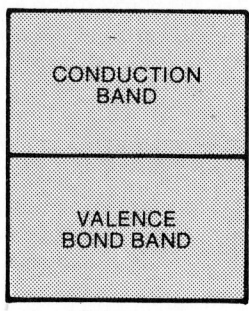

Fig. 1-12: Energy level diagrams.

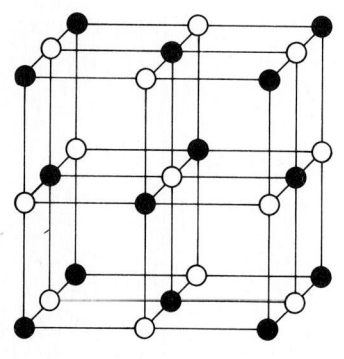

● SODIUM ION   ○ CHLORINE ION

Fig. 1-13: Atomic crystal lattice structure of salt.

While all materials have some degree of conductance, metals are commonly associated with electrical conduction. The best conductors are silver, copper, gold, and aluminum, in that order. However, copper and aluminum are used more extensively because they are the least expensive.

Insulators. An insulator is a material, or combination of materials, in which the atomic structure opposes the movement of electrons from atom to atom. In other words, an insulator is a material that has few loosely bound electrons. Insulators usually have 5, 6, 7, or 8 electrons in their valence shell.

No known material is a perfect insulator, but there are materials which are such poor conductors that for all practical purposes they are classed as perfect insulators. Glass, dry wood, rubber, mica, and certain plastics such as polystyrene are good insulators.

Semiconductors. Between the extremes of good conductors and good insulators are a number of materials which are neither good conductors nor good insulators. Germanium, silicon, carbon, and selenium are substances falling in this category. These materials, due to their peculiar crystalline structure, may, under certain conditions such as an increase in temperature, act as conductors; under other conditions, as insulators. Semiconductors are frequently called *tetravalent;* that is, they have valence shells of ± 4 electrons (Fig. 1-10B).

Energy Levels

The energy level diagram shown in Fig. 1-12 helps explain the conductive properties of conductors, semiconductors, and insulators. We should keep in mind that most solids are made up of atoms that combine to form a definite and repeated geometric shape called a *crystal.* In this crystal lattice condition (Fig. 1-13), the shells around the atoms interact and broaden into bands. The electrons now may exist only in the bands, but not in between. They may move from band to band, but only if they receive or lose the necessary and specific quantum of energy—the greater the distance between bands, the larger the quantum of energy necessary for the jump. As mentioned earlier, the spaces between the bands where no electrons may exist are called the *forbidden bands.*

In the atom, the valence shell is of great importance because its electrons are most readily removed and because it may capture other electrons. In the crystal, the outermost valence band is similarly of great importance. It is formed by the interaction of the valence shells of the various atoms.

Farther out, adjacent to the valence band, is another forbidden band. The latter is then followed by another band that may receive electrons. However, this outside band is empty and is called the *conduction band.* Electrons that are raised to the conduction band (or energy levels) are free to be made to flow or move as an electric current. In comparing the energy level diagrams for an insulator and a conductor, the conductor is seen to have little or no forbidden gap. Since this is true, under normal conditions the conduction band for a conductor contains a sufficient number of electrons to make it a good conductor of electricity.

The semiconductor, being neither a good conductor nor a good insulator, has a gap energy between that of a conductor and that of an insulator.

In the following chapters, the role of the conductor, semiconductor, and insulator will assume greater importance as the various electronic devices are developed and discussed. In fact, in the final analysis, all electronic phenomena are based on the electrical nature of matter.

SUMMARY

Electricity is produced by the movement of small atomic particles known as electrons. All matter is electrical in structure and is largely composed of a combination of positive and negative particles. Matter is anything which has mass (weight) and occupies space. There are three forms or states of matter: gases, liquids, and solids. One fundamental property of matter is its ability to change from one state to another under an external influence.

Matter is composed of elements and compounds. Elements are basic substances that cannot be decomposed into simpler substances. At this time there are 106 known elements. Compounds are formed when two or more elements chemically combine through a process called synthesis. When elements combine to form compounds, they lose their individual chemical and physical properties.

An atom is the smallest particle of an element that still retains the chemical properties of that element. A combination of two or more atoms which make up the smallest particle of a compound is called a molecule. The structure of the atom consists of a nucleus made up of protons and neutrons surrounded by orbiting electrons. Protons are positively charged particles. Neutrons are particles with no electrical charge. Electrons are negatively charged particles which normally exist in numbers equal to the number of protons in the nucleus.

Electrons orbit the nucleus in fixed orbits or shells with each shell only holding a fixed number of electrons. The number of electrons in the outermost shell, or valence shell, determines the valence of the atom, its ability to gain or lose electrons. A negative valence indicates a tendency to gain electrons, while a positive valence indicates the atom will give up electrons. Atoms which have lost or gained electrons take on a positive or negative charge and are called ions. The amount of energy required to free electrons from an atom is called the ionization potential of that atom.

All atoms enter into chemical reactions in the form of positive or negative ions. When atoms combine they form bonds through the interaction of their valence electrons. These atomic bonds constitute the binding force which holds all matter together.

In accordance with their electrical properties, all elements and compounds can be classified as conductors, insulators, and semiconductors. Conductors are substances that permit the free motion of a large number of electrons. Insulators are materials in which the atomic structure opposes the movement of electrons from atom to atom. Semiconductors, between the two extremes, are materials which are neither good conductors nor good insulators. The conductive properties of materials determine how they are best used in electronic devices.

CHECK YOURSELF (Answers at back of book)

Fill in the blanks with the appropriate word or words.

1. Matter can exist in three states: _____, _____, and _____ .
2. A/An _____ contains the smallest known unit of positive electricity.
3. Atoms that combine to form definite and repeated geometric shapes are called _____ .
4. _____ gases are those that will not combine chemically with other substances.
5. Freeing electrons from an atom requires raising the internal energy of the atom. The amount of energy required is called the _____ _____ .
6. Electrons are positioned within the atom in groups of orbits called _____ .
7. The atomic number is equal to the number of _____ found in the nucleus.
8. _____ are formed when two or more different elements chemically combine by the process called _____ .
9. _____ may under certain conditions act as conductors and under other conditions act as insulators.
10. When an atom has more than its normal count of electrons, it is called a/an _____ ion, while an atom giving up electrons is called a/an _____ ion.
11. Matter in the _____ state possesses neither a fixed shape nor a fixed volume.
12. Pauli's Exclusion Principle can only be used to determine the number of electrons in the _____ shells closest to the nucleus of an atom.
13. _____ are substances that permit the free motion of a large number of electrons.
14. The combination of two or more atoms is known as a/an _____ .
15. A negative valence indicates an atom's tendency to _____ electrons, while a positive valence indicates a tendency to _____ electrons.
16. The nucleus of an atom consists of _____ and _____ .
17. Light energy exists in the form of _____ and vibrational energy exists in the form of _____ .
18. Atomic weights are determined by using the element _____ as a reference.

CHAPTER 2

Static and Dynamic Electricity

In a natural, or neutral, state each atom in a body of matter will have the proper number of electrons in orbit around it. Consequently, the whole body of matter comprised of the neutral atoms will also be electrically neutral. In this state, it is said to have a "zero charge" and will neither attract nor repel other matter in its vicinity. Electrons will neither leave nor enter the neutrally charged body should it come in contact with other neutral bodies. If, however, any number of electrons are removed from the atoms of a body of matter, there will remain more protons than electrons, and the whole body of matter will become electrically positive. Should the positively charged body come in contact with another body having a normal charge or having a negative (too many electrons) charge, an electric current will flow between them. Electrons will leave the more negative body and enter the positive body. This electron flow will continue until both bodies have equal charges.

When two bodies of matter have unequal charges and are near one another, an electric force is exerted between them because of their unequal charges. However, since they are not in contact, their charges cannot equalize. The existence of such an electric force, where current cannot flow, is referred to as *static electricity*. "Static" means "not moving." This is also referred to as an *electrostatic force*.

One of the easiest ways to create a static charge is by the friction method. With the friction method, two pieces of matter are rubbed together and electrons are "wiped off" one onto the other. If materials that are good conductors are used, it is quite difficult to obtain a detectable charge on either. The reason for this is that equalizing currents will flow easily in and between the conducting materials. These currents equalize the charges almost as fast as they are created. A static charge is easier to obtain by rubbing a hard, nonconducting material against a soft, fluffy nonconductor. Electrons are rubbed off one material and onto the other material. This is illustrated in Fig. 2-1.

When a hard rubber rod is rubbed in fur, the rod accumulates electrons. Since both fur and rubber are poor conductors, little equalizing current can flow and an electrostatic charge is built up. When the charge is great enough, equalizing currents will flow regardless of the material's poor conductivity. These currents will cause a crackling sound and if viewed in darkness will produce visible sparks.

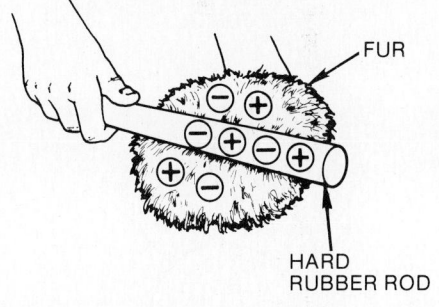

+ CHARGES AND ELECTRONS ARE PRESENT IN EQUAL QUANTITIES IN THE ROD AND FUR

FUR

HARD RUBBER ROD

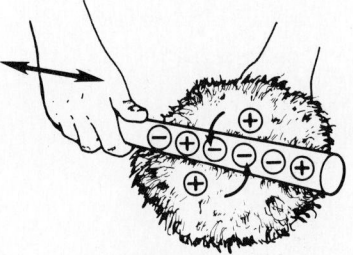

ELECTRONS ARE TRANSFERRED FROM THE FUR TO THE ROD

Fig. 2-1: Producing static electricity by friction.

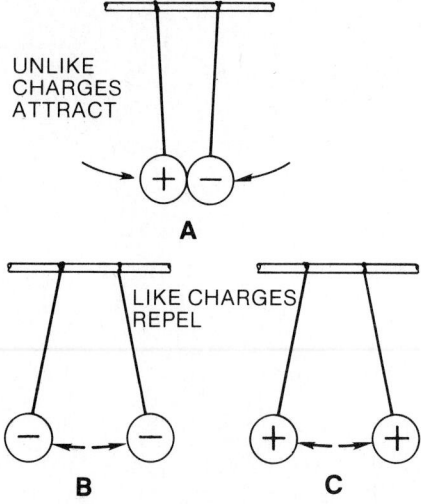

UNLIKE CHARGES ATTRACT

A

LIKE CHARGES REPEL

B   **C**

Fig. 2-2: Reaction between two charged bodies.

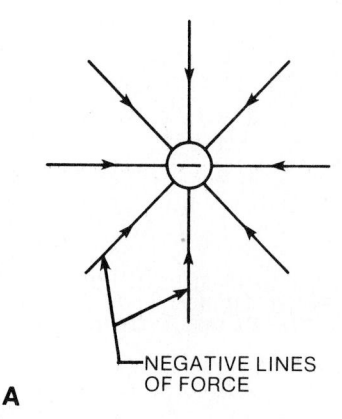

NEGATIVE LINES OF FORCE

A

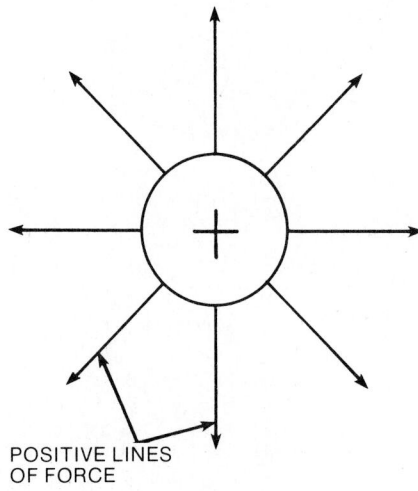

POSITIVE LINES OF FORCE

B

Fig. 2-3: Electrostatic lines of force about a negatively charged body (A), and a positively charged body (B).

Charged Bodies

One of the fundamental laws of electricity mentioned in Chapter 1 is that like charges repel each other and unlike charges attract each other. A positive charge and negative charge, being unlike, tend to move toward each other. In the atom the negative electrons are drawn toward the positive protons in the nucleus. This attractive force is balanced by the electron's centrifugal force caused by its rotation about the nucleus. As a result, the electrons remain in orbit and are not drawn into the nucleus. Electrons repel each other because of their like negative charges, and protons repel each other because of their like positive charges.

The law of charged bodies may be demonstrated by a simple experiment. Two pith (paper pulp) balls are suspended near one another by threads, as shown in Fig. 2-2.

If the hard rubber rod is rubbed to give it a negative charge and then held against the right-hand ball, the rod will impart a negative charge to the ball. The right-hand ball will be charged negative with respect to the left-hand ball. When released, the two balls will be drawn together, as shown in Fig. 2-2A. They will touch and remain in contact until the left-hand ball acquires a portion of the negative charge of the right-hand ball, at which time they will swing apart as shown in Fig. 2-2B. If, positive charges are placed on both balls (Fig. 2-2C), the balls will also be repelled from each other.

Electric Fields

The space between and around charged bodies in which their influence is felt is called an *electric field of force*. The electric field is always terminated on material objects and extends between positive and negative charges. It can exist in air, glass, paper, or a vacuum. *Electrostatic fields* and *dielectric fields* are other names used to refer to this region of force.

Fields of force spread out in the space surrounding their point of origin (Fig. 2-3) and, in general, diminish in proportion to the square of the distance from their source.

The field about a charged body is generally represented by lines which are referred to as *electrostatic lines of force*. These lines are imaginary and are used merely to represent the direction and strength of the field. To avoid confusion, the lines of force exerted by a positive charge are always shown leaving the charge, and for a negative charge they are shown as entering. Figures 2-4 and 2-5 illustrate the use of lines to represent the field about charged bodies.

Figure 2-4 represents the repulsion of like-charged bodies and their associated fields. Figure 2-5 represents the attraction between unlike-charged bodies and their associated fields.

Coulomb's Law of Charges

The amount of attracting or repelling force which acts between two electrically charged bodies in free space depends on two things: their charges and the distance between them. The relationship of charge and distance to electrostatic force was first discovered and written by a French scientist named Charles A. Coulomb. Coulomb's Law states that *charged bodies attract or repel each other with a force that is directly proportional to the product of their charges and is inversely proportional to the square of the distance between them.* This is expressed mathematically as:

$$F = \frac{kQ_1Q_2}{d^2}$$

where F is in newtons, d is in meters, k is 9×10^9 (in free space), and Q is the unit of charge in coulombs.

Example: A body which has a positive charge of 2 coulombs is brought to a point 2 meters from a body possessing a positive charge of 4 coulombs. What is the repelling force between the charged bodies?

Given:
$$Q_1 = 2 \text{ coulombs}$$
$$Q_2 = 4 \text{ coulombs}$$
$$d = 2 \text{ meters}$$
$$k = 9 \times 10^9$$

Find F:
Solution:

$$F = 9 \times 10^9 \times \frac{Q_1Q_2}{d^2}$$

$$F = 9 \times 10^9 \times \frac{2 \times 4}{2^2}$$

$$F = 9 \times 10^9 \times \frac{8}{4}$$

$$F = 9 \times 10^9 \times \frac{2}{1}$$

$$F = 18 \times 10^9 \text{ newtons}$$

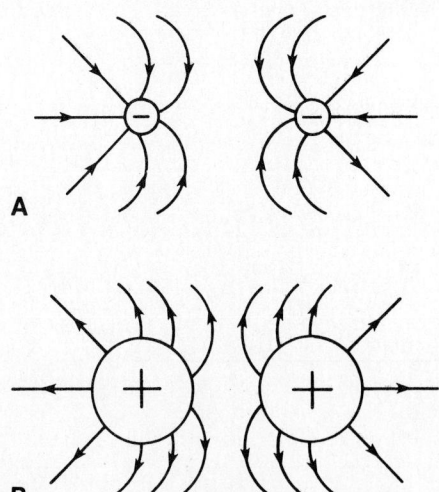

Fig. 2-4: Lines of force of like charges repel one another.

CLOSE TOGETHER, ALL LINES INTERLINK

FAR APART, THERE IS LITTLE FIELD INTERLINKAGE

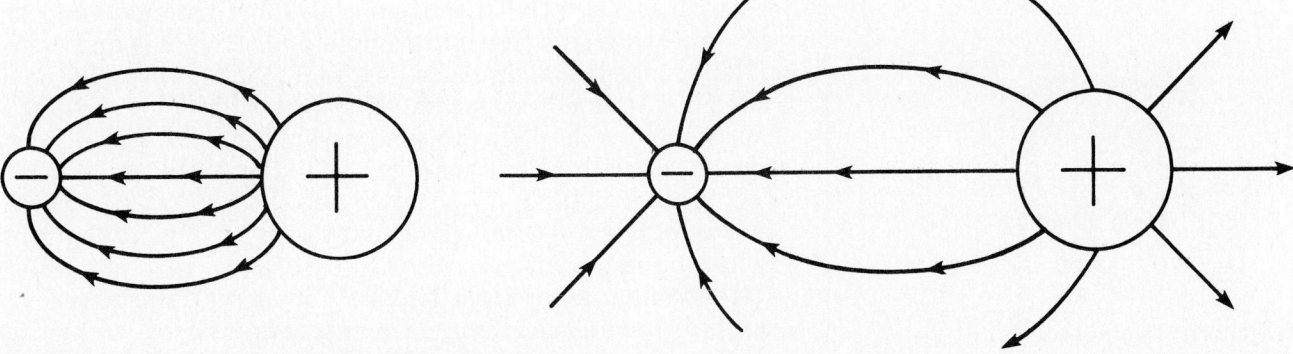

Fig. 2-5: Lines of force from unlike charges.

17

It is interesting to note that experimental measurements often result in values such as 6.28 billion billion electrons and are usually written, 6.28×10^{18} electrons. This means that when a body has a positive charge of 1 coulomb, it has a deficiency of 6.28×10^{18} electrons. If a body has a negative charge of 1 coulomb, it has an excess of 6.28×10^{18} electrons.

The newton (named in honor of Sir Isaac Newton) is the force required to accelerate a mass of 1 kilogram at the rate of 1 meter per second per second. A force of 1 newton is equal to a force of 0.2248 pound.

CURRENT OR CHARGE FLOW

Static electricity, as mentioned earlier, is electricity at rest. The science of electrostatics is concerned with the study of static electricity. It has useful applications, but in the field of electronics, *dynamic electricity* is of utmost importance.

As the name implies, dynamic electricity is electricity in motion. The name is not of great importance, but it is descriptive. The heart of the matter is movement. Actually, electric current is the movement of charged particles in a specified direction. The charged particle may be an electron, a positive ion, or a negative ion. Charged particles can move through a solid, liquid, gas, or vacuum. In a solid, such as a copper wire, ions are held rigidly in place by the atomic structure of the material and electrons serve as the charged particles. In liquids and gases ions are not held firmly in place and can move about and become current carriers.

The speed of electron motion is approximately the speed of light, 186,000 miles per second, but the individual electrons move at a much slower rate. The concept of the effect of current being instantaneous and the individual electrons moving at a slower rate is illustrated in Fig. 2-6. Suppose you had a very long rubber hose with a diameter just large enough to fit a golf ball into the hose. If you were to fill the hose completely with golf balls and then push an extra ball into one end of the hose, another ball would immediately come out the other end of the hose. If you did not know that the hose was full of golf balls, you might think that the golf ball you pushed in one end of the hose traveled very quickly down the hose and out the other end. The effect (the movement of the balls from one end to the other end) is very fast. Yet each ball moves only a short distance. Suppose you stacked up six hoses filled with balls (Fig. 2-7) and alternately pushed a ball into one hose and then another. Now you could have a steady stream of balls appearing at one end of the hoses, yet only the balls in one hose would be moving at any one time. Even within that hose each individual ball would move only a short distance. This is comparable to the way in which current carriers (electrons) move through a wire when current is flowing in the wire.

Now let's take a closer look at how actual electrons flow through a solid conductor, such as a copper wire. The flow of charge through a conductor always takes place between points

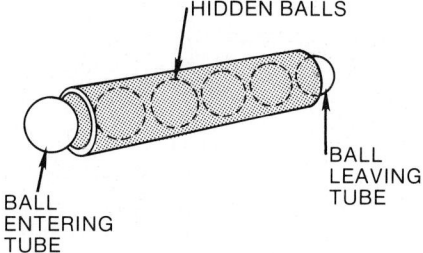

Fig. 2-6: A ball exits the instant another ball enters the hose.

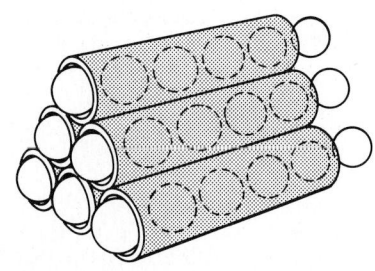

Fig. 2-7: Typical electron movement.

18

of different potentials. The point of negative potential has an excess of electrons which will flow toward the point of positive potential which has a deficiency of electrons. This phenomenon is often illustrated by connecting the positive electrode of a battery to the right end of a copper wire and the negative electrode of the battery to the left end of the wire (Fig. 2-8). The positive electrode, being deficient of electrons, will draw electrons from the atoms in the right end of the wire. These atoms suddenly become positive ions, short of electrons. They, in turn, attract electrons from the atoms to their left, which continue this chain reaction throughout the length of the wire to the negative electrode (Fig. 2-9). The electrons move from the negative electrode to the positive electrode. Disconnecting the copper wire from the electrodes eliminates the electromotive force and the movement of electrons ceases.

The symbol for current is *I*. The symbol *I* was chosen because early scientists talked about the *intensity* of the electricity in a wire.

Direct and Alternating Current

Electric current is generally classified into two *general* types: direct current (DC) and alternating current (AC). The electron flow in a DC system is always in a constant, uniform direction (Fig. 2-10A); that is, the electrical current flows from the source, through the load, and back to the source.

With an AC system, the electrical current starts, rises to a maximum value, and then drops to zero. Then the current reverses direction and repeats the same cycle, rising to a maximum value and then dropping to a no-current-flow condition (Fig. 2-10B). After one complete reversal, the cycle is repeated and continues in this manner. Actually, this cycle of AC is repeated a given number of times per second. The rate of this repetition is called the frequency of the electrical current. This term has been related to a standard reference of time, the second. The frequency of an electrical current is described in terms of "cycles per second." This term has become obsolete in recent years. International agreement in the scientific world has accepted the term *hertz* to describe cycles per second. The correct way of stating frequency is illustrated by the sentence, "The frequency of the electrical current is 60 hertz."

Current Rate of Flow

In order to determine the amount (number) of electrons flowing in a given conductor, it is necessary to adopt a unit of measurement of current flow. The term *ampere* is used to define the unit of measurement of the rate at which current flows (electron flow). Current flow is measured in amperes. The abbreviation for ampere is A. One ampere may be defined as the flow of 6.28×10^{18} electrons per second past a fixed point in a conductor.

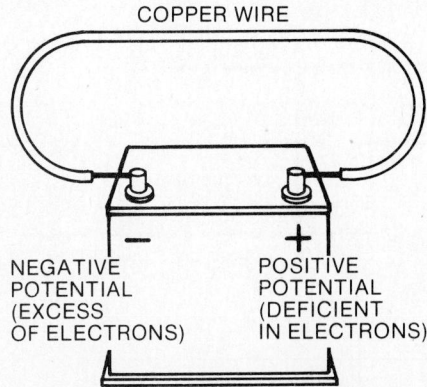

Fig. 2-8: In a battery, a difference of potential exists between the positive terminal and the negative terminal. This is because the positive terminal is deficient in electrons while the negative terminal has an excess of electrons.

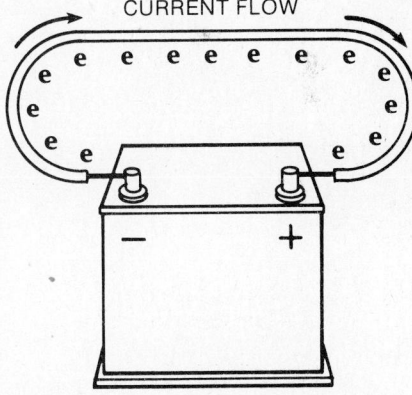

Fig. 2-9: When the terminals of a battery are connected with a copper wire, a chain reaction occurs in which the positive terminal deficient in electrons draws electrons from the copper wire. In turn, the copper wire replenishes its electrons by drawing electrons from the negative terminal. This creates a current flow from the negative terminal through the wire to the positive terminal of the battery.

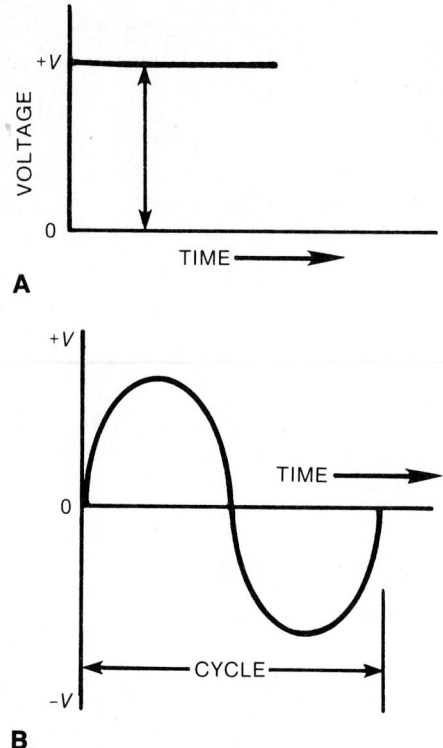

A

B

Fig. 2-10: Current flow in (A) DC voltage and (B) AC voltage.

A unit quantity of electricity is moved through an electric circuit when 1 ampere of current flows for 1 second of time. This unit is a coulomb (Q) and, as we know, is equivalent to 6.28×10^{18} electrons. The rate of flow of current in amperes and the quantity of electricity moved through a circuit are related by the common factor of time. Thus, the quantity of electric charge, in coulombs, moved through a circuit is equal to the product of the current in amperes, I, and the duration of flow in seconds, t. Expressed as an equation, $Q = It$.

For example, if a current of 2A flows through a circuit for 10 seconds, the quantity of electricity moved through the circuit is 2×10, or 20C. Conversely, current flow may be expressed in terms of coulombs and time in seconds. Thus, if 20C are moved through a circuit in 10 seconds, the average current flow is $\frac{20}{10}$, or 2A. Note that the current flow in amperes implies the rate of flow of coulombs per second without indicating either coulombs or seconds. Thus a current flow of 2A is equivalent to a rate of flow of 2C per second.

THE FORCE THAT CAUSES CURRENT FLOW: VOLTAGE

The force that causes loosely bound electrons to move in a conductor as an electric current may be referred to as follows:
1. Electromotive force (emf)
2. Voltage (V)
3. Difference in potential

When a difference in potential exists between two charged bodies that are connected by a conductor, electrons will flow along the conductor. This flow is from the negatively charged body to the positively charged body until the two charges are equalized and potential difference no longer exists.

An analogy of this action is shown in the two water tanks connected by a pipe and valve in Fig. 2-11. At first the valve is closed and all the water is in tank A. Thus, the water pressure across the valve is at maximum. When the valve is opened, the water flows through the pipe from A to B until the water level becomes the same in both tanks. The water then stops flowing

TANK A   TANK B

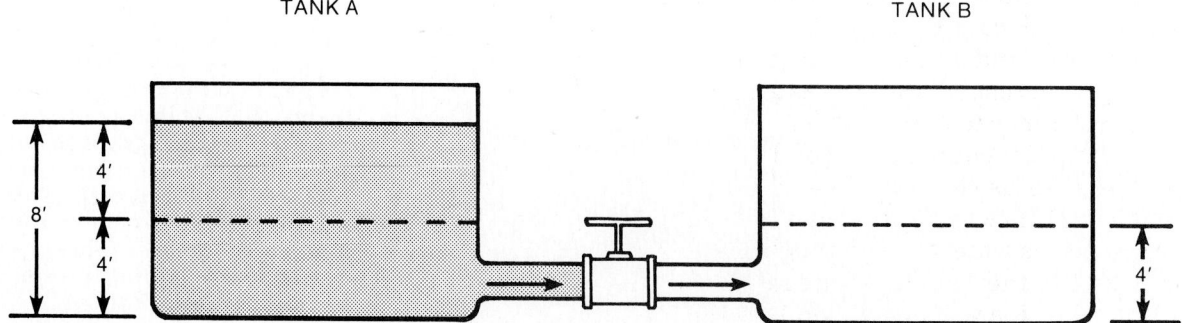

Fig. 2-11: Water analogy of electric difference in potential.

in the pipe, because there is no longer a difference in water pressure between the two tanks.

Current flow through an electric circuit is directly proportional to the difference in potential across the circuit, just as the flow of water through the pipe in Fig. 2-11 is directly proportional to the difference in water level in the two tanks.

A fundamental law of electricity is that the current is directly proportional to the applied voltage; that is, if the voltage is increased, the current is increased. If the voltage is decreased, the current is decreased.

Primary Methods of Producing a Voltage

Presently, there are seven commonly used methods of producing electromotive force (emf). Some of these methods are much more widely used than others. The following is a list of the seven most common methods of producing an emf.

1. *Friction.* Voltage produced by rubbing two materials together.

2. *Pressure.* (Piezoelectricity). Voltage produced by squeezing crystals of certain substances.

3. *Heat.* (Thermoelectricity). Voltage produced by heating the joint (junction) where two unlike metals are joined.

4. *Light.* (Photoelectricity). Voltage produced by light striking photosensitive (light sensitive) substances.

5. *Chemical Action.* Voltage produced by chemical reaction in a battery cell.

6. *Magnetism.* Voltage produced in a conductor when the conductor moves through a magnetic field or a magnetic field moves through the conductor in such a manner as to cut the magnetic lines of force of the field.

7. *Radioactivity.* Electricity based on various direct actions of radioactivity.

Of the seven sources of emf, the only ones that are generally of concern in electronic activities are: chemical action (see Chapter 4); magnetism (see Chapter 9); and pressure (see Chapter 24).

RESISTANCE

Every material offers some resistance, or opposition, to the flow of electric current through it. Good conductors, such as copper, silver, and aluminum, offer very little resistance. Poor conductors, or insulators, such as glass, wood, and paper, offer a high resistance to current flow.

The size and type of material of the wires in an electric circuit are chosen so as to keep the electrical resistance as low as possible. In this way, current can flow easily through the conductors, just as water flows through the pipe between the tanks in Fig. 2-11. If the water pressure remains constant, the flow of water in the pipe will depend on how far the valve is opened. The smaller the opening, the greater the opposition to the flow and the smaller will be the rate of flow in gallons per second.

In the electric circuit, the larger the diameter of the wires, the lower will be their electrical resistance (opposition) to the flow of current through them. In the water analogy, pipe friction opposes the flow of water between the tanks. This friction is similar to electrical resistance. The resistance of the pipe to the flow of water through it depends upon the length of the pipe, the diameter of the pipe, and the nature of the inside walls (rough or smooth). Similarly, the electrical resistance of the conductors depends upon the length of the wires, the diameter of the wires, and the material of the wires (copper, aluminum, etc.).

Temperature also affects the resistance of electrical conductors to some extent. In most conductors (copper, aluminum, iron, etc.), the resistance increases with temperature. Carbon is an exception. In carbon the resistance decreases as temperature increases. Certain alloys of metals (manganin and constantan) have a resistance that does not change appreciably with temperature.

The relative resistance of several conductors of the same length and cross section is given in the following list with silver as a standard of 1 and the remaining metals arranged in an order of ascending resistance:

Silver	1.0
Copper	1.08
Gold	1.4
Aluminum	1.8
Platinum	7.0
Lead	13.5

The resistance in an electrical circuit is expressed by the symbol R. Manufactured circuit parts containing definite amounts of resistance are called resistors (see Chapter 4). *Resistance* (R) is measured in *ohms*. One ohm is the resistance of a circuit element, or circuit, that permits a steady current of 1 A (1 C per second) to flow when a steady emf of 1 V is applied to the circuit.

CONDUCTANCE

Electricity is a study that is frequently explained in terms of opposites. The term that is exactly the opposite of resistance is conductance. Conductance (G) is the ability of a material to pass electrons. The unit of conductance is the siemens (S). The relationship that exists between resistance and conductance is the reciprocal. A reciprocal of a number is obtained by dividing the number into 1. In terms of resistance and conductance:

$$R = \frac{1}{G}$$

$$G = \frac{1}{R}$$

If the resistance of a material is known, dividing its value into 1 will give its conductance. Similarly, if the conductance is known, dividing its value into 1 will give its resistance.

Table 2-1 gives the complete electrical characteristics of items covered in this chapter.

Table 2-1: Electrical Characteristics

Characteristic	Symbol	Unit	Description
Charge	Q or q^1	Coulomb (C)	Quantity of electrons or protons: $Q = I \times t$
Current	I or i^1	Ampere (A)	Charge in motion: $I = Q/t$
Voltage	V or $v^{1,2}$	Volt (V)	Potential difference between two unlike charges; makes charge move to produce I
Resistance	R or r^3	Ohm (Ω)	Opposition that reduces amount of current
Conductance	G or $g^{2,3}$	Siemens (S)	Reciprocal of R, or $G = 1/R$

Note: [1] Small letter q, *i,* or *v* is used for an instantaneous value of a varying charge, current, or voltage.
[2] *E* or *e* is sometimes used for a generated emf, but the standard symbol is *V* or *v* for any potential difference in the international system of units (SI). Siemens is the SI standard unit for conductance, not mho.
[3] Small letter *r* or *g* is used for internal resistance or conductance of transistors and tubes.

SUMMARY

Between two bodies of matter that are charged unequally and separated by a nonconductor, there exists an electrostatic force, sometimes called static electricity. The electrostatic or dielectric field is the region between and around charged bodies in which their influence is felt. Coulomb's Law states that the force of this field is directly proportional to the product of the charges and is inversely proportional to the square of the distance between them.

$$F = k \frac{Q_1 Q_2}{d^2}$$

Dynamic electricity is electricity in motion. The movement of charged particles in a specified direction is called current (I). If a conductor is connected between two bodies with different charge levels, current will flow until both bodies have equal charges. Direct current (DC) maintains a constant magnitude and a uniform direction. Alternating current (AC) periodically switches directions of current flow. The frequency of alternations is measured in hertz (cycles per second).

The ampere is the unit measure of current and is equal to 1 coulomb per second. The coulomb is a unit of electrical charge equal to 6.28×10^{18} electrons.

The difference in potential, called voltage (V) or electromotive force (emf), that exists between two bodies, is the force that causes current to flow. There are several methods of producing this electromotive force: friction, pressure, heat, light, chemical action, magnetism, and radioactivity.

Resistance (R), measured in ohms, is the opposition to current flow. One ohm is the amount of resistance that will allow 1 ampere of current to flow in a circuit with 1 volt of electromotive force applied.

Conductance (*G*), measured in siemens (*S*), is the ability of a material to pass electrons. Conductance is the reciprocal of resistance.

CHECK YOURSELF (Answers at back of book)

Questions

Determine whether the following statements are true or false.

1. A static charge can be developed by rubbing two conductive pieces of matter together.
2. A voltage can be produced by moving a conductor through a magnetic field.
3. One ampere is equivalent to 6.28×10^{18} electrons.
4. When a body has a positive charge of 1 coulomb, it has an excess of 6.28×10^{18} electrons.
5. In most conductors, such as copper and iron, resistance increases as temperature increases.
6. The flow of current through a conductor always takes place between points of different potential.
7. If an electrostatic charge is great enough, current will flow between two charged bodies even when they are not in contact.
8. A dielectric field cannot exist in a vacuum.
9. Copper offers more resistance to current than silver does.
10. Current is directly proportional to the applied voltage.
11. The coulomb is a unit of charge.
12. Coulomb's Law determines the amount of force which exists between two charged bodies.
13. It is necessary to have a negatively charged body and a positively charged body to produce a potential difference.
14. An insulator is a material that readily allows the flow of charges.
15. Dynamic electricity is electricity in motion.

Problems

1. Two metal spheres, separated by a distance of 10cm, are both charged to 2.5×10^{-8} coulomb. Calculate the force in newtons between them.
2. If 400 coulombs are moved through a circuit in 25 seconds, what is the average current flow?
3. If a carbon resistor has a value of resistance equal to 4Ω, what is its value of conductance?
4. How much is the charge in coulombs for a surplus of 25×10^{18} electrons?
5. With the same voltage applied, which resistance will allow more current to flow, 5Ω or 500Ω?

CHAPTER 3

Circuit Laws

Over the years various people have contributed laws and rules that help us to understand electrical phenomena. Some of these people were Volta, Ampere, and Coulomb. Others, such as Ohm, Watt, Joules, and Kirchhoff, have developed laws that show how the various concepts of voltage, current, resistance, and power are interrelated. The most significant of these rules are called Ohm's Law and Joule's Law and are discussed in this chapter. Kirchhoff's laws are covered fully in Chapters 6 and 7.

Persons involved in electrical or electronic work need rules with which to work. Each electronic circuit works in a certain way. Knowledge of the way the circuit works and how to identify each of the concepts in the circuit is required if one is to be successful. There are very few "new" electronic circuits. Most of the "new" circuits are really modifications of another circuit. The rules identified by Ohm, Watt, Joules, and Kirchhoff apply to every electrical and electronic circuit. Quantities of voltage, current, resistance, and power are identified by applying the basic circuit laws developed by these men.

Once the electronic circuit is broken down into its component parts, it is better understood. The rules about to be presented are really very simple. They should be learned well by everyone planning to spend any amount of time working with electricity and electronics. Most circuit analysis is based upon an understanding of these laws. It is paramount that they be clearly understood. Not only should these laws be understood, but you must be able to apply the laws in practical work.

SIMPLE ELECTRICAL CIRCUIT

Before taking a careful look at two of the laws that govern electricity and electronics, certain principles must be reviewed. As was mentioned in Chapter 2, whenever two unequal charges are connected by a conductor, a complete pathway for current flow exists. Current will flow from the negative to the positive charge.

An electrical circuit is a completed conducting pathway, consisting not only of the conductor, but including the path

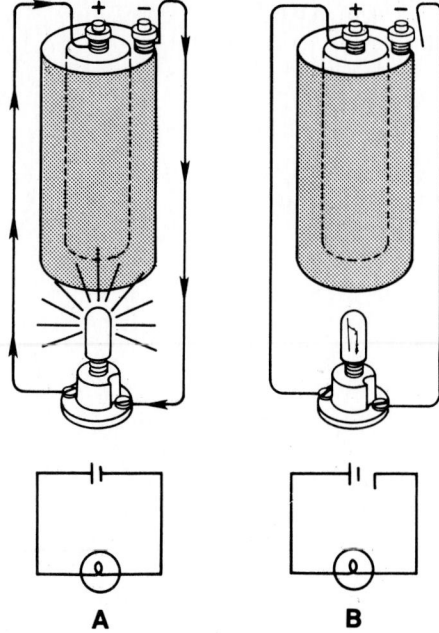

Fig. 3-1: (A) Simple electric circuit (closed); (B) simple electric circuit (open).

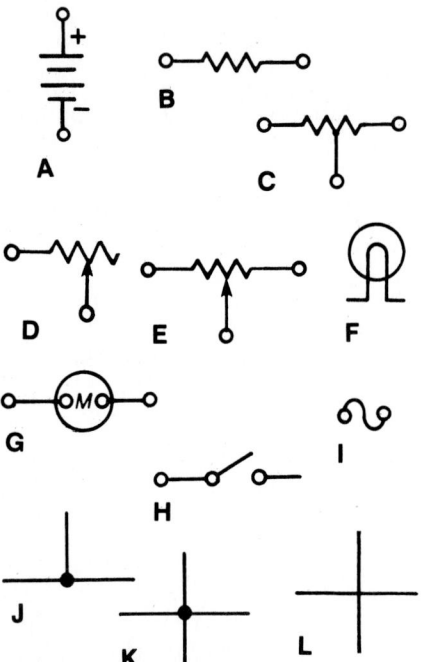

Fig. 3-2: Schematic circuit symbols: (A) battery, (B) resistor, (C) tapped resistor, (C) rheostat, (E) potentiometer, (F) lamp, (G) meter, (H) switch, (I) fuse, (J) junction, (K) junction, and (L) cross-over.

through the voltage source. Current flows from the positive terminal through the source, emerging at the negative terminal. As an example, a lamp connected by conductors across a dry cell forms a simple electrical circuit (Fig. 3-1).

Current flows from the negative (-) terminal of the battery through the lamp to the positive (+) battery terminal and continues by going through the battery from the positive (+) terminal to the negative (-) terminal. As long as this pathway is unbroken, it is a closed circuit and current will flow. However, if the path is broken at *any* point, it is an open circuit and no current flows (Fig. 3-1B).

Current flow in the external circuit is the movement of electrons (Fig. 3-1A) in the direction indicated by the arrows (from the negative terminal through the lamp to the positive terminal). Current flow in the internal battery circuit is the simultaneous movement in opposite directions of positive ions toward the positive terminal of the battery and negative ions toward the negative terminal. Current flow through a battery is discussed in more detail in Chapter 4.

Schematic Representation

A *schematic* is a diagram in which symbols are used for the various components instead of pictures. These symbols are used in an effort to make the diagrams easier to draw and easier to understand. In this respect, schematic symbols aid you in the same way that shorthand aids the stenographer. A complete listing of all schematic electronic symbols is given in Appendix F. The symbols used in this chapter can be found illustrated in Fig. 3-2.

A schematic diagram of a simple circuit is shown in Fig. 3-3. The battery source voltage is designated by the letter symbols V_s; the light bulb in the circuit is labeled R_1. Since, in reality, the light bulb element is nothing more than a wirewound resistor, the conventional resistor symbol is used for the bulb in this discussion. It should be noted, however, that the light bulb has its own specific schematic symbol and is not normally drawn as a resistor.

In studies of electricity and electronics many circuits are analyzed which consist mainly of specially designed resistive components. As previously stated, these components are called *resistors*. Throughout the remaining analysis of the basic circuit, the resistive component will be a physical resistor. However, the resistive component could be any one of several electrical devices.

A closed loop of wire (conductor) is not necessarily a circuit. A source of voltage must be included to make it an electrical circuit. In any electrical circuit where electrons move around a closed loop, current, voltage, and resistance are present. The physical pathway for current flow is actually the circuit. By knowing any two of the three quantities, such as voltage and current, the third (resistance) may be determined. This is done mathematically by the use of *Ohm's Law*.

OHM'S LAW

In the early part of the 19th century Georg Simon Ohm, a German scientist, proved by experiment that a precise relationship exists between current, voltage, and resistance. This relationship is called Ohm's Law and is stated as follows:

The current in a circuit is *directly* proportional to the applied voltage and *inversely* proportional to the circuit resistance. Ohm's Law may be expressed as an equation:

$$I = \frac{V}{R}$$

Where:
I = current in amperes
V = voltage in volts
R = resistance in ohms

If any two of the quantities in this equation are known, the third may be easily found. For example, Fig. 3-4 shows a circuit containing a resistance of 1.5Ω and a source voltage of 1.5V. How much current flows in the circuit?

Given:
$V = 1.5V$
$R = 1.5\Omega$
$I = ?$

Solution:
$$I = \frac{V}{R}$$

$$I = \frac{1.5V}{1.5\Omega}$$

$$I = 1A$$

To observe the effect of source voltage on circuit current, the above problem will be solved again using double the previous source voltage.

Given:
$V = 3V$
$R = 1.5\Omega$
$I = ?$

Solution:
$$I = \frac{V}{R}$$

$$I = \frac{3V}{1.5\Omega}$$

$$I = 2A$$

Notice that as the source voltage doubles, the circuit current also doubles. Circuit current is directly proportional to applied voltage and will change by the same factor that the voltage changes.

To verify the statement that current is inversely proportional to resistance, assume the resistor in Fig. 3-4 to have a value of 3Ω.

Given:
$V = 1.5V$
$R = 3\Omega$
$I = ?$

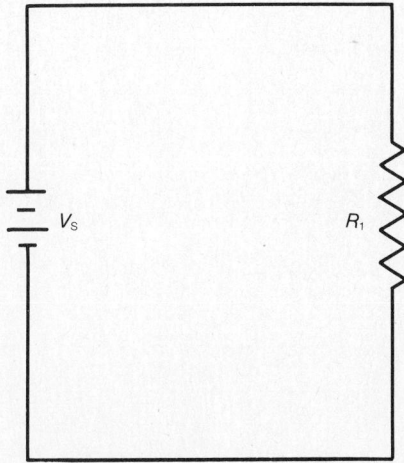

Fig. 3-3: Schematic diagram of a simple circuit.

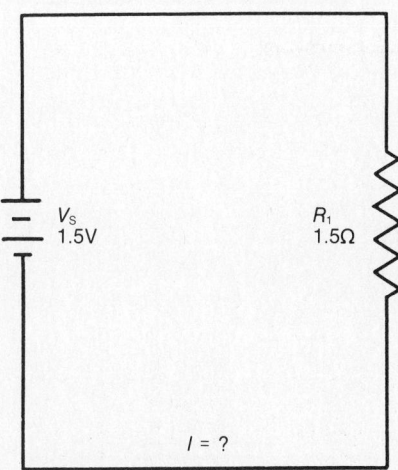

Fig. 3-4: Determining current in a simple circuit.

27

Solution:

$$I = \frac{V}{R}$$

$$I = \frac{1.5\text{V}}{3\Omega}$$

$$I = 0.5\text{A}$$

Comparing this current of 0.5A for the 3Ω resistor to the 1A of current obtained with the 1.5Ω resistor shows that doubling the resistance will reduce the current to one-half the original value. Circuit current is inversely proportional to the circuit resistance.

In many circuit applications current is known and either the voltage or the resistance will be the unknown quantity. To solve a problem in which current and resistance are known, the basic formula for Ohm's Law must be transposed to solve for V as follows:

Basic equation:

$$I = \frac{V}{R}$$

Multiply both sides of the equation by R.

$$IR = \frac{V}{R} R$$

$$IR = V$$

$$V = IR$$

To transpose the basic formula when resistance is unknown:

Basic equation:

$$I = \frac{V}{R}$$

Multiply both sides of the equation by R.

$$IR = \frac{V}{R} R$$

$$IR = V$$

Divide both sides of the equation by I.

$$\frac{IR}{I} = \frac{V}{I}$$

$$R = \frac{V}{I}$$

Example: What voltage is required to properly light a lamp having a resistance of 10Ω and a current rating of 1A?

First draw a circuit like Fig. 3-5, including all the given information.

Given:
$R = 10\Omega$
$I = 1\text{A}$
$V = ?$

Solution:
$V = IR$
$V = 1\text{A} \times 10\Omega$
$V = 10\text{V}$

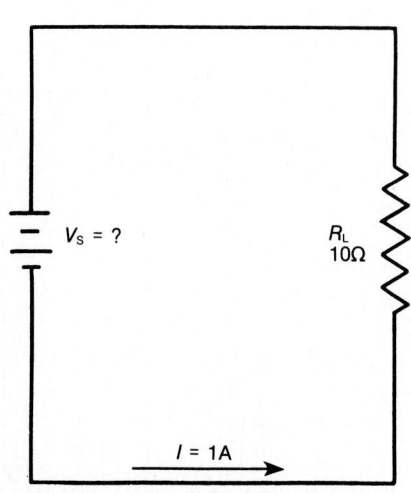

$V_s = ?$ R_L 10Ω $I = 1\text{A}$

Fig. 3-5: Determining voltage in a simple circuit.

Example: When a 10V source is connected to a circuit, the circuit draws 5A of current from the source. How much resistance is contained in the circuit?

Given:
$$V = 10V$$
$$I = 5A$$
$$R = ?$$

Draw and label a circuit like Fig. 3-6.

Solution:
$$R = \frac{V}{I}$$
$$R = \frac{10V}{5A}$$
$$R = 2\Omega$$

Although the three equations representing Ohm's Law are fairly simple, they are perhaps the most important of all electrical and electronic equations. These three equations and the law they represent must be thoroughly understood before continuing on to more advanced theory.

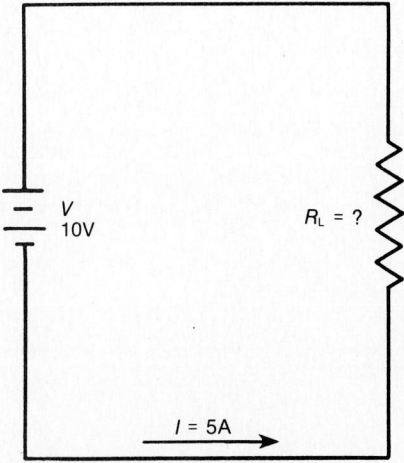

Fig. 3-6: Determining resistance in a simple circuit.

Graphical Analysis

One of the most valuable methods of inquiry available to you is that of graphical analysis. No other method provides a more convenient or more rapid way to observe the characteristics of an electrical device.

The first step in constructing a graph consists of obtaining a table of data from which the graph will evolve. The information in the table can be obtained experimentally by taking laboratory measurements on the device under examination or can be obtained theoretically through a series of computations. The latter method will be used here.

Let us assume that the characteristics of the circuit shown in Fig. 3-7 are to be investigated using Ohm's Law and graphical methods. Since there are three variables (V, I, and R) under consideration, there are three unique graphs that may be constructed.

In constructing any graph of electrical quantities, it is standard practice to vary one quantity in a specified way and note the changes which occur in a second quantity. The quantity which is intentionally varied is called the independent variable and is plotted on the *X-axis*. The second quantity which changes as a result of changes in the first quantity is called the dependent variable and is plotted on the *Y-axis*. Any other quantities involved are held constant.

In the circuit of Fig. 3-7, the resistance will remain fixed (constant) and the voltage (independent variable) will be varied. The resulting changes in current (dependent variable) will then be graphed.

To aid in compiling the data, a table of values is completed as shown in Fig. 3-8. This table shows R to be held constant at 10Ω as V is varied from 0 to 20V in 5V steps. Through the use of

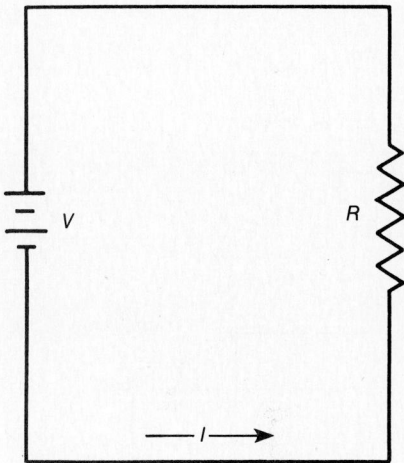

Fig. 3-7: The three variables in a simple circuit.

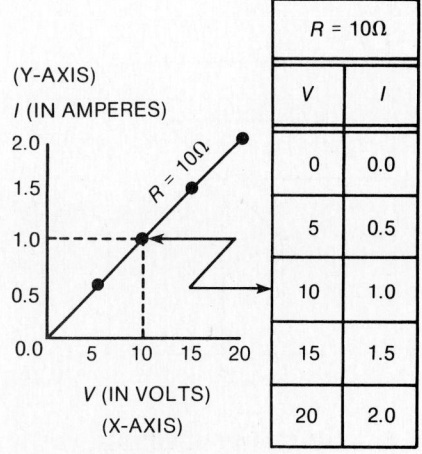

R = 10Ω	
V	I
0	0.0
5	0.5
10	1.0
15	1.5
20	2.0

Fig. 3-8: Volt-ampere characteristic.

Ohm's Law, the value of current in column two of the table can be calculated for each value of voltage in column one. When the table is complete, the information it contains can be used to construct the graph in Fig. 3-8. For example, when the voltage applied to the 10Ω resistor is 10V, the current is 1A. These values of current and voltage determine a point on the graph. When all the points have been plotted, a smooth curve is drawn through the points. This curve is called the volt-ampere characteristic for the 10Ω resistor.

Through the use of this curve, the value of current through the resistor can be quickly determined for any value of voltage between 0 and 20V.

A very important characteristic of a fixed resistor is illustrated by the graph in Fig. 3-8. Since the volt-ampere characteristic curve is a straight line, it shows that equal changes of voltage across a resistor produce equal changes in current through the resistor. Because of this straight line characteristic, the fixed resistor is called a *linear device*. Actually, this graph illustrates an important characteristic of the basic law; namely, the current varies directly with the applied voltage if the resistance is constant.

If the voltage across a load is maintained at a constant value, the current through the load will depend solely upon the effective resistance of the load. For example, with a constant voltage of 12V and a resistance of 12Ω the current will be $\frac{12V}{12\Omega}$, or 1A. If the resistance is halved, the current will be doubled; if the resistance is doubled, the current will be halved. In other words, the current will vary inversely with the resistance.

If the resistance of the load is reduced in steps of 2Ω starting at 12Ω and continuing to 2Ω, the current through the load becomes $\frac{12V}{10\Omega}$ = 1.2A; $\frac{12V}{8\Omega}$ = 1.5A; $\frac{12V}{6\Omega}$ = 2A; and so forth. The relationship between current and resistance in this example is expressed as a graph (Fig. 3-9), whose equation is $I = \frac{12V}{R}$. The numerator of the fraction represents a constant value of 12V in this example. As R approaches a small value, the current approaches a very large value. The example illustrates a second equally important relation in Ohm's Law; namely, the current varies inversely with the resistance.

If the current through the load is maintained constant at 5A, the voltage across the load will depend upon the resistance of the load and will vary directly with it. The relationship between voltage and resistance is shown in the graph of Fig. 3-10. Values of resistance are plotted horizontally along the X-axis to the right of the origin, and corresponding values of voltage are plotted vertically along the Y-axis above the origin. The graph is a straight line having the equation $V = 5R$. The coefficient 5 represents the assumed current of 5A which is constant in this

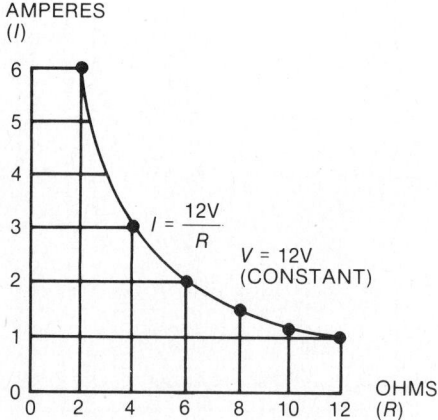

Fig. 3-9: Relation between current and resistance.

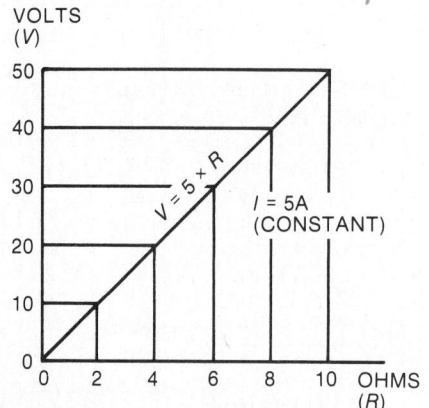

Fig. 3-10: Graph of voltage vs. resistance with constant current.

30

example. Thus, a third important relationship is illustrated; namely, that the voltage across a device varies directly with the effective resistance of the device provided the current through the device is maintained constant.

Applying Ohm's Law

The basic Ohm's Law formula may be transposed to solve for the resistance if the current and voltage are known or to solve for the voltage if the current and resistance are known; thus, $R = \dfrac{V}{I}$ and $V = IR$. For example, if the voltage across a device is 50V and the current through it is 2A, the resistance of the device will be $\dfrac{50}{2}$, or 25Ω. Also, if the current through a wire is 3A and the resistance of the wire is 0.5Ω, the voltage drop across the wire will be 3 × 0.5, or 1.5V.

The basic Ohm's Law formula and its transpositions may be obtained readily with the aid of Fig. 3-11. The circle containing V, I, and R is divided into two parts with V above the line and IR below it. To determine the unknown quantity, first cover that quantity with a finger. The location of the remaining uncovered letters in the circle will indicate the mathematical operation to be performed. For example, to find I, cover I with a finger. The uncovered letters indicate that V is to be divided by R, or $I = \dfrac{V}{R}$. To find V, cover V. The result indicates that I is to be multiplied by R, or $V = IR$. To find R, cover R. The result indicates that V is to be divided by I, or $R = \dfrac{V}{I}$.

CONDUCTANCE

Ohm's Law can also be expressed in terms of conductance, G, in which G is the reciprocal of R:

$$V = \frac{I}{G}$$

$$I = VG$$

and

$$G = \frac{I}{V}$$

For example, if the current in a 120V circuit is 6A, the conductance of the circuit is

$$G = \frac{I}{V} = \frac{6\text{A}}{120\Omega} = 0.05S$$

The siemens (S) is the unit formed by taking the reciprocal of 1Ω.

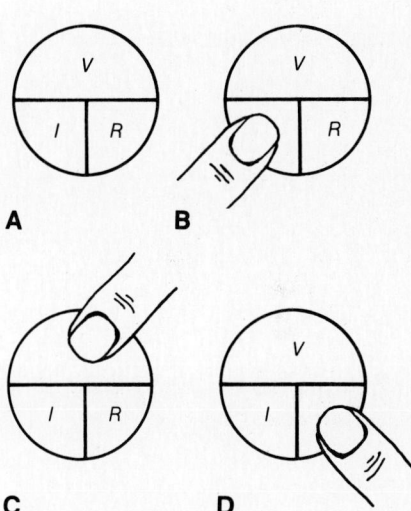

A   B

C   D

Fig. 3-11: Memory aid for learning Ohm's Law. Part A of this figure shows a circle with the three symbols arranged in separate spaces. If any one of the symbols is covered, the arrangement of the other two symbols forms the right-hand side of the formula for determining the value of the covered symbol. Thus, (B) if a finger is placed over the I, V/R remains and indicates that $I = V/R$; (C) if V is covered, IR remains and indicates that $V = IR$; and (D) if R is covered, V/I remains and indicates that $R = V/I$.

31

When the voltage is fixed, a high resistance indicates a low current, whereas when the resistance is small, the current is large. The converse is true of conductance. A low conductance value indicates a small current. A high conductance indicates a large current.

An open circuit is the electrical condition in which no current flows; that is, the numerical value of current is zero (0). The current in a circuit is given by Ohm's Law:

$$I = \frac{V}{R} = VG$$

When V is fixed, the current becomes smaller and smaller as R becomes larger and larger. The limiting condition for R as R becomes larger and larger is infinity. Thus, we say that an open circuit has a resistance of infinite (∞) ohms. The conductance of the open circuit is $0S$.

As R in Ohm's Law becomes smaller and smaller, the current becomes larger and larger. Theoretically, the current becomes infinite as R approaches zero. This is a short circuit. The corresponding value of G is infinite under short-circuit conditions. Actually, the current in a short circuit is limited by the resistance of the conductors and connection contacts in the circuit and by the limitations of the power supply itself.

JOULE'S LAW OF POWER AND ENERGY

Joule's Law of electrical power and energy—based on James Watt's Laws for mechanical power—is important to the electronics field. Power, as defined by Watt, is the rate of doing work per unit of time. Work results from a force acting on a mass over a distance. The operation of electrical circuits involves a potential acting on a mass (electrons) over a distance. The amount of time required to perform a given amount of work will determine the power expended. Expressed as an equation, the relationship between power, work, and time is:

$$P = \frac{W}{t}$$

where:
P = power in watts
W = work in joules
t = time in seconds

Since energy is the capacity to do work, power can also be defined as the time rate of developing or expending energy. In every electrical circuit, electrical energy is transformed into heat energy. Following the law of conservation of energy, the heat energy will be equal in value to the electrical energy causing it. Therefore, by measuring the amount of heat energy given off by an electrical circuit in a given amount of time, the amount of electrical power consumed in the circuit can be determined.

An experiment measuring the heat given off by an electric circuit was performed by an English physicist, James Joule, in 1843. He experimentally proved that the amount of heat produced by an electrical circuit was dependent upon current and resistance. This proportional relationship is known as Joule's Law and is stated as follows: The amount of heat produced by a circuit element is directly proportional to resistance, the square of the current, and time. Expressed as an equation:

$$\text{Heat} = I^2Rt$$

The amount of heat energy produced is equal to the amount of electrical energy consumed, or the amount of work performed.

Therefore: $$\text{Work} = I^2Rt$$

Since:
$$P = \frac{W}{t}$$

$$P = \frac{I^2R\cancel{t}}{\cancel{t}}$$

$$P = I^2R$$

By substituting Ohm's Law values into the power formula developed from Joule's Law, other equations can be derived that are useful in determining power.

$$P = I^2R$$

Since:
$$I = \frac{V}{R} \qquad P = (\frac{V}{R})^2R$$

$$P = \frac{V^2}{R^2} \times R$$

$$P = \frac{V^2}{R}$$

The resultant equation is useful when the resistance and voltage are known.

The power formula can also be expressed as an equation in terms of current and voltage.

$$P = I^2R$$

Since:
$$R = \frac{V}{I} \qquad P = I^2(\frac{V}{I})$$

$$P = I^2\frac{V}{I}$$

$$P = IV$$

The unit of measure for electrical power is the *watt*. With the watt unit for power, 1 watt used during 1 second equals the work of 1 joule. Or 1 watt is 1 joule per second ($1W = J/s$). The joule is a basic practical unit of work or energy. One joule is the energy required to transport 1 coulomb between two points having a potential difference of 1 volt.

In each of the three derived equations for power, power will be in watts when: V is in volts, I is in amperes, and R is in ohms. As an example: When 1 volt of potential difference produces a current of 1 ampere, the power expended is 1 watt. The watt

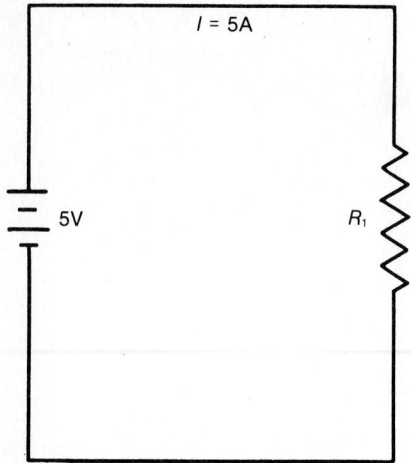

Fig. 3-12: Determining power in a simple circuit.

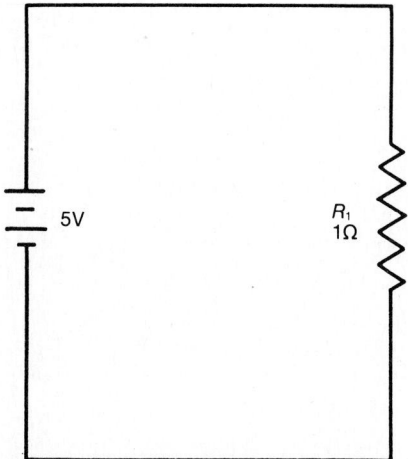

Fig. 3-13: Computing power in a simple circuit.

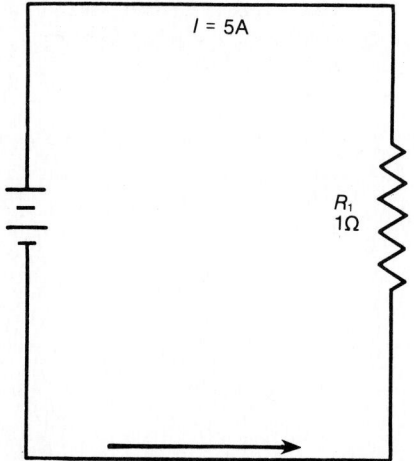

Fig. 3-14: Solving for power in a simple circuit.

represents the rate at any given instant at which work is being done in moving electrons through a circuit.

Example: What is the power expended in a circuit when a voltage of 5V causes a current of 5A as indicated in Fig. 3-12?

Given:

$$I = 5A$$
$$V = 5V$$
$$P = ?$$

Solution:

$$P = IV$$
$$P = 5A \times 5V$$
$$P = 25W$$

If voltage and resistance are known, as in Fig. 3-13, the formula containing voltage and resistance is best suited to compute power.

Given:

$$V = 5V$$
$$R = 1\Omega$$
$$P = ?$$

Solution:

$$P = \frac{V^2}{R}$$

$$P = \frac{25V}{1\Omega}$$

$$P = 25W$$

Had the current and resistance been known, as in Fig. 3-14, the formula, $P = I^2R$, would apply.

Given:

$$I = 5A$$
$$R = 1\,\Omega$$
$$P = ?$$

Solution:

$$P = I^2R$$
$$P = (5A)^2 1\Omega$$
$$P = 25W$$

The previous examples proved that any form of the power formula can be used to find power in a circuit. Likewise, if the power dissipated in a simple circuit is known and the value of any one of the other circuit quantities is known (V, I, or R), the value of the remaining quantities can be found.

Example: The power dissipated by the 1Ω resistor in Fig. 3-15 is 25W. What is the value of current and voltage in the circuit?

Given:

$$P = 25W$$
$$R = 1\Omega$$
$$V = ?$$

Solve for V:

$$P = \frac{V^2}{R}$$

$$V^2 = PR$$

$$V = \sqrt{PR}$$

$$V = \sqrt{25 \times 1}$$

$$V = \sqrt{25} \text{ or } 5V$$

Solve for I:

$$P = I^2R$$

Divide by R:

$$\frac{P}{R} = \frac{I^2 R}{R}$$

$$\frac{P}{R} = I^2$$

Take the square root of both sides:

$$\sqrt{\frac{P}{R}} = I$$

$$I = \sqrt{\frac{P}{R}}$$

$$I = \sqrt{\frac{25}{1}}$$

$$I = \sqrt{25}$$

$$I = 5A$$

With a knowledge of transposition of equations, the solution of a simple circuit can be found when any two values are known. Circuits having known values of power and current or power and resistance can be solved similar to the circuit of Fig. 3-15 by the correct usage of Joule's Law and Ohm's Law. Always begin a circuit analysis by choosing a formula containing two known values and an unknown that you wish to find. In fact, all the interrelationships between these two laws are shown in Fig. 3-16. Note that this wheel is divided into four sections. Each section shows three formulas. Each formula may be used to solve for the quantity shown near the center of the circle. To use this, find the section that has the formula for the two known factors. Use this formula to solve for the unknown.

Power Rating

Electrical lamps, soldering irons, and motors are examples of electrical devices that are rated in watts. The wattage rating of a device indicates the rate at which the device converts electrical energy into another form of energy, such as light, heat, or motion.

For example, a 100W lamp will produce a brighter light than a 75W lamp, because it converts more electrical energy into light.

Electric soldering irons are of various wattage ratings, with the high-wattage irons changing more electrical energy to heat than those of low-wattage ratings.

If the normal wattage rating is exceeded, the equipment or device will overheat and probably be damaged. For example, if a lamp is rated 100W at 110V and is connected to a source of 220V, the current through the lamp will double. This will cause the lamp to use four times the wattage for which it is rated, and it will burn out quickly.

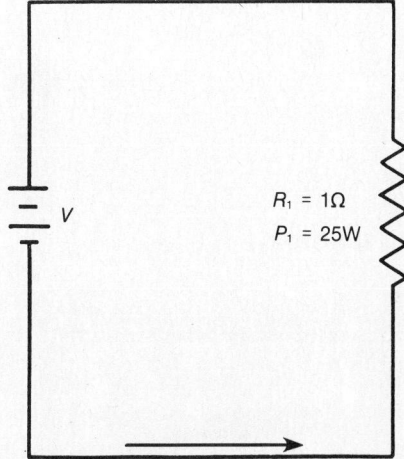

Fig. 3-15: Solving for current and voltage in a simple circuit.

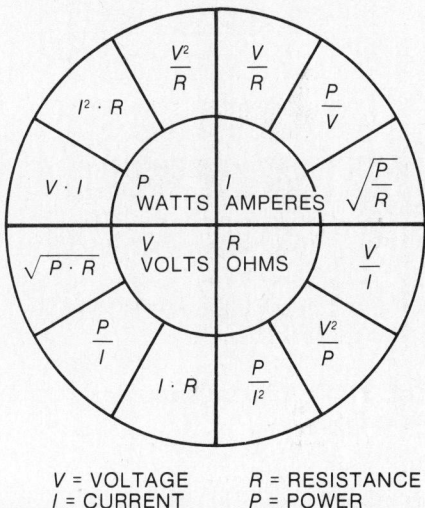

V = VOLTAGE   R = RESISTANCE
I = CURRENT   P = POWER

Fig. 3-16: Wheel used to show the interaction between Ohm's and Joule's Laws.

Fig. 3-17: Typical engineer's calculator.

UNITS OF MEASUREMENT

As you have probably noticed, base units of electricity and electronics are often expressed in very large or very small quantities; for example, 1kWh, the base unit used by electric companies to measure electrical use, is equal to 3,600,000J. In contrast, the current passing through a solid-state device may be less than 0.0000001A. Writing out such long numerical expressions is slow and cumbersome. There is also a great risk of error. Large numerical expressions are easier with which to work if they are condensed into a simpler system of notation. Two such systems are often used in electronic work: scientific notation and multiple and submultiple units.

Scientific Notation: Working With Powers of 10

Scientific notation is commonly used in electronics and other sciences to simplify the expression of large numbers. Scientific notation uses the powers of 10 as its base system. The powers of 10 refer to numbers that are exponents of the number 10. The exponent (power) indicates the number of times the digit 1 must be multiplied or divided by 10. For example, 10^2 means that 1 will be multiplied by 10 two times ($1 \times 10 \times 10$) and is equivalent to 100; 10^{-2} means that 1 will be divided by 10 twice ($1 \div 10 \div 10$) and is equivalent to 0.01.

Pocket calculators use scientific notation to display larger numbers. For example, the calculator in Fig. 3-17 reads 1.2×10^3. This represents the number $1.2 \times 10 \times 10 \times 10 = 1,200$. Notice that the exponent of "3" indicates that the decimal point is moved three places to the right to determine the actual number represented. Table 3-1 lists the commonly used powers of 10 and their numerical equivalents.

Converting to Powers of 10. Converting a number to a power of 10 is simply a matter of moving the decimal point and multiplying by some power of 10. For example, 1,200 can be written as a coefficient times a power of 10 by moving the decimal place three places to the left and multiplying by 10 to the third power. The expression is written in this way:

$$1,200 = 1.2 \times 10^3$$

When converting to scientific notation, always leave at least one number on the left of the decimal.

When converting a number smaller than 1, move the decimal point to the right of the first digit larger than 0 (zero). Then, multiply by 10 with a negative exponent equal to the number of places you moved the decimal point. For example, 0.000042 is equal to 4.2×10^{-5}. As with positive exponents, leave one digit on the left of the decimal point.

Expressing numbers in powers of 10 not only simplifies numerical expressions, but it also simplifies arithmetic involving large numbers.

Table 3-1: Powers of 10 and Numerical Equivalents

Power of 10	Numerical Equivalent
10^{12}	1,000,000,000,000
10^{11}	100,000,000,000
10^{10}	10,000,000,000
10^{9}	1,000,000,000
10^{8}	100,000,000
10^{7}	10,000,000
10^{6}	1,000,000
10^{5}	100,000
10^{4}	10,000
10^{3}	1,000
10^{2}	100
10^{1}	10
10^{0}	1
10^{-1}	0.1
10^{-2}	0.01
10^{-3}	0.001
10^{-4}	0.0001
10^{-5}	0.00001
10^{-6}	0.000001
10^{-7}	0.0000001
10^{-8}	0.00000001
10^{-9}	0.000000001
10^{-10}	0.0000000001
10^{-11}	0.00000000001
10^{-12}	0.000000000001

Addition and Subtraction. To add or subtract numbers expressed as powers of 10 and having equal exponents, simply add the coefficients. For example,

$$(3.4 \times 10^5) + (2.5 \times 10^6) = (0.34 \times 10^6) + (2.5 \times 10^6) = 2.84 \times 10^6$$

Multiplication. To multiply two numbers expressed in powers of 10, multiply the coefficients and add the powers. The result will be a coefficient multiplied by a new power of 10.

For example:

$$(3.3 \times 10^7) \times (1.8 \times 10^2) =$$

$$(3.3 \times 1.8) \times (10^7 \times 10^2) =$$

$$5.94 \times 10^{7+2} =$$

$$5.94 \times 10^9$$

Division. To divide numbers expressed in powers of 10, first divide the coefficients; then subtract the exponent in the dividend from the exponent in the divisor.

For example:

$$(1.4 \times 10^4) \div (2 \times 10^2) =$$

$$(1.4 \div 2) \times (10^4 \div 10^2) =$$

$$0.7 \times 10^{4-2} =$$

$$0.7 \times 10^2$$

Multiple and Submultiple Units

Large numbers may be expressed in terms even simpler than the powers of 10. For example, 1,800,000A, even when expressed as a power of 10 (1.8×10^6), is cumbersome to say: "one point eight times ten to the sixth amperes." To shorten such long expressions, prefixes are used to express units larger than (multiples) or smaller than (submultiples) a single digit.

Table 3-2 lists the prefixes and their symbols commonly used in electronics. Their numerical equivalents and equivalent powers of 10 are also given.

Multiple and submultiple units are designated by adding the appropriate prefix to the given unit of measurement. For example, when measuring electrical current in amperes, we can speak of kiloamperes (1,000 amperes) or microamperes (0.000001 amperes).

Convert numbers into multiple or submultiple units by moving the decimal point the appropriate number of places. For example, 1,500 ohms is expressed as kiloohms by moving the decimal point three places to the left and using the appropriate symbol: 1,500 ohms = 1.5kΩ. Also, 0.000035 amperes is converted to microamperes by moving the decimal point six places to the right and using the appropriate symbol: 0.000035 = 35μA.

Expressing numbers in multiple units or submultiple units makes multiplication and division as well as addition and subtraction much simpler due to the fact that you must only perform the major operation on the significant units. For example, 45kV + 20kV + 15kV = 80kV.

Table 3-2: Multiple and Submultiple Units

Prefix	Symbol	Value	Power of 10
tera	T	1,000,000,000,000	10^{12}
giga	G	1,000,000,000	10^{9}
mega	M	1,000,000	10^{6}
kilo	k	1,000	10^{3}
hecto	h	100	10^{2}
deka	da	10	10^{1}
deci	d	0.1	10^{-1}
centi	c	0.01	10^{-2}
milli	m	0.001	10^{-3}
micro	μ	0.000 001	10^{-6}
nano	n	0.000 000 001	10^{-9}
pico	p	0.000 000 000 001	10^{-12}

Significant Figures

Most scientific calculators are capable of displaying numbers having either 8 or 10 digits. However, many of the calculations that must be made in the field of electronics are based on measurements that have been made using meters having tolerances of as much as 10% to 20% and, as a result, 8 to 10 digit accuracy is unnecessary.

The accuracy of a number is represented by the significant figures contained in that number. A *significant figure* can be defined as any digit that is considered reliable as a result of measurement or mathematical computation.

The following guidelines may be used to determine the number of significant figures in a decimal number:

1. The significant figures of a number are those figures that are not zero, unless the zero is used as a place holder between two other non-zero digits; that is, the zeros in 0.024 and 0.0038 are *not* significant, while the zeros in 30.7, 208.6, and 8.90 *are* significant.

2. The only other time zero is considered to be a significant figure is when zero is to the right of the decimal point, unless it is used only as a place holder between the decimal point and the first non-zero number. Thus, each of the numbers 9.570, 0.009570, and 957.0 has four significant figures.

3. Use judgment in forming the answer to a calculation. Most of the numbers used in electronics are significant only to two or three figures.

4. Use powers of 10 or scientific notation to indicate the significant digits of very large and very small quantities. For example, writing 470,000Ω to indicate the resistance of a resistor might indicate six-figure significance, but the manufacturer's tolerance of the component is only significant to three figures. By expressing the number in scientific notation as 4.70×10^5 or as 470×10^3 in powers of 10 notation, only three significant figures are indicated.

SUMMARY

Current, voltage, and resistance are present in any electrical circuit where electrons move around a closed loop path. The relationship between these three elements is stated in Ohm's Law:

$$I = V/R$$

Since conductance is the reciprocal of resistance, Ohm's Law can also be expressed in terms of conductance:

$$I = VG$$

One joule is the energy required to transport 1 coulomb between two points having a potential difference of 1 volt. One joule per second equals 1 watt, the unit of measure for electrical power.

Power, mathematically expressed as:

$$P = W/t,$$

is the rate at which work is done. By substituting Joule's Law ($W = I^2Rt$) into the power equation, the following equation can be developed:

$$P = I^2R$$

Other valuable equations can be derived by substituting Ohm's Law into this power formula. Thus, any value ($I, R, V,$ or P) can be found in a simple circuit if two values are known.

Graphical analysis, which was used in this chapter to show equal changes in voltage across a fixed resistor result in equal changes in current, is a convenient method for showing the relationship of electrical quantities in any circuit.

Scientific notation and multiple/submultiple units are two systems of notation used in electronics work to simplify calculations and to reduce error.

A schematic is a diagram in which electronic symbols are used in an effort to make circuit understanding easier.

CHECK YOURSELF (Answers at back of book)

Questions

Fill in the blanks with the appropriate word or words.

1. Electric power is measured in _____ .
2. The prefix _____ is equivalent to 10^3.
3. The prefix *milli* is equivalent to _____ .
4. One _____ per second equals one watt.
5. Current is _____ proportional to voltage and _____ proportional to resistance.

Problems

For problems 1-6, fill in missing values using the values given.

	V	I	R	P
1.	6V	?	5kΩ	?
2.	?	25mA	400Ω	?
3.	20V	125mA	?	?
4.	10V	?	?	50mW
5.	?	1μA	?	3μW
6.	?	?	120Ω	60mW

7. Use scientific notation to express the following numbers in simpler terms: 15,600; 0.00000023; 1,000,000; 0.03; 1,560; 0.002.
8. Use multiple/submultiple prefixes to express the following numbers in simpler terms: 3,000; 0.050; 13,000,000; 0.00000001; 0.0004; 27,000.

9. Solve: $\dfrac{(1.4 \times 10^2)\ (3.9 \times 10^3)\ (5.1 \times 10^1)}{(4.2 \times 10^5)}$

10. If given the value of current and resistance in a circuit, what equation should be used to calculate the voltage? the power?
11. If given the value of power and resistance in a circuit, what equation should be used to calculate the voltage? the current?

CHAPTER 4

DC Circuit Components

Circuit components are the items that comprise a circuit. In the previous two chapters, voltage, current, and resistance were thoroughly investigated. Also described was the fundamental structure of all electrical and electronic devices—the *circuit*.

In the simplest form, a DC circuit consists of three basic parts, which are: the *source,* the *load,* and the *conductor* (Fig. 4-1). The source is a device, such as a battery, which supplies electrical energy to the circuit. The electrical energy is carried from the source to the load by the conductors, which are usually in the form of wire. The load, which may be a device such as an electric light bulb or other form of resistance, is the receiver of electrical energy.

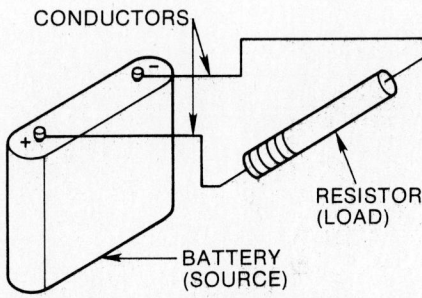

Fig. 4-1: Electric circuit.

VOLTAGE CELLS AND BATTERIES

As stated in Chapter 2, one of the seven means that creates an electromotive force (emf) is the release of energy during a chemical reaction. In this process, chemical energy is converted into electrical energy in a device known as a *voltaic* or *voltage* cell. The cell is defined as a single unit capable of producing electrical energy. Cells may be, and often are, wired together to increase the amount of electrical energy available for use. When this is done, the combination of cells is called a *battery*. Wiring combinations and resulting energy levels are discussed in Chapters 6 and 7.

Cells used to produce electrical energy are often described as either *wet* or *dry* cells. In either case they work in the same basic way, as shown in Fig. 4-2. The cell contains three different units. These units are an electrolyte material and two electrodes. The electrodes are always made of two different materials. One electrode will collect positive (+) charges from the interaction of the electrode and the electrolyte. The second electrode will collect negative (–) charges.

In the study of electricity, there are two accepted methods of presenting material related to the movement of an electrical current. One method is to consider the movement of the negative charges in the system. Electron movement in an electrical circuit is from the negative connection on the power source, through the external circuit and load, and back to the positive

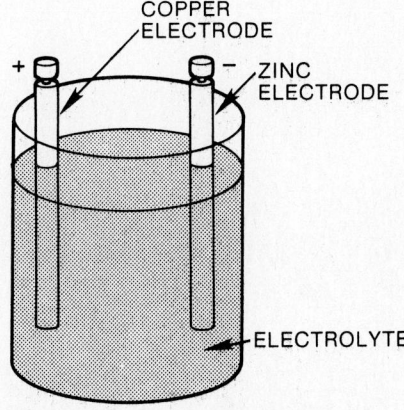

Fig. 4-2: Chemical cell consisting of two electrodes and an electrolyte material.

41

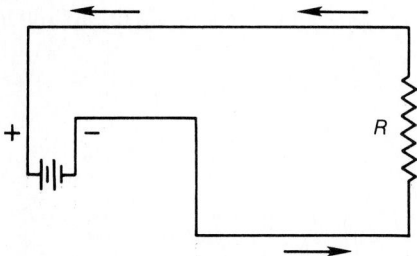

Fig. 4-3: The flow of negatively charged electrons in the circuit. This flow is from the negative terminal of the battery, through the load (R), and to the positive terminal.

connection on the power source. This is illustrated in Fig. 4-3. When these electrons are moving, they form an electric current. When they are moving in a circuit, work is being done in the circuit.

The second method of describing the flow of electrical charges, as mentioned in Chapter 3, is to look at the movement of the positive charges in the circuit. The positive charges move from the positive connection on the power source, through the external circuit, and then back to the negative terminal of the power source. This charge movement produces the same results as does the flow of electrons. It seems to be a matter of personal choice as to which of the two charge paths one wishes to follow. This book uses the movement of the negative charges as a means of explaining how electrical circuits work. This is called *electron current flow*. Electron current flow is from the high concentration of negative-charged electrons found at the minus (–) terminal of the power source, out through the external circuit and load, and then to the positive terminal (+) of the power source. The positive terminal has a shortage of electrons. It readily accepts all that it is capable of handling from the negative terminal of the power source.

Dry Cells

Several different kinds of dry cells are available at this time. The oldest of the group is described as a *carbon-zinc cell*. The carbon-zinc cell is illustrated in Fig. 4-4. It consists of a carbon

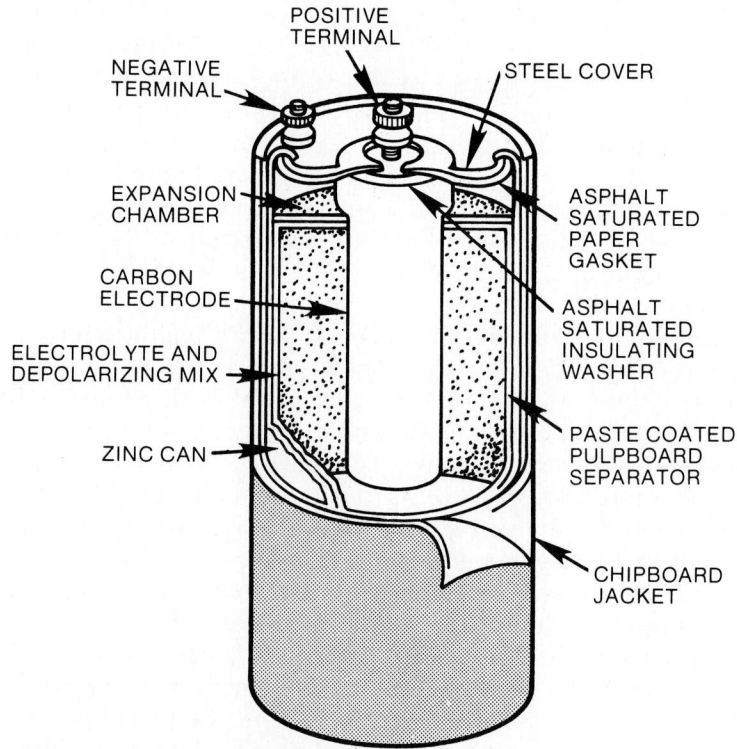

Fig. 4-4: Cutaway view of a dry cell.

rod housed in a zinc can. The rod is surrounded by an electrolyte material. This cell is capable of producing 1.55V of electricity when it is fresh. Electrical current capacity of the cell depends upon the size of the two electrodes. This cell has a shelf life of about 1 year if it has not been used. It is not considered to be rechargeble.

Another common cell is the *alkaline* type (Fig. 4-5). This cell uses electrodes of manganese dioxide and zinc. This cell is capable of handling heavier electrical currents over longer periods of time than the carbon-zinc cell. Its terminal voltage is also 1.55V. This cell is also not normally rechargeable.

A third type of cell used today is the *mercury cell* (Fig. 4-6). One of its electrodes is made of mercuric oxide and graphite. The other electrode is zinc and mercury. The terminal voltage of this cell is 1.34V. This cell is the most expensive of the three cells discussed so far. It has the ability to provide a steady voltage and current over its operational life. It also has the characteristic of holding its rated voltage until it is almost fully discharged. Again, it is not considered a rechargeable type of cell.

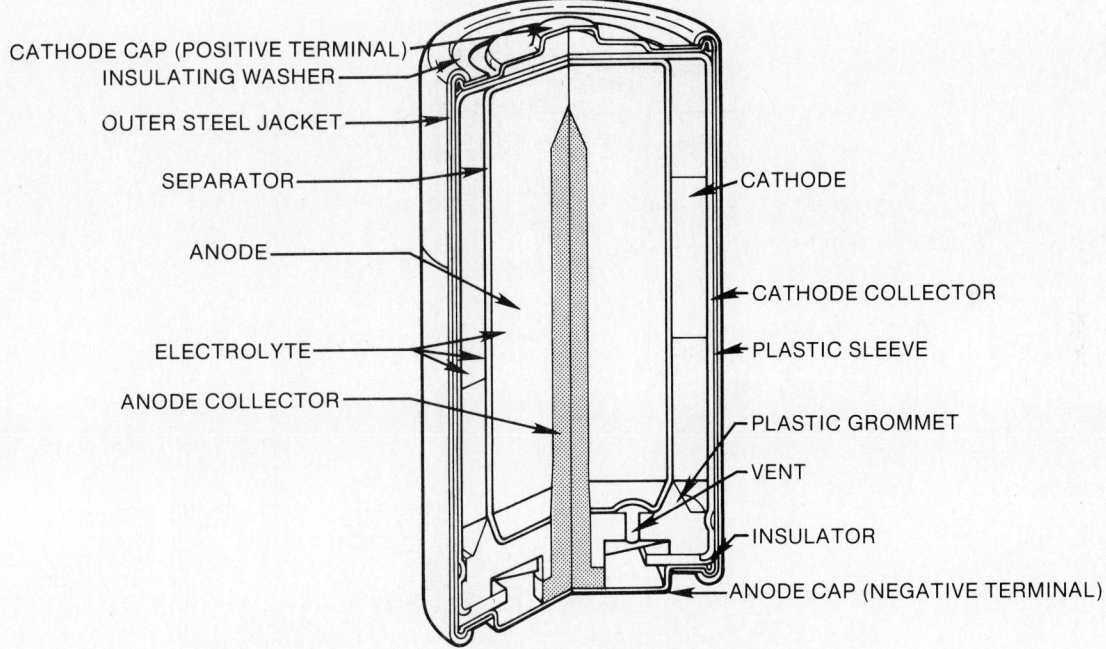

Fig. 4-5: Cutaway view of an alkaline manganese cell.

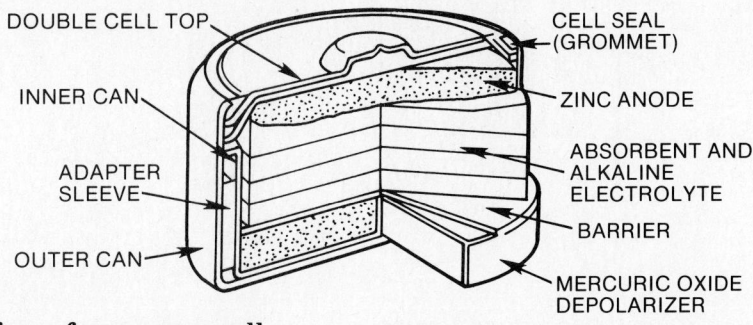

Fig. 4-6: Cutaway view of a mercury cell.

Wet Cells

A major disadvantage of most dry cells is that they are not rechargeable. Once the chemical energy is exhausted, the cell must be replaced. The discovery of the wet cell overcame this drawback. Wet cells are often called *secondary cells* but operate on the same basic chemical principles as primary cells. They differ mainly in that they may be recharged, whereas the primary cell is not normally recharged. (Some primary cells have been developed that may be recharged.) Some of the materials of a primary cell are consumed in the process of changing chemical energy to electrical energy. In the secondary cell, the materials are merely transferred from one electrode to the other as the cell discharges. Discharged secondary cells may be restored (charged) to their original state by forcing an electric current from some other source through the cell in the direction opposite to that of discharge.

The storage battery (Fig. 4-7) consists of a number of secondary cells connected in series. Properly speaking, this battery

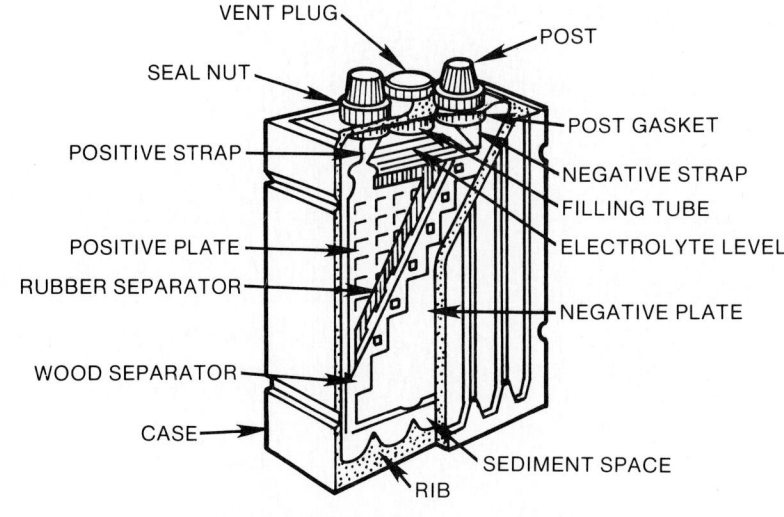

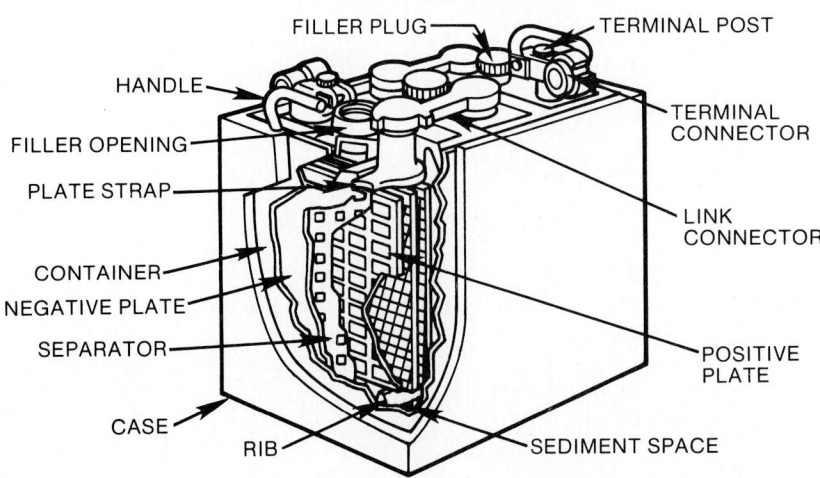

Fig. 4-7: Cutaway view of a lead-acid cell and battery.

44

does *not* store electrical energy but is a source of chemical energy which produces electrical energy. There are various types of storage cells—the lead-acid type, which has an emf of 2.1V per cell; the nickel-iron alkali type, with an emf of 1.2V per cell; the silver-zinc type, with an emf of 1.5V per cell; and the nickel-cadmium type, which has an emf of 1.25V per cell. The latter is also available in the so-called "dry" cell. These small cylindrically shaped rechargeable batteries have one electrode of nickel and another of cadmium.

While most of these cells use potassium hydroxide as an electrolyte, the cells are sealed, so they do not ordinarily spill or leak the electrolyte. However, they are usually equipped with a vent to release the pressure in case the cell overheats (Fig. 4-8). This can happen if the battery is improperly charged at an excessively high rate or if a defective battery—one that is internally shorted—is placed on charge.

One additional type of cell is the gelled electrolyte cell. These cells are rechargeable and have the advantage of being lower in cost than the nickel-cadmium battery. It has another advan-

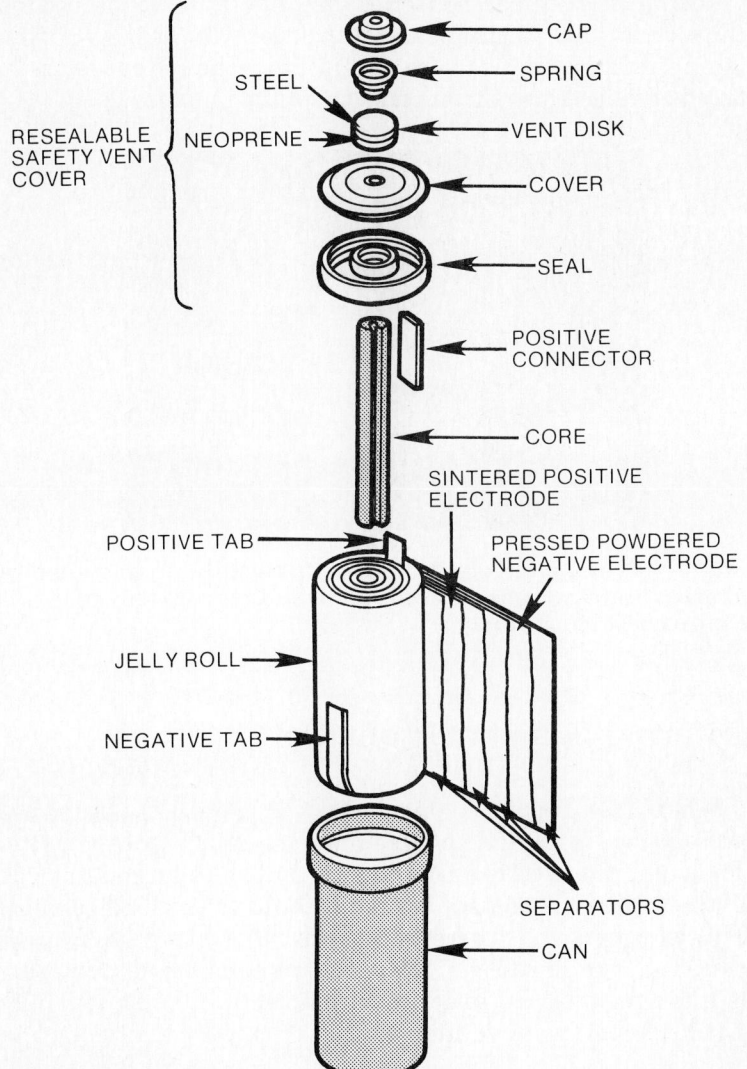

Fig. 4-8: Cutaway view of nickel-cadmium (ni-cad) "dry" cell.

tage in that its cells cannot be permanently reversed in polarity, as is the case with a nickel-cadmium battery. The disadvantage of this type of cell is its size and weight. If one required a lightweight, longer-life battery with a high discharge rate and rapid recharging characteristic, then the nickel-cadmium battery should be considered. The gelled electrolyte cell battery is illustrated in Fig. 4-9.

Battery life is a consideration in the selection of any battery. Battery life is directly related to the amount of electron current and the length of time that the maximum current will flow from the battery to the load. Batteries are rated in units of the ampere-hour (Ah). This figure tells the user the amount of current available, usually on the basis of 1 hour of activity. For example, a 100Ah rating will provide 100A of current for a period of 1 hour. This same battery will provide 10A of current for 10 hours or any other combination that will produce the 100Ah figure.

A second factor to consider regarding battery life for a rechargeable battery is the number of times the battery is capable of being recharged. Batteries are not able to be operated through a charge and discharge cycle forever. The law of diminishing returns applies here. Eventually, the output of the battery is reduced and the battery has to be replaced.

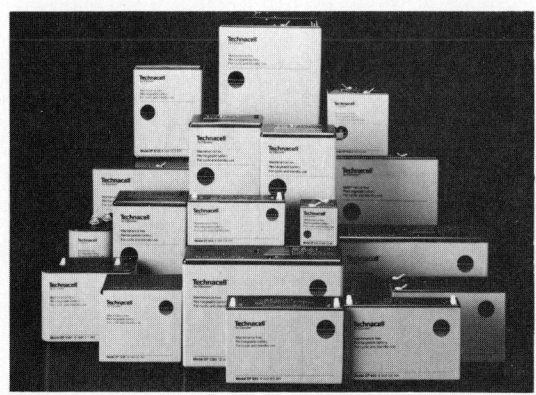

Fig. 4-9: The gelled electrolyte cell is available in a variety of sizes and voltages. (Photo furnished by courtesy of Elpower Corporation.)

Internal Resistance of a Cell

When a load is placed on a battery (Fig. 4-10A), the terminal voltage, V_T, is found to be less than the emf, V_S, of the battery. The amount of decrease in the terminal voltage from the value of the emf of the battery is proportional to the load current, I. This suggests that there is an *internal resistance, R_{in}*, associated with the battery. If the battery is considered to be equivalent to a resistor, R_{in}, in series with an emf, V_S, as shown in Fig. 4-10B, the terminal voltage is

$$V_T = V_S - IR_{in}$$

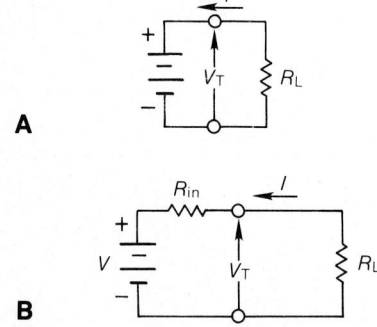

Fig. 4-10: (A) A battery under load and (B) the internal resistance.

If this equation is multiplied through by I, a power equation is obtained:

$$V_T I = V_S I - I^2 R_{in}$$

The term, $V_T I$, is the useful power which is delivered to the load. The term, $V_S I$, represents the conversion of chemical energy into electrical energy, and the term $I^2 R_{in}$ is the power lost within the battery as internal heating.

The internal resistance, R_{in}, of a fresh, fully charged battery is low. As the battery approaches the end of its life, R_{in} increases. While the emf of the battery does not change significantly over its life, as R_{in} increases, the terminal voltage, V_T, decreases under load to the point where it is insufficient to perform efficiently. Details on how to compute internal resistance of a cell are given in Chapter 8.

CONDUCTORS

All electronic circuits utilize conductors to provide a path on which a current flows. In nearly all instances, it is desirable to keep the resistance of the conductor as small as possible. To accomplish this, it is important to know the factors affecting the resistance of a conductor.

Type of Conductor Material

As stated in Chapter 1, any substance that permits the free motion of a large number of electrons is classed as a conductor. Dependent upon their atomic structures, different materials will have different values of resistance.

The atoms of materials such as silver, copper, and aluminum are so close together that the electrons in the outer shell of the atom are associated with one atom as much as with a neighbor (Fig. 4-11A). As a result, the force of attachment of an outer electron with any individual atom is practically zero. Depending on the metal, at least one, sometimes two, and, in a few cases, three electrons per atom exist in this state. In such a case, a relatively small amount of additional electron energy would free the outer electrons from the attraction of the nucleus. Materials of this type are good conductors. Good conductors, as already mentioned, will have a low resistance.

In materials such as glass, rubber, porcelain, mica, quartz, etc., the atoms are farther apart (Fig. 4-11B). The electrons in the outer shells will not be equally attached to several atoms as they orbit the nucleus. They will be attracted by the nucleus of the parent atom; therefore, a great amount of energy is required to free any of these electrons. Materials of this type are poor conductors or insulators and, therefore, have a high resistance.

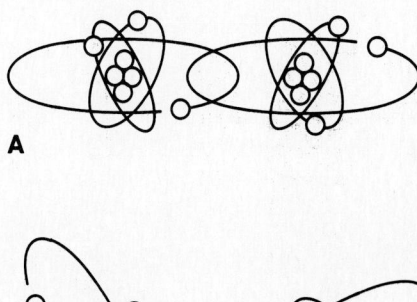

A

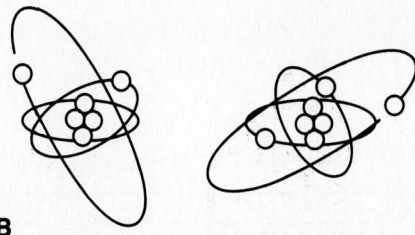

B

Fig. 4-11: Atomic spacing in (A) conductors and (B) insulators.

47

Effect of Cross-Sectional Area

The cross-sectional area of a material greatly affects the amount of resistance it has. The rule is: *The resistance of a material is inversely proportional to its cross-sectional area.* For example, if the cross-sectional area of a conductor is increased, a greater quantity of electrons are available for movement through the conductor; therefore, a larger current will flow for a given amount of applied voltage. The increase in current tells us that when the cross-sectional area of a conductor increases, its resistance decreases. On the other hand, if the cross-sectional area of a conductor decreases, the number of electrons available decreases. For a given applied voltage, the current through the conductor decreases. The decrease in current flow indicates that when the cross-sectional area of a conductor decreases, its resistance increases.

The diameters of conductors used in electronics are often only a fraction of an inch. Because they are so small, the diameter of a wire is stated in mils (0.001″). It is also standard practice to use the unit circular mil to state the cross-sectional area of a conductor. The circular mil is found by simply squaring the diameter when the diameter is expressed in mils (Fig. 4-12). For example, if the diameter of a wire is 35 mils (0.035″), its circular mil area is equal to $(35)^2$ or 1,225 circular mils.

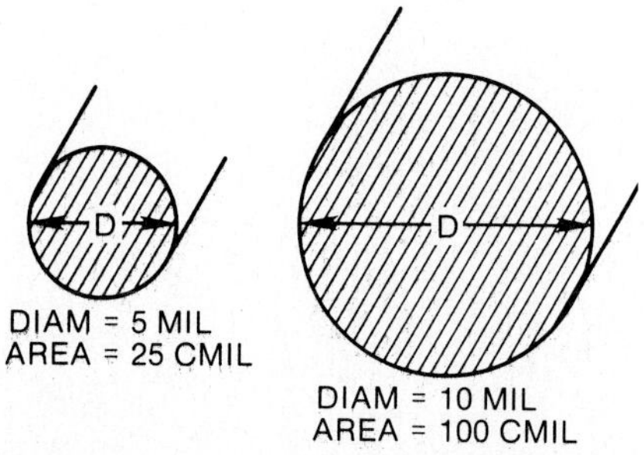

Circular-Mil-Foot. A circular-mil-foot, as shown in Fig. 4-13, is actually a unit of volume. It is a unit conductor 1′ in length and having a cross-sectional area of 1 circular mil. Because it is considered a unit conductor, the circular-mil-foot is useful in making comparisons between wires that are made of different metals. For example, a basis of comparison of the resistivity of various substances may be made by determining the resistance of a circular-mil-foot of each of the substances.

Specific Resistance or Resistivity

Specific resistance, or resistivity, is the resistance in ohms offered by the unit volume (the circular-mil-foot) of a substance to the flow of electric current. Resistivity is the reciprocal of conductivity. A substance that has a high resistivity will have a low conductivity, and vice versa.

Thus, the specific resistance of a substance is the resistance of a unit volume of that substance. Many tables of specific resistance are based on the resistance in ohms of a volume of the substance 1′ long and 1 circular mil in cross-sectional area. The temperature at which the resistance measurement is made is also specified. If the kind of metal of which a conductor is made is known, the specific resistance of the metal may be obtained from a table. The specific resistances of some common substances are given in Table 4-1.

The resistance of a conductor of uniform cross section varies directly as the product of the length and the specific resistance of the conductor and inversely as the cross-sectional area of the conductor. Therefore, the resistance of a conductor may be

DIAM = 5 MIL
AREA = 25 CMIL

DIAM = 10 MIL
AREA = 100 CMIL

Fig. 4-12: Cross-sectional areas for circular wire. Double the diameter equals four times the circular area.

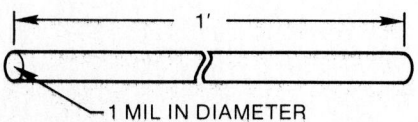

1′

1 MIL IN DIAMETER

Fig. 4-13: The circular-mil-foot.

Table 4-1: Specific Resistance

Substance	Specific Resistance at 20°C	
	Centimeter Cube (microohms)	Circular-Mil-Foot (ohms)
Silver	1.629	9.8
Copper (drawn)	1.724	10.37
Gold	2.44	14.7
Aluminum	2.828	17.02
Carbon (amorphous)	3.8 to 4.1	—
Tungsten	5.51	33.2
Brass	7.0	42.1
Steel (soft)	15.9	95.8
Nichrome	109.0	660.0

calculated if the length, cross-sectional area, and specific resistance of the substance are known. Expressed as an equation, the resistance, R in ohms, of a conductor is

$$R = \rho \frac{L}{A}$$

where ρ (Greek rho) is the specific resistance in ohms per circular-mil-foot, L the length in feet, and A the cross-sectional area in circular mils.

For example, what is the resistance of a 1,000′ copper wire having a cross-sectional area of 10,400 circular mils (No. 10 wire), the wire temperature being 20°C? The specific resistance, from Table 4-1, is 10.37. Substituting the known values in the preceding equation, the resistance, R, is determined as

$$R = \rho \frac{L}{A} = 10.37 \times \frac{1,000}{10,400} = 1\,\Omega \text{ (approximately)}$$

If R, ρ, and A are known, the length may be determined by a simple mathematical transposition.

As has been mentioned in preceding chapters, conductance, G, is the reciprocal of resistance. When R is in ohms, the conductance is expressed in siemens (S). Where resistance is opposition to flow, conductance is the ease with which the current flows. Conductance in siemens is equivalent to the number of amperes flowing in a conductor per volt of applied emf. Expressed in terms of the specific resistance, length, and cross section of a conductor,

$$G = \frac{A}{\rho L}$$

The conductance, G, varies directly as the cross-sectional area, A, and inversely as the specific resistance, ρ, and the length, L. When A is in circular mils, ρ is in ohms per circular-mil-foot, L is in feet, and G is in siemens.

The relative conductance of several substances is given in Table 4-2.

Table 4-2: Relative Conductance	
Substance	Relative Conductance (Silver = 100%)
Silver	100
Copper	98
Gold	78
Aluminum	61
Tungsten	32
Zinc	30
Platinum	17
Iron	16
Lead	15
Tin	9
Nickel	7
Mercury	1
Carbon	0.05

Length of Conductor

The length of a conductor also affects its resistance. If the length of a conductor is increased, the number of electron collisions that occur throughout the conductor increases proportionally. As a result of the greater number of collisions, more energy is given up in the form of heat. This additional energy loss subtracts from the energy being transferred through the conductor, resulting in a decrease in current flow for a given applied voltage. A decrease in current flow indicates an increase in resistance; therefore, if the length of a conductor is increased, its resistance is increased; and, if its length is decreased, its resistance is decreased. The resistance of a conductor is directly proportional to its length.

Effects of Temperature

Temperature changes affect the resistance of material in different ways. In some materials an increase in temperature causes an increase in resistance; in others, an increase in temperature causes a decrease in resistance. The amount of change of resistance per degree change in temperature is known as a temperature coefficient. If the resistance of a material increases when its temperature increases, the material is said to have a *positive temperature coefficient*. A material whose resistance decreases with an increase in temperature has a *negative temperature coefficient*. Most conductors have positive temperature coefficients. However, carbon and certain semiconductor materials used in transistors have negative temperature coefficients.

With positive temperature coefficients, for example, a temperature reduction can continue to a point where the material would have zero resistance. For copper, our most common conductor, this point would occur at -234.5°C.

Resistance variations above this temperature (–234.5°C) are approximately linear. From this relationship, the following equation for copper results:

$$\frac{R_2}{R_1} = \frac{234.5 + T_2}{234.5 + T_1}$$

where R_1 represents the resistance in ohms at temperature T_1, and R_2 is the resistance in ohms at temperature T_2. Each of these temperatures is in °C. Remember that the equation above is only for copper. Other materials have a different zero resistance temperature and thus the equation above would have to be modified accordingly.

Example: A copper conductor has a resistance of 15.68 Ω at 32°C. What would its resistance be at 63°C?

Solution:

$$\frac{R_2}{R_1} = \frac{234.5 + T_2}{234.5 + T_1}$$

Given: $T_1 = 32°C$,
$\quad\quad\quad R_1 = 15.68\,\Omega$,
$\quad\quad\quad T_2 = 63°C$

$$\frac{R_2}{15.68} = \frac{234.5 + 63}{234.5 + 32} = \frac{297.5}{266.5}$$

$$R_2 = 15.68 \times 1.12 = 17.5\,\Omega$$

WIRE SIZE AND TYPES

Various electronic applications demand different conductor sizes. Some wires are large, and others are almost as fine as a human hair. All wire is designated by definite gauge sizes. Each number designates a wire of specific diameter. As the diameter of the wire decreases, the gauge number increases. Table 4-3 illustrates some various wire sizes, their comparative areas, and resistance per 1,000'. The resistance values apply only to copper conductors. A complete breakdown of gauge sizes is given in Appendix L.

The ratio of the diameter of one gauge number to the diameter of the next higher gauge number is a constant, 1.123. The cross-sectional area varies as the square of the diameter; therefore, the ratio of the cross section of one gauge number to that of the next higher gauge number is the square of 1.123, or 1.261. Because the cube of 1.261 is very nearly 2, the cross-sectional area is approximately halved, or doubled, every three gauge numbers. Also, because 1.261 raised to the 10th power is very nearly equal to 10, the cross-sectional area is increased or decreased 10 times every 10 gauge numbers.

A No. 10 wire has a diameter of approximately 102 mils, a cross-sectional area of approximately 10,400 circular mils, and a resistance of approximately 1 Ω per 1,000'. From these facts it is possible to estimate quickly the cross-sectional area and the resistance of any size copper wire without referring directly to a wire table.

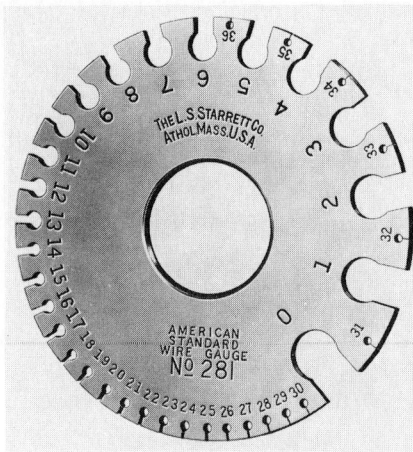

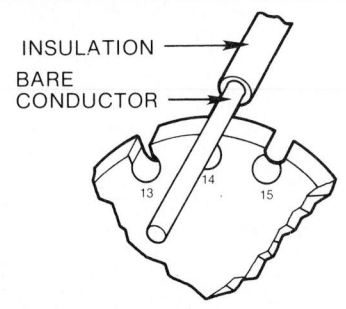

INSULATION

BARE
CONDUCTOR

Fig. 4-14: Wire gauge and how it is used. (Photo furnished by courtesy of L.S. Starrett Company.)

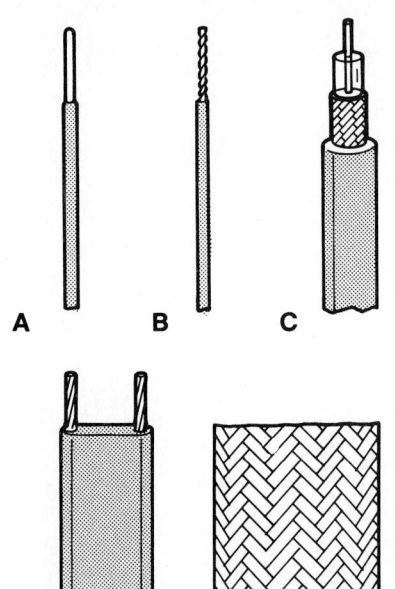

A B C

D E

Fig. 4-15: Common types of wire conductors: (A) solid, (B) stranded, (C) coaxial cable, (D) twin-lead cable, and (E) braided.

Table 4-3: Standard Annealed Solid Copper Wire			
Gauge Number	Diameter (mils)*	Cross-Sectional Circular Mils*	Ohms per 1,000' 25°C (=77°F)
0000	460.0	212,000.0	.0500
2	258.0	66,400.0	.159
6	162.0	26,300.0	.403
10	102.0	10,400.0	1.02
14	64.0	4,110.0	2.58
17	45.0	2,050.0	5.16
18	40.0	1,620.0	6.51
20	32.0	1,020.0	10.4
22	25.3	642.0	16.5
26	15.9	254.0	41.6
30	10.0	101.0	105.0
36	5.0	25.0	423.0
38	4.0	15.7	673.0
40	3.1	9.9	1,070.0

*Conversion table to metric figures can be found in Appendix L.

For example, to estimate the cross-sectional area and the resistance of 1,000' of No. 17 wire, the following reasoning might be employed. A No. 17 wire is three sizes removed from a No. 20 wire and, therefore, has twice the cross-sectional area of a No. 20 wire. A No. 20 wire is ten sizes removed from a No. 10 wire and, therefore, has one-tenth the cross section of a No. 10 wire. Therefore, the cross-sectional area of a No. 17 wire is $2 \times 0.01 \times 10,000 = 2,000$ circular mils. Since resistance varies inversely with the cross-sectional area, the resistance of a No. 17 wire is $10 \times 1 \times 0.5 = 5\Omega$ per 1,000'.

A wire gauge (Fig. 4-14) can be used to give both gauge number and the diameter. The width of the slot leading to the hole in the gauge indicates the gauge size. The diameter is generally stamped on the reverse side of the gauge.

Wire Types

A *wire* is a slender rod or filament of drawn metal. The definition restricts the term to what would ordinarily be understood as "solid wire." The word "slender" is used because the length of a wire is usually large in comparison with the diameter. If a wire is covered with insulation, it is properly called an insulated wire.

Figure 4-15 illustrates some of the more common types of wire used in electronic circuits. Stranded wire, as its name suggests, is made up of a number of strands of solid wire which are twisted together in order to form a wire with a size equivalent to the sum of the areas for the individual strands. For example, two strands of No. 30 wire are equivalent to one piece of No. 27 wire. Stranded wire is more flexible and less likely to break open than solid wire.

52

Heavier wires (those with lower gauge numbers) are generally enclosed in an insulating sleeve made of rubber, cotton, or one of many plastics. General-purpose hook-up wire about No. 20 gauge (either solid or stranded) is usually for connecting electronic components (Figs. 4-15A and B). Hollow insulating sleeves, called *spaghetti,* are frequently used to enclose or insulate bare, solid hook-up wire.

Very thin wire (No. 30 gauge and higher) is frequently coated with an insulating material such as shellac or enamel. This coating may look like copper, but where useful electrical connections are required, it must be scraped off.

When two or more conductors are enclosed in a common covering, they are referred to as a *cable.* Coaxial cable is shown in Fig. 4-15C. It consists of a center conductor surrounded by a concentric conducting shell. Either or both of the conductors may be solid or stranded, depending on both the electrical and mechanical application. The two conductors are separated by an insulator such as teflon, polyethylene, or air. The outer conductor, which is connected to ground, shields the inner conductor from any external magnetic interference. Thus, coaxial cable is an ideal *transmission line.*

A transmission line is any pair of conductors with constant spacing between them. The *twin-lead* transmission line shown in Fig. 4-15D is generally made of No. 20 gauge wires separated by rubber or plastic. Twin-lead is commonly used for connecting external television antennas to their receivers.

The braided wire illustrated in Fig. 4-15E is basically a form of stranded wire. It is commonly used for the outer conductor of a coaxial cable when mechanical flexibility is desired.

The so-called *flat ribbon cable* is used in digital electronic systems because it can be easily connected to printed circuit boards (Fig. 4-16).

Copper is used for most wire conductors, although aluminum and silver are sometimes used. In many cases the copper is tinned with a thin layer of solder. Solder is employed to provide protection against oxidation as well as to make connections into a circuit easier.

Printed Circuit Boards

Printed circuit boards (Fig. 4-17), often called *PC boards,* are found in a great deal of today's sophisticated electronic equipment. One side of the insulated plastic PC board contains the circuit components—resistors, capacitors, coils, diodes, transistors, etc. On the other side is the printed wiring conductive paths. These printed wiring circuits are commonly of silver, copper, and aluminum. The components are fastened to printed wiring by sockets, metal eyelets, or holes in the board.

Small tears or breaks in the printed wiring may be corrected by creating a "bridge" of solder, thus connecting the open path. Larger breaks may be repaired by soldering a short length of bare wire over the open hole or soldering a length of hook-up wire between end terminals of the printed wiring.

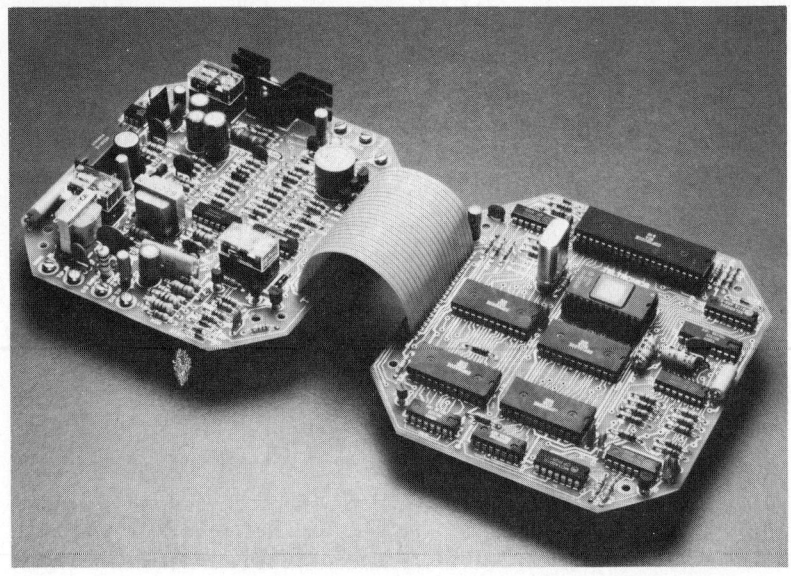

Fig. 4-16: Flat cable and how it can be easily connected to printed circuit boards. (Photo furnished by courtesy of the Brand-Rex Company.)

Fig. 4-17: Typical printed circuit board. (Photo furnished by courtesy of Diagnostic Testing Laboratory.)

Where faulty components are found on a PC board, the "bad" part can be removed by desoldering it from the printed wiring and soldering in a new component. But remember that soldering to a PC board must be done with caution. If too much heat is applied to the conductive side, the thin strips may lift or, in some cases, tear. A 25 to 35W rated soldering iron is best for this application. Remember to always use a heat sink when soldering semiconductor components to a board. The short distance between the solder connection and the semiconductor junction does not lend itself to the conduction of the heat away from the component.

Printed circuit boards serve a very useful function in that they reduce the space needed for "hard wiring" a circuit, as well as the weight associated with many wires. Also, the safe placement of integrated circuit (IC) chips requires a PC board to reduce the chance of destruction from static electricity.

More information on the use of conductors, insulators, and semiconductors can be found in Chapter 19.

RESISTORS

Commercial devices which provide a specific amount of resistance to a circuit are called resistors. They are categorized in several different ways. Each adds a certain dimension to the total description. When each category is described, then, and only then, is the description complete.

Ohmic Value

Ohmic value relates to the amount of resistance in a specific resistor. Each resistor produced has a design value of resistance. Very often, resistors are manufactured that vary from the design value. Each resistor type has a tolerance rating. This may vary anywhere from 0.5% to 20%. This means that the exact value of resistance of the specific resistor may not be as marked on the resistor. A 100Ω resistor with a $\pm 10\%$ tolerance may range between 90 and 110Ω in value and still be within the tolerance range. This resistor is acceptable for use in a circuit where the tolerance is not critical. Tolerance does not mean that the resistor is off its design value by 10% or so. All it means is that it could be off and still be acceptable. The only way to be certain is to measure the ohmic value with a resistance meter. Conversely, a resistor having a $\pm 20\%$ tolerance whose ohmic value was exactly 100Ω could be used to replace a $100\Omega \pm 1\%$ resistor successfully as long as temperature is not a factor. However, 1% resistors are designed for greater stability as well as critical tolerance. Generally, the closer tolerance resistors are more expensive than those with high tolerances.

In order to provide uniformity in resistor values, electronic manufacturers have established a number of preferred values. For example, resistors with a $\pm 10\%$ tolerance are manufactured with resistance values in ohms of

1.0	2.7	5.6
1.2	3.3	6.8
1.5	3.9	8.2
1.8	4.7	10.0
2.2		

and multiples of 10 of these values.

That is, resistors are available in the sizes indicated, as well as in multiples of 10 of these sizes, up to 22MΩ. For example, resistors are available in sizes of 1.8, 18, 180, and 1,800Ω, and so on.

Resistors with ± 20% tolerance are available in fewer resistance values. Resistors with ± 5% tolerance are available in a greater variety of resistance values than those available with ± 10% tolerance. Remember that most electronics do not require resistors with low tolerance values. Usually, resistors with ± 10% tolerance, or even ± 20% tolerance, are satisfactory.

Power (Wattage) Rating

When a current is passed through a resistor, heat is developed within the resistor. The resistor must be capable of dissipating this heat into the surrounding air; otherwise, the temperature of the resistor rises, causing a change in resistance or possibly causing the resistor to burn out.

The ability of a resistor to dissipate heat depends on the amount of its surface which is exposed to the air. A resistor designed to dissipate a large amount of heat must, therefore, have a large physical size.

The heat dissipation capability of a resistor is measured in watts and gives the maximum power handling rating for the resistor. Additional amounts of power tend to destroy the resistor. The lowest standard rating is 1/10W. The physical size of the resistor container is related to its power-handling rating. As the power rating increases, so does the physical dimension of the unit.

In many cases a larger power resistor may be substituted in a circuit for a lower power rating resistor. This is true only when the circuit design permits it. The ohmic value of both resistors must be the same. Never substitute a lower wattage-rated resistor for one of higher ratings. The substitute resistor in this situation will overheat. It may catch on fire, causing circuit damage or damage to adjacent parts, the device in which it is placed, or even the building in which it is housed.

Temperature Coefficient of Resistance

As mentioned earlier for conductors, the coefficient of resistance indicates the variation of resistance with temperature and is designated in parts per million per degree centigrade, ppm/°C, with 25°C used as a reference temperature. A resistor with a positive temperature coefficient of 100 ppm/°C would be designated as P100 (P indicating a positive coefficient). This means that such a resistor would show an increase in resistance of 100 ppm for each °C increase in temperature above 25°C.

Temperature Conditions

Resistors are not usually a limiting factor as far as operating ambient temperatures are concerned; that is, other circuit components have temperature-range limitations that restrict their use. For example, most resistor types have at least a normal operation range of from –55° to +125°C.

RESISTOR TYPES

There are two common kinds of resistors available. They are generally classified as either fixed or variable.

Fixed Resistors

Most resistors are of the fixed type; that is, their resistance cannot be varied (Table 4-4). There are two common types of fixed resistors: composition and wirewound.

Composition Resistors. The most common type of composition resistors are those made principally of the element carbon (Fig. 4-18). Their specific resistance of carbon lies between the range of 20Ω to 27Ω per circular-mil-foot. Carbon has a conductance that is approximately 1/500 that of silver. In the manufacture of carbon resistors, insulating fillers or binders, such as clay, rubber, talc, or phenolic materials, are added to the carbon to obtain various resistor values. For instance, a high value of resistance would have a greater proportion of insulating filler material than would a low resistance value.

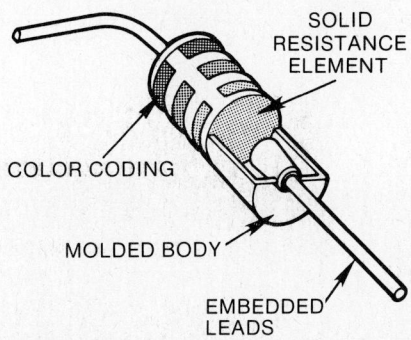

Fig. 4-18: Carbon composition resistor.

Table 4-4: Summary of the Range of Typical Specifications for Selected Fixed Resistors

Resistor Type	Resistance Range	Power Range (W)	Tolerance (%)	Temp. Coef. (ppm/°C)	Ambient Temperature Range (°C)
Carbon Composition	1 to 22MΩ	1/10 to 2	± 5 or ± 10	± 300	–55 to 125
Carbon film	0.5 to 33MΩ	1/10 to 1	± 2 or ± 5	± 300	–55 to 150
Metal film	0.5 to 100MΩ	1/10 to 1/4	± 0.25 to ± 2	± 100	–55 to 150
Cermet film	10 to 10MΩ	1/10 to 1/2	± 0.5 or ± 2	± 50	–55 to 165
Wirewound					
Power	1 to 100kΩ	3 to 225	± 5	± 30	–55 to 300
Precision	0.001 to 150kΩ	1/2 to 25	0.01 to 1	± 10	–55 to 250

Carbon resistors are commonly available in resistance values from 1Ω to about 22MΩ. Their power ratings are usually 1/10, 1/8, 1/4, 1/2, 1, and 2W (Fig. 4-19). Remember that the physical size of carbon resistors has nothing to do with the ohmic value of the resistor. Physical size of any resistor basically relates to its power-handling capability.

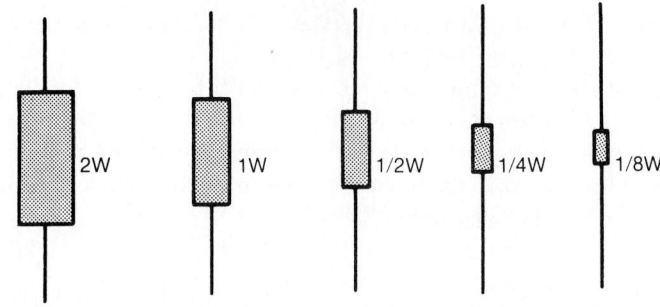

Fig. 4-19: Full-size drawing of carbon composition resistor of various power ratings.

Another type of composition resistor finding increased application in industry is the cermet resistor. This resistor uses a metallized ceramic material instead of carbon composition for its resistive element. The reason for the use of this resistor is the higher reliability of the wire lead connection to the resistance element. Cermet resistors also have closer tolerance to the ohmic value than do carbon composition resistors.

The cermet film resistance element is spirally wound around an aluminum core material (Fig. 4-20). The wire leads are connected to the cermet element. The ohmic value of this type of resistor, which ranges up to $10M\Omega$ at tolerances of $\pm 0.5\%$ to $\pm 2\%$, depends wholly on the size of the cermet element.

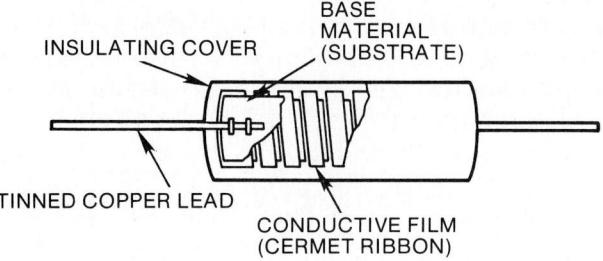

Fig. 4-20: Cermet film resistor.

Other film coatings are employed to form composition resistors similar to the cermet type. For example, metal film-fixed resistors are made by distributing a very thin metal film onto an insulating rod. The metal film is spiral cut to the desired resistance. Tolerances ranging from $\pm 0.025\%$ to $\pm 2\%$ of the nominal value are commercially available.

A thin carbon coating is sometimes formed around an insulator to produce the so-called carbon film resistor. These resistors are characterized by good stability, low noise, and a wide frequency range. They cost less than hold-molded carbon composition resistors.

There are two thick film resistors in common use today. They are the metal-oxide film resistors and bulk-property film resistors. Metal-oxide film resistors are manufactured by oxidizing tin chloride on a heated glass base material. This base material is called a substrate. The resulting resistors have very low-noise characteristics and good temperature stability. Units with ultrahigh voltage (8kV/in., 3kV/cm) and very high resistance (up to $10^6 M\Omega$) are available.

Bulk-property film resistors are made of photoetched metal film and provide close resistance tolerances of $\pm 0.1\%$ to $\pm 1\%$. They are extremely low noise, practically noninductive, high-frequency resistors with a very low temperature coefficient and extremely good long-term stability.

Resistor networks (Fig. 4-21) composed of many film resistors on a single substrate are available in single in-line packages (SIP) or dual in-line packages (DIP). Fabricated from thin film and thick film metals, often laser-trimmed to close tolerances, these packages may contain eight fully isolated resistors of equal value typically from 100Ω to $100k\Omega$ or some resistor network combination. They are especially useful where matched resistor pairs within $\pm 0.1\%$ to $\pm 0.5\%$ are required.

Wirewound Resistors. Wirewound resistors (Fig. 4-22) are generally for low-resistance, high-power applications and are are available in power ratings from 5W to several hundred watts with a resistance range of less than 1Ω to several thousand ohms. Though quite costly when compared to other resistor types, the wirewound resistor has outstanding electrical properties, including low noise, very good time stability, and good overload (high current) characteristics. However, the wirewound resistor cannot be used in high frequency circuits due to the inductance and capacitance inherent in its methods of manufacture.

In the construction of these resistors, resistance wire such as Nichrome, constantan, German silver, or manganin is wrapped around an insulating core made of porcelain, cement, phenolic materials, or just plain paper. It is the length and specific resistivity of the wire that determine the resistance of the device.

Of all the resistor types, wirewound is the least standardized and is available in a wider variety of shapes and sizes. Ultra-precision wound resistors have tolerances as exacting as $\pm 0.002\%$.

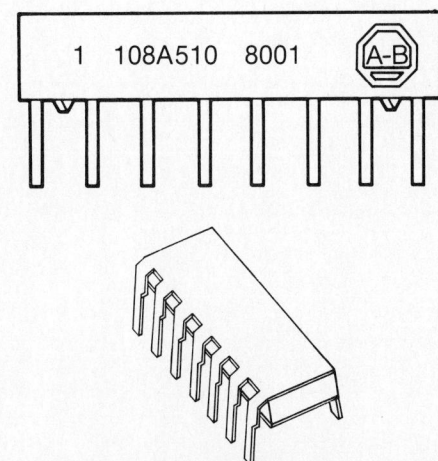

Fig. 4-21: Resistors in single in-line package (SIP) and dual in-line package (DIP). These resistor networks are used primarily with electronic circuits on printed circuit boards.

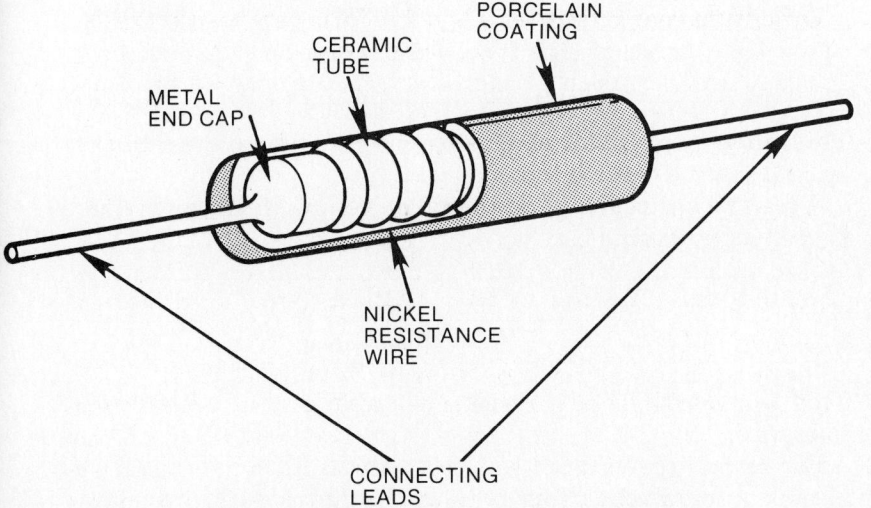

Fig. 4-22: Construction of a wirewound resistor.

Resistor Color Code

There are two systems to designate the ohmic value of resistors. One is called the standard system (most common), and the other is called the body-end-dot system. Both systems make use of a color code. The position of the color and the color code indicate the value of the resistor.

The molded carbon composition resistor is the one most commonly used in electronics circuits. Its size is usually such that printing its resistance and tolerance value on its case would create difficulties. As a result, a system of color-coded stripes is used. These stripes are readily seen in any position the resistor is placed, which would not always be the case if printed numerals were used. The color code for carbon composition resistors is shown in Table 4-5.

Table 4-5: Color Code for Carbon Composition Resistors					
Color	Number	Color	Number	Color	Tolerance (%)
Black	0	Green	5	Gold	5
Brown	1	Blue	6	Silver	10
Red	2	Violet	7	**Color**	**Multiplier**
Orange	3	Gray	8	Gold	0.1
Yellow	4	White	9	Silver	0.01

The use of the color code will be shown in identifying carbon composition resistors with axial and radial leads.

Resistors—Axial Leads. This type of resistor is always held so that the color stripes are on the left. This is shown in Fig. 4-23. The first color band designates the first digit of the resistance; the second color band, the second digit of the resistance. The third band, if other than gold or silver, designates the number of zeros to be added. The fourth band indicates the resistor tolerance. A resistor with bands of green, blue, orange, and gold is used as an example.

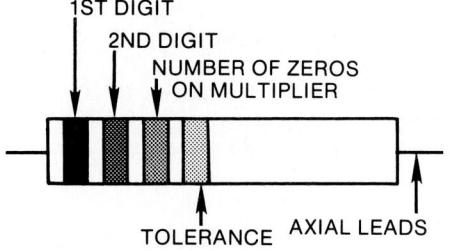

Fig. 4-23: Color-coded resistor—axial leads.

Green	Blue	Orange	Gold
First digit	Second digit	No. zeros	Tolerance
5	6	3	5%

This, then, represents a resistor with a resistance value of 56,000 Ω with a 5% tolerance.

A resistor with color bands of violet, yellow, black, and silver is used as an example.

Violet	Yellow	Black	Silver
7	4	0 (no zeros)	10%

This would represent a resistance value of 74 Ω with a 10% tolerance.

For resistance values below 10 Ω, the third band would be either gold or silver. In this instance the third band represents a decimal multiplier, 0.1 if gold, 0.01 if silver. This multiplier is used to multiply the first two digits. The fourth band, as before, designates the tolerances.

For example, a resistor with bands of red, red, gold, and gold is used.

Red	Red	Gold	Gold
First digit 2	Second digit 2	Decimal multiplier 0.1	Tolerance 5%
	$22 \times 0.1 = 2.2\,\Omega$		

The resistance value represented is 2.2Ω with a 5% tolerance.

Resistors with axial leads often have a fifth color stripe. This indicates the expected percentage of failures that will occur in a test period of 100 hours; thus, an indication of the reliability of the device is provided. The color code for the fifth stripe is shown in Table 4-6.

Table 4-6: Color Code—Fifth Stripe, 100-Hour Test	
Color	**Percent Failure**
Brown	1.0
Red	0.1
Orange	0.01
Yellow	0.001

Resistors—Radial Leads. This type of resistor is always held so that the leads are toward the observer. The use of the color code is exactly the same as with axial lead resistors. With this type of resistor, however, the color locations are different, as shown in Fig. 4-24.

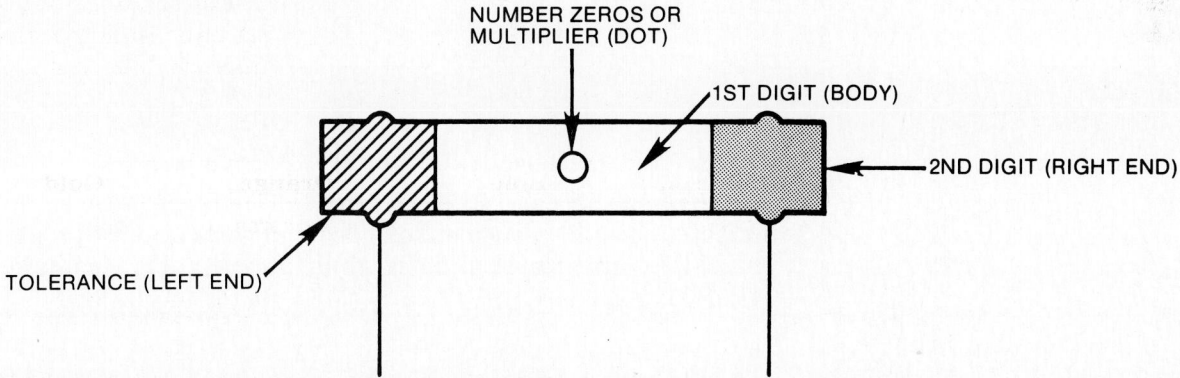

Fig. 4-24: Color-coded resistor— radial leads.

Wirewound Resistors—Axial Leads. Wirewound resistors, similar in form to carbon composition resistors, use the same color code for identification. They are identified as wirewound resistors by the first-digit stripe, which is double the width of that of a carbon composition resistor (Fig. 4-25).

Industrial-Commercial Precision Resistor Coding. In industry, an alternative method of coding is sometimes used. This system can be briefly explained as follows:

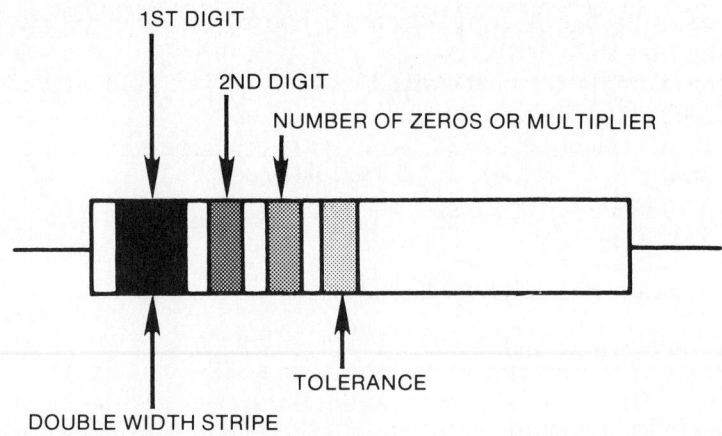

1ST DIGIT

2ND DIGIT

NUMBER OF ZEROS OR MULTIPLIER

TOLERANCE

DOUBLE WIDTH STRIPE

Fig. 4-25: Wirewound resistor.

If the ohms symbol (Ω) appears after the last digit, retain the preceeding number as is. The tolerance, if not printed, is understood to be ± 1%. For example, "560Ω" will read as 560Ω± 1%.

If no "Ω" sign appears next to the last digit, the numbers preceeding the last digit are retained and the last digit indicates the number of zeros to be added. For example, "9312" will read as 93,100Ω or 93.1kΩ.

The tolerance, if not printed, is understood to be ± 1%. If a tolerance is stamped on the resistor, read as is. For example, "1270± 2%" is read as 1270Ω± 2% or 1.27kΩ± 2%.

In certain cases, a letter is utilized to indicate the tolerance. This letter follows the (4) numbers. For example, "1101F" is read as 1100Ω± 1%. A "1270G" would read as 127Ω± 2%. (Note that in the above example the last digit indicates that no zeros are to be added.) Listings for letter coded types are as follows:

A	± .05%	F	± 1%
B	± .1%	G	± 2%
C	± .25%	J	± 5%
D	± .5%	K	± 10%
		M	± 20%

In some special cases, the placement of a letter between digits indicates both the placement of a decimal point and the proper unit. "R" indicates Ω and "K" indicates kΩ. For example, "4R5F" is read as 4.5Ω± 1% and "20K4" is read as 20.4kΩ.

Variable Resistors

Variable resistors may be of either the carbon composition, carbon film, cermet, or wirewound types. Like fixed resistors, this style is used to control current flow and provide desired voltage drops (Table 4-7). Variable resistors may be familiar to most of us in the form of a light dimmer. This device allows the intensity of the lamp to be changed from full brightness to dark. Such devices are also used to regulate the gain of an amplifier; used in this manner, we refer to them as volume controls.

Table 4-7: Summary of the Range of Selected Specifications for Several Variable Resistors

Resistor Type	Resistance	Power (W)	Rotation
Potentiometers			
Power	0.5 to 10kΩ	7 to 500	300°
Low power	50 to 5MΩ	1/2 to 2	280°
High resolution	10 to 1MΩ	1/2 to 1	Single turn Multiturn
Trimmers	50 to 5MΩ	1/4 to 3/4	Single turn Multiturn
Variable wirewound	1 to 150kΩ	3 to 200	Slide

There are two main categories of variable resistors: potentiometers and adjustable wirewound resistors (Fig. 4-26).

Potentiometers. These resistors are usually round with a moveable wiper arm which makes contact with the resistive variation (Fig. 4-27). Single-turn potentiometers, or "pots" as they are frequently called, with a rotation of about 300°, usually have linear resistance over the degrees of rotation; that is, one degree of shaft rotation results in the same variation of resistance regardless of the shaft position. These types of potentiometers are commonly called *linear-tapers*. This movement of the shaft is called the *taper* of the potentiometer. It is also possible to have a nonlinear potentiometer taper. In this case the resistance is not linear and usually varies on a logarithmic basis. This type of potentiometer is often used for audio purposes and is commonly referred to as an *audio-taper*.

Figure 4-28 shows how a potentiometer operates. The numbers 1, 2, and 3 represent the terminals of the device. The total resistance is always found between the ends of the resistance element. A certain amount of resistance is present between one end, terminal 1, and the variable, terminal 2. This is less than the total resistance of the device. Identify this as resistance A. A certain amount of resistance is also present between the adjustable terminal 2 and the other end, terminal 3. Identify

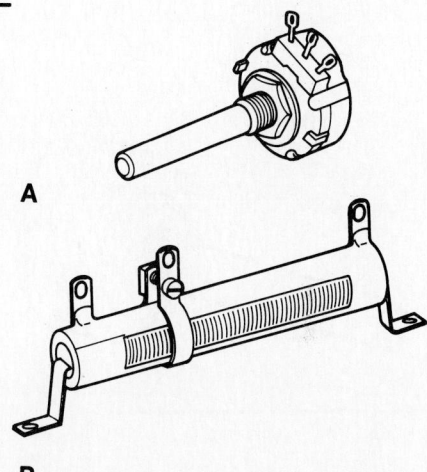

A

B

Fig. 4-26: (A) Typical variable potentiometer and (B) adjustable wirewound resistor.

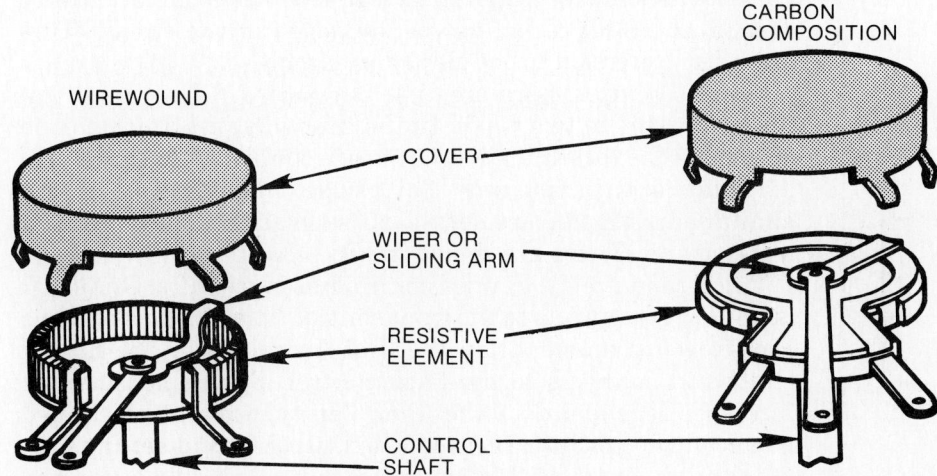

Fig. 4-27: Construction of potentiometer.

63

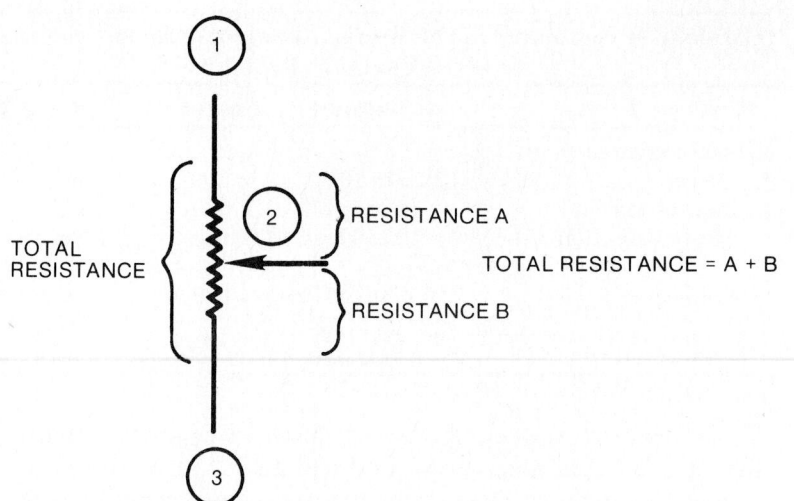

Fig. 4-28: The variable resistor has a fixed total value of resistance between its outer terminals. The arm is used to vary the resistance between it and one of the outer terminals.

this as resistance B. The exact amount of resistance between terminal 2 and the ends will vary depending on the position of the adjustable arm of the pot. In any case, resistance A and resistance B will add to equal the total resistance between the adjustable arm and the ends.

A second way of using the same pot is to utilize two of the three terminals. One end terminal and the adjustable terminal are used. This is illustrated schematically in Fig. 4-29. In this type of connection the varying resistance between terminals 1 and 2 is utilized to change the resistance in a circuit. The third terminal is not always used. In some cases the third terminal and the adjustable terminal are connected together, as shown in Fig. 4-29B. The amount of resistance between terminals 1 and 2 still varies as the control shaft is turned. The resistance between these terminals decreases as the moveable arm approaches the end terminal. If the arm section should fail for any reason, the total resistance of the pot is in the circuit. This provides protection to the electronic circuit.

Some potentiometers are "gang mounted." This is accomplished in one of two ways. In the first way, the unit has one shaft. All the rotating slider arms are connected to this shaft. All rotate at the same time. This is used in devices where two similar adjustments are necessary, such as the tone control on a stereo unit. These are shown in Fig. 4-30.

The second type of ganged control has two shafts. One shaft is hollow. It connects to the front control. The second shaft has a narrower diameter than the first. It fits inside the hollow shaft and connects to the back control. Each shaft may be rotated independently of the other. This type of control is called a *concentric* control. It is used when two different settings are required. A control of this type saves panel space. One common use is the contrast and brightness control on certain television sets.

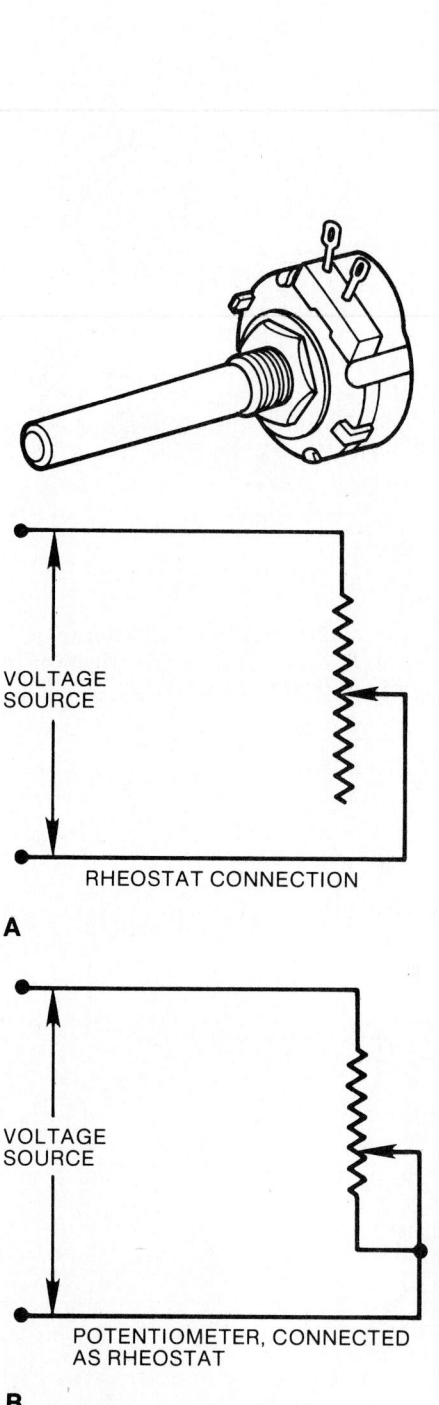

Fig. 4-29: Typical rheostat connections.

Fig. 4-30: Variable resistors may be "gang mounted" in one of two ways. These are either with one common shaft or two concentric shafts. (Photo furnished by courtesy of CTS Corporation.)

Another description for potentiometers is the designation of whether the pot is mounted in such a way that the user has access to it and may adjust it. Many pots are mounted inside control panels or on printed circuit boards. These are used to make minor adjustments in circuit values. In a sense they "trim up" the circuit. Because of this they are called *trim pots* (Fig. 4-31).

Adjustable Wirewound Resistors. Construction of these resistors involves winding nichrome wire around a ceramic core and covering the unit with an insulative coating. A window (Fig. 4-32A) or uninsulated section is preserved, exposing the resistive wire. An adjustable tap rides along the exposed wire and makes electrical contact. A nut and bolt with a screwdriver slot is used to adjust and secure the tap at the desired resistance. Tapped adjustable wirewound resistors (Fig. 4-32B) are also available.

Like the fixed wirewound resistors, these devices find applications in power supplies as voltage divider resistors and in low-frequency AC circuits. Because of their inductive and capacitive properties, they are not suited to high-frequency applications. A wide range of resistances, tolerances, and power ratings are available.

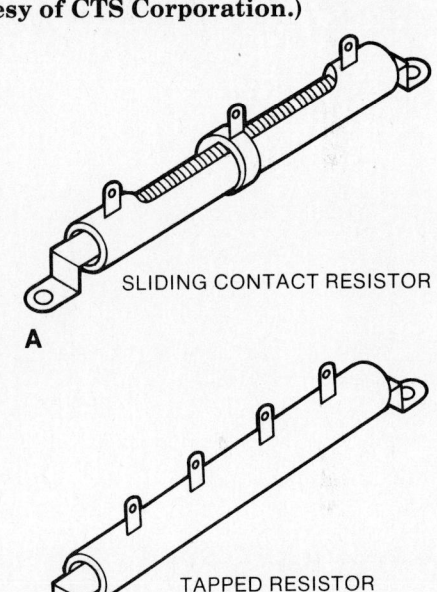

SLIDING CONTACT RESISTOR

A

TAPPED RESISTOR

B

Fig. 4-32: Typical tapped resistors.

Fig. 4-31: Typical trim pots. (Photo furnished by courtesy of CTS Corporation.)

SPECIAL-PURPOSE RESISTORS

There are a number of special resistors available that are specifically designed to have a high temperature coefficient so that their resistance varies drastically and nonlinearly with temperature. The more common of these nonlinear resistors are discussed below.

1. *Ballast Resistors.* These resistors are often employed to maintain a constant current flow in spite of variations in line voltage. When the line voltage increases, the current would normally increase. Because of the ballast resistor's high positive temperature coefficient, even a slight increase in current will cause an increase in its resistance; thus, the current tends to remain constant at its original value.

2. *Voltage-Dependent Resistors.* These resistors are used to protect circuits of electronic communications equipment from very sudden surges of voltage. Their resistance value is high when cold (room temperature) and drops as they heat up, allowing the current to rise to normal operating value. While they are available under a variety of trade names, all voltage-dependent resistors have a high negative temperature coefficient.

3. *Thyrites.* These nonlinear resistors are made of silicon-carbide particles bonded in a ceramic matrix. It is a characteristic of this material that its resistance decreases sharply as the voltage across it is increased. Thyrite resistors are frequently used to protect electronic equipment against line voltage surges as might occur from lightning discharges.

4. *Thermistors.* These thermally sensitive resistors are made of solid semiconductor material (usually some variety of metallic oxide) and have a very high negative temperature coefficient. Thermistor elements are available in bead, rod, disk, probe, or washer forms. Because thermistors produce large changes in resistance for small changes in temperature, they are used in temperature-sensing and temperature-measuring devices and in a variety of applications wherein changes of temperature are an indirect result.

OTHER CIRCUIT COMPONENTS

There are three "other" DC circuit components that should be considered (Fig. 4-33). They are switches, fuses, and miniature lamps.

Switches

Fundamentally a "switch" is a mechanical device for completing or interrupting an electronic circuit. There are many types of switches available, but they all perform the same basic function of opening or closing circuits.

Switch Types and Uses. A few of the more common types of small switches are shown in Fig. 4-34. These are fairly simple switches which usually control only one or two circuit paths.

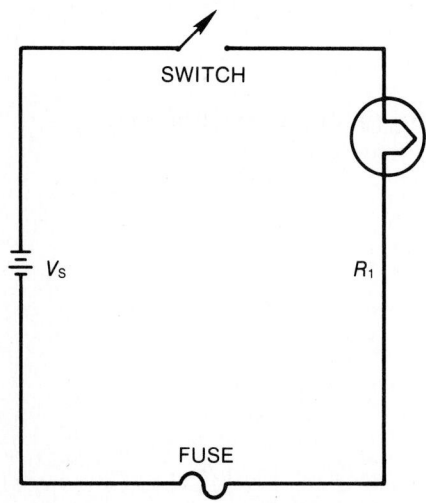

Fig. 4-33: Circuit showing the use of a fuse, switch, and lamp.

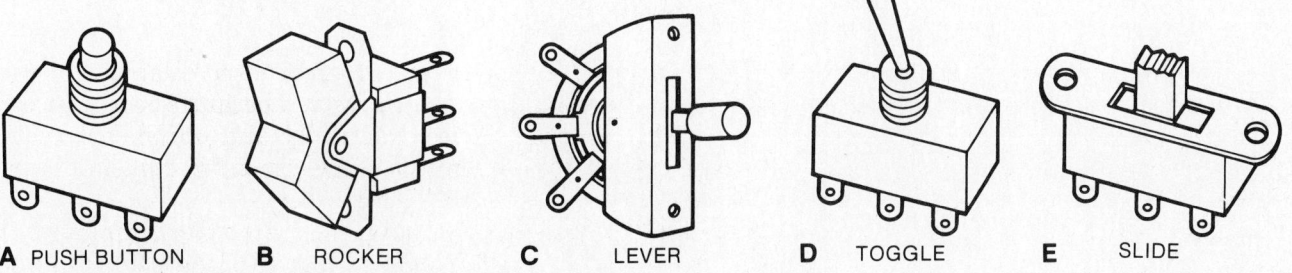

A PUSH BUTTON   **B** ROCKER   **C** LEVER   **D** TOGGLE   **E** SLIDE

Fig. 4-34: Common types of switches.

The schematic symbols of these switches are shown in Fig. 4-35. The number of poles represents the number of separate circuit connecting points, and the number of throws represents the number of different contact points to which each pole can connect. A typical on-off switch has a circuit connection (single pole) and one contact point (single throw). The dotted line in Fig. 4-35 indicates that the two poles are mechanically connected, but not electrically connected.

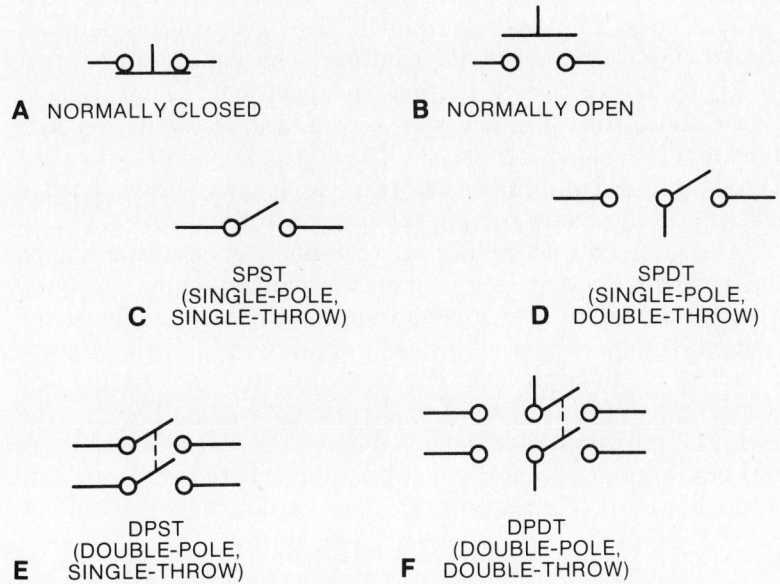

A NORMALLY CLOSED   **B** NORMALLY OPEN

C SPST (SINGLE-POLE, SINGLE-THROW)   **D** SPDT (SINGLE-POLE, DOUBLE-THROW)

E DPST (DOUBLE-POLE, SINGLE-THROW)   **F** DPDT (DOUBLE-POLE, DOUBLE-THROW)

Fig. 4-35: Switch symbols and names.

Some switches are manufactured so that they always return to the same condition after being released by the operator. These are known as momentary contact switches. They can be either "normally open" (NO) or "normally closed" (NC). Another name given to these switches is a "spring-loaded" switch since it is a spring that returns it to its normal position.

Rotary switches are illustrated in Fig. 4-36. There are many different types of rotary switches available, ranging from a simple on-off switch to a very complex multideck switch such as the tuners in a television. Figure 4-37 illustrates the schematic diagrams used for rotary switches.

Fig. 4-36: Various types of rotary switches. (Photo furnished by courtesy of Oak Industries, Inc.)

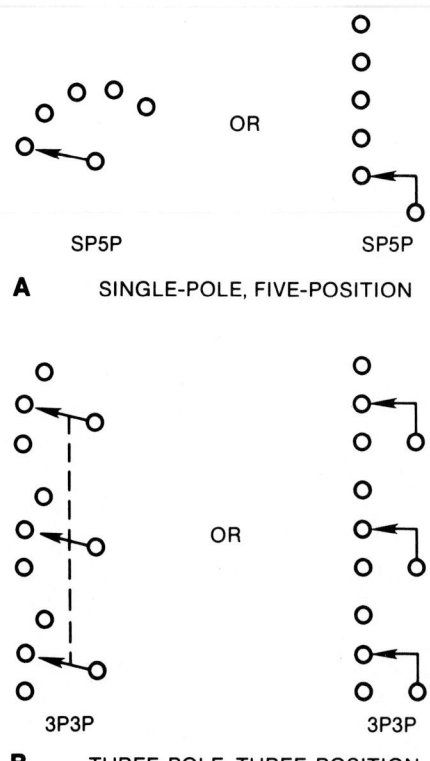

A   SINGLE-POLE, FIVE-POSITION

B   THREE-POLE, THREE-POSITION

Fig. 4-37: Rotary switch symbols.

There are both "shorting-type" and "nonshorting-type" rotary switches (Fig. 4-38). The shorting-type is sometimes referred to as a "make-before-break" switch. In this type of switch, the pole makes contact with the new position before it breaks contact with the old position. The wiper will contact position 2 before it loses contact with position 1.

Construction. The material, size, and shape of the contacts have much to do with their life. The contact pressure when the switch is closed and the speed with which the switch is opened are other major construction factors controlling switch life.

The switch contact points are commonly made from a good electrical-conducting, high-corrosion-resistant metal, usually silver or platinum. The mechanism that opens and closes the contacts is generally a copper or brass alloy, which allows good electrical conduction from the terminals to the contacts and offers high resistance to atmospheric corrosion. The moving portion is spring-loaded in such a way to ensure a fast making and breaking action and enough distance between the on and off position to discourage any electrical arcing between the con-

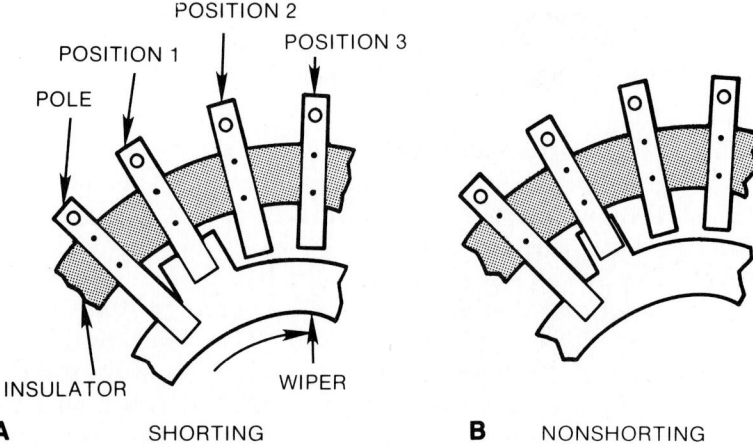

A   SHORTING   **B**   NONSHORTING

Fig. 4-38: Shorting and nonshorting rotary switches.

tacts. These parts are contained in an insulating case designed to keep dust, moisture, and other foreign matter out of the mechanism and to protect the user from electrical shock.

Ratings. Switches have both a voltage and a current rating. The switch manufacturer will rate a switch for the type of service and the maximum current allowable for a certain design, but he has no control over its usage. Exceeding the manufacturer's ratings shortens the life of the switch and can also be dangerous to the operator. If the ratings are exceeded enough to cause arcing, an arc can char and break down the insulation in the switch, putting the operator in contact with the circuit.

Switches are often given multiple current and voltage ratings; a switch may be rated for 3A at $250V_{AC}$, 6A at $125V_{AC}$, and 1A at $120V_{DC}$. Generally, switches can handle higher alternating currents than direct currents.

Fuses

When using an electronic device, care must be taken that the electric current passing through the device does not become excessive. All electronic equipment contains a certain amount of resistance, and when electric current passes through a resistance, electrical energy is tranformed into heat energy. If faulty circuit operation causes an excess amount of current flow through an electronic device, the resulting increased temperature could cause considerable damage. For protection from such occurrences, it is necessary to have an inexpensive electrical component that instantly opens the circuit when subjected to excessive current. Such a device is called a *fuse*. Several types of fuses and holders are illustrated in Fig. 4-39.

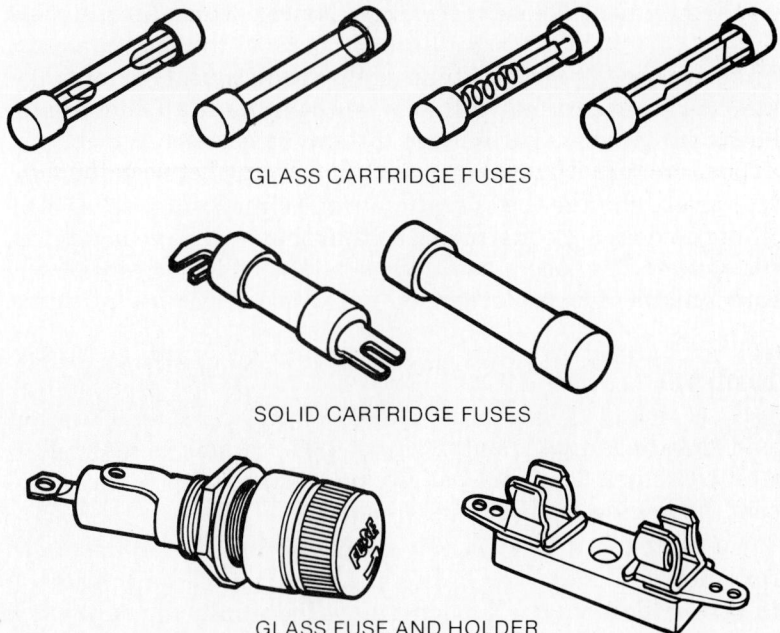

GLASS CARTRIDGE FUSES

SOLID CARTRIDGE FUSES

GLASS FUSE AND HOLDER

Fig. 4-39: Typical fuse types and holders.

69

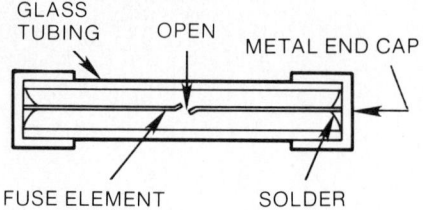

GLASS TUBING OPEN METAL END CAP

FUSE ELEMENT SOLDER

Fig. 4-40: Open fuse.

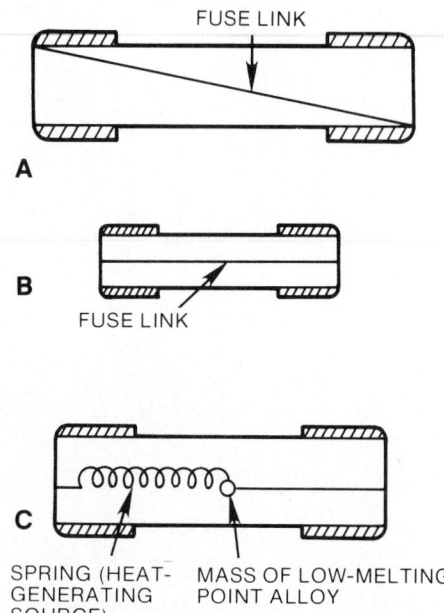

FUSE LINK

A

FUSE LINK

B

SPRING (HEAT-GENERATING SOURCE) MASS OF LOW-MELTING POINT ALLOY

C

Fig. 4-41: Types of fuses: (A) fast blow fuse, (B) medium blow fuse, and (C) slow blow fuse.

Fuses are fabricated from wire made of zinc or similar metals having a low resistance value and a low melting point. When the current in a circuit is less than or equal to the current rating of the fuse, the link element of the fuse is below its melting temperature. However, as soon as current flow in the circuit exceeds the current rating of the fuse, the fuse element melts rapidly due to the increased heat. Thus, when the fuse "blows" (Fig. 4-40), the circuit will open and the circuit components are protected. Fuses are available in current ratings from 1/200A to hundreds of amperes. The current rating indicates the maximum amount of current that can pass through the fuse link without melting the link.

Fuses are also frequently rated by "blow-time" or by how rapidly they blow (open) when subjected to specified overloads. The three most widely used categories (Fig. 4-41) are *fast blow, medium blow,* and *slow blow.* All three categories react rapidly (about 1ms) to heavy overloads (more than 10 times the rated current). All three categories react approximately the same to very small overloads. At 1.35 times (135%) the rated current, they all take a minute or two to open. In between these two extremes, the three categories differ substantially. For instance, at five times rated current, a slow blow may take up to 2s to blow, while a fast blow will usually open in less than 1ms. Under the same conditions, a medium blow would take approximately 10ms.

Medium blow fuses are the general-purpose fuses used in most electronic circuits, while fast blow fuses, which have the same physical appearance as the medium blow fuses, are employed to protect sensitive equipment such as meters and test gear. Slow blow fuses are used wherever short duration current surges are anticipated.

Fuse resistors are sometimes found in electronic communications equipment. These wirewound, low resistance (usually less than 200Ω) resistors have limited current carrying capacity. They are used in circuits that require high initial currents but low operating currents. If the current remains high, the resistor burns out (opens) and protects the circuit.

The insertion of a switch, a fuse, or both into a circuit will not effectively change the operation of the circuit since these electronic devices are constructed so that they will have negligible resistance. The fuse must always be placed in series with the components to be protected.

Lamps

Miniature lamps or bulbs (Fig. 4-42) are found in many electronic circuits. They are usually one of three types: incandescent, neon glow, or light-emitting diode (LED).

Incandescent Lamps. Small incandescent lamps are used for pilot and indicator lights or instrument illumination. A tungsten filament enclosed in a glass bulb emits the light when heated by an electric current. Miniature incandescent lamps

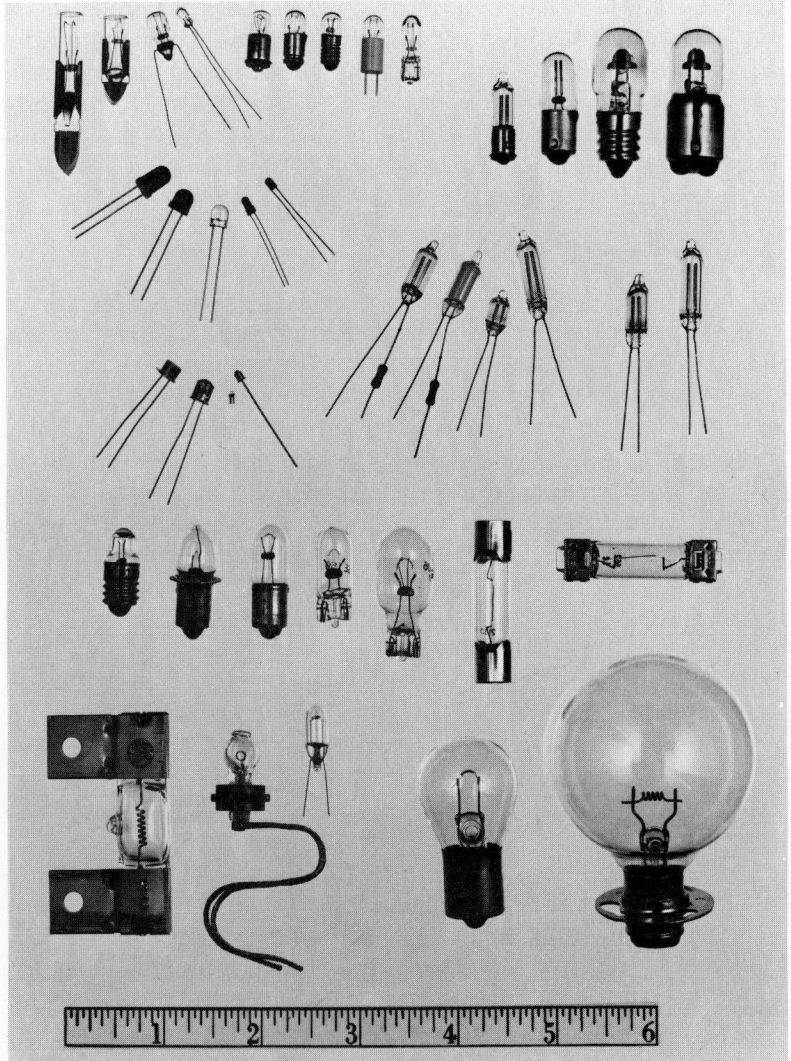

Fig. 4-42: Typical miniature lamps. (Photo furnished by courtesy of General Electric Company.)

are available in a wide range of voltage, current, and wattage ratings. They come with either a screw-type base or a bayonet base. The former is screwed into a lamp holder in the same manner as the common household light bulb. The bayonet base lamp has one or two small pins which slide into grooves in the bulb holder. To insert or remove the lamp, press down on the bulb and twist slightly to set the pins.

Neon Glow Lamps. Neon lamps are used in electronic circuits in the same way as the incandescent types and are available in a number of different bulb and base styles. The major advantage of neon glow lamps over the incandescent type is that they require very little power to operate. Most will operate at less than 1/2W and require less than 2mA.

Light-Emitting Diodes. In addition to being used for many of the same purposes as incandescent and neon glow lamps, LED lamps are also employed for numeric readouts on meters, calculators, and other digital devices. They are available in either a wire-terminal or socket-type base. Operating only on DC, LEDs use anywhere from 3 to 40mA of current. While most LED lamps are red when lighted, they are also available in yellow and green.

SUMMARY

A DC circuit consists of three basic elements: the source, the load, and the conductor.

The voltage cell is one type of source used to produce electrical energy. If a number of these cells are wired together to increase the amount of available electrical energy, a battery is created. Dry cells, or primary cells, such as the carbon-zinc, alkaline, and the mercury cell, are usually not rechargeable. Wet cells, sometimes called secondary cells, are capable of being recharged numerous times. There are a variety of wet cells; namely, lead-acid, nickel-iron alkali, and the silver-zinc types. Every battery has a small amount of internal resistance when placed in circuit.

Every electrical circuit requires conductors with minimum resistance to provide a path for current flow. The actual resistance of a conductor may be calculated, if its specific resistance, length, and cross-sectional area are known, by using the formula:

$$R = \rho \, \frac{L}{A}$$

Most conductors have a positive temperature coefficient (resistance rises as temperature rises).

There are a variety of wire sizes and types required to meet the demands of various electronic applications. A pair of conductors with constant spacing between them is called a transmission line; twin-lead and coaxial cable are two examples. Flat ribbon cable is used in digital circuitry. PC boards reduce the space needed to hard wire a circuit.

Commercial devices which provide a specific amount of resistance to a circuit are called resistors. They are rated by their ohmic value and their power rating. There are two basic categories of resistors: fixed and variable resistors. Composition and wirewound resistors are the two common types of variable resistors. The taper of a potentiometer can be linear or logarithmic. Ballast resistors, voltage dependent resistors, thyrites, and thermistors are all special resistors designed with high temperature coefficients. Each has its special purpose.

There are three other DC circuit components: the switch, a mechanical device used for completing or interrupting an electrical circuit; the fuse, a device used to prevent the flow of excess currents; and miniature lamps, used for pilot and indicator lights or instrument illumination.

CHECK YOURSELF (Answers at back of book)

Determine whether the following statements are true or false.

1. A dry cell is not considered rechargeable.
2. Fast blow fuses are used wherever short duration current surges are expected.
3. The coaxial cable is considered an ideal transmission line. Because the outer conductor is connected to ground, it shields the inner conductor from any external magnetic interference.
4. The number of poles represents the number of different contact points, and the number of throws represents the number of separate circuit connecting points.
5. The ballast resistor has a high negative temperature coefficient.
6. The resistance of a conductor is directly proportional to its cross-sectional area.
7. Wirewound resistors are generally used for high power, low-resistance applications.
8. Specific resistance is measured in ohms per circular-mil-foot of a substance.
9. The use of a printed circuit board will reduce the chance of destruction of an IC chip due to static electricity.
10. Every battery has a certain amount of internal resistance.
11. As the battery ages its internal resistance decreases, and the terminal voltage decreases under load to the point where it is insufficient to perform efficiently.
12. Most conductors have a positive temperature coefficient.
13. Thyrites are used in temperature-sensing and temperature-measuring devices.
14. Generally, low-tolerance resistors are less expensive than high-tolerance resistors.
15. Conductance is inversely proportional to the cross-sectional area of the conductor.
16. Solder helps to prevent the oxidation of a copper wire.
17. Although the metal-oxide film resistor is noted for its low-noise characteristics, it has poor temperature stability.
18. The metal film resistor has tolerances ranging from $\pm 0.0025\%$ to $\pm 2\%$.
19. The electrodes of a voltage cell are always made of two different materials.
20. It is possible for a switch to have multiple voltage and current ratings.

Fill in the blanks with the appropriate word or words.

21. A storage battery consists of a number of secondary cells connected in _____ .
    A. series
    B. parallel
    C. series-parallel

22. In materials such as glass, rubber, and mica, the electrons in the outer shells of these atoms _____ .
   A. are equally attracted to several atoms in the area
   B. are more strongly attracted to their parent atom
   C. require relatively small amounts of energy to be freed from their orbits.
23. If the resistance of a material increases when its temperature increases, the material is said to have a _____ .
   A. negative temperature coefficient
   B. positive temperature coefficient
   C. biased temperature coefficient
24. The mercury dry cell produces a terminal voltage of _____ .
   A. 1.55V
   B. 1.43V
   C. 1.34V
25. Resistors employed to maintain a constant current flow in spite of variations in line voltage are called _____ .
   A. thyrite resistors
   B. ballast resistors
   C. voltage-dependent resistors

CHAPTER 5

DC Measuring Instruments

A

B

Fig. 5-1: Major types of meters in use today: (A) analog with a needle and scale and (B) digital readout. (Photo A furnished by courtesy of Triplett Corporation and photo B by courtesy of Simpson Electric Company.)

In the field of electronics, as in all the other physical sciences, accurate quantitative measurements are essential. This involves two important items—numbers and units. Simple arithmetic is used in most cases and the units are well defined and easily understood. The standard units of current, voltage, and resistance, as well as other units, are defined by the National Bureau of Standards. At the factory various instruments are calibrated by comparing them with established standards.

There are several different types of measuring devices in common use today. For most DC measurement, however, the ammeter, voltmeter, and ohmmeter are employed. It is interesting to note that the measuring industry is going through a change of state at the present time. For years analog types of measuring devices were used. These devices converted the electrical quantity being measured into a reading found on a pre-calibrated scale of a meter. These types of measuring devices are still in use. They are gradually being replaced by digital readout types of measuring devices. Both analog and digital style meters are shown in Fig. 5-1.

To use measuring instruments properly, a basic understanding of their construction, operation, and limitation, coupled with the theory of circuit operation, is essential.

BASIC METER MOVEMENTS

One of the most popular analog meter movements still in use is the d'Arsonval type. This meter movement, named after its inventor, Arsene d'Arsonval, operates by virtue of the magnetic field associated with current flow. A sketch of this type of meter movement is shown in Fig. 5-2. The movement consists of a permanent magnet, an armature, and a pointer. A set of coil springs is attached to the armature. The springs hold the armature in the rest position when the meter is not connected to a circuit.

When the armature is connected to a test circuit, an electric current flows through its winding. This forms an electromagnet. The magnetic field of the armature reacts with the magnetic field established by the permanent magnet's pole pieces. The armature's field causes it to rotate. The rotation of

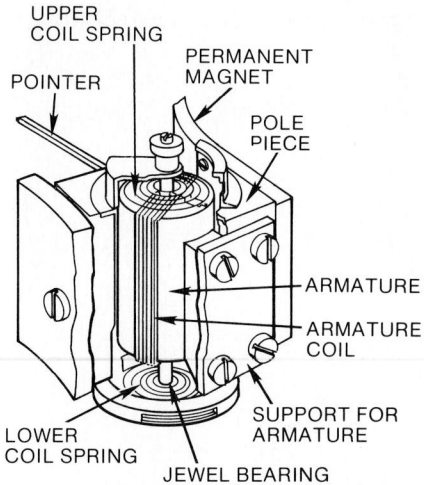

UPPER
COIL SPRING
POINTER
PERMANENT
MAGNET
POLE
PIECE
ARMATURE
ARMATURE
COIL
SUPPORT FOR
ARMATURE
LOWER
COIL SPRING
JEWEL BEARING

Fig. 5-2: Detailed view of basic d'Arsonval meter movement.

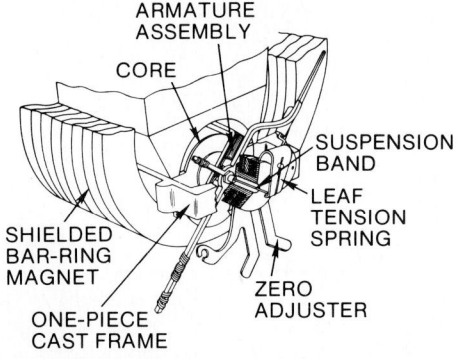

ARMATURE
ASSEMBLY
CORE
SUSPENSION
BAND
LEAF
TENSION
SPRING
SHIELDED
BAR-RING
MAGNET
ZERO
ADJUSTER
ONE-PIECE
CAST FRAME

Fig. 5-3: Taut-band movement. (Photo furnished by courtesy of Triplett Corporation.)

the armature swings the pointer needle to some position other than the "rest" position on the left side of the scale. The amount of current flowing through the armature will move the pointer to any point on the scale. As more current flows, the pointer swings closer to the right side of the scale. At maximum current flow, the pointer is on the right side of the scale. The amount of current flowing through the armature coil in this type of meter determines the amount of deflection. Armatures are produced with various current ratings.

This type of meter movement has some limitations. These include: (1) the overall current limitation which is based on the size of the wire in the armature, (2) the deflection of the needle which is limited to only one direction, and (3) only a very small amount of voltage will cause full-scale deflection of the needle. That is, any amount of voltage greater than this very small voltage will pin the needle and, as a result, will not be able to be measured. There are many methods in use today that modify the above limitations. These methods extend the capability of the meter movement.

Taut-Band Meters. No matter how well designed, the jewel bearing of a d'Arsonval will introduce frictional losses. In a movement drawing high power, these friction losses are insignificant. However, in high-sensitivity movements, pivot friction can be serious. The taut-band suspension eliminates this friction loss by eliminating the pivots. Instead, the moving coil is held in place by means of tightly stretched metal ribbons at the top and bottom. The instrument shown in Fig. 5-3 uses a taut-band suspension. The taut band serves three purposes:

1. Provides frictionless suspension for the moving coil
2. Provides current path to and from the coil
3. Provides the mechanical restoring torque, formerly supplied by the spiral springs

Due to the lack of friction, less driving and restoring torque is necessary. As a result, for a given full-scale current, the taut-band movement will have fewer turns and a lower resistance. Conversely, for a given coil resistance, these mechanisms can measure smaller currents. Taut-band instruments are available with sensitivities up to 1μA.

It should not be concluded that the taut-band design is always preferable. It is recommended where driving powers of less than 0.05mW are desired. On the other hand, vibrational stresses can interfere with their operation and extreme shock can make them permanently inoperative.

DC AMMETER

The ammeter is used to measure current. It may be calibrated in amperes (A), milliamperes (mA), or microamperes (μA). It must be inserted as part of the circuit so that all of the current being measured flows through the meter. A typical ammeter is shown in Fig. 5-4.

As mentioned previously, armature wire size will limit current flow in the meter. A meter movement's range is extended when a portion of the current is passed around the meter armature. This is illustrated in Fig. 5-5. One or more resistance (R) values are wired so that they are in parallel with the meter movement. If the value of R_1 is equal to the ohmic value of the armature, the current will divide equally between the two resistances. If the value of R_2 is smaller than the ohmic value of the meter movement, a greater percentage of the total circuit current will flow through R_2 than flows through the meter. R_3, being even a smaller value than R_1 or R_2, will allow even more current to flow in the circuit. These resistors are called *shunt* resistors. They permit a calculated percentage of current to bypass, or shunt, the meter movement. In this way the current capacity of the meter is extended.

For example, suppose it is desired to use a $100\,\mu A$ d'Arsonval meter having an internal coil resistance of $100\,\Omega$ to measure line currents up to 1A. The meter will deflect to its full-scale position when the current through the deflection coil is $100\,\mu A$. As the coil resistance is $100\,\Omega$, the coil's voltage is easily calculated by Ohm's Law, $V = I \times R = 0.0001 \times 100 = 0.01V$, where V is voltage in volts, I is current in amperes, and R is resistance in ohms. When the pointer is deflected to full scale, there will be $100\,\mu A$ of current flowing through the coil and $0.01V$ dropped across it. It should be remembered that $100\,\mu A$ is the maximum safe current for the meter movement. More than this value of current will damage the meter. It is, therefore, required that the shunt carry any additional load current.

There will be a $0.01V$ drop across the meter coil, and because the shunt and coil are in parallel, the shunt must also have a voltage drop of $0.01V$. The current that flows through the shunt must be the difference between the full-scale meter current and the line current being fed in. In this case the meter current is $100\,\mu A$. It is desired that this current cause full-scale deflection only when the total current is 1A; therefore, the shunt current must equal 1A minus $100\,\mu A$ or $0.9999A$. Ohm's Law now provides the value of required shunt resistance. $R = V/I = 0.01/0.9999 = 0.01\,\Omega$ (approximately).

The range of the $100\,\mu A$ meter can be increased to 1A (full-scale deflection) by placing a $0.01\,\Omega$ shunt in parallel with the meter movement.

The $100\,\mu A$ instrument may also be converted to a 10A meter by the use of a proper shunt. For full-scale deflection of the meter, the voltage drop, V, across the coil (and the shunt) is still $0.01V$. The meter current is still $100\,\mu A$. The shunt current must therefore be $9.9999A$ under full-scale deflection conditions. The required shunt resistance is $R = V/I = 0.01/9.9999 = 0.001\,\Omega$. Again the shunt resistance value is an approximation.

The same instrument can likewise be converted to a 50A meter by the use of a proper shunt. Calculating: $R = V/I = 0.01/49.9999 = 0.0002\,\Omega$ (approximately).

The above method of computing the shunt resistance is satisfactory in most cases when the line current is in the ampere range and the meter current is relatively small compared to the load current. In such cases it is allowable to use an approximate value of resistance for the shunt as was done above. However,

Fig. 5-4: Typical multirange ammeter. (Photo furnished by courtesy of Electrical Instrument Service.)

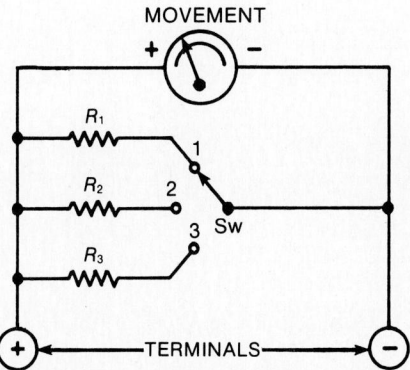

Fig. 5-5: Current ranges are increased by use of a shunt to bypass part of the current.

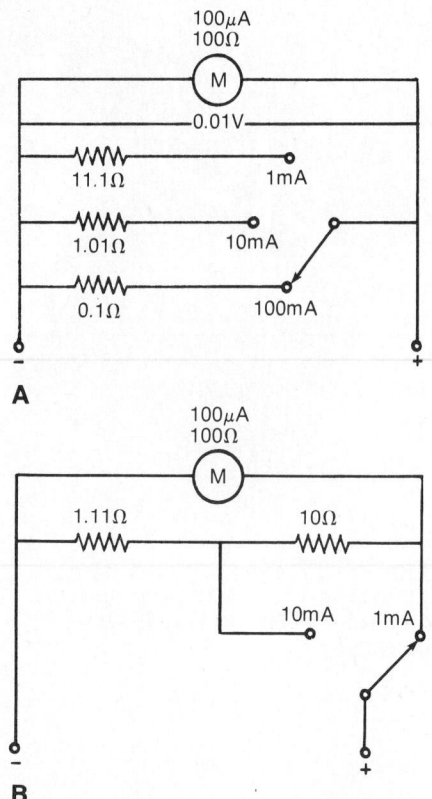

A

B

Fig. 5-6: Two methods of connecting internal meter shunts: (A) individual shunt and (B) Ayrton or universal shunt.

when the total current is in the milliampere range and the coil current becomes an appreciable percentage of the line current, a more accurate calculation must be made.

Example: It is desired to use a meter movement which has a full-scale deflection of 1mA and a coil resistance of 50Ω to measure currents up to 10mA. Using Ohm's Law the voltage across the meter coil (and the shunt) at full-scale deflection is: $V = I \times R = 0.001 \times 50 = 50mV$. The current that flows through the shunt is the difference between the line current and the meter current. $I = 10mA - 1mA = 9mA$. The shunt resistance must be $R = V/I = 0.05/0.009 = 5.55\Omega$. Notice that in this case the exact value of shunt resistance has been used rather than an approximation.

A formula for determining the resistance of the shunt is given by $R_S = I_m/(I_S \times R_m)$ where R_S is the shunt resistance in ohms, I_m is the meter current at full-scale deflection, I_S is the shunt current at full-scale deflection, and R_m is the resistance of the meter coil. If the values given in the previous example are used in this equation, it will yield 5.55Ω, the value previously calculated.

Various values of shunt resistance may be used, by means of a suitable switching arrangement, to increase the number of current ranges that may be covered by the meter. Two switching arrangements are shown in Fig. 5-6.

Figure 5-6A is the simpler of the two arrangements from the point of view of calculating the value of the shunt resistors when a number of shunts are used. However, it has two disadvantages:

1. When the switch is moved from one shunt resistor to another, the shunt is momentarily removed from the meter and the line current then flows through the meter coil. Even a momentary surge of current could easily damage the coil.

2. The contact resistance—that is, the resistance between the blades of the switch when they are in contact—is in series with the shunt, but not with the meter coil. In shunts that must pass high currents, the contact resistance becomes an appreciable part of the total shunt resistance. Because the contact resistance is of a variable nature, the ammeter indication may not be accurate.

A more generally accepted method of range switching is shown in Fig. 5-6B. Although only three ranges are shown, as many ranges as are needed can be used. In this type of circuit, known as the Ayrton or universal shunt, the range selector switch contact resistance is external to the shunt and meter in each range position and, therefore, has no effect on the accuracy of the current measurement.

Ammeter Connections

Current measuring instruments must always be connected in series with a circuit, *never* in parallel with it. If an ammeter is connected across a constant potential source of appreciable voltage, the low internal resistance of the meter bypasses the circuit resistance. This results in source voltage (or a good portion of it) being applied directly to the meter terminals and the resulting excessive current burning up the meter.

If the approximate value of current in a circuit is not known, it is best to start with the *highest range* of the ammeter and switch to progressively lower ranges until a suitable reading is obtained.

Most ammeter needles indicate the magnitude of the current being deflected from left to right. If the meter is connected with reversed polarity, the needle will be deflected backwards, an action that may damage the movement; therefore, the proper polarity should be observed in connecting the meter in the circuit. The meter should always be connected so that the current flow will be into the negative terminal and out of the positive terminal. Figure 5-7 shows various circuit arrangements and the proper ammeter connection methods to measure current in various portions of the circuits.

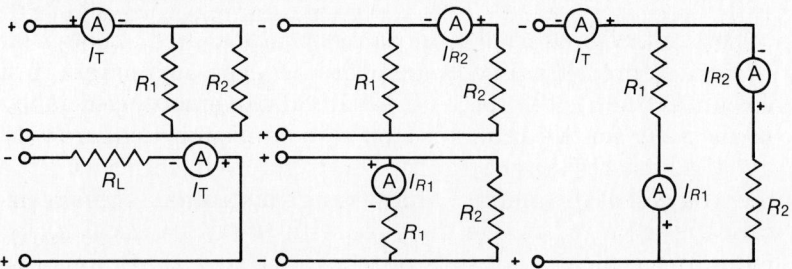

Fig. 5-7: Proper ammeter connections.

Ammeter Sensitivity

Meter sensitivity is determined by the amount of current required by the meter coil to produce full-scale deflection of the pointer. The smaller the current required to produce this deflection, the better the sensitivity of the meter. Thus, a meter movement which required only $100\mu A$ for full-scale deflection would have a better sensitivity than a meter movement which requires 1mA for the same deflection.

Good meter sensitivity is especially important in ammeters that are to be used in circuits where small currents are flowing. As the meter is connected in series with the load device, the current flows through the meter. If the internal resistance of the meter is an appreciable portion of the load resistance, a meter loading effect will occur. Meter loading is defined as "the condition which exists when the insertion of a meter into a circuit changes the operation of that circuit." This condition is not desired. When a meter is inserted into a circuit, its use is intended to allow the measurement of circuit current in the normal operating condition. If the meter alters the circuit operation and changes the amount of current flow, the reading obtained must be erroneous. An example of this problem is shown in Fig. 5-8.

In Fig. 5-8A the circuit to be tested is shown. There is an applied voltage of 100mV and a resistance of $100\,\Omega$. The current

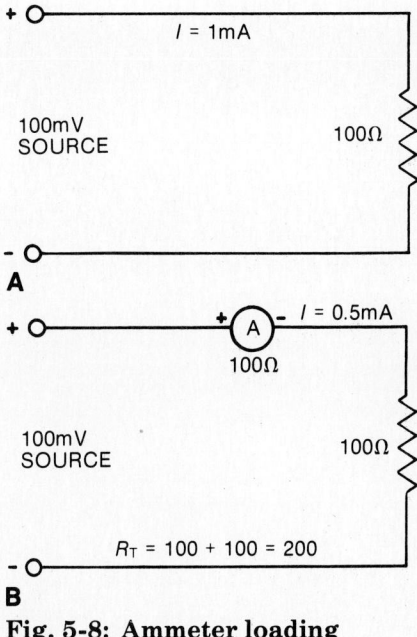

Fig. 5-8: Ammeter loading effect.

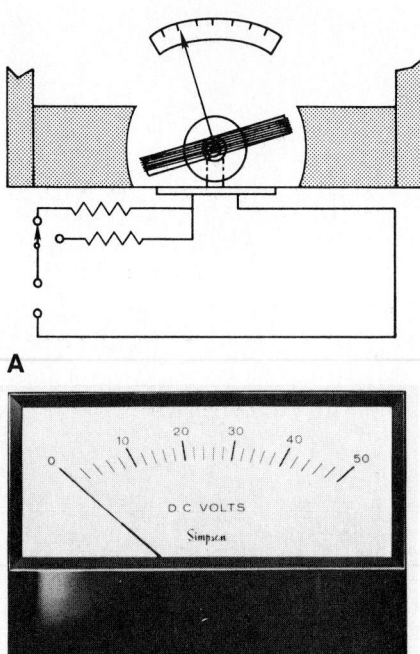

A

B

Fig. 5-9: (A) Internal construction and circuit of simplified voltmeter. (B) Typical single-range voltmeter. (Photo furnished by courtesy of Simpson Electric Company.)

normally flowing in this circuit would then be 1mA. In Fig. 5-8B, we insert an ammeter which requires 1mA for full-scale deflection and has an internal resistance of 100Ω. As there is 1mA of current flow in Fig. 5-8A, there is the tendency to say that with the meter inserted, a full-scale deflection will occur and the 1mA of circuit current will be measured. *This does not happen.* In Fig. 5-8B with the ammeter inserted into the circuit, the total resistance of the circuit now becomes 200Ω. With an applied voltage of 100mV, Ohm's Law yields the current as $I = V/R = 100 \times 10^{-3}/200 = 0.5 + 10^{-3} = 0.0005A$ or 0.5mA instead of the desired 1mA. Since the meter will read 0.5mA instead of the normal value of current, the meter has a definite loading effect. In such cases as this, it is desirable to use ammeters which have a better current sensitivity and a lower internal resistance.

DC VOLTMETER

The voltmeter is used to measure the difference of potential (electrical pressure or voltage drop) between two points in a circuit. Actually, the $100\mu A$ d'Arsonval movement used as the basic meter for the ammeter may also be used to measure voltage if a high resistance (multiplier) is placed in series with the moving coil of the meter. For low-range instruments this resistance is mounted inside the case with the d'Arsonval movement and typically consists of resistance wire that has a low temperature coefficient and is wound on either spools or card frames. For the higher voltage ranges, the series resistance may be connected externally. A simplified diagram of a voltmeter is shown in Fig. 5-9.

It must be remembered that the d'Arsonval meter movement, like most analog types, operates by use of a current flow to produce a magnetic field which is proportional to the current. This meter movement is, therefore, an indicator of current flow rather than voltage. The addition of a series resistance allows the meter to be calibrated in terms of voltage. It still operates because of the current flow through the meter, but the scale may be marked off in volts.

The meter movement being used in Fig. 5-10, for an example, has an internal resistance of 100Ω, requires $100\mu A$ for full-scale deflection, and will have a voltage drop of 10mV when full-scale deflection is achieved. If this meter were placed directly across a 10V source, an excessive current would flow ($I = V/R = 10/100 = 100$mA or 99.9mA too many) and the meter movement would be destroyed.

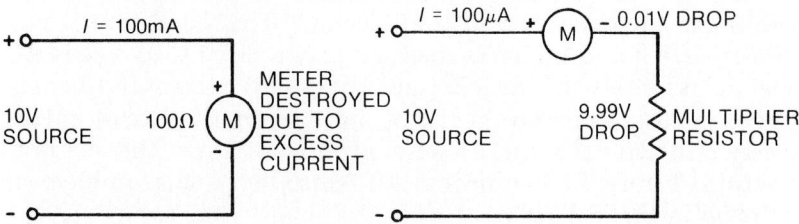

Fig. 5-10: Usage of multiplier resistors with a d'Arsonval movement.

80

Since the normal meter voltage drop is 10mV for full-scale deflection condition, some means is required to drop 9.99V without applying it directly to the meter. This is the function of the *multiplier resistor* shown in Fig. 5-10.

Extending Voltmeter Ranges

The value of the necessary series resistance is determined by the current required for full-scale deflection of the meter and by the range of voltages to be measured. Because the current through the meter circuits is directly proportional to the applied voltage, the meter scale can be calibrated directly in volts for a fixed value of series resistance.

For example, assume that the basic meter is to be made into a voltmeter with a full-scale reading of 1V. The coil resistance of the basic meter is 100Ω, and $100\mu A$ of current causes a full-scale deflection. The total resistance which is required of the circuit (to limit total current to $100\mu A$) is given by Ohm's Law as $R = V/I = 1V/100\mu A = 10k\Omega$ (for a full-scale 1V range). As the meter coil already contains 100Ω, the series resistance is therefore required to be $10k\Omega - 100\Omega = 9.9k\Omega$.

Multirange voltmeters utilize one meter movement with the required resistances connected in series with the meter by a conventional switching arrangement. A schematic diagram of a multirange voltmeter with three ranges is shown in Fig. 5-11. The total resistances for each of the three ranges, beginning with the 1V range, are given by Ohm's Law as:

> For 1V range—$R = 1/0.0001 = 10k\Omega$
> For 100V range—$R = 100/0.0001 = 1M\Omega$
> For 1,000V range—$R = 1,000/0.0001 = 10M\Omega$

The multiplying series resistor (R_m) for each of these circuits is 100Ω less than the total resistance.

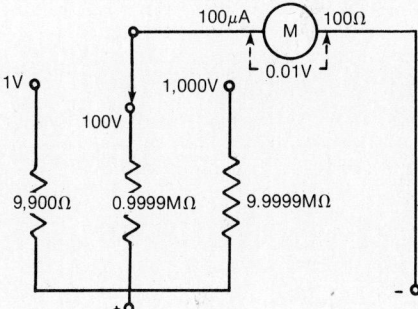

Fig. 5-11: Multirange voltmeter.

Voltmeter Connection

Voltage measuring instruments are connected across (in parallel with) the circuit component to be measured. If the voltage to be measured is not known, it is best to start with the highest range of the voltmeter and progressively lower the range until a suitable reading is obtained.

In many cases, the voltmeter is not a center-zero indicating instrument. In a center-zero indicating instrument, the zero-current mark is in the exact center of the scale. Current flowing in one direction will deflect the pointer to the right, while current flowing in the other direction will deflect the pointer to the right. When the polarity markings on the input terminals are observed, a positive current deflects the pointer to the right and a negative current deflects it to the left. Thus, it is necessary to observe the proper polarity when connecting the instrument to the circuit. The voltmeter is connected so that current will flow into the negative terminal of the meter.

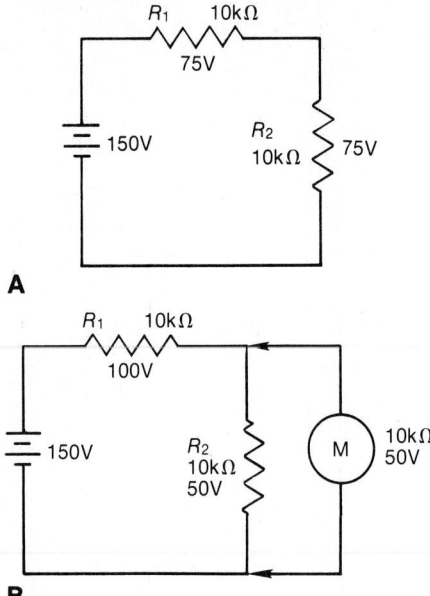

A

B

Fig. 5-12: Shunting action caused by a voltmeter.

Influence in a Circuit

The function of a voltmeter is to indicate the potential difference between two points in a circuit. When the voltmeter is connected across a circuit, it shunts the circuit. If the voltmeter has a low resistance, it will draw an appreciable amount of current. The effective resistance of the circuit will be lowered, and the voltage reading will consequently be changed. When voltage measurements are made in high-resistance circuits, it is necessary to use a high-resistance voltmeter to prevent this shunting action of the meter. The effect is less noticeable in low-resistance circuits because the shunting effect is less. The problem of voltmeter shunting is illustrated in Fig. 5-12.

In Fig. 5-12A, a source voltage of 150V is applied to a series circuit consisting of two 10kΩ resistors. As shown, the voltage drop across each resistor is 75V. It is now assumed that in the 150V range the voltmeter to be used has a total internal resistance of 10kΩ. Figure 5-12B shows the voltmeter connected. It is in parallel with R_2. The parallel combination of R_2 and the meter now exhibit a total resistance of 5kΩ. As a result the voltage drops change to 100V across R_1 and 50V across R_2, which is in parallel with the meter. It is easily seen that this is not the normal voltage drop across R_2.

Sensitivity

The sensitivity of a voltmeter is given in ohms per volt (Ω/V) and may be determined by dividing the resistance of the meter, R_m, plus the series resistance, R_S, by the full-scale reading in volts. Thus, in equation form, sensitivity = $(R_m + R_S)/V$. This is the same as saying that the sensitivity is equal to the reciprocal of the full-scale deflection current. In equation form then,

$$\text{Sensitivity} = \frac{\Omega}{V} = \frac{1}{\dfrac{V}{\Omega}} = \frac{1}{A}$$

Thus, the sensitivity of a 100μA movement is the reciprocal of 0.0001A, or 10,000Ω per volt.

Accuracy. Test meters are available with accuracies ranging from as high as ±0.1% to a low of ±5%. A point that is not often realized is that this accuracy figure applies to full-scale deflection. For example, a 10A meter with ±2% accuracy may have an error of ±2% of full scale or 0.2A anywhere on its scale. So, when an indication of 1.0A is read on this meter, the actual current may be 0.8 or 1.2A, and the actual error at this point on the scale may be as high as ±20%. It is desirable, for good accuracy, to use a range that will give readings in the upper half of the meter scale.

Accuracy of any indicating instrument (ammeter and voltmeter) depends on the quality of the components used. The tolerance of the resistors used as shunts (for ammeters) or multipliers (for voltmeters) contributes to the meter accuracy.

Precision resistors must be used for high accuracy. The magnets, bearings, and springs that supply the restoring torque also contribute to the accuracy of the finished instrument. Finally, where extreme accuracy is required, the scale is hand calibrated to match the actual deflection of the moving coil.

When the manufacturer has done all that can be done to produce an accurate instrument, there is still another source of potential error—the user. To prevent sticking of the needle against the face of the scale, the needle must of necessity be raised a short distance above the scale. Because of this spacing, an exact reading can be made only if the eye is directly over the needle. If the eye is to the right, the indication would be read to the left—giving a lower reading. The converse is true if the eye is to the left. This effect is known as *parallax*. A little experiment will show how serious parallax error can be. Close one eye and hold up one finger approximately 1' in front of the other eye. Now sight on some mark on a wall. Keeping your finger stationary, move your head just a small distance to either side and notice how far off the mark your sighting has shifted. Of course, this experiment exaggerates the error, but when you are dealing with accuracies of ±1% or better, meter parallax cannot be ignored (Fig. 5-13). To avoid parallax errors, high quality instruments use mirror scales. When reading such a meter, the eye is properly aligned when the reflection of the needle in the mirror is directly below the needle itself.

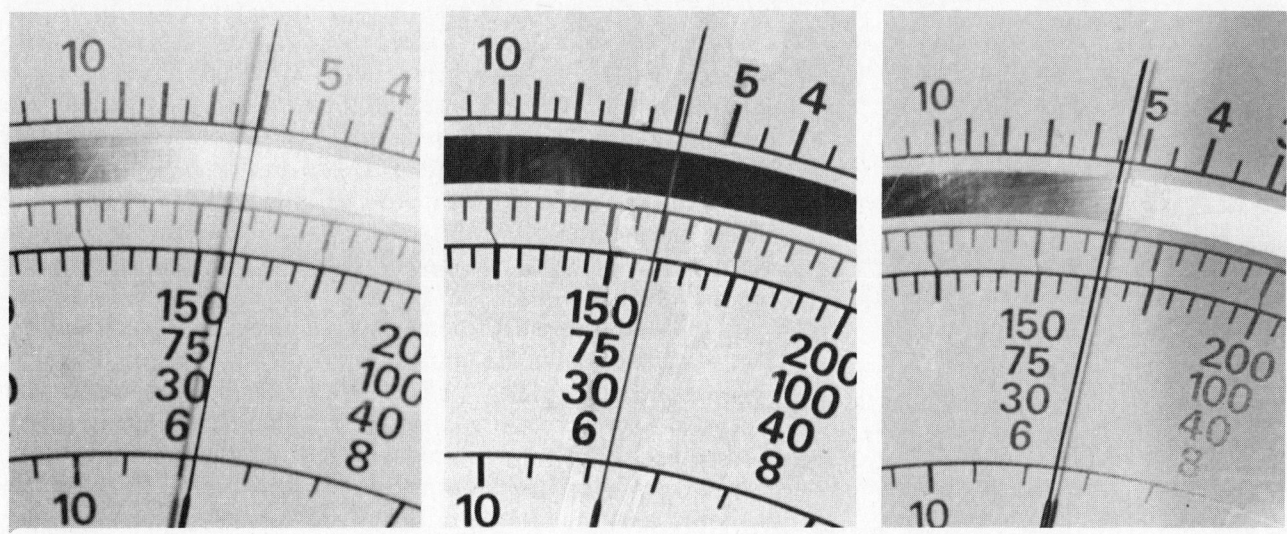

Fig. 5-13: Parallax can be a problem if needle position is not carefully read.

THE OHMMETER

The ohmmeter consists of a DC milliammeter with a few added features. The added features are:
1. A DC source of potential.
2. One or more resistors (one of which is variable).

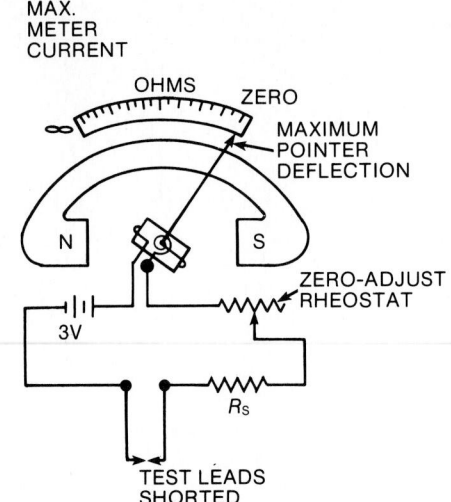

Fig. 5-14: Simple series ohmmeter circuit.

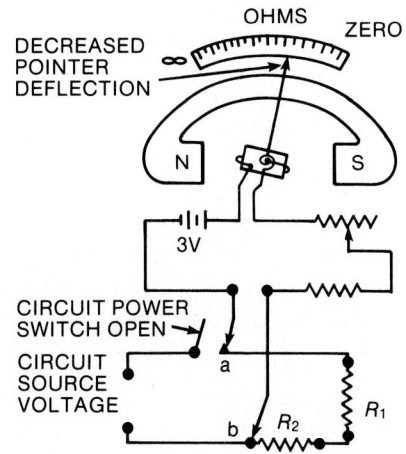

Fig. 5-15: Measuring circuit resistance with a series ohmmeter.

Series Type. A simple series ohmmeter circuit is shown in Fig. 5-14. The ohmmeter's pointer deflection is controlled by the amount of battery current passing through the moving coil. Before measuring the resistance of an unknown resistor or electrical circuit, the ohmmeter must be calibrated. If the value of resistance to be measured can be estimated within reasonable limits, a range is selected which will give approximately half-scale deflection when this resistance is inserted between the probes. If the resistance is unknown, the selector switch should be set on the highest scale. Whatever range is selected, the meter must be calibrated to read zero before the unknown resistance is measured.

Calibration is accomplished by first shorting the test leads together as shown in Fig. 5-14. With the test leads shorted, there will be a complete series circuit consisting of the 3V source, the resistance of the meter coil, R_m, the resistance of the zero-adjust rheostat, and the series multiplying resistor, R_S. Current will flow and the meter pointer will be deflected. The zero point on the ohmmeter scale (as opposed to the zero for voltage and current) is located at the extreme right side of the scale. With the test leads shorted, the zero-adjust potentiometer is set so that the pointer rests on the zero mark; therefore, full-scale deflection indicates zero resistance between the test leads.

If the range is changed, the meter must be "zeroed" again to obtain an accurate reading. When the test leads of an ohmmeter are separated, the pointer of the meter will return to the left side of the scale, due to the spring tension acting on the movable coil assembly. This reading indicates an infinite resistance.

After the ohmmeter is adjusted for zero reading, it is ready to be connected in a circuit to measure resistance. A typical circuit and ohmmeter arrangement is shown in Fig. 5-15. The power switch of the circuit to be measured should always be in the OFF position. This prevents the circuit's source voltage from being applied across the meter, which could cause damage to the meter movement.

As indicated above, the ohmmeter is an open circuit when the test leads are separated. In order to be capable of taking a resistance reading, the path for current produced by the meter's battery must be completed. In Fig. 5-15, this is accomplished by connecting the meter at points a and b (putting resistors R_1 and R_2 in series with the resistance of the meter coil, zero-adjust rheostat, and the series multiplying resistor). Since the meter has been preadjusted (zeroed), the amount of coil movement now depends solely on the resistance of R_1 and R_2. The inclusion of R_1 and R_2 raised the total series resistance, decreased the current, and thus decreased the pointer deflection. The pointer will now come to rest at a scale figure, indicating the combined resistance of R_1 and R_2. If R_1 or R_2, or both, were replaced with a resistor having a larger ohmic value, the current flow in the moving coil of the meter would be decreased still more. The deflection would also be further decreased and the scale indication would read a still higher circuit resistance. Movement of the moving coil is proportional to the intensity of current flow. The scale reading of the meter, in ohms, is inversely proportional to current flow in the moving coil.

The amount of circuit resistance to be measured may vary over a wide range. In some cases it may be only a few ohms, and in other cases it may be as great as $1M\Omega$. To enable the meter to indicate any value being measured, with the least error, scale multiplication features are incorporated in most ohmmeters. Most ohmmeters are equipped with a selector switch for selecting the multiplication scale desired. For example, a typical meter may have a six position switch, marked as follows: $R \times 1$, $R \times 10$, $R \times 100$, $R \times 1K$, $R \times 10K$, and $R \times 100K$. This is shown in Fig. 5-16.

The range to be used in measuring any particular unknown resistance (R_x in Fig. 5-16) depends on the approximate ohmic value of the unknown resistance. For instance, assume the ohmmeter scale is calibrated in divisions from zero to 1,000. If R_x is greater than $1,000\Omega$, and the $R \times 1$ range is being used, the ohmmeter cannot measure it. This occurs because the combined series resistor $R \times 1$ and R_x is too great to allow sufficient battery current to flow to deflect the pointer away from infinity (∞). The switch would have to be turned to the next range, $R \times 10$.

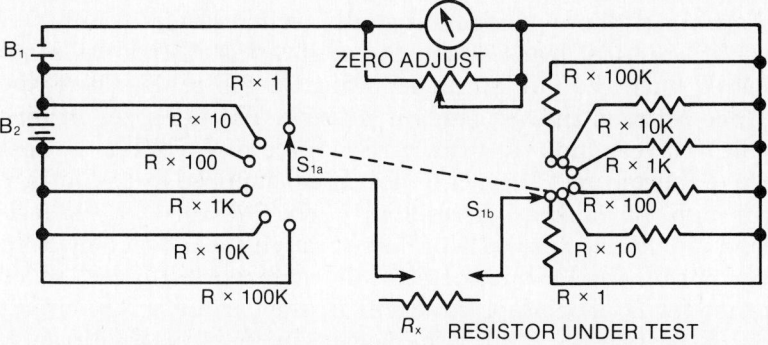

Fig. 5-16: Series ohmmeter with multiplication switch.

When this is done, assume the pointer deflects to indicate 375. This would indicate that R_x has $375 \times 10 = 3,750\Omega$ of resistance. The change of range caused the deflection because resistor $R \times 10$ has only one-tenth the resistance of resistor $R \times 1$. Thus, selecting the smaller series resistance allowed a battery current of sufficient value to cause a useful pointer deflection. If the $R \times 100$ range were used to measure the same $3,750\Omega$ resistor, the pointer would deflect still further, to the 37.5Ω position. This increased deflection would occur because resistor $R \times 100$ has only one-tenth the resistance of resistor $R \times 10$.

The foregoing circuit arrangement allows the same amount of current to flow through the meter moving coil whether the meter measures $10,000\Omega$ on the $R \times 1$ scale, or $100,000\Omega$ on the $R \times 10$ scale, or $1M\Omega$ on the $R \times 100$ scale.

It always takes the same amount of current to deflect the pointer to a certain position on the scale (midscale position for

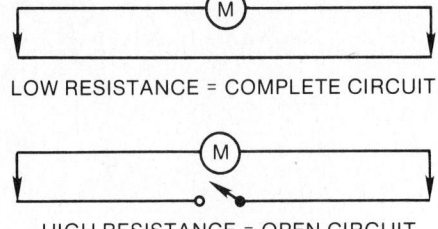

LOW RESISTANCE = COMPLETE CIRCUIT

HIGH RESISTANCE = OPEN CIRCUIT

Fig. 5-17: Continuity measurements are determined as shown.

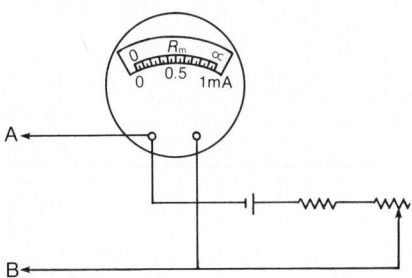

Fig. 5-18: Shunt-type ohmmeter.

example), regardless of the multiplication factor being used. Since the multiplier resistors are of different values, it is necessary to always "zero" adjust the meter for each multiplication scale selected. The operator of the ohmmeter should select the range that will result in the pointer coming to rest as near midpoint of the scale as possible. This enables the operator to read the resistance more accurately, because the scale readings are more easily interpreted at or near midpoint.

There are instances when one wishes to check for a continuous circuit. This is called a *continuity check*. The ohmmeter may be used to do this. The check for continuity is actually a resistance measurement. The amount of resistance measured in the circuit will give an excellent indication of the quality of the circuit. If the meter shows a very low resistance, as illustrated in Fig. 5-17, then the circuit has continuity. If, on the other hand, the circuit has a break in it, the resistance value is very high. Wires and switches are tested in this manner. If the test indicated that there was some resistance, but the value was much higher than expected, some component or wire would probably be failing.

Shunt Type. The shunt ohmmeter is used to measure low values of resistance. Instead of shunting the movement and connecting R_x in series with the calibrating resistor and battery as before, the unknown resistor itself is used as the shunt. A diagram of this type of ohmmeter is seen in Fig. 5-18.

With the test leads apart, the calibrating resistance is adjusted until the milliammeter indicates full scale. The resistance between the test leads is infinite. The full-scale current point corresponds to infinite resistance of R_x. With the test leads shorted, $R_x = 0$, and all the current supplied by the battery through the calibrating resistance will flow through the test-lead circuit. No current will flow through the movement. Zero current on the milliammeter scale corresponds to zero resistance for R_x. For other values of R_x the current in the circuit divides between R_x and the movement inversely as their respective resistances. When R_x is equal to the resistance of the movement, the current division will be 50% in each. The half-way point on the current scale should, therefore, be marked with a resistance value equal to the resistance of the moving coil. For a 1mA movement, this value may be as low as 27Ω. Good accuracy on low resistance can be obtained with this type of ohmmeter.

MULTIMETERS

The typical VOM (volt-ohm-milliampere) multimeter (Fig. 5-19) is capable of measuring DC current, DC resistance, DC voltage, as well as reading AC voltage quantities. Selection of the proper function switch permits this versatility. The voltage ranges cover low values of 1 or 2V up to a high value of 1,000V. This type of meter is desired as a portable test meter because of its versatility. One unit has the capability of making a wide variety of measurements.

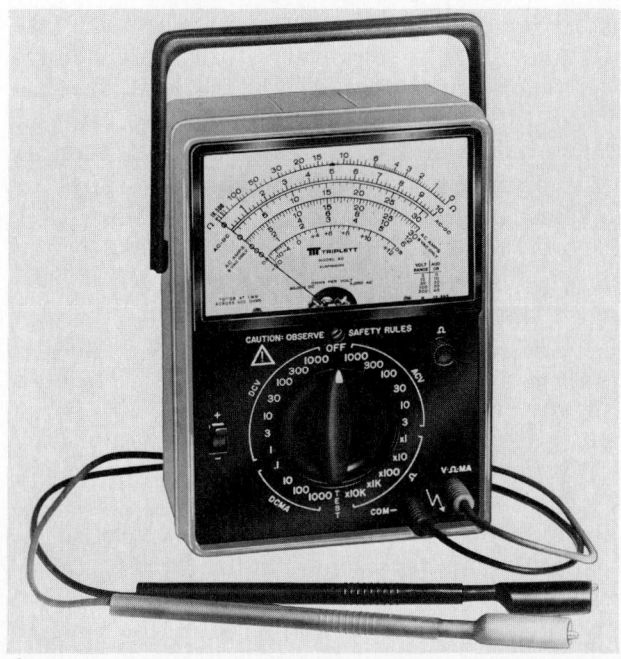
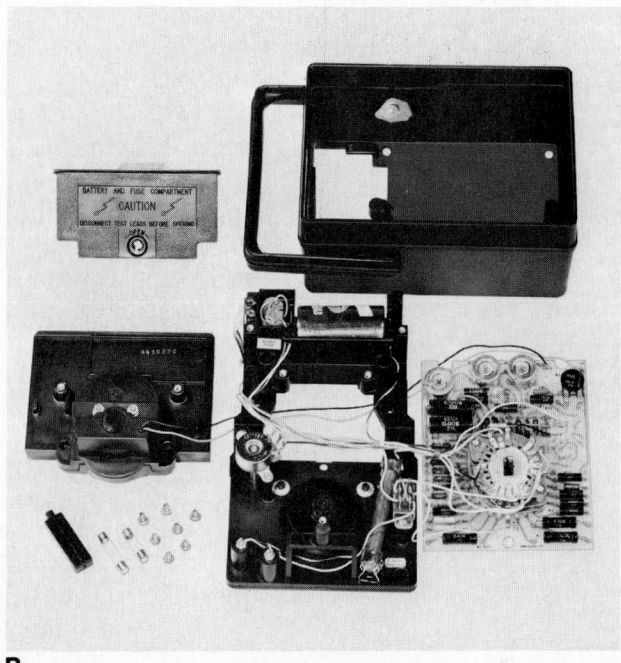

A   **B**

Fig. 5-19: Front and internal view of typical VOM. (Photos furnished by courtesy of Triplett Corporation.)

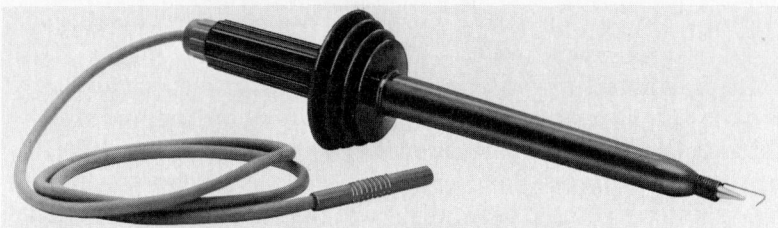

Fig. 5-20: High-voltage probe accessory for VOM. (Photo furnished by courtesy of Triplett Corporation.)

Better quality multimeters have a much higher ohms per volt rating. Most of these meters use a value of either 20 or 30kΩ per volt. This value is called the *input sensitivity* of the meter.

Many multimeters have an accessory probe (Fig. 5-20) available that can be employed to measure DC voltage up to 30kV. One application is measuring the anode voltage of 20 to 30kV for the picture tube in a TV receiver. Actually, this probe is only an external multiplier resistance for the voltmeter. The required R for a 30kV probe is 580MΩ with a 20kΩ/V meter on the 1,000V range.

ELECTRONIC METERS

Meter input sensitivities that are constantly high are desired for work on low-voltage and low-current circuits. Meters with an input sensitivity of 1 or 2MΩ for each range are used for measuring a wide variety of circuit values. These meters are

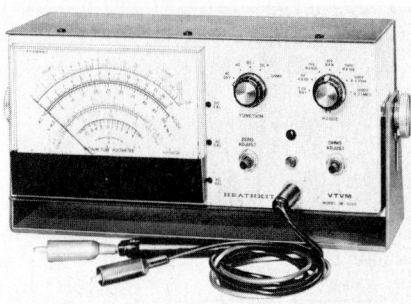

Fig. 5-21: Typical VTVM. (Photo furnished by courtesy of Heath Company.)

Fig. 5-22: Solid-state FET (field effect transistor) meter. (Photo furnished by courtesy of Simpson Electric Company.)

classified as electronic multimeters. Older versions of these meters used vacuum tubes as amplifiers to raise the input resistance value. These are called VTVMs (Fig. 5-21). This stands for vacuum tube voltmeters. Many of these types of meters in current production use solid-state devices such as transistors or integrated circuits instead of vacuum tubes for the amplifiers. These units are called EVM, TRVM, or FET meters. These letters indicate an electronic or a transistorized voltmeter. A picture of this type of meter is shown in Fig. 5-22.

Some electronic meters do not measure current (Fig. 5-23). This is a design characteristic. Many of the newer electronic meters have a digital readout instead of the more traditional analog meter movement. The digital type of readout offers a more precise reading than can be obtained from an analog dial. Like analog multimeters, the digital type has a function switch, a range switch, and two terminals. This type of instrument also has an external supply of electricity or is battery operated.

The quantity measured is represented in the form of a four-digit display with a properly placed decimal point. The polarity is symbolized by means of a + or − sign preceding the number when DC quantities are measured. Figure 5-24 shows the displays obtained when measuring −1,000V. The most accurate measurements with such meters, as with the analog, are achieved when the lowest possible range is used. Some digital meters do not have a range switch but use "automatic ranging circuits" that automatically vary the range to produce the best display for a given input.

Digital meters are obviously much easier to use, because they reduce the human error that often can occur in reading the different scales of an analog meter. They are also less likely to be damaged by overload conditions. Digital readout meters, having more complex circuitry, are normally more expensive than their analog counterparts.

There are times when an analog type of meter movement is mandated. In some circuitry it is necessary to adjust components for either a maximum or a minimum value. The fairly slow response of the digital meter makes it difficult to use in these cases. The analog type of meter movement displays changes more conveniently. Some digital meters have an accessory analog display just for this purpose.

SUMMARY

The three most commonly used instruments for measuring DC circuit values are the ammeter (a device for measuring current), the voltmeter (for measuring voltage), and the ohmmeter (a device for measuring resistance).

These measuring instruments may be of either the analog or digital types. Analog meters employ basic meter movements that use the electrical quantity being measured to deflect a needle, or pointer, over a precalibrated scale. The operator of the device must then interpolate the location of the pointer in respect to the scale to attain an accurate reading. In a digital meter, the quantity measured is represented in a four-digit display with a properly placed decimal point and a + or – symbol to indicate polarity. Because digital meters are so much more easy to use, they are gradually replacing their analog predecessors.

There are two basic types of meter movements: the d'Arsonval type and the taut-band type. Due to the major differences in the construction of these two types of meter movements, each type has definite advantages and disadvantages.

The ammeter is used to measure current; however, the size of the wire in the armature of the meter movement limits the amount of current allowed to flow through the meter. Therefore, shunt resistors must be added to extend the usefulness of the meter. Various values of shunt resistance may be employed by means of a switching arrangement to increase the number of current ranges that may be covered by the meter.

Ammeters must always be connected in series with a circuit, never in parallel with it. If an approximate value of the current to be measured is not known, measurements should be made with the highest range of the ammeter. The range switch can then be lowered to obtain a more accurate reading.

Ammeters must be connected with the right polarity. Current should flow into the negative terminal and out of the positive terminal of the device.

Meter sensitivity is determined by the amount of current required by the meter to cause full-scale deflection. The smaller the current required to produce this deflection, the better the sensitivity of the meter.

Voltmeters are used to measure the difference of potential between two points in a circuit. A voltmeter is similar to an ammeter in construction in that it, too, uses a d'Arsonval meter movement and current limiting resistors. However, in a voltmeter, these resistors (multipliers) are connected in series with the meter movement. A group of these multiplier resistors can be connected by means of a switching arrangement to produce a multirange voltmeter.

To make a voltage measurement, the voltmeter must be attached in parallel with the circuit component(s) to be measured.

When the voltmeter is connected across a circuit, it shunts the circuit. If the voltmeter has a low resistance, it will draw an appreciable amount of current from the circuit; thus, the effective resistance of the circuit will be lowered and the voltage reading will be inaccurate. In such a case, it is said that the voltmeter is loading the circuit. High-resistance voltmeters must be used in high-resistance circuits in order to prevent circuit loading.

The sensitivity of a voltmeter is given in ohms per volt (Ω/V) and may be determined by dividing the resistance of the meter, plus the series resistance, by the full-scale reading in volts. In other words, sensitivity equals the reciprocal of the full-scale deflection current.

Accuracy of any measuring instrument depends mostly on the quality of the components used in making the instruments. Precision resistors and quality magnets, bearings, and springs

Fig. 5-23: Digital volt-ohm-meter. (Photo furnished by courtesy of Simpson Electric Company.)

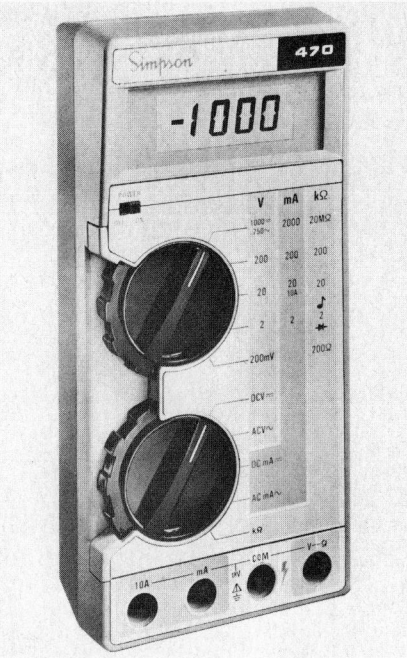

Fig. 5-24: Electronic meter that utilizes a digital readout. (Photo furnished by courtesy of Simpson Electric Company.)

all contribute to the accuracy of the finished instrument. Test meters are available with accuracies ranging from ±0.1% to 5%.

Parallax must be considered in making meter readings. An ohmmeter is nothing more than an ammeter with a DC source of potential and one or more resistors (one of which is variable). It must be calibrated before resistance measurements are made. Resistance measurements are made with the resistors to be measured in series with the ohmmeter.

Shunt type ohmmeters are available for measuring low values of resistance. Instead of shunting the movement and connecting the unknown resistance in series with the calibrating resistor and battery, the unknown resistor itself is used as the shunt.

Multipurpose meters (multimeters) are also available. They are capable of measuring DC current, resistance, and voltage as well as reading AC quantities.

CHECK YOURSELF (Answers at back of book)

Questions

Determine whether the following statements are true or false.

1. The range of an ammeter can be extended by using a shunt resistor.
2. A 50mA meter movement will make a more sensitive voltmeter than a 20mA meter movement.
3. An ohmmeter is calibrated with its leads shorted together.
4. Ammeter accuracy is expressed as a percentage of full-scale value.
5. A voltmeter can be made from a DC current meter movement if a multiplier resistor is connected in parallel with the meter.
6. The current meter should have a resistance that is high when compared to circuit resistances.
7. The jewel bearing of a taut-band meter movement introduces frictional losses which can be serious in high-frequency applications of the meter.
8. Full-scale deflection on an ohmmeter scale represents infinite resistance.

Problems

1. A basic meter movement with a full-scale deflection current of 75mA and a coil resistance of 900 Ω is to be used as a voltmeter. Calculate the values of the individual multiplier resistances required to give a full-scale deflection of (A) 100V, (B) 30V, and (C) 5V.
2. What is the sensitivity of a voltmeter constructed from a 30mA movement?
3. If a 200mA meter with a specified accuracy of ±3% indicates 130mA, what are the possible minimum and maximum values of the actual current?

CHAPTER 6

DC Circuits— Series

There are two basic circuits found in electrical and electronic work. These are the *series circuit* and the *parallel circuit*. Any other circuit that is used is a combination of these two circuits. Each one of these has its own set of rules. The rules apply to current, resistance, voltage, and power in the circuit. Circuit analysis is based upon recognition of the type of circuit and application of the related rules for that circuit. The circuit explained in this chapter is the series circuit. It is illustrated in Fig. 6-1. It consists of a source, two or more loads, and connecting lines.

Identification of the Series Circuit

The series circuit is easy to identify. The major means of identification in this circuit is the number of paths in which current can flow. The series circuit has only one path. It may be drawn as illustrated in Fig. 6-2A. Often this circuit is simplified. The power source is not illustrated. A number representing the voltage value is shown instead. This number is placed at one end of the components making up the circuit. Usually, this is the positive end of the circuit. This is shown in Fig. 6-2B. When this type of circuit is shown, the connections to the power source are assumed.

RESISTANCE IN THE SERIES CIRCUIT

The total resistance in the series circuit depends upon the number of resistive loads in the circuit and the resistance value of each of these loads. Resistance values are added in the series circuit. Total circuit resistance is equal to the sum of each resistance in the circuit.

$$R_T = R_1 + R_2 + R_3 + \dots R_n$$

Example: Three resistors of 10Ω, 15Ω, and 30Ω are connected in series across a battery whose emf is 110V (Fig. 6-3). What is the total resistance?

Given: $R_1 = 10\Omega$
              $R_2 = 15\Omega$
              $R_3 = 30\Omega$
              $R_T = ?$

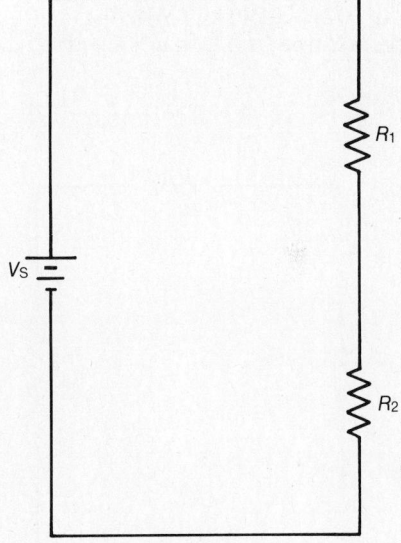

Fig. 6-1: The basic series circuit with one source (V_S) and two loads (R_1 and R_2).

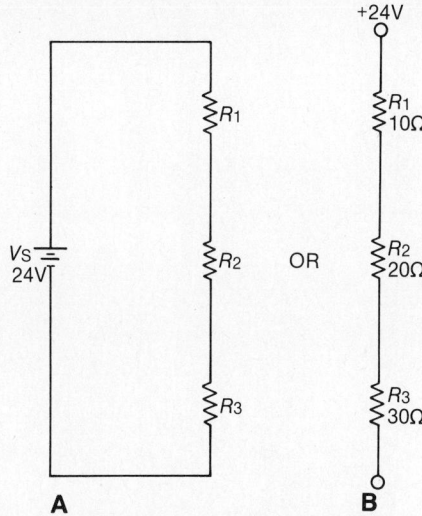

Fig. 6-2: Two methods of showing the same circuit. One shows the source. The other shows a value that represents the source.

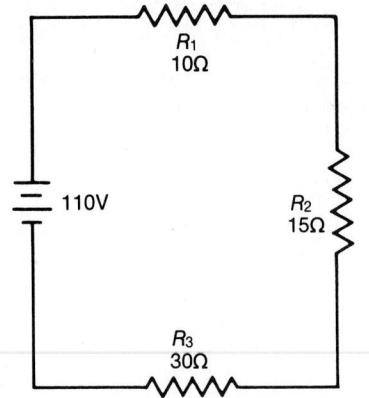

Fig. 6-3: Solving for total resistance (R_T) in a series circuit.

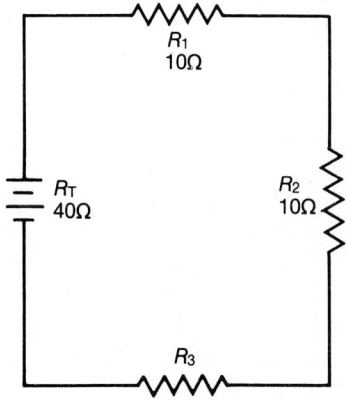

Fig. 6-4: Calculating the value of one resistance in a series circuit.

Solution:
$$R_T = R_1 + R_2 + R_3$$
$$R_T = 10\Omega + 15\Omega + 30\Omega$$
$$R_T = 55\Omega$$

In some circuit applications, the total resistance is known and the value of a circuit resistor has to be determined. The total circuit resistance equation can be transposed to solve for the value of the unknown resistance.

Example: The total resistance of a circuit containing three resistors is 40Ω (Fig. 6-4). Two of the circuit resistors are 10Ω each. Calculate the value of the third resistor.

Given:
$$R_T = 40\Omega$$
$$R_1 = 10\Omega$$
$$R_2 = 10\Omega$$
$$R_3 = ?$$

Solution:
$$R_T = R_1 + R_2 + R_3$$

Subtracting ($R_1 + R_2$) from both sides of the equation:
$$R_3 = R_T - R_1 - R_2$$
$$R_3 = 40\Omega - 10\Omega - 10\Omega$$
$$R_3 = 20\Omega$$

CURRENT IN THE SERIES CIRCUIT

Current in a series circuit depends upon two factors. These are the total circuit resistance and the source voltage. There is but one path for current flow in the series circuit. This being the case, the circuit current is the same throughout the circuit. Figure 6-5 illustrates a series circuit with three load resistances. If one inserts a current-measuring device, or ammeter, at each of the points identified as A through F, the measured current is the same. An increase in the resistance of any resistor(s) in a series circuit will increase the total resistance, thus decreasing the current.

Circuit current is determined by using Ohm's Law. In order to find the R for this formula, total resistance is determined by adding each resistance value.

Example: In Fig. 6-5,
$$R_T = R_1 + R_2 + R_3 = 10\Omega + 20\Omega + 30\Omega = 60\Omega$$

The total resistance value is inserted into the formula $I = V/R$. Substituting known values for the letters given,

$$I = \frac{12V}{60\Omega} = 0.2A$$

The current flowing through each resistor is 0.2A. The amount of current is determined by the total resistance.

VOLTAGE IN THE SERIES CIRCUIT

The total voltage across a series circuit is called the source voltage and consists of the sum of two or more individual voltage drops. The voltage developed across an individual resistor is called a voltage drop. In any series circuit the *sum* of the resistor voltage drops must equal the source voltage. This statement can be proven by an examination of the circuit shown in Fig. 6-6. In this circuit a source potential (V_T) of 20V is impressed across a series circuit consisting of two 5Ω resistors. The total resistance of the circuit is equal to the sum of the two individual resistances, or 10Ω. Using Ohm's Law, the circuit current may be calculated as follows:

$$I = \frac{V_T}{R_T}$$

$$I = \frac{20V}{10\,\Omega}$$

$$I = 2A$$

Knowing the size of the resistors to be 5Ω each and the current through the resistors to be 2A, the voltage drops across the resistors can be calculated. The voltage (V_1) across R_1 is therefore:

$$V_1 = IR_1$$

$$V_1 = 2A \times 5\,\Omega$$

$$V_1 = 10V$$

Since R_2 is the same ohmic value as R_1 and carries the same current, the voltage drop across R_2 is also equal to 10V. Adding these two 10V drops together gives a total voltage of 20V (exactly equal to the applied voltage). For a series circuit then:

$$V_T = V_1 + V_2 + V_3 + \dots V_n$$

Example: A series circuit consists of three resistors having values of 20Ω, 30Ω, and 50Ω respectively. Find the applied voltage if the current through the 30Ω resistor is 100mA.

To solve the problem, a circuit diagram is first drawn and labeled (Fig. 6-7).

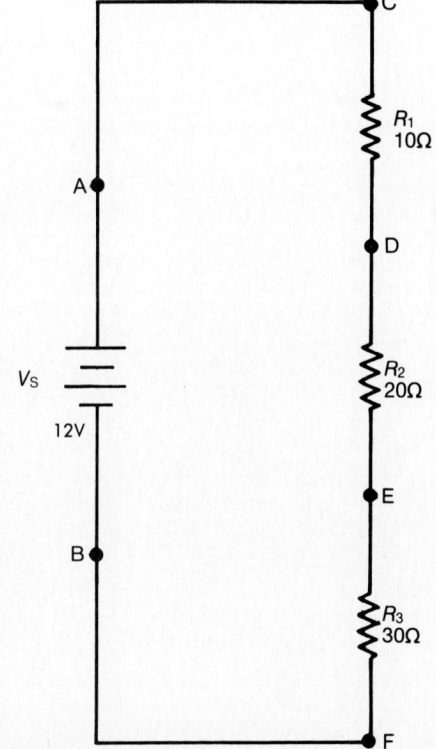

Fig. 6-5: Series circuit used as an example in the text.

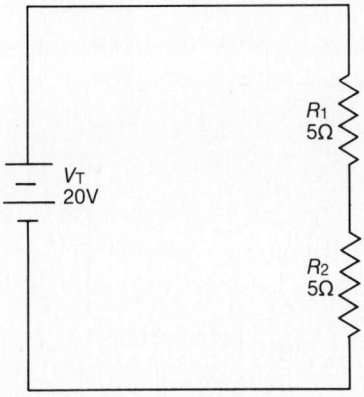

Fig. 6-6: Circuit used to prove $V_T = V_1 + V_2$.

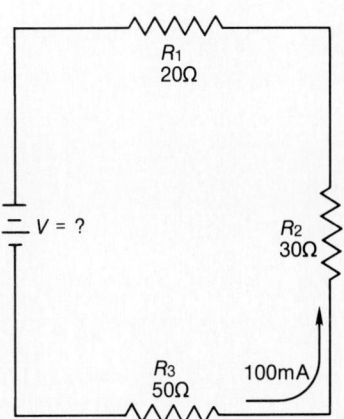

Fig. 6-7: Solving for applied voltage (V_T) in a series circuit.

93

Given:
$$R_1 = 20\,\Omega$$
$$R_2 = 30\,\Omega$$
$$R_3 = 50\,\Omega$$
$$I = 100\text{mA}$$

Solution: Since the circuit involved is a series circuit, the same 100mA of current flows through each resistor. Using Ohm's Law, the voltage drops across each of the three resistors can be calculated and are:

$$V_1 = 2\text{V}$$
$$V_2 = 3\text{V}$$
$$V_3 = 5\text{V}$$

Once the individual drops are known, they can be added to find the total or applied voltage:

$$V_T = V_1 + V_2 + V_3$$
$$V_T = 2\text{V} + 3\text{V} + 5\text{V}$$
$$V_T = 10\text{V}$$

Note: In using Ohm's Law, the quantities used in the equation *must* be taken from the *same* part of the circuit. In the above example, the voltage across R_2 was computed using the current through R_2 and the resistance of R_2.

As long as the source produces electrical energy as rapidly as it is consumed by a resistance, the voltage drop across resistance will remain at a constant value. The value of this voltage is determined by the applied voltage and the proportional relationship of circuit resistances. The voltage drops that occur in a series circuit are in direct proportions to the resistance across which they appear. This is a result of having the same current flow through each resistor. Thus, the larger the resistor, the larger will be the voltage drop across it.

Cells in Series. When cells are wired in series, each individual cell voltage is added in order to determine the total voltage at the terminals of the package (Fig. 6-8). In the case of a dry or wet type, each cell is connected inside the battery to

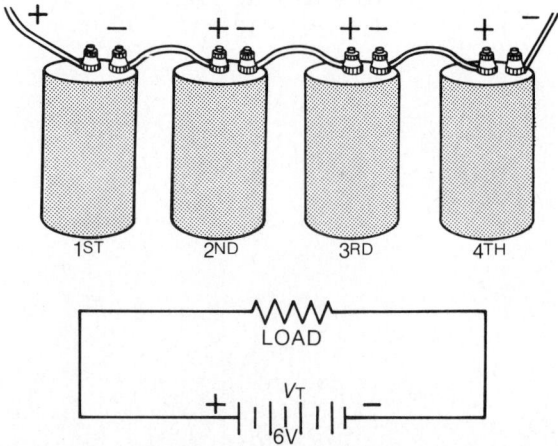

Fig. 6-8: Individual cells may be connected in series as shown. The voltage produced by each cell adds together to provide the total battery voltage (V_T).

produce the terminal voltage. Automobile batteries contain six cells. Each lead-acid battery cell produces 2.1V. Adding the voltage from each cell gives a terminal voltage of 12.6V.

The current-producing capability of the series-wired battery is limited to the current produced by a single cell. This is due to the way in which the cells are wired and to the way in which the current flows through the battery. Current flow must travel through each cell in turn in order to get from one terminal in the battery to the other terminal. The maximum current flow is limited to the capability of the smallest cell in the chain. If six 2.1V cells, each with a capacity of 1.5A, were wired in series, the terminal voltage would be 12.6V. The current capacity of the battery would be limited to 1.5A in this circuit. If one cell in this battery were to fail, the whole battery would fail.

POWER IN THE SERIES CIRCUIT

Each of the resistors in a series circuit consumes power which is dissipated in the form of heat. Since this power must come from the source, the total power must be equal in amount to the power consumed by the circuit resistances. In a series circuit, the total power is equal to the *sum* of the powers dissipated by the individual resistors. Total power (P_T) is thus equal to:

$$P_T = P_1 + P_2 + P_3 + \dots P_n$$

Example: A series circuit consists of three resistors having values of 5Ω, 10Ω, and 15Ω. Find the total power dissipation when 9V is applied to the circuit (Fig. 6-9).

Given:
$$R_1 = 5\Omega$$
$$R_2 = 10\Omega$$
$$R_3 = 15\Omega$$
$$V_T = 9V$$

Solution: The total resistance is found first.
$$R_T = R_1 + R_2 + R_3$$
$$R_T = 5\Omega + 10\Omega + 15\Omega$$
$$R_T = 30\Omega$$

Using total resistance and the applied voltage, the circuit current is calculated.

$$I = \frac{V_T}{R_T}$$

$$I = \frac{9V}{30\Omega}$$

$$I = 0.3A = 300mA$$

Using the power formulas, the individual power dissipations can be calculated. For resistor R_1:

$$P_1 = I^2 R_1$$
$$P_1 = (0.3A)^2 5\Omega$$
$$P_1 = 450mW$$

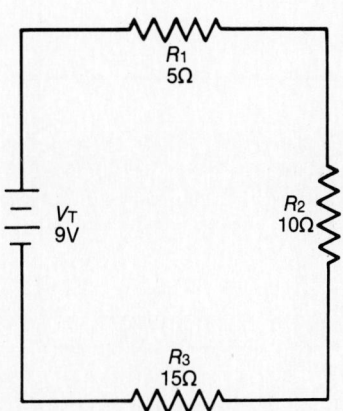

Fig. 6-9: Solving for total power (P_T) in a series circuit.

95

For R_2:

$$P_2 = I^2 R_2$$
$$P_2 = (0.3A)^2 10\Omega$$
$$P_2 = 900mW$$

For R_3:

$$P_3 = I^2 R_3$$
$$P_3 = (0.3A)^2 15\Omega$$
$$P_3 = 1,350mW = 1.35W$$

To obtain total power:

$$P_T = P_1 + P_2 + P_3$$
$$P_T = 450mW + 900mW + 1,350mW$$
$$P_T = 2,700mW = 2.7W$$

To check the answer, the total power delivered by the source can be calculated:

$$P_{source} = I_{source} \times V_{source}$$
$$P_{source} = 300mA \times 9V$$
$$P_{source} = 2,700mW = 2.7W$$

Thus, the total power is equal to the sum of the individual power dissipations.

GENERAL CIRCUIT ANALYSIS

To establish a procedure for solving series circuits, the following sample problem will be solved.

Example: Three resistors of 5kΩ, 10kΩ, and 15kΩ are connected in series across a 25V battery. Completely solve the circuit (Fig. 6-10).

In solving the circuit, the total resistance will be found first. Next, the circuit current will be calculated. Once the current is known, the voltage drops and power dissipations can be calculated.

The total resistance is:

$$R_T = R_1 + R_2 + R_3$$
$$R_T = 5k\Omega + 10k\Omega + 15k\Omega$$
$$R_T = 30k\Omega$$

By Ohm's Law the current is:

$$I = \frac{V_T}{R_T}$$
$$I = \frac{25V}{30,000\Omega}$$
$$I = 0.000833A = 833\mu A$$

The voltage (V_1) across R_1 is:

$$V_1 = IR_1$$
$$V_1 = 0.000833A \times 5,000\Omega$$
$$V_1 = 4.17V$$

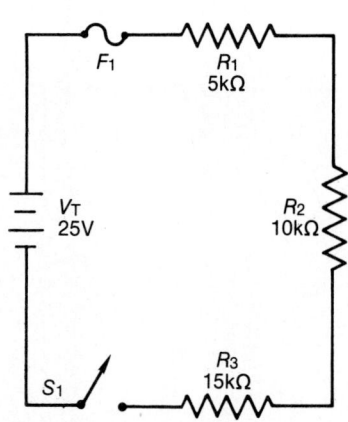

Fig. 6-10: Solving for various values in a series circuit.

The voltage (V_2) across R_2 is:

$$V_2 = IR_2$$
$$V_2 = 0.000833A \times 10,000\Omega$$
$$V_2 = 8.33V$$

The voltage (V_3) across R_3 is:

$$V_3 = IR_3$$
$$V_3 = 0.000833A \times 15,000\Omega$$
$$V_3 = 12.5V$$

The power dissipated in R_1 is:

$$P_1 = IV_1$$
$$P_1 = 0.000833A \times 4.17V$$
$$P_1 = 0.00347W = 3.47mW$$

The power dissipated in R_2 is:

$$P_2 = IV_2$$
$$P_2 = 0.000833A \times 8.33V$$
$$P_2 = 0.00694W = 6.94mW$$

The power dissipated in R_3 is:

$$P_3 = IV_3$$
$$P_3 = 0.000833A \times 12.5V$$
$$P_3 = 0.0104W = 10.4mW$$

The total power dissipated is:

$$P_T = IV_T$$
$$P_T = 0.000833A \times 25V$$
$$P_T = 0.0208W = 20.8mW$$

An important fact to keep in mind when applying Ohm's Law to a series circuit is to consider whether the values used are component values or total values. When the information available enables the use of Ohm's Law to find total resistance, total voltage, and total current, total values must be inserted into the formula.

To find total resistance:

$$R_T = \frac{V_T}{I_T}$$

To find total voltage:

$$V_T = I_T \times R_T$$

To find total current:

$$I_T = \frac{V_T}{R_T}$$

Note: I_T is equal to I in a series circuit. However, the distinction between I_T and I in the formula should be noted, the reason being that future circuits may have several currents, and it will be necessary to differentiate between I_T and other currents.

To compute any quantity (V, I, R, or P) associated with a single given resistor, the values used in the formula must be obtained from that particular resistor. For example, to find the value of an unknown resistance, the voltage across and the current through that particular resistor must be used.

To find the value of a resistor:

$$R = \frac{V_R}{I_R}$$

To find the voltage drop across a resistor:

$$V_R = I_R \times R$$

To find current through a resistor:

$$I_R = \frac{V_R}{R}$$

KIRCHHOFF'S VOLTAGE LAW

In 1847, Gustav R. Kirchhoff extended the use of Ohm's Law by developing a simple concept concerning the voltages contained in a series circuit loop. Kirchhoff's Law is stated as follows: The algebraic sum of the voltage drops including the source around any closed circuit loop is zero.

Through the use of Kirchhoff's Law, circuit problems can be solved which would be difficult and often impossible with only a knowledge of Ohm's Law. When the law is properly applied, an equation can be set up for a closed loop and the unknown circuit values may be calculated. Variations of these *loop equations* will be applied later in future circuit computations.

Polarity of Voltage

To apply Kirchhoff's Voltage Law, the meaning of voltage *polarity* must be understood. In the circuit shown in Fig. 6-11, the current is seen to be flowing in a clockwise direction due to the arrangement of the battery source V_T. Notice that the end of resistor R_1 into which the current flows is marked negative (–). The end of R_1 at which the current leaves is marked positive (+). These polarity markings are used to show that the end of R_1 into which the current flows is at a higher negative potential than is the end of the resistor at which the current leaves. Point A is thus more negative than point B.

Point C (which is the same point as point B) is labeled negative. This is to indicate that point C, though positive with respect to point A, is more negative than point D. To say a point is positive (or negative) without stating what it is positive or negative *in respect to* has no meaning.

Kirchhoff's Voltage Law can be written as an equation as shown below:

$$V_a + V_b + V_c + ...V_n = 0$$

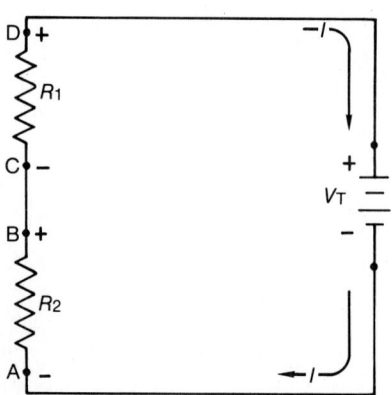

Fig. 6-11: Voltage polarities.

where V_a, V_b, etc. are the voltage drops and emf's around any closed circuit loop. To set up the equation for an actual circuit, the following procedure is used.

1. Assume a direction of current through the circuit. (Correct direction desirable but not necessary.)

2. Using assumed direction of current, assign polarities to all resistors through which the current flows.

3. Place correct polarities on any source included in the circuit.

4. Starting at any points in the circuit, trace around the circuit, writing down the magnitude and polarity of the voltage across each component in succession. Use the polarity sign previous to the component or voltage source being traced. Stop when reaching the point at which the trace was started.

5. Place these voltages with their polarities into the loop equation and solve for the desired quantity.

Example: Three resistors are connected across a 50V source. What is the voltage across the third resistor if the voltage drops across the first two resistors are 25V and 15V?

Solution: A diagram is first drawn as shown in Fig. 6-12. Next, a direction of current is assumed as shown. Using this current, the polarity markings are placed at each end of each resistor and also on the terminals of the source. Starting at point A, trace around the circuit in the direction of current flow recording the voltage and polarity of each component. Starting at point A these voltages would be as follows:

Basic formula:

$$V_a + V_b + V_c + ... V_n = 0$$

From the circuit:

$$(-V_x) + (-V_2) + (-V_1) + (+V_A) = 0$$

Substituting values from circuit:

$$-V_x - 15V - 25V + 50V = 0$$
$$-V_x + 10V = 0$$
$$-V_x = -10V$$
$$V_x = 10V$$

Thus, the unknown voltage (V_x) is found to be 10V.

Using the same idea as above, a problem can be solved in which the current is the unknown quantity.

Example: A circuit having a source voltage of 60V contains three resistors of 5Ω, 10Ω, and 15Ω. Find the circuit current.

Solution: Draw and label the circuit (Fig. 6-13). Establish a direction of current flow and assign polarities. Next, starting at any point (point A will be chosen in this example), write out the loop equation.

Basic equation:

$$V_a + V_b + V_c + ... V_n = 0$$
$$- V_2 - V_1 + V_A - V_3 = 0$$

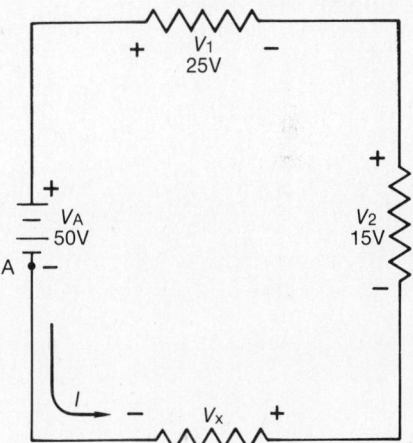

Fig. 6-12: Determining unknown voltage in a series circuit.

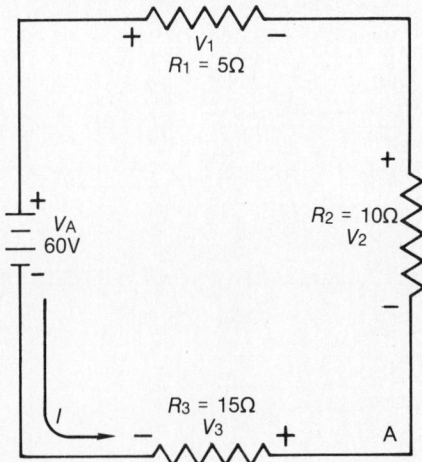

Fig. 6-13: Correct direction of assumed current.

99

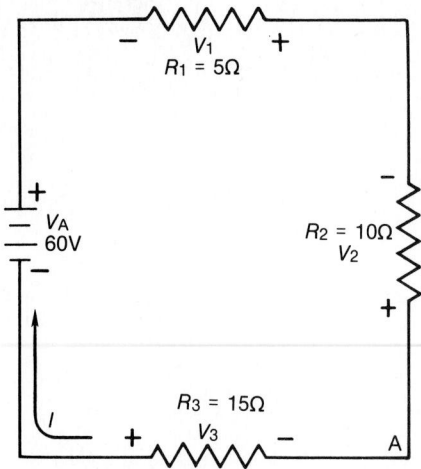

Fig. 6-14: Incorrect direction of assumed current.

Since $V = IR$, by substitution:

$$-IR_2 - IR_1 + V_A - IR_3 = 0$$

Substituting values:

$$-10I - 5I + 60A - 15I = 0$$

Combining like terms:

$$-30I + 60A = 0A$$
$$-30I = -60A$$
$$-I = -2A$$
$$I = 2A$$

Since the current obtained in the above calculations is a positive 2A, the assumed direction of current was correct. To show what happens if the incorrect direction of current is assumed, the problem will be solved as before but with the opposite direction of current. The circuit is redrawn showing the new direction of current and new polarities in Fig. 6-14.

Solution:

$$V_a + V_b + V_c + ... V_n = 0$$

Starting at point A:

$$-V_3 - V_A - V_1 - V_2 = 0$$
$$-IR_3 - V_A - IR_1 - IR_2 = 0$$
$$-15I - 60A - 5I - 10I = 0$$
$$-30I = 60A$$
$$I = -2A$$

Notice that the *amount* of current is the *same* as before. Its polarity, however, is *negative.* The negative polarity simply indicates the wrong direction of current was assumed. Should it be necessary to use this current in further calculations on the circuit, the negative polarity should be retained in the calculations.

Series Aiding and Opposing Sources

In many practical applications a circuit may contain more than one source. Sources of emf that cause current to flow in the same direction are considered to be *series aiding* and their voltages add. Sources of emf that would tend to force current in opposite directions are said to be *series opposing,* and the effective source voltage is the difference between the opposing voltages. When two opposing sources are inserted into a circuit, current flow would be in a direction determined by the larger source. Examples of series aiding and opposing sources are shown in Fig. 6-15.

Multiple Source Solutions

A simple solution may be obtained for a multiple source circuit through the use of Kirchhoff's Voltage Law. In applying

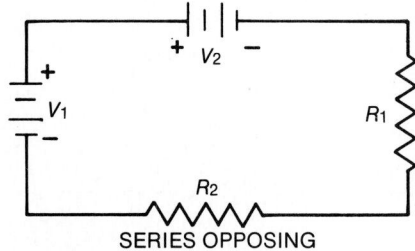

Fig. 6-15: Aiding and opposing sources.

this method, the exact same procedure is used for the multiple source as was used above for the single source circuit. This is demonstrated by the following problem.

Example: Using Kirchhoff's Voltage equation, find the amount of current in the circuit shown in Fig. 6-16.

Solution: As before, a direction of current flow is assumed and polarity signs are placed on the drawing. The loop equation will be starting at point A.

Basic equation:

$$V_a + V_b + V_c + ... V_n = 0$$

From the circuit:

$$V_{S2} + V_1 - V_{S1} + V_{S3} + V_2 = 0$$
$$-20 - 60I + 180 - 40 - 20I = 0$$

Combining like terms:

$$-80I + 120A = 0$$
$$-80I = -120A$$
$$-I = -1.5A$$
$$I = 1.5A$$

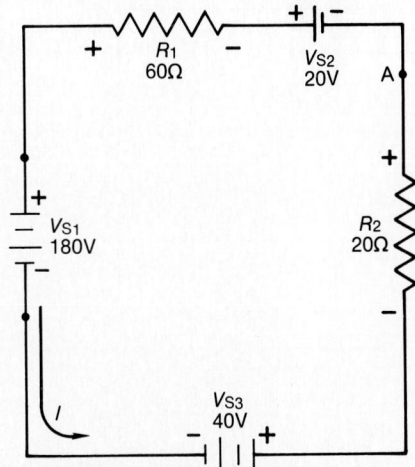

Fig. 6-16: Solving for circuit current using Kirchhoff's voltage equation.

VOLTAGE REFERENCES

A voltage reference point is an arbitrarily chosen point to which all other points are compared. In series circuits, any point can be chosen as a reference and the electrical potential at all other points can be determined in reference to the initial point. In the example of Fig. 6-17, point A shall be considered as the reference. Each series resistor in the illustrated circuit is of equal value; therefore, the applied voltage is equally distributed across each resistor. The potential at point B is 25V more positive than A. Points C and D are 50V and 75V respectively more positive than point A.

If point B is used as the reference as in Fig. 6-18, point D would be positive 50V in respect to the new reference point B. The former reference point A is 25V negative in respect to point B.

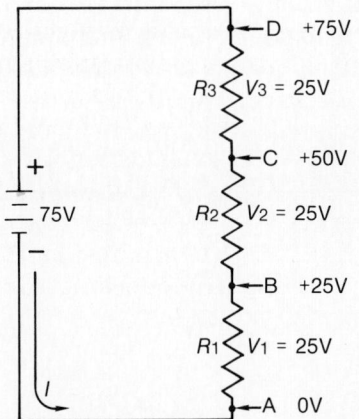

Fig. 6-17: Reference points in a series circuit.

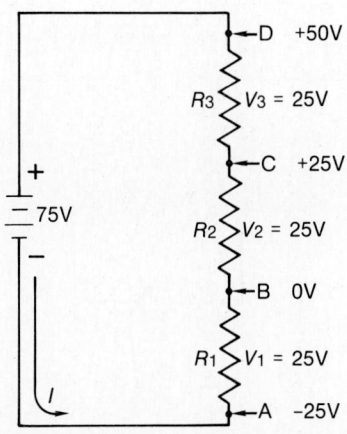

Fig. 6-18: Determining potentials with respect to a reference point.

101

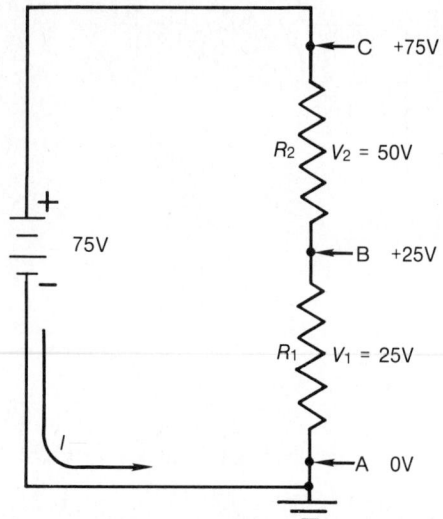

Fig. 6-19: Use of ground symbol.

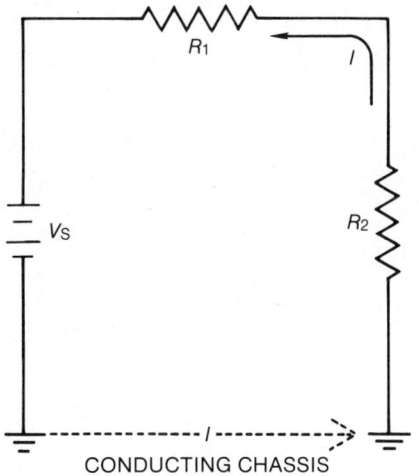

Fig. 6-20: Ground used as a conductor.

Ground

As in the previous circuit illustration, the reference point of a circuit is always considered to be at zero potential. Since the earth (ground) is said to be at a zero potential, the term *ground* is used to denote a common electrical point of zero potential. In Fig. 6-19, point A is the zero reference or ground and is symbolized as such. Point C is 75V positive and point B is 25V positive with respect to ground.

In much electrical/electronic equipment, the metal chassis is the common ground for the many electrical circuits. The value of ground is noted when considering its contribution to economy, simplification of schematics, and ease of measurement. When completing each electrical circuit, common points of a circuit at zero potential are connected directly to the metal chassis, thereby eliminating a large amount of connecting wire. The electrons pass through the metal chassis (conductor) to reach other points of the circuit. An example of a grounded circuit is illustrated in Fig. 6-20.

Most voltage measurements used to check proper circuit operation in electronic equipment are taken in respect to ground. One meter lead is attached to ground and the other meter lead is moved to various test points.

OPEN AND SHORT CIRCUITS

A circuit is said to be *open* when a break exists in a complete conducting pathway. Although an open occurs any time a switch is thrown to deenergize a circuit, an open may also develop accidentally due to abnormal circuit conditions. To restore a circuit to proper operation, the open must be located and its cause determined.

Sometimes an open can be located visually by a close inspection of the circuit components. Defective components, such as open resistors and fuses, can usually be discovered by this method. Others, such as a break in wire covered by insulation or the melted element of an enclosed fuse, are not visible to the eye. Under such conditions, the understanding of an open's effect on circuit conditions enables a technician to make use of a voltmeter or ohmmeter to locate the open component.

In Fig. 6-21, the series circuit consists of two resistors and a fuse. Notice the effects on circuit conditions when the fuse opens.

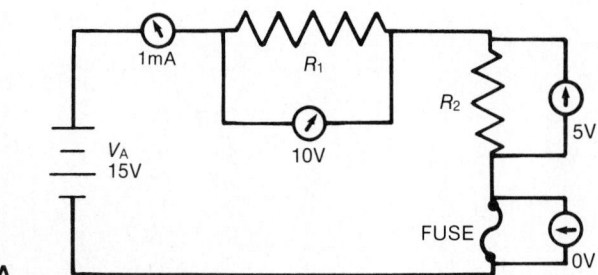
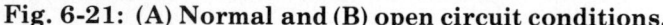
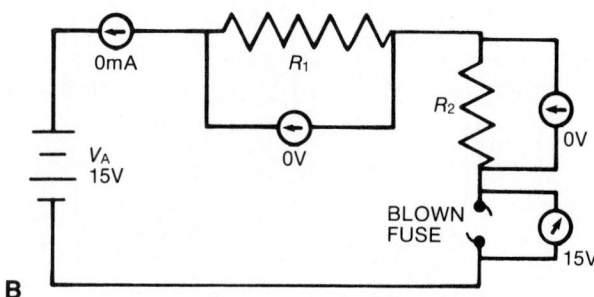

A   B

Fig. 6-21: (A) Normal and (B) open circuit conditions.

102

Current ceases to flow; therefore, there is no longer a voltage drop across the resistors. Each end of the open conducting path becomes an extension of the battery terminals and the voltage felt across the open is equal to the applied voltage.

An open circuit, such as found in Fig. 6-21, could also have been located with an ohmmeter. However, when using an ohmmeter to check a circuit, it is important to first deenergize the circuit. The reason for this is that an ohmmeter has its own power source and would be damaged if connected to an energized circuit.

The ohmmeter used to check a series circuit would indicate the ohmic value of each resistance across which it is connected. The open circuit, due to its almost infinite resistance, would cause no deflection on the ohmmeter as indicated by Fig. 6-22.

A *short circuit* is an accidental path of low resistance which passes an abnormal amount of current. A short circuit exists whenever the resistance of the circuit or the resistance of a part of a circuit drops in value to almost $0\,\Omega$. A short often occurs as a result of improper wiring or broken insulation.

In Fig. 6-23, a short is caused by improper wiring. Note the effect on current flow. Since the resistor has in effect been replaced with a piece of wire, practically all the current flows through the short and very little current flows through the resistor. Electrons flow through the short, a path of almost zero resistance, and complete the circuit by passing through the $10\,\Omega$ resistor and the battery. The amount of current flow increases greatly because its resistive path has decreased from $10{,}010\,\Omega$ to $10\,\Omega$. Due to the excessive current through the $10\,\Omega$ resistor, the increased heat dissipated by the resistor will destroy the component.

ELECTRICAL SHOCK

Every electrical circuit is dangerous. Electric shock may cause burns of varying degree, cessation of breathing and unconsciousness, ventricular fibrillation or cardiac arrest, and death. If a current is passed through a person from hand to hand or from hand to foot, the effects when current is gradually increased from zero are as follows:

1. At about 1mA the shock will be felt.
2. At about 10mA the shock is severe enough to paralyze muscles and a person may be unable to release the conductor.
3. At about 100mA the shock is usually fatal if it lasts for 1 second or more. **It is important to remember that funda-**

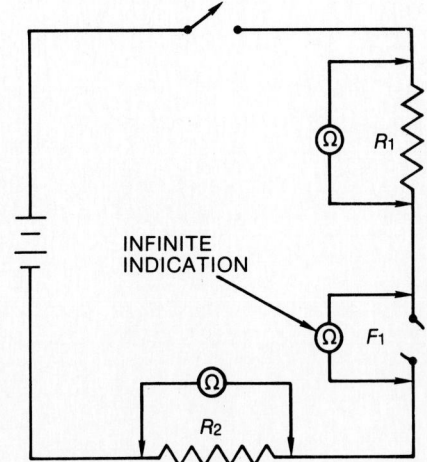

Fig. 6-22: Ohmmeter reading in an open series circuit.

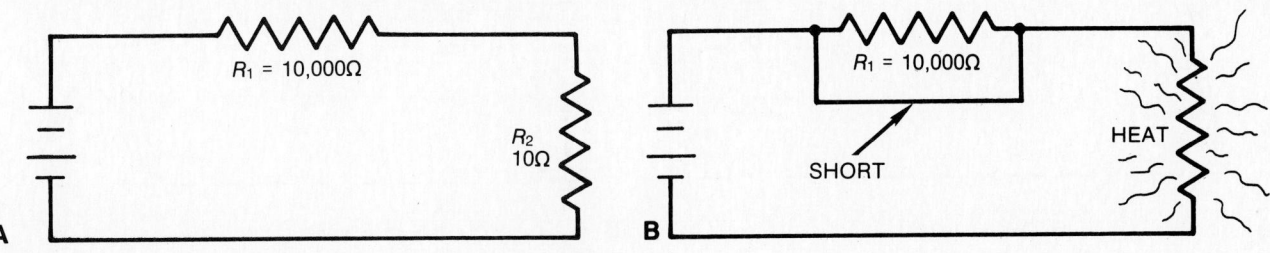

Fig. 6-23: (A) Normal and (B) short circuit conditions.

mentally, current, rather than voltage, is the criterion of shock intensity.

It should be clearly understood that resistance of body will vary; that is, if the skin is dry and unbroken, body resistance will be quite high, on the order of 300,000 to 500,000 Ω. However, if the skin becomes moist or broken, body resistance may drop to as low as 300 Ω. Thus, a potential as low as 30V could cause a fatal current flow. Therefore, any circuit with a potential in excess of this value must be considered dangerous. Also keep in mind that points of negative potential are as dangerous as positive potential. Let us consider Fig. 6-24. In Fig. 6-24A, the negative terminal is grounded. Examination of the illustration shows clearly that the *figure* is standing on the ground with one hand on the positive terminal of the power source which is connected directly across it. In Fig. 6-24B, the positive terminal of the power source is connected to ground. This does not mean that it is safe to handle the negative terminal. Comparison of Figs. 6-24A and B shows that, in both illustrations, the figure is connected directly across the power source. The negative terminal in Fig. 6-24B is just as hot as the positive terminal in Fig. 6-24A. The only difference is the direction of current through the figure, and current from head to toes can be just as fatal as current from toes to head. Remember that current through the body is limited only by the resistance of the body. And to repeat again, under certain conditions the body resistance may be very low and a very low voltage may produce a fatal current.

SUMMARY

There are two basic circuit types, series circuits and parallel circuits. In a series circuit there is only one path in which the current can flow. A series circuit consists of two or more components and a power source.

The total resistance in a series circuit is equal to the sum of each of the individual resistances in the circuit; that is:

$$R_T = R_1 + R_2 + R_3 + ...R_n$$

Total current in a series circuit is dependent upon the total resistance of the circuit and the applied voltage of the circuit and is constant throughout the circuit; that is:

$$I_T = I_{R_1} = I_{R2} = I_{R3} = I_{R_n}$$

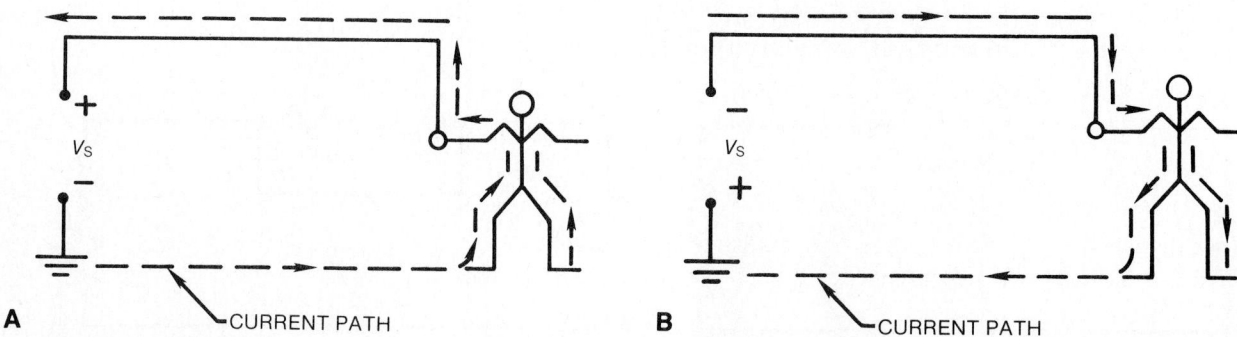

A　　CURRENT PATH　　　　　　　　　　**B**　　CURRENT PATH

Fig. 6-24: Points of negative and positive potential are equally as dangerous.

Voltage is the term used to identify a difference in electrical potential found in a circuit. A voltage drop is produced across a resistance when current flows through that resistance. The sum of each of the voltage drops in a series circuit will always equal the applied voltage; that is:

$$V_T = V_{R1} + V_{R2} + V_{R3} + ...V_{Rn}$$

When cells are connected in series, each individual cell voltage is added in order to determine the total potential voltage at the terminal of the package. The current-producing capability of cells wired in series is limited to the current capabilities of one of the single cells.

Each of the resistors in a series circuit consumes power which is dissipated in the form of heat. The total power in a series circuit is equal to the sum of the powers dissipated by the individual resistors; thus,

$$P_T = P_{R1} + P_{R2} + P_{R3} + ...P_{Rn}$$

Kirchhoff's Voltage Law states that the algebraic sum of the voltage drops including the source around any closed circuit loop is zero; that is:

$$V_1 + V_2 + V_3 + ...V_n = 0$$

To apply Kirchhoff's Law an understanding of voltage polarity and reference points is required.

Many circuits contain more than one power source. These sources can be either series aiding or series opposing. The effective voltage in a circuit with series aiding sources is equal to the sum of the individual voltages. In a series opposing circuit, the effective voltage is equal to the difference between the opposing voltages, and the polarity is determined by the larger source.

An open circuit is one that has a very high resistance which is often due to a break in the current path. No current can flow in an open circuit.

A short circuit is one that has very low resistance (usually less than 1Ω). A short often occurs as the result of broken insulation or improper wiring. Current flow in a short circuit is very high.

Care should be taken when working on electric circuits. One hand should be kept in your pocket in order to avoid the possibility of connecting yourself to two points between which there is a potential difference. Under certain conditions the body resistance may be very low and a very low voltage may produce a fatal current.

CHECK YOURSELF (Answers at back of book)

Questions

1. In a series circuit, the _____ is the same at any point.
2. The resistance of a circuit which has an I = 20A and V = 100V is _____ .
3. A circuit has an I = 3mA and R = 50kΩ. The circuit voltage is _____ .

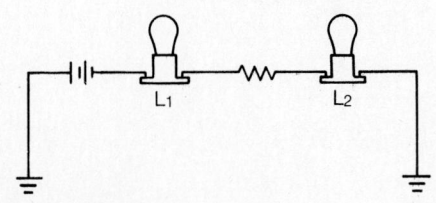

Fig. 6-25:

105

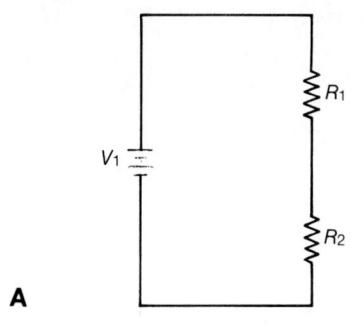

A

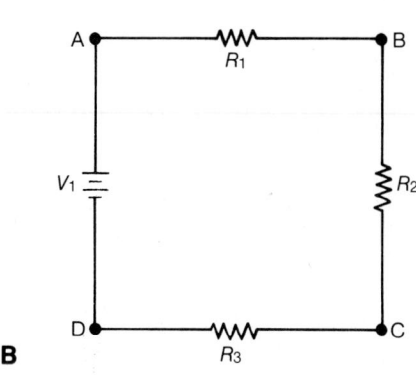

B

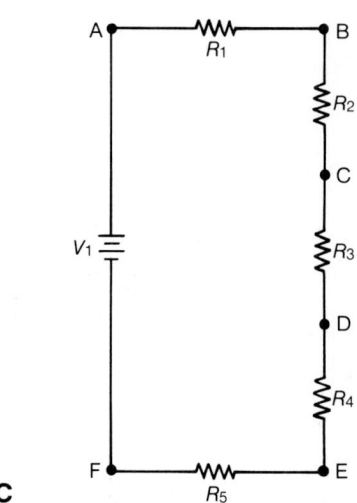

C

Fig. 6-26:

4. A series circuit has $R_1 = 5\Omega$, $R_2 = 20\Omega$, and $R_3 = 50\Omega$. $R_T =$ _____ .

5. In a series circuit, if $V_{R_1} = 20V$ and $V_{R_2} = 20V$, $V_T =$ _____ .

6. A circuit consists of three equal series resistances. If $V_{R_1} = 50V$ and $I_{R_3} = 5A$, $R_T =$ _____ .

7. In a series circuit, if $V_T = 200V$, $V_{R_1} = 50V$, and $V_{R_2} = 78V$, $V_{R_3} =$ _____ .

8. A circuit consists of R_1, R_2, and R_3 in series. If $V_T = 800V$, $R_1 = 50\Omega$, $R_2 = 60\Omega$, and $I = 2A$, the size of $R_3 =$ _____ .

9. If you remove L_2 from its socket in the circuit of Fig. 6-25, the current flow stops and the voltage drop across the resistor will be _____ .

10. To find the total resistance of a series circuit, you must _____ the individual _____ . To find the applied voltage, you must _____ the individual _____ . The current in all parts of a series circuit is the _____ .

Problems

Using Fig. 6-26A, solve each problem for the unknowns.

1. $R_1 = 56\Omega$, $R_2 = 47\Omega$, $V_1 = 2.0V$; find I_T and V_{R_1}.
2. $R_1 = 10k\Omega$, $R_2 = 500\Omega$, $V_1 = 15V$; find V_{R_1}, V_{R_2}, I_T, and P_T.

Use Fig. 6-26B for these problems:

3. $R_1 = 470\Omega$, $R_2 = 330\Omega$, $R_3 = 220\Omega$, $V_1 = 51V$; find I_T, and P_{R_1}, P_{R_2}, P_{R_3}, and P_T.
4. $R_1 = 30\Omega$, $R_2 = 10\Omega$, $I_T = 0.1A$, $V_{R_3} = 2V$; find V_1 and R_3.
5. In the circuit of Fig. 6-26C all resistors are valued at 500Ω. $V_T = 50V$. Find current, voltage drop, and power values for each component in the circuit.
6. In an electrical measuring device, five resistors with the following values were in series: $1.2k\Omega$, $1.5M\Omega$, $250k\Omega$, 10Ω, and $500.4k\Omega$. What is the total resistance of this series string?
7. The output from a transistor cassette player power supply provided 12V, but a lower voltage was required. A 200Ω resistor was placed in series with a 600Ω resistor. What voltage was available from across the 600Ω, with the 12V applied across the series string?
8. In an electronic test instrument there are three series resistors with the following values: 500Ω, 450Ω, and 50Ω. If there is 50V across this series circuit, what current flows? What is the voltage drop across each resistor? What is the power dissipated in each resistor?

CHAPTER 7

DC Circuits— Parallel

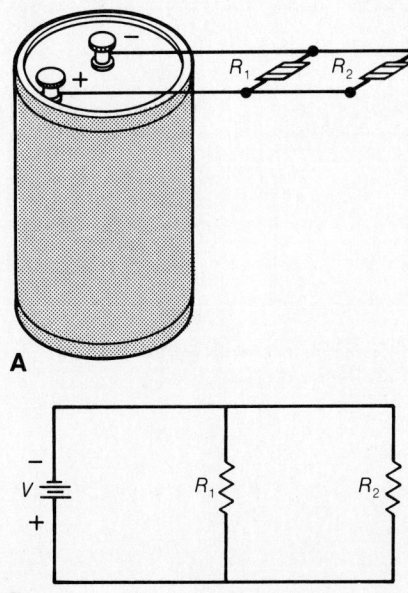

Fig. 7-1: (A) Actual parallel circuit; (B) schematic circuit.

The second major DC circuit configuration is called the *parallel circuit*. This circuit differs from the series circuit in that it always has multiple paths for current flow. The series circuit has only one current flow path. The parallel circuit consists of a source, two or more loads, and connecting lines. An example of a basic parallel circuit is shown in Fig. 7-1.

Commencing at the voltage source (V_S) and tracing clockwise around the circuit, two complete and separate paths can be identified in which current can flow (Fig. 7-2). One path is traced from the source through resistance R_1 and back to the source; the other, from the source through resistance R_2 and back to the source.

VOLTAGE IN PARALLEL CIRCUITS

One of the easiest parts to understand in the parallel circuit is the voltage across each of the components. This is simple because it is all the same. Each load is wired directly across the source. The formula for voltage in the parallel circuit is:

$$V_T = V_{R_1} = V_{R_2} = V_{R_3} = V_{R_n}$$

Voltage measurements taken across the resistors of a parallel circuit, as illustrated by Fig. 7-3, verify the above equation. Each voltmeter indicates the same amount of voltage. Notice that the voltage across each resistor is the same as the applied or total voltage of the circuit.

Example: Assume that the current through a resistor of a parallel circuit is known to be 4.5mA and the value of the

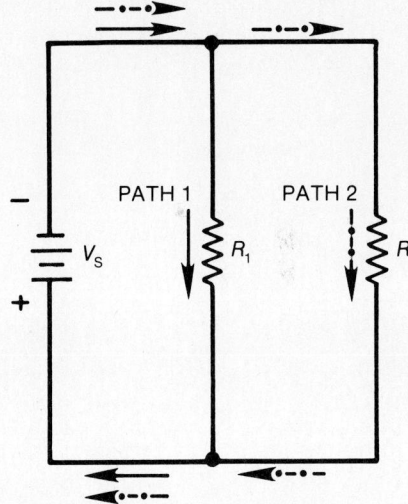

Fig. 7-2: Example of a basic parallel circuit.

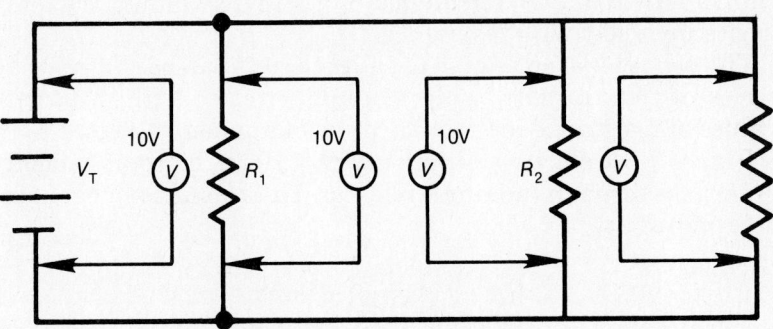

Fig. 7-3: Voltage comparison in a parallel circuit.

107

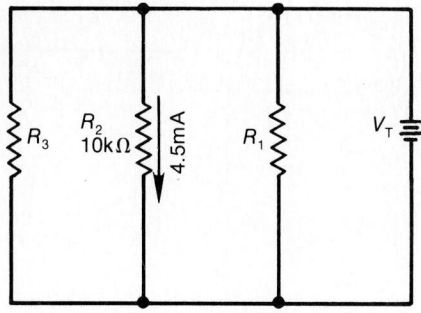

Fig. 7-4: Example problem of parallel circuit.

resistor is 10kΩ. Determine the potential across the resistor. The circuit is shown in Fig. 7-4.

Given:

$$R_2 = 10k\Omega$$
$$I_{R2} = 4.5mA$$

Find:

$$V_{R2} = ?$$
$$V_T = ?$$

Solution: Select proper equation.

$$V = IR$$

Substitute known values:

$$V_{R2} = I_{R2} \times R_2$$
$$V_{R2} = 4.5mA \times 10k\Omega$$

Express in powers of ten:

$$V_{R2} = (4.5 \times 10^{-3}) \times (10 \times 10^3)$$
$$V_{R2} = 4.5 \times 10$$

Resultant:

$$V_{R2} = 45V$$

Therefore:

$$V_T = 45V$$

Having determined the voltage across one resistor (R_2) in a parallel circuit, the value of the source voltage (V_T) and the potentials across any other resistors that may be connected in parallel with it are known.

CURRENT IN PARALLEL CIRCUITS

The current in a circuit is inversely proportional to the circuit resistance. This fact, obtained from Ohm's Law, establishes the relationship upon which the following discussion is developed.

A single current flows in a series circuit. Its value is determined in part by the total resistance of the circuit. However, the source current in a parallel circuit divides among the available paths in relation to the value of the resistors in the circuit. Ohm's Law remains unchanged. For a given voltage, current varies inversely with resistance.

The behavior of current in parallel circuits will be shown by a series of illustrations using example circuits with different values of resistance for a given value of applied voltage.

Figure 7-5A shows a simple circuit. Here the total current must pass through the single resistor. The amount of current is determined as

$$I_T = \frac{V}{R_1} = \frac{50V}{10k\Omega} = 5mA$$

Figure 7-5B shows the same resistor (R_1) with a second resistor (R_2) of equal value connected in parallel across the voltage source. Applying the proper equation from Ohm's Law, the current flow through each resistor is seen to be the same as through the single resistor in Fig. 7-5A. These individual currents are determined as follows:

$$I_{R_1} = \frac{V}{R_1} = \frac{50V}{10k\Omega} = 5mA$$

$$I_{R_2} = \frac{V}{R_2} = \frac{50V}{10k\Omega} = 5mA$$

However, it is apparent that if 5mA of current flows through each of the two resistors, there must be a total current of 10mA drawn from the source. The distribution of current in the simple parallel circuit is shown in Fig. 7-5B.

The total current of 10mA leaves the negative terminal of the battery and flows to point X. Since point X is a connecting point for the two resistors, it is called a junction. At junction X the total current divides into two smaller currents of 5mA each. These two currents flow through their respective resistors and rejoin at junction Y. The total current then flows from junction Y back to the positive terminal of the source. Thus, the source supplies a total current of 10mA and each of the two equal resistors carries one-half the total current.

Each individual current path in the circuit of Fig. 7-5B is referred to as a branch. Each branch will carry a current that is a portion of the total current. Two or more branches form a network.

From the foregoing observations, the characteristics of current in a parallel circuit can be expressed in terms of the following general equation:

$$I_T = I_1 + I_2 + \dots I_n$$

The analysis of current in parallel circuits is continued with the use of the following example circuits.

Compare Fig. 7-6A with Fig. 7-5B. Notice that doubling the value of the second branch resistor (R_2) has no effect on the current in the first branch (I_{R_1}), but does reduce its own branch current (I_{R_2}) to one-half its original value. The total circuit current drops to a value equal to the sum of the branch currents. These facts are verified as follows:

$$I_1 = \frac{V}{R_1} = \frac{50V}{10k\Omega} = 5mA$$

$$I_2 = \frac{V}{R_2} = \frac{50V}{20k\Omega} = 2.5mA$$

$$I_T = I_1 + I_2$$

$$I_T = 5mA + 2.5mA = 7.5mA$$

Now compare the two circuits of Fig. 7-6. Notice that the sum of the ohmic values of the resistors in both circuits is equal and

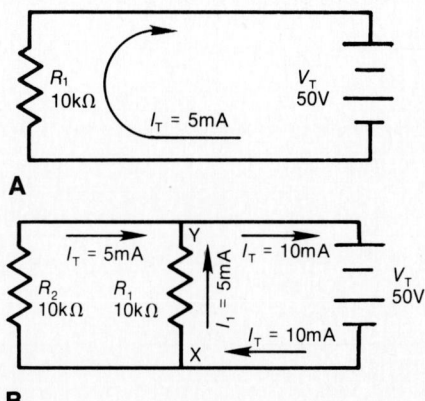

Fig. 7-5: Current in a series and a parallel circuit.

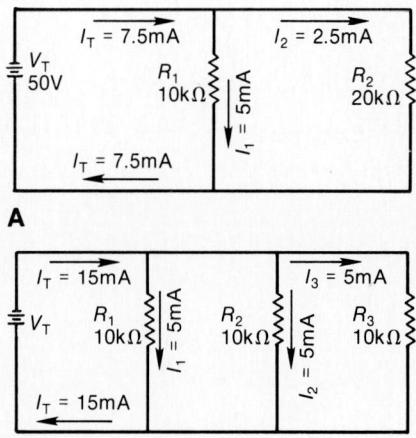

Fig. 7-6: Current behavior in parallel circuits.

109

that the applied voltage is the same value. However, the total current in Fig. 7-6B is twice the amount in Fig. 7-6A. It is apparent, therefore, that the manner in which resistors are connected in a circuit, as well as their actual ohmic value, affects the total current flow. This phenomenon will be illustrated in more detail in the discussion of resistance. The amount of current flow in the branch circuits and the total current in the circuit (Fig. 7-6B) are determined as follows:

$$I_1 = \frac{V}{R_1} = \frac{50V}{10k\Omega} = 5mA$$

$$I_2 = \frac{V}{R_2} = \frac{50V}{10k\Omega} = 5mA$$

$$I_3 = \frac{V}{R_3} = \frac{50V}{10k\Omega} = 5mA$$

$$I_T = I_1 + I_2 + I_3$$

$$I_T = \frac{V}{R_1} + \frac{V}{R_2} + \frac{V}{R_3}$$

$$I_T = \frac{50V}{10k\Omega} + \frac{50V}{10k\Omega} + \frac{50V}{10k\Omega} = 5mA + 5mA + 5mA = 15mA$$

Cells in Parallel. When identical cells are connected in parallel, they are capable of providing a total current equal to the sum of each cell's current capability. However, the voltage of this battery is the same as for one cell. For example, if four 1.5V cells, each capable of producing 1.5A of current, were wired in parallel (Fig. 7-7), the resulting battery would produce 1.5V with a maximum current capability of 6A.

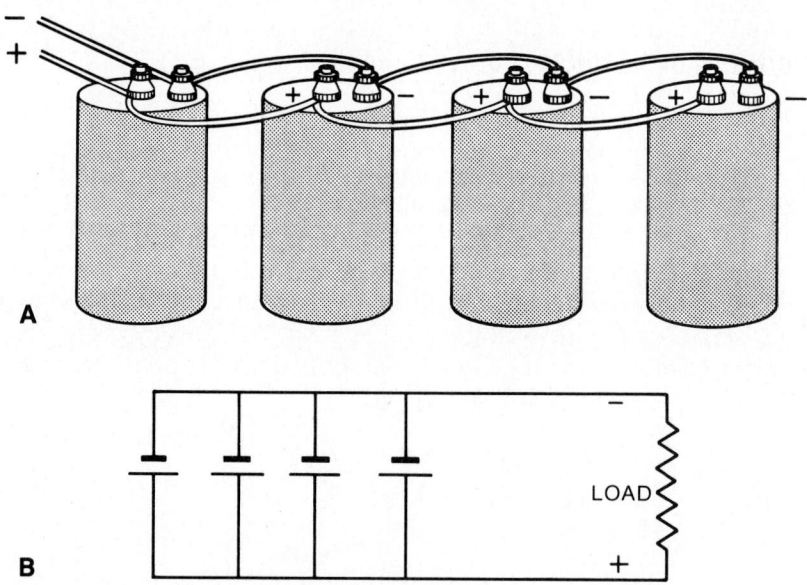

Fig. 7-7: (A) Pictorial view of parallel connected cells; (B) schematic of parallel connection.

Kirchhoff's Current Law

The division of current in a parallel network follows a definite pattern. This pattern is described by Kirchhoff's current law which is stated as follows: *The algebraic sum of the currents entering and leaving any junction of conductors is equal to zero.* This law can be stated mathematically as:

$$I_a + I_b + \ldots\ldots I_n = 0$$

where I_a, I_b, etc., are the currents entering and leaving the junction. Currents entering the junction are assumed to be positive, and currents leaving the junction are considered negative. When solving a problem using this Kirchhoff equation, the currents must be placed into the equation with the proper polarity signs attached.

Example: Solve for the value of I_3 in Fig. 7-8.

Solution: First the currents are given proper signs.

$$I_1 = +10A$$
$$I_2 = -3A$$
$$I_3 = \ ? \text{ amperes}$$
$$I_4 = -5A$$

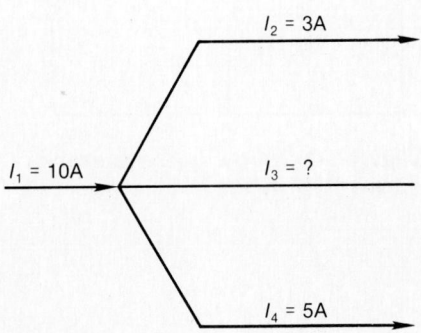

Fig. 7-8: Applying Kirchhoff's current law.

These currents are placed into the Kirchhoff equation with the proper signs as follows:

Basic equation:

$$I_a + I_b + \ldots.. I_n = 0$$

Substitution:

$$I_1 + I_2 + I_3 + I_4 = 0$$
$$(+10A) + (-3A) + (I_3) + (-5A) = 0$$

Combining like terms:

$$I_3 + 2A = 0$$
$$I_3 = -2A$$

Thus, I_3 has a value of 2 amperes, and the negative sign shows it to be a current *leaving* the junction.

Example: Using Fig. 7-9, solve for the magnitude and direction of I_3:

Solution:

$$I_a + I_b + \ldots.. I_n = 0$$
$$I_1 + I_2 + I_3 + I_4 = 0$$
$$(+6A) + (-3A) + (I_3) + (-5A) = 0$$
$$I_3 - 2A = 0$$
$$I_3 = 2A$$

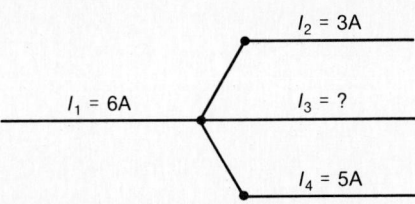

Fig. 7-9: Finding an unknown current.

Thus, I_3 is 2A and its positive sign shows it to be a current *entering* the junction.

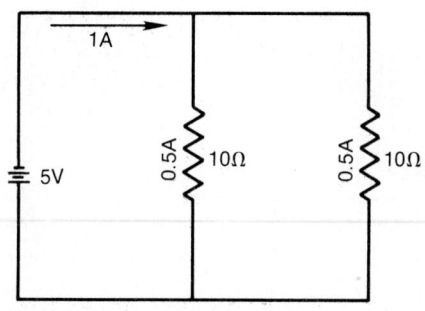

Fig. 7-10: Two equal resistors connected in parallel.

RESISTANCE IN PARALLEL CIRCUITS

The preceding discussion of current introduced certain principles involving the characteristics and effects of resistance in parallel circuits. A detailed explanation of the characteristics of parallel resistances will be considered in this section. The explanation will commence with a simple parallel circuit. Various methods used to determine the total resistance in parallel circuits will be described.

Equal Resistors Method

In Fig. 7-10, two resistors having a resistance value of 10Ω each are connected across a 5V source. A complete circuit consisting of two parallel paths is formed and current will flow as shown.

Computing the individual currents shows that there is 0.5A of current flowing through each resistance. Accordingly, the total current flowing from the battery to the junction of the resistors and returning from the resistors to the battery is equal to 1A. The total resistance of the circuit can be determined by substituting total values of voltage and current into the following equation. This equation is derived from Ohm's Law.

$$R_T = \frac{V}{I_T}$$

$$R_T = \frac{5}{1} = 5\Omega$$

This computation shows the total resistance to be 5Ω, one-half the value of either of the two resistors.

Since the total resistance of any parallel circuit is smaller than the smallest resistance in the circuit, the term "total resistance" does not mean the sum of the individual resistor values. The total resistance of resistors in parallel is also referred to as equivalent resistance. In many texts the terms total resistance and equivalent resistances are used interchangeably.

There are several methods used to determine the equivalent resistance of parallel circuits. The most appropriate method for a particular circuit depends on the number and value of the resistors. For the circuit described above, the following simple equation is used:

$$R_{eq} = \frac{R}{N}$$

where

R_{eq} = equivalent parallel resistance

R = ohmic value of one resistor

N = number of resistors

This equation is valid for any number of *equal value* parallel resistors.

112

Example: Four 40Ω resistors are connected in parallel. What is their equivalent resistance?

Solution:

$$R_{eq} = \frac{R}{N} = \frac{40\Omega}{4} = 10\Omega$$

Circuits containing parallel resistance of unequal value will now be considered. Refer to the example circuit in Fig. 7-11.

Given:

$$R_1 = 3k\Omega,\ R_2 = 6k\Omega,\ V_T = 30V$$

Known:

$$I_1 = 10mA,\ I_2 = 5mA,\ I_T = 15mA.$$

Determine:

$$R_{eq} = ?$$

Solution:

$$R_{eq} = \frac{V}{I_T}$$

$$R_{eq} = \frac{30V}{15mA} = 2k\Omega$$

Notice that the equivalent resistance of $2k\Omega$ is less than the value of either branch resistor. In parallel circuits the equivalent resistance will always be smaller than the resistance of any branch.

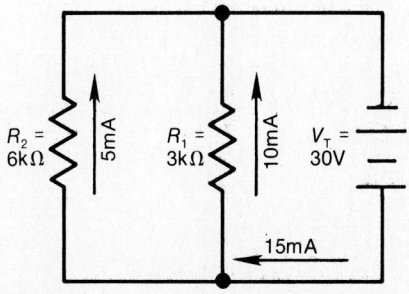

Fig. 7-11: Two unequal parallel resistors.

Reciprocal Method

Many circuits are encountered in which resistors of unequal value are connected in parallel. It is, therefore, desirable to develop a formula which can be used to compute the equivalent resistance of two or more unequal parallel resistors. This equation can be derived as follows:

Given:

$$I_T = I_1 + I_2 + \dots\dots I_n$$

Substituting $\dfrac{V}{R}$ for I gives:

$$\frac{V_T}{R_T} = \frac{V_1}{R_1} + \frac{V_2}{R_2} + \dots\dots \frac{V_n}{R_n}$$

Since in a parallel circuit $V_T = V_1 = V_2 = V_n$

$$\frac{V}{R_T} = \frac{V}{R_1} + \frac{V}{R_2} + \dots\dots \frac{V}{R_n}$$

Dividing both sides by V:

$$\frac{V}{V\,R_\mathrm{T}} = \frac{V}{V\,R_1} + \frac{V}{V\,R_2} + \ldots\ldots \frac{V}{V\,R_\mathrm{n}}$$

$$\frac{1}{R_\mathrm{T}} = \frac{1}{R_1} + \frac{1}{R_2} + \ldots\ldots \frac{1}{R_\mathrm{n}}$$

Taking the reciprocal of both sides:

$$\frac{1}{\dfrac{1}{R_\mathrm{T}}} = \frac{1}{\dfrac{1}{R_1} + \dfrac{1}{R_2} + \ldots \dfrac{1}{R_\mathrm{n}}}$$

Simplifying:

$$R_\mathrm{T} = \frac{1}{\dfrac{1}{R_1} + \dfrac{1}{R_2} + \ldots \dfrac{1}{R_\mathrm{n}}}$$

This formula is called "the reciprocal of the sum of the reciprocals" and is the one normally used to solve for the equivalent resistance of a number of parallel resistors.

Example: Given three parallel resistors of 20Ω, 30Ω, and 40Ω, find the equivalent resistance using the reciprocal equation (Fig. 7-12).

Solution:

Select the proper equation:

$$R_\mathrm{eq} = \frac{1}{\dfrac{1}{R_1} + \dfrac{1}{R_2} + \dfrac{1}{R_3}}$$

Substitute:

$$R_\mathrm{eq} = \frac{1}{\dfrac{1}{20\Omega} + \dfrac{1}{30\Omega} + \dfrac{1}{40\Omega}}$$

Find LCD:

$$R_\mathrm{eq} = \frac{1}{\dfrac{6}{120} + \dfrac{4}{120} + \dfrac{3}{120}} = \frac{1}{\dfrac{13}{120}}$$

Invert:

$$R_\mathrm{eq} = \frac{120}{13} = 9.23\Omega$$

Keep in mind that the total or equivalent resistance of a parallel circuit is always smaller than that of the smallest branch. This relationship provides a quick check of parallel circuit calculations. If the calculations result in a total resistance larger than the smallest branch resistance, the results are in error and must be recalculated.

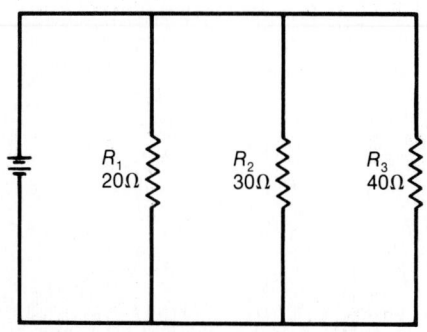

Fig. 7-12: Example of parallel circuits with unequal branch resistors.

114

Conductance Method

Some parallel circuit problems can be solved more conveniently by considering the ease with which current can flow. The degree to which a circuit permits or conducts current, as mentioned in Chapter 3, is called the conductance (G) of the circuit and the unit of conductance is the siemens (S). The conductance of a circuit is the reciprocal of the resistance. The conductance can therefore be found using the following formula:

$$G = \frac{1}{R}$$

also:

$$R = \frac{1}{G}$$

In a parallel circuit, the total conductance is equal to the sum of the individual branch conductances. As an equation:

$$G_T = G_1 + G_2 + \ldots\ldots G_n$$

Example: Determine the equivalent (total) resistance of the circuit shown in the preceding example (Fig. 7-12), using the conductance method.

Solution:

$$G_1 = \frac{1}{R_1} = \frac{1}{20\Omega} = 0.050S$$

$$G_2 = \frac{1}{R_2} = \frac{1}{30\Omega} = 0.033S$$

$$G_3 = \frac{1}{R_3} = \frac{1}{40\Omega} = 0.025S$$

$$G_T = G_1 + G_2 + G_3$$

$$G_T = 0.050S + 0.033S + 0.025S = 0.108S$$

Since:

$$R_T = \frac{1}{G_T}$$

$$R_T = \frac{1}{0.108} = 9.25\Omega$$

The value of equivalent resistance determined by the conductance method is almost identical to the value determined by the reciprocal of the sum of the reciprocals method.

Product Over the Sum Method

A convenient formula for finding the equivalent resistance of *two* parallel resistors can be derived from the following equation:

$$R_T = \cfrac{1}{\cfrac{1}{R_1} + \cfrac{1}{R_2}}$$

Finding the LCD:

$$R_T = \cfrac{1}{\cfrac{R_2 + R_1}{R_1 R_2}}$$

Taking the reciprocal:

$$R_T = \frac{R_1 R_2}{R_1 + R_2}$$

This equation is called the product over the sum formula.

Example: What is the equivalent resistance of a 20Ω and a 30Ω resistor connected in parallel?

Given:

$$R_1 = 20\Omega$$
$$R_2 = 30\Omega$$

Find:

$$R_{eq} = ?$$

Solution:

$$R_T = \frac{R_1 R_2}{R_1 + R_2}$$

$$R_T = \frac{20\Omega \times 30\Omega}{20\Omega + 30\Omega}$$

$$R_T = 12\Omega$$

COMPARISON—SERIES versus PARALLEL CIRCUIT

A comparison of series and parallel circuits is given in Table 7-1 and Fig. 7-13.

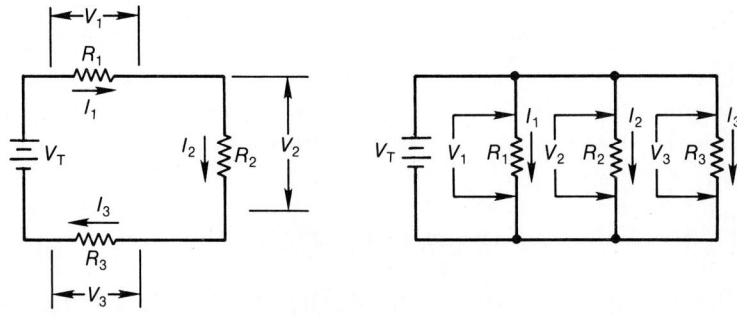

Fig. 7-13: Comparison—series vs. parallel circuit.

Table 7-1: Series versus Parallel Circuit

	Series	Parallel
Resistance	The total resistance (R_T): 1. Is greater than any component resistance 2. Is equal to the sum of the individual component resistances 3. $R_T = R_1 + R_2 + R_3 + ...R_n$	The equivalent resistance of the total circuit (R_T): 1. Is *less* than the lowest branch resistance 2. Its reciprocal is equal to the sum of the reciprocals of the individual branch resistances
Current	1. Only one path for current to flow 2. The current through each component is the same, regardless of its resistance value 3. $I_T = I_1 = I_2 = I_3 = ...I_n$	1. Has as many paths for current flow as there are branches 2. The branch with the lowest resistance has a proportionately higher current flow 3. $I_T = I_1 + I_2 + I_3 + ...I_n$
Voltage	1. The component with the highest resistance has a proportionately higher voltage drop across it 2. The sum of the voltage drops across each component is equal to the supply voltage 3. $V_T = V_1 + V_2 + V_3 + ...V_n$	1. The voltage across each branch (load) is the same as the supply voltage, regardless of the individual branch resistances 2. $V_T = V_1 = V_2 = V_3 = ...V_n$

POWER IN PARALLEL CIRCUITS

Power computations in a parallel circuit are essentially the same as those used for the series circuit. Since power dissipation in resistors consists of a heat loss, power dissipations are additive regardless of how the resistors are connected in the circuit. The total power dissipated is equal to the sum of the powers dissipated by the individual resistors. Like the series circuit, the total power consumed by the parallel circuit is:

$$P_T = P_1 + P_2 + ... P_n$$

Example: Find the total power consumed by the circuit in Fig. 7-14.

Solution:

$$P_{R1} = V \times I_{R1}$$
$$P_{R1} = 50V \times 5mA$$
$$P_{R1} = 250mW$$
$$P_{R2} = V \times I_{R2}$$
$$P_{R2} = 50V \times 2mA$$
$$P_{R2} = 100mW$$

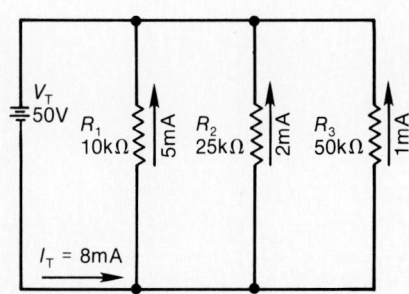

Fig. 7-14: Example of a parallel circuit.

117

$$P_{R3} = V \times I_{R3}$$
$$P_{R3} = 50\text{V} \times 1\text{mA}$$
$$P_{R3} = 50\text{mW}$$
$$P_T = P_1 + P_2 + P_3$$
$$P_T = 250\text{mW} + 100\text{mW} + 50\text{mW}$$
$$P_T = 400\text{mW}$$

Note that the power dissipated in the branch circuits is determined in the same manner as the power dissipated by individual resistors in a series circuit. The total power (P_T) is then obtained by adding the powers dissipated in the branch resistors using the equation mentioned.

Since, in the example shown in Fig. 7-14, the total current is known, the total power could be determined by the following method:

$$P_T = V \times I_T$$
$$P_T = 50\text{V} \times 8\text{mA}$$
$$P_T = 400\text{mW}$$

OPENS AND SHORTS IN PARALLEL CIRCUITS

In Chapter 6 the series circuit was examined under open-circuit and short-circuit conditions. The parallel circuit will now be examined under similar conditions of operation.

In Fig. 7-15, a parallel circuit composed of four vacuum tube filaments is shown under conditions of an open circuit. An open circuit is seen to have occurred in tube 4. In a parallel circuit such as this, the only effect would be a single inoperative vacuum tube. Thus, an open circuit in a parallel circuit only affects the device in that branch. All other branches continue uninterrupted operation and are completely unaffected by the open-circuit condition. The circuit current would decrease because the open-circuit branch is no longer drawing current.

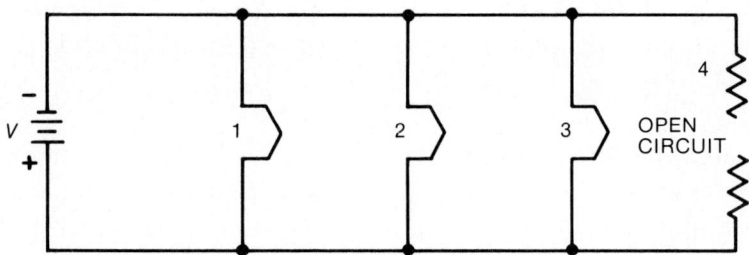

Fig. 7-15: Parallel circuit—open branch.

In Fig. 7-16, a parallel circuit composed of three resistances is shown with a short circuit across one branch. A short circuit is

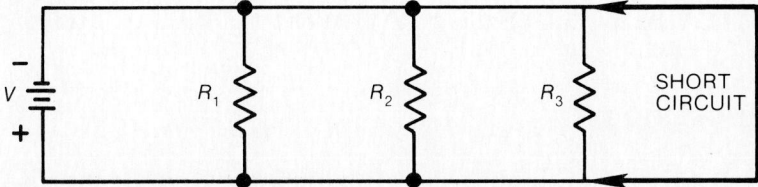

Fig. 7-16: Parallel circuit—short circuit.

considered to have zero resistance. This means that we are placing in parallel with R_1, R_2, and R_3 a resistance of $0\,\Omega$. Since the current in a parallel circuit divides according to the resistance of each branch, the current, if one branch has a resistance of $0\,\Omega$, of that branch would be

$$I = \frac{V}{R} = \frac{V}{0}$$

Effectively, then, all the current would flow through the short-circuited branch. That is, dividing a number by zero results in an infinite value; in this case, an infinite value of current. In reality, the power supply is limited in the amount of current that it can deliver. In addition, the circuit conductors are also limited in the amount of current they may carry. In practical operations, circuits are protected from such conditions by over-current devices such as fuses (see Chapter 4). These devices merely "interrupt" the supply circuit when a short-circuit or overload condition occurs on any branch. This is designed to occur before harmful levels of current are reached.

SUMMARY

Rules for solving parallel DC circuits are as follows:
1. The same voltage exists across each branch of a parallel circuit and is equal to the source voltage.
2. The current through a branch of a parallel network is inversely proportional to the amount of resistance of the branch.
3. The total current of a parallel circuit is equal to the sum of the currents of the individual branches of the circuit.
4. The total resistance of a parallel circuit is equal to the reciprocal of the sum of the reciprocals of the individual resistances of the circuit. For two resistances, the product over the sum method may be used.
5. The total circuit conductance is equal to the sum of the branch conductance.
6. Adding additional parallel branches decreases the circuit resistance and results in an increase in line current.
7. The equivalent resistance of a parallel circuit is always smaller than the smallest branch resistance.
8. The total power consumed in a parallel circuit is equal to the sum of the power consumption of the individual resistances.

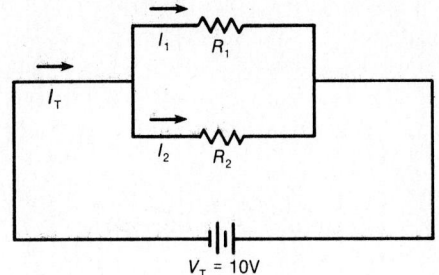

Fig. 7-17

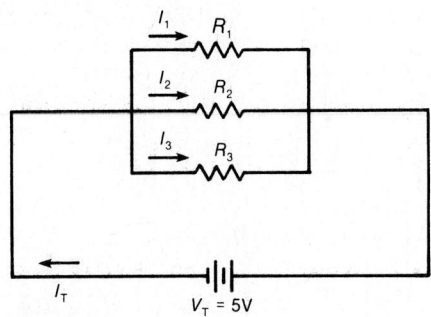

Fig. 7-18

CHECK YOURSELF (Answers at back of book)

Questions

Fill in the blanks with the appropriate word or words.

1. A(n) _____ in a branch of a parallel circuit only affects that branch and I_T.
2. A(n) _____ in a branch of a parallel circuit will affect the entire circuit.
3. The _____ of identical cells in parallel is the same as for one cell.
4. Cells connected in parallel are capable of producing a total _____ equal to the sum of each cell's _____ capability.
5. The algebraic sum of the currents entering and leaving any junction of conductors is _____ .

Problems

1. See Fig. 7-17. If $R_1 = R_2 = 1\text{k}\Omega$, determine the R_{eq} for the circuit using the equal resistors method.
2. Using the R_{eq} from problem 1, calculate I_T.
3. See Fig. 7-18. If $I_T = 100\text{mA}$, $I_1 = 20\text{mA}$, and $I_2 = 60\text{mA}$, calculate I_3 using Kirchhoff's Current Law.
4. See Fig. 7-18. Using the reciprocal method, determine R_{eq} if $R_1 = 800\Omega$, $R_2 = 10\text{k}\Omega$, and $R_3 = 1\text{k}\Omega$.
5. Determine the current flowing through each branch using the values of resistance from the previous problem. Also determine I_T.
6. Using values from the previous problem, calculate the total amount of power consumed by this circuit.
7. See Fig. 7-17. $R_1 = 20\text{k}\Omega$, $R_2 = 30\text{k}\Omega$. Use the product over the sum method to find R_{eq}.
8. Use the conductance method to determine R_{eq} in Fig. 7-18, if $R_1 = 700\Omega$, $R_2 = 900\Omega$, and $R_3 = 850\Omega$.
9. Use the values from the previous example to find I_1, I_2, I_3, and I_T.
10. Now determine the amount of power dissipated by this circuit (using data from problems 8 and 9).

CHAPTER 8

DC Circuits— Combination

In Chapters 6 and 7, series and parallel DC circuits have been considered separately. However, there are few occasions when an electronic circuit will be considered solely of one type or the other. In fact, most electronic circuits consist of both series and parallel elements. A circuit of this type will be referred to as a *combination circuit* (Fig. 8-1).

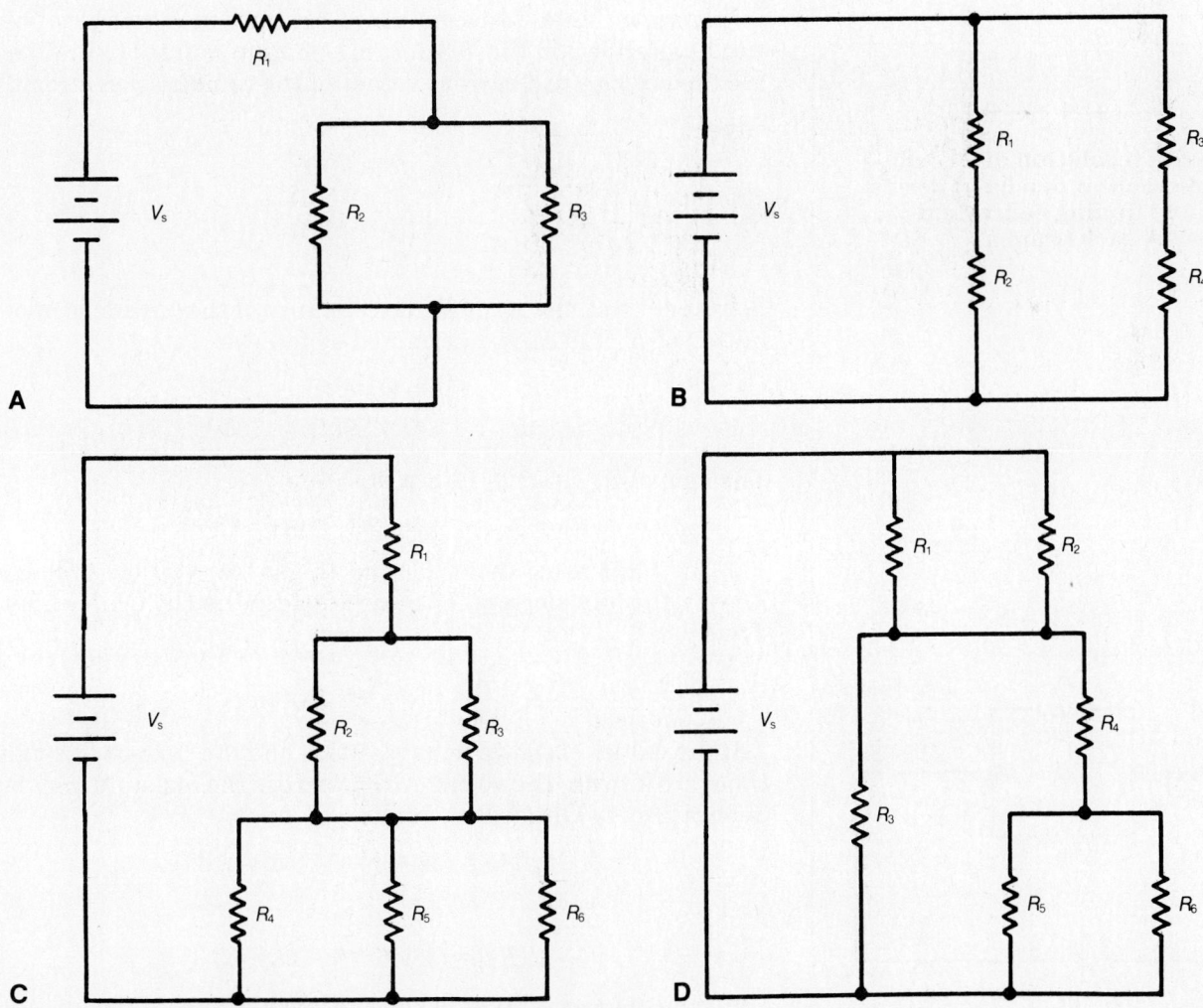

Fig. 8-1: Typical combination circuit configurations.

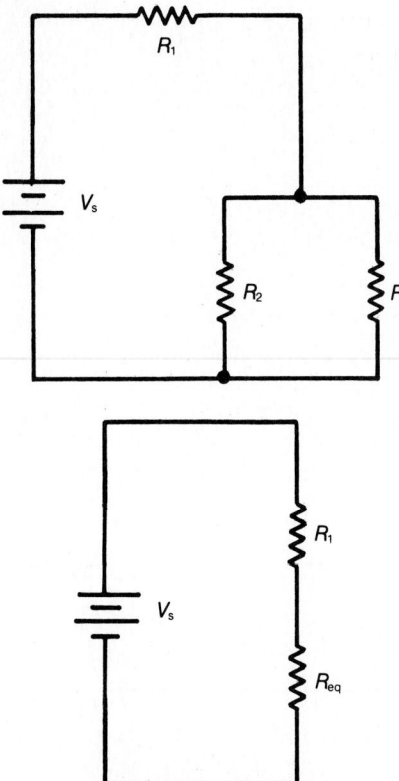

Fig. 8-2: Resolution of a combination circuit into a basic series circuit by finding equivalent values for each branch.

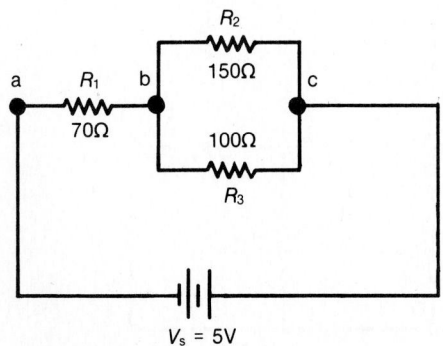

Fig. 8-3: R_1 in series with parallel combination of R_2 and R_3.

SOLVING A COMBINATION CIRCUIT

In the study of electronics, it is often necessary to resolve a complex circuit into a simpler form. Any complex circuit consisting of resistances can be reduced to a basic equivalent circuit containing the source and total resistance (Fig. 8-2). The process is called reduction to an *equivalent circuit*.

Most combination equivalent circuits can be broken down into either a *series-parallel circuit* or *parallel-series circuit*, then finally into a simple series or parallel circuit.

Equivalent Circuits—Series-Parallel

The two-step equivalent is a breakdown of a series circuit. Resistor R_1 is wired in series with the parallel combination of R_2 and R_3. This circuit is solved by first finding the equivalent resistance of R_2 and R_3. The first step is to apply the laws relating to two parallel resistances. Change these into a single equivalent resistance, R_{eq}. Now the circuit values are used to identify specific voltages, currents, resistances, and power values.

Example: Three resistors are connected in a combination circuit as shown in Fig. 8-3 across a voltage source of 5V. What are the voltage and current values of the combination circuit?

Given:

$$R_1 = 70\Omega$$
$$R_2 = 150\Omega$$
$$R_3 = 100\Omega$$
$$R_T = ?$$

Solution: First, the equivalent resistance of the parallel combination of R_2 and R_3 is determined as follows:

$$R_{2,3} = \frac{R_2 R_3}{R_2 + R_3} = \frac{(150\Omega)(100\Omega)}{150\Omega + 100\Omega} = \frac{15,000}{250} = 60\Omega$$

The sum of $R_{2,3}$ and R_1,—that is R_T—is

$$R_T = R_{2,3} + R_1 = 60\Omega + 70\Omega = 130\Omega.$$

If the total resistance (R_T) and the source voltage (V_S) are known, the total current (I_T) may be determined by Ohm's Law. Thus,

$$I_T = \frac{V_S}{R_T} = \frac{5V}{130\Omega} = 38.46\text{mA}$$

If the values of the various resistors and the current through them are known, the voltage drops across the resistors may be determined by Ohm's Law. Thus,

$$V_{ab} = I_T R_1 = (38.46\text{mA})(70\Omega) = 2.69V$$

and

$$V_{bc} = I_T R_{2,3} = (38.46\text{mA})(60\Omega) = 2.31V$$

According to Kirchhoff's voltage law, the sum of the voltage drops around the closed circuit is equal to the source voltage. Thus,

$$V_{ab} + V_{bc} = V_T$$

or

$$2.69V + 2.31V = 5V$$

If the voltage drop (V_{bc}) across $R_{2,3}$—that is, the drop between points b and c—is known, the current through the individual branches may be determined as follows:

$$I_2 = \frac{V_{bc}}{R_2} = \frac{2.31V}{150\Omega} = 15.4mA$$

$$I_3 = \frac{V_{bc}}{R_3} = \frac{2.31V}{100\Omega} = 23.1mA$$

According to Kirchhoff's current law, the sum of the currents flowing in the individual parallel branches is equal to the total current. Thus,

$$I_2 + I_3 = I_T$$

or

$$15.4mA + 23.1mA = 38.46mA \text{ (approx.)}$$

The total current flows through R_1; and at point b it divides between the two branches in inverse proportion to the resistance of each branch.

Equivalent Circuits—Parallel-Series

The basic design of the circuit shown in Fig. 8-4 is a parallel combination. The circuit has one resistor (R_1) in parallel with two series resistors (R_2 and R_3). The parallel-series combination is connected in parallel to the voltage source (V_S). The first step in the solution of this circuit is to simplify the series-wired resistances (R_2 and R_3). Once the equivalent resistance of the series branch is found, the circuit is treated as a parallel circuit.

Example: Three resistors are connected in a combination circuit as shown in Fig. 8-5 across a voltage source of 5V. What are the voltage and current of the combination circuit?

Given:

$$R_1 = 60\Omega$$
$$R_2 = 20\Omega$$
$$R_3 = 100\Omega$$
$$R_T = ?$$

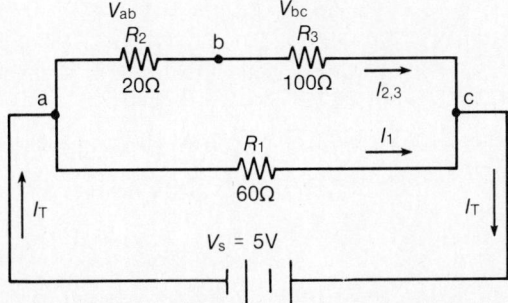

Fig. 8-5: R_1 in parallel with the series combination of R_2 and R_3.

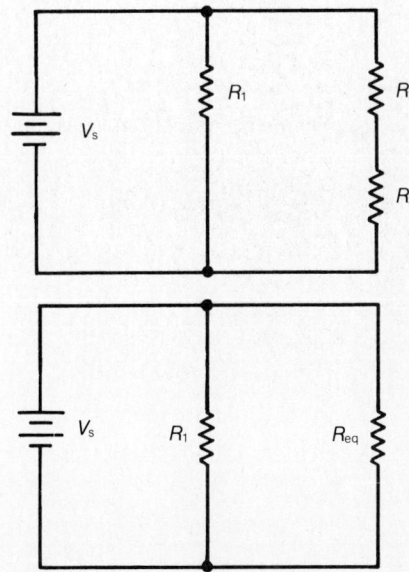

Fig. 8-4: Resolve each series branch into an equivalent value.

123

Solution: The total resistance (R_T) is determined in two steps as just mentioned. First, the sum of the resistance of R_2 and R_3—that is, $R_{2,3}$—is determined as follows:

$$R_{2,3} = R_2 + R_3 = 20\Omega + 100\Omega = 120\Omega$$

Second, the total resistance (R_T) is the result of combining $R_{2,3}$ in parallel with R_1, or

$$R_T = \frac{R_{2,3}R_1}{R_{2,3} + R_1} = \frac{(120\Omega)(60\Omega)}{(120\Omega + 60\Omega)} = 40\Omega$$

If the total resistance (R_T) and the source voltage (V_S) are known, the total current (I_T) may be determined by Ohm's Law. Thus, in Fig. 8-5,

$$I_T = \frac{V_T}{R_T} = \frac{5V}{40\Omega} = 125mA$$

A portion of the total current flows through the series combination of R_2 and R_3 and the remainder flows through R_1. Because current varies inversely with the resistance, two-thirds of the total current flows through R_1 and one-third flows through the series combination of R_2 and R_3, since R_1 is one-half of $R_2 + R_3$.

The source voltage (V_S) is applied between points a and c and, therefore, the current I_1 through R_1 is

$$I_1 = \frac{V_S}{R_1} = \frac{5V}{60\Omega} = 83.3mA$$

and the current, $I_{2,3}$, through $R_{2,3}$ is

$$I_{2,3} = \frac{V_S}{R_{2,3}} = \frac{5V}{120\Omega} = 41.6mA$$

According to Kirchhoff's current law the sum of the individual branch currents is equal to the total current, or

$$I_T = I_1 + I_{2,3}$$
$$I_T = 83.3mA + 41.6mA = 125mA \quad (approx.)$$

Knowing the current $I_{2,3}$ through $R_{2,3}$, voltage drops V_2 and V_3 can be found as follows:

$$V_2 = I_2R_2$$
$$= (41.6mA)(20\Omega)$$
$$= 832mV$$
$$V_3 = I_3R_3$$
$$= (41.6mA)(100\Omega)$$
$$= 4.16V$$

Note that the $V_2 + V_3$ branch is equal to the voltage developed across R_1 and also V_S. As a check:

$$V_S = V_2 + V_3$$
$$= 0.832V + 4.16V$$
$$= 4.99V \text{ or } 5V$$

124

Compound Combination Circuits

Combination circuits may be made up of a number of resistors arranged in numerous series and parallel arrangements. The easiest way to simplify complicated circuits is to decide whether they are basically series or parallel. For example, the circuit shown in Fig. 8-6A is more of a series circuit than it is a parallel circuit. Thus, the total resistance (R_T) is obtained in the following three logical steps:

Since R_3, R_4, and R_5 are in series (there is only one path for current), they may be combined in Fig. 8-6B to give the resistance, R_S, of the three resistors. Thus,

$$R_S = R_3 + R_4 + R_5 = 50\Omega + 90\Omega + 100\Omega = 240\Omega$$

and it is now in parallel with R_2 (because they both receive the same voltage).

The combined resistance of R_S in parallel with R_2 is

$$R_{S,2} = \frac{R_2 R_S}{R_2 + R_S} = \frac{80\Omega \times 240\Omega}{80\Omega + 240\Omega} = 60\Omega$$

as in Fig. 8-6C.

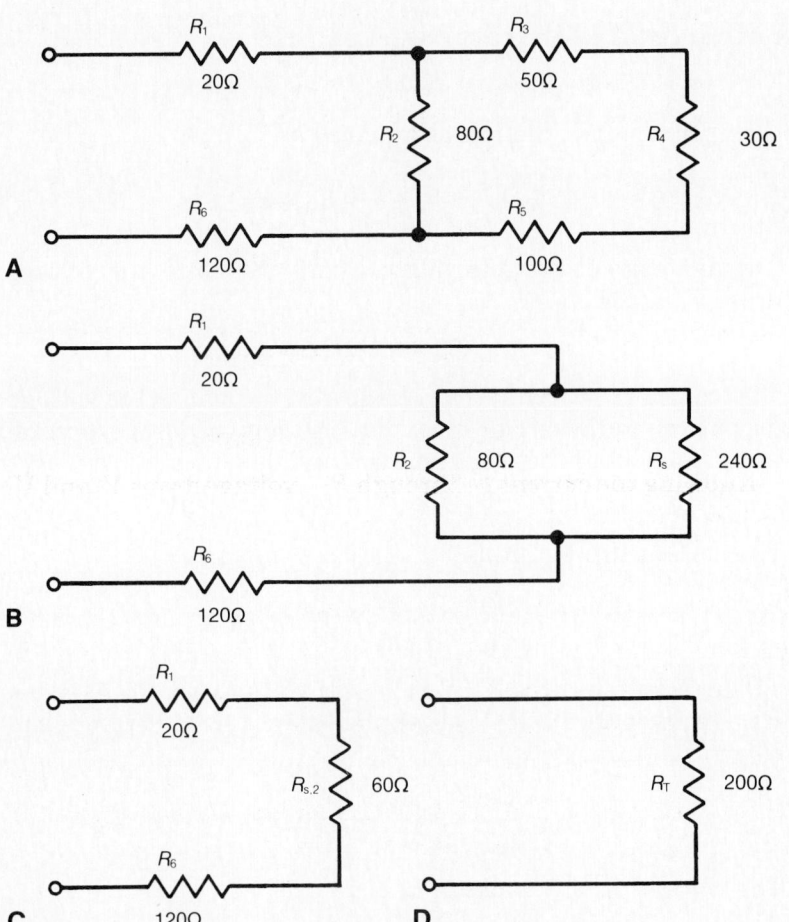

Fig. 8-6: Solving total resistance in a compound circuit.

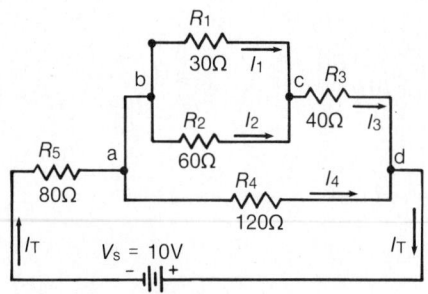

Fig. 8-7: Compound circuit for solving resistance, voltage, current, and power.

Third, the total resistance (R_T) is determined by combining resistors R_1 and R_6 with $R_{S,2}$ as

$$R_T = R_1 + R_6 + R_{S,2} = 20\Omega + 120\Omega + 60\Omega = 200\Omega$$

Other compound circuits may be solved in a similar manner. For example, in Fig. 8-7, the total resistance (R_T) may be found by simplifying the circuit in successive steps beginning with the resistance, R_1 and R_2. Thus,

$$R_{1,2} = \frac{R_1 R_2}{R_1 + R_2} = \frac{30\Omega \times 60\Omega}{30\Omega + 60\Omega} = \frac{1800\Omega}{90\Omega} = 20\Omega$$

and it is in series with R_3.

The resistances, $R_{1,2}$ and R_3, are added to give the resultant resistance, $R_{1,2,3}$. Thus,

$$R_{1,2,3} = R_{1,2} + R_3 = 20\Omega + 40\Omega = 60\Omega$$

$R_{1,2,3}$ is in parallel with R_4. The combined resistance, $R_{1,2,3,4}$, is determined as follows:

$$R_{1,2,3,4} = \frac{R_{1,2,3} R_4}{R_{1,2,3} + R_4} = \frac{60\Omega \times 120\Omega}{60\Omega + 120\Omega} = \frac{720\Omega}{180\Omega} = 40\Omega$$

This equivalent resistance is in series with R_5. Thus, the total resistance (R_T) of the circuit is

$$R_T = R_{1,2,3,4} + R_5 = 40\Omega + 80\Omega = 120\Omega$$

By Ohm's Law, the line current (I_T) is

$$I_T = \frac{V_S}{R_T} = \frac{10V}{120\Omega} = 83.3mA$$

The line current flows through R_5 and, therefore, the voltage drop, V_5, across R_5 is

$$V_5 = I_T R_5 = (83.3mA)(80\Omega) = 6.666V$$

According to Kirchhoff's voltage law, the sum of the voltage drops around the circuit is equal to the source voltage; accordingly, the voltage between points a and d is

$$V_{ad} = V_S - V_5 = 10V - 6.666V = 3.333V$$

The current through R_4 is

$$I_4 = \frac{V_4}{R_4} = \frac{3.33V}{120\Omega} = 27.77mA$$

The resistance, $R_{1,2,3}$, of parallel resistors R_1 and R_2 in series with resistor R_3 is 60Ω. V_{ad} is applied across 60Ω; therefore, the current, I_3, through R_3 is

$$I_3 = \frac{V_{ad}}{R_{1,2,3}} = \frac{3.333V}{60\Omega} = 55.55mA$$

The voltage drop, V_3, across R_3 is

$$V_3 = I_3 R_3 = (55.55mA)(40\Omega) = 2.222V$$

126

and the voltage across the parallel combination of R_1 and R_2—that is, V_{bc}—is

$$V_{bc} = I_{1,2}R_{1,2} = (55.55\text{mA})(20\Omega) = 1.111\text{V}$$

where $I_{1,2}$ is the current through the parallel combinations of R_1 and R_2. By Kirchhoff's current law, $I_{1,2}$ is equal to I_3. The current, I_1, through R_1 is

$$I_1 = \frac{V_{bc}}{R_1} = \frac{1.111\text{V}}{30\Omega} = 37.03\text{mA}$$

and the current, I_2, through R_2 is

$$I_2 = \frac{V_{bc}}{R_2} = \frac{1.111\text{V}}{60\Omega} = 18.518\text{mA}$$

The preceding computations may be checked by the application of Kirchhoff's voltage and current law to the entire circuit. Briefly, the sum of the voltage drops around the circuit is equal to the source voltage. Voltage V_5 across R_5 is 6.666V and voltage V_{ad} across R_4 is 3.333V—that is,

$$V_S = V_5 + V_{ad}$$

or

$$10\text{V} = 6.666\text{V} + 3.333\text{V}$$

Likewise, the voltage drop, V_{bc}, across the parallel combination of R_1 and R_2 plus the voltage drop, V_3, across R_3 should be equal to the voltage across points a and d. V_{bc} is 1.111V and V_3 is 2.222V; therefore,

$$V_{ad} = V_{bc} + V_3 = 1.111\text{V} + 2.222\text{V} = 3.333\text{V}$$

Kirchhoff's current law says in effect that the sum of the branch currents is equal to the line current, I_T. The line current is 83.3mA; therefore, the sum of I_4 and I_3 should be 83.3mA or

$$I_T = I_4 + I_3 = 27.77\text{mA} + 55.55\text{mA} = 83.3\text{mA}$$

The power consumed in a circuit element is determined by one of the three power formulas. For example, in Fig. 8-7 the power, P_1, consumed in R_1 is

$$P_1 = I_1 V_{bc} = (37.03\text{mA})(1.111\text{V}) = 41.14\text{mW}$$

the power, P_2, consumed in R_2 is

$$P_2 = I_2 V_{bc} = (18.518\text{mA})(1.111\text{V}) = 20.57\text{mW}$$

the power, P_3, consumed in R_3 is

$$P_3 = I_3 V_3 = (55.55\text{mA})(2.222\text{V}) = 123.4\text{mW}$$

the power, P_4, consumed in R_4 is

$$P_4 = I_4 V_4 = (27.77\text{mA})(3.33\text{V}) = 92.5\text{mW}$$

and the power, P_5, consumed in R_5 is

$$P_5 = I_5 V_5 = (83.3\text{mA})(6.666\text{V}) = 555\text{mW}$$

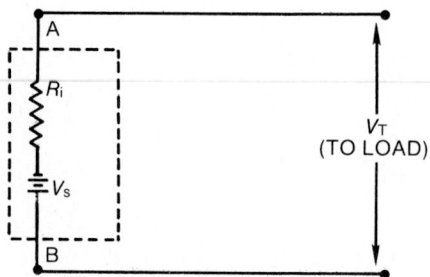

Fig. 8-8: Battery with internal resistance.

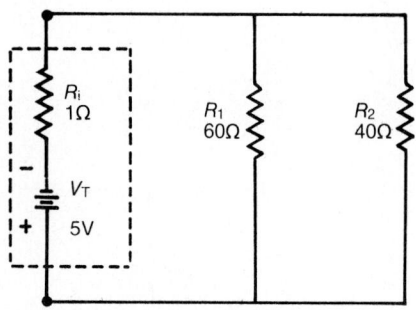

Fig. 8-9: Effect of source resistance on a parallel circuit.

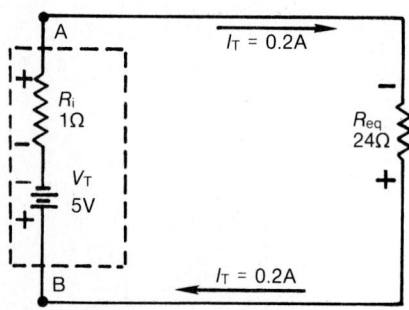

Fig. 8-10: Source resistance in an equivalent circuit.

The total power, P_T, consumed is

$$P_T = P_1 + P_2 + P_3 + P_4 + P_5$$
$$= 41.14\text{mW} + 20.57\text{mW} + 123.4\text{mW} + 92.5\text{mW} + 555\text{mW}$$
$$= 833\text{mW}$$

The total power is also equal to the total current multiplied by the source voltage or

$$P_T = I_T V_S = (83.3\text{mA})(10\text{V}) = 833\text{mW}$$

EFFECTS OF SOURCE RESISTANCE

The combination circuits discussed up to this point have been explained and solved without considering the internal resistance of the source. Every known source possesses resistance. In a battery, as mentioned in Chapter 4, the resistance is partially due to the opposition offered to the movement of current through the electrolyte. A schematic representation of source resistance is shown in Fig. 8-8.

The internal resistance of the battery is labeled R_i and is always shown schematically connected in series with the source. Under load conditions this internal resistance will have a voltage drop across it and must be considered as part of the external circuit. The voltage at battery terminals A and B will always be less than the generated voltage of the battery since a portion of the generated voltage will be dropped across the internal resistance of the battery.

The presence of internal resistance results in (1) a diminished voltage supplied to the components that comprise the load, (2) a decrease in total current, and (3) an increase in total resistance. The power dissipated by the circuit is also affected. The effect of internal resistance on the circuit is analyzed using the example circuit shown in Fig. 8-9. This circuit can no longer be classified as a parallel circuit because there is a series resistance to be considered. The circuit is solved in the following manner.

Determine R_{eq} for the parallel network:

$$R_{eq} = \frac{R_1 R_2}{R_1 + R_2} = \frac{60 \times 40}{60 + 40} = \frac{2,400}{100} = 24\Omega$$

Reduce to an equivalent circuit (Fig. 8-10).

Compute the total series resistance:

$$R_T = R_i + R_{eq}$$
$$R_T = 1\Omega + 24\Omega = 25\Omega$$

Compute total current:

$$I_T = \frac{V_T}{R_T} = \frac{5\text{V}}{25\Omega} = 0.2\text{A}$$

Determine voltage drop across R_{eq}:

$$V_{R_{eq}} = I_T \times R_{eq}$$
$$V_{R_{eq}} = 0.2\text{A} \times 24\Omega$$
$$V_{R_{eq}} = 4.8\text{V}$$

Find voltage drop across R_i:

$$V_{R_i} = I_T \times R_i$$
$$V_{R_i} = 0.2A \times 1\Omega$$
$$V_{R_i} = 0.2V$$

Determine power dissipated by load resistors:

$$P_{R_{eq}} = I_T \times V_{R_{eq}}$$
$$P_{R_{eq}} = 0.2A \times 4.8V$$
$$P_{R_{eq}} = 0.96W$$

Determine power dissipated by source resistance:

$$P_{R_i} = I_T \times V_{R_i}$$
$$P_{R_i} = 0.2A \times 0.2V$$
$$P_{R_i} = 0.04W$$

Determine total power dissipation:

$$P_T = P_{R_{eq}} + P_{R_i}$$
$$P_T = 0.96W + 0.04W$$
$$P_T = 1.00W$$

Circuit efficiency is determined by the following formula:

$$\text{Percent Eff } (\eta) = \frac{P_O}{P_{in}} \times 100$$

(The letter symbol for efficiency is the Greek letter η [eta].) where

$\qquad$ Percent Eff = percent of efficiency

$\qquad$ P_O = power in W supplied to load device

$\qquad$ P_{in} = power in W supplied by the source

For the circuit of Fig. 8-10 the percent efficiency is:

$$\text{Percent Eff} = \frac{P_O}{P_{in}} \times 100$$

$$\text{Percent Eff} = \frac{0.96W}{0.96W + 0.04W} \times 100$$

$$\text{Percent Eff} = \frac{0.96W}{1.00W} \times 100 = 96\%$$

From this efficiency relationship, we may conclude that the source resistance does affect the total power dissipated by the equivalent (load) resistance. The source resistance also affects the transfer of power.

OPEN AND SHORT COMBINATION CIRCUITS

In comparing the effects of an open in series and parallel circuits, as discussed in Chapters 6 and 7, the major difference to be noted is that an open in a parallel circuit would not necessarily disable the entire circuit, i.e., the current flow would not be reduced to zero unless the open condition existed at some point electronically common to all other parts of the circuit.

A short circuit in a parallel network has an effect similar to a short in a series circuit. In general, the short will cause an increase in current and the possibility of component damage regardless of the type of circuit involved.

Opens and shorts alike, if occurring in a branch circuit of a combination network, will result in an overall change in the equivalent resistance. This can cause undesirable effects in other parts of the circuit due to the corresponding change in the total current flow.

VOLTAGE DIVIDER

In practically all electronic devices, certain design requirements recur again and again. For instance, many electronic units require a number of different voltages at various points in their circuitry. In addition, all the various voltages must be derived from a single primary power supply. The most common method of meeting these requirements is by the use of a voltage-divider network. A typical voltage divider consists of two or more resistors connected in series across the primary power supply. The primary voltage, V_s, must be as high or higher than any of the individual voltages it is to supply. As the primary voltage is dropped by successive steps through the series resistors, any desired fraction of the original voltage may be "tapped off" to supply individual requirements. The values of the series resistors to be used are dictated by the voltage drops required.

If the total current flowing in the divider circuit is affected by the loads placed on it, then the voltage drops of each divider resistor will also be affected. When a voltage divider is being designed, the maximum current drawn by the loads will determine the value of the resistors that form the voltage divider. Normally, the resistance values chosen for the divider will permit a current equal to 10% of the total current drawn by the external loads. This current which does not flow through any of the load devices is called *bleeder current*.

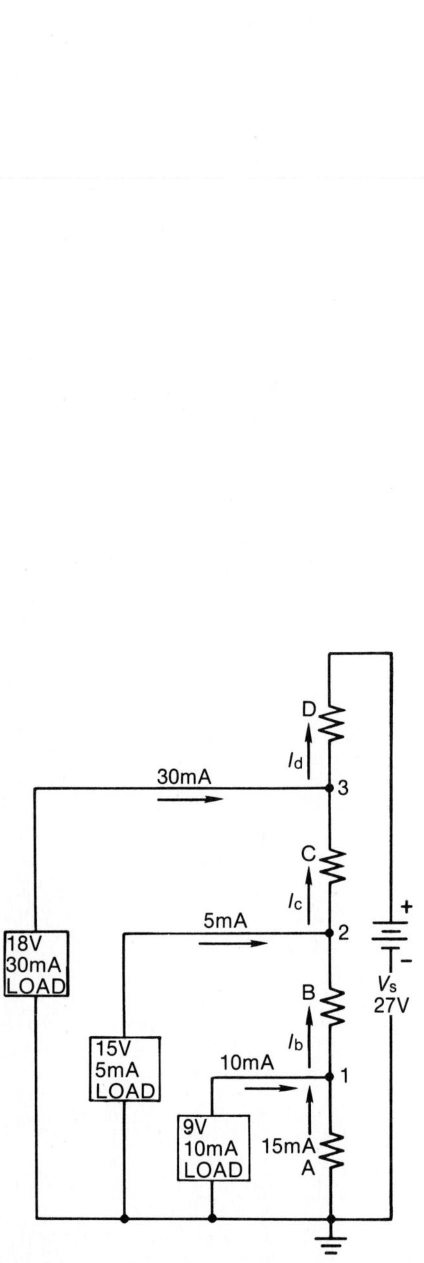

A voltage-divider circuit is shown in Fig. 8-11. The divider is connected across a 27V source and supplies three loads simultaneously—10mA at 9V between terminal 1 and ground; 5mA at 15V between terminal 2 and ground; and 30mA at 18V between terminal 3 and ground. The current in resistor A (bleeder current) is 15mA. The current, voltage, resistance, and power of the four resistors are to be determined.

Kirchhoff's law of currents applied to terminal 1 indicates that the current in resistor B is equal to the sum of 15mA from resistor A and 10mA from the 9V load. Thus,

$$I_b = 15\text{mA} + 10\text{mA} = 25\text{mA}$$

Similarly,

$$I_c = 25\text{mA} + 5\text{mA} = 30\text{mA}$$

and

$$I_d = 30\text{mA} + 30\text{mA} = 60\text{mA}$$

Fig. 8-11: Typical voltage divider, to determine R and P.

Kirchhoff's voltage law indicates that the voltage across resistor A is 9V; the voltage across B is

$$V_b = 15V - 9V = 6V$$

the voltage across C is

$$V_c = 18V - 15V = 3V$$

and the voltage across D is

$$V_d = 27V - 18V = 9V$$

Applying Ohm's Law to determine the resistances:

$$\text{resistance of A if } R_a = \frac{V_a}{I_a} = \frac{9V}{15mA} = 600\Omega$$

$$\text{resistance of B if } R_b = \frac{V_b}{I_b} = \frac{6V}{25mA} = 240\Omega$$

$$\text{resistance of C if } R_c = \frac{V_c}{I_c} = \frac{3V}{30mA} = 100\Omega$$

$$\text{resistance of D if } R_d = \frac{V_d}{I_d} = \frac{9V}{60mA} = 150\Omega$$

The power dissipated by

resistor A is $P_a = V_a I_a = 9V \times 0.015A = 135mW$
resistor B is $P_b = V_b I_b = 6V \times 0.025A = 150mW$
resistor C is $P_c = V_c I_c = 3V \times 0.030A = 90mW$
resistor D is $P_d = V_d I_d = 9V \times 0.060A = 540mW$

The total power dissipated by the four resistors is

$$135mW + 150mW + 90mW + 540mW = 915mW$$

The power dissipated by the load connected to

terminal 1 is $P_1 = V_1 I_1 = 9V \times 0.010A = 90mW$
terminal 2 is $P_2 = V_2 I_2 = 15V \times 0.005A = 75mW$
terminal 3 is $P_3 = V_3 I_3 = 18V \times 0.030A = 540mW$

The total power supplied to the three loads is

$$90mW + 75mW + 540mW = 705mW$$

The total power supplied to the entire circuit including the voltage divider and the three loads is

$$915mW + 705mW = 1.62W$$

This value is checked as

$$P_T = V_s \times I_T = 27V \times 0.060A = 1.62W$$

In Fig. 8-12, the voltage-divider resistances are given and the current in R_5 is to be found. The load current in R_1 is 6mA; the current in R_2 is 4mA; and the current in R_3 is 10mA. The source voltage is 510V. Kirchhoff's current law may be applied at junctions a, b, c, and d to determine expressions for the current in resistors R_4, R_5, R_6, and R_7. Accordingly, the current in R_4 is $I + 6 + 4 + 10$, or $I + 20$; the current in R_5 is I; the current in R_6 is $I + 6$; the current in R_7 is $I + 6 + 4$, or $I + 10$.

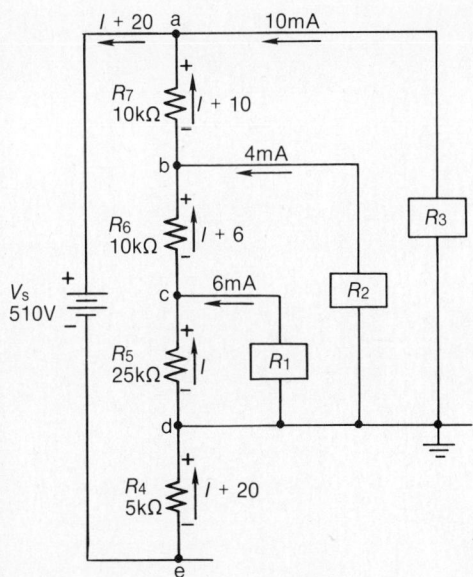

Fig. 8-12: Typical voltage divider, to determine V and R.

131

The voltage across R_4 may be expressed in terms of the resistance in kΩ and the current in mA as $5(I + 20)$V. Similarly, the voltage across R_5 is equal to $25I$; the voltage across R_6 is $10(I + 6)$ and the voltage across R_7 is $10(I + 10)$. Kirchhoff's law of voltages may be applied to the voltage divider to solve for the unknown current, I, by expressing the source voltage in terms of the given values of voltage, resistance, and current (both known and unknown values). The sum of the voltages across R_4, R_5, R_6, and R_7 is equal to the source voltage as follows:

$$V_4 + V_5 + V_6 + V_7 = V_6$$
$$5(I + 20) + 25I + 10(I + 6) + 10(I + 10) = 510$$
$$5I + 100 + 25I + 10I + 60 + 10I + 100 = 510$$
$$50I + 260 = 510$$
$$50I = 510 - 260$$
$$50I = 250$$
$$I = 5\text{mA}$$

The current of 5mA through R_5 produces a voltage drop across R_5 of 5mA $\times$ 25kΩ, or 125V. Since R_1 is in parallel with R_5, the voltage across load R_1 is 125V. The current through R_4 is $5 + 20$, or 25mA and the corresponding voltage is 5×25, or 125V. Since point d is at ground potential, point c is 125V positive with respect to ground, whereas point e is 125V negative with respect to ground. The current in R_6 is $5 + 6$ or 11mA and the voltage drop across R_6 is 11×10, or 110V. The current in R_7 is $5 + 10$, or 15mA and the voltage drop is 15×10 or 150V. The total voltage is the sum of the voltages across the divider. Thus,

$$125\text{V} + 125\text{V} + 110\text{V} + 150\text{V} = 510\text{V}$$

The power dissipated by each resistor in the voltage divider may be found by multiplying the voltage across the resistor by the current in the resistor. If the current is expressed in amperes and the emf in volts, the power will be expressed in watts. Thus, the power in R_4 is

$$P_4 = V_4 I_4 = 125\text{V} \times 0.025\text{A} = 3.125\text{W}$$

Similarly, the power in R_5 is 125V $\times$ 0.005A = 0.625W; the power in R_6 is 110V $\times$ 0.011A = 1.21W; and in R_7 is 150V $\times$ 0.015A = 2.25W. The total power in the divider is

$$3.125\text{W} + 0.625\text{W} + 1.21\text{W} + 2.25\text{W} = 7.21\text{W}$$

The voltage across load R_1 is the voltage across R_5, or 125V. The power in R_1 is

$$P_1 = V_1 I_1 = 125\text{V} \times 0.006\text{A} = 0.750\text{W}$$

The voltage across load R_2 is equal to the sum of the voltages across R_5 and R_6. Thus,

$$V_2 = V_5 + V_6 = 125\text{V} + 110\text{V} = 235\text{V}$$

The power in load R_2 is

$$P_2 = V_2 I_2 = 235\text{W} \times 0.004\text{W} = 0.940\text{W}$$

The voltage across load R_3 is equal to the sum of the voltages across R_5, R_6, and R_7. Thus,

$$V_3 = V_5 + V_6 + V_7 = 125V + 110V + 150V = 385V$$

The power in load R_3 is

$$P_3 = V_3 I_3 = 385V \times 0.010A = 3.85W$$

The total power in the three loads is

$$0.75W + 0.94W + 3.85W = 5.54W$$

and the total power supplied by the source is equal to the sum of the power dissipated by the voltage divider and the three loads, or

$$7.21W + 5.54W = 12.75W$$

The total power may be checked by

$$P_T = V_T I_T = 510V \times 0.025A = 12.75W$$

The resistances of load resistors R_1, R_2, and R_3 are determined by means of Ohm's Law as follows:

$$R_1 = \frac{V_1}{I_1} = \frac{125V}{6A} = 20.83k\Omega$$

$$R_2 = \frac{V_2}{I_2} = \frac{235V}{4A} = 58.75k\Omega$$

and

$$R_3 = \frac{V_3}{I_3} = \frac{385V}{10A} = 38.5k\Omega$$

The variation of voltages and currents found in the previous examples are undesirable in a voltage divider. It must be designed to provide voltages that are as stable as possible. A voltage divider consisting of two resistors will be designed using the circuit configuration shown in Fig. 8-13. The supply voltage is 20V. It is desired to furnish voltages of 5 and 20V to two loads drawing 6 and 18mA respectively. Assume bleeder current to be 10% of the required load current.

Total load current is specified as 24mA. The bleeder current, therefore, should be

$$I_b = 10\% I_L$$
$$I_b = 10\% \times 24mA$$
$$I_b = 2.4mA$$

The bleeder current and the current through resistor R_3 combine and both currents flow through R_1. This current value may be computed as follows:

$$I_{R_1} = I_b + I_{R_3}$$
$$I_{R_1} = 2.4mA + 6mA$$
$$I_{R_1} = 8.4mA$$

The total current may also be determined:

$$I_T = 8.4mA + 18mA$$
$$I_T = 26.4mA$$

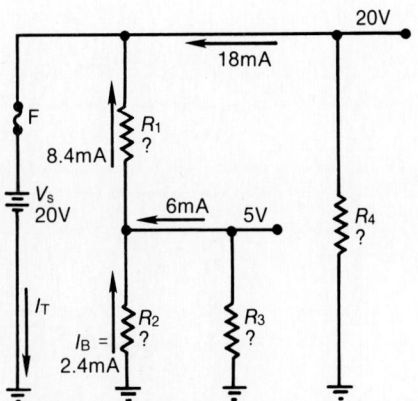

Fig. 8-13: Example circuit for proposed voltage divider.

133

The resistance values of R_3 and R_4 must be as follows:

$$R_3 = \frac{V_{R3}}{I_{R3}} = \frac{5\text{V}}{6\text{mA}} = 833\,\Omega$$

$$R_4 = \frac{V_{R4}}{I_{R4}} = \frac{20\text{V}}{18\text{mA}} = 1,111\,\Omega$$

Computing for R_1 and R_2:

$$R_1 = \frac{V_{R1}}{I_{R1}} = \frac{35\text{V}}{8.4\text{mA}} = 4.17\text{k}\Omega$$

$$R_2 = \frac{V_{R2}}{I_{R2}} = \frac{5\text{V}}{2.4\text{mA}} = 2.082\text{k}\Omega$$

BRIDGE CIRCUITS

A resistance bridge circuit in its simplest form is shown in Fig. 8-14. Two identical resistors, R_1 and R_2, are connected in parallel across a 20V power source. The network becomes a bridge circuit when a cross connection or "bridge" is placed between the two resistors.

The voltage across both resistors is dropped at the same rate, because the resistors are identical; therefore, points a-a', b-b', c-c', and d-d' are at equal potentials. If the bridge is connected between points of equal potential, as in Fig. 8-14A, no current will flow through the bridge. However, if the bridge is connected between points of unequal potential, current will flow from the more negative to the less negative end as shown in Figs. 8-14B and C. In Fig. 8-14B, current flows from right to left from b' to d. In Fig. 8-14C, current flows from left to right from a to c'. Thus, you see that the direction of bridge current is controlled by the

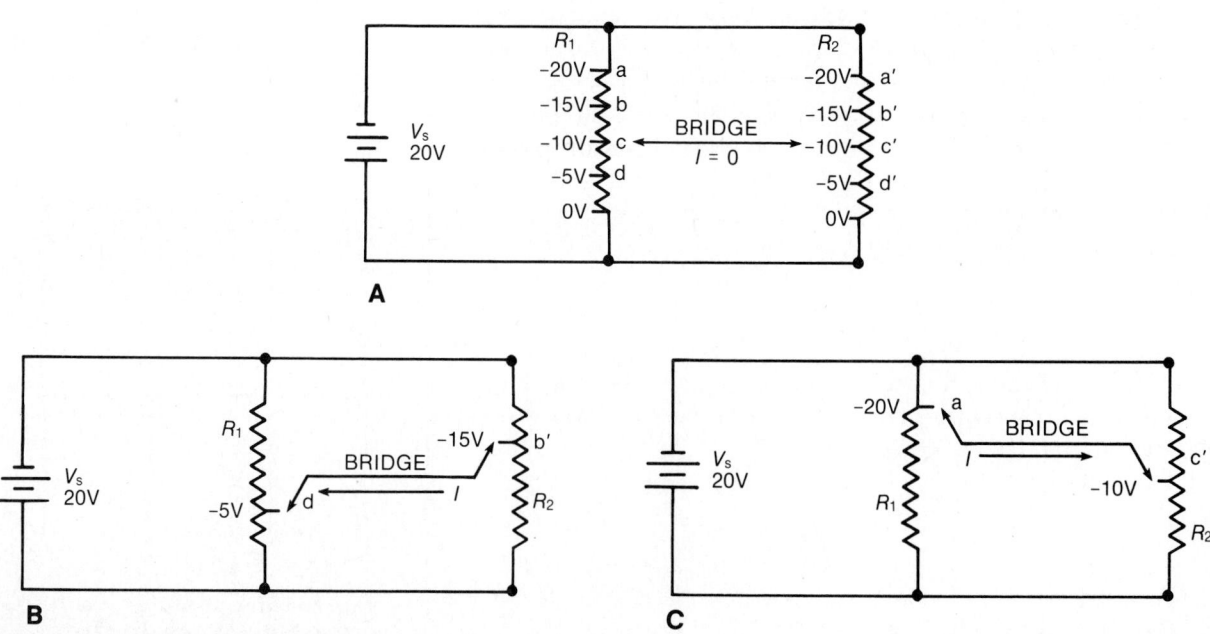

Fig. 8-14: Simple resistance bridges.

difference in potential between the two ends of the bridge resistor. When the bridge resistor is across points of equal potential, no current flows through the bridge resistor and the bridge is said to be *balanced*. When it is across unequal potentials, current flows through the bridge resistor and the bridge is said to be *unbalanced*. The bridge may be unbalanced in either or both of two ways: (1) by connecting the bridge to unequal potentials or (2) by using resistors of unequal value.

Unbalanced Resistance Bridge

Figure 8-15 shows an unbalanced bridge using unequal resistors. The two parallel legs are R_1-R_4 and R_3-R_5. R_2 is the bridge. The current and voltage drop of each resistor and the source voltage (V_S) are to be determined.

The current of 0.1A flowing into junction a divides into two parts. The part flowing through R_1 is indicated as I_1 and the part through R_3 is 0.1 - I_1. Similarly, at junction b, I_1 divides, part flowing through R_2 and the remainder through R_4. The part through R_2 is designated I_2, and the part through R_4 is I_1 - I_2. The direction of current through R_2 may be assumed arbitrarily.

If the solution indicates a positive value for I_2, the assumed direction is proved to be correct. The current through R_4 is I_1 - I_2. At junction d the currents may be analyzed in a similar manner. Current I_2 through R_2 joins current 0.1 - I_1, from R_3; and the current through R_5 is 0.1 - I_1 + I_2.

The unknown currents, I_1 and I_2, may be determined by establishing two voltage equations in which they both appear. These equations are solved for I_1 and I_2 in terms of the given values of current and resistance. The first voltage equation is developed by tracing clockwise around the closed circuit containing resistors R_1, R_2, and R_3. The trace starts at junction a and proceeds to b, to d, and back to a. The algebraic sum of the voltages around this circuit is zero. These voltages are expressed in terms of resistance and current. Going from a to b, the voltage drop is in the direction of the arrow and is equal to $-50I_1$; the drop across R_2, going from b to d, is $-100I_2$; and the voltage from d to a (in opposite direction to the arrow) is + 70(0.1 - I_1). Thus, the first equation is

$$-50I_1 - 100I_2 + 70(0.1 - I_1) = 0$$

Multiplying both sides by –1,

$$50I_1 + 100I_2 - 70(0.1 - I_1) = 0$$

and transposing and simplifying,

$$120I_1 + 100I_2 = 7$$

The second voltage equation is established by tracing clockwise around the circuit, which includes resistors R_4, R_5, and R_2. Starting at junction b, the trace proceeds to c, to d, and returns to b. The voltage across R_4, from b to c, is $-250(I_1 - I_2)$; the

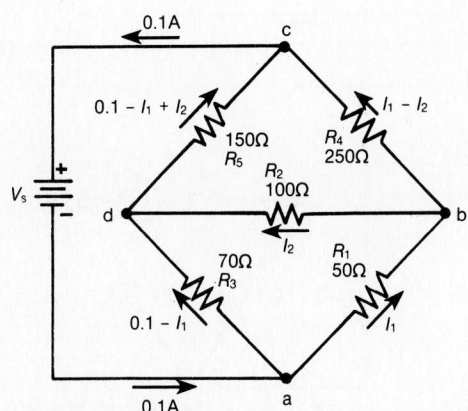

Fig. 8-15: Unbalanced resistance bridge.

135

voltage across R_5, from c to d, is $+ 150(0.1 - I_1 + I_2)$; and the voltage across R_2, from d to b, is $+ 100I_2$. Thus,

$$-250(I_1 - I_2) + 150(0.1 - I_1 + I_2) + 100I_2 = 0$$

from which,

$$400I_1 - 500I_2 = 15$$

The two equations may be solved simultaneously by multiplying the first equation by the factor 5 and then adding the second equation to eliminate I_2 as follows:

$$400I_1 - 500I_2 = 15$$
$$\underline{600I_1 + 500I_2 = 35}$$
$$1{,}000I_1 = 50$$
$$I_1 = 0.05\text{A}$$

Substituting the value of 0.05 for I_1 in the first equation and solving for I_2,

$$120(0.05) + 100I_2 = 7$$
$$100I_2 = 1$$
$$I_2 = 0.01\text{A}$$

Thus the current in R_1 is $I_1 = 0.05\text{A}$. The current in R_2 is $I_2 = 0.01\text{A}$. The current in R_3 is $0.1 - I_1 = 0.1 - 0.05$, or 0.05A. The current in R_4 is $I_1 - I_2 = 0.05 - 0.01$, or 0.04A. The current in R_5 is $0.1 - I_1 + I_2$ or $0.1 - 0.05 + 0.01 = 0.06\text{A}$. The voltages V_1, V_2, V_3, V_4, and V_5 are as follows:

V_1 across R_1 is $I_1 R_1 = 0.05\text{A} \times 50\Omega = 2.5\text{V}$.

V_2 across R_2 is $I_2 R_2 = 0.01\text{A} \times 100\Omega = 1.0\text{V}$.

V_3 across R_3 is $(0.1 - I_1)R_3 = 0.05\text{A} \times 70\Omega = 3.5\text{V}$.

V_4 across R_4 is $(I_1 - I_2)R_4 = 0.04\text{A} \times 250\Omega = 10\text{V}$.

V_5 across R_5 is $(0.1 - I_1 + I_2)R_5 = 0.06\text{A} \times 150\Omega = 9.0\text{V}$.

The source voltage, V_S, is equal to the sum of voltages across R_3 and R_5 or R_1 and R_4. Thus,

$$V_S = V_1 + V_4$$
$$= 2.5\text{V} + 10\text{V} = 12.5\text{V}$$

and

$$V_S = V_3 + V_5$$
$$= 3.5\text{V} + 9\text{V} = 12.5\text{V}$$

The voltage across R_2 is the difference in the voltages across R_3 and R_1. It is also the difference in the voltages across R_4 and R_5.

Wheatstone Bridge

A type of circuit that is widely used for precision measurements of resistance is the Wheatstone bridge. The circuit diagram of a Wheatstone bridge is shown in Fig. 8-16. R_1, R_2, and R_3 are precision variable resistors, and R_x is the resistor whose unknown value of resistance is to be determined. After the

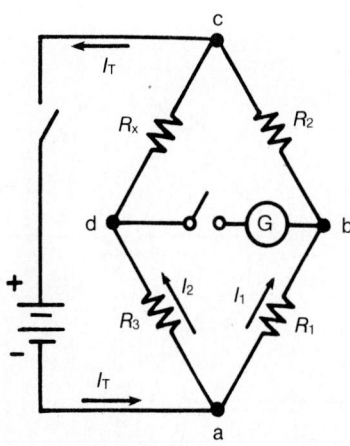

Fig. 8-16: Schematic of a Wheatstone bridge circuit.

136

bridge has been properly balanced, the unknown resistance may be determined by means of a simple formula. The galvanometer (G) is inserted across terminals b and d to indicate the condition of balance. (A galvanometer is a meter device that indicates very small amounts or relative amounts of current or voltage.) When the bridge is properly balanced, there is no difference in potential across terminals bd and the galvanometer deflection, when the switch is closed, will be zero.

The operation of the bridge is explained in a few logical steps. When the switch to the battery is closed, electrons flow from the negative terminal of the battery to point a. Here the current divides, as it would in any parallel circuit, a part of it passing through R_1 and R_2 and the remainder passing through R_3 and R_x. The two currents, labeled I_1 and I_2, unite at point c and return to the positive terminal of the battery. The value of I_1 depends on the sum of resistances R_1 and R_2, and the value of I_2 depends on the sum of resistances R_3 and R_x. In each case, according to Ohm's Law, the current is inversely proportional to the resistance.

R_1, R_2, and R_3 are adjusted so that when the galvanometer switch is closed there will be no deflection of the needle. When the galvanometer shows no deflection, there is no difference of potential between points b and d. This means that the voltage drop (V_1) across R_1, between points a and b, is the same as the voltage drop (V_3) across R_3, between points a and d. By similar reasoning, the voltage drops across R_2 and R_x—that is, V_2 and V_x—are also equal. Expressed algebraically,

$$V_1 = V_3$$

or

$$I_1 R_1 = I_2 R_3$$

and

$$V_2 = V_x$$

or

$$I_1 R_2 = I_2 R_x$$

Dividing the voltage drops across R_1 and R_3 by the respective voltage drops across R_2 and R_x,

$$\frac{I_1 R_1}{I_1 R_2} = \frac{I_2 R_3}{I_2 R_x}$$

Simplifying,

$$\frac{R_1}{R_2} = \frac{R_3}{R_x}$$

Therefore,

$$R_x = \frac{R_2 R_3}{R_1}$$

The resistance values of R_1, R_2, and R_3 are readily determined from the markings on the standard resistors or from the calibrated dials if a dial-type bridge is used.

The Wheatstone bridge may be of the slide-wire type, as shown in Fig. 8-17. In the slide-wire circuit, the slide-wire (b to d) corresponds to R_1 and R_3 of Fig. 8-16. The wire may be an alloy of uniform cross section; for example, German silver or nichrome, having a resistance of about 100Ω. Point a is established where the slider contacts the wire. The bridge is balanced by moving the slider along the wire.

The equation for solving for R_x in the slide-wire bridge of Fig. 8-17 is similar to the one used for solving for R_x in Fig. 8-16. However, in the slide-wire bridge the length, L_1, corresponds to the resistance, R_1, and the length, L_2, corresponds to the resistance, R_3. Therefore, L_1 and L_2 may be substituted for R_1 and R_3 in the equation. The resistance of L_1 and L_2 varies uniformly with slider movement because in a wire of uniform cross section, the resistance varies directly with the length; therefore, the ratio of the resistances equals the corresponding ratio of the lengths. Substituting L_1 and L_2 for R_1 and R_3

$$R_x = \frac{L_2 R_2}{L_1}$$

A meter stick is mounted underneath the slide-wire and L_1 and L_2 are easily read in centimeters. For example, if a balance is obtained when $R_2 = 150\Omega$, $L_1 = 25cm$, and $L_2 = 75cm$, the unknown resistance is

$$R_x = \frac{75}{25} \times 150 = 450\Omega$$

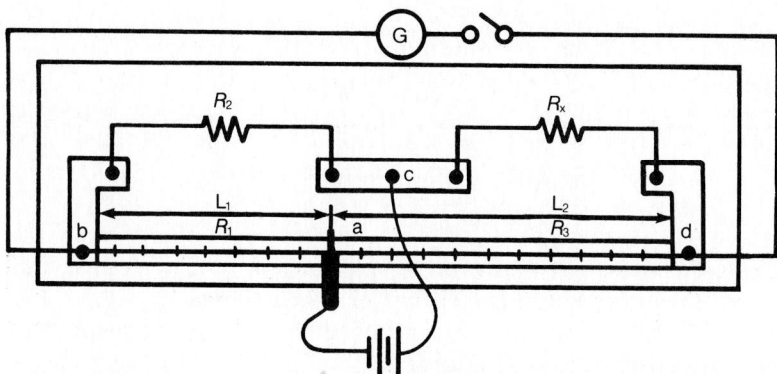

Fig. 8-17: Typical Wheatstone bridge circuit.

MULTIPLE SOURCE CIRCUITS

Quite frequently, networks containing more than one source must be solved. Although a circuit of this type may look complicated, the solution is no more difficult than the one discussed previously. In fact, the same method of solution is used in both single and multiple source circuits.

The circuit shown in Fig. 8-18 illustrates two sources of emf, V_{S_1} and V_{S_2}, having internal resistances of 2Ω and 2.5Ω respectively, connected in parallel, and supplying a 500Ω load.

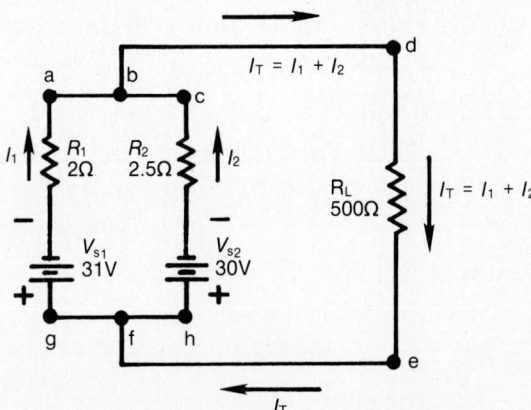

Fig. 8-18: Parallel sources supply a common load.

Neglecting the resistance of the lead wires, it is desired to determine the current delivered by each source to the load, the load current, and the load voltage.

The problem may be solved by establishing two voltage equations in which the voltages are expressed in terms of the unknown currents I_1 and I_2, the known resistances, and the known voltages. The equations are then solved simultaneously as in previous examples, to eliminate one of the unknown currents. The other unknown current is solved by substitution.

The first voltage equation is established by starting at point g and tracing clockwise around circuit *gabdefg*. The total load current is equal to the sum of the source currents, $I_1 + I_2$. The first voltage equation is,

$$V_{S1} - R_1 I_1 - R_L I_T = 0$$

from which,

$$31V - 2\Omega(I_1) - 500\Omega(I_1 + I_2) = 0$$
$$31V - 502\Omega(I_1) - 500\Omega(I_2) = 0$$
$$502\Omega(I_1) + 500\Omega(I_2) = 31V$$

The second voltage equation is established by starting at point h and tracing around the circuit *hcbdefh*. Thus,

$$V_{S2} - R_2 I_2 - R_L I_T = 0$$
$$30V - 2.5\Omega(I_2) - 500\Omega(I_1 + I_2) = 0$$
$$30V - 500\Omega(I_1) - 502.5\Omega(I_2) = 0$$
$$500\Omega(I_1) + 502.5\Omega(I_2) = 30V$$

I_2 is eliminated by multiplying the first voltage equation by 1.005 and subtracting the second voltage equation from the result, as follows:

$$1.005[502\Omega(I_1) + 500\Omega(I_2) = 31V] = 504.5\Omega(I_1) + 502.5\Omega(I_2) = 31.155V$$

$$504.5\Omega(I_1) + 502.5\Omega(I_2) = 31.155V$$
$$\underline{500.0\Omega(I_1) + 502.5\Omega(I_2) = 30.000V}$$
$$4.5\Omega(I_1) + \quad 0\Omega(I_2) = \quad 1.155V$$

$$4.5\Omega(I_1) = 1.155V$$

$$I_1 = 256.66mA$$

139

Substituting this value in the second voltage equation:

$$500\,\Omega(I_1) + 502.5\,\Omega(I_2) = 30V$$
$$500\,\Omega(256.66mA) + 502.5\,\Omega(I_2) = 30V$$
$$128.33V + 502.5\,\Omega(I_2) = 30V$$
$$502.5\,\Omega(I_2) = -98.33V$$
$$I_2 = -195.69mA$$

The load current is,

$$I_T = I_1 + I_2$$
$$I_T = 256.66mA + (-)195.69mA$$
$$I_T = 60.978mA$$

The load voltage is equal to the voltage developed across terminals f and b and is equal to the difference in a given source voltage and the internal voltage absorbed across the corresponding source resistance.

The statement applies equally to either source since both are in parallel with the load. In terms of source V_{S_1}

$$V_{fb} = V_{S_1} - I_1R_1$$
$$= 31V - 0.512V$$
$$= 30.488V$$

and in terms of V_{S2}

$$V_{fb} = V_{S2} - I_2R_2$$
$$= 30V - (-)0.489V$$
$$= 30.489V$$

A further check on the load voltage is to express this value in terms of the load current (I_T) and the load resistance (R_L) as follows:

$$V_{fb} = (I_1 + I_2)R_L$$
$$= (60.978mA)(500\,\Omega)$$
$$= 30.489V$$

Multiple Cell Circuits. Figure 8-19 depicts a battery network supplying power to a load requiring both a voltage and current greater than one cell can provide. To provide the required 4.5V, groups of three 1.5V cells are connected in series. To provide the required 0.5A of current, four series groups are connected in parallel, each supplying 0.125A of current.

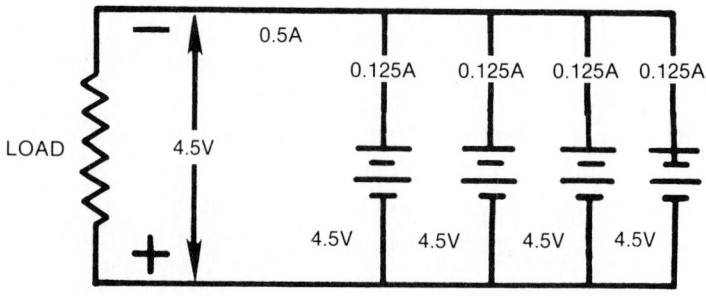

Fig. 8-19: Combination connected battery cells.

140

SUMMARY

A circuit consisting of both series and parallel elements is referred to as a combination or complex circuit. Most combination circuits can be broken down into either series-parallel or parallel-series circuits. These basic circuit types can be further reduced to form basic equivalent circuits containing the source and total resistance.

When analyzing a complex circuit, it is necessary to consider the internal resistance of the source (R_i). The presence of internal resistance results in (1) a diminished voltage supplied to the components that comprise the load, (2) a decrease in total current, and (3) an increase in total resistance. The power dissipated by the circuit is also affected.

The effect of an open or short circuit in a combination circuit, as in parallel circuits, depends on just where in the circuit the fault occurs.

It is often necessary in many electronic devices for a single voltage source to supply a number of different voltages at various points of the circuitry. The most common method of meeting these requirements is by the use of a voltage-divider network. The typical resistance values chosen, when designing a voltage-divider, will permit a current equal to 10% of the total current drawn by the external loads. This current, which does not flow through any of the load devices, is called bleeder current.

Another type of combination circuit which is widely used is the bridge circuit. A bridge circuit may be either balanced or unbalanced. The bridge is said to be balanced when the bridge resistor is across points of equal potential and no current flows through it. When the bridge resistor is across unequal potentials, current flows through it and the bridge is said to be unbalanced. Another method of attaining an unbalanced bridge is to use resistors of unequal value. The Wheatstone bridge is the most common of the bridge circuits. It is widely used for precision resistance measurements.

Many complex circuits employ more than one source. These circuits can be solved by establishing two voltage equations in which the voltages are expressed in terms of the unknown currents, the known resistances, and the known voltages. The two equations are then solved simultaneously to eliminate one of the unknown currents. The other unknown current is solved by substitution.

CHECK YOURSELF (Answers at back of book)

Questions

Fill in the blanks with the appropriate word or words.

1. In the combination circuit of Fig. 8-20, _____ .
   A. R_1 is in series with R_3
   B. R_2 is in series with R_3
   C. R_4 is in parallel with R_3
   D. R_1 is in parallel with R_3

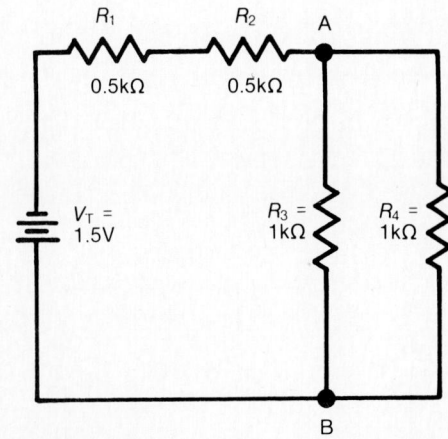

Fig. 8-20

141

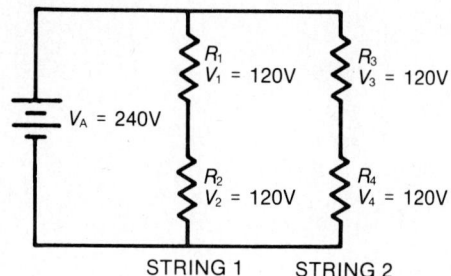

Fig. 8-21

2. In the parallel-series circuit of Fig. 8-21, _____ .
   A. R_1 is in series with R_2
   B. R_1 is in parallel with R_3
   C. R_2 is in parallel with R_4
   D. R_2 is in series with R_4

3. In the Wheatstone bridge of Fig. 8-22, at balance, _____ .
   A. $I_B = 0$
   B. $V_2 = 0$
   C. $V_{AB} = 0$
   D. $I_A = 0$

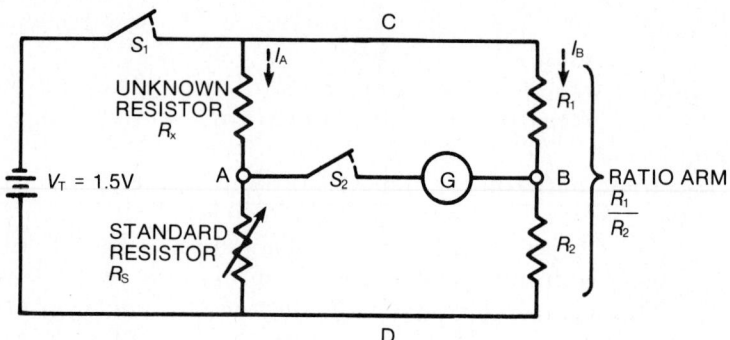

Fig. 8-22

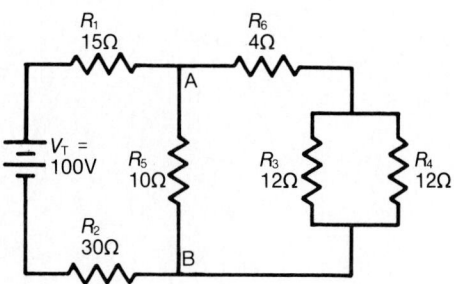

Fig. 8-23

4. In the complex circuit of Fig. 8-23, the total current flowing into branch point A and out of branch point B is equal to _____ .
   A. 2A
   B. 4A
   C. 1/2A
   D. 1A

5. In Fig. 8-24, with the switch closed, R_T is equal to _____ .
   A. 100Ω
   B. 50Ω
   C. 97Ω
   D. 150Ω

6. In Fig. 8-1A, $R_2 = 200\Omega$ and $R_3 = 20\Omega$. Current through R_2 is _____ R_3.
   A. less than
   B. equal to
   C. greater than
   D. independent of

7. In Fig. 8-1B, the voltage drop across $R_1 + R_2$ is _____ the voltage drop across $R_3 + R_4$.
   A. less than
   B. equal to
   C. more than
   D. independent of

8. In Fig. 8-1C, the voltage drop across R_4 and R_5 is _____ the voltage drop across R_6.
   A. less than
   B. equal to
   C. more than
   D. independent of

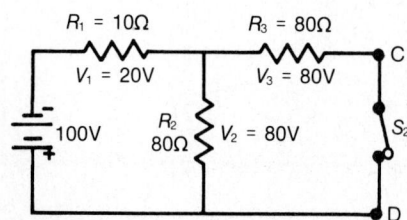

Fig. 8-24

142

9. In Fig. 8-1D, the voltage drop across R_1 is _____ the voltage drop across R_2.
   A. less than
   B. equal to
   C. more than
   D. independent of
10. In Fig. 8-1D, current through R_1 is _____ the current through R_3.
    A. less than
    B. equal to
    C. more than
    D. independent of
11. If the voltage across a series-parallel circuit is 200V and the current through the series branch is 3mA, the total circuit resistance, R_T, is _____ .
12. A circuit consists of a parallel and series branch. If $V = 100V$ and $V_{parallel} = 96V$, $V_{series} =$ _____ .
13. $I_{R2} = 3.2mA$ and $I_{R3} = 5.6mA$ in the parallel branches of a series-parallel circuit. The current through the series part of the circuit is $I_{series} =$ _____ and $I_T =$ _____ .
14. If the parallel branches of a series-parallel circuit consumes 1,520W and the circuit power supplied to the circuit is 2.62kW, the power consumed by the series circuit is $P_{series} =$ _____ .
15. If a series-parallel circuit has an $R_{series} = 52.9\Omega$ and $R_{parallel} = 15.6\Omega$, the value of $R_T =$ _____ .

Problems

1. Calculate R_T for Fig. 8-25.
2. Refer to Fig. 8-26 and find:

   $R_T =$ _____ $V_{R2} =$ _____
   $I_T =$ _____ $V_{R3} =$ _____
   $P_T =$ _____ $P_1 =$ _____
   $I_2 =$ _____ $P_2 =$ _____
   $I_3 =$ _____ $P_3 =$ _____
   $V_{R1} =$ _____

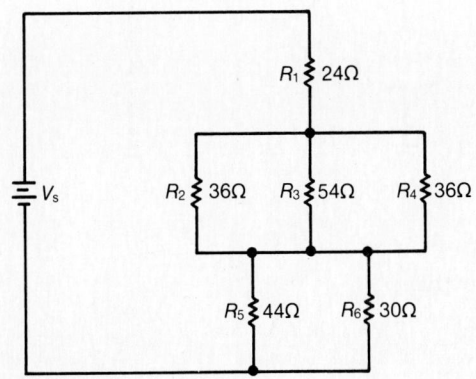

Fig. 8-25

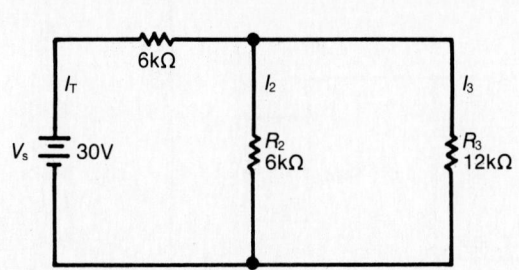

Fig. 8-26

143

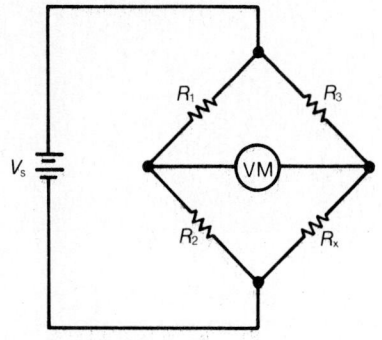

Fig. 8-27

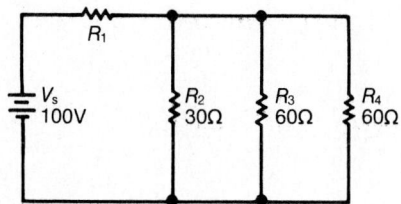

Fig. 8-29

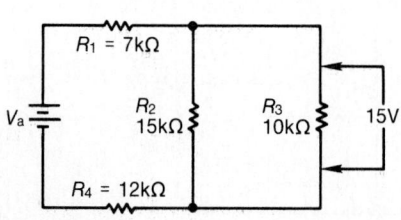

Fig. 8-31

3. Refer to the balanced Wheatstone bridge in Fig. 8-27 and determine the value of R_X when $R_1 = 100\,\Omega$, $R_2 = 200\,\Omega$, and $R_3 = 450\,\Omega$.

4. Refer to the balanced Wheatstone bridge in Fig. 8-27 and calculate R_2 when $R_1 = 1\,\Omega$, $R_3 = 900\,\Omega$, and $R_X = 340\,\Omega$.

5. Find the resistance between points A and B in Fig. 8-28.

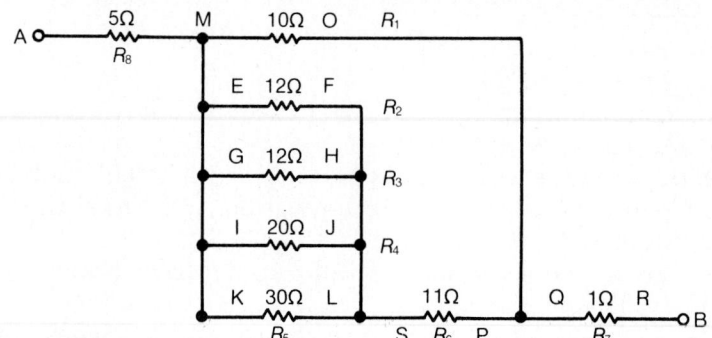

Fig. 8-28

6. The total current in Fig. 8-29 is 5A. Find the value of R_1 and calculate I_{R4}.

7. Calculate the values of R_1, R_2, and R_3 for the voltage divider circuit in Fig. 8-30 with loads as shown. Allow 10% bleeder current. Do not calculate the value of the loads R_{L1}, R_{L2} and R_{L3}.

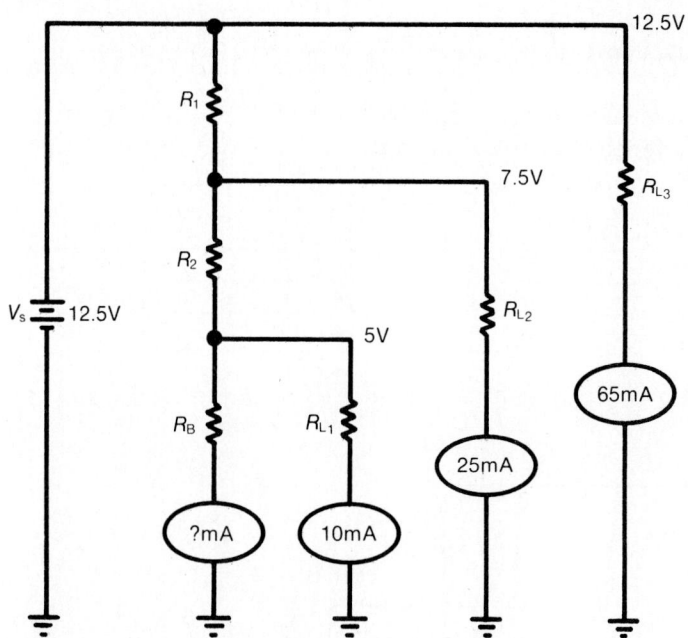

Fig. 8-30

8. Refer to Fig. 8-31 and find:

$I_1 = $ _____

$I_T = $ _____

$V_a = $ _____

$R_T = $ _____

144

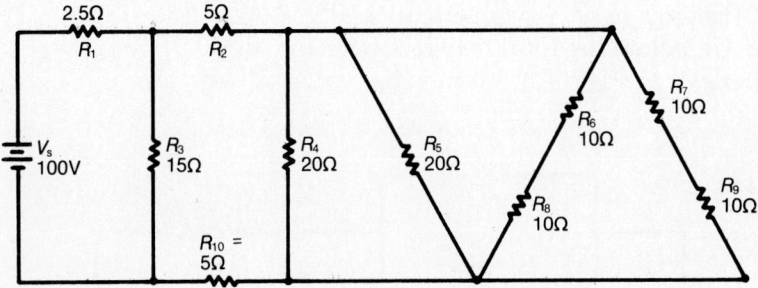

Fig. 8-32

9. Calculate R_T, I_T, and P_T for the circuit of Fig. 8-32.
10. Refer to Fig. 8-33 and calculate:

$R_T =$ _____
$I_T =$ _____
$V_{R_1} =$ _____
$V_{R_4} =$ _____
$V_{R_6} =$ _____
$P_{R_1} =$ _____
$P_{R_5} =$ _____
$P_{R_2} =$ _____

11. Calculate R_T, I_T, and I_{R_2} for the circuit of Fig. 8-34.
12. What is the value of I_T in Fig. 8-35?

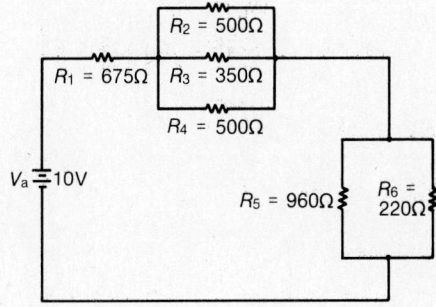

Fig. 8-33

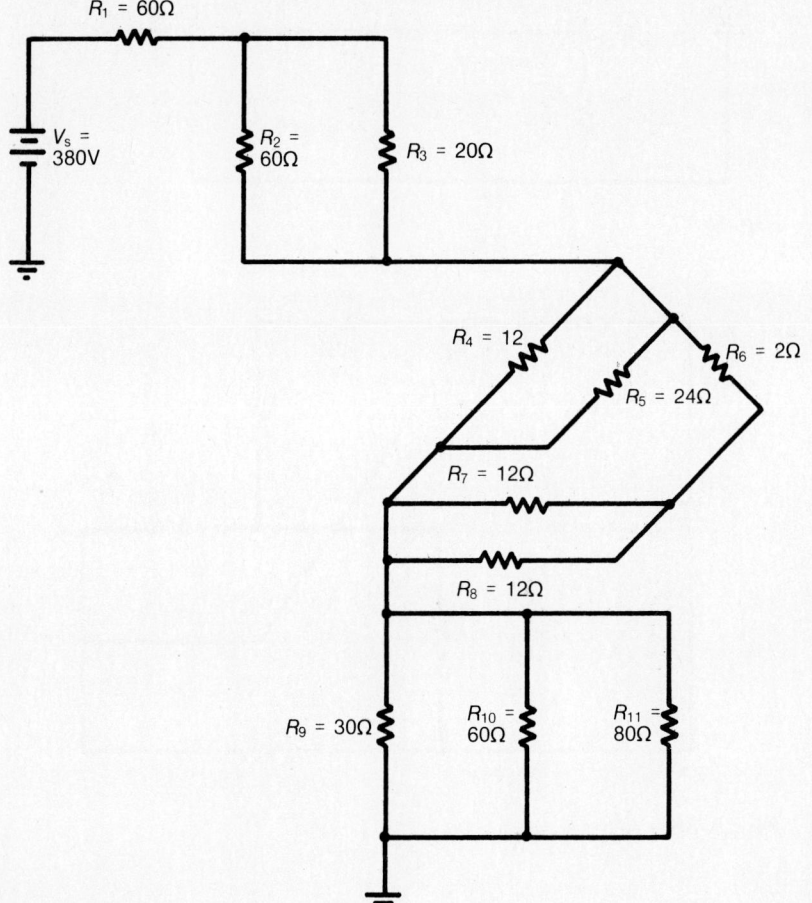

Fig. 8-34

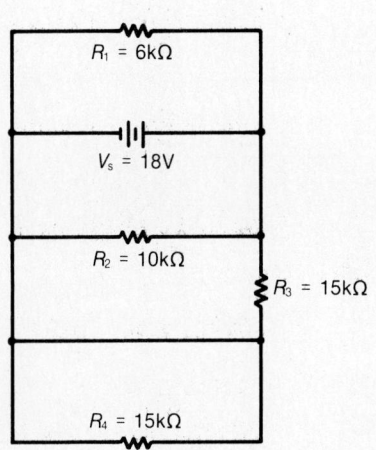

Fig. 8-35

13. Determine R_T for the circuit of Fig. 8-36.
14. Calculate the total resistance for Fig. 8-37.
15. Refer to Fig. 8-38. What is the value of R_T?

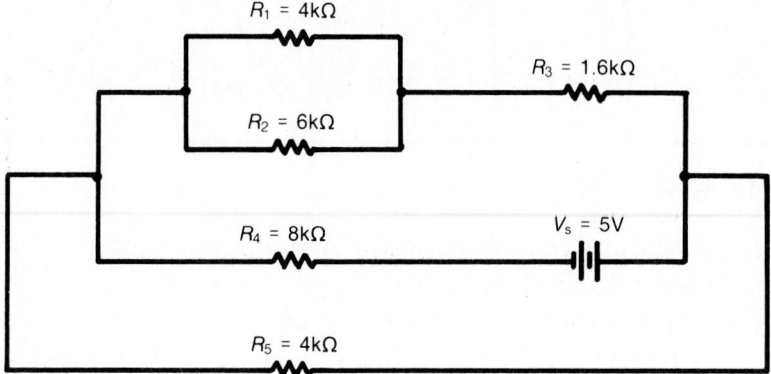

Fig. 8-36

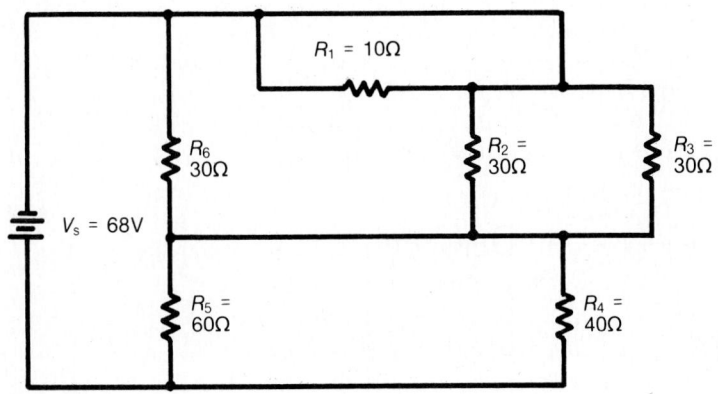

Fig. 8-37

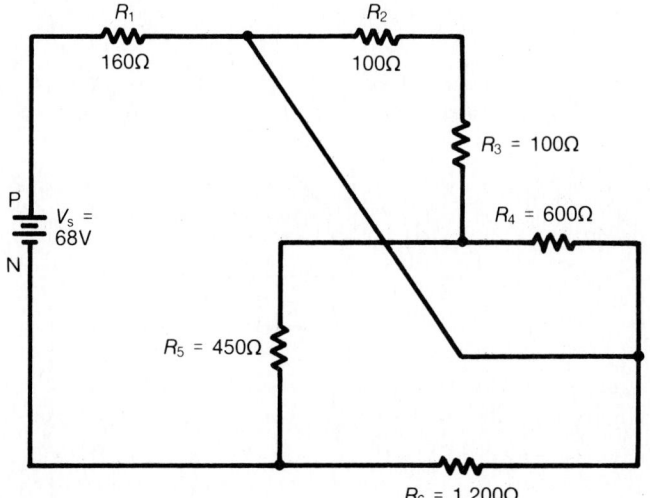

Fig. 8-38

CHAPTER 9

Principles of Magnetism

Magnetism is another invisible force which has been known to civilized people for many centuries. However, its relationship to an electric force was not realized for some two thousand years. Early students of science believed that magnetism was a completely unrelated phenomenon. In modern science the connection between the two forces is quite apparent, for without magnetism very few of our modern devices would be possible. Magnets or magnetic effects can be found in numerous electric and electronic devices. Some of these devices in which the principles of magnetism are employed are:

- *Microphones and speakers,* used in radios, recorders, public address systems, etc.
- *Phonographs,* employing magnetic type cartridges.
- *Tape recorders,* used in both audio and video systems.
- *Meters,* employed in most electronic test equipment.
- *Transformers,* found in numerous types of electronic equipment where magnetic lines of force must interact for proper operation.
- *Relays,* used in automation and many other branches of electricity and electronics.
- *Magnetic amplifiers,* employed in industrial control systems.

The principles of magnetism are applied by industry in electric motors and generators, as well as electromagnets used for hoisting metal objects.

PROPERTY OF MAGNETISM

Magnetism is generally defined as that property of a material which enables it to attract pieces of iron. A material possessing this property is known as a *magnet.* The word originated from the ancient Greeks who found stones possessing this characteristic. Materials that are attracted by a magnet such as iron, steel, nickel, and cobalt have the ability to become magnetized; these materials are called *magnetic.* Materials such as paper, wood, glass, or tin which are not attracted by magnets are considered *nonmagnetic.* Nonmagnetic materials are not able to become magnetized.

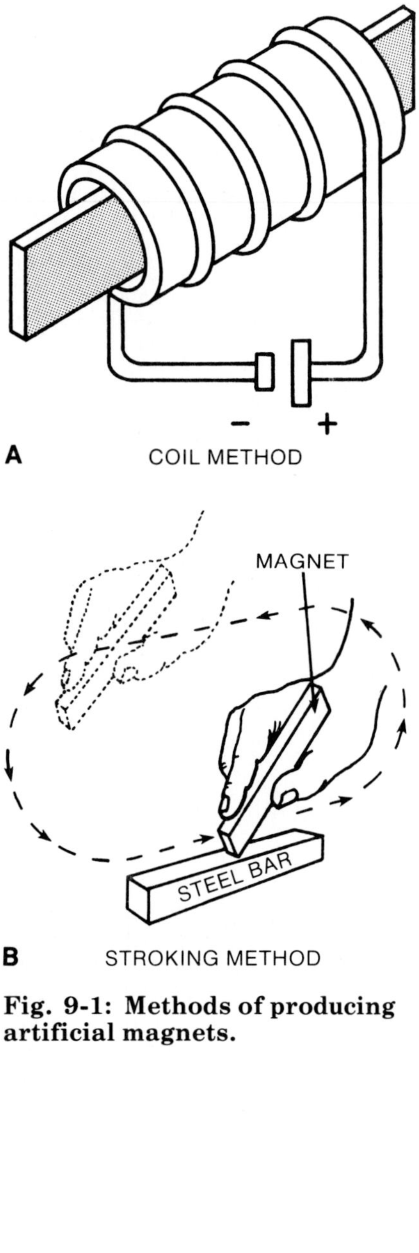

Classification of Magnets

Magnets may be conveniently divided into two groups: natural and artificial.

1. *Natural Magnets.* The lodestone (or magnetite) is a *natural magnet*. It exists in a natural state. Lodestone can presently be found in the United States, Norway, and Sweden. These natural magnets are crude, generally weak, and serve little purpose in the modern world. The earth itself is a huge natural magnet, and it is an exception to the rule of a natural magnet serving no useful purpose. The magnetic properties of the earth are used constantly.

2. *Artificial Magnets.* Magnets produced from magnetic materials are called *artificial magnets*. They can be made in a wide variety of shapes, sizes, and materials. For example, many commercial magnets are made from special steels and alloys— for example, alnico, made principally of aluminum, nickel, and cobalt. The name is derived from the first two letters of the three principal elements of which it is composed.

An iron, steel, or alloy bar can be magnetized by inserting the bar into a coil of insulated wire and passing a heavy direct current through the coil, as shown in Fig. 9-1A. This aspect of magnetism is treated in Chapter 10. The same bar may also be magnetized if it is stroked with a bar magnet, as shown in Fig. 9-1B. It will then have the same magnetic property that the magnet used to induce the magnetism has; namely, there will be two poles of attraction, one at either end. This process produces a permanent magnet by *induction;* that is, the magnetism is induced in the bar by the influence of the stroking magnet.

Artificial magnets may be classified as "permanent" or "temporary" depending on their ability to retain their magnetic strength after the magnetizing force has been removed. Hardened steel and certain alloys are relatively difficult to magnetize and are said to have a *low permeability* because the magnetic lines of force do not easily permeate or distribute themselves readily through the steel. Once magnetized, however, these materials retain a large part of their magnetic strength and are called *permanent magnets.* Permanent magnets are used extensively in electric instruments, meters, telephone receivers, permanent-magnet loudspeakers, and magnetos. Conversely, substances that are relatively easy to magnetize, such as soft iron and annealed silicon steel, are said to have a *high permeability.* (*Permeability* is a measure of the ease with which a substance passes magnetic lines of force.) Such substances retain only a small part of their magnetism after the magnetizing force is removed and are called *temporary magnets.* The magnetism that remains in a temporary magnet after the magnetizing force is removed is called *residual magnetism.* The fact that temporary magnets retain even a small amount of magnetism is an important factor in some electrical and electronic equipment.

A COIL METHOD

B STROKING METHOD

MAGNET

STEEL BAR

Fig. 9-1: Methods of producing artificial magnets.

NATURE OF MAGNETISM

The earth is a permanent magnet (Fig. 9-2); hence, any permanent magnet, such as a compass needle, will align itself with the earth's magnetic field. The end of the compass needle that points to the geographical north pole of the earth is itself called the magnetic north-seeking pole or the *north pole*. The south-seeking end is called the *south pole*. It should be noted that the magnetic axis of the earth is located about 15° from its geographical axis, thereby locating the magnetic poles some distance from the geographical poles.

The ability of the north pole of the compass needle to point toward the north geographical pole is due to the presence of the magnetic pole nearby. This magnetic pole is named the magnetic north pole. However, in actuality, it must have the polarity of a south magnetic pole since it attracts the north pole of a compass needle. The reason for this conflict in terminology can be traced to the early users of the compass. Knowing little about magnetic effects, they called the end of the compass needle that pointed towards the north geographical pole, the north pole of a compass. With our present knowledge of magnetism, the north pole of a compass needle (a small bar magnet) can only be attracted by an unlike magnetic pole or a pole of south magnetic polarity.

Weber Theory

One of the first theories about the nature of magnetism was put forth by the German physicist, Wilhelm Weber. Known as the Weber Theory, it assumes all magnetic substances to be composed of tiny molecular magnets. All unmagnetized materials have the magnetic forces of their molecular magnets neutralized by adjacent molecular magnets, thereby eliminating any magnetic effect. A magnetized material will have most of its molecular magnets lined up so that the north-seeking pole of each molecule points in one direction and the south-seeking pole faces the opposite direction. A material with its molecules thus aligned will then have one effective north pole and one effective south pole. An illustration of Weber's Theory is shown in Fig. 9-3A, where a steel bar is magnetized by stroking. When a steel bar is stroked several times in the same direction by a magnet, the magnetic force from the north pole of the magnet causes the molecules to align themselves. The polarity of the magnet formed is dependent upon the direction of the magnetizing force as it is brought over the random magnetic molecules.

Some justification of Weber's Theory occurs when a magnet is split in half. It is found that each half possesses both a north and a south magnetic pole as shown in Fig. 9-3B. The polarities of the poles are in the same respective directions as the poles of the original magnet. If a magnet is further divided into small

Fig. 9-2: The earth is a permanent magnet.

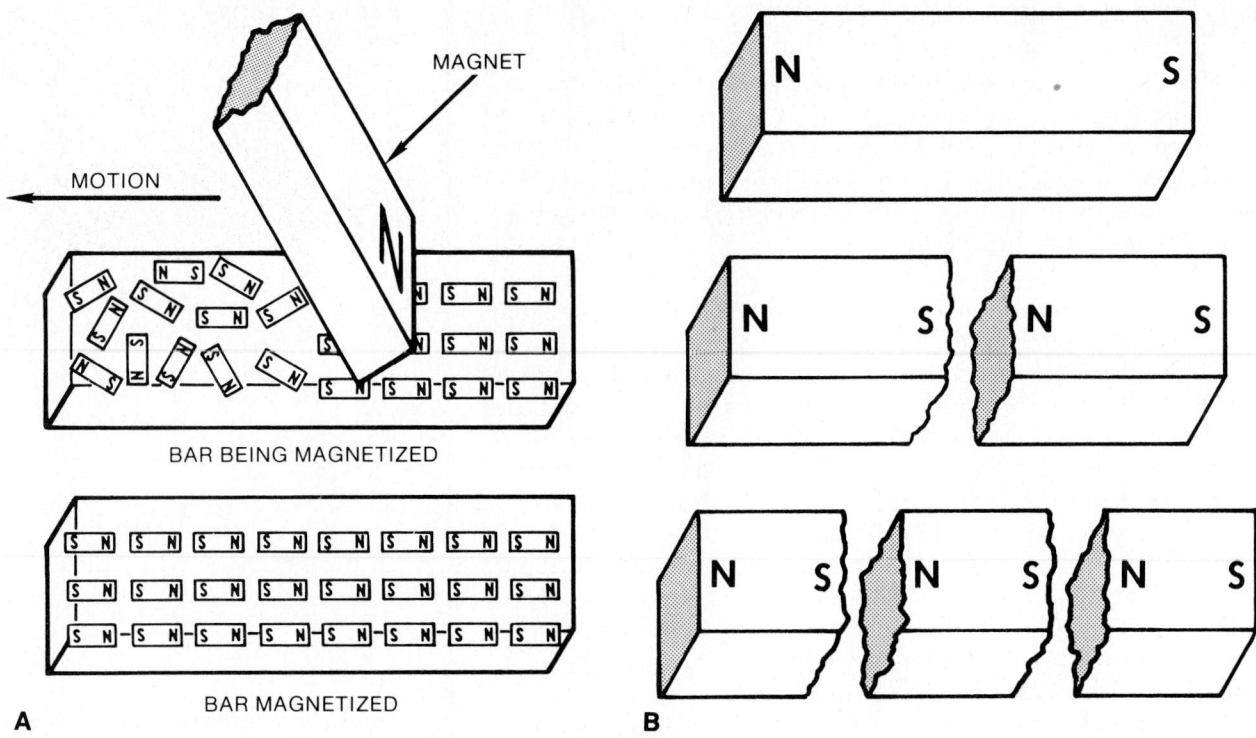

MAGNET

MOTION

BAR BEING MAGNETIZED

BAR MAGNETIZED

A

B

Fig. 9-3: (A) Molecular magnets; (B) broken magnets.

parts, it will be found that each part, down to its last molecule, will all have similar north and south poles. Each part would exhibit its own magnetic properties.

Further support of Weber's Theory comes from the fact that when a bar magnet is held out of alignment with the earth's magnetic field and repeatedly jarred or heated, the molecular alignment is disarranged and the material becomes demagnetized. For example, measuring devices which make use of permanent magnets become inaccurate when subjected to severe jarring or exposure to opposing magnetic fields.

Domain Theory

A more modern theory of magnetism is based on the electron spin principle. From the study of atomic structure it is known that all matter is composed of vast quantities of atoms, each atom containing one or more orbital electrons. The electrons are considered to orbit in various shells and subshells depending upon their distance from the nucleus. The structure of the atom has previously been compared to the solar system, wherein the electrons orbiting the nucleus correspond to the planets orbiting the sun. Along with their orbital motion about the sun, these planets also revolve on their axes. It is believed that the electron also revolves on its axis as it orbits the nucleus of an atom.

It has been experimentally proven that an electron has a magnetic field about it along with an electric field. The effectiveness of the magnetic field of an atom is determined by the number of electrons spinning in each direction. If an atom has equal numbers of electrons spinning in opposite directions, the magnetic fields surrounding the electrons cancel one another, and the atom is unmagnetized. However, if more electrons spin in one direction than another, the atom is magnetized. An atom such as iron with an atomic number of 26 has 26 protons in the nucleus and 26 revolving electrons orbiting its nucleus. If 13 electrons are spinning in a clockwise direction and 13 electrons are spinning in a counterclockwise direction, the opposing magnetic fields will be neutralized. When more than 13 electrons spin in either direction, the atom is magnetized. An example of a magnetized atom of iron is shown in Fig. 9-4. Note that in this specific illustration the electrons' magnetic fields in all except the M shell neutralize each other. As illustrated in the diagram, there exist 15 electrons spinning in one direction and only 11 electrons spinning in an opposite direction; therefore, the unopposed magnetic fields of four electrons will cause this iron atom to become an extremely small magnet.

When a number of such atoms are grouped together to form an iron bar, there is an interaction between the magnetic forces of various atoms. The small magnetic force of the field surrounding an atom affects adjacent atoms, thus producing a small group of atoms with parallel magnetic fields. This group of from 10^{14} to 10^{15} magnetic atoms, having their magnetic poles orientated in the same direction, is known as a *domain*. Throughout a domain there is an intense magnetic field without the influence of any external magnetic field. Since about 10 million tiny domains can be contained in 1 cubic millimeter, it is apparent that every magnetic material is made up of a large

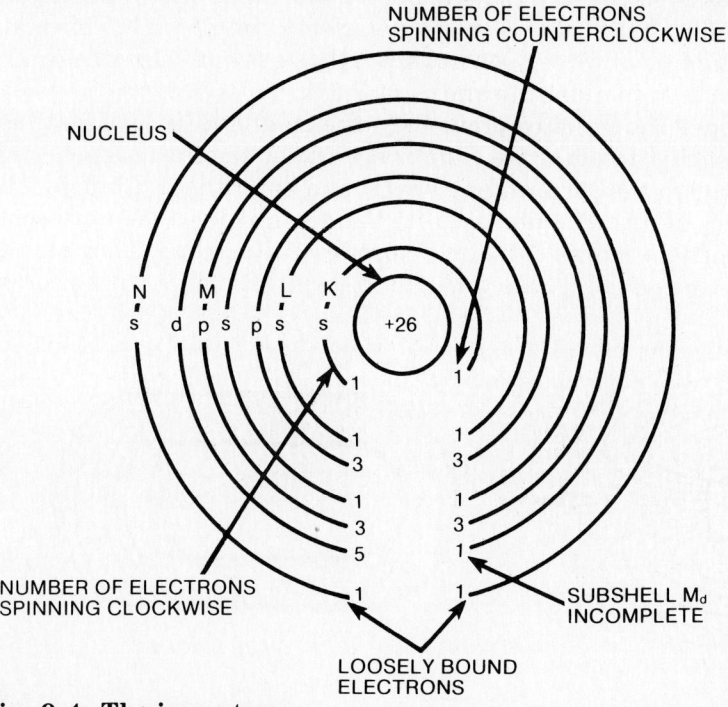

Fig. 9-4: The iron atom.

number of domains. The domains in any substance are always magnetized to saturation but are randomly orientated throughout a material; thus, the strong magnetic field of each domain is neutralized by opposing magnetic forces of other domains. When an external field is applied to a magnetic substance, the domains will line up with the external field. Since the domains themselves are naturally magnetized to saturation, the magnetic strength of magnetized material is determined by the number of domains aligned by the magnetizing force.

MAGNETIC FIELDS AND LINES OF FORCE

If a bar magnet is dropped into iron filings, many of the filings are attracted to the ends of the magnet, but none are attracted to the center of the magnet. As mentioned previously, the ends of the magnet where the attractive force is the greatest are called the *poles* of the magnet. By using a compass, the line of direction of the magnetic force at various points near the magnet may be observed. The compass needle itself is a magnet. The north end of the compass needle always points toward the south pole, S, as shown in Fig. 9-5A and thus the sense of direction (with respect to the polarity of the bar magnet) is also indicated. At the center, the compass needle points in a direction that is parallel to the bar magnet.

When the compass is placed successively at several points in the vicinity of the bar magnet, the compass needle aligns itself with the field at each position. The direction of the field is indicated by the arrows and represents the direction in which the north pole of the compass needle will point when the compass is placed in this field. Such a line along which a compass needle aligns itself is called a *magnetic line of force*. This magnetic line of force does not actually exist but is an imaginary line used to illustrate and describe the pattern of the magnetic field. As mentioned previously, the magnetic lines of force are assumed to emanate from the north pole of a magnet, pass through the surrounding space, and enter the south pole. The lines of force then pass from the south pole to the north pole inside the magnet to form a closed loop. Each line of force forms an independent closed loop and does not merge with or cross

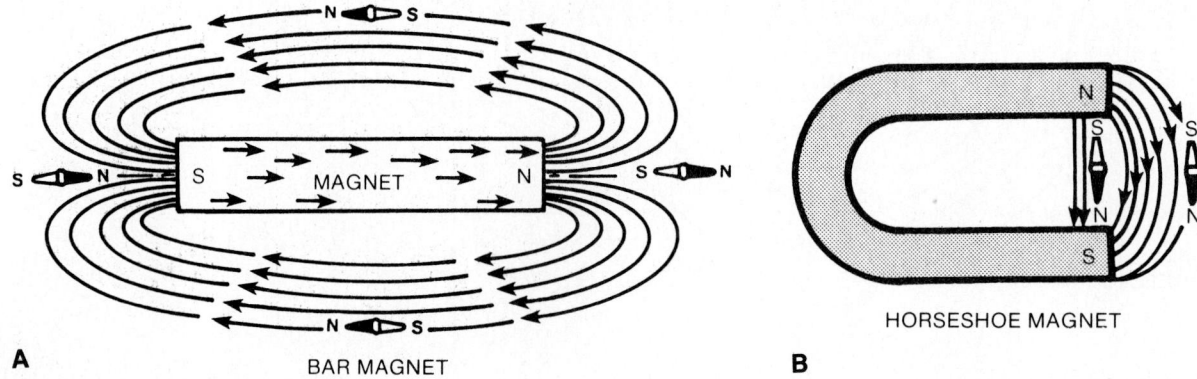

A   BAR MAGNET   B   HORSESHOE MAGNET

Fig. 9-5: Magnetic lines of force.

other lines of force. The lines of force between the poles of a horseshoe magnet are shown in Fig. 9-5B.

Although magnetic lines of force are imaginary, a simplified version of many magnetic phenomena can be explained by assuming the magnetic lines to have certain real properties. The lines of force can be compared to rubberbands which stretch outward when a force is exerted upon them and contract when the force is removed. The characteristics of magnetic lines of force can be described as follows:

1. Magnetic lines of force are continuous and will always form closed loops.

2. Magnetic lines are elastic or flexible.

3. Magnetic lines of force will never cross one another.

4. Parallel magnetic lines of force traveling in the same direction repel one another. Parallel magnetic lines of force traveling in opposite directions tend to unite with each other and form into single lines traveling in a direction determined by the magnetic poles creating the lines of force.

5. Magnetic lines of force tend to shorten themselves; therefore, the magnetic lines of force existing between two unlike poles cause the poles to be pulled together.

6. Magnetic lines of force pass through all materials. They will take the path of least *reluctance*. Reluctance is the opposition offered to the passage of magnetic lines.

7. They cannot be insulated against, but can be shielded or deflected.

The space surrounding a magnet in which the magnetic force exists is called a *magnetic field*. Michael Faraday, a British physicist and chemist, was the first scientist to visualize the magnet field as being in a state of stress and consisting of uniformly distributed lines of force. The entire quantity of magnetic lines surrounding a magnet is called *magnetic flux*. Unlike an electric current, magnetic flux is not thought to be a stream of motion. Instead, magnetism is just simply a field of force exerted in space.

The weber (Wb) is the unit used to represent the flux of the magnet. This term is very large; one weber is equal to 100,000,000 or 1×10^8 magnetic lines of force. It is seldom used when calculating flux. A more common method uses a fraction of the weber. The microweber (μWb) is most often used. One microweber represents 100 lines of flux in the magnetic field.

The number of lines of force per unit area is called *flux density*. In the CGS (centimeter, gram, second) system, the unit of measurement is the gauss. One gauss is one maxwell per square centimeter. One maxwell is equal to one magnetic line.

In the MKS (meter, kilogram, second) system, the unit of measurement is the tesla. A tesla (named for Nikola Tesla, a Yugoslav-born American scientist) is equal to one weber per square meter. Flux density is expressed by the equation:

$$B = \frac{\Phi}{A},$$

where B (flux density) is in teslas, Φ (total number of lines of flux) is in webers, and A is in meters (Fig. 9-6). Thus, the unit of

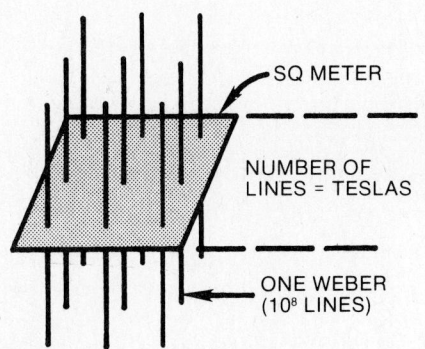

Fig. 9-6: Webers and teslas defined.

153

flux density, B, is expressed in webers per square meter (Wb/m²).

A visual representation of the magnetic field around a magnet can be obtained by placing a plate of glass over a magnet and sprinkling iron filings onto the glass. The filings arrange themselves in definite paths between the poles. This arrangement of the filings shows the pattern of the magnetic field around the magnet, as in Fig. 9-7.

The magnetic field surrounding a symmetrically shaped magnet has the following properties:

1. The field is symmetrical unless disturbed by another magnetic substance.

2. The lines of force have direction and are represented as emanating from the north pole and entering the south pole.

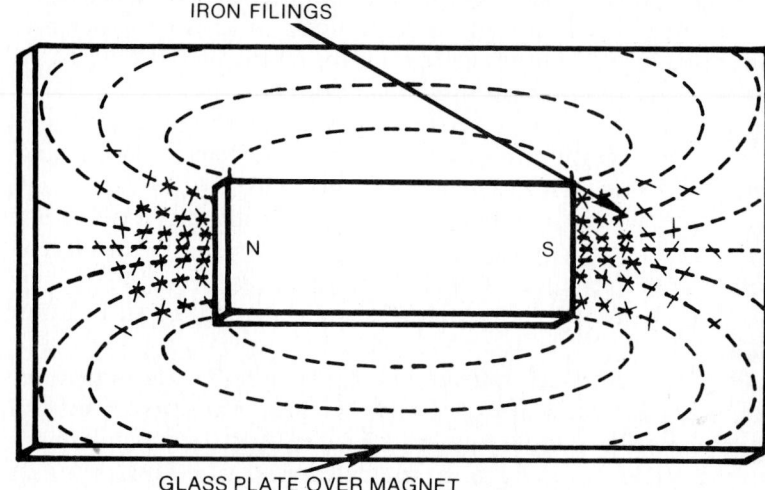

IRON FILINGS

GLASS PLATE OVER MAGNET

Fig. 9-7: Magnetic field pattern around a magnet.

Laws of Attraction and Repulsion

If a magnetized needle is suspended near a bar magnet, as in Fig. 9-8, it will be seen that a north pole repels a north pole and a south pole repels a south pole. Opposite poles, however, will attract each other. Thus, the first two laws of magnetic attraction and repulsion are:

1. *Like* magnetic poles *repel* each other.

2. *Unlike* magnetic poles *attract* each other.

The flux patterns between adjacent *unlike* poles of bar magnets, as indicated by lines, are shown in Fig. 9-9A. Similar patterns for adjacent *like* poles are shown in Fig. 9-9B. The lines do not cross at any point and they act as if they repel each other.

Figure 9-10 shows the flux pattern (indicated by lines) around two bar magnets placed close together and parallel with each other. Figure 9-10A shows the flux pattern when opposite poles are adjacent; and Fig. 9-10B shows the flux pattern when like poles are adjacent.

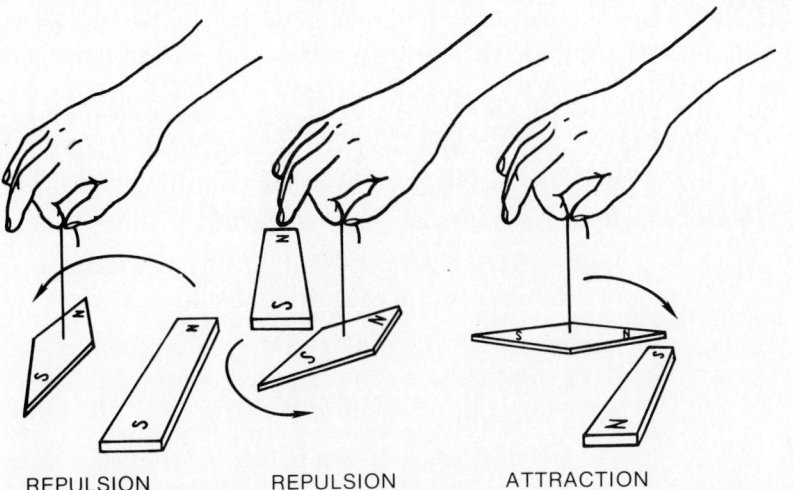

REPULSION   REPULSION   ATTRACTION

Fig. 9-8: Laws of attraction and repulsion.

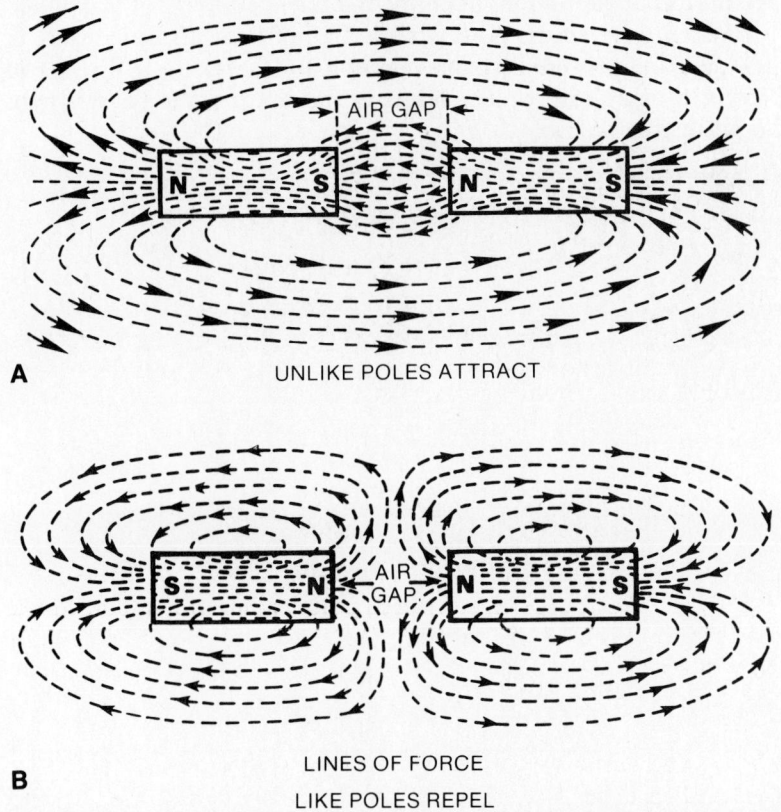

A   UNLIKE POLES ATTRACT

B   LINES OF FORCE
LIKE POLES REPEL

Fig. 9-9: Lines of force between unlike and like poles.

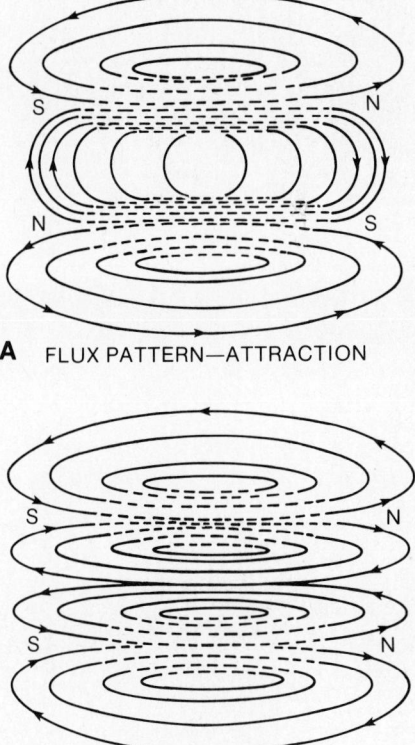

A   FLUX PATTERN—ATTRACTION

B   FLUX PATTERN—REPULSION

Fig. 9-10: Flux patterns of adjacent parallel bar magnets.

The *third law* of magnetic attraction and repulsion states in effect that the force of attraction or repulsion existing between two magnetic poles decreases rapidly as the poles are separated from each other. Actually the force of attraction or repulsion follows Coulomb's Law with respect to the force between two charged bodies. Thus, the magnetic force is inversely propor-

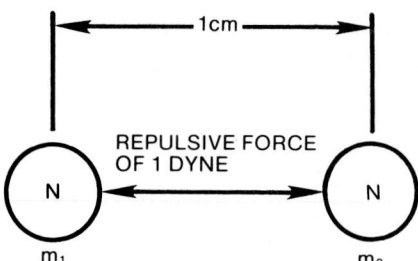

Fig. 9-11: A unit pole.

tional to the square of the distance between the two magnets and directly proportional to the product of the strength of the two forces:

$$F = \frac{m_1 m_2}{\mu d^2}$$

where: F = force in dynes between two poles

m_1 = magnetic strength of first pole in unit poles

m_2 = magnetic strength of second pole in unit poles

d = distance between poles in cm

μ = permeability of the medium through which the force acts (air or vacuum μ = 1)

The magnetic strength of each magnetic pole is measured in *unit poles*. A unit pole has a strength such that when placed 1 centimeter from an equal pole (in air or vacuum), there will be a force exerted of 1 dyne (Fig. 9-11). This can be a force of either attraction or repulsion, depending on the polarity of the poles.

Example: A south pole with a strength of 40 unit poles is placed 10 centimeters from a north pole having a magnetic strength of 20 unit poles. What is the magnitude of the force between the two poles?

Given:

$$m_1 = -40 \text{ unit poles}$$
$$m_2 = 20 \text{ unit poles}$$
$$d = 10 \text{ centimeters}$$
$$\mu = 1$$

Find F:

Solution:

$$F = \frac{(m_1)(m_2)}{\mu d^2}$$

$$F = \frac{-40 \times 20}{1 \times 10^2}$$

$$F = \frac{-800}{100}$$

$$F = -8$$

$$F = 8 \text{ dynes of attraction}$$

Note: Unit pole strengths are expressed as positive quantities for north poles and negative quantities for south poles. Resultant forces of a positive value indicate a force of repulsion; a negative value force indicates attraction.

The air space between poles of a magnet is known as an *air gap* (Fig. 9-9). The shorter or smaller the air gap, the stronger the magnetic field in the gap. Keep in mind that since air is nonmagnetic and cannot concentrate magnetic lines, a larger air gap only provides additional space for the magnetic lines to expand out.

MAGNETIC INDUCTION

Earlier in the chapter, it was mentioned that it was possible to induce magnetism by stroking with a magnet. It is also possible to induce magnetism by placing a substance in a magnetic field. A bar magnet, such as shown in Fig. 9-12, has uniform and symmetrical fields unless these fields are disturbed by falling under the influence of the fields of another magnetic material. If, for instance, a soft (unmagnetized) iron bar is brought into close proximity to the bar magnet, some of the lines of force of the latter will diverge from their normal region and pass through the soft iron bar. The iron bar then becomes a magnet, with its south pole near the north pole of the bar magnet and the north pole near the south pole of the original magnet. The magnetism of the soft iron bar was induced into it by the original magnet when the two were brought close together.

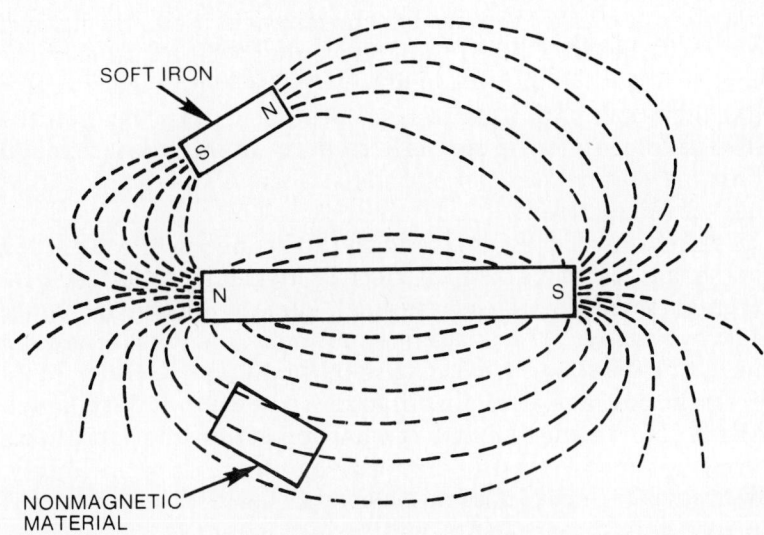

Fig. 9-12: Effects of magnetic and nonmagnetic substances on a magnetic field.

When the soft iron is removed from the vicinity of the bar magnet, it will retain some magnetism. The strength of the newly formed magnet, however, depends on the composition of the material and its ability to retain the induced magnetism. This ability is known as *retentivity*, and this varies with the type of material. Steel, for instance, has a much greater retentivity than iron, and some alloys are even superior to steel. "Nonmagnetic" materials have a low degree of retentivity even if powerful magnetic lines of force are used for the magnetic induction process. The retentivity factor depends on the ability of the electron spins of the material to maintain their new domain order after the magnetizing force has been removed.

MAGNETIC SHIELDING

There is not a known insulator for magnetic flux. If a non-magnetic material is placed in a magnetic field, there is no appreciable change in flux; that is, the flux penetrates the nonmagnetic material. For example, as illustrated previously in Fig. 9-7, a glass plate placed between the poles of a horseshoe magnet will have no appreciable effect on the field although glass itself is a good insulator in an electric circuit. If a magnetic material (for example, soft iron) is placed in a magnetic field, it then becomes a magnet with its south pole near the north pole of the magnet and its north pole near the south pole of the inducing magnet (Fig. 9-12). The insertion of the smaller soft iron bar also bends the lines of force from the larger magnet, thus changing the shape of the force field. Insertion of a non-magnetic material will have no effect on the shape of the magnetic force. Removal of the smaller soft iron bar, of course, returns the larger magnet's force field to its original shape. The soft iron bar will lose its magnetic properties at the same time.

If an iron washer is placed in the magnetic field, the lines of force will be directed around the hole in the middle of the washer. This is illustrated in Fig. 9-13. The lines of force tend to take the easier path. In this case, as in most, it is easier for the lines of force to travel through the iron washer than through the air; therefore, the center of the washer will have few, if any, lines of force present.

Application of the above information is necessary to protect the sensitive mechanism of electric instruments as such instruments become inaccurate when subjected to the influence of stray magnetic fields. Because an instrument's mechanism cannot be insulated from magnetic flux, it is necessary to redirect the passage of the flux lines. It is known that the magnetic lines of force take the path of least opposition; therefore, if we surround an object with a material having high permeability, the magnetic lines taking the easiest path will flow through the surrounding material. A sensitive instrument is protected by enclosing it in a soft iron case called a *magnetic screen* or *shield*, as shown in Fig. 9-14. It must be emphasized again that there is no insulator for magnetic lines of force; but by placing an instrument inside the iron shield, an insulating effect occurs.

TYPES OF MAGNETIC MATERIALS

Early magnetic studies classified materials merely as being magnetic and nonmagnetic. Present studies classify materials into one of the following groups:

Paramagnetic materials are those that become only slightly magnetized even though under the influence of a strong magnetic field. This slight magnetization is in the same direction as the magnetizing field. Materials of this type are aluminum, chromium, platinum, and air.

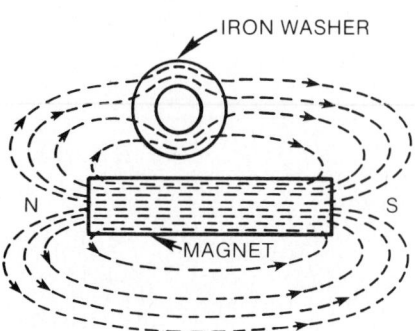

Fig. 9-13: A metal iron washer will direct the forces around a specific area in the field.

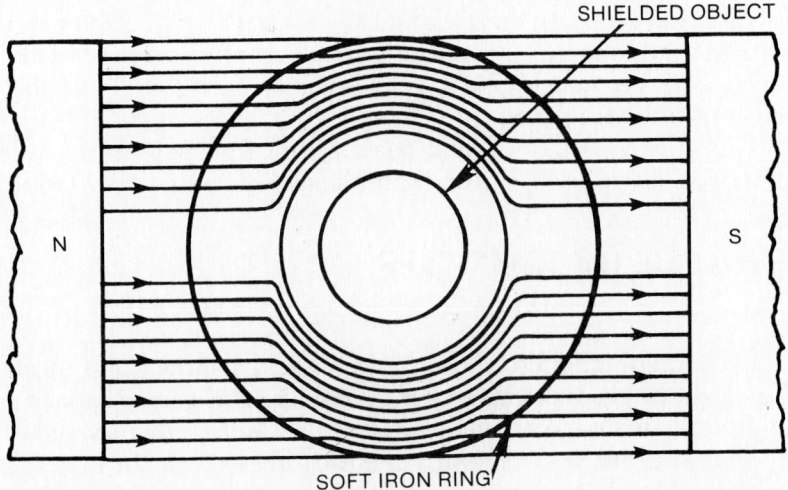

N

S

SOFT IRON RING

Fig. 9-14: Magnetic shielding.

Diamagnetic materials can also be only slightly magnetized when under the influence of a very strong field. These materials, when slightly magnetized, are magnetized in a direction opposite to the external field. Some diamagnetic materials are copper, silver, gold, and mercury.

Paramagnetic and diamagnetic materials have a very low permeability. Paramagnetic materials have a permeability slightly greater than one; diamagnetic materials have a permeability less than one. Because of the difficulty in obtaining some magnetization of paramagnetic and diamagnetic materials, these materials are considered for all practical purposes as nonmagnetic materials.

The most important group of materials for applications of electricity and electronics are the *ferromagnetic materials*. Ferromagnetic materials are those which are relatively easy to magnetize such as iron, steel, cobalt, Alnico, and Permalloy, the latter two being alloys. Alnico consists primarily of aluminum, nickel, and cobalt. These new alloys can be very strongly magnetized with Alnico capable of obtaining a magnetic strength great enough to lift five hundred times its own weight.

Ferromagnetic materials all have a high permeability, typically from 50 to 5,000. However, as previously discussed, a material such as steel used to make a permanent magnet is considered to have a relatively low permeability in comparison to other ferromagnetic materials.

Ferrites. Ferrites are a group of powdered, compressed magnetic materials having a high resistivity and consisting of ferric oxide with ceramic as a binder. They display the same properties as ferromagnetic iron. Unlike iron, however, ferrites are poor electrical conductors. They are also a great deal more efficient than iron when the current alternates at high frequencies, thus making them ideal for RF (radio frequency) transformers.

Another use of ferrites employing their efficiency at high frequencies is as RF chokes. For this application, a ferrite bead

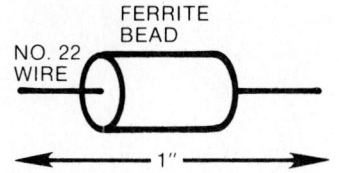

Fig. 9-15: Typical ferrite bead choke.

or beads are molded on a wire as shown in Fig. 9-15. Since the ferrite bead concentrates the magnetic field of the current in the wire, it serves as a choke to reduce the current at only an undesired radio frequency.

SHAPES OF MAGNETS

Permanent magnets are manufactured in a variety of shapes. Each shape produces its own magnetic field. Because opposite magnetic fields attract each other, the closer together the poles of a magnet are made, the stronger the force outside the magnet will be. This is because the lines of magnetic force are more concentrated. The four most common magnet shape classifications are bar, horseshoe, ring, and round magnets (Fig. 9-16).

Bar Magnets. While the bar magnet (Fig. 9-16A) was useful in demonstrating magnetic effects in the preceding text, it is seldom used in electronic equipment.

Horseshoe Magnets. This is the magnet shape most frequently used in electrical or electronic equipment. It is similar to a bar magnet but is bent in the shape of a horseshoe. The horseshoe magnet provides much more magnetic strength than a bar magnet of the same size and material because of the closeness of the magnetic poles. The magnetic strength from one pole to the other, as illustrated in Fig. 9-16B, is greatly increased due to the concentration of the magnetic field in a smaller area. Electrical measuring devices quite frequently use horseshoe type magnets.

Ring Magnets. This type of magnet has uniform lines of force throughout its circular frame (Fig. 9-16C). The ring magnet lends itself well to uses in meters as well as computer memory cores.

Frequently, ring magnets form a closed magnetic loop to concentrate the magnetic lines within the magnet itself. Because the loop has no open ends, there are no poles or air gap. To better understand how a closed magnetic loop functions, look at Fig. 9-16D where two horseshoe magnets are placed together to form such a loop. The north and south poles of each magnet cancel as opposite poles touch. However, while the magnetic lines outside the magnets cancel because they are in opposite directions, each magnet has its own lines inside. This is in addition to the magnetic lines of the other magnet. The effect of the closed magnetic loop, therefore, is maximum concentration of magnetic lines within the magnet, but with a minimum number of lines outside.

Round Magnets. Magnets of this type (Fig. 9-16E) are found in loudspeakers (see Chapter 22). As mentioned earlier, the magnets used today are found in a variety of physical shapes. The final form depends upon the application. All of these magnetic forms except ring magnets have one thing in common, the pair of poles.

160

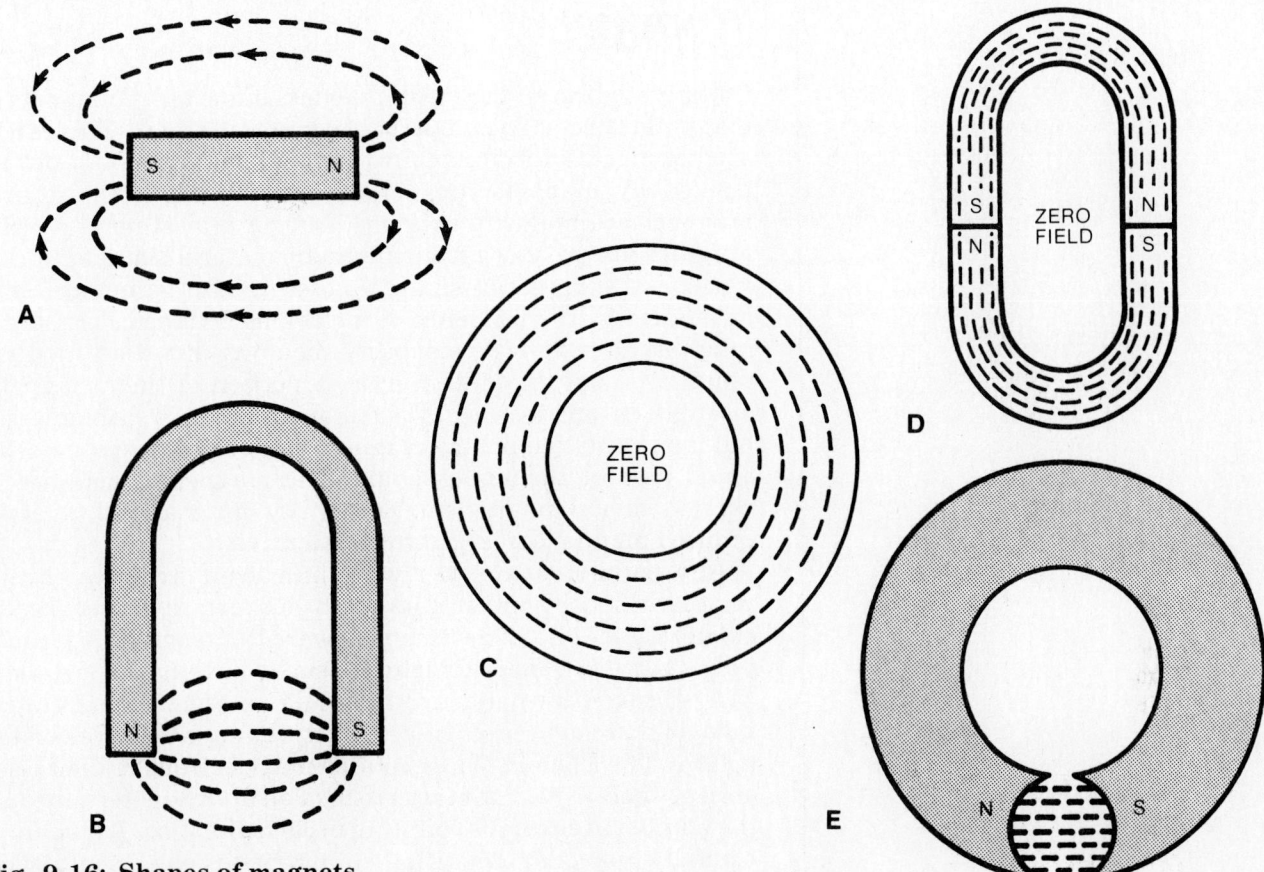

Fig. 9-16: Shapes of magnets.

CARE OF MAGNETS

A piece of steel that has been magnetized can lose much of its magnetism by improper handling. If it is jarred or heated, there will be a disalignment of its domains, resulting in the loss of some of its effective magnetism. Had this steel formed the horseshoe magnet of a meter, the meter would no longer be operable or would give inaccurate readings; therefore, care must be taken when handling instruments containing permanent magnets. Severe jarring or subjecting the instrument to high temperature will damage the device.

A magnet may also become weakened from loss of flux. Thus, when storing magnets, one should always try to avoid excess leakage of magnetic flux. Figures 9-17A and B show how to store permanent magnets in a closed loop to prevent loss of flux. A horseshoe magnet should always be stored with a *keeper* (Fig. 9-17C), a soft iron bar used to join its magnetic poles. By use of the keeper while the magnet is being stored, the magnetic flux will continuously circulate through the magnet and not leak off into space.

When storing bar magnets, the same principle must be kept in mind; therefore, bar magnets should always be stored in pairs with a north pole and a south pole placed together. This provides a complete path for the magnetic flux without any flux leakage.

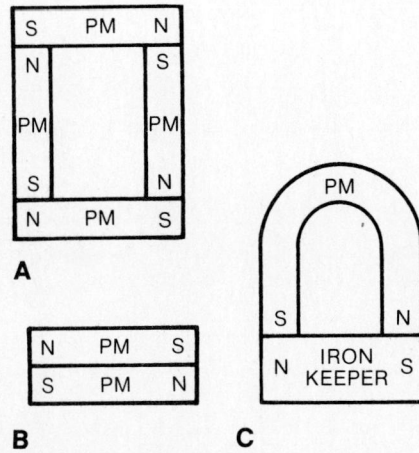

Fig. 9-17: Methods of storing permanent magnets.

SUMMARY

Many electric and electronic devices make use of magnets or magnetic effects. The property of a material which enables it to attract pieces of iron is called magnetism. Materials not attracted by magnets are considered nonmagnetic. Natural magnets are magnets existing in a natural state. Artificial magnets are produced from magnetic materials and exist in a variety of shapes and sizes. Magnetism can be induced in a substance by the influence of a stroking magnet. Permanent magnets have a low permeability, meaning they are difficult to magnetize, but they retain a large portion of their magnetic strength. Highly permeable substances are easy to magnetize, but make only temporary magnets because they retain only a small part of their magnetism. The magnetism retained is called residual magnetism. The ability of a material to retain induced magnetism is known as retentivity.

Any permanent magnet will align itself with the earth's magnetic field by pointing its north pole toward the geographical north pole and its south pole toward the geographical south pole. The Weber Theory of magnetism holds that all substances can be broken down into tiny molecular magnets. The arrangement of these molecular magnets determines a material's magnetism. The Domain Theory states that electrons in their orbit spin on their axes. The magnetism of an atom is determined by the number of electrons spinning in each direction. A domain is a small group of atoms with their magnetic poles orientated in the same direction.

Magnetic lines of force are continuous imaginary lines emanating from the north pole of a magnet and entering the south pole. They show the pattern of a magnetic field, which is the space surrounding a magnet influenced by the magnetic force. All of the lines together are called magnetic flux. The flux density, or number of lines of force per unit area, can be measured in lines per square centimeters or webers per square meter. Flux patterns show us that like magnetic poles repel each other while unlike magnetic poles attract. As the distance between magnetic poles increases, the force between them decreases. Unit poles measure the magnetic strength of a pole. The strength of the air gap, or space between poles of a magnet, is determined by its size. The uniform lines of a magnetic field can be disturbed by the influence of another material that has been brought in contact with them. Magnetism can be induced when some of the lines of force pass through the material being introduced to the magnetic field. Because there is no known insulator for magnetic flux, a magnetic screen or shield made from highly permeable material should be placed around sensitive instruments that must be protected from stray magnetic fields.

Materials can be classified as paramagnetic, diamagnetic, or ferromagnetic. Paramagnetic materials become only slightly magnetized and in the same direction as the magnetizing field. Diamagnetic materials also become slightly magnetized but in the opposite direction from the magnetizing field. Ferromag-

162

netic materials become strongly magnetized in the same direction as the magnetizing field. Ferromagnetic materials are easy to magnetize. Ferrites display the same properties as ferromagnetic iron but are more efficient at high frequencies. Just as there are different magnetic materials, there are a variety of shapes in which magnets are manufactured. Magnets shouldn't be jarred or heated and must be properly stored utilizing a keeper to prevent loss of magnetism.

CHECK YOURSELF (Answers at back of book)

Fill in the blanks with the appropriate word or words.

1. A material possessing the property which enables it to attract pieces of iron is known as a _____ .
2. Stroking an iron, steel, or alloy bar with a bar magnet will produce a permanent magnet through the process of _____ .
3. _____ magnets are produced from magnetic materials and can be made in a wide variety of shapes, sizes, and materials.
4. Substances that are relatively easy to magnetize are said to have a _____ permeability, while substances relatively difficult to magnetize are said to have a _____ permeability.
5. The end of the compass that points toward the geographical north pole of the earth is called the _____ , and the end pointing toward the south geographical pole is called the _____ .
6. The _____ Theory assumes all magnetic substances to be composed of tiny molecular magnets which line up with most of their north-seeking poles pointing in one direction and their south-seeking poles facing the opposite direction.
7. A small group of atoms with their magnetic poles orientated in the same direction is known as a _____ .
8. According to the Domain Theory, an atom is _____ when more of its electrons spin in one direction than another and _____ when equal numbers of electrons spin in opposite directions.
9. The attractive force is greatest at the ends, or _____ , of a magnet.
10. The line along which a compass needle aligns itself when placed in the vicinity of a bar magnet is called a _____ .
11. The area surrounding a magnet in which the magnetic force exists is called the _____ .
12. Together the magnetic lines surrounding a magnet are called the magnetic _____ .
13. _____ is the number of magnetic lines of force per unit area.
14. Like magnetic poles _____ each other, while unlike magnetic poles _____ each other.
15. The force of attraction or repulsion existing between two magnetic poles _____ as the poles are separated from each other.

16. The size of the _____ determines the strength of the magnetic field in the air space between the poles of a magnet.
17. When a soft (unmagnetized) iron bar is brought near a bar magnet, some of the lines of force will diverge from the bar magnet and pass through the soft iron bar, thus inducing _____ in the soft iron bar.
18. The ability of a newly formed magnet to retain induced magnetism is known as _____ .
19. In order to protect a sensitive electric instrument from magnetic lines of force, enclose it in a soft iron case called a _____ .
20. Magnetic lines of force will _____ a nonmagnetic material placed in a magnetic field.
21. A magnetic material placed in the magnetic field of another substance will change the _____ of the force field of that substance.
22. _____ materials become slightly magnetized in the same direction as the magnetizing field.
23. _____ materials become slightly magnetized in a direction opposite to the magnetizing field.
24. _____ materials, used in electricity and electronics, are highly permeable.
25. _____ are a group of powdered, compressed magnetic materials having a high resistivity and consisting of ferric oxide with ceramic as a binder.
26. Because opposite magnetic fields attract each other, the closer together the poles of a magnet are made, the _____ the force outside the magnet will be.
27. The _____ is the magnet shape most frequently used in electrical or electronic equipment.
28. A ring magnet's magnetic lines are concentrated _____ the magnet.

CHAPTER 10

Electromagnetism and Magnetic Circuits

Magnetism and basic electricity/electronics are so closely related that one cannot be studied at length without involving the other. To become proficient in electricity and electronics, one must become familiar with such general relationships that exist between magnetism and electricity as follows:

1. Electric current flow will always produce some form of magnetism.

2. Magnetism is by far the most commonly used means for producing or using electricity.

3. The peculiar behavior of electricity under certain conditions is caused by magnetic influences.

ELECTROMAGNETISM

The magnetic strength of a permanent magnet is limited by its size and the cost of producing this kind of magnet. A permanent magnet has many limitations. For example, one cannot turn off a permanent magnet. Also, the strength of the magnetic field cannot be varied unless the magnet is moved from one position to another position. These limitations, and others, are overcome when an electromagnet is used.

Magnetic Field Around a Current-Carrying Conductor

An electrical current moving through a conductive material, such as a wire, will produce, as stated earlier, magnetic lines of force (Fig. 10-1). These lines of force surround the conductor. They also form at right angles to the conductor. The strength of the magnetic field produced in this manner is directly proportional to the amount of electric current flowing through the wire. The magnetic field size increases as more current flows. A reduction in current makes the magnetic field smaller. It is very difficult to show, but picture the magnetic field completely surrounding the conducting wire. This magnetic field is very strong close to the surface of the wire. It diminishes in strength as the distance from the wire increases.

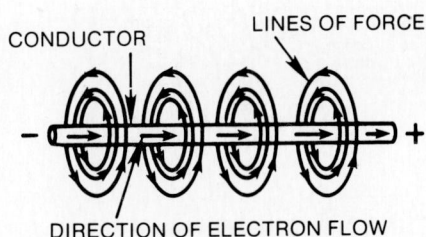

Fig. 10-1: Magnetic lines surrounding wire carrying current.

Magnetic lines of force have direction. In 1819, Hans Christian Oersted, a Danish physicist, discovered that an electric current is accompanied by certain magnetic effects and that these effects obey definite laws. If a compass is placed in the vicinity of a current-carrying conductor, the needle aligns itself at right angles to the conductor, thus indicating the presence of a magnetic force. The presence of this force can be demonstrated by passing an electric current through a vertical conductor which passes through a horizontal piece of cardboard, as illustrated in Fig. 10-2. The magnitude and direction of the force are determined by setting a compass at various points on the cardboard and noting the deflection.

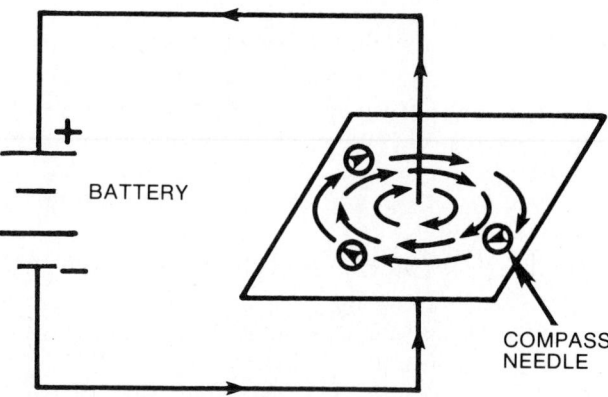

BATTERY

COMPASS
NEEDLE

Fig. 10-2: Oersted showed how the magnetic field around a conductor affects a compass needle.

The direction of the force is assumed to be the direction in which the north pole of the compass points. These deflections show that a magnetic field exists in circular form around the conductor. When the current flows upward, the field direction is clockwise, as viewed from the top, but if the polarity of the supply is reversed so that the current flows downward, the direction of the field is counterclockwise.

The relation between the direction of the magnetic lines of force around a conductor and the direction of current flow along the conductor may be determined by means of the *left-hand rule for a conductor*. If the conductor is grasped in the left hand with the thumb extended in the direction of electron flow (− to +), the fingers will point in the direction of the magnetic lines of force (Fig. 10-3). This is the same direction in which the north pole of a compass would point if the compass were placed in the magnetic field.

Arrows generally are used in electric diagrams to denote the direction of current flow along the length of wire. Where cross sections of wire are shown, a special view of the arrow is used. A cross-sectional view of a conductor that is carrying current toward the observer is illustrated in Fig. 10-4A. The direction of current is indicated by a dot, which represents the head of the arrow. A conductor that is carrying current away from the

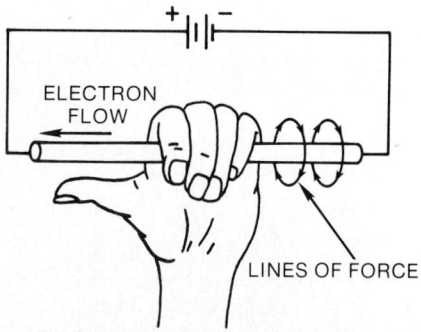

ELECTRON
FLOW

LINES OF FORCE

Fig. 10-3: The left-hand rule for determining the direction of magnetic lines.

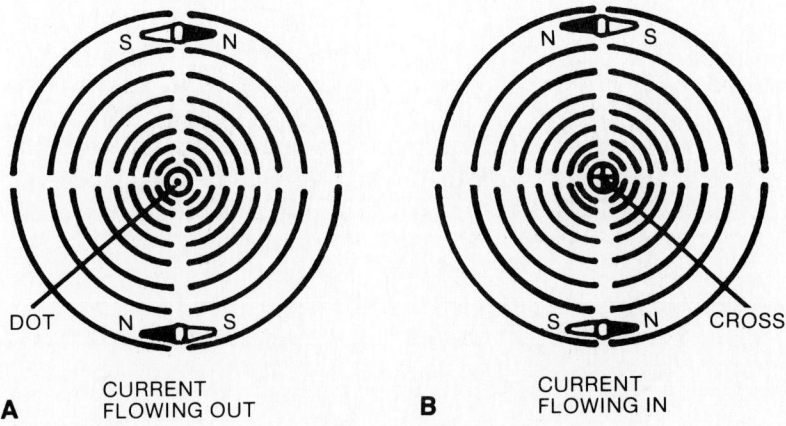

A   CURRENT
FLOWING OUT

B   CURRENT
FLOWING IN

Fig. 10-4: Magnetic field around a current-carrying conductor.

observer is illustrated in Fig. 10-4B. The direction of current is indicated by a cross, which represents the tail of the arrow.

When two parallel conductors carry current in the same direction, the magnetic fields tend to encircle both conductors, drawing them together with a force of attraction, as shown in Fig. 10-5A. Two parallel conductors carrying currents in opposite directions are shown in Fig. 10-5B. The field around one conductor is opposite in direction to the field around the other conductor. The resulting lines of force are crowded together in the space between the wires and tend to push the wires apart. Therefore, two parallel adjacent conductors carrying currents in the same direction attract each other, and two parallel conductors carrying currents in opposite directions repel each other.

MAGNETIC FIELD OF A COIL

The magnetic field around a current-carrying wire exists at all points along its length. As previously mentioned, the field consists of concentric circles in a plane perpendicular to the wire. When this straight wire is wound around a core, as shown in Fig. 10-6A, it becomes a coil and the magnetic field assumes a different shape. This drawing is a partial cutaway view which shows the construction of a simple coil. Figure 10-6B is a three-dimensional view of the same coil. As can be noted, the coil current sets up circular magnetic lines of force around each coil turn. Consequently, the magnetic lines of force encircle the upper part of each coil turn in a clockwise direction. In the center of the coil, all the magnetic lines of force run in the same direction, thereby aiding each other to produce a net positive effect. Between adjacent coil turns, the magnetic lines of force cancel each other. Consequently, magnetic lines of force travel the entire length of the coil in order to complete their loops. The combined influence of all the turns produces a two-pole field similar to that of a simple bar magnet. One end of the coil will be a north pole and the other end will be a south pole.

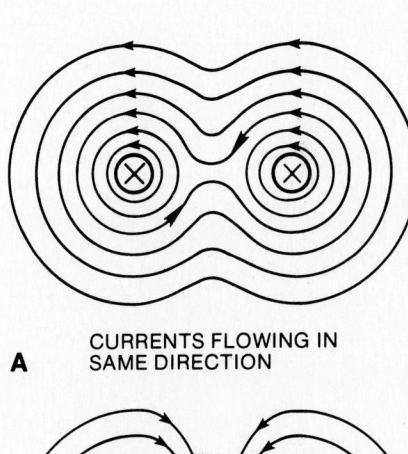

A   CURRENTS FLOWING IN
SAME DIRECTION

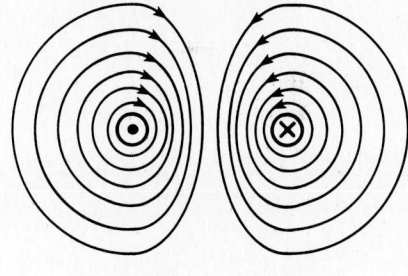

B   CURRENTS FLOWING IN
OPPOSITE DIRECTION

Fig. 10-5: Magnetic field around two parallel conductors.

167

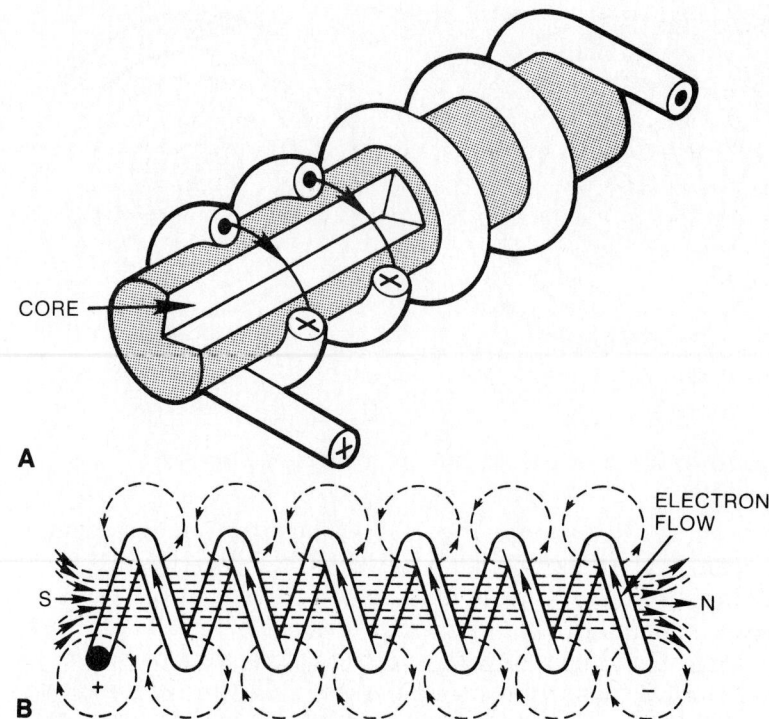

CORE

A

ELECTRON FLOW

S   N

+   −

B

Fig. 10-6: **Magnetic field produced by a current-carrying coil.**

Polarity of an Electromagnetic Coil

In Fig. 10-4, it was shown that the direction of the magnetic field around a straight conductor depends on the direction of current flow through that conductor. Thus, a reversal of current flow through a conductor causes a reversal in the direction of the magnetic field that is produced. It follows that a reversal of the current flow through a coil also causes a reversal of its two-pole field. This is true because that field is the product of the linkage between the individual turns of wire on the coil. Therefore, if the field of each turn is reversed, it follows that the total field (coil's field) is also reversed.

When the direction of electron flow through a coil is known, its polarity may be determined by use of the left-hand rule for coils. This rule is illustrated in Fig. 10-7 and is stated as follows: Grasping the coil in the left hand, with the fingers "wrapped around" in the direction of electron flow, the thumb will point toward the north pole.

Strength of an Electromagnetic Field

The strength, or intensity, of a coil's field depends on a number of factors. The major factors are listed below:
1. The number of turns of conductor.
2. The amount of current flow through the coil.
3. The physical dimensions of the coil.
4. The permeability of the core.

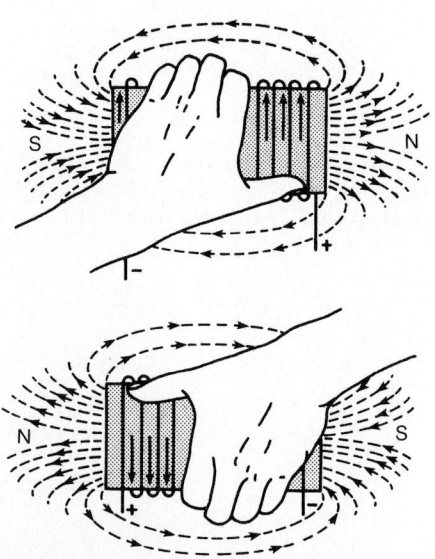

Fig. 10-7: **Left-hand rule for coil polarity.**

MAGNETIC CIRCUITS

Many electrical devices depend upon magnetism in one or more forms for their operation. To have these devices function efficiently, engineers work out intricate designs for the required magnetic conditions. The magnets designed to do a particular job must have the required strength and must be provided with paths, or circuits, of suitable shapes and materials.

A magnetic circuit is defined as the path (or paths) taken by the magnetic lines of force leaving a north pole, passing through the entire circuit, and returning to the south pole. A magnetic circuit may be a series or parallel circuit or any combination.

Ohm's Law Equivalent for Magnetic Circuits

The law of current flow in the electric circuit is similar to the law for the establishing of flux in the magnetic circuit. Ohm's Law for electric circuits states, as already detailed, that the current is directly proportional to the applied voltage and inversely proportional to the resistance offered by the circuit. Expressed mathematically,

$$I = \frac{V}{R}$$

Rowland's Law for magnetic circuits, formulated by Henry Rowland, an American physicist, states, in effect, that the number of lines of magnetic flux in webers (Φ) is directly proportional to the magnetomotive force (mmf) and inversely proportional to the reluctance ($\mathfrak{R}$) offered by the circuit. Expressed mathematically:

$$\Phi = \frac{\text{mmf}}{\mathfrak{R}}$$

The similarity of Ohm's Law and Rowland's Law is apparent; however, the units used in the expression for Rowland's Law need to be explained.

Ampere-Turns. The magnetomotive force, F, or mmf, comparable to voltage or electromotive force in the Ohm's Law formula, is the force that produces the flux in the magnetic circuit. The practical unit of magnetomotive force is the *ampere-turn*. Stated as a formula,

$$\text{ampere-turns} = NI$$

where N is the number of turns, multiplied by the current, *I*, in amperes. The practical unit for N*I* is the ampere-turn. The standard abbreviation for ampere-turn is A, the same as for the ampere. For this reason, ampere is spelled out in this chapter.

One of the circuits shown in Fig. 10-8 is a coil that has 100 turns of wire. The measured electrical current is 1 ampere. The magnetomotive force for this coil is 100 × 1, or 100A. Figure 10-8B shows a coil that has 500 turns of wire. The measured electrical current flowing through the coil is 0.15 ampere. The

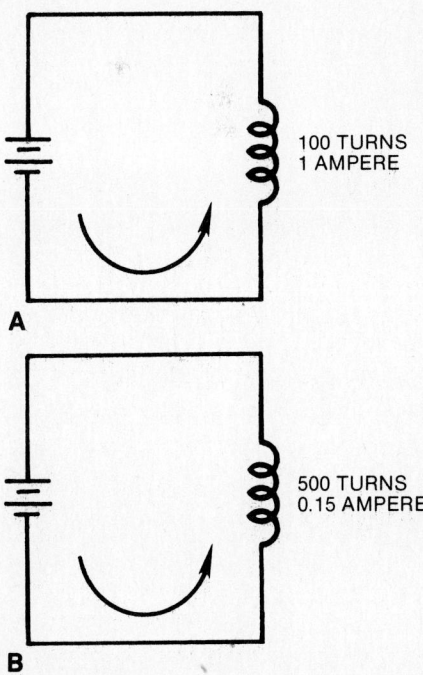

100 TURNS
1 AMPERE

A

500 TURNS
0.15 AMPERE

B

Fig. 10-8: Ampere-turns are determined by the number of turns of wire on the form and the current flowing through these turns of wire.

169

mmf of this coil is 500 × 0.15, or 22.5A. In summary, one-ampere-turn is equal to one turn of wire on the coil and 1 ampere of current flowing through the coil.

Magnetic Flux. As described in Chapter 9, magnetic flux (Φ) is similar to current in the Ohm's Law formula and comprises the total number of lines of force existing in the magnetic circuit. Keep in mind that 1 weber (Wb) equals 1×10^8 lines. Since the weber is a large unit for most typical fields used in electronics, the microweber unit can be employed:

$$1\mu Wb = 10^{-6}Wb.$$

Flux Density. Flux density of a magnetic field is the number of magnetic lines of force in a given area. Stated as a formula,

$$\beta = \frac{\Phi}{A}$$

where β is in teslas, Φ (phi) is in webers, and A is in square meters.

Magnetic Field Strength. Magnetic field strength relates to the magnetic strength (force) exerted by the field and is measured in units of ampere-turns per meter. Stated as a formula,

$$H = \frac{mmf}{l}$$

where H is the magnetic field strength, mmf is the magnetomotive force in ampere-turns, and l represents the length in meters of the coil form. The answer is given in amperes per meter.

Example: Suppose a coil wire has 50 turns and carries a current of 0.75 ampere. Its length is 25 cm. What is the magnetomotive force?
Solution:

$$mmf = NI$$
$$= 50 \text{ turns} \times 0.75 \text{ amperes}$$
$$= 37.5A$$

Use this information to determine the magnetic field strength of the coil.

$$H = \frac{mmf}{l}$$
$$= \frac{37.5A}{25 \times 10^{-2}m}$$
$$= 150A/m$$

It is interesting to be able to draw some conclusions from the use of these formulas. Consider the effect of the number of turns of wire on the amount of magnetic field. As the number of turns of wire increases, the magnetic field also increases. The same is true about the amount of current flowing. If this increases owing to a larger supply source or to the size of the wires used to wind the coil, the size of the magnetic field also increases. The

170

overall length of any coil also affects its magnetic strength. As the length is increased, the coil has a lower magnetic force per unit of measure. All these factors are used in determining the magnetic field strength of any electromagnetic device.

Reluctance. Reluctance is the opposition offered by a material to magnetic flux and is also referred to as *reluctivity*. Generally, it corresponds to the resistance to current flow in electric circuits, although more strictly it relates to the proportionate factor between the applied magnetic force and the magnetic flux that results from it. Thus, reluctance can be thought of as a measure of the limiting factors of a material with respect to the setting up of magnetic lines of force. The symbol for reluctance is the script letter $\mathcal{R}$. The unit for reluctance is ampere per weber (A/Wb) and the formula is:

$$\mathcal{R} = \frac{mmf}{\Phi}$$

Permeance. This is a seldom-used term and refers to the property that determines the magnitude of the flux in a material and is equal to the reciprocal of reluctance. The symbol of permeance is the script capital letter $\mathcal{P}$ and the unit is weber per ampere (Wb/A). Thus, permeance, much like conductance in electric circuits, can be considered a measure of how well a given material permits the setting up of magnetic lines of force and is related to the discussion on permeability that follows. Mathematically, permeance is thus indicated as follows:

$$\mathcal{P} = \frac{1}{\mathcal{R}}$$

Retentivity. Retentivity is the measure of how well a material is able to hold its magnetic properties after it is exposed to a magnetizing force. This is measured after the magnetizing force is removed. The terms *hard* and *soft* are used to describe this. A magnetically hard material has excellent retention capabilities. The magnetically soft material loses its magnetic properties quickly after the magnetizing force is removed. Another way of stating this is that permanent magnets have a high retentivity value and temporary magnets have a low retentivity value.

Residual Magnetism. This term relates to the amount of magnetic field strength left in any magnetic material after the magnetizing force is removed. Magnetically hard materials have a high residual magnetism. Residual magnetism is low in magnetically soft materials. The amount of residual magnetism is measured in terms of the flux density at the poles of the magnet.

Permeability. Permeability, designated by the Greek letter mu, μ, is treated to some degree in Chapter 9. However, it is defined here to permit a fuller interpretation of Rowland's Law and also a practical application of this law. Permeability is a measure of the relative ability of a substance to conduct magnetic lines of force as compared with air. The permeability of air is taken as 1. Permeability is indicated as the ratio of the flux density in webers per square meter (teslas) to the intensity of the magnetizing force in ampere-turns per meter of length, indicated by H. Expressed mathematically,

$$\mu = \frac{\beta}{H} \quad \text{or} \quad \beta = \mu \times H$$

Keep in mind that a good magnetic material with high relative permeability can concentrate flux and produce a large value of flux density, β, for a specified H.

Table 10-1 lists the meter-kilogram-second (mks) system or SI units and symbols relating to magnetism.

A comparison of the units, symbols, and equations used in applying Ohm's Law to electric circuits and Rowland's Law to magnetic circuits is given in Table 10-2.

Table 10-1: Magnetic Quantities

Quantity	Symbol	Unit	Unit Abbreviation
Magnetic field strength	H	ampere per meter	A/m
Magnetic flux	Φ	weber	Wb
Magnetic flux density	β	tesla	T
Magnetomotive force	$\mathcal{F}$	ampere (ampere-turns)	A
Permeability	μ	henry per meter	H/m
Permeance	$\mathcal{P}$	weber per ampere	Wb/A
Reluctance	$\mathcal{R}$	ampere per weber	A/Wb

Table 10-2: Comparison of Electric and Magnetic Circuits

	Electric Circuit	Magnetic Circuit
Force	Volt, V, or emf	mmf
Flow	Ampere, I	Flux, Φ, in webers
Opposition	Ohms, R	Reluctance, $\mathcal{R}$
Law	Ohm's Law, $I = \dfrac{V}{R}$	Rowland's law, $\Phi = \dfrac{mmf}{\mathcal{R}}$
Intensity of force	Volts per cm of length	$H = \dfrac{NI}{l}$, amperes per meter of length.
Density	Current density—for example, amperes per cm².	Flux density, β—for example, lines or webers per meter or tesla.

MAGNETIZATION CURVES

The density of the flux that can be created by an electromagnet depends on the number of ampere-turns of the winding as well as the current through it. Hence, the flux density in a given electromagnet depends on the magnetizing force established by the current that flows through the coil. For a graphic representation, the relationship between the field strength, H, and the flux density, β, can be plotted as shown in Fig. 10-9 to form what is known as a β-H magnetization curve. Curves of this type are often used to show how much flux density, β, results from increasing the amount of field strength, H.

It is interesting to note that as the magnetizing force (H) is increased, the flux density (β) increases linearly up to a point (Fig. 10-10). This is called the *saturation point*. At this point most of the magnetic domains have been aligned and little further change in β occurs with increases in the magnetizing force, H. The material is considered saturated beyond this point. The nonlinear relationship of the curve for lower values of H is due to initial effects in aligning the magnetic domains.

β-H curves provide important information to design engineers regarding the characteristics of core material. These curves enable engineers to anticipate core dimensions, number of required turns of wire, and other factors necessary for specific applications.

Hysteresis

There is a certain degree of residual magnetism in any magnetic field. Figure 10-11 exhibits the different magnetic qualities of three typical materials: (A) soft iron, (B) hard steel, and (C) ferrite. The vertical lines on the chart represent the flux density of the magnet. The symbols + and – are used to indicate the ability of applying a magnetizing force in either direction. This means that the polarity of the force may be reversed. The horizontal line represents the field strength of the magnet

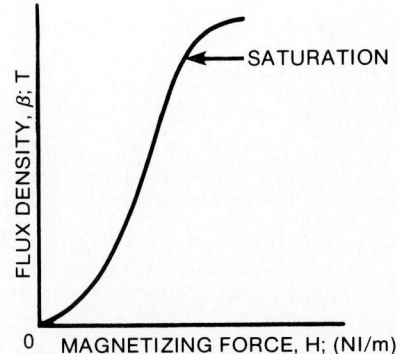

Fig. 10-10: Magnetization curves for iron.

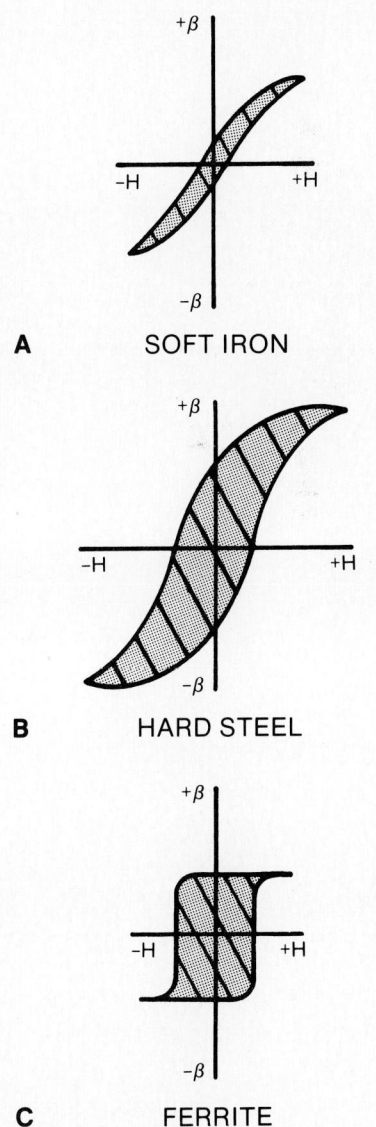

Fig. 10-11: Magnetization curves for typical magnetic materials.

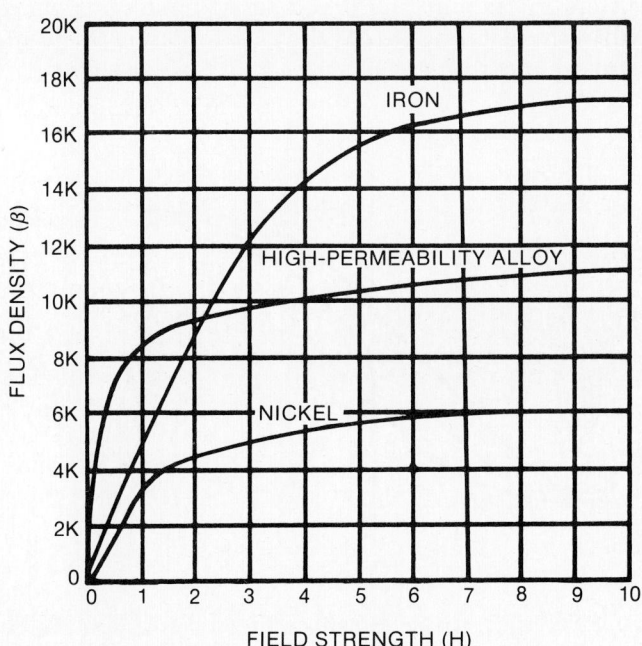

Fig. 10-9: Magnetization curves for several metals.

created in this manner. Here, too, the symbols + and − are used to show a polar reversal of the magnet due to the effects of the applied force.

These three illustrations show that the forces of magnetism are not instantaneous. They require a little time to magnetize each of the materials. The left side of each curve represents the amount of magnetism produced in the core when the magnetizing force is applied to the core. The curve on the right side of each figure shows the time delay that occurs when the magnetizing force is removed. Note that each curve is symmetrical. Also consider the shaded area in each of the curves. Soft iron has the smallest shaded area. This indicates that soft iron cores maintain magnetic properties for the shortest period of time.

Also note the curve of the ferrite core. This core has rather clearly defined vertical lines. This means that the magnetic effect is turned "on" quickly. It is also turned "off" quickly. This capability is useful in computers and other data-processing devices where a hangover of the magnetic field will affect the subsequent operation of the device. Computers require a very positive magnetic field with little or no time lag. Ferrite core materials in electric magnets provide this type of reaction. Iron cores have a relatively long carryover time and cannot be used successfully for this purpose.

This hysteresis curve information relates to frequency of any electrical waves being processed by the coils. A gradual hysteresis curve, such as illustrated in Fig. 10-12, responds slowly to changes in electrical waves. These types of cores are best used at lower frequencies or changes. These are best illustrated by identifying power line frequencies as low frequencies. Faster changes occur as the rate of change, or frequency, increases. Radio, television, and data processing require fast changes at much higher frequencies. These systems would use ferrite core materials for any coils associated with these higher frequencies.

Each magnetic material has a set of hysteresis curves and from all of them it is apparent that *hysteresis is the property of a magnetic substance that causes the magnetization to lag*

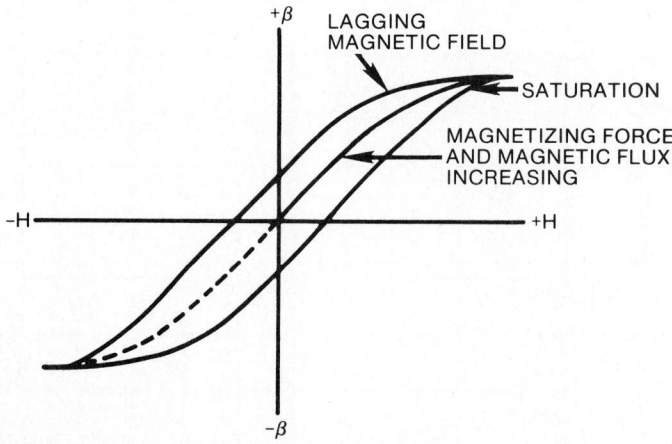

Fig. 10-12: **Magnetic hysteresis causes a delay in the time required to either magnetize or demagnetize a material.**

174

behind the force that produces it. The lag of magnetization behind the force that produces it is caused by molecular friction. Energy is needed to move the molecules or (domains) through a cycle of magnetization. If the magnetization is reversed slowly, the energy loss may be negligible. However, if the magnetization is reversed rapidly, as when commerical alternating current is used, considerable energy may be dissipated. If the molecular friction is great, as when hard steel is used, the losses may be very great. Another factor that determines hysteresis loss is the maximum density of the flux established in the magnetic material.

A comparison of the hysteresis loops for annealed steel and hard steel is shown in Fig. 10-13. The area within each loop is a measure of the hysteresis energy loss per cycle of operation. Thus, as shown in the figure, more energy is dissipated in molecular friction in hard steel than in annealed steel. It is, therefore, important that substances having low hysteresis loss be used for transformer cores and similar AC applications.

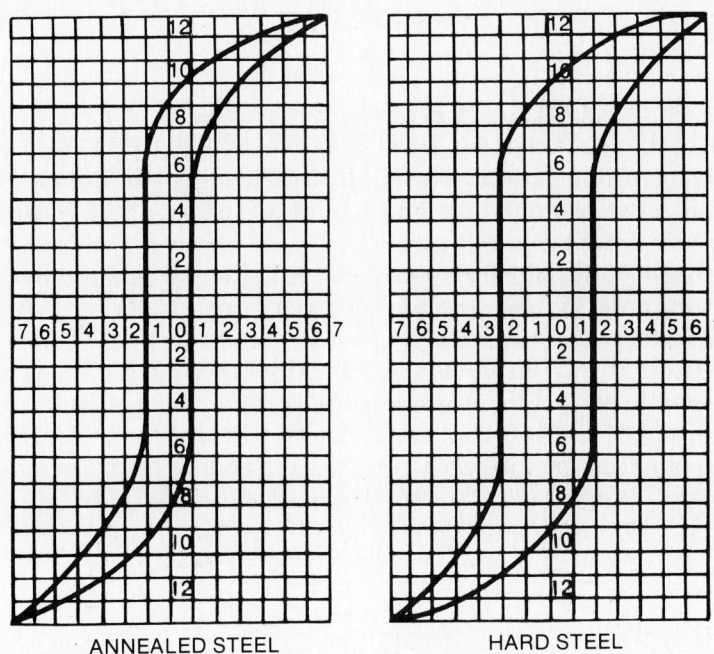

ANNEALED STEEL   HARD STEEL

Fig. 10-13: Comparison of hysteresis loops.

Demagnetization

Magnetically hard materials are very difficult to demagnetize. One method that has proved successful is to apply a rapidly changing magnetic field to the magnet. This is done with a device called a degaussing coil. This coil consists of many turns of wire. It is powered by an alternating polarity electrical current such as used in the home. The coil is placed near the magnetic device and then the coil is slowly taken away. The effect on the magnetically hard material of this changing electromagnetic field neutralizes the magnet. This minimizes any residual magnetism in the material.

Common use of this principle is found in any color television set. The color picture tube is affected by magnetic fields. The steel frame supporting the picture tube acts like a magnet. It is magnetically hard. Most color television sets have a built-in degaussing coil that is used to neutralize this magnetic effect. The circuit only works during the first minute or so that the set is turned on. Television technicians use a portable degaussing coil to neutralize the color picture tube's magnetic field on a new installation.

Another application of the degaussing coil is the removal of residual magnetism on a tape recording head. This is required on industrial, broadcast, and home quality tape recorders. The magnetically hard tape heads build up a magnetic field. This field affects the quality of the recording. It must be neutralized periodically to provide true fidelity of the information recorded on the tape.

The term *gauss* is used in older systems of measurement to indicate the flux density of a magnet. This term is replaced in the SI system by the tesla. The term *degaussing* is a carryover from the earlier use of this term.

ELECTROMAGNETS

As mentioned in Chapter 9, an electromagnet is composed of a coil of wire wound around a core of soft iron. When direct current flows through the coil, the core will become magnetized with the same polarity that the coil (solenoid) would have without the core. If the current is reversed, the polarity of both the coil and the soft iron core is reversed.

Figure 10-14 illustrates the effect that an iron core has on the magnetic lines of force surrounding a current coil. Actually the addition of the soft iron core does two things for the current-carrying coil, or solenoid. First, the magnetic flux is increased because the soft iron core is more permeable than the air core;

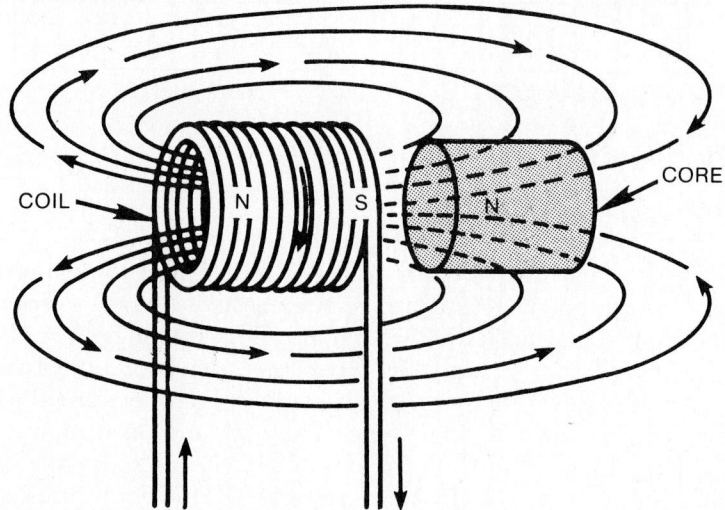

Fig. 10-14: Solenoid with soft iron core.

second, the flux is more highly concentrated. The permeability of soft iron is many times that of air and, therefore, the flux density is increased considerably when a soft iron core is inserted in the coil.

The magnetic field around the turns of wire making up the coil influences the molecules in the iron bar causing, in effect, the individual molecular magnets, or domains, to line up in the direction of the field established by the coil. Essentially the same effect is produced in a soft iron bar when it is under the influence of a permanent magnet.

The magnetomotive force resulting from the current flow in the coil does not increase the magnetism that is inherent in the iron core; it merely reorientates the "atomic" magnets that were present before the magnetizing force was applied. If substantial numbers of the tiny magnets are orientated in the same direction, the core is said to be magnetized.

When soft iron is used, most of the atomic magnets return to what amounts to a miscellaneous orientation upon removal of the magnetizing current and the iron is said to be demagnetized. If hard steel is used, more of them will remain in alignment with the direction of the flux produced by the flow of current through the coil and the metal is said to be a permanent magnet. Soft iron and other magnetic materials having high permeability and low retentivity are generally used in electromagnets.

Figure 10-15 shows another interesting effect of the iron core. In Fig. 10-15A, notice that the lines of force passing through the coil are confined to the iron core. If the iron core is pulled partially out of the coil, as shown in Fig. 10-15B, the magnetic lines of force will be extended to enter the end of the iron core that is outside the coil. Once the lines have established themselves in the core, they tend to shorten, thereby exerting a force on the core. This force tends to pull the core until its center coincides with the center of the coil, as shown by the dashed lines in Fig. 10-15B. This action has many practical applications in the various types of electrical/electronic controlling devices used in industry and in the home.

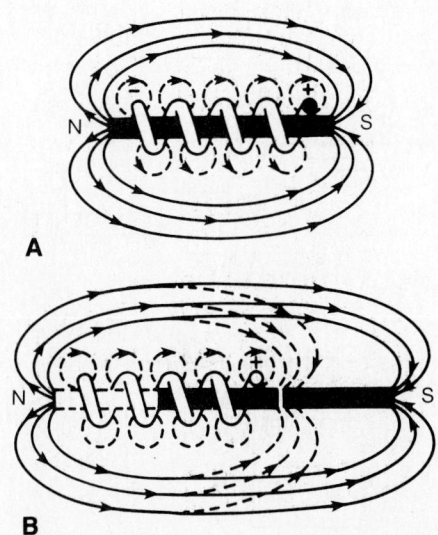

Fig. 10-15: **Effect of an iron core on the magnetic field of a coil.**

ELECTROMAGNETIC INDUCTION

One of the remarkable features of magnetism (from the electronic viewpoint) is its ability to *induce* electric power into certain circuits. If, for instance, the units shown in Fig. 10-16A are set up, electromagnetic induction can be demonstrated. Here, a zero-center sensitive meter is employed; attached to it is a loop of wire to form a closed circuit. When this wire is passed through the magnetic fields at the end of a magnet, a current will flow in the circuit and cause a needle deflection to the left as shown. This current is referred to as *induced current*.

The principle of electromagnetic induction was discovered by Faraday at the beginning of the last century. He demonstrated that when a conductor is moved through a magnetic field, a voltage differential is established between the ends of the con-

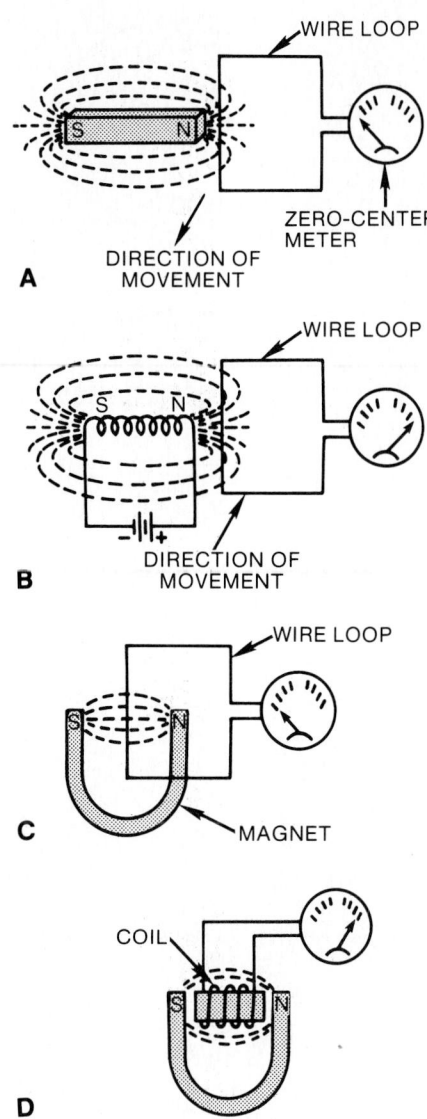

A

DIRECTION OF MOVEMENT

WIRE LOOP

ZERO-CENTER METER

B

WIRE LOOP

DIRECTION OF MOVEMENT

C

WIRE LOOP

MAGNET

D

COIL

Fig. 10-16: Magnetic induction.

ductor and an electromotive force is induced into the latter. The induced current flow and induced voltage exist only while the conductor is in motion with respect to the magnetic field. For this reason, the conductor can be kept stationary and the magnetic field moved to cut through the conductor. With the wire loop moving in one direction through the magnetic field, the needle deflection will be to the left; for a reverse movement, the meter needle will deflect to the right.

For inducing a voltage into another circuit, an electromagnet can be employed instead of the bar magnet initially shown in Fig. 10-16A. The electromagnetic induction is shown in Fig. 10-16B. Here, a battery is connected to a coil (shown schematically) so that magnetic lines are formed around the coil, as shown. Again, a movement through the magnetic lines of force will cause an induced current to flow, which is indicated by the meter indicator movement. A greater concentration of magnetic lines of force is achieved when a horseshoe-type magnet is employed, as shown in Fig. 10-16C. Here the flux lines of force are highly concentrated between the two pole pieces, and the meter needle is deflected more when the conductor passes through the electromagnetic fields. A much greater induced current can be realized by utilizing the method shown in Fig. 10-16D. Here a coil is placed within the concentrated magnetic lines of force. When the coil is rotated, many conductor sections cut magnetic lines of force to produce a greater induced current. This is the principle employed for the generation of electric potentials, which will be explained more fully in Chapter 11.

By use of the coil, a bar magnet can be employed, as shown in Fig. 10-17A. When the north pole end of the bar magnet is inserted into the coil, as shown, the meter needle will deflect to the left; if the bar magnet is removed, the needle deflects in the opposite direction. A needle deflection in the opposite direction would also be accomplished by pushing the bar magnet through the coil so that it emerges from the other end as shown in Fig. 10-17B. If the bar magnet is stopped in its movement through the inductor, the meter will drop to its center zero setting. The relationship between the magnetic lines of force

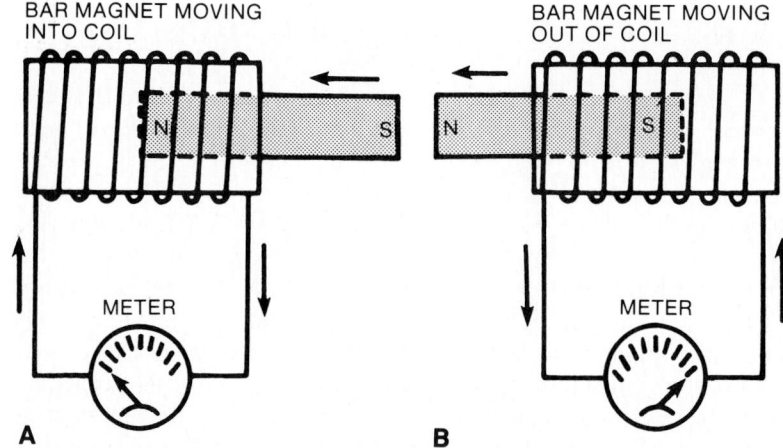

BAR MAGNET MOVING INTO COIL

BAR MAGNET MOVING OUT OF COIL

METER

METER

A

B

Fig. 10-17: Magnetic induction with stationary coil.

surrounding the conductor and those of the magnetic field that induces the voltage were first verified by Heinrich Lenz, who lived in the first half of the last century. This German physicist established a law that bears his name, *Lenz's Law: An induced current set up by the relative motion of a conductor and a magnetic field always flows in such a direction as to form a magnetic field that opposes the motion.*

As shown in Fig. 10-17, the direction in which the needle of the meter moves indicates the electron-flow direction, which is shown by the arrows. From the left-hand rule, we ascertain that the electromagnet formed by the coil has its north pole facing the north pole of the bar magnet. Since two like poles repel each other, the repulsion that exists between the two poles must be overcome by the torque of the magnet movement. Thus, by our motion in pushing the magnet through the coil we provide the necessary mechanical energy. Once the magnet is within the coil, there will no longer be an induced current and the meter needle will drop to zero. If the bar magnet is pushed through the coil, its magnetic lines of force, again, cut across the conductor wires while the field is in motion and an induced current again flows. The current reverses direction after the magnet has been pushed through the coil. If the magnet had been inserted and then pulled out again, without passing through the coil, the left end of the coil would become a south pole and thus have an attraction for the north pole of the magnet. This holding characteristic must also be overcome by mechanical torque energy manually applied to the magnet as it is withdrawn.

A magnet that has a stronger magnetic field will make it possible for a greater number of lines of force to be cut within a certain time interval. Thus, the magnitude of the induced voltage is proportional to the number of magnetic lines of force that are cut by a conductor per second. If the number of turns of wire in a coil is increased, the induced voltage will be greater. Also, if we move the bar through the coil at a greater speed, we will increase voltage. The rate at which the magnetic flux is changed thus has a direct bearing on the production of induced electromotive force.

In the SI system, the following equation applies for solving induced voltage (v):

$$v = \frac{N d\Phi}{dt}$$

where

      N = number of turns

     $d\Phi$ = the change in the number of flux lines (in webers) cutting N

     dt = the change in time (in seconds)

Thus, the amount of induced voltage is determined by the following three factors:

1. *Amount of flux.* The stronger the magnetic field, the greater the induced voltage.

2. *Number of turns.* The greater the number of conductors cutting the magnetic field, the greater the induced voltage.

3. *Time rate of cutting.* The greater the speed of relative motion between the magnetic field and conductors, the greater the induced voltage.

It has been found that if a conductor cuts across 10^8 lines of force in a second, a voltage of 1V will be induced between its ends. The polarity of the induced voltage is determined by the direction of current flow. The current flow can be determined using the method established by the British scientist, John A. Fleming. This method is commonly referred to as Fleming's right-hand rule (Fig. 10-18). Assume that the conductor is moving down between the poles of the magnet. Extend the thumb, the forefinger, and the middle finger of the right hand so that they are at right angles to one another. Let the thumb point in the direction in which the conductor is moving (up). Now let the forefinger point in the direction of the magnetic lines of force (recall that it is assumed they go from the north to the south pole). The middle finger will point in the direction of the induced electromotive force (namely toward the positive terminal).

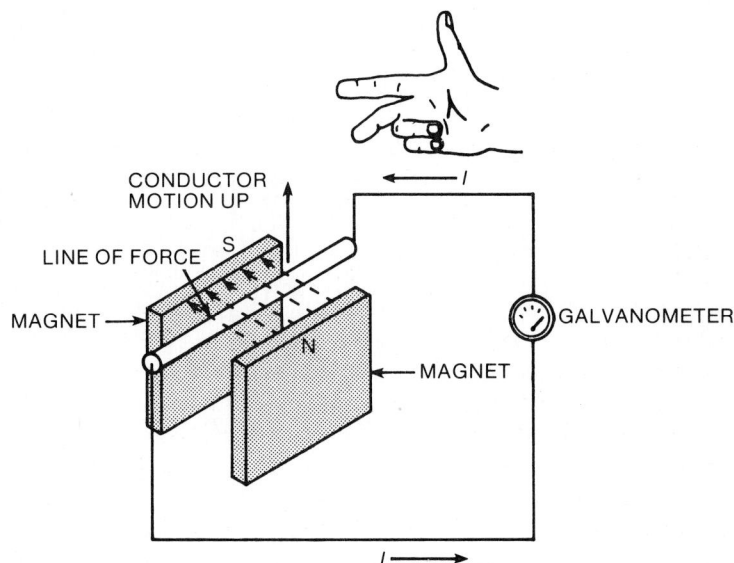

Fig. 10-18: Right-hand rule for determining the direction of induced current flowing in a conductor.

DC GENERATORS

Mechanical devices that utilize the principles of electromagnetic induction for producing DC (or AC as described in Chapter 11) are known as *generators*. The basic components that make up a DC generator are shown in Fig. 10-19. Here, a framework composed of laminated iron sheets or other ferromagnetic metal has a coil wound on it to form an electromagnet. When current flows through this coil, magnetic fields are created between the pole pieces as shown. Permanent magnets could also be employed instead of the electromagnet.

To simplify the initial explanation, a single wire loop is shown between the south and north pole pieces. When this wire loop is turned within the magnetic fields, it cuts the lines of

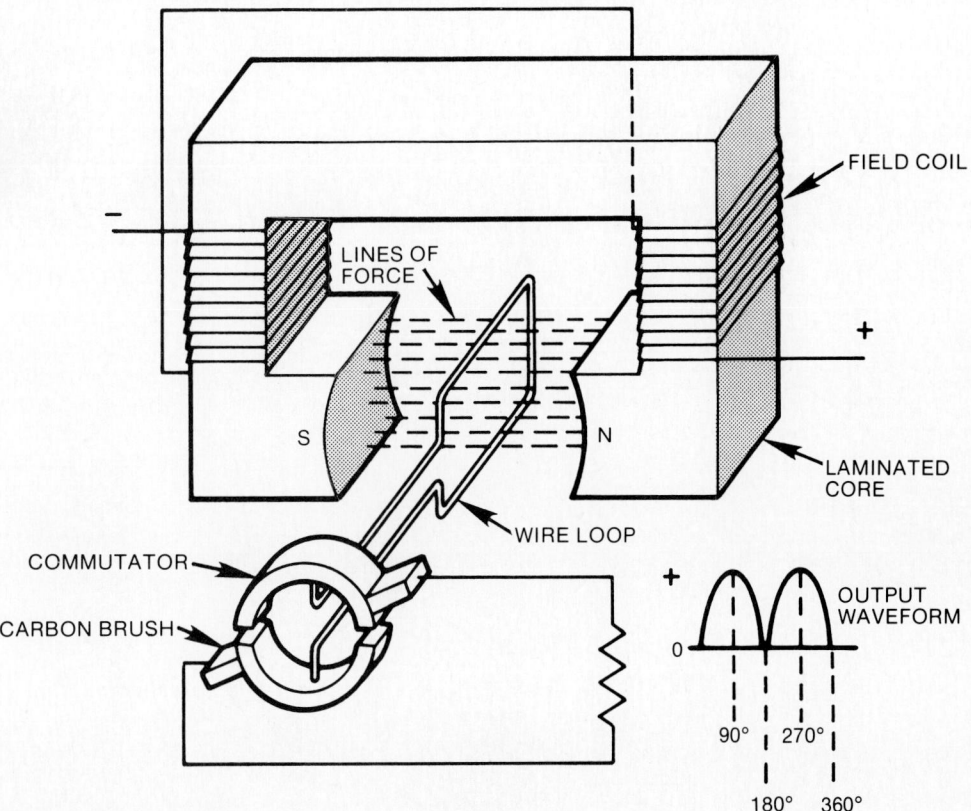

Fig. 10-19: Basic DC generator.

force and a voltage is induced. If there is a complete circuit from the wire loop, current will flow. In Fig. 10-19, the wire loop is connected to a split ring known as a *commutator,* and carbon brushes pick off the electric energy as the commutator rotates. Connecting wires from the carbon brushes transfer the energy to the load circuit.

When the wire loop makes a half-turn, the energy generated rises to a maximum level and drops to zero, as shown in Fig. 10-19. As the wire loop completes a full rotation, the induced voltage would reverse itself and the current would flow in the opposite direction then for the initial half-turn. To provide for an output having a single polarity, the split-ring commutator is used. Thus, for the second half-turn, the carbon brushes engage commutator segments opposite to those over which they slid for the first half-turn. The output waveform is not a steady-level DC, but rises and falls to form a pattern referred to as *pulsating* DC. Thus, for a complete 360° turn of the wire loop, two waveforms are produced, as shown in Fig. 10-20. (Chapter 11 gives a step-by-step explanation of the loop rotation for the production of AC.)

In practical generators, an increased output and a smoother DC is produced by using a number of coils instead of a single wire loop. The individual coils are wound around a central core to form what is known as an *armature.* Instead of a two-segment commutator, a multisection commutator is used.

181

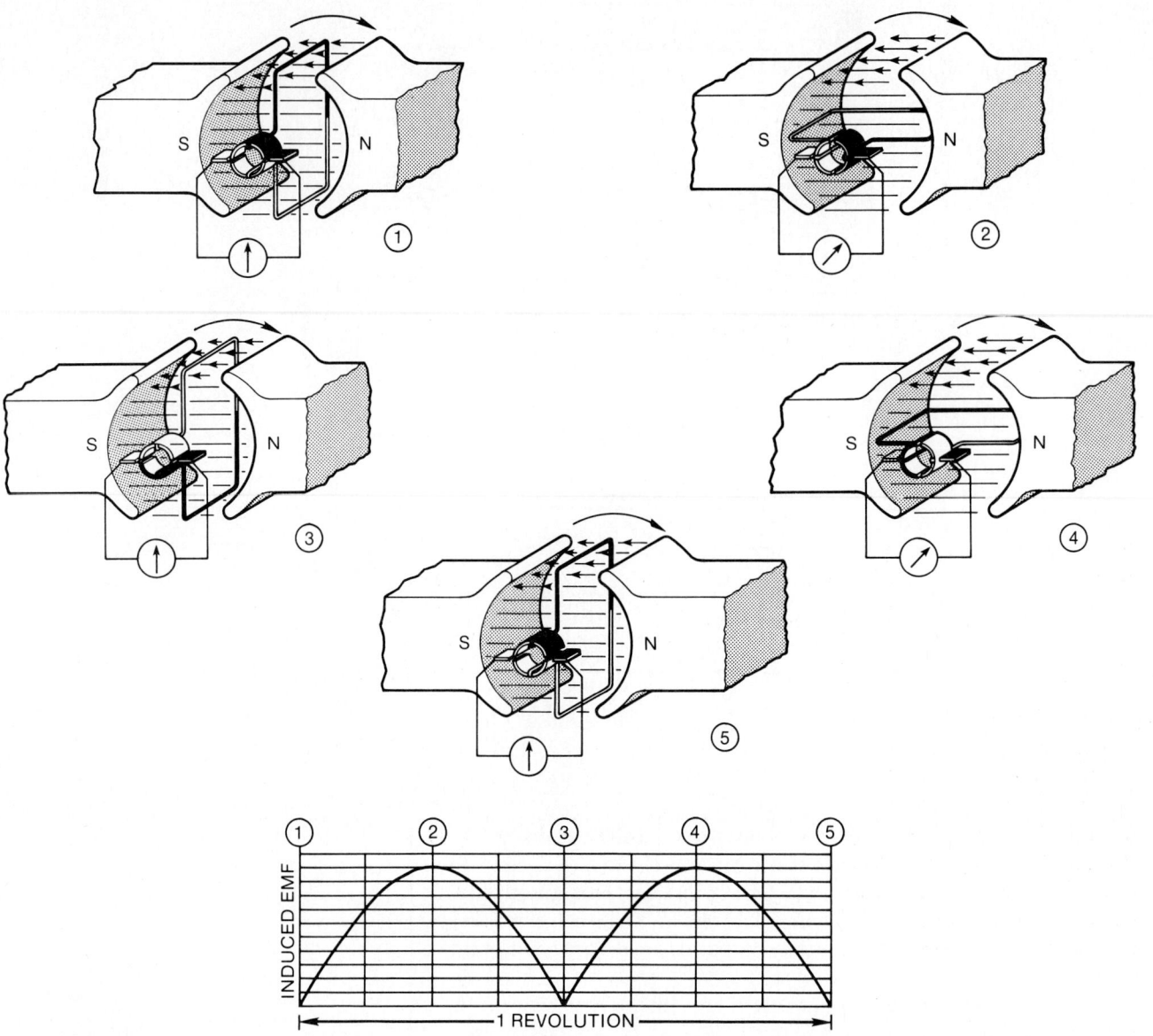

Fig. 10-20: Loop of wire connected to a split ring, rotating in a magnetic field.

There are as many segments to the commutator as there are coils, and each coil is connected to two segments, as shown in Fig. 10-21. Note that all the coils are connected in *series,* with the carbon brushes connected to opposite sides of the commutator. Thus, they provide a plus and minus output, as was the case with the split-ring commutator shown in Fig. 10-20. As the armature rotates, the brushes slide over the commutator segments and pick off the energy generated. The brushes, as shown in Fig. 10-21, are placed vertically in relation to the north and south pole pieces, so that the top and bottom coils are in a neutral magnetic plane. This is necessary to prevent a current short-circuit across the coils while the brushes engage two segments at one time, as shown in Fig. 10-21. With the coils in the neutral plane, there is no induced voltage and hence no current flows through the coil and short-circuited segments.

182

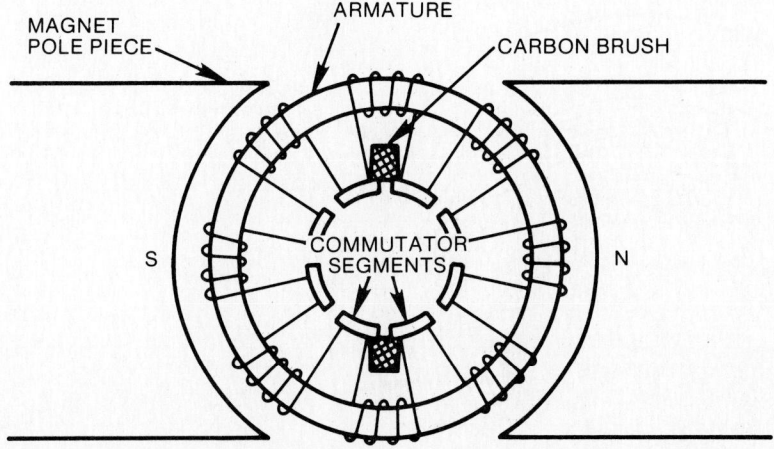

Fig. 10-21: Multicoil generator.

In the multicoil generator, the DC produced is smoother than the pulsating DC shown earlier. As more and more coils are employed, the pulsating DC from the individual coils combines, as shown in Fig. 10-22. Note that as the peak DC from one coil starts its decline, the DC peak is reached by the next coil. The addition of the various voltages existing at any moment produces the resultant shown above the waveforms in Fig. 10-22.

In commercial generators, the coils are not wound around the armature frame, but are actually wired into slots in the armature, as shown in greater detail in the cross-sectional view in Fig. 10-23. Here, the generator has four poles instead of the two shown in Fig. 10-21. With four (or more) poles, the magnetic fields are more uniform around the periphery of the armature. Hence, the four-pole generator has a lower internal resistance than the two-pole generator and is generally more efficient. To simplify our drawing, the carbon brushes and commutator segments are not shown. A generator having four poles can be operated at only half the rotation speed of the two-pole generator and still produce the same output voltage. As mentioned earlier, the amount of voltage induced and generated is dependent on the rate at which the armature coils cut through the lines of force. With a four-pole generator, each wire of the armature coil cuts through four sets of lines of force for each complete armature turn. Since the increase in the number of poles makes for lower speeds, many practical generators employ a dozen or more poles, plus as many brushes. Usually, each brush is lo-

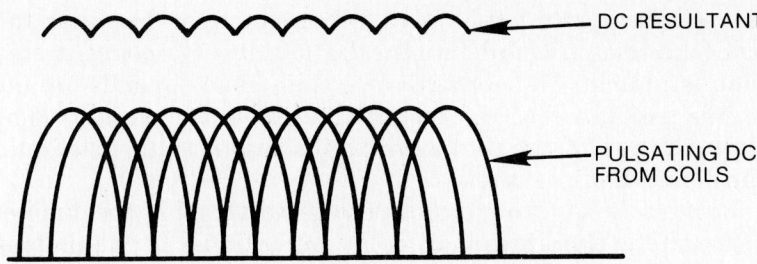

Fig. 10-22: Output from multicoil DC generator.

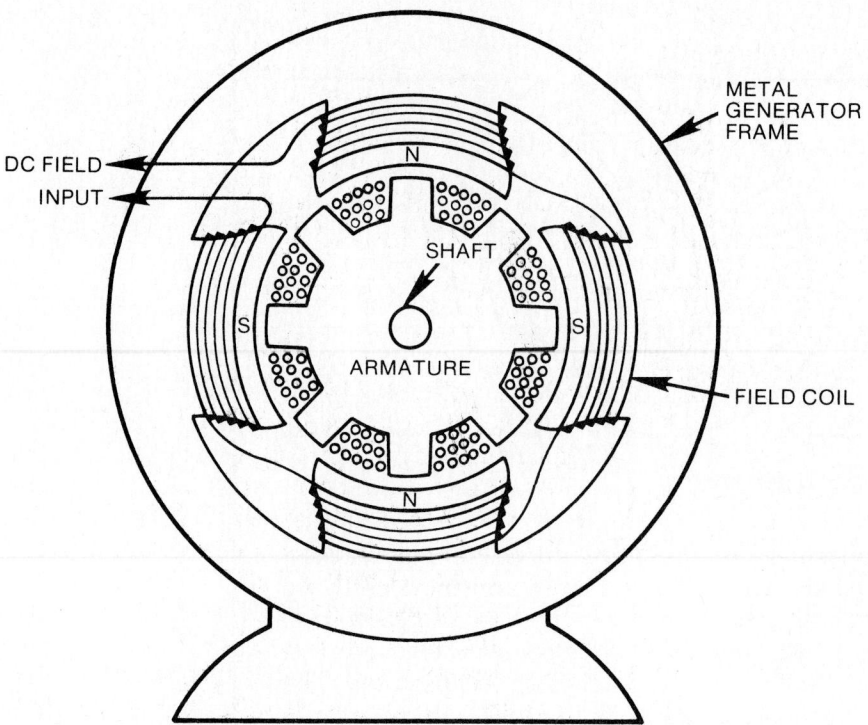

Fig. 10-23: Four-pole generator.

cated halfway between two poles. Half of the brushes are connected together to form the positive output potential, and the other half are connected together to form the negative output potential.

As shown in Fig. 10-23, the field coils are wound around the pole pieces in series. When current flows through the field coil, a high degree of magnetic density is established between the pole pieces adjacent to each other. The metal generator frame has sufficient permeance to carry the magnetic lines of force around and through the pole pieces and across the armature area for a complete magnetic circuit.

SUMMARY

An electrical current flowing through a conductor will produce a magnetic field around that conductor. The direction of the magnetic lines of force around a conductor may be determined by using the left-hand rule for a conductor. The left-hand rule for coils is used to determine the polarity of its concentrated magnetic field. The strength of a coil's field depends on: the number of turns of conductor; the amount of current flow through the coil; the ratio of the coil's length to its width; and the material of the core.

Rowland's Law for magnetic circuits states that the number of magnetic flux lines is directly proportional to the magnetomotive force and inversely proportional to the reluctance of the circuit. The practical unit for magnetomotive force is the ampere-turn. The number of flux is described in webers.

The relationship between flux density and field intensity can be shown on a β-H magnetization curve.

Hysteresis is the lag of magnetization behind the force that produces it. It is caused by molecular friction.

Demagnetization, or degaussing, can be achieved by applying a rapidly changing magnetic field to a magnet.

When a wire is passed through the magnetic field of a magnet, a current will flow through the wire. This is called electromagnetic induction. The magnitude of the induced current is determined by the amount of flux, number of turns, and the time rate at which the flux lines are cut.

Mechanical devices utilizing the principles of electromagnetic induction to produce electricity are known as generators.

CHECK YOURSELF (Answers at back of book)

Questions

Fill in the blanks with the appropriate word or words.

1. _____ is by far the most commonly used means for producing electricity.
2. The amount of induced voltage is determined by three factors: _____, _____, and _____ .
3. Reluctance is the opposition offered by a material to magnetic _____ .
4. The _____ is the unit measurement for magnetomotive force.
5. A generator having _____ poles can be operated at half the rotation speed of a _____ generator and still produce the same voltage.
6. The strength of an electromagnetic field depends on four factors: _____, _____, _____, and _____ .
7. _____ is the property of a magnetic substance that causes the magnetization to lag behind the force that produces it.
8. _____ is another term for demagnetizing.
9. Rowland's Law states that the number of flux lines is directly proportional to the _____ and inversely proportional to the _____ offered by the circuit.
10. _____ is a measure of how well a given substance permits the setting up of magnetic lines of force.
11. As the magnetizing force (H) is increased, the flux density increases linearly up to a point called _____ .
12. The _____ of soft iron is many times that of air.
13. High _____ and low _____ are usually desired for electromagnets.
14. The direction of magnetic lines of force around a current-carrying conductor can be determined by the _____ .
15. _____ of a magnetic field is the number of magnetic lines of force in a given area.
16. The number of magnetic flux is measured in _____ .
17. Flux density is measured in _____ .

18. _____ is a measure of the relative ability of a substance to conduct magnetic lines of force as compared with air.
19. _____ has a relatively high level of retentivity compared to soft iron.
20. The output waveform of a generator is a _____ .

Problems

1. If a coil has 1,000 turns and 500mA of current flowing through it, what is the magnetomotive force of the coil?
2. What is the reluctance of a magnetic circuit in which a total flux of 2×10^{-3}Wb is set up by a 6A current flowing through a coil consisting of 300 turns of wire?
3. Determine the magnetic field strength of a coil that has 200 turns and 200mA flowing through it. The coil is 10cm long.
4. The amount of flux in a 75 turn coil changes by 471μWb in 0.1s. What is the amount of voltage induced on the coil?
5. A coil, 5cm in length, has a magnetomotive force of 760 ampere-turns. Determine the magnetic field strength.

CHAPTER 11

AC Voltage and Current

The first faltering steps of scientific achievement in the field of electricity were performed with crude, and for the most part, homemade apparatus. Great men, such as George Simon Ohm, had to fabricate nearly all the laboratory equipment used in their experiments. The only convenient source of electrical energy available to these early scientists was the voltaic cell, invented some years earlier. Due to the fact that cells and batteries were the only sources of power available, some of the early electrical devices were designed to operate from direct current (DC).

When the use of electricity became widespread, certain disadvantages in the use of direct current became apparent. In a direct-current system the supply voltage must be generated at the level required by the load. To operate a 240V lamp for example, the generator must deliver 240V. A 120V lamp could not be operated from this generator by any convenient means. A resistor could be placed in series with the 120V lamp to drop the extra 120V, but the resistor would waste an amount of power equal to that consumed by the lamp.

Another disadvantage of direct-current systems is the large amount of power lost due to the resistance of the transmission wires used to carry current from the generating station to the consumer. This loss could be greatly reduced by operating the transmission line at very high voltage and low current. This is not a practical solution in a DC system, however, since the load would also have to operate at high voltage. As a result of the difficulties encountered with direct current, practically all modern power distribution systems use a type of current known as alternating current (AC). In an alternating-current system, the current flows first in one direction, then reverses and flows in the opposite direction.

As far as electronics is concerned, a major disadvantage is that direct-current power cannot be radiated from an antenna. The whole art of radio communication as it is known today depends in large measure on the capability of AC to radiate its energy into space from an antenna and to project itself for long distances across the surface of the earth.

Because of many uses and advantages, AC gradually became recognized as a more suitable and versatile source of

187

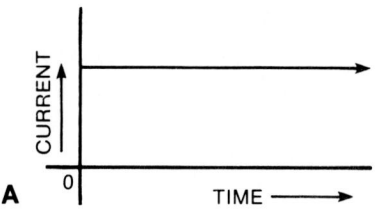

A

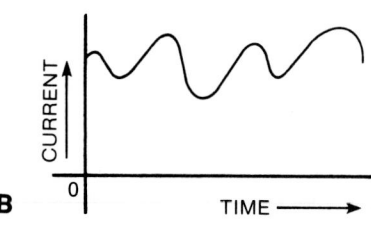

B

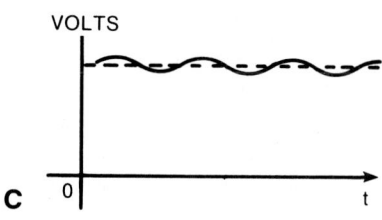

C

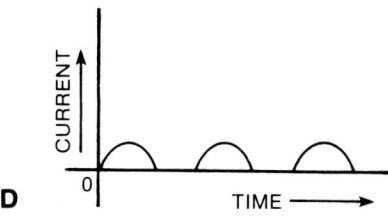

D

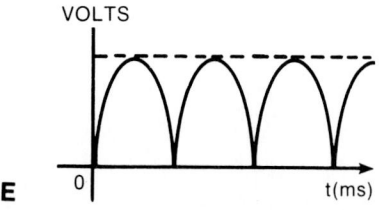

E

Fig. 11-1: Types of DC: (A) pure DC; (B) varying DC; (C, D, and E) pulsating DC. The latter is often the output of a DC generator.

power than direct current, particularly since alternating current has the overall advantage of being converted with relative ease to direct current. On the other hand, the conversion of DC to AC, though possible, is often neither convenient nor easy. This does not mean that DC is outmoded or useless; there are many situations in which direct current is the proper if not the only source of power.

Waveforms. As has already been discussed, a direct current is a current which flows in one direction only. Pure DC has a constant magnitude (value) with respect to time (Fig. 11-1A). Varying DC varies in magnitude with respect to time (Fig. 11-1B). When such variations occur at regular intervals, the DC is called a "pulsating" direct current (Figs. 11-1C, 11-1D and 11-1E). As shown in these illustrations, when these changes in voltage amplitude with respect to time are plotted on graph paper or viewed on an oscilloscope, the pattern is known as a *waveform*.

Alternating current, as mentioned in Chapter 2, is an electric current which moves first in one direction for a fixed period of time and then in the opposite direction for an equal period of time, constantly changing in magnitude. From a zero value, alternating current builds up to a maximum in a positive direction, then falls off to zero value again before building up to a maximum in the opposite or negative direction, and then finally returning to zero. For this reason, alternating current may be further defined as a current which is constantly changing in magnitude (either building up or falling off) and periodically (at set intervals of time) changing direction. The shape of such alternating flow is that of a wave.

The most common and most important alternating current wave shape is the sine wave, so named after the trigonometric function whose graphic pattern it follows. However, due to the presence of electrical noise (random voltages generated within an electric circuit), sine waves are often distorted. Consequently, many different kinds of alternating wave shapes can occur; these alternating wave shapes are referred to as complex waves. Figure 11-2 illustrates a sine wave and two types of complex waves.

In each instance, one cycle or complete pattern from positive to negative is shown (Fig. 11-2). Assuming that each wave shape starts at the same instant of time, it can be seen that the wave shape of Fig. 11-2C completes its cycle in half the time of either Figs. 11-2A or 11-2B; therefore, the frequency of Fig. 11-2C is twice that of Figs. 11-2A and 11-2B.

It is apparent, then, that the proper recognition and definition of an alternating current depends upon:
1. The wave shape of one cycle.
2. The value at some specified point in the cycle.
3. The length of time to complete one cycle (the period).

BASIC AC GENERATOR

An AC generator converts mechanical energy into electrical energy. It does this by utilizing the principle of electromagnetic induction. In the study of magnetism, it was shown that a current-carrying conductor produces a magnetic field around itself. It is also true that a changing magnetic field may produce an emf in a conductor. If a conductor lies in a magnetic field and either the field or conductor moves, an emf is induced in the conductor. This effect is called electromagnetic induction.

Cycle

Figure 11-3 shows a suspended loop of wire (conductor) being rotated (moved) in a counterclockwise direction through the magnetic field between the poles of a permanent magnet. For ease of explanation, the loop has been divided into a dark and a light half (Fig. 11-4). Notice that in Fig. 11-4A, the dark half is moving along (parallel to) the lines of force. Consequently, it is cutting none of these lines. The same is true of the light half, moving in the opposite direction. Since the conductors are cutting no lines of force, no emf is induced. As the loop rotates toward the position shown in Fig. 11-4B, it cuts more

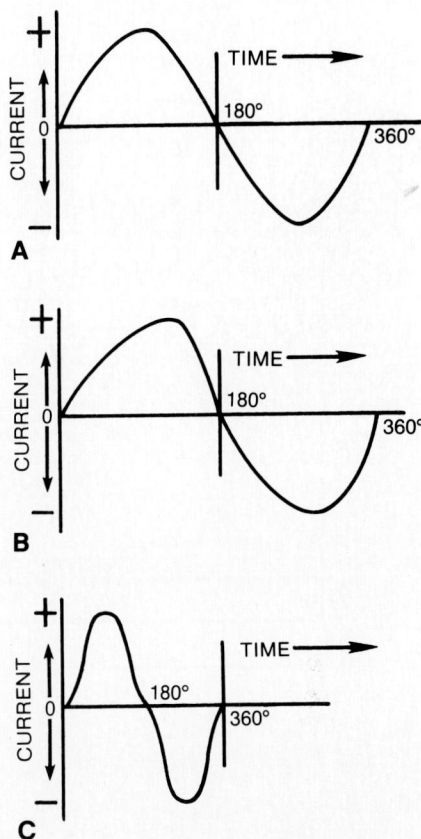

Fig. 11-2: AC waveshapes: (A) sine wave—one complete cycle; (B) complex wave—one complete cycle; and (C) complex wave—one-half the period of A and B.

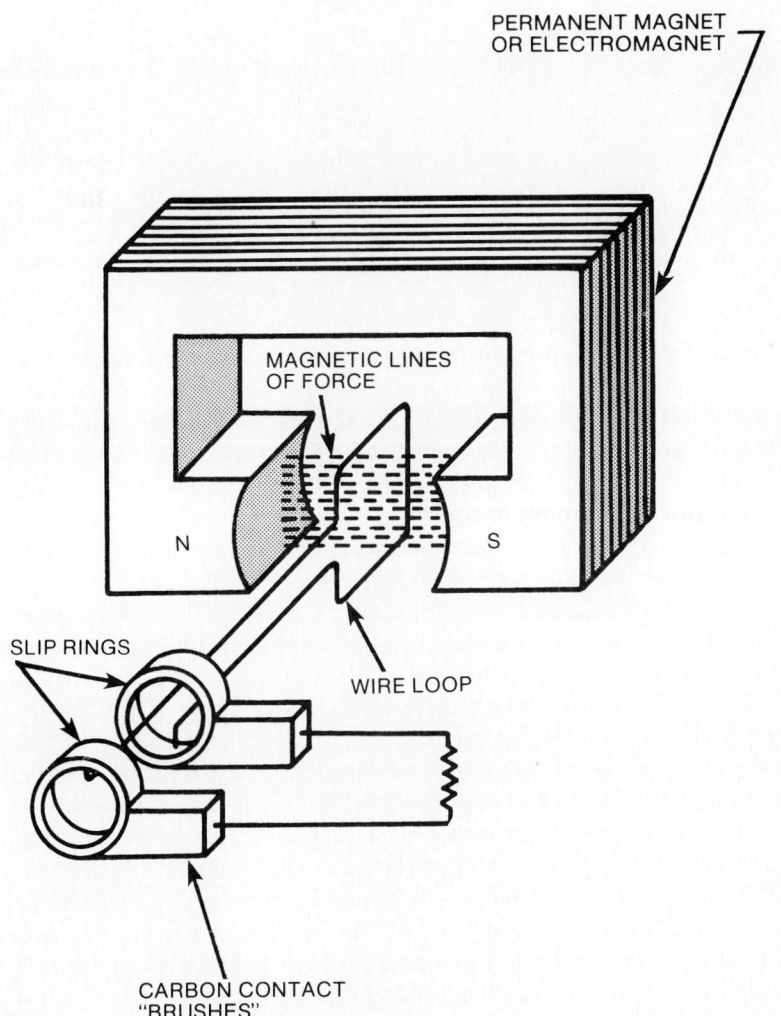

Fig. 11-3: Basic AC generator.

189

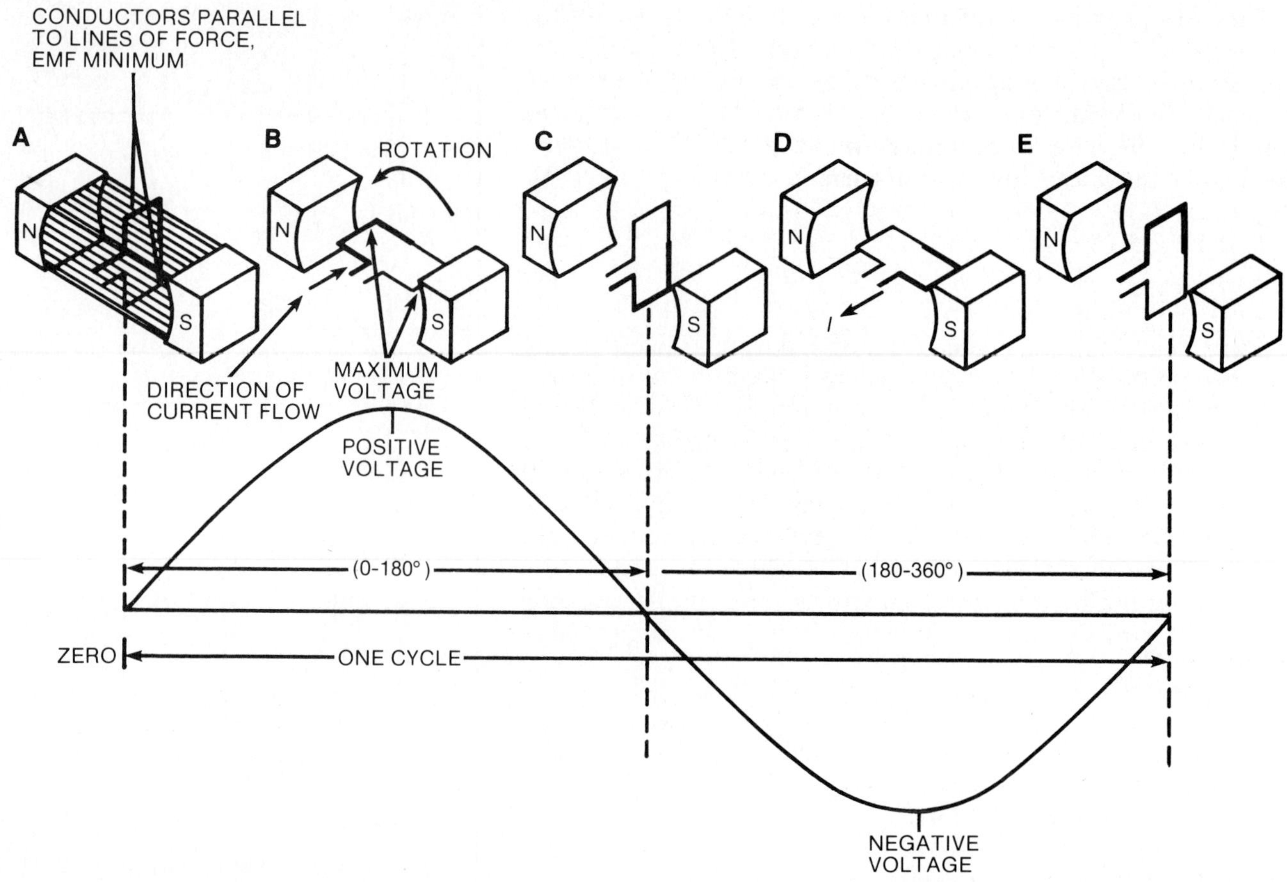

CONDUCTORS PARALLEL
TO LINES OF FORCE,
EMF MINIMUM

A   B   ROTATION   C   D   E

N   N   N   N   N

S   S   S   S   S

DIRECTION OF
CURRENT FLOW

MAXIMUM
VOLTAGE

POSITIVE
VOLTAGE

(0-180°)   (180-360°)

ZERO   ONE CYCLE

NEGATIVE
VOLTAGE

Fig. 11-4: Basic AC generator in action.

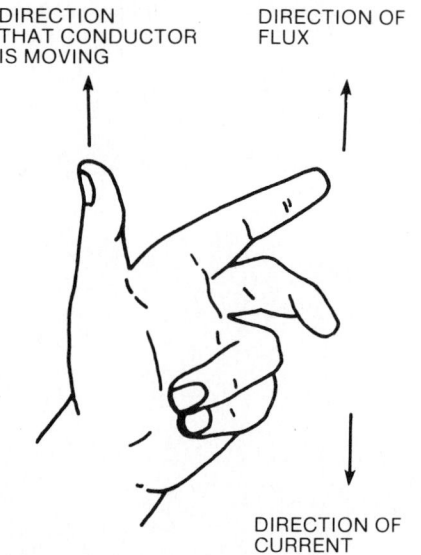

DIRECTION
THAT CONDUCTOR
IS MOVING

DIRECTION OF
FLUX

DIRECTION OF
CURRENT

Fig. 11-5: Determining the direction of current using the left-hand rule for generators.

and more lines of force per second because it is cutting more directly across the field (lines of force) as it approaches the position shown in Fig. 11-4B. At the position shown in Fig. 11-4B, the induced voltage is greatest because the conductor is cutting directly across the field.

As the loop continues to be rotated toward the position shown in Fig. 11-4C, it cuts fewer and fewer lines of force per second. The induced voltage decreases from its peak value. Eventually, the loop is once again moving in a plane parallel to the magnetic field, and no voltage (zero voltage) is induced. The loop has now been rotated through half a circle (one alternation, or 180°). The sine curve shown in the lower part of the figure shows the induced voltage at every instant of rotation of the loop. Notice that this curve contains 360°, or two alternations. Two alternations represent one complete cycle of rotation.

The direction of current flow during the rotation from Figs. 11-4A to 11-4C, when a closed path is provided across the ends of the conductor loop, can be determined by using the *left-hand rule for generators* (Fig. 11-5). The left-hand rule is applied as follows: Extend the left hand so that the thumb points in the direction of conductor movement and the forefinger points in the direction of magnetic flux (north to south). By pointing the middle finger 90° from the forefinger, it will point in the direction of current flow within the conductor.

Applying the left-hand rule to the dark half of the loop in Fig. 11-4B, the direction of current flow can be determined and is indicated. Similarly, direction of current flow through the light half of the loop can be determined. The two induced voltages add together to form one total emf. When the loop is further rotated to the position shown in Fig. 11-4D, the action is reversed. The dark half is moving up instead of down, and the light half is moving down instead of up. By applying the left-hand rule once again, it is readily apparent that the direction of the induced emf and its resulting current have reversed. The voltage builds up to maximum in this new direction, as shown by the sine wave tracing. The loop finally returns to its original position (Fig. 11-4E), at which point voltage is again zero. The wave of induced voltage has gone through one complete cycle.

If the loop is rotated at a steady rate and if the strength of the magnetic field is uniform, the number of hertz and the voltage will remain at fixed values. Continuous rotation will produce a series of sine wave voltage cycles or, in other words, an AC voltage. In this way mechanical energy is converted into electrical energy.

The rotating loop in Fig. 11-4 is called an *armature*. The armature may have any number of loops or coils.

Frequency

The frequency (*f*) of an alternating current or voltage is the number of complete cycles occurring in each second of time. Hence, the speed of rotation of the loop determines the frequency. For a single loop rotating in a two-pole field you can see that each time the loop makes one complete revolution, the current reverses direction twice. A single hertz will result if the loop makes one revolution each second. If it makes two revolutions per second, the output frequency will be 2Hz. In other words, the frequency of a two-pole generator happens to be the same as the number of revolutions per second. As the speed is increased, the frequency is increased.

If an alternating-current generator has four pole pieces, as in Fig. 11-6, every complete mechanical revolution of the armature will produce two AC cycles. When the dark half of the loop passes between poles S_1 and N_2, a voltage is induced which causes current to flow into the slip ring attached to the dark end of the loop.

When the dark half passes between N_2 and S_2, the induced voltage reverses direction. Another reversal occurs when it passes between S_2 and N_1. The voltage at the slip rings reverses direction *four times* during each revolution. In other words, two cycles of an AC voltage are generated for each mechanical revolution. If each revolution lasts 1 second, the frequency of the output is 2Hz. The more poles that are added, the higher the fre-

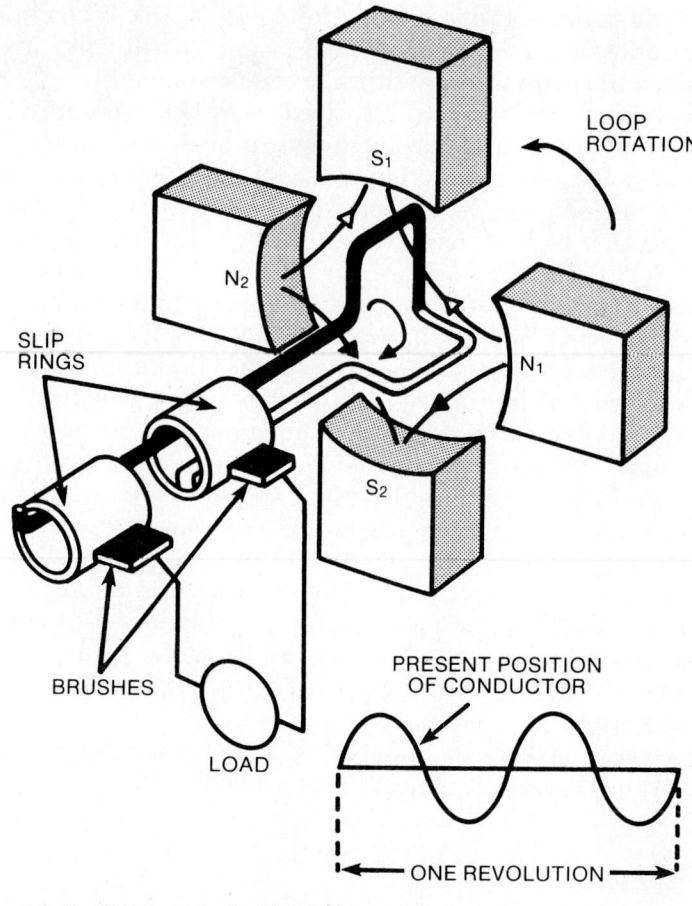

Fig. 11-6: Four-pole basic AC generator.

quency per revolution becomes. To find the output frequency of any AC generator, the following formula can be used:

$$f = \frac{P \times \text{rpm}}{120}$$

where f is frequency in Hz, rpm is revolutions per minute, and P is the number of poles.

A generator made to deliver 60Hz, having two field poles, would need an armature designed to rotate at 3,600rpm. If it had four field poles, it would need an armature designed to rotate at 1,800rpm. In either case, frequency would be the same. In actual practice, a generator designed for low-speed operation generally has a greater number of pole pieces, while a high-speed machine will have relatively fewer pole pieces, if both are to deliver power at the same frequency.

As mentioned earlier in the chapter, the more cycles per second, the higher the frequency and the less time for one cycle. That is, the changes in values are faster for higher frequencies. But, when comparing sine waves, remember that the amplitude has *no* relationship to the frequency. Two waveforms can have the same frequency with different amplitudes (Fig. 11-7), the same amplitude with different frequencies (Fig. 11-8), or different frequencies and amplitudes. Keep in

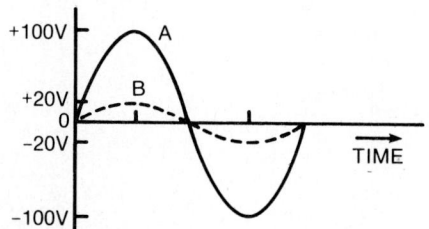

Fig. 11-7: Waveform A and B have different amplitudes, but both are sine waves.

192

mind that amplitude indicates how much the current or voltage is, while the frequency determines the time rate of change of the amplitude variations, in cycles per second.

The word *hertz* (Hz), as already noted, has been adopted internationally to designate the number of cycles per second. It was named for Heinrich R. Hertz, a German physicist of the latter 19th century. The common designations of Hz are as follows:

$$1Hz = 1 \text{ cycle per second(c/s)}$$
$$1kHz = 1 \text{ kilohertz} = 1,000c/s$$
$$1MHz = 1 \text{ megahertz} = 10^6c/s$$
$$1GHz = 1 \text{ gigahertz} = 10^9c/s$$
$$1THz = 1 \text{ terahertz} = 10^{12}c/s$$

The use of Hz as a universal term eliminates the many different translations of the cycles-per-second expression in the various languages.

Electrical signals are divided into two major groups: audio frequencies (AF) and radio frequencies (RF). Audio frequencies range from about 10Hz to about 20,000Hz. These frequencies may be heard by most human ears. Information carried by radio or tape recorders is produced in the AF range.

Electronic equipment operates on frequencies ranging from 20,000Hz through frequencies in the visible light range (Fig. 11-9). These frequencies are generally divided into eight bands as shown in Table 11-1. Full information on RF frequencies can be found in Appendix B.

Velocity. Sound waves, AC waves from power stations, or RF transmitted waves are all represented by the sine wave

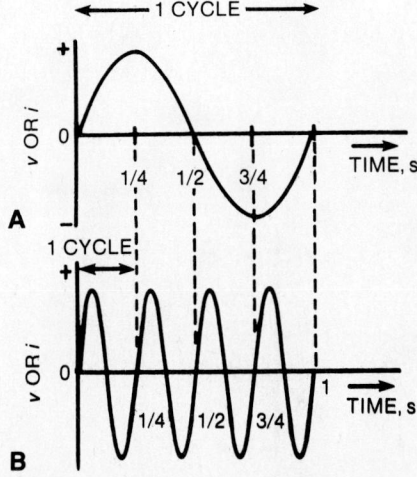

Fig. 11-8: Number of cycles per second (Hz) is the frequency (f). (A) f = 1Hz and (B) f = 4Hz.

Table 11-1: Bands of Frequencies

Abbreviation	Frequency Band	Frequency Range
VLF	Very low frequency	below 30kHz
LF	Low frequency	30–300kHz
MF	Medium frequency	300–3,000kHz
HF	High frequency	3,000–30,000kHz
VHF	Very high frequency	30–300MHz
UHF	Ultrahigh frequency	300–3,000MHz
SHF	Superhigh frequency	3,000–30,000MHz
EHF	Extremely high frequency	30–300GHz

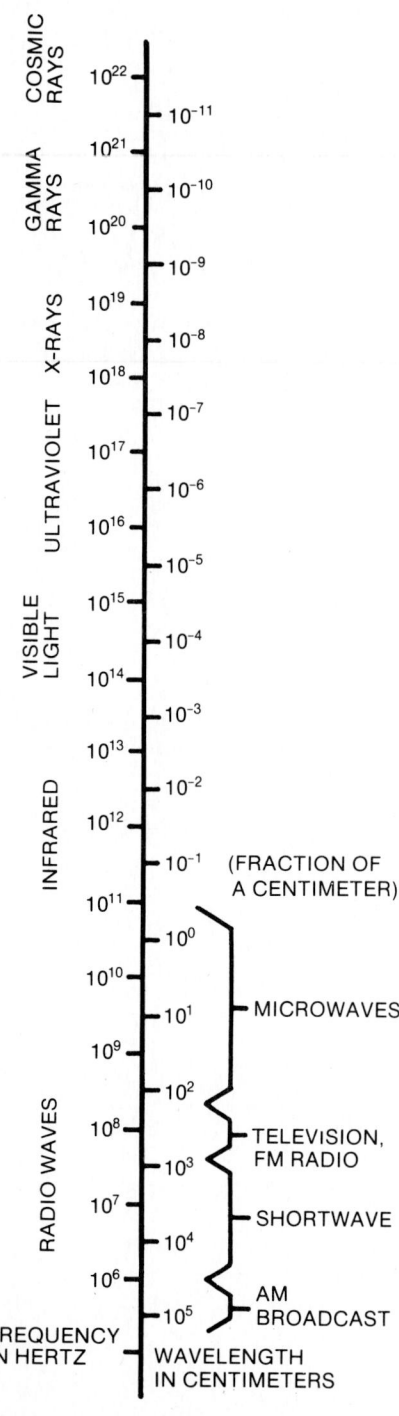

COSMIC RAYS	10^{22}
	10^{-11}
GAMMA RAYS	10^{21}
	10^{-10}
	10^{20}
	10^{-9}
X-RAYS	10^{19}
	10^{-8}
	10^{18}
	10^{-7}
ULTRAVIOLET	10^{17}
	10^{-6}
	10^{16}
	10^{-5}
VISIBLE LIGHT	10^{15}
	10^{-4}
	10^{14}
	10^{-3}
	10^{13}
	10^{-2}
INFRARED	10^{12}
	10^{-1} (FRACTION OF A CENTIMETER)
	10^{11}
	10^{0}
	10^{10}
	10^{1} — MICROWAVES
	10^{9}
	10^{2}
RADIO WAVES	10^{8} — TELEVISION, FM RADIO
	10^{3}
	10^{7} — SHORTWAVE
	10^{4}
	10^{6} — AM BROADCAST
	10^{5}
FREQUENCY IN HERTZ	WAVELENGTH IN CENTIMETERS

Fig. 11-9: Frequency spectrum.

type signal as shown in Fig. 11-10, where a positive and negative alternation make up a complete cycle. *Velocity* is the time rate of motional change of position in a specific direction. Thus, when radio frequency (RF) waves are propagated, they span a given distance in a time interval dependent on velocity factors. The word velocity is often used loosely as a synonym for speed, though actually the latter is more accurately defined as the time rate of change of position in a given direction.

The velocity of sound is determined by the medium (air, water, etc.) through which sound travels. The velocity is dependent on temperature as well. In air, the velocity of sound is approximately 1,100 feet per second (ft/s), or 750 miles per hour (mph). At 32°F, the velocity is 1,088ft/s. In the metric system the velocity at 0°C is 349.8 meters per second (m/s).

Electromagnetic radio waves traveling through space have the same velocity as light—299,792.5 kilometers per second (km/s), or 186,282 miles per second in a vacuum. Most commonly used formulas for light waves and radio waves use a rounded-off velocity figure (300,000,000m/s or 186,000 miles per second).

Wavelength. The distance spanned by one cycle of a propagated waveform in a given period is termed the *wavelength*. Hence, if a sound wave has a velocity of 331 meters per second, and one cycle is produced per second, the wavelength is 331 meters. Thus, for the single cycle shown in Fig. 11-10A, the wavelength, represented by the Greek *lambda* (λ), is the span between the beginning and end of the cycle and represents a distance of 331 meters. As the frequency (Hz) is increased, the wavelength shortens as shown in Fig. 11-10B. For a frequency of 3Hz, the wavelength would be one-third of 331, or 110 meters. Wavelength (λ), velocity (v), and frequency (f) are related as follows:

$$v = f\lambda$$

$$f = \frac{v}{\lambda}$$

$$\lambda = \frac{v}{f}$$

Example: Calculate the wavelength (λ) in centimeters for a radio wave with a frequency (f) of 92MHz. The velocity (v) of electromagnetic radio waves, in air or vacuum, is 3×10^{10}m/s

Solution:

$$\lambda = \frac{v}{f} = \frac{3 \times 10^{10}\text{m/s}}{92 \times 10^{6}} = \frac{3}{92} \times 10^{4}$$

$$\lambda = 0.0326 \times 10^{4} = 326\text{cm}$$

Example: What is the wavelength (λ) in inches for a radio wave with a frequency (f) of 62MHz? The velocity (v) of electromagnetic radio waves, in air or vacuum, is 186,000mi/s.

194

Solution:

$$\lambda = \frac{v}{f} = \frac{186 \times 10^3}{62 \times 10^6} = \frac{186}{62} \times 10^{-3}$$

$\lambda = 3 \times 10^{-3}$ mi or

$\lambda = 3 \times 10^{-3} \times 5{,}280^* = 15.84$ft

$\lambda = 15.84 \times 12 = 190$in.

*5,280ft = 1 mi

Example: Calculate the corresponding frequency for a wavelength of 100m.

Solution:

$$f = \frac{v}{\lambda} = \frac{3 \times 10^{10}\text{m/s}}{1 \times 10^2\text{m}} = 3 \times 10^8\text{Hz}$$

$f = 300 \times 10^6$Hz = 300MHz

Example: A radio wave has a wavelength of 314.14m with a corresponding frequency of 95.5MHz. What is its velocity?

Solution:

$v = f\lambda = 314.14$m $\times$ 95.5MHz (or 1/s)

$v = 3 \times 10^{10}$m/s

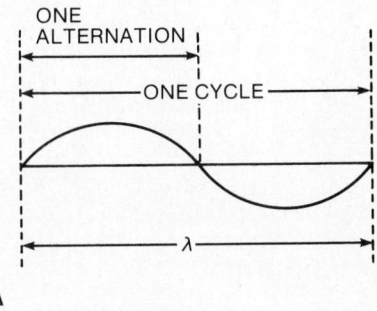

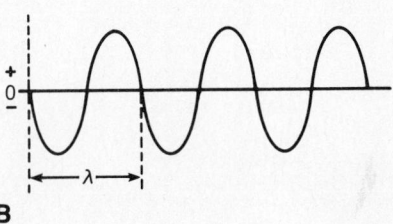

Fig. 11-10: Wavelength of sine wave signal.

Period

An individual cycle of any sine wave represents a finite amount of *time*. Figure 11-11 shows two cycles of a sine wave which has a frequency of 2Hz. Since two cycles occur each second, one cycle must require one-half second of time. The time required to complete one cycle of a waveform is called the *period* of the wave. In this example the period is one-half second.

Each cycle of the waveform in Fig. 11-11 is seen to consist of two pulse shaped variations in voltage. The pulse which occurs during the time the voltage is positive is called the *positive alternation*. The pulse which occurs during the time the voltage is negative is called the *negative alternation*. For a sine wave these two alternations will be identical in size and shape and opposite in polarity.

The period of a wave is inversely proportional to its frequency. Thus, the higher the frequency (greater number of Hz), the shorter the period (Table 11-2). In terms of an equation:

$$t = \frac{1}{f}$$

where t = period in the number of cycles occurring in 1s.

f = frequency in the time in seconds to make 1 cycle.

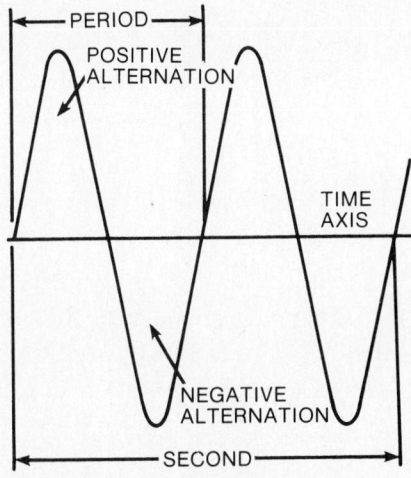

Fig. 11-11: Period of a sine wave.

195

Frequency	Period of 1 Cycle			
	ms(10^{-3})	μs(10^{-6})	ns(10^{-9})	ps(10^{-12})
25kHz	0.04	40	40,000	
25MHz		0.04	40	40,000
25GHz			0.04	40
25THz				0.04

Table 11-2: Frequency versus Time

Generated Voltage

Alternating-current generators are usually constant-potential machines because they are driven at constant speed and have fixed magnetic field strength for a given load. The effective voltage, V, generated by a single-winding (phase) AC generator is related to the total field strength per pole, Φ; the frequency, f; and the total number of active conductors, N, in the armature winding; as indicated in the following equation:

$$V = 2.22\Phi f N 10^{-8}$$

For example, if $\Phi = 2.5 \times 10^6$, $f = 60$Hz, and N = 96 conductors, the voltage generated is

$$V = 2.22 \times 2.5 \times 10^6 \times 60 \times 96 \times 10^{-8}$$
$$V = 320V$$

The length of active conductor extending under a pole does not appear in the equation directly because it is included in the factor of total magnetic flux per pole. The longer the active conductor, the more flux there will be for each pole, since the pole length and conductor length are assumed to be the same. For example, if an active conductor length is doubled, the pole length is doubled, the flux per pole is doubled, and the generated voltage is doubled.

ANALYSIS OF A SINE WAVE

An alternating current or voltage is one in which the magnitude and direction change periodically. The electron movement is first in one direction, then in the other. The variation is of sine waveform. Straight lines drawn to scale, called *phasors,* are used in solving problems involving sine wave currents and voltages.

A simple phasor is a straight line used to denote the magnitude and direction of a given quantity. Magnitude is denoted by the length of the line, drawn to scale, and direction is indicated by an arrow at one end of the line, together with the angle that the phasor makes with a horizontal reference phasor.

For example, if a certain point B (Fig. 11-12) lies 1 mile east of point A, the direction and distance from A to B can be shown as phasor e by using a scale of approximately 1/2″ = 1 mile.

Phasors may be rotated like the spokes of a wheel to generate angles. Positive rotation is counterclockwise and generates positive angles. Negative rotation is clockwise and generates negative angles.

The vertical projection (dotted line in Fig. 11-13) of a *rotating phasor* may be used to represent the voltage at any instant. Phasor V_m represents the maximum voltage inducted in a conductor rotating at uniform speed in a two-pole field (points 3 and 9). The phasor is rotated counterclockwise through one complete revolution (360°). The point of the phasor describes a circle. A line drawn from the point of the phasor perpendicular to the horizontal diameter of the circle is the vertical projection of the phasor.

The circle also describes the path of the conductor rotating in the bipolar field. The vertical projection of the phasor represents the voltage generated in the conductor at any instant corresponding to the position of the rotation phasor as indicated by angle θ. Angle θ represents selected instants at which the generated voltage is plotted. The sine curve plotted at the right of the figure represents successive values of the AC voltage induced in the conductor as it moves at uniform speed through the two-pole field because the instantaneous values of rotationally induced voltage are proportional to the sine of the angle θ that the rotating phasor makes with the horizontal.

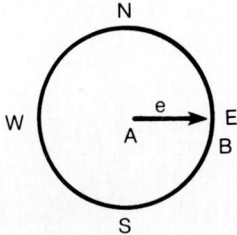

Fig. 11-12: Designating direction and distance by phasors.

Equation of Sine Wave Voltage

The sine curve is a graph of the equation

$$v = V_m \sin \theta,$$

where v is the instantaneous voltage, V_m is the maximum voltage, θ is the angle of the generator armature, and "sine" is one of the trigonometric functions shown in Fig. 11-13.

The instantaneous voltage, v, depends on the sine of the angle. It rises to a maximum positive value as the angle reaches 90°. This occurs because the conductor cuts directly across the flux at 90°. It falls to zero at 180°, because the conductor cuts no lines of flux at 180°. It reaches a negative peak at 270° and becomes zero again at 360°. For example, when $\theta = 60°$ and $V_m = 100V$, $v = 100 \times \sin 60° = 86.6V$. When $\theta = 240°$, $v = 100 \sin 240° = -86.6V$.

Another form of the trigonometric equation for a sine wave of voltage involves the angular velocity of its rotating phasor. The term "angular velocity" refers to the number of degrees (angles) through which a voltage phasor rotates per second. Angular

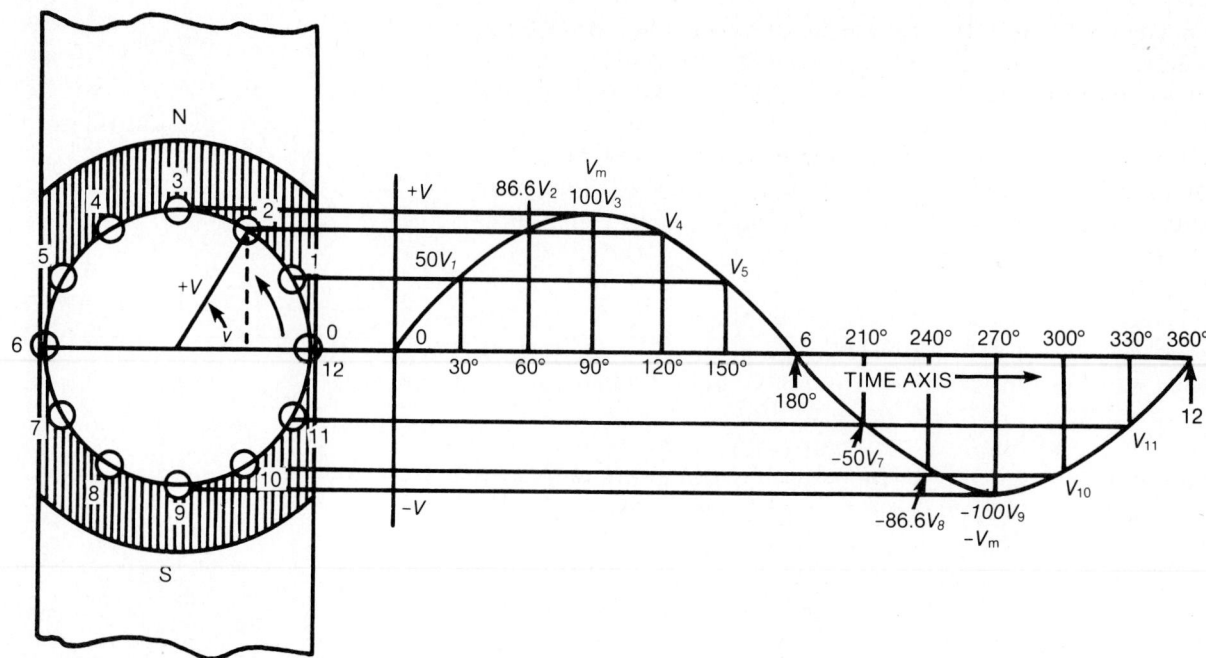

Fig. 11-13: Generation of sine wave voltage.

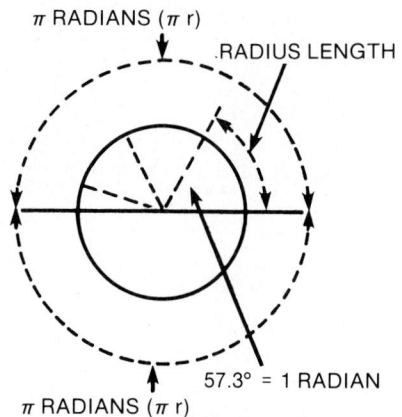

Fig. 11-14: Circle divided into radians. 2π radians = 360°.

velocity is symbolized by the Greek letter omega (ω). In practice, ω is generally given in terms of *radians per second*, rather than degrees per second. A radian is a segment of the circumference of a circle. This segment is always exactly equal in length to the *radius* of that circle. There are π (3.14) radians in half a circle, and 2π (6.28) radians in the circumference of a complete circle (Fig. 11-14). Therefore, when a voltage phasor makes one complete revolution, describing one complete circle, it traverses 2π or 6.28 radians. In terms of degrees of rotation, one radian is 360°/6.28 or 57.32°.

The number of radians per second traversed by an alternating-voltage phasor is closely related to its frequency, because either may be used to express the other. Since one phasor revolution equals one complete cycle, then each cycle equals 6.28 radians of phasor travel; that is, an alternating voltage whose frequency is 60Hz would be said to have an angular velocity of 6.28 × 60, or roughly 377 radians per second. Written as a formula involving angular velocity, the term $2\pi f$ may be replaced by the simpler symbol ω, if convenient. As previously stated, another form of the trigonometric equation for a sine wave of voltage involves the angular velocity of the generating phasor. The equation is:

$$v = V_m \times \sin \omega t.$$

It is used to determine the voltage of a rotating phasor at some given instant of time. The starting reference or "time zero," is usually when the voltage phasor is at zero. The equation time factor t is the elapsed time from time zero and is the exact instant at which the voltage is to be determined. To determine the exact angular position at any instant, multiply the angular velocity (ω) of the phasor by the time elapsed (t). By dividing elapsed time, t, by the period for one cycle and multiplying by 360°, the angular position of the voltage phasor may be determined for any instant. Multiplying V_m by the sine of that instantaneous angle will yield the instantaneous voltage v.

For example, consider a 60Hz voltage whose peak value is 100V. To determine the voltage 0.00139 second from the zero point, the equation would be:

$$\sin \omega t = \sin 2\pi ft = \sin 360(60)(0.00139) = \sin 30° = 0.5$$

The instantaneous voltage v is

$$v = V_m \times \sin 30° = 100 \times 0.5$$
$$v = 50\text{V}.$$

There are four important values associated with sine waves of voltage or current: instantaneous—designated as v or i; maximum—designated as V_m or I_m; average—designated as V_{avg} or I_{avg}; and effective—designated as V or I.

The *instantaneous value* may be any value between zero and maximum depending on the instant chosen, as indicated by the equation $v = V_m \sin \omega t$.

The *maximum value* of voltage is reached twice each cycle and is the greatest value of instantaneous voltage generated during each cycle.

The *ratio of the instantaneous value of voltage to the maximum value* is equal to the sine of the angle corresponding to that instant.

Peak Amplitude

One of the most frequently measured characteristics of a sine wave is its amplitude. Unlike DC measurement, the amount of alternating current or voltage present in a circuit can be measured in various ways. In one method of measurement, the maximum amplitude of either the positive or the negative alternation is measured. The value of current or voltage obtained is called the *peak voltage* or the *peak current*. To measure the peak value of current or voltage, an oscilloscope or a special meter (peak reading meter) must be used. The peak value of a sine wave is illustrated in Fig. 11-15.

Peak-to-Peak Amplitude

A second method of indicating the amplitude of a sine wave consists of determining the total voltage or current between the positive and negative peaks. This value of current or voltage is called the *peak-to-peak value* (Fig. 11-15). Since both alternations of a pure sine wave are identical, the peak-to-peak value is twice the peak value. Peak-to-peak voltage is usually measured with an oscilloscope, although some voltmeters have a special scale calibrated in peak-to-peak volts.

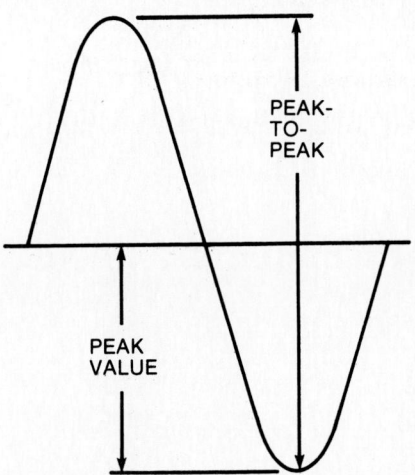

Fig. 11-15: Peak and peak-to-peak values.

199

Instantaneous Amplitude

The instantaneous value of a sine wave of voltage for any angle of rotation is expressed by the formula:

$$v = V_m \times \sin \theta$$

where

v = the instantaneous voltage

V_m = the maximum or peak voltage

$\sin \theta$ = the sine of angle at which v is desired.

Similarly, the equation for the instantaneous value of a sine wave of current would be:

$$i = I_m \times \sin \theta$$

where

i = the instantaneous current

I_m = the maximum or peak current

$\sin \theta$ = the sine of the angle at which i is desired.

Effective or RMS Value

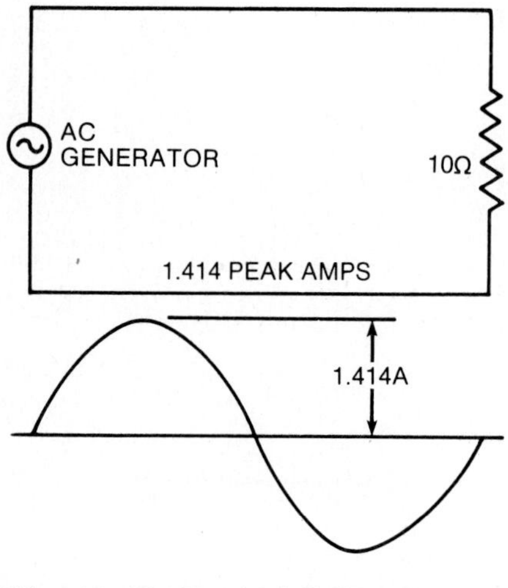

As the use of alternating current gained popularity, it became increasingly apparent that some common basis was needed on which AC and DC could be compared. A 100-watt light bulb, for example, should work just as well on 120V AC as it does on 120V DC. It can be seen, however, that a sine wave of voltage having a peak value of 120V would not supply the lamp with as much power as a steady value of 120V DC.

Since the power dissipated by the lamp is a result of current flow through the lamp, the problem resolves to one of finding a *mean* alternating current ampere which is equivalent to a steady ampere of direct current.

Figure 11-16 shows a circuit in which the peak alternating current through the 10Ω resistor is 1.414A. Since the current through the resistor is changing continuously, the power dissipated by the resistor will also vary. It will be minimum when the current is zero.

The variations in power throughout the cycle can best be analyzed by plotting a curve showing the instantaneous power at each point in the cycle. In the procedure to follow, the instantaneous current, the square of the instantaneous current, and the instantaneous power will be calculated in 10° steps for the first quarter of the cycle. These values are shown in Table 11-3.

Notice that at 0° the instantaneous current (i) is zero, causing the power dissipated by the resistor to be zero. At 10° the instantaneous current is 0.245A, the current squared is 0.060, and the power is 0.60W. At 90° the current has reached its maximum value of 1.414A, the square of the current is 2.000, and the power dissipated is 20.00W.

Fig. 11-16: Basic AC circuit.

Degrees	i	i^2	P
0°	.000	.000	.00
10°	.245	.060	.60
20°	.486	.236	2.36
30°	.707	.500	5.00
40°	.909	.826	8.26
50°	1.083	1.173	11.73
60°	1.225	1.500	15.00
70°	1.329	1.766	17.66
80°	1.393	1.940	19.40
90°	1.414	2.000	20.00

Table 11-3: Instantaneous Values of Current and Power

During the part of the sine wave of current from 90° to 180° the same values could be used as before but in a reverse order. Thus, at 100° the values of current and power would be identical to those at 80°.

Using the values of i and P from Table 11-3, a graph can be constructed showing the way in which power varies throughout the cycle. This graph is shown in Fig. 11-17.

In this graph a sine wave of current is plotted first, using the instantaneous values from Table 11-3. Next the curve representing i^2 and power is constructed.

Notice that the power curve has twice the frequency of the current curve and that *all power is positive*. This is due to the fact that heat is dissipated regardless of which way the current flows through the resistor.

Since all the alternations of the power curve are identical, the *mean* or *average power* is the value half-way between the max-

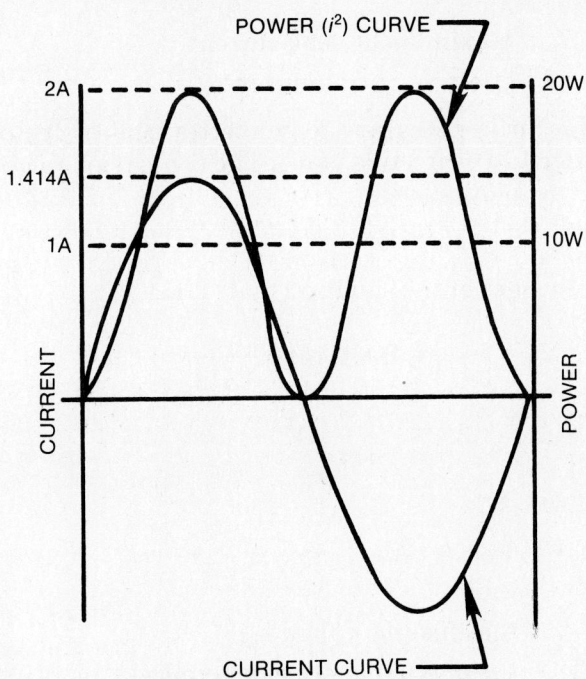

Fig. 11-17: Current and power curves.

imum and minimum values of power. Thus, the average power dissipated by the 10Ω resistor is 10W, one-half the peak power. Since the curve representing power also represents current squared (i^2), the average or mean of the curve also lies halfway between the maximum and minimum values of i^2. As power is proportional to i^2, a DC current having a value equal to the square root of the mean of the i^2 values would produce the same average power as the original sine wave of current. This mean current is called the *root mean square* (rms) current. One rms ampere of alternating current is as effective in producing heat as one steady ampere of direct current. For this reason an rms ampere is also called an effective ampere. In Fig. 11-17, the peak current of 1.414A produces the same amount of average power as 1A of effective (rms) current.

Anytime an alternating voltage or current is stated without any qualifications, it is assumed to be an effective value. Since effective values of AC are the ones generally used, most meters are calibrated to indicate effective values of voltage and current.

In many instances it is necessary to convert from effective to peak or vice-versa. Figure 11-17 shows that the peak value of a sine wave is 1.414 times the effective value and therefore:

$$V_m = V \times 1.414$$

where

$$V_m = \text{maximum or peak voltage}$$
$$V = \text{effective or rms voltage}$$

and

$$I_m = I \times 1.141$$

where

$$I_m = \text{maximum or peak current}$$
$$I = \text{effective or rms current.}$$

Upon occasion, it is necessary to convert a peak value of current or voltage to an effective value. The conversion factor may be derived as follows:

$$V_m = V \times 1.414$$

Multiplying both sides of the equation by 1/1.414

$$V_m \times \frac{1}{1.414} = V \times 1.414 \times \frac{1}{1.414}$$

$$V_m \times \frac{1}{1.414} = V$$

Dividing 1 by 1.414

$$V = V_m \times 0.707$$

where

$$V = \text{the effective voltage}$$
$$V_m = \text{the maximum or peak voltage}$$

Similarly for current

$$I = I_m \times 0.707$$

where

I = the effective current

I_m = the maximum or peak current.

Example: The rms value of voltage is 120V. What is its peak value? If the peak value of a current is 1.62A, what is its rms value?

Solution:

$$V = 1.41 \times \text{rms voltage}$$
$$= 1.41 \times 120$$
$$= 169.2\text{V}$$
$$I_{rms} = I_{peak} \times 0.707$$
$$= 1.62 \times 0.707$$
$$= 1.145\text{A}$$

Average Value

The average value of a complete cycle of a sine wave is zero, since the positive alternation is identical to the negative alternation. In certain types of circuits, however, it is necessary to compute the average value of one alternation. This could be accomplished by adding together a series of instantaneous values of the wave between 0°, and 180°, and then dividing the sum by the number of instantaneous values used. Such a computation would show one alternation of a sine wave to have an average value equal to 0.637 of the peak value. In terms of an equation:

$$V_{avg} = V_m \times 0.637$$

where

V_{avg} = the average voltage of one alternation

V_m = the maximum or peak voltage

Similarly,

$$I_{avg} = I_m \times 0.637$$

where

I_{avg} = the average current in one alternation

I_m = the maximum or peak current.

Figure 11-18 shows a comparison between the various values that are used to indicate the amplitude of a sine wave. Table 11-4 gives the computation data for peak, rms, or average voltage values.

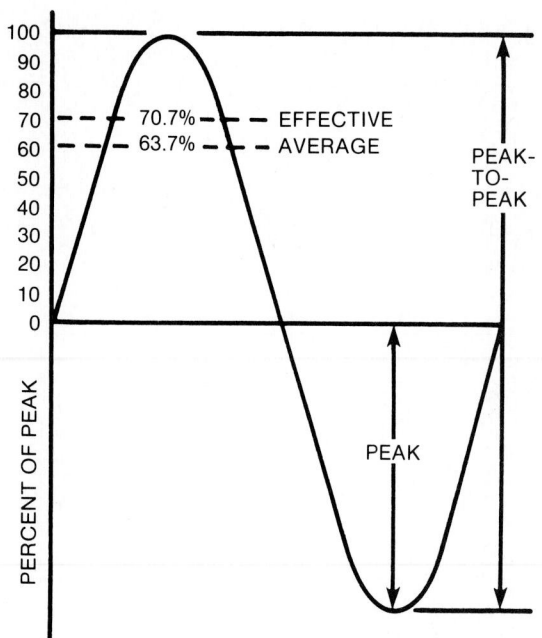

Fig. 11-18: **Various values used to indicate sine wave amplitude.**

Table 11-4: Computation of Peak, rms, or Average Voltage Values		
Given	**To find**	**Multiply by**
V_{peak}	V_{rms}	0.707
V_{peak}	V_{avg}	0.637
V_{rms}	V_{peak}	1.414
V_{rms}	V_{avg}	0.899
V_{avg}	V_{peak}	1.57
V_{avg}	V_{rms}	1.11

The same formulas are used for current as are used for voltage:

$$\text{rms current} = 0.707 I_p$$

$$\text{peak current} = 1.414 \text{rms current}$$

$$\text{peak-to-peak current} = 2.828 \text{rms current}$$

$$\text{average current} = 0.637 I_p \text{ (full-wave)}$$

Sine Waves in Phase

If a sine wave of voltage is applied to a resistance, the resulting current will also be a sine wave. This follows Ohm's Law which states that the current is directly proportional to the applied voltage. Figure 11-19 shows a sine wave of voltage and the resulting sine wave of current superimposed on the same axis. Notice that as the voltage increases in a positive direction, the current increases along with it. When the voltage reverses direction, the current reverses direction. At all times the voltage and current pass through the same relative parts of their respective cycles at the same time. When two waves, such as those in Fig. 11-19, are precisely in step with one another, they are said to be *in phase*. To be in phase, the two waves must go through their maximum and minimum points at the same time and in the same direction.

In some circuits, several sine waves can be in phase with each other. Thus, it is possible to have two or more voltage drops in phase with each other and also in phase with the circuit current.

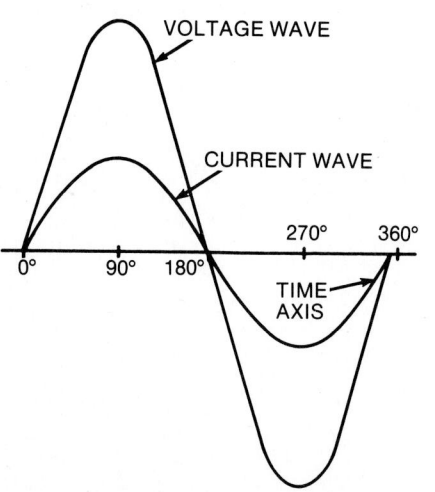

Fig. 11-19: **Voltage and current waves in phase.**

Sine Waves Out of Phase

Figure 11-20 shows a voltage wave, V_1, considered to start at 0° (time one). As voltage wave V_1 reaches its positive peak, a second voltage wave, V_2, starts its rise (time two). Since these waves do not go through their maximum and minimum points at the same instant of time, a *phase difference* exists between the two waves. The two waves are said to be out of phase. For the two waves in Fig. 11-20, this phase difference is 90°.

To further describe the phase relationship between two waves, the terms *lead* and *lag* are used. The amount by which one wave leads or lags another is measured in degrees. Referring again to Fig. 11-20, wave V_2 is seen to start 90° later in time than wave V_1, thus wave V_2 lags wave V_1 by 90°. This relationship could also be described by stating that wave V_1 leads wave V_2 by 90°.

It is possible for one wave to lead or lag another by any number of degrees, except 0° or 360°, in which condition the two waves are in phase. Thus, two waves may differ in phase by 45°, but two waves differing by 360° would be considered as in phase.

A phase relationship that is quite common is the one shown in Fig. 11-21. The two waves illustrated have a phase difference of 180°. Notice that although the waves pass through their maximum and minimum values at the same time, their instantaneous voltages are always of opposite polarity. If two such waves existed across the same component, they would have a cancelling effect on each other. If the two waves are equal in amplitude, the resultant wave would be zero. However, if they have different amplitudes, the resultant wave would have the polarity of the larger and be the difference of the two.

To determine the phase difference between two sine waves, locate the points on the time axis where the two waves cross the time axis traveling in the same direction. The number of degrees between the crossing points is the phase difference. The wave that crosses the axis at the later time (to the right on the time axis) lags the other.

COMBINING AC VOLTAGES

Phasors may be used to combine AC voltages of sine waveform and of the same frequency. The angle between the phasors indicates the time difference between their positive maximum values. The length of the phasors represents either the effective value or the positive maximum values as desired.

The sine wave voltages generated in coils a and b of the simple generators (Fig. 11-22A) are 90° out of phase because the coils are located 90° apart on the two-pole armatures. The armatures are on a common shaft. When coil a is cutting squarely across the field, coil b is moving parallel to the field and not cutting through it. Thus, the voltage in coil a is maximum when the voltage in coil b is zero. If the frequency is 60Hz, the time

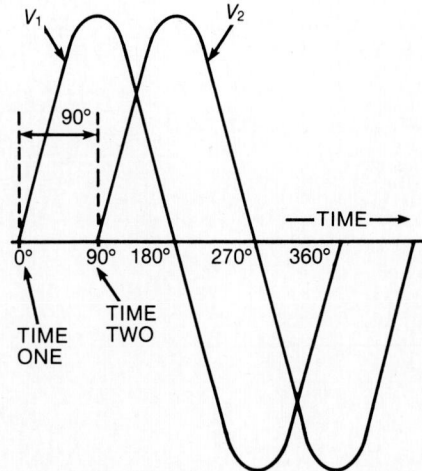

Fig. 11-20: Voltage waves 90° out of phase.

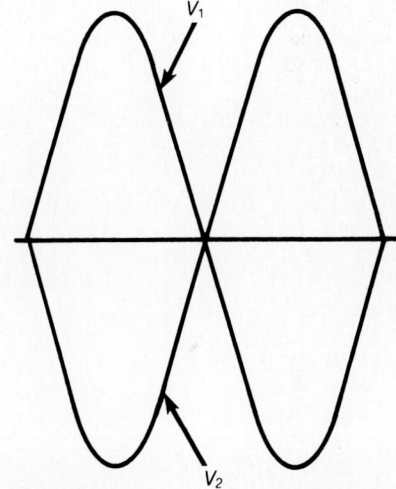

Fig. 11-21: Two waves 180° out of phase.

difference between the positive maximum values of these voltages is $\dfrac{90}{360} \times \dfrac{1}{60} = 0.00416$s.

Phasor V_a leads phasor V_b by 90° in Fig. 11-22B and sine wave a leads sine wave b by 90° in Fig. 11-22C. The curves are shown on separate axes to identify them with their respective generators; they are also projected on a common axis to show the relation between their instantaneous values. If coils a and b are connected in series and the maximum voltage generated in each coil is 10V, the total voltage is not 20V because the two maximum values of voltage do not occur at the same instant, but are separated one-fourth of a cycle.

The voltages cannot be added arithmetically because they are out of phase. These values, however, can be added phasorally.

V_c in Fig. 11-22B is the phasor sum of V_a and V_b and is the diagonal of the parallelogram, the sides of which are V_a and V_b. The effective voltage in each coil is 0.707 × 10 or 7.07V (where 10V is the maximum) and represents the length of the sides of the parallelogram. The effective voltage of the series combination is 7.07 × $\sqrt{2}$ or 10V and represents the length of the diagonal of the parallelogram.

Curve c in Fig. 11-22C represents the sine wave variations of the total voltage V_c developed in the series circuit connecting coils a and b. Voltage V_a leads V_b by 90°. Voltage V_c lags V_a by 45° and leads V_b by 45°.

Counterclockwise rotation of the phasors is considered positive rotation, giving the sense of lead or lag. Therefore, if V_a and V_b (Fig. 11-22B) are rotated counterclockwise and their movement observed from a fixed position, V_a will pass this position first; then 90° later V_b will pass the position. Thus V_b lags V_a by 90°. If the maximum voltage in each coil is 10V, the maximum value of the combined voltage will be 10 × $\sqrt{2}$ or 14.4V (Fig. 11-22C).

The important point to be made in this discussion is that two out of phase voltages may be resolved into a single resultant value by the use of phasors. Though generated voltages are used to illustrate the point in this case, the same rules apply to voltage drops as well. Any number of out of phase voltages may be combined phasorally, as long as they all have the same frequency; that is, as long as they remain a fixed number of degrees apart, like the two generator loops. Their magnitudes (length of their phasors) may be different.

NONSINUSOIDAL AC WAVEFORMS

The AC waveform also may be nonsinusoidal and symmetrical. Three common types of such waveforms are illustrated in Fig. 11-23. In Fig. 11-23A is shown the square, or rectangular, waveform. This type of AC waveform is used extensively in computer circuits. With a square wave the magnitude of the current (or voltage) is not continuously varying. However, both amplitude and direction do periodically change.

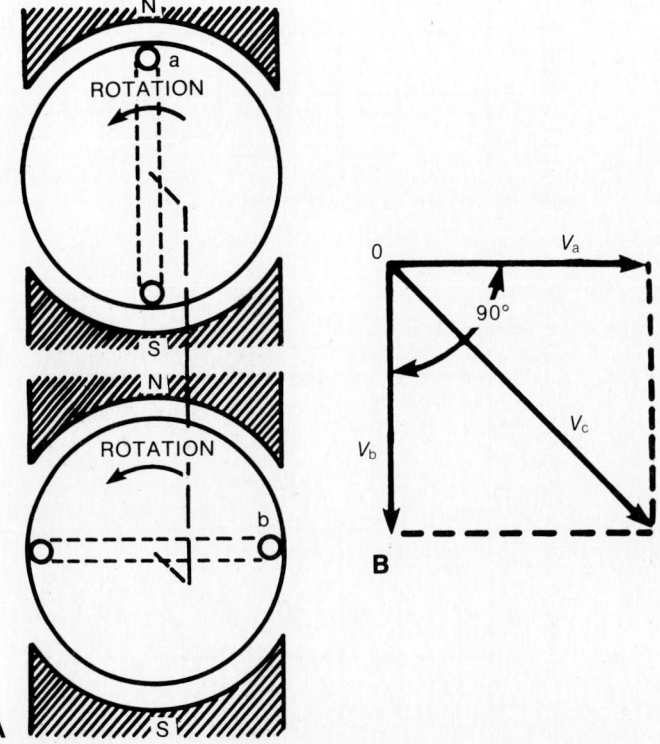

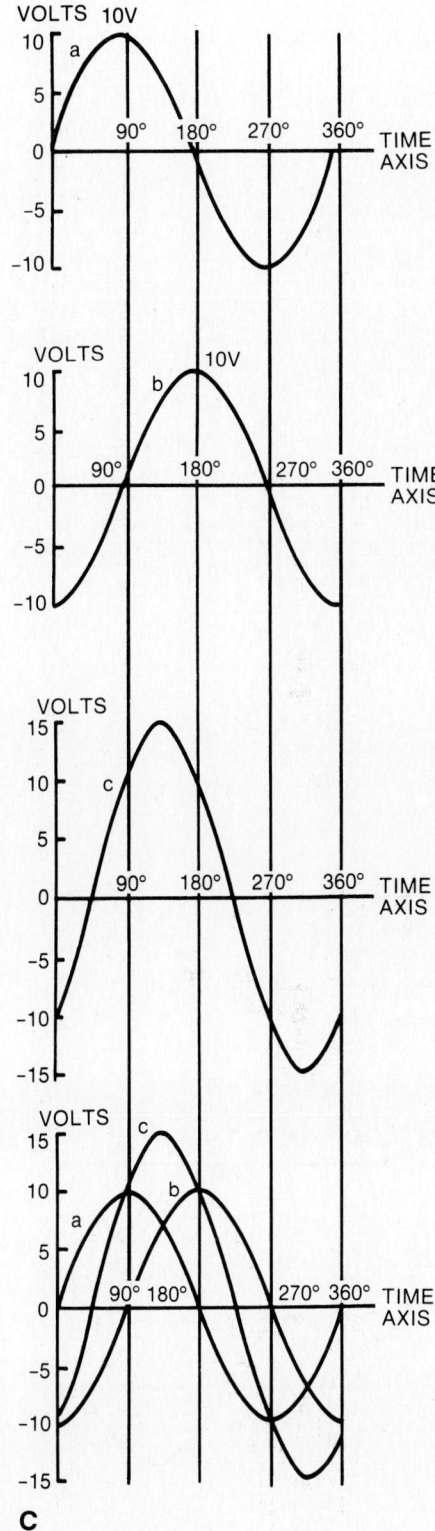

Fig. 11-22: Combining AC voltages.

The triangular waveform (Fig. 11-23B) and sawtooth waveform of Fig. 11-23C are used in television receivers, radar receivers, and other electronic devices. Sawtooth voltages and currents are used in the circuits which produce the picture on a television screen.

Figure 11-24A shows a distorted sine wave signal where the signal has a dual peak for each alternation with a dip at the center. Such a waveform is actually composed of more than one signal: the fundamental signal shown in Fig. 11-24B and its third harmonic shown in Fig. 11-24C. A *harmonic* is a higher-frequency signal made up of multiples of a fundamental signal. Harmonics can be even (second, fourth, sixth, etc.) or odd (third, fifth, etc.) For Fig. 11-24A, if the fundamental frequency were 200Hz, the harmonic component would be 600Hz with a much lower amplitude than the fundamental signal. A pure sine wave, on the other hand, has only one frequency, and if it deviates from one or more of the following three essentials, harmonics are present: (1) The negative amplitude must be the same as the positive amplitude; (2) the duration of each alternation must be the same; (3) the incline and decline of amplitudes must be gradual and identical for both alternations of the cycle.

Note that the waveform shown in Fig. 11-24A resembles the waveform in Fig. 11-24C. If additional odd harmonics were added, the result would be the formation of square waves. Thus, square waves are made up of a fundamental frequency, a third harmonic having one-third the amplitude of the fundamental, a fifth harmonic having one-fifth the amplitude of the fundamental, and successive odd harmonics with decreasing ampli-

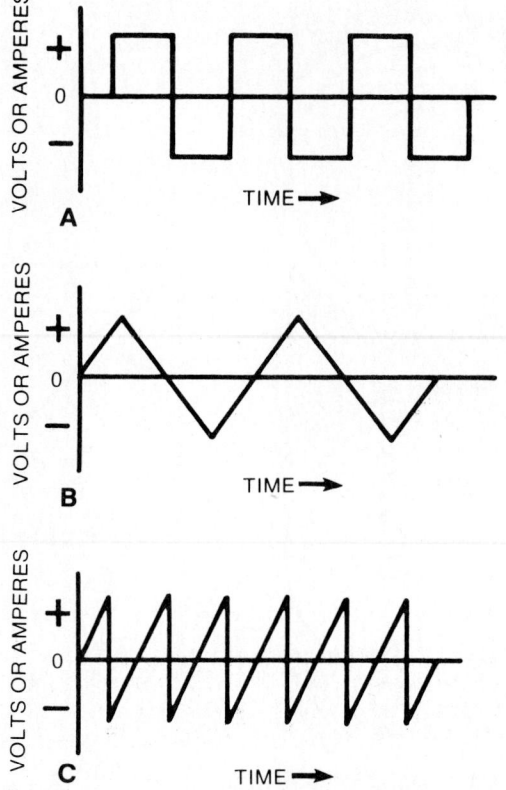

Fig. 11-23: Several types of nonsinusoidal waveforms.

tudes. This is important in the design of circuits, since many circuits must be capable of processing the high-frequency signal components of the square wave so that the wave shape is not distorted.

Some interesting results may be obtained if a steady direct current (or voltage) is added to an alternating current (or voltage). In Fig. 11-25A is shown the resultant of the graphical addition of a steady direct current and a symmetrical, sinusoidal alternating current, where the amplitude of the former is greater than that of the latter. That the resultant current is direct is indicated by the fact that its waveform lies entirely in one-half of the graph. But, since its amplitude varies continuously, it is a fluctuating direct current.

If the amplitude of the alternating current is greater than that of the direct current (Fig. 11-25B), the resultant is an alternating current. This is indicated by the fact that the waveform lies in both the positive and negative halves of the graph. But note that the two halves of the cycle are no longer equal in amplitude; hence, the resultant is a nonsymmetrical alternating current.

It is clear, then, that a fluctuating direct current or a nonsymmetrical alternating current contains both steady DC and AC components. Note that the shape of the waveform of the resultant current (or voltage) is determined by the waveform of the AC component. The steady DC component merely affects

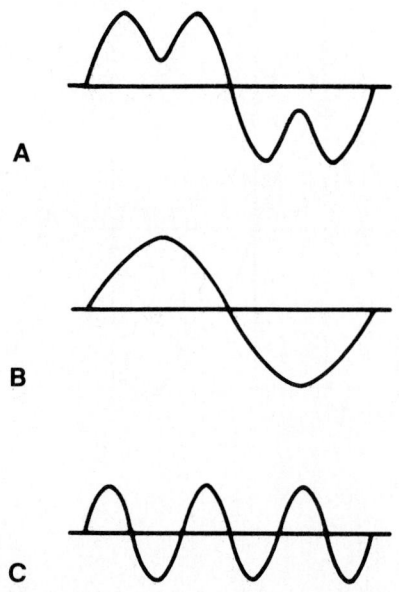

Fig. 11-24: Signal waveforms.

208

the vertical positioning of the resultant curve of the graph. If the steady DC component is more positive, the curve is moved higher up the graph. If the steady DC component is more negative (or, what is the same thing, less positive), the curve is moved lower down the graph.

Similar results may be obtained from the addition of steady DC and nonsinusoidal AC components. Here, too, the shape of the waveform of the resultant current (or voltage) is determined by the waveform of the AC component, and the vertical position of the curve of the graph is affected by the steady DC component.

Under certain circumstances, composite currents (or voltages) may be separated into their various components. The alternating currents (or voltages) discussed so far are *all continuous*—that is, a cycle starts where the previous one has left off and is repeated over and over. However, the currents (or voltages) need not be continuous. In Fig. 11-26 are shown the waveforms of several types of bursts, or pulses, of current or voltage. These waveforms may be sinusoidal, spiked, triangular, or rectangular, as illustrated. Or they may be composites of two or more waveforms.

Note in Fig. 11-26A that the pulse starts at some reference value of current (or voltage), rises to a peak value, and falls back to its original reference value. Such a pulse is said to be *positive-going*. In Fig. 11-26B the pulse starts at its reference value, falls to a negative (or less positive) peak value, and rises to its original reference value. Such a pulse is said to be *negative-going*.

The pulses may be *repetitive*, recurring at regular intervals of time, or they may be *transient*, appearing at irregular intervals.

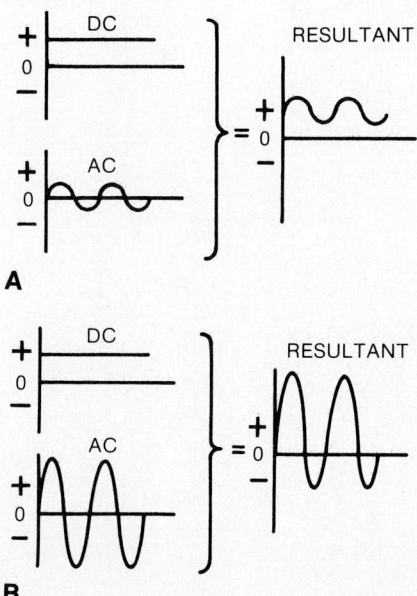

Fig. 11-25: (A) How a pure DC and a symmetrical AC are added to produce a fluctuating DC. (B) How a pure DC and a symmetrical AC are added to produce a nonsymmetrical AC.

SUMMARY

Unlike direct current, which flows in only one direction and has a steady value, an alternating current flows in one direction and then the other. The value or magnitude of an alternating current varies as it flows in each direction. Alternating current is used more extensively than direct current because it is more versatile. It is easier and cheaper to produce, and it may also be easily converted into DC.

An AC generator is able to produce an alternating current through a process known as electromagnetic induction. The voltage induced within a generator is affected by the strength of the magnetic field, the speed of conductor movement, the length of the conductor in the field, and the total number of active conductors in the armature winding. The polarity of the induced voltage is determined by the direction of conductor motion and the direction of the magnetic field.

One cycle of output voltage is produced during one complete revolution of the armature in a two-pole generator. The frequency of an alternating voltage (or current) is the number of cycles occurring in each second of time. The period of a sine wave is the time required to produce one complete cycle.

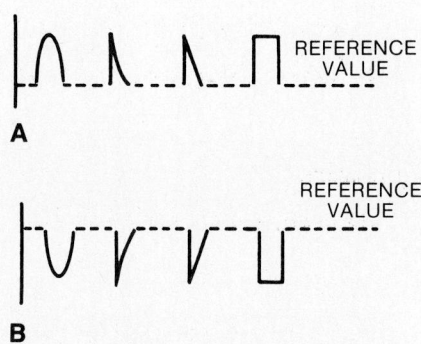

Fig. 11-26: Waveforms of various types of pulses: (A) positive-going pulses; (B) negative-going pulses.

The amount of voltage induced on the armature at any moment in time can be determined by the formula $v = V_m \sin \theta$. The maximum voltage reached during each alternation is called the peak voltage. The total value of a waveform, between its peak values, is called its peak-to-peak value. The average value of one alternation is equal to 0.637 times the peak value. However, the average value of one complete cycle is zero. The effective, or rms, value of an AC sine wave is equal to 0.707 times its peak value.

A phase lead or lag refers to the difference in starting time of one wave compared to the starting time of another wave. Phasors may be used to combine sine waves of the same frequency.

A variety of nonsinusoidal waveforms, such as the square wave, triangular wave, and sawtooth wave, are also used in electronic applications.

CHECK YOURSELF (Answers at back of book)

Questions

Fill in the blanks with the appropriate word or words.

1. An eight-pole alternation would develop one-half the frequency of a _____ -pole alternator.
2. One-half of a sine wave is called an _____ .
3. V_{avg} = _____ × V_{peak}.
4. I_{peak} = _____ × I_{avg}.
5. V_{rms} = _____ × V_{peak}.
6. I_{rms} = _____ × I_{peak}.
7. V_{peak} = 1.414 _____ .
8. _____ and _____ are examples of nonsinusoidal waveforms.
9. Adding odd harmonics to a fundamental results in the formation of a _____ .
10. If a voltage leads the current, the current is said to _____ the voltage.
11. The rotating loop in an AC generator is called an _____.
12. The phasor is rotated _____ through one complete revolution (360°).
13. The difference in rotational degrees of starting of two electrical waves is called the _____ .
    A. phase angle
    B. rotation delay
    C. starting difference
    D. none of the above
14. The shape of the electrical wave produced by an AC generator is called a _____ .
    A. square wave
    B. triangular wave
    C. signal wave
    D. sine wave

15. One complete revolution of the generator armature is called _____ .
    A. a hertz
    B. an alternation
    C. a frequency
    D. a cycle
16. Rms values are _____ of peak values.
    A. 1.414
    B. 2.828
    C. 0.707
    D. 0.634
17. Voltages used in the electric power industry are almost always _____ .
    A. DC
    B. RF
    C. AC
    D. signal
18. A(n) _____ is used to show wave shapes, amplitude, and deviation in waveforms.
    A. oscilloscope
    B. frequency generator
    C. AC meter
    D. DC meter
19. The angular velocity of a 60Hz waveform is _____ .
    A. 240rad/s
    B. 377rad/s
    C. 120rad/s
    D. 60rad/s
20. The root-mean-square value of a sine wave having a peak-to-peak value of 10V is _____ .
    A. 7.07V
    B. 6.37V
    C. 3.535V
    D. 5.0V
21. Given the formula $v = 50\sin 4,000\pi t$, the amplitude and frequency are _____ .
    A. 100V, 2kHz
    B. 50V, 2kHz
    C. 50V, 4kHz
    D. 100V, 4kHz
22. The AC current through a 5Ω resistor is 2mA. The rms voltage developed across the resistor is _____ .
    A. 10mV
    B. 7.07mV
    C. 14.14mV
    D. 3.535mV
23. A peak current of 3A is flowing through a 10Ω resistor. The peak and average power are _____ .
    A. 9W, 30W
    B. 45W, 25W
    C. 90W, 45W
    D. 4.5W, 22.5W

24. A period of $0.01\,\mu s$ corresponds to the frequency of _____ MHz.
25. For an rms value of 70.7V, the peak value of voltage is _____ .

Problems

1. Calculate the period for the following frequencies:
   A. 1,000Hz
   B. 60Hz
   C. 2MHz
   D. 15kHz
2. What is the average value of an AC wave with a 120V peak?
3. A sine wave has a V_{peak} = 500V. What is the instantaneous value of voltage at:
   A. 63°
   B. 90°
   C. 178°
   D. 240°
4. A sine wave has an rms value of 300V. What is V_{avg} and V_{peak}?
5. The peak value of a sine wave is 200V. What is the V_{rms} and V_{avg}?
6. Calculate the frequency for each of the following periods:
   A. 5s
   B. 0.0002s
   C. $1\,\mu s$
   D. 0.05s

CHAPTER 12

Inductance

Inductance is the characteristic of an electrical circuit that makes itself evident by opposing the starting, stopping, or changing of current flow. The above statement is of such importance to the study of inductance that it bears repeating in a simplified form. Inductance is the characteristic of an electrical conductor which opposes a *change* in current flow.

In the analysis of AC and DC circuits previously considered, any opposition to the flow of current was termed *resistance*. The current in such a typical DC circuit may be compared to an object in motion, such as an automobile, which is retarded only by the friction or resistance of the surface on which it moves. If, however, a DC circuit is broken suddenly, by opening a switch, for instance, a considerable spark jumps across the contacts of the switch as it opens. It may be said then that opening a DC circuit is like suddenly stopping an object in motion. But from Isaac Newton's *first* law of motion it is known that an object in motion tends to remain in motion and that a considerable force must be exerted to bring it to a stop. In the case of a speeding automobile stopped suddenly by a stone wall, this inertia or momentum which tends to keep the car moving will smash the car and dissipate itself as heat. In the case of a DC circuit suddenly opened, particularly one carrying heavy current, the inertia or momentum of the current meeting the very high resistance of the open circuit produces a high voltage and dissipates itself as heat in a blue spark.

Furthermore, Newton's first law of motion states that an object at rest tends to remain at rest unless acted on by an external force. For example, an automobile must exert considerable power in order to start; after reaching speed, the only power necessary to keep it moving is the power used to overcome friction. In like manner, an electric current cannot be started instantaneously; there is a delay in time between the application of the voltage and the rise of the current to its maximum amount. This effect of inertia in a DC circuit generally is not noticed, since the starting delay is slight. Also, except in circuits carrying large amounts of current, the spark resulting from stopping the current, or opening the circuit, is not dangerous. It should be noted, however, that in DC circuits carrying heavy power, provision must be made to open the circuit by gradually increasing resistance until current reaches a safe value.

213

In an AC circuit, however, this effect of electrical inertia, or inductance, is ever present, since by definition an alternating current is one constantly changing in magnitude and periodically changing direction. Thus, if the current in a circuit is increasing, the inductance of the circuit is defined as that property of the circuit which tends to prevent the increase; if the current is decreasing, the inductance of the circuit tends to prevent the decrease. The greater the inductance of the circuit, the greater the opposition to a *change* in current.

COUNTER OR BACK ELECTROMOTIVE FORCE

In AC circuits there is ever present an opposition to the flow of current other than the DC resistance of the circuit. This additional opposition is caused by a *counter* or *back electromotive force,* so named to distinguish it from the applied voltage, which is the original force tending to set up a current flow. Back emf or counter emf is analogous to the force which opposes a change in motion of a mass. Stated generally, it may be said that the velocity of a mass cannot be changed instantaneously, whether that mass be an automobile, an electron, or an electrostatic charge.

In an electric circuit, the cemf (counter electromotive force) is an *induced* emf or voltage. This voltage is induced in conductors of the circuit not by means of an external magnetic field, as in the case of the simple two-pole generator, but by means of the magnetic field already surrounding any conductor carrying a current. Any change in current changes the intensity of this magnetic field, and the resultant emf induced, the counter emf, is a *self-induced* voltage. Thus, the property of a circuit which produces such an emf is called *self-inductance*. Actually, all elements in a circuit, including connecting wires, show some self-inductance; but for all practical purposes, only those elements designed to make use of this property to advantage are known as *inductances* or *inductors*. Moreover, it may be said that counter emf is present in any AC circuit, but its effect is negligible in a circuit of moderate power, such as an electric lamp, which uses almost pure resistance as a load. But the effect of counter emf is considerable in circuits (even of very low power) which use an inductance as part of the load, such as the primary of the power transformer in an ordinary radio receiver.

Generation of Counter Emf

It will be remembered that in the case of the simple generator, *motion* either of the conductor or of the magnet was necessary to an induced emf. In self-inductance, the equivalent of motion, that is—a change in flux density of the magnetic lines of force about a conductor—is caused by the rise or fall of the current, since, as was seen in the study of electromagnetism, the intensity of a magnetic field about a conductor is directly proportional to the current through the conductor. The force setting up the flux lines is equal to $0.4\pi NI$. Since 0.4π is a constant, the factor NI is called the *ampere-turns*. Any change in current changes the factor NI, or ampere-turns, and, accordingly, the flux density. Thus, self-inductance is present constantly in an AC circuit because current is constantly changing; but it is present in a DC circuit only at the moments of closing or opening the circuit.

That the emf self-induced in a conductor carrying current is a *counter emf* (cemf) was deduced by Lenz from the principle of the conservation of energy. If the emf self-induced was not a counter emf, then an increase of current would aid the applied voltage, and this increase in applied voltage would, in turn, tend to increase the current. This process would continue, of course, until current reached an infinite amount, a condition not possible in the physical universe. As described in Chapter 10, Lenz' Law, in effect, states: *An induced emf always has such a direction as to oppose the action that produces it.* Thus, when a current flowing through a circuit is varying in magnitude, it produces a varying magnetic field which sets up an induced emf that opposes the current change producing it. Or, it may be said that when the current in a circuit is increasing, the induced emf opposes the applied voltage and tends to keep the current from increasing; and when the current is decreasing, the induced emf aids the line voltage and tends to keep the current from decreasing (Fig. 12-1).

The effect of counter emf may be observed experimentally in that an alternating current through an inductor is opposed by a force much greater than its simple DC resistance. For example, the DC resistance of the primary of an ordinary power transformer used in a typical radio receiver is approximately 6Ω. As in Fig. 12-2, this primary is connected directly to the 120V, AC outlet in the home. From Ohm's Law:

$$I = \frac{V}{R} = \frac{120}{6}$$

$$I = 20\text{A}$$

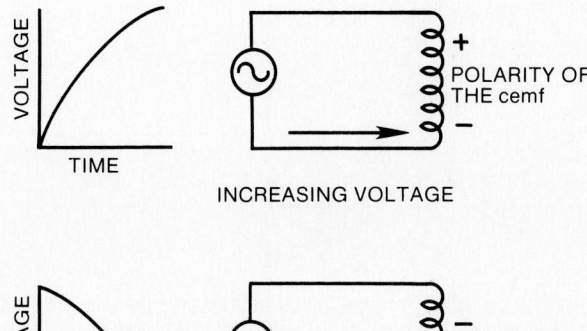

INCREASING VOLTAGE

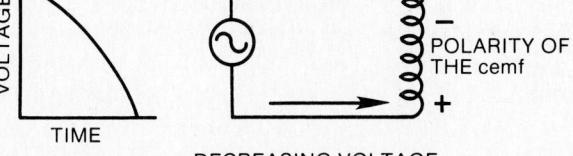

DECREASING VOLTAGE

Fig. 12-1: Polarity of counter emf.

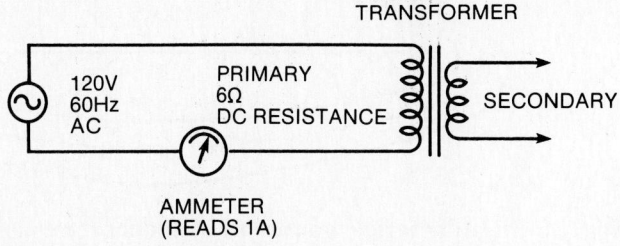

Fig. 12-2: Effect of cemf on current flow.

Thus, the current is calculated to be 20A; but when actually measured, the current is found to be approximately 1A. It may be seen that some opposition other than the 6Ω resistance is present in an AC circuit. This opposition is the counter emf. If by mistake such a radio set is connected to a 120V, DC line, the current through the primary of the transformer is 20A, and the transformer burns up. Hence, Ohm's Law as stated for DC circuits must be modified to include this effect of electrical inertia present in AC circuits.

The magnitude of a counter emf depends upon the same factors that govern any induced emf. In the analysis of an induced emf it was shown that the magnitude of the emf induced in a conductor of unit length depended on the number of flux lines cut per second. This principle may be restated as Faraday's Law of electromagnetic induction: *The emf induced in any circuit is dependent upon the rate of change of the flux linking the circuit.* Since there is no physical movement of the conductor or of the lines of force in self-inductance, the rate of change of the flux density is equivalent to movement. But, as was shown here and in Chapter 10, the flux density about a conductor is directly proportional to the current in the conductor. Therefore, the magnitude of self-induced emf depends directly upon the rate of change of the current in the circuit. Thus, a rapidly changing current induces a greater counter emf than a slowly changing current. But for any AC, the rate of change of current depends on the number of Hz or the frequency. The counter emf then depends directly upon frequency.

The total magnitude of an induced emf depends also on the length of the conductor, since in the simple generator the length of the conductor and the number of conductors in a coil side determine the total emf induced. Thus, a long conductor has greater counter emf induced, or has more self-inductance, than a short one. If, however, a long conductor is wound on itself in the form of a coil, its self-inductance is increased because of the increase in total flux density. Such a coil, or inductance, is a solenoid, and, as was shown in the study of electromagnetism, the flux density about it may also be increased by the addition of a core material of high permeability, such as soft iron. Figure 12-3 illustrates this type of inductance.

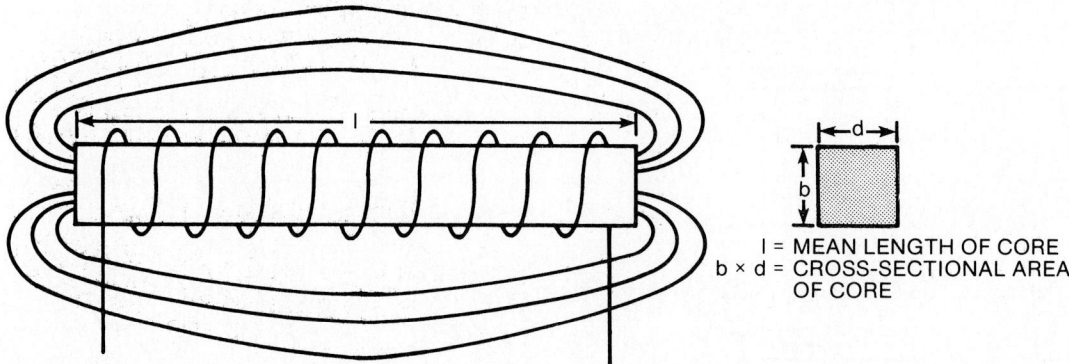

l = MEAN LENGTH OF CORE
b × d = CROSS-SECTIONAL AREA OF CORE

Fig. 12-3: Inductance wound on iron core.

216

UNIT OF INDUCTANCE

A coil has an *inductance* of 1H if 1V is induced into the coil when the current in it is changing at the rate of 1A/s (Fig. 12-4).

$$\text{inductance} = \frac{\text{induced voltage}}{\text{rate of current change/second}}$$

$$L = \frac{-v_L}{\dfrac{di}{dt}}$$

from which induced voltage

$$-v_L = \frac{di}{dt} L$$

where

$\quad v_L \quad$ is induced voltage

$\quad L \quad$ is inductance in H

$\quad \dfrac{di}{dt} \quad$ is $\dfrac{A}{s}$

Note: The symbol d (delta, Δ) is used to indicate a change. The factor di/dt for the current variation with respect to time really specifies how fast the current's associated magnetic flux is cutting the conductor to produce v_L.

The negative sign indicates that the induced voltage is in a direction opposite to the voltage that produced the current change. It is not used in calculations.

Example: The current in an inductor changes from 12A to 16A in 2s. How much is the inductance of the coil if the induced voltage is 40V?

Solution:

$$L = \frac{v_L}{\dfrac{di}{dt}}$$

$$= \frac{40V}{\dfrac{16A - 12A}{2s}}$$

$$= \frac{40V}{2}$$

$$= 20H$$

Example: What is the v_L across an inductance of 10H when the current through it is changing from 6mA to 14mA in 8μs?

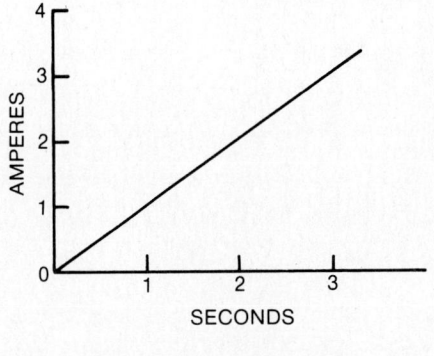

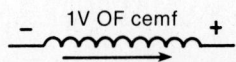

Fig. 12-4: A 1H inductor produces 1V of cemf when the current changes at a rate of 1A/s.

Solution:

$$L = \frac{v_L}{\frac{di}{dt}}$$

$$v_L = \frac{di}{dt} L$$

$$v_L = \frac{(14 - 6) \times 10^{-3} \times 10}{8 \times 10^{-6}}$$

$$= \frac{80 \times 10^{-3}}{8 \times 10^{-6}}$$

$$= 10{,}000V$$

$$= 10kV$$

Example: What value of inductance is necessary to induce a voltage of 250V if the current changes at the rate of 5A in 10ms?

Solution:

$$L = \frac{v_L}{\frac{di}{dt}} = \frac{250}{\frac{5}{10 \times 10^{-3}}}$$

$$= \frac{250}{\frac{5}{10^{-2}}}$$

$$= 50 \times 10^{-2}$$

$$= 0.5H$$

MEASUREMENT OF INDUCTANCE

The unit for measuring inductance (L) is the *henry* (H), named after Joseph Henry, an American physicist known for his extensive work in electromagnetism. An inductor has an inductance of 1H if an emf of 1V is induced in the inductor when the current through the inductor is changing at the rate of 1A per second. The relation between the induced voltage, inductance, and rate of change of current with respect to time is stated mathematically as:

$$v_L = L \frac{di}{dt}$$

where v_L is in V, L in H, and di/dt in A/s.

Example: How much is the self-induced voltage across a 3H inductance produced by a current change of 12A/s?

Solution:

$$v_L = L \frac{di}{dt} = 3 \times 12$$

$$= 36V$$

The henry is a large unit of inductance and is used with relatively large inductors. The unit employed with small inductors is the millihenry, mH. For still smaller inductors the unit of

218

inductance is the microhenry, μH. In fact, since the henry is defined in terms of practical units, the factor 10^{-8} must be used if the cemf is to be read in volts and the rate of change of the current in amperes per second.

Then

$$L = \frac{0.4\pi N^2 \mu A}{1 \times 10^8}$$

where

L = self-inductance of solenoid in H

N = number of turns of coil

μ = permeability of core in electromagnetic units

A = cross-sectional area of core in cm^2

l = mean length of core in cm

Example: An air core coil has 1,200 turns, is 6cm long, and has a 2cm cross-sectional area. What is the inductance?

Solution:

$$L = \frac{0.4\pi N^2 \mu A}{1 \times 10^8} = \frac{1.256 \times (1,200)^2 \times 1 \times 2}{10^8 \times 6} = 6\text{mH}$$

The above formula reveals the following important relationships:

1. The inductance of a coil is proportional to the square of the number of turns.

2. The inductance of a coil increases directly as the permeability of the material making up the core increases.

3. The inductance of a coil increases directly as the cross-sectional area of the core increases.

4. The inductance of a coil decreases as its length increases. Figure 12-5A shows two coils of a fixed number of turns with different cross-sectional areas. The larger coil has a greater total flux, or less reluctance, and therefore greater inductance. Figure 12-5B shows two coils of a fixed number of turns and the same cross-sectional area, but of different lengths. The longer coil has less total flux, or greater reluctance, and therefore less inductance.

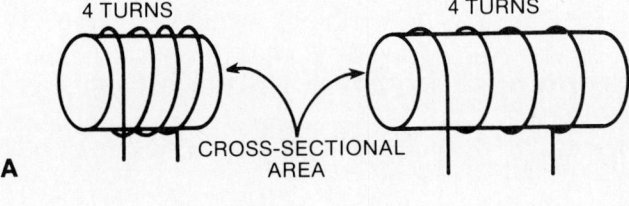

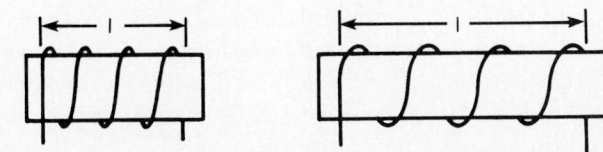

Fig. 12-5: **Variation of inductance with size of solenoid.**

It should also be noted that for ferromagnetic (iron core) inductances, the permeability, μ, of the core material is not a constant, but depends on the magnitude of the magnetizing current. In AC circuits, the current is constantly changing in magnitude and periodically in direction; and, accordingly, an error is introduced in calculations of the magnitude of the inductance. In Fig. 12-6A, the relationship between the flux density, B, and the field intensity, H, is shown in the form of a hysteresis loop graph. But the ratio B/H is the definition of permeability. Therefore, it may be seen from this graph that the value of μ varies as the ratio of B to H varies for different points on the loop.

In Fig. 12-6B, a permeability curve for cast steel is shown. Permeability for this material increases to a maximum value at approximately 7,000 lines per square centimeter and then falls off as the flux density increases. This variation of μ invalidates calculations of inductance based on the formula given previously.

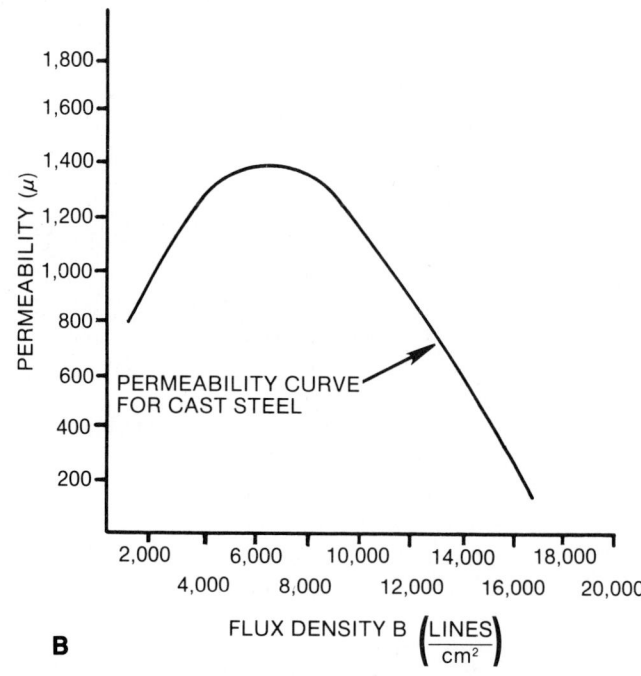

Fig. 12-6: Variation of μ for iron core materials.

Measurement of Counter Emf: Voltage

The formula for the magnitude of a cemf is as follows:

$$\text{cemf} = -L\frac{di}{dt}$$

An examination of this formula reveals that the greater the inductance, or the faster the rate of change of the current, the greater is the cemf induced in the circuit. For example, a coil of 1H inductance has a current of 1A flowing through it. If this current changes to 2A in 1s, the cemf will be:

220

$$\text{cemf} = -1 \times \frac{(2-1)}{1} = -1\text{V}$$

If the current change remains the same but the coil used has an inductance of 10H, then:

$$\text{cemf} = -10 \times \frac{(2-1)}{1} = -10\text{V}$$

If the inductance remains, as in the first instance, at 1H and the current change from 1A to 2A takes place in one-tenth of a second, then:

$$\text{cemf} = -1 \times \frac{(2-1)}{1/10} = -10\text{V}$$

From these examples it may be seen that a high value of opposition to the flow of current may be obtained either by increasing the inductance or the speed of the change of current in a circuit, or both. Thus, low-frequency AC circuits, because of the slow speed of change of the current, generally employ high values of inductance (iron cores) to obtain a high cemf. High-frequency AC circuits, because of the great speed of change of the current, often may generate sufficient cemf with small air core inductances. Table 12-1 illustrates the rise in cemf as the rate of change of the current increases.

Table 12-1: Relationship of cemf and Frequency			
L henrys	di amperes	dt seconds	cemf in volts
1	1	1	–1
1	1	1/2	–2
1	1	1/4	–4
1	1	1/10	–10
1	1	1/20	–20
1	1	1/50	–50
1	1	1/100	–100
1	1	1/500	–500
1	1	1/1,000	–1,000
1	1	1/1,000,000	–1,000,000

From Table 12-1 it is apparent that if a change of 1A were to take place instantaneously, that is, if dt = 0, the induced voltage, v, would become infinitely large. This would violate Kirchhoff's first law, which states that at any instant the applied voltage in a circuit must equal the sum of the voltage drops around the circuit. Certainly, if dt = 0, the voltage drop across the inductance would be greater than any applied voltage could be. And by extension, it may be seen that at any instant, no matter how fast the change of current or how great the value of inductance, the induced voltage cannot be greater than the applied voltage. On the other hand, if there were no change of current, that is, if dt were equal to infinity, the circuit would be a DC circuit and v would be zero.

MUTUAL INDUCTANCE

Whenever two coils are located so that the flux from one coil links with the turns of the other, a change of flux in one coil will cause an emf to be induced in the other coil. The two coils have *mutual inductance*. The amount of mutual inductance depends on the relative position of the two coils. If the coils are separated a considerable distance, the amount of flux common to both coils is small and mutual inductance is low. Conversely, if the coils are close together so that nearly all the flux of one coil links the turns of the other, mutual inductance is high. The mutual inductance can be increased greatly by mounting the coils on a common iron core.

Two coils placed close together with their axes in the same plane are shown in Fig. 12-7. Coil A is connected to a battery through switch S, and coil B is connected to galvanometer G. When the switch is closed (Fig. 12-7A), the current that flows in coil A sets up a magnetic field that links coil B, causing an induced current and a momentary deflection of galvanometer G. When the current in coil A reaches a steady value, the galvanometer returns to zero. If the switch is opened (Fig. 12-7B), the galvanometer deflects momentarily in the opposite direction, indicating a momentary flow of current in the opposite direction in coil B. This flow of current in coil B is produced by the collapsing flux of coil A.

When current flows through coil A, the flux expands in coil A, producing a north pole nearest coil B. Some of the flux, that expands from left to right, cuts the turns of coil B. This flux induces an emf and current in coil B that opposes the growth of current and flux in coil A. Thus the current in B tries to establish a north pole nearest coil A (like poles repel).

When the switch is opened, the magnetic field produced by coil A collapses. The collapse of the flux cuts the turns of coil B in the opposite direction and produces a south pole nearest coil A (unlike poles attract). This polarity aids the magnetism of coil A, tending to prevent the collapse of its field.

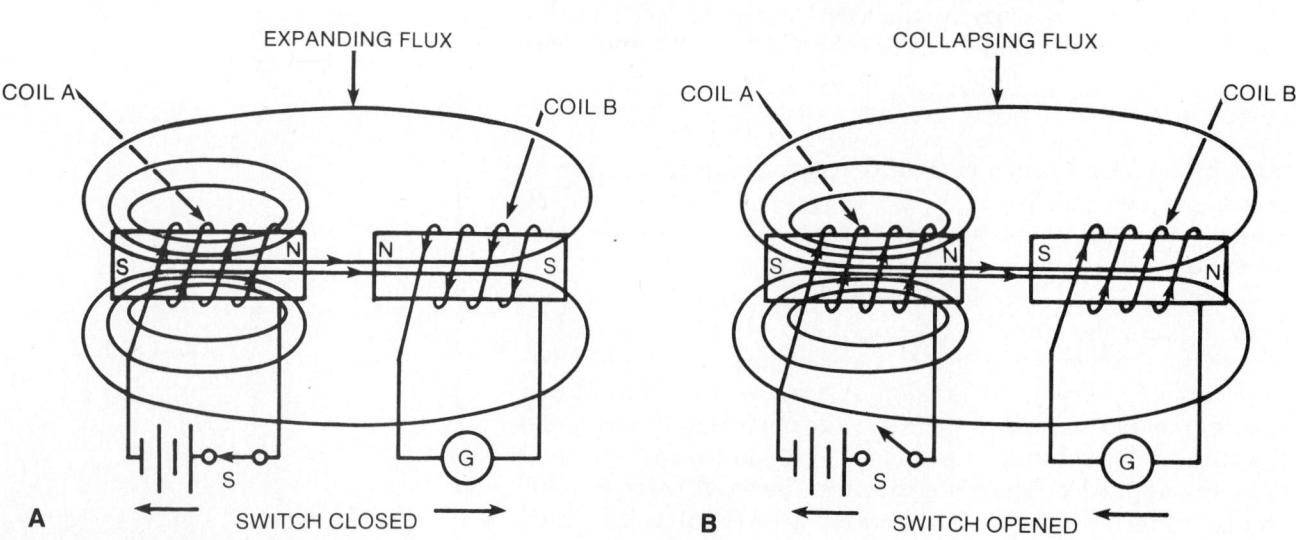

Fig. 12-7: Mutual inductance.

Factors Affecting Mutual Inductance

The mutual inductance of two adjacent coils is dependent upon the (1) physical dimensions of the two coils, (2) number of turns in each coil, (3) distance between the two coils, (4) relative positions of the axes of the two coils, and (5) the permeability of the cores.

If the two coils are positioned so that all the flux of one coil cuts all the turns of the other, the mutual inductance can be expressed as follows:

$$L_M = \frac{0.4\pi\mu SN_1N_2}{10^8 l}$$

where

L_M equals mutual inductance of coils in H

N_1 and N_2 equal the number of turns in coils 1 and 2 respectively

S   equals area of core in square cm

μ equals permeability of core

l   equals length of core in cm

Coefficient of Coupling. The degree or amount of magnetic coupling between two coils is known as the *coefficient of coupling (K)*. The maximum coefficient of coupling occurs when every line of flux in the first coil cuts the turns of the second coil. This degree of closeness of the coils is known as *unity coupling*.

One-hundred percent coupling (theoretical only) is expressed as 1. Coupling of less than 100% is always expressed in decimal values; for example, 75% coupling is expressed as 0.75. High percentages of coupling are commonly referred to as *tight coupling*. Loose coupling produces a small value of coupling; for example, if only 1% of the lines of force from coil 1 were cutting coil 2, then the coefficient of coupling would be 0.01.

Coefficient of coupling is dependent not only upon the closeness of one coil to another, but also upon the angle between the two. Coils placed parallel to one another (Figs. 12-8A and 12-8B) produce the highest value of coupling for a specific distance apart. At right angles (perpendicular), minimal coupling occurs regardless of the closeness (Fig. 12-8C). This is because the expanding magnetic lines of flux in the first coil move parallel to the turns in the second coil and no cutting can take place.

Coefficient of coupling is expressed by the letter K. K is a ratio and it has no units.

$$K = \frac{\text{lines of flux in coil 2}}{\text{lines of flux in coil 1}}$$

$$K = 1 = 100\% \text{ coupling}$$

$$K = 0 = \text{no coupling}$$

The relationship between two coils which are coupled, and their coefficient of coupling, is

$$L_M = K\sqrt{L_1 \times L_2}$$

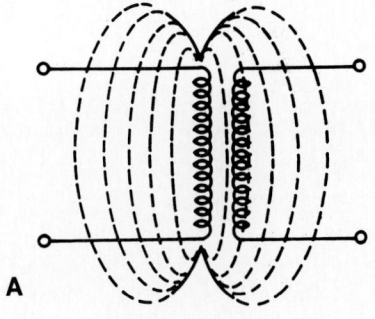

A

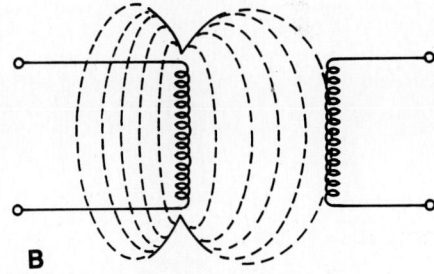

B

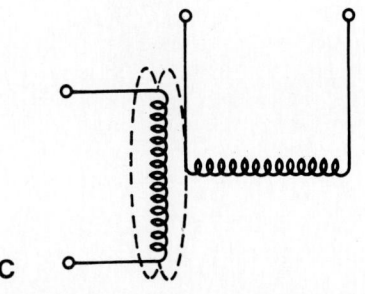

C

Fig. 12-8: (A) Tight coupling—K is high; (B) loose coupling—K is low; and (C) no coupling—$K = 0$.

223

where

L_M is the mutual inductance of the two coils in H

L_1 is the inductance of the first coil in H

L_2 is the inductance of the second coil in H

K is the coefficient of coupling in decimal values

Example: Calculate the mutual inductance of a 2H coil coupled to a 6H coil. The coefficient of coupling is 0.8.

Solution:

$$L_M = K \sqrt{L_1 \times L_2}$$
$$= 0.8 \sqrt{2 \times 6}$$
$$= 2.77H$$

Example: What is the coefficient of coupling between two coils if one has an inductance of 12mH, the other has an inductance of 36mH, and the mutual inductance of the two is 18mH?

Solution:

$$L_M = K \sqrt{L_1 \times L_2}$$
$$K = \frac{L_M}{\sqrt{L_1 \times L_2}}$$
$$= \frac{18 \times 10^{-3}}{\sqrt{(12 \times 10^{-3}) \times (36 \times 10^{-3})}}$$
$$= \frac{18 \times 10^{-3}}{\sqrt{432 \times 10^{-6}}}$$
$$= \frac{18 \times 10^{-3}}{20.78 \times 10^{-3}}$$
$$= 0.86$$

Series Inductors Without Magnet Coupling

When inductors are well shielded, placed at right angles to each other, or located far enough apart to make the effects of mutual inductance negligible, the inductance of the various inductors are added in the same manner that the resistances of resistors in series are added. For example,

$$L_T = L_1 + L_2 + L_3 + ...L_n$$

where L_T is the total inductance; L_1, L_2, and L_3 are the inductances of L_1, L_2, and L_3; and L_n means that any number (n) of inductors may be used.

Example: Three inductances are present in a circuit but separated so that their magnetic fields do not interact with each other. What is the total inductance if one coil is rated at 0.02H, another at 39mH, and the third at 460μH?

Solution:

$$L_T = L_1 + L_2 + L_3$$
$$L_T = 0.02 + 0.039 + 0.00046 = 0.05946H$$
$$= 59.46mH$$

Series Inductors With Magnetic Coupling

When two inductors in series are so arranged that the field of one links the other, the combined inductance is determined as

$$L_T = L_1 + L_2 \pm 2L_M$$

where L_T is the total inductance, L_1 and L_2 are the self-inductances of L_1 and L_2 respectively, and L_M is the mutual inductance between the two inductors. The plus sign is used with L_M when the magnetomotive forces of the two inductors are aiding each other. The minus sign is used with L_M when the mmf's of the two inductors oppose each other. The factor 2 accounts for the influence of L_1 on L_2 and L_2 on L_1.

Example: A 10H and a 12H coil are connected series-aiding magnetically and have a mutual inductance of 2H. What is the total inductance?

Solution:

$$L_T = L_1 + L_2 + 2L_M$$
$$= 10 + 12 + (2 \times 2)$$
$$= 26H$$

If the above coils were connected series-opposing, what would be the total inductance?

$$L_T = L_1 + L_2 - 2L_M$$
$$= 10 + 12 - (2 \times 2)$$
$$= 18H$$

Example: A 50mH coil is connected series-aiding magnetically to a 75mH coil. The coefficient of coupling (K) is 0.8. What is the total inductance?

Solution:

$$L_M = K \sqrt{L_1 \times L_2}$$
$$= 0.8 \sqrt{(50 \times 10^{-3}) \times (75 \times 10^{-3})}$$
$$= 0.8 \sqrt{3,750 \times 10^{-6}}$$
$$= 0.8 \times (61.24 \times 10^{-3})$$
$$= 49 \times 10^{-3}H$$
$$= 49mH$$
$$L_T = L_1 + L_2 + 2L_M$$
$$= (50 \times 10^{-3}) + (75 \times 10^{-3}) + 2(49 \times 10^{-3})$$
$$= (50 \times 10^{-3}) + (75 \times 10^{-3}) + (98 \times 10^{-3})$$
$$= 223 \times 10^{-3}H$$
$$= 223mH$$

If the coils are arranged so that one can be rotated relative to the other to cause a variation in the coefficient of coupling, the mutual inductance between them can be varied. The total inductance, L_T, may be varied by an amount equal to $4L_M$.

225

Parallel Inductors

The total inductance of a circuit containing more than one inductance connected in parallel is calculated in the same manner as that used for resistors in parallel. The total inductance is the reciprocal of the sum of the reciprocals of the individual inductances (provided the coils are shielded from each other, L_M equal to zero). Then:

$$L_T = \frac{1}{\dfrac{1}{L_1} + \dfrac{1}{L_2} + \dfrac{1}{L_3} \cdots + \dfrac{1}{L_n}}$$

where L_1, L_2, and L_3 are the respective inductances of inductors L_1, L_2, and L_3; and L_n means that any number of inductors may be used.

If the coils are of equal inductance, total inductance may be obtained immediately by dividing the inductance of one coil by the number of inductances in the circuit. Thus, four 1H inductors total 0.25H when connected in parallel.

For any two inductances, the simplified formula used with resistances may also be employed:

$$L_T = \frac{L_1 \times L_2}{L_1 + L_2}$$

Example: Two shielded coils—one rated at 30mH, the other at 15mH—are placed in parallel. What is the total inductance?

Solution:

$$L_T = \frac{L_1 \times L_2}{L_1 + L_2}$$

$$L_T = \frac{30 \times 15}{30 + 15} = \frac{450}{45}$$

$$= 10\text{mH}$$

As noted above, these formulas for parallel inductors hold true only when the coils are shielded from each other. Any mutual inductance existing between inductances in parallel tends to reduce the total inductance. Thus, there can be no gain in total inductance and, hence, little practical use for unshielded inductances in parallel. However, if by chance this type of mutual inductance occurs, the total inductance may be found by the formula:

$$L_T = \frac{(L_1 \pm L_M)(L_2 \pm L_M)}{(L_1 \pm L_M) + (L_2 \pm L_M)}$$

TYPES OF INDUCTORS

There are several types of inductors or coils used in electronic circuits. They are generally classified by their function and/or the type of material used for the *core* of the inductor. The core refers to the substance occupying the space enclosed by the turns of the inductor.

The material used for the inductor core may be either *magnetic* or *nonmagnetic*. Inductors may also be shielded in magnetic cases to prevent magnetic fields or *stray inductance* from influencing the inductance of the circuit.

Air Core Inductors. Air core inductors are usually found in high-frequency circuits and are seldom larger than 5mH. These inductors may be either wound on nonmagnetic forms made of ceramic or phenolic materials or may be *self-supporting*, meaning it requires no coil form. The physical appearance of an air core, self-supporting inductor is shown in Fig. 12-9.

Powdered Iron (Ferrite) Core Inductors. Ferrite or powdered iron core inductors are primarily used in circuits operating at frequencies about the audio range and usually have less than 200mH of inductance. Because metallic core has a high inductance, it can be obtained with a smaller number of turns, resulting in lower resistance, smaller coil size, and a higher quality.

When it is required that the inductance of a unit be adjustable, the iron core of the inductor is arranged so that its position can be varied (Fig. 12-10). The inductance of the coil is then controlled, within fixed limits, by varying the amount of powdered iron or ferrite slug inserted in the coil. When the core is all the way in the coil, the inductance will be at its maximum value; and when the core is withdrawn from the coil, the inductance will be at its minimum value.

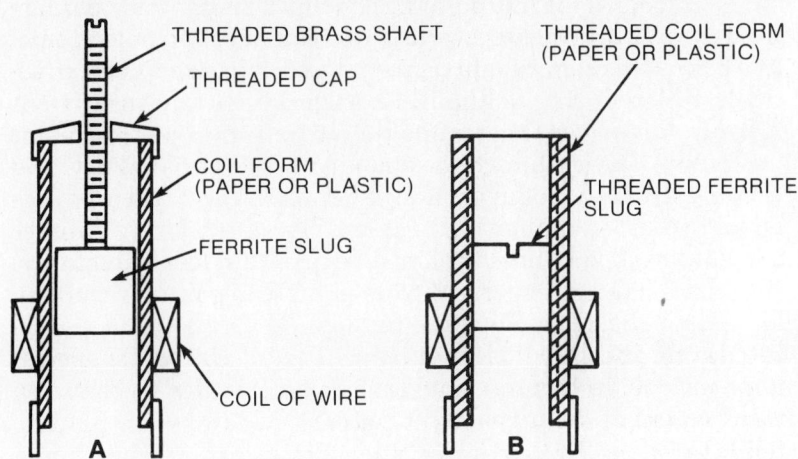

Fig. 12-10: Two methods of adjusting inductance: (A) threaded shaft and (B) threaded slug.

Toroid Inductors. One type of inductor used in electronic equipment has its coil of wire wound around the entire length of a doughnut-shaped powdered iron, or ferrite core (Fig. 12-11). These inductors are generally called *toroids* or *toroidal inductors*. Advantages of toroids are (1) high qualities can be obtained; (2) except for minor stray losses, their entire magnetic field is contained within the coil; and (3) inductance values can be kept to within 1% tolerance. Since all magnetic flux lines exist within the core, a toroidal inductor has a high inductance value for its physical size.

Molded Inductors. Molded inductors (Fig. 12-12) are encased in an insulating material in order to protect the inductive coil. These inductors can have cores of air, powdered iron, or ferrite. Some molded inductors are also available in a shield

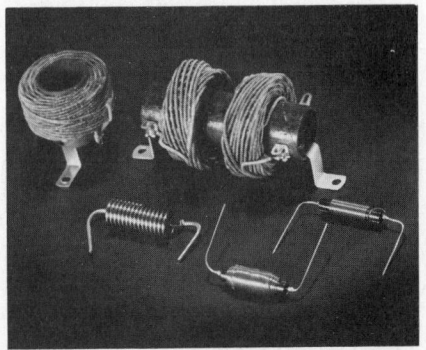

Fig. 12-9: Air core inductor.

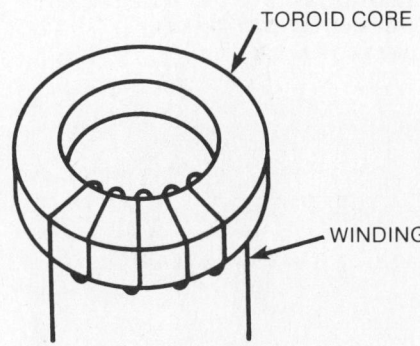

Fig. 12-11: Toroid inductor. All of the flux produced by the coil is concentrated in the low-reluctance core.

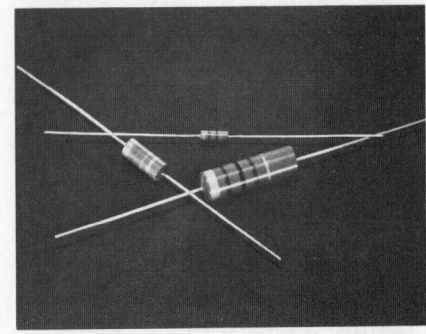

Fig. 12-12: Molded inductors.

227

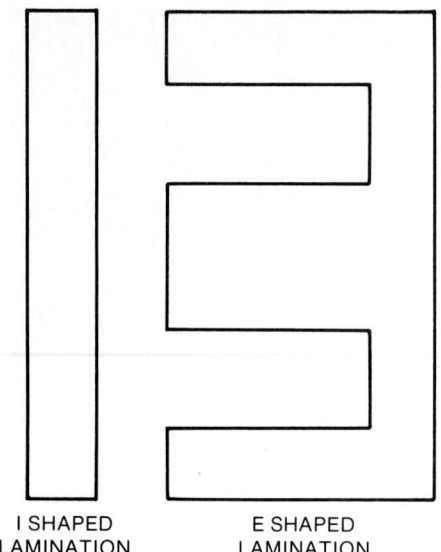

I SHAPED LAMINATION   E SHAPED LAMINATION

Fig. 12-13: I and E shaped laminations are stacked in various configurations for iron cores.

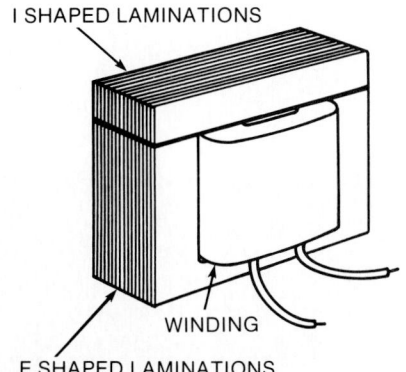

I SHAPED LAMINATIONS

WINDING

E SHAPED LAMINATIONS

Fig. 12-14: Laminated iron core inductor. The coil fits over the center leg of the E shaped laminations.

form; their shields are enclosed underneath the outside molding. Molded inductors in general are available that range from $0.1\mu H$ to 100mH.

Laminated Iron Core Inductors. Laminated iron cores are used for virtually all large inductors operated at power frequencies (for instance 60Hz). The inductances of such inductors range from approximately 0.1H to 100H.

The typical laminated inductor uses I and E shaped laminations like those illustrated in Fig. 12-13. Both the I and E shaped laminations are stacked together to the desired thickness and the winding is placed on the center extension of the E shaped lamination. The I shaped lamination is then situated across the open end of the E shaped lamination (Fig. 12-14).

Figure 12-15 shows the two parallel paths for flux formed by the I and E shaped laminations. The center extension of the E shaped lamination is twice as thick as the outside extensions because it must carry twice as much flux. The laminated iron core inductor produces more flux than other types of inductors for a given amount and rate of current change. This changing flux, in turn, produces a greater counter electromotive force.

It is the amount of current flowing through the winding that controls the magnetic field strength of the inductor. For example, in Fig. 12-16, suppose the current through an inductor changes from point A to point B. The flux density would change from A′ to B′ and create a certain amount of counter electromotive force. This cemf would represent a certain amount of inductance. Still referring to Fig. 12-16, assume now that the current in the inductor was greater and the current changes from point C to point D. Even though the amount of change between C and D is equal to the amount of change between A and B, it produces a much smaller change (C′ to D′) in flux. Therefore, at higher current levels, the inductor has less cemf and less inductance. Because a change in current from point E to point F results in almost no change in flux, the inductor is said to be in a saturated state and very little inductance occurs. Except for special applications, inductors should never be operated in the saturated region of the permeability curve.

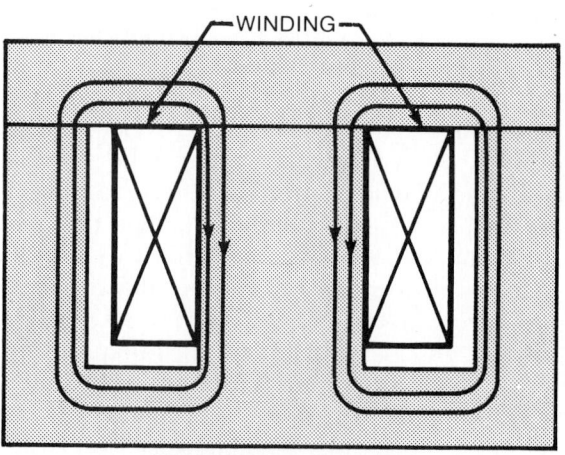

WINDING

CROSS-SECTIONAL VIEW

Fig. 12-15: Flux paths in a laminated core.

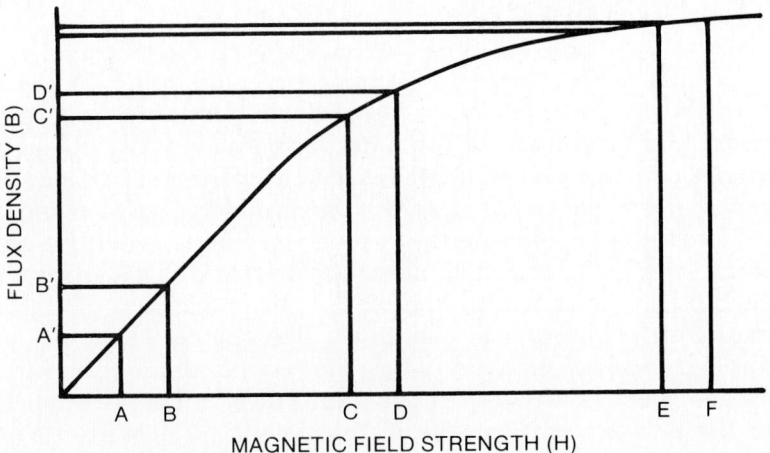

Fig. 12-16: Permeability decreases as the flux density and magnetic field strength increase.

Filter Chokes. Laminated iron core inductors are often called filter chokes (Fig. 12-17). While filter chokes are used in a wide variety of electrical and electronic equipment, they are seen most often in power supply filtering circuits. Their function in such a filter circuit is to smooth out the AC or fluctuating DC until it is nearly pure DC.

There are two basic types of filter chokes: the *swinging choke* and the *smoothing choke*. The swinging choke is one in which the I shaped lamination and the E shaped lamination are closely butted together; thus, there is a minimum air gap between them. This makes the amount of inductance change with the amount of current. A common swinging choke may be rated 20H at 50mA and 5H at 200mA.

The smoothing choke usually has a very small air gap (about 0.1mm thick) between the I and E shaped laminations. Since air does not saturate as easily as iron, this makes the inductance less dependent on the amount of current.

RF Chokes. These choke coils are operated at frequencies above the audio range (Fig. 12-18). They are used in RF circuits to provide a high impedance to RF currents and a low impedance to DC. RF chokes are available at ratings from a fraction of 1μH to hundreds of mH and have low values of DC resistance compared with their impedance. These air core type inductors are usually wound, nonmetallic cores of ceramic or phenolic materials.

Schematic symbols for both fixed and variable inductors are given in Fig. 12-19.

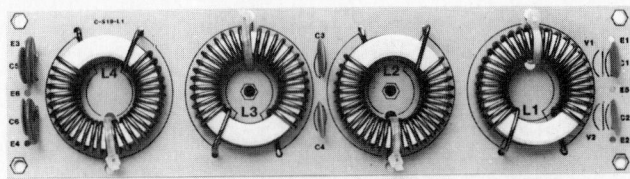

Fig. 12-17: Typical filter choke. (Photo furnished by courtesy of J.W. Miller Division/Bell Industries.)

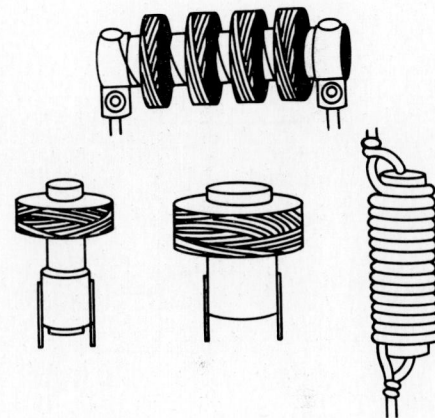

Fig. 12-18: Typical RF chokes.

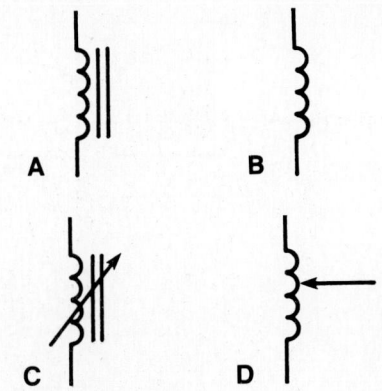

Fig. 12-19: Schematic symbols for fixed and variable inductors: (A) fixed iron core (magnetic); (B) fixed air core (nonmagnetic); (C) variable iron core (magnetic); and (D) variable air core (nonmagnetic).

TRANSFORMERS

One of the best working examples of mutual inductance is a *transformer*. They are extensively used in electronics to couple one circuit to another, to alter impedances so that two unequal values can be matched, to increase or decrease voltage amplitudes, and to isolate DC circuits while transferring AC signals. A transformer is constructed by placing coils sufficiently close to one another so that their magnetic lines and fields link with each other. As shown in Fig. 12-20A, the basic structure of a transformer consists of a primary winding to which AC is applied and a secondary winding that produces an output with an amplitude related to the transformer design. The AC applied to the primary can consist of any potential encountered in electronics, including 120V, 60Hz current, as well as audio signals, radio and television signals, square waves, and pulses. Transformers, however, cannot be operated on DC voltage because the magnetic field in the primary circuit must be moving in order to induce a voltage into the secondary circuit.

Transformers used at power line frequencies and audio frequencies are constructed with laminated high-permeability

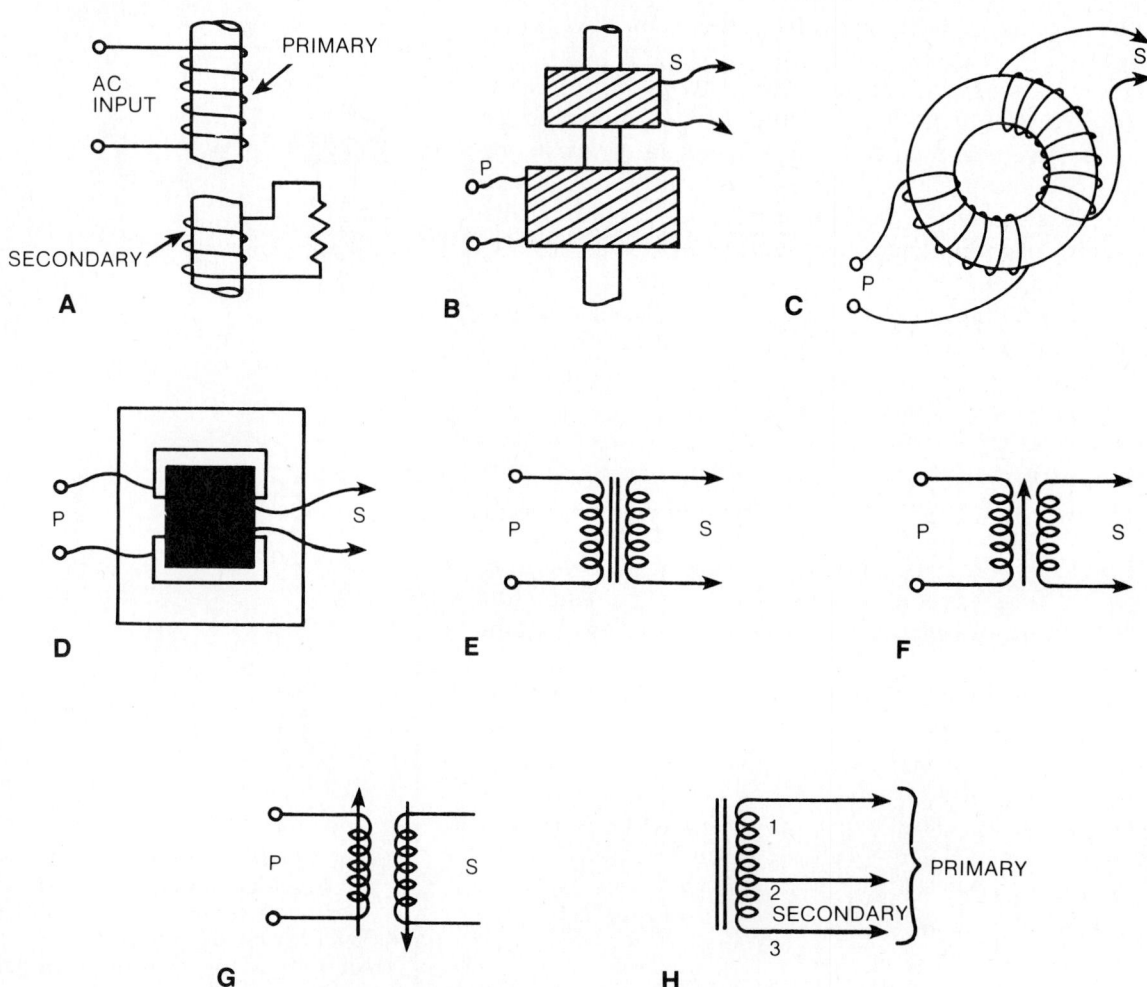

Fig. 12-20: Transformer representations.

230

soft iron cores. The coefficient of coupling is as high as possible, approaching the value of 1. Radio frequency transformers, on the other hand, are wound on cores of low permeability and low coefficient of coupling, from as low as 0.002 to 0.75. RF transformers are used in tuned circuit applications.

The arrangement shown in Fig. 12-20A is a transformer design often used for RF signals, with the actual component shown in Fig. 12-20B, where the windings are mounted on a plastic tube. A transformer may also be wound around a powdered iron core or a ferrite ring, as shown in Fig. 12-20C. (Ferrite has a higher permeability than air and increases transformer efficiency.) Another arrangement is shown in Fig. 12-20D; the primary and secondary windings are wound on the center arm of a laminated metal core. Laminations are made of thin metal sheets often stamped out in E formation so that the open E portions can be inserted through the coil cores to form a surrounding metal section, as shown. Laminations are used instead of a solid core to reduce eddy-current losses, which are caused by circulating currents within core material. Sometimes the laminations are treated so that they oxidize to produce insulating characteristics. Varnish is also used.

A transformer is commonly shown in schematics as in Fig. 12-20E, with the core material represented by the straight lines between the primary and secondary coils. If an arrow is shown, as in Fig. 12-20F, it indicates a movable core that can be adjusted for tuning. Often, compressed metallic particles are used for this. The schematic representation shown in Fig. 12-20G, a double variable core, may also be used. The upward-pointing arrow denotes that the tuning slot for this coil is available from the top of the chassis; a downward-pointing arrow indicates that the tuning slot is available from the bottom of the chassis.

Figure 12-20H shows the schematic for a transformer utilizing a continuous single coil tapped as shown. The windings between terminals 1 and 3 form the primary winding, and the windings between terminals 2 and 3 form the secondary winding. This is called an *autotransformer*. (If leads 2 and 3 formed the primary and leads 1 and 2 formed the secondary, a true autotransformer would not be formed, since it would be equal to a separate primary and secondary with common connections at lead 2.)

Turns or Voltage Ratio

The input side of a transformer (L_1) is called the *primary winding* and the output winding (L_2), which is cut by the magnetic lines of force from the primary winding, is called the *secondary winding*. Note that the primary winding receives energy from the source and the secondary winding delivers energy to the load(s). When AC is applied to the primary, the changing currents induce a voltage across the secondary winding. If the secondary winding were connected to a resistor or other component forming a closed circuit, current would flow in the secondary. The energy utilized by a resistor in the secondary is obtained from the primary by induction. The voltage is

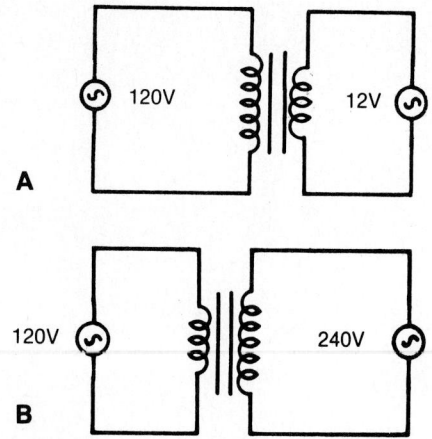

Fig. 12-21: (A) Step-down transformer; (B) step-up transformer.

proportional to the number of turns; hence, for a secondary winding with fewer turns than the primary, the voltage is stepped down across the transformer (Fig. 12-21A). For a secondary winding with more turns than the primary, voltage is stepped up (Fig. 12-21B). In equation form, these factors are expressed as follows:

$$\frac{V_p}{V_s} = \frac{N_p}{N_s} = a$$

where

V_p and V_s are the primary and secondary voltages,

N_p and N_s are the primary and secondary number of turns,

V_p/V_s is the voltage ratio between the primary and the secondary,

N_p/N_s is the turns ratio between the primary and the secondary,

a is the transformation ratio.

As an example, assume that the secondary of a transformer is to deliver 30V. The primary has 400 turns and the AC source is 120V. How many turns must the secondary winding have? Using the turns ratio formula, the following is obtained:

$$a = 120V/30V = 4$$
$$N_s = 400/4 = 100 \text{ turns}$$

The secondary output of a transformer may be in phase or 180° out of phase with the primary voltage (Fig. 12-22), depending upon the direction of the windings. Restating Lenz' Law, the polarity of the induced voltage will be opposite to the voltage producing it.

Current Ratio

The current available from the secondary of a transformer is inversely proportional to the number of turns. For a given primary winding, there is a limit to the power that it can handle. As available current is increased in the secondary, the available voltage is reduced. Similarly, as available voltage is increased, the available current is reduced. The power drawn from the secondary must be obtained from the primary. If it is

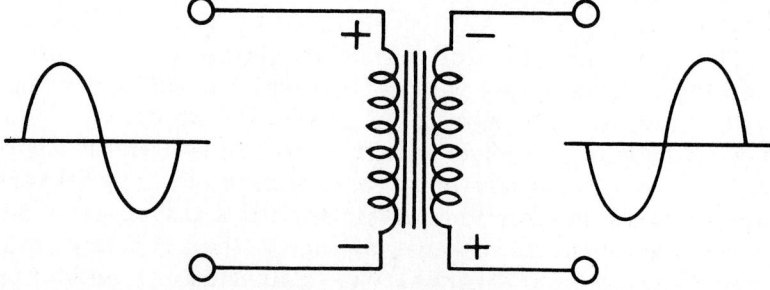

Fig. 12-22: The normal phase relationship between the primary and secondary of the transformer.

assumed that the transformer is 100% efficient, then the power in the secondary must equal the power in the primary winding; that is:

$$I_p \times V_p = I_s \times V_s$$

Suppose a step-up transformer were to produce 500V in the secondary when 100V is applied to the primary. A current of 1A would flow if a 500Ω load resistance were connected across the secondary.

$$I = \frac{500}{500}$$
$$I = 1A$$

Thus, 500W of power would be used in the secondary:

$$P_s = I_s \times V_s$$
$$P_s = 1 \times 500$$
$$P_s = 500W$$

Since the power in the secondary must be supplied by the primary, the primary current equals:

$$I_p = P_p/V_p$$
$$I_p = \frac{500}{500}$$
$$I_p = 1A$$

Efficiency of Transformers

The maximum transfer of power between the primary and secondary occurs when both windings are wound so that the turns overlap, or when one winding is wound over the other. Such *tight coupling* is also called *overcoupling,* and it assures that virtually all the lines of force of both windings interact fully. These transformers are highly efficient, and losses can be held to less than 5%. The ratio of power out to power in determines the efficiency of a transformer:

$$\% \text{ Efficiency} = \frac{P_{out}}{P_{in}} \times 100 \text{ or } \frac{P_s}{P_p} \times 100$$

There are four types of losses associated with transformer construction: *hysteresis losses, eddy current losses, copper losses,* and losses resulting from *skin effect.*

Hysteresis Losses. The molecular friction which results when magnetic particles change polarity in step with an induced voltage is called a *hysteresis loss.* Heat energy is released every time the magnetic polarity is switched and the residual magnetism of the last polarity has to be overcome.

A typical hysteresis loop is shown in Fig. 12-6. Hysteresis losses occur in the area between the level of residual magnetism and the X-axis, which is the level of zero flux density. Therefore, the narrower the hysteresis loop, the smaller the hysteresis losses will be. By using a special alloy, such as a silicon steel, hysteresis losses can be kept quite low. This type of laminated core has a narrow hysteresis loop *and* high permeability.

The laminated iron alloy core transformer is not generally used above the audio frequency range. This is because hysteresis losses will increase when the frequency of the primary current increases.

Eddy Current Losses. Because the iron core of a transformer is a conductor, a voltage will be induced into the core by the alternating current on the primary winding. This voltage will cause an eddy current to flow in a circle through the cross section of the core. This current flow will result in power dissipated in the form of heat. The amount of power dissipated equals:

$$P = IV$$

I = eddy current

V = induced voltage on the core

If the value of eddy current must be reduced, the resistance of the core must be increased. (Remember, $I = V/R$ and the induced voltage is a fixed value.) This can be done by breaking up the core structure into thin layers of metal called *laminations*. Each of these layers is insulated from the next by an oxide coating. The resistance which opposes the flow of eddy current can be increased by making the laminations thinner. At the same time, however, the thinner laminations decrease the permeability of the core. All of these factors must be considered in the design of the transformer. The most efficient lamination shape is shown in Fig. 12-23.

Copper Losses. Copper loss is the power dissipated by the DC resistance of the transformer windings. It is sometimes called the I^2R losses. Copper losses can be minimized by using as large a wire size as is practical for the size of the transformer.

Skin Effect. Radio-frequency resistance is called *skin effect*. Radio-frequency currents have a tendency to flow near the surface of a conductor instead of through it. The use of fine multistrand conductors, called *Litz wire,* for coils increases the total surface area of the conductor in the coil and reduces the RF resistance in the same way that a large diameter wire reduces DC resistance.

Inductors used in radio-frequency applications such as antenna coils and radio-frequency transformers are wound of Litz wire which is very small in diameter and thus has a relatively large DC resistance per foot but a low RF resistance. Litz wire consists of a number of very fine strands of wire, each insulated by a thin fabric material, twisted together to form one conductor. The advantage of Litz wire is its low resistance to RF currents.

Impedance Matching

The current in the primary winding of an unloaded transformer is nearly 90° out of phase with the voltage. The copper

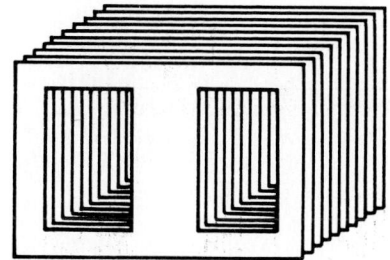

Fig. 12-23: Efficient lamination shape.

and core losses are the reasons why the current is not quite 90° out of phase with the voltage.

Both the primary current and its phase angle change when a resistive load is connected to the transformer's secondary. As more power is supplied to the secondary load, the transformer appears more resistive and less inductive. This change in phase angle can be observed for different loads in Fig. 12-24.

Also notice in Fig. 12-24 that the magnitude of primary current increases as the load resistance increases. This is caused when the current in the secondary coil creates a magnetizing force which opposes the magnetizing force of the primary. Because the flux created by the primary produces the secondary voltage and current, the primary current must increase to provide a greater magnetizing force. The primary current flowing due to the load on the secondary will be in phase with the primary voltage; therefore, the total primary current will consist of the reflected resistive current and the energizing current (Fig. 12-25). The more the transformer is loaded, the more resistive the transformer appears to the power source.

One of the most important features of a transformer in an electronic circuit is its ability to make a small load impedance appear to be large and equal to the large internal impedance of a voltage source, or to make the small impedance of a voltage source appear large and equal to a large load impedance. This transformation is called *impedance matching*. Impedance matching is one of the most common uses of transformers other than providing power for power supplies. Some examples where transformers are used for impedance matching are:

1. Matching loudspeakers to audio amplifiers or stereo equipment.

2. Matching deflection coils to power amplifiers or transistors in television receivers, television cameras, and radar receivers.

3. Matching any low-resistance device to a high-resistance device, and vice versa.

The transformer, therefore, has the ability to effectively transform impedance; that is, to step up or step down impedance as demonstrated in the following development of the impedance matching equation for transformers.

As stated earlier:

$$\frac{V_p}{V_s} = \frac{I_s}{I_p} = \frac{N_p}{N_s} = a \text{ (turns ratio)}$$

Therefore:

$$V_p = \frac{N_p}{N_s} V_s = a V_s$$

$$\text{and } I_p = \frac{N_s}{N_p} I_s = I_s / a$$

Because

$$Z_p = \frac{V_p}{I_p}$$

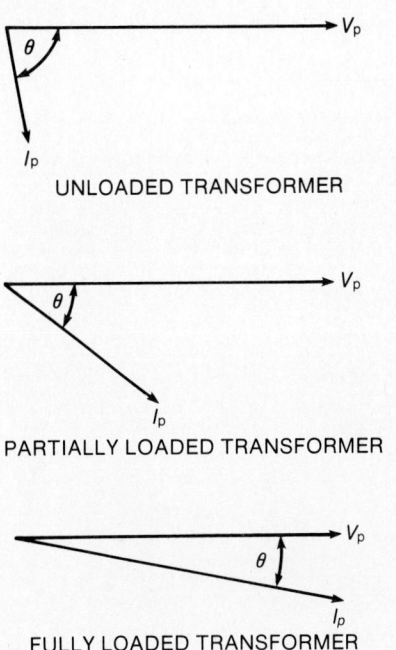

Fig. 12-24: **Primary current and voltage phase relationships. When fully loaded, a transformer appears to be close to a resistive load.**

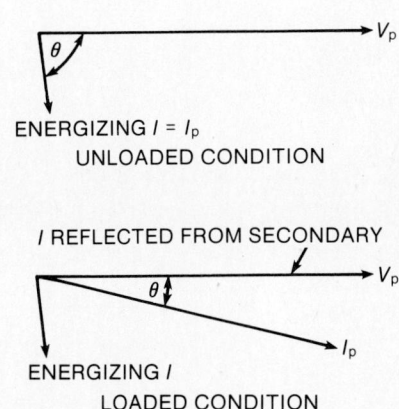

Fig. 12-25: **Reflected load current.**

235

can also be expressed as

$$Z_p = \frac{\dfrac{N_p}{N_s} V_s}{\dfrac{N_s}{N_p} I_s}$$

$$Z_p = \frac{\dfrac{N_p}{N_s}}{\dfrac{N_p}{N_s}} Z_s$$

$$Z_p = Z_s \left(\frac{N_p}{N_s} \times \frac{N_p}{N_s} \right)$$

$$Z_p = Z_s \left(\frac{N_p}{N_s} \right)^2$$

it can be said that the load connected across the secondary reflects its impedance across the transformer to the primary (since Z_s consists mainly of the load impedance). Hence, Z_p is often referred to as the *reflected impedance*. This reflected impedance can also be expressed as:

$$Z_p = a^2 Z_s$$

Therefore,

$$a^2 = \frac{Z_p}{Z_s}$$

The turns ratio necessary to match two impedances can be expressed as:

$$a = \sqrt{\frac{Z_p}{Z_s}}$$

Example: A voltage source has an internal impedance of $3{,}600\Omega$. The load is 100Ω. What transformer turns ratio will make the two impedances appear equal?

Solution:

$$a = \sqrt{\frac{Z_p}{Z_s}} = \sqrt{\frac{3{,}600}{100}}$$

$$a = 6{:}1$$

That is, the two impedances will appear equal if the primary winding has 6 turns for every 1 turn in the secondary winding.

Example: A transistor amplifier requires a $2{,}400\Omega$ load impedance. The speaker has an impedance of 8Ω. The primary has 865 turns. How many turns must the secondary winding contain?

Solution:

$$a = \sqrt{\frac{2{,}400}{8}} = \sqrt{300} = 17.3{:}1$$

$$N_s = \frac{N_p}{a} = \frac{865}{17.3} = 50 \text{ turns}$$

Example: The turns ratio of a transformer was tested by applying 10V to the secondary and reading the primary voltage to find the voltage ratio. The primary voltage reads 200V. If a 4,000Ω generator is to be connected to the primary, what is the load impedance for a match?

Solution:

$$\frac{200}{10} = 20:1 \text{ turns ratio}$$

turns ratio squared (a^2) = 20 × 20 = 400

$$a^2 = \frac{Z_p}{Z_L} = \frac{4,000}{Z_L}$$

Therefore,

$$400 = \frac{4,000}{Z_L}$$

$$Z_L = \frac{4,000}{400} = 10Ω$$

Up to this point, only transformers that have a single secondary winding have been discussed. Many transformers, however, have a number of secondary windings; in this case, calculations for turns ratio, impedances, and other values must be applied to each secondary winding and then compared to the primary; that is, each secondary winding must be considered individually as though no other windings existed. On occasion, a secondary winding may be tapped for voltage division or impedance separation. Typical of such applications is the power supply transformer shown in Fig. 12-26. This type found extensive applications in tube-type television receivers and older audio systems that used vacuum tubes. A 5V secondary winding supplies the rectifier tube filament, while a 6.3V winding supplies energy to tube filaments. The high-voltage secondary is tapped to deliver 300V on each side of the common

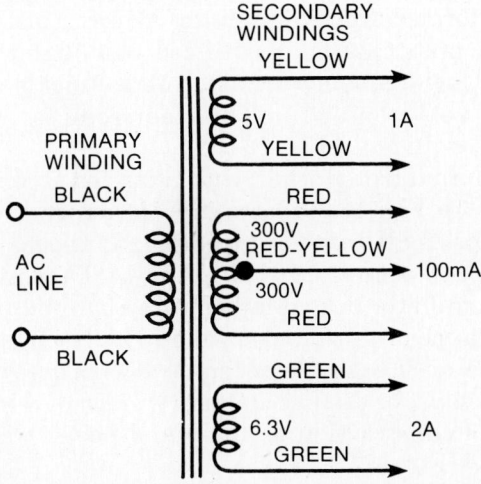

Fig. 12-26: Power supply transformer with typical current and voltage.

center when used in full-wave rectification circuitry. Current ratings are the maximum values that the transformer can deliver without overheating. Actual currents drawn during operation, however, depend on the resistances of the applied load circuitry. The color code for power transformers is given in Fig. 12-26.

Most transformers that handle RF signals (as well as individual inductors) are surrounded by a metal enclosure known as a *shield* (Fig. 12-27). The shield minimizes the effect of magnetic fields on adjacent circuits and components. Unshielded coils can cause interference signals in sensitive circuits. Often power transformers use a metal container as a shield so that interfering fields will not cause an audible hum in high-fidelity systems. A shield is usually connected to the metal chassis or a grounding connection. Since a shield surrounds an inductor, it intercepts magnetic fields, and an induced current is established that opposes the magnetic field of the coil. Thus, the effect of the field on nearby sections of the circuit is minimized.

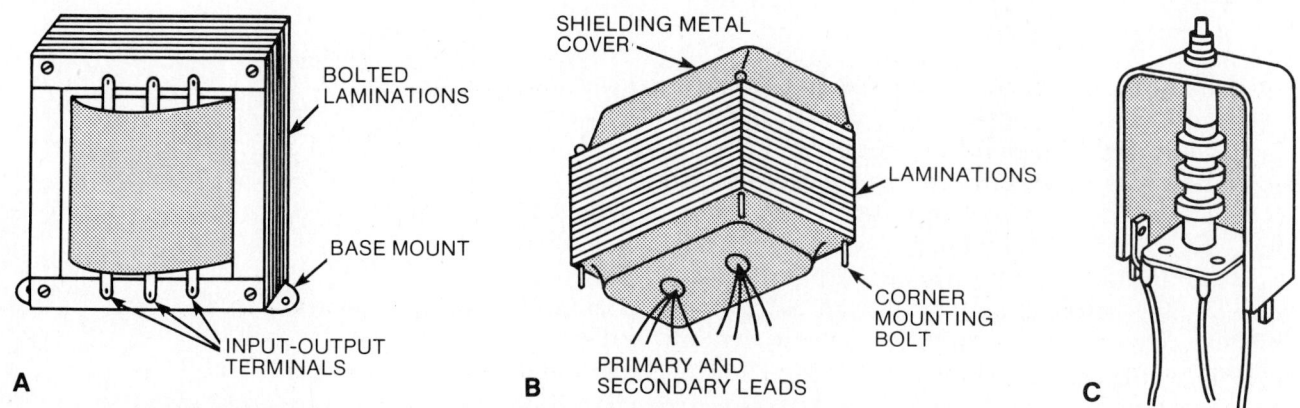

Fig. 12-27: (A) Unshielded-type transformer; (B) shielded chassis mount transformer; and (C) shielded RF transformer.

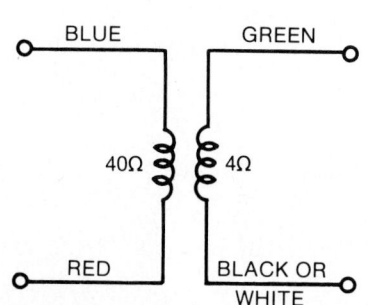

Fig. 12-28: Color code and typical DC resistance for RF transformer.

Since a shield intercepts some energy, some slight signal loss usually occurs. Most shields for high-frequency transformers or coils are formed from aluminum. Copper, brass, or other nonmagnetic materials can be utilized, although these are expensive. For low-frequency transformers, shields are usually composed of soft iron or tin. The color code for RF air core transformers is given in Fig. 12-28.

An audio output transformer with a tapped secondary winding is shown in Fig. 12-29. Here, the transformer will reflect either a 16Ω or an 8Ω speaker impedance to appear as a 9,000Ω load to the circuit connected to the primary. Thus, such a transformer can match the output impedance of a tube or transistor to the impedance of the speaker voice coil. Some audio transformers have only a single secondary winding; others may have several taps, or dual secondary windings. There are two basic types of construction of audio and power transformers: core and shell (Fig. 12-30).

Transformers are also used as isolation devices for DC and AC, for they will pass AC but keep DC confined to the primary

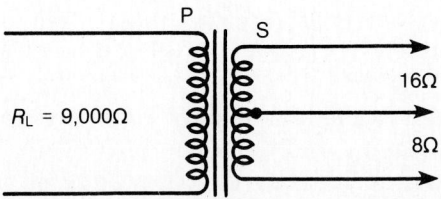

Fig. 12-29: Audio output transformer with typical DC resistances.

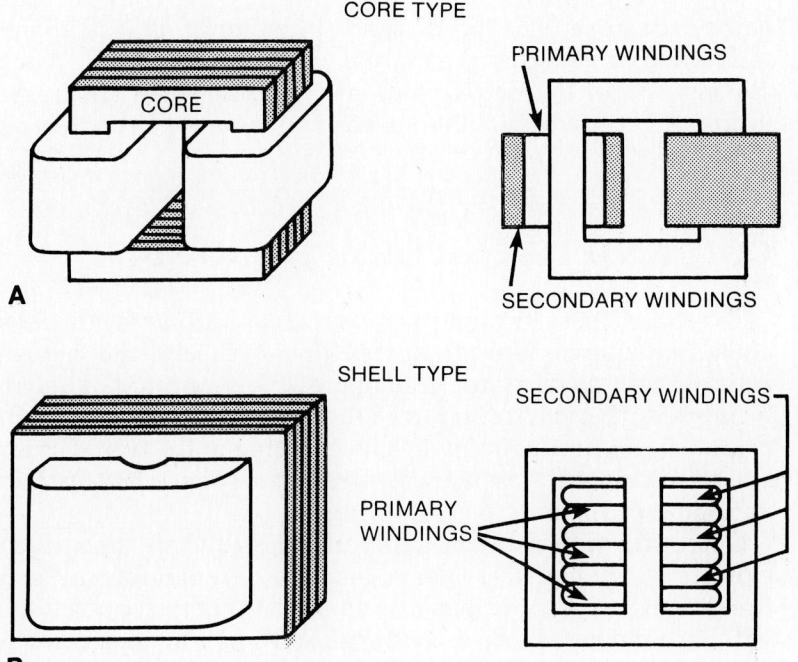

CORE TYPE

PRIMARY WINDINGS

CORE

SECONDARY WINDINGS

SHELL TYPE

SECONDARY WINDINGS

PRIMARY WINDINGS

Fig. 12-30: The construction of (A) core and (B) shell type transformers.

winding. Transformers with primary and secondary windings also minimize the danger of shock by separating the AC power mains (and hence the electrical grounding system) from the apparatus connected to the secondary. Where this consideration is not a factor, the autotransformer is sometimes employed. This transformer uses a single winding tapped as shown in Fig. 12-31. Thus, the line AC is applied to the lower two terminals and the load circuit to the upper two. In such an instance, the center terminal is common to both the load and the source AC. The load current flows through the entire winding. The autotransformer shown in Fig. 12-31A is a step-up type, and the one in Fig. 12-31B is a step-down unit.

Variable transformers—generally of the autotransformer type—are available with adjustable secondaries. The voltage on the secondary can usually be adjusted from 0% to 110% of the primary voltage.

Dot Notation. The transformers already discussed have all had the secondary induced voltage 180° out of phase with the primary voltage. However, it is possible to have the voltage on

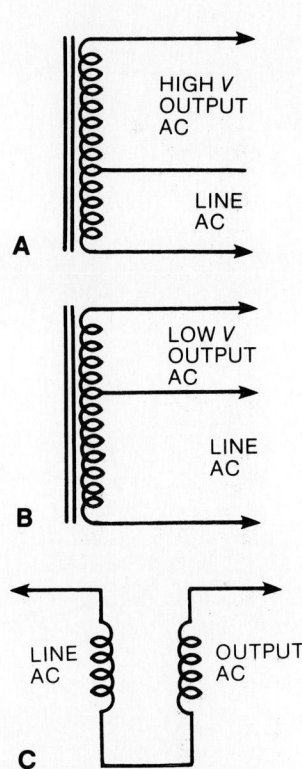

HIGH V OUTPUT AC

LINE AC

A

LOW V OUTPUT AC

LINE AC

B

LINE AC

OUTPUT AC

C

Fig. 12-31: Autotransformer arrangements: (A) step-up, (B) step-down, and (C) alternate drawing.

239

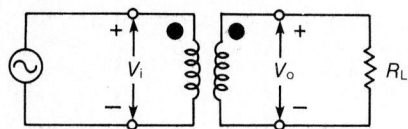

Fig. 12-32: Dot notation as used with transformers.

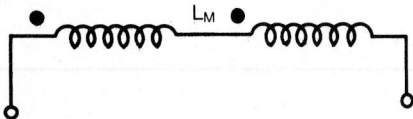

Fig. 12-33: Dot notation as used with coils that are aiding.

both sides of the transformer in phase. This can be accomplished simply by reversing the direction of the secondary windings.

Dot notation is a method of showing the phase relationship between the primary and the secondary. A dot is used at one end of each coil to identify in-phase ends of each coil (Fig. 12-32). Thus, when the primary voltage has the instantaneous polarity shown (dotted at the positive end of the coil), the secondary voltage is also positive at the dotted end of the transformer.

The dot notation method is also used in coils to show that the same ends have the same direction of winding. This is important when considering the mutual coupling of a circuit. When current enters the dotted ends of two coils, their fields are *aiding* and mutual inductance has the same sense as the coil inductance (Fig. 12-33).

COIL AND TRANSFORMER TROUBLES

To properly use a transformer, its *voltage* and *current ratings* should be known. Manufacturers always specify the voltage ratings of the primary and secondary windings, but they usually only specify the current rating for the secondary winding. Of course, from these ratings the power rating of the transformer can be calculated. Most transformers are also specified by their voltampere (apparent power) ratings.

Operating transformers and coils higher than their rated capacities can cause problems such as open windings. When the primary of a transformer is open, for example, no voltage will be found across the secondary windings. An open secondary winding will be indicated by no voltage across its load. Remember that if a transformer has several secondary windings, the open winding in one secondary generally does not affect transformer action for the other secondary windings that are operating properly. A short across the secondary will often cause the primary to burn out and result in an open.

The DC resistance of both coils and transformer can be measured with an ohmmeter. The amount of DC resistance will depend on the current rating, the inductance of the coil (number of turns), the diameter of the wire, and the application in which the inductor is used. For example, the RF inductors, employing only 10 to 100 turns of fine wire, may vary from a fraction of an ohm to about $25\,\Omega$, while those used for audio frequencies and 60Hz power circuits with several hundred turns may have DC resistances ranging from $10\,\Omega$ to over $500\,\Omega$, depending on the wire size. (The heavier the wire, the less the resistance.) An open winding will give an infinite resistance indicated on the ohmmeter.

When a coil or transformer is defective, it is best to replace the entire unit with one of the same rating. If, however, a single unit of the desired value is unobtainable, the correct total value can be procured by using the series or parallel combinations (without interacting fields).

SUMMARY

Inductance is the characteristic of an electrical circuit that opposes any change in the amount of current flow. The counter electromotive force is the self-induced voltage that opposes any change in current. This induced emf always has such a direction as to oppose the action that produces it.

The unit for measuring inductance (L) is the henry (H). An inductor has an inductance of 1H if 1V is induced in the inductor when the current through the inductor is changing at the rate of 1A per second.

Mutual inductance occurs when the changing flux of one coil induces a voltage on a neighboring coil. The degree of mutual inductance between two coils is known as the coefficient of coupling (K).

The sum of series inductors determines the total inductance for that circuit. If two series inductors are magnetically coupled, this mutual inductance must also be accounted for: $L_T = L_1 + L_2 \pm 2Lm$. The total inductance of parallel inductors is determined in the same manner as that used for resistors in parallel. If two parallel inductors are magnetically coupled, this mutual inductance must also be accounted for:

$$L_T = \frac{(L_1 \pm L_M)(L_2 \pm L_M)}{(L_1 \pm L_M) + (L_2 \pm L_M)}$$

There are several types of inductors: air core, powdered iron core, toroid, molded, laminated iron core, filter chokes, and RF chokes. Each has its specific purpose.

The transformer is a device that utilizes the effects of mutual inductance. The input side of a transformer is called the primary winding and the output winding, which is cut by the flux of the primary, is called the secondary. The transformer ratio (a) equals N_p/N_s (or V_p/V_s).

Hysteresis losses, eddy current losses, copper losses, and the skin effect are all losses in power due to the construction of the transformer.

The transformer has the ability to step up or step down voltages or currents as well as to match impedance. It is also a device used for isolation.

Operating transformers and coils higher than their rated capacitors can cause problems such as open windings.

CHECK YOURSELF (Answers at back of book)

Questions

Fill in the blanks with the appropriate word or words.

1. When the fields of two series coils are aiding, total inductance is increased by a factor of twice the _____ inductance.
2. The greater the _____ of the circuit, the greater the opposition to a change in current.

3. A short across the _____ of a transformer will often cause the _____ to burn out and result in an open.
4. _____ is a method of showing the phase relationship between the primary and secondary of a transformer.
5. An induced emf always has such a direction as to _____ the action that produces it.
6. A _____ inductor has its coil of wire wound around the entire parameter of a doughnut-shaped iron core.
7. Tight coupling is also called _____ .
8. _____ losses occur every time the magnetic polarity is switched and the residual magnetism of the last polarity has to be overcome.
9. Z_p is often referred to as the _____ impedance.
10. Transformers will pass _____ but keep _____ confined to the primary winding.
11. _____ first law of motion states that an object in motion tends to remain in motion and that a considerable _____ must be exerted to bring it to a stop.
12. The inductance of a coil is directly proportional to the square of the _____ .
13. The inductance of a coil decreases as its _____ increases.
14. _____ inductors are usually found in high-frequency circuits and are seldom larger than 5mH.
15. _____ are used in RF circuits to provide high impedance to RF currents and low impedance to DC.
16. An inductance of one _____ will induce 1V on the coil when the current in it is changing at the rate of 1A/s.

Problems

1. Three inductors whose inductances are 20mH, 40mH, and 50mH are located so that there is no mutual induction between them. What is the effective inductance when the three coils are connected in series?
2. What is the effective inductance when the above three coils are connected in parallel?
3. What voltage will be induced on a 10H inductor in which current changes from 1VA to 7A in 90ms?
4. A voltage of 50mV is induced in an inductor when the current changes at the rate of 200mA/s. What is the inductance of the coil?
5. What is the transformer ratio of a transformer having 40 turns on the secondary winding and 680 turns on the primary?
6. What is the coefficient of coupling between two coils if one has an inductance of 24H, the other has an inductance of 9H, and the mutual inductance of the two is 8.18H?
7. An impedance matching transformer is connected between 4,200Ω in the primary and 8Ω in the secondary. What must be the turns ratio to properly match the two impedances?

242

CHAPTER 13

Inductive Circuits

Inductors have differing effects on DC and AC circuits. It is important that these effects are understood since electronic circuits use both direct and alternating currents.

INDUCTORS IN DC CIRCUITS

An inductor has absolutely no effect upon direct current unless the DC is changing. In most DC circuits, the current flowing in the circuit is a constant value. When an inductor is used in such a circuit, only the resistance of the wire affects the current. The property of inductance depends upon a changing current which produces a self-induced voltage. When the current in a DC circuit changes, the inductance affects it. The current in a DC circuit usually changes only when the voltage is applied or removed.

Figure 13-1 shows an inductor, L, connected to a DC source through a switch. When the switch is in position A, the inductor is disabled and no DC is applied to it. When the switch is moved to position B, the DC voltage from the battery is connected to the inductor. When the switch is moved from position A to position B, the DC source voltage will cause current to flow. The instant current begins to flow, a magnetic field will be developed. As the magnetic field expands outward, it cuts the turns of the inductor and induces a counter emf. The polarity of the induced voltage opposes the applied voltage; therefore, the amount of current flowing in the circuit is initially limited. Because of the opposition of the induced voltage, the current in the circuit will not rise instantaneously. Instead, it takes a finite period of time for the current to rise to its maximum value. The graph in Fig. 13-2A shows how the current varies with time when a DC voltage is first applied to the inductor. Once the magnetic field stops expanding, no further counter emf will be induced. It is at this time that the total current in the circuit will be the maximum value as determined by the amplitude of the applied voltage and the resistance of the inductor.

When the switch is now moved from position B to position A, the connection to the DC source is broken. This means that the magnetic field in the inductor collapses. As it collapses, it induces a voltage as the magnetic field cuts the turns of the coil.

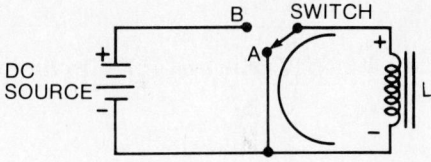

Fig. 13-1: Inductance in a DC circuit.

When the DC source voltage is removed, current does not drop to zero instantaneously. Instead the collapsing magnetic field induces a voltage which causes the current flow to continue in the same direction. When the magnetic field collapses completely, no further voltage will be induced and the current flow will drop to zero. Figure 13-2B shows the decay in current when the power is removed from the inductor.

As can be seen from the graphs in Fig. 13-2, the inductor clearly opposes changes in current in the DC circuit. When a voltage is applied to the inductor, the inductor itself opposes the rise of current in the circuit. When the voltage is removed from the circuit, the inductor again opposes the change in current flow. This time it tends to keep the current flowing.

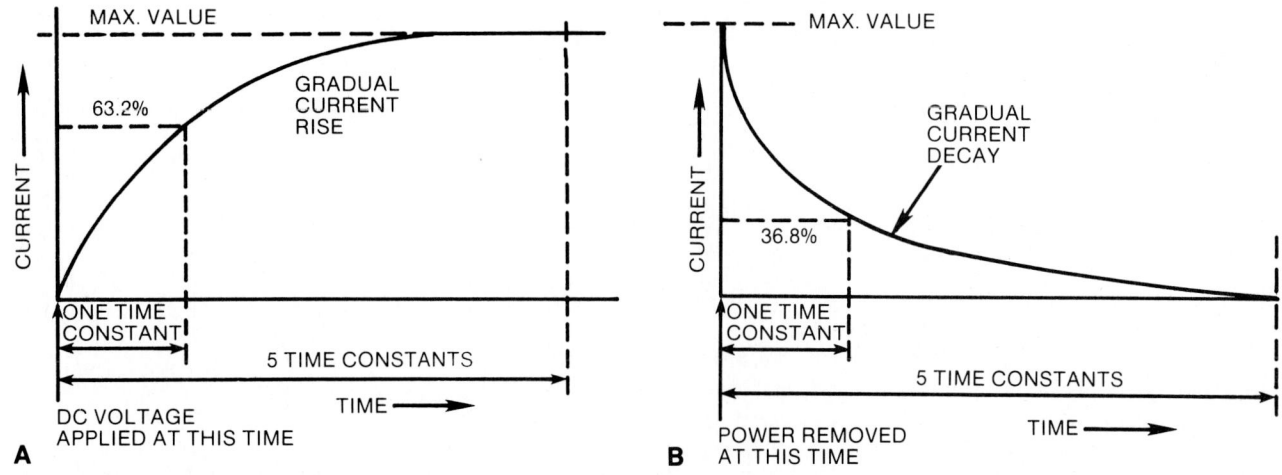

Fig. 13-2: **Variation of DC in an inductor when (A) power is applied and (B) power is removed.**

INDUCTORS IN AC CIRCUITS

As stated in Chapter 12, the opposition that an inductance offers to a changing current was called self-induced voltage or counter emf and was measured in volts. However, opposition to current flow is normally measured in ohms, not in volts. Since a coil reacts to a current change by generating a counter emf, a coil is said to be reactive. The opposition of a coil is, therefore, called reactance (X) and is measured in ohms. Since more than one kind of reactance exists, the subscript L is added to denote *inductive reactance*. Thus, the opposition offered by a coil to alternating current is termed inductive reactance, designated X_L.

Inductive Reactance

In the preceding chapter, it was shown that when an alternating voltage is applied to an inductor, the inductor reacts in such a way as to oppose alternating current flow. This opposition is thus referred to in general as an inductive reactance. The amount of opposition exhibited by a coil depends on the magni-

244

tude of its self-induced voltage, V_{ind}. The effective value of this self-induced voltage, or counter emf in V, is

$$V_{ind} = 2\pi fLI$$

where 2π is a constant, I is in effective A, L is in H, and f is in Hz. The induced voltage varies directly with the frequency, the inductance, and the current.

The amount of opposition to current flow offered by an inductor is referred to as its inductive reactance. This reactance is expressed as the ratio of the inductor's counter emf to the current through the inductor, or

$$X_L = \frac{V_{ind}}{I}$$

where X_L is the inductor's opposition to current flow measured in Ω, V_{ind} is the inductor's counter emf in V, and I is the inductor current in A. Assuming the current and counter voltage have sine waveforms, the inductive reactance X_L, in Ω, is

$$X_L = 2\pi fL$$

Note: The expression $2\pi f$ or $6.28 \times f$ is also known as the *angular velocity* and its symbol is the lowercase Greek letter *omega* (ω). Thus, the equation $X_L = 2\pi fL$ can also be expressed as $X_L = \omega L$.

From this formula it may be seen that the higher the frequency, or the greater the inductance, the greater the inductive reactance. Figure 13-3A is a graph of X_L plotted against frequency, and Fig. 13-3B is a graph of X_L plotted against inductance. The graphs are similar and illustrate that inductive reactance increases directly, or linearly, with frequency and inductance: the greater the frequency for any L, or the greater the inductance for any f, the greater the inductive reactance of the circuit.

At the high frequencies, even small amounts of inductance may offer very great inductive reactance. Figure 13-3C is a graph of current in an inductive circuit plotted against frequency. It may be said, then, that at high frequencies an induc-

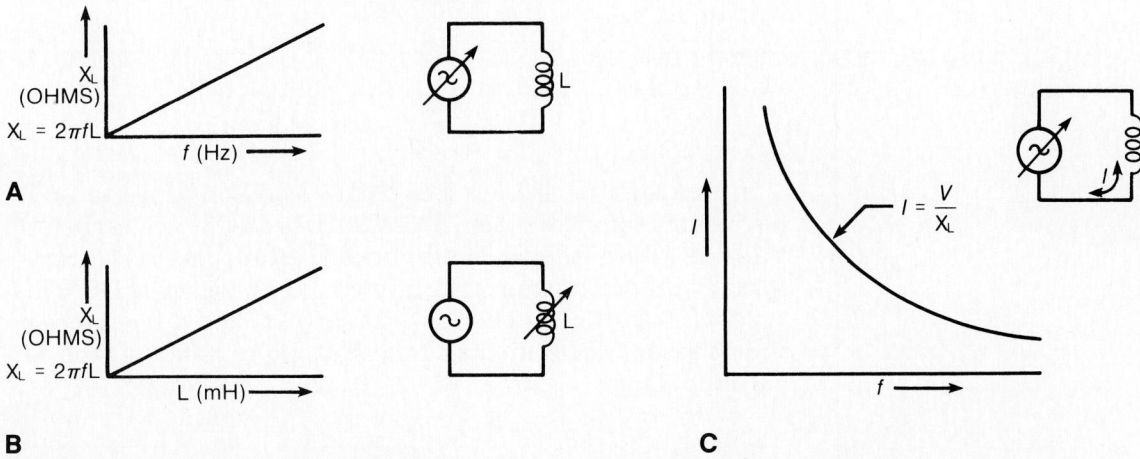

Fig. 13-3: Variation of inductive reactance in an AC circuit.

245

tance tends to act as an open circuit, since very little current flows in the circuit; and at low frequencies, an inductance tends to act as a simple conductor—a short circuit—since high current flows. Thus, inductances designed for use at low frequencies are generally large iron core inductances. They are measured in full units of H. Those used at high frequencies are generally small air core inductances, which are measured in mH (1/1,000 of a henry) or in μH (1/1,000,000 of a henry).

For example, the opposition offered by a 20H inductance to a 120V, 60Hz AC source may be determined by use of the formula for inductive reactance:

$$X_L = 2\pi fL$$
$$X_L = 6.28 \times 60 \times 20$$
$$X_L = 7,536\Omega$$

Ohm's Law applies to inductive AC circuits as it does to resistive circuits. The current flowing in an inductive AC circuit (I) is directly proportional to the applied voltage (V) and inversely proportional to the inductive reactance (X_L). This relationship is represented mathematically by the expression:

$$I_L = \frac{V_L}{X_L}$$

In this expression the current is in A, the voltage in V, and the reactance in Ω. Increasing the voltage or decreasing the reactance will cause an increase in the current. Decreasing the applied voltage or increasing the reactance will cause a decrease in the current. The current in the AC circuit given above is:

$$I_L = \frac{V_L}{X_L} = \frac{120}{7,536}$$
$$I_L = 0.016A$$

If the frequency of this same circuit is increased to 400Hz, the inductive reactance is increased:

$$X_L = 6.28 \times 400 \times 20$$
$$X_L = 50,240\Omega$$

Current falls to:

$$I_L = \frac{120}{50,240} = 0.0024A$$

In a transmitter operating on 5MHz (5 million Hz per second), a small inductance of 2.5mH (0.0025H) is used to keep substantially all alternating currents out of a certain part of the transmitter while at the same time allowing the passage of DC. This type of inductance is called a *choke* since inductive reactance at that frequency is quite high and effectively chokes off the AC currents. Thus:

$$X_L = 2\pi fL$$
$$X_L = 6.28 \times (5 \times 10^{+6}) \times (2.5 \times 10^{-3})$$
$$X_L = 78,500\Omega \text{ at 5MHz.}$$

Example: What is the inductive reactance of a 40mH coil connected across a 10kHz signal?
Solution:

$$X_L = 2\pi f L$$
$$= 6.28 \times (10 \times 10^3) \times (40 \times 10^{-3})$$
$$= 2,512\Omega$$

Example: The current in an inductive circuit is 100mA when 40V and 400Hz are connected across it. What is the inductance of the circuit?
Solution:

$$X_L = \frac{V_L}{I_L} = \frac{40}{(100 \times 10^{-3})}$$
$$= 400\Omega$$

From $X_L = 2\pi f L$
then,

$$L = \frac{X_L}{2\pi f} = \frac{400}{6.28 \times 400} = 0.159H$$
$$= 159mH$$

Example: What is the value of current which will flow through a 75mH coil when a 3,200Hz, 25V signal is applied to it?
Solution:

$$X_L = 2\pi f L$$
$$= 6.28 \times 3,200 \times (75 \times 10^{-3})$$
$$= 1,507.2\Omega$$

$$\text{current } (I_L) = \frac{V_L}{X_L} = \frac{25}{1,507.2}$$
$$= 16.587mA$$

Voltage and Current of an Inductance

Inductive reactance, as has been shown, not only limits the current flowing in an AC circuit, but also tends to limit the rising and the falling of current. In Fig. 13-4A, a sine wave of voltage is applied to a pure inductance. The current in the circuit also follows the form of the sine wave; but it is necessary to determine precisely the delay in time between the application of the maximum voltage and the moment the current reaches maximum value—that is, the phase shift between the voltage and current.

Figure 13-4B illustrates the voltage and current sine curves for this circuit. At a time corresponding to point a, the voltage, v, is positive, and the current is increasing with time. At point b, the voltage is negative, and current is decreasing with time. Thus, at point A (180° on the time axis), the curve for current must flatten out (neither increasing nor decreasing) as it passes through its maximum value. In like manner, the rest of the

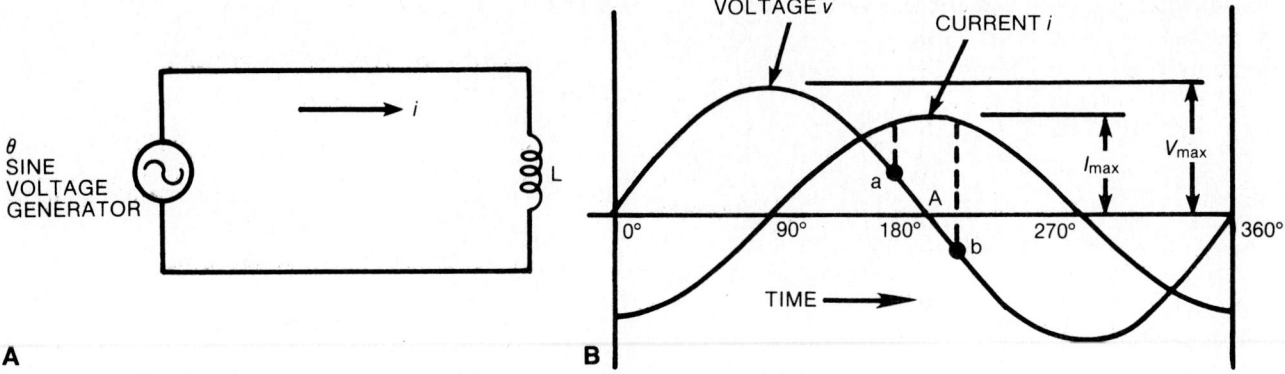

Fig. 13-4: Current and voltage in an inductive AC circuit.

current curve in B may be drawn. It may be seen, then, that current is zero when the applied voltage is maximum or maximum when the applied voltage is zero. Consequently, the current is said to *lag* the applied voltage by one-quarter cycle, or by 90°.

Voltage and Current in an RL Circuit: Phase Angle

In any resistive circuit without inductance (L equal to zero), the voltage and current are said to be *in phase*. But it should be noted that in no circuit is the value of the inductance ever actually zero, since, by definition, it would then be possible to change the value of the current instantaneously. However, for all practical purposes, those circuits which do not contain inductors, or appreciable amounts of inductance, can be considered as pure resistive circuits. The time lag of the current in such a circuit is so small as to be negligible. Figure 13-5 shows this circuit and illustrates graphically the in-phase relationship of a sine wave of voltage and current across a resistance. From the graph, it may be seen that the current and the voltage are alternating in polarity at the same frequency. Current rises as the applied voltage rises and is maximum when the applied voltage is maximum; current falls off as the voltage falls and is zero when the voltage is zero. The magnitude of the current may be determined by Ohm's Law for maximum, effective, or average value, since the frequency of the current change has no effect on the in-phase relationship. Thus:

$$I_{max} = \frac{V_{max}}{R}$$

$$I_{rms} = \frac{V_{rms}}{R}$$

$$I_{av} = \frac{V_{av}}{R}$$

In any circuit containing both inductance and resistance, there is a 90° phase shift of voltage across and current through the inductance alone and no phase shift across the resistance.

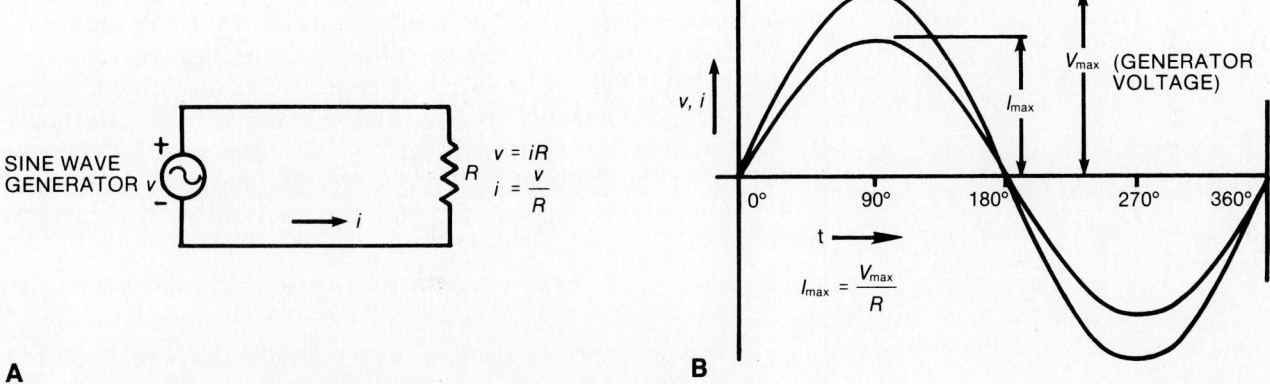

Fig. 13-5: Current and voltage in a resistive AC circuit.

But it must be emphasized that in any series circuit, current is the same throughout the circuit. Since current is the line of reference for both the inductance and the resistance, it may be seen that the voltage developed across the resistance (by the current through it) will be 90° out of phase with the voltage across the inductance. Figure 13-6A shows an RL circuit and Fig. 13-6B shows the relationships among the current, the voltage across the inductance, and the voltage across the resistance. Thus, it may be seen that the presence of resistance in an inductive circuit results in two separate voltage drops 90° out of phase with each other. The resultant voltage of these two voltages is the voltage drop in the whole circuit and is, by Kirchhoff's Law, equal to the applied voltage. The amount of phase shift of the current in such a circuit is measured, not with relation to the voltage across the inductance alone (always 90°), but with relation to the resultant voltage, which is the applied voltage.

The resultant voltage and the phase angle (generally ascribed the symbol θ) of any RL circuit may be determined by

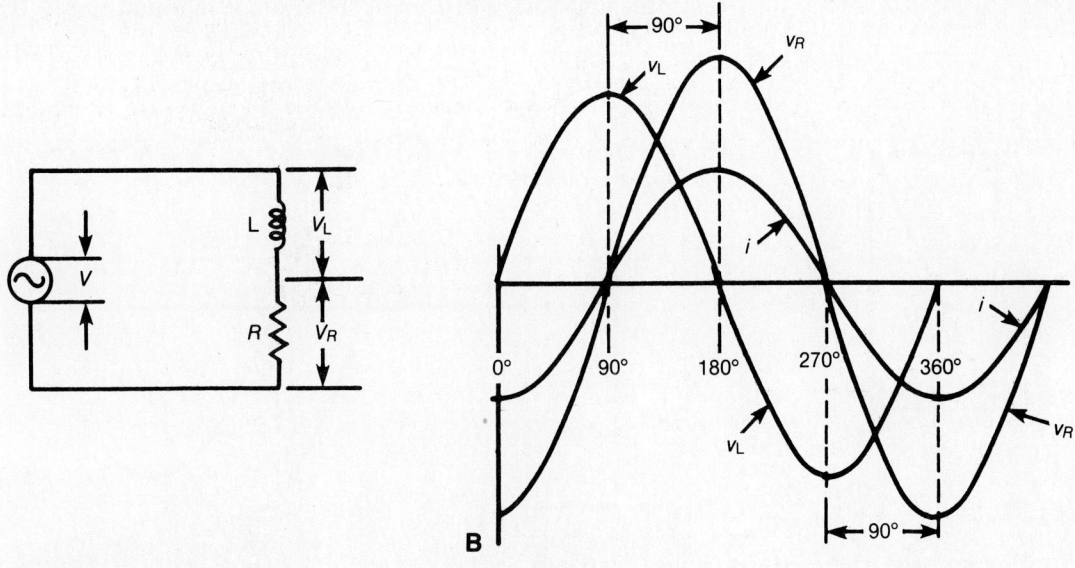

Fig. 13-6: Current and voltages in a series RL circuit.

249

means of phasors. In Fig. 13-7A, the voltage across the resistance is laid off on the horizontal phasor and the voltage across the inductance on the vertical phasor. Since these two voltages are 90° out of phase, the angle between them is a right angle. By drawing in a parallelogram based on these two sides, the resultant phasor V is seen to be the hypotenuse of a right triangle. Then, by the theorem of Pythagoras, the square on the hypotenuse is equal to the sum of the squares on the other two sides, or:

$$V^2 = V_R{}^2 + V_L{}^2$$
$$V = \sqrt{V_R{}^2 + V_L{}^2}$$

From the previous discussion, it is known that the current in the circuit is in phase with the voltage across the resistance; therefore, the position of the current with relation to the applied voltage is the same as the phasor V_R, the voltage across the resistance. The phase angle θ, then, is the angle that the applied voltage phasor V makes with the phasor V_R, as shown in Fig. 13-7A. The angle θ may be measured in terms of any of its trigonometric functions, depending on the values known. If the voltage across the resistance is large with relation to that across the inductance, the resultant phasor will approach the horizontal and the phase angle will be small; and in like manner, if the voltage across the resistance is small, the resultant phasor will approach the vertical and the phase angle will approach 90°. Hence, the presence of resistance in an inductive circuit causes current to lag the applied voltage by some angle *less* than 90°. Figure 13-7B illustrates in graph form the relative positions of voltage and current and the phase angle θ.

Inductive Reactance and Resistance: Impedance

In any series circuit containing both inductance and resistance, the total opposition offered by the circuit is not the simple arithmetical sum of the inductive reactance X_L and the

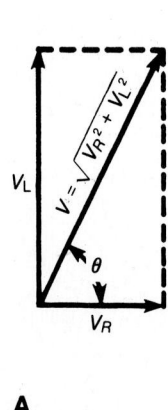

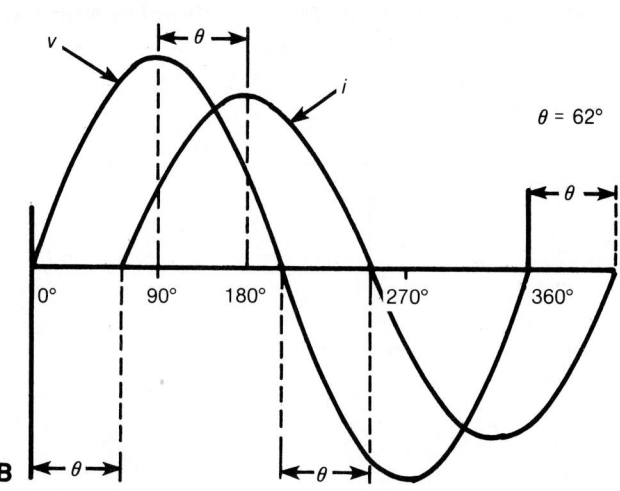

A   **B**

Fig. 13-7: Applied voltage and current in an *R*L circuit.

resistance R. The inductive reactance must be added to the resistance in such a manner as to take into account the 90° phase difference between the two voltages in the circuit. This total opposition is termed *impedance* and assigned the symbol Z. Since the voltage across the inductance is determined by the inductive reactance and the current through the inductance, then:

$$V_L = IX_L$$

The voltage across the resistance is determined by the resistance and the current through it:

$$V_R = IR$$

Then the resultant voltage of these two, or the applied voltage, is determined by the current and the total opposition of the circuit:

$$V = IZ$$

But as previously shown:

$$V = \sqrt{V_R^2 + V_L^2}$$

or

$$V = \sqrt{(IR)^2 + (IX_L)^2}$$

Then:

$$IZ = \sqrt{I^2 (R^2 + X_L^2)}$$
$$IZ = I \sqrt{R^2 + X_L^2}$$
$$Z = \sqrt{R^2 + X_L^2}$$

Thus, *the impedance of an RL series circuit is equal to the square root of the sum of the squares of the resistance and the inductive reactance.*

The same result may be obtained more readily by means of phasors. The voltage across the resistance V_R is equal to IR, and the voltage across the inductance V_L to IX_L. Since each phasor represents a product of which current is a common factor, the phasors may be laid off proportional to R and X_L and separated by 90°. Figure 13-8 shows these phasors. The resultant phasor Z is the hypotenuse of a right triangle and represents the impedance of the circuit. Then:

$$Z = \sqrt{R^2 + X_L^2}$$

It will be seen also that the angle θ is the phase angle because the direction of the impedance phasor is actually the same as that of the applied voltage phasor. This angle is generally determined in terms of its tangent, X_L/R, or in terms of its cosine, R/Z.

From the phasor diagram of Fig. 13-8, it may be seen that if the resistance is large with relation to the inductive reactance, the circuit tends to act as a pure resistive circuit, the phase angle approaches 0°, and the impedance approaches R. If the inductive reactance is large with relation to the resistance, the

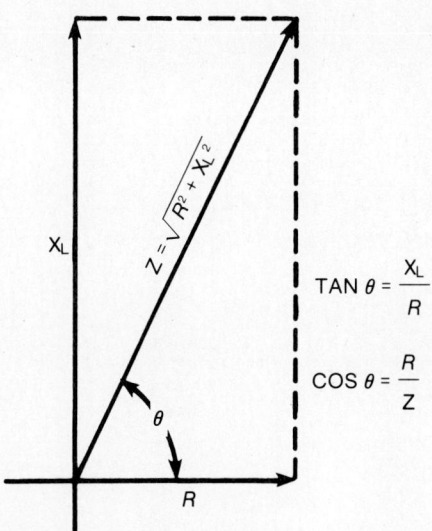

$$\text{TAN } \theta = \frac{X_L}{R}$$

$$\text{COS } \theta = \frac{R}{Z}$$

Fig. 13-8: Impedance of an RL circuit. Trigonometry tables can be found in Appendix Q.

251

circuit acts as a pure inductive circuit, the phase angle approaches 90°, and the impedance approaches X_L. For practical purposes, then, the impedance of the circuit may be taken to be substantially equal to the larger quantity if the ratio of the reactance to the resistance, or of the resistance to the reactance, is 10 to 1 or greater.

The current in an AC circuit containing inductance and resistance may be determined by substituting the impedance Z for the resistance R used in the formula applied to DC circuits. Thus:

$$I = \frac{V}{Z}$$
$$V = IZ$$

and

$$Z = \frac{V}{I}$$

These formulas are called Ohm's Law for AC circuits and they apply to solutions for maximum, effective, or average values but not to any instantaneous value. Therefore:

$$I_{max} = \frac{V_{max}}{Z}$$

$$I_{rms} = \frac{V_{rms}}{Z}$$

$$I_{av} = \frac{V_{av}}{Z}$$

Example: A 120V, 60Hz AC line is connected across an inductance of 5H and a resistance of 1,000Ω in series. The inductive reactance alone is:

$$X_L = 2\pi fL = 6.28 \times 60 \times 5$$
$$X_L = 1,884\,\Omega$$

and as given:

$$= 1,000\,\Omega$$

Then the total impedance Z of the circuit is:

$$Z = \sqrt{R^2 + X_L{}^2}$$
$$Z = \sqrt{(1,000)^2 + (1,884)^2}$$
$$Z = \sqrt{1,000,000 + 3,549,456}$$
$$Z = \sqrt{4,549,456}$$
$$Z = 2,133\,\Omega$$

The effective or rms current in this circuit is:

$$I = \frac{V}{Z} = 120/2,133$$
$$I = 0.056A$$

The phase angle θ of the lag of the current behind the applied voltage is:

$$\text{Cos } \theta = \frac{R}{Z} = 1{,}000/2{,}133 = 0.469$$

$$\theta = 62°$$

Example: What is the impedance of a 1.5H choke with 200Ω of resistance at 60Hz?
Solution:

$$X_L = 2\pi fL$$
$$X_L = 6.28(60)(1.5) = 565.2\Omega$$
$$Z = \sqrt{(R)^2 + (X_L)^2}$$
$$Z = \sqrt{(200)^2 + (565.2)^2}$$
$$Z = \sqrt{40{,}000 + 319{,}451}$$
$$Z = \sqrt{359{,}451} = 599.5\Omega$$

Example: In a series RL circuit, V_R is 20V and the applied V is 32V at 400Hz. The resistance is 1,000Ω. What is the inductance of the coil?
Solution:

$$I = \frac{V_R}{R} = \frac{20}{1{,}000} = 0.02\text{A}$$

$$Z = \frac{V}{I} = \frac{32}{0.02} = 1{,}600\Omega$$

$$X_L = \sqrt{(Z)^2 - (R)^2}$$
$$X_L = \sqrt{(1{,}600)^2 - (1{,}000)^2}$$
$$X_L = \sqrt{2{,}560{,}000 - 1{,}000{,}000}$$
$$X_L = \sqrt{1{,}560{,}000} = 1{,}249\Omega$$

If $X_L = 6.28fL$, then:

$$L = \frac{X_L}{2\pi f}$$

$$L = \frac{1{,}249}{6.28(400)} = \frac{1{,}249}{2{,}512}$$

$$L = 0.497\text{H}$$

Admittance. The reciprocal of impedance is *admittance,* and its symbol is Y; hence, it equals 1/Z. Admittance is a measure of how easily a circuit passes alternating current. As for conductance and susceptance, the unit value is the *siemens.* The reciprocal form is sometimes used in circuit calculations.

PARALLEL RL CIRCUITS

Figure 13-9A shows an inductance L and a resistance R connected in parallel across an AC source. In a circuit of this type it will be seen that, by Kirchhoff's Law for parallel circuits, the voltage across the inductance is equal to the voltage across

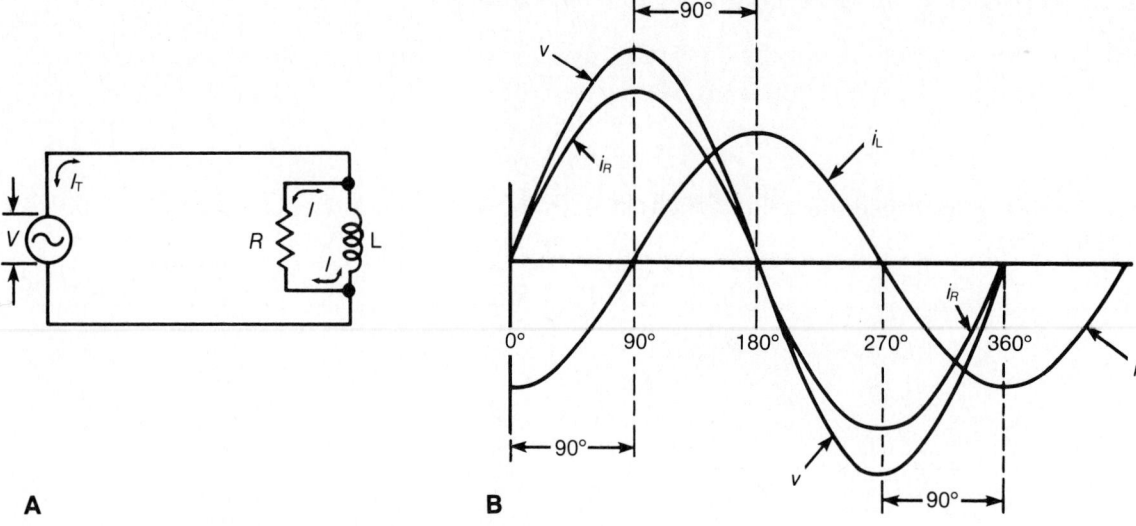

Fig. 13-9: Voltage and currents in a parallel RL circuit.

the resistance and that this voltage is the same as the applied voltage. Therefore, all voltages in this circuit, being the same voltage, are in phase with each other. However, the current through the inductance lags the applied voltage by 90°, and the current through the resistance is in phase with the applied voltage (Fig. 13-9B). Then the current in the inductance lags the current in the resistance by 90°. The resultant current, or line current, is the phasor sum of these two currents.

In Fig. 13-10, the current through the resistance I_R is laid off on the horizontal phasor and the current through the inductance I_L on the vertical phasor. The I_L phasor is laid off in the negative direction because this current lags the current in the resistance, which is taken as the reference phasor, since it is in phase with the applied voltage and represents also the direction of the applied voltage. Now, as in any parallel circuit, the current in the resistance is equal to the voltage divided by the resistance:

$$I_R = \frac{V}{R}$$

The current in the inductance is equal to the voltage divided by the X_L, the inductive reactance:

$$I_L = \frac{V}{X_L}$$

The resultant phasor I_T represents the total current in the circuit, and the angle this phasor makes with the horizontal is the phase angle θ. Therefore, the angle θ is the angle the line current makes with relation to the applied voltage, since, as shown above, the direction of the applied voltage is the same as that of the phasor I_R. By convention, phasors are rotated in a counterclockwise direction; since the I_T phasor follows the V phasor, the line current is said to *lag* the applied voltage by the

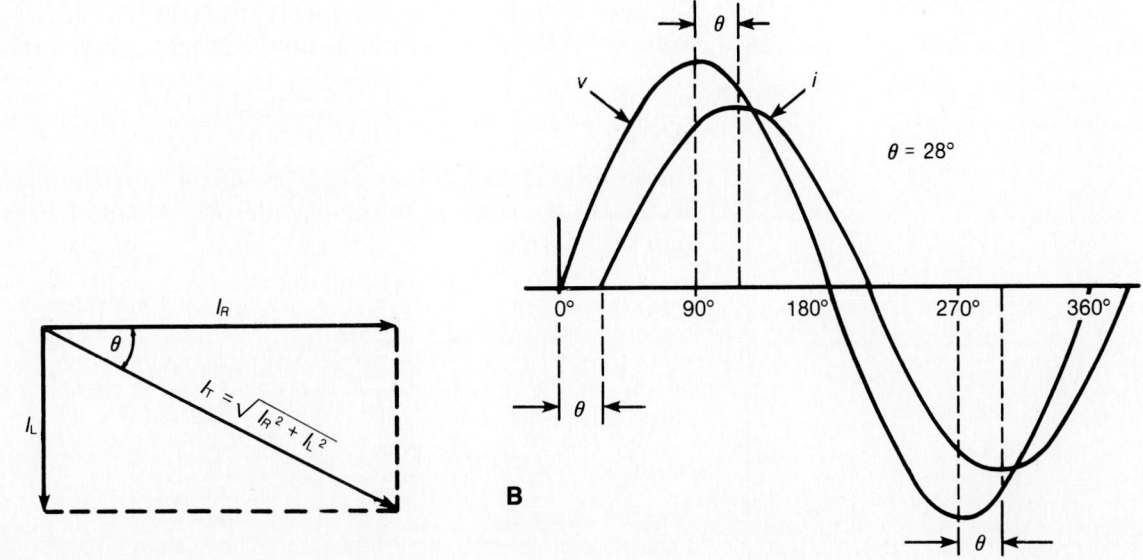

Fig. 13-10: Total current and voltage in a parallel RL circuit.

angle θ. The tangent of this angle is then I_L/I_R. But I_L is equal to V/X_L and I_R is equal to V/R. By substitution and cancellation:

$$\text{Tan } \theta = \frac{R}{X_L}$$

$$\text{Cos } \theta = \frac{Z}{R}$$

The magnitude of the line current phasor I_T must always be greater than either I_R or I_L because it is the hypotenuse of a right triangle.
Then:

$$I_T = \sqrt{I_R^2 + I_L^2}$$

Thus, it may be seen that, as in DC circuits, total current in a parallel RL circuit is always greater than the current in either branch; and, by extension, the impedance of the circuit is less than the opposition of either branch. By Ohm's Law:

$$Z = \frac{V}{I_T}$$

The impedance of a parallel RL circuit may also be obtained by the use of a formula similar to that used for resistances in parallel. It will be remembered that the total resistance of two resistances in parallel is equal to their product divided by their sum:

$$R_T = \frac{R_1 \times R_2}{R_1 + R_2}$$

Then, by analogy:

$$Z = \frac{R \times X_L}{R + X_L}$$

255

But, as has been shown previously, the addition of two vector quantities ($R + X_L$) may not be made directly; therefore:

$$Z = \frac{RX_L}{\sqrt{R^2 + X_L{}^2}}$$

Example: If the values used previously for the series RL circuit are transferred to the parallel RL circuit of Fig. 13-11, then as before:

$$R = 1,000\,\Omega$$
$$X_L = 1,884\,\Omega$$
$$Z = \frac{RX_L}{\sqrt{R^2 + X_L{}^2}} = \frac{1,884,000}{2,132}$$
$$Z = 884\,\Omega$$

The line current $I_T = \dfrac{V}{Z} = \dfrac{120}{884}$

$$I_T = 0.136\,\text{A}$$

The line current lags the applied voltage by the angle θ:

$$\text{Cos } \theta = \frac{Z}{R} = \frac{884}{1,000} = 0.884$$
$$\theta = 28°$$

The following comparisons may now be made between series and parallel RL circuits:

1. The tangent of the phase angle of the series circuit θ_s is the reciprocal of the tangent of the phase angle of the parallel circuit θ_p.

$$\text{Tan } \theta_s = \frac{1}{\text{Tan } \theta_p} = \frac{1}{\dfrac{R}{X_L}} = \frac{X_L}{R}$$

2. If the values of inductance and resistance for the two circuits are the same, the two angles are complementary:

$$\theta_s + \theta_p = 90°$$

3. The *greater* the resistance added to a series circuit, the more resistive the circuit becomes; θ_s approaches $0°$.

4. The greater the resistance added across a parallel circuit, the more inductive the circuit becomes; θ_p approaches $90°$.

TIME CONSTANT OF AN RL CIRCUIT

The property of inductance, as has been shown, introduces a delay in time between the applied voltage and the current produced by this voltage in any circuit in which a change of current occurs. In AC sine wave circuits, the current is a sine wave also, following the applied voltage by the phase angle θ. In such a circuit, θ depends on the ratio X_L/R—that is, on the frequency, inductance, and resistance of the circuit. In DC circuits, how-

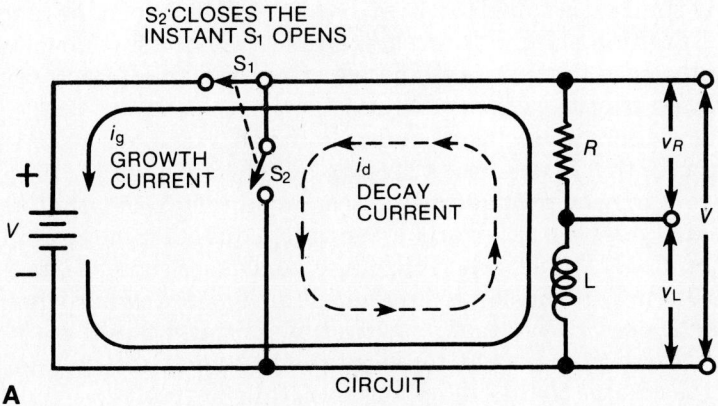

CIRCUIT

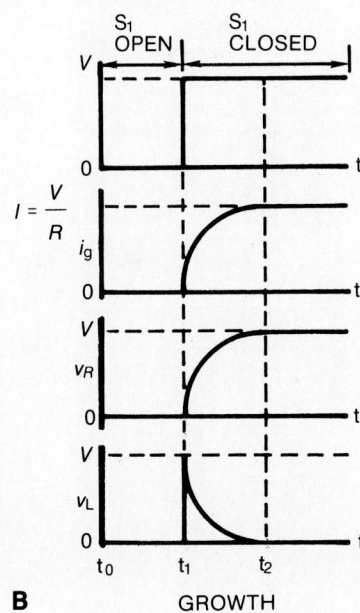

B GROWTH

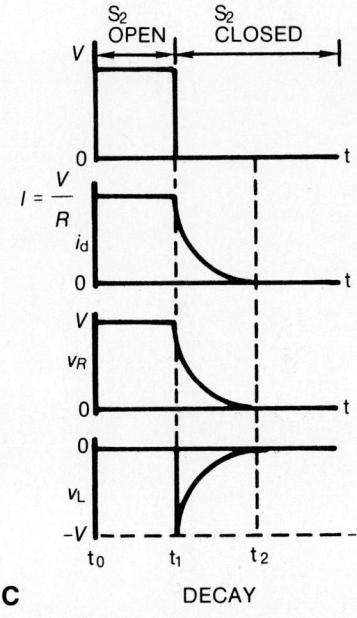

C DECAY

Fig. 13-11: Growth and decay of current in an RL series circuit.

ever, the time lag resulting from the starting and stopping of the voltage does not depend on the frequency, since frequency is zero, but on the inductance and the resistance of the circuit.

Thus, when a DC is applied to a pure resistance, the current is said to rise immediately to its V/R or maximum value. But if an inductance is added to the circuit, the current is held back by the counter emf and does not rise to its V/R value immediately. As shown in Fig. 13-11A, a voltage divider containing resistance and inductance may be connected in a circuit by means of a special switch arrangement. If switch S_1 is closed (as shown), a voltage (V) appears across the divider. A current attempts to flow, but the inductor opposes this current by building up a back emf that, at the initial instant, exactly equals the input voltage (V). Because no current can flow under this condition, there is no voltage across resistor R. Figure 13-11B shows that all of the voltage is impressed across L and no voltage appears across R at the instant switch S_1 is closed.

As current starts to flow, a voltage, v_R, appears across R, and v_L is reduced by the same amount. The fact that the voltage across L is reduced means that the growth current, i_g, is increasing and consequently v_R is increasing. Figure 13-11B shows that v_L finally becomes zero when i_g stops increasing, while v_R builds up to the input voltage, V, as i_g reaches its maximum value. Under steady-state conditions, only the resistor limits the size of the current.

When the battery switch, S_1, in the R_L circuit of Fig. 13-11A is closed, the rate of current increase is maximum in the inductive circuit. At this instant all the battery voltage is used in overcoming the emf of self-induction which is a maximum because the rate of change of current is maximum. Thus, the battery voltage is equal to the drop across the inductor and the voltage across the resistor is zero. As time goes on, more of the battery voltage appears across the resistor and less across the inductor. The rate of change of current is less and the induced emf is less. As the steady-state condition of the current flow is approached, the drop across the inductor approaches zero and all of the battery voltage is used to overcome the resistance of the circuit.

The voltages across the inductor and resistor change in magnitudes during the period of growth of current the same way the force applied to a boat divides itself between the inertia and friction effects. In both examples, the force is developed first across the inertia-inductive effect and finally across the friction-resistive effect.

If switch S_2 is closed (source voltage V removed from the circuit), the flux that has been established around L collapses through the windings and induces a voltage, v_L, in L that has a polarity opposite to V and essentially equal to it in magnitude (Fig. 13-11C). The induced voltage, v_L, causes current i_d to flow through R in the same direction that it was flowing when S_1 was closed. A voltage, v_R, that is initially equal to V, is developed across R. It rapidly falls to zero as the voltage v_L, across L, due to the collapsing flux, falls to zero.

RL Time Constant

Figure 13-12A shows the graph of the rise of current in a DC circuit containing resistance and inductance. This curve is an exponential curve; that is, the current rises rapidly at first and then gradually tapers off to its maximum value, thus describing the characteristic curve shown. An analysis of this curve reveals that in regular units of time, the current rises in decreasing amounts. Thus, in the first unit of time, current rises to 63.2% of its final value, in the second unit to 63.2% of the 36.8% remaining (86.4% of maximum), and in like manner for each succeeding unit of time. Theoretically, in such a progression, the current would never reach a maximum because there would always be a remainder; but for practical purposes, current is considered at a maximum after five units of time.

Conversely, the current in a DC inductive circuit cannot fall immediately to zero, as it would in a pure resistive circuit from which the voltage was removed. Figure 13-12B shows the graph of the fall of the current. This curve is the same as that described for current rise, but is turned inside out; that is, current falls rapidly at first and then tapers gradually to a minimum. In the first unit of time, it falls 63.2% from its maximum V/R value, or to 36.8% of its maximum value; in the next unit of time, it falls an additional 63.2% of the remainder. This process continues, as described above, until, after five units of time, the current is considered to be at zero.

This recurring 63.2% of the maximum rise or fall of current in a fixed unit of time is called the *time constant* (τ) of a circuit. The more inductance in a circuit, the longer the unit of time required to reach this initial 63.2% of the V/R value; and the more resistance in a circuit, the shorter the time required to reach this value, since the greater the resistance of an RL circuit, the less the effect of inductance. Therefore, it may be said that the time constant is equal to inductance divided by resistance, or:

$$\tau \text{ (time constant)} = \frac{L}{R}$$

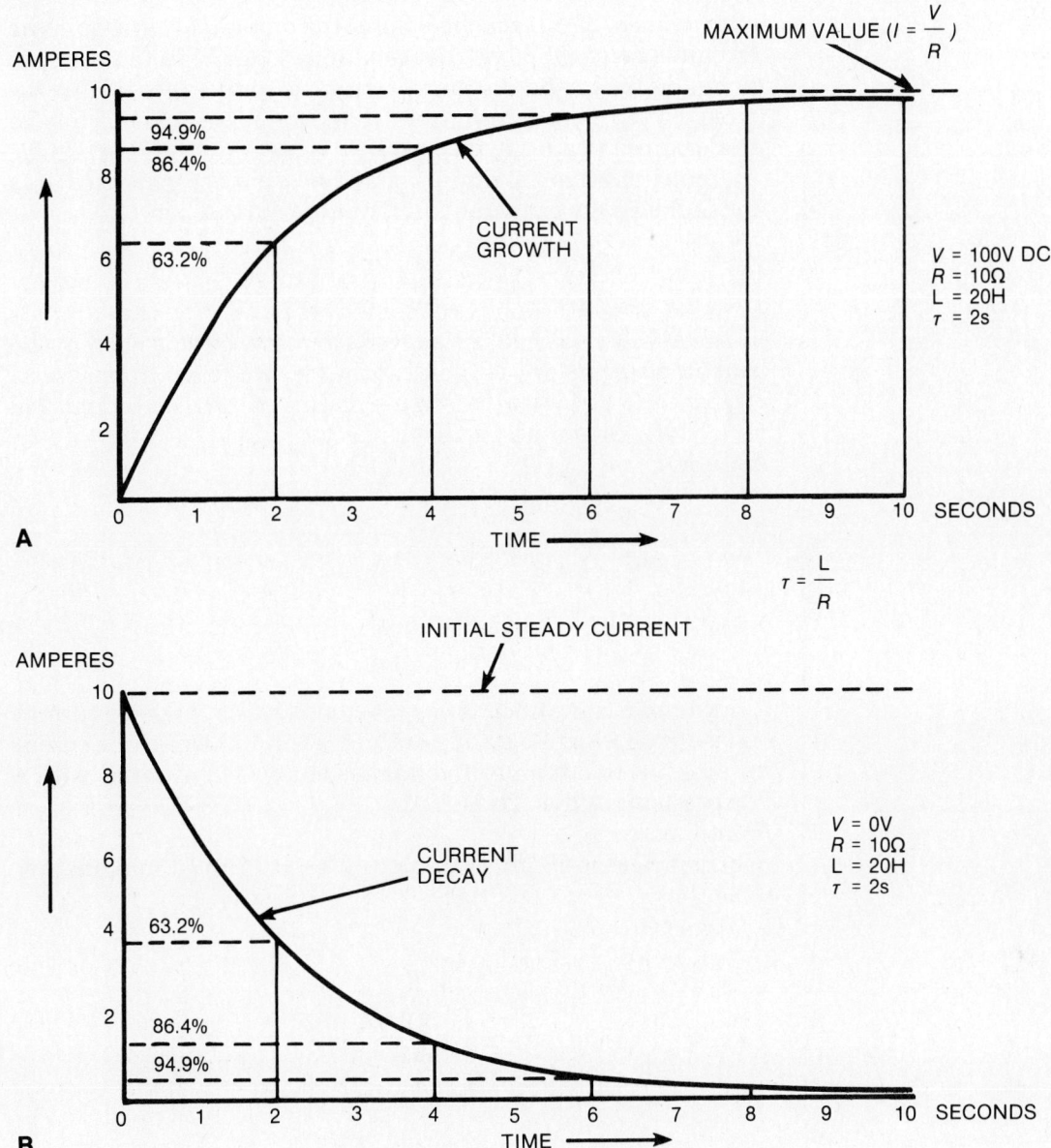

MAXIMUM VALUE ($I = \dfrac{V}{R}$)

AMPERES

CURRENT
GROWTH

V = 100V DC
R = 10Ω
L = 20H
τ = 2s

A

TIME ⟶

SECONDS

$$\tau = \frac{L}{R}$$

INITIAL STEADY CURRENT

AMPERES

CURRENT
DECAY

V = 0V
R = 10Ω
L = 20H
τ = 2s

B

TIME ⟶

SECONDS

Fig. 13-12: Time constant in RL circuit.

If L is in H and R in Ω, then τ is in s and is the time required for the current to reach 63.2% of its DC V/R value.

For example, if 100V DC is applied to an RL circuit containing a 20H inductance and a 10Ω resistance, the V/R value of the current is 10A. The time constant is L/R, 20/10 or 2s. Therefore, it will require 2s for the current to reach 6.32A, and an additional 8s before it reaches approximately its maximum of 10A. If the resistance is lowered, the time constant is increased. In a circuit of 20H inductance and 4Ω resistance, the time constant is 5s. However, it must be remembered that since resistance is lowered, the maximum current in the circuit is now 25A. At the end of the first 5s, the current will have reached a value of 63.2% of 25, or 15.8A; and 20 more seconds will elapse before the

259

maximum of 25 A is reached. Since the amount of current in the circuit is determined by the resistance alone, it is necessary to vary the time constant by varying the inductance, if it is desired to keep maximum current at a certain fixed value. Thus, in the original example, if resistance is kept at 10Ω, current remains at 10A and the time constant of 2s may be increased to 4s by doubling the inductance (L equal to 40H). Then:

$$\tau = \frac{L}{R} = \frac{40}{10} = 4s$$

If the applied voltage is removed from this circuit, then at the end of 4s current is 3.68A and at the end of 20s it is almost zero.

Example: Calculate the time constant of a 10H coil connected in a 500Ω resistor and a 20V battery.
Solution:

$$\tau = \frac{L}{R}$$

$$= \frac{10}{500}$$

$$= 20ms$$

Example: How much time, in seconds, will it take for current to rise to 63.2% of its steady-state (maximum) value in a circuit in which a 16V source of voltage is connected in series with a 4kΩ resistor and a 12mH coil?
Solution:
Current rises to 63.2% of its maximum value in 1 time constant (τ).

$$\tau = \frac{L}{R}$$

$$= \frac{12 \times 10^{-3}}{4 \times 10^{3}} s$$

$$= 3 \times 10^{-6}s$$

$$\tau = 3\mu s$$

Example: A series circuit with a 50Ω resistor would require what size inductor to produce a time constant of 20ms?
Solution:

$$\tau = \frac{L}{R}$$

$$L = \tau R = (0.020)(50)$$

$$L = 1H$$

In the examples above, the circuit values and the time constant were made deliberately large for purposes of explanation, but it should be understood that in most electronics work, the time constant may be measured in milliseconds (ms) or even in microseconds (μs). Small inductances and high resistances are generally used in such cases. Even though these circuits are primarily AC circuits, the time constant determines the level of the DC in those circuits containing both AC and DC. In addi-

260

tion, it is often necessary to use waveforms other than sine waves in AC circuits. These complex waveforms are shaped in various ways. Figure 13-13A shows a current sawtooth waveform which was shaped by first applying a DC voltage to an *RL* circuit and then removing the voltage. In Fig. 13-13B, the voltage has not been removed until after the current reached a maximum and remained there for a given time, resulting in the flat-top form shown.

More details of the *RL* time constant and a comparison with the *RC* time constant for capacitive circuits are given in Chapter 15.

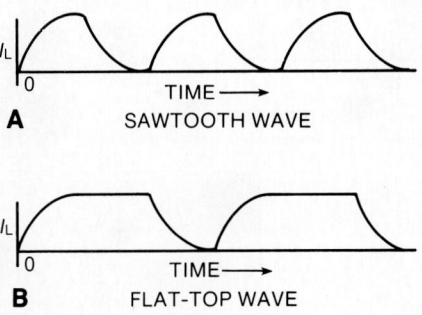

Fig. 13-13: Complex waveforms based on time constant.

POWER IN AN INDUCTIVE CIRCUIT

The power in a DC circuit is equal to the product of volts and amperes, but in an AC circuit this is true only when the load is resistive and has no reactance.

In a circuit possessing inductance only, the true power is zero (Fig. 13-14). The current lags the applied voltage by 90°. The *true power* is the average power actually consumed by the circuit, the average being taken over one complete cycle of alternating current. The *apparent power* is the product of rms volts and rms amperes. Thus, in Fig. 13-14A, the apparent power is $100 \times 10 = 1,000\,\text{V/A}$. However, the power absorbed by the coil during the time the current is rising (Fig. 13-14B) is returned to the source during the time the current is falling so that the average power is zero.

The product of instantaneous values of current and voltage yield the double frequency power curve *P* (Fig. 13-14B). The shaded areas under this curve above the X axis represent positive energy and the shaded areas below the X axis represent negative energy.

From 0° to 90° current is negative and falling; the magnetic field is collapsing and the energy in the field is being returned to the source (negative energy). The product of negative current and positive voltage is negative power.

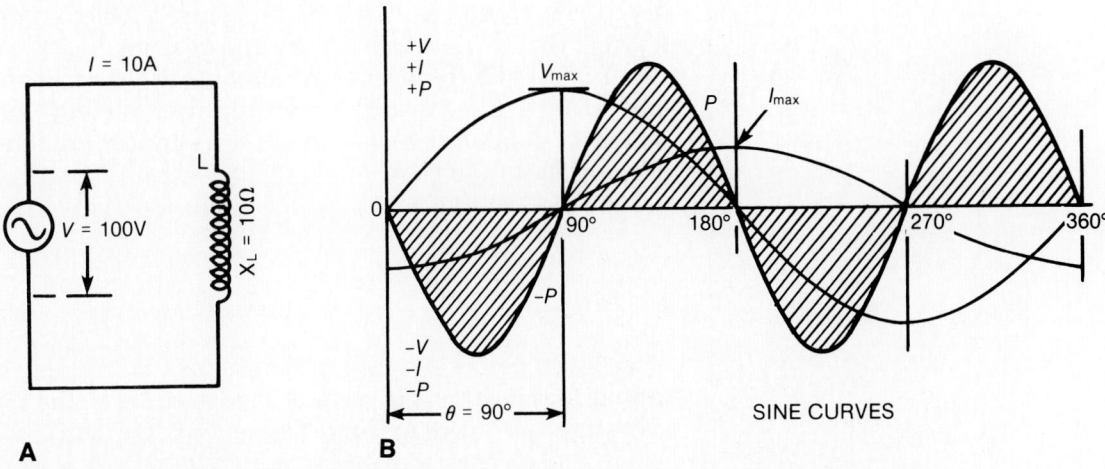

Fig. 13-14: Power in an inductive circuit.

261

From 90° to 180° the current is positive and rising; energy from the source is being stored in the rising magnetic field (positive energy). The product of positive current and positive voltage is equal to positive power.

From 180° to 270° current is positive and falling. Again energy is being returned to the source; the product of positive current and negative voltage is negative power.

From 270° to 360° current is negative and rising. Energy (positive) is being supplied by the source and stored in the magnetic field. The product of negative current and negative voltage is positive power.

Thus when current is rising, power is being supplied by the source and stored in the magnetic field; when current is falling, power is being returned to the source from the collapsing magnetic field. In the theoretically pure inductance shown in Fig. 13-14, the supplied power is equal to the returned power; thus, average power used (true power) is zero.

The ratio of the true power to the apparent power in an AC circuit is called the *power factor*. It may be expressed as a percentage or as a decimal. In the inductor of Fig. 13-14A, the power factor is $\frac{0}{1,000}$ = 0. The power factor is also equal to the cos θ, where θ is the phase angle between the current and voltage. The phase angle between V and I in the inductor is $\theta = 90°$; thus, the power factor of an inductor of negligible losses is cos 90° = 0. The apparent power in a purely inductive circuit is called *reactive power* and the unit of reactive power is called the *VAR*. This unit is derived from the first letters of the words volt-ampere-reactive.

Resistance in an AC Circuit

In an AC circuit containing only resistance, the current and voltage are always in phase (Fig. 13-15). The true power dissipated in heat in a resistor in an AC circuit when sine waveforms of voltage and current are applied is equal to the product of the rms volts and the rms amperes. In the circuit of Fig. 13-15A, the power absorbed by the resistor is $P = VI = 100 \times 10 = 1,000$W. The product of the instantaneous values of current and voltage (Fig. 13-15B) gives the power curve, P, the axis of which is displaced above the X axis by an amount that is proportional to 1,000W.

The true average power is

$$P = \frac{V_{max} \times I_{max}}{2} = V_{eff} \times I_{eff}$$

The power factor of a resistive circuit is cos 0° = 1, or 100%. The apparent power in the resistor is also equal to the true power. The reactive power in the resistive circuit is zero.

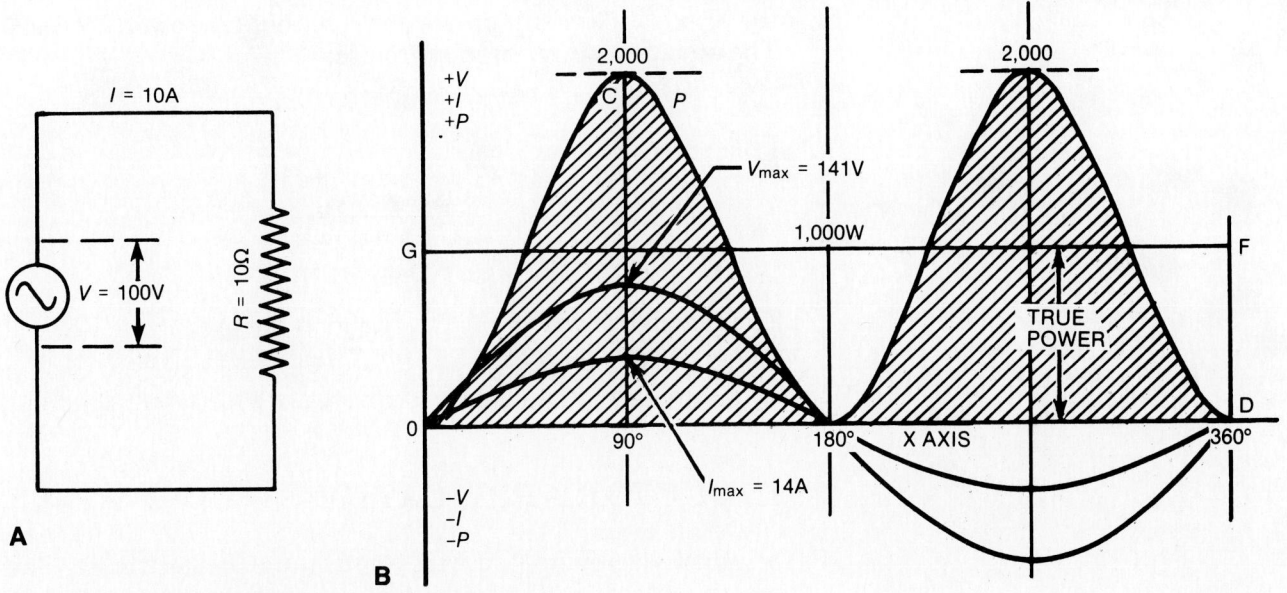

Fig. 13-15: Power in a resistive circuit.

Power in a Circuit Containing RL in Series

The true power in any circuit is the product of the applied voltage, the circuit current, and the cosine of the phase angle between them. Thus, in Fig. 13-16 the true power is

$$P = VI \cos \theta = 100 \times 7.07 \ (\cos 45° = 0.707)$$
$$= 500W$$

The power curve is partly above the X axis (Fig. 13-16B) and partly below it. The axis of the power curve is displaced above

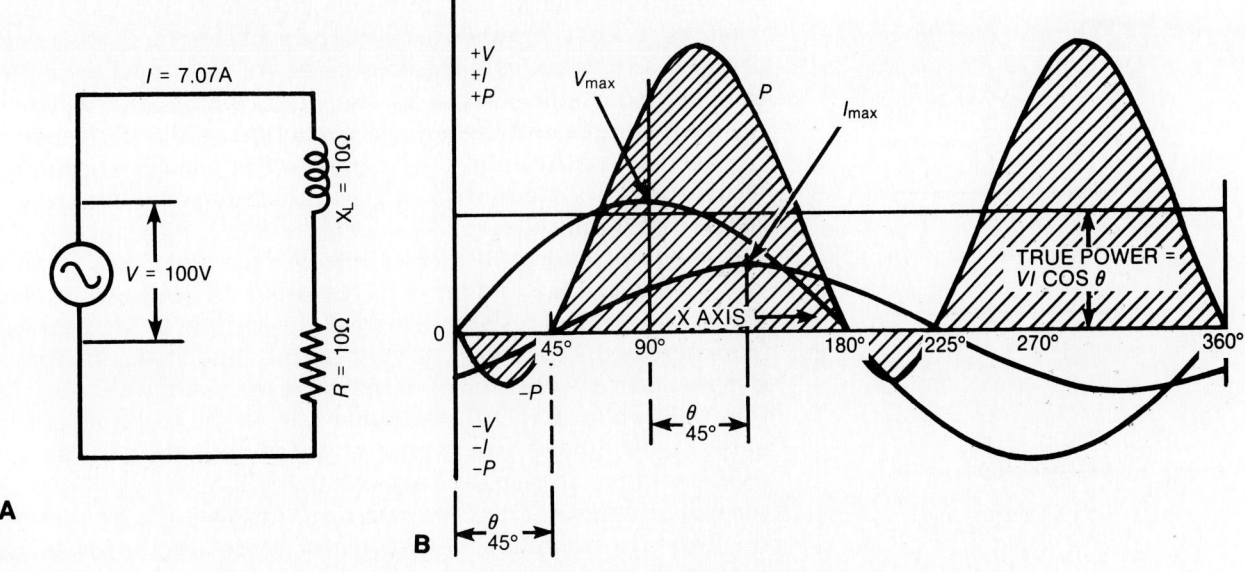

Fig. 13-16: Power in a circuit containing L and R in series.

263

the X axis an amount proportional to the true power, $VI \cos \theta$. The apparent power in this circuit is

$$100 \times 7.07 = 707\text{VA}$$

The power factor is

$$\cos \theta = \frac{\text{true power}}{\text{apparent power}} = \frac{500}{707}$$
$$= 0.707, \text{ or } 70.7\%$$

The reactive power in the RL circuit is the product of $VI \sin \theta$ where $\sin \theta$ is the reactive factor. Thus the reactive power is

$$100 \times 7.07 \,(\sin 45° = 0.707) = 500 \text{ VARs (lagging)}$$

INDUCTOR APPLICATIONS

Inductive circuits find wide applications in electronics. The reactive effect of an inductor in an AC circuit makes the inductor valuable in filtering and phase shift applications.

RL Filters

Series RL networks can be used as simple low and high-pass filters (Fig. 13-17). These circuits are resistor-inductor voltage dividers.

The circuit in Fig. 13-17A is a *low-pass filter*. The input signal is applied across the coil and resistor in series and the output voltage is taken from across the resistor. At low frequencies, the reactance of the coil is low; therefore, very little voltage is dropped across it. Most of the voltage is dropped across the output resistor. As the input frequency increases, the inductive reactance increases. The inductive reactance becomes larger with respect to the resistance and more of the input voltage is dropped across the coil; therefore, less voltage appears across the output resistor. As you can see, increasing the frequency causes an increase in the inductive reactance and a corresponding decrease in the output voltage. Low frequencies are passed with little or no attenuation, while high frequencies are greatly reduced in amplitude.

The circuit in Fig. 13-17B is a *high-pass filter*. Again this circuit is a voltage divider with the input AC voltage applied across the coil and resistor in series. The output voltage is taken from across the inductor. At very high frequencies, the inductive reactance is very high compared to the resistance. Most of the input voltage will then appear across the coil and at the output terminals. As the input frequency decreases, the inductive reactance decreases. Less voltage will be dropped across the coil and more across the resistor. The lower the frequency, the lower the inductive reactance and, therefore, the lower the output voltage. High frequencies are passed with little or no attenuation. Low frequencies are greatly attenuated due to the voltage divider effect.

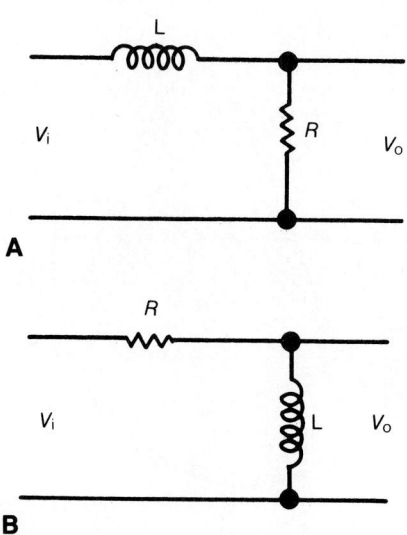

Fig. 13-17: Single sections of RL filters: (A) low-pass filter and (B) high-pass filter.

The cutoff frequency (f_{co}) of the RL network shown in Fig. 13-17 is given by the expression below:

$$f_{co} = \frac{R}{2\pi L} = \frac{R}{6.28L}$$

The *cutoff frequency* is that frequency above or below which frequencies are passed or attenuated. In the expression above, R is expressed in Ω, L is expressed in H, and f is expressed in Hz. At the cutoff frequency, X_L equals R and the phase shift is 45°. At the cutoff frequency, the output voltage (V_o) will be approximately 70% of the input voltage (V_i) or $V_o = 0.707 V_i$. For example, if the input voltage is 8V, the output voltage at the cutoff frequency will be $8 \times 0.707 = 5.656$V.

Phase Shifting Networks

In many electronic circuits, the phase of the input signal is a critical factor. It may not be sufficient to have an input signal of a specified frequency, shape, and amplitude, but it may be necessary to start the input at some specified point along its time axis. If the input were to arrive in just the right phase, it would be purely accidental. *Phase shifting networks* are used to rearrange the phase of a signal to suit the demands of the circuit.

RL circuits are based on the fact that the current lags the voltage in an RL circuit. With a proper arrangement of resistors and inductors, the output can be made to assume any desired phase angle in respect to the input. An RL negative or lagging phase shifting network is shown in Fig. 13-18A.

Assuming that the input is a sine wave, the output will also be a sine wave, but it will lag the input between 0° and 90°. The input signal represents the applied voltage. The inductor opposes the change of current. The output is developed across the resistor as a result of the circuit current. With components of equal ohmic value, the phase shift is exactly 45° (Fig. 13-18B). The equal size components also provide an output with an amplitude of about seven-tenths of the input amplitude.

To obtain phase shifts greater than 90°, the RL network can be cascaded. As shown in Fig. 13-19, each of these sections provides a 45° phase shift and reduces the input signal by about 70%. A 10V signal at the input produces an output of about 3V. The output lags the input by 135°.

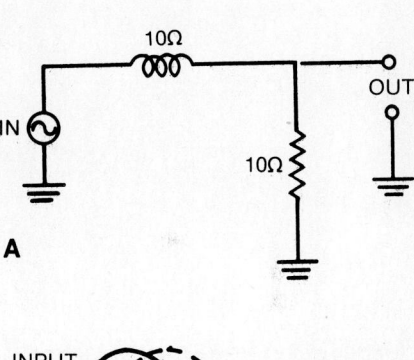

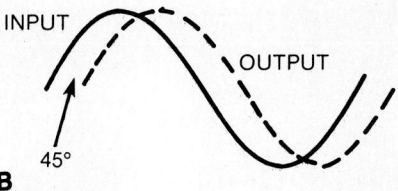

Fig. 13-18: Negative 45° phase shift.

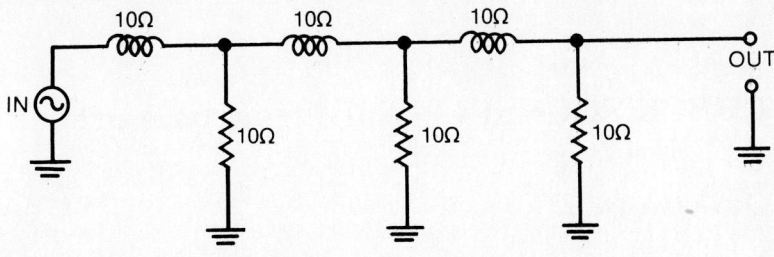

Fig. 13-19: Negative 135° phase shift.

All RL phase shifting circuits need not produce exact multiples of $45°$. For a given section to have a greater phase shift, the section must have a higher X_L to R ratio. For instance, if the ratio of X_L to R were 4 to 3, the phase shift would be $53.1°$. Reversing this ratio ($X_L = 3$, $R = 4$) would reduce the phase shift to $36.9°$. Actually, the amount of phase shift produced by these circuits is a function of the inductance, the resistance, and the input frequency. The phase shift is given by the formula below:

$$\theta = \arctan\frac{X_L}{R}$$

This formula assumes that the resistance of the inductor is negligible or small compared to the external resistor.

In the RL phase shifting circuit, the phase is shifted clockwise with respect to the phase of the input signal. The term, "output lags input," is equivalent to assigning a negative phase angle to the output.

SUMMARY

In a DC circuit, only the resistance in the wire of the inductor affects the current flow. The opposition offered by a coil to AC is termed inductive reactance (X_L). The value of this reactance is determined by the formula $X_L = 2\pi f_L$. The current is said to lag the applied voltage by $90°$.

The impedance of an RL circuit is equal to the square root of the sum of the squares of the resistance and the inductive reactance: $Z = \sqrt{R^2 + X_L^2}$. The formula $I_T = \sqrt{I_R^2 + I_L^2}$ is used to determine the total current in a parallel RL circuit. The cosine of the current phase angle equals Z/R.

The current of a circuit will rise (or fall) 63.2% of the maximum in one time constant (τ). The time constant of an inductive circuit is equal to L/R.

True power is the power actually consumed by the RL circuit. Apparent power is the product of rms voltage and rms amperes. The ratio of true power to apparent power is called the power factor. The apparent power in a purely inductive circuit is called reactive power.

The reactive effect of an inductor in an AC circuit makes the inductor valuable in filtering and phase shift applications.

CHECK YOURSELF (Answers at back of book)

Questions

Fill in the blanks with the appropriate word or words.

1. An inductor stores energy in a _____ .

2. When a voltage is first applied to an RL circuit, all the voltage appears across the _____ .
3. The current in an RL circuit will rise to the maximum value in _____ time constants.
4. Opposition to changing circuit values is called _____ .
5. Inductive reactance is directly proportional to _____ and _____ .
6. An RL time constant is equivalent to the _____ divided by the _____ .
7. In a low-pass filter, the reactance of the coil is _____ , and the output is taken across the _____ .
8. The _____ frequency is that frequency above or below which frequencies are passed or attenuated.
9. _____ is the reciprocal of inductive reactance.
10. The impedance of an RL circuit is equal to the square root of the sum of the squares of the _____ and the _____ .
11. An increase in DC resistance will _____ the phase angle of an inductive circuit.
12. When an inductor is used in a DC circuit, only the _____ affects the current.
13. The true power in any circuit is the product of the _____ , the _____ , and the _____ of the phase angle between them.
14. At _____ frequencies, even small amounts of inductance may offer a large amount of inductive reactance.
15. An inductor having a value of 2H and negligible resistance is used at a frequency of 400Hz. Its inductive reactance is

    _____ .
    A. $1,600\pi\Omega$
    B. $5.1k\Omega$
    C. $2.513k\Omega$
    D. $1,600\Omega$
16. If the inductor in Question 15 is connected to a 240V, 400Hz supply, the current is _____ .
    A. 12.06mA
    B. 4.8mA
    C. 15mA
    D. 47.7mA
17. A series RL circuit has an inductive reactance of 40Ω and a resistance of 30Ω. The impedance of the circuit is _____ .
    A. 35Ω
    B. 50Ω
    C. 70Ω
    D. 10Ω

Problems

1. At what frequency will a 0.5H inductor have a reactance of $2,000\Omega$?
2. What is the impedance of a series circuit containing $1,500\Omega$ of resistance and $1,000\Omega$ of inductance?
3. What is the phase angle θ of the lag of the current behind the applied voltage in problem 2?
4. How long will it take a 500mH coil to fully charge if it is connected in series with a Ik Ω?

5. What is the true power dissipated by a circuit having an apparent power of 5W and a phase angle of 30° between voltage and current?
6. What is the reactance of a 250mH inductor at a frequency of 4.5MHz?
7. If a 500Ω resistor is connected in series with a 50mH coil, what is the total impedance at 10kHz?
8. At what phase angle θ will the current lag the voltage in the previous problem?
9. Three inductors having values of 30mH, 50mH, and 80mH are connected in series with a 10kΩ resistor. What is the time constant for this circuit?
10. What is the total reactance of the circuit in problem 9 at a frequency of 100kHz?
11. What is the true power dissipated by a circuit having a phase angle θ of 70° and an apparent power dissipation of 20W?
12. What is the reactance of a 98mH coil at 20kHz? at 5kHz?

CHAPTER 14

Capacitance

Every electrical/electronic circuit, no matter how complex, is composed of no more than three basic electrical properties: resistance, inductance, and capacitance. Therefore, a thorough understanding of each of these basic properties is a necessary step toward the understanding of electrical equipment. Since resistance and inductance have been covered, the last of the basic three, capacitance, will now be discussed.

Two conductors separated by a nonconductor (Fig. 14-1) exhibit the property called *capacitance,* because this combination can store an electric charge. Whereas inductance was defined as a property of a circuit which opposes a change in *current,* capacitance is a property of a circuit which opposes a change in *voltage.* Where inductance stored energy in an electromagnetic field, capacitance stores energy in an *electrostatic field.*

Electrostatic Field

When one charged body is brought close to another charged body, as discussed in Chapters 1 and 2, there is a force that either causes the charged bodies to attract or repel each other. If these charged bodies have the same sign of charge, they will tend to repel each other. If they have unlike signs, there will be a force of attraction between them. This force of attraction or repulsion is caused by the electrostatic field that surrounds every charged body. If the material is charged positively, it has a lack of electrons. If it is charged negatively, it has an excess of electrons.

Look at Fig. 14-2. Here two charged bodies exist in close proximity to each other. One is negatively charged, and one is positively charged. Between these two charged bodies there is an *electrostatic force field* existing due to natural laws of attraction and repulsion between differently charged bodies. If we could see the lines of force, we would see them emanating from the positive body and going towards the negative body. Remember, these are lines of force, not electrons. An electron will move from the negative to the positive material *against* the lines of force.

The force between charged bodies is very aptly explained by a common electrical law: The force existing between two charged bodies is directly proportional to the product of the charges and

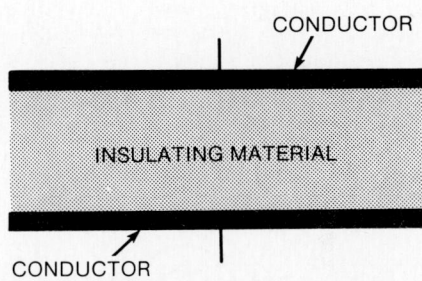

Fig. 14-1: Cross section of a basic capacitor.

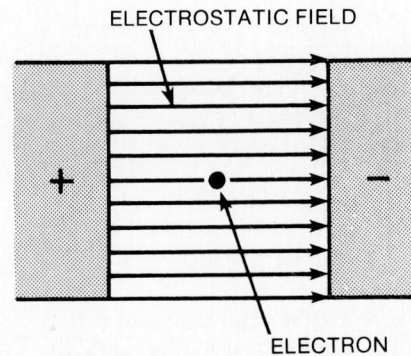

Fig. 14-2: An electrostatic force field.

inversely proportional to the square of the distance separating them. In other words, the greater the charge on the charged bodies, the greater will be the force field; and the greater the distance between the charged bodies, the less the force of attraction. An interesting characteristic of electrostatic lines of force is that they have the ability to pass through any known material.

THE CAPACITOR

Capacitance is defined as the property of an electrical device or circuit that tends to oppose a change in voltage. Capacitance is also a measure of the ability of two conducting surfaces, separated by some form of nonconductor, to store an electric charge. For the present time air will be used as the insulating material between the conducting surfaces.

The device used in electrical circuits to store a charge by virtue of an electrostatic field is called a *capacitor*. (The larger the capacitor, the larger the charge that can be stored.)

The simplest type of capacitor consists of two metal plates separated by air. It has been established that a loosely bound electron inserted in an electrostatic field will move. The same is true, with qualifications, if the electron is in a bound state. The material between the two charged surfaces of Fig. 14-2 (air in this case) is composed of atoms containing bound orbital electrons. Since the electrons are bound, they cannot travel to the positively charged surface. Therefore, the resultant effect will be a distorting of the electron orbits. The bound electrons will be attracted toward the positive surface and repelled from the negative surface. This effect is illustrated in Fig. 14-3. In Fig. 14-3A, there is no difference in charge placed across the plates; and the structure of the atom's orbits is undisturbed. If there is a difference in charge across the plates, as shown in Fig. 14-3B, the orbits will be elongated in the direction of the positive charge.

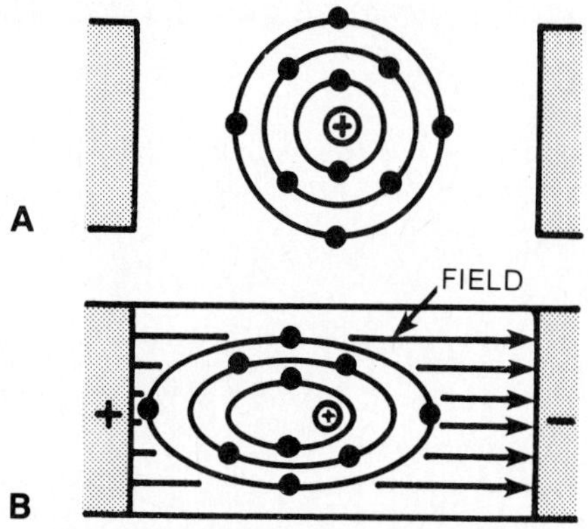

An energy is required to distort the orbits. Energy is transferred from the electrostatic field to the electrons of each atom between the charged plates. Since energy cannot be destroyed, the energy required to distort the orbits can be recovered when the electron orbits are permitted to return to their normal positions. This effect is analogous to the storage of energy in a stretched spring. A capacitor can thus "store" electrical energy.

An illustration of a simple capacitor and its schematic symbol is shown in Fig. 14-4. The conductors that form the capacitor are called *plates*. The material between the plates is called the *dielectric*. In Fig. 14-4B, the two vertical lines represent the connecting leads. The two horizontal lines represent the capacitor plates. Notice that the schematic symbol (Fig. 14-4B) and the simple capacitor diagram (Fig. 14-4A) are similar in appearance. In a practical capacitor, the parallel plates may be constructed in various configurations (circular, rectangular, etc.); but the cross-sectional area of the capacitor plates is tremendously large in comparison to the cross-sectional area of

Fig. 14-3: Electron orbits with and without the presence of an electron field.

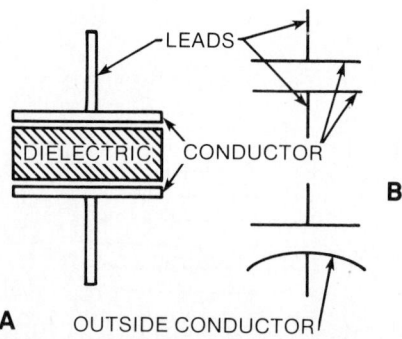

Fig. 14-4: Capacitor and its schematic symbol. The outside conductor is frequently the ground connect for a capacitor.

270

the connecting conductor. This means that there is an abundance of electrons available in each plate of the capacitor. If the cross-sectional area and plate material of the capacitor plates are the same, the number of electrons in each plate must be approximately the same. It should be noted that there is a possibility of the difference in charge becoming so large as to cause ionization of the insulating material to occur (cause bound electrons to be freed). This places a limit on the amount of charge that can be stored in the capacitor.

Figure 14-4A depicts a capacitor in its simplest form. It consists of two metal plates separated by a thin layer of insulating material (dielectric). When connected to a voltage source (battery), the voltage forces electrons onto one plate, making it negative, and pulls them off the other, making it positive. Electrons cannot flow through the dielectric. Since it takes a definite quantity of electrons to "fill up," or charge, a capacitor, it is said to have a *capacity*. This characteristic is referred to as capacitance.

Dielectric Materials

Various materials differ in their ability to support electric flux or to serve as dielectric material for capacitors. This phenomenon is somewhat similar to permeability in magnetic circuits. Dielectric materials or insulators are rated in their ability to support electric flux in terms of a figure called the *dielectric constant*. The higher the value of dielectric constant (other factors being equal), the better is the dielectric material.

A vacuum is the standard dielectric for purposes of reference and is assigned the value of unity (or one). The dielectric constant of a dielectric material is also defined as the ratio of the capacitance of a capacitor having that particular material as the dielectric to the capacitance of the same capacitor having air as the dielectric. By way of comparison, the dielectric constant of pure water is 81; flint glass, 9.9; and paraffin paper, 3.5. The range of dielectric constants is much more restricted than is the range of permeabilities. Dielectric constants for some common materials are given in the following list:

Material	Constant
Vacuum	1.0000
Air	1.0006
Paraffin paper	3.5
Glass	5–10
Mica	5–6
Rubber	2.5–35
Wood	2.5–8
Glycerine (15°C)	56
Petroleum	2
Pure water	81

Notice the dielectric constant for a vacuum. Since a vacuum is the standard of reference, it is assigned a constant of 1; and

271

the dielectric constants of all materials are compared to that of a vacuum. Since the dielectric constant of air has been determined experimentally to be approximately the same as that of a vacuum, the dielectric constant of *air* is also considered to be equal to 1. The formula used to compute the value of capacitance using the physical factors just described is:

$$C = 0.2249 \left(\frac{KA}{d} \right)$$

Where

C = capacitance, in picofarads (10^{-12})

A = area of one plate, in square inches

d = distance between plates, in inches

K = dielectric constant of insulating materials

0.2249 = a constant resulting from conversion from Metric to British units

Example: Find the capacitance of a parallel plate capacitor with paraffin paper as the dielectric.

Given:

K = 3.5

d = 0.05″

A = 12sq in.

Solution:

$$C = 0.2249 \left(\frac{KA}{d} \right)$$

$$C = 0.2249 \left(\frac{3.5 \times 12}{0.05} \right)$$

$$C = 189pF$$

Using the above equation, it is easy to visualize the effects on capacitance of the physical factors involved. It can be seen that capacitance is a direct function of the dielectric constant and the area of the capacitor plates and an inverse function of the distance between the plates.

Unit of Capacitance

Capacitance is measured in a unit called the *farad*. This unit is a tribute to the memory of Michael Faraday, a scientist who performed many early experiments with electrostatics and magnetism.

It was discovered that for a given value of capacitance, the ratio of charge deposited on one plate to the voltage producing the movement of charge is a constant value. This constant value is a measure of the amount of capacitance present. The symbol used to designate a capacitor is C. The capacitance is equal to 1F when a voltage changing at the rate of 1V/per second causes an average current of 1A to flow. This is expressed by the equation:

272

$$C = \frac{i}{\frac{\Delta v}{\Delta t}}$$

Where

C = capacitance, in farads

i = average charging current in A

$\frac{\Delta v}{\Delta t}$ = rate of change of voltage, in V, with time, in s

This equation may be clearer if expressed as follows:

Capacitance equals 1F when =

$$\frac{\text{Average charging current of 1A flows}}{\text{When voltage changes 1V in 1s}}$$

The farad can also be defined in terms of charge and voltage. A capacitor has a capacitance of 1F if it will store 1C of charge when connected across a potential of 1V. This relationship can be expressed mathematically as:

$$C = \frac{Q}{V}$$

Where

C = capacitance in farads

Q = charge in coulombs (C)

V = applied potential in volts

Example: What is the capacitance of two metal plates separated by 1cm of air, if 0.001C of charge is stored when a potential of 200V is applied to the capacitor?

Given:

$$Q = 0.001C$$
$$V = 200V$$

Solution:

$$C = \frac{Q}{V}$$

Converting to powers of ten:

$$C = \frac{10 \times 10^{-4}}{2 \times 10^2}$$

$$C = 5 \times 10^{-6}$$

$$C = 0.000005 \text{ farads or } 5\mu F$$

Although this capacitance might appear rather small (five millionths of a F), many electronic circuits require capacitors of much smaller value. Consequently the F is a cumbersome unit, far too large for most applications. The *microfarad,* which is one millionth of a F (1×10^{-6} farad), is a more convenient unit. The symbol used to designate microfarad is μF. In high-frequency circuits even the microfarad becomes too large, and the unit picofarad (one millionth of a microfarad) is used. The symbol for picofarad is pF. (The older term for pF was micromicrofarad or $\mu\mu f$).

In using the previous equation, one must not deduce the mistaken idea that capacitance is dependent upon charge and voltage. Capacitance is determined entirely by physical factors such as plate area, plate spacing, etc.

Factors Affecting the Value of Capacitance

The capacitance of a capacitor depends on the three following factors:
1. The opposing plate area.
2. The distance between the plates.
3. The dielectric constant of the material between the plates.

These three factors are related to the capacitance of a parallel-plate capacitor consisting of two plates by the formula

$$C = 0.2249 \left(\frac{KA}{d} \right)$$

where C is in pF, A is the opposing plate area in square inches, d is the distance between the plates in inches, and K is the dielectric constant of the insulator separating the plates.

For example, the capacitance of a parallel-plate capacitor with an air dielectric and spacing of 0.0394″ between the plates, each of which has an area of 15.5 square inches, is approximately

$$C = 0.225 \left(\frac{1 \times 15.5}{0.0394} \right) = 88.5 pF$$

From this formula it may seem that the capacitance increases when the plates are increased in area; it decreases if the spacing of the plates is increased; and it increases if the K-value is increased.

As previously mentioned, the dielectric constant, K, expresses the relative capacitance when materials other than air are used as the insulating material between the plates. For example, if mica is substituted for air as the dielectric, the capacitance increases from five to six times because the dielectric constant of mica is from five to six times greater than air (1).

If the capacitor is composed of more than two parallel plates, the capacitance is calculated by multiplying the preceding formula by N – 1, where N is the number of plates. The plates are interlaced as shown in Fig. 14-5, and the effect is that of increasing the capacitance of the two-plate capacitor by the factor N – 1. In Fig. 14-5 there are 11 plates and the capacitance is 10 times that of a two-plate capacitor of the same plate area, spacing, and dielectric material.

Voltage Rating of Capacitors

In selecting or substituting a capacitor for use in a particular circuit, consideration must be given to (1) the value of capaci-

274

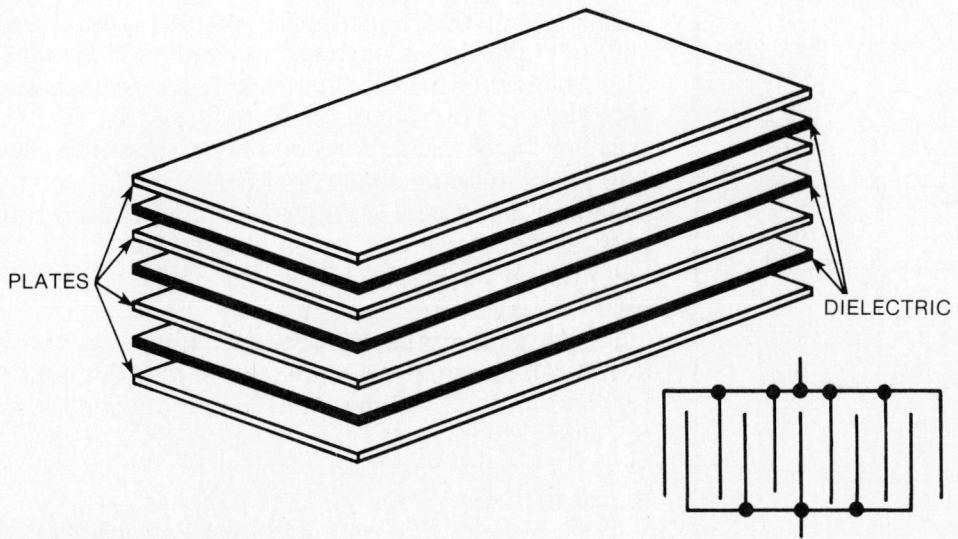

Fig. 14-5: Multiple-plate capacitor.

tance desired and (2) the amount of voltage to which the capacitor is to be subjected. If the voltage applied across the plates is too great, the dielectric will break down and arcing will occur between the plates. The capacitor is then short-circuited and the possible flow of current through it can cause damage to other parts of the equipment. Capacitors have a voltage rating that should not be exceeded.

The working voltage of the capacitor is the maximum voltage that can be steadily applied without danger of arc-over. The working voltage depends on (1) the type of material used as the dielectric and (2) the thickness of the dielectric.

The voltage rating of the capacitor is a factor in determining the capacitance because capacitance decreases as the thickness of the dielectric increases. A high-voltage capacitor that has a thick dielectric must have a larger plate area in order to have the same capacitance as a similar low-voltage capacitor having a thin dielectric. The voltage rating also depends on frequency because the losses and the resultant heating effect increase as the frequency increases.

A capacitor that may be safely charged to 500V DC cannot be safely subjected to alternating or pulsating direct voltages whose effective values are 500V. An alternating voltage of 500V (rms) has a peak voltage of 707V, and a capacitor to which it is applied should have a working voltage of at least 750V. The capacitor should be selected so that its working voltage is at least 50% greater than the highest voltage to be applied to it. Effective (rms) voltage and the action of capacitors in AC circuits are described in Chapter 15.

CHARGING AND DISCHARGING A CAPACITOR

In order to better understand the action of a capacitor in conjunction with other components, the charge and discharge

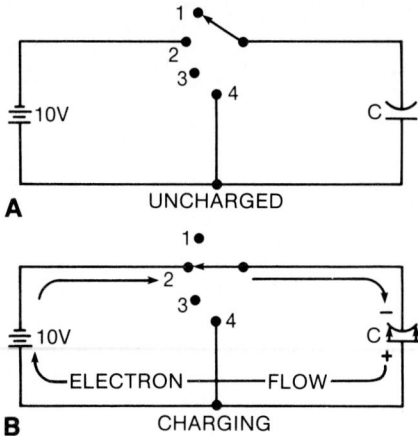

A   UNCHARGED

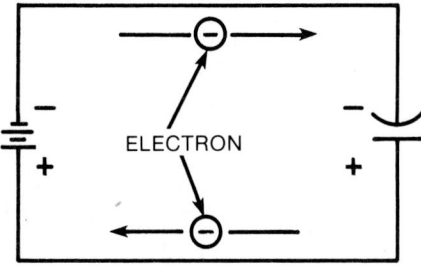

B   CHARGING

Fig. 14-6: Charging a capacitor.

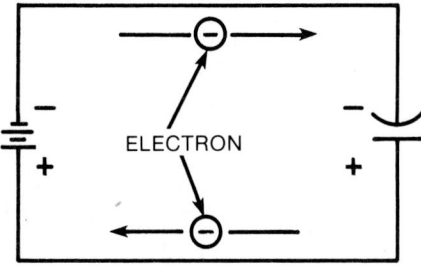

Fig. 14-7: Electron motion during the charge.

action of a purely capacitive circuit will be analyzed first. For ease of explanation the capacitor and voltage source used in Fig. 14-6 will be assumed to be perfect (no internal resistance, etc.) although this is impossible in practice.

In Fig. 14-6A an uncharged capacitor is shown connected to a four-position switch. With the switch in position 1, the circuit is open and no voltage is applied to the capacitor. Initially each plate of the capacitor is a neutral body, and until a difference of potential is impressed across the capacitor, no electrostatic field can exist between the plates.

To *charge* the capacitor, the switch must be thrown to position 2, which places the capacitor across the terminals of the battery. Under the given conditions the capacitor would reach full charge instantaneously; however, the charging action will be spread out over a period of time in the following discussion so that a step-by-step analysis can be made.

At the instant the switch is thrown to position 2 (Fig. 14-6B), a displacement of electrons will occur simultaneously in all parts of the circuit. This electron displacement is directed away from the negative terminal and toward the positive terminal of the source. An ammeter connected in series with the source will indicate a brief surge of current as the capacitor charges.

If it were possible to analyze the motion of individual electrons in this surge of charging current, the following action would be observed (Fig. 14-7).

At the instant the switch is closed, the positive terminal of the battery extracts an electron from the bottom conductor and the negative terminal of the battery forces an electron into the top conductor. At this same instant an electron is forced into the top plate of the capacitor and another is pulled from the bottom plate. Thus, in every part of the circuit a clockwise displacement of electrons occurs in the manner of a chain reaction.

As electrons accumulate on the top plate of the capacitor and others depart from the bottom plate, a difference of potential develops across the capacitor. Each electron forced onto the top plate makes that plate more negative, while each electron removed from the bottom causes the bottom plate to become more positive. Notice that the polarity of the voltage which builds up across the capacitor is such as to oppose the source voltage. The source forces current around the circuit of Fig. 14-7 in a clockwise direction. The emf developed across the capacitor, however, has a tendency to force the current in a counterclockwise direction, opposing the source. As the capacitor continues to charge, the voltage across the capacitor rises until it is equal in amount to the source voltage. Once the capacitor voltage equals the source voltage, the two voltages balance one another and current ceases to flow in the circuit.

In studying the charging process of a capacitor, it must be emphasized that *no* current flows *through* the capacitor. The material between the plates of the capacitor must be an insulator.

To an observer stationed at the source or along one of the circuit conductors, the action has all the appearances of a true flow of current even though the insulating material between

276

the plates of the capacitor prevents having a complete path. The current which appears to flow in a capacitive circuit is called *displacement current*.

To provide a better understanding of charging action, a capacitor can be compared to the mechanical system in Fig. 14-8. The diagram in Fig. 14-8A shows a metal cylinder containing a flexible rubber membrane which blocks off the cylinder. The cylinder is then filled with round balls as shown. If an additional ball is now pushed into the left-hand side of the tube, the membrane will stretch and a ball will be forced out of the right-hand end of the tube. To an observer who could not see inside the tube, the ball would have the appearance of traveling all the way through the tube. For each ball inserted into the left-hand side, one ball would leave the right-hand side, although no balls actually pass all the way through the tube.

As more balls are forced into the tube, it becomes increasingly difficult to force in additional balls, due to the tendency of the membrane to spring back to its original position.

If too many balls are forced into the tube, the membrane will rupture, and any number of balls can then be forced all the way through the tube.

A similar effect occurs in a capacitor when the voltage applied to the capacitor is too high. If an excessive amount of voltage is applied to a capacitor, the insulating material between the plates will break down and allow a current flow through the capacitor. In most cases this destroys the capacitor, necessitating its replacement.

When a capacitor is fully charged and the source voltage is equalled by the counter electromotive force (cemf) across the capacitor, the electrostatic field between the plates of the capacitor will be maximum. Since the electrostatic field is maximum, the energy stored in the dielectric will be maximum.

If the switch is now opened as shown in Fig. 14-9A, the electrons on the upper plate are isolated. Due to the intense repelling effect of these electrons, no electrons will return to the positive plate. Thus, with the switch in position 3, the capacitor will remain charged indefinitely. At this point it should be noted that the insulating dielectric material in a practical capacitor is not perfect and small leakage current will flow through the dielectric. This current will eventually dissipate the charge. A high quality capacitor may hold its charge for a month or more, however.

To review briefly, when the capacitor is connected across a source, a surge of charging current will flow. This charging current develops a cemf across the capacitor which opposes the applied voltage. When the capacitor is fully charged, the cemf will be equal to the applied voltage and charging current will cease. At full charge the electrostatic field between the plates is at maximum intensity and the energy stored in the dielectric is maximum. If the charged capacitor is disconnected from the source, the charge will be retained for some period of time. The length of time the charge is retained depends on the amount of leakage current present. Since electrical energy is stored in the capacitor, a charged capacitor can act as a source.

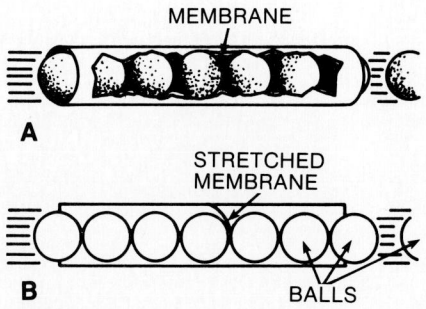

Fig. 14-8: Mechanical equivalent of capacitor action.

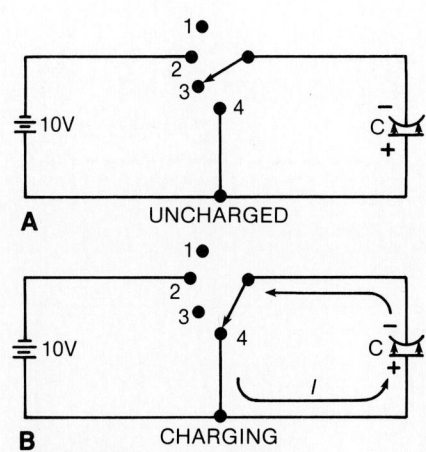

Fig. 14-9: Discharging a capacitor.

Discharging

To *discharge* a capacitor, the charges on the two plates must be neutralized. This is accomplished by providing a conducting path between the two plates (Fig. 14-9B). With the switch in position 4, the excess electrons on the negative plate can flow to the positive plate and neutralize its charge. When the capacitor is discharged, the distorted orbits of the electrons in the dielectric return to their normal positions and the stored energy is returned to the circuit. It is important to note that a capacitor does not consume power. The energy the capacitor draws from the source is recovered when the capacitor is discharged.

The discharge of large capacitors must be carried out with care. This process must be performed using a shorting device large enough to handle the current flow involved and must be carried out such that the operator is insulated from the circuit. Also, a sufficient time must be allowed for the capacitor to become fully discharged.

CAPACITORS IN SERIES AND PARALLEL

Capacitors may be connected in series or parallel to give resultant values, which may be either the sum of the individual values (in parallel) or a value less than that of the smallest capacitance (in series).

Capacitors in Series

A circuit consisting of a number of capacitors in series is similar in some respects to one containing several resistors in series. In a series capacitive circuit, the same *displacement current* flows through each part of the circuit and the applied voltage will divide across the individual capacitors.

Figure 14-10 shows a circuit containing a source and two series capacitors. When the switch is closed, current will flow in the direction indicated by the arrows on the diagram. Since there is only one path for current, the amount of charge current in motion is the same in all parts of the circuit. This current is of brief duration and will flow only until the total voltage across the capacitors is equal to the source voltage. Since the charge (Q) is the same in all parts of the circuit:

$$Q_T = Q_1 = Q_2 = \ldots\ldots Q_n$$

also:

$$C = \frac{Q}{V}$$

transposing:

$$V = \frac{Q}{C}$$

Since the sum of the capacitor voltages must equal the source voltage (Kirchhoff's Law):

$$V_T = V_1 + V_2 + \ldots V_n$$

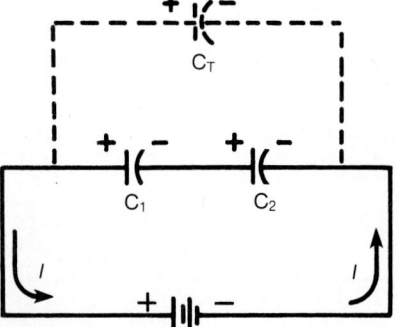

Fig. 14-10: Series capacitive circuit.

278

Substituting the equation $V = \dfrac{Q}{C}$ into $V_T = V_1 + V_2 + \ldots V_n$,

$$\frac{Q_T}{C_T} = \frac{Q_1}{C_1} + \frac{Q_2}{C_2} + \ldots\ldots \frac{Q_n}{C_n}$$

Since, by the equation $Q_T = Q_1 = Q_2 = \ldots Q_n$, all the charges are the same, dividing each term of $\dfrac{Q_T}{C_T} = \dfrac{Q_1}{C_1} + \dfrac{Q_2}{C_2} + \ldots\ldots \dfrac{Q_n}{C_n}$ by Q_T yields:

$$\frac{1}{C_T} = \frac{1}{C_1} + \frac{1}{C_2} + \ldots\ldots \frac{1}{C_n}$$

Taking the reciprocal of both sides:

$$C_T = \cfrac{1}{\dfrac{1}{C_1} + \dfrac{1}{C_2} + \ldots\ldots \dfrac{1}{C_n}}$$

Where C_T, C_1, etc., are in F. The equation above is the general equation used to compute the total capacitance of capacitors connected in series. Notice the similarity between this equation and the one used to find equivalent resistance of parallel resistors. If the circuit contains only two capacitors (Fig. 14-11), the product over the sum formula can be used.

$$C_T = \frac{C_1 C_2}{C_1 + C_2}$$

Where: C_T, C_1, etc., are in F.

As might be anticipated from the above equations, the total capacitance of series connected capacitors will always be smaller than the smallest of the individual capacitors (Fig. 14-12).

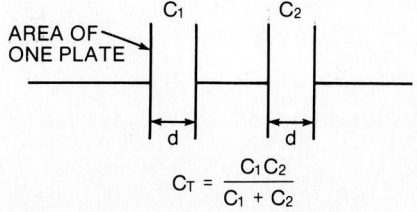

$$C_T = \frac{C_1 C_2}{C_1 + C_2}$$

Fig. 14-11: Two capacitors in series.

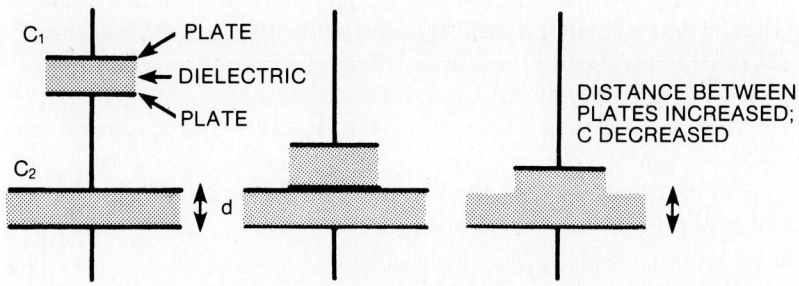

Fig. 14-12: Capacitors in series. The equivalent capacitance in C is less than the lower of the two capacitances in A.

Example: Determine the total capacitance of a series circuit containing three capacitors of $0.01\,\mu\text{F}$, $0.25\,\mu\text{F}$, and $50,000\,\text{pF}$ respectively.

Given:

$$C_1 = 0.01\,\mu\text{F}$$
$$C_2 = 0.25\,\mu\text{F}$$
$$C_3 = 50,000\,\text{pF}$$

279

Solution:

$$C_T = \cfrac{1}{\cfrac{1}{C_1} + \cfrac{1}{C_2} + \cfrac{1}{C_3}}$$

$$C_T = \cfrac{1}{\cfrac{1}{.01\mu F} + \cfrac{1}{.25\mu F} + \cfrac{1}{50,000 pF}}$$

Convert to powers of ten.

$$C_T = \cfrac{1}{\cfrac{1}{1 \times 10^{-8}} + \cfrac{1}{25 \times 10^{-8}} + \cfrac{1}{5 \times 10^{-8}}}$$

$$C_T = \cfrac{1}{100 \times 10^6 + 4 \times 10^6 + 20 \times 10^6}$$

$$C_T = \cfrac{1}{124 \times 10^6}$$

$$C_T = 0.008\mu F$$

The total capacitance of $0.008\mu F$ is slightly smaller than the smallest capacitor $(0.01\mu F)$.

Capacitors in Parallel

When capacitors are connected in parallel, one plate of each capacitor is connected directly to one terminal of the source, while the other plate of each capacitor is connected to the other terminal of the source. In Fig. 14-13, since all the negative plates of the capacitors are connected together and all the positive plates are connected together, C_T appears as a capacitor with a plate area equal to the sum of all the individual plate areas (Fig. 14-14). It should be noted that the thickness of the dielectric remains unchanged, thus not affecting the value of capacitance. As previously mentioned, capacitance is a direct function of plate area. Connecting capacitors in parallel effectively increases plate area and thereby the capacitance (Fig. 14-15).

For capacitors connected in parallel, the total charge is the sum of all the individual charges.

$$Q_T = Q_1 + Q_2 + Q_3 + ... Q_n$$

Transposing the formula $C = \dfrac{Q}{V}$:

$$Q = CV$$

Substitute $Q = CV$ into $Q_T = Q_1 + Q_2 + Q_3 + ... Q_n$.

$$C_T V = C_1 V + C_2 V + C_3 V$$

Divide both sides by V:

$$C_T = C_1 + C_2 + C_3 + ... C_n$$

where all capacitances are in the same units.

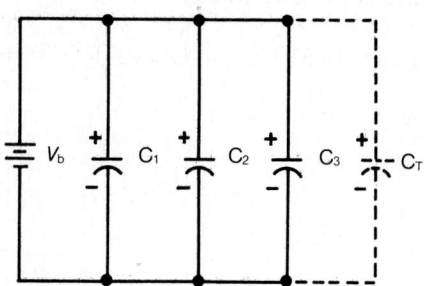

Fig. 14-13: Parallel capacitive circuit.

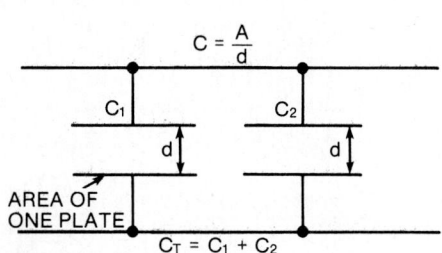

Fig. 14-14: Two capacitors in parallel.

280

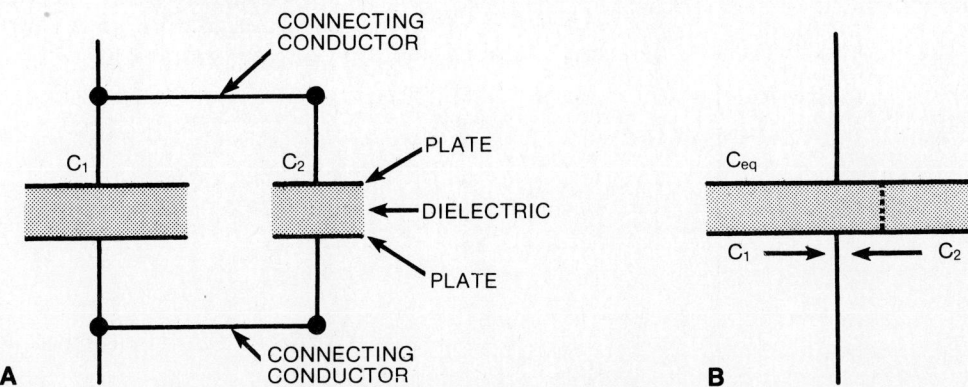

Fig. 14-15: Capacitance in parallel. The equivalent capacitance in B is equal to the sum of the capacitances in A.

Example: Determine the total capacitance in a parallel capacitive circuit:

Given:

$$C_1 = 0.03\mu F$$
$$C_2 = 2\mu F$$
$$C_3 = 0.25\mu F$$

Solution:

$$C_T = C_1 + C_2 + C_3$$
$$C_T = 0.03 + 2 + 0.25$$
$$C_T = 2.28\mu F$$

Series Parallel Configuration

If capacitors are connected in a combination of series and parallel, the total capacitance is found by applying the equations $C_T = \dfrac{1}{\dfrac{1}{C_1} + \dfrac{1}{C_2} + ... \dfrac{1}{C_n}}$ and $C_T V = C_1 V + C_2 V + ... C_n V$ to the individual branches.

Example: Determine the total capitance of the circuit in Fig. 14-16.

Given:

$$C_1 = 0.06\mu F$$
$$C_2 = 0.002\mu F$$
$$C_3 = 3\mu F$$
$$C_4 = 0.005\mu F$$
$$C_5 = 0.001\mu F$$

Find:

$$C_T = ?$$

Solution: Simplify the circuit into separate branches.

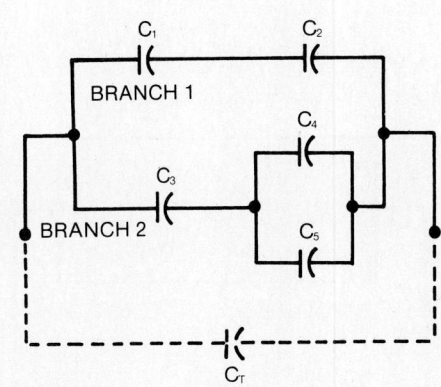

Fig. 14-16: Series-parallel capacitance configuration.

281

BRANCH 1: Consists of a series combination C_1 and C_2. To determine total capacitance of this branch only:

Branch 1: $\quad C_{T_1} = \dfrac{C_1 C_2}{C_1 + C_2}$

$$C_{T_1} = \dfrac{0.06 \times 0.002}{0.06 + 0.002}$$

$$C_{T_1} = \dfrac{0.00012}{0.062}$$

$$C_{T_1} = 0.00193 \mu F \text{ or } 1.93 nF$$

BRANCH 2: Consists of a series-parallel combination. Solve for total capacitance of branch 2 only. The equivalent capacitance (C_{eq}) of the parallel combination of C_4 and C_5 must be determined first by the use of the equation for parallel configurations:

$$C_{eq} = C_4 + C_5$$
$$C_{eq} = 0.005 + 0.001$$
$$C_{eq} = 0.006 \mu F \text{ or } 6 nF$$

Branch 2 is now reduced to an equivalent series circuit consisting of C_3 and the equivalent capacitance of C_4 and C_5.

The circuit in Fig. 14-16 has been simplified considerably and appears as the circuit in Fig. 14-17. To illustrate the use of the reciprocal method, the total capacitance of branch 2 is determined using the following equation:

$$C_{T_2} = \dfrac{1}{\dfrac{1}{C_3} + \dfrac{1}{C_{eq}}}$$

$$C_{T_2} = \dfrac{1}{\dfrac{1}{3} + \dfrac{1}{0.006}}$$

$$C_{T_2} = \dfrac{1}{0.333 + 166.666}$$

$$C_{T_2} = 0.00598 \mu F \text{ or } 5.98 nF$$

The entire problem is now reduced to a simple parallel combination consisting of C_{T_1} (capacity branch 1) and C_{T_2} (capacity branch 2).

$$C_T = C_{T_1} + C_{T_2}$$
$$C_T = 0.00193 + 0.00598$$
$$C_T = 0.00791 \mu F \text{ or } 7.91 nF$$

DISTRIBUTED CAPACITANCE

Stray or random capacitance caused by the intricate circuit wiring and the physical construction of components is present everywhere throughout AC circuits. Such stray capacitance is called the *distributed capacitance* of the circuit as distinguished from the *lumped capacitance* of the capacitor elements.

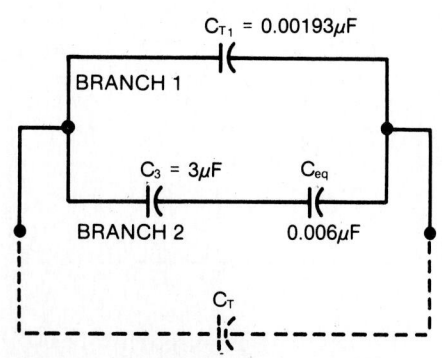

Fig. 14-17: Simplified series-parallel capacitance configuration.

Since a capacitor is, by definition, any two conductors separated by a dielectric, small amounts of capacitance exist between connecting wires (Fig. 14-18), between components located physically near each other, and between different parts of a given component. Generally, such distributed capacitance is undesired, and at low frequencies its effect is negligible. At high frequencies, however, the effects of distributed capacitance must be considered in circuit design. In some instances, the distributed capacitance is used as the capacitance of the circuit; but for the most part, every effort is made to cut down its total by careful wiring, positioning of components, and special physical construction of certain components.

Since the voltage drop across an inductance is distributed proportionately along its length and therefore exists between turns, turns adjacent to each other appear as plates of a capacitor. Figure 14-19A illustrates this effect across an inductance, the dotted line capacitors representing the effective distributed capacitance. It will be noticed that the capacitors as shown are effectively in series from one end of the inductance to the other, resulting in a total capacitance less than the smallest. In large, tightly wound inductances, however, the turns fall one upon the other in layers in such a manner that the small distributed capacitances are effectively in parallel and the total amount is considerable. Figure 14-19B shows a cross section of such an inductance and the accumulated distributed capacitance. Inductances using a large number of turns are often specially constructed to avoid the effect of accumulated distributed capacitance.

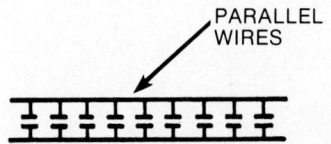

Fig. 14-18: Capacitance exists between two parallel wires.

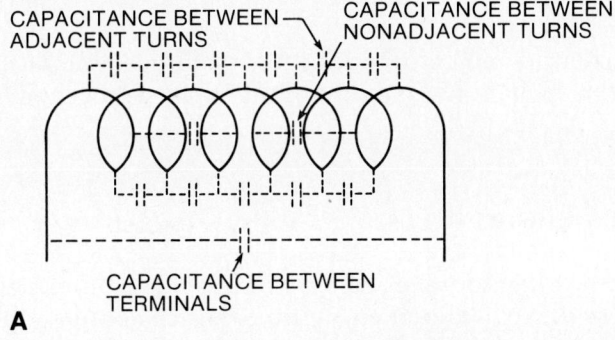

A

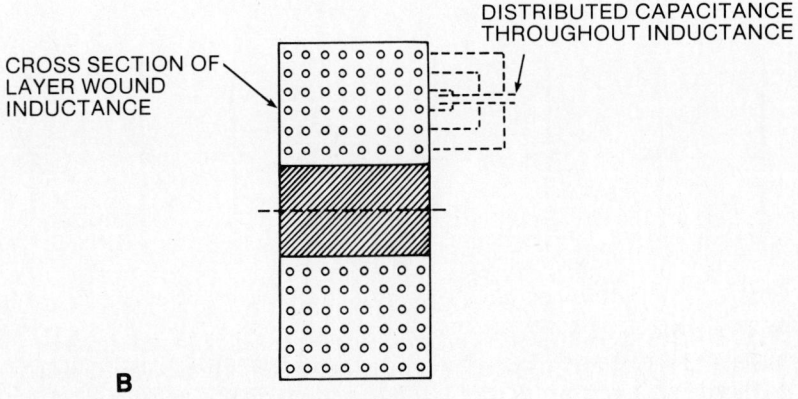

B

Fig. 14-19: Distributed capacitance.

CAPACITOR LOSSES

Losses occurring in capacitors may be attributed to either dielectric leakage or dielectric absorption (dielectric hysteresis). Dielectric absorption may be defined as an effect in a dielectric material similar to the hysteresis found in magnetic materials. It is a result of the changes in orientation of electron orbits in the dielectric because of the rapid reversals of the polarity of the line voltage. The amount of loss due to hysteresis depends upon the type of dielectric used. Vacuum shows the smallest loss.

Dielectric leakage occurs in a capacitor as the result of leakage of current through the dielectric. Normally, it is assumed that the dielectric will effectively prevent the flow of current through the capacitor. However, while the resistance of the dielectric is extremely high, a minute amount of current does flow. Ordinarily, this current is so small that for all practical purposes it is considered to be of no consequence. However, if the leakage through the dielectric is abnormally high, there will be a rapid loss of charge and an overheating of the capacitor.

The power factor of a capacitor is determined by dielectric losses. If the losses are negligible and the capacitor returns the total charge to the circuit, it is considered to be a perfect capacitor with a power factor of zero; that is, the lower the numerical value of the power factor, the better the quality of the capacitor. Therefore, the power factor of a capacitor is a measurement of its efficiency.

CAPACITOR TYPES

Capacitors may be divided into two major groups: fixed and variable. Schematic symbols of the two types of capacitors are shown in Fig. 14-20.

Fixed Capacitors

Fixed capacitors are constructed in such a manner that they possess a fixed value of capacitance which cannot be adjusted. They may be classified according to the type of material used as

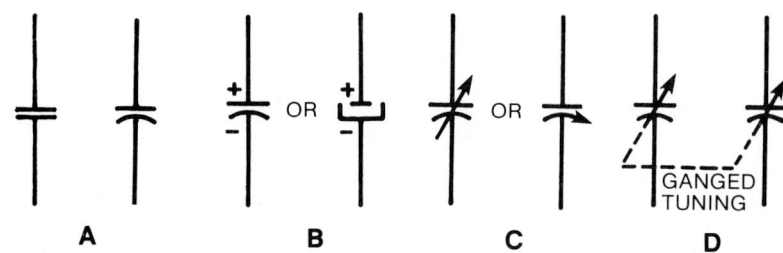

Fig. 14-20: Schematic symbols for common types of capacitors: (A) fixed air, paper, ceramic, mica; (B) electrolytic, with polarity; (C) variable; and (D) ganged variable on one shaft.

the dielectric, such as paper, oil, mica, and electrolyte. Capacitance tolerance generally depends upon the type of dielectric employed. The ± tolerance frequently is usually less on the minus side to provide enough capacitance.

Paper Capacitors. A paper capacitor is one that uses paper as its dielectric. It consists of flat, thin strips of metal foil conductors, separated by the dielectric material. In this capacitor the dielectric used is waxed paper. Paper capacitors usually range in value from about 300pF to about 4μF at tolerances of ±10%. Normally, the voltage limit across the plates rarely exceeds 600V. Paper capacitors are sealed with wax to prevent the harmful effects of moisture and to prevent corrosion and leakage.

Many different kinds of outer covering are used for paper capacitors, the simplest being a tubular cardboard. Some types of paper capacitors are encased in a mold of very hard plastic; these types are very rugged and may be used over a much wider temperature range than the cardboard-case type. Figure 14-21A shows the construction of a tubular paper capacitor; Fig. 14-21B shows a completed cardboard encased capacitor. Figure 14-21C illustrates the encapsulated paper type used on printed circuit boards.

In general, only electrolytic capacitors show polarity; all other capacitors may be connected across AC or across pulsating DC without regard to polarity. However, it will be observed that those paper capacitors in cardboard containers and designed for use in relatively high-frequency AC circuits carry a notation such as *negative* or *outside foil* for a particular lead or have a single black band near one end. This notation is to insure that the outside foil is connected to ground, since it then becomes a shield for the entire capacitor. This notation is, of course, unnecessary for a paper capacitor in a metal container, which itself acts as an effective shield.

Synthetic Film Capacitors. Synthetic film capacitors are similar to paper capacitors; however, the plastic or synthetic film type (Fig. 14-22) uses a dielectric of polystyrene, Mylar, polyethylene, or Teflon rather than paper. The improved dielectric properties of the synthetic film over the paper produce high leakage resistances and low dissipation factors, even at elevated temperatures. Synthetic film capacitors are more expensive and are used only where their improved properties are required. Some examples of their uses are polystyrene in high-Q tuned circuits and as integrating capacitors in operational amplifiers, and mylar and Teflon in high-temperature applications, up to 150° and 200°C, respectively.

Mica Capacitors. A mica capacitor is made of metal foil plates that are separated by sheets of mica, which form the dielectric. The whole assembly is covered in molded plastic. Figure 14-23 shows a cutaway view of a mica capacitor. By molding the capacitor parts into a plastic case, corrosion and damage to the plates and dielectric are prevented in addition to making the capacitor mechanically strong. Various types of terminals are used to connect mica capacitors into circuits; these are also molded into the plastic case.

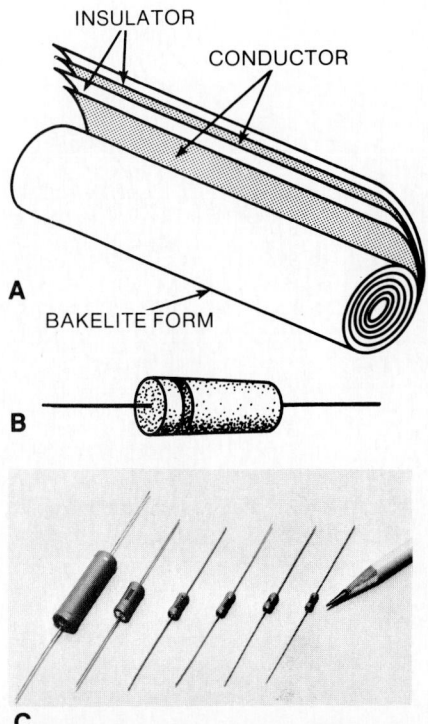

Fig. 14-21: Two types of paper capacitors. (Photo furnished by courtesy of Sprague Electric Company.)

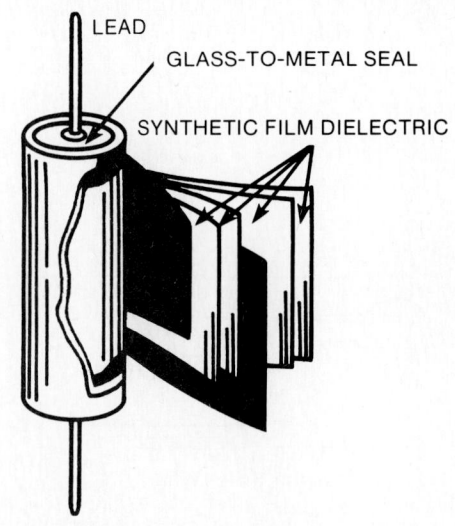

Fig. 14-22: Plastic film capacitor.

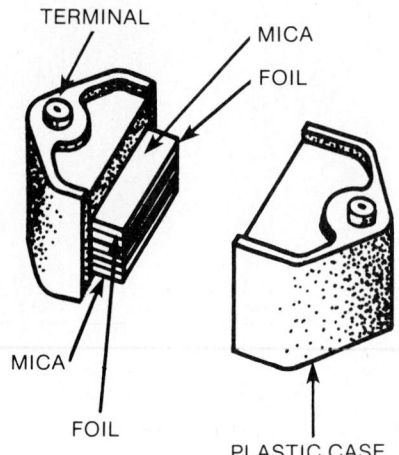

Fig. 14-23: Typical construction of a mica capacitor.

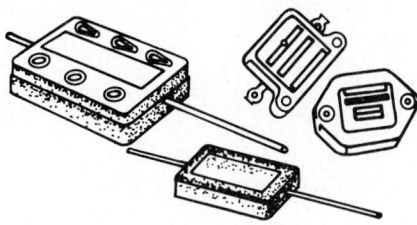

Fig. 14-24: Shapes and sizes of mica capacitors.

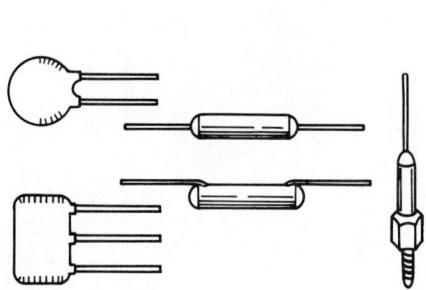

Fig. 14-25: Ceramic capacitors.

Mica capacitors are used generally for capacities between 1pF and 0.05μF at tolerances ranging from ±2% to 20%. These relatively low values of capacitors are used for higher frequency circuits. Since the dielectric constant of mica is more than five or six times that of air and its breakdown voltage much higher (about 2,000V per 0.001″), this relatively expensive material is used for high-frequency, high-voltage capacitors (up to 7,500V) which are physically small in size in comparison to the same capacity and breakdown voltage of a paper capacitor.

One variation of this type of capacitor is called *silvered mica*. The mica dielectric has silver deposited on it instead of using metalized foil for the plates. This produces capacitors with excellent stability, close tolerances (as low as ±1%), and repeatable, linear temperature-capacitance variations. Various typical mica capacitors are shown in Fig. 14-24.

Ceramic Capacitors. A ceramic capacitor (Fig. 14-25) is so named because of the use of ceramic dielectric. When titanium dioxide or one of several types of silicates are used, very high values of dielectric constant can be obtained.

One type of ceramic capacitor uses a hollow ceramic cylinder as both the form on which to construct the capacitor and as the dielectric material. The plates consist of thin films of metal deposited on the ceramic cylinder. They range in value between 1pF and 0.05μF and tolerances of ±2% to ±20%.

A second type of ceramic capacitor is manufactured in the shape of a disk (Fig. 14-26). After leads are attached to each side of the capacitor, the capacitor is completely covered with an insulating moistureproof coating. Ceramic capacitors usually range in value between 1pF and 0.01μF and may be used with voltages as high as 30,000V. The tolerances of the disk type ceramic capacitors are usually ±20%.

Ceramic capacitors are frequently employed for temperature compensation, to decrease or increase capacitance with a rise in temperature. This temperature coefficient of ceramic capacitors is usually rated in parts per million (ppm) per degree C,

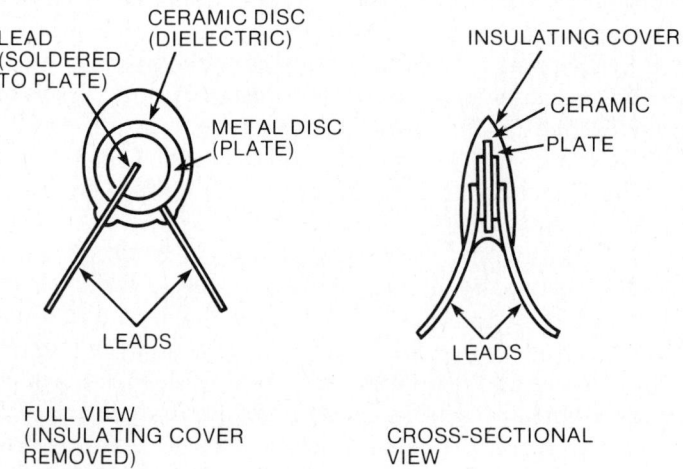

Fig. 14-26: Construction of ceramic disk capacitor.

with the reference point being 25°C. For instance, a positive 750ppm ceramic capacitor is designated as P750. A negative temperature coefficient of the same value would be labeled as N750. Capacitors that do not change in capacitance for a specific temperature range are noted as *NPO*.

Electrolytic Capacitors. These capacitors are used where a large amount of capacitance is required. As the name implies, electrolytic capacitors contain an electrolyte. This electrolyte can be in the form of either a liquid (wet electrolytic capacitor) or a paste (dry electrolytic capacitor). Wet electrolytic capacitors are no longer in popular use due to the care needed to prevent spilling of the electrolyte.

Dry electrolytic capacitors consist essentially of two metal plates between which is placed the electrolyte. In most cases the capacitor is housed in a cylindrical aluminum container which acts as the negative terminal of the capacitor (Fig. 14-27A). The positive terminal (or terminals if the capacitor is of the multisection type) is in the form of a lug on the bottom end of the container. The size and voltage rating of the capacitor is generally printed on the side of the aluminum case.

An example of a multisection type of electrolytic capacitor is depicted in Fig. 14-28. The cylindrical aluminum container will normally enclose four electrolytic capacitors into one can. Each section of the capacitor is electrically independent of the other sections, and one section may be defective while the other sections are still good (Fig. 14-27B). The can is the common negative connection with separate terminals for the positive connections identified by an embossed mark. The common identifying marks on electrolytic capacitors are the half moon, triangle, square, and no identifying mark. By looking at the bottom of the container and the identifying sheet pasted to the side of the container, it is easy to identify each section. *Note:* Correct polarity must be observed when connecting an electrolytic capacitor into a circuit. If the correct polarity is not observed and re-

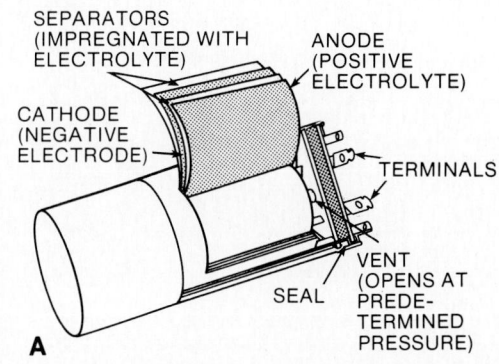

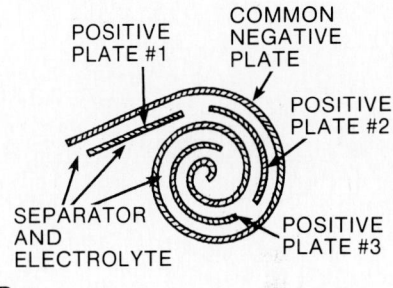

Fig. 14-27: **(A) Construction of aluminum dry electrolytic capacitor. (B) A capacitor block with a common negative and three positive plates.**

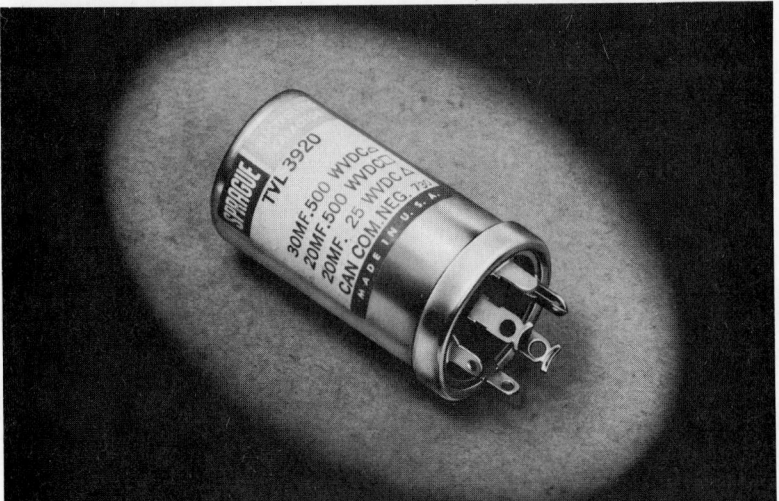

Fig. 14-28: Can type electrolytic capacitor. (Photo furnished by courtesy of Sprague Electric Company.)

287

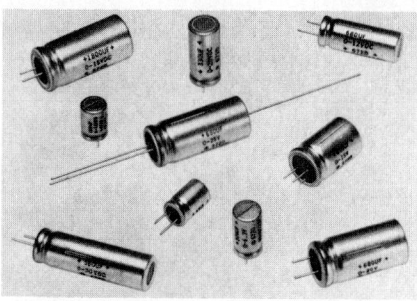

Fig. 14-29: Other types of electrolytic capacitors. (Photo furnished by courtesy of Sprague Electric Company.)

verse polarity connections are made, chemical reactions occur, causing the thin oxide film dielectric to dissolve. With no dielectric, the capacitor is effectively a short-circuit. In some cases, the chemicals in the capacitor will form a gas when the capacitor is connected in reverse polarity. This gas expands in the container and, since it has nowhere else to go, it explodes the container.

Internally, the electrolytic capacitor is constructed similarly to the paper capacitor. The positive plate consists of aluminum foil covered with an extremely thin film of oxide which is formed by an electrochemical process. This thin oxide film acts as the dielectric of the capacitor. Next to, and in contact with, the oxide is placed a strip of paper or gauze which has been impregnated with a paste-like electrolyte. The electrolyte acts as the negative plate of the capacitor. A second strip of aluminum foil is then placed against the electrolyte to provide electrical contact to the negative electrode (electrolyte). When the three layers are in place, they are rolled up into a cylinder as shown in Fig. 14-28. Other types of electrolytic capacitors are shown in Fig. 14-29.

Electrolytic capacitors have two primary disadvantages in that they are *polarized* and they have a *low leakage resistance*. Should the positive plate be accidentally connected to the negative terminal of the source, the thin oxide film dielectric will dissolve and the capacitor will become a conductor (i.e., it will short). The polarity of the terminals is normally marked on the case of the capacitor. Since electrolytic capacitors are polarity sensitive, their use is ordinarily restricted to DC circuits or circuits where a small AC voltage is superimposed on a DC voltage. Special electrolytic capacitors are available for certain AC applications, such as motor starting capacitors. Dry electrolytic capacitors vary in size from about $4\,\mu F$ to several thousand μF and have a voltage limit of approximately 500V. Electrolytic capacitors have a wide tolerance range—as much as –10% to +50%.

The type of dielectric used and its thickness govern the amount of voltage that can safely be applied to a capacitor. If the voltage applied to a capacitor is high enough to cause the atoms of the dielectric material to become ionized, an arc-over will take place between the plates. If the capacitor is not self-healing, its effectiveness will be impaired. The maximum safe voltage of a capacitor is called its *working voltage* and is indicated on the body of the capacitor. The working voltage of a capacitor is determined by the type and thickness of the dielectric. If the thickness of the dielectric is increased, the distance between the plates is also increased and the working voltage will be increased. Any change in the distance between the plates will cause a change in the capacitance of a capacitor. Because of the possibility of voltage surges (brief high amplitude pulses), a margin of safety should be allowed between the circuit voltage and the working voltage of a capacitor. The working voltage should always be higher than the maximum circuit voltage.

Tantalum Capacitors. Electrolyte capacitors that employ tantalum plates instead of aluminum foil are called *tantalum capacitors,* or *tantalytics.* This type of electrolyte capacitor has a dielectric which is a better insulator than the aluminum electrolytics. Therefore, the amount of leakage current allowed to pass through the dielectric is much smaller compared to the aluminum electrolytic. The tantalytics, while having a lower voltage rating than the aluminum electrolytics, are superior in terms of higher capacitance in a smaller size. Typical tantalytics range in size from 0.01 to 400 μF, with breakdown voltages in the 5 to 50V range. The tantalytics are mainly used in lower-voltage, solid-state devices, where size is of importance.

Oil Capacitors. This type of capacitor (Fig. 14-30) is often used in radio transmitters where high output power is desired. Oil-filled capacitors are nothing more than paper capacitors that are immersed in oil. The oil impregnated paper has a high dielectric constant which lends itself well to the production of capacitors that have a high value. Many capacitors will use oil with another dielectric material to prevent arcing between the plates. If an arc should occur between the plates of an oil-filled capacitor, the oil will tend to reseal the hole caused by the arc. These types of capacitors are often called *self-healing* capacitors.

Variable Capacitors

Variable capacitors are constructed in such a manner that their value of capacitance can be varied. Since capacity depends both on area of the plates and distance between them, either specification may be varied. Variable capacitors generally vary the effective area of the plates and adjustable capacitors vary the distance between them. A typical variable capacitor consists of two sets of plates which move in and out of mesh to vary capacitance. One set of plates is stationary, called the *stator,* and the other set is movable, called the *rotor* (Fig. 14-31). The rotor is usually grounded to reduce *hand effect.* The rotor plates usually number one less than the stator plates and are mounted on a shaft for rotation into and out of mesh with the stator. Both sets of plates are rigid and made of aluminum to prevent corrosion.

Since the thickness of the plates does not affect capacitance, they are made as thin as possible while remaining suitably rigid. Air is the dielectric, and care must be taken to prevent the rotor from touching the stator at any point. The distance between any rotor and stator plate varies with the use to which the capacitor is put. Thus, a receiving circuit of very low voltage may use variable capacitors with the plates as close as is mechanically feasible, but transmitting capacitors with high voltages must space plates a considerable distance apart to prevent arcing.

The rotor plates of variable tuning capacitors are often placed off center on the shaft in order to produce a nonlinear change of capacitance with relation to shaft rotation. Thus, as

Fig. 14-30: The oil-filled capacitor. (Photo furnished by courtesy of Sprague Electric Company.)

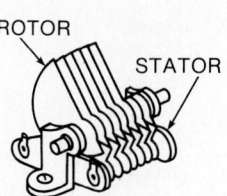

Fig. 14-31: Rotor-stator type variable capacitor.

289

the shaft moves through 180° mechanically, the plates mesh very slowly at the minimum capacitance end (high-frequency end) and move rapidly as the full mesh or maximum capacitance (low-frequency end) is reached. This distribution of capacitance change tends to spread out the crowded higher frequencies and bring together the spread of the lower frequencies, thus achieving a more uniform or linear frequency change compared to mechanical movement of the shaft. Other variable capacitors have split-stator arrangements; that is, half of the stator plates are insulated from the other half and both rotors are mounted on a common shaft so that the capacitor becomes, in effect, two balanced capacitors if connected in series (stator-rotor, rotor-stator) or one capacitor if connected in parallel. The multisection or ganged capacitor is shown in Fig. 14-32. Essentially, this type consists of two or more variable capacitors mounted in such a manner on a common shaft that all are varied simultaneously. Single-dial tuning of a number of circuits is possible.

Low-frequency variable capacitors for broadcast band transmission and reception generally range from 250pF to 500pF in fully meshed position. For higher-frequency applications, capacitors having ranges between 25pF to 150pF are used. These smaller variable capacitors are specially designed for use in high-frequency circuits. Other small variable capacitors with a maximum capacity of 5pF, called *micro* capacitors, are also used in certain very high-frequency applications.

Adjustable capacitors are another type of variable capacitor. They are so designed that they can be set to a desired capacitance and then left at that capacitance. Adjustable capacitors are often mica capacitors and consist of a spring-metal plate which is screwed into position against another plate from which it is insulated by the mica. These adjustable capacitors are generally subdivided into *trimmers* and *padders*. It should be noted that there is no electrical difference between the two types. The trimmer is a small adjustable capacitor generally placed in parallel with some larger variable capacitor so that a

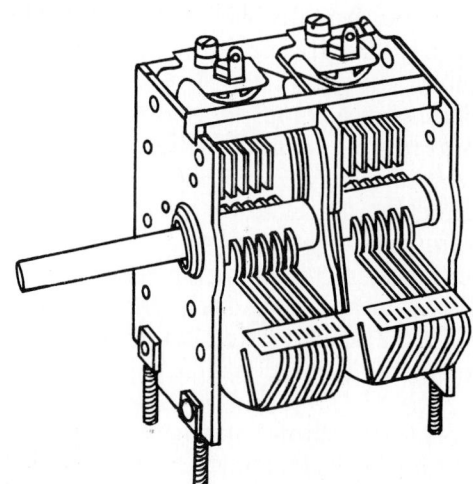

Fig. 14-32: A typical ganged capacitor.

certain small amount of capacitance is added to the larger unit; the padder is a physically small, high-capacitance capacitor placed in series with a variable capacitor to lessen its capacitance. Trimmer adjustments range from 0.5pF to 10pF, others from 10pF to 50pF. Padder adjustments range anywhere between 50pF to 100pF (minimum) and 500pF to 1,000pF (maximum). These smaller capacitors are shown in Fig. 14-33. *Note:* When making adjustments on trimmer and padder capacitors, the screw should be turned with a special tool or plastic screwdriver called an *alignment tool*. The capacitive effect of a steel screwdriver, if used, would result in inaccurate adjustment.

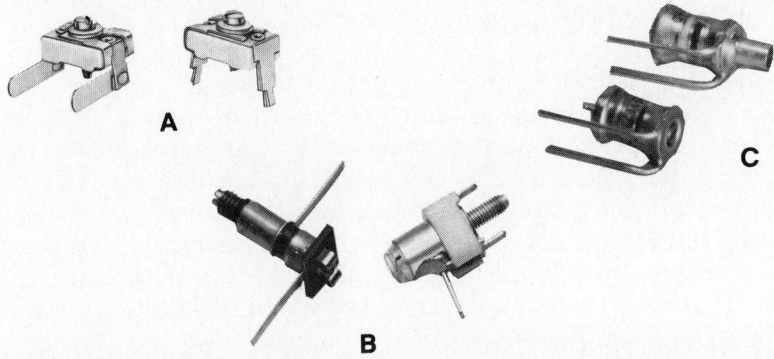

Fig. 14-33: Several types of trimmer capacitors. (Photo furnished by courtesy of Sprague Electric Company.)

Color Codes for Capacitors

Although the value of a capacitor may be indicated by printing on the body of a capacitor, many capacitive values are indicated by the use of a color code. The colors used to represent the numerical value of a capacitor are the same as those used to identify resistance values. There are two color coding systems that are currently in popular use: the Joint Army-Navy (JAN) and the Radio Manufacturers' Association (RMA) systems.

In each of these systems a series of colored dots (sometimes bands) is used to denote the value of the capacitor. Mica capacitors are marked with either three dots or six dots. Both systems are similar, but the six dot system (Fig. 14-34) contains more information about the capacitor, such as working voltage, temperature coefficient, etc. Capacitors are manufactured in various sizes and shapes. Some are small, tubular, resistor-like devices and others are molded, rectangular, flat components.

An explanation of capacitor codes is provided in Appendix E.

CAPACITOR PROBLEMS

Capacitors generally short or open as a result of deterioration of the dielectric. The most common causes of this deterioration are high temperatures, stress of charging voltage, punctured dielectric, and drying up of electrolyte. A leaky capacitor is usually the result of dielectric gradually losing its insulating

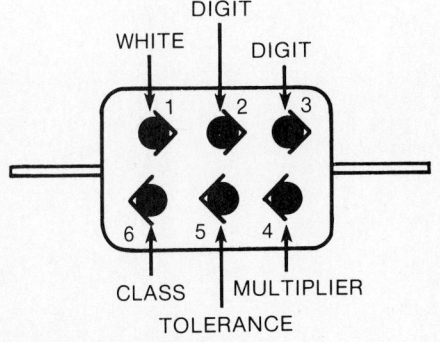

Fig. 14-34: Six-dot color code for mica capacitors.

Fig. 14-35: How to discharge a large capacitor.

properties, thus lowering its resistance. Paper and electrolytic capacitors have the highest failure rate.

Most capacitors can be tested for open or shorted conditions with an ohmmeter. Before testing, however, the capacitor should be disconnected and discharged. To safely discharge the capacitor, connect a 20,000 Ω, 2W resistor across the terminals (Fig. 14-35).

Set the ohmmeter on the highest possible range, such as R×1MΩ. When making the reading, do not touch the connections, since body resistance will lower the reading, giving a false reading.

For a low value of capacitance or when a large value of capacitance is open, the needle will not deflect from the infinite or at-rest position. When shorted, the needle will deflect and remain at or near the zero-resistance point. For a large value of capacitance charging is indicated when the needle deflects toward zero, then returns to the near infinite position. Keep in mind that a good capacitor has a very high resistance. In fact, for ceramic, mica, and paper capacitors, the resistance reading could range from 500 to 1,000MΩ or more. Electrolytic capacitors usually have less resistance (about 0.5MΩ) than other capacitor types. Leaky capacitors of any type will have lower resistances than they should.

When a capacitor is defective, the replacement should have the same rating in microfarads and in the voltage it can withstand. If, however, a single unit of the desired value is unobtainable, the correct total value can be had by using the series or parallel combinations necessary. Two 8μF capacitors in parallel, for instance, will have a total capacity of 16μF.

SUMMARY

Capacitance is one of the three basic factors that affect electronic circuits. The capacitor is constructed of a dielectric medium sandwiched between two metallic plates. There are many different dielectric materials used in capacitors. These include: air, mica, paper, synthetic film, oil, and ceramic.

The capacitor tends to oppose any change in the amount of voltage charge on its plates. In other words, it will give up charges as the circuit voltage drops and absorb charges as the circuit voltage rises.

Charging a capacitor is achieved by applying a potential, such as a battery, to the plates of the capacitor. A positive charge is established on the plate connected to the positive terminal of the battery, while a negative charge is established on the plate connected to the negative terminal of the battery. When the charge difference between the plates equals the voltage of the battery, the capacitor is said to be charged.

A capacitor stores charge through molecular distortion of the dielectric medium by electrostatic stress provided by the battery voltage.

If the dielectric is subjected to excessive potential, electrostatic stress, dielectric breakdown will occur.

292

Discharging a capacitor may be accomplished by shorting its terminals together. This permits the accumulated negative charges on one plate to return to the other plate where there is a deficiency of electrons, thus establishing neutral equilibrium.

The size of one unit of capacitance (the farad) is too large for most electronic applications; therefore, capacitor values are usually given in micro- or picofarads.

The charge of a capacitor is measured in coulombs and the charging potential is measured in volts. These factors are related as

$$Q = CV$$

Increasing the plate area of a capacitor increases the charge (Q) and the capacitance (C), and vice versa.

Increasing the thickness of the dielectric decreases the charge (Q) and the capacitance (C), and vice versa.

Capacitances in series combine as resistances in parallel:

$$C_T = \cfrac{1}{\cfrac{1}{C_1} + \cfrac{1}{C_2} + \cfrac{1}{C_3} + \ldots \cfrac{1}{C_n}}$$

Capacitances in parallel combine as resistances in series:

$$C_T = C_1 + C_2 + C_3 + \ldots C_n$$

Most capacitors are not polarity sensitive; however, correct polarity must be observed when using electrolytic capacitors. Electrolytic capacitors provide a large amount of capacitance in a small size unit.

Not all capacitors have fixed values. Some are made so that the amount of capacitance may be varied. Capacitors of this nature are called variable or adjustable capacitors. In either case, the amount of capacitance is varied by moving one plate with respect to a fixed plate. As the plates move away from each other, the amount of capacitance decreases.

Open-circuit faults appear on a VOM as infinite resistance and a "no-charging" indication. A capacitor with a short-circuit condition has a zero resistance reading when checked with a VOM. Excessive leakage in a large capacitor has some charging indication and a lower-than-normal dielectric resistance, indicating a higher-than-normal leakage current.

CHECK YOURSELF (Answers at back of book)

Questions

Determine whether the following statements are true or false.

1. The property of a capacitor to store electrical charge and energy is known as capacitance.
2. The charge stored on one plate of a capacitor is identical to the charge on the other plate.
3. The larger the physical size of a capacitor, the larger the charge it can hold.

4. Two identical capacitors connected in series have a greater capacitance than the same two connected in parallel.
5. Plastic capacitors have paper dielectric, which is encased in plastic.
6. A tantalum capacitor is not polarity sensitive.
7. The charge on a capacitor is expressed as $Q = CV$.

Fill in the blanks with the appropriate word or words.

8. The unit of capacitance is the _____ .
   A. coulomb
   B. ohm
   C. volt
   D. farad
   E. None of the above
9. If a charge of 6C raises the potential difference of a capacitor from 0 to 3V, the value of capacitance is _____ .
   A. 2pF
   B. $2\mu F$
   C. $0.5\mu F$
   D. $18\mu F$
   E. None of the above
10. The insulating material used in the construction of a capacitor is called the _____ .
    A. nonconductive layer
    B. insulating layer
    C. dielectric
    D. charge separator
    E. None of the above

Problems

1. An $8\mu F$ filter capacitor in a power supply circuit holds 1.6×10^{-2} coulombs. What is the applied voltage?
2. A special-purpose capacitor has no unit value marked on it. The two plates measure $3'' \times 3''$ and are separated by air for a distance of $0.001''$. Determine the capacitance.
3. What value capacitor must be connected in series with a $40\mu F$ capacitor to give a total capacitance of $24\mu F$?
4. A capacitor is produced by placing two sheets of brass, each $20''$ long $\times 15''$ wide, at either side of a sheet of glass that is $0.25''$ thick. Given the dielectric constant of the glass as 7, what is the capacitance of the capacitor thus formed?
5. How many coulombs of charge are stored in a $12\mu F$ capacitor which has 500V developed across it?
6. Determine the capacity of a two-plate capacitor if the plates are separated by a polystyrene dielectric $1.25 \times 10^{-3}''$ thick. The dielectric constant of polystyrene is 2.6 and the area of each plate is 64.56 sq in.

CHAPTER 15

Capacitive Circuits

When capacitors are used in circuits involving DC voltages, they exhibit certain properties and operational characteristics. When used in AC circuits, the capacitors operate in a different manner. The major part of this chapter is devoted to a discussion of the operation of a capacitor involving AC voltages; however, many AC circuits also involve the use of DC voltages. For that reason, it is desirable to cover the operation of a capacitor in a DC circuit before discussing AC capacitive circuits.

CAPACITORS IN DC CIRCUITS

Because there is no direct electrical connection between the plates of a capacitor, a capacitor represents essentially an open circuit or infinite resistance to a DC voltage. There is no direct conducting path for electrons to flow through the capacitor. However, when a capacitor is charged or discharged, a movement of electrons does take place within the circuit. This movement of electrons does constitute a current flow, but this charging or discharging action is transient in nature. This means that it occurs for only a brief period of time. Once the capacitor is charged or discharged, no current flows. While no electrons cross between the plates of a capacitor, the charging and discharging action does cause a movement of electrons in the external circuit. A component connected in series with the capacitor, such as a resistor, will have a voltage drop developed across it as the charging or discharging current occurs.

Figure 15-1A illustrates a DC circuit containing only capacitance. When the switch is closed, a surge of current charges the capacitor. Once the surge is over, current flow stops (Fig. 15-1B). The capacitor, at this point, appears to the battery like an open switch. We can assume that it has infinite impedance for all practical purposes.

The amplitude of the surge current is controlled by the resistance in the circuit, which, in this case, consists of the lead resistance and the internal resistances of the battery and the capacitor. These resistances are generally quite low and, as a result, the surge current can be quite high. However, this surge current lasts for only an instant.

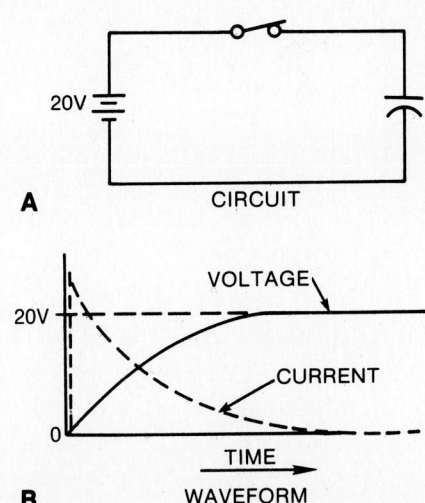

Fig. 15-1: Capacitor charging in the DC circuit. The current reaches its maximum value almost the instant the switch is closed. The voltage developed across the capacitor increases more slowly.

295

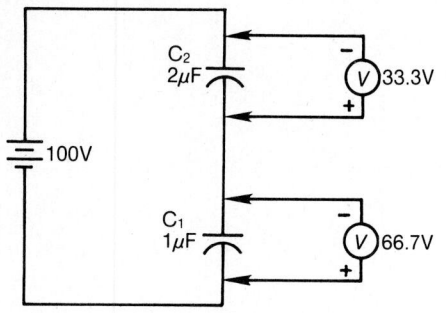

Fig. 15-2: DC voltage distribution with series capacitors. The smallest capacitor develops the most voltage.

Notice that in Fig. 15-1B the current is at peak value while the voltage is just leaving zero. By the time the voltage reaches its maximum value, the current has fallen to zero. This is characteristic of capacitors in both AC and DC circuits.

Voltage Distribution

Figure 15-2 shows a circuit consisting of two capacitors and a battery. The DC supply voltage will divide among these series capacitors in inverse proportion to their capacitance; that is, a voltmeter connected across C_1 (1μF) would indicate twice as much voltage as a voltmeter connected across C_2 (2μF). This is because capacitors in series must receive the same quantity of charge. The same amount of charge produces more voltage in a small capacitor than it does in a large capacitor. Remember that $V = Q/C$. If Q is the same for C_1 and C_2, then the higher the capacitance is, the lower the voltage will be.

RC CIRCUITS

Ohm's Law states that the voltage across a resistance is equal to the current through it times the value of the resistance. This means that a voltage will be developed across a resistance *only when current flows through it.*

A capacitor is capable of storing or holding a charge of electrons. When uncharged, both plates are in a neutral state. When charged, one plate is negatively charged and the other is positively charged. The difference in the number of electrons is a measure of the charge on the capacitor. The accumulation of this charge builds up a voltage across the terminals of the capacitor, and the charge continues to increase until this voltage equals the applied voltage. The charge on a capacitor is related to the capacitance and voltage as follows:

$$Q = CV,$$

in which Q is the charge in coulombs, C the capacitance in farads, and *V* the difference of potential in volts. Thus, the greater the voltage, the greater the charge on the capacitor. Unless a discharge path is provided, a capacitor keeps its charge indefinitely. Any practical capacitor, however, has some leakage through the dielectric so that the charge will gradually leak off.

A voltage divider containing resistance and capacitance may be connected in a circuit by means of a switch, as shown in Fig. 15-3A. Such a series arrangement is called an *RC* series circuit.

If S_1 is closed, electrons flow counterclockwise around the circuit containing the battery, capacitor, and resistor. This flow of electrons ceases when C is charged to the battery voltage. At the instant current begins to flow, there is no voltage on the capacitor and the drop across R is equal to the battery voltage.

The initial charging current, *I,* is therefore equal to $\dfrac{V_s}{R}$. Figure

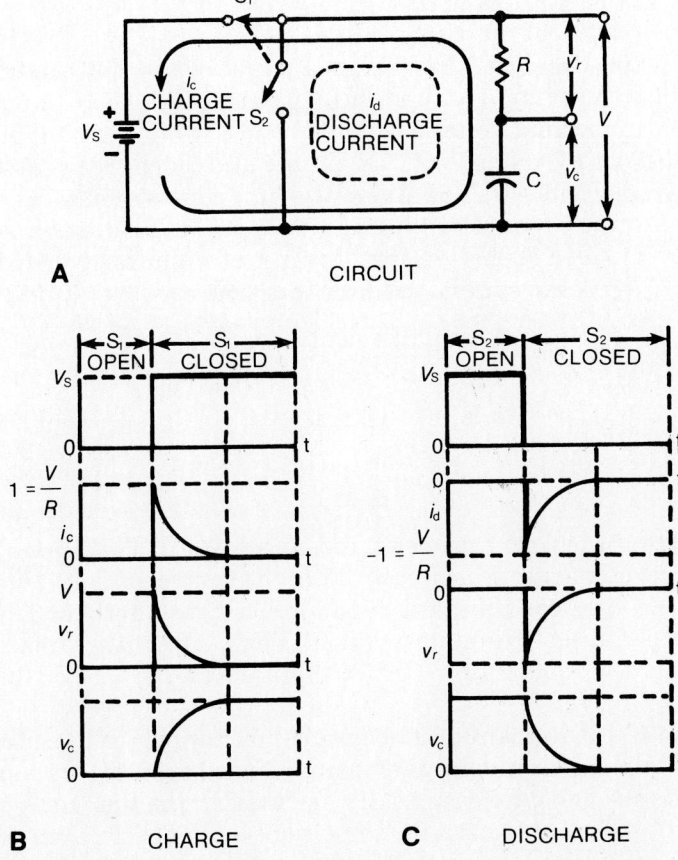

A   CIRCUIT

B   CHARGE   **C**   DISCHARGE

Fig. 15-3: Charge and discharge of an RC series circuit.

15-3B shows that at the instant the switch is closed, the entire input voltage, V_s, appears across R and that the voltage across C is zero.

The current flowing in the circuit soon charges the capacitor. Because the voltage on the capacitor is proportional to its charge, a voltage, v_c, will appear across the capacitor. This voltage opposes the battery voltage—that is, these two voltages buck each other. As a result, the voltage v_r across the resistor is $V_s - v_c$, and this is equal to the voltage drop ($i_c R$) across the resistor. Because V_s is fixed, i_c decreases as v_c increases.

The charging process continues until the capacitor is fully charged and the voltage across it is equal to the battery voltage. At this instant, the voltage across R is zero and no current flows through it. Figure 15-3B shows the division of the battery voltage, V_s, between the resistance and capacitance at all times during the charging process.

If S_2 is closed (S_1 opened) in Fig. 15-3A, a discharge current, i_d, will discharge the capacitor. Because i_d is opposite in direction to i_c, the voltage across the resistor will have a polarity opposite to the polarity during the charging time. However, this voltage will have the same magnitude and will vary in the same manner. During discharge the voltage across the capacitor is equal and opposite to the drop across the resistor, as shown in Fig. 15-3C. The voltage drops rapidly from its initial value and then approaches zero slowly, as indicated in the figure.

RC Time Constant

The time required to charge a capacitor to 63% (actually 63.2%) of maximum voltage or to discharge it to 37% (actually 36.8%) of its final voltage is known as the *time constant* of the circuit. An *RC* circuit with its charge and discharge graphs is shown in Fig. 15-4. The value of the time constant in seconds is equal to the product of the circuit resistance in ohms and its capacitance in farads, one set of values of which is given in Fig. 15-4A. *RC* is the symbol used for this time constant; that is,

$$1 \text{ time constant } (\tau) = R \times C$$

$$1 \text{ TC } (\tau) = RC$$

where τ is in s

R is in Ω

C is in F

During the first time constant, as shown in Fig. 15-4A, the capacitor charges from 0 to 63.2% of the source voltage. During the next time constant, the capacitor charges another 63.2% of the remaining available voltage. Thus, after two time constants, the capacitor is 86.5% charged ($63.2 = 0.632 \times 36.8 = 86.5$).

The actual amount of time needed to charge the capacitor to 86.5% of the source voltage is entirely based upon the amount of resistance and capacitance. By calculating the time for a time constant, we can determine the exact amount of time required.

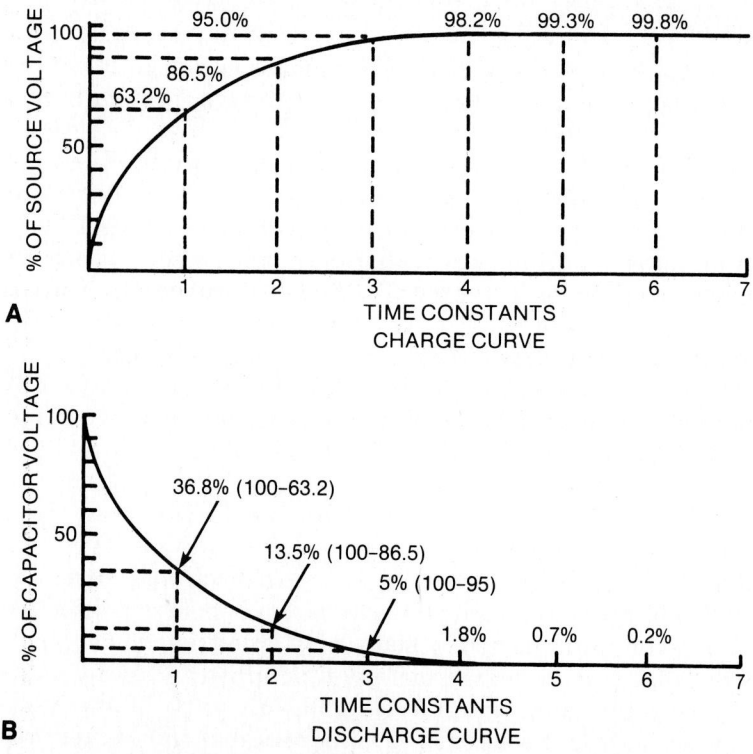

Fig. 15-4: Typical *RC* charge-discharge curves. After five or six time constants, the capacitor is considered to be fully charged or discharged.

298

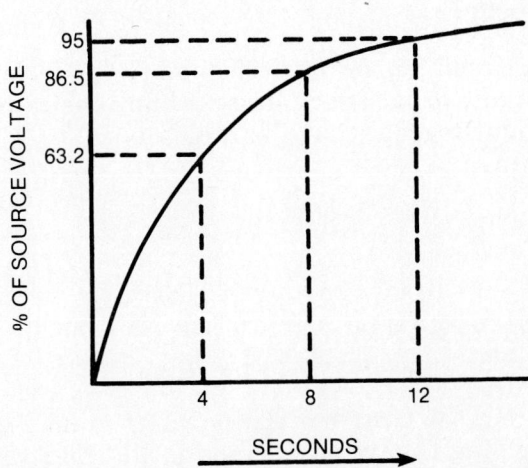

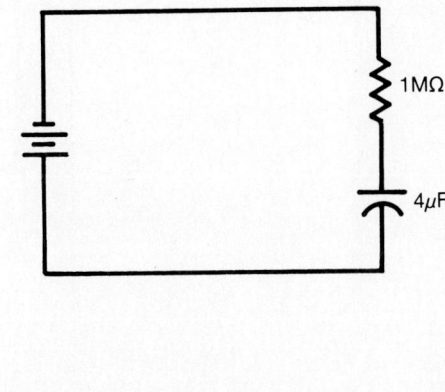

Fig. 15-5: Charging time for a capacitor. After 12s (three time constants), the capacitor is 95% charged.

For example, the time constant for the RC circuit in Fig. 15-5 would be

$$\tau = RC = 1\text{M}\Omega \times 4\mu\text{F}$$
$$= 1 \times 10^6\,\Omega \times 4 \times 10^{-6}\text{F}$$
$$= 4\text{s}$$

It requires two time constants for the capacitor to reach 86.5% of the source voltage; therefore, it takes 8 seconds for the capacitor to charge to 86.5% of the supply.

The exact amount of voltage on the capacitor after two time constants depends upon the value of the source voltage. If, in Fig. 15-5, the source voltage is 100V, it will be 86.5V. If the source is 10V, it will be 8.65V.

The discharge of a capacitor is represented in Fig. 15-6. Notice that the values of the resistance and the capacitance are the same as those used in Fig. 15-5. The time constant is still 4 seconds. When the switch is closed, the capacitor will charge to 100V. When the switch is opened, the capacitor discharges through the resistor. After one time constant, it will have discharged 63.2% of its voltage; therefore, the voltage across the capacitor will have decreased 63.2V. (36.8V will be left.) During the next time constant, the capacitor will discharge 63.2% of the remaining 36.8V. Thus, it will discharge 23.3V ($0.632 \times 36.8 = 23.3$). There will be about 13.5V left on the capacitor after two time constants, as shown in Fig. 15-6.

Referring back to Fig. 15-4, notice that a capacitor is essentially charged or discharged after a period of approximately five time constants.

Example: A circuit has a time constant of 0.001s. What amount of capacitance was involved, if $R = 2.2\text{M}\Omega$?
Solution:

$$\tau = RC$$

$$C = \frac{\tau}{R} = \frac{0.001}{2.2 \times 10^6} = 0.0454\mu\text{F} = 454\text{pF}$$

Example: Determine the time constant of an RC circuit, where $R = 500\text{k}\Omega$ and C $= 0.001\mu\text{F}$.
Solution:

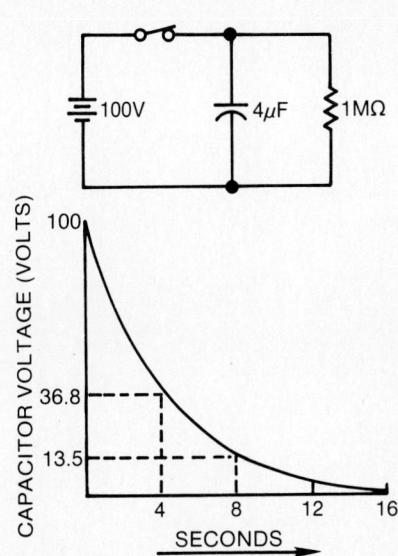

Fig. 15-6: Discharging time for a capacitor.

$$\tau = RC = 500 \times 10^3 \times 0.001 \times 10^{-6} = 0.5 \times 10^{-3} = 0.5 \text{ms}$$

Example: A circuit has an $R = 3,000$ and $C = 10\mu\text{F}$. What is the time constant? What effect does doubling the resistance have on τ? What effect does halving R have on τ?
Solution:

$$\tau = RC = 3,000 \times 10 \times 10^{-6} = 0.03\text{s}$$
For $R = 6,000\Omega$
$$\tau = 6,000 \times 10 \times 10^{-6} = 0.06\text{s}$$
For $R = 1,500\Omega$
$$\tau = 1,500 \times 10 \times 10^{-6} = 0.015\text{s}$$

Example: From the circuit of the previous example, what effect does doubling C have on τ? Halving C? The resistance remains as $R = 3,000\Omega$.
Solution:

For $C = 20\mu\text{F}$:
$$\tau = RC = 3,000 \times 20 \times 10^{-6} = 0.06\text{s}$$
For $C = 5\mu\text{F}$:
$$\tau = 3,000 \times 5 \times 10^{-6} = 0.015\text{s}$$

Universal Time Constant Chart

Because the impressed voltage and the values of R and C or R and L usually will be known, a universal time constant chart (Fig. 15-7) can be used. Curve A is a graph of the voltage across the resistor in series with the inductor on the growth of current. Curve B is a graph of capacitor voltage on discharge, capacitor

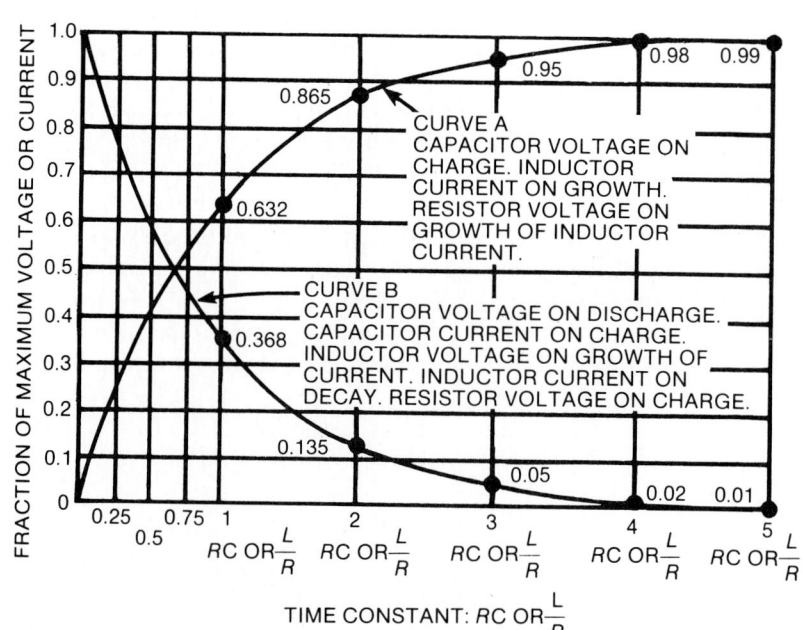

Fig. 15-7: Universal time constant chart for RC and RL circuits.

300

current on charge, inductor current on decay, or the voltage across the resistor in series with the capacitor on charge. The graphs of resistor voltage and current in the capacitor discharge circuit and the inductor voltage when the field is decaying are not shown because negative values would be involved.

The time scale (horizontal scale) is graduated in terms of the RC or L/R time constants so that the curves may be used for any value of R and C or L and R. The voltage and current scales (vertical scales) are graduated in terms of the fraction of the maximum voltage or current so that the curves may be used for any value of voltage or current. If the time constant and the initial or final voltage for the circuit in question are known, the voltages across the various parts of the circuit can be obtained from the curves for any time after the switch is closed, either on charge or discharge. The same reasoning is true of the current in the circuit.

For example, an RL circuit may have a time constant equal to 0.02s. This means that 1τ, on the universal time constant curve in Fig. 15-7, represents 0.02s, 2τ represents 0.04s, 3τ represents 0.06s, etc. This means, according to the definition of a time constant, that at 0.02s the current has risen to 63.2% of the full value. At 0.06s the current has decayed to a level of 4.9% of its maximum value. At 0.025s, the current decay may be read off the curve as 26%.

Now consider an RC circuit with a time constant of $1\mu s$. Each time constant on the curve would then represent $1\mu s$. Thus, at a time of $3\mu s$, the capacitor voltage would have reached a level of 95.1% of its maximum. Given a source voltage of 200V, at this time,

$$v_c = 200 \times 95.1\% = 190.2V$$

at $2.5\mu s$, the percentage of voltage discharge as read from the curve is approximately 9.5%. Thus, the capacitor voltage on discharge would be $v_c = 200 \times 9.5\% = 19V$.

Example: An RL circuit has an $R = 1k\Omega$ and $L = 0.03mH$. The source voltage is 9V. Using the universal time constant curves, determine the value of i_L on rise at $0.05\mu s$ and i_L on decay at $0.02\mu s$.

Solution:

$$\tau = \frac{L}{R} = \frac{0.03 \times 10^{-3}}{1 \times 10^3} = 0.03\mu s$$

The full current level has a value equal to

$$I = V/R = 9/1 \times 10^3 = 9 \times 10^{-3}A = 9mA$$

The full voltage is the supply voltage of 9V. From curve A (Fig. 15-7) at $0.05\mu s$, i_L on rise is about 82%, or

$$i_L = 9 \times 0.82 = 7.38mA$$

At $0.02\mu s$, i_L on decay is about 50%, from curve B, or

$$i_L = 9 \times 0.50 = 4.5mA$$

Example: An RC circuit has a supply voltage of 30V. The circuit R is $0.068M\Omega$, while the capacitance is 250pF. What is v_c on charge at 68s and v_c on decay at $23.8\mu s$?

Solution:

$$\tau = RC = 0.068 \times 10^6 \times 250 \times 10^{-12} = 17 \times 10^{-6} = 17\mu s$$

From curve A (Fig. 15-7), as each time constant on the curve represents $17\mu s$, 1τ would be $17\mu s$, 2τ would be $34\mu s$, and 4τ would be $68\mu s$. Thus, at this point, v_c would reach 98.1% of its full value. Thus, at $68\mu s$,

$$v_c = 0.981 \times 30 = 29.43V$$

On decay, at $23.8\mu s$, v_c will reach a level of about 25% of its full value. Thus, at $23.8\mu s$,

$$v_c = 0.25 \times 30 = 7.5V$$

AC IN A CAPACITIVE CIRCUIT

As already noted, a capacitor that is initially uncharged tends to draw a large current when a DC voltage is first applied. During the charging period, the capacitor voltage rises. After the capacitor has received sufficient charge, the capacitor voltage equals the applied voltage and the current flow ceases. If a sine wave AC voltage is applied to a pure capacitance, the current is a maximum when the voltage begins to rise from zero, and the current is zero when the voltage across the capacitor is a maximum (Fig. 15-8A). The current leads the applied voltage by 90°, as indicated in the phasor diagram of Fig. 15-8B.

Factors that Control Charging Current

Because a capacitor of large capacitance can store more energy than one of small capacitance, a larger current must flow to charge a large capacitor than to charge a small one, assuming the same time interval in both cases. Also, because the current flow depends on the rate of charge and discharge, the higher the frequency, the greater is the current flow per unit time.

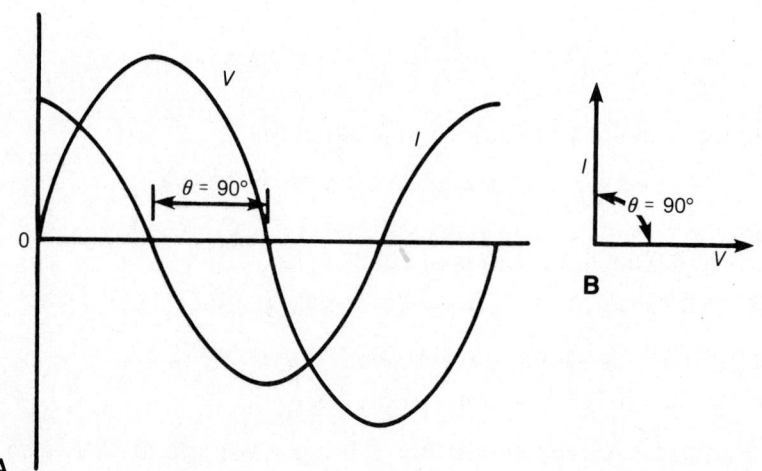

Fig. 15-8: Phase relation between V and I in a capacitive circuit.

302

The charging current in a purely capacitive circuit varies directly with the capacitance, voltage, and frequency;

$$I = 2\pi fCV$$

where I is effective current in A, f is in Hz, C is the capacitance in F, and V is in effective volts.

Formula for Capacitive Reactance

The ratio of the effective voltage across the capacitor to the effective current is called the *capacitive reactance*, X_c, and represents the opposition to current flow in a capacitive circuit of zero losses.

$$X_c = \frac{1}{2\pi fC}$$

when f is in Hz and C is in F, then X_c is in Ω.

From this equation it can be seen that X_c is inversely proportional to f and C; that is, increasing f or C will result in a decrease in X_c, and decreasing f or C will result in an increase in X_c. The application of this equation is shown in the following examples.

Example: Find the capacitive reactance of a capacitor of C = 0.5μF at a frequency of 5kHz. What is X_c if f = 10kHz? if f = 20kHz? if f = 1kHz?

Solution:

At f = 5kHz:

$$X_c = \frac{1}{2\pi fC} = \frac{1}{6.28 \times 5 \times 10^3 \times 0.5 \times 10^{-6}} = 63.7\Omega$$

At f = 10kHz:

$$X_c = \frac{1}{2\pi fC} = \frac{1}{6.28 \times 10 \times 10^3 \times 0.5 \times 10^{-6}} = 31.8\Omega$$

At f = 20kHz:

$$X_c = \frac{1}{2\pi fC} = \frac{1}{6.28 \times 20 \times 10^3 \times 0.5 \times 10^{-6}} = 15.9\Omega$$

At f = 1kHz:

$$X_c = \frac{1}{2\pi fC} = \frac{1}{6.28 \times 1 \times 10^3 \times 0.5 \times 10^{-6}} = 318.4\Omega$$

Transposing the equation for capacitive reactance provides useful relationships with the frequency, f, the capacitor, C, and the capacitive reactance, X_c. These are shown in the following examples.

Example: What value capacitor is needed to develop a capacitive reactance of 5kΩ at a frequency of 2kHz?

Solution:

$$X_c = \frac{1}{2\pi fC}$$

$$C = \frac{1}{2\pi fX_c} = \frac{1}{6.28 \times 2,000 \times 5,000} = 0.016F$$

Example: What frequency will provide a capacitive reactance of 5.26MΩ in a circuit with a capacitor of 25pF?
Solution:

$$X_c = \frac{1}{2\pi f C}$$

$$f = \frac{1}{2\pi C X_c} = \frac{1}{6.28 \times 25 \times 10^{-12} \times 5.26 \times 10^6}$$

$$= 0.00121 \times 10^6 = 1,210 \text{Hz}$$

Power in Capacitive Circuits

With no voltage or charge, the electrons in the dielectric between the capacitor plates rotate around their respective nuclei in normally circular orbits. When the capacitor receives a charge, the positive plate repels the positive nuclei and, at the same time, the electrons in the dielectric are strained toward the positive plate and repelled away from the negative plate. This distorts the orbits of the electrons in the direction of the positive charge. During the time the electron orbits are changing from normal to the strained position, there is a movement of electrons in the direction of the positive charge. This movement constitutes the *displacement current* in the dielectric. When the polarity of the plates reverses, the electron strain is reversed. If a sine wave voltage is applied across the capacitor plates, the electrons will oscillate back and forth in a direction parallel to the electrostatic lines of force. *Displacement current* is a result of the movement of bound electrons, whereas *conduction current* represents the movement of free electrons.

Figure 15-9A shows a capacitive circuit and Fig. 15-9B indicates the sine waveform of charging current, applied voltage, and instantaneous power. The effective voltage is 70.7V. The effective current is 7.07A. Because the losses are neglected, the phase angle between current and voltage is assumed to be 90°. The true power is zero, as indicated by the expression

$$P = VI \cos \theta$$

$$= 70.7 \times 7.07 \ (\cos 90° = 0) = 0\text{W}$$

Multiplying instantaneous values of current and voltage over one cycle, or 360°, gives the power curve, *P*. During the first quarter cycle (from 0° to 90°) the applied voltage rises from zero to a maximum and the capacitor is receiving a charge. The power curve is positive during this period and represents energy stored in the capacitor. From 90° to 180° the applied voltage is falling from maximum to zero and the capacitor is discharging. The corresponding power curve is negative and represents energy returned to the circuit during this interval. The third quarter cycle represents a period of charging the capacitor and the fourth quarter cycle represents a discharge period. Thus, the average power absorbed by the capacitor is zero. The action is like the elasticity of a spring. Storing a charge in the capacitor is like compressing the spring. Dis-

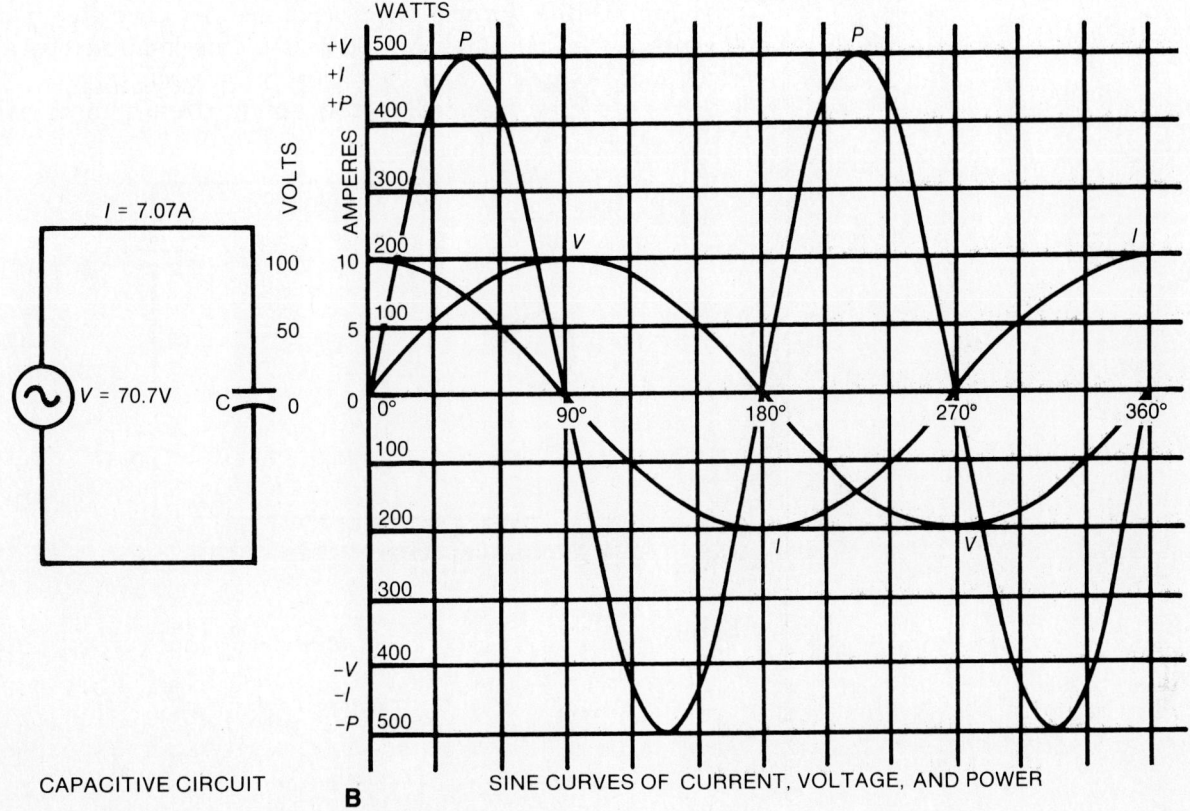

WATTS

VOLTS

AMPERES

A CAPACITIVE CIRCUIT

B SINE CURVES OF CURRENT, VOLTAGE, AND POWER

Fig. 15-9: Power in a capacitive circuit.

charging the capacitor is like releasing the pressure on the spring, thus allowing it to return the energy that was stored within it on compression.

The apparent power in the capacitor is $VI = 70.7 \times 7.07 = 500 V/A$. The reactive power in the capacitor is

$$VI \sin \theta = 70.7 \times 7.07 \ (\sin 90° = 1)$$
$$= 500 \ \text{VARs (leading)}$$

Series or Parallel Capacitive Reactances

Series or parallel reactances are combined in the same way as resistances. This is because capacitive reactance, like resistance, is an opposition in Ω to the circuit current.

Figure 15-10A shows the addition of two series reactances of $100\,\Omega$ and $200\,\Omega$. Their sum equals $300\,\Omega$, as shown. The formula for adding capacitive reactances in series is

$$X_{cT} = X_{c1} + X_{c2} + X_{c3} + ... X_{cn}$$

For parallel capacitive reactances, the combined reactance is calculated by the reciprocal formula, as shown in Fig. 15-10B.

$$\frac{1}{X_{cT}} = \frac{1}{X_{c1}} + \frac{1}{X_{c2}} + \frac{1}{X_{c3}} + ... \frac{1}{X_{cn}}$$

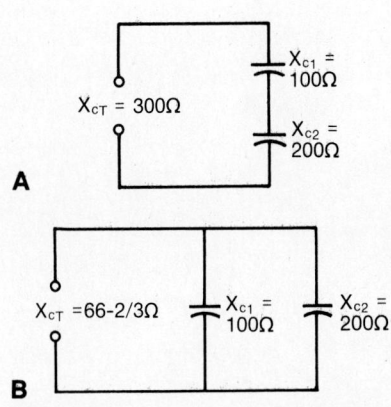

Fig. 15-10: Typical (A) capacitive series circuit and (B) parallel circuit.

305

In Fig. 15-10B, the combined reactance of X_{c1} (100Ω) in parallel with X_{c2} (200Ω) equals X_{cT} (66-2/3Ω). This combined parallel reactance is smaller than the lowest branch reactance.

Example: Find the total reactance of the circuit shown in Fig. 15-11.

Fig. 15-11: Typical capacitive series circuit.

Solution:

$$X_{c1} = \frac{1}{2\pi f C_1} = \frac{1}{6.28 \times 100 \times 2 \times 10^{-6}} = 796\,\Omega$$

$$X_{c2} = \frac{1}{2\pi f C_2} = \frac{1}{6.28 \times 100 \times 4 \times 10^{-6}} = 398\,\Omega$$

$$X_{cT} = X_{c1} + X_{c2} = 1{,}194\,\Omega$$

The answer to the preceding example could also be found by first calculating the total capacitance and then determining the total reactance; that is,

$$C_T = \frac{C_1 C_2}{C_1 + C_2} = \frac{2 \times 4}{2 + 4} = \frac{8}{6} = 1.33\,\mu F$$

$$X_{cT} = \frac{1}{2\pi f C_T} = \frac{1}{6.28 \times 100 \times 1.33 = 10^{-6}} = 1{,}194\,\Omega$$

Ohm's Law is also applicable in capacitor circuits. Just substitute R with X_c.

Example: Find the circuit current and voltage across each capacitor in Fig. 15-11.
Solution:

$$I_T = \frac{V_T}{X_{cT}} = \frac{40V}{1{,}194\,\Omega} = 0.0335A = 33.5mA$$

$$V_{c1} = 0.0335A \times 796\,\Omega = 26.7V$$

$$V_{c2} = 0.0335A \times 398\,\Omega = 13.3V$$

Note that in the preceding example, the largest capacitor dropped the least amount of voltage. This is because X_c and C are inversely proportional.

Example: Given the circuit of Fig. 15-12, find I_T, I_{c1}, C_T, and X_{cT}.

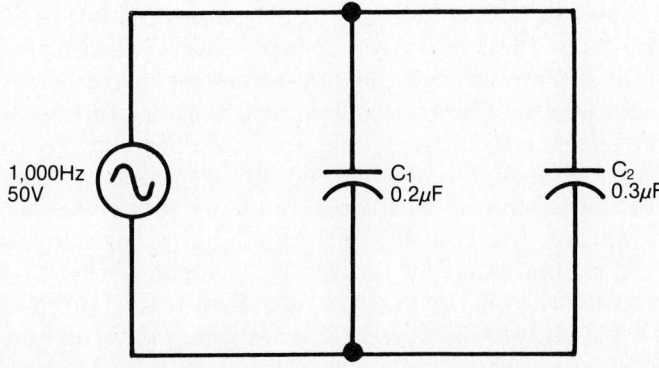

Fig. 15-12: Typical capacitive parallel circuit.

Solution:

$$C_T = 0.2\mu F + 0.3\mu F = 0.5\mu F$$

$$X_{cT} = \frac{1}{6.28 \times 1 \times 10^3 \times 0.5 \times 10^{-6}} = 318$$

$$I_T = \frac{50V}{318} = 0.157A = 157mA$$

$$X_{c1} = \frac{1}{6.28 \times 1 \times 10^3 \times 0.2 \times 10^{-6}} = 796$$

$$I_{c1} = \frac{50V}{796} = 0.0628A = 62.8mA$$

CURRENT-VOLTAGE RELATIONSHIPS IN CAPACITIVE AC CIRCUITS

The relationship between the current and the applied voltage in a capacitive circuit is different from purely resistive AC circuits. Because of the way that a capacitor works, the current and voltage in a capacitive AC circuit are not in step with one another. In a circuit where AC voltage is applied to a resistance, current through the resistor follows the voltage applied to it. We say that the current and voltage in such a circuit are in phase. The positive and negative half cycles of voltage and current in a resistive AC circuit are in step with one another.

In a capacitive AC circuit, the capacitor constantly charges and discharges with a change in the applied voltage. Once the capacitor is initially charged, the voltage across it acts as a voltage source. Its effect is to oppose changes in the external supply voltage. Since the capacitor must charge or discharge to follow the changes in the applied voltage, the resulting current flow is out of step with the changes in the applied voltage; that is, there is a *phase shift* between the voltage and current in the circuit.

To understand the relationship between the current and voltage in a capacitive AC circuit, it is best to go back to the basic principles of capacitor operation with a DC voltage. Figure

307

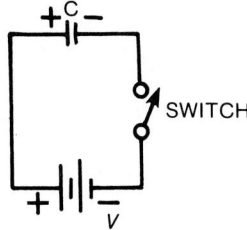

Fig. 15-13: A capacitor being charged by a DC source.

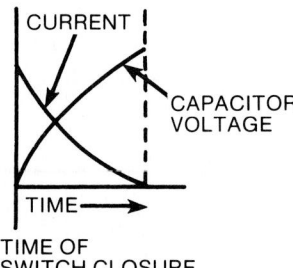

Fig. 15-14: Relationship between voltage and current in a charging capacitor.

15-13 shows the DC voltage source (*V*) connected to a capacitor (C). The capacitor is initially discharged and the switch is open. When the switch is closed, the instantaneous voltage across the capacitor is zero. While it would appear that the voltage across the capacitor is initially equal to the applied voltage, the instant the switch is closed, it is zero, since electrons have not had time to flow to and from the plates of the capacitor. In other words, on the initial closure of the switch, the capacitor is still uncharged. Immediately thereafter, electrons begin to flow. Electrons flow from the negative terminal of the battery to the right-hand plate of the capacitor, giving that plate an excess of electrons and a negative charge. At the same time, electrons are drawn from the left-hand plate of the capacitor to the positive terminal of the battery, giving the left-hand plate a positive charge. As electrons begin to flow, a voltage builds up across the capacitor. The polarity of voltage across the capacitor is as indicated in Fig. 15-13. Note that this voltage is in direct opposition to the applied voltage. When the capacitor is fully charged to the applied voltage, the two voltages equal one another and their effects cancel. The effective voltage in the circuits is zero, so no further current flows.

Figure 15-14 illustrates the relationship between the current and capacitor voltage in the circuit of Fig. 15-13. When the switch is initially closed, the capacitor voltage is zero while the current in the circuit is maximum. As the electron flow charges the capacitor, the capacitor voltage begins to build up. The capacitor voltage opposes the applied voltage, thereby reducing the amount of current flowing in the circuit. When the capacitor becomes fully charged, the current in the circuit is reduced to zero.

Now, using the relationship shown in Fig. 15-14, we can show the actual current and voltage relationships in a capacitor circuit when an AC signal is applied. The current flowing in the circuit is out of step with the voltage as indicated in Fig. 15-14.

The exact relationship between the current and voltage is a capacitive circuit when a sine wave AC signal is applied as shown in Fig. 15-15. Note that when the current is at maximum, the voltage across the capacitor is zero. As can be seen, there is

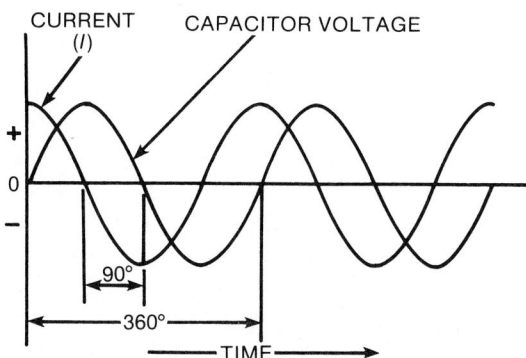

Fig. 15-15: Current and voltage relationship in a purely capacitive AC circuit.

308

a phase shift between the current and voltage in the circuit. This phase shift is expressed in terms of degrees. Remember that one complete cycle of an AC sine wave contains 360°. The amount of phase shift in the capacitor circuit is one-fourth of this or 90°. It can be said that the current and voltage in a purely capacitive circuit are 90° out of phase with one another. Another important fact to note is that the change in current *leads* the change in voltage. Looking at Fig. 15-15, it can be seen that the capacitor voltage change *follows* the current change in time; thus, the current leads the voltage in a capacitive circuit. In other words, there is a 90° leading phase shift in a purely capacitive circuit.

SERIES AC RC CIRCUITS

Figure 15-16 illustrates the simplest form of RC circuit connected to a source of AC voltage. The source voltage produces a sine wave designated V, which causes current to flow in the circuit. The capacitor charges and discharges as the AC voltage varies. This causes a movement of electrons in the circuit, but no electrons flow through the capacitor. The charge on the capacitor causes a voltage to be developed across it. This voltage is designated V_c. The voltage across the capacitor is a sine wave that lags the circuit current by 90°.

A voltage, V_R, is developed across the resistor, R, as the circuit current flows through it. This voltage is in phase with the circuit current. The voltage across the resistor is determined by the resistance value and the current ($V_R = IR$). The voltage across the capacitor is a function of the current flowing in the circuit and the capacitor reactance ($V_c = IX_c$).

The phase relationships of the various voltages and currents in this series circuit are shown in Fig. 15-17. One of the fundamental characteristics of a series circuit is that the current is common to all components of the circuit; therefore, the current sine wave (I) is used as the reference. The voltage drop across the resistor, V_R, is in phase with the current; that is, the maximum, minimum, and zero crossing points for the voltage wave-

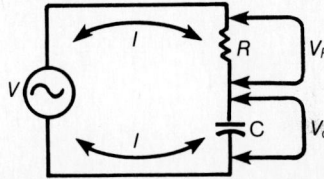

Fig. 15-16: Series *R*C circuit.

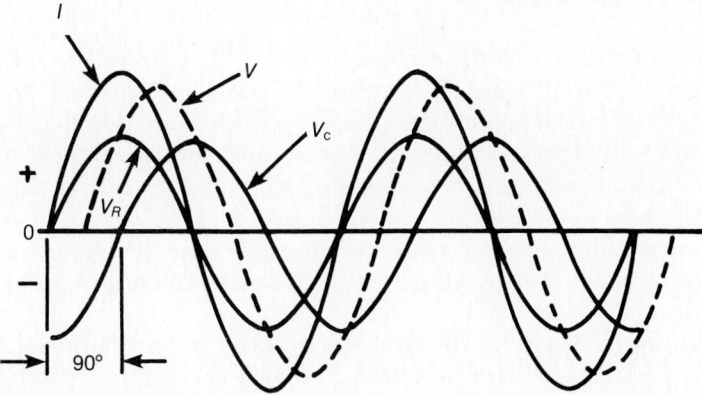

Fig. 15-17: Phase relationships between the current and voltage in a series *R*C circuit.

form coincide with those of the current waveform. Note also that the current through a capacitor leads the voltage across it by 90°. Thus, the voltage across the capacitor lags (occurs later in time than) the current.

Kirchhoff's Voltage Law also applies to AC capacitive circuits. As stated previously, the sum of the voltage drops across the components in a series circuit equals the applied voltage. Referring to Fig. 15-17, if V_R and V_c are added by summing the amplitudes of the two sine waves at multiple points and plotting the resulting curve, the sine wave obtained is represented by the dashed line. This waveform represents the applied voltage, V. Observe that the amplitude is higher than either the capacitor voltage or the resistor voltage. Another important point to keep in mind is that the applied voltage is not in phase with the current, capacitor voltage, or resistor voltage. The circuit current leads the applied voltage as it will in any capacitive current; but since the circuit is not purely capacitive (because of the series resistor), the current will lead the voltage by some phase angle less than 90°. The phase shift between the current and the applied voltage is approximately 45° in this example. In a purely capacitive circuit, the current and the applied voltage will be out of phase by 90°. In a purely resistive circuit, the current and the voltage will be in phase. When both capacitance and resistance are in a circuit, the phase difference between the current and the applied voltage will be some value greater than 0° but less than 90°. The exact amount of phase shift is determined by the size of the resistance and the capacitive reactance.

It is important to note that when using Kirchhoff's Law to analyze an AC RC circuit, it is not possible to *algebraically* add the numerical values of the resistor and capacitor voltages in order to obtain the applied voltage; that is, the relationship $V = V_R + V_c$ is incorrect. This is due to the capacitance in the circuit. The capacitance causes a phase difference between the voltage of the capacitor and the voltage across the resistor. To obtain the correct sum, this phase difference must be taken into consideration. This can be accomplished by vector or phasor addition.

Phasor Diagrams

A *phasor,* or *vector,* as mentioned earlier, is a line whose length represents the peak value of an AC voltage or current. The direction in which the phasor points indicates its phase relationship to other phasors. Various phasors are combined to form a phasor diagram which shows the phase and amplitude relationships of currents and voltages in a reactive AC circuit. A phasor diagram can be drawn to represent the current and voltages in the series AC circuit that has been analyzed to this point.

Figure 15-18 shows the basic format for drawing phasor diagrams. Here a phasor labeled I is drawn starting at the origin (designated O) in the center of the diagram. The point of the

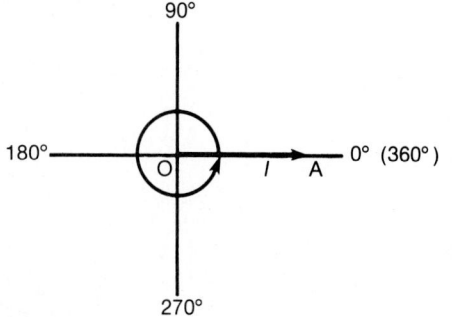

Fig. 15-18: Phasor diagram format.

arrow is designated A. The distance between O and A represents the peak value of an alternating current (I). Note that the current phasor is pointing to the right, which is designated as the 0° position. The phasor is assumed to be rotating in the counterclockwise direction. One complete 360° rotation corresponds to one cycle of the AC sine wave represented by the phasor. The position of the phasor anywhere during its rotation is representative of the particular point in the occurrence of one cycle of the sine wave. By adding other phasors to the diagram, a complete picture of the currents and voltages in the AC circuit can be obtained.

Figure 15-19 shows the phasor diagram for a series RC circuit. A current phasor (I) is shown pointing to the right. This phasor represents the peak value of the current flowing in the series RC circuit. Coinciding with the current phasor is another phasor labeled V_R. The length of the phasor from the origin (O) to the point of the arrow represents the peak value of the voltage across the resistor. The resistive voltage phasor overlaps or coincides with the current phasor because these two signals are in phase.

The voltage across the capacitor V_c is 90° out of phase with the current. This is illustrated by the phasor labeled V_c (Fig. 15-19). The direction of V_c is shifted 90° from the direction of the current phasor. The length of this phasor will represent the peak value of the capacitor voltage.

To find the applied voltage (V), add the capacitor and resistor voltages. Since they are out of phase with one another, the peak values cannot be added directly as indicated earlier. Instead, the addition must take into consideration both the magnitudes of the voltages and their directions.

To accomplish the addition of the resistor and capacitor voltages in graphical manner, a rectangle can be formed by using the capacitor and resistor voltage phasors as shown in Fig. 15-19. The dashed lines in the diagram indicate how the rectangle is completed. The magnitude of the applied voltage is the distance between the origin and the far corner formed by completing the rectangle. In other words, the applied voltage in the circuit is represented by the diagonal line drawn from the origin. The angle formed by the applied voltage phasor (V) and the current phasor (I) represents the amount of phase shift in the circuit. This phase angle is designated theta (θ). Keep in mind that it is some value less than 90°.

The magnitude of the applied voltage can be determined in a more direct manner than by this graphical means. If the resistor and capacitor voltages are known, a simple mathematical formula (Pythagorean's Theorem) can be employed; that is, consider the triangle formed by the resistor voltage (V_R), the applied voltage (V), and the dashed line connecting these two phasors in Fig. 15-19. The dotted line has the same length as the phasor representing the capacitor voltage (V_c) and can be considered as an equivalent phasor. Then use Pythagorean's Theorem (the square of two legs of a right triangle are equal to the hypotenuse).

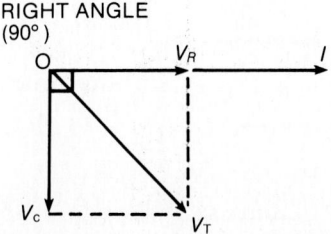

Fig. 15-19: Phasor diagram for a series RC circuit.

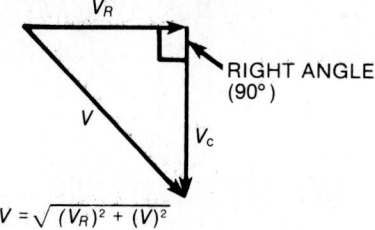

$$V = \sqrt{(V_R)^2 + (V)^2}$$

Fig. 15-20: Voltage phasor tri-angle for a series *RC* circuit.

Referring to Fig. 15-20, the side of the triangle directly oppo-site the 90° angle (the hypotenuse) represents the phasor for the applied voltage (*V*). To find *V* (the hypotenuse), proceed as follows:

$$V = \sqrt{(V_R)^2 + (V_c)^2}$$

This formula can also be employed to determine the applied voltage when the resistor and capacitor voltages are known.

$$V_R = \sqrt{(V)^2 - (V_c)^2}$$
$$V_c = \sqrt{(V)^2 - (V_R)^2}$$

Example: What is the value of the applied voltage in a series *RC* circuit where the capacitor voltage is 12V and the resistor voltage is 16V?
Solution:

$$V = \sqrt{(V_R)^2 + (V_c)^2}$$
$$= \sqrt{(16)^2 + (12)^2}$$
$$= \sqrt{256 + 144}$$
$$= \sqrt{400}$$
$$= 20V$$

Example: 50V is the voltage applied to a series *RC* circuit. The circuit current is 100mA. The resistor value is 270Ω. De-termine the value of the capacitor voltage.
Solution:

$$V_R = IR$$
$$= (0.1A)(270Ω)$$
$$= 27V$$
$$V_c = \sqrt{(V)^2 - (V_R)^2}$$
$$V_c = \sqrt{(50V)^2 - (27V)^2}$$
$$= \sqrt{2,500 - 729}$$
$$= \sqrt{1,771}$$
$$= 42.08V$$

Impedance in Series RC Circuits

As stated earlier, Ohm's Law applies in AC circuits if Z is substituted for *R* in the following formula:

$$Z = \frac{V}{I}$$

Because of the phase shift caused by the capacitor in a series *RC* circuit, the voltage drops across the capacitor and resistor cannot be added directly to obtain the desired value. The volt-age drops across the resistor and capacitor in a series *RC* cir-cuit, however, are directly proportional to the current. Since the current through a series circuit is the same in all elements, it can be said that the voltage drops across the circuit compo-

312

nents are directly proportional to their resistance or reactance. For this reason, it is possible to draw a diagram in the exact same manner as the voltage phasor diagram described earlier to obtain the total impedance (Z_T) of the circuit (Fig. 15-21). Here the current phasor (I) is used as the reference. The resistance phasor coincides with the current phasor since the resistive voltage drop is in phase with the current. In this case, the length of the phasor is proportional to resistance. Another phasor representing the magnitude of the capacitive reactance (X_c) is drawn 90° out of phase with the resistance phasor to take into account the 90° phase shift produced by the capacitor. Completing the rectangle formed by the resistance and reactance phasors and drawing the diagonal line represents the total opposition to current flow. Using Pythagorean's Theorem, an expression for the total impedance of the circuit can be written in terms of the resistance and reactance:

$$Z_T = \sqrt{R^2 + X_c^2}$$

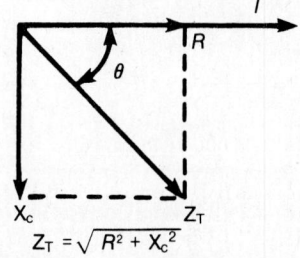

Fig. 15-21: Impedance phasor diagram of a series RC circuit.

Thus, the impedance is equal to the square root of the sum of the resistance squared and the capacitive reactance squared. The formula above can be rearranged so that the circuit resistance in terms of the impedance and reactance, or the capacitive reactance in terms of the impedance and resistance, can be calculated. These formulas are as follows:

$$R = \sqrt{Z^2 - X_c^2}$$
$$X_c = \sqrt{Z^2 - R^2}$$

Example: Determine the impedance of a series RC circuit with a resistance of 25Ω and a reactance of 50Ω.
Solution:

$$\begin{aligned} Z_T &= \sqrt{R^2 + X_c^2} \\ &= \sqrt{25^2 + 50^2} \\ &= \sqrt{625 + 2,500} \\ &= \sqrt{3,125} \\ &= 55.90 \end{aligned}$$

Example: A series RC circuit has 120V applied to it. The current is 20mA. The circuit resistance is 4,000Ω. Find the capacitive reactance.
Solution:

$$\begin{aligned} Z_T &= \frac{V}{I} \\ &= \frac{120V}{20mA} \\ &= 6,000 \\ X_c &= \sqrt{Z^2 - R^2} \\ &= \sqrt{6,000^2 - 4,000^2} \\ &= \sqrt{36,000,000 - 16,000,000} \\ &= \sqrt{20,000,000} \\ &= 4,472Ω \end{aligned}$$

313

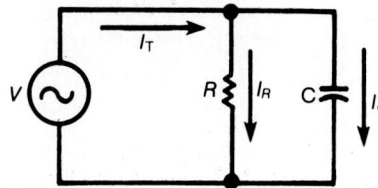

Fig. 15-22: Typical parallel *RC* circuit.

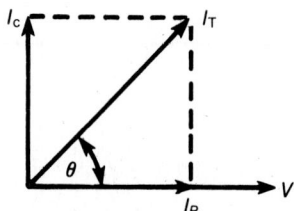

Fig. 15-23: Current phasor diagram of a parallel *RC* circuit.

Parallel RC Circuits

Figure 15-22 illustrates an AC parallel *RC* circuit. The applied voltage is common to both the capacitor and the resistor. The applied voltage and the resistance determine the amount of current (I_R) flowing in the resistor:

$$I_R = \frac{V}{R}$$

The applied voltage and the capacitance determine the current (I_c) in the capacitor:

$$I_c = \frac{V}{X_c}$$

The total current (I_T) drawn from the source is the sum of the resistor and capacitor currents ($I_R + I_c$). However, these currents must be added phasorally in order to account for the phase shift of the capacitor. (The current through the resistor is in phase with the applied voltage, but the current through the capacitor leads the applied voltage by 90°.) This is illustrated by the phasor diagram shown in Fig. 15-23. Since the applied voltage (V) is common to both circuit elements, it is used as the reference phasor as shown. The resistor current (I_R) is represented by the phasor coinciding with the applied voltage phasor. It is in phase with the applied voltage. The capacitor current is drawn shifted from the applied voltage by 90° in order to show that the current leads the applied voltage. The total current flowing in the circuit then can be determined by phasor addition:

$$I_T = \sqrt{(I_R)^2 + (I_c)^2}$$

Since the circuit is capacitive, the current will always lead the applied voltage. However, since it is not purely capacitive, the current leads by some angle less than 90°.

The impedance of a parallel *RC* circuit can be expressed in terms of Ohm's Law as follows:

$$Z_T = \frac{V_T}{I_T}$$

The general formula for finding the total resistance of two resistors connected in parallel can also be used to calculate the impedance of a parallel *RC* circuit. This formula is:

$$R_T = \frac{R_1 R_2}{R_1 + R_2}$$

The following expression can be used to determine the impedance if the resistance and capacitive reactance are known:

$$Z_T = \frac{RX_c}{\sqrt{R^2 + X_c^2}}$$

314

CAPACITOR APPLICATIONS

Capacitors are used in many ways in electronic circuits. A few of the more common applications include AC voltage dividers, filters, coupling networks, phase shifters, and wave formers.

Capacitive AC Voltage Dividers

Like the capacitive DC voltage divider discussed earlier in this chapter, the AC type is a series capacitive circuit whose output voltage is equal to some fraction of its input voltage. Figure 15-24 illustrates a simple voltage divider composed of two connected capacitors, C_1 and C_2 in series. The input voltage (V_i) is connected to both C_1 and C_2. The output voltage (V_o) is taken from across C_2. The circuit current produces a voltage drop across each capacitor. Kirchhoff's Voltage Law applies as the sum of the voltage drops is equal to the input voltage:

$$V_i = V_{c1} + V_{c2}$$

In this application the output voltage equals the voltage across capacitor C_2:

$$V_o = V_{c2}$$

The exact amount of voltage developed across each capacitor is a function of the circuit current and the capacitive reactance. Therefore, according to Ohm's Law, the output voltage will be:

$$V_o = V_{c2} = IX_{c2}$$

The voltage drops across the capacitors in the circuit divide in proportion to their capacitive reactance. The greater the capacitive reactance, the greater the voltage drop across the capacitor.

$$V_o = \frac{X_{c2}}{X_{c1} + X_{c2}} (V_i)$$

That is, the output voltage is equal to the input voltage multiplied by the ratio of the reactance of C_2 to the sum of the individual reactances and is called the *voltage division ratio*.

The output voltage can also be expressed in terms of the capacitor values rather than the reactances:

$$V_o = \frac{C_1}{C_1 + C_2} (V_i)$$

Note that the output voltage is directly proportional to the voltage division ratio $\left(\dfrac{X_{c2}}{X_{c1} + X_{c2}}\right)$ or in this relationship, $\dfrac{C_1}{C_1 + C_2}$.

The voltage division ratio is not affected by frequency. Even though a change in frequency causes a change in the reactances, the reactances change together and the voltage division ratio remains constant. Also, no phase shift occurs between the

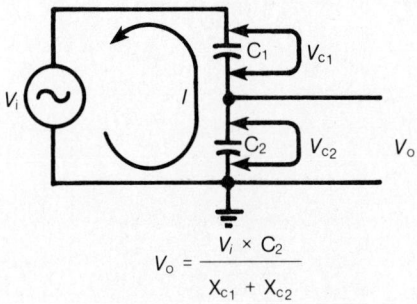

$$V_o = \frac{V_i \times C_2}{X_{c1} + X_{c2}}$$

Fig. 15-24: Capacitive voltage divider.

input and the output of a purely capacitive voltage divider. The circuit current leads the applied voltage, but the voltage across the output capacitor is in phase with the input voltage. Again, changing frequency has no effect on this characteristic.

Capacitive voltage dividers are frequently found in certain types of oscillator circuits as well as in high-frequency amplifier circuits. The output voltage of a capacitive voltage divider can be varied by changing either C_1 or C_2. Using a variable capacitor for this task, it is possible to change the value of capacitance, thus allowing the output voltage to be adjusted to a desired value.

RC Filters

A filter is a frequency discriminating circuit; that is, a filter will greatly attenuate some frequencies while allowing others to pass with virtually no opposition.

An *RC* series circuit, since it contains reactance, will not respond equally to all frequencies. Therefore, an *RC* circuit will exhibit frequency discrimination similar in many respects to that encountered in the *RL* circuits.

The terms cutoff frequency, half-power point, and critical frequency have the same meanings as previously defined. Series circuits containing reactances are AC voltage dividers. The voltage developed across a reactive component depends on the reactance of the component which in turn depends on the frequency. X_c is an inverse function of frequency. Therefore, at a frequency of zero hertz per second (DC) the opposition of the capacitor will be maximum, and all of the applied voltage will be dropped across the capacitor. As frequency is increased, the reactance of the capacitor will decrease and the voltage will divide between the resistor and capacitor. The cutoff frequency is reached when the voltage divides equally between the R and C.

Cutoff Frequency of an RC Circuit. A formula can be developed for determining the cutoff frequency (f_{co}) of an *RC* series circuit in the following manner:

Since f_{co} occurs when:

$$R = X_c$$

Substituting the equations for X_c:

$$R = \frac{1}{2\pi fC}$$

Substituting f_{co} for f and transposing yields:

$$f_{co} = \frac{1}{2\pi RC}$$

Example: Determine the cutoff frequency of a series *RC* circuit consisting of an $80\,\Omega$ resistor and a $1.99\,\mu F$ capacitor.

316

Solution:

$$f_{co} = \frac{1}{2\pi RC}$$

$$f_{co} = \frac{1}{6.28 \times 80 \times 1.99 \times 10^{-6}}$$

$$f_{co} = 1\text{kHz}$$

Therefore, at 1kHz, the resistance will be equal to the X_c and V_R will equal V_c.

Low-Pass Filter. While there are many types of filters used in electronic circuits, two of the more common are the *low-pass filter* and *high-pass filter*.

An RC series circuit connected as a low-pass filter will pass frequencies below the f_{co} and attenuate the frequencies above the f_{co}. Since the capacitor develops the most voltage at the low frequencies, the output of an RC low-pass filter (high-frequency discriminator) will be taken across the capacitor as shown in Fig. 15-25.

The frequency response curve (Fig. 15-26) shows the amount of output voltage (V_o) with respect to frequency (f). On the left-hand side of the curve, at very low frequencies, the output voltage is nearly equal to the input voltage. In fact, with a frequency of 0Hz or DC, the capacitor offers maximum opposition and the output voltage will be equal to the input voltage. As the frequency increases, the capacitive reactance begins to decrease. The output voltage then begins to drop off. After (to the right of) the cutoff frequency (f_{co}), the output voltage drops off at a constant rate. At the cutoff frequency, the output voltage is equal to approximately 70% of the input voltage or $V_o = 0.707 V_i$.

High-Pass Filter. An RC series circuit connected as a high-pass filter will pass frequencies above the f_{co} and attenuate the frequencies below the f_{co}. Since the resistor develops the most voltage at the high frequencies, the output of an RC high-pass filter (low-frequency discriminator) will be taken across the resistor as shown in Fig. 15-27.

The frequency response curve of an RC high-pass filter is shown in Fig. 15-28. Notice that at high frequencies, the output voltage is approximately equal to the input voltage. Note also that as the frequency decreases, the output voltage decreases. At the cutoff frequency, the output voltage is about 70% of the input voltage. Below the cutoff frequency, the opposition increases and the output voltage drops.

Circuits Combining AC and DC

Many electronic circuits operate with a combination of both AC and DC voltages. The applied voltage to these circuits appears as a DC voltage with an AC signal superimposed on it; that is, the zero line of the sine wave signal coincides with the DC voltage level. RC networks are often used with such AC/DC voltage circuits.

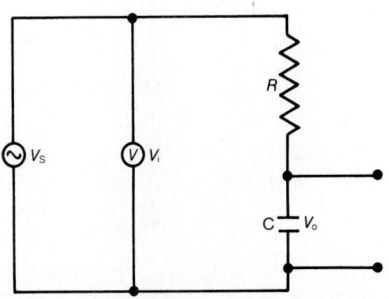

Fig. 15-25: RC low-pass filter.

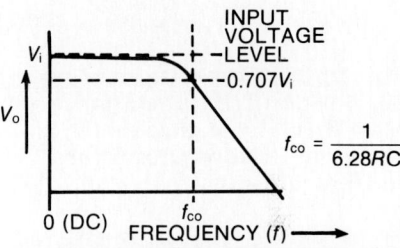

Fig. 15-26: Frequency response of an RC low-pass filter.

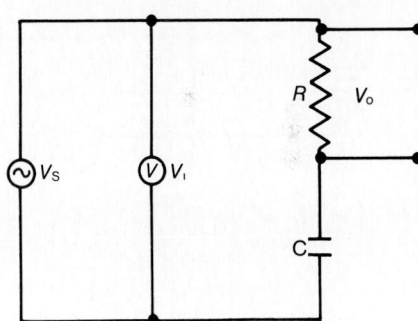

Fig. 15-27: RC high-pass filter.

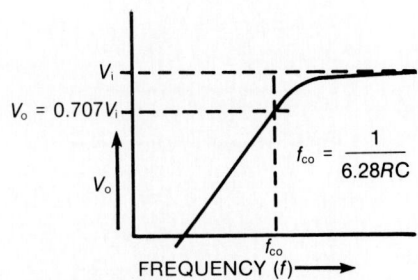

Fig. 15-28: Frequency response of an RC high-pass filter.

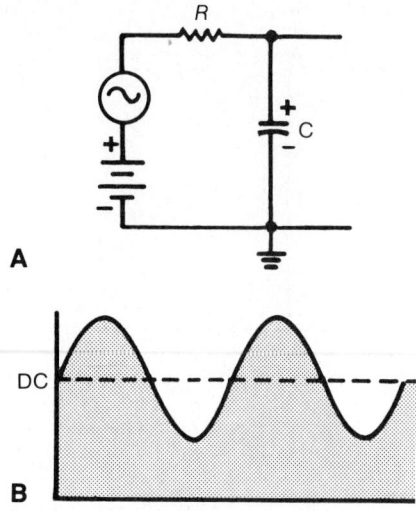

A

B

Fig. 15-29: (A) *RC* decoupling network and (B) the output waveform. Note that this is actually an *RC* low-pass filter configuration.

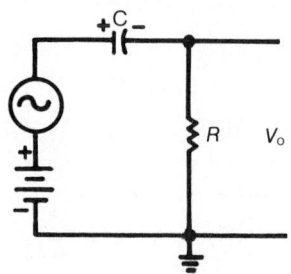

Fig. 15-30: *RC* coupling network.

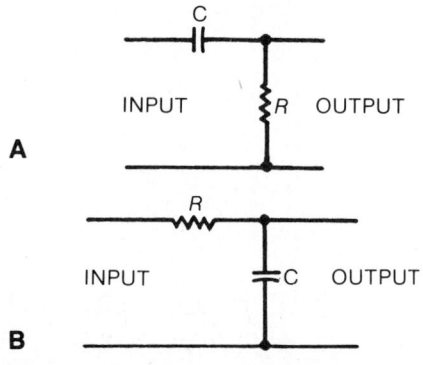

A

B

Fig. 15-31: Basic *RC* phase shift networks: (A) leading output; (B) lagging output.

Figure 15-29A illustrates one type of *RC* circuit which has both AC and DC voltages applied. The capacitor will charge to the value of the DC voltages as indicated by the polarity shown. The sine wave voltage will cause the charge on the capacitor to vary above and below the DC value by an amount equal to the peak value of the AC. Figure 15-29B shows the voltage across the capacitor. The combination voltage is often called pulsating DC. The composite voltage is still DC because it never goes negative, but it does vary or pulsate. In Fig. 15-29B, the frequency of the AC voltage is assumed to be lower than the cutoff frequency of the *RC* low-pass network. If the frequency of the AC is higher than the cutoff frequency of the *RC* network, its amplitude will be greatly attenuated.

Another application of an *RC* circuit with AC/DC is illustrated in Fig. 15-30. This circuit is referred to as a *coupling network*. A coupling network is used to couple an AC signal from one point to another while blocking any DC. The source is a combination of both AC and DC voltages and is applied across a series *RC* network. The output is taken from across the resistor. The capacitor charges to the applied DC voltage. Once it charges, however, no further DC flows in the circuit; therefore, no DC appears across the output resistance. Instead, only the AC voltage will appear across the output. The sine wave source will cause the capacitor to charge and discharge at the AC rate, thereby creating current flow in the resistor which appears at the output in the form of an AC voltage. In this application, only the AC signal appears at the output. The cutoff frequency of the *RC* network is adjusted so that it is low enough to permit the AC input to pass without noticeable attenuation.

Phase Shift Networks

Phase shifting is another application of *RC* networks. This process is employed to correct undesirable phase shift in a circuit, or to produce special desired effects. *RC* networks are ideal for these purposes because the capacitor causes the circuit current to lead the applied voltage. Figure 15-31 shows the two most commonly used configurations. In Fig. 15-31A, the input signal is applied across the *RC* combination and the output is taken from across the resistor. The circuit current leads the applied voltage by some phase angle less than 90° but greater than 0° depending on the values of the resistor and capacitor. The voltage across the resistor will be in phase with the current flowing through it; therefore, the output voltage will lead the input voltage (Fig. 15-32A).

The output is taken from across the capacitor in the phase shift circuit of Fig. 15-31B. Again the circuit current leads the applied voltage by some phase angle between 0° and 90°, but the voltage across the capacitor lags the applied voltage (Fig. 15-32B).

These simple *RC* phase shift networks can be used only where small amounts of phase shift are required. When phase

318

shifts of greater than about 60° are required, other phase shifting techniques are used. This is because as the phase shift approaches 90°, the output voltage drops to a very low level. (Remember that these *RC* networks are also voltage dividers; therefore, at 90°, the output voltage is theoretically zero.)

When phase shifts of greater than approximately 60° are required, it is often possible to cascade the simple *RC* network shown in Fig. 15-31. The output of one network can be connected to the input of another, thereby creating a total phase shift equal to the sum of the individual phase shifts. For example, if three 45° phase shift networks are cascaded as illustrated in Fig. 15-33A, the total phase shift would be 3 × 45° = 135°. In this case, the output voltage would lead the input voltage by 135°.

To achieve a lagging phase shift greater than that obtainable with a single *RC* network, the circuit of Fig. 15-33B can be employed. Here two networks are cascaded. For example, if each network provided a phase shift of 45°, the output would lag the input by a total of 90°.

RC Circuits and Waveforms

Earlier in this chapter, the method of charging and discharging *RC* constant circuits was by means of a switch operation resulting in an instantaneous increase or decrease in voltage across the circuit. In these switching operations, the voltage waveform was shown as a single, or one-time, procedure which did not repeat. This type of waveform is called a *step function*. A positive step function is the portion of the waveform where the voltage rises; a negative step function is the portion where the voltage decreases.

When positive and negative step functions are combined and repeated at regular intervals, the result is referred to as a *square wave*. The waveform in Fig. 15-34A is an example of a square wave. This representation shows the square wave to be always above the zero reference axis. In most cases, however, a square wave has a reference level such that a positive and negative segment of the square wave result (Fig. 15-34B). Observe in these figures the repeating nature of the waveform. Observe, too, that the voltage wave occurs in pulses. The rate at which these pulses recur is known as the *pulse repetition frequency* (prf).

Effect of Short and Long Time Constants on Waveforms. In Figs. 15-3 through 15-7, the output voltage and current waveforms were the result of step function voltage being applied to the input. In Fig. 15-35, the repeating nature of the output waveforms was observed when a square wave was used. The distortion of the input waveform (represented by the output waveform) is not always undesirable. In fact, it is often extremely useful for many electronics tasks.

A long or short time constant is determined on the basis of the width of the applied voltage pulse. The width of the voltage pulse is measured in seconds. The time constant of an *RC*

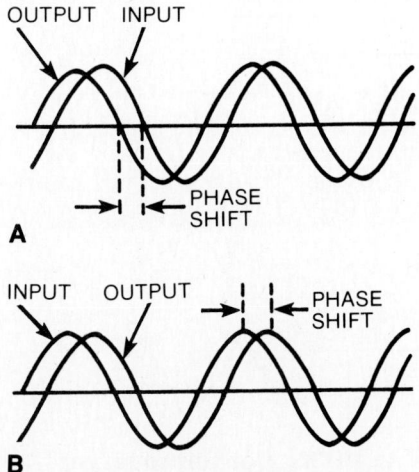

Fig. 15-32: Phase shift in a series *RC* circuit: (A) leading output; (B) lagging output.

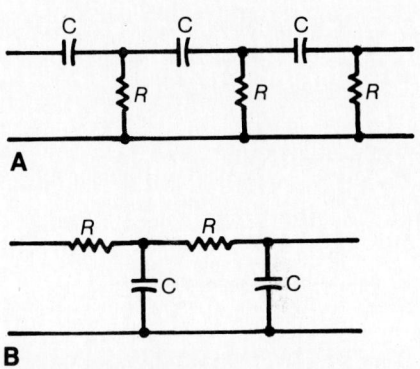

Fig. 15-33: Cascade *RC* phase shift networks.

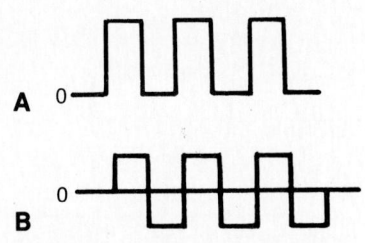

Fig. 15-34: Square wave representation.

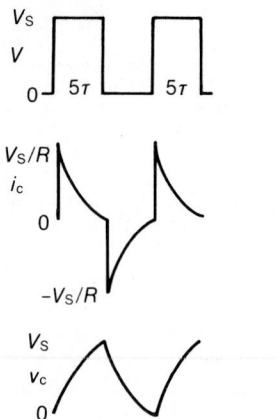

Fig. 15-35: Conventional representation of waveforms for an RC circuit.

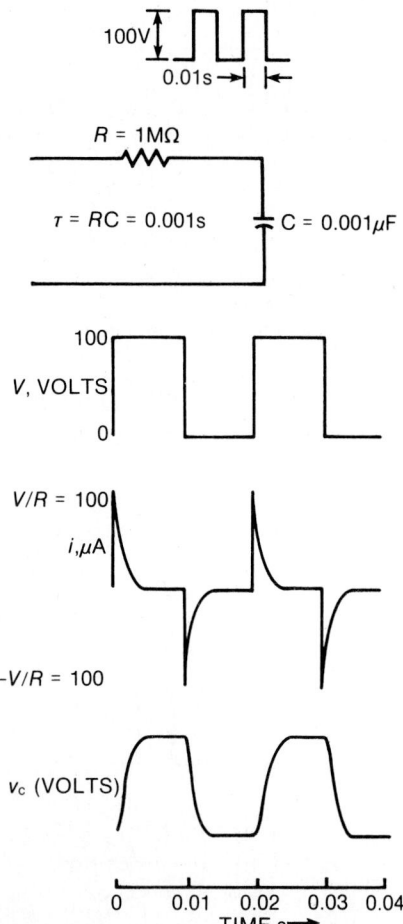

Fig. 15-36: Short time constant waveforms.

circuit is also measured in seconds. Thus, we see that a long or short time constant is a relationship between the time and the duration of a voltage pulse to that of the time constant of the circuit.

A long time constant is usually considered to be one that is 10 or more times greater than the duration of the input voltage pulse. For example, if $RC = 0.001$s and the duration of the voltage pulse is 0.00001s, the relation between the two is:

$$\text{time constant/pulse duration} = \frac{0.001}{0.00001} \text{ or } 100.$$

A short time constant is usually considered to be one that is no more than one-tenth of the time of the duration of the pulse. For example, if $RC = 0.001$s and the duration of the voltage pulse is 0.01s, the relation between the two is:

$$\text{time constant/pulse duration} = \frac{0.001}{0.01} \text{ or } \frac{1}{10}$$

In a circuit with a short time constant, the time constant of the circuit is seen to be much shorter than that of the pulse duration. Thus, the voltage charge on the capacitor builds up to its peak voltage quickly and remains at this level for the duration of the pulse. The current quickly falls from its peak V/R level to zero and remains for the duration of the pulse. Figure 15-36 illustrates these waveforms.

Figure 15-37 shows the waveforms for a circuit with a long time constant. In a circuit of this nature, the voltage charge on the capacitor does not have enough time to build to anywhere near the input voltage. The current waveform falls only slightly from its maximum V/R value.

SUMMARY

When the switch of a DC circuit containing only capacitance is closed, a surge of current charges the capacitor. Once the surge is over, current flow stops. At this point, the capacitor appears to the battery like an open switch.

A DC supply voltage will divide among series capacitors in inverse proportion to their capacitances. In other words, a voltmeter connected across a 1μF capacitor would indicate twice as much as a voltmeter connected across a 2μF capacitor. This is because $V = Q/C$, and capacitors in series must receive the same quantity of charge.

In a circuit composed of a capacitor (C) in series with a resistor (R) in series with a battery (V_s) and a switch, at the instant the switch is closed, all of V_s appears across R and the voltage across C is zero. The initial charging current, I, is therefore equal to V_s/R. This current soon charges the capacitor and creates a voltage (V_c) across it. When $V_c = V_s$, the capacitor is fully charged. At this point, the voltage across R is zero and no current flows through it.

In a capacitive circuit, there is a time lag for the voltage to rise or fall to its full or final value. This time lag is called the

320

time constant of the circuit. Voltage rises 63.2% in one time constant. Five time constants are required for a full rise. The same holds true when the voltage falls.

A universal time constant chart may be employed to determine the voltages across the various parts of a circuit for any time after the switch is closed, either on charge or discharge, if the time constant and the initial or final voltage for the circuit in question are known.

If an AC voltage is applied to a pure capacitance, the current is maximum when the voltage begins to rise from zero, and the current is zero when the voltage across the capacitor is maximum. The current leads the applied voltage by 90°.

An AC circuit containing capacitance is said to be reactive. The charging current in a purely capacitive circuit varies directly with the capacitance, voltage, and frequency:

$$I = 2\pi f C V$$

where I is effective current in A, f is in Hz, C is the capacitance in F, and V is in effective volts.

The ratio of the effective voltage across the capacitor to the effective current is called the capacitive reactance, X_c, and represents the opposition to current flow in a capacitive circuit of zero losses.

$$X_c = \frac{1}{2\pi f C}$$

where f is in Hz, C is in F, and X_c is in Ω.

Series or parallel reactances are combined in the same manner as resistances; that is,

$$X_{CT(series)} = X_{C1} + X_{C2} + X_{C3} + ...X_{Cn}$$

and

$$\frac{1}{X_{CT(parallel)}} = \frac{1}{X_{C1}} + \frac{1}{X_{C2}} + \frac{1}{X_{C3}} + \cdots \frac{1}{X_{Cn}}$$

The voltage across a capacitor in an AC circuit is a function of the current flowing in the circuit and the capacitive reactance; that is,

$$V_c = IX_c$$

A series AC RC circuit has a total voltage given by:

$$V_T = V_R{}^2 + V_c{}^2$$

and a phase angle between the current and applied voltage given by:

$$\theta = -\tan^{-1}\frac{X_c}{R}$$

The impedance of a series RC circuit is given by:

$$Z = \sqrt{R^2 + X_c{}^2}$$

In a parallel AC RC circuit, the applied voltage is common to both the capacitor and resistor. The applied voltage and the resistance determine the amount of current flowing in the resis-

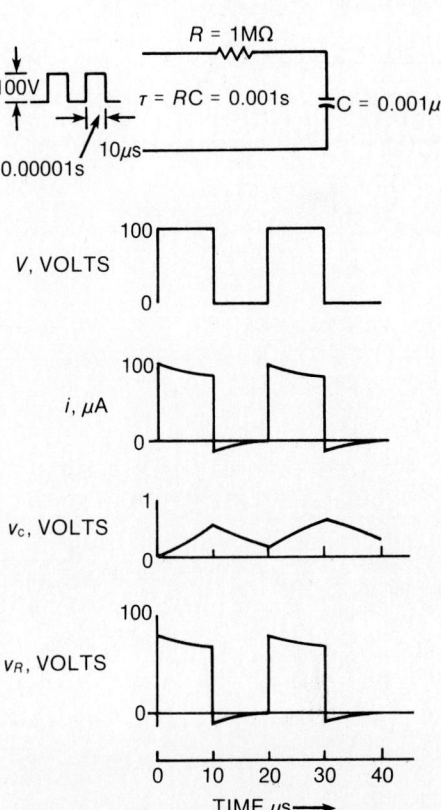

Fig. 15-37: Long time constant waveforms.

321

tor; the applied voltage and the capacitive reactance determine the current in the capacitor. The total current is the phasor sum of the resistor and capacitor currents:

$$I_T = \sqrt{I_R^2 + I_c^2}$$

The total impedance can be expressed as:

$$Z_T = \sqrt{\frac{RX_c}{R^2 + X_c^2}}$$

Capacitors are often employed in electronic circuits as AC voltage dividers, filters, coupling networks, and phase shifters.

CHECK YOURSELF (Answers at back of book)

Questions

Fill in the blanks with the appropriate word or words.

1. In a capacitive circuit, _____ .
   A. current and voltage start together
   B. current lags voltage
   C. voltage lags current
   D. voltage leads current
   E. none of the above
2. Opposition to current in an AC capacitive circuit is called

   _____ .
   A. capacitive resistance
   B. ohmic holdback
   C. reluctance
   D. capacitive reactance
   E. none of the above
3. A $1\mu F$ capacitor, at 500Hz, has a capacitive reactance approximately equal to _____ .
   A. $1,000\,\Omega$
   B. $1,000\pi\,\Omega$
   C. $1,000/\pi\,\Omega$
   D. $640\,\Omega$
   E. $320\,\Omega$
4. If the capacitor of Question 3 were connected to a 10V, 500Hz supply, the current would be approximately equal to

   _____ .
   A. 31.4mA
   B. 10mA
   C. 3.14mA
   D. 62.8mA
   E. none of the above
5. If a series RC circuit has $X_c = 2R$, the phase angle between the applied voltage and current is _____ .
   A. 30°
   B. 26.6°
   C. 60°
   D. 45°
   E. 63.4°

6. The first instant after voltage is applied, all the voltage appears across the _____ .
7. When the capacitor reaches full charge, all the voltage appears across the _____ .
8. The X_c of a capacitor will increase when frequency _____ .
9. The capacitor changes its charge by _____ % during _____ time constant(s).
10. Capacitors _____ DC and effectively _____ AC.
11. The current through a capacitor _____ voltage by _____ degrees.
12. An RC time constant is the product of _____ and _____ .

Determine whether the following statements are true or false.

13. A phasor is a line that represents a quantity that varies with time.
14. The capacitor current, in a sinusoidal AC circuit, is maximum when the capacitor voltage is zero.
15. The use of a capacitor to couple AC from one stage to another depends on the capacitor's ability to block alternating current (AC) and pass direct current (DC).
16. When a 120V source is applied to two series capacitors, the voltage across each capacitor is 60V.
17. Two phasors drawn at right angles can represent the voltage and current waveforms in a purely capacitive circuit.

Problems

1. What value of capacitance is required if its capacitive reactance is to be 50Ω at 400Hz?
2. Calculate the capacitive reactance of the following capacitors when a 60Hz voltage source is applied:
   A. 0.002μF
   B. 10μF
   C. 270pF
   D. 3,000nF
3. A 759kΩ resistor is connected in series with a 500pF capacitor and a 150V battery. How long will it take, in seconds, to fully charge the capacitor?
4. A 0.1μF capacitor is connected in series with a 250kΩ resistor and a 160V battery as shown in Fig. 15-38.
   A. Determine the voltage developed across C_1 after the switch S_1 is closed to position A for 2τ.
   B. How many seconds is the switch left closed to position A if it is closed for 2τ?
5. In designing a signal-filtering network circuit, it was found that a time constant of 4ms was necessary with a resistance value of 2kΩ. What value capacitor is required for such a time constant?
6. At what frequency will a 0.002μF capacitor have a reactance of 80Ω?

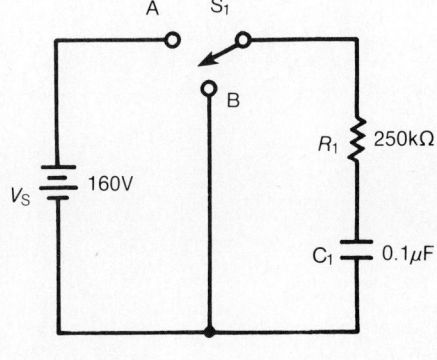

Fig. 15-38

323

7. The input resistance to an audio amplifier is 47kΩ. What value coupling capacitor is required if the reactance is to equal the input resistance at 20Hz?

8. A. What value of current will flow in a pure capacitive circuit when a $0.01\,\mu F$ capacitor is connected in series with a 120V, 60Hz voltage source?

   B. What value of current would flow if the 120V, 60Hz generator in part A were replaced by a 240V DC generator?

9. Determine the voltage across a $0.1\,\mu F$ capacitor when the current is 5mA at 400Hz.

10. Three capacitors ($0.1\,\mu F$, $0.15\,\mu F$, and $0.2\,\mu F$) are parallel connected across a 48V, 60Hz voltage source. Determine:

    A. The total capacitive reactance.

    B. The current drawn by each capacitor.

    C. The total current drawn by the capacitors.

    D. The total capacitance of the circuit.

CHAPTER 16

RLC Circuits

RLC SERIES CIRCUIT

When resistive, inductive, and capacitive elements are connected in series, their individual characteristics are unchanged; that is, the current through and the voltage drop across the resistor are in phase while the voltage drops across the reactive components (assuming pure reactances) and the current through them are 90° out of phase. However, a new relation must be recognized with the introduction of the three-element circuit. This pertains to the effect on total line voltage and current when connecting reactive elements in series, whose individual characteristics are opposite in nature, such as inductance and capacitance. Such a circuit is shown in Fig. 16-1.

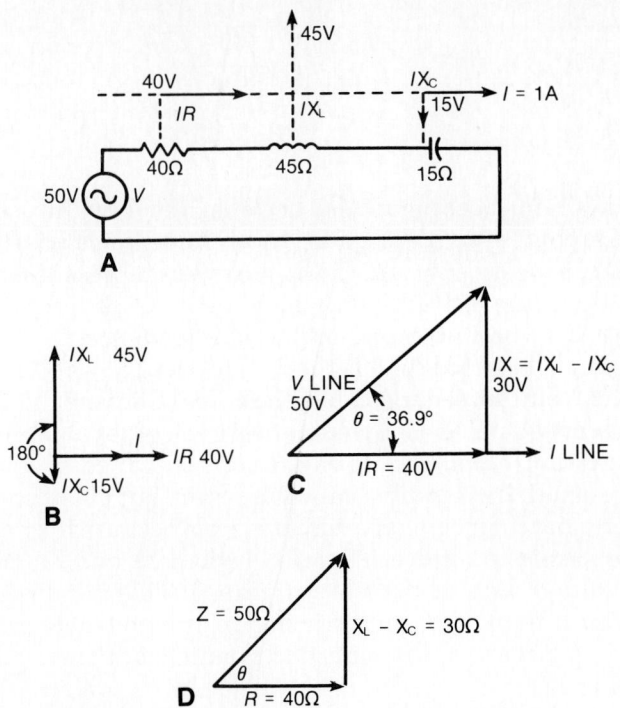

Fig. 16-1: Resistance, inductance, and capacitance connected in series.

In the figure, note first that current is the common reference for all three element voltages, because there is only one current in a series circuit and it is common to all elements. The common series current is represented by the dashed line in Fig. 16-1A. The voltage phasor for each element, showing its individual relation to the common current, is drawn above each respective element. The total source voltage V is the phasor sum of the individual voltages of IR, IX_L and IX_C.

The three element voltages are arranged for summation in Fig. 16-1B. Since IX_L and IX_C are each 90° away from I, they are therefore 180° from each other. Phasors in direct opposition (180° out of phase) may be subtracted directly. The total reactive voltage, VX, is the difference of IX_L and IX_C. Or, $V_X = IX_L - IX_C$ = 45 – 15 = 30V. The final relationship of line voltage and current, as seen from the source, is shown in Fig. 16-1C. Had X_C been larger than X_L, the voltage would lag, rather than lead. When X_C and X_L are of equal values, line voltage and current will be in phase.

The impedance of an RLC (three-element) series circuit is computed in exactly the same manner described for the two-element circuits in the preceding chapters; however, there is one additional operation to be performed. The difference of X_L and X_C must be determined prior to computing total impedance. When employing the Pythagorean Theorem-based formula for determining series impedance, the net reactance of the circuit is represented by the quantity in parentheses $(X_L - X_C)$. Applying this formula to the circuit in Fig. 16-1, the impedance is

$$
\begin{aligned}
Z &= \sqrt{R^2 + (X_L - X_C)^2} \\
&= \sqrt{40^2 + (45 - 15)^2} \\
&= \sqrt{1{,}600 + 900} \\
&= \sqrt{2{,}500} \\
Z &= 50\,\Omega
\end{aligned}
$$

Series impedance may also be determined by the use of phasoral layout, or triangulation. In an impedance triangle (Fig. 16-1D) for a series circuit, the base always represents the series resistance, the altitude represents the *net* reactance $(X_L - X_C)$, and the hypotenuse represents total impedance.

It should be noted that as the difference of X_L and X_C becomes greater, total impedance also increases. Conversely, when X_L and X_C are equal, their effects cancel each other and impedance is minimum, equal only to the series resistance. When X_L and X_C are equal, their individual voltages are 90° out of phase with current, but their collective effect is zero, because they are equal and opposite in nature. Therefore, when X_L and X_C are equal, line voltage and current are in phase. This condition is the same as if there were only resistance and no reactances in the circuit. A circuit in this condition is said to be at *resonance* (see Chapter 17).

Example: Figure 16-2 shows a series circuit combining resistance, inductance, and capacity. What is the total impedance? What is the phase angle? What is the power factor?

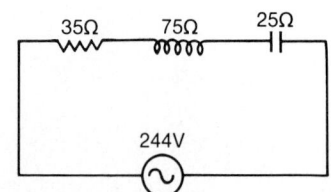

35Ω   75Ω   25Ω

244V

Fig. 16-2

Solution:

$$X_T = (X_L - X_C) = (75 - 25)$$
$$= 50\,\Omega$$
$$Z_T = \sqrt{R^2 + X_T{}^2} = \sqrt{35^2 + 50^2}$$
$$= \sqrt{3{,}725} = 61\,\Omega$$
$$\theta = \cos^{-1}\frac{R}{Z_T} = \cos^{-1}\frac{35}{61} = 55°$$
$$\text{power factor} = \frac{R}{Z_T} \text{ or } \cos\theta = 0.573$$

Example: If, in Fig. 16-2, 244V is applied to the circuit, what would be the total current? What are the voltage drops across each component?

Solution:

$$I_T = \frac{V}{Z_T} = \frac{244}{61} = 4\text{A}$$
$$V_R = I \times R = 4 \times 35$$
$$= 140\text{V}$$
$$V_L = I \times X_L = 4 \times 75$$
$$= 300\text{V}$$
$$V_C = I \times X_C = 4 \times 25$$
$$= 100\text{V}$$

proof

$$V_T = \sqrt{V_R{}^2 + (V_L - V_C)^2} = \sqrt{V_R{}^2 + V_X}$$
$$= \sqrt{140^2 + (300 - 100)^2} = \sqrt{140^2 + 200^2}$$
$$= \sqrt{59{,}600} = 244\text{V}$$

Power in an RLC Series Circuit

Total true power in an *RLC* series circuit is the product of line voltage and current times the cosine of the angle between them. When X_L and X_C are equal, total impedance is at a minimum and, thus, current is maximum. When maximum current flows, the series resistor dissipates maximum power. When X_L and X_C are made unequal, total impedance increases and line current decreases and moves out of phase with line voltage. Both the decrease in current and the creation of a phase difference cause a decrease in true power.

Multiple Component Series RLC Circuits

When series circuits contain a variety of resistors, capacitors, and inductors, the individual values of the inductive reactances are combined, as are the values of capacitive reactance. The difference between the two then becomes the total reactance. After adding the various resistance values, the phasor sum of

327

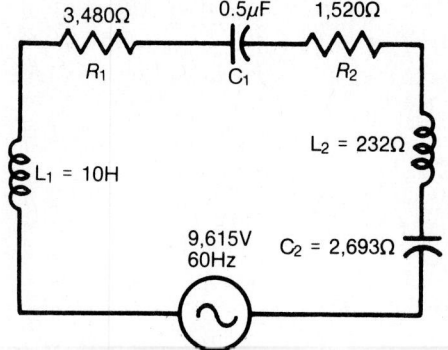

Fig. 16-3: Multiple component series *RLC* circuit.

combined resistance and reactance is again taken as before. An example of such a circuit is shown in Fig. 16-3, and the accompanying calculations are:

For L_1,

$$X_L = 6.28 \times 10 \times 60 = 3,768\,\Omega$$

For L_2,

$$X_L \qquad\qquad = \underline{232\,\Omega}$$
$$X_L \text{ (total)} \qquad = 4,000\,\Omega$$

For C_1,

$$X_C = \frac{1}{6.28 \times 0.5 \times 10^{-6} \times 60} = 5,307\,\Omega$$

For C_2,

$$X_C = 2,693\,\Omega$$
$$X_C \text{ (total)} = 5,307 + 2,693 = 8,000\,\Omega$$
$$R_T = 3,480 + 1,520 = 5,000\,\Omega$$
$$Z = \sqrt{5,000^2 + (8,000 - 4,000)^2} = 6,403\,\Omega$$
$$\theta = \cos^{-1}\frac{R}{Z} = \cos^{-1}\frac{5,000}{6,403} = 38.65°$$
$$\text{power factor} = \frac{R}{Z} \text{ or } \cos\theta = 0.78$$
$$I = \frac{V_T}{Z_T} = \frac{9,615\text{V}}{6,403\,\Omega} = 1.5\text{A}$$
$$\text{true power} = VI\cos\theta = 14,422.5 \times 0.78 = 11,249.5\text{W}$$

RLC PARALLEL CIRCUIT

In the discussion of inductance in Chapter 13 and of capacitance in Chapter 15, it was shown that a parallel *RL* circuit or a parallel *RC* circuit differed from its respective series circuits in that the *voltage* in the circuit is everywhere the same, the *current* is 90° out of phase, and the resultant line current is either lagging (*RL*) or leading (*RC*) the applied voltage by some angle less than 90°. In addition, it was shown that the line current is greater than the current in either branch and that, therefore, the total impedance is less than the impedance of either branch.

Unlike the series reactive circuits, an increase in the resistance of a parallel circuit lessens the current through that branch and increases the relative effectiveness of the inductance or capacitance, resulting in a more reactive circuit as the phase angle θ approaches 90°. The limiting condition in this instance would be a resistance of infinite ohms, effectively opening that branch and making current and voltage across the reactive element 90° out of phase. On the other hand, a decrease in the value of the resistance in a parallel circuit causes an increase in current in that branch and decreases the relative effectiveness of the inductance or capacitance, resulting in a more resistive circuit as θ approaches $0°$. The limiting condition in this instance would be a resistance of 0Ω, directly short-circuiting the reactive element so that current and voltage are in phase in this short circuit.

In bringing together the parallel RL and RC circuits into a parallel circuit, the *basic* procedures are similar to those outlined before for these circuits. When both inductance and capacitance are present, however, the capacitive current is subtracted from the inductive current (or inductive from capacitive, whichever is larger). The net reactive current is then combined by phasor addition with the resistive current to obtain circuit current.

The circuit in Fig. 16-4 shows the ohmic values of the three parallel components (R, C, and L). Solving for individual currents,

$$I_R = \frac{V}{R} = \frac{30}{500} = 0.06\text{A or }60\text{mA}$$

$$I_C = \frac{V}{X_C} = \frac{30}{600} = 0.05\text{A or }50\text{mA}$$

$$I_L = \frac{V}{X_L} = \frac{30}{250} = 0.12\text{A or }120\text{mA}$$

Now solve for total current using the phasor equation:

$$I_T = \sqrt{0.06 + (0.12 - 0.05)^2} = \sqrt{0.06 + 0.07^2}$$
$$= \sqrt{0.0085} = 0.092\text{A} = 92\text{mA}$$

Now solve for true and apparent powers:

$$P_T = (I_R)^2 R = (0.06)^2 (500) = 1.8\text{W}$$

$$P_A = I_T V_T = (0.092)(30) = 2.76\text{W}$$

$$PF = \frac{P_T}{P_A} = \frac{1.8}{2.76} = 0.65$$

$$\theta = PF \times \cos^{-1} = (0.65)(\cos^{-1}) = 49.3°$$

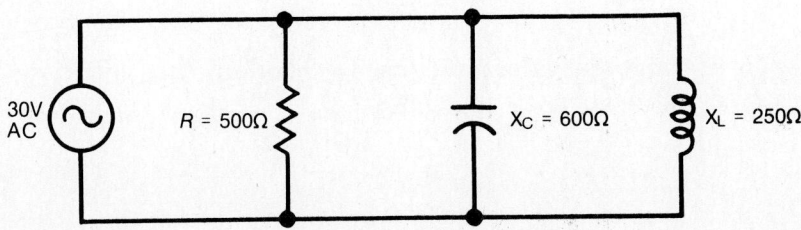

Fig. 16-4: Typical parallel RLC circuit.

Example: Figure 16-5 shows a typical parallel circuit, where two individual inductances shunt a resistor and a capacitor. Find the I_T and true power.

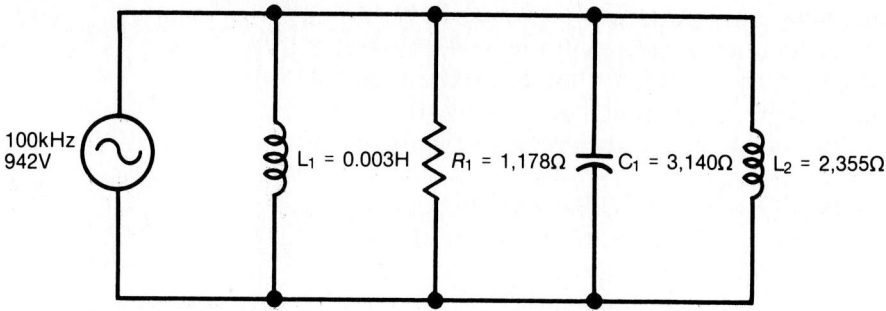

Fig. 16-5

Solution:

For L_1,

$$X_L = 6.28 \times 0.003 \times 10^3 = 1,884\,\Omega$$

For L_1,

$$I = \frac{942}{1,884} = 0.5A$$

$$I_R = \frac{942}{1,178} = 0.8A$$

For L_2,

$$I = \frac{942}{2,355} = 0.4A$$

For L_1 and L_2,

$$I_T = 0.9A$$

For C_1,

$$I = \frac{942}{3,140} = 0.3A$$

The foregoing gives the necessary values for inclusion in the equation for finding total current. The formula is an extension of the previous formula for total current:

$$I_T = \sqrt{I_R^2 + (I_L - I_C)^2}$$

Substituting the values previously found in this equation solves for total current as follows:

$$I_T = \sqrt{0.8^2 + (0.9 - 0.3)^2}$$
$$= \sqrt{0.64 + 0.36} = 1A$$

Solving for true and apparent powers:

$$P_T = \frac{V^2}{R} = \frac{(942)^2}{1,178} = 753.3W$$

$$P_A = I_T V_T = (1)(942) = 942W$$

$$PF = \frac{P_T}{P_A} = \frac{753.3}{942} = 0.80$$

$$\theta = PF \times \cos^{-1} = (0.80)(\cos^{-1}) = 36.9°$$

In a circuit composed of both L and C, the opposing effects of each have a decreasing influence on power as their relative values approach equality. This is exemplified in the circuit shown in Fig. 16-6, where it is necessary to combine the dual components in each series branch. Solving for branch currents,

$$I_R = \frac{V}{R_1 + R_2} = \frac{45}{230} = 0.19565A = 0.2A$$

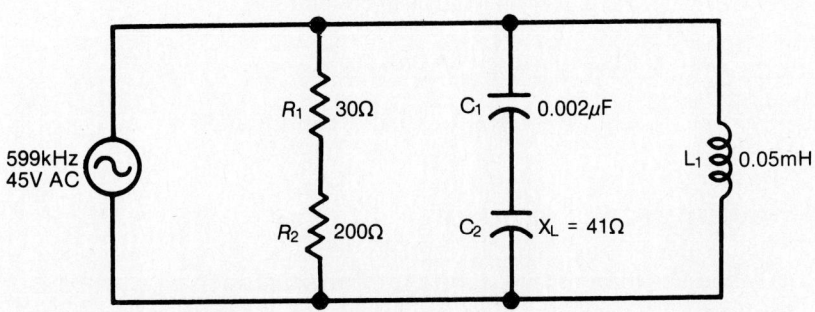

Fig. 16-6: Complex parallel *R*LC circuit.

For convenience, 0.19565A is converted to 0.2A. In some circuits, however, more accurate expressions may be needed where critical-value components must be used. It is a common practice to round off numbers such as 0.15 to 0.2 or to drop the last number if below 5, such as 0.14 to 0.1. When expressing current values in milliamperes, however, it is apparent that a noticeable difference exists for fractional values. Thus 0.19565A is actually 195.65mA; this would establish specific characteristics in a critical circuit that would be significantly different from those produced by an alternating current of 200mA.

In Fig. 16-6, the reactance for C_1 is not given; hence, it is solved as follows:

$$X_C = \frac{1}{6.28 \times 500,000 \times 0.002 \times 10^{-6}}$$

$$= \frac{1}{0.00628} = 159\Omega$$

Now combine 159Ω with the 41Ω given for C_2:

$$I_X = \frac{V}{X_1 + X_2} = \frac{45}{200} = 0.225A$$

Because the reactance of the inductor is also not given, it must be solved before finding total current:

$$X_L = 6.28fL = 6.28 \times 500,000 \times 0.00005 = 157\Omega$$

Now solve for current through the inductor, again rounding off the numbers for convenience:

$$I_X = \frac{45}{157} = 0.2866A = 0.3A$$

Subtracting I_C from I_L yields 0.300 − 0.225 = 0.075 A.

Total current can now be found by using the phasor equation:

$$I_T = \sqrt{0.2^2 + 0.075^2} = \sqrt{0.045625} = 0.2136A \text{ or } 213.6mA$$

Note how close this value is to what it would be if only the resistance were in the circuit (200mA).

$$P_T = \frac{V^2}{R} = \frac{(45)^2}{230} \times 8.80W$$

$$P_A = I_T V_T = (0.2136)(45) = 9.61W$$

$$PF = \frac{P_T}{P_A} = \frac{8.80}{9.61} = 0.92$$

$$\theta = PF \times \cos^{-1} = (0.92)(\cos^{-1}) = 23.8°$$

Rectangular Notation

When the sum of two in phase quantities such as a 5Ω resistor and a 10Ω resistor is desired, the total is written as:

$$R_T = R_1 + R_2$$
$$R_T = 5 + 10$$
$$R_T = 15Ω$$

This is simply an algebraic addition. When the sum of a non-reactive component and a reactive component is desired, the equation cannot be written in the form used above because algebraic addition is no longer valid. Out of phase quantities must be combined by some method of addition other than algebraic. Using the impedance formula ($Z = \sqrt{R^2 + X^2}$), a resistance of 15Ω and a reactance of 20Ω were added, by phasor addition, to obtain an impedance of 25Ω. In Fig. 16-7, the resistance is plotted in standard position while the reactive component (X_L) is plotted 90° counterclockwise from the standard position. The phasor position of X_L may be indicated by the j operator; that is, if the 20Ω of X_L were multiplied by j, it would indicate that X_L was 90° counterclockwise from the standard position. Therefore, by use of the j operator, the phasor addition of two quantities at right angles may be expressed mathematically as:

$$Z = R \pm jX$$

where:

Z = impedance in Ω

R = resistance in Ω

$\pm jX$ = reactance in Ω. Plus j is used for X_L and minus j is used for X_C.

Inserting values from Fig. 16-7 in the equation:

$$Z = R \pm jX$$
$$= 15 + j20Ω$$

Thus, the impedance may be read as: "Impedance" equals the resistive component of 15Ω added phasorally to the reactive

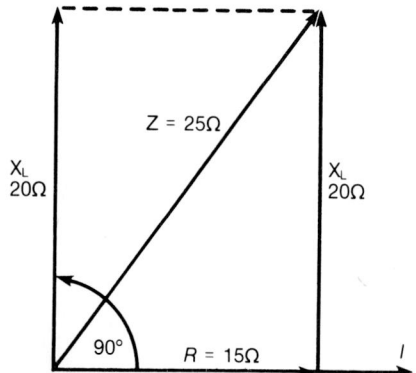

Fig. 16-7: Impedance phasor diagram.

component of $20\,\Omega$. Note that the right side of the equation is in the proper form to be inserted in the equation $Z = \sqrt{R^2 + X^2}$ to solve for impedance.

The equation $Z = R \pm j\text{X}$ is a *complex number*. In electronics, a complex number is an expression containing both reactive and nonreactive components. Quantities that are at right angles to each other cannot be added or subtracted in the usual sense of the word. These quantities are always added phasorally. The plus or minus sign simply indicates phasor quadrant. When a quantity is written in the form of this equation, it is said to be written in *rectangular notation,* or rectangular form. Complex numbers may be added or subtracted directly when in rectangular form.

Example: Three impedances are to be added to find their total.

Z_1 consists of a $20\,\Omega$ resistor in series with $60\,\Omega$ of inductive reactance.

Z_2 consists of a $10\,\Omega$ resistor and $10\,\Omega$ of inductive reactance in series.

Z_3 consists of a $50\,\Omega$ resistor and $20\,\Omega$ of capacitive reactance in series.

Solution: Write all impedances in rectangular form:

$$Z_T = R \pm j\text{X}$$
$$Z_1 = 20 + j60$$
$$Z_2 = 10 + j10$$
$$\underline{Z_3 = 50 - j20 \text{ (note the } -j \text{ due to } X_C)}$$
$$Z_T = 80 + j50$$

The resistive components are added arithmetically:

$$20 + 10 + 50 = 80$$

The reactive components are added algebraically with due regard for the signs of the j operators:

$$(+j60) + (+j10) + (-j20) = +j50$$

The three original impedances are now described by a total impedance of $80\,\Omega$ of resistance added to $50\,\Omega$ of inductive reactance.

$$Z_T = 80 + j50 \text{ (rectangular notation)}$$

In order to determine the actual magnitude of the resultant phasor quantity Z_T, either phasor addition or the Pythagorean Theorem can be used. By substitution of the known quantities into the impedance formula:

$$Z_T = \sqrt{R^2 + X_L^2}$$
$$Z_T = \sqrt{80^2 + 50^2}$$
$$Z_T = 94.3\,\Omega$$

In rectangular form the phasor (Z_T) is described in terms of the two sides of a right triangle, the hypotenuse of which is the phasor (Fig. 16-7). Thus, rectangular notation describes the

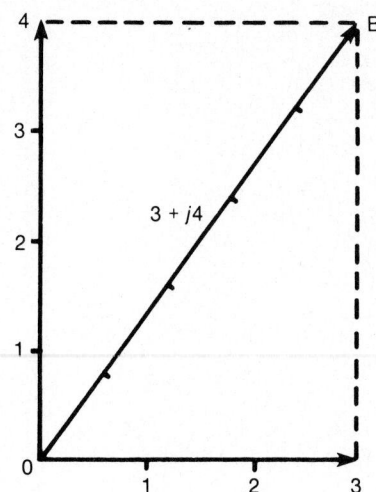

Fig. 16-8: Phasor diagram, rectangular form.

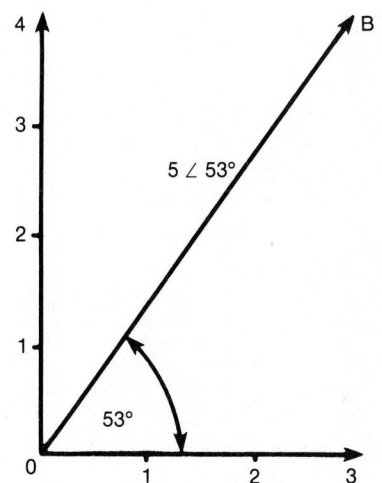

Fig. 16-9: Phasor diagram, polar form.

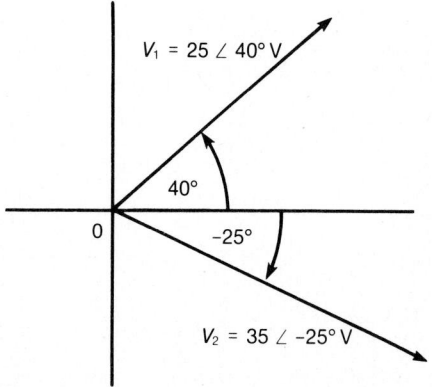

Fig. 16-10: Phasor diagram, polar notation.

resultant phasor quantity to a limited degree. It supplies information as to how much of the resultant phasor is due to resistance and how much is due to reactance, but it does not directly indicate the actual magnitude of the resultant phasor nor the angle of the resultant phasor with respect to the reference line.

Polar Notation

Up to this point methods have been discussed by which a resultant phasor could be determined. Now a system of notation is needed that will express the information required to accurately describe the resultant phasor. The phasor 0B (Fig. 16-8) is described in rectangular form by the complex number $3 + j4$. In other words, the information contained in the rectangular form is sufficient to plot the phasor diagram (Fig. 16-8). From this diagram, the length and angle of the resultant phasor with respect to the reference line may be determined by physical measurement.

The phasor 0B (Fig. 16-8) can be described in a combined form if its length and angle of rotation are given. When a phasor is described by means of its magnitude and angle, it is said to be expressed in *polar notation* or polar form. Figure 16-9 illustrates a phasor with a magnitude of 5 units and an angle of 53°.

A phasor expressed in polar form may be graphed directly (without using the parallelogram method).

Example: Graph the two phasors described in polar notation as:

$$\text{Phasor one } (V_1) = 25 \angle 40° \text{V}$$
$$\text{Phasor two } (V_2) = 35 \angle -25° \text{V}$$

Notice that phasor two has an angle of -25°. This means that the rotation of the phasor is clockwise or 25° below the reference line. Figure 16-10 illustrates the two phasors graphed from the information supplied in the polar notation form.

Note that the rectangular and polar notations are simply convenient methods of describing circuit conditions from mathematical and electrical viewpoints. An individual phasor may be described in either form. For example, the phasor 0B (Fig. 16-8) is:

$$5 \angle 53° = 3 + j4$$

Multiplication and division may be easily accomplished using polar notation. The product of two polar phasors is calculated by multiplying their magnitudes and adding their angles algebraically.

Example:

$$2.88 \angle -33° \times 7.56 \angle 72° = (2.88 \times 7.56) \angle (-33° + 72°)$$
$$= 21.77 \angle 39°$$

The quotient of two phasors is calculated by dividing their magnitudes and subtracting the angle of the divisor from the angle of the dividend.

334

Example:

$$\frac{7.56 \angle 72°}{2.88 \angle -33°} = \left(\frac{7.56}{2.88}\right) \angle (72° - [-33°])$$
$$= 2.625 \angle 105°$$

Conversion From Rectangular to Polar Form

The process of converting a phasor described in rectangular form to a polar form utilizes trigonometric functions. As shown in Fig. 16-11, the original rectangular form was

$$Z = 4 + j3\Omega$$
$$\tan \theta = \frac{opp}{adj} = \frac{3}{4} = 0.7500$$

Therefore,

$$\theta = \tan^{-1} 0.7500$$
$$\theta = 36.8°$$

The impedance can be found by using the series impedance formula:

$$Z = \sqrt{R^2 + X_L^2}$$
$$Z = \sqrt{4^2 + 3^2} = 5$$

Knowing the phase angle, however, it is possible to use (as an alternative) the trigonometric function for finding the hypotenuse (Z) as follows:

$$Z = \frac{R}{\cos \theta}$$
$$= \frac{4}{\cos 36.8°} = \frac{4}{0.8} = 5\Omega \angle 36.8°$$

The same calculations would have prevailed for problems involving phasor voltages or currents, in similar fashion to the example previously described.

Example: The total impedance of the circuit in Fig. 16-12 is described in rectangular form as $Z_T = 20 + j50$. Find Z and state in polar form.

Solution: Determine the phase angle:

$$\theta = \tan^{-1} \frac{X_L}{R}$$

Insert values from rectangular notation:

$$\theta = \tan^{-1} \frac{50}{20}$$
$$\theta = \tan^{-1} 2.5$$
$$\theta = 68.2°$$

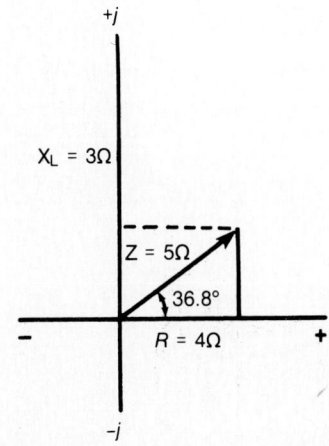

Fig. 16-11: Original rectangular form.

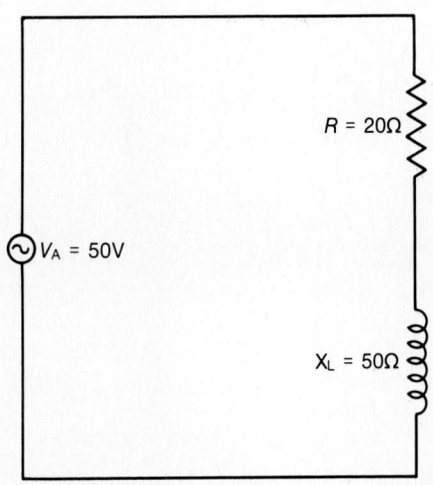

Fig. 16-12

335

Determine total impedance;

$$\sin \theta = \frac{X}{Z}$$

Transpose:

$$Z = \frac{X}{\sin \theta}$$

$$Z = \frac{50}{\sin 68.2°}$$

$$Z = \frac{50}{0.9285}$$

$$Z = 53.8\Omega$$

Therefore, $Z = 20 + j50\Omega$ written in polar form is:

$$Z = 53.8 \angle 68.2°$$

Conversion from Polar to Rectangular Form

The conversion of a quantity described in polar form involves the determination of its rectangular quantities through use of the trigonometric functions. Given the circuit impedance in polar form ($Z \angle n°$), an equation can be derived for conversion into rectangular form ($Z = R \pm jX$) as follows:

As a trigonometric function:

$$\cos \theta = \frac{R}{Z}$$

Transposed:

$$R = Z \cos \theta$$

As a trigonometric function:

$$\sin \theta = \frac{X}{Z}$$

Transposed:

$$X = Z \sin \theta$$

Also:

$$Z = R \pm jX$$

Substituting the above trigonometric equations for R and X into this equation yields:

$$Z = R \pm jX = \mathbb{Z} \cos \theta \pm j\mathbb{Z} \sin \theta$$

where $\mathbb{Z}$ is a special symbol explained below. Factoring the $\mathbb{Z}$ from the right side:

$$Z = \mathbb{Z} (\cos \theta \pm j \sin \theta)$$

where

Z = the circuit impedance in rectangular form in Ω

$\mathbb{Z}$ = the absolute magnitude of the polar impedance in Ω

θ = the phase angle of the impedance in degrees.

336

Example: It is desired to determine the value of the resistive and reactive quantities in a series circuit whose impedance is described in polar form as 2,411 $\angle$ 51.3°.

Given:

$$Z = 2,411 \angle 51.3°$$

Solution: Convert to rectangular form:

$$Z = \mathbb{Z} (\cos \theta + j \sin \theta)$$
$$Z = 2,411 (\cos 51.3° + j \sin 51.3°)$$
$$Z = 2,411(0.6252 + j0.7804)$$
$$Z = (2,411 \times 0.6252) + (2,411 \times j0.7804)$$
$$Z = 1,507 + j1,881.5\Omega$$

Therefore, the resistive quantity of the circuit is equal to 1,507Ω and the reactive quantity is inductive (+j) and is equal to 1,881.5Ω.

Example: What is the rectangular form of $I_T = 61\mu A \angle 235°$?

Solution: A phasor diagram is often convenient to help visualize the angle and phasor position. For this problem, use the diagram as shown in Fig. 16-13A for reference. The actual angle for the right-angle triangle is:

$$235 - 180 = 55°$$
$$I_T = -61 \cos 55° - j61 \sin 55°$$
$$= -35 - j50 \mu A$$

Example: What is the rectangular form of expression of $Z = 23.3\Omega \angle -59°$?

Solution: The negative sign for the 59° angle indicates a phasor diagram, as shown in Fig. 16-13B. Expressed in rectangular form this becomes:

$$23.3 \cos 59° - j23.2 \sin 59°$$
$$= 23.3 \times 0.5150 - j23.3 \times 0.8572$$
$$= 12 - j20$$

Circuit Solution with *j*-Factors

Each of the two methods of notation (polar and rectangular) has its own particular advantages. Final circuit solutions are usually expressed in polar form, because the actual impedance values, total currents, and total voltages are obtained. The rectangular form, because it is possible to apply ordinary algebraic processes, becomes convenient in circuit solution processes and also simplifies problems of circuits that contain two or more impedances which must be added phasorally for total impedance.

Example: Determine the total impedance and total current for the circuit of Fig. 16-14.

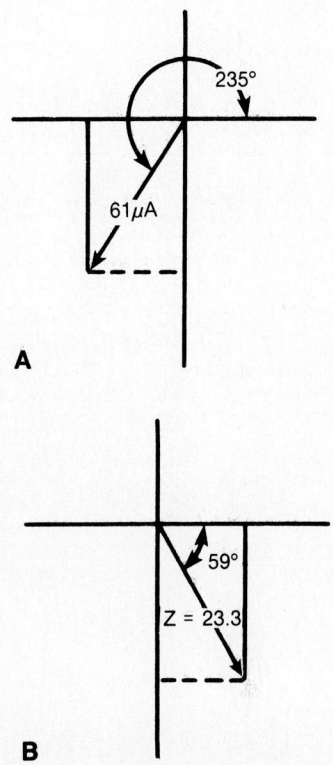

A

B

Fig. 16-13: Phasor diagram with – *j* values.

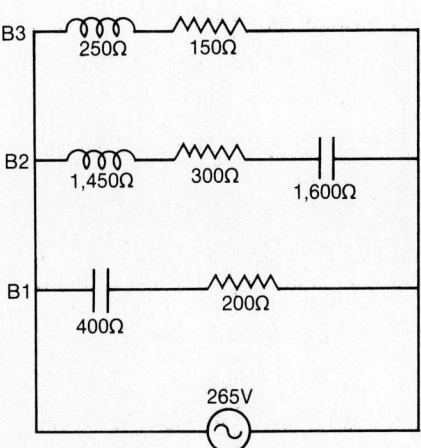

Fig. 16-14

337

Solution:

$$Z_1 = 200 - j400 = 447.214 \angle -63.435°$$

$$Z_2 = 300 - j150 = 335.410 \angle -26.565°$$

$$Z_3 = 150 + j250 = 291.548 \angle 59.036°$$

$$I_1 = \frac{265 \angle 0°}{447.214 \angle -63.435°} = 0.592 \angle 63.435° = 0.265 + j0.53$$

$$I_2 = \frac{265 \angle 0°}{335.41 \angle -26.565°} = 0.790 \angle 26.565° = 0.707 + j0.353$$

$$I_3 = \frac{265 \angle 0°}{291.548 \angle 59.036°} = 0.909 \angle -59.036° = 0.468 - j0.779$$

$$I_T = \begin{array}{r} 0.265 + j0.53 \\ 0.707 + j0.353 \\ +0.468 - j0.779 \\ \hline 1.44 \ + j0.104 \end{array}$$

$$I_T = 1.444 \angle 4.131°$$

$$Z_T = \frac{265 \angle 0°}{1.444 \angle 4.131°} = 184\,\Omega \angle -4.131° = 183.522 - j13.255$$

The product-over-the-sum method can be used to verify the preceeding calculations:

$$Z_{eq} = \frac{Z_1 Z_2}{Z_1 + Z_2}$$

$$= \frac{(447.214 \angle -63.435°)(335.41 \angle -26.565°)}{500 - j550}$$

$$= \frac{150,000 \angle -90}{743.303 \angle -47.726}$$

$$= 201.802 \angle -42.274° \text{ or } 149.321 - j135.748$$

$$Z_T = \frac{Z_{eq} Z_3}{Z_{eq} + Z_3}$$

$$= \frac{(201.802 \angle -42.274°)(291.548 \angle 59.036°)}{299.321 + j114.252}$$

$$= \frac{58,834.9695 \angle 16.762°}{320.385 \angle 20.892°}$$

$$= 184\,\Omega \angle -4.13° = 183.522 - j13.252$$

SUMMARY

In a series RLC circuit, the total voltage is the phasor sum of the individual voltages of I_R, IX_L, and IX_C. The total reactance voltage V_X is equal to the difference between IX_L and IX_C.

The impedance of an RLC series circuit can be calculated as $Z = \sqrt{R^2 + (X_L - X_C)^2}$. The phase angle can be expressed as $\theta = \cos^{-1} \frac{R}{Z_T}$. To find I, V_T is divided by Z.

True power in an RLC circuit is the product of line voltage and current times the cosine of the angle between them.

In a parallel RLC circuit, the total current is equal to $I_T = \sqrt{I_R{}^2 + (IX_L - IX_C)^2}$, and its phase angle is $\theta = PF \times \cos^{-1}$. To find Z, V_T is divided by I_T.

Complex numbers may be described in either polar or rectangular notation. In rectangular notation a positive j operator indicates a 90° phase angle and a negative j operator indicates a negative 90° phase angle.

Rectangular notation must be used to add or subtract complex numbers. A practical application for rectangular notation is the addition of branch currents, containing both reactive and resistive elements, in a parallel circuit. Final circuit solutions are usually given in polar notation. It is also easier to multiply and divide in polar notation.

In RLC circuit calculations, it is often necessary to convert from rectangular to polar notation or from polar to rectangular notation.

CHECK YOURSELF (Answers at back of book)

Questions

Match the values in the column at the left with those in the column at the right.

1. $25 + j15 + 10 + j20$
2. $23 - j4 + 17 - j6$
3. $1.5k\Omega$ of $R + 800\Omega$ of X_C
4. $6A$ of $I_R + 5A$ of I_C
5. $45V$ of $V_R + 30V$ of V_L
6. $12 \angle 38° \times \angle 32°$
7. $12 \angle 38° \div 2 \angle 32°$
8. $16 \angle -75° \times 4 \angle 30°$
9. $100 \angle 45° \div 10 \angle -45°$
10. $50 \angle -70° \div 5 \angle 9°$

A. $1,500 - j800$
B. $12 \angle 70°$
C. $35 + j35$
D. $64 \angle -45°$
E. $6 + j5A$
F. $45 + j30V$
G. $10 \angle 90°$
H. $40 - j10$
I. $6 \angle 6°$
J. $10 \angle -79°$

Determine whether the following statements are true or false.

11. An oscilloscope having accurately calibrated horizontal and vertical inputs is required to determine the phase angle of a circuit.
12. The impedance of a series circuit must always be less than its resistance or reactance.
13. If a capacitor of the correct reactance is connected in series with a series RL circuit, a purely resistive circuit may result.
14. Because it obeys Kirchhoff's Current Law, the total line current in a parallel RLC circuit must always be greater than any one of the branch currents.
15. When resistance is increased in a series impedance, Z increases but θ decreases.
16. The phase angle, θ, of a series RLC circuit is determined only by the values of R, L, and C.

17. The total impedance of a parallel RLC circuit can be larger than the smallest branch reactance or resistance.
18. If a capacitor is connected in series with a series RL circuit, a decrease in impedance will always occur.

Problems

1. Determine the total impedance for the circuit of Fig. 16-15.

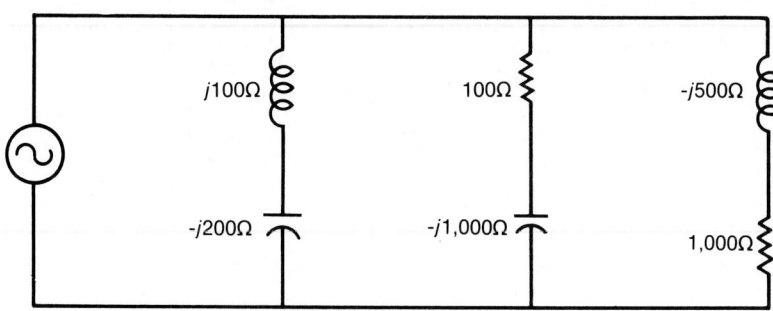

Fig. 16-15

2. Determine the total impedance and current for the circuit of Fig. 16-16.

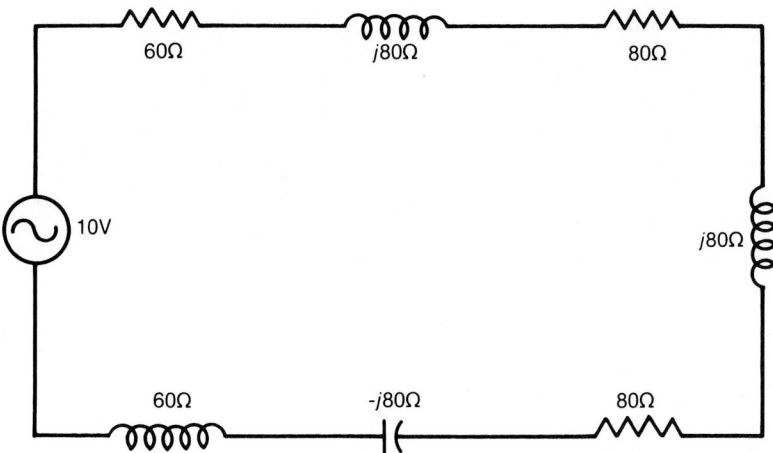

Fig. 16-16

340

CHAPTER 17

Resonance

Resonant circuits make it possible to select and remove one particular signal from the thousands of frequencies which exist in space. Resonant circuits make possible radio and television, two-way radio communication systems of all types, radar and navigation systems, and the latest applications of satellite transmission and reception.

In general, resonant circuits are used to select desired signals and to reject unwanted signals. To see how this is accomplished, take a look at Fig. 17-1, a parallel circuit composed of a capacitor and an inductor. If a voltage is applied between points a and b, the operation of the circuit will depend upon the frequency of the applied voltage. If the applied voltage is DC, the capacitor will act as an open while the inductor will act as a short; that is, X_C will be infinite while X_L will be zero.

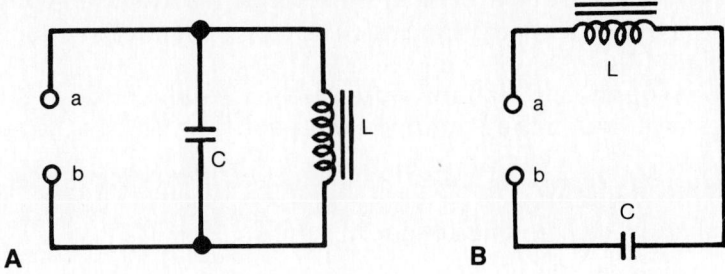

Fig. 17-1: LC circuits: (A) parallel LC circuit and (B) series LC circuit.

On the other hand, suppose the applied voltage is not DC, but rather a very low frequency AC voltage. In such a case, X_L will be very low and X_C will be high. Their exact values are dependent on the circuit values of L and C and the frequency of the applied voltage. If the frequency is gradually increased, the X_L will gradually increase, while the X_C will gradually decrease. As the frequency is increased further, a point is eventually reached at which the value of X_L is the same as the value of X_C; that is, for any combination of L and C, there is some frequency at which X_L equals X_C. This is true whether the two components are connected in parallel as shown in Fig. 17-1A or in series as shown in Fig. 17-1B. The condition at which X_L is equal to X_C is called *resonance*. Also, the frequency at which X_L is equal to X_C is called the *resonant frequency* and is abbreviated f_r.

341

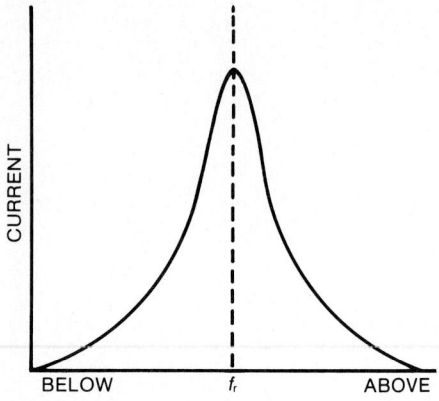

Fig. 17-2: Typical series resonant response curve.

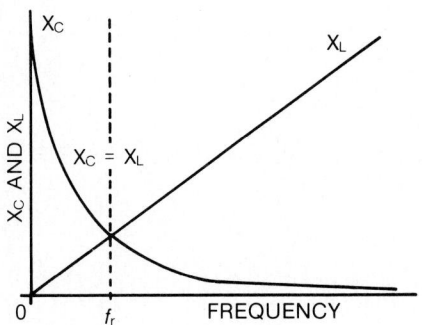

Fig. 17-3: Reactance curves for series RLC circuit.

SERIES CIRCUIT RESONANCE

When the reactances in a series circuit cancel and, as far as the source is concerned, the circuit appears to contain only resistance, the circuit is said to be in a condition of resonance. When resonance is established in a series circuit certain conditions will prevail.

1. The inductive reactance will be equal to the capacitive reactance.
2. The circuit impedance will be minimum.
3. The circuit current will be maximum.

Current decreases on either side of resonance. Actually, the current values can be plotted against their respective frequencies on linear graph paper. The result is a series resonant response curve similar to the one shown in Fig. 17-2.

Resonant Frequency

The reactance of capacitors and inductors is determined by their physical construction and the applied frequency. X_L varies directly with frequency and X_C varies inversely with frequency. Due to this relationship, any combination of inductance and capacitance will have a specific frequency at which the reactances will be equal. The relationship of X_L, X_C, frequency, and the resonant frequency point (f_r) is shown in Fig. 17-3. In a series RLC circuit, the largest reactance value determines the appearance and phase angle of the circuit. It can be seen that below the resonant frequency point X_C is the larger reactance and above f_r the X_L is the larger reactance. Therefore, the circuit will appear capacitive below f_r and inductive above f_r (Fig. 17-4).

An equation for determination of the resonant frequency can be developed in the following manner:

At resonance $X_L = X_C$.

Substituting the equation for X_L and X_C:

$$2\pi f L = \frac{1}{2\pi f C}$$

Transposing:

$$1 = 4\pi^2 f^2 LC$$

$$f^2 = \frac{1}{4\pi^2 LC}$$

Solving for f:

$$f_r = \frac{1}{2\pi \sqrt{LC}}$$

where:

f_r = resonant frequency in hertz (Hz)

L = inductance in henrys (H)

C = capacitance in farads (F).

342

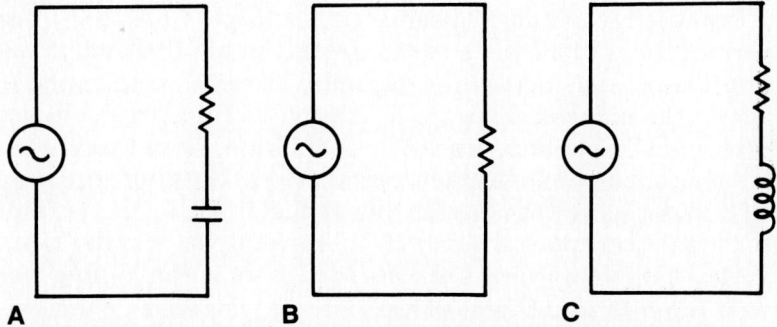

Fig. 17-4: Equivalent circuits: (A) below f_r, resistive capacitive; (B) resonance, resistive; and (C) above f_r, resistive-inductive.

Example: Determine the resonant frequency of a series RLC circuit consisting of an $80\mu\text{F}$ capacitor, $10\mu\text{H}$ coil, and a 10Ω resistor.

Solution:

$$f_r = \frac{1}{2\pi\sqrt{LC}}$$

$$f_r = \frac{1}{6.28 \times \sqrt{80 \times 10 \times 10^{-12}}}$$

$$f_r \approx \frac{1 \times 10^6}{177.5}$$

$$f_r \approx 5.63\text{kHz}$$

Thus, when a frequency of 5.63kHz is applied to the circuit in the example, the capacitive and inductive reactances will be equal.

Resonant Series Circuit Analysis

In order for the series circuit (Fig. 17-5A) to be in resonance, the frequency of the applied voltage must be such that $X_L = X_C$.

When a series circuit contains resistance, inductive reactance, and capacitive reactance, the total impedance for any frequency is:

$$Z = R + j(X_L - X_C)$$

Because X_L increases and X_C decreases with an increase in frequency, at a certain frequency (the resonant frequency) X_L will equal X_C, they will cancel, the j term will drop out, and Z will equal R. Furthermore, because the total impedance is now only the resistance, R, of the circuit, the circuit current is maximum. In other words, at resonance the generator is looking into a pure resistance.

At frequencies below resonance, X_C is greater than X_L and the circuit contains resistance and capacitive reactance; at frequencies above resonance, X_L is greater than X_C and the circuit contains resistance and inductive reactance. At resonance, the current is limited only by the relatively low value of resistance.

Because the circuit is a series circuit (Fig. 17-5A), the same current flows in all parts of the circuit (Fig. 17-5B) and, therefore, the voltage across the capacitor is equal to the voltage across the inductor, because X_L is equal to X_C. These voltages (Fig. 17-5C), however, are 180° out of phase, since the voltage across a capacitor lags the current through it by approximately 90°, and the voltage across the inductor leads the current through it by approximately 90°. The total value of the input voltage, V_T, then appears across R and is shown as V_R, in phase with the current, I (Fig. 17-5C).

Assume that at a given instant corresponding to angle 0°, the current through the circuit is a maximum as indicated in Fig. 17-5B. During the first quarter cycle (from 0° to 90°), the circuit current falls from maximum to zero. The capacitor is receiving a charge, as is indicated by the rising voltage, v_c, across it. The product of the instantaneous values of v_c and i for this interval indicates a positive power curve (p). The shaded area under this curve represents the energy stored in the capacitor during this time it is receiving a charge.

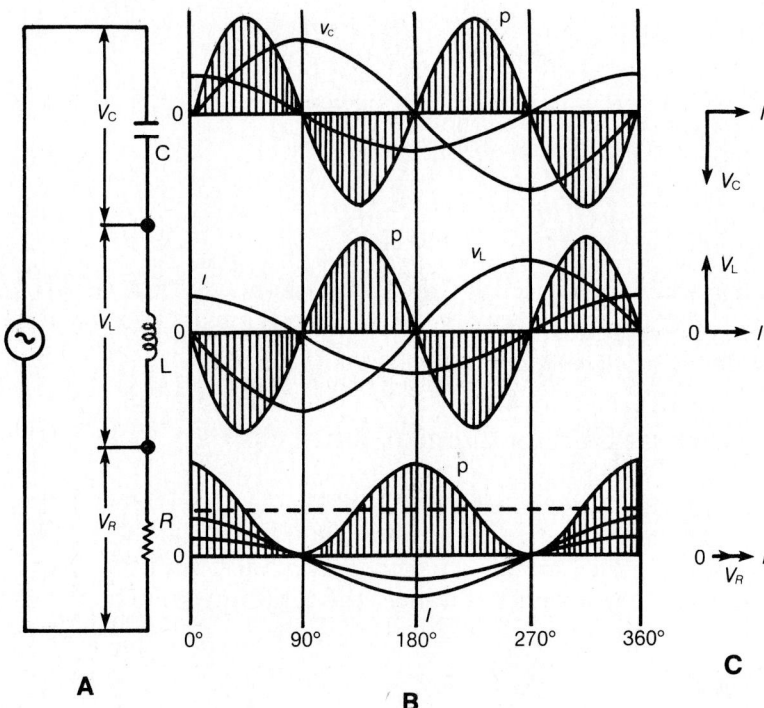

Fig. 17-5: Series resonance: (A) the circuit; (B) voltage sine wave curves; and (C) phasors.

During the first quarter of a cycle (0° to 90°), when the capacitor is receiving a charge, the magnetic field about the inductor is collapsing because the circuit current is falling, and the inductor acts like a source of power that supplies the charging energy to the capacitor. The voltage, v_L, across the coil is opposite in phase to the voltage building up across the capacitor and is shown below the line. Therefore, the product of the instan-

344

taneous values of the current and voltage across the inductor indicates a negative power curve for the coil between 0° and 90°.

During the second quarter cycle (90° to 180°), the capacitor discharges from maximum to zero, as indicated by the capacitor voltage curve, v_c, and the coil reverses its function and acts like a load on the capacitor. Thus, the capacitor now acts as a source of power. The product of a negative current and a positive voltage (v_c) indicates a negative power curve for the capacitor for this interval. During the same quarter cycle the current is rising through the inductor (in the opposite direction), and energy is being stored in the magnetic field. The product of the negative current and negative voltage, v_L, for the second quarter cycle indicates a positive power curve for the inductor.

A similar interchange of energy between the capacitor and inductor takes place in the third and fourth quarter cycles. Therefore, the average power supplied to the inductor and capacitor by an external source is essentially zero. All circuit losses are assumed to be in the resistor, R. The voltage across the resistor and the current through it are in phase. The product of the voltage and current curves associated with the resistor indicates a power curve that has its axis displaced above the X axis. The displacement is proportional to the true average power which is equal to the product, VI (where V and I are effective values). Whatever power is dissipated in R is supplied by the source.

Circuit Q

The *quality* or Q of a circuit is an important consideration in determining the actual merit or efficiency of an inductor, capacitor, or a combination of these components.

The Q of an Inductor. The ratio of the energy stored in an inductor during the time the magnetic field is being established to the losses in the inductor during the same time is called the *quality,* or Q, of the inductor. It is also called the *figure of merit* of the inductor. This ratio is:

$$Q = \frac{I^2 X_L}{I^2 R}$$

Cancellation yields:

$$Q = \frac{X_L}{R}$$

where:

Q = a number representing the quality of the inductor

X_L = inductive reactance of the coil in ohms

R = combined DC and AC resistances of the coil in ohms.

The Q of the inductor is, therefore, equal to the ratio of the inductive reactance to the effective resistance in series with it, and it approaches a high value as R approaches a low value.

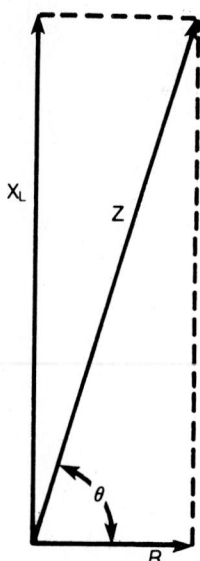

Fig. 17-6: Impedance triangle.

Thus, the more efficient the inductor, the lower the losses in it and the higher is the Q.

In terms of the impedance triangle (Fig. 17-6):

$$Q = \frac{X_L}{R}$$

$$Q = \tan \theta$$

where θ is the phase angle between the hypotenuse, Z, and the base, R. As θ approaches 90°, $\tan \theta$ approaches infinity, and the coil losses approach zero.

The Q of a coil does not vary extensively within the operating limits of a circuit. It would seem that since X_L is a direct function of frequency, Q also must be a direct function of frequency. Such is not the case. It is true that as frequency increases, the X_L will increase; but as frequency increases, the effective resistance of the coil also increases. Since Q is an inverse function of the effective resistance, the net effect of a frequency increase is to leave Q relatively unchanged.

The Q of a Capacitor. A capacitor's Q is a measure of the ratio of the energy stored to the energy dissipated in heat within the capacitor for equal intervals of time. This ratio is reduced by algebraic manipulation to the equation expressed below:

$$Q = \frac{X_C}{R}$$

where:

  R = the effective resistance of the capacitor dielectric (losses)

  Q = a number representing the quality of the capacitor

  X_C = capacitive reactance.

The effective resistance is low with respect to the capacitive reactance and is such that, when multiplied by the square of the effective capacitor current, it equals the true power dissipated in heat within the capacitor.

Since most of the losses in a solid-dielectric capacitor occur within the dielectric rather than in the plates, the Q of a low-dielectric-loss capacitor is high. The losses of an air-dielectric capacitor are negligible and, thus, the Q of such a capacitor may have a very high value.

The Q of a series resonant circuit is the ratio of the energy stored to the energy lost in equal intervals of time. The expression becomes:

$$Q = \frac{X_L}{R} = \frac{X_C}{R}$$

where R represents the total effective series resistance of the entire circuit. Since the capacitor has negligible losses, the circuit Q becomes equivalent to the Q of the coil. The circuit Q may be maintained satisfactorily high by keeping the circuit resistance to a minimum. This may be expressed mathematically in the following manner. The inductive voltage drop is:

346

$$V_{\mathrm{L}} = I\mathrm{X_L}$$

Since

$$I = \frac{V}{R}$$

then

$$V_{\mathrm{L}} = \frac{V\mathrm{X_L}}{R}$$

and

$$Q = \frac{\mathrm{X_L}}{R}$$

then

$$V_{\mathrm{L}} = Q V$$

transposed

$$Q = \frac{V_{\mathrm{L}}}{V}$$

where

    Q = a number representing the quality of the circuit

    V_{L} = the voltage across the inductor (can be V_{C}, voltage across the capacitor)

    V = the voltage across the effective series resistance.

Therefore, the Q of the circuit is the ratio of the voltage across either the inductor or capacitor to that across the effective series resistance. In other words, the voltage gain (VG) of the series resonant circuit depends on the circuit Q. Expressed mathematically:

$$VG = \frac{V_{\mathrm{L}}}{V} = \frac{I\mathrm{X_L}}{IR} = \frac{\mathrm{X_L}}{R} = \frac{I\mathrm{X_C}}{IR} = \frac{\mathrm{X_C}}{R} = Q$$

Series RLC Resonant Circuit Analysis

Figure 17-7 shows the relation between the effective current and frequency in the vicinity of resonance for a series circuit containing a 159μH coil, a 159pF capacitor, and an effective series resistance of either 10Ω or 20Ω.

The resonant frequency, f_r, is:

$$f_r = \frac{1}{2\pi \sqrt{LC}}$$

$$f_r = \frac{1}{6.28 \sqrt{159 \times 10^{-6} \times 159 \times 10^{-12}}}$$

$$f_r = 1 \times 10^6 \mathrm{Hz} \text{ or } 1{,}000\mathrm{kHz}.$$

The reactances and impedance at resonance may likewise be determined. Thus:

$$\mathrm{X_{L_r}} = 2\pi f L = 6.28 \times 10^6 \times 159 \times 10^{-6}$$

$$\mathrm{X_{L_r}} = 1{,}000 \angle +90°\Omega$$

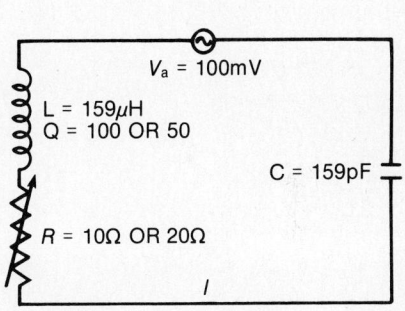

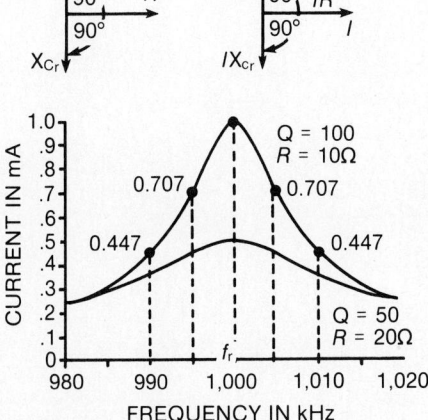

Fig. 17-7: Resonance curves of a series _R_LC circuit.

347

where X_{L_r} is the inductive reactance at resonance. The $+90°$ angle indicates that the IX_{L_r} and X_{L_r} phasors are plotted vertically upward because the current phasor is horizontal and extends to the right. The current phasor thus lags the voltage, IX_{L_r}, across the coil by $90°$ (counterclockwise rotation is positive, Fig. 17-7).

$$X_{C_r} = \frac{1}{2\pi f C} = \frac{1}{6.28 \times 10^6 \times 159 \times 10^{-12}}$$
$$X_{C_r} = 1{,}000 \angle -90° \Omega$$

where X_{C_r} is the capacitive reactance at resonance. The $-90°$ angle indicates that the phasors X_{C_r} and IX_{C_r} are plotted vertically downward because the current phasor is the horizontal reference phasor extending to the right and leads the voltage drop, IX_{C_r}, across the capacitor by $90°$ (Fig. 17-7). Note that current is a common factor to both voltage and impedance phasors. When the effective resistance of the circuit is equal to 10Ω, the impedance at resonance (Z_r) will be:

$$Z_r = R + j(X_L - X_C)$$
$$Z_r = 10 + j(1{,}000 - 1{,}000)$$
$$Z_r = 10 + j0\,\Omega$$
$$Z_r = 10 \angle 0° \Omega$$

If the applied voltage is assumed to be 10mV at a frequency of 1,000kHz, the circuit current is:

$$I = \frac{V}{Z}$$
$$I = \frac{0.01}{10}$$
$$I = 1\text{mA}$$

At the resonant frequency, the voltage across the inductor is:

$$V_L = IX_L$$
$$V_L = 0.001 \times 1{,}000$$
$$V_L = 1\text{V}$$

The voltage across the capacitor is the same, except it is $180°$ out of phase with the voltage across the coil. The losses in the coil and capacitor are assumed to be lumped in the effective series resistance. The circuit Q is:

$$Q = \frac{X_L}{R}$$
$$Q = \frac{1{,}000}{10}$$
$$Q = 100$$

The voltage gain at resonance is:

$$VG = \frac{V_L}{V}$$

$$VG = \frac{1.00}{0.01}$$

$$VG = 100$$

The Q of a tuned circuit can be found by this equation:

$$Q = \frac{1}{R} \times \sqrt{\frac{L}{C}}$$

This can be proven by substituting the values from Fig. 17-7 into the equation and solving for the results. It is important to note that the higher the resistance in a tuned circuit, the lower will be the circuit Q.

The resonance curves of current vs. frequency are symmetrical about a vertical line (f_r) extending through the point of maximum current (Fig. 17-7). The shape of the resonance curve may be approximated in the vicinity of resonance by applying the following rules that can be derived from the resonant circuit equations. (The derivation is not given because of its length.)

RULE 1. If the frequency of the applied voltage is decreased by an amount $\frac{1}{2Q}$ times the resonant frequency, f_r, the current in the tuned circuit decreases to 0.707 of its value at the resonant frequency and leads the applied voltage by 45°.

Example: The input frequency of the circuit (Fig. 17-7) is decreased from the resonant frequency (f_r) of 1,000kHz by an amount equal to:

$$f_{\text{decrease}} = \frac{1}{2Q} \times f_r$$

$$f_{\text{decrease}} = \frac{1}{2 \times 100} \times 1,000\text{kHz}$$

$$f_{\text{decrease}} = 5\text{kHz}$$

The new applied frequency is then:

$$f_{\text{new}} = f_r - f_{\text{decrease}}$$
$$f_{\text{new}} = 1,000 - 5$$
$$f_{\text{new}} = 995\text{kHz}$$

The X_L at the new applied frequency is:

$$X_L = 2\pi f L$$
$$X_L = 6.28 \times 0.995 \times 10^6 \times 159 \times 10^{-6}$$
$$X_L \approx 995 \angle +90°\,\Omega$$

The X_C at the new applied frequency is:

$$X_C = \frac{1}{2\pi f C}$$

$$X_C = \frac{1}{6.28 \times 9.995 \times 10^6 \times 159 \times 10^{-12}}$$

$$X_C \approx 1,005 \angle - 90°\,\Omega$$

The circuit impedance at 995kHz is:

$$Z_T = R + j(X_L - X_C)$$
$$Z_T = 10 + j(995 - 1,005)$$
$$Z_T = 10 - j10\,\Omega$$

Converting to polar form:

$$Z_T = 14.14 \angle - 45°$$

The circuit current at 995kHz is:

$$I_T = \frac{V_T}{Z_T}$$

$$I_T = \frac{0.01 \angle \ 0°}{14.14 \angle - 45°}$$

$$I_T \approx 0.707 \angle \ 45°\,\text{mA.}$$

At this frequency (which is the cutoff frequency), the voltage across the coil, or the capacitor, is reduced to approximately 70% of its value at resonance.

The voltage across the coil is:

$$V_L = I X_L$$
$$V_L = 0.707 \times 995$$
$$V_L \approx 705\text{mV.}$$

RULE 2. If the frequency of the applied voltage is decreased by an amount $1/Q$ times the resonant frequency, the current decreases to 0.447 of its value at resonance and leads the applied voltage by 63.4°.

Example: The input frequency of the circuit (Fig. 17-7) is decreased from the resonant frequency at an amount equal to:

$$f_{\text{decrease}} = \frac{1}{Q} \times f_r$$

$$f_{\text{decrease}} = \frac{1}{100} \times 1,000$$

$$f_{\text{decrease}} = 10\text{kHz}$$

The new applied frequency is then:

$$f_{\text{new}} = f_r - f_{\text{decrease}}$$
$$f_{\text{new}} = 1,000 - 10$$
$$f_{\text{new}} = 990\text{kHz}$$

The X_L at the new applied frequency of 990kHz is:

$$X_L = 2\pi f L$$
$$X_L = 6.28 \times 0.990 \times 10^6 \times 159 \times 10^{-6}$$
$$X_L \approx 990 \angle +90° \Omega$$

The X_C at the new applied frequency of 990kHz is:

$$X_C = \frac{1}{2\pi f C}$$
$$X_C = \frac{1}{6.28 \times 0.990 \times 10^6 \times 159 \times 10^{-12}}$$
$$X_C \approx 1,010 \angle -90° \Omega$$

The impedance of the series circuit at 990kHz is:

$$Z_T = R + j(X_L - X_C)$$
$$Z_T = 10 + j(990 - 1,010)$$
$$Z_T = 10 - j20$$

Converting to polar form yields:

$$Z_T \approx 22.4 \angle -63.4° \Omega$$

At this frequency the circuit current is:

$$I_T = \frac{V_T}{Z_T}$$
$$I_T = \frac{0.10 \angle 0°}{22.4 \angle -63.4°}$$
$$I_L \approx 0.447 \angle +63.4° \text{mA}$$

The voltage across the coil is:

$$V_L = I X_L$$
$$V_L = 0.447 \times 990$$
$$V_L \approx 444 \text{mV}.$$

Decreasing the applied frequency is seen (by the positive phase angles of I_T) to cause the total current to lead the applied voltage. Corresponding increases in the frequency of the applied voltage above the resonant frequency will produce the same reductions in circuit current and voltage across the reactive portions of the circuit. In this case, however, the current lags the applied voltage instead of leading it. Thus, the resonance curve is symmetrical about the resonant frequency in the vicinity of resonance.

The series resonant circuit increases the voltage gain at the resonant frequency. If the circuit losses are low, the circuit Q will be high and the voltage gain relatively large. For resonant circuits involving iron-core coils, the Q may range from 20 to 100. In practice, because nearly all of the resistance of a circuit is in the coil, the ratio of the inductive reactance to the resistance is especially important. The higher the Q of the coil, the better is the coil and the more effective is the series resonant circuit that utilizes it.

Power in Resonant Series Circuits

Total true power in an RLC resonant series circuit is the product of line voltage and current times the cosine of the angle between them. When X_L and X_C are equal, total impedance is at a minimum and, thus, current is maximum. When maximum current flows, the series resistor dissipates maximum power. When X_L and X_C are made unequal, total impedance increases, and line current decreases and moves out of phase with line voltage. Both the decrease in current and the creation of a phase difference cause a decrease in true power.

In the previous example, decreasing the applied frequency caused a leading current, therefore, a positive phase angle. A positive phase angle causes a leading power factor and vice versa. Thus, the pF is leading when the frequency is decreased and lagging when the frequency is increased (with respect to the resonant frequency).

Bandwidth

If the circuit Q is low, the gain at resonance is relatively small and the circuit does not discriminate sharply between the resonant frequency and the frequencies on either side of resonance, as is shown by the lower curve in Fig. 17-7. The range of frequencies included between the two frequencies at which the current drops to 70% of its value at resonance is called the *bandwidth* for 70% response.

Frequencies beyond (outside of) the 70% area, referred to as the *half-power* points on the curve, are considered to produce no usable output. The series resonant circuit is seen to have two half-power points, one above the resonant frequency point and one below. The two points are designated upper f_r and lower f_r or simply f_1 and f_2. The range of frequencies between these two points comprises the bandwidth. Figure 17-8 illustrates the bandwidths for high and low Q series resonant circuits. The bandwidth may be determined by the equation:

$$BW = \frac{f_r}{Q} = f_2 - f_1$$

where:

   BW = bandwidth of a series resonant circuit in units of frequency

   f_r = resonant frequency

   f_2 = the highest frequency the circuit will pass

   f_1 = the lowest frequency the circuit will pass

   Q = as defined previously.

Example: Determine the bandwidth for the curve shown in Fig. 17-8B.

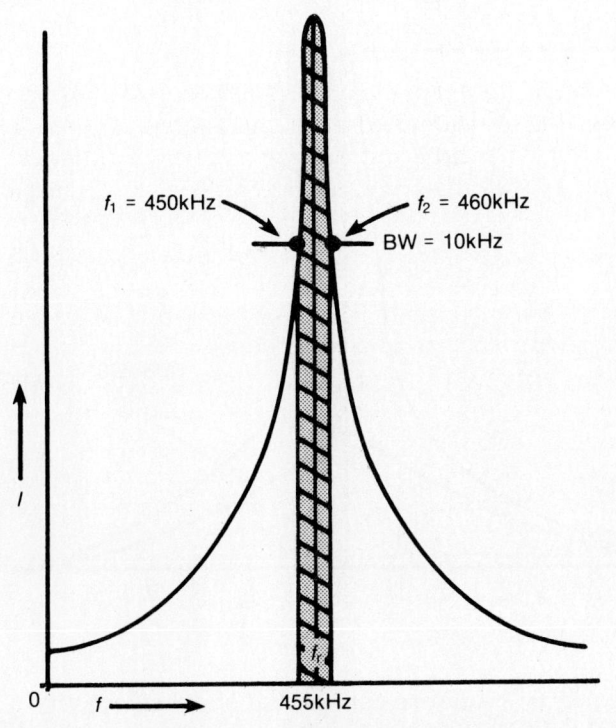

A

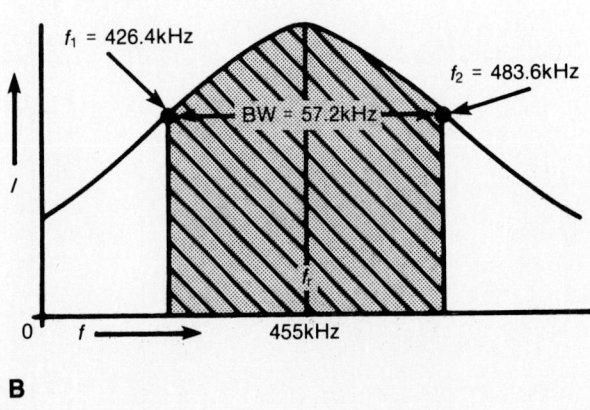

B

Fig. 17-8: Bandwidth for (A) high and (B) low Q series resonant circuits.

Solution:

$$BW = f_2 - f_1$$
$$BW = 483.6\text{kHz} - 426.4\text{kHz}$$
$$BW = 57.2\text{kHz}$$

The Q of the curve in Fig. 17-8A is given as 45.5; determine the bandwidth.

$$BW = \frac{f_r}{Q}$$
$$BW = \frac{455\text{kHz}}{45.5}$$
$$BW = 10\text{kHz}$$

Selectivity. Curve A in Fig. 17-9 compared to curve B is a very selective curve because it responds sharply to the resonant frequency but responds very poorly to frequencies on either side of resonance.

Selectivity is the ability of a circuit to accept or pass currents of one frequency and to exclude all others. Curve A satisfies this definition, as it responds to a narrow band of frequencies around 1,500kHz but not to the other frequencies. Curve B responds equally well to a wide range of frequencies around 1,500kHz and thus it has very poor selectivity. The selectivity of a tuned circuit depends upon the Q of the tuned circuit.

353

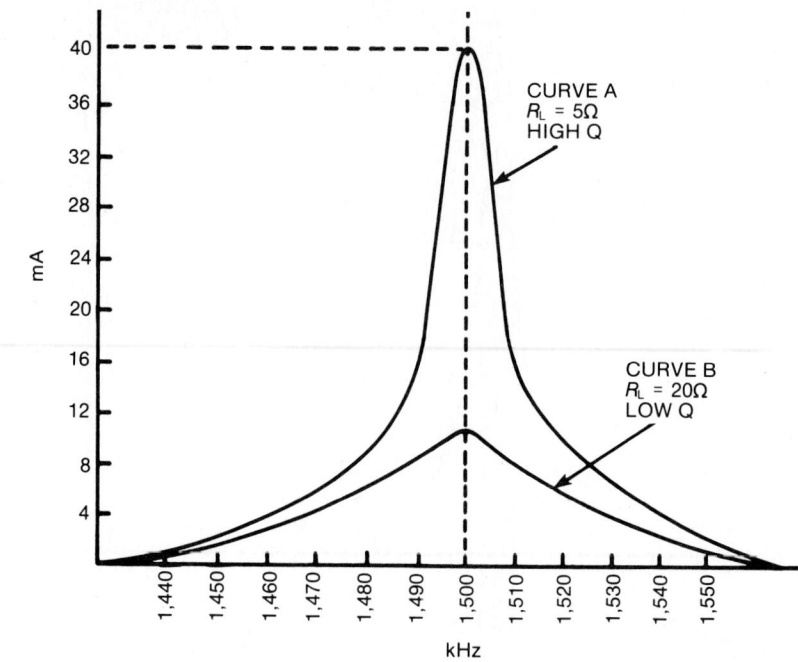

Fig. 17-9: Frequency-response curves that show selectivity of a circuit.

The selectivity of a series resonant circuit is dependent upon the Q of the circuit which, in turn, is determined chiefly by the Q of the coil. Q is often considered a measure of the circuit's ability to store energy. The energy stored alternately in the reactance is reactive power (I^2X_L) and the energy dissipated in the form of heat is dissipated power (I^2R). In other words, Q is the ratio of reactive power to dissipated power:

$$Q = \frac{I^2X_L}{I^2R}$$

The I^2 cancel. Therefore

$$Q = \frac{X_L}{R} = \frac{\text{energy stored}}{\text{energy dissipated}} = \text{figure of merit}$$

where

X_L is inductance reactance at resonance

R is resistance of the circuit

Another definition of Q is that it is a measure of quality or purity of an inductance. The resistance of a coil is considered to be in series with its inductive reactance. Curve A in Fig. 17-9 has a coil resistance of $5\,\Omega$; whereas, curve B has a coil resistance of $20\,\Omega$. Curve A has high selectivity because of the low value of coil resistance. The coil in curve A is said to have a high Q. For example, the Q's of the coils in Fig. 17-9 are:

Curve A: $\quad Q = \dfrac{X_L}{R_1} = \dfrac{1,000}{5} = 200$

Curve B: $\quad Q = \dfrac{X_L}{R_2} = \dfrac{1,000}{20} = 50$

354

As seen in Fig. 17-9, the higher the Q—due to a lower coil resistance—the larger the circuit current at resonance and the higher the circuit selectivity. This is because the ratio of the amount of energy stored in the magnetic field in the coil to the amount of heat energy lost in the DC resistance is greater.

L/C Ratio. The L/C ratio also has an effect on selectivity. As noted earlier, the Q of a circuit can also be increased by increasing the inductance of a coil, by winding the coil with a greater number of turns of wire, or by using a coil core of a higher permeability. In any case, because the resistance of the coil should be kept constant, the coil wire used to rewind the coil should be of a larger diameter.

If the coil inductance is increased, the value of capacitance used to resonate at the same frequency would have to be reduced. In other words, the L/C ratio of the combination is increased. As can be seen from Table 17-1, there are many combinations of L and C values which will produce a resonance frequency of 1,500kHz, even though for each combination there is only one resonance frequency.

Table 17-1: Some L and C Combinations Which Resonate at 1,500kHz

Equation	L in μH	C in pF	X_L	X_C
A	26.5	424	250	250
B	53	212	500	500
C	106	106	1,000	1,000
D	212	53	2,000	2,000
E	424	26.5	4,000	4,000

The combination of L in μH and C in pF in Eq. C produces an L/C ratio of

$$L/C = 106\mu/106p = 1,000,000/1$$

The combination of L in μH and C in pF in Eq. E produces an L/C ratio of

$$L/C = 424\mu/26.5p = 16,000,000/1$$

The L and C combination in Eq. E (Table 17-1) will produce the largest Q, if the DC resistance has not increased. This is because the X_L is larger in Eq. E than in Eq. C. The X_L in Eq. E is 4,000Ω but is only 1,000Ω in Eq. C. This means that the X_L is four times larger in Eq. E than the X_L in Eq. C. The result is that the $Q(X_L/R)$ of the coil in Eq. E will be four times larger than that in Eq. C if the DC resistance of the 424μH coil is kept at the same value as in the 106μH coil. Theoretically, the Q of a coil could be made to approach infinity by increasing the value of inductance with respect to the capacitance. In practical situations, however, a value of inductance is reached beyond which the Q can no longer be increased, because the DC resistance of the wire increases in proportion to the inductance increase. The optimum value of X_L for best selectivity in radio frequencies is 1,000Ω.

Figure 17-10 shows the frequency response curves for three different L/C ratios. Curve C has the highest L/C ratio and is the most selective.

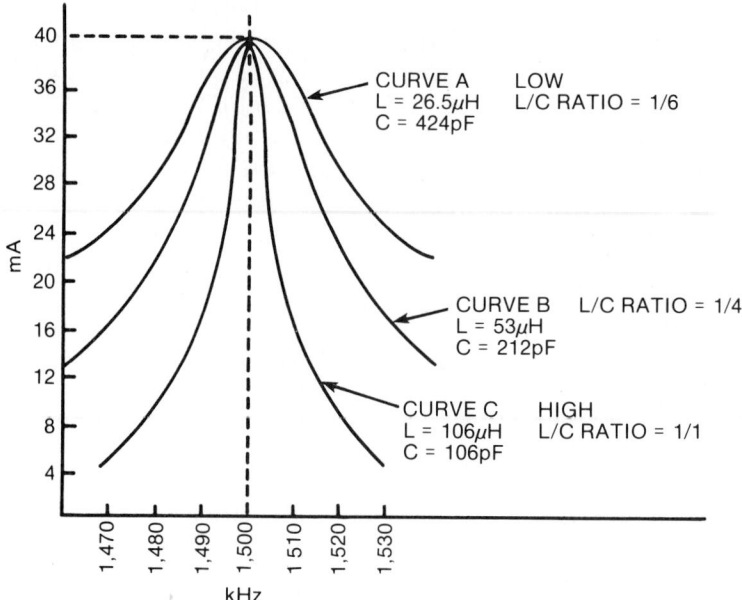

Fig. 17-10: Series resonant response curves showing the effect of L/C ratios on selectivity.

Applications of Series Resonant Circuits

Series resonant circuits are used largely as filters (to be treated later in this chapter) for audio and radio frequencies. With proportionately larger component values, the series circuit may be used as a power-supply filter. For example, assume that a DC generator has a ripple frequency of 400 Hz. A series resonant circuit tuned to 400 Hz may be connected across the terminals of the generator and thus effectively short-circuit the ripple voltage. The coil and capacitor insulation must be able to withstand the relatively high AC voltage caused by the series resonant action.

The series-tuned circuit may also be used to give an indication of frequency if the capacitor is calibrated for the appropriate frequency range. The capacitor and the inductor are connected in series with a current-indicating device across the source of the unknown frequency. At resonance, the current as indicated by the device will be maximum; therefore, if the value of L and C are known, the value of the unknown frequency may be calculated.

Summary of Resonant Conditions for Series RLC Circuits

The major characteristics of series RLC circuits at resonance are summarized in Table 17-2.

Table 17-2: Major Characteristics of Series RLC Circuits at Resonance

Quantity	Series Circuit
At resonance: Reactance ($X_L - X_C$)	Zero, because $X_L = X_C$
Resonant frequency	$\dfrac{1}{2\pi\sqrt{LC}}$
Impedance	Minimum; $Z = R$
I_{line}	Maximum value
I_L	I_{line}
I_C	I_{line}
V_L	$Q \times V_{line}$
V_C	$Q \times V_{line}$
Phase angle between V_{line} and I_{line}	$0°$
Angle between V_L & V_C	$180°$
Angle between I_L & I_C	$0°$
Desired value of Q	10 or more
Desired value of R	Low
Highest selectivity	High Q, low R, high L/C ratio
When f is greater than f_r: Reactance	Inductive (lagging current)
When f is less than f_r: Reactance	Capacitive (leading current)

PARALLEL RESONANT CIRCUITS

The ideal parallel resonant circuit is one that contains only inductance and capacitance. Resistance and its effects are not considered in an ideal parallel resonant circuit. One condition for parallel resonance is the application of that frequency which will cause the inductive reactance to equal the capacitive reactance. The formula used to determine the resonant frequency of a parallel LC circuit is the same as the one used for a series circuit.

$$f_r = \frac{1}{2\pi\sqrt{LC}}$$

where:

f_r = resonant frequency in Hz

L = inductance in H

C = capacitance in F

357

If the circuit values are those shown in Fig. 17-11A, the resonant frequency may be computed as follows:

$$f_r = \frac{1}{2\pi \sqrt{LC}}$$

$$f_r = \frac{1}{6.28 \sqrt{(3.5 \times 10^{-3})(2.5 \times 10^{-6})}}$$

$$f_r = 1{,}700\,\text{Hz}$$

At the resonant frequency:

Determine X_L:

$$X_L = 2\pi fL$$
$$X_L = (6.28)(1.7 \times 10^3)(3.5 \times 10^{-3})$$
$$X_L \approx 37.4\,\Omega$$

Determine X_C:

$$X_C = \frac{1}{2\pi fC}$$

$$X_C = \frac{1}{(6.28)(1.7 \times 10^3)(2.5 \times 10^{-6})}$$

$$X_C \approx 37.4\,\Omega$$

Determine I_C:

$$I_C = \frac{V_a}{-jX_C}$$

$$I_C = \frac{60}{-j37.4}$$

$$I_C \approx 0 + j1.605\,\text{A}$$

Determine I_L:

$$I_L = \frac{V_a}{jX_L}$$

$$I_L = \frac{60}{j37.4}$$

$$I_L = 0 - j1.605\,\text{A}$$

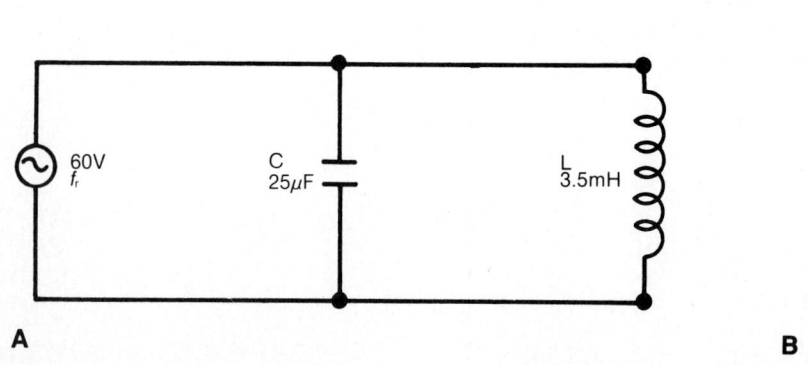
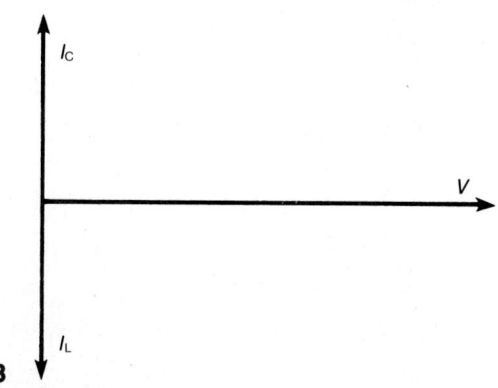

Fig. 17-11: (A) Parallel LC circuit at resonance; (B) current-voltage phase relationships.

The total current is determined by addition of the two currents in rectangular form:

$$I_C \approx 0 + j1.605$$
$$I_L \approx 0 - j1.605$$
$$\overline{I_T \approx 0}$$

Therefore, in an ideal resonant parallel circuit, the line current (I_T) is zero. If total current is zero, then:

$$Z = \frac{V_{app}}{I_T} = \frac{60}{0} = \text{undefined}$$

Or it may be said that the impedance approaches infinity.

At frequencies other than the natural resonant frequency of the circuit, X_C will not be equal to X_L and some amount of current will be drawn from the source. If the applied frequency is lower than the resonant frequency of the circuit, X_L will be smaller than X_C and a lagging source current will result. When the applied frequency is above the resonant frequency, X_C is smaller than X_L, and the source current leads the source voltage.

To obtain an overall view of the operation of a parallel LC circuit, a graph can be constructed in which impedance and current are plotted as a function of frequency. To obtain the information for the graph, the capacitive and inductive reactances, impedance, and total current are computed for a group of frequencies centered about the resonant frequency. These computations have been performed for a parallel circuit containing a $2.5\,\mu F$ capacitor and a $3.5mH$ inductor and have been tabulated in Table 17-3.

Table 17-3: Reactance, Impedance, and Current as a Function of Frequency

f (Hz)	X_L (ohms)	X_C (ohms)	Z (ohms)	I_T (amps)
700	15.39	90.95	18.53	3.238
800	17.59	79.58	22.59	2.657
900	19.79	70.74	27.48	2.183
1,000	21.99	63.66	33.60	1.786
1,100	24.19	57.88	41.56	1.444
1,200	26.39	53.05	52.51	1.142
1,300	28.59	48.97	68.69	0.874
1,400	30.79	45.47	95.34	0.629
1,500	32.99	42.44	148.08	0.405
1,600	35.19	39.79	314.25	0.197
*1,700	37.39	37.45	22,221.00	Zero
1,800	39.58	35.34	332.18	0.181
1,900	41.78	33.51	169.17	0.355
2,000	43.98	31.83	115.22	0.521
2,100	46.18	30.32	88.27	0.680
2,200	48.38	28.94	72.01	0.833
2,300	50.58	27.68	61.13	0.982
2,400	52.78	26.53	53.33	1.125
2,500	54.98	25.47	47.44	1.265
2,600	57.18	24.49	42.83	1.401
2,700	59.38	23.58	39.11	1.534

*Resonant frequency

At 700Hz, for example, the inductive and capacitive reactances are found to be 15.39Ω and 90.95Ω respectively. The impedance of the circuit can be computed by the product over the sum method as follows:

$$Z = \frac{(jX_L)(-jX_C)}{(jX_L) + (-jX_C)}$$

$$Z = \frac{(0 + j15.39)(0 - j90.95)}{0 - j75.56}$$

$$Z = \frac{1,339.72 \angle 0°}{75.56 \angle -90°}$$

$$Z = 18.53 \angle 90° \Omega$$

If the applied voltage is 60V the total current is:

$$I_T = \frac{V_a}{Z}$$

$$I_T = \frac{60 \angle 0°}{18.52 \angle 90°}$$

$$I_T \approx 3.24 \angle -90° A$$

Notice that at 700Hz the impedance and current have phase angles of 90°, indicating that the circuit appears as a pure reactance. At all frequencies below the resonant frequency of 1,700Hz, the ideal circuit appears as a pure inductance and the source current lags the source voltage by exactly -90°.

As the frequency is increased above 700Hz, the impedance rises and the source current decreases. At 1,700Hz the inductive branch current becomes equal to the capacitive branch current causing the total (line) current to diminish to zero. Since no current is drawn from the source, the impedance of the circuit moves toward infinity.

To investigate the operation of the circuit for applied frequencies above the natural resonant frequency of the circuit, the impedance and current will be computed for a frequency of 2,100Hz.

$$Z = \frac{(jX_L)(-jX_C)}{(jX_L) + (-jX_C)}$$

$$Z = \frac{(0 + j46.18)(0 - j30.32)}{0 + j15.86}$$

$$Z = \frac{1,400 \angle 0°}{15.86 \angle 90°}$$

$$Z \approx 88.27 \angle -90° \Omega$$

The current at 2,100Hz is:

$$I_T = \frac{V_a}{Z}$$

$$I_T = \frac{60 \angle 0°}{88.27 \angle -90°}$$

$$I_T \approx 0.68 \angle 90° A$$

The phase angles of the impedance and current at 2,100Hz show that the LC circuit appears as a pure capacitive reactance to the source. This condition exists for all applied frequencies which are higher than the resonant frequency, since for these frequencies I_C is greater than I_L.

At this point certain conclusions can be drawn concerning the characteristics of a parallel LC circuit. These are as follows:

1. For every possible parallel combination of inductance and capacitance, a frequency exists which will make X_C equal to X_L. This frequency is called the resonant frequency of the circuit.

2. When the circuit is operated at its resonant frequency, I_C is equal to I_L.

3. At resonance the source current is minimum (zero current in the ideal LC circuit).

4. At resonance the impedance of the circuit is maximum (infinite in the ideal circuit).

5. At resonance the circuit appears resistive to the source, the phase angle is zero, and the power factor is unity.

6. When the applied frequency is below the resonant frequency of the circuit, I_L is greater than I_C and the circuit appears inductive.

7. When the applied frequency is above the resonant frequency of the circuit, I_C is greater than I_L and the circuit appears capacitive.

Practical Resonant Parallel Circuits

The primary difference that exists between the ideal parallel resonant circuit and the practical parallel resonant circuit is that the practical series LC circuit contains resistance. This resistance exists throughout the circuit; however, most of it is located in the inductive branch of the circuit. For purposes of analysis all of the circuit resistance will be represented by a single resistor placed in series with the inductive branch. This resistance will be assumed to account for all of the circuit losses, both AC and DC. The LC circuit is shown in Fig. 17-12. In this circuit the impedance of the capacitive branch is equal to X_C, while the impedance of the inductive branch is equal to the phasor sum of X_L and R. If the source is adjusted to the frequency at which X_L is equal to X_C, the current through the inductive branch will be smaller than the current through the capacitive branch. The resulting phasor cancellation of I_C and I_L yields a low value of I_C (the total amount being determined by the size of the resistance, R). Therefore, the total current will lead the applied voltage by a small angle, making the circuit appear slightly capacitive to the source.

If the applied frequency is reduced slightly, the current through the capacitive branch decreases and the current through the inductive branch increases. The current through the inductive branch will cancel a greater percentage of the capacitive branch current, and the total current will attain its minimum value.

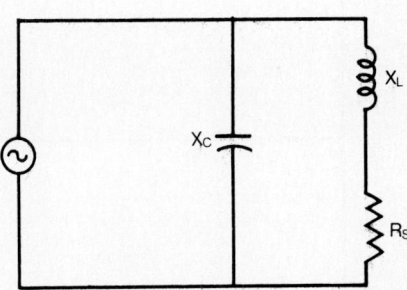

Fig. 17-12: Two branch parallel resonant circuit.

361

Due to the energy losses which occur in a practical parallel LC circuit, some current must be drawn from the source. Because of this current, the circuit will have a finite impedance at the resonant frequency.

Effects on Q Caused by Loading a Parallel Resonant Circuit

As for series RLC resonant circuits, the Q of the parallel resonant circuit shown in Fig. 17-12 is determined by the ratio of X_L/R_s (or X_C/R_s). The parallel circuit impedance at resonance varies inversely with the resistance in the coil branch. This (series) resistance may act as the total load on the parallel circuit. Hence, an increase in series resistance inherent in the coil or produced by increasing the physical resistance in the coil branch of the circuit represents an increased load. This lowers the Q and the total impedance of the circuit.

The total or line current (I_T) is equal to the sum of the branch currents I_C and I_L. These currents must be converted to rectangular form before they can be added.

The impedance is found by dividing the source voltage by the total current. At resonance, the phase angle of the impedance is zero and the circuit is purely resistive to the source.

The combination of L, C, and R in Fig. 17-13A forms a parallel resonant circuit in which a parallel resistance has been added. Notice that there is no resistance shown in series with the inductive branch. At resonance, the phasor diagram of the branch currents will be as shown in Fig. 17-13B. As before, if the lengths of the phasors representing the reactive currents are equal, the circuit will appear to be a purely resistive circuit having a resistance value that is equal to R.

The circuit shown in Fig. 17-13A is to be used to solve for current, impedance, circuit Q, and other characteristics in the following manner:

Find:

$$I_C = ? \qquad\qquad I_{tank} = ?$$
$$I_L = ? \qquad\qquad I_{line} = ?$$
$$Q = ?$$

Solution: Determine I_C.

$$I_C = \frac{V_a}{X_C}$$

$$I_C = \frac{25}{51}$$

$$I_C \approx 490\text{mA or } 0 + j490\text{mA}$$

Determine I_L:

$$I_L = \frac{V_a}{X_L}$$

$$I_L = \frac{25}{51}$$

$$I_L \approx 490\text{mA or } 0 - j490\text{mA}$$

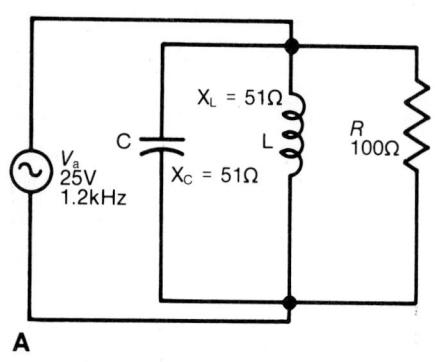

A

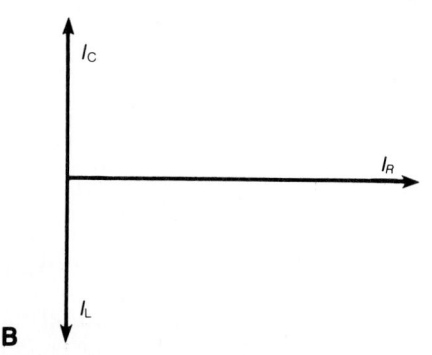

B

Fig. 17-13: Parallel resonant circuit with parallel resistance.

Determine I_R:

$$I_R = \frac{V_a}{R}$$

$$I_R = \frac{25}{100}$$

$$I_R \approx 250 + j0 \text{mA}.$$

To determine the current drawn from the source (line current), insure all currents are in rectangular form and add:

$$
\begin{aligned}
I_C &= 0 +j490 \\
I_L &= 0 -j490 \\
I_R &= 250 +j 0 \\
\hline
I_{line} &= 250 +j 0
\end{aligned}
$$

Since the reactive currents cancel, the line current is equal to the current drawn by the resistive component. At resonance, in a three branch circuit:

$$I_{line} = I_R$$

Since the circulating current of the tank is the same for either reactive component, then tank current may be found by determining current flow through the capacitor or inductor.

$$I_{tank} = \frac{V_a}{X_C}$$

$$I_{tank} = \frac{25}{51}$$

$$T_{tank} \approx 490 \text{mA}$$

or at resonance:

$$I_{tank} = I_C = I_L$$

For the purpose of discussion, it can be assumed that the LC combination in Fig. 17-13A forms an ideal parallel resonant network. The impedance of such a circuit would be higher than any preassigned value. Thus, the equivalent resistance (impedance) offered to the source by the parallel arrangement of the extremely high impedance of the tank circuit in parallel with a known value of resistance is equal to the value of the known resistance. The circuit impedance can therefore be determined in the following manner:

$$Z = \frac{V_a}{I_{line}}$$

$$Z = \frac{25}{0.25}$$

$$Z = 100\Omega$$

Determining the value of Q for a three branch circuit:

$$I_{line} = \frac{I_{tank}}{Q}$$

transposing:

$$Q = \frac{I_{\text{tank}}}{I_{\text{line}}}$$

$$Q = \frac{490}{250}$$

$$Q \approx 1.96$$

Q may be determined by:

$$Q = \frac{I_{\text{tank}}}{I_{\text{line}}}$$

$$I_{\text{tank}} = \frac{V_a}{X_L}$$

$$I_{\text{line}} = \frac{V_a}{R}$$

Therefore, by substitution:

$$Q = \frac{\dfrac{V_a}{X_L}}{\dfrac{V_a}{R}}$$

To divide, invert denominator and multiply:

$$Q = \frac{V_a}{X_L} \times \frac{R}{V_a}$$

Cancelling V_a:

$$Q = \frac{R}{X_L}$$

$$Q = \frac{100}{51}$$

$$Q \approx 1.96$$

The bandpass of a three branch RLC circuit such as that shown in Fig. 17-13A undergoes slight modification, i.e., it will not be the same as for a two branch circuit if expressed as a function of R, L, and C. The shape of the current versus frequency curve is the same as for the series resonant circuit since as the resistance, R, is increased, the line current decreases. Since:

$$BW = \frac{f_r}{Q}$$

and: $Q = \dfrac{R}{X_L}$ in the three branch circuit,

$$BW = \frac{f_r \times X_L}{R}$$

If the parallel resistance is increased, the line current goes down. As the shunting resistance approaches infinity, the line current approaches zero. As resistance is increased, the bandwidth becomes narrower and the selectivity increases. Therefore, it can be seen how regulation of the bandwidth may be accomplished by variation of the "shunt" resistance. This shunt resistance is sometimes referred to as a "swamping" resistance.

The inverse relationship between resistance and bandwidth may be seen by examination of the BW (bandwidth) formula above or the impedance and current versus frequency curves in Fig. 17-14, where f_{cr} defines the maximum and minimum cutoff frequencies.

Applications of Parallel Resonant Circuits

The parallel resonant circuit is one of the most important circuits used in electronic transmitters, receivers, and frequency-measuring equipment.

The IF transformers of radio and television receivers employ parallel-tuned circuits. These are transformers used at the input and output of each intermediate frequency amplifier stage for coupling purposes and to provide selectivity. They are enclosed in a shield and provided with openings at the top through which screwdriver adjustments may be made when the set is being aligned.

Parallel-tuned circuits are also used in the driver and power stages of transmitters, as well as in the oscillator stages of transmitters, receivers, and frequency-measuring equipment.

Various types of filter circuits employ parallel-tuned circuits as well as series-tuned circuits.

Summary of Resonant Conditions for Parallel RLC Circuits

The major characteristics of parallel RLC circuits at resonance are summarized in Table 17-4.

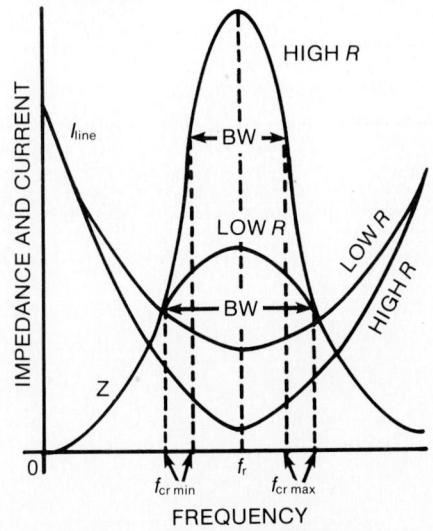

Fig. 17-14: Effect of shunt resistance on bandwidth, I_{line}, and Z.

Table 17-4: Major Characteristics of Parallel RLC Circuits at Resonance

Quantity	Parallel Circuit
At resonance: Reactance ($X_L - X_C$)	Infinite; because $I_{line} = 0$
Resonance frequency	$f_r = \dfrac{1}{2\pi \sqrt{LC}}$
Impedance	Maximum; $Z = \dfrac{L}{CR}$
I_{line}	Minimum value
I_L	$Q \times I_{line}$

365

Table 17-4: Major Characteristics of Parallel RLC Circuits at Resonance (Continued)

Quantity	Parallel Circuit
I_C	$Q \times I_{line}$
V_L	V_{line}
V_C	V_{line}
Phase angle between V_{line} and I_{line}	$0°$
Angle between V_L and V_C	$0°$
Angle between I_L and I_C	$180°$
Desired value of Q	10 or more
Desired value of R	Low
Highest selectivity	High Q, low R_s, high C/L ratio
When f is greater than f_r: Reactance	Capacitive (leading current)
When f is less than f_r: Reactance	Inductive (lagging current)

TUNED CIRCUITS AS FILTERS

Tuned circuits are employed as filters for the passage or rejection of specific frequencies. Bandpass filters and band-rejection filters are examples of this type. Tuned circuits have certain characteristics that make them ideal for certain types of filters, especially where high selectivity is desired. A series-tuned circuit offers a low impedance to currents of the particular frequency to which it is tuned and a relatively high impedance to currents of all other frequencies. A parallel-tuned circuit, on the other hand, offers a very high impedance to currents of its natural, or resonant, frequency and a relatively low impedance to others.

One of the most common electronic uses of tuned circuits is as a filter. As already mentioned, a filter can be described as a circuit designed to separate a specific signal or band of signals from other signals. A filter can also be defined as a circuit designed to separate voltage or current of a particular frequency from other voltages or currents.

Resonant Filters

Resonant filters are frequency-sensitive circuits comprised of at least one capacitor and one inductance, either in series or parallel arrangement. They are merely applications of series or parallel resonant circuits as discussed earlier in this chapter. Resonant filters depend upon a resonance frequency for their maximum response and the Q of the tuned circuit for the width of their response. The lower the Q, the wider the response. All the characteristics and parameters which apply to series or parallel resonant circuits also apply to resonant filters. Resonant filters are usually referred to as *wave traps,* or just as *traps.* The expression is derived from the fact that a filter effectively captures a specific frequency and its side bands and removes them from the circuit in which they are present.

The side band frequencies in bandpass filter circuits range from a minimum of 70.7% of the resonant frequency voltage to the maximum value of the frequency at resonance.

In bandstop or band-elimination filters, the side band voltages extend from a maximum of 70.7% of the resonant frequency voltage to a theoretical minimum of 0V at the resonant frequency.

Bandpass Filters. A bandpass filter, as its name implies, is designed to pass currents of frequencies within a continuous band, limited by an upper and lower cutoff frequency, and substantially to reduce, or attenuate, all frequencies above and below that band. A simple bandpass filter is shown in Fig. 17-15A.

The curves of current versus frequency are shown in Fig. 17-15B. The high Q circuit gives a steeper current curve; the low Q circuit gives a much flatter current curve.

The series- and parallel-tuned circuits are tuned to the center frequency of the band to be passed by the filter. The parallel-tuned circuit offers a high impedance to the frequencies within this band, while the series-tuned circuit offers very little impedance. Thus, the desired frequencies within the band will travel on to the load without being affected; but the currents of unwanted frequencies, that is, frequencies outside the desired band, will meet with a high series impedance and a low shunt impedance so that they are in a greatly attentuated form at the load.

There are many circuit arrangements for both bandpass and band-elimination filters. For the purpose of a brief analysis the bandpass circuit shown in Fig. 17-15A will be considered. Let it be assumed that a band of frequencies extending from 90kHz to

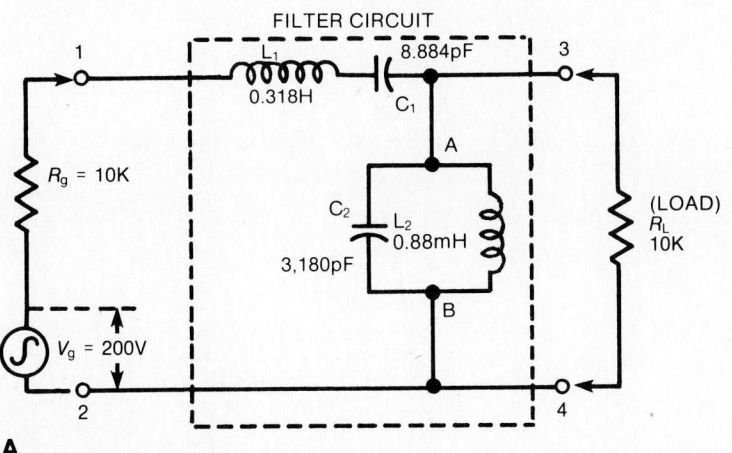

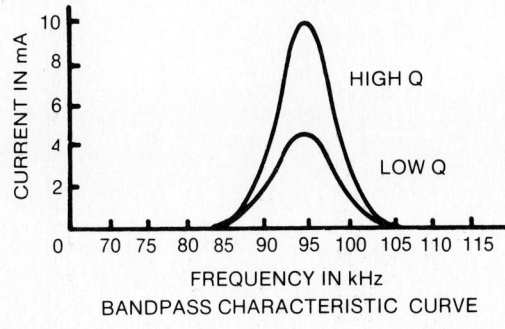

A

B

Fig. 17-15: Bandpass filter.

367

100kHz is to be passed by the filter. For an input and output resistance of 10,000 Ω, the values of inductance and capacitance are as indicated in the figure. The formulas by which these values are obtained may be found in handbooks on the subject.

The resonant frequency of the series circuit, L_1C_1, is:

$$f_r = \frac{1}{2\pi \sqrt{LC}}$$

$$f_r = \frac{1}{6.28 \sqrt{0.318 \times 8.884 \times 10^{-12}}} \approx \frac{0.159 \times 10^{+6}}{\sqrt{2.825}}$$

$$\approx \frac{159,000}{1.68} \approx 95,000 \text{Hz},$$

and for the parallel circuit, L_2C_2,

$$f_r \approx \frac{0.159}{\sqrt{3,180 \times 10^{-12} \times 8.8 \times 10^{-4}}} \approx 95,000 \text{Hz}.$$

Thus, both circuits are resonant at the center frequency of the bandpass filter, the upper limit of which is 100kHz and the lower limit 90kHz.

At resonance the impedances of L_1 and C_1 cancel and maximum current flows through the load, R_L; also, the parallel circuit, C_2L_2, offers almost infinite impedance and may be considered an open circuit. The inherent resistances associated with the filter components are neglected. Thus, at resonance, with an assumed source voltage of 200V and a total impedance of 20,000 Ω (10,000 Ω at the source and 10,000 Ω at the load), the current through the load is approximately

$$I = \frac{V}{R} = \frac{200}{20,000} = 0.01\text{A, or 10mA}$$

Below resonance—for example, at 90kHz—the impedances of L_1C_1 and C_2L_2 are such that with the assumed source voltage of 200V only about 4mA flow through the load. The calculations are as follows:

At 90kHz:

$$X_{L_1} = 2\pi fL_1 = 6.28 \times 90 \times 10^3 \times 0.318$$
$$\approx 180,000\Omega$$

$$X_{C_1} = \frac{1}{2\pi fC_1} = \frac{1}{6.28 \times 90 \times 10^3 \times 8.884 \times 10^{-12}}$$
$$\approx 200,000\Omega$$

$$X_{C_1} - X_{L_1} = j20,000\Omega$$

$$X_{L_2} = 2\pi fL_2 = 6.28 \times 90 \times 10^3 \times 0.88 \times 10^{-3}$$
$$\approx 497\Omega \angle 90°$$

$$X_{C_2} = \frac{1}{2\pi fC_2} = \frac{1}{6.28 \times 90 \times 10^3 \times 3,180 \times 10^{-12}}$$
$$\approx 556\Omega \angle -90°$$

The impedance offered to the flow of current entering and leaving terminals AB is equal to the product of L_2C_2 impedances divided by their sum.

$$Z_{AB} = \frac{X_{L_2}X_{C_2}}{+jX_{L_2} - jX_{C_2}} = \frac{(497 \angle 90°)(556 \angle -90°)}{+j497 - j556}$$

$$= \frac{(497 \times 556) \angle 0°}{59 \angle -90°} \approx 4{,}700 \angle 90°\,\Omega$$

The parallel impedance, Z_{AB34}, of R_{LOAD} and Z_{AB} is equal to their product divided by their sum.

$$Z_{AB34} = \frac{Z_{AB}R_L}{R_L + Z_{AB}} = \frac{(4{,}700 \angle 90°)(10{,}000 \angle 0°)}{10{,}000 + j4{,}700}$$

$$= \frac{(4{,}700 \times 10{,}000) \angle 90°}{11{,}049 \angle 25.2°} \approx 4{,}254 \angle 64.8°$$

$$\approx 1{,}811 + j3{,}849\,\Omega$$

The total impedance, Z_T, of the circuit is equal to the phasor sum of R_g, X_{L1}, X_{C1}, and Z_{AB34}.

$$Z_T = 10{,}000 + j180{,}000 - j200{,}000 + 1{,}811 + j3{,}849$$

$$= 11{,}811 - j16{,}150$$

$$\approx 20{,}000\,\Omega \angle -53.8°$$

$$I_T = \frac{V_g}{Z_T} = \frac{200}{20{,}000} = 0.010\text{A}$$

$$V_{34} = I_T Z_{AB34}$$

$$= 0.010 \times 4{,}254$$

$$= 42.5\text{V}$$

$$I_{load} = \frac{V_{AB34}}{R_{load}}$$

$$= \frac{42.4}{10{,}000} = 0.00425\text{A, or } 4.25\text{mA}$$

Further below resonance the current through the load is even less; and at 75kHz the load current drops to approximately 0.0024mA. The same relative decrease in current occurs through the load with a corresponding increase in frequency. Figure 17-15B is a graph of the load current versus frequency characteristic of the filter shown in Fig. 17-15A.

Figure 17-16 is called the *pi filter* because of the resemblance to the Greek letter π. It employs two parallel resonant circuits in junction with a series resonant circuit. The pi-type arrangement gives a more critical filtering action.

Bandstop Filter. A bandstop filter (or band-elimination filter) is designed to suppress current of all frequencies within a continuous band, limited by the lower and upper cutoff frequencies, and to pass all frequencies below or above that band. A simple band-suppression filter is shown in Fig. 17-17A. This type of filter is just the opposite of the bandpass filter; currents of frequencies within the band are greatly attenuated or weak-

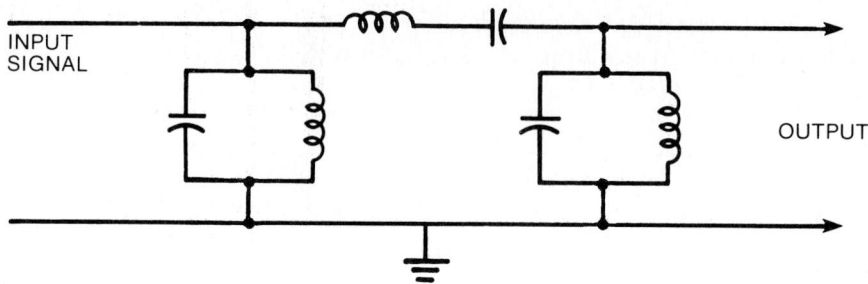

Fig. 17-16: Pi-type bandpass filter.

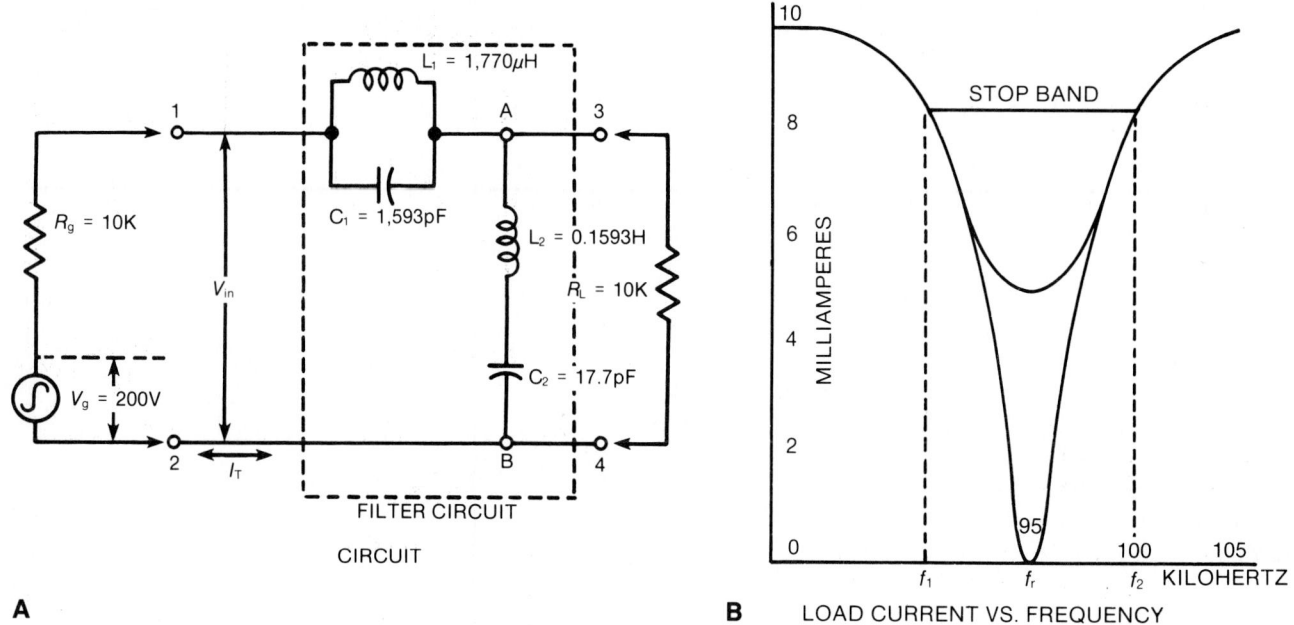

A

B LOAD CURRENT VS. FREQUENCY

Fig. 17-17: Band-elimination or bandstop filter.

ened. The series- and parallel-tuned circuits are tuned to the center of the band to be eliminated. The parallel-tuned circuit in series with the source offers a high impedance to this band of frequencies, and the series-tuned circuit in shunt with the load offers very low impedance; therefore, the signals are blocked and diverted from the load. All other currents, that is, currents at all frequencies outside the stop band (Fig. 17-17B), pass through the parallel circuit with very little opposition unaffected by the series-tuned circuit since it acts as an open circuit at these frequencies.

Representative voltage and component values are shown in Fig. 17-17A. Proof of the selectivity or rejectivity of the circuit, however, is left to the reader. The current versus frequency characteristics of the filter are shown in Fig. 17-17B.

The pi-type bandstop filters can be used where critical band-stopping action is required. In this type of circuit, as in all bandstop filters, the parallel-resonant section is in series between the source of signals and the output, hence it offers a high impedance for the resonant frequency signals. For other sig-

370

nals the parallel circuit presents a low impedance. In shunt with the incoming signals are two series resonant circuits, as shown in Fig. 17-18.

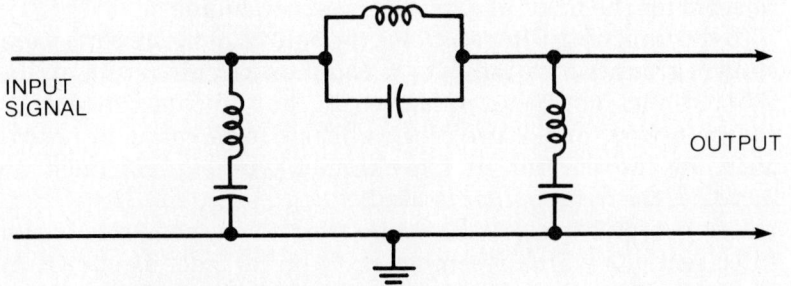

Fig. 17-18: Pi-type bandstop filter.

T- and H-Type Filters. Another design of the resonant filter results in what are termed *T-types,* as shown in Fig. 17-19A and B. In Fig. 17-19A, the T-type bandpass filter is illustrated. Instead of the conventional single series bandpass filter with two shunt parallel types (Fig. 17-16), the T-type uses two series circuits in series with a single parallel resonant circuit in shunt to form a T configuration. Here, the band of signals traveling through the circuit is filtered twice for better series selectivity. The parallel branch has a low impedance for signals above and below resonant frequency just as with the bandpass type previously described. Again, all resonant circuits are tuned to the same frequency for the signals to be passed.

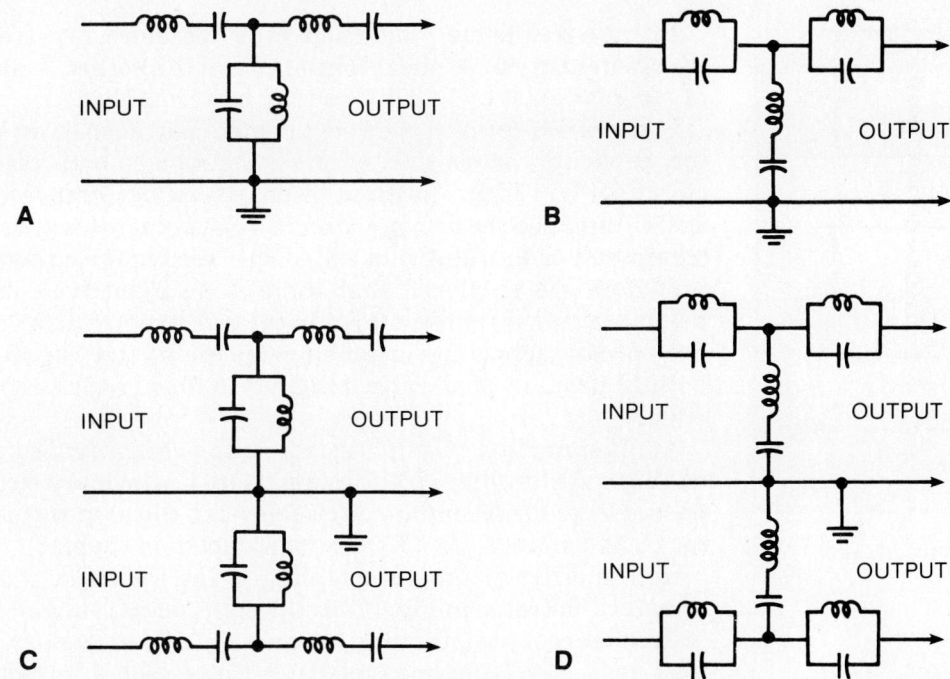

Fig. 17-19: T-type resonant filters and balanced H-types.

371

The band-elimination or bandstop filter in the T-type is illustrated in Fig. 17-19B. Here two parallel resonant circuits are in series, with only a single series resonant circuit in shunt. Again, each circuit is tuned to resonance for the signal to be rejected (or the band of signals around resonance).

In the foregoing filter sections, the bottom of the circuits were shown grounded to indicate a common circuit return path. Ground does not mean actual earth, but an interconnecting common wire, or the use of the chassis for a common return path for the current. If two-channel type signals must be handled, the H-type filter is used.

The H-type filter gets its name from the appearance of the filter sections which resemble an H on its side, as shown in Figs. 17-19C and D. Here, duplicate T sections are used; the lower section is inverted as compared to the upper. This makes the ground terminals common, as shown. The circuit in Fig. 17-19C is thus a bandpass type filter, and the same arrangement could be formed with the bandstop, as shown in Fig. 17-19D. Also, similar H-type filters could be formed with the pi-type filters. The H type shown is also referred to as a *balanced-type filter* because the upper circuit is above ground to the same degree as the lower circuit, thus each is balanced above ground. The T types in Figs. 17-19A and B (and the pi-types shown in Figs. 17-16 and 17-18) are *unbalanced filters* because one signal path is above ground with respect to the other.

Coupling of Electronic Circuits

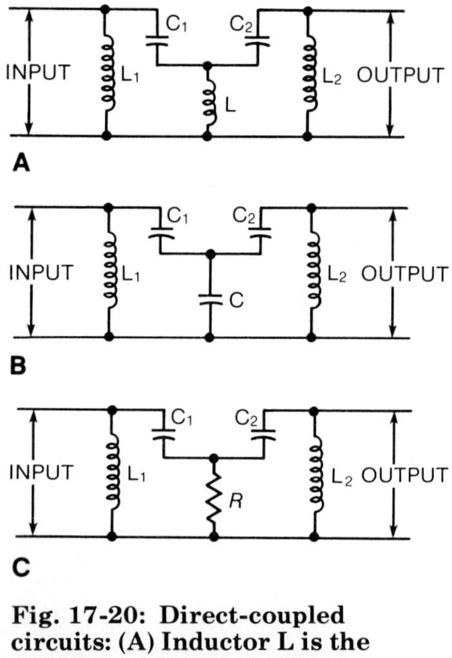

In high-frequency circuits, it often becomes necessary to transfer energy from one resonant circuit to another. A number of methods exist by which this may be accomplished.

In the *direct-coupled method,* a component, such as an inductor, capacitor, or resistor, is made common to both resonant circuits (Fig. 17-20). The input resonant circuit (composed of L_1 and C_1) is called the *primary circuit*. The output resonant circuit (composed of L_2 and C_2) is called the *secondary circuit*. The inductors are so placed that there is no inductive coupling between them. As current flows in the primary circuit, a voltage drop occurs across the common component (L, C, or R). This voltage drop, in turn, causes current to flow in the secondary circuit.

Another method of coupling is the *capacitive-coupling method* illustrated in Fig. 17-21. Energy from the primary circuit is passed on to the secondary circuit through the coupling capacitor C. As before, L_1 and L_2 are not inductively coupled.

Still another method of coupling is the *inductive-coupling method* illustrated in Fig. 17-22. In this case, coils L_1 and L_2 are inductively coupled, forming the primary and secondary windings, respectively, of a transformer. Since such a transformer operates at high frequencies, it is of the air-core or powdered iron-core type. As such, the coupling between the primary and secondary windings is considerably less than unity, and the

Fig. 17-20: Direct-coupled circuits: (A) Inductor L is the common component; (B) capacitor C is the common component; and (C) resistor *R* is the common component.

step-up or step-down effects arising out of turn ratios do not hold. Coupling between the inductors can be varied by placing the windings closer or farther apart.

The reflected impedances or resistances from one circuit to the other become quite important. Assume that both the primary and secondary circuits are tuned to the same resonant frequency and that the coils are placed a considerable distance apart. As current flows in the primary circuit, a certain amount of voltage will be induced in the secondary circuit. Because the coupling is very small, the energy transferred will be small, too, and so will be the voltage output of the secondary circuit.

Since, at resonance, the inductive reactance cancels out the capacitive reactance, both the primary and secondary circuits are resistive in nature. But at all frequencies above and below the resonant frequency (f_r), X_L does not cancel out X_C, and these circuits, therefore, become reactive in nature.

At the resonant frequency the reflected resistance from each circuit reduces the selectivity of the other. At frequencies other than the resonant, the reflected impedance from one circuit to the other upsets the tuning and makes each circuit resonant for some other frequency than the original resonant frequency. Thus, the voltage output becomes maximum for these new frequencies, whereas it drops off at the original resonant frequency because of the effect of the reflected resistance.

With most high-frequency tuned transformer circuits, the coupling is rarely such that all the magnetic lines of the primary cut the secondary. When loose coupling is employed in transformers rather than tight coupling, the mutual inductance formula must be modified to take into consideration the *coefficient of coupling*. The latter is usually designated as *K* and the formula now becomes

$$M = K \sqrt{L_1 L_2}$$

As an example, assume that the coefficient of coupling is 0.5. This actually represents the percentage of coupling, and hence it means that only one-half the lines of force of the primary winding are intercepting the secondary; hence, the coefficient of coupling is really 50%.

When both the primary and secondary of a transformer are tuned, the transfer characteristics of the device depend considerably on the coefficient of coupling. The degree of coupling will affect the bandpass characteristics of the resonant circuits as well as the amplitude of the signals transferred across the transformer. Consider, for instance, the circuit shown in Fig. 17-22, which is typical for the transformer arrangement of an RF amplifier. Here, the generator could be a transistor and the load could be another amplifier stage or a transmission line feeding an antenna. The primary of the transformer consists of L_1, with C_1 forming the primary resonant circuit. Inductance L_2 forms the transformer secondary, with C_2 in shunt for the secondary parallel resonant circuit. The secondary is applied to a load circuit which consumes RF power and can be pictured as a pure resistance. If any reactive components were present in the load, they would affect the resonance and the tuning of the two circuits.

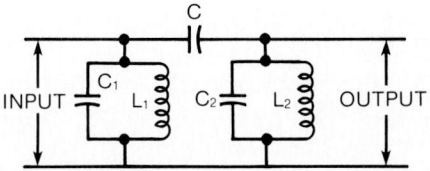

Fig. 17-21: Capacitive-coupled circuit.

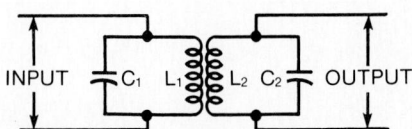

Fig. 17-22: Inductive- or transformer-coupled circuit.

373

Since RF energy is transferred via the inductances only, the induced voltage is working into a series circuit consisting of the inductance and the shunt capacitor. Within the inductor, current and voltage are out of phase; although for the entire secondary circuit, reactive equality will prevail at resonance. With a resistive load, the Z_L will be reflected to the primary and appear as a shunting resistor consuming energy. Thus, the primary current will rise in proportion to the wattage being consumed. The amount of such impedance which is reflected, however, also depends on the degree of coupling (assuming a fixed turns ratio in the transformer). For analytical purposes, three types of coupling exist: *loose coupling, critical coupling,* and *tight coupling*. The latter is sometimes referred to as *overcoupling*.

The effect that each of these degrees of coupling has on the response curve is shown in Fig. 17-23, where frequency is plotted against primary current. Curves for frequency versus secondary current are shown in Fig. 17-24, and these are particularly important since they are the transfer characteristics of the bandpass and represent the signal magnitude felt at the load circuit.

If the coefficient of coupling is very small, the curve for the primary current in Fig. 17-23 (loose coupling) is almost unaffected and appears to be highly selective, with a high amplitude. The secondary current curve in Fig. 17-24 shows that the loose coupling produces a current in the secondary that is of low amplitude and sharply peaked. As the primary and secondary are brought closer together, thus increasing the coefficient of coupling, the curve of the primary current broadens as a greater percentage of the load resistance reflects back into the primary. The increase of resistance which is felt here tends to reduce the circuit current. The secondary current peak, however, becomes greater in amplitude and broader. The changes in both the primary and secondary currents continue as the coupling is increased.

When the coefficient of coupling has been increased to the point where the reflected resistance is equal to the primary resistance, the condition of critical coupling has been reached.

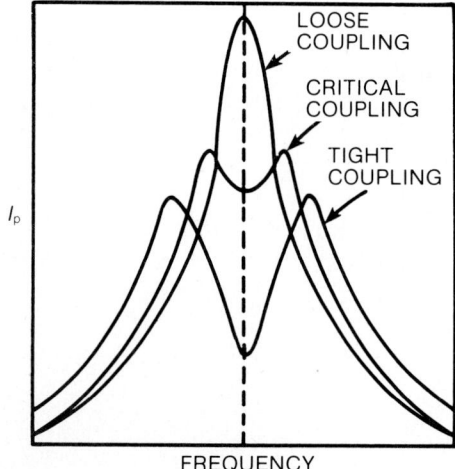

Fig. 17-23: Coupling effects on I_p response curve.

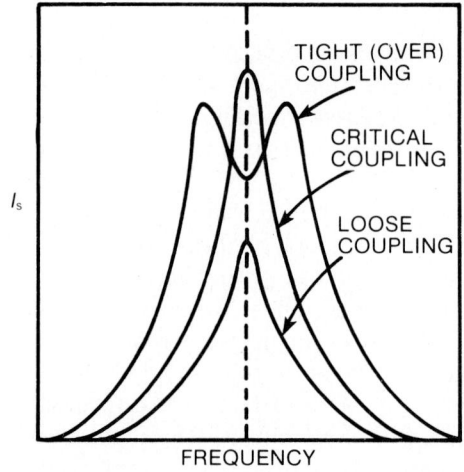

Fig. 17-24: Coupling effects on I_s response curve.

374

Now the secondary current will be at its maximum, as shown in Fig. 17-24. Note that the primary current has two peaks (Fig. 17-23—critical coupling curve) because the increased reflected resistance raises primary impedance and decreases current.

When the coupling is increased beyond the critical point, it becomes tight coupling, or overcoupling. Now the double peaks of the primary current not only decrease in amplitude, but shift off resonance to a greater degree, as shown in Fig. 17-23. Because of the increased reflected resistance added to the primary resistance, the current curve undergoes a pronounced dip at the resonant points. This significant current drop in the primary around the resonant frequency decreases the magnetic flux lines, and less electromotive force is induced into the secondary. Thus, the secondary current curve undergoes a dip also around the resonant frequency, as shown for the tight coupling curve in Fig. 17-24. Obviously this has poor selectivity for the resonant frequency signals, because the circuit has resonant peaks above and below actual resonance. For maximum energy transfer with optimum selectivity, the coefficient of coupling should be critical.

There are occasions where the broad selectivity of the tight coupling circuit is combined with the sharp selectivity of the critical coupling circuit to produce a wide resonance curve with a substantially flat top for even response to frequencies of the passband signals. Such a broad response is required in FM and television systems, where wide bands of resonance are employed. Two or more amplifier stages are so designed that their resonant curves combine to produce the required result. Thus, one amplifier would have tight coupling, and this would be followed by a second amplifier with critical coupling. The two curves are as shown in Fig. 17-25A. The combined curve for the two is shown in Fig. 17-25B, and the slight dips in the flat-top portion produce signal-amplitude variations that are negligible if the entire curve amplitude is sufficiently high.

While the critical coupling for tuned primary and secondary circuits produces maximum current and voltage output, the circuit Q will affect current magnitude and bandwidth. This

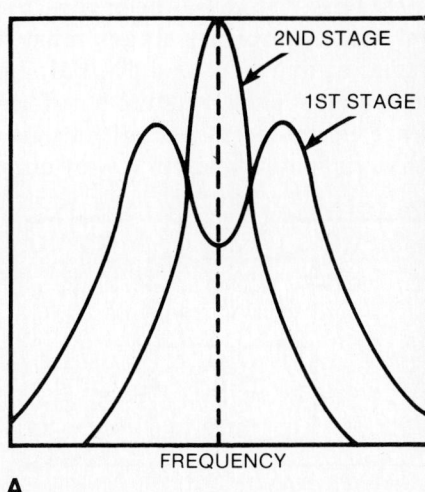

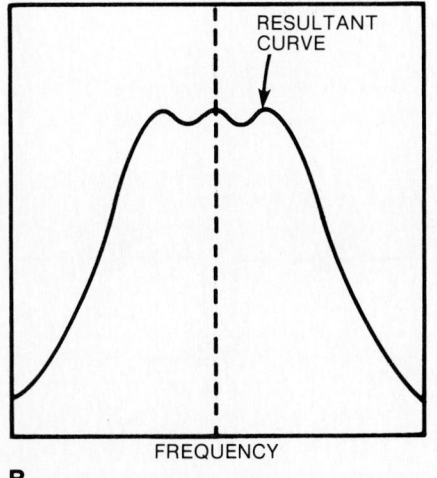

Fig. 17-25: Curves for double-tuned circuits.

conforms to the principles of the Q factor related for single series and parallel resonant circuits discussed earlier. With both the primary and secondary circuits tuned to the resonant frequency, an increase in Q will provide increased selectivity in the transfer characteristics of the two circuits. As with the circuits discussed previously, Q is affected by circuit resistance, which should be kept at a minimum for a high Q. If a low Q is desired, such as in wide-band amplification, the parallel resonant circuits may be loaded with a shunt resistor of sufficient value to drop the Q to the desired level.

At critical coupling the coefficient (K_c) related to the individual circuit Q of the primary (Q_p) and secondary (Q_s) becomes

$$K_c = \frac{1}{\sqrt{Q_p Q_s}}$$

Other Coupling Types. In addition to the transformer coupling methods that employ single parallel resonant circuits in the primary and secondary sides, other methods are sometimes employed for achieving certain specific results. On occasion, for instance, series resonant circuits may be employed, as shown in Fig. 17-26A, where it is necessary to transfer energy from one low impedance source to another. This can be accomplished by utilizing low impedance series-resonant circuits, using the inductors of each for primaries and secondaries.

As shown in Fig. 17-26A, a capacitor is placed in each side of the line to maintain a balanced circuit. This procedure is sometimes used to feed RF energy to a transmission line of the two-wire type. Because of series resonance, current flow is high in each series resonant circuit. Assuming that C_1 and C_2 are connected to the power source, the circulating current builds up the necessary magnetic lines of force around L_1, which then induces a voltage across L_2. The principles of resonance, circuit Q, and magnitude of signal transfer follow the principles previously discussed.

There are occasions where it is convenient to use a double-transformer arrangement, such as shown in Fig. 17-26B. Such a circuit is useful where two distinct types of signals must be handled at different times. This system has been widely used in AM-FM radio receivers, where amplifier stages must handle the transmitted AM signal at one time and the FM signal at another. Instead of two systems, each tuned to a different signal frequency, the two are combined in a single circuit which functions virtually as though separate circuits were employed.

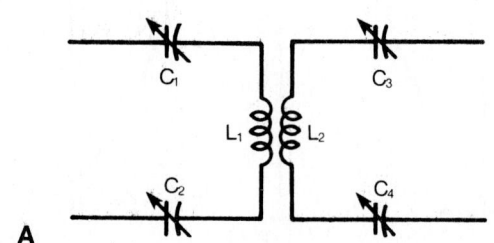

A

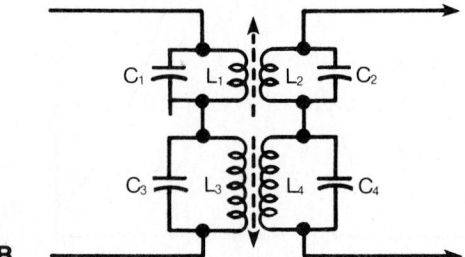

B

Fig. 17-26: Types of resonant circuit couplings.

Assume that the FM signal has a frequency of 10.7MHz. Resonant circuits at the top are tuned to this frequency. If the AM signal is 455kHz, the lower circuits are tuned to this frequency. Now, if the incoming signal is 10.7MHz, the upper resonant circuits have a high impedance and transfer the signal to the next stage. The lower circuits, however, have a very low impedance for this frequency because they are off resonance and, therefore, act as low-resistance paths and do not affect the 10.7MHz signal.

When the 455kHz signal enters the circuit, the lower resonant units have a high impedance and transfer the signal to the succeeding stage. The upper resonant circuits, however, are not tuned to this signal and have a low impedance and are ineffectual with respect to the 455kHz signals. Thus, the system automatically switches its high impedance selective characteristics to the particular resonant circuits required for the signal coming in.

Capacitors C_1 and C_2 as well as C_3 and C_4 are fixed, and tuning is done when necessary by adjustment of the metallic cores of the transformers (shown by the dashed arrows). The coils are mounted in a metal can for shielding purposes, with one tuning core at the top and the other at the bottom. Since these are fixed-frequency transformers, tuning is only necessary on rare occasions. Actual tuning to the various AM and FM stations is done in the tuner circuits with either variable capacitors or inductors.

SUMMARY

The condition at which X_L and X_C of a circuit are equal is called resonance. The frequency at which this condition exists is called the resonant frequency.

The major characteristics of series RLC circuits at resonance are listed in Table 17-2. The major characteristics of parallel RLC circuits at resonance are presented in Table 17-4.

The quality or Q of a circuit is calculated as $Q = X_L/R$. The voltage gain of a series resonant circuit depends on the circuit Q.

The range of frequencies included between the frequencies at which the current drops to 70% of its value at resonance is called the bandwidth. The bandwidth is calculated as $BW = f_r/Q$. Selectivity is the ability of a circuit to accept or pass currents of one frequency and to exclude all others.

Series resonant circuits are most frequently used in filter circuits. Various types of filter circuits use parallel-tuned circuits as well as series-tuned circuits. Parallel-tuned circuits are also in the driver and power stages of transmitters and in the oscillator stage of transmitters, receivers, and frequency-measuring equipment.

There are a number of methods utilized to transfer energy from one resonant circuit to another: direct-coupled method, capacitive-coupling method, and the inductive-coupling method. For the inductive-coupling method, three types of coupling exist: loose coupling, critical coupling, and tight coupling.

377

CHECK YOURSELF (Answers at back of book)

Questions

Fill in the blank with the appropriate word or words.

1. For a given Q, increasing f_r will _____ the BW.
2. If the frequency of the applied voltage is decreased by an amount $1/2Q$ times the resonant frequency, the current in the tuned circuit decreases to _____ of its value at the resonant frequency and leads the applied voltage by _____ .
3. In a parallel RLC circuit, the reactance is _____ if the frequency is greater than f_r.
4. Current is at a _____ in a series resonant circuit at the resonant frequency.
5. The H-type filter is also referred to as a _____ .
6. There are three types of inductor coupling: _____ , _____ , and _____ .
7. Resonant filters are often referred to as _____ .
8. Current is at a _____ in a parallel resonant circuit at the resonant frequency.
9. The selectivity of a tuned circuit depends upon the _____ of the tuned circuit.
10. In a series RLC circuit, the reactance is _____ when the frequency is greater than the resonant frequency.
11. Regulation of a bandwidth may be accomplished by variations of the _____ .
12. The IF transformers of radio and television receivers employ _____ .
13. A _____ filter is designed to pass currents of frequencies within a continuous band and to reject all other frequencies.
14. In a series RLC circuit, the reactance is _____ when the frequency is less than the resonant frequency.

Problems

1. A series circuit has an inductance of 20mH and a capacitance of $2\mu F$. Calculate the resonant frequency of the circuit.
2. Find the quality of a circuit with L = $200\mu H$, C = 50pF, and R = 10Ω.
3. Calculate the bandwidth for the circuit in the previous example.
4. The resonant frequency of a circuit is 15MHz (Q = 10). At what lower frequency will the current in the tuned circuit decrease to 0.707 of its resonant value?
5. What is the resonant frequency of a parallel circuit with a $10\mu H$ coil, a 10pF capacitor, and 5Ω of resistance?

CHAPTER 18

Electron Tubes

Vacuum or electron tubes were *once* the heart of all electronic equipment. Today, however, solid-state devices have replaced them in most modern electronic circuits. Since electron tubes behave similarly to many of the solid-state devices, it is a good idea to have some knowledge of how they function. In addition, electron tubes still perform important tasks as amplifier/rectifiers in communications, stereo equipment, and navigation systems.

ELECTRON TUBE DIODES

The diode or two-element tube is the simplest type of electron tube. It consists of an emitter of electrons (called the *cathode*) and a positively charged collector (known as the *anode*) sealed in a tube or bulb called the *envelope*. These envelopes are made of glass, metal, ceramic, or a combination. They are constructed in many different sizes and shapes (Fig. 18-1).

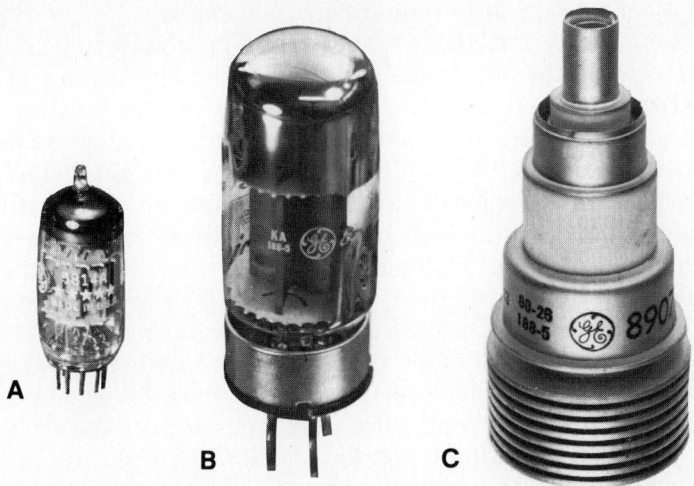

Fig. 18-1: Various types of electron tubes: (A) miniature, glass-envelope type; (B) octal-base, metal-envelope type; and (C) metal-ceramic power tube. (Photos furnished by courtesy of General Electric.)

The cathodes of most diodes are formed of compounds or oxides which lend themselves to rapid and easily accomplished emission of electrons. Anode material does not readily lend itself to such behavior.

Electron Emission. When the cathodes of most of the electron tubes discussed in this chapter are heated by an electric current, they immediately start to emit electrons. (This emission of electrons from solid materials is known as *thermionic emission* or the *Edison effect*.) Although this emission is low, the potential between the anode and cathode can be measured (before an external potential difference is applied between them) if adequately sensitive equipment is used. Such an indication is achieved by comparing the stable or reference state of the anode material to the electron emission from the cathode material.

Another method of emission is to supply an external electric field strong enough to exert an attraction upon the electrons near the surface of the metal sufficient to overcome the surface barrier potential. The electrons are literally pulled away. This method is called *field emission.*

When certain metals are exposed to light, enough light energy or flux may be absorbed by the surface electrons so that they overcome the barrier potential and escape into space. This method of freeing electrons is called *photoelectric emission.*

Finally, if the metal is bombarded by particles with sufficient force, some of the surface electrons may absorb sufficient kinetic energy to escape into space. This method is called *secondary emission.* Any or all of these methods may be employed to free electrons so that they may flow through the electron tube.

Electron Flow. At the instant a difference of potential is applied externally between anode (positive) and cathode (negative), an electric field exists between the electrodes of the diode. Electrons, now under the directed influence of the electric field, flow toward the anode through the source to cathode and in the complete cycle path as long as the field exists.

When the electrons flow through a conductor, such as a copper wire, they are confined within the boundaries of the conductor and, therefore, control over them is limited. But when electrons move through a vacuum or gas, they are much freer and can be more easily controlled by external means.

In electron tube diodes, the intensity of current flow is directly proportional to the level of potential difference between the cathode and anode. This potential difference may also be described as the electric field which exists between the cathode and anode or such field's strength and/or density.

Since the material of an electron tube's anode is not conducive to the emission of electrons and this material is not heated to prepare it for such emission, current flow from the anode to cathode does not normally occur. Hence, the electron tube diode performs as a unidirectional valve or switch.

Should a difference of potential high in value and reversed in polarity be placed across an electron tube diode, the "dielectric" between anode and cathode may be "broken down," and an excessive quantity of reverse current flow will naturally occur.

Electron tube reverse current breakdown may be closely compared to the breakdown of a capacitor's dielectric. Each is due to excessive electric field strength or extraordinarily high difference of potential across device elements. Note that in each case of breakdown there exists some physical means of current conduction between the poles of potential difference. In capacitors, this medium is the dielectric; in electron tubes it is also a form of dielectric. Specifically, in electron tubes, the evacuated envelope is not an absolute void or "perfect" vacuum.

To improve the vacuum qualities of some electron tubes, a "getter" is employed; that is, chemicals are placed on a small platform inside the tube and are exploded after the tube is sealed in order to absorb any remaining gases left in the envelope.

Operation of the Electron Tube Diode

As has been stated, the evacuated envelope surrounding the electron tube diode is not a perfect vacuum; therefore, some matter remains in the existing space. When the cathode is heated, it begins to emit electrons. Upon application of an electric field, these electrons are directed through the material of the "vacuum" to the anode, through the source, and back to the cathode.

A question which frequently arises concerning the operation of electron tube diodes is the one of improving its efficiency by not evacuating the envelope. Such thinking results from the fact that residual matter in the evacuated tube provides the current conduction path from cathode to anode. However, increasing this matter by leaving air in the device has two fundamental obstacles to effective operation. One, the oxygen in air would encourage the rapid burnout of the heater and, two, an abundance of matter in the tube would allow excessive electron dissipation in the forms of ionization and heat, even should problem one be overcome. Ionization, incurred in the process just discussed, develops whenever electron flow occurs. Visible evidence of ionization is manifested when the ionized matter is principally composed of an inert gas or gases.

Cold Cathode Electron Tube Diodes

In some cases electron tube diodes are constructed without facilities for heating the cathode. Such diodes are frequently used as voltage regulators as well as conventional rectifiers.

Operation of cold cathode diodes occurs when a difference of potential is placed across the diode anode (positive) and cathode (negative) causing an electric field to exist between these elements. Since cathode material is of a relatively high electron emission level, some of these electrons are started on their path to the anode. Usually, the medium of conduction is an inert gas such as neon, argon, krypton, or xenon. When electron flow in these gases occurs, the gases glow in one of several colors and are said to have become ionized.

Since some energy is dissipated in the form of heat while cold cathode electron tubes are operating, the cathode temperature does rise; therefore, the term "cold" cathode is not completely accurate and at best only indicates the lack of a cathode heating element in such electron tubes.

As current flow through an inert gas diode increases, more energy is dissipated in ionizing the gas. Within the limits established by the quantity and type of gas, such diodes will maintain a constant voltage drop across their elements, though current flow through the devices varies.

Gas Electron Tube Diodes

A problem frequently occurring with diodes is *space charge*. This is the repelling action caused by the cloud of electrons around the cathode on electrons flowing to the anode (plate). A relatively high plate voltage is required to remove the electrons of the space charge and permit the emitted electrons a free path to the plate. If it were not for the space charge, any small plate voltage could produce the maximum space current.

One method for overcoming the space charge is to insert a very small amount of some inert gas, such as nitrogen, argon, helium, neon, or mercury vapor, into the tube after the air has been removed. Such tubes are known as *gas electron tubes*. As the electrons emitted by the cathode travel to the plate, they collide with the gas atoms. Since the velocity of these electrons is high, the collisions knock electrons off the gas atoms and thus form positive ions. In turn, the positive ions tend to neutralize the electrons of the space charge, thereby dispelling it. The emitted electrons now can reach the plate more easily.

In manufacture, the pressure of the gas is carefully controlled. The atoms of gas will not form ions (that is, ionize) unless they are struck by electrons with sufficient velocity. Since the velocity of the electrons moving from the cathode to the plate depends upon the potential difference between the two electrodes, ionization cannot take place until this potential difference reaches a certain minimum (called the *firing, striking,* or *ignition potential*). This firing potential depends upon the type of gas employed and its pressure. For typical tubes it ranges from about 10V to 25V.

Once started, ionization maintains itself at a voltage considerably lower than the firing potential of the gas; however, a certain minimum voltage is required to maintain the ionization. Should the voltage across the tube fall below this minimum value, the gas becomes deionized, or extinguished, and the current flowing through the tube drops to practically zero.

Cathode Construction

This chapter, for the most part, will be concerned with electron tubes which employ thermionic emission. Thermionic cathodes are heated in one of two ways—directly or indirectly.

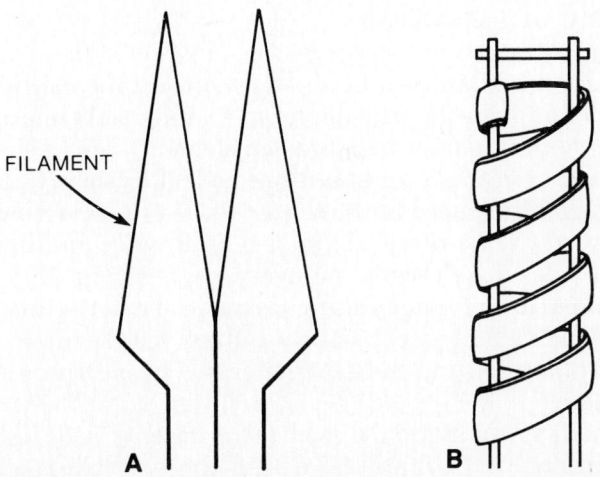

FILAMENT

Fig. 18-2: Filament emitters: (A) thin-wire type and (B) heavy-ribbon type.

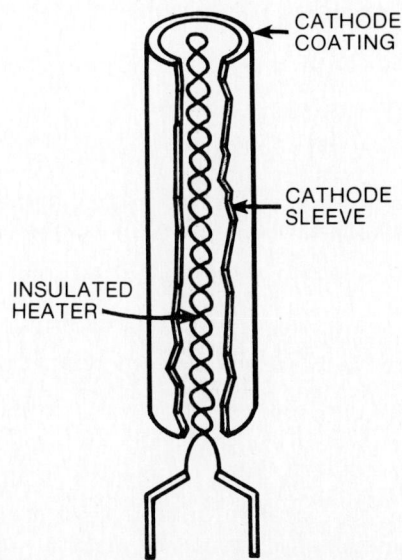

CATHODE COATING

CATHODE SLEEVE

INSULATED HEATER

Fig. 18-3: Cutaway view of the indirectly heated cathode.

A directly heated cathode is one in which the current used to supply the heat flows directly through the emitting electrode (Fig. 18-2). In an indirectly heated cathode, the heating current does not flow through the emitting electrode (Fig. 18-3).

The directly heated cathode, commonly called a filament, has the advantage of being fairly efficient and capable of emitting large amounts of energy. However, due to the small mass of the filament wire, the filament temperature fluctuates with changes in current flow. If an AC source is used to heat the filament, undesired hum may be introduced into the circuit. This is especially evident in low level signal circuits.

A relatively constant rate of emission under mildly fluctuating current conditions may be obtained with an indirectly heated cathode. This type of cathode is in the form of a cylinder, in the center of which is a twisted, electrically insulated wire called the heater. The cathode is maintained at the correct temperature by heat radiated from the heater. The emitting material in this type of cathode remains at a relatively constant temperature, even with an alternating heater current. Figure 18-4 illustrates the schematic symbols for directly and indirectly heated cathodes, as well as a cold cathode.

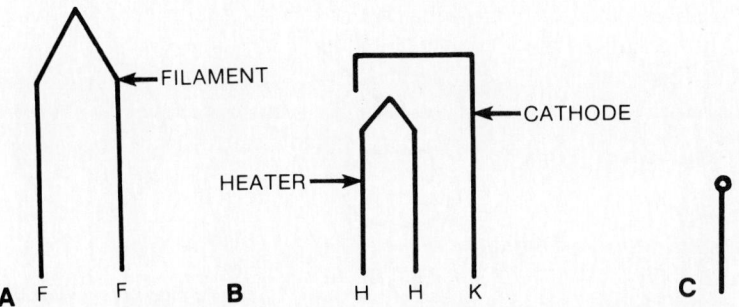

FILAMENT

HEATER

CATHODE

A F   F   B   H H K   C

Fig. 18-4: Schematic symbols for (A) directly and (B) indirectly heated cathodes. The symbol for a cold cathode is shown in C.

Diode Construction

As the name implies, the diode electron tube contains two electrodes or elements (the heater in an indirectly heated cathode type of tube is not considered an element). These electrodes are called the cathode and plate or anode. Physically, the cathode is normally placed in the center of the tube structure and is surrounded by the plate. Examples of directly and indirectly heated diodes are shown in cutaway views in Fig. 18-5.

The plate, as previously stated, is made of a material which is not conducive to the emission of electrons. It must also be capable of dissipating the heat generated by electrons striking the plate.

The schematic symbols used to represent diode tubes are shown in Fig. 18-6. Figure 18-7 details the symbolic representation of several types of gas-filled diodes. In many cases, two or more diodes may be included in the same envelope in order to conserve space. A tube which contains two plates and one or two cathodes is called a duo-diode or twin diode. Figure 18-8 illustrates the construction and schematic symbols of duo-diode electron tubes.

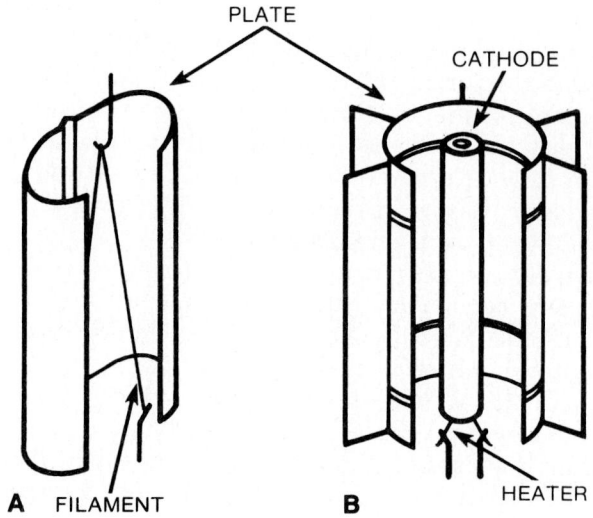

Fig. 18-5: Cutaway views of (A) directly and (B) indirectly heated diodes.

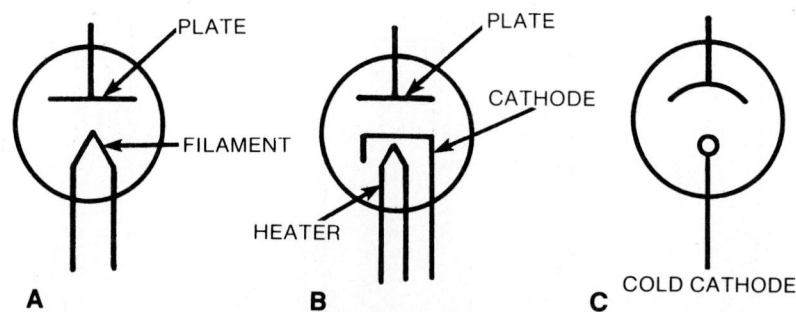

Fig. 18-6: Schematic symbols used to represent vacuum diode tubes: (A) directly heated; (B) indirectly heated; and (C) cold cathode.

384

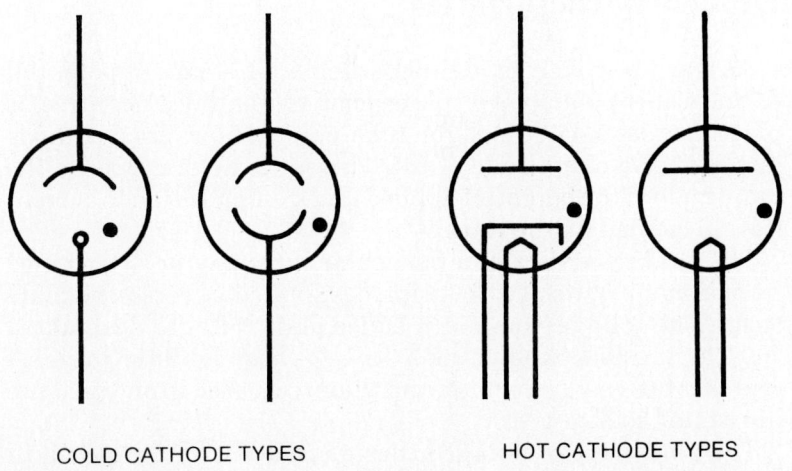

COLD CATHODE TYPES   HOT CATHODE TYPES

Fig. 18-7: Schematic symbols used to represent gas-filled diode tubes.

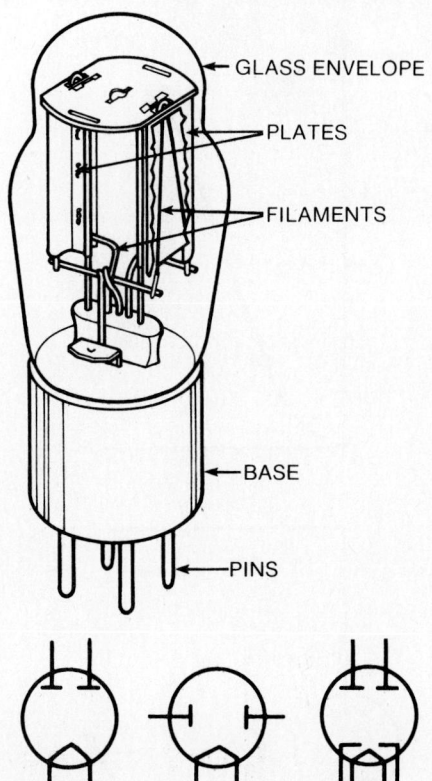

GLASS ENVELOPE

PLATES

FILAMENTS

BASE

PINS

As a rule, electron tubes do not have as long a life expectancy as resistors, capacitors, or other circuit components. This is due, in large part, to wearing out or breaking down of the oxide coated filament or cathode, resulting in a weak or defective tube which must then be replaced. To provide for easy removal and replacement of the tube, the base of the tube (Fig. 18-8) is constructed in the form of a plug, which is inserted into a socket on the chassis. The electrical connections between the tube elements and the circuit are completed through the plug terminals, called *pins* (Fig. 18-8).

There are various types of tube bases, containing different numbers and sizes of pins. Each type of tube base has some kind of guide or key to prevent the tube from being plugged into the socket improperly. For ease of circuit tracing, the tube pins are assigned numbers. At the bottom of the tube or socket, the pins are numbered in a clockwise direction, beginning with pin number one at the key or guide. The pin numbering systems for several types of tubes are shown in Fig. 18-9.

Fig. 18-8: Construction and schematic symbols of duo-diodes.

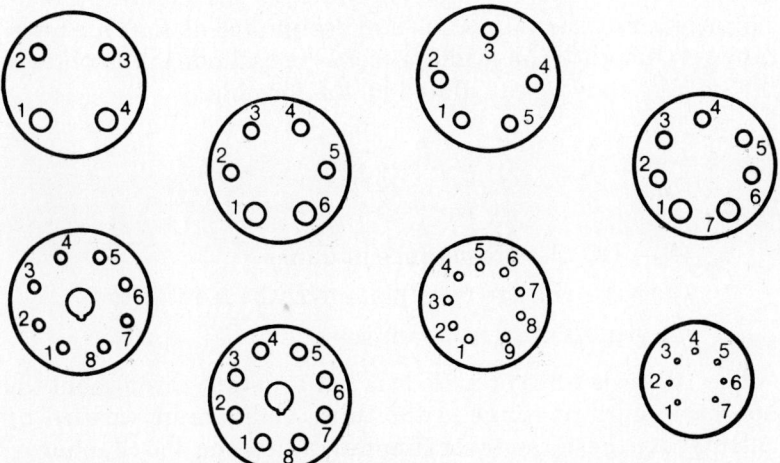

Fig. 18-9: Pin numbering systems for several tube types.

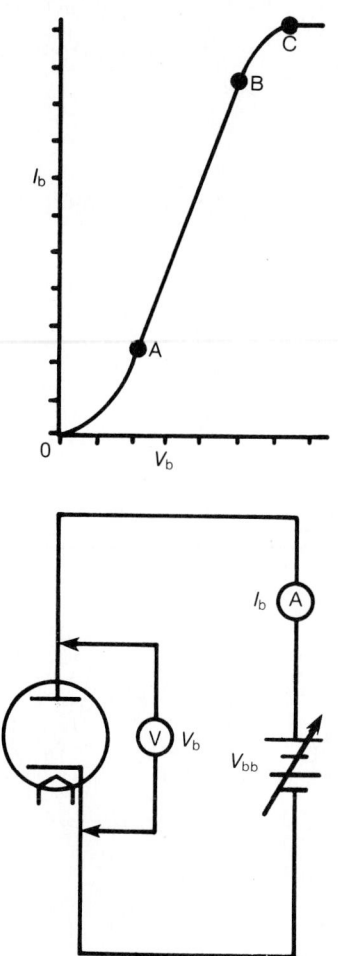

Fig. 18-10: Representative
V_b - I_b **curve and the circuit used to obtain it.**

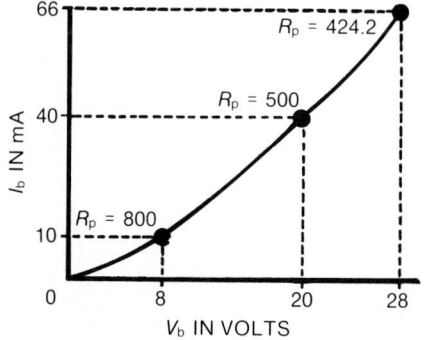

Fig. 18-11: Calculating R_p from the V_b - I_b characteristic curve.

Diode Characteristics

As was previously explained, when a difference of potential of the proper polarity (i.e., plate positive—cathode negative) is placed across a diode electron tube, current flows from cathode to plate. The magnitude of this current is determined by the amplitude of the potential applied and the opposition to current flow presented by the tube.

The characteristics of a particular tube may be determined from a graph of its plate current, designated I_b, versus its plate to cathode voltage (commonly called plate voltage), designated V_b. The graph is known as a V_b - I_b characteristic curve. A representative V_b - I_b curve and the circuit used to obtain it are illustrated in Fig. 18-10.

The plate supply voltage is designated V_{bb}. In the circuit in Fig. 18-10, V_b is equal to V_{bb}. However, in practical circuits, V_b will not equal V_{bb}, and the two voltages should not be confused. These terminologies were derived from the now obsolete use of batteries to supply the electrode potentials. The plate to cathode potential was supplied by what was called a "B" battery, hence the designations V_{bb} and V_b are now used.

To obtain a V_b - I_b characteristic curve, the plate voltage (V_b) is increased from zero in steps, and the corresponding increases in plate current (I_b) are plotted as shown in Fig. 18-10.

The curve obtained from the circuit in Fig. 18-10 consists of three distinct regions: O to A, A to B, and B to C. Note that in the regions of O to A and B to C appreciable nonlinearity exists. The region from A to B is essentially linear.

As the curve is plotted, a point is finally reached (point C) where a further increase in plate voltage no longer produces an increase in plate current. At this point the plate is attracting all the electrons the cathode is capable of emitting. This condition is known as saturation, and at this point the upper limit of the tube's conduction capabilities have been reached.

Note that the current flow through a tube is some finite value. This indicates that the tube is effectively offering opposition to current flow. The opposition offered to the flow of direct current, by an electron tube, is called *DC plate resistance*. DC plate resistance is measured in ohms and designated as R_p. This resistance is thought of as existing from the cathode to the plate of the tube. R_p may be calculated by the formula

$$R_p = \frac{V_b}{I_b}$$

Where:

R_p = DC plate resistance in ohms

V_b = the voltage from plate to cathode in volts

I_b = plate current in amperes.

The DC resistance of a diode is not constant throughout the operating current range of the tube. This can be verified by calculating the R_p at several points along the V_b - I_b characteristic curve, as shown in Fig. 18-11. However, it is practically constant throughout the linear portion of the curve.

386

DC plate resistance is a "static" characteristic. A static characteristic is one which is obtained with DC potentials applied to the tube. Another characteristic, AC plate resistance, is a dynamic characteristic; that is, it is indicative of tube performance under signal, or AC, conditions.

The term AC plate resistance is defined as the opposition offered by an electron tube to the flow of alternating current. The symbol designating AC plate resistance is r_p. The r_p of a diode electron tube may be found with the use of the formula

$$r_p = \frac{\Delta v_b}{\Delta i_b}$$

Where:

r_p = AC plate resistance in ohms

Δv_b = a small change in plate voltage

Δi_b = a small change in plate current.

The values required to calculate r_p may be obtained from the V_b - I_b characteristic curve of the tube concerned. This is illustrated in Fig. 18-12. AC plate resistance also varies throughout the operating range of the tube; however, it, too, is essentially constant in the linear portion of the curve.

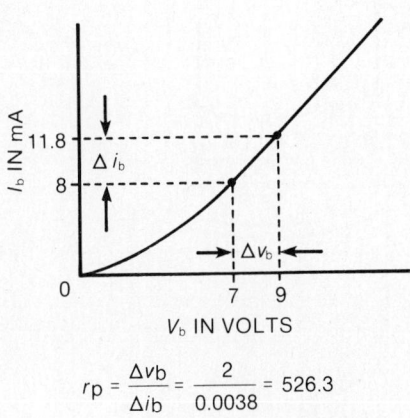

$$r_p = \frac{\Delta v_b}{\Delta i_b} = \frac{2}{0.0038} = 526.3$$

Fig. 18-12: Calculating r_p from the V_b - I_b characteristic curve.

Diode Ratings

Each diode has certain voltage, current, and power values which should not be exceeded in normal operation. These values are called ratings. The following are the most important diode electron tube ratings:

• *Plate dissipation* is the maximum average power, in the form of heat, which the plate may safely dissipate.

• *Maximum average current* is the highest average plate current which may be handled continuously. It is based on the tube's permissible plate dissipation.

• *Maximum peak plate current* is the highest instantaneous plate current that a tube can safely carry, recurrently, in the direction of normal current flow.

• *Peak Inverse Voltage (PIV)* is the highest instantaneous plate voltage which the tube can withstand recurrently, acting in a direction opposite to that in which the tube is designed to pass current (i.e., plate negative—cathode positive).

It should also be noted that the correct filament or heater voltage and current are required for proper operation of an electron tube. If heater current is too low, the cathode will not emit sufficient electrons. This will result in low emission, and the tube will be incapable of proper operation. Excessive heater or filament current may reduce the life of the tube or destroy the heater or filament.

Where small diodes (Fig. 18-13) are used, sufficient heat can be removed by radiation into the surrounding air. For some of the larger tubes that pass relatively large quantities of current, however, cooling by forced air or circulating water is employed

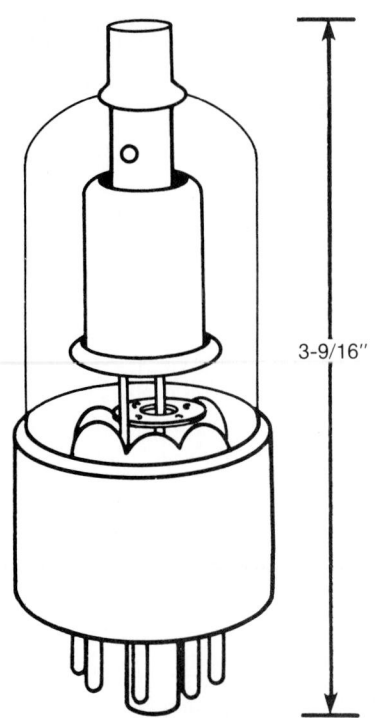

Fig. 18-13: Typical vacuum diodes.

(Fig. 18-14). In the case of the air-cooled tube, a finned radiator helps remove the heat from the plate. In the water-cooled tube, the plate is surrounded by a water jacket.

Voltage Regulator (*VR*) Tubes. Cold cathode gas tubes are frequently used in voltage regulator circuits (Fig. 18-15). The diode is connected in series with a limiting resistor across a power supply. The load is connected across the *VR* tube. When the power supply voltage changes, the degree of ionization and the *VR* tube resistance will change, which, in turn, maintains a constant voltage across the tube.

Voltage regulator tubes are usually designed to operate within limited ranges, generally at voltages from about 50V to 150V and currents in the order of 30mA to 40mA. It gets its name (*glow-discharge*) from the light glow produced as positive gas ions meet with electrons within the tube and become neutral atoms, giving off light in the process.

Since, within the limits determined by the tube, the *IR* drop remains constant despite fluctuations of the current, there is a constant source of voltage. Hence the glow-discharge tube may also act as a source of reference voltage required for certain applications. A tube used in this manner is called a *voltage-reference tube*.

A

B

Fig. 18-14: (A) Forced-air-cooled power tube. (Photo furnished by courtesy of Raytheon Company.) (B) Water-cooled power tube. (Photo furnished by courtesy of Amperex Electronic Corporation.)

Mercury Vapor Rectifiers. These tubes (Fig. 18-16) are used in communication transmitters because they possess the following characteristics:

1. Low constant voltage drop (approximately 15V).
2. Use oxide coated cathode.
3. High current rating.
4. Low filament power.
5. Economical.
6. Low inverse peak voltage (exceeding it causes arc-back or flashback).
7. Filament may be damaged if not preheated before plate voltage is applied.
8. When mercury vapor rectifiers are connected in parallel, small resistors are connected in series with the plate to ensure ionization of both tubes.

In the operation of a mercury vapor rectifier, the positive potential on the plate causes electrons to flow. These electrons on their way to the plate strike the gas molecules, ionizing the gas.

TRIODES

If a third electrode, called a control grid, is placed between the cathode and plate of an electron tube, a device known as a triode (three element tube) is created. Figure 18-17 illustrates a cutaway view and schematic symbol of a triode tube.

The addition of a control grid makes possible the use of an electron tube as an amplifier. The grid usually consists of fine wire wound on two support rods, extending the length of the cathode. Two types of grid structure are illustrated in Fig. 18-18.

The grid wire is made of a material which has a low level of electron emission. The spacing between the turns of the grid wire is large compared to the size of the wire. This allows the passage of electrons from the cathode to the plate to be practically unobstructed by the grid structure.

Control of Plate Current

The purpose of the control grid is to control the flow of plate current. The grid acts as an electric shield between cathode and plate; therefore, both grid and plate potentials are effective in controlling plate current. However, the grid voltage has much more control over plate current than does the plate voltage. It is this fact which accounts for the triode's ability to amplify.

SPECIAL-PURPOSE ELECTRON TUBES

There are a number of electron tubes used in the electronic industry, each designed to serve some special function. Although most of these do not resemble the tubes already discussed, they all operate upon the same basic principle—electrons emitted by a cathode flow in a one-way path to a positively charged anode.

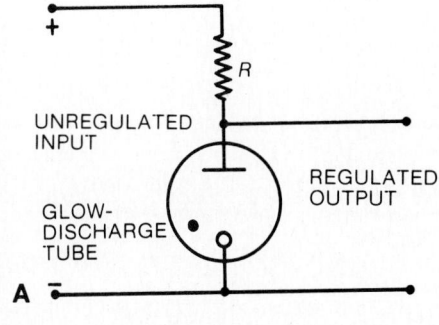

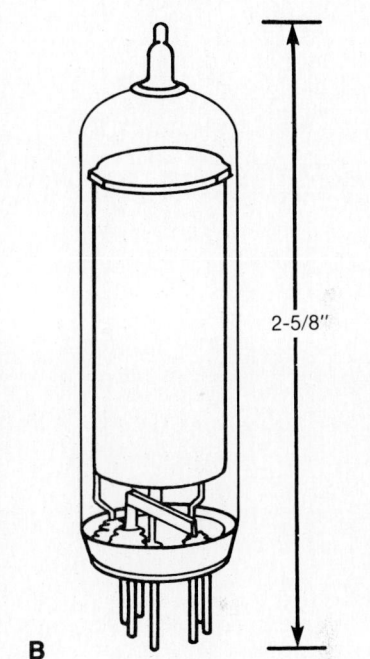

Fig. 18-15: (A) Diagram of a cold cathode gas diode used as a voltage regulator. (B) Typical glow-discharge tube.

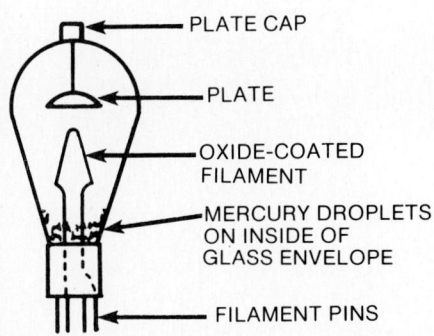

Fig. 18-16: Transmitter type mercury vapor rectifier tube.

389

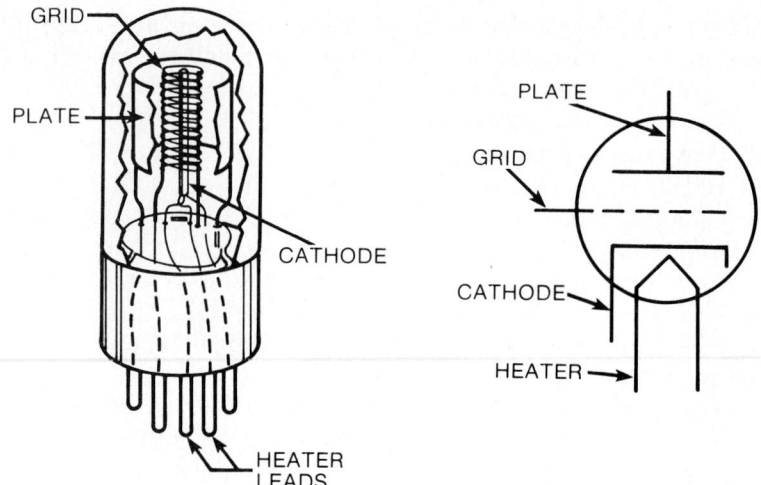

Fig. 18-17: Cutaway view and schematic symbol of a triode tube.

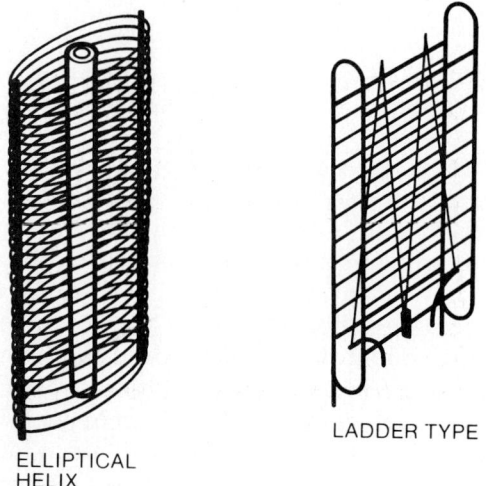

ELLIPTICAL HELIX

LADDER TYPE

Fig. 18-18: Typical grid structures.

Photoelectric Tubes

A photoelectric cell is an electron tube that causes light energy to be converted into electrical energy. Any change in light diversity, color, or wavelength will produce corresponding changes in electric current and voltage.

There are three general types of photoelectric cells or tubes:

1. *Photo conductive*—resistance varies with light flux.
2. *Photo voltaic*—voltage is generated with light flux.
3. *Photo emission*—light flux causes electrons to be emitted from a cold cathode. The photo emission type is the most commonly used in electronic circuits.

The cathode of the photo emission type electron tube generally consists of a half-cylinder of metal coated on its inner surface with a photosensitive material such as potassium, sodium, calcium (alkaline material), barium, and rare earths. The cathode is not heated and the emission of electrons depends solely upon the amount and wavelength of the light flux. The

390

shorter the wavelength, the greater the energy imparted to the electrons. Also, the photosensitive materials used for coating the cathodes of photo emission type tubes are each sensitive to a different range of wavelengths.

The anode usually is a thin metal rod lying along the central axis of the cathode (Fig. 18-19). To keep out unwanted light, the inner surface of the envelope may be masked, except for a small, circular window facing the inner surface of the cathode through which the light may enter, or else a shield with a similar window may be placed over the entire tube. Both the cathode and anode are enclosed in the glass envelope. If the envelope contains a vacuum, the tube is known as a *vacuum photoelectric tube*. If, instead of a vacuum, a little gas is added, it is called a *gas photoelectric tube*. The characteristic curve for a typical vacuum and gas photoelectric cell is given in Fig. 18-20.

Cathode-Ray Electron Tube

One of the most ingenious of all electron tubes is the cathode-ray tube (CRT). It is used in television (Fig. 18-21), radar receivers, and in the cathode-ray oscilloscope.

The cathode-ray tube (Fig. 18-22A) is a special type of electron tube in which electrons emitted by a cathode are focused and accelerated to form a narrow beam having high velocity. The direction of this beam is then controlled and allowed to strike fluorescent screen, whereupon light is emitted at the point of impact to produce a visual indication of the beam position. The electronic process of forming, focusing, accelerating, controlling, and deflecting the electron beam is accomplished by the following principal elements of the cathode-ray tube:

1. The electron-gun.
2. Deflection plates.
3. A fluorescent screen.

A simplified form of the electron gun, shown in Fig. 18-22B, provides a concentrated beam of high-velocity electrons. The cathode, when properly heated, emits electrons. These electrons are attracted toward the accelerating and focusing anodes because of their high positive potential with respect to the cathode. In order to reach these anodes, the electrons are forced to pass through a cylindrical control grid, closed at one end except for a tiny circular opening, which concentrates the electrons and starts the formation of a beam. Electrons leaving this small aperture are strongly attracted by the positive potential on the focusing anode (anode No. 1) and accelerating anode (anode No. 2), which are also cylindrical in shape and have small openings to permit beam passage.

The electron beam is deflected after leaving the electron gun proper through the use of two pairs of parallel plates located on each side, above and below the beam. These plates, called the vertical and horizontal deflection plates, are oriented such that the electron beam must pass between them. If no electric potential exists between these two plates, the electron beam will

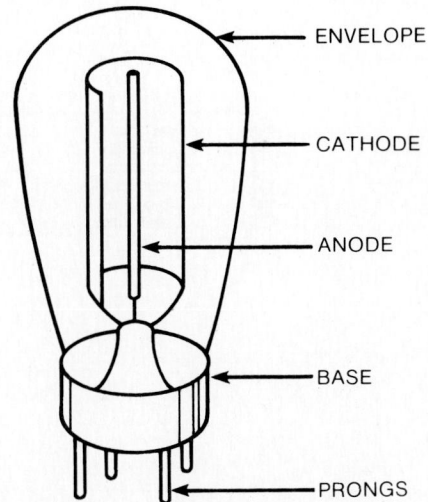

Fig. 18-19: The photoelectric tube.

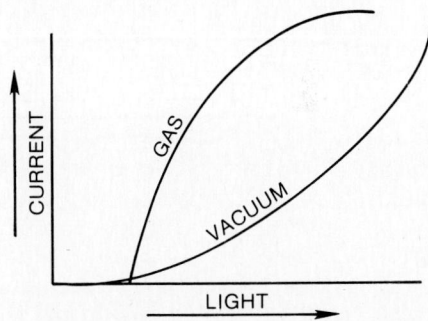

Fig. 18-20: Characteristic curve of a typical vacuum and gas photoelectric cell.

Fig. 18-21: Cathode-ray tube used in the television receiver. (Photo furnished by courtesy of Raytheon Company.)

391

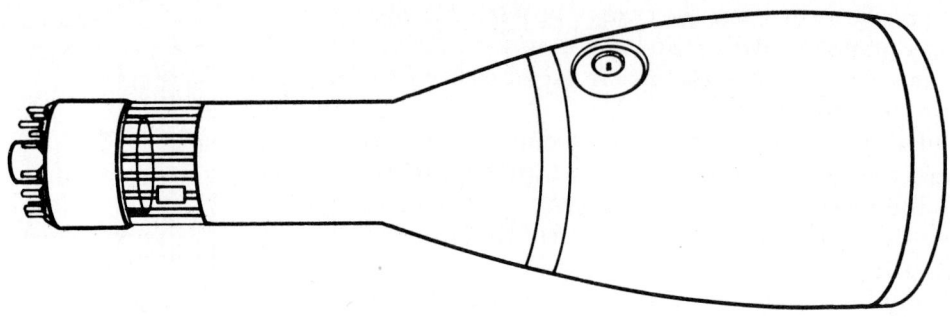

A

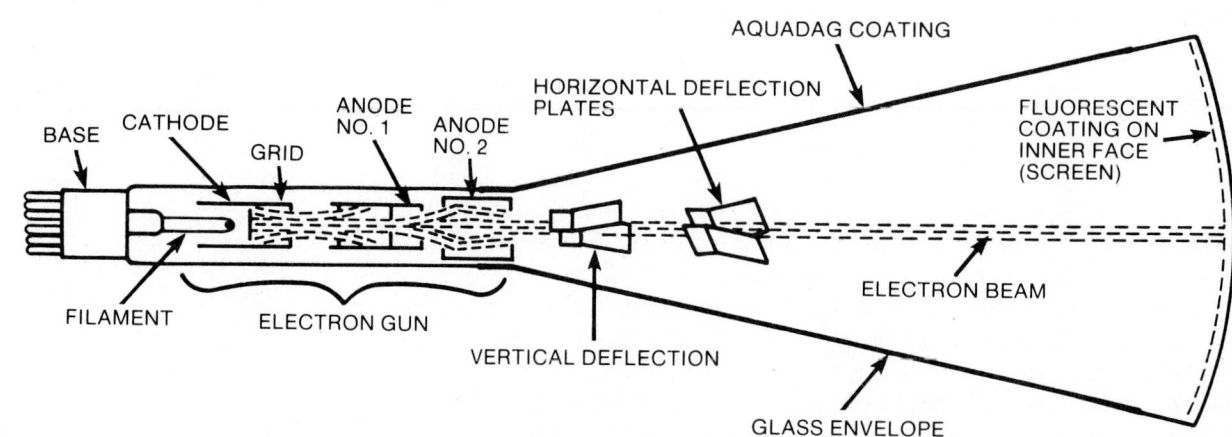

B

Fig. 18-22: (A) Typical cathode-ray tube; (B) cross-sectional view of cathode-ray tube employing an electrostatic deflection system.

continue moving forward and strike the screen at or near the center. If a source of potential is applied to one or both sets of plates, the electron beam will "veer," or be bent away, from the center of the screen. Therefore, the electron beam is shaped and accelerated by the electron gun and bent, or deflected, by the deflection plates (Fig. 18-23).

In order to convert the energy of the electron beam into a visible light, that area where the beam strikes, the screen, is coated with a phosphor chemical, which, when bombarded by electrons, has the property of emitting light. This property is known as "fluorescence." The intensity of the lighted spot on the screen depends upon two factors:

1. The speed of the electrons in the beam, and
2. The number of electrons that strike the screen at a given point.

The amount of light per unit area which the phosphor is capable of emitting is limited; and once the maximum has been reached, any further increase in the electron bombardment has no further effect on the intensity of the light.

Oscilloscopes. The cathode-ray oscilloscope (CRO) is a piece of test equipment that gives a visual presentation of the

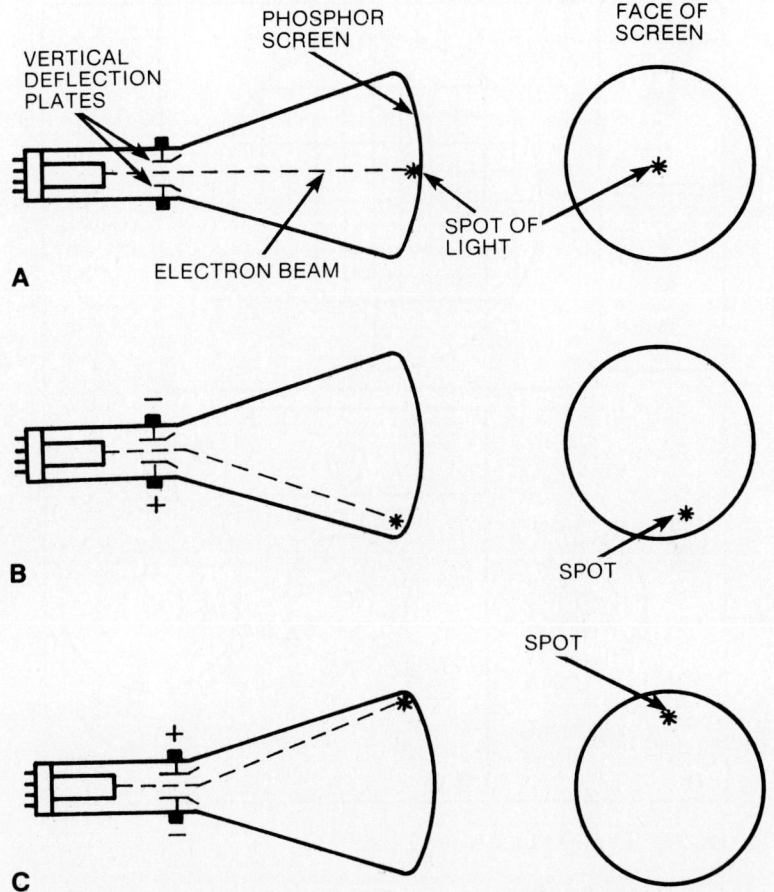

Fig. 18-23: **How the electrostatic deflection system operates: (A) No charge on the vertical deflection plates; (B) negative charge on top plate, positive charge on bottom plate; and (C) positive charge on top plate, negative charge on bottom plate.**

measured signal on a front-panel screen, much like that contained in a television set. A typical basic oscilloscope is shown in Fig. 18-24. The "screen" is shown in the upper middle of the figure.

The oscilloscope (sometimes called a *scope*) is the most versatile piece of test equipment that a technician/engineer has at his/her disposal. Most test equipment is designed to measure amplitude or quantity. The oscilloscope permits many characteristics of a circuit to be observed and measured. Some of these are:

1. Frequency.
2. Duration (or time) of one or more hertz.
3. Phase relationships between waveforms.
4. Shape of waveforms.
5. Amplitude of waveforms.

The oscilloscope may be used in preventive maintenance procedures for alignment of receivers and transmitters, frequency comparisons, percentage modulation checks, and equipment calibration.

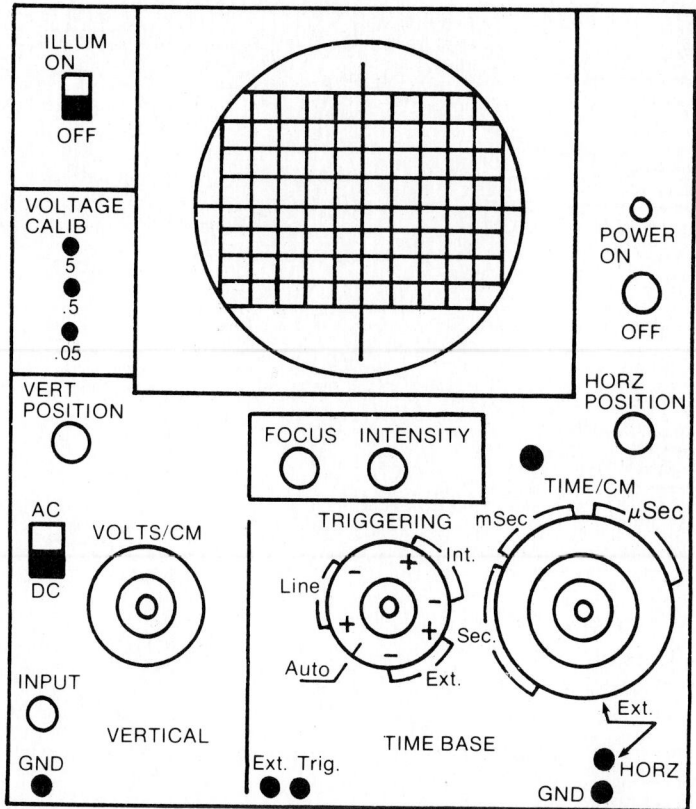

Fig. 18-24: Typical basic oscilloscope.

A waveform may be considered as a pictorial representation of a varying signal. Distortion of a waveform is an undesired change or deviation in its shape. This is especially noticeable in amplifiers, when their components begin to deteriorate, break down, or change in value. When this happens, the output waveform will change in amplitude, frequency, phase, etc. This change (or distortion) can be more readily detected with an oscilloscope as part of preventive or corrective checks. Naturally, the usefulness of an oscilloscope in any preventive or corrective technique depends upon the operator's knowledge of front-panel controls and operation and how to interpret the resultant pictorial display on the screen.

High-voltage and low-voltage power supplies are required for the operation of the oscilloscope. The high-voltage power supply is used to provide operating potentials to the cathode-ray tube. The low-voltage power supply is used to supply operating potentials to the associated oscilloscope circuitry, i.e., amplifiers, oscillators, etc. The output of the high-voltage power supply is usually over 1,000V DC depending upon the size of the cathode-ray tube. The output of the low-voltage power supply is usually in the 250V to 400V range for tube types and 5V to 50V for modern solid-state types.

Figure 18-24 shows a typical basic oscilloscope with square gridwork on the front of the screen. It is through this gridwork that we are able to accurately measure circuit outputs. This gridwork is normally detachable and can be replaced with grids of different sizes in order to measure different quantities. These detachable grids are called *graticules* and are normally provided on separate plates of glass or plastic that are accurately marked and can be mounted on the oscilloscope in front of the CRT.

Before using any oscilloscope, be completely familiar with the associated operating instructions. A lack of knowledge of operating instructions can lead to personal injury when you are using a piece of test equipment containing voltages as high as that of an oscilloscope. Always consult the oscilloscope technical manual for proper operation and applicable safety precautions.

SUMMARY

Vacuum tubes were once the heart of all electronic equipment. They have largely been replaced by solid-state devices.

Electron emission is the freeing of electrons from the surface of a body due to some form of energy being added to the atomic structure. There are several types: thermionic emission, field emission, photoelectric emission, and secondary emission.

The electron tube diode is a simple two element tube, consisting of a cathode and anode. The cathode readily emits electrons when heated. If a positive voltage is placed on the anode (or plate) in respect to the cathode, the free electrons will flow to the anode. The magnitude of the current flowing from cathode to anode is determined by the amplitude of the potential applied and the opposition to current flow by the tube.

If a third electrode, called a control grid, is added between the cathode and anode, a triode is created. The control grid controls the flow of plate current.

There are a number of special purpose tubes. A photoelectric cell is an electron tube which converts light energy into electrical energy. The cathode-ray tube is a tube that uses a concentrated beam of light velocity electrons to produce a visible trace on its screen.

CHECK YOURSELF (Answers at back of book)

Fill in the blanks with the appropriate word or words for the following sentences.

1. The diode has two elements, the _____ and _____, sealed in a tube or bulb called the _____ .
2. Cold cathode gas tubes are frequently used in _____ circuits.
3. _____ may produce undesired coupling between input and output circuitry in high gain RF amplifiers.
4. The intensity of the lighted spot on a CRT screen depends on _____ and _____ .
5. A _____ is used to improve the vacuum qualities of some electron tubes.

Choose the most correct answer for each of the following:

6. The _____ is an emitter of electrons.
   A. anode
   B. cathode
   C. heater
   D. control grid

7. In the photoconductive tube, _____ varies with light flux.
   A. current
   B. voltage
   C. resistance
   D. oxidation

8. Ionization cannot take place in a gas tube until the _____ has been reached.
   A. firing potential
   B. striking potential
   C. ignition potential
   D. All of the above

9. _____ is the point at which the plate is attracting all the electrons the cathode is capable of emitting.
   A. Ionization
   B. Cutoff
   C. Emitting potential
   D. Saturation

Determine whether each of the following statements is true or false.

10. The photo emission tube is the most commonly used photo-electric tube.
11. The photoelectric tube has a small output at low levels of illumination.
12. The electron gun of a CRT is used to deflect the electron beam.
13. The electron tube diode performs as a unidirectional switch.
14. The emission of electrons from solid materials is known as thermionic emission or the Edison defect.
15. The evacuated envelope of a tube is a perfect vacuum.

CHAPTER 19

Semiconductor Theory and Diodes

With the introduction of semiconductors in 1948, the science of electronic technology has rapidly expanded. Prior to this time, vacuum tubes had been used in the processing of electronic signals: amplification, switching, generation, waveshaping, and so on. Today, semiconductors perform these same electronic functions with greater efficiency and, as stated in Chapter 18, have therefore virtually replaced vacuum tubes. In addition to greater efficiency, semiconductors offer such other advantages over vacuum tubes as small physical size, mechanical ruggedness, and long life.

INTRINSIC SEMICONDUCTOR MATERIALS

In Chapter 1, it was stated that materials are classified as *semiconductors* if their electrical conductivity is intermediate between metallic *conductors*, which have a large number of loosely bound electrons available as charge carriers, and non-metallic *insulators*, which have practically no loosely bound electrons available to conduct current. While there are a variety of semiconductors available, the two most widely used in electronics are *germanium* (Ge) and *silicon* (Si).

A silicon atom is illustrated in Fig. 19-1A. An orderly, three-dimensional pattern called a *crystal* is formed through the process of covalent bonding (electron sharing) when individual silicon atoms are combined to form a solid.

Each individual silicon atom, as is true for all semiconductor atoms, possesses four valence electrons (tetravalent). If the valence shell shares four additional electrons (a total of eight), this outermost shell is complete and a strong covalent electron bond is produced. Figure 19-1B shows a central silicon atom *sharing* four electrons from four neighboring atoms. Each of these neighboring atoms, in turn, is also sharing an electron from the central atom.

The closer an electron is to the nucleus, the stronger are the forces that bind it. Each shell has an *energy level* associated with it which represents the amount of energy that would have to be supplied to extract an electron from the shell. Since the electrons in the valence shell are furthest from the nucleus, they

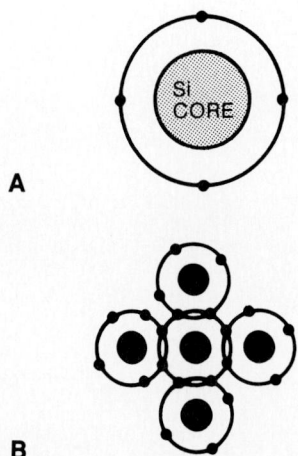

Fig. 19-1: (A) A silicon atom; (B) a silicon atom sharing four electrons from neighboring atoms.

397

require the least amount of energy to extract them from the atom. Conversely, those electrons closest to the nucleus require the greatest energy application to extract them from the atom.

Remember that electrons of an isolated atom are acted upon only by the forces within that atom. However, when atoms are brought closer together, as in a solid, the electrons come under the influence of forces from other atoms. The energy levels that may be occupied by electrons merge into bands of energy levels. As mentioned in Chapter 1, within any given material there are two distinct *energy bands* in which electrons may exist, the *valence band* and the *conduction band*. Separating these two bands is an *energy gap* in which no electrons can normally exist. This gap is termed the *forbidden gap*. The valence band, conduction band, and forbidden gap are shown diagrammatically in Fig. 19-2.

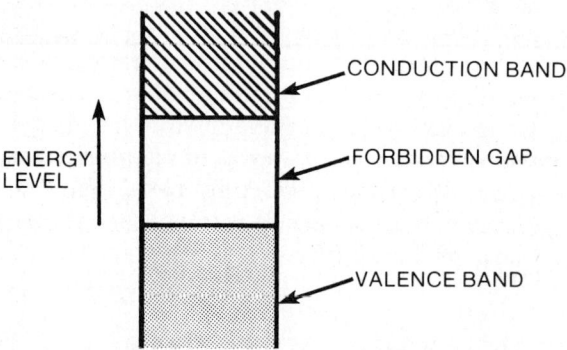

Fig. 19-2: Energy band diagram.

Electrons in the conduction band have escaped from their atoms or are only weakly held to the nucleus. Conduction band electrons may be easily moved around within the material by the application of relatively small amounts of energy. Much larger amounts of energy must be applied to extract an electron from the valence band or to move it around within the valence band. Electrons in the valence band are usually in normal orbit around a nucleus. For any given type of material, the forbidden gap may be large, small, or nonexistent. The distinction between conductors, insulators, and semiconductors is largely concerned with the relative widths of the forbidden gap (Fig. 19-3); that is, the degree of difficulty in dislodging valence electrons of an atom determines whether the material is a conductor, a semiconductor, or an insulator.

When an electron is forced loose in a block of a pure semiconductor material, it creates a "hole" which acts as a positively charged current carrier. Thus, electron liberation creates two currents, known as *electron current* and *hole current*.

Holes and electrons do not necessarily travel at the same velocity, and when an electric field is applied, they are accelerated in opposite directions. The life spans (time until recombination) of a hole and a free electron in a given semiconductor sample are not necessarily the same. Hole conduction (Fig. 19-4) may be thought of as the unfilled tracks of a moving electron. Because the hole is a region of net positive charge, the

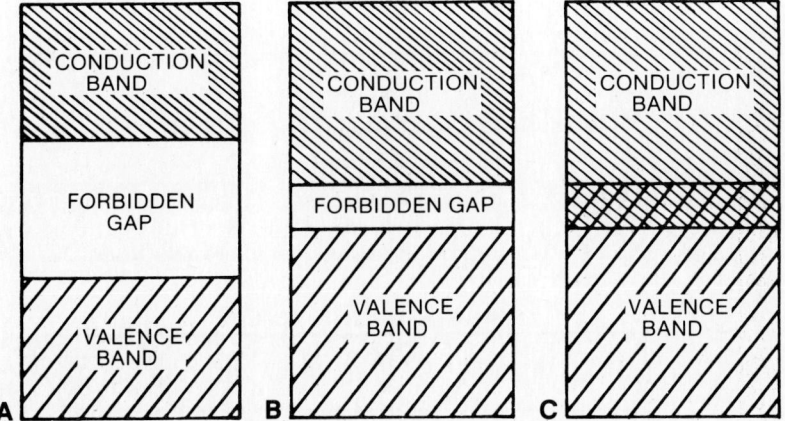

Fig. 19-3: Energy band diagrams for (A) insulator, (B) semiconductor, and (C) conductor.

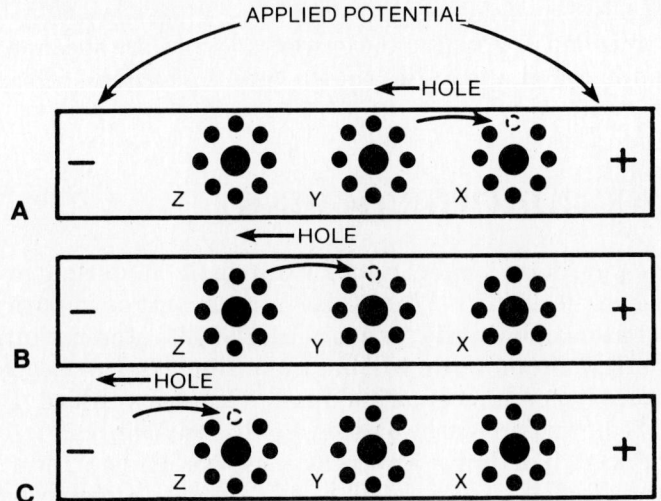

Fig. 19-4: Conduction by hole transfer. (A) Electron jumps from atom y to atom x. (B) It fills the hole in atom x and leaves a hole in atom y. (C) If an electron jumps from atom z to atom y, it will leave a hole in atom z.

apparent motion is like the flow of particles having a positive charge. An analogy of hole motion is the movement of the hole as balls are moved through a tube (Fig. 19-5). When ball number 1 is removed from the tube, a space is left. This space is then filled by ball number 2. Ball number 3 then moves into the space left by ball number 2. This action continues until all balls have moved one space to the left at which time there is a space left by ball number 8 at the right-hand end of the tube.

A pure specimen of semiconductor material will have an equal number of free electrons and holes, the number depending on the temperature of the material and the type and size of the specimen. Such a specimen is called an *intrinsic* semiconductor; the current, which is born equally by hole conduction and electron conduction, is called intrinsic conduction.

If a suitable "impurity" is added to the semiconductor, the resulting mixture can be made to have either an excess of electrons, thus causing more electron current, or an excess of

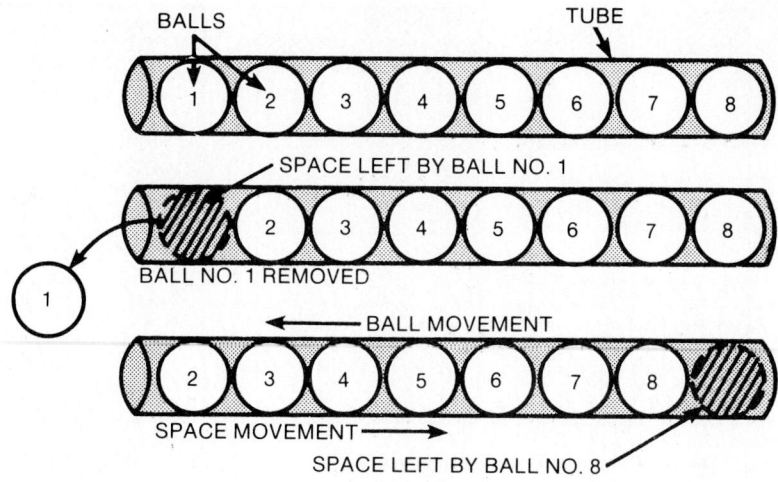

Fig. 19-5: Analogy of hole movement.

holes, thus causing more hole current. An impure specimen of semiconductor material is known as an *extrinsic* semiconductor.

SEMICONDUCTOR DOPING

In the pure form, semiconductor (intrinsic) materials are of little use in electronics. When a certain amount of impurity is added (1 atom of impurity for every 100,000,000 atoms of intrinsic material), the material will have more (or less) free electrons than holes depending upon the kind of impurity added. Both forms of conduction will be present, but the majority carrier will be dominant. The holes are called positive carriers, and the electrons negative carriers. The one present in the greatest quantity is called the *majority carrier;* the other is called the *minority carrier.* The quality and quantity of the impurity are carefully controlled by a process known as *doping.*

Two different types of doping are possible, *donor* doping and *acceptor* doping. Donor doping generates free electrons in the conduction band (i.e., electrons that are not tied to an atom). Acceptor doping produces valence band holes or a shortage of valence electrons in the material.

Donor doping is affected by adding impurity atoms which have five electrons in their valence shells. The impurity atoms form covalent bonds with the silicon or germanium atoms, but since semiconductor atoms have only four electrons (tetravalent) in their valence shells, one spare valence shell electron is produced for each impurity atom added. Each spare electron produced in this way enters the conduction band as a free electron. In Fig. 19-6 there is no hole for the fifth electron from the outer shell of the impurity atom; therefore, it becomes a free electron. Since free electrons have negative charges, donor doped materials are known as *N-type* semiconductor materials.

Free electrons in the conduction band are easily moved around under the influence of an electric field. Therefore, conduction occurs largely by electron motion in donor doped semi-

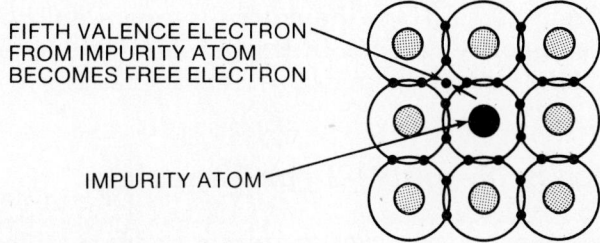

FIFTH VALENCE ELECTRON
FROM IMPURITY ATOM
BECOMES FREE ELECTRON

IMPURITY ATOM

Fig. 19-6: Donor doping.

conductor material. The doped material remains electrically neutral because the total number of electrons (including the free electrons) is still equal to the total number of protons in the atomic nuclei. The term *donor doping* comes from the fact that an electron is donated to the conduction band by each impurity atom. Typical donor impurities are antimony, phosphorous, and arsenic. Since these atoms have five valence electrons, they are referred to as *pentavalent* atoms.

In acceptor doping, impurity atoms are added with outer shells containing three electrons. Suitable atoms with three valence electrons (which are called *trivalent*) are boron, aluminum, indium, and gallium. These atoms form bonds with the semiconductor atoms, but the bonds lack one electron for a complete outer shell of eight. In Fig. 19-7 the impurity atom has *robbed* an electron from a neighboring silicon bond. The net result of this robbery is a hole in the crystal structure. Thus in acceptor doping, holes are introduced into the valence band so that conduction may occur by the process of hole transfer.

Since holes can be said to have a *positive* charge, acceptor doped semiconductor materials are referred to as *P-type* materials. As with N-type material, the material remains electrically neutral because the total number of electrons in the structure is equal to the total number of protons in the structure. Holes can accept free electrons, hence the term *acceptor* doping.

Even in intrinsic (undoped) semiconductor material at room temperature there are a number of free electrons and holes. These are due to thermal energy causing some electrons to break the bonds with their atoms and enter the conduction band, so creating pairs of holes and electrons. The process is termed *hole-electron-pair generation,* and its converse is a process called *recombination.* As the name implies, recombination occurs when an electron falls into a hole in the valence band. Since there are many more electrons than holes in N-type mate-

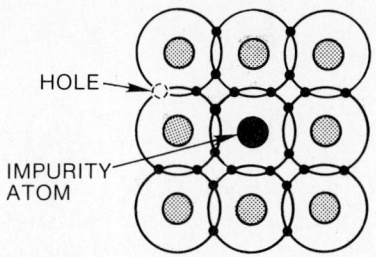

HOLE

IMPURITY
ATOM

Fig. 19-7: Acceptor doping.

rial, electrons are said to be the majority carriers, and holes are said to be minority carriers. In P-type material holes are the majority carriers and electrons are minority carriers.

Charges in N- and P-Type Materials

When a donor material such as arsenic is added to silicon, the fifth electron in the outer ring of the arsenic atom does not become a part of a covalent bond. This extra electron drifts away from the arsenic atom and moves randomly through the structure.

The arsenic atom has a positive charge of five units on the inner circle, as shown in Fig. 19-8A. When the electron moves away from the arsenic atom, there will be only four electrons to neutralize the positive charge and as a result there will be a region of positive charge around the arsenic atom. However, the net charge on the N-type crystal is zero because the positive charge around the arsenic atom is neutralized by the negatively charged free electron.

In a P-type material having an impurity such as indium added to it, a similar situation may exist. Indium has only three electrons in its outer ring. Three electrons are all that are needed to neutralize the net positive charge of three units on the inner circle (Fig. 19-8B). When the indium atom robs an electron from a neighboring silicon bond (in order to complete its four electron pair bond), it becomes negatively charged. However, the net charge on the P-type crystal is zero because the negative charge around the indium atom is neutralized by the positively charged hole.

The charged atoms produced in both N- and P-type silicon are not concentrated in any one part of the crystal, but instead are spread uniformly throughout the crystal. If any region within

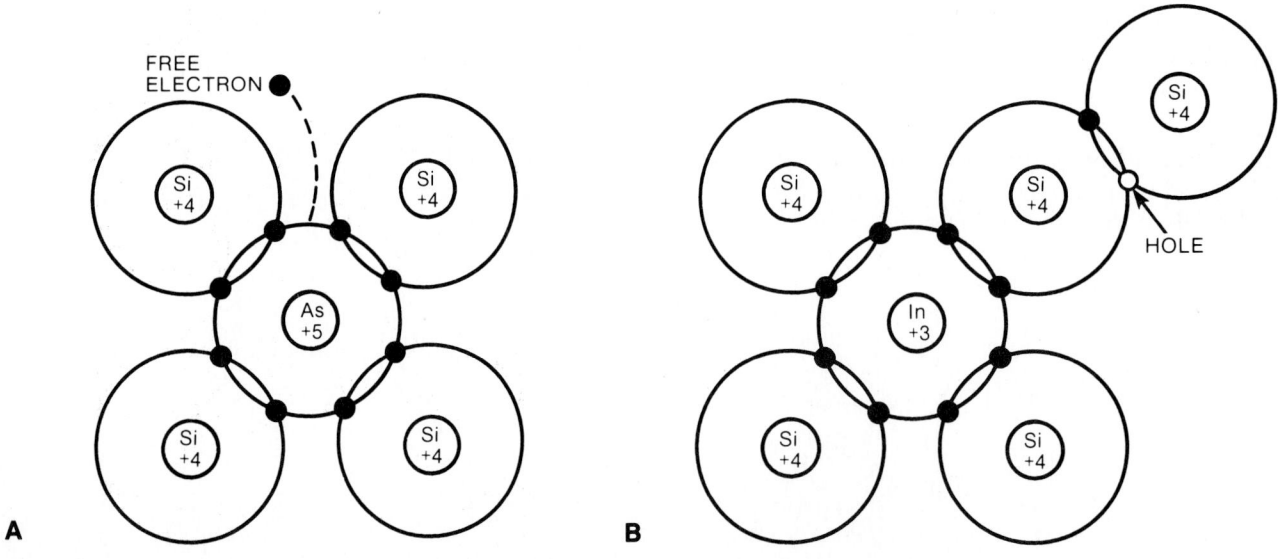

Fig. 19-8: (A) Atomic structure of N-type material; (B) atomic structure of P-type material.

402

the crystal were to have a very large number of positively charged atoms, these atoms would attract free electrons from other parts of the crystal to neutralize part of the charged atoms so that the charge would spread uniformly throughout the crystal. Similarly, if a large number of atoms within a small region had an excess of electrons, these electrons would repel each other and spread throughout the crystal.

As stated previously, both holes and electrons are involved in conduction. In N-type material the electrons are the majority carriers and holes the minority carriers. In P-type material the holes are the majority carriers and electrons the minority carriers.

Current Flow in N-Type Materials

Current flow through an N-type material is illustrated in Fig. 19-9. The application of voltage across the material will cause the free electron to be repelled by the negative potential point and attracted toward the positive potential point.

The semiconductor resistance decreases with temperature increase because more carriers are made available at higher temperatures. Increasing the temperature releases electrons from more of the impurity atoms in the lattice causing increased conductivity (decreased resistance).

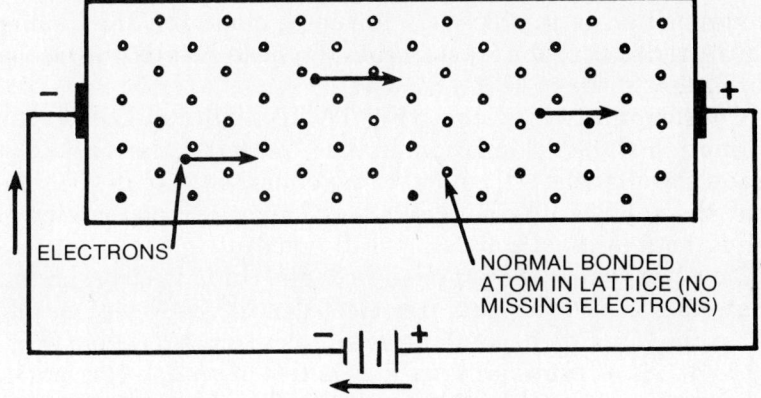

ELECTRONS

NORMAL BONDED
ATOM IN LATTICE (NO
MISSING ELECTRONS)

Fig. 19-9: Current flow in N-type material.

Current Flow in P-Type Materials

Current flow through a P-type material is illustrated in Fig. 19-10. Conduction in this material is by positive carriers (holes) from the positive to the negative terminal. Electrons from the negative terminal cancel holes in the vicinity of the terminal; while at the positive terminal, electrons are being removed from the covalent bonds, thus creating new holes. The new holes then move toward the negative terminal (the electrons shifting to the positive terminal) and are canceled by more electrons emitted from the negative terminal. This process continues as a steady stream of holes (hole current) moving toward the negative terminal.

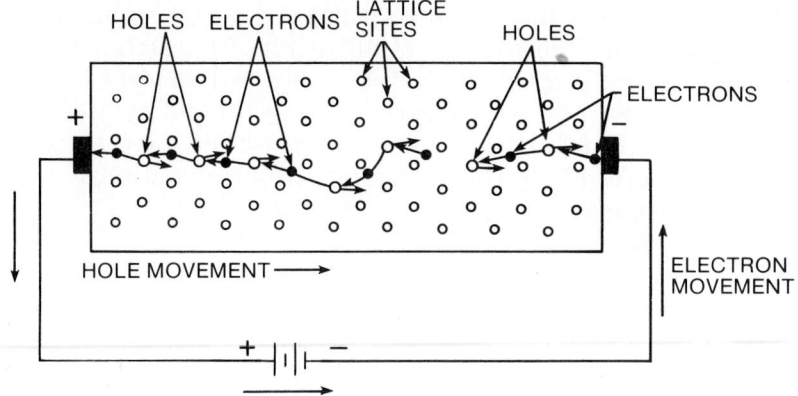

Fig. 19-10: Current flow in P-type material.

In both N-type and P-type materials, current flow in the external circuit is out of the negative terminal of the battery and into the positive terminal.

The Basic PN Junction

A PN junction is created when a P region and an N region are formed side by side in the same semiconductor crystal. This juxtaposition is produced by abruptly changing the doping characteristics of the crystal from P-type to N-type during the formation of the crystal.

Consider the two regions in Fig. 19-11A separately, before the formation of the PN junction. In the N material, the number of free electrons equals the number of donor atoms. In the P material, the number of holes equals the number of acceptor atoms. Thus, both materials are electrically neutral.

However, when the junction between these two dissimilar materials is first formed, free electrons in the N region will cross the junction to combine with holes in the P region (Fig. 19-11B). As a result, an area that is free of charge carriers is produced on either side of the junction. This area, shown in Fig. 19-11C, is called the depletion region.

Observe that once the free electrons near the junction combine with holes, there are no longer any free charges to neutralize the charges around the donor and acceptor atoms. Thus, a potential difference, called the potential barrier (or potential hill), is created across the junction, as shown in Fig. 19-11D. This potential is sufficiently large to oppose any further movement of charges across the junction, and so, in effect, it limits the size of the depletion region.

It is found that at room temperature, the potential barrier is approximately 0.3V for a germanium PN junction and 0.7V for a silicon PN junction. The potential barrier of a PN junction can play an important role in the performance of a circuit, especially in low-voltage applications.

The minority carriers (not shown) present in both halves of the PN junction diffuse through the crystal in a random

404

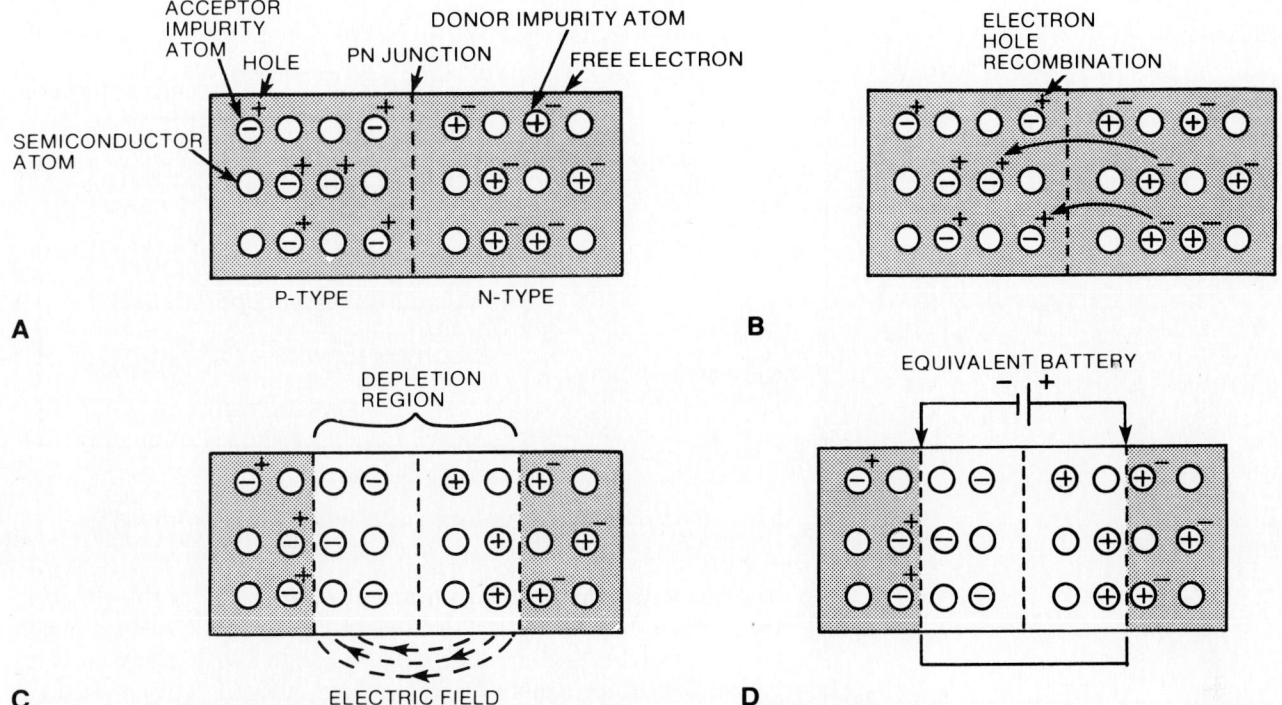

Fig. 19-11: (A) P and N regions before the formation of the PN junction; (B) electron hole recombination; (C) the area empty of free charge carriers called the depletion region; and (D) the potential barrier across the junction.

motion. When they approach the depletion region, they are swept across the junction under the influence of the potential barrier. At equilibrium of the junction, however, the minority electron flow from P to N is balanced by an equal but opposite minority hole flow from N to P, and there is no net flow of charge carriers across the junction.

Forward Voltage

If a forward voltage is applied to the PN junction—that is, a positive battery terminal connected to the P region and a negative terminal to the N region—a drift of free charge carriers will occur. Free electrons in the N region, repelled by the negative terminal, will move toward the junction. At the same time holes in the P region, repelled by the positive terminal, will move toward the junction. If the voltage applied is large enough to overcome the barrier potential, these charge carriers will combine at the junction. This action is illustrated in Fig. 19-12. As an electron from the N region combines with a hole in the P region, an electron will enter the crystal from the negative terminal of the battery. Similarly, as a hole combines with an electron, an electron from an electron hole pair will leave the P region to enter the positive battery terminal. As a result, a new hole in the P region is created. Thus, the two regions outside of the depletion region are able to remain electrically neutral.

The constant movement of holes toward the negative terminal and electrons toward the positive terminal will result in a

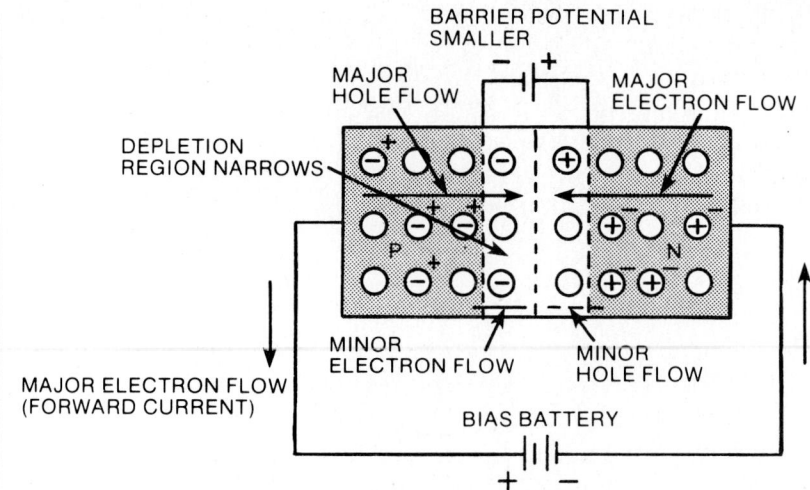

Fig. 19-12: Forward voltage applied to the PN junction.

high forward current. A fundamental property of the PN junction, therefore, is that it conducts under forward voltage conditions (positive battery to P region, negative battery to N region). The magnitude of this forward current is exponentially related to the magnitude of the applied forward voltage.

Reverse Voltage

If the reverse voltage is applied to the PN crystal, as shown in Fig. 19-13, electrons in the N region will be attracted away from the junction by the positive terminal of the battery. At the same time holes in the P region will be attracted away from the junction by the negative terminal of the battery. In this way, the size of the depletion region is enlarged and, thus, the potential barrier is increased. If the potential barrier is increased, there can be no movement of charge carriers across the junction and, therefore, practically no current will flow.

The small amount of current that does flow, which is called *reverse current* or *leakage current*, is due to the minority carri-

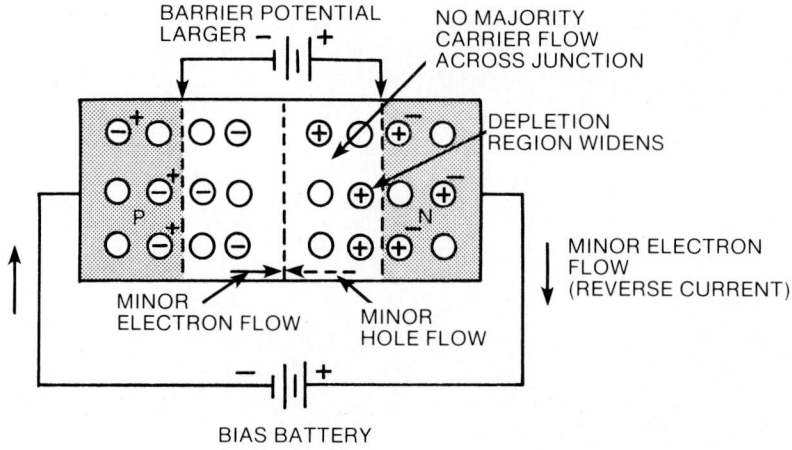

Fig. 19-13: Reverse voltage applied to the PN junction.

ers present in the PN junction. The reverse voltage sweeps these carriers across the junction; but since relatively few minority carriers are available, the leakage current is small (in the microampere range). We recall, however, that the number of minority carriers increases with temperature; therefore, the leakage current of the PN junction increases with an increase in temperature.

Characteristic Curve

A PN junction (semiconductor) diode allows appreciable current flow in one direction, while restricting current flow to an almost negligible value in the other direction. The N material section of the device is called the cathode, and the P material section is called the anode. The device permits current flow from cathode to anode and restricts current flow from anode to cathode.

Figure 19-14 is a graph of the current flow through a semiconductor diode, plotted by values of anode to cathode voltage. Note that the forward current increases slowly at low values of forward voltage. As the forward voltage is increased, the barrier is neutralized and current increased rapidly for further increases in applied voltage. It should be noted that excessive forward voltage could destroy the device through excessive forward current.

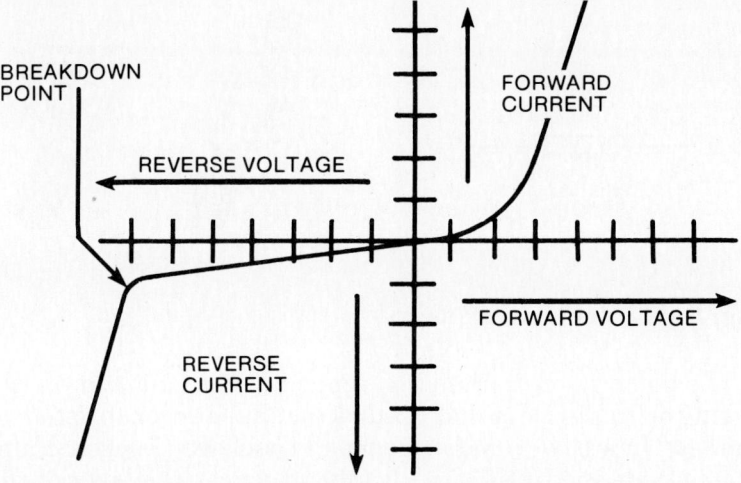

Fig. 19-14: Semiconductor diode characteristic curve.

With reverse voltage applied, a very small reverse current exists. This reverse current increases minutely, with an increase in reverse voltage. However, if an excessive reverse voltage is applied, the structure of the semiconductor material may be broken down by the resulting high electric stress, and the device may be permanently damaged. The value of reverse voltage at which breakdown occurs is called the breakdown voltage. At this voltage, current increases rapidly with small increases in reverse voltage. Certain semiconductor devices are designed and doped to operate in the breakdown region without harm.

407

SEMICONDUCTOR DIODES

Semiconductor materials treated to form PN junctions are used extensively in electronic circuitry. Variations in doping agent concentrations and physical size of the substrate produce diodes which are suited for different applications. There are rectifier diodes, signal diodes, zener diodes, reference diodes, varactors, and others.

Various types of diodes are shown in Fig. 19-15. This is but a very limited representation of the wide assortment in case design. However, the shape of characteristic curves of these diodes is very similar; primarily, current and voltage limits and relationships are different. The schematic symbol for a semiconductor diode is illustrated in Fig. 19-16.

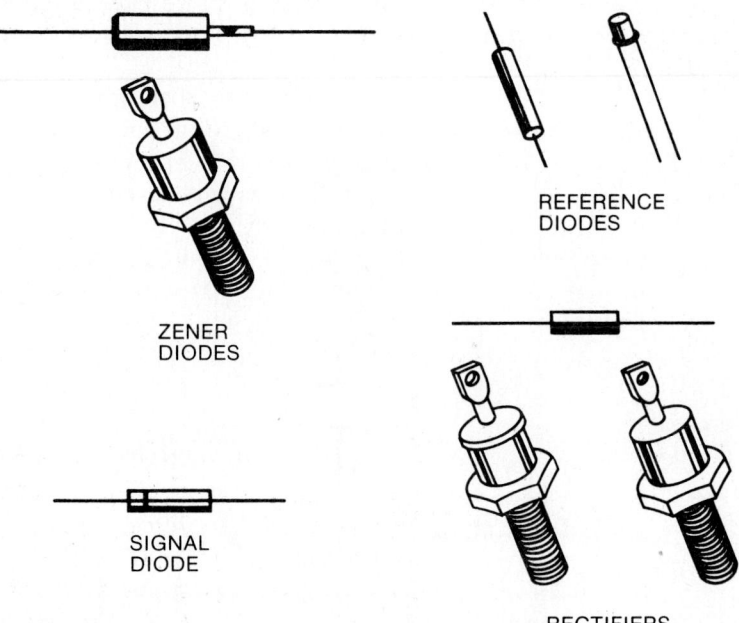

REFERENCE DIODES

ZENER DIODES

SIGNAL DIODE

RECTIFIERS

Fig. 19-15: Junction diodes.

The polarity of the junction diode is either marked on the casing of the device or indicated by the shape or construction of the case. Low power diodes are usually packaged in plastic, and the cathode is identified with a band across the body of the device, close to the cathode lead. In some cases, plus and minus signs are used to indicate polarity. High power diodes are usually packaged in a metal container. The case of the device is then the cathode, and the insulated lead emerging from the top of the case is the anode. It is recommended practice, however, to consult the diode manual or the data sheet to verify the polarity before inserting the diode in a circuit.

PN JUNCTION DIODES

The PN junction diode is an excellent switch. Figure 19-17 shows how an ordinary PN junction diode might be used as a

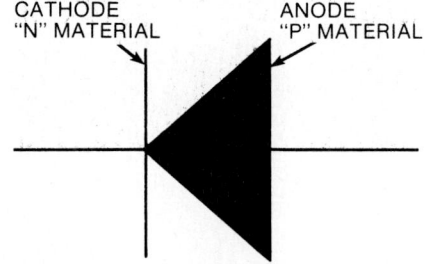

CATHODE "N" MATERIAL

ANODE "P" MATERIAL

Fig. 19-16: Schematic symbol for a semiconductor diode.

408

switch to control the frequency of an LC tank circuit. Although a switch (SW_1) and a battery are used for ease of explanation, the control voltage may be taken from other voltage sources. Therefore, the battery and switch can be replaced by control lines. The combination of C_2 and CR_1 in parallel with C_1 and L_1 will be resonant at some specific frequency according to the values of capacitance and inductance present. With SW_1 open the value of capacitance presented by the PN junction diode is much lower than it is when the switch is closed. This change in capacitance occurs when the PN junction diode conducts heavily (SW_1 closed) due to the forward bias and is an effective short. The value of capacitance presented in parallel with C_1 and L_1 is now approximately equal to the value of C_2. When the switch is open, the value of capacitance in parallel with C_1 and L_1 is the series value of C_2 and the junction capacitance of CR_1. The junction capacitance of the diode chosen in this application will be inconsequential and the resonant frequency will be determined primarily by L_1 and C_1. The high resonant frequency condition of this circuit is determined by the nonconducting state of CR_1, and the low resonant frequency occurs when CR_1 is conducting heavily.

One application of a PN junction diode is as a switch. Remember that a conducting PN junction diode acts as a very low resistance and that a reverse biased PN junction diode presents a very high resistance. This switching characteristic has many other applications. The PN junction diode, as a switch, will be encountered in various equipment.

Although it is a very good switch, a PN junction diode is *not* perfect. However, when a simplified analysis of a diode circuit is desired, the ideal diode can be used for calculations. The *ideal diode* is a one-way switch with absolutely zero resistance to the flow of current in one direction and infinite resistance in the other direction. That is, the transition from conducting to blocking occurs abruptly at exactly zero bias voltage. The volts-ampere relationship of such a switch is shown in Fig. 19-18.

Example: A 20Vp sine wave is applied to the circuit shown in Fig. 19-19. Graphically depict the voltage drop across the load resistor. What is the maximum reverse voltage the diode must endure?

Solution: During the positive half cycle, the diode is forward-biased and current is flowing. Figure 19-20A shows the circuit at the positive peak. The output across the load resistor has a positive peak of 20V because the diode acts as a closed switch. During the negative half cycle, the diode is reverse-biased, thus the diode appears to be an open switch (Fig. 19-20B). No current

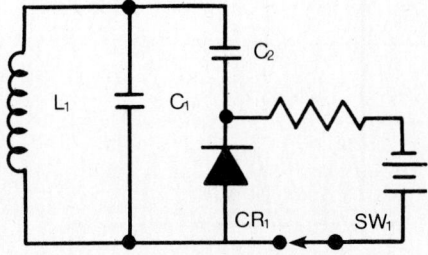

Fig. 19-17: Changing resonant frequency with a PN junction diode.

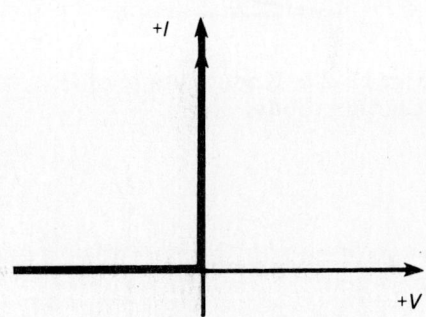

Fig. 19-18: *I-V* **characteristic of a perfect switch.**

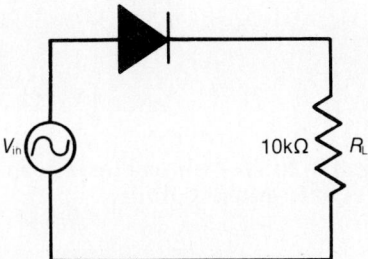

Fig. 19-19

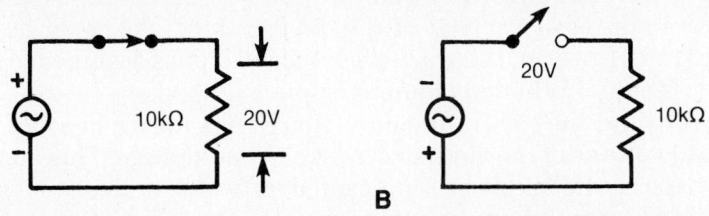

A   **B**

Fig. 19-20: (A) Equivalent circuit at positive peak. (B) Equivalent circuit when reverse-biased.

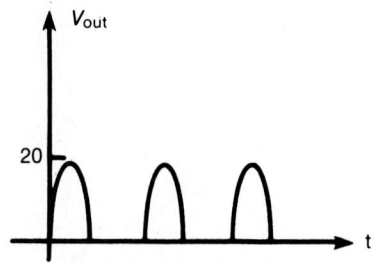

Fig. 19-21: Voltage output.

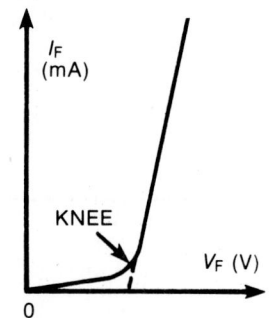

Fig. 19-22: Knee voltage of PN junction diode.

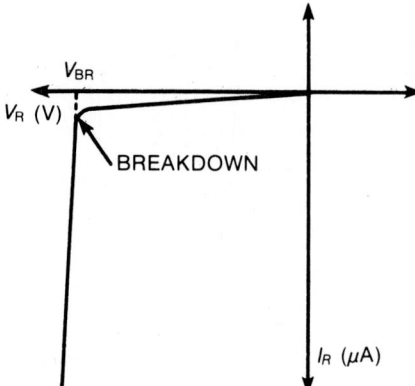

Fig. 19-23: *I-V* characteristics of a reverse-biased diode.

can flow and the 20V is dropped across the diode. Figure 19-21 shows the voltage output dropped across R_L. The *peak inverse voltage* (PIV) is the maximum reverse voltage a diode must withstand during the cycle. The maximum negative peak in this example is 20V.

As stated earlier, the magnitude of the forward current is exponentially related to the forward voltage. The diode current equals zero when the diode voltage equals zero. As the forward voltage is gradually increased, the forward current starts to increase slowly. Once the diode voltage overcomes the barrier potential of the PN junction, current flow increases significantly, and the volts-amps characteristic curve rises sharply. This is depicted in Fig. 19-22. The diode voltage at which this steep increase in current occurs is called the *knee voltage*. This voltage is approximately equal to the barrier potential of the diode (around 0.7V for silicon diodes; around 0.3V for germanium diodes).

Breakdown. The voltage-current characteristic of the reverse-biased diode is shown in Fig. 19-23. Initially there is a minute reverse current I_R, caused by minority carriers crossing the PN junction. Since the number of minority carriers in a crystal depends solely on temperature, the reverse current should remain constant as the reverse voltage is increased. Because of surface effects, however, the reverse current increases slightly with increasing reverse voltage. The reverse current is negligibly small and for most practical purposes the reverse-biased diode is regarded as an open circuit.

Once the reverse voltage reaches the breakdown value V_{BR}, the voltage remains virtually constant as the current increases rapidly. This breakdown of the diode can be caused by either of two physical processes: *zener* breakdown or *avalanche* breakdown.

Zener breakdown occurs when an electrical field is sufficiently strong enough to pull numerous electrons out of their covalent bonds simultaneously. Once these electrons are pulled from these bonds, a sudden increase in reverse current is realized.

Avalanche breakdown is also caused by a high reverse voltage. Minority carriers gain sufficient energy to knock electrons out of the covalent bonds, thereby creating a number of electron-hole pairs. These extra carriers are swept across the junction where they, in turn, collide with and liberate additional electrons. As this process builds, the resulting avalanche of electrons produces a rapid increase in reverse current that cannot be stopped.

If the reverse voltage applied across a diode exceeds that diode's breakdown value and the reverse current surpasses the maximum power rating of the device, the diode could be destroyed. However, there are a few special diodes designed to run specifically in the breakdown region.

Consider the fact that conduction starts only when the forward voltage of the diode exceeds the knee voltage. This means the actual diode can be represented as an ideal diode in series with a battery whose voltage is equal to the knee voltage of the actual diode. In other words, the voltage across the diode (during conduction) is 0.7V or 0.3V, depending on whether it is a silicon or germanium diode.

Example: Calculate the maximum forward current of the circuit shown in Fig. 19-24. Also graph the output waveform and determine the peak inverse voltage.

Solution: The silicon diode conducts only after overcoming the 0.7V barrier potential. Figure 19-25 shows the equivalent circuit at the positive voltage peak. The output voltage peak equals the supply voltage minus the barrier potential:

$$2V - 0.7V = 1.3V$$

The maximum current peak occurs at the output voltage peak:

$$I_{max} = \frac{1.3V}{10k\Omega} = 130\mu A$$

Because the diode acts as an open switch on the negative half cycle, there is no current flow and no voltage drop across R_L. The output waveform is shown in Fig. 19-26. This graph shows the output of a single *half-wave rectifier*. The entire voltage (2V) is dropped across the diode on the negative half cycle, as stated earlier; therefore, 2V is the peak inverse voltage.

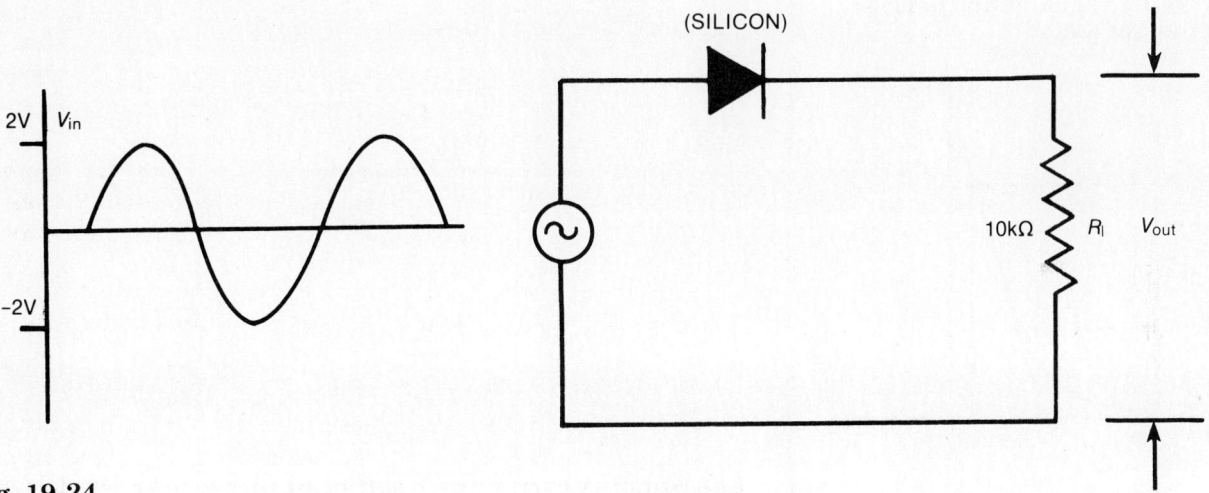

Fig. 19-24

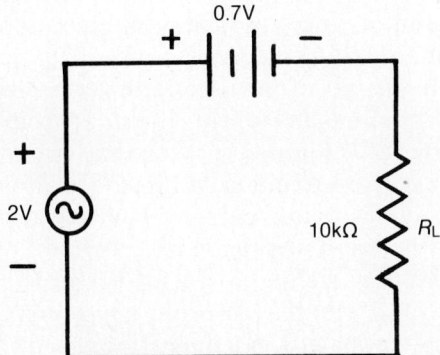

Fig. 19-25: Equivalent circuit at positive voltage peak.

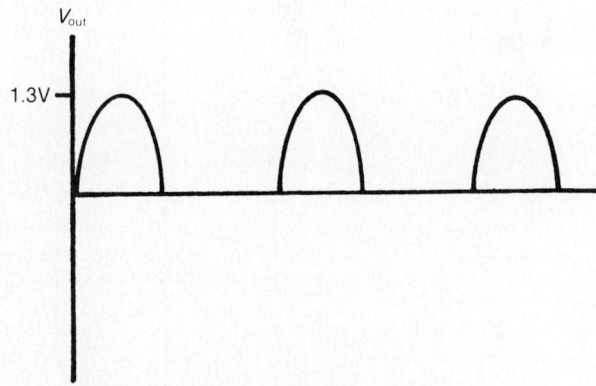

Fig. 19-26: Output waveform.

411

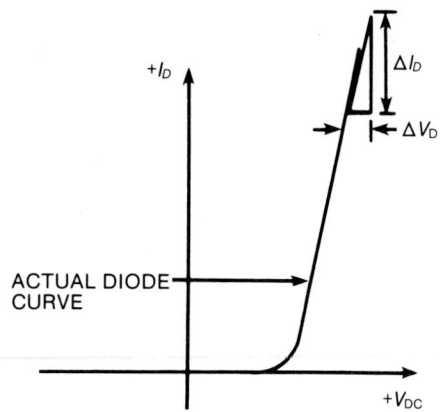

Fig. 19-27: Voltage change vs. current change (above knee voltage).

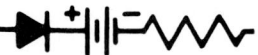

Fig. 19-28: Equivalent circuit showing the forward resistance and voltage equal to the barrier potential of a diode.

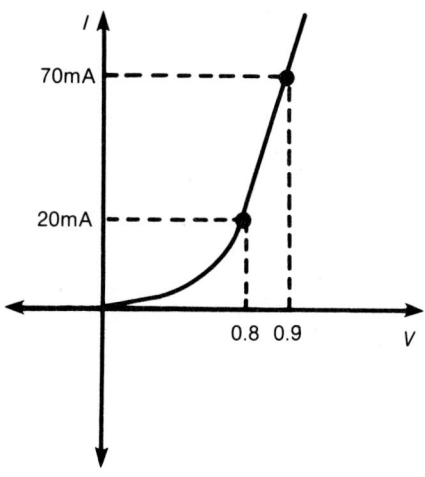

Fig. 19-29

As a rule, the forward resistance of a silicon diode is usually less than 10Ω. Since circuit resistance is usually much greater than the forward resistance of a diode, the forward resistance of a diode can usually be ignored. However, if the circuit resistance in series with the diode is not much larger than the diode's forward resistance, the forward resistance can be calculated.

In Fig. 19-27, it is apparent that the change in voltage is nearly proportional to the change in current. So it can be said that once a diode starts conducting, it acts like a small resistor. The equivalent circuit for this diode is illustrated in Fig. 19-28.

The *forward resistance* of a diode can be determined by selecting any two points well above the knee of the volts-amps characteristic curve. Using the changes in voltage and current between these two points, the forward resistance of the diode is calculated as follows:

$$r_d = \frac{\Delta V_d}{\Delta I_d}$$

Example: Referring to Fig. 19-29, calculate the forward resistance. This is also the value of forward resistance for the diode in Fig. 19-30.

Solution: The forward resistance equals:

$$r_d = \frac{\Delta V_d}{\Delta I_d} = \frac{.1V}{50mA} = 2\Omega$$

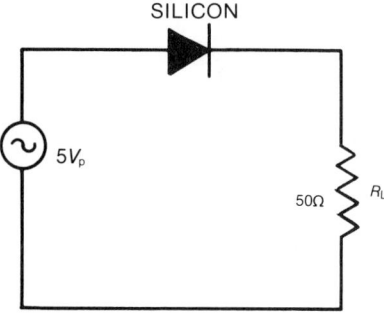

Fig. 19-30

Rectifier Diodes

One of the most common and simplest of semiconductor PN junction devices is the rectifier diode. The use of PN junction rectifier diodes in the design of power supplies for electronic equipment is increasing. Reasons for this include these characteristics: no requirement for filament power, immediate operation without need for warm-up time, low internal voltage drop substantially independent of load current, low operating temperature, and generally small physical size. PN junction rectifier diodes are particularly well-adapted for use in the power supplies of portable and small electronic equipment where weight and space are important considerations (see Chapter 21).

Rectifier diodes are primarily of the silicon type because of this material's inherent reliability and higher overall performance compared to other materials. Silicon allows higher maximum forward current, lower reverse leakage current, and operation at higher temperatures compared to other materials.

The major electrical characteristics of rectifier diodes are as follows:

1. Peak Reverse Voltage (PRV). Maximum reverse DC voltage which will not cause breakdown.

2. Average Forward Voltage Drop (V_F). Average forward voltage drop across the rectifier given at a specified forward current and temperature, usually specified for rectified forward current at 60 Hz.

3. Average Rectifier Forward Current (I_F). Average rectified forward current at a specified temperature, usually at 60 Hz with a resistive load. The temperature is normally specified for a range, typically –65 to +175°C.

4. Average Reverse Current (I_R). Average reverse current at a specified temperature, usually at 60 Hz.

5. Peak Surge Current (I_{surge}). Peak current specified for a given number of cycles or portion of a cycle (for example, 1/2 cycle at 60 Hz).

Signal Diodes

Signal diodes fall into various categories, such as general purpose, high-speed switch, parametric amplifiers, etc. These devices are used as mixers, detectors, and switches, as well as in many other applications.

Signal diodes' major electrical characteristics are:

1. Peak Reverse Voltage (PRV). Maximum reverse voltage which can be applied before reaching the breakdown point.

2. Reverse Current (I_R) or Leakage Current. Small value of direct current that flows when a semiconductor diode has reverse bias.

3. Maximum Forward Voltage Drop at Indicated Forward Current (V_F @ I_F). Maximum forward voltage drop across the diode at the indicated forward current.

4. Reverse Recovery Time (t_{rr}). Time required for reverse current to decrease from a value equal to the forward current to a value equal to I_R when a step function of voltage is applied.

The schematic diagram for the rectifier and signal diode is shown in Fig. 19-31. Forward current flows into the point of the arrow and reverse current is with the arrow.

Zener Diodes

The zener diode is unique compared to other diodes in that it is designed to operate reverse biased in the breakdown region. The device is used mainly as a voltage regulator or clipper.

The major electrical characteristics of zener diodes are:

1. Nominal Zener Breakdown ($V_{Z(nom)}$). Sometimes a $V_{Z(max)}$ and $V_{Z(min)}$ are used to set absolute limits between which breakdown will occur.

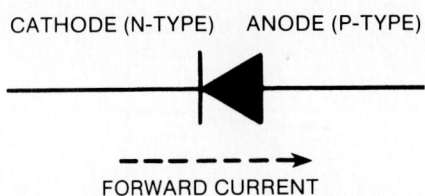

CATHODE (N-TYPE)   ANODE (P-TYPE)

FORWARD CURRENT

Fig. 19-31: Schematic diagram for the rectifier and signal diode.

413

Fig. 19-32: Zener diode schematic symbol.

2. Maximum Power Dissipation (P_D). Maximum power the device is capable of handling. Since voltage is a constant, here is a corresponding current maximum (I_{ZM}).

Virtually all zener diodes are made of silicon. Silicon, in comparison to germanium, has superior thermal qualities and lower reverse current below breakdown. The breakdown region of this diode is designed, during the manufacturing process, to be as close to vertical as possible. A schematic symbol of the zener is shown in Fig. 19-32. Zener current flows in the direction of the arrow. In many schematics a distinction is not made for this diode and a signal diode symbol is used.

As shown in Fig. 19-33, the zener diode behaves much like an ordinary diode under forward-biasing voltage. When reverse-biased, the diode first exhibits a relatively high resistance until the breakdown or zener voltage is reached. Here it breaks down and the voltage across the device remains virtually constant over a wide range of reverse current. They also are available in a breakdown voltage range from approximately 1.4V to several hundred volts.

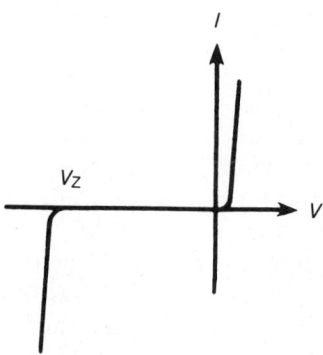

Fig. 19-33: *I-V* characteristic curve for zener diode.

As a practical consideration, a zener diode operating in the breakdown region can be considered a battery (constant voltage source). A more accurate description of the zener diode (operating in breakdown region) is a battery in series with a small resistor. The small resistor represents the small change in voltage as current increases. This is called the *dynamic slope resistance.*

Example: Given the values in Fig. 19-34, calculate the dynamic slope resistance, r_Z.

Solution:

$$r_Z = \frac{V_z}{I_z}$$

$$r_Z = \frac{.3V}{20mA}$$

$$r_Z = 15\Omega$$

The smaller this value of resistance is, the closer the breakdown region resembles a vertical line and, thus, approaches the ideal voltage regulator.

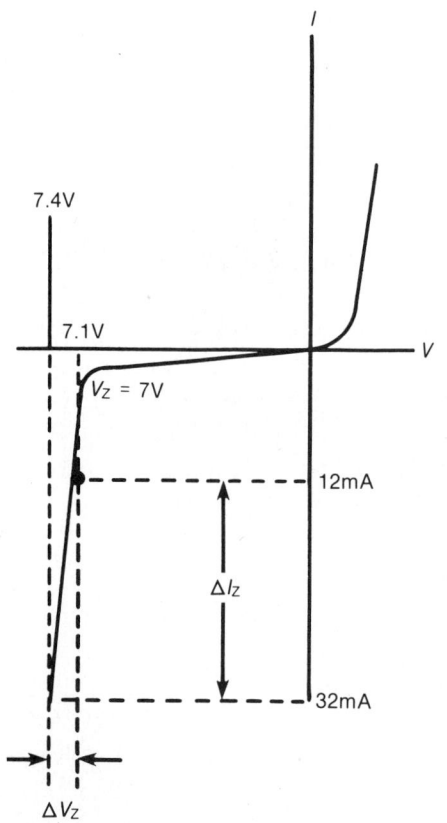

Fig. 19-34

Schottky Diodes

The Schottky diode is a silicone-to-metal semiconductor junction diode used for high speed operations. Gold, titanium, aluminum, or platinum is used to form a junction with N-type silicon. When unbiased, the free electrons in the metal travel in larger orbit than the free electrons in the N material. This difference in energy levels is called the *Schottky barrier*. When forward biased, the free electrons in the N material gain enough energy to travel in larger orbits, cross the junction, and enter the metal. Thus, a large current is produced.

In the Schottky diode the holes have been eliminated. This effectively eliminates transition capacitance, which, in turn, eliminates reverse current flow during the early part of the "off" half cycle of the diode's switching action. Thus, distortion at high frequencies is suppressed.

One major disadvantage of the Schottky diode was its relatively low reverse breakdown voltage. However, this problem has been reduced by the use of a P-type guard ring around the periphery of the metal-to-N-type silicon, and repetitive peak inverse working voltages of 30V to 45V are now common.

The advantages of the Schottky diode are: current flow is entirely due to the majority carrier (electrons); it is relatively unaffected by reverse recovery transients caused by the transition capacitance and minority carrier flow; the reverse recovery time is very small; the forward voltage drop is small, which results in a low forward conduction power loss; and it has relatively low reverse leakage current.

In the power electronic field, the major applications of these devices are in low voltage, high current DC power supplies and in high frequency power supply applications. Other applications of the Schottky principle are in transistor-transistor logic (TTL), photodiodes, and power thyristors. The current symbol for a Schottky diode is shown in Fig. 19-35.

Tunnel Diodes

The tunnel diode (also called the Esaki diode after its discoverer Leo Esaki, a Japanese scientist) is one of the most significant solid-state devices to emerge from the research laboratory since the transistor. It is smaller and faster in operation than either the transistor or the electron tube. It also offers a host of additional outstanding features. The high switching speed of tunnel diodes, coupled with their simplicity and stability, makes them particularly suitable for high speed operation. They operate effectively as amplifiers, oscillators, and converters at microwave frequencies. In addition, tunnel diodes have extremely low power consumption and are relatively unaffected by radiation, surface effects, or temperature variations.

The two terminal nature of the tunnel diode is something which should be considered. On one hand, this feature allows the construction of circuits which are very simple and consequently provide a savings in size and weight. This also pro-

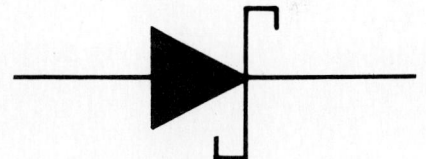

Fig. 19-35: Schematic symbol for Schottky diode.

415

vides a significant improvement in reliability. On the other hand, the lack of isolation between the input and the output can be a serious problem in some applications. Below the microwave frequency region, transistors are usually more practical and economical; however, at microwave frequencies, tunnel diodes have several advantages and are highly competitive with other high frequency devices.

A tunnel diode is a small two-lead device having a single PN junction which is formed from very heavily doped semiconductor materials. It differs from other PN junction diodes in that the doping levels are from one hundred to several thousand times higher in the tunnel diode. The high impurity levels in both the N- and P-type materials result in an extremely thin barrier region at the junction. The effects which occur at this junction produce the unusual current-voltage characteristics and high frequency capabilities of the tunnel diode.

There are various schematic symbols used to represent the tunnel diode as shown in Fig. 19-36. Figure 19-37 compares the current voltage characteristic curve of the tunnel diode with that of a conventional PN diode. The broken line shows the curve for the conventional diode and the solid line indicates the characteristic curve of the tunnel diode.

In the standard PN diode, the forward current does not begin to flow freely until the forward voltage reaches a value on the order of half a volt. This corresponds to point Y in Fig. 19-37. This forward voltage is sometimes referred to as the offset voltage. In the reverse direction, the conventional diode has a high resistance to current flow until the breakdown region is reached. The breakdown region is shown at point X in Fig. 19-37.

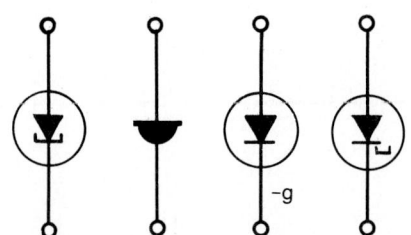

Fig. 19-36: Schematic symbols for tunnel diode.

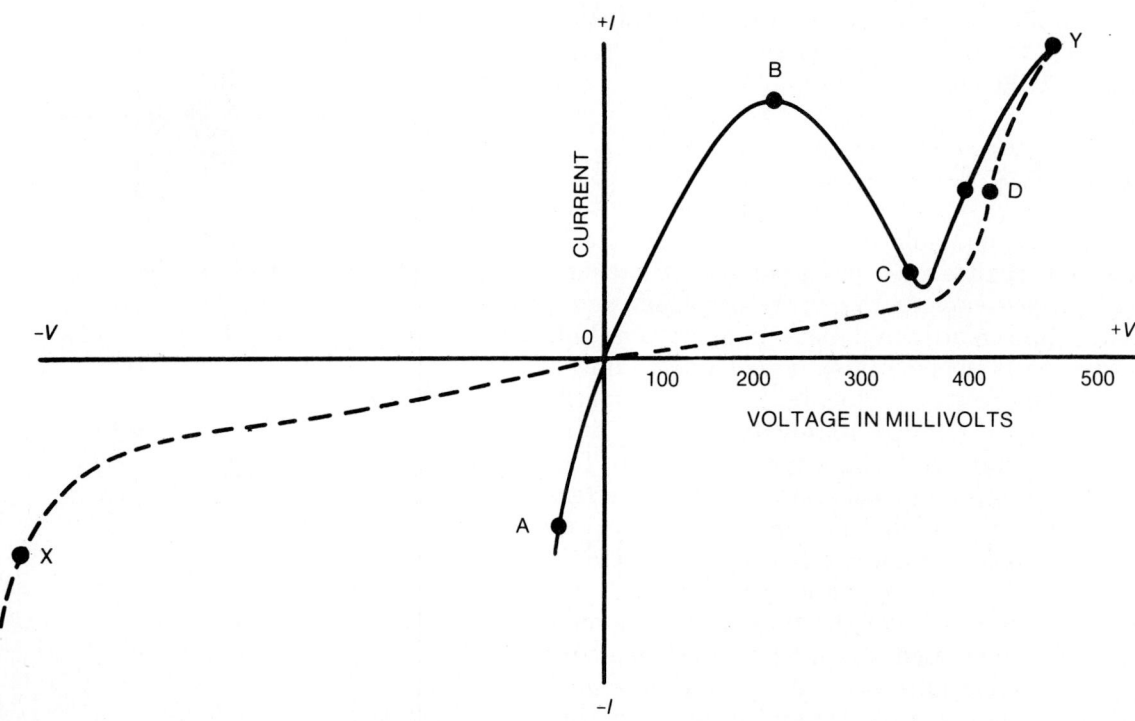

Fig. 19-37: Characteristic of a tunnel diode compared to a PN junction diode.

416

The tunnel diode is much more conductive near zero voltage. Appreciable current flows when a small bias is applied in either the forward or the reverse direction. Because the active region of the tunnel diode is at a much lower voltage than the standard semiconductor devices, it is an extremely low power device.

As the forward bias on the tunnel diode is increased, the current reaches a sharp maximum as shown at point B in Fig. 19-37. This point is referred to as peak voltage and a peak current point. The curve then drops to a deep minimum at point C. This is called *valley point voltage* and *valley point current* respectively. The curve now increases exponentially with the applied voltage and finally coincides with the characteristic curve of a conventional rectifier. This is shown at point Y on the characteristic curve. The drop in current with an increasingly positive voltage (area B to C) gives the tunnel diode the property of negative resistance in this region. This negative resistance enables the tunnel diode to convert DC power supply current into AC circuit current, and thus permits its use as an oscillator (see Chapter 25).

Varactor Diodes

PN junctions exhibit capacitance properties because the depletion area represents a dielectric and the adjacent semiconductor material represents two conductive plates. Increasing reverse bias decreases this capacitance while increasing forward bias increases it. When forward bias is large enough to overcome the barrier potential, high forward conduction destroys the capacitance effect, except at very high frequencies; therefore, the effective capacitance is a function of external applied voltage. This characteristic is undesirable in conventional diode operation, but it is enhanced by special doping in the varactor or variable-capacitance (varicap) diodes. Application categories of the varactor can be divided into two main types, tuning and harmonic generation. Different characteristics are required by the two types, but both use the voltage dependent junction capacitance effect. Figure 19-38 shows the voltage capacitance relationships.

As a variable capacitor, the varactor is rugged and small, is not affected by dust or moisture, and is ideal for remote control and precision fine tuning. The current uses of tuning diodes

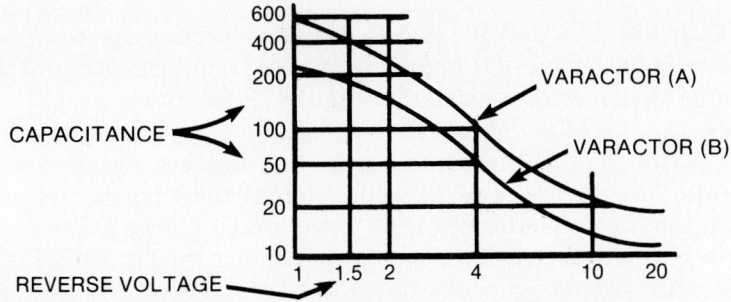

Fig. 19-38: Typical voltage capacitance relationship in varactors.

span the spectrum from AM radio to the microwave region. The most significant parameters of a tuning diode are the capacitance ratio, Q series resistance, nominal capacitance, leakage current, and breakdown voltage.

The capacitance ratio, which defines the tuning range, is the amount of capacitance variation over the bias voltage range. It is normally expressed as the ratio of the lower voltage capacitance divided by the high voltage capacitance. For example, a typical specification which reads $C_4/C_{60} = 3$ indicates that the capacitance value at 4V is three times the capacitance value at 60V. The high voltage in the ratio is usually the minimum breakdown voltage specification. A 4V lower limit is quite common since it describes the approximate lower limit of linear operation for most devices. The capacitance ratio of tuning diodes varies in accordance with construction.

Q is inversely proportional to frequency, nominal capacitance, and series resistance. Ideally, tuning diodes should have high Q, low series resistance, low reverse leakage, and high breakdown voltage at any desired capacitance ratio; however, as might be expected, these parameters are not unrelated and improving one degrades another so that often a compromise must be reached. As a rule, diodes with low capacitance values have the highest Q.

Various schematic symbols are used to designate varactor diodes, as shown in Fig. 19-39. The application of the varactor as a control device in LC tanks will be described. The resonant frequency of any tank circuit containing L and C is found by the formula $f_r = \dfrac{1}{2\pi \sqrt{LC}}$. By changing the values of either L or C, the resonant frequency of the tank changes. The resonant frequency of a tank can be changed mechanically (i.e., the use of a movable slug in the coil or the movement of plates on a variable capacitor). However, the varactor provides a means of obtaining electronic control of the tuned tank.

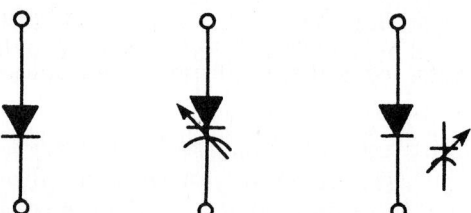

Fig. 19-39: Schematic symbols of varactors.

Capacitors in parallel add so as to give an increased capacitance value. If one of the capacitors is variable, the range of the combination is also variable. Figure 19-40 shows a varactor that is controlled by a variable voltage supply; the junction capacitance of the device is part of the tank's reactive components. The degree of reverse bias across the varactor will determine the capacitance of the varactor. C_2 blocks DC current flow through L_1. As the varactor's capacitance is varied, the resonant frequency of the tuned tank composed of L_1, C_1, and the junction capacitance will change. If C_2 is large in comparison to the junction capacitance, C_2 will have a minimal effect in

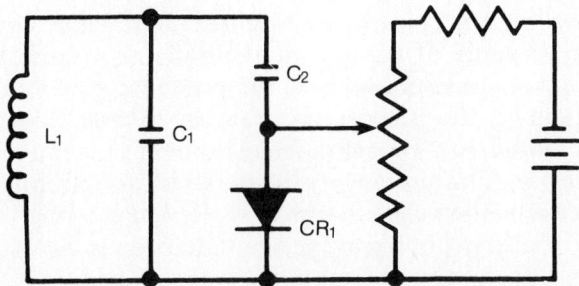

Fig. 19-40: Changing resonant frequency with a varactor.

determining the resonant frequency. A decrease in the reverse bias on the varactor will cause an increase in its capacitance. The increase in capacitance causes the overall capacitance (C_1 and varactor) to increase with a resultant decrease in resonant frequency of the circuit. An increase in reverse bias of the diode causes an opposite effect on resonant frequency.

Light-Emitting Diodes

The light-emitting diode (LED) has a fast on-off switch speed, typically 300ns and 200ns respectively and a high life expectancy as well as low power consumption. Among a wide range of applications, LEDs are used in smoke detectors, optical encoders, optical switches, security systems, level indication, and remote control.

The symbol for a light-emitting diode is given in Fig. 19-41. As already mentioned, when a conventional diode is forward-biased, free electrons cross the junction and recombine with the holes. When these electrons recombine, energy is released in the form of heat. In the LED, electron hole recombination produces a proton of light.

Through the use of various elements, namely gallium, arsenic, nitrogen, and phosphorus, the LED can be manufactured to produce a wide spectrum of colored light. Gallium arsenide phosphide emits red light. Gallium phosphide emits yellow and/or green light. Gallium nitride emits blue light. It should be noted that an LED cannot glow white.

Photoelectric Cells

Photoelectric cells (also called *photocells*) are devices which produce electrical variations in response to changes in light intensity. Any change in light diversity or wavelength will produce a corresponding change in current, voltage, or resistance. A photocell may be classified as either *photoconductive* or *photovoltaic*.

A photoconductive cell changes resistance in response to a change in light intensity. As the intensity goes up, the resistance goes down, and vice versa. Figure 19-42 shows both the schematic symbol and internal structure of a photoconductive cell.

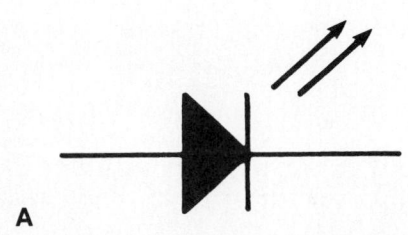

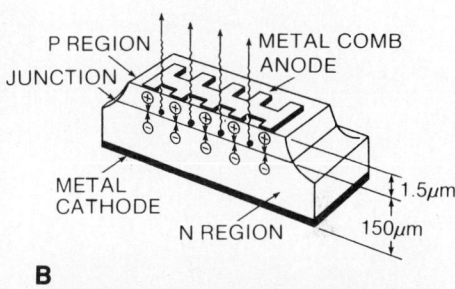

Fig. 19-41: (A) LED schematic symbol. (B) LED construction. The comb-type anode gives even current distribution while blocking less than 25% of the available light.

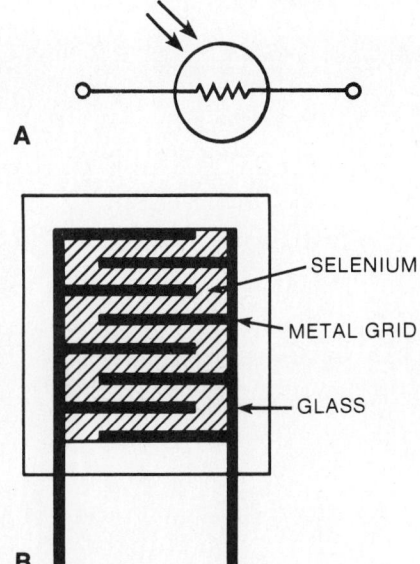

Fig. 19-42: (A) Schematic symbol for a typical photoconductive cell. (B) Structure of a selenium photoconductive cell.

Light energy, when applied to a semiconductor, causes the breaking of some of the covalent bonds. For each bond so broken, a free electron and hole are produced. The greater the light intensity, the greater the number of free electron hole pairs so created. As a result, the resistance of the semiconductor will be lower. The property of a material to exhibit a lower resistance when exposed to light is called *photoconduction*.

One of the first photoconductive materials to be discovered was the semiconductor selenium. Two metal grids, which are to serve as the terminals, are cemented to a sheet of glass or some nonconductive transparent plastic. A thin layer of selenium is deposited between the grids. The whole then is covered by another sheet of glass or plastic.

Light striking the selenium through the glass causes the resistance of the cell to drop sharply. The more intense the light rays, the greater will be the drop in resistance. Hence, the photoconductive cell can act as a control which permits more current to flow through it the more strongly it is illuminated.

For use in photoconductive cells, selenium has, for the most part, been replaced by another semiconductor, *cadmium sulfide* (CdS). When in the dark, the cadmium sulfide electrons are firmly bonded; therefore, the resistance of the cell is very high. Light striking the semiconductor imparts sufficient energy to a number of the electrons and enables them to break their bonds and so become free electrons. As a result, the resistance of the cell drops to a low value. When the light is removed, the electrons are recaptured and the resistance of the cell is restored to its original high value. (As the electrons return to their normal states, they surrender the extra energy they had absorbed in the form of a reddish glow.)

The photoconductive cell has good color sensitivity—especially at short wavelengths; that is, as the wavelength of light (color) changes, the resistance varies accordingly. If placed in a circuit with a source voltage, the variations in resistance (due to variations in light) will cause variations in the current flowing through the circuit.

A photovoltaic cell (or solar cell) is a device which produces a difference of potential when exposed to light. This device is composed of a single crystal of P-type silicon into whose top surface a thin layer of N-type silicon has been diffused. The schematic symbol and structural diagram for a photovoltaic cell are shown in Fig. 19-43.

When light strikes the N layer of the crystal, sufficient energy is supplied to break a number of covalent bonds, thus releasing free electrons and holes. Normally, the potential barrier at the junction between the layers would keep the free electrons from migrating into the P layer. However, the energy furnished by the light enables some of the electrons to overcome the barrier and enter into the P layer, thus making it negative. The holes, left behind, make the N layer positive. The resulting potential difference across the device is typically 0.45V. If an external path is provided, an electron current will flow through it from the P layer to the N layer.

420

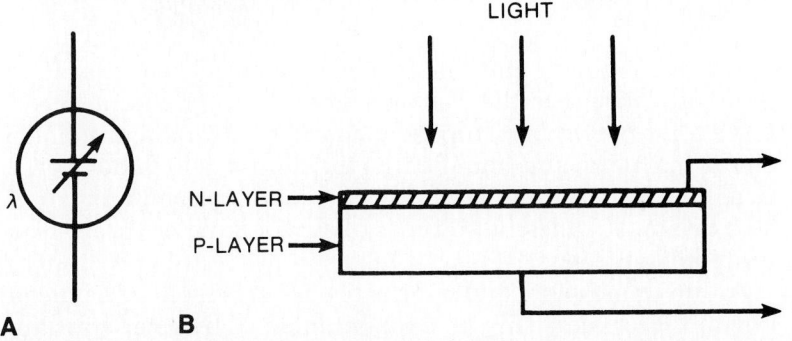

Fig. 19-43: (A) Schematic symbol for a typical photovoltaic cell; (B) structure of a solar cell.

Photovoltaic cells, having the capability of detecting high frequency (GHz) variations in light intensity, are commonly used in light metering devices.

In recent applications, highly developed solar cells are used in combinations or banks to supply electrical energy for charging batteries at remote locations and for powering portable equipment.

Recent consumer applications include solar powered watches and calculators.

DIODE TESTING

The general condition of a PN junction diode may be tested for a good/bad indication through the use of an ohmmeter. Since an ohmmeter uses a battery for its operation, polarity of the voltage appearing at the leads must be known.

Remember that a forward biased PN junction diode should exhibit a very low resistance when the negative (battery) lead of an ohmmeter is connected to the cathode (N material) and the positive (battery) lead of the meter makes contact with the anode (P material). Reversing the leads should result in a very high resistance reading since reverse bias causes a high resistance across the junction.

By using a known good PN junction diode, polarity of an ohmmeter's leads can be ascertained as follows: Connect the ohmmeter leads to the diode terminals, and note the resistance reading. If a low resistance reading is obtained, the diode is forward biased, indicating that the negative battery lead is connected to the cathode and the positive battery lead is connected to the anode. A high resistance reading would indicate opposite polarities.

The ohmmeter being used should be set to the $R \times 100$ scale in order to assure the best reading. A bad diode will be one that either reads extremely high in both directions (indicating an open diode) or extremely low in both directions (indicating a shorted diode). Care must be taken to ensure that the battery voltage and series resistance contained in the meter are such that the current and/or voltage rating of the diode under test are not exceeded.

SUMMARY

A semiconductor material is one whose electrical characteristics are intermediate between metallic conductors, which have a large number of loosely bound electrons available as charge carriers, and nonmetallic insulators, which have practically no loosely bound electrons available to conduct current. The two most widely used types of semiconductors are germanium and silicon.

An intrinsic semiconductor is a pure specimen of semiconductor material having an equal number of free electrons and holes, the number depending on the temperature of the material and the type and size of the specimen.

If an impurity is added to the semiconductor, the resulting mixture can be made to have either an excess of electrons, thus causing more electron current, or an excess of holes, thus causing more hole current. Such an impure specimen is called an extrinsic semiconductor.

An N-type semiconductor is formed by a process called doping in which an impure material such as arsenic (a pentavalent or donor material) is added, causing an increase in free electrons and a reduction in resistivity. In an N-type semiconductor, the majority carriers are electrons and the minority carriers are holes.

Trivalent (acceptor) material, such as gallium or aluminum, is used to form P-type semiconductors. In a P-type semiconductor the majority carriers are holes and the minority carriers are electrons.

P material and N material can be combined to form a PN junction, or diode. A diode acts as a two-state resistance when connected in a circuit. The state of the diode is dependent upon the polarity of the voltage applied to its elements. Its forward-voltage state is its ON or low resistance state. When the reverse voltage is applied to the diode, the diode is in its OFF or high-resistance state.

The diode VI characteristics show that a turn-on voltage of more than 0.7V is required for the forward current to flow through a silicon diode (0.3V for a germanium diode).

Every diode has a peak inverse voltage above which the diode breaks down with an increase in reverse current.

There are many types of diodes each of which has its own characteristics and applications. Among these are rectifier diodes, signal diodes, zener diodes, Schottky diodes, tunnel diodes, varactor diodes, light-emitting diodes, and photoelectric cells.

The general (good/bad) condition of a PN junction diode may be determined by using an ohmmeter.

CHECK YOURSELF (Answers at back of book)

Questions

Determine whether the following statements are true or false.

1. An N-type semiconductor is negatively charged.

2. Both silicon and germanium semiconductors have a valence of four.
3. A hole may be considered as a positive charge carrier.
4. Since free electrons and holes flow in the same directions, the total current is the sum of the two.
5. If an ohmmeter indicates zero when the positive terminal is connected to a diode's anode and the negative side is connected to the cathode, the diode is good.
6. The application of a reverse bias reduces the potential barrier while a forward bias increases the barrier.
7. Although it has a large number of positive holes, a P-type material is still neutral.
8. An acceptor impurity, being trivalent, is so called because it has an incomplete bond that can accept a valence electron and create a hole.
9. The peak inverse voltage is the maximum reverse voltage a diode can withstand before an appreciable current begins to flow.
10. The 0.7V turn-on voltage of a silicon diode is a direct result of the potential barrier across a PN junction.
11. If the temperature of pure silicon rises, the number of free electrons increases and the number of holes decreases.
12. Forward bias involves connecting the positive to the N-type and the negative to the P-type.
13. In pure silicon there are just as many free electrons as there are holes.
14. Absolutely zero current flows through a reverse-biased diode since the barrier is made so large.
15. A donor impurity must be trivalent to form an N-type semiconductor.
16. The voltage-current characteristics can be approximated by a closed switch when forward-biased and an open switch when reverse-biased.
17. If the majority carriers in a semiconductor are electrons, the material must be P-type.
18. The potential barrier makes the P side positive and the N side negative.
19. The resistance of a forward-biased diode is high because majority carriers can easily cross the PN junction.
20. When current flows in pure germanium, there are two components: one due to hole flow, the other to free electron flow.

Fill in the blanks with the appropriate word or words.

21. The two most widely used types of semiconductor materials are _____ and _____ . Individual atoms of these materials have _____ electrons in the valence shell.
22. The simplest approximation of a rectifier diode is the ideal diode. There is no_____ current, no breakdown, and no forward_____ drop. The equivalent circuit is a switch.
23. The second and third approximations utilize the_____ voltage drop across a diode. The second approximation uses _____ V for silicon diodes and _____ V for germanium diodes. The third approximation takes into ac-

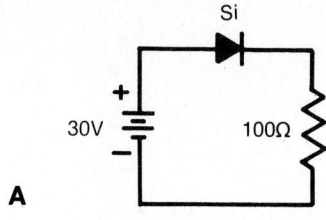

A

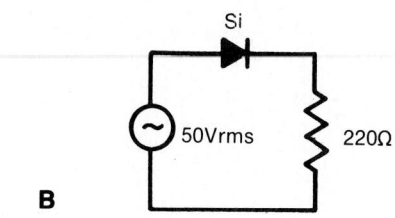

B

Fig. 19-44

Fig. 19-45

count an additional voltage caused by the _____ resistance.

24. A _____ is a vacancy in the outer shell. It can attract and hold a nearby electron. This fusion is referred to as _____.

25. The process of adding trivalent or pentavalent impurities to a pure semiconductor is called _____ . The P-type semiconductor has an excess of holes, and the N-type has an excess of _____ .

26. The lack of _____ means the Schottky diode can switch off faster than an ordinary PN junction diode.

27. The region between the peak and valley points of a tunnel diode is known as a _____ region. When operating along this part of the curve, the tunnel diode converts DC input power to AC output power.

28. Diodes that are optimized for variable capacitance are called _____ . In many applications, they are replacing mechanically tuned capacitors.

29. In a forward biased PN junction diode, free electrons cross the junction and fall into holes. When they _____ , these electrons radiate energy. In an LED, this energy is radiated in the form of _____ instead of heat.

30. A _____ diode operating in the breakdown region is ideally equal to a battery and a small resistance. The main use for this type of diode is in voltage _____ , holding voltage constant.

Problems

1. Calculate the current through the diode in Fig. 19-44. Consider the diode to be an ideal switch.
2. Calculate the maximum current in Figs. 19-44A and B, taking into account the barrier potential.
3. The light-emitting-diode of Fig. 19-45 has 1.8V across it. What is the current?

CHAPTER 20

DC Power Supplies

Electric power is generally available in the United States in the form of an alternating supply at a fixed frequency of 60Hz. The nominal voltage is generally $120V_{rms}$. The question of why a change of AC to DC is necessary probably arises. The answer, simply, is that for proper operation many electronic circuits depend upon DC. As already pointed out, the PN junction diode conducts more easily in one direction than in the other. Transistors and electron tubes are also unidirectional and a constantly alternating source voltage would be undesirable.

Figure 20-1 is a block diagram of a power supply showing an AC input to and a DC output from a block labeled positive power supply and filter-regulator network. Although Fig. 20-1 shows a power supply that provides a unidirectional current which causes a positive voltage output, it might well be designed to furnish a negative voltage output.

The typical DC power supply (Fig. 20-2) can be divided into four basic stages: power transformation, rectification, filtering, and regulation. But, before describing how an AC input is

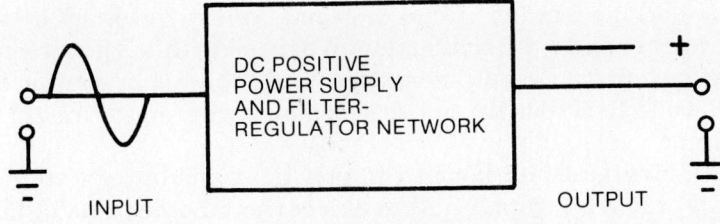

Fig. 20-1: Positive voltage output from an AC input.

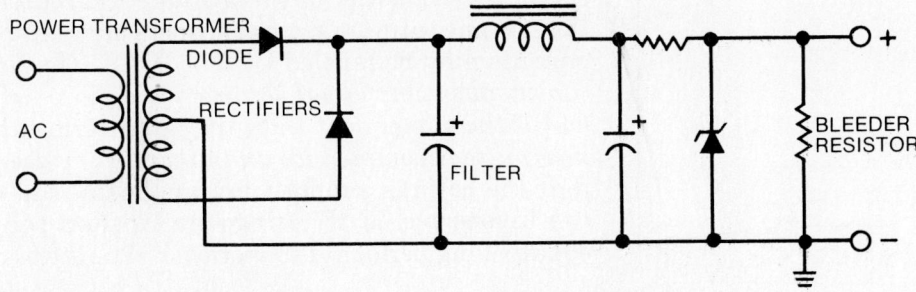

Fig. 20-2: Stages of DC power supply.

converted into a DC output the definition "load" as it applies to power supplies must be understood. Load is the current supplied to the power consuming device or devices connected to the power supply. The power consuming device needs voltage and current for proper operation and this voltage and current is supplied by the power supply. The power consuming device may be a simple resistor or one or more electronic circuits using resistors, capacitors, coils, and active devices.

POWER TRANSFORMERS

Power transformers are used in power supply circuits because of the efficiency and ease with which they transfer energy. As described in Chapter 12, the power transformer is capable of receiving a voltage at one level and delivering it at the same level, some higher level, or some lower level. In fact, many power-supply transformers will have one primary winding and several secondary windings that provide different amounts of voltage and have different current ratings. The transformer may have one relatively high voltage winding, a medium-voltage winding, and still another with a relatively low voltage output. When using a full-wave center-tapped rectifier configuration, the transformer secondary must have a secondary winding that is tapped at its center. Technically, the secondary is composed of but one winding, which is tapped at the center with a lug or conductor that can be utilized in a full-wave center-tapped rectifier arrangement. The amount of voltage from the center tap to either the upper or lower end of the transformer is identical, and the DC output voltage will be decided by this AC voltage across one-half of the entire secondary winding. This same transformer could be used in a full-wave bridge rectifier circuit or even a half-wave circuit by not using the center-tap connection. When using this type of circuit configuration, DC output voltage would be calculated from the AC voltage across the entire secondary winding of the transformer.

As mentioned earlier in Chapter 12, transformers are normally rated by voltage output and current. Some transformers may be rated in volt-amperes or watts, with individual ratings compiled for both the primary and secondary windings. This is especially true of transformers of the multipurpose variety that can be used in both full-wave bridge and center-tap configurations. Some ratings on transformers and other electronic components may be labeled CCS or ICAS. The first term means *continuous commercial service*, which implies a more rugged use of the component than the latter term, which stands for *intermittent commercial and amateur service*. CCS indicates that the ratings supplied are applicable for round-the-clock continuous use; ICAS ratings are for short periods of use with equally long periods of inoperation. The latter ratings are normally higher for a specific transformer because intermittent use with periods of inoperation usually allows for better cooling of the transformer core and windings.

RECTIFICATION

Rectification is described as the changing of an alternating current to a unidirectional or direct current. The normal PN junction diode is well suited for this purpose as it conducts very heavily when forward biased (low resistance direction) and only slightly when reverse biased (high resistance direction).

There are three basic types of rectifier circuits: *half-wave rectifiers, full-wave rectifiers,* and *full-wave bridge rectifiers.*

Half-Wave Rectifiers

Figure 20-3 shows the PN junction diode functioning as a half-wave rectifier. A half-wave rectifier is one that uses only half of the input cycle to produce an output.

The induced voltage across L_2 (the transformer secondary) will be as shown in Fig. 20-3. The dots on the transformer indicate points of the same polarity. During that portion of the input cycle which is going positive (solid line), CR_1, the PN junction diode, will be forward biased and current will flow through the circuit. L_2, acting as the source voltage, will have current flowing from the top to the bottom. This current then flows up through R_L causing a voltage drop across R_L equal to the value of current flowing times the value of R_L. This voltage drop will be positive at the top of R_L with respect to its other side and the output will, therefore, be a positive voltage with respect to ground. It is common practice for the end of a resistor receiving current to be given a sign representing a negative polarity of voltage, and the end of the resistor through which current leaves is assigned a positive polarity of voltage. The voltage drop across R_L, plus the voltage drops across the conducting diode and L_2, will equal the applied voltage. Although the peak

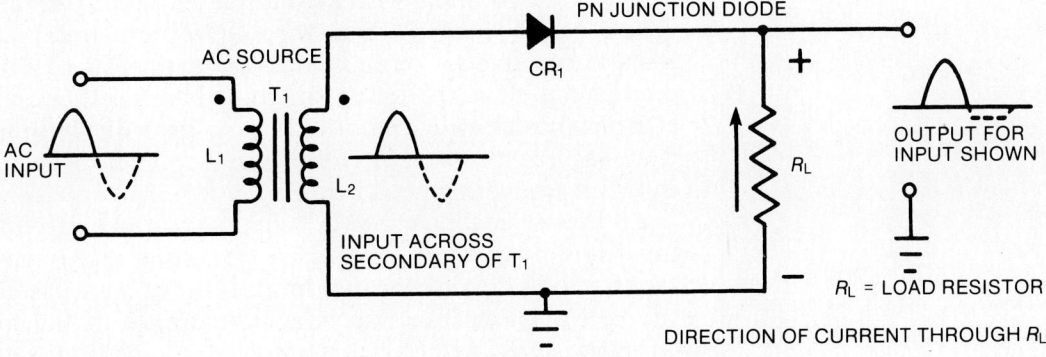

Fig. 20-3: **Positive voltage output half-wave PN junction diode rectifier.**

427

output voltage will nearly equal the peak input voltage, it cannot reach this value due to the voltage drops, no matter how small, across CR_1 and L_2.

The negative half cycle of the input is illustrated by the broken line. When the negative half cycle is felt on CR_1, the PN junction diode is reverse biased. The reverse current will be very small—but it will exist. The voltage resulting from the reverse current, as shown below the line in the output, is exaggerated in Fig. 20-3 to bring out the point of its existence. Although only one cycle of input is shown in Fig. 20-3, it should be realized that the action described above continually repeats itself, as long as there is an input.

By reversing the diode connection in Fig. 20-3—having the anode on the right instead of the left—the output would now become a negative voltage. The current would be going from the top of R_L toward the bottom—making the output at the top of R_L negative in respect to the bottom or ground side.

The same negative output can be obtained from Fig. 20-3 if the reference point (ground) is changed from the bottom (where it is shown) to the top, or cathode connected end of the resistor. The bottom of R_L is shown as being negative in respect to the top, and reading the output voltage from the "hot" side of the resistor to ground would result in a negative voltage output.

The half-wave rectifier will normally indicate improper functioning in one of two manners—there is no output or the output is low. The *no output* condition can be caused by no input—the fuse has blown, the transformer primary or secondary winding is open, the PN junction diode is open.

The low output condition might be caused by an aged diode. A check of both forward and reverse resistance of the diode may reveal the condition of the diode. Low output can be the result of an increased forward resistance or a decreased reverse resistance of the diode.

It is necessary to check the AC input voltage to see if it is of the correct value—a low input voltage will result in a low output voltage. A check of the transformer secondary voltage should be made to see if it is of the correct value also, as a low secondary voltage will also result in a low output voltage.

By removing the input voltage, resistance checks of the components can be made with an ohmmeter. Is the primary open? The secondary open? Has the diode increased in forward resistance value or decreased in reverse resistance value? Is the diode open? Has the load resistance become shorted? Do the components show signs of excessive heat dissipation—have they become discolored? Does a measurement of load current indicate an overload condition?

All these questions should be answered when troubleshooting the half-wave rectifier. If a trouble is discovered in the rectifier, one should then determine if the *cause* is a local one (in the rectifier itself) or due to some changes in the following circuitry, such as the power supply filter components or changing load impedance. While it is important that the trouble be repaired, elimination of the cause of the trouble is of greater importance.

428

Full-Wave Rectifier

The PN junction diode works just as well in a full-wave rectifier circuit as shown in Fig. 20-4. The circuit shown has a negative voltage output; however, it might just as well have a positive voltage output. This can be accomplished by either changing the reference point (ground side of R_L) or by reversing the diodes in the circuit.

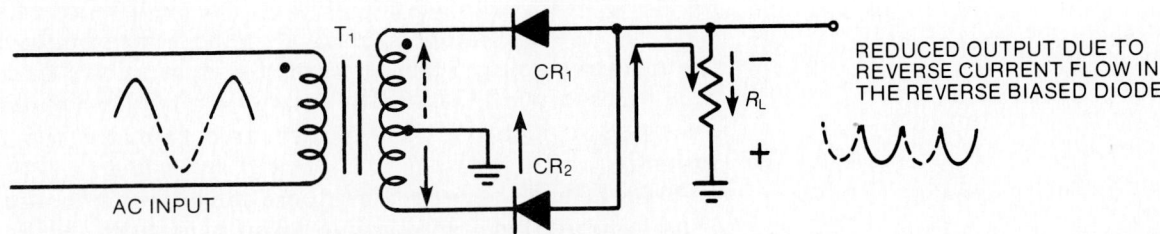

Fig. 20-4: Unfiltered negative output full-wave rectifier.

The AC input is felt across the secondary winding of T_1. This winding is center tapped as shown; the center of the secondary is at ground potential. This seems to be a good time to define ground as a reference point which is of no particular polarity. When the polarity is such that the top of T_1 secondary is negative, the bottom is positive. At this time, the center tap, as shown, has two polarities, positive with respect to the top half of the winding and negative with respect to the bottom half of the winding. When the secondary winding is positive at the top, the bottom is negative and the center tap is negative with respect to the top and positive with respect to the bottom. What is the polarity of the reference point (ground)? The answer must be in terms of "with respect to." For each alternation of the input, one of the diodes will be forward biased and the other one reverse biased.

For ease of explanation, the negative alternation will be considered when the rectifier circuit is initially energized by the AC source. CR_1 will be forward biased—negative voltage felt on its cathode—and CR_2 will be reverse biased—a positive voltage felt on its cathode. Therefore, the top of T_1 secondary must be negative with respect to the bottom. When forward bias is applied to CR_1, it conducts heavily from cathode to anode (dashed arrow), down through R_L—this current flow creates a voltage drop across R_L—negative at the top with respect to the bottom or ground side of R_L. The current passing through R_L is returned to CR_1 by going through the grounded center tap and up the upper section of the center tapped secondary winding of T_1. This completes the first alternation of the input cycle. The second alternation of the input now is of such polarity as to forward bias CR_2—a negative voltage at the bottom of T_1 secondary winding with respect to ground. CR_1 is now reverse biased. CR_2 conducts, current moves in the same direction through R_L (solid arrow)—top to bottom—back through the lower half of the center tapped secondary to CR_2. One may

429

wonder why current does not flow from the anode of one of the diodes through the anode to cathode of the other diode. The answer is simple: Current flow through a reverse biased diode is very slight due to the high resistance of the diode when reverse biased. This rectifier has a slightly reduced output (Fig. 20-4) because of the reverse current flow.

As can be seen in the output waveform of Fig. 20-4, there are two pulses of DC out for every cycle of AC in; this is full-wave rectification. Current flow through R_L is in the same direction, no matter which diode is conducting. The positive going alternation of the input allows one diode to be forward biased and the negative going alternation of the input allows the other diode to be forward biased. The output for the full-wave rectifier shown is a negative voltage measured from the top R_L to ground.

As in the half-wave rectifier, there can be two indications of trouble: no output or low output. No output conditions are indications of no input, shorted load circuits, open primary winding, open or shorted secondary winding, or defective diodes. Low output conditions are possible indications of aging diodes, open diodes, or opens in either half of the secondary winding (allowing the circuit to act as a half-wave rectifier).

The method for troubleshooting the full-wave rectifier is the same as used for the half-wave rectifier. Check voltages of both primary and secondary windings, check current flow, and when the circuit is deenergized, take resistance measurements. Shorted turns in the secondary windings give a lower voltage output and possibly shorted turns in the primary winding will produce a lower voltage input. (Shorted turns are hard to detect with an ohmmeter; they are more easily detected by taking a voltage reading across various terminals of the energized transformer).

Full-Wave Bridge Rectifier

Figure 20-5 shows a full-wave bridge rectifier circuit capable of producing a positive output voltage. When the AC input is applied across the secondary winding of T_1, it will forward bias diodes CR_1 and CR_3 or CR_2 and CR_4. When the top of the transformer is positive with respect to the bottom, as illustrated in Fig. 20-5 by the designation number 1, both CR_1 and CR_2 will feel this positive voltage. CR_1 will have a positive voltage on its cathode, a reverse bias condition, and CR_2 will have a positive voltage on its anode, a forward bias condition. At this same time the bottom of the secondary winding will be negative with respect to the top, placing a negative voltage on the anode of CR_3 (a reverse bias condition) and on the cathode of CR_4 (a forward bias condition).

During the half cycle of the input designated by the number 1 in Fig. 20-5, CR_2 and CR_4 are forward biased and will, therefore, conduct heavily. The conducting path is shown by the solid arrows, from the source (the secondary winding of T_1) through CR_4 to ground, up through R_L making the top of R_L positive with respect to the grounded end, to the junction of CR_2 and CR_3.

430

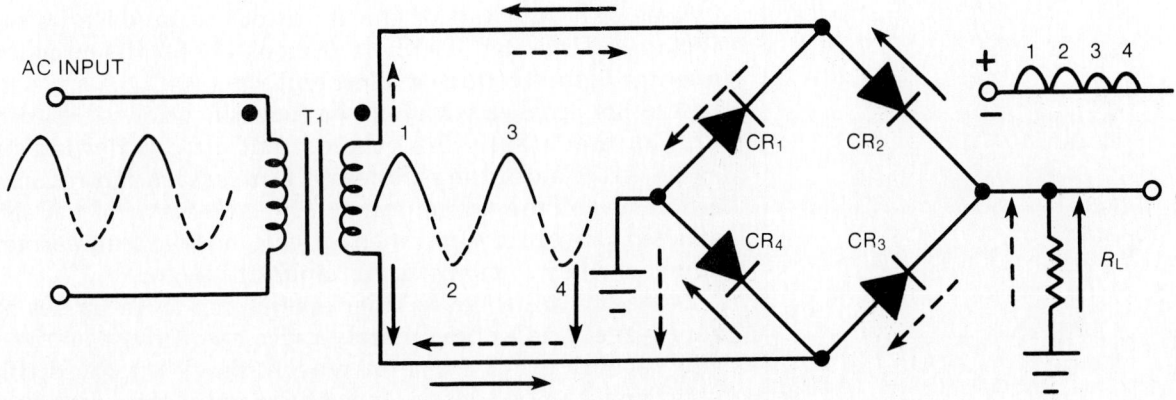

Fig. 20-5: Typical bridge rectifier circuit.

CR₂, being forward biased, offers the path of least resistance to current flow and this is the path current will take to get back to the source.

During the alternation designated by the number 2 in Fig. 20-5, shown by the dashed arrows, the top of the secondary winding is going negative while the bottom is going positive. The negative voltage at the top is felt by both CR_1 and CR_2, forward biasing CR_1 and reverse biasing CR_2. The positive voltage on the bottom of T_1 secondary is felt by CR_3 and CR_4, forward biasing CR_3 and reverse biasing CR_4. Current flow starting at the source (T_1 secondary winding) is through CR_1 to ground, up through R_L (this is the same direction as it was when CR_2 and CR_4 were conducting, making the top of R_L positive with respect to its grounded end) to the junction of CR_2 and CR_3. This time CR_3 is forward biased and offers the least opposition to current flow, and current takes this path to return to its source.

As can be seen, the diodes in the bridge circuit operate in pairs; first one pair—CR_1 and CR_3—conduct heavily and then the other pair—CR_2 and CR_4—conduct heavily. As shown in the output waveform, we get one pulse out for every half cycle of the input or two pulses out for every cycle in. This is the same as for the full-wave rectifier circuit explained previously.

The bridge circuit will also indicate a malfunction in one of two manners: It has no output or a low output. The causes for both conditions are the same as they were for the half- or full-wave rectifier. If any one of the diodes opens, the circuit will act as a half-wave rectifier with a resultant lower output voltage.

Comparison of Rectifiers

Each of the three basic rectifier circuits discussed has individual advantages and disadvantages. For example, a *half-wave rectifier* is very simple to construct; it requires only one diode and may be used with or without a transformer. Its major disadvantage, however, is that it is inefficient. Remember that

431

it uses only one-half of the AC input to produce its output voltages. This inefficiency is responsible for the circuit's low average output voltage and current.

Another problem with half-wave rectification is DC *core saturation* that results from the current *always* flowing in the same direction in the secondary. Core saturation reduces the efficiency of the transformer. The half-wave is usually restricted to use in low current applications or in transformerless supplies when economy is an important factor.

Since the full-wave rectifier operates on both halves of the sine wave, it is approximately twice as efficient as the half-wave rectifier. For the same reason, the full-wave rectifier is also capable of producing twice the current for a given output voltage. When using this type of system, the efficiency of the power transformer is very high; approximately 90% of the input power is useable in the output.

Its greatest disadvantage is that it requires a center-tapped transformer and *matched diodes*. The potential barriers of matched diodes are identical. If these voltages were not equal, the output waveform would resemble the one shown in Fig. 20-6. Also, for a given output voltage, the peak inverse voltages of these diodes must be two times greater than the peak inverse voltage of the diode in the half-wave rectifier. For example, the diodes in a full-wave rectifier with an output voltage of 20V must have a peak inverse voltage (PIV) rating of at least 40V. Remember, all of the voltage produced on the secondary winding is felt across the reverse-biased diode, but only half of this induced voltage is developed across the output. On the other hand, the diode in the half-wave rectifier only needs a PIV rating of at least 20V. It should be noted here that the diode(s) in both the half-wave and full-wave rectifiers will require higher PIV ratings once filtering capacitors have been added.

The *full-wave bridge rectifier* has several advantages over the full-wave rectifier. It can be designed with or without a transformer. If a transformer is desired for isolation or a step-up or step-down in voltage, no center tap is necessary. Also, for

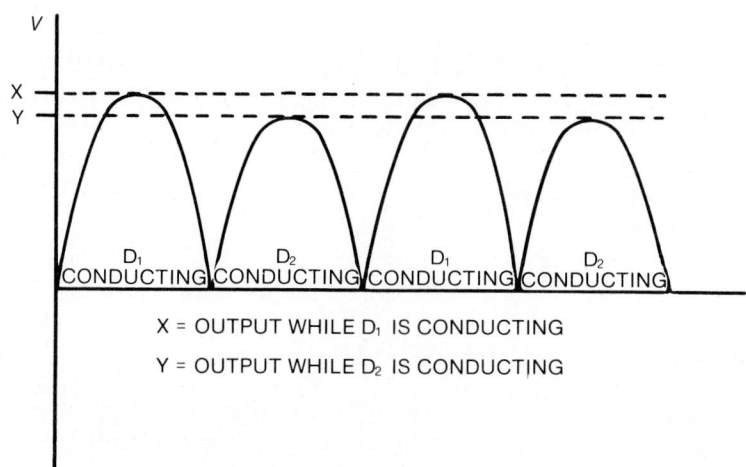

X = OUTPUT WHILE D₁ IS CONDUCTING

Y = OUTPUT WHILE D₂ IS CONDUCTING

Fig. 20-6: Waveform when voltages are not equal.

432

a given transformer, the output voltage from the bridge is nearly twice the height of the output voltage from the full-wave rectifier.

Unlike the other two types of rectifiers, the reverse-biased diodes in the bridge rectifier are never exposed to more than the peak voltage induced on the secondary winding, even when filter capacitors are added. While this circuit does require four matched diodes, its advantages make the bridge rectifier preferable for most electronic tasks.

Junction Diode Considerations

The junction rectifier diode has four important ratings that must be taken into consideration when designing a power supply. They are the maximum
- average forward current
- repetitive peak reverse voltage
- surge current
- repetitive forward current

These ratings are important when it becomes necessary to troubleshoot a power supply or when selecting junction diodes for replacement when the desired one is not readily available.

The maximum average forward current is the maximum amount of average current that can be permitted to flow in the forward direction. This rating is usually given for a specified ambient temperature and should not be exceeded for any length of time as damage to the diode will occur. The repetitive peak reverse voltage (PRV) is that value of reverse bias voltage that can be applied to the diode without causing it to break down.

The maximum surge current is that amount of current allowed to flow in the forward direction in nonrepetitive pulses. Current should not be allowed to exceed this value at any time and should only equal this value for a period not to exceed one cycle of the input. The maximum repetitive forward current is the maximum value of current that may flow in the forward direction in repetitive pulses.

All of the ratings mentioned above are subject to change with temperature variations. If the temperature increases, the ratings given on the specification sheet should all be lowered or damage to the diode will result.

Diode Protection

A silicon diode allows current to flow for only one-half of the AC cycle, so when it does conduct, it may pass at least twice the average DC. With a capacitive-input filter, it may pass peak current of 10 or more times the average direct current. This peak current, I_{rep}, is also known as the peak repetitive current, and, as explained earlier, is much less than the I_s or surge current rating, which applies to the surge current that flows through the rectifier diode when the circuit is initially switched on. One other current rating for solid-state rectifiers is the I_0 rating, or the average DC current rating. This sets the limit on the amount of current the diode may safely handle for continuous periods of operating time.

To sum up these current ratings, I_0 sets the limits on average current through the diode, I_{rep} limits the instantaneous peaks of current that may occur at intervals throughout the operating cycle, and I_s limits the instantaneous surge of current during initial circuit activation. The duration of time involved in the surge-current rating is for one AC cycle at 60Hz, or about sixteen thousandths of a second.

Comparing these current ratings, a ratio of approximately 4:1 is obtained for I_{rep} and I_0. Thus if the I_0 rating is 1A, the I_{rep} rating will be about 4A. The ratio of I_s to I_0 is about 12:1, which would be approximately 12A in a rectifier with an I_0 rating of 1A.

Thermal protection of diodes is necessary in certain applications owing to the small physical size of these components. Fortunately, solid-state rectifier diodes have a low internal resistance, and heat problems do not occur in units rated for less than 2A if operated within their ratings; but rectifiers rated for 2A or more usually must depend on external heat sinks in order for maximum ratings to apply.

A heat sink is an external, usually metal, device that attaches directly to the solid-state component. It is a solid piece of metal that often has metal fins or vanes protruding from each side in order to release more heat into the air. The heat sink effectively increases the physical size of the solid-state rectifier, thus allowing a more efficient method of channeling heat away from the delicate crystal interior. Heat sinks are usually insulated from other portions of the circuit because, in most instances, the full DC voltage will be present on their exterior surfaces. If this is not practical, the rectifier diode will be insulated electrically from the heat sink but not thermally. A small piece of plastic or mica sheet will be placed between the rectifier and the heat sink for the electrical insulation. To ensure proper conductance of heat from the rectifier to the sink, silicone grease is sometimes applied to both units at the insulating material where they meet. Silicone grease is an excellent conductor of heat, and at the same time maintains a good electrical insulation property. The heat is conducted to the heat sink while the current is limited to the surface of the rectifier only. The heat sink may now be grounded to the common ground or chassis without danger of an electrical short circuit.

Transient Protection. A frequent cause of rectifier failure in DC power-supply circuits is transient voltage or spikes in the AC power line. These are quick surges of voltage that cause the transformer output voltage to be much higher than that normally applied to the rectifier. Spikes can be caused by high-current devices such as motors operating on the same AC line being switched on and off or even by distant lightning strokes; whatever their cause, transient voltage problems can cause permanent damage to solid-state rectifier diodes.

434

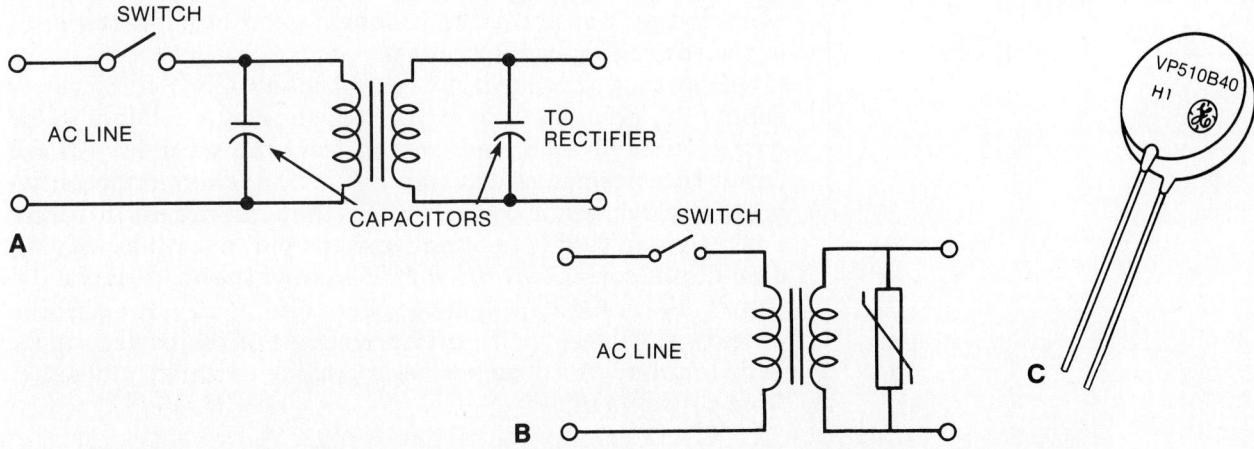

Fig. 20-7: (A) Suppressing line transients using capacitors; (B) suppressing line transients using a GE-MOV (metal oxide varistor); and (C) a typical metal oxide varistor.

Suppressing line transients is easily accomplished. Figure 20-7 shows two ways. In Fig. 20-7A, two capacitors are used, one in the primary input to the secondary, the other across the secondary. At 60Hz, the primary winding capacitor looks like a very high resistance and has almost no effect on operation of the circuit; however, when a spike occurs, its high-frequency component sees the capacitor as a short circuit and cancels itself before entering the transformer primary. The capacitor in the transformer secondary is an additional protection device, working in much the same manner as the primary capacitor.

The circuit shown in Fig. 20-7B utilizes a suppression device called a GE-MOV (metal oxide varistor). Metal oxide varistors (Fig. 20-7C) are voltage dependent, symmetrical resistors which operate in a manner similar to back-to-back zener diodes. The impedance of the varistor changes from a very high standby value to a very low conducting value (thus clamping the line voltage to a safe level) when exposed to high energy voltage transients. The excessive energy of the incoming high voltage pulse is absorbed by the GE-MOV, thus protecting the voltage sensitive circuit components.

The two types of transient voltage protection shown do an adequate job of eliminating spikes from the DC power circuits, but neither is 100% effective in all cases. If persistent problems recur, rectifiers with double the PIV rating normally called for should eliminate further problems if used with these forms of spike protection.

Surge Protection. As was mentioned earlier, each time the power supply is turned on, the rectifier diode sees what appears to be a short circuit for the instant when the filter capacitor is charging. During this fraction of a second, large amounts of current are drawn through the diodes. In many cases and applications, this current may exceed the I_s rating of even the finest rectifiers and cause eventual failure. Surge protection is usually necessary to protect the diodes until the capacitor has

435

been charged and normal continuous operation characteristics of the power supply are in effect.

One method of reducing the surge potential when the power supply is activated is to install resistors of a relatively low ohmic value in each lead coming from the secondary of the power transformer, as depicted in Fig. 20-8. This circuit configuration works well in power supplies that are required to deliver a continuous value of voltage and current; but when varying amounts of current are drawn throughout the operation of the supply, the resistors will cause poor voltage regulation. Surge protection devices in the primary circuit of the power supply will eliminate this drawback and provide adequate protection from current surges.

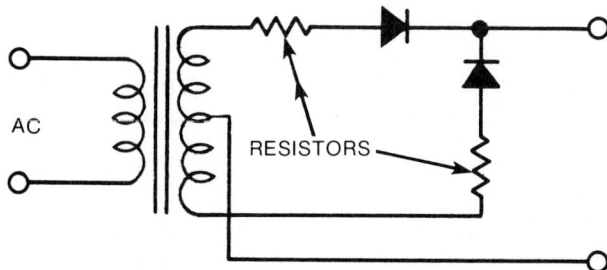

Fig. 20-8: Surge protection using series resistors.

Figure 20-9 depicts a typical primary surge protection circuit. The surge-limiting resistor is located in the AC line to the power transformer and is electrically removed from the circuit a fraction of a second after the supply is activated and the filter capacitor is fully charged. The line resistor causes the input AC voltage to drop when the circuit is initially activated until the filter capacitor is nearly charged. When this capacitor approaches a full charge, the surge current drops and allows the relay in the primary circuit to activate. When this happens, the relay contacts short out the surge resistor, and for the rest of the operating cycle, the entire circuit functions as if the resistor were not there at all. This entire process takes place in a fraction of a second, and the delay of the relay in activating cannot usually be detected. A relay with contacts rated to handle the AC current drawn by the supply must be chosen because the full current demand is conducted across these contacts. The

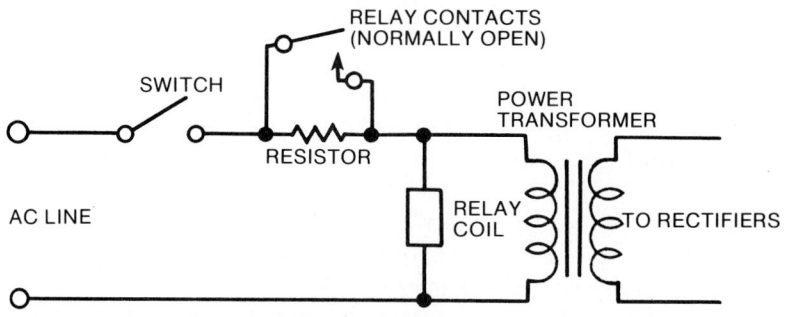

Fig. 20-9: Surge protection using series resistors in the primary line and a line-activated relay.

436

value of the surge resistor will vary depending on the type of supply and the primary voltage, but resistance will usually be between $10\,\Omega$ and $75\,\Omega$ with a power rating of 5W to 50W.

Diode Stacks

As was previously discussed, solid-state rectifier diodes have specific limits in voltage and current ratings per unit. Owing to cost and physical space factors, it is often convenient to stack diodes in series or parallel configurations for increased ratings. Two diodes in parallel exhibit the same PIV or PRV ratings, but forward current may be almost doubled. Diodes in series still maintain the same current ratings as would apply with a single component, but the voltage ratings may be almost doubled. Owing to differences in internal resistance, special circuitry should be provided to make certain that one rectifier does not pass the majority of the forward current or is not subjected to the majority of the PIV. When internal resistances differ, either of the two conditions could exist, resulting in one or more diodes handling the majority of the work and eventually failing in conditions beyond normal operating limits.

Diodes in Series. When two or more solid-state rectifiers are placed in series, a capacitor and resistor should be placed in parallel across each component in the stack to equalize the PIV drops and as further protection against voltage spikes. When two resistors are combined in parallel, the resulting total resistance is less than the value of either resistor. By thinking of the internal resistance of the rectifier as a resistor, the added external resistor, which is of a much higher resistance value, will equalize the internal resistance to a value almost equal to the next diode-resistor combination in the string or stack (Fig. 20-10A). The value of the external resistor is not critical, but a means of calculating the resistance can be had by multiplying the PIV rating of the diode by 500, with the resulting figure being the ohmic value of the external resistor. A rectifier with a PIV rating of 1,000 would require a shunting resistor of approximately $500,000\,\Omega$ or $500\,k\Omega$. This calculation holds true whether there are 2 or even 20 diodes in the stack.

The capacitors aid the rectifiers in the string to conduct and to block current flow at the same time and should be of a value of about $0.01\,\mu F$. Capacitors used for this purpose should be of the noninductive ceramic-disc type, with a voltage rating equivalent to the voltage dropped at each rectifier. If two diodes are

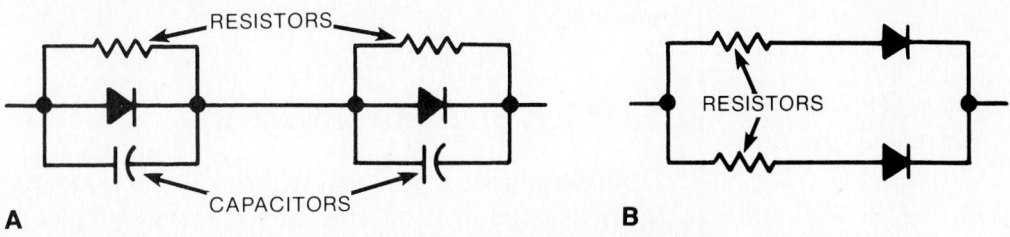

Fig. 20-10: Protective circuits for diodes in series and in parallel.

437

used in a power supply with a 500V DC output, the protection capacitors should be rated higher than half of this amount, since the total output is divided by two. Doubling the ratings of these capacitors will assure adequate protection from voltage breakdown. Capacitors should be of the same value and manufacture, with 10% tolerances usually termed as adequate. All diodes in the string should also be of the same type and manufacture to assure proper equalization.

Diodes in Parallel. For greater current-handling capabilities, diodes may be placed in parallel provided adequate equalization procedures are taken. Figure 20-10B shows how two equal-value resistors are placed in series with the rectifiers and in parallel with each other. If the resistors are omitted from the circuit, one diode may draw most of the current and malfunction after a brief period of time. The resistors equalize the internal resistance of the stack and are of a very low ohmic value. Resistor values are chosen to deliver a voltage drop of about 1V at peak current demand from the supply using Ohm's Law of $V = IR$, where V is the voltage drop of 1V and I is the peak current. R will then be the required value in ohms of the equalization resistors.

Other Combinations. Figure 20-11 shows a combined series-parallel configuration using diodes and equalization-protection components. Here the stack contains paralleled diodes and equalization resistors for greater current-handling capability. These individual stacks connect into a larger series string. This is a combination that can occupy a larger physical space and is not very practical, but it can be used effectively in a situation where the correct components are not available. Examine this schematic closely. It can be seen that both forms of diode combination, series and parallel, have been used with proper protection circuitry.

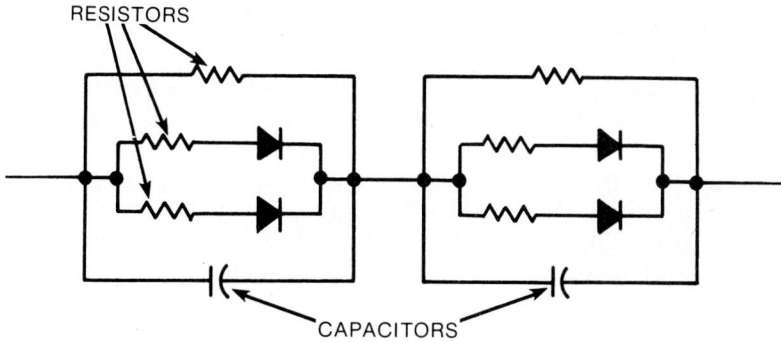

Fig. 20-11: Series-parallel combinations and associated circuitry.

POWER SUPPLY FILTERS

As previously indicated the operation of most electronic circuits is dependent upon a direct current source. It has been illustrated how alternating current can be changed into a pul-

438

sating direct current; that is, a current that is always positive or negative with respect to ground although it is not of a steady value (it has "ripple").

Ripple can be defined as the departure of the waveform of a rectifier from pure DC. It is the amplitude excursions, positive and negative, of a waveform from the pure DC value—the alternating component of the rectifier voltage. Ripple contains two factors which must be considered: frequency and amplitude. Ripple frequency in the rectifiers that have been presented is either the same as line frequency for the half-wave rectifier or twice the line frequency for the full-wave rectifiers.

In the half-wave rectifier, one pulse of DC output was generated for one cycle of AC input; the ripple frequency is the same as the input frequency. In the full-wave rectifiers (center tapped and bridge), two pulses of DC output were produced for each cycle of AC input—the ripple frequency is twice that of the line frequency. With a 60Hz input frequency there will be a 60Hz ripple frequency in the output of the half-wave rectifier and a 120Hz ripple frequency in the output of the full-wave rectifier.

The amplitude of the ripples in the output of a rectifier circuit will give us a measure of the effectiveness of the filter being used—the ripple factor. The ripple factor is defined as the ratio of the rms value of the AC component to the average DC value, $r = \dfrac{V_{rms}}{V_{DC}}$. The lower the ripple factor the more effective the filter. The term "percent of ripple" may be used. This is different from the ripple factor only because the figure arrived at in the ripple factor formula is multiplied by 100 to give us a percent figure, % Ripple = $\dfrac{V_{rms}}{V_{DC}} \times 100$.

In both formulas given, V_{rms} is the rms value of ripple voltage and V_{DC} is the DC value (average value) of the output voltage.

Filter circuits used in power supplies are usually low pass filters. (As described in Chapter 15, a low pass filter is a network which passes all frequencies below a specified frequency with little or no loss but is highly discriminate against all higher frequencies.) The filtering is done through the use of resistors or inductors and capacitors. The purpose of power supply filters is to smooth out the ripple contained in the pulses of DC obtained from the rectifier circuit while increasing the average output voltage or current.

Filter circuits used in power supplies are of two general types: capacitor input and choke input. There are several combinations that may be used, although they are referred to by different names (Pi, RC, L section, etc.). The closest element electrically to the rectifier determines the basic type of filter being used.

Figure 20-12 depicts the basic types. In Fig. 20-12A a capacitor shunts the load resistor, therein bypassing the majority of ripple current which passes through the series elements. In Fig. 20-12B an inductor (choke) in series with the load resistor opposes any change in current in the circuit. The capacitor input filter will keep the output voltage at a higher level compared to

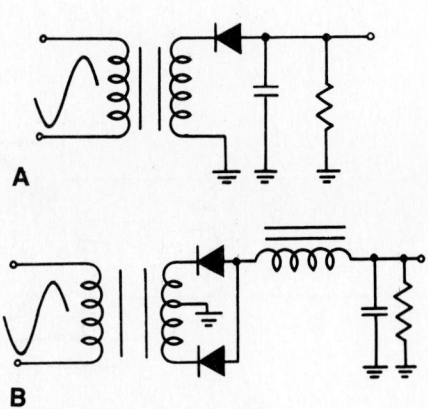

Fig. 20-12: Filter circuits.

a choke input. The choke input will provide a steadier current under changing load conditions. From this it can be seen that a capacitor input filter would be used where voltage is the prime factor and the choke input filter is used where a steady flow of current is required.

Capacitor Input Filter

First, an analysis will be made of the simple capacitor input filter depicted in Fig. 20-13A. The output of the rectifier, without filtering, is shown in Fig. 20-13B and the output, after filtering, is illustrated in Fig. 20-13C. Without the capacitor, the output across R_L will be pulses as previously described. The average value of these pulses would be the V_{DC} output of the rectifier.

With the addition of the capacitor, the majority of the pulse changes are bypassed through the capacitor and around R_L. As the first pulse appears across the capacitor, changing it from negative to positive, bottom to top, the peak voltage is developed across the capacitor. When the first half cycle has reached its peak and starts its negative going excursion, the capacitor will start to discharge through R_L maintaining the current through R_L in its original direction, thereby holding the voltage across R_L at a higher value than its unfiltered load. Before the capacitor can fully discharge, the positive excursion of the next half cycle is nearing its peak, recharging the capacitor. As the pulse again starts to go negative, the capacitor starts to discharge once again. The positive going excursion of the next half cycle comes in and recharges the capacitor; this action continues as long as the circuit is in operation.

The charge path for the capacitor is through the transformer secondary and the conducting diodes, and the discharge path is through the load resistor. The reactance of the capacitor, at the line frequency, is small compared to R_L, which allows the changes to bypass R_L and, effectively, only pure DC appears across R_L.

Figures 20-13B and C illustrate the use of an RC time constant. If the value of C_1 or R_L were such that the discharge time was the same or less than that of the charge time, we would have no filtering action. The larger the values of C_1 and R_L, the longer the discharge time constant and the lower the ripple factor. The charge time of the capacitor is the RC values of the

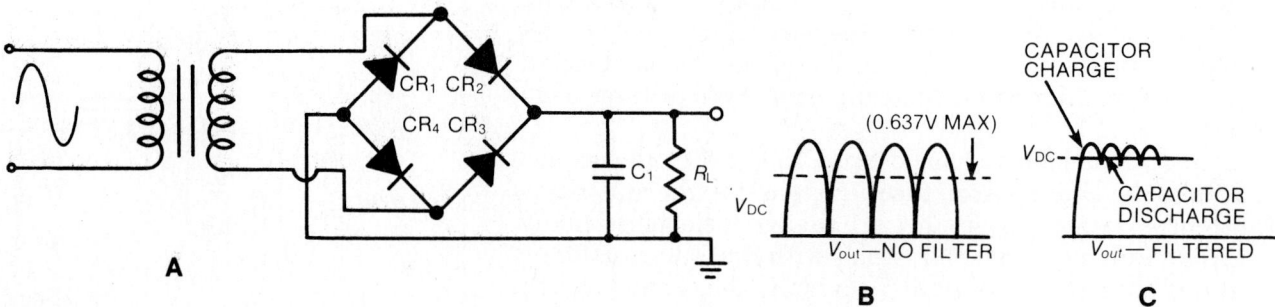

Fig. 20-13: Low voltage power supply with a simple capacitor filter.

440

capacitor, the conducting diodes, and the transformer secondary. The impedance offered by these elements is very small when compared to the impedance in the discharge path of the capacitor—the value of R_L. The output voltage is practically the peak value of the input voltage. This circuit provides very good filtering action for low currents, but results in little filtering in higher current power supplies due to the smaller resistance of the load.

It is important to keep in mind that when a rectifier diode is immediately followed by a capacitor, an additional stress is placed on the diodes in that rectifier. Figure 20-14A shows a half-wave rectifier with a capacitor as a filter. C_1 charges to the peak of the secondary voltage during the half-cycle in which the diode conducts. The capacitor is large enough to hold the voltage across the load at approximately this same level throughout the cycle. The voltage across the secondary reverses during the next half cycle. At the peak of the negative cycle, the anode of the diode is at –20V with respect to ground. At this point, the difference of potential across the diode is twice the peak value of the secondary, or 40V. A diode must be selected which can withstand this voltage.

The diode used in this example must have a PIV of at least 40V. However, a good rule of thumb is to operate the diode at no more than 80% of its rated value. Thus, the PIV rating should be no lower than:

$$PIV = \frac{40}{0.8} = 50V$$

Figure 20-14B illustrates that a similar condition exists in the full-wave rectifier. When C_1 is charged to the positive peak and the top of the secondary is charged to the negative peak, D_1 will experience twice the peak voltage.

This situation, however, does not hold true for the bridge rectifier. Here the diodes are never exposed to more than the peak of the secondary voltage (Fig. 20-14C). It might appear that D_2 is being subjected to twice the peak value, but actually D_3 is conducting. Thus, point A is effectively at ground potential and the voltage across D_2 is equal to the peak of the secondary voltage. You will find that no diode in the bridge rectifier is ever exposed to more than the peak secondary voltage. Thus, diodes with lower PIV ratings can be used for a bridge rectifier.

Choke Input Filter

The next filter to be analyzed is the choke input filter, or the L-section filter. Figure 20-15 shows this filter and the resultant output of the rectifier after filtering has taken place. The series inductor, L, (Fig. 20-15) will oppose rapid changes in current. The output voltage of this filter is less than that of the capacitor input filter since the choke is in series with the output impedance. The parallel combination of R_L and C in connection with L smooths out the peaks of the pulses and results in a steady, although reduced, output.

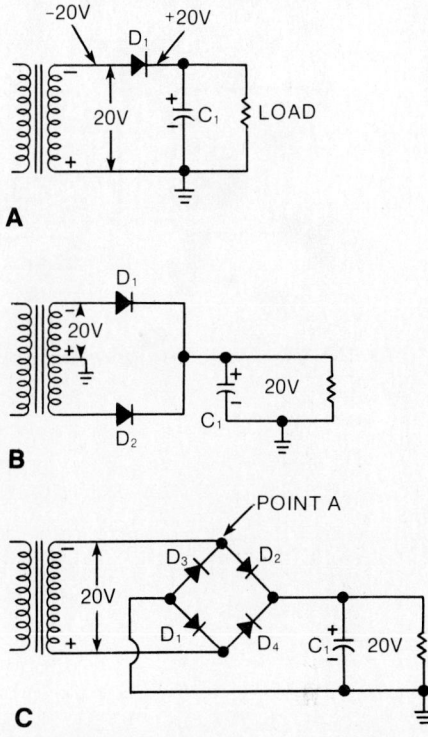

Fig. 20-14: **Effect of a capacitor on the circuit.**

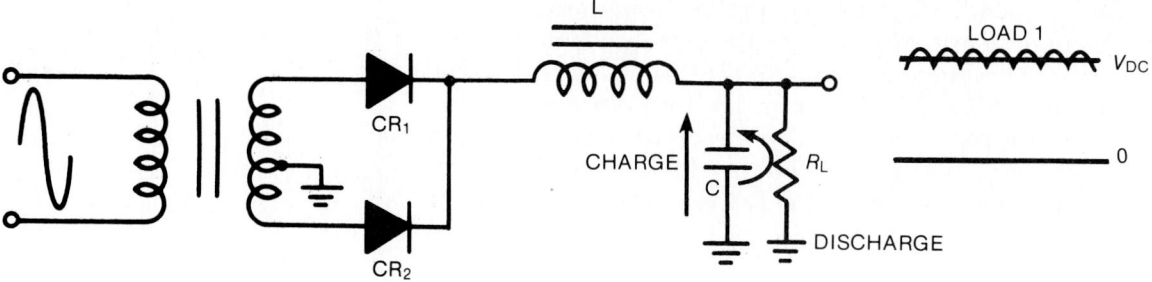

Fig. 20-15: L-section (choke input) filter showing representative waveforms.

The inductance chokes off the peaks of the alternating components of the rectified waveform and the DC voltage is the average, or DC value, of the rectified wave. The choke input filter allows a continuous flow of current from the rectifier diodes rather than the pulsating current flow as seen in the capacitor input filter. The X_L of the choke reduces the ripple voltage by opposing any change in current during either the rapid increases in current during the positive excursions of the pulses or decreases in current during the negative excursions. This keeps a steady current flowing to the load throughout the entire cycle. The voltage developed across the capacitor is maintained at a relatively constant value approaching the average value of input voltage because of this steady current flow.

Multiple Section Choke Input Filter. The filtering action provided by the choke input filter can be enhanced by using more than one such section. Figure 20-16 shows two sections with representative waveforms approximating the shape of the voltage with respect to ground at different points in the filter networks.

While Fig. 20-16 shows two choke input sections being used as a multiple section filter, more sections may be added as desired. While the multiple section filter does reduce the ripple content—and they are found in applications where only a minimum ripple content can be tolerated in the output voltage—they also result in reduced regulation. The additional sections add more resistance in series with the power supply which results in increased voltage variations in the output when the load current varies.

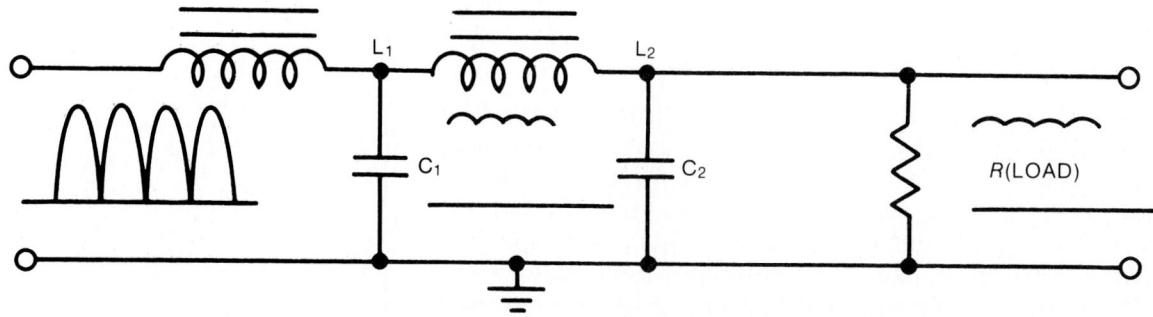

Fig. 20-16: Multiple section choke input filter with representative waveforms.

442

Pi Filter

As mentioned in Chapter 17, the pi filter is a combination of the simple capacitor input filter and the choke input filter. This filter is shown in Fig. 20-17.

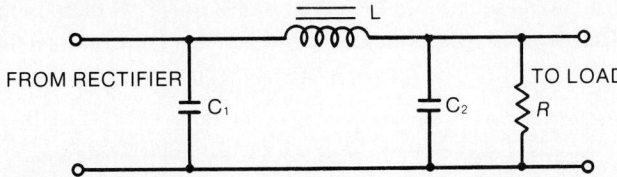

Fig. 20-17: Pi filter.

The resistor, R, is known as a bleeder resistor and is found in practically all power supplies. The purpose of this resistor is two-fold: When the equipment has been working and is then turned off it provides a discharge path for the capacitors, preventing a possible shock to maintenance personnel; it also provides a fixed load, no matter what equipment is connected to the power supply. It is also possible to use this resistor as a voltage dividing network through the use of appropriate taps. More on this will be subsequently discussed.

The pi filter is basically a capacitor input filter with the addition of an L-section filter. The majority of the filtering action takes place across C_1 which charges through the conducting diode(s) and discharges through R, L, and C_2. As in the simple capacitor input filter, the charge time is very fast compared to the discharge time. The inductor smooths out the peaks of the current pulses felt across C_2, thereby providing additional filtering action. The voltage across C_2, since C_2 is in parallel with the output, is the output voltage of the power supply. Although the voltage output is lower in this filter than it would be if taken across C_1 and the load, the amount of ripple is greatly reduced.

Even though C_1 will charge to the peak voltage of the input when the diodes are conducting and discharge through R when they are cut off, the inductor is also in the discharge path and opposes any changes in load current. The voltage dividing action of L and C_2 is responsible for the lower output voltage in the pi filter when compared to the voltage available across C_1.

In Fig. 20-17, the charge path for both C_1 and C_2 is through the transformer secondary, through the capacitors, and, in the case of C_2, through L. Both charge paths are through the conducting diodes. However, the discharge path for C_1 is through R and L while the discharge path for C_2 is through R only. How fast the input capacitor, C_1, discharges is mainly determined by the ohmic value of R. The discharge time of the capacitors is directly proportional to the value of R. If C_1 has very little chance to discharge, the output voltage will be high. For lower values of R the discharge rate is faster, and the output voltage will decrease. With a lower value of resistance, the current will be greater and the capacitor will discharge further. The V output is the average value of DC; and the faster the discharge time, the lower the average value of DC and the lower the V_{out}.

RC Capacitor Input Filter. While the pi filter previously discussed had an inductor placed between two capacitors, the inductor can be replaced by a resistor, as shown in Fig. 20-18. The main difference in operation between this pi filter and the one previously discussed is the reaction of an inductor to AC when compared to the resistor. In the former filter the combination of the reactances of L and C_2 to AC was such as to provide better filtering, giving a relatively smooth DC output.

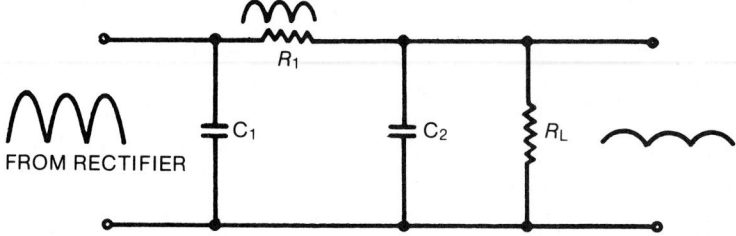

Fig. 20-18: Capacitor input filter and associated waveforms.

In Fig. 20-18 both the AC and DC components of rectified current pass through R_1. The output voltage is reduced due to the voltage drop across R_1 and the higher the current, the greater this voltage drop. This filter is effective in high voltage-low current applications. As in choke input filters, the capacitor input filters shown may be multiplied; for example, identical sections may be added in series.

The choice of a filter for a particular use is a design problem, but the purpose and operation of filters should be understood because of their importance to the proper operation of equipment following the power supply.

Bleeder Resistor Voltage Divider

Figure 20-19 shows how the bleeder resistor may function as a voltage divider network. Terminal 3 is grounded. Current is flowing as indicated, from the bottom to the top, making terminal 4 negative with respect to ground (terminal 3). Terminals 1 and 2, on the other hand, are positive with respect to ground.

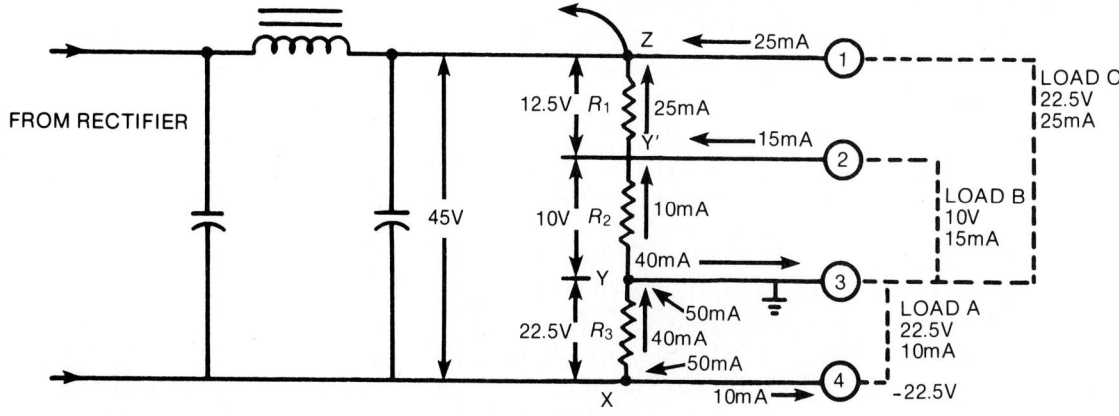

Fig. 20-19: Voltage divider network with bleeder resistor.

444

Since collector voltage from NPN and PNP transistors used in amplifier circuits needs positive and negative values, respectively, a voltage dividing network such as that shown in Fig. 20-19 is typical. At point X, 50mA of current enters the junction of R_3 and terminal 4 (load A). At point X this current divides; 40mA flows through R_3 and 10mA through load A. They both have a voltage drop of 22.5V across them.

At point Y the current again divides into the parallel paths of R_2, load B, and load C. As indicated, 40mA flows through loads B and C, and 10mA flows up through R_2. The current through R_2 causes a voltage drop of 10V. The top of R_2 is positive 10V with respect to the bottom, or ground side, of R_2. Load B, being in parallel with R_2, is also a positive 10V, although there are 15mA of current flowing through load B. It should be evident that the value of R_2 is greater than the impedance of load B by a ratio of 1.5:1. This can be proven by using the formula $R(Z) = \dfrac{V}{I}$, since you have the values of both V and I. At point Y' the currents through R_2 and load B rejoin at terminal 2 and flow up through R_1. The 25mA flowing through R_1 make the voltage drop across R_1 a positive 12.5V, with respect to its lower end. The voltage measured from point Z to ground, however, will be 22.5V since the voltage drops across R_1 (12.5V) and R_2 (10V) are between point Z and ground. (The voltage across load C is in parallel with the series combination of R_1 and R_2, with load B in parallel with R_2.) At point Z the 25mA flowing in load C combines with the 25mA through R_1, which satisfies Kirchhoff's current law.

Figure 20-20 shows this division and joining of current in line form with the arrowheads indicating the direction of current.

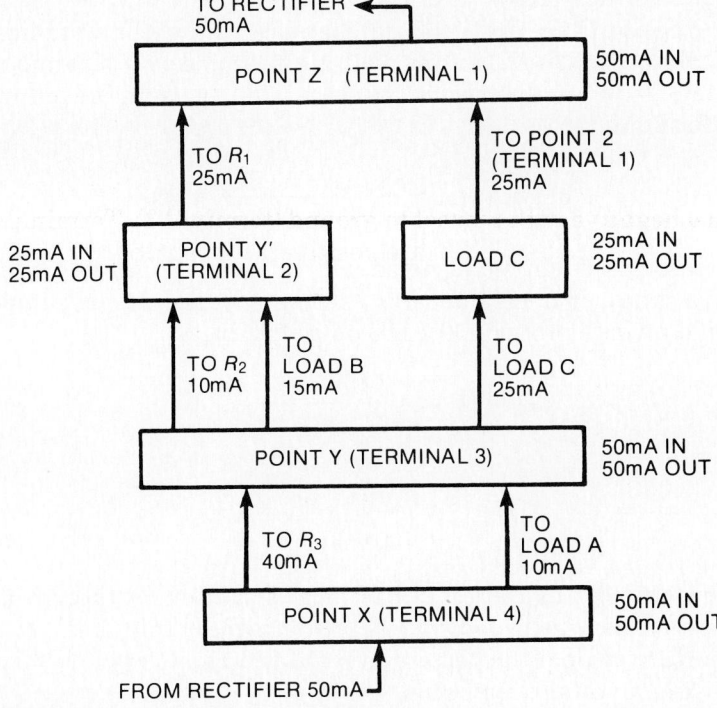

Fig. 20-20: Flowchart of Fig. 20-19.

445

While it may seem that current is flowing in two directions at one time, it will be evident that it is not. Just think in terms of Kirchhoff's current law—"The algebraic sum of currents entering and leaving a junction of conductors is zero."

VOLTAGE REGULATORS

It is often required that the output voltage from a power supply be maintained at a constant value regardless of input voltage or load variations. The device used to give us this control is the voltage regulator. Regardless of the specific operating device used, the action is basically the same—that of providing a constant value of output voltage from the power supply irrespective of reasonable variations in input voltage or load current.

A voltage regulator, then, is an electronic device connected in the output of a power supply to maintain the output voltage at its constant rated value. It reacts automatically within its rated limits to any variations in the output voltage. Should the output voltage rise or fall, the voltage regulator automatically compensates for the change and maintains the output voltage at the required value. Although there may be large changes in load current drawn from a power supply or changes in the applied voltage, the voltage regulator maintains a constant output voltage.

The regulating action of the voltage regulator is, in effect, that of a variable resistor which responds to any changes in the current flowing through it. This "variable resistance" may be in series or in shunt with the load.

Percent of Regulation. Percent of regulation is an indication of how much the output voltage changes over a range of load resistance values. It can be calculated by using the following formula:

$$\% \text{ Regulation} = \frac{V_{\text{no-load}} - V_{\text{full-load}}}{V_{\text{full-load}}} \times 100$$

Ideally, there should be no change in output voltage over the entire range of operation. For example, a 10V power supply should always produce 10V DC output. Thus,

$$\text{Regulation} = \frac{10V - 10V}{10V} \times 100$$

$$= \frac{0}{10V} \times 100$$

$$= 0\%$$

Therefore, 0% regulation is the ideal situation. Although all circuits must settle for something less than this ideal, it is important to hold the percent of regulation to a very low value by using a voltage regulator.

446

Series Voltage Regulator

A simple series voltage regulator is shown in Fig. 20-21. Here a variable resistor (R) is connected in series with a load resistor (R_L).

As in any series circuit the current through R and R_L will depend upon the value of source voltage and the total amount of resistance ($R + R_L$) in the circuit. It can be seen that the amount of current through R_L and, therefore, the amount of voltage drop across R_L will be dependent on the setting of R. If the value of R is increased, the current will decrease and the voltage drop across R_L will also decrease. If the value of R is decreased, the value of total resistance also decreases and current will increase causing a larger voltage drop to be felt across R_L. If a fixed value for the voltage across R_L is set, the regulator will insure that this value remains constant by properly varying the resistor R to compensate for circuit changes.

If the input voltage increases, without changing R, the output voltage would increase. However, if R is increased in value, a larger voltage drop across R results. The increase in the value of R and the voltage drop across it are proportional to the rise in input voltage (a 5V increase in the input results in a 5V increase in the drop across R) so that the output voltage (the drop across R_L) remains exactly what it was before the increased input voltage. Conversely, a decrease in the input voltage requires a decrease in the value of R so that the voltage drop across R is proportional to the decreased input voltage (a 5V decrease in the input results in a 5V decrease in the voltage drop across R) and the output voltage remains constant.

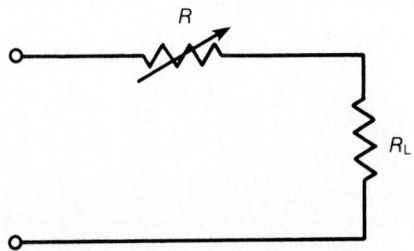

Fig. 20-21: Series voltage regulating circuit.

Shunt Voltage Regulator

Figure 20-22 depicts a simple shunt voltage regulator. As in the simple series voltage regulator, a voltage dividing action is used to obtain regulation. The amount of current through R_s is now determined by the shunt regulating device R_v. The larger the current through R_s, the higher its voltage drop and the lower the voltage across R_L.

The shunt voltage regulator operates in the following manner. R_s (the series resistance) is in series with R_v (the shunt regulating device) and R_L (which represents the load impedance). R_v and R_L, being in parallel, will have the same voltage drop

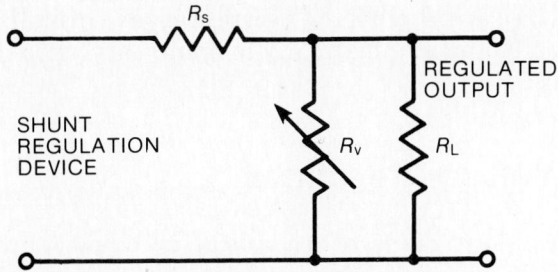

Fig. 20-22: Shunt voltage regulating circuit.

447

across them. R_s, being in series, will have the total current flowing through R_v and R_L flowing through it. To keep the voltage constant, R_v must be adjusted to compensate for changes in the input voltage as well as for changes caused by load impedance changes.

Analysis of an increasing source voltage will exemplify the regulating action. The voltage across R_v and R_L would tend to increase with this increase of source voltage. To compensate for this undesired change, the value of R_v must be decreased. This results in more current flowing through the entire circuit (the total resistance has been decreased) and the increase in current through R_s causes a larger voltage drop across R_s; consequently, the output voltage remains constant.

Conversely, when the source voltage decreases, the voltage across the parallel leg, R_v and R_L, would tend to decrease. To compensate for this change, the value of R_v must be increased. This increases the total resistance and decreases total current. Less current through R_s means less voltage drop across R_s and more voltage available across R_v and R_L. Once again the voltage is held constant. The amount of voltage increase or decrease from the source is the same amount of increase or decrease in the voltage drop across R_s, resulting in the desired amount of voltage for the load.

If the value of R_L were to decrease (the equivalent of adding another resistor in parallel with R_v and R_L), the total resistance would decrease. Current would increase, causing a greater voltage drop across R_s. This would tend to cause the output voltage to decrease. To compensate for this change, the value of R_v must be increased, reducing the amount of current through R_s and bringing the output voltage to the desired amount.

If the value of R_L were to increase (the equivalent of removing a circuit from the system), the total resistance would increase. Total current would decrease and the decrease in current through R_s would result in a lower voltage drop across the element. This would tend to cause the output voltage to rise. To compensate for this change the value of R_v must be decreased, increasing the amount of current through R_s, increasing the voltage drop across R_s, and decreasing the voltage across the load to the desired amount.

Both the voltage regulators discussed rely upon a mechanical adjustment and would require continuous monitoring of the equipment to obtain, at best, barely satisfactory results. These voltage regulators were used for explanation purposes. Electronic devices accomplish the function of the variable resistors electronically and automatically to immediately provide excellent results.

Zener Voltage Regulator

The breakdown diode, or zener, is an excellent source of variable resistance. Zener diodes come in voltage ratings ranging from 2.4V to 200V, with tolerances of 5%, 10%, and 20% and with power dissipation ratings as high as 50W.

The zener diode will regulate to its rated voltage with changes in load current or input voltage. Referring to the zener diode shunt-type regulator in Fig. 20-23, the zener diode VR_1 is in series with fixed resistor R_s. The voltage across the zener is constant, thus holding the voltage across the parallel load R_L constant. Although the circuit shown depicts a positive voltage output, it is a simple matter to have a negative output voltage—reverse the zener and the polarities shown in Fig. 20-23.

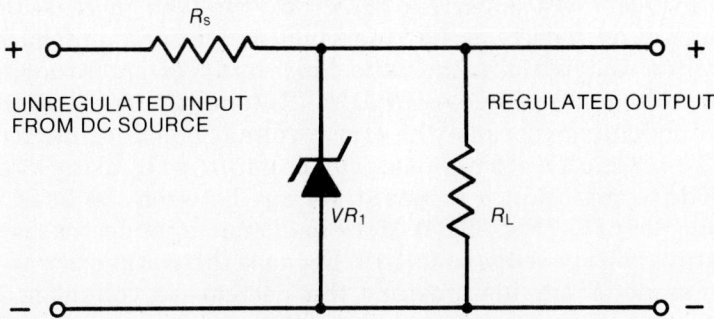

Fig. 20-23: Zener diode shunt type regulator circuit.

The value of R_s in Fig. 20-23 has been fixed so that it can handle the combined currents of the diode and the load and still allow the diode to conduct well within the breakdown region. R_s stabilizes the load voltage by dropping the difference between the diode operating voltage and the unregulated input voltage.

The zener diode is a PN junction that has been specially doped during its manufacture so that when reverse biased it will operate at a specific breakdown voltage level. It operates well within its rated tolerance over a considerable range of reverse current.

Zener Diode Shunt Regulator. The zener diode regulator shown in Fig. 20-23 is, as already mentioned, of the shunt type. If the input voltage across the zener diode must decrease, the zener current will also decrease. The total current in the circuit decreases, much the same as when the value of R_v was increased in the simple shunt regulator of Fig. 20-22. The current through R_s, having decreased, results in a lower voltage drop across R_s. This results in the voltage drop across the zener and the load returning to the desired voltage. The zener diode has replaced the variable resistor of Fig. 20-22 and makes the necessary adjustment automatically with the change in input voltage or load current. Effectively, the variable resistor that required manual adjustment has been replaced with an electronic equivalent—the zener diode.

When input voltage increases, the change is immediately felt across the zener. This effectively biases the zener so that there is an increase in zener current. The increase in zener current means an increase in total circuit current. R_s, in series with the source, will have an increase in current through it, resulting in a larger voltage drop across it. With the larger drop across R_s, the voltage across the zener and, therefore, the load are reduced to the desired output voltage. The zener, in this instance, has a lower resistance.

449

For changes in load current the zener makes the adjustment so that source current remains constant and the output voltage will also be constant. For instance, if the current drawn by the load decreases, the zener current will increase by a corresponding amount. The total current remains the same. If the voltage across R_s is the same as it was before the load current decrease and if source voltage has not changed, then it is logical to infer that the output voltage is at the desired regulated amount. Conversely, if the current drawn by the load increases, the zener current decreases by the same amount and total current in the circuit remains the same. Since total current is the same, the voltage drops across R_s and VR_1 (and, therefore, R_L also) cannot change because the source voltage has not changed.

The operation of a regulator can be improved by using a zener diode to maintain a constant voltage between the base and collector of Q_1 (Fig. 20-24). If the load resistance decreases, the output voltage tends to fall off. Because the voltage across the zener diode remains constant, this decrease in voltage is seen across R_1; thus, the forward-bias on Q_1 (and Q_2) decreases. This causes the collector current of Q_2 to decrease. The voltage drop across R_s is reduced, and the output voltage rises to its former value.

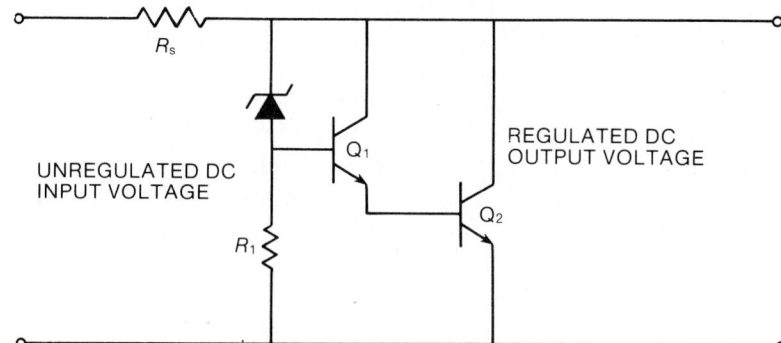

Fig. 20-24: **Improved diode shunt type regulator circuit.**

One advantage of a shunt regulator is its ability to withstand a short circuit without serious damage (the series resistor could get quite hot). But, some words of caution about the shunt type voltage regulator are in order. Do not operate this circuit without a load. If a no-load condition exists, the transistor (Q_2) must dissipate more power than usual—the load power as well as its normal power. If this condition occurs, it is quite possible that the maximum ratings will be exceeded and damage may result.

If there is a failure in the voltage regulator circuit, the following checks should help in locating the trouble:

1. Check to find out if there is a load on the regulator circuit. The lack of a load might indicate a damaged transistor as the source of trouble.

2. Check the DC voltage measurements at the input to determine whether voltage is applied and whether it is within tolerance.

3. Since the operation of this regulator is based upon the voltage divider principle, measurements of the voltages across

the output terminals and across R_s might be necessary to determine if the output voltage is within tolerance or if the drop across series resistor R_s is excessive (be sure to observe correct voltage polarity when making these checks).

4. If the load is shorted or if R_s is open, a voltage measurement across R_s will indicate source voltage. In both cases there will be no output. It will now be necessary to check the value of R_s. (Disconnect R_s from the circuit when making the measurement.)

5. If the zener diode becomes open, there will be no regulation and the output voltage should be higher than normal.

6. If the zener diode becomes shorted there will be no output.

In general, if the output voltage is above normal, it is an indication of an open circuit in the shunt elements, either VR_1 or the load, or an increase in impedance of these same elements. An output voltage that is below normal is an indication of an increased value of R_s, a low input voltage, or an excessive load current due to a decrease in load impedance.

Zener Diode Series Regulator. In a series regulator circuit, the control device is placed in series with the load resistance. The resistance of the control device is adjusted automatically to ensure that the voltage across the load remains constant. A simple series regulator is shown in Fig. 20-25.

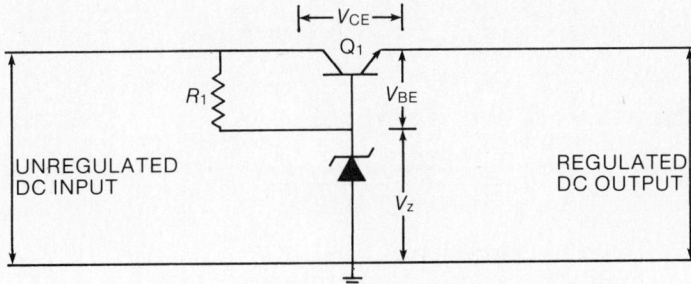

Fig. 20-25: Zener diode series type regulator circuit.

The resistor (R_1) is chosen to allow the necessary reverse current to flow for the zener diode to operate within its breakdown region. This guarantees the voltage at the base of Q_1 will remain constant. The regulated output will equal the zener voltage (V_z) minus the base-emitter junction voltage (V_{BE}). If the output voltage tries to fall due to a decrease in load resistance, the difference in potential between the emitter of Q_1 and its base will increase. (Remember the voltage at the base remains constant.) When V_{BE} increases, the transistor will conduct harder and V_{CE} will decrease. Consequently, less of the input voltage is dropped across the transistor, and the output voltage can return to its former value. The circuit will respond similarly to any changes in the unregulated input voltage.

It is an easy matter to design a simple series regulator circuit (Fig. 20-26). Assume, for example, the desired output is 9.3V, the unregulated DC input will vary between 18V and 25V and the load current can vary from 0 to 100mA. Beta (β—ratio of DC collector current to DC base current) of Q_1 equals 30. For this circuit, a V_z value of 10V is selected, since V_z minus V_{BE} equals the desired output voltage:

$$10V - 0.7V = 9.3V$$

451

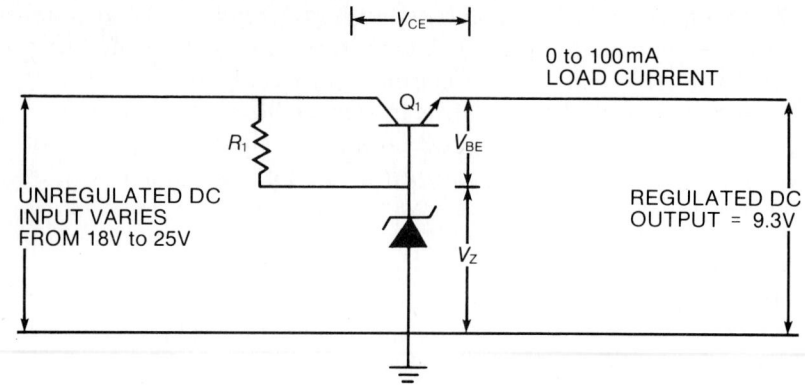

Fig. 20-26.

First, the value of R_1 must be determined. The *maximum* value of R_1 is approximated as:

$$R_1 = \frac{V_{in}\,(min) - V_z}{I_B\,(max)}$$

The numerator represents minimum voltage across R_1. $I_B\,(max)$ is the minimum base current of Q_1. Because:

$$I_B\,(max) = \frac{I_L\,(max)}{\beta}$$

The formula for R_1 can be rewritten:

$$R_1 = \frac{V_{in}\,(min) - V_z}{I_B\,(max)} \times \beta$$

Substituting the values listed here:

$$R_1 = \frac{18 - 10}{0.1} \times 30$$

$$R_1 = 2{,}400\,\Omega$$

Now, the absolute maximum current that would ever flow through the zener diode can be determined:

$$I_z\,(max) = \frac{V_{in}\,(max) - V_z}{R_1}$$

In this example:

$$I_z\,(max) = \frac{25 - 10}{2{,}400\,\Omega}$$

$$= 6.25\text{mA}$$

The zener diode should have a power rating somewhat higher than:

$$10V \times 6.25\text{mA} = 62.5\text{mW}$$

The transistor may be required to drop a voltage as large as:

$$V_{in}\,(max) - V_{out}$$

452

in this case:

$$25V - 9.3V = 15.75V$$

Therefore, the transistor may have to dissipate as much as:

$$P = 15.7V \times 100mA$$
$$P = 1.57W$$

Series Regulation With Darlington Configuration

In the series regulator circuit just examined, the power dissipated by the zener diode will become quite large when higher output voltage and load currents are desired. However, this amount of power dissipated can be reduced by adding another transistor to the circuit. Figure 20-27 illustrates a series regulator with a Darlington configuration.

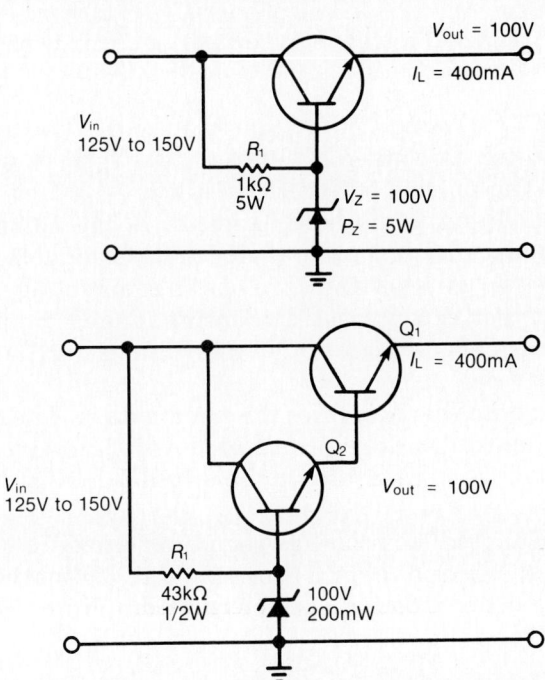

Fig. 20-27: The Darlington configuration permits series voltage regulation with a lower wattage zener diode.

Q_1 has a β of 30 and Q_2 has a β of 50, thus the overall β is 1,500 (30 × 50). Because Q_2 carries only the base current of Q_1, Q_2 can have a much lower power dissipation than Q_1. Notice that R_1 only carries the base current Q_2 now; therefore, R_1 can have a higher resistance value and a lower power rating. Since the maximum current through R_1 is lower, the zener diode can have a lower current and power rating.

Because there are now two base-emitter junctions between the zener and the load, this circuit is more sensitive to changes in temperature than the single transistor version.

453

Feedback Regulator

Figure 20-28 shows the block diagram of a feedback regulator. It consists of five basic circuits.

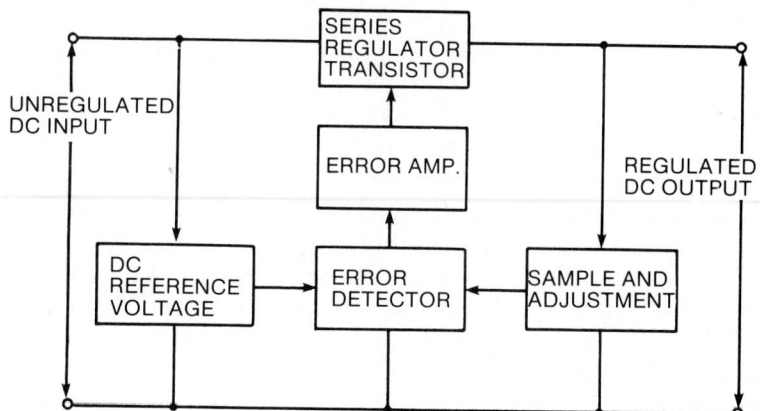

Fig. 20-28: Block diagram of a series regulator with feedback.

A *sampling circuit*, which is nothing more than a voltage divide network, is connected across the output. A voltage is tapped off and applied to the *error detector*.

The other input to the error detector is the *DC reference voltage*, produced across a zener diode. The error detector compares these two voltages and develops an error voltage which is applied to the *error amplifier*. The error voltage is proportional to the difference between the reference voltage and the sampled voltage.

The error amplifier amplifies the error voltage. The output of this amplifier controls the conduction level of the *series regulator transistor* to compensate for the original change in output voltage.

A practical feedback voltage regulator circuit is shown in Fig. 20-29. R_1, R_2, and R_3 form the sample and adjust circuit. Q_1 acts as the error detector and the error amplifier. The zener

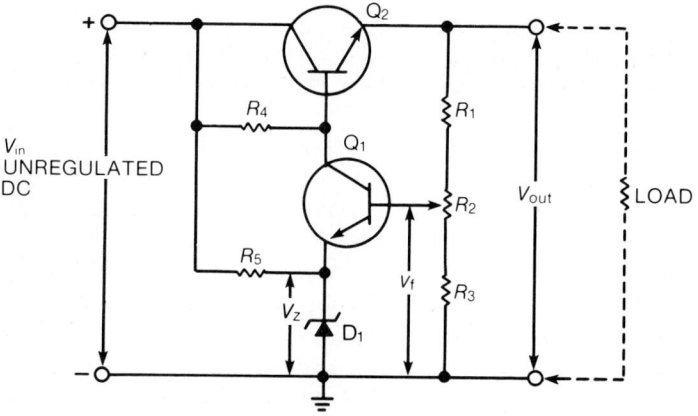

Fig. 20-29: Feedback voltage regulator.

diode and R_5 are used to produce the reference voltage. R_4 is the collector load resistor for Q_1 and a base biasing resistor for Q_2. Q_2 is the series regulator transistor.

To see how this circuit operates, assume that V_{out} attempts to increase. When V_{out} increases, feedback voltage (V_f) increases proportionately. This increases the voltage at the base of Q_1. The emitter voltage of Q_1 is held constant by the zener; therefore, the forward bias on Q_1 increases, causing the transistor to conduct harder. As Q_1 forces more current through R_4, the voltage at the collector of Q_1 and at the base of Q_2 decreases. This decreases the forward bias on Q_2, causing the series regulator transistor to conduct less. If Q_2 conducts less, it draws less current through the load; consequently, the voltage across the load (V_{out}) tends to decrease and the output is restored to its previous level.

The output voltage is adjustable over a range of voltages from slightly above the zener voltage to about three-fourths of the unregulated input voltage. This is done by adjusting variable resistor R_2. For example, to decrease the value of V_{out}, simply move the arm of R_2 up. This increases the value of V_f at the base of Q_1. The forward bias of Q_1 increases and so Q_1 conducts harder. This causes the collector voltage of Q_1 and the base voltage of Q_2 to decrease. Thus, the forward bias on Q_2 decreases and the transistor conducts less, which ultimately decreases the output voltage. The voltage adjust is an easy way of compensating for component tolerances and aging.

The feedback regulator is relatively inexpensive and fairly sensitive. The circuit can be made even more sensitive by increasing the gain of the error amplifier.

Short Circuit Protection. One disadvantage of a series regulator is that the series transistor could easily be destroyed by a large overload current if a short develops in the load. Fortunately there are several methods available to protect this circuit from excessive currents. A fuse, overload relay, or thermal relay could be added to the circuit. In many instances, however, these devices do not react fast enough to protect the series regulator transistor. An automatic electronic overload protection circuit has been added to the series regulator shown in Fig. 20-30. Q_3 and R_6 form the current limiting circuit.

Remember that the transistor will not conduct unless the base to emitter voltage exceeds 0.7V. Resistor R_6 will develop 0.7V when the load current equals:

$$I_L = \frac{.7V}{1\Omega}$$

$$I_L = 700mA$$

When the current is below 700mA, Q_3 will be cut off and the circuit will act exactly like the regulator previously examined.

If the current tries to rise above 700mA, the voltage dropped across R_6 rises above 0.7V, which causes Q_3 to conduct through R_4. This decreases the forward-bias of Q_2, causing the series regulator transistor to conduct less; hence, the current cannot rise above 700mA.

455

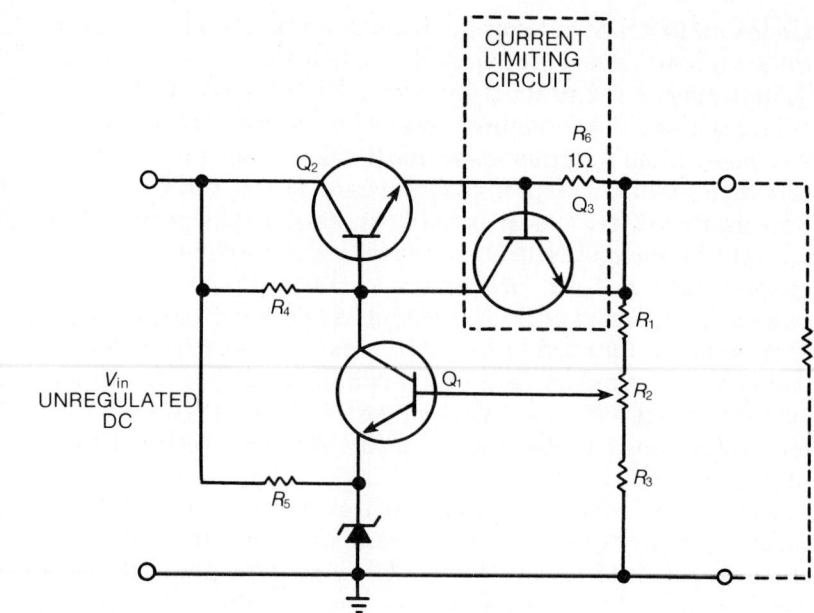

Fig. 20-30: Automatic electronic overload protection circuit.

The current can be limited to a lower value by using a larger resistance for R_6. For example a 10Ω resistor will limit the current to 70mA (0.7V/10Ω = 70mA).

Overvoltage Protection. Suppose the series regulator transistor of a power supply becomes shorted. The entire output voltage of the rectifier would be felt across the load, and that could destroy many of the solid-state devices in the load. Some power supplies use overvoltage protection (OVP) circuitry to guard the load from such excessive voltages.

A protection circuit called a *crow bar* is illustrated in Fig. 20-31. An SCR is connected directly across the load. As long as the power supply output remains at 5V, the zener diode cannot conduct and there is no current for the SCR. The SCR will remain off and the 5V will be delivered to the load. If the output voltage attempts to rise to 8V due to a short in the power supply, the zener diode will start to conduct as soon as its breakdown voltage is exceeded, and a voltage drop will appear across R_1. When the voltage drop across R_1 is sufficient to trigger the SCR (0.7V), current will flow through the device, causing it to latch immediately, and the load will be shorted out. The power supply current will bypass the load and flow through the SCR. The resulting short-circuit current will cause the fuse to blow or the circuit breaker to trip, and the power supply will be shut down.

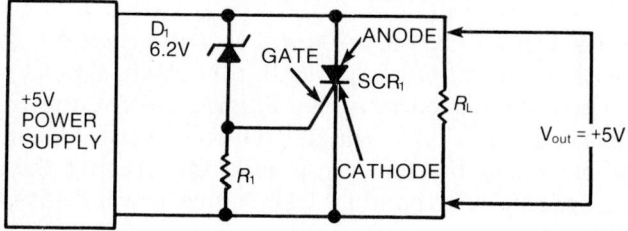

Fig. 20-31: "Crow bar" circuit.

456

COMBINATION VOLTAGE SUPPLIES

Power supplies found in modern electronics equipment must supply a great variety of voltages and current; for example a typical oscilloscope requires voltages in ranges of –5V, –15V, –55V, +5V, +120V, +2,400V, and –580V.

The necessity of having a power supply capable of delivering a high voltage-low current output, a low voltage-high current output, and a high voltage-high current output, simultaneously, is the reason for having combinational power supplies.

It is not the purpose of this section to present all possible varieties of combinational power supplies. It has been described how both negative and positive voltages can be obtained from a voltage dividing network connected to a power supply. In the section dealing with voltage multipliers, it will be shown how aiding voltages are combined to increase the output. It should be realized that, in addition to the sum of the voltages available, any of the voltages comprising that sum might also be used. Two basic combinations will be presented here—the full-wave bridge and the full-wave full-wave.

Full-Wave Bridge

Figure 20-32 is an example of the full-wave bridge combinational power supply. Figure 20-32A shows a simplified schematic drawing and Fig. 20-32B shows the entire schematic,

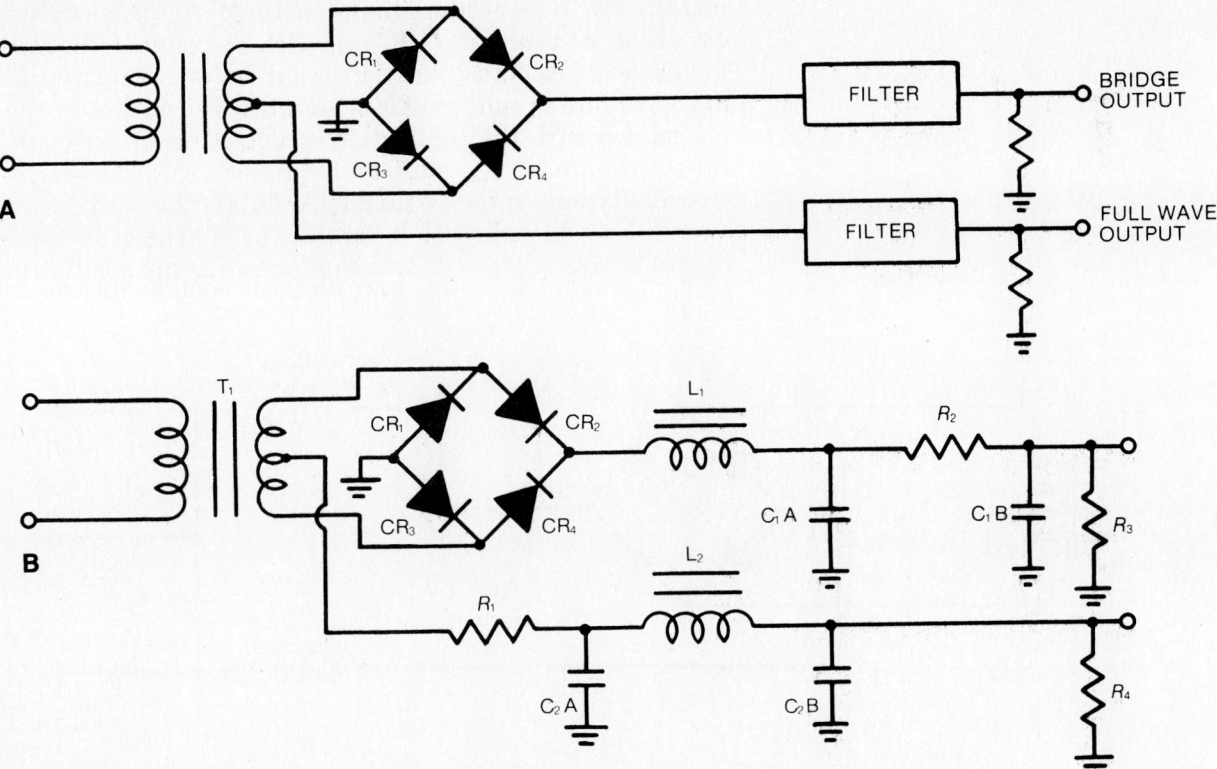

Fig. 20-32: Full-wave bridge combination power supply.

457

including the filtering network. It is a typical arrangement of the full-wave bridge combination supply, quite often called the "economy" power supply.

In Fig. 20-32B, CR_1 and CR_3 form the full-wave rectifier circuit and C_2A, C_2B, and L_2 form the filter network. R_1 is a current-limiting resistor used to protect the diodes from surge currents. R_4 is the bleeder resistor and also assures that the power supply has a minimum load at all times. CR_1, CR_2, CR_3, and CR_4 form the bridge rectifier circuit with L_1, R_2, C_1A and C_1B doing the filtering; R_3 is a bleeder resistor and assures that the bridge always has a minimum load. Each circuit, of itself, works in the conventional manner. Troubleshooting will be the same as for the other power supplies covered, entailing the *no output-low output* factors.

Full-Wave Full-Wave

Figure 20-33 illustrates a full-wave full-wave combinational power supply with positive and negative outputs. It has one primary distinguishing feature compared to a bridge circuit— the center tapped transformer secondary. The components associated with the negative voltage output are CR_1, CR_3, L_2, C_2, and R_2, while CR_2, CR_4, L_1, C_1, and R_1 are the components in the positive voltage output. Transformer T_1 is a component common to both supplies.

The operation of each full-wave rectifier is identical. As a refresher, the operation will be reiterated. When point A is negative with respect to ground, point B will be positive. This condition causes both CR_1 and CR_4 to conduct. Both these diodes are associated with different full-wave rectifiers. The negative power supply will be described first; this is the one associated with the conduction of CR_1. Current through CR_1 flows through L_2, C_2, and R_2 and completes the path via ground to transformer centertap then to A. This is the path during the first half cycle while CR_1 is conducting. At the same time this action is taking place, CR_4 is conducting in the positive power

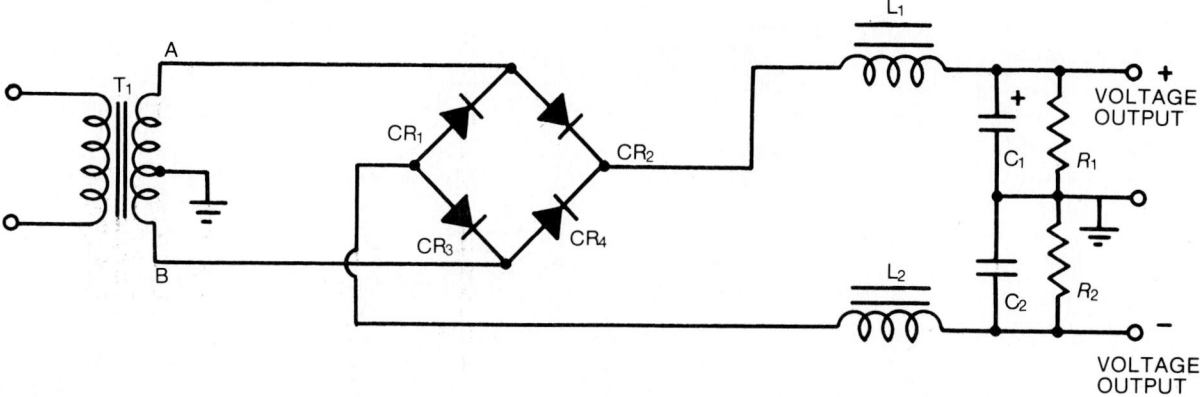

Fig. 20-33: Full-wave full-wave combination power supply with positive and negative output voltages.

supply. Since point B is positive and the lower half of the transformer secondary acts as a source, current from point B to ground flows up through C_1 (charging it as shown), L_1, CR_4, and back to the source. When the polarities at point A and point B are reversed, point A now being positive and point B being negative, with respect to ground, the conducting diodes are now CR_3 in the negative supply and CR_2 in the positive supply.

Again, taking the negative supply first, CR_3 conducts and the current path (from point B and back to point B) is through CR_3, L_2, and up through C_2 in the same manner as when CR_1 was conducting (to ground and through the center tap of the transformer to point B). This completes the full cycle of operation for the negative supply and CR_1 and CR_3 will be conducting alternately as long as there is an input.

At the time CR_3 is conducting in the negative power supply, CR_2 is conducting in the positive power supply. Current, from point A back to point A, is going down the upper half of the transformer secondary to ground, up through C_1 and L_1, through CR_2, and back to point A. Again, the charge path for C_1 is the same as it was when CR_4 was conducting.

While it has been said that when troubleshooting the combinational power supply, the troubles to look for are of the no output-low output type, it should be realized that in the full-wave bridge combinational power supply that a low output can occur in either section or the voltage out of the full-wave rectifier can be normal, while the bridge rectifier might indicate a low output. If the load currents are not excessive and the filter components have checked satisfactorily, the defective diodes, in the first case, would be assumed to be CR_1 or CR_3 and in the second case CR_2 or CR_4.

VOLTAGE MULTIPLIERS

Figure 20-34 depicts a simple half-wave rectifier circuit that is capable of delivering a voltage increase (more voltage output than voltage input), providing the current being drawn is low. It is shown that it is possible to get a larger voltage out of a simple half-wave rectifier as long as the current is low. If the current demand increases, the output voltage will decrease. This can best be explained by the use of RC time constant. The charge time for the circuit in Fig. 20-34 is very fast since the

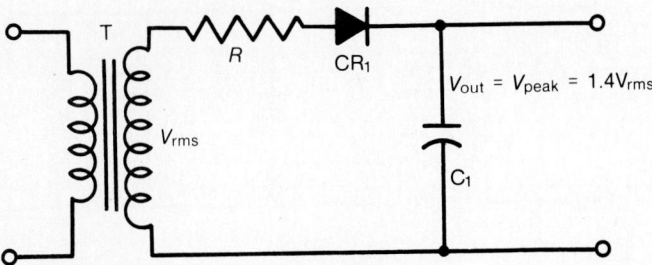

Fig. 20-34: Half-wave rectifier used to deliver an increased voltage output.

459

circuit elements in the capacitor's charge path are the diode, CR_1, the surge resistor, R, and the secondary of the transformer. These elements combine to form a very low impedance, since CR_1 is conducting during the charge time and the value of R is about 20Ω. In comparison, the discharge path for the capacitor is through the load which offers an impedance several hundred times higher than that of the charge path. The lower the load impedance, the greater the current. If the discharge path for the capacitor offers a lower impedance, the capacitor will discharge further, lowering the output voltage.

Rectifier circuits that can be used to double, triple, and quadruple the input voltage will now be discussed. All these circuits have one thing in common—they use the charge stored on capacitors to increase the output voltage. Figure 20-35 is a block diagram of a voltage multiplier circuit. The input is AC and the output is a DC multiple.

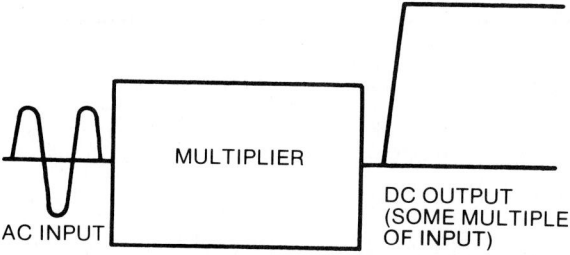

Fig. 20-35: Block diagram of a voltage multiplier circuit.

Half-Wave Voltage Doubler

The first voltage multiplier circuit is the half-wave voltage doubler. As the name implies, this circuit gives a DC output that is approximately twice that obtained from the equivalent half-wave rectifier circuit.

Figure 20-36 shows a typical half-wave voltage doubler circuit. While the circuit shown uses a transformer and the output voltage is positive with respect to ground, it could just as well operate as a negative voltage output by reversing the diodes. The transformer, which may be used to step up secondary voltage or as an isolation transformer, may also be eliminated with the proper choice of circuit elements.

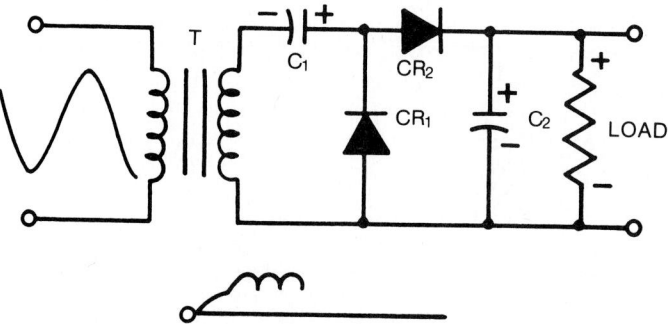

Fig. 20-36: Half-wave voltage doubler circuit.

460

When the top of the transformer secondary in Fig. 20-36 is negative, C_1 will charge through conducting CR_1 to approximately the peak of the secondary voltage. The direction of charge is indicated by the polarity signs. At this time there is no output. On the next alternation, when the top of the transformer secondary is positive with respect to the bottom, C_1 will discharge through the transformer and CR_2, which is now conducting. C_2, which is also in this discharge path, is charged to approximately twice the peak of the secondary voltage because the charge on C_1 is in series with the applied AC and, therefore, adds to the voltage applied to C_2.

Since C_2 receives only one charge for every cycle of operation, the ripple frequency, as in the half-wave rectifier, is the same as the input frequency. Also, as in the half-wave rectifier, C_2 will discharge slightly between charging cycles so that filtering is required to smooth the output and give us a relatively pure DC.

The procedures for troubleshooting the half-wave voltage doubler are the same as those used for the half-wave rectifier. No output conditions might be caused by a defective transformer, defective rectifiers, an open C_1, or short circuited C_2. The low output condition might be caused by a low input, rectifier aging, or excessive load current (caused by a decrease in load impedance).

The voltage multiplier circuits which follow all have one thing in common with the circuit just described—they use the charge stored on a capacitor to increase the output voltage. As the voltage across C_1 is added to the input voltage to approximately double the charge applied to C_2, so will the charges on other capacitors add to the charge applied to an input capacitor to double, triple, or quadruple the output voltage.

Full-Wave Voltage Doubler

The full-wave rectifier circuit can also be adapted to a voltage doubling circuit. Figure 20-37 depicts a basic full-wave voltage doubler circuit. Depending upon the circuit application, it may or may not use a power or isolation transformer. The resistor R_s is a surge resistor that is used to limit the charge current and protect the diode; it might not be necessary in some equipment

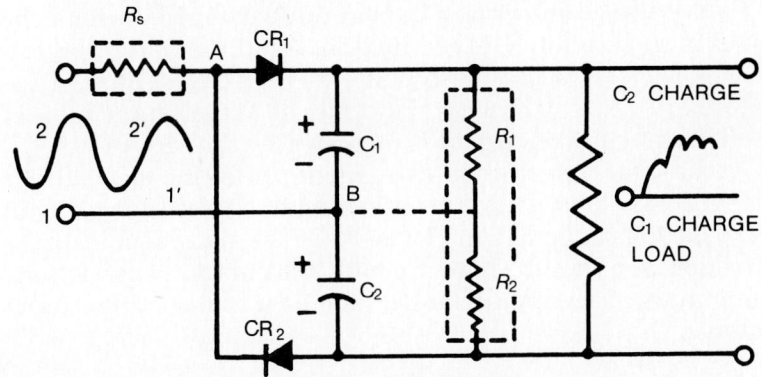

Fig. 20-37: Basic full-wave voltage doubler.

461

and when used it is placed in series with the AC source. Resistors R_1 and R_2 are not necessary for circuit operation but may be used to act as bleeder resistors to discharge their associated capacitors when the circuit is deenergized. When used they also tend to equalize the voltages across C_1 and C_2.

The circuit operates much the same as the full-wave rectifier previously discussed with the exception that now two capacitors are employed, each one charging to approximately the peak voltage of the input, adding their charges to provide an output. When point A is positive with respect to point B, C_1 will charge through the conducting diode, CR_1, and the source. It will charge to approximately the peak of the incoming voltage. On the next half cycle of the input, point A is now negative with respect to point B, CR_2 conducts charging C_2 in the direction indicated. The voltage across the load will be the total of the voltages across C_1 and C_2. C_1 and C_2 will be equal value capacitors and R_1 and R_2 will also be equal value resistors. The value of R_s will be small, probably in the 20Ω to 500Ω range.

VOLTAGE TRIPLER

Figure 20-38 depicts a typical voltage tripler circuit with waveforms and circuit operation. Figure 20-38A shows the complete circuit. In Fig. 20-38B, C_1 is shown charging as CR_2 is conducting. In Fig. 20-38C, C_3 is illustrated charging as CR_3 is conducting. Figure 20-38D reveals the charge path for C_2 while CR_1 is conducting. In Figure 20-38E a comparison is made of the input signal and its effects on the voltages felt across C_1, C_2, C_3, and the load. The following explanation uses Fig. 20-38 as the operating device.

Close inspection of Fig. 20-38A should reveal that removal of CR_3, C_3, R_2, and the load resistor results in the voltage doubler circuit previously described. The connection of circuit elements CR_3 and the parallel network C_3 and R_2 to the basic doubler circuit is arranged so that they are in series across the load. The combination provides approximately three times more voltage in the output than is felt across the input. Fundamentally this circuit is a combination of a half-wave voltage doubler and a half-wave rectifier circuit arranged so that the output voltage of one circuit is in series with the output voltage of the other.

Figure 20-38B shows how C_1 is initially charged. Assume the input is such that CR_2 is conducting. A path for charging current is from the right-hand plate of C_1 through CR_2 and the secondary winding of the transformer, to the left-hand plate of C_1. The direction of current is indicated by the arrows.

At the same time that the above action is taking place, CR_3 is also forward biased and is conducting and C_3 is charged with the polarities indicated in Fig. 20-38C. The arrows indicate the direction of current. There are now two energized capacitors, each charged to approximately the peak value of the input voltage.

462

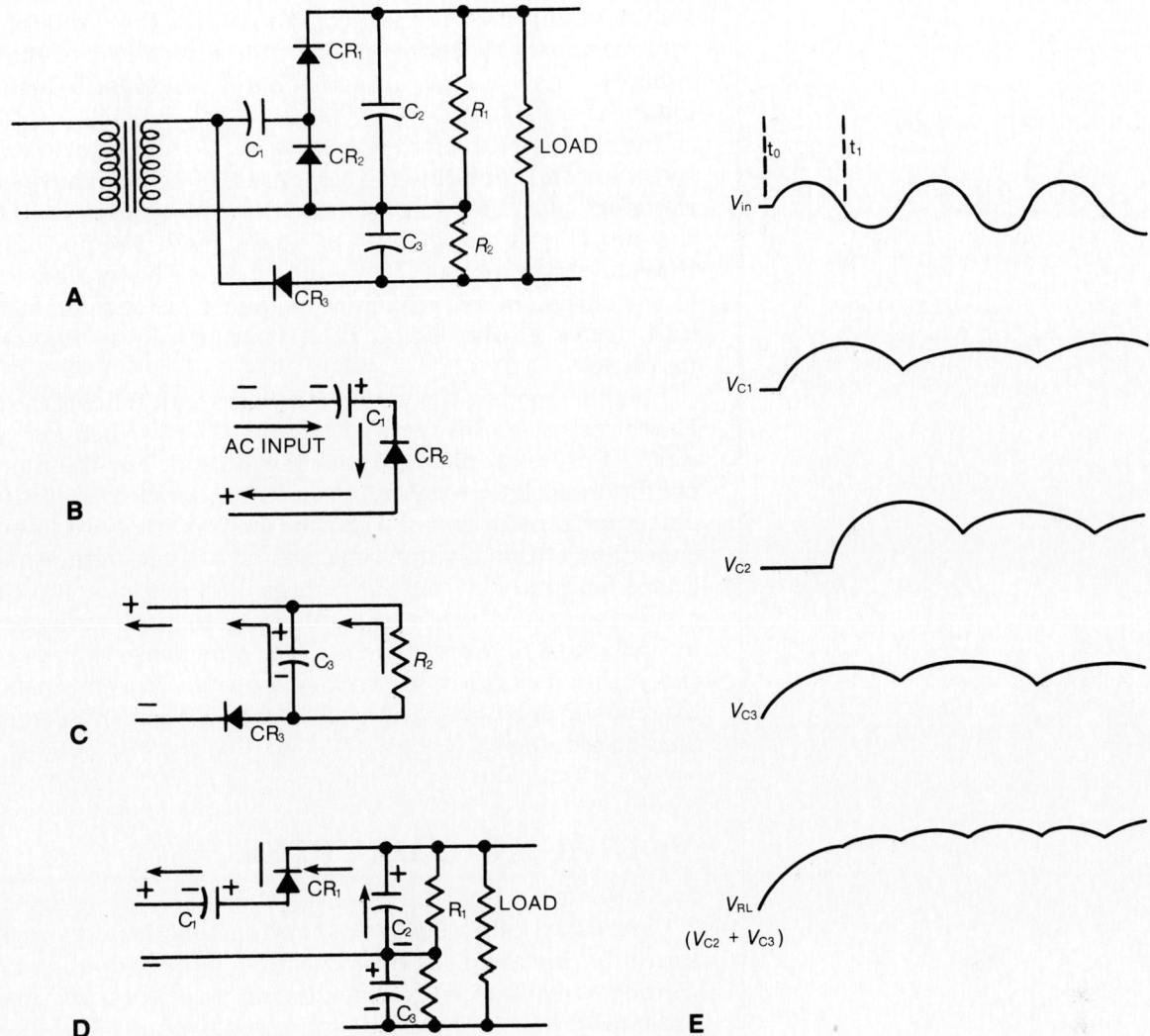

Fig. 20-38: Typical voltage tripler circuit operation and waveforms.

On the next half cycle of the input, the polarities change so that CR_1 is now the conducting diode. Figure 20-38D indicates how capacitor C_1, now in series with the applied voltage, adds its potential to the applied voltage. Capacitor C_2 charges to approximately twice the peak value of the incoming voltage ($V_{in} + V_{C1}$). As can be seen C_2 and C_3 are in series and the load resistor is in parallel with this combination. The output voltage then will be the total voltage felt across C_2 and C_3, or approximately three times the peak voltage of the input.

Figure 20-38E indicates the action taking place using time and the incoming voltage. At time zero (t_0) the AC input is starting on its positive excursion. At this time the voltage on C_1 and C_3 is increasing. When the input starts to go into its negative excursion, t_1, both C_1 and C_3 start to discharge and the voltage across C_2 is increasing. The discharge of C_1 adds to the source voltage when charging C_2 so that the value of V_{C2} is approximately twice the value of the peak value of the input. Since V_{C2} and V_{C3} are in series across the load resistor, the output is their sum.

The charge paths for capacitors C_1, C_2, and C_3 are comparatively low impedance when compared to their discharge paths; therefore, even though there is some ripple voltage variation in the output voltage, the output voltage will be approximately three times the value of the input voltage. The ripple frequency of the output, since capacitors C_2 and C_3 charge on alternate half cycles of the input, is twice that of the input ripple frequency.

Troubleshooting the voltage tripler circuit follows the general practice given for rectifier circuits. The two general categories of failure are no output or low output. For the no output condition, look for a no input condition, a lack of applied AC, a defective transformer, or a shorted load circuit. For a low output condition, check the input voltage. Low input voltage means a correspondingly low output voltage. Low output voltage might also be a result of any of the following: rectifier aging causing an increased forward resistance or a decreased reverse resistance, any leakage in the capacitors or decrease in their effective capacitance, or an increase in the load current (decrease in load impedance).

VOLTAGE QUADRUPLER

Figure 20-39 shows a typical voltage quadrupler circuit. Essentially this circuit is two half-wave voltage doubler circuits connected back to back and sharing a common AC input. In order to show how the voltage quadrupler works, Fig. 20-39 has been shown as two voltage doublers. The counterparts of the one circuit are shown as a prime (') in the second circuit. (C_1 in the first circuit is the same as C_1' in the other circuit; CR_1 and CR_1' conduct at the same time, etc.) When the circuit is first turned on, it will be assumed that the top of the secondary winding, point A, is negative with respect to B. At this instance CR_1 is forward biased and conducts allowing C_1 to charge. On the next alternation point B is negative with respect to point A.

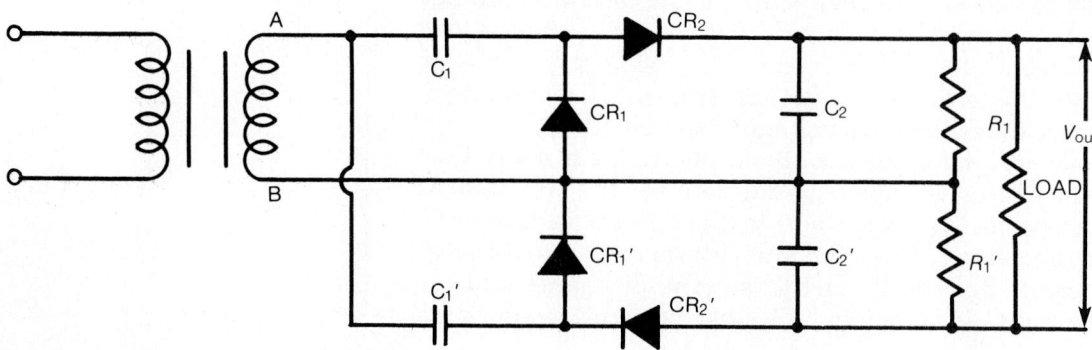

Fig. 20-39: Voltage quadrupler circuit.

464

At this time two things are going to occur: (1) C_1, which was charged to approximately the peak voltage across the secondary winding, will aid the source and, since CR_2 is now conducting, C_2 will be charged to approximately twice the incoming voltage; and (2) CR_1' will conduct charging C_1' to approximately the peak voltage of the input. During the following alternation C_1' adds to the input which allows CR_2' to conduct, charging C_2' to twice the input voltage.

When CR_2 conducts, C_1 will aid the input voltage in charging C_2 to approximately twice the peak voltage of the secondary and CR_2'; conducting at the same time charges C_1'. On the next alternation CR_1' conducts, and since C_1 is in series with the input, it aids in charging C_2' to twice the peak of the secondary. The voltages across C_2 and C_2' add to provide four times the peak secondary voltage in the output.

VOLTAGE DIVIDER DESIGN

When intermediate voltages are desired, a resistor network can be employed. A voltage divider design (see Chapter 6), such as the one illustrated in Fig. 20-40, can be used with any of the power supplies discussed earlier.

To determine the resistor values in Fig. 20-40, proceed as follows. The first amplifier has a 600Ω load resistance, and the second amplifier has a 300Ω load resistance; thus, amplifier 1 would have a current demand of $I = V/R = 30/600 = 50$mA. Amplifier 2 would have $12/300 = 40$mA. Resistor R_3 could be any value, but to initiate the calculations, assume a value of 600Ω. With this ohmic value, the current through R_3 would then be $12/600 = 20$mA. Since amplifier 2 shunts R_3, the total current drawn by both would be 40mA $+ 20$mA $= 60$mA; this current must flow through R_2. The resistance of R_3 can be solved if the voltage across it is known. For R_2 to drop from 30V to 12V, it would require a drop of 30V – 12V = 18V. Hence $R_2 = V \div I = 18 \div 0.06 = 200\Omega$. Resistor R_1 must reduce the 60V to 30V;

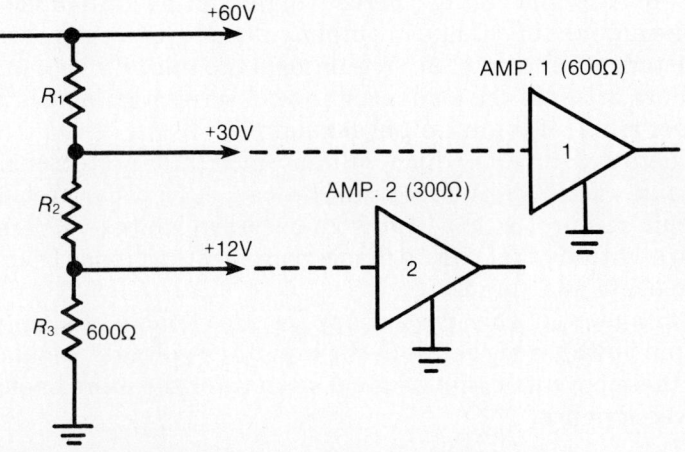

Fig. 20-40: Voltage-divider calculation factors.

465

hence, there would be a 30V drop across it. The current through R_1 would consist of the current of amplifier 1 (50mA) and the current through R_2. R_2 passes the current for amplifier 2 (40mA) and for R_4 (20mA) for a total of 60mA, which was found earlier; thus, 50mA + 60mA = 110mA flowing through R_2. Since there is a 30V drop across R_2, the resistance would be V/I = 30 ÷ 0.11 = 272.7Ω. Hence, for the voltages shown plus the amplifiers' resistances, only these values for the voltage-divider resistor network would be correct. If different load resistances were applied to the 30V and 12V output lines, all the resistor values for R_1, R_2, and R_3 would have to be changed.

SUMMARY

DC power supplies are used to convert an AC source into a DC voltage. There are four basic stages to a DC power supply.

The first stage is the power transformation stage. A transformer is used to provide isolation and/or a step-up or step-down in the AC voltage. This is not an essential stage in some designs.

The second stage is the rectification stage. This most important stage converts the AC sine wave to a pulsating DC voltage. There are three basic types of rectifier circuits. The simplest in construction is the half-wave rectifier. The full-wave rectifier is approximately twice as efficient as the half-wave rectifier. The full-wave bridge rectifier, while just as efficient as the other full-wave rectifier, does not require a center-tapped transformer.

The third stage, the filtering stage, removes the ripple component from the rectified output. Because the ripple frequency of a full-wave rectifier is twice as high as that of a half-wave rectifier, filtering of a full-wave rectifier is more easily accomplished. The degree of filtering is expressed in a figure of merit called percent ripple.

Regulation is the final stage. Its purpose is to keep the DC output constant regardless of variations in either load resistance or AC input voltage. Percent of regulation is an indication of the amount of change in output voltage over a range of load resistance values. A shunt regulator is capable of withstanding a short circuit in the load resistance. A series regulator is more power efficient than a shunt regulator.

There are circuits which will automatically protect a series regulator from short circuits in the load. A crow bar protection circuit will protect the load from excessive voltage. There are also additions that can be made to protect the diodes from line transients and surges.

Voltage multiplier power supplies are capable of supplying output voltages higher than the input line voltage. Regulation for these circuits is not as good as that for the more basic DC power supplies.

CHECK YOURSELF (Answers at back of book)

Questions

Fill in the blanks with the appropriate word or words.

1. The greatest disadvantage of a full-wave rectifier is its construction. It requires _____ and _____ diodes.
2. A _____ filter supplies a high degree of filtering for medium current loads.
3. _____ % regulation is the highest degree of regulation possible.
4. In a crow bar protection circuit, an _____ is connected directly across the load.
5. In the design of a voltage divider power supply, both _____ and _____ must be taken into consideration.
6. A 1:1 transformer would be used to provide _____ .
7. The reverse biased diodes in a _____ rectifier are never exposed to more than the peak voltage induced on the secondary winding.
8. A capacitive filter reduces the level of _____ and raises the _____ level of _____ voltage.
9. The feedback regulator consists of five basic blocks: _____ , _____ , _____ , _____ , and _____ .
10. A voltage multiplier utilizes the accumulated _____ of _____ to obtain a final voltage that equals the sum of these charges.
11. The maximum _____ current is the maximum amount of average current that can be permitted to flow in the forward direction through a diode.
12. The maximum _____ current is that amount of current allowed to flow in the forward direction through a diode in nonrepetitive pulses.
13. _____ can be suppressed by using two capacitors, one across the primary, the other across the secondary.
14. Diodes can be _____ in series or parallel configurations for increased ratings.
15. To insure proper conductance of heat from the rectifier to the sink, _____ is sometimes applied to both units at the insulating material where they meet.
16. The diode should be operated at no more than _____ % of its rated value.
17. One advantage of a shunt regulator is its ability to withstand a _____ without series damage.

Problems

1. The output of a power supply is 50V when the load current is 0 and 45V when the load current is at its maximum. Determine the percent of regulation.

2. The input voltage to a zener diode regulator varies from 105V to 120V. The load current varies from 175mA to 200mA. If the desired output is 100V, determine the resistance value of the series resistance.
3. Determine the value of each of the four resistors in Fig. 20-41.
4. A filter stage has a peak ripple of 2V for an output voltage of 40V. What is the percent ripple?

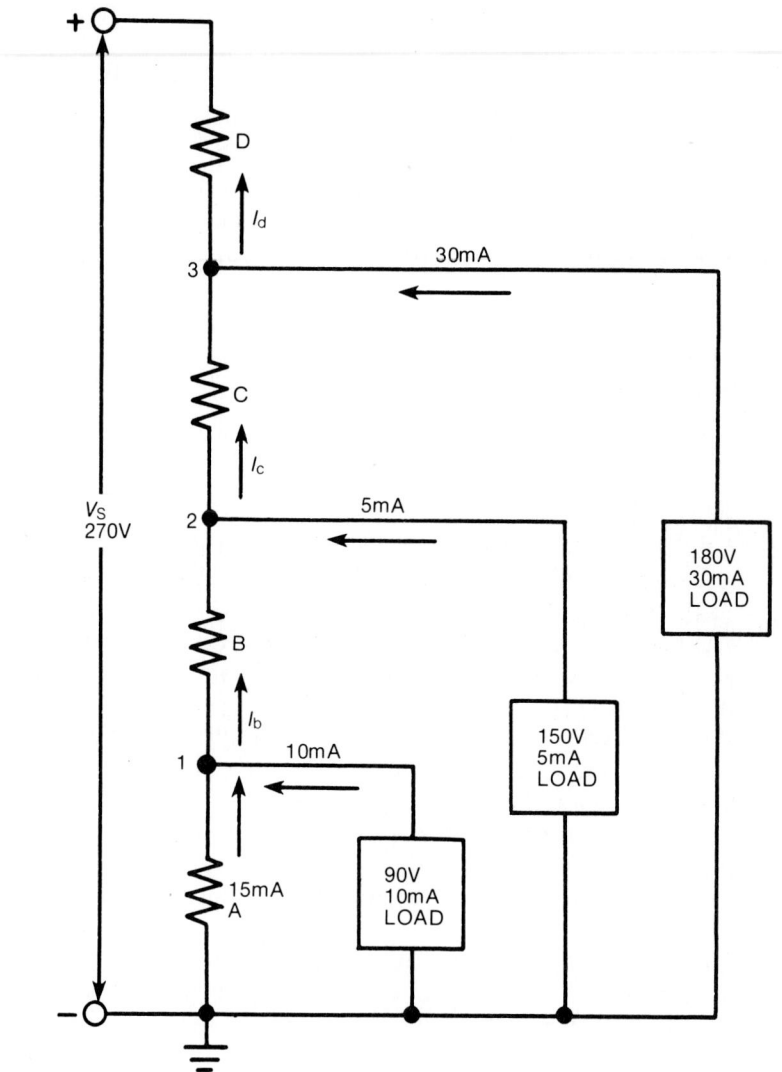

Fig. 20-41.

CHAPTER 21

Transistors

A transistor is a device that utilizes a small change in current to produce a large change in voltage, current, or power. The transistor, therefore, may function as an amplifier or an electronic switch. An amplifier is a device that increases the voltage, current, or power level of a signal applied to that device. The output of an amplifier may or may not be an exact replica of the input wave shape. An electronic switch is a device which utilizes a small voltage or current to turn on or turn off a large current flow with great rapidity.

BIPOLAR JUNCTION TRANSISTORS

As mentioned in Chapter 19, a diode is a device formed by a single junction between N-type material and P-type material. A *bipolar junction transistor* (BJT) is formed by making two such diodes in a single piece of material. A piece of N-type material sandwiched between two pieces of P-type material becomes a *PNP transistor*. A piece of P-type material sandwiched between two pieces of N-type material becomes an *NPN transistor*. The most common case styles used to hold the semiconductor materials are shown in Fig. 21-1.

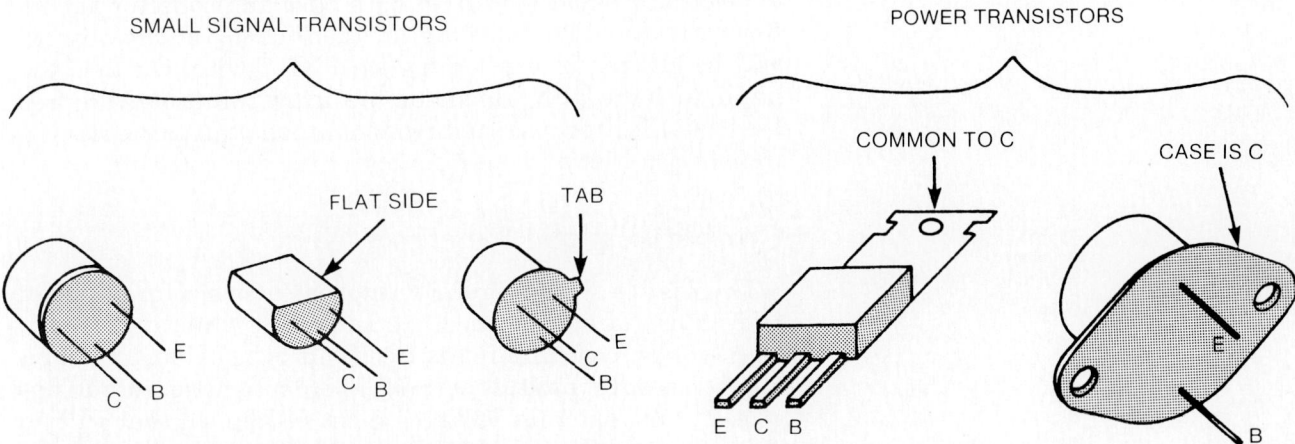

Fig. 21-1: Bipolar junction transistor case styles.

Each of the three pieces of material has an assigned name. The center material is always the base, and it is much thinner than the two outer layers. The two outer materials are called *emitter* and *collector*. The emitter and collector are of the same type of material; the only difference is in the physical size. The collector is generally much larger than the emitter. These names, terminals, and schematic symbols are illustrated in Fig. 21-2.

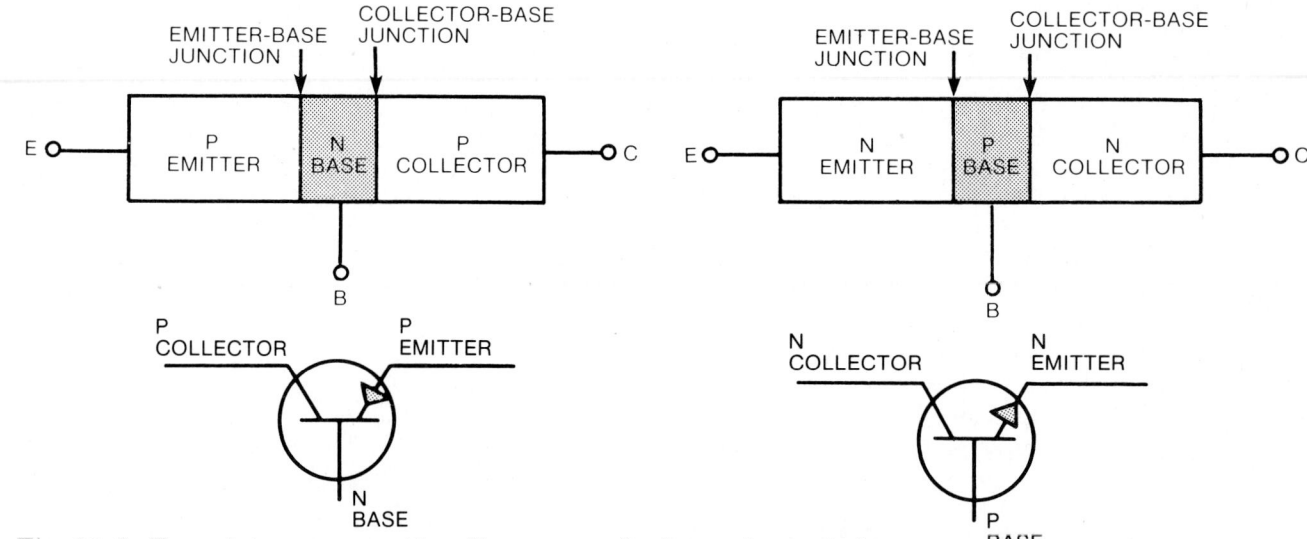

Fig. 21-2: Transistor cross section diagrams and schematic symbols.

The potential on the base controls the current between collector and emitter. The two junctions are the *emitter-base junction* and the *collector-base junction*. Each of these junctions has a depletion region and a barrier potential as described for diode junctions in Chapter 19.

Notice in the schematic symbol that the emitter contains an arrow. In a PNP transistor, this arrow points toward the base. Electron current through a transistor always moves against the emitter arrow. Since this is a PNP transistor, the direction of electron current is *in* from both base and collector and *out* through the emitter. For simplification of analysis, hole current will be utilized in explaining the PNP device. On the other hand, with the NPN transistor, the arrow points outward from the base. Electron current direction through this transistor is *in* from the emitter and *out* from both base and collector.

Controlling the Electron Flow

There is always some random movement of electrons inside a transistor. A free electron drops into a hole while another electron breaks loose and floats free. This type of random movement does not constitute current. In order to have a useful flow of electrons, external voltages must be applied that will not only direct the electron flow, but will enhance the development of current carriers as well.

Basic Biasing Principles. A potential difference across a PN junction constitutes bias. When the voltage is connected with negative to the N-type material and positive to the P-type material, the junction has *forward bias*. When the polarity is reversed, positive to the N-type material and negative to the P-type material, the junction has *reverse bias*. In most applications, a transistor conducts *only* when the emitter-base junction is forward-biased and the collector-base junction is reverse-biased.

The depletion regions, barrier potentials, and electric fields at the junctions of NPN and PNP transistors are illustrated in Figs. 21-3A and B. As shown in the illustrations, the center

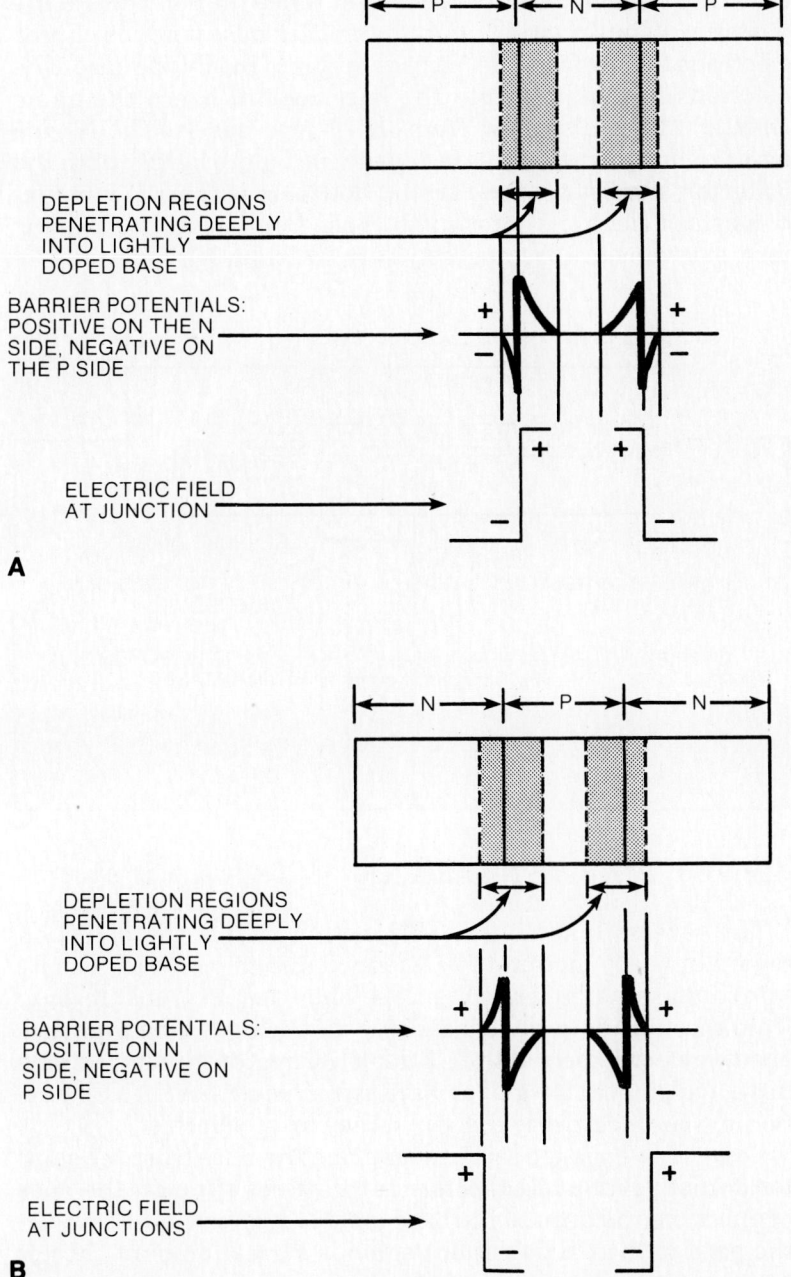

Fig. 21-3: Barrier potentials and depletion regions for un-biased (A) PNP transistors and (B) NPN transistors.

layer in each case is made a great deal narrower than the two outer layers. Also, the outer layers are doped much more heavily than the center layer. This causes the depletion regions to penetrate deeply into the base, thus the distance between the emitter-base (EB) and collector-base (CB) depletion regions is minimized. Note that the barrier potentials and electric fields for the PNP device are positive on the base and negative on the emitter and collector, and they are negative on the base and positive on the emitter and collector for the NPN device.

Refer to the NPN transistor shown in Fig. 21-4. For normal (linear) transistor operation, the EB junction is forward-biased and the CB junction is reverse-biased. (Note the polarities of the batteries.) The forward bias at the emitter-base junction causes electrons to flow from the N-type emitter to the P-type base. The electrons are emitted into the base region, hence the name *emitter*. Holes also flow from the P-type base to the N-type emitter, but since the base is doped much more lightly than the collector, almost all the current flow across the EB junction consists of electrons entering the base from the emitter. Therefore, electrons are the majority carriers in an NPN device.

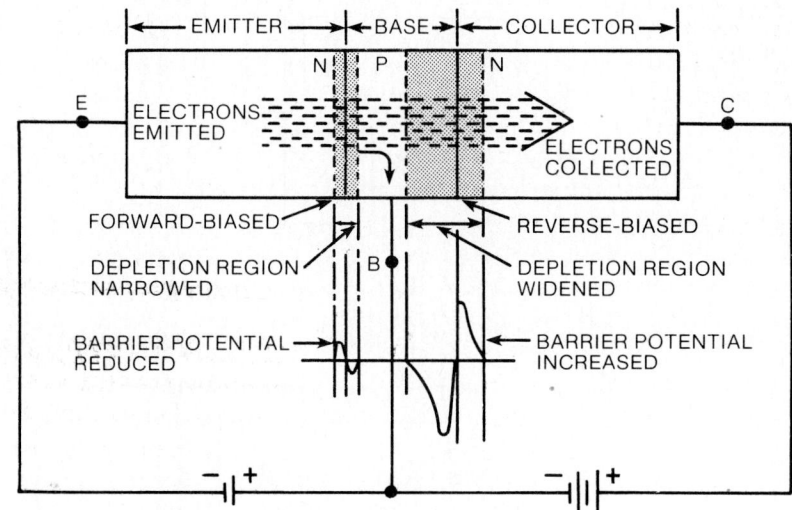

Fig. 21-4: Biased NPN transistor.

The reverse bias at the CB junction causes the depletion region at that junction to be widened and to penetrate deeply into the base, as shown in Fig. 21-4. Thus, the electrons crossing from the emitter to the base arrive quite close to the negative-positive electric field at the CB depletion region. Since electrons have a negative charge, they are attracted across the CB junction by this electric field; that is, they are "collected."

Some of the charge carriers entering the base from the emitter do not reach the collector, but flow out through the base connection and around the base-emitter bias circuit. However, the path to the CB depletion region is a great amount shorter than that to the base terminal, so that only a very small percentage of charge carriers flows out of the base terminal. Also, since the base region is doped very lightly, there are few holes available in the base to recombine with the charge carriers from the emitter. The result is that about 98% of the charge carriers from the emitter are collected at the CB junction and flow through the collector circuit by way of the bias batteries back to the emitter.

Another way of looking at the effect of the reverse-biased collector-base junction is from the viewpoint of majority and minority charge carriers. It has already been shown that a reverse-biased junction opposes the flow of majority carriers and aids the flow of minority carriers. Majority carriers are holes coming from the P-side of a junction and electrons coming from the N-side. Conversely, minority carriers are holes coming from the N-side and electrons coming from the P-side. In the case of the NPN transistor, the charge carriers arriving at the CB junction are electrons from the emitter traveling through the P-type base. Consequently, to the CB junction they appear as minority charge carriers, and the reverse bias helps them to cross the junction.

Since the emitter-base junction is forward-biased, it has the characteristics of a forward-biased diode. Substantial current will not flow until the forward bias is approximately 0.7V for a silicon device or approximately 0.3V for a germanium device. Reducing the level of EB bias voltage in effect reduces the PN-junction forward bias and thus reduces the current that flows from the emitter through the base to the collector. Increasing the EB bias voltage increases this current. Reducing the bias voltage to zero, or reversing it, cuts the current off completely. Thus, variations of the small forward-bias voltage on the EB junction control the emitter and collector currents, and the EB controlling voltage source has to supply only the small base current.

The PNP transistor operates exactly the same as an NPN transistor, with the exception that the majority charge carriers are holes. As shown in Fig. 21-5, holes are emitted from the P-type emitter across the forward-biased EB junction into the base. In the lightly doped N-type base, the holes find few elec-

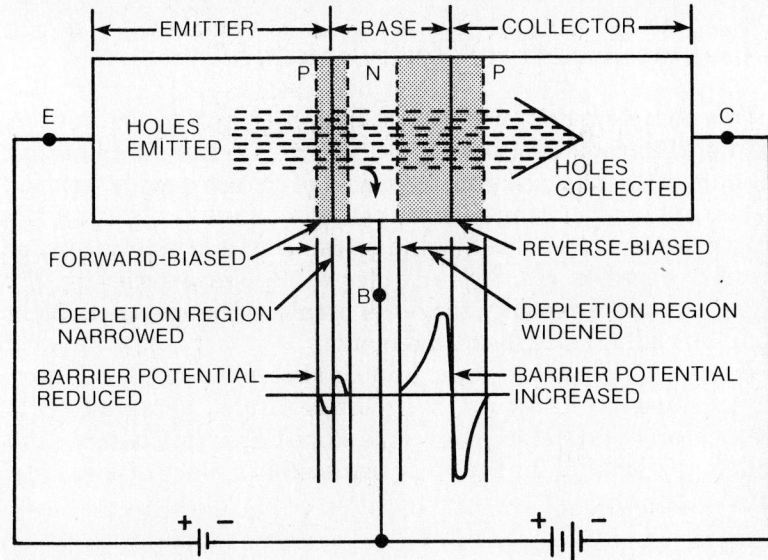

Fig. 21-5: Biased PNP transistor.

473

trons with which to recombine. Some of the holes flow out through the base terminal, but most are drawn across to the collector by the positive-negative electric field at the reverse-biased CB junction. As in the case of the NPN device, the forward bias at the EB junction controls the collector and emitter currents.

AMPLIFIER CONFIGURATIONS

When a transistor is used as the principal controlling element in an amplifier circuit, it can be connected in basically three different ways and still perform its amplifying function. Each amplifier must have two input leads and two output leads, but a transistor has only three connections or leads (emitter, base, and collector). As a result, one of the transistor's leads must be common to both the input and the output of the circuit.

The three possible circuit configurations are, therefore, formed by using the emitter, base, or collector lead as the common lead and the two remaining leads as input and output leads. For this reason, the three configurations are commonly called *common-emitter, common-base,* and *common-collector* circuits.

The common lead is often connected to circuit ground so that the input and output signals are referenced to ground. However, when considering AC signals, the common lead does not have to be connected directly to ground to be at ground potential. For example, the common lead can be connected to ground through a battery or a resistor that is bypassed with a capacitor and still be at AC ground potential. The battery and the bypassed resistor are short circuits as far as the AC signal is concerned. When the common lead is connected directly to circuit ground, it is then considered to be at both AC and DC ground. The common leads in the circuits which will now be described are considered to be at AC ground only.

Common-Emitter Configuration

In a common-emitter (CE) circuit configuration, the emitter lead of the transistor is common to both the input and output signals, while the transistor's base and collector leads serve as input and output terminals, respectively. Both NPN and PNP transistors may be used in the forming of a common-emitter circuit. Because of its many desirable characteristics, the common-emitter configuration is employed more extensively than any other circuit arrangement.

A basic CE circuit configuration is shown in Fig. 21-6. Figure 21-6A illustrates how a CE circuit is formed using an NPN transistor. Figure 21-6B shows the same CE circuit, but uses the schematic symbol for the NPN transistor in place of the structural diagram.

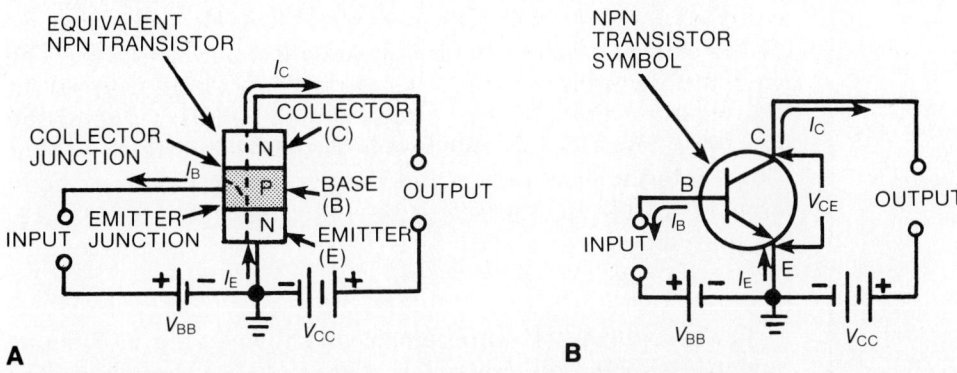

Fig. 21-6: An NPN transistor connected in the common-emitter configuration.

The input signal is applied between the base (B) and emitter (E) of the transistor, and the output signal appears between the collector (C) and emitter (E) of the transistor; thus, the emitter is common to both the input and output signals.

The transistor's emitter-base junction must be forward-biased under normal operating conditions; that is, it must be subjected to a voltage which will allow current to flow through it. The emitter junction is forward-biased when the P-type base is made positive with respect to the N-type emitter. This forward-biased voltage is provided by voltage source V_{BB} as illustrated. When an input signal is connected to the amplifier, V_{BB} is effectively connected across the emitter junction through the internal resistance of the signal source. This causes a small current to flow through the emitter junction, the signal source, and V_{BB}. This current, which flows out of the base region, is referred to as the *base current,* I_B. If the polarity of voltage source V_{BB} were reversed, the emitter junction would be reverse-biased and essentially no base current would flow.

The collector-base junction of the transistor is normally reverse-biased; that is, the N-type collector region must be made positive in respect to the P-type base region. This reverse-biased voltage is provided by voltage source V_{CC}. When an external load is connected to the output terminals, the positive side of V_{CC} is applied through the load to the N-type collector. The negative side of V_{CC} is effectively applied through the forward-biased emitter junction to the P-type base. The emitter junction exhibits a low resistance because it is forward-biased and is capable of conducting a small base current (I_B).

It might be reasonable to assume that no current could possibly flow through the collector junction, since it is reverse-biased; however, this is not the case. The internal action of the transistor allows current to flow through the reverse-biased collector-base junction. The forward-biased emitter-base junction allows a small I_B value to flow from the emitter to base, but most of the electrons that flow into the base do not flow out of the base to produce I_B. In fact, only about 2% of these electrons flow out of the base. As stated earlier, the majority (the remaining 98%) are attracted by the positive charge placed on the collector region by voltage source V_{CC}. These electrons flow out of the collector region to produce a *collector current* (I_C). The current flowing into the emitter region of the transistor (from the negative side of each voltage source) is referred to as the *emitter current* (I_E); therefore, it is the emitter current (I_E) that flows into the base region and then divides into two paths to create I_B and I_C. In other words, I_E is equal to the sum of I_B and I_C or, mathematically stated:

$$I_E = I_B + I_C$$

The same basic CE amplifier circuit shown in Fig. 21-6 can be formed using a PNP transistor. Like the NPN transistor, the PNP device must operate with its emitter junction forward-biased and its collector junction reverse-biased. This means that the polarities of bias batteries V_{BB} and V_{CC} must be reversed as shown in Fig. 21-7. This PNP circuit arrangement functions in basically the same manner as its NPN counterpart even though its bias voltages and currents are opposite to those in the NPN circuit. The exact same relationship between I_E, I_B, and I_C still exists.

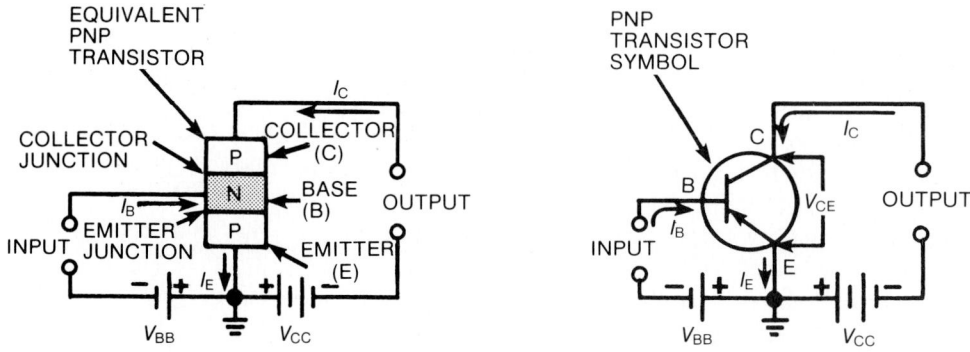

Fig. 21-7: A PNP transistor connected in the common-emitter configuration.

Current Gain. In a CE configuration, the very small base current (I_B) causes a much larger collector current (I_C) to flow through the transistor. Since I_B serves as the input current and I_C is essentially the output current, it can be said that an increase in current, or a *current gain*, is provided by the overall circuit. Furthermore, any change in I_B will produce a proportional change in I_C; that is, as I_B increases, I_C increases by a proportional amount. When I_B decreases, I_C will decrease proportionally. If I_B decreases to zero, I_C will essentially decrease to zero (although a very small leakage current may still flow due to impurities within the semiconductor material). There is also an operative limit to the maximum value of I_B and I_C. If I_B is increased too far, the increase in I_C will reach a *saturation point* where it begins to taper off. The values of I_B and I_C must be held to safe operating limits so that the internal structure of the transistor is not damaged by an excessive amount of internally generated heat.

The current gain of a transistor in the CE configuration is often referred to as the transistor's *beta* (β). Beta is simply the

476

ratio of the transistor's output current (I_C) to the transistor's input current (I_B). If the ratio of the DC or steady-state input and output currents is determined, the end result will be the transistor's DC beta. It is expressed mathematically as:

$$\text{DC beta} = \frac{I_C}{I_B}$$

However, as explained earlier, a small change in I_B (ΔI_B) will be accompanied by a corresponding change in I_C (ΔI_C). If the ratio of a small change in I_C to a corresponding change in I_B is determined, the end result will be the AC beta of the transistor. The AC beta may be expressed mathematically as:

$$\text{AC beta} = \frac{\Delta I_C}{\Delta I_B}$$

As was the DC beta previously described, the AC beta is simply a ratio of output current to input current. However, the AC beta represents the gain obtained when a changing (AC) signal is applied to the common-emitter transistor, whereas the DC beta represents the gain under steady-state or no-signal conditions. The AC and DC beta values are roughly the same for any given transistor. They may range from 10 to more than 200 with a typical value being from 80 to 150. Beta values are always determined while the collector-to-emitter voltage, V_{CE}, is held constant (i.e., no load resistor is included in the circuit). Two basic circuit configurations which can be used to determine the beta of a transistor are shown in Fig. 21-8. Figure 21-8A illustrates how steady DC values are used to determine the DC beta and Fig. 21-8B illustrates how changing values are used to determine the AC beta. In both circuits, the collector-to-emitter voltage is held constant by the source, V_{CE}. Note that in Fig. 21-8B the change in I_B is the result of varying V_{BE}. Although NPN transistors are shown here, the same basic techniques are used to determine the AC or DC beta of a PNP transistor.

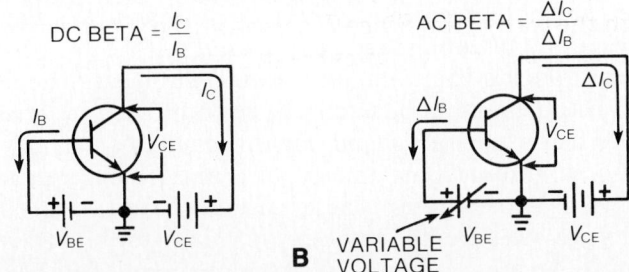

Fig. 21-8: Determining the current gain of a transistor in the common-emitter configuration.

The current gain or beta of a CE transistor is sometimes called the *common-emitter, forward-current transfer ratio* of the transistor. The DC beta (forward-current transfer ratio) is often designated by the symbol h_{FE} and the AC beta (forward-current transfer ratio) is represented by h_{fe}.

Common-Base Configuration

The transistor's base is common to both the input and the output signals in the common-base circuit configuration. The input signal is applied across the transistor's emitter and base, and the output signal appears across the transistor's collector and base. The transistor's emitter junction is still forward-biased and its collector junction is still reverse-biased in the same manner as the common-emitter circuit. Although it is not used as extensively as the common-emitter configuration, the common-base configuration has certain characteristics that make it useful in a number of applications.

Figure 21-9A shows a simple common-base circuit using an NPN transistor. The voltage source, V_{EE}, provides the forward-bias voltage necessary for the transistor's EB junction and the voltage source, V_{CC}, supplies the reverse-bias voltage needed across the transistor's CB junction.

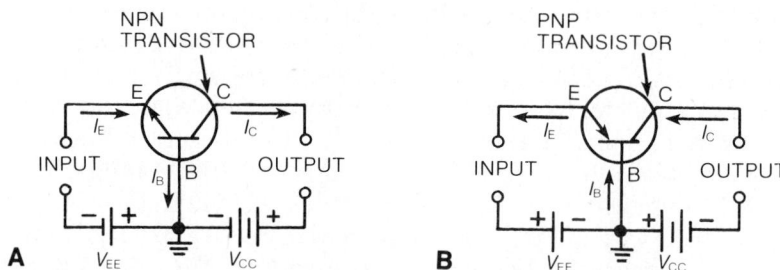

Fig. 21-9: A common-base circuit using (A) an NPN transistor and (B) a PNP transistor.

As explained previously, the transistor's emitter current, I_E, must equal the sum of its base current, I_B, and collector current, I_C. Stated in another way, the emitter current must divide into two separate currents (I_B and I_C). Typically, the amount of I_C is at least 95% of the I_E total. The remaining 5% of current, I_B, flows through the base. The transistor's emitter, base, and collector currents are directly related so that a change in I_E produces a proportional change in I_B and I_C.

I_E serves as the input current and I_C serves as the output current in the common-base circuit configuration. Variations in I_E, controlled by the input signal, cause I_C to vary by a proportional amount. The amount of I_E and I_C, under no-signal or steady-state conditions, is determined by V_{EE} and V_{CC}. These two voltages are adjusted to keep the transistor biased within its linear operating region so that proportional changes in I_E and I_C result.

As shown in Fig. 21-9B, the common-base circuit configuration may also be formed with a PNP transistor. This circuit is basically the same as the NPN circuit, except the polarities of the voltage sources have been reversed to provide the necessary forward and reverse-bias voltage for the PNP transistor. Even though its bias voltages and currents are reversed, the PNP circuit operates in basically the same manner as the NPN circuit.

478

Current Gain. In the common-base configuration, the current gain of a transistor, called alpha, is represented by the symbol α. The current gain of a common-base circuit is determined by dividing the output current, I_C, by the input current, I_E. Current gain is usually calculated for the transistor alone without any additional components in the circuit.

The transistor's alpha can be obtained by using either fixed or changing values of I_E and I_C. When fixed values are used, the DC alpha of the transistor is determined; and when changing values are used, the AC alpha of the transistor is determined.

The DC alpha is equal to the ratio of I_C to I_E under steady-state or no-signal conditions. It can be expressed mathematically as:

$$\text{DC alpha} = \frac{I_C}{I_E}$$

By using a simple circuit like the one shown in Fig. 21-10A, the DC alpha of a transistor can be determined. V_{BE} and V_{CE} are adjusted to obtain fixed values of I_E and I_C, and then I_C is divided by I_E to obtain the DC alpha.

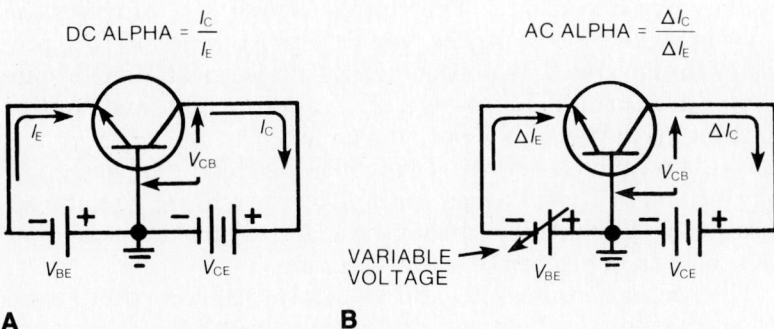

Fig. 21-10: Determining the current gain of a transistor in the common-base configuration.

The transistor's DC alpha must always be less than unity or 1, because I_C is slightly lower than I_E. However, alpha may be very close to 1 since I_C is typically 95% to 99% of I_E. Therefore, typical DC alpha values will range from 0.95 to 0.99. What this means, essentially, is that the common-base circuit configuration does not provide an increase in current. In fact, there is a minute loss in current at the output.

The AC alpha, which represents the current gain seen by an AC input signal, can be determined by varying I_E a small amount and observing the corresponding change in I_C. The change in I_C is then divided by the change in I_E, as shown below:

$$\text{AC alpha} = \frac{\Delta I_C}{\Delta I_E}$$

The AC alpha can be determined by using a simple circuit like the one shown in Fig. 21-10B. V_{BE} is varied to produce changes in I_E, which results in changes for I_C. The transistor's AC alpha is represented by the ratio of these changes.

479

The DC and AC alpha values, for any given transistor, are nearly the same. They are always just slightly less than 1. It should be noted that these DC and AC alpha values are always determined while the transistor's collector-base voltage, V_{CB}, is held constant. A constant V_{CB} can be obtained by connecting V_{CE} directly across the transistor's collector and base leads, as shown in Fig. 21-10.

Alpha is sometimes referred to as the *common-base, forward current transfer ratio*. Also, the DC and AC alpha values are sometimes represented by the symbols h_{FB} and h_{fb} respectively.

Common-Collector Circuits

The transistor's collector lead is common to both the input and the output signals in the common-collector circuit configuration. The input signal is applied across the base and collector, and the output signal appears across the collector and emitter. Once again, the emitter junction remains forward-biased while the collector junction is reverse-biased.

A simple common-collector circuit using an NPN transistor is shown in Fig. 21-11A. The voltage source, V_{BB}, provides the forward-bias voltage for the transistor's EB junction. V_{CC} provides the reverse bias needed across the transistor's CB junction. Base current, I_B, serves as the input current, and emitter current, I_E, serves as the output current.

The level of base current, I_B, is controlled by the input signal. Variations in I_B will cause the emitter current, I_E, to vary accordingly. I_E must flow through a load resistor or some other type of load to perform a useful function.

The source voltages, V_{BB} and V_{CC}, are adjusted so that I_B and I_C are within the transistor's linear operating region. As a result, any change in I_B will produce a proportional change in I_C which, in turn, will produce a proportional change in the output voltage.

As shown in Fig. 21-11B, the common-collector circuit configuration may also be formed using a PNP transistor. This circuit is basically the same as the NPN circuit, except for the difference in polarity of the source voltages and the direction of the input and output currents.

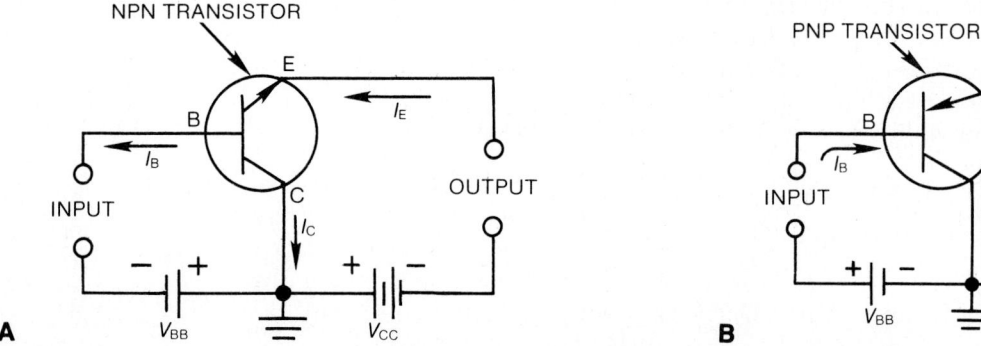

Fig. 21-11: (A) An NPN transistor connected in the common-collector configuration. (B) A PNP transistor connected in the common-collector configuration.

480

Current Gain. The CG circuit produces a substantial current gain since the output current, I_E, is much higher than the input current, I_B. In fact, the current gain in this circuit is even slightly higher than the current gain in a common-emitter configuration. This is due to the slightly higher output current in a common-collector configuration, assuming that the same transistor is used in each configuration.

The current gain of a common-collector configuration is equal to beta (of the transistor) plus 1. For example, if a transistor used in a CC circuit has a beta of 30 (which is current gain of 30 when connected in CE configuration), the current gain of this CC circuit will equal 30 + 1, or 31. Thus, the amplitude of the output signal current will be 31 times greater than the input signal current amplitude.

Transistor Characteristics

A *family of curves* showing current-voltage characteristics for a particular transistor is shown in Fig. 21-12. This graph is a plot of emitter current and collector current for all possible values of collector voltage. Since the collector voltages are represented as positive values, it may be assumed that the transistor is an NPN. However, the same graph applies equally well to a PNP transistor if the collector voltages are referred to as negative values. Complete current-voltage characteristics may be obtained from the *static characteristic curves* supplied by the manufacturer. The static curves show only the direct current and direct voltage conditions. As will be discussed later in the book, a load line added to this graph converts it into a dynamic characteristics graph.

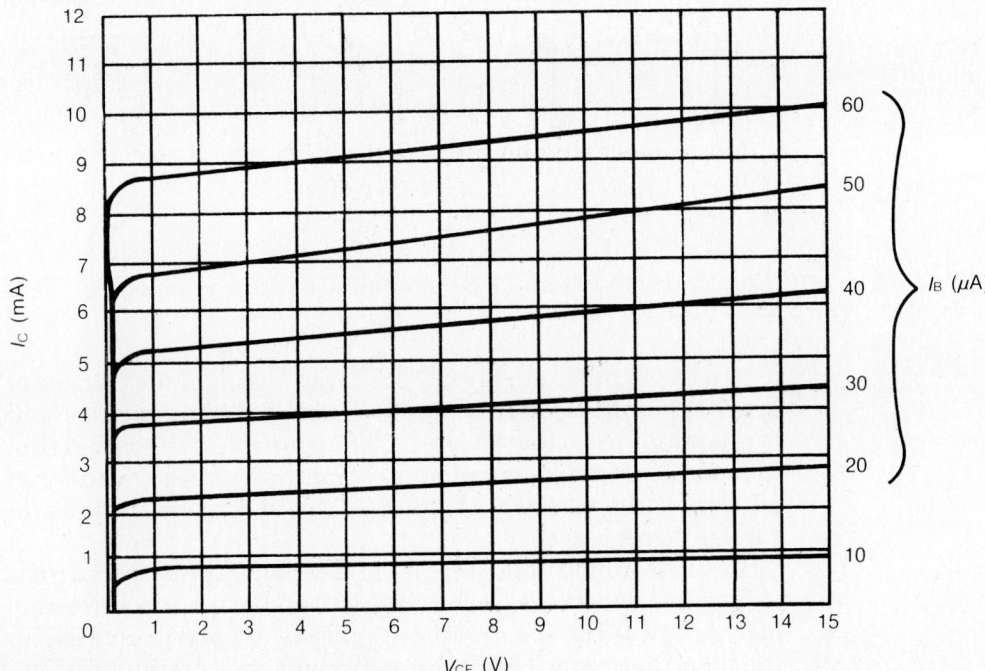

Fig. 21-12: Static characteristic curves.

481

Common-Emitter Transistor Amplifier

The NPN and PNP common-emitter transistor amplifiers are shown in Figs. 21-13A and B, respectively. Bias (the average difference of potential across the emitter-base junction) of the transistors shown in Fig. 21-13 is in the forward, or low resistance, direction. This means that direct current is flowing continuously, with or without an input signal, throughout the entire circuit. The bias current, flowing through the emitter-base junction, is supplied by bias battery V_{BB}. The average base voltage, as measured with respect to the emitter, depends upon the magnitude of the bias voltage supply in the emitter-base circuit and the base resistor. The current and the amplitude of the input signal determine the class of operation of the transistors. The average collector voltage depends upon the magnitude of the collector supply voltage, V_{CC}.

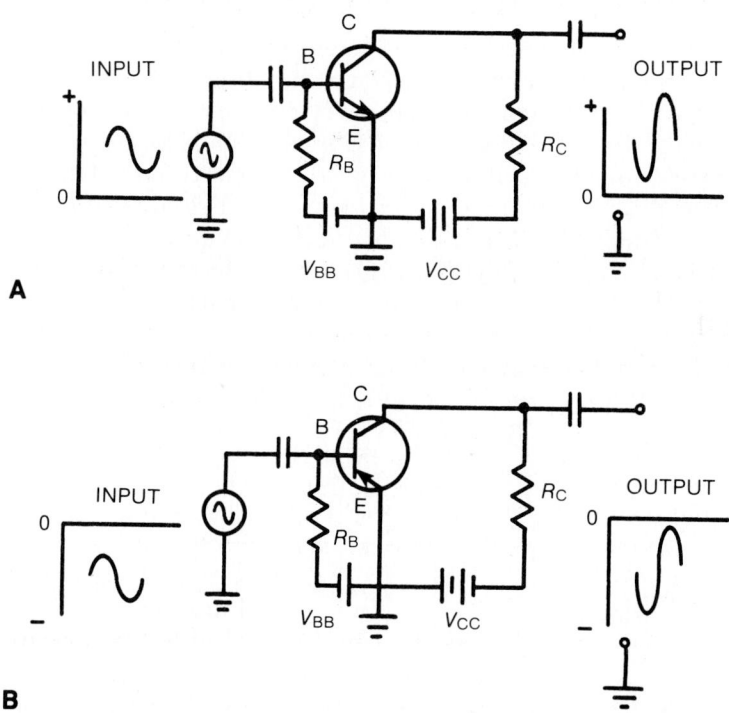

A

B

Fig. 21-13: NPN and PNP common-emitter transistor amplifiers.

When a signal is supplied to the input circuit of the transistor (the base-emitter junction here), the bias current varies about the average, or no-signal, value. That portion of the signal that aids the bias causes transistor current to increase, while that portion of the signal that opposes the bias causes transistor current to decrease.

Since the input signal is capable of causing transistor current to increase or decrease, according to its polarity, it is also capable of controlling the greater amplitude of current flowing in the collector circuit. This, as already explained, is amplification.

The input to the common-emitter amplifier is between the base and the emitter as shown in Fig. 21-13. The output voltage for this type circuit is that voltage felt across the collector to emitter terminals. The emitter is common to both the input and output circuits, hence the designation of the common-emitter configuration.

The output voltage will vary with collector current. This is the same as stating that the voltage on the collector will vary with collector current. As collector current increases, collector voltage decreases and vice versa due to the drop across the collector resistor R_C.

It is possible to visualize the transistor acting as a variable resistance. When the transistor is allowing more current to flow, its resistance is lower than when it is causing less current to flow. Figure 21-14 will be referred to for further explanation. That portion of the input signal indicated by point 1 in the input circuit effectively aids the bias, thereby reducing resistance of the transistor to current flow. Current flowing in the transistor increases so that current in the entire circuit is increased. This increased current must flow through R_C and thereby increases the voltage drop across R_C. The increased voltage drop across R_C leaves less voltage to be dropped across the transistor. Since the output is taken from the collector terminal to the emitter terminal (ground), it is evident that collector voltage (V_C) is less than it had been at its no-signal value of +4V. At point 1 of the output waveform (corresponding to point 1 of the input signal), the collector voltage is at +3V.

As the input signal goes from point 1 to point 2, the transistor is increasing its resistance until, at point 2, it offers its maximum resistance to current flow. Less current flowing in the transistor means less current is flowing in the circuit. This reduction in current flowing through R_C results in a reduced voltage drop across R_C. The reduction in the voltage drop across R_C means that the collector voltage is higher. This increase is shown as +5V at point 2 on the output signal. Again point 2 on the input signal and point 2 on the output signal occur essentially at the same time.

In succeeding cycles, as the input signal aids the forward bias, transistor current increases, causing an increased voltage drop across R_C, thereby decreasing the output voltage. As the input signal opposes the forward bias, transistor current de-

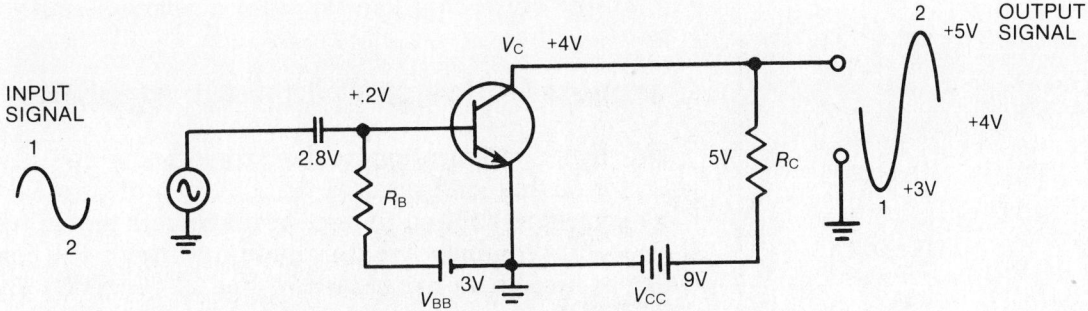

Fig. 21-14: Basic NPN common-emitter transistor amplifier.

creases. This decrease in current through R_C means a reduced voltage drop across R_C and more voltage available in the output.

In any of the amplifier circuits which are presented, moving the base potential towards the collector voltage (V_C) increases transistor current, and moving the base voltage away from collector voltage decreases transistor current. As shown in Figs. 21-13 and 21-14, the input and output signals are 180° out of phase. The common-emitter amplifier, whether it be an NPN or PNP circuit, provides a phase inversion.

Some typical values from common-emitter amplifiers are given below:

1. Input resistance: $500\,\Omega$ to $1{,}500\,\Omega$.
2. Output resistance: $30\mathrm{k}\Omega$ to $50\mathrm{k}\Omega$.
3. Voltage gain: 300 to 1,000.
4. Current gain: 25 to 50.
5. Power gain: 25dB to 40dB (a gain of roughly 200 to 10,000).

The manufacturer also provides static characteristics graphs for common-emitter amplifiers. A family of curves that composes such a graph reflects base current and collector current for all possible values of collector voltage. Such a graph is shown in Fig. 21-15.

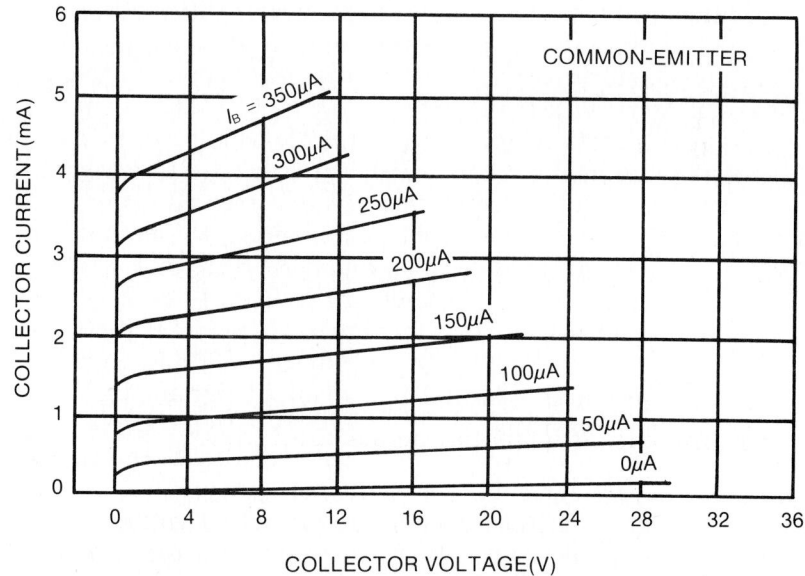

Fig. 21-15: Static characteristics for a common-emitter.

Common-Collector Transistor Amplifier

The last configuration to be presented is the common-collector. In this configuration the input signal is applied to the base element and the output will be taken from the emitter with the collector common to both input and output. The common-collector amplifiers are shown in Fig. 21-16 (NPN) and Fig. 21-17 (PNP).

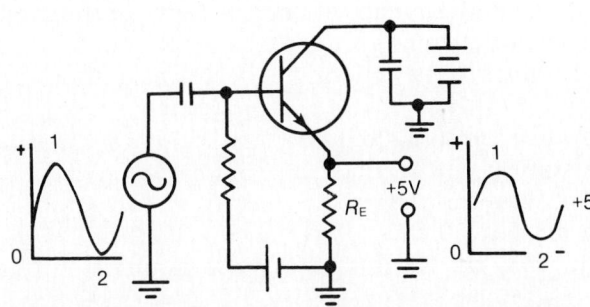

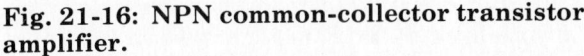

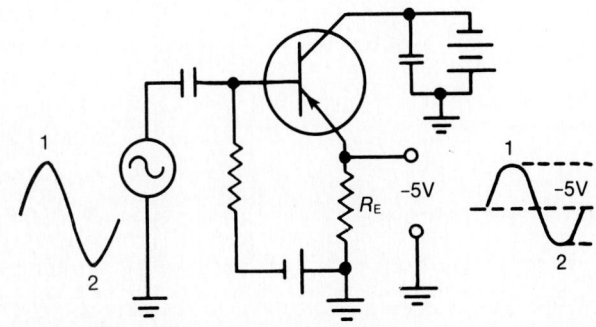

Fig. 21-16: NPN common-collector transistor amplifier.

Fig. 21-17: PNP common-collector transistor amplifier.

The common-collector transistor amplifier is often referred to as an emitter-follower circuit. Since the input signal is across the base-collector junction—the reverse-biased, high-resistance direction—the input impedance of the common-collector configuration is high. On the other hand, the output impedance of this circuit is low, since it is taken in the forward-bias, low-resistance direction of the base-emitter junction. The high input impedance and low output impedance of the common-collector circuit make it useful as an impedance matching device. Figure 21-17 will be used to explain the circuit operation of the basic common-collector transistor amplifier circuit.

As shown in Fig. 21-17, the transistor is forward-biased. When the input signal is moving the base potential away from the collector voltage (point 1 on the input signal), forward bias on the emitter-base junction is opposed and less current flows in the circuit. Less current flowing through R_E means that the negative potential, with respect to ground on the output end of R_E, is something less than its quiescent value of –5V. The output signal is moving in a positive direction as shown by point 1 on the output signal. Again points 1 and 2 on the input and output waveforms occur at approximately the same period in time. The input signal at point 2 is of the polarity necessary to increase the difference of potential across the emitter-base junction. This aids the forward bias, causing an increase in transistor current. Increased current flow through R_E results in a larger voltage drop across R_E, and the output voltage becomes more negative as shown by point 2 on the output waveform.

The reason that the common or grounded collector configuration is called an emitter-follower is because the output voltage on the emitter is practically a replica of the input voltage; that is, it has the same waveform and amplitude. In a practical circuit, the output amplitude will be slightly less than the magnitude of the input. It was stated that the input is across the base-collector junction; however, a closer examination will reveal that the input is also across the series combination consisting of the base-emitter junction and the output load resistor. Since the input is across two resistors and the output is across only one of these, the output voltage must be less than the input.

485

However, since all currents in the circuit flow through the load resistor and only a small amount of current flows in the input circuit, a considerable current gain may be obtained.

Some typical values from common-collector amplifiers are given below:

1. Input resistance: 20kΩ to 500kΩ.
2. Output resistance: 50Ω to 1,000Ω.
3. Voltage gain: less than unity.
4. Current gain: 25 to 50.
5. Power gain: 14dB to 17dB (roughly the same as current gain).

In summary, the transistor behaves like a variable resistance. When it is so biased as to cause decreased resistance, current flow increases and vice versa. A comparison chart for the three configurations is shown in Table 21-1.

Table 21-1: Transistor Amplifier Comparison Chart

Amplifier Type	Common-Base	Common-Emitter	Common-Collector
input/output phase relationship	0°	180°	0°
voltage gain	high	medium	low
current gain	low (α)	medium (β)	high (γ)
power gain	low	high	medium
input resistance	low	medium	high
output resistance	high	medium	low

Gain in Transistor Circuits

Gain is the ratio of output voltage, power, or current with respect to an input reading in the same unit of measurement. Each transistor configuration will give a different value of gain even though the same transistor is being used. The configuration used is a matter of design consideration; however, it is interesting to note the ratio of output with respect to the input. This is important in determining whether or not the transistor is performing its circuit application function. Transistor specifications usually state the normal current gain for a given transistor at a particular voltage.

Generally, if the output circuit has a low impedance, the transistor provides a large current or power gain. Conversely, a high output impedance is an indication of a voltage gain. Some transistor circuits have a low input impedance; others have high input impedances. Some transistor circuits have low output impedances and others high output impedances.

Operating Limits

Some of the limits imposed on operating transistors are: maximum collector current, maximum collector voltage, maximum collector power dissipation, and collector cutoff (leakage) current. The first three limitations listed are transistor ratings, and the last item is a characteristic of the transistor. A rating is defined as a limiting value which, if exceeded, may result in permanent damage to the device. On the other hand, a characteristic is a measurable property of the device under specific operating conditions.

Collector cutoff current, also called *collector-base leakage current,* or I_{CBO} for the common-base configuration (I_{CEO} for the common-emitter configuration), is determined primarily by two factors: temperature and voltage.

The rating of I_{CBO} or I_{CEO} is given as some maximum at a specified collector-to-base or collector-to-emitter voltage and a particular temperature. If either the temperature or the collector voltage increases, cutoff current will increase. The rating must be proportionally increased for voltages or temperatures which are higher than those specified. For example, with most transistors, the I_{CBO} doubles for each 8° to 10°C increase. Because I_{CBO} adds to the collector current and subtracts from the base current, the control of I_B over I_C is reduced. Excessive leakage current can produce excessive temperatures, which can build up to *thermal runaway.*

The peak collector power dissipation rating, P_C, is that value of power that can safely be dissipated by the collector. P_C represents the limit of the product of the DC quantities V_C and I_C. If the maximum power dissipation rating is 5W, the product of collector voltage times collector current flow must not exceed 5W.

The maximum collector voltage rating is the maximum DC potential that can be applied to the collector with safety. The maximum collector voltage rating is usually given in terms of the collector-emitter voltage (V_{CEO}). The value of this voltage, in part, determines the quantity of reverse bias across the collector-base junction and the quantity of forward bias across the emitter-base junction. For a specified value of I_B, both I_C and I_B will remain relatively constant until the maximum value of V_{CEO} is reached. When this value, V_{CEO}, is exceeded, both I_C and I_B begin increasing. When the breakdown voltage (BV_{CEO}), which is just above the maximum value of V_{CEO}, is reached, I_C and I_B increase very rapidly, and the device breaks down.

The maximum collector current rating is the maximum value of DC that can flow through the collector. As may be seen from the above information, permanent damage to the device will occur if this value is exceeded.

Base Lead Current

Transistor current is dependent upon many variables. Some of these variables are: (1) the type of bias, i.e., fixed bias, self bias; (2) the magnitude of the input signal; (3) the configuration being used; (4) the type of transistor being used; and (5) the direction of current flow in the base lead.

Of the variables mentioned, the direction of current flow in the base lead deserves further examination. Whether the majority of current flow is in the emitter or the collector is determined by this variable. If the base lead current is flowing into the base material in the PNP (Fig. 21-18A) or out of the base material into the base lead in an NPN (Fig. 21-18B), the majority of current is flowing in the emitter. If the reverse is true (Figs. 21-18C and D), then the majority of current is flowing in the collector.

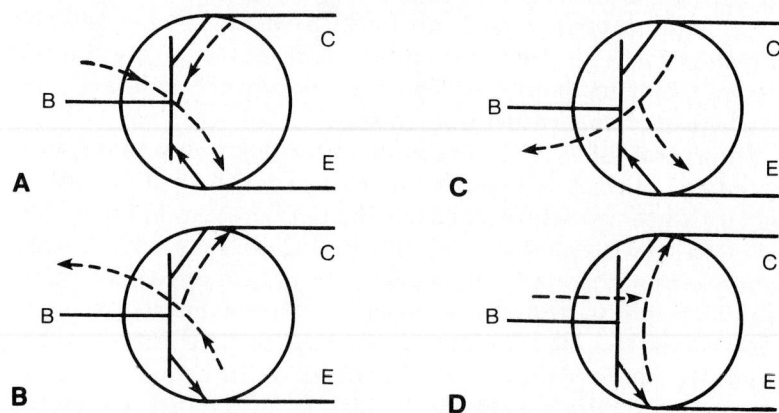

Fig. 21-18: Base lead current flowing into, or out of, the base material in NPN and PNP transistors.

Reverse Current. Reverse current in transistors as in the PN junction diode is that portion of electron flow taking place across a reverse-biased junction. Normally, the magnitude of this reverse current is very low (in the microampere range) and does not subtract any significant quantity from the base current (current in the base lead or I_B). However, since this reverse current increases rapidly as temperature increases, it might reach a magnitude where damage to the transistor or malfunctioning of the circuit would occur.

In order to explain how this reverse current may cause trouble by reducing the magnitude of I_B, or even reversing the normal flow of I_B, the block diagram of an NPN transistor shown in Fig. 21-19 will be used. The solid lines indicate the normal flow of current, while the dotted lines indicate the direction of reverse current when the transistor is biased as shown.

In Fig. 21-19, when S_1 is closed, current flow is as indicated by the solid lines: 100% of the current flows in the emitter, some

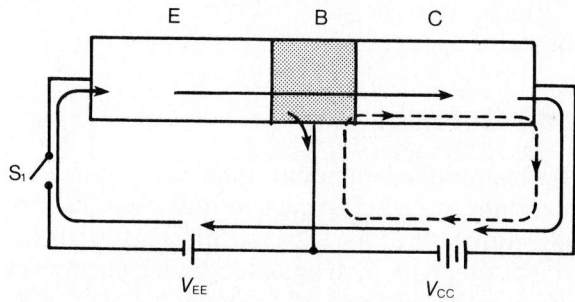

Fig. 21-19: Normal and reverse current in an NPN transistor.

488

small percentage flows out of the base lead, and the remaining current flows in the collector. With the switch open there is a small amount of current flowing into the base material from the base lead and the collector electrode, continuing through the collector back to the source, V_{CC}.

Transistor Lead Identification

Identification of the leads or terminals of a semiconductor device is necessary before it can be connected into a circuit. As there is no standard method of identifying transistor leads or terminals, it would be quite possible to mistake one lead for another.

In Fig. 21-20, the bases of four categories of transistors are shown. Each set, while similar in appearance, has different elements connected to the leads. In the first set, the leads are emitter, base, collector, reading from left to right, for the top transistor, while in the bottom transistor the leads are emitter, collector, and base. If one of these transistors were connected into a circuit as a replacement for the other, the circuit would not function properly or the transistor might be destroyed. The same general results apply for the other sets of transistors. Note that in the right-hand type, the case can be used for collector connections.

To further complicate matters, field effect transistors also have the same casing as those shown. It is emphasized that you *cannot trust* to shape in replacing one transistor with another. Be sure the leads are where they should be and that the transistor chosen as a replacement is suitable for performing the circuit application. If there is any doubt, consult the equipment manual or a transistor manual showing the specifications for the transistor being used.

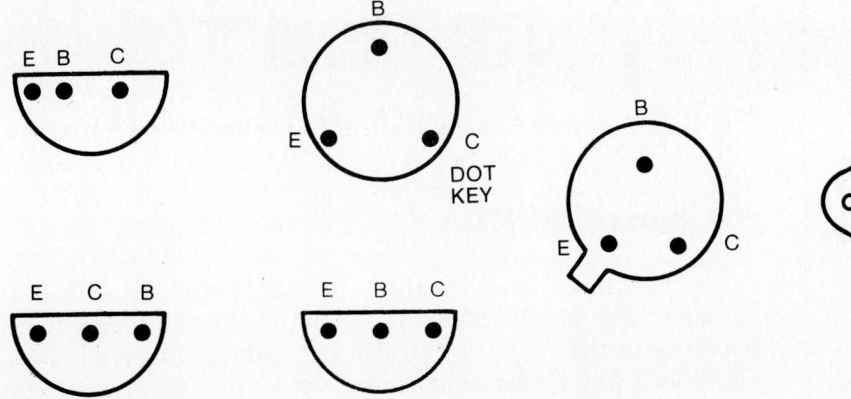 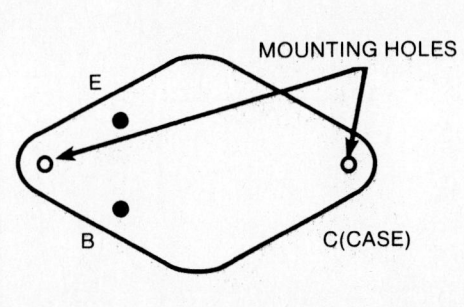

Fig. 21-20: Bottom view of some common transistor cases.

FIELD EFFECT TRANSISTORS (FETs)

The bipolar junction transistor is called bipolar because it has both loosely bound electrons and holes as charge carriers. In contrast to the BJT, the junction field effect transistor is a unipolar, voltage-sensitive device that uses an electrostatic field to control conduction. It combines the inherent advantages of small size, low power, and mechanical ruggedness with a high input impedance. In addition, it has a square-law transfer characteristic that makes it ideally suited for low cross modulation. It is in many ways superior to a bipolar transistor, although both types still have a definite place in electronic design. There are two general types of field effect transistors: junction (JFET) and insulated gate (IGFET). The operation is similar in the two types, but the structure is different.

JUNCTION FIELD EFFECT TRANSISTORS

Basically, a JFET is a bar of doped silicon, called a channel, that behaves as a resistor. The doping may be N or P-type, creating either an N-channel or P-channel JFET. There is a terminal at each end of the channel. One terminal is called the source, and the other is called the drain. Current flow between source and drain, for a given drain-source voltage, is dependent upon the resistance of the channel. This resistance is controlled by a gate. The gate consists of P-type regions diffused into an N-type channel or N-type regions diffused into a P-type channel. Cutaway views of both types are shown in Fig. 21-21.

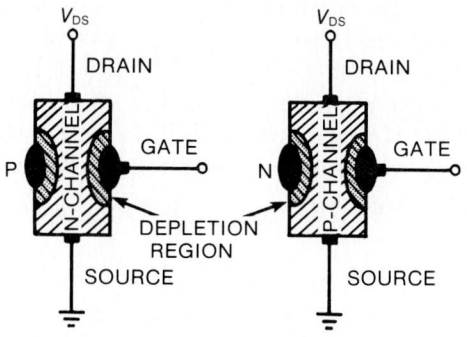

Fig. 21-21: Cutaway view of N- and P-channel JFETs.

N-Channel JFET

The two most used symbols for an N-channel JFET are shown in Fig. 21-22. The arrow is now on the gate element, and it points to the N-type channel. The only difference between symbols A and B is the position of the arrow. The symbol used in any specific instance is a matter of individual choice.

Figure 21-23 shows the proper voltage polarities and the effect of different bias levels. A relatively large positive potential is connected between the drain and the source. This voltage is designated as V_{DS} (voltage drain to source). A smaller negative potential, V_{GS}, is connected between the gate and the source. This arrangement places the drain as the most positive element and the gate as the most negative element.

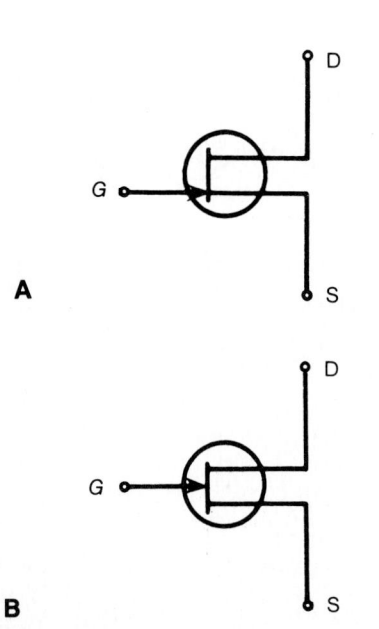

Fig. 21-22: Schematic symbols for an N-channel JFET.

490

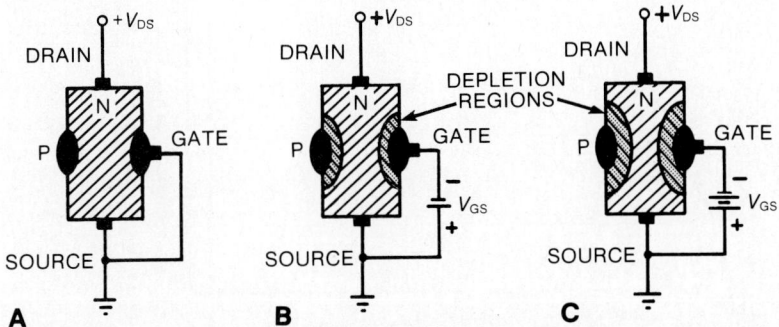

Fig. 21-23: Biasing of an N-channel JFET: (A) zero gate bias, (B) small negative gate bias, and (C) large negative gate bias.

Increasing the negative potential on the gate widens the depletion region around the junction and decreases the current (Figs. 21-23B and C). The gate control of current is complete, all the way from saturation to cutoff.

Characteristics of an N-Channel JFET with Zero Gate Bias. In this configuration, the gate is connected directly to the source as shown in Fig. 21-23A. The drain current-voltage characteristics for zero gate bias ($V_{GS} = 0$) are shown in the diagram of Fig. 21-24. When the drain-to-source voltage, V_{DS}, is zero, the drain current is zero. As V_{DS} increases, I_D starts and increases in a linear fashion until the current nears saturation. Now the drain current increase slows down until *pinch-off voltage, V_p*, is reached. When V_{DS} reaches the V_p point, the drain current flattens out. This value of drain current is known as the *drain-source saturation current, or I_{DSS}*. This I_{DSS} is normally considered to be the maximum I_D with the gate shorted to the source. This is the same as saying that $V_{GS} = 0$. When V_{DS} is increased beyond the V_p point, there is little or no change in drain current. However, if V_{DS} is increased to the breakdown point, the JFET goes into avalanche conduction with a sudden large increase in drain current.

The area on Fig. 21-24 from point A to point B is the *ohmic channel* or triode region. The reason for these names is the fact that the channel is behaving like a resistor, similar to the action of a triode vacuum tube. The area from point B to point C is the pinch-off region. For all values of V_{DS} in the pinch-off region, the drain current remains at the saturation level. The JFET should be operated at V_{DS} levels well below the breakdown potential. When driven to breakdown, the avalanche effect may destroy the transistor. Normal operating values of V_{DS} fall anywhere within the pinch-off region.

Characteristics of an N-Channel JFET with Gate Bias. A family of current-voltage characteristics curves is produced by plotting a separate curve for each of several values of gate voltage. The diagram in Fig. 21-25 is such a family of curves. Now it can be seen that the controlling factor is the potential applied to the gate. For each different level of V_{GS}, the areas on the diagram change slightly. Maximum I_D occurs at different values of V_{GS}, pinch-off voltage changes, and the point of breakdown changes. Notice that both V_p and breakdown poten-

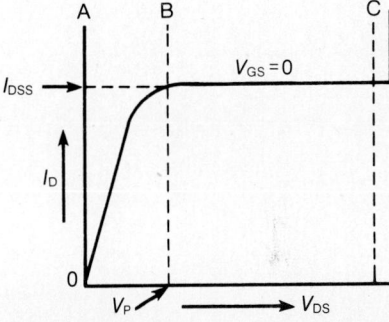

Fig. 21-24: Drain characteristics ($V_{GS} = 0$).

491

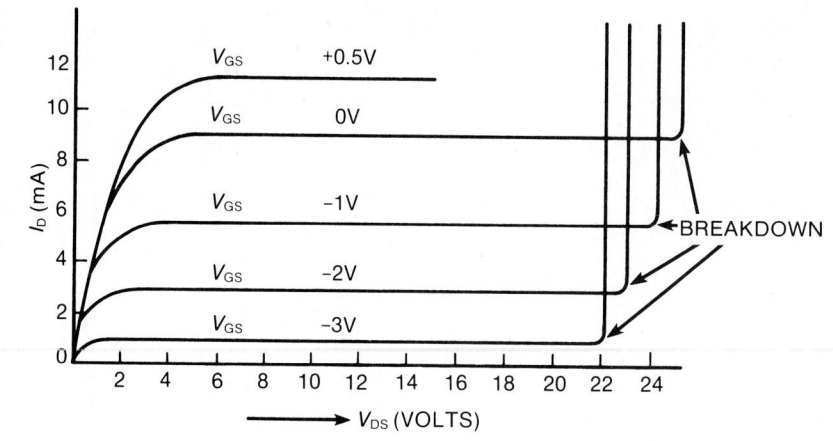

Fig. 21-25: **Drain characteristics with gate bias.**

tial occur at lower values of V_{DS} as the gate potential becomes more negative. Such a family of curves presents only the (static) direct current and voltage characteristics. Load lines can be constructed on these diagrams to indicate dynamic conditions in a particular circuit.

P-Channel JFET

As illustrated in Fig. 21-21, the P-channel JFET is constructed by diffusing a ring of N-type material around a piece of P-type material. In this case, the channel is composed of P-type material and the majority current carriers are the holes. The controlling element, the gate, is of N-type material. This reversal of material causes a reversal of all biasing potentials. The direction of electron movement is from drain to source, so V_{DS} is a negative potential. The gate-to-source junction must have reverse bias, so the V_{GS} is a positive potential.

The two most used schematic symbols for the P-channel JFET are exactly the same as those for the N-channel JFET except for the direction of the gate arrow. The arrow now points away from the P-type channel. Bias and symbols are illustrated in Fig. 21-26.

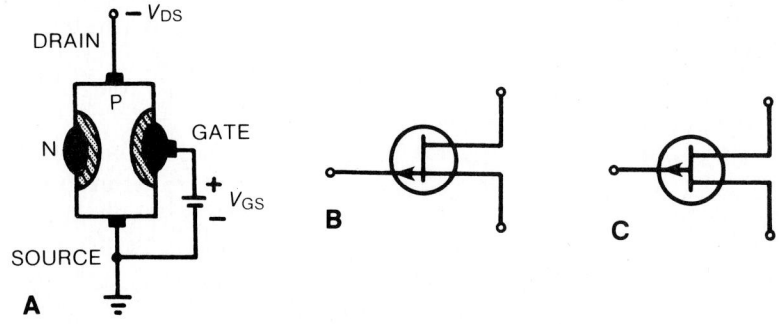

Fig. 21-26: **Bias and schematic symbols for a P-channel JFET.**

Characteristics of a P-Channel JFET. There are only two differences between N-channel characteristics and P-channel characteristics: the current direction and voltage polarities are reversed. Compare the diagram in Fig. 21-27 with that in Fig. 21-25.

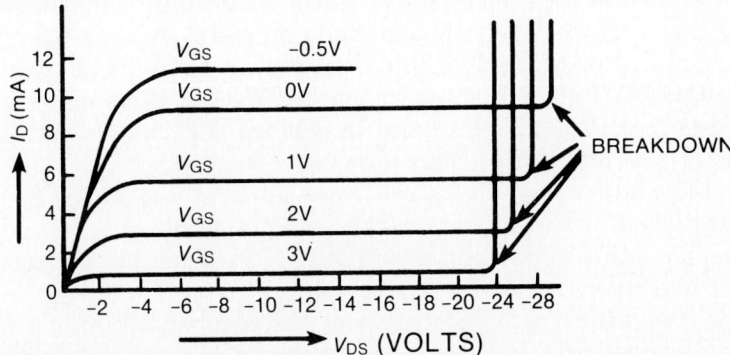

Fig. 21-27: Drain characteristics of a P-channel JFET with a gate bias.

BJTs Versus JFETs

It is important at this point, however, to know the theory of operation of JFETs and the major differences between JFETs and BJTs. The operation of a JFET can be summarized as follows: Free electrons moving between the drain and the source must flow through the narrow channel between the depletion regions. The size of these depletion regions defines the width of the conducting channel. The more negative the gate voltage, the narrower the conducting channel. In other words, the gate voltage controls the current between the source and drain. The greater the negative gate voltage, the smaller the current.

The main difference between a JFET and a BJT is that the gate of a JFET is reverse-biased; whereas, the base of a BJT is forward-biased. As a result, the JFET is a voltage-controlled device; input voltage alone controls output current. The BJT is a current-controlled device, where input current controls output current.

Another major difference between the JFET and BJT is the input impedance of each device. The input resistance of a JFET is much greater than the input impedance of a BJT; however, smaller control over output current is the price paid for higher input resistance. In other words, the JFET is less sensitive to variations in input voltages than the BJT. That is, in virtually any JFET, an input voltage variation of 0.1V results in a change in output current of less than 10mA. In a BJT, however, a 0.1V input variation easily results in an output variation of more than 10mA. As will be discussed later, this means that a JFET has less voltage gain than a BJT.

INSULATED GATE FIELD EFFECT TRANSISTOR (IGFET)

Insulated gate is a descriptive name for a type of FET that actually has no junction between the channel and the gate. In fact, the conducting channel between source and drain (in one type) must be induced by the potential on the gate. This insulated gate FET (IGFET) relies heavily on metal oxide in the construction process, and for this reason it is commonly referred to as MOSFET, which means *metal oxide semiconductor field effect transistor*. Two of the important differences between JFET and MOSFET are as follows:

1. The extremely high input resistance of the MOSFET is unaffected by the polarity of the gate potential.

2. The leakage current in a MOSFET is largely independent of the temperature.

The MOSFET is particularly well suited for such applications as RF amplifiers, voltage control circuits, and voltage amplifiers. There are two general types of MOSFET: the channel-enhancement and the channel-depletion types.

Structure of a Channel-Enhancement MOSFET

The channel-enhancement MOSFET takes its name from the fact that the conduction channel between source and drain is nonexistent until the proper bias is applied. The channel must be induced by applying forward bias between gate and source. The enhancement-channel MOSFET is divided into the N-channel and the P-channel types. The N-channel type is discussed here, but reversing the P- and N-type materials will cause the information to apply to the P-channel type.

The N-channel MOSFET begins with a lightly doped P-type substrate of highly resistant semiconductor material. This substrate is usually in the form of a thin silicon wafer. The wafer is polished, then oxidized in a furnace with a thin layer of insulating metal oxide. Spots are then etched through the oxide layer for the source and the drain. The wafer now goes into a furnace with a gas of N-type material. The N-type material diffuses into the two spots of exposed P-type substrate. Another oxidation process insulates the areas of exposed silicon; then small spots are etched for attaching the source and drain contacts. Metal is now evaporated over the entire wafer surface. Another etching process removes the excess metal and separates the drain, source, and gate contacts.

Connector wires are bonded to the metal contacts to form the electrode leads. Each unit is hermetically sealed in a case in an inert atmosphere. The transistor is then tested, and all external leads are shorted together. These leads should remain shorted together until the MOSFET is soldered into its circuit. The short is intended to prevent static charges from damaging the gate insulation.

The drawing in Fig. 21-28 is greatly exaggerated to illustrate the mechanical structure. Notice that there is no conduction channel between the two spots of N-type material that constitute source and drain. This conduction channel must be induced by placing a positive potential on the gate. This positive will attract electrons in the substrate and cause them to collect in the area between source and drain. The concentration of electrons in this area has the same effect as a strip of N-type material. The gate-positive potential has enhanced the negative carriers and created a conduction channel. In the absence of a positive potential on the gate, there can be no useful current between source and drain. Even with zero gate bias, there is no current. So the enhancement MOSFET is normally off, and the gate bias turns it on.

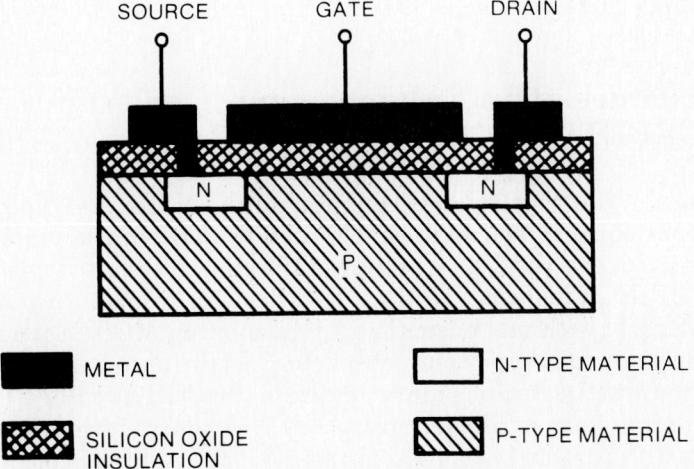

Fig. 21-28: **Structure of an N-channel-enhancement MOSFET.**

The schematic symbols for the enhancement MOSFET are illustrated in Fig. 21-29. The lead with the arrow is the substrate. It is sometimes labeled with an SS and sometimes with a U. If there is such an external lead, it should be connected to the circuit ground. If this fourth external lead is not present, it means that the substrate is internally connected to the source. The direction of the substrate arrow indicates the type of channel. The arrow points toward the N-type channel and away from the P-type channel. The broken line between source and drain identifies the enhancement MOSFET.

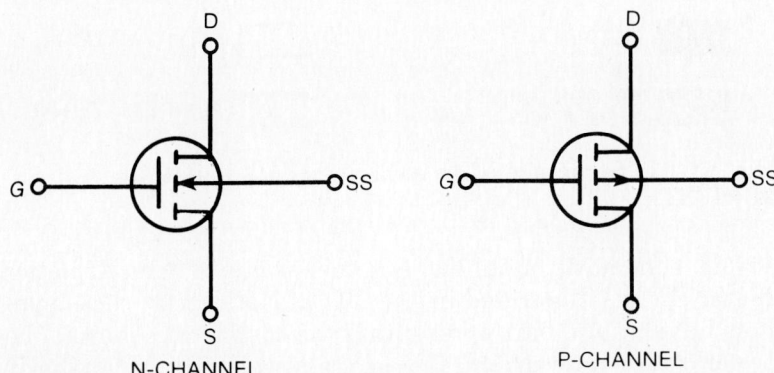

Fig. 21-29: **Schematic symbols for a channel-enhancement MOSFET.**

495

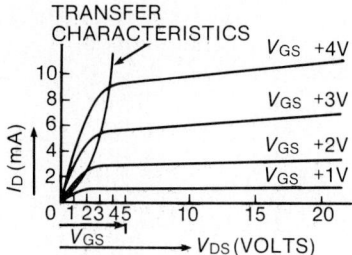

Fig. 21-30: Characteristics of a channel-enhancement MOSFET.

Characteristics of an Enhancement MOSFET

The diagram in Fig. 21-30 represents the drain and transfer characteristics of an N-channel enhancement MOSFET. The N-channel type is implied by the fact that all voltages on the diagram are positive. The same diagram with negative voltages represents a P-channel enhancement MOSFET.

This diagram shows that the conductivity of the channel is enhanced by the positive potential on the gate. Notice that the drain current increases as the V_{GS} becomes more positive. The fact that the gate is insulated from the channel accounts for the extremely high input resistance and the virtual absence of leakage current.

Structure of an Enhancement/Depletion MOSFET

Again, the discussion is centered on the N-channel type but applies equally well to the P-channel type when the material types are reversed. The structure of the enhancement/depletion MOSFET is identical to that described for the enhancement MOSFET, with one exception. The depletion MOSFET has a lightly doped N-type conduction channel diffused between the source and the drain. This structure is illustrated in Fig. 21-31. The enhancement/depletion MOSFET has current from source to drain with zero bias on the gate. In fact, it requires a considerable reverse bias to deplete the electron carriers in the channel and cut it off. The enhancement/depletion MOSFET is normally on.

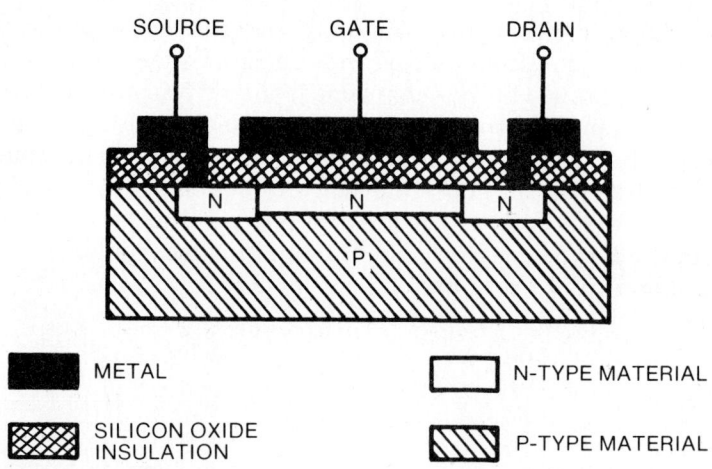

Fig. 21-31: Structure of an enhancement/depletion MOSFET.

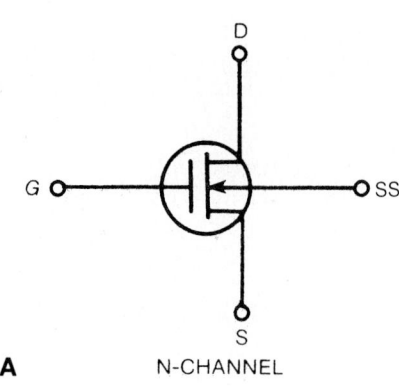

A   N-CHANNEL

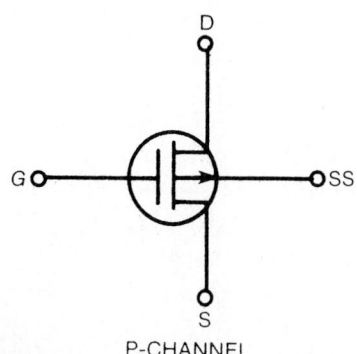

B   P-CHANNEL

Fig. 21-32: Schematic symbols for an enhancement/depletion MOSFET.

The schematic symbols for the enhancement/depletion MOSFET are illustrated in Fig. 21-32. Notice that these symbols have a solid bar representing the conduction channel between source and drain. The arrow still points in for the N-channel type and out for the P-channel type. The lead marked

SS is the substrate, and it is handled as explained under enhancement-channel MOSFET. The gate is separated from the channel as before. This signifies that the gate is insulated.

Characteristics of an Enhancement/ Depletion MOSFET

The enhancement/depletion MOSFET can be operated with the gate either positive, negative, or zero with respect to the source. When operated with zero bias on the gate, the diffused strip of doped material between source and drain functions reasonably well as a conduction channel. Applying a positive to the gate sets up a field that attracts more electron carriers into the channel area. These electrons enhance the N-channel carriers, and the drain current increases as the positive gate potential increases. When the gate is made negative, the resulting field attracts positive (holes) carriers into the channel area while repelling the negative electron carriers. This action depletes the N-channel carriers, and the drain current decreases as the gate becomes more negative. So the resistance of the channel depends upon the polarity and level of the applied gate voltage. The characteristics are summed up by saying that the enhancement/depletion MOSFET has two modes of operation: the depletion mode and the enhancement mode. When operated with a positive on the gate, it is in the enhancement mode. When it has a negative on the gate, it is in the depletion mode. The two modes are illustrated in Fig. 21-33.

As always, the drain is positive with respect to the source. In the depletion mode, V_{GS} is negative, and this depletes the carriers in the N-type channel. There is less conduction because the channel resistance has gone up. When $-V_{GS}$ reaches approximately 4V, the transistor will cut off. This complete stoppage of drain current indicates that for all practical purposes, the channel has ceased to exist.

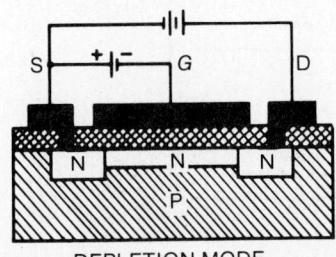

DEPLETION MODE   ENHANCEMENT MODE

Fig. 21-33: Modes of operation.

In the enhancement mode, V_{GS} is positive, and this attracts electrons from the substrate to increase the channel area. The N-type channel is enhanced, and drain current is increased. The characteristics are further illustrated in Fig. 21-34. This figure shows that in the depletion mode, I_D is largely independent of the value of V_{DS}. The only requirement here is that V_{DS}

497

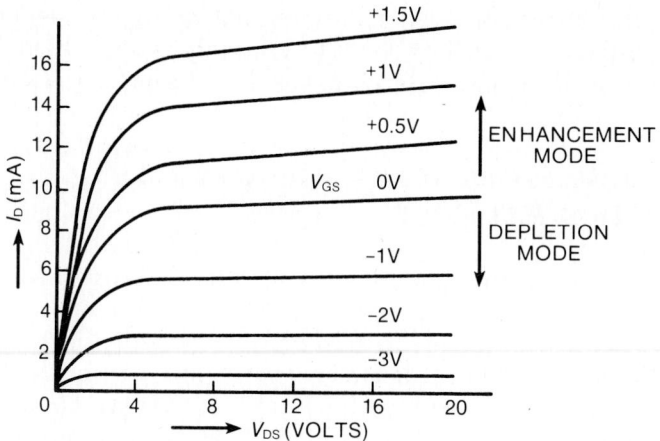

Fig. 21-34: Characteristics of an enhancement/depletion MOSFET.

be some positive potential. In the enhancement mode, I_D increases slightly with an increase of V_{DS}. This graph is for an N-channel enhancement/depletion MOSFET. For the P-channel type, the polarity of V_{DS} becomes negative and the V_{GS} polarities are reversed. These differences are shown in Fig. 21-35. Since the gate potential may be either positive or negative, the transfer characteristics for the N-channel enhancement/depletion MOSFET are approximately as depicted in the diagram of Fig. 21-36.

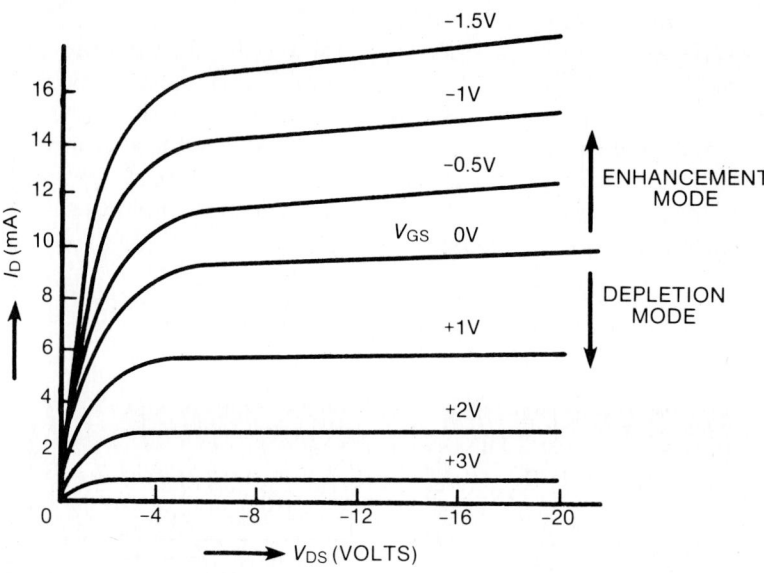

Fig. 21-35: Characteristics of a P-channel-enhancement/ depletion MOSFET.

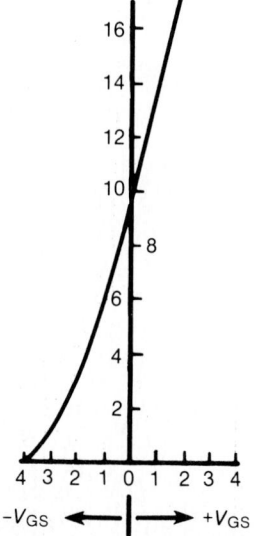

Fig. 21-36: Transfer characteristics of an enhancement/ depletion MOSFET.

BASIC FET CIRCUITS

The FET, like the BJT, is used in three basic circuit configurations. These are *common gate, common source,* and *common drain.* The name is taken from the element that is placed in a position to be *common* to both input and output circuits. Since this common element is generally grounded, as mentioned earlier, the word *grounded* is sometimes used instead of common. The common-gate configuration finds popular application in high-frequency voltage amplifiers. The common-drain configuration provides high power gain and makes a good buffer amplifier. The common-source configuration lends itself to a variety of uses because it amplifies voltage, current, and power. The common source is used more often than the other two.

Common-Gate Circuit

The common-gate circuit, as the name implies, has the gate as the element that is common to both input and output. This configuration is illustrated in Fig. 21-37. This is an N-channel enhancement MOSFET. The substrate arrow and the voltage polarities are reversed for the P-channel type. In this arrangement, the common element is readily apparent. The gate-to-source voltage places the gate slightly positive with respect to the source. The drain supply voltage places the drain positive with respect to both source and gate, with the source being the most negative. The positive V_{GS} attracts electrons from the substrate to create an N-type channel between source and drain. The input signal is applied across R_S and the output is taken across the drain load resistor, R_D.

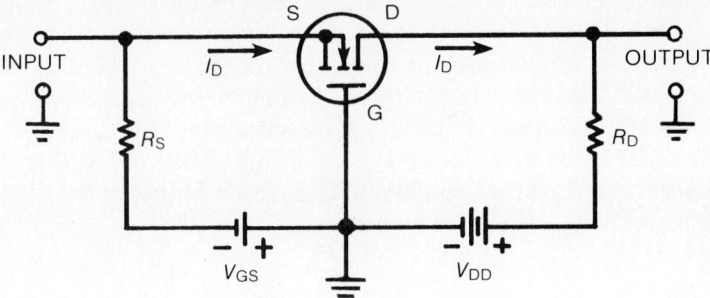

Fig. 21-37: Common gate.

The common gate effectively has an input impedance equal to the value of R_S. Since R_S is in series with R_D, the drain current I_D passes through both R_S and R_D. The value of R_D must be at least 10 times as large as that of R_S to realize a reasonable gain. Even with a large ohmic value for R_S, this is considered a low input impedance for a MOSFET. This characteristic makes the circuit useful for matching a low impedance to a high impedance.

The common gate has a low voltage amplification, which may be considered a disadvantage of the circuit. However, the low gain has some advantage in circuit design because it allows the neutralization circuits to be eliminated in many applications.

Common-Source Circuit

The common-source configuration is the most popular MOSFET circuit because it provides good voltage amplification together with reasonable gain in both current and power. An N-channel enhancement MOSFET is used in Fig. 21-38 to illustrate the common-source circuit. The grounded source is the most negative point in the circuit, and the drain is the most positive. The small voltage, V_{GS}, keeps the gate slightly more positive than the source. The input is applied across R_G to the gate. The output is taken across the drain load resistor, R_D.

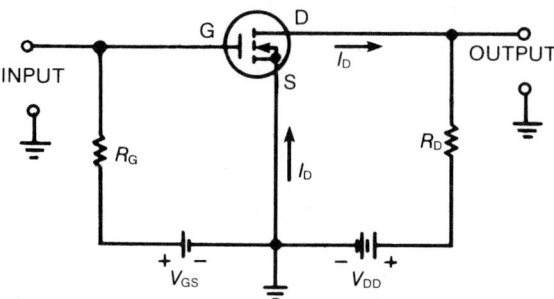

Fig. 21-38: Common source.

The common-source circuit combines a very high input impedance with a reasonably high output impedance. The value of R_G is on the order of $1M\Omega$, which gives the circuit a high input impedance. The output impedance of the device is high enough that it has little effect on the circuit output impedance. The circuit output impedance is then approximately equal to the value of R_D. The value of R_D has a direct bearing on the voltage gain; the voltage gain is directly proportional to the value of R_D.

The common source does have a voltage phase inversion. As the signal on the gate swings positive, the voltage on the drain swings negative, and vice versa.

Common-Drain Circuit

The common-drain circuit has a voltage gain less than unity, but it is frequently used as a power amplifier. An N-channel enhancement MOSFET is used in Fig. 21-39 to illustrate the common-drain configuration. Even though the drain is grounded, it is still the most positive point in the circuit. The small voltage, V_{GD}, keeps the gate potential less positive than the drain but still more positive than the source. The input is applied across R_G to the gate. The output is taken across R_S.

JFET Biasing

The biasing arrangements for a JFET resemble electron-tube biasing circuits and have little association with biasing of bipolar transistors. The reason for this is the fact that both

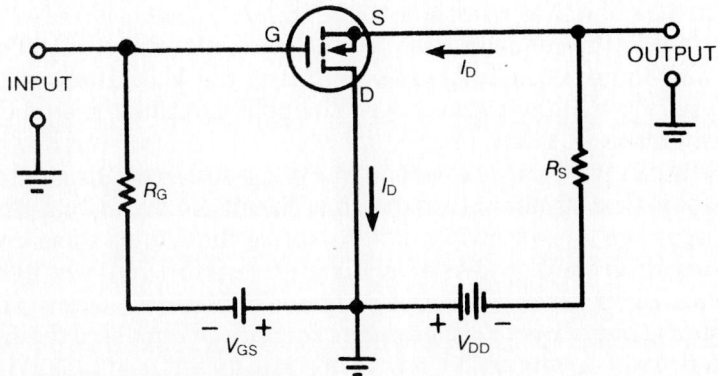

Fig. 21-39: Common drain.

electron tubes and field effect transistors are voltage-controlled devices. The JFET has no thermal-runaway problem. The two principal types of bias for the JFET are fixed bias and self-bias.

JFET Fixed Bias. The first step in designing any FET bias circuit is to determine the desired levels of V_{DS} and I_D under quiescent conditions. These are the conditions that exist when there is no input signal applied. The intersection of V_{DS} and I_D on the applicable characteristic graph reveals the level of V_{GS} that will produce these conditions. Consider the circuit in Fig. 21-40.

With the +24V drain supply voltage indicated, maximum undistorted amplification can be obtained if the static V_{DS} is +12V, just half of the total supply voltage. So the desired static V_{DS} is +12V. The remaining 12V must be dropped across R_D by the static value of I_D. Since R_D is 2.2kΩ, the static I_D must be

$$I_D = \frac{12V}{2.2k\Omega}$$

$$= 5.5mA$$

Now refer to Fig. 21-41 and determine what value of V_{GS} will produce an I_D of 5.5mA with 12V of V_{DS}.

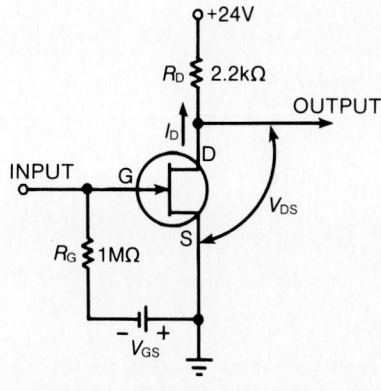

Fig. 21-40: JFET fixed bias.

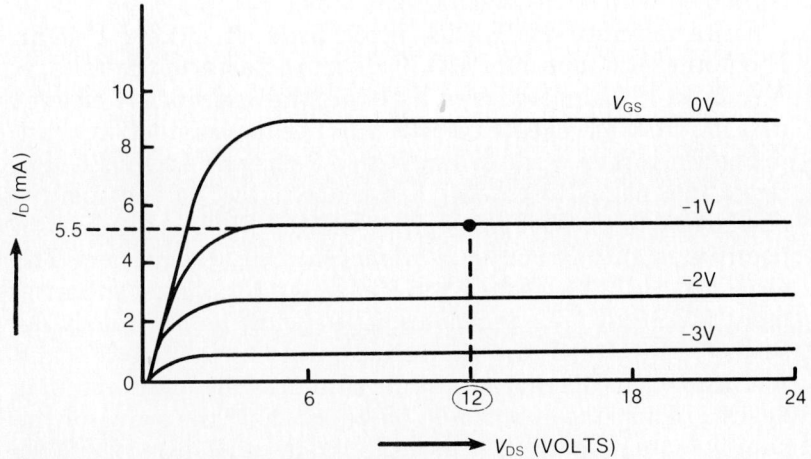

Fig. 21-41: V_{GS} for V_{DS} = 12V and I_D = 5.5mA.

501

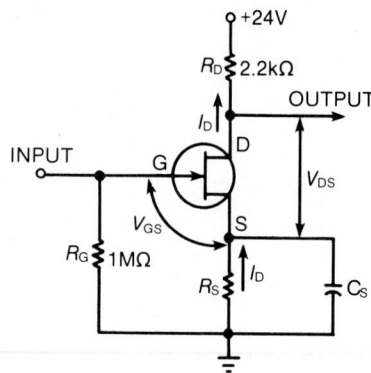

Fig. 21-42: JFET self-bias.

The 12V V_{DS} line intersects the 5.5mA I_D line at a V_{GS} of –1V. The –1V is the required bias for the circuit in Fig. 21-40. The biasing circuit is completed by replacing the V_{GS} battery with any steady voltage equal to 1V. The polarity must be as indicated on the diagram.

JFET Self-Bias. The circuit of Fig. 21-40 can be changed to a self-biasing circuit with only a few circuit modifications. The changes are shown in Fig. 21-42. Notice that R_G has been returned to ground and that a biasing resistor, R_S, has been connected in series with the source. Capacitor C_S across R_S keeps a steady direct voltage on the source and bypasses the AC signal. This arrangement provides a steady value of bias, V_{GS}, but prevents signal degeneration.

Assuming the same desired conditions as in the previous example (V_{DS} = 12V, I_D = 5.5mA, and V_{GS} = –1V), the only problem is to determine what value of R_S will produce a voltage drop of 1V.

$$R_S = \frac{V_{GS}}{I_D}$$
$$= \frac{-1V}{5.5mA}$$
$$= 180\Omega$$

When the 180Ω resistor is inserted as R_S, I_D becomes 5.5mA and V_{GS} becomes –1V. There is still a 12V drop across R_D, and this leaves 12V from drain to ground. The V_{DS}, of course, is 12V less the 1V dropped across R_S. So there is a V_{DS} of 11V instead of the desired 12V. This makes no substantial difference in the circuit operation. There is now a maximum swing of 11V in both directions instead of 12V.

MOSFET Biasing

Biasing for a MOSFET is exactly the same as for a JFET as long as fixed bias is concerned. However, pure self-bias for an enhancement MOSFET is impossible. If self-bias is used, it is combined with some form of fixed bias.

Enhancement-MOSFET Fixed Bias. In either a P- or an N-channel enhancement MOSFET, the polarity of the voltage, V_{GS}, must be such that it will attract the appropriate current carriers from the substrate and enhance the conduction channel between source and drain. With a P-channel MOSFET, V_{GS} must be a negative potential in order to attract the positive current carriers. With an N-channel MOSFET, V_{GS} must be a positive potential in order to attract the negative carriers. The circuit in Fig. 21-43 is for fixed bias on an N-channel enhancement MOSFET. The procedure is the same as previously described. First determine the desired quiescent values of V_{DS} and I_D. Assume, in this case, that a maximum swing of the output is desired. This fixes the value of V_{DS} as half the value of the supply voltage. With +24V as a supply, R_D must drop 12V. With an R_D of 2kΩ, I_D must be

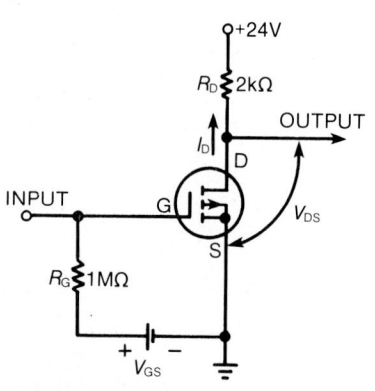

Fig. 21-43: Enhancement-MOSFET fixed bias.

$$I_D = \frac{V}{R_D}$$

$$= \frac{12V}{2k\Omega}$$

$$= 6mA$$

So the quiescent conditions are $V_{DS} = 12V$ and $I_D = 6mA$. The graph in Fig. 21-44 reveals the value of V_{GS} that will produce these desired conditions. The intersection of $V_{DS} = 12V$ and $I_D = 6mA$ is on the +3V V_{GS} line. So +3V of V_{GS} will produce the desired conditions. The V_{GS} battery is now replaced by any steady voltage that causes the gate to be 3V more positive than the source.

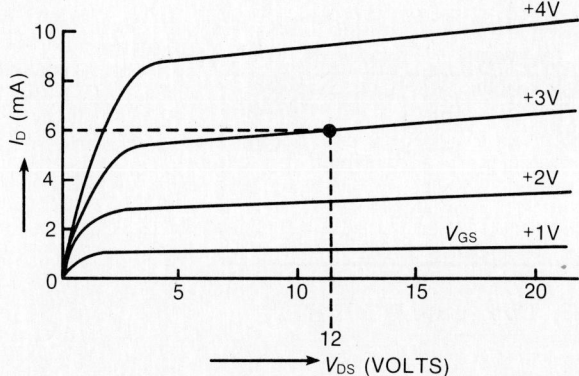

Fig. 21-44: V_{GS} for $V_{DS} = 12V$ and $I_D = 6mA$.

Enhancement-MOSFET Combination Bias. When self-bias is used with an enhancement MOSFET, it is combined with some form of fixed bias. One arrangement of this combination bias is illustrated in Fig. 21-45. The fixed bias is provided by R_1 in series with R_G. The self-bias is produced by I_D through the source resistor, R_S. Current through R_G from ground to $+V_{DD}$ produces a positive potential between gate and ground. Current, I_D, from ground through R_S produces a positive potential between source and ground. The voltage across R_G tends to give V_{GS} a positive potential, which it must have. The voltage across R_S tends to give V_{GS} a negative potential. So the effective V_{GS} is the difference between the voltage across R_G and the voltage across R_S. For proper operation this difference value must be equal to the predetermined V_{GS}, and the gate must be positive with respect to the source.

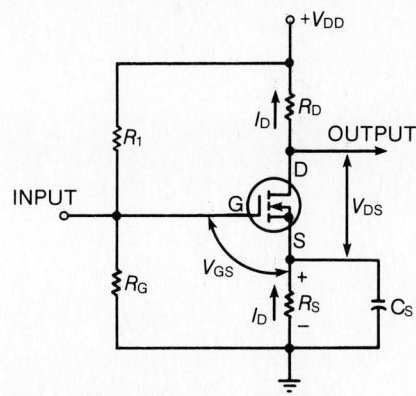

Fig. 21-45: Enhancement-MOSFET combination bias.

Dual-Gate MOSFET

With MOSFET, the combination of metal, insulating layer, and semiconductor forms an equivalent parallel plate capacitance. This equivalent capacitance between drain and gate has a large effect on the MOSFET's performance at high frequencies; as a result, it is an unwanted effect. To minimize the effect, a second gate region is introduced between the control gate and the drain as shown in Fig. 21-46. The resulting device is called a *dual-gate* MOSFET. The second gate is normally biased positively in order to reduce the resistance of the second channel and to break up the effective drain-to-gate capacitance. In all other respects, the operation of the dual-gate MOSFET is the same as that of the enhancement/depletion mode MOSFET, using gate 1 as the input. Figure 21-47 shows the schematic symbols for dual-gate MOSFETs.

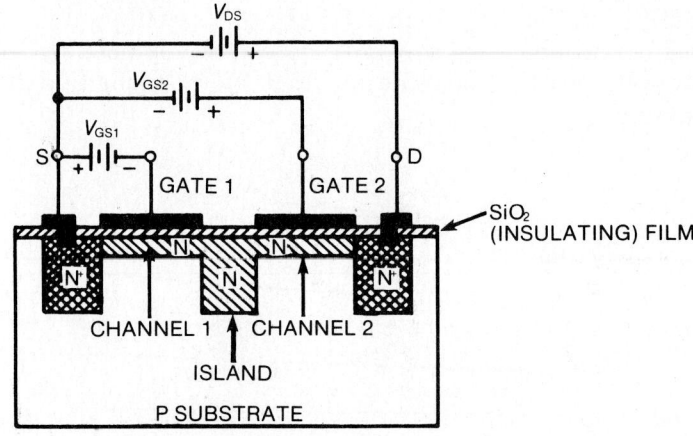

Fig. 21-46: Dual-gate MOSFET.

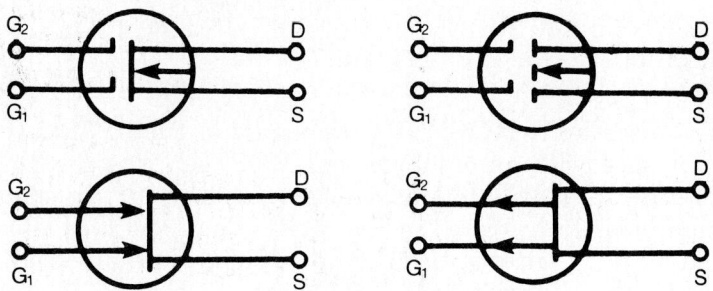

Fig. 21-47: Schematic symbols for dual-gate MOSFET.

Power MOSFETs

MOSFETs have been used commonly in large scale integration (LSI) and very large scale integration (VLSI) circuit applications for a number of years, though limited in their application because of their limited current carrying capability and the 10V to 15V drain-source breakdown voltage. Now there is a new type of MOSFET, the *power* MOSFET, with drain-to-source voltages ranging up to 500V and continuous drain currents up to 28A (up to 70A in pulse applications).

Power MOSFETs make use of a new technology called *vertical metal-oxide semiconductor* (VMOS) which employs a diffused channel and vertical current flow to achieve high voltage and current ratings comparable to the power transistor and power Darlington. Figure 21-48 illustrates the structure of the VMOS.

504

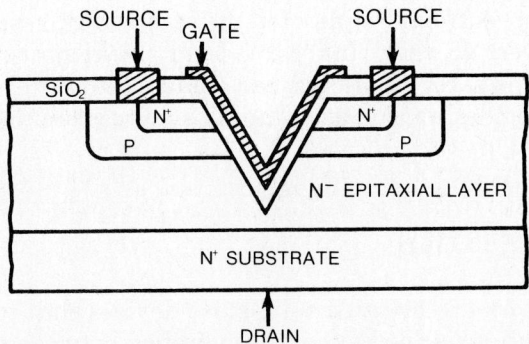

Fig. 21-48: Structure of the VMOS

The function of the N+ substrate is to provide the drain connection and a low resistance path. The N- epitaxial layer is provided to increase the drain-to-source voltage breakdown capability by absorbing the depletion region from the reverse-biased drain body junction. An additional benefit of the N- epitaxial layer is that it reduces the feedback capacitance. The P body and N+ source are diffused into the epitaxial layer. At this point the V groove is etched into the epitaxial layer, the silicon dioxide layer is formed, and finally the aluminum source and gate connections are made.

When a positive gate voltage is applied, the surface of the P base region is inverted and a channel is formed under the gate electrode surface, which in turn permits current to flow from the drain to source.

The advantages of these devices as compared to the bipolar power transistors are: (1) faster switching speed, and (2) no thermal runaway.

In a bipolar power transistor, the charge storage (due to the effective capacitance of the junction) prevents fast turn-off. That is, until these stored charges have been removed, the transistor continues to conduct. With the VMOS, however, this is not so. Because it has no charge storage, it can turn off much more rapidly than a bipolar transistor. A VMOS transistor can shut off amperes of current in tens of nanoseconds. This is approximately 10 to 100 times faster than a comparable bipolar transistor.

We have already discussed the phenomenon of thermal runaway in respect to transistors. Because the VMOS transistor has no gate current, which is analogous to base current, it is free of thermal runaway and therefore will not self-destruct like the bipolar transistors.

Handling MOSFETs

Some MOSFETs have protective diodes inside the device and may be handled in the same manner as any other device, but many MOSFETs have no such internal protection. These MOSFETs are usually supplied with a ring which shorts together all of the leads of the device. The purpose of the ring is to prevent static charges from building up on the gate. Even static

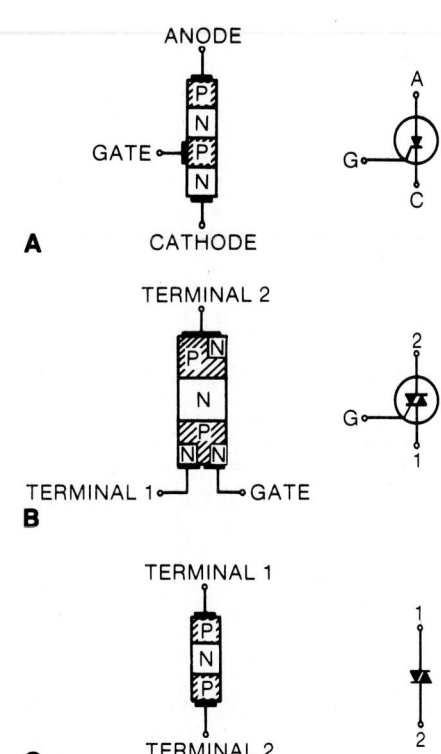

Fig. 21-49: Junction arrangements and schematic symbols for (A) SCR, (B) the triac, and (C) the diac.

charges are sufficient to destroy the insulating properties of the oxide layer, so this ring should not be removed until the MOSFET is safely installed and soldered into the circuit. The tip of the soldering iron used should also be grounded as another precaution.

THYRISTORS

Thyristors are bistable solid-state devices that have two or more junctions and that can be switched between conducting states (from OFF to ON, or from ON to OFF). Thyristors generally are heavy-current devices (to several hundred amperes) and are extensively used in applications involving power rectification, control, and switching.

1. The *silicon-controlled rectifier* (SCR) is a four-layer device with three electrodes: an anode, a cathode, and a control gate. The basic junction diagram and schematic symbol are shown in Fig. 21-49A. The SCR is a *unidirectional* reverse-blocking thyristor, providing current conduction in only one direction. It can be switched from the OFF state to the ON state by a positive trigger pulse applied to the gate.

2. The *triac* is a four-layer device with three electrodes: terminal 1, terminal 2, and a control gate. The basic junction diagram and schematic symbol of a triac are shown in Fig. 21-49B. It is a *bidirectional* device which can block voltages of either polarity and conduct current in either direction. It can be switched from the OFF state to the ON state, in either direction, by a positive or negative gate pulse. The triac is often considered analogous to two SCRs connected in parallel but oriented in opposite directions (inverse-parallel connection).

3. The *diac* is a two-electrode, three-layer *bidirectional* avalanche diode, which can be switched from the OFF state to the ON state in either direction. The basic junction diagram and schematic symbol of a diac are shown in Fig. 21-49C. It does not have a control gate. It is switched on, for either polarity of applied voltage, simply by exceeding the avalanche breakdown voltage.

SILICON-CONTROLLED RECTIFIER (SCR)

Figure 21-50A shows the typical voltage-current characteristic of an SCR. In the forward-bias region (anode positive with respect to cathode) the SCR has two distinct operating states. As the forward bias is initially increased from zero volts, the SCR allows only a small forward current and exhibits a high forward resistance. This region of operation is called the *forward blocking region* or OFF state. As the forward bias is further increased, the OFF-state current increases very slowly until the *breakover voltage*, $V_{(BO)}$, is reached. At $V_{(BO)}$, the SCR suddenly converts to the *high conductance region*, or ON state, where the anode current is limited mainly by the resistance in the external circuit.

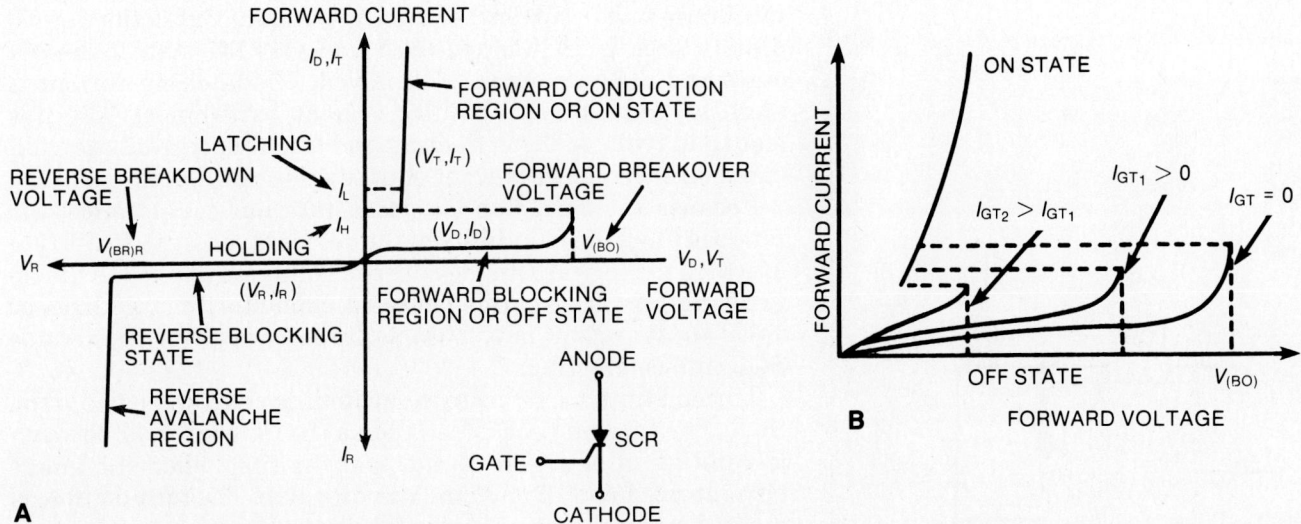

Fig. 21-50: (A) SCR voltage-current characteristic. (B) SCR breakover voltage decreases with increasing values of gate current.

The breakover voltage can be varied by applying a current pulse of the right character to the gate electrode, causing the SCR to switch from the OFF state to the ON state at a lower forward voltage, as indicated in Fig. 21-50B. As the amplitude of the gate pulse is increased, the breakover point moves toward the origin until the curve resembles that of a conventional rectifier diode. Normally, the SCR is operated well below $V_{(BO)}$ and is then made to switch ON by a gate signal of sufficient amplitude. This assures that the SCR starts forward conduction at exactly the right instant.

After the SCR is switched ON, a certain minimum anode current, called the holding current, I_H, is required to maintain the device in the ON state. If the anode current drops below I_H, the SCR reverts to the OFF state. In fact, to switch the SCR from ON to OFF, the anode current must be reduced to a value below I_H.

In the reverse-bias region (anode negative with respect to cathode), the SCR presents a very high impedance and allows only a small reverse current. This is the *reverse-blocking region*. The SCR remains blocked until the reverse bias exceeds the reverse breakdown voltage $V_{(BR)R}$. At this avalanche breakdown point the SCR conducts heavily and is subject to the usual problems of thermal runaway. Thermal runaway means destruction of the device, and $V_{(BR)R}$ should, therefore, never be exceeded.

In addition to the usual power rectifier nomenclature used in the ratings, detailed information on the dynamic gating and switching parameters is given in the electrical characteristics. Gate-triggering requirements are supplied in terms of the gate-to-cathode voltage, V_{GT}, and the gate current, I_{GT}; typical values are specified in the characteristics.

After an SCR has been switched to the ON state, a certain minimum anode current is required to maintain the SCR in the ON state. This holding current I_H is specified at room temperature (T=25°C) with the gate open. Latching current I_L is the

507

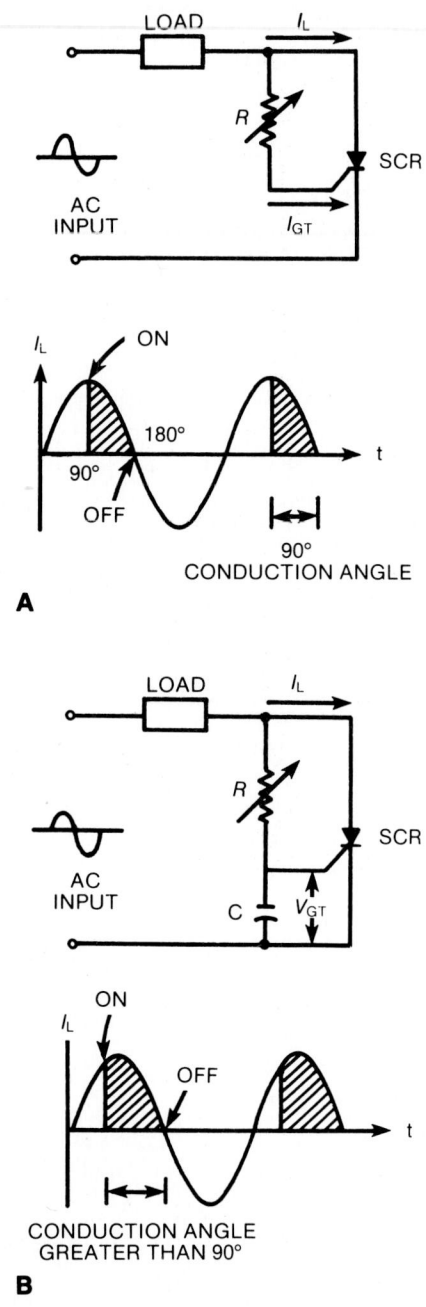

A

B

Fig. 21-51: Basic gate trigger circuits, controlling the conduction angle of the SCR.

minimum anode current required to sustain conduction immediately after the SCR is switched from the OFF state to the ON state and the gate signal is removed. The latching current is slightly larger than the holding current, but for most SCRs it is less than twice I_H. Both holding and latching current are indicated on the voltage-current characteristic of Fig. 21-50A.

Because the SCR has a certain internal capacitance, the forward blocking capability of the device is sensitive to the rate at which the forward anode-cathode voltage is applied. A quickly rising forward voltage causes a capacitive charge current ($i = C dV_D/dt$) which may become large enough to trigger the SCR into conduction.

Turn-on time, t_{on}, refers to the switching characteristics of the SCR. The turn-on time is defined as the time interval between the initiation of a gate signal and the time when the anode current reaches 90% of its maximum value. The turn-on time is dependent on the magnitude of the gate trigger pulse, decreasing with increasing gate current magnitude.

The commutated turn-off time, t_q, of an SCR is the time interval between the cessation of forward current and the reapplication of forward blocking voltage. The turn-off time depends on several circuit parameters, in particular the SCR junction temperature and the ON-state current prior to switching.

Control of the Phase Angle

Many AC power control applications require that the thyristor be switched full ON or full OFF, an action similar to that of a relay. In cases like this, the SCR handles the heavy load current, while only a small gate current is required to trigger the SCR into conduction. Figure 21-51A shows a simple method of controlling the AC load power by gating the SCR into the ON state.

With maximum resistance, R, in the gate circuit, the SCR is in the OFF state. As R is reduced, a point is reached where sufficient gate current I_{GT} is provided at the positive peak of the AC input voltage waveform (90°) to trigger the SCR on. The SCR then conducts from the 90° point to the 180° point (where the anode current drops to zero), for a total *conduction angle* of 90°. This conduction angle can be increased to almost 180° by reducing the resistance in the gate circuit and thereby increasing I_{GT}. In the half-wave circuit of Fig. 21-51A, the AC power delivered to the load can be controlled over a considerable range simply by adjusting the conduction angle of the SCR.

A different method of obtaining conduction angle control greater than 90° for half-wave operation is shown in the RC trigger network of Fig. 21-51B. Here the SCR is in series with the AC load and in parallel with the RC network. At the start of each positive half-cycle, the SCR is in the OFF state (zero anode current) and the rising AC input voltage drives current through R and C to charge the capacitor. When the capacitor voltage reaches the SCR breakover voltage, the capacitor discharges through the gate terminal and turns the SCR on. Full AC power is then transferred to the load for the remainder of the half-cycle. When the potentiometer resistance is decreased, the capacitor charges more rapidly and the SCR breakover voltage is reached earlier in the cycle. In this case the AC power delivered to the load is increased.

Many triggering circuits can be used to meet the needs of the SCR. In the typical AC power control circuits of Fig. 21-52, the trigger circuits are labeled CONTROL.

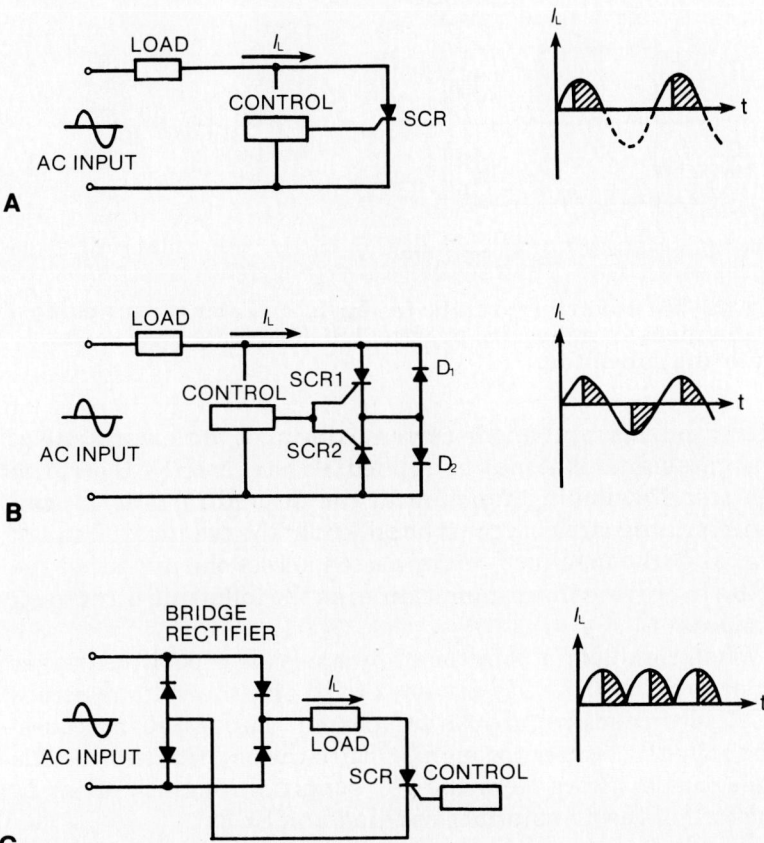

Fig. 21-52: SCR AC phase control in power applications: (A) controlled half-wave alternating current; (B) controlled full-wave alternating current; and (C) controlled full-wave rectified current.

SCR Applications

SCR *inverters* provide an efficient and economical way of converting direct current or voltage into alternating current or voltage. In this application the SCR serves as a controlled switch, alternately opening and closing a DC circuit. Consider the basic inverter circuit of Fig. 21-53A, where an AC output is generated by alternately closing and opening switches S_1 and S_2. Replacing the mechanical switches with SCRs, whose gates are triggered by an external pulse generator, we obtain the practical SCR inverter of Fig. 21-53B.

It should be realized that the SCR is basically a *latching* device, which means that the gate loses control as soon as the

509

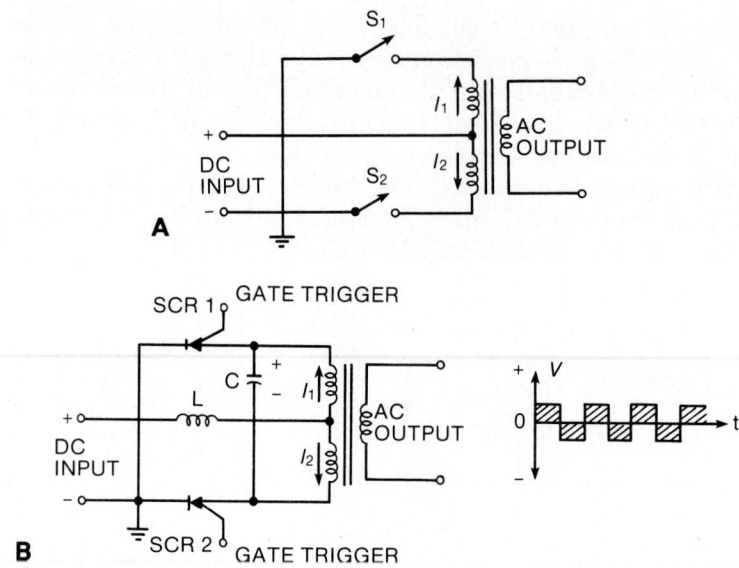

Fig. 21-53: Inverter circuits: (A) basic inverter circuit using mechanical switches; (B) practical SCR inverter with commutating capacitor.

SCR conducts, and anode current continues to flow in spite of any gate signal that may be applied. To have each SCR perform the on-off switching function at the desired instant, special commutating circuitry must be added. In the practical circuit of Fig. 21-53B capacitor C is connected across the anodes of the SCRs to provide this commutation, as the following discussion explains:

When conduction is initiated by applying a positive trigger pulse to SCR1 (SCR2 is assumed to be off), the voltage across SCR1 decreases rapidly as the current through it increases. Capacitor C charges through SCR1 in the polarity shown. The load current flows from the DC supply through inductor L, one-half of the transformer winding and SCR1.

When a firing pulse is now applied to the gate of SCR2, this SCR turns on and conducts. At this instant, capacitor C begins to discharge through SCR1 and SCR2. This discharge current flows through SCR1 in a *reverse* direction. Conduction of SCR1 ceases as soon as the charge stored in the SCR has been removed by the reverse recovery current. At this time, with SCR1 turned OFF, the capacitor voltage (approximately – 2V) appears across SCR1 as a reverse voltage, long enough to allow this SCR to recover for forward blocking.

Simultaneously during this interval, conducting SCR2 allows the capacitor to discharge through the transformer winding and inductor L. The function of L is to control the discharge rate of C to allow sufficient time for SCR1 to turn OFF. Capacitor C discharges rapidly from -2V to zero and then charges up in the opposite direction to +2V. The load current is now carried through the second half of the transformer winding and SCR2.

When the trigger current is applied again to the gate of SCR1, this device conducts and the process just described repeats.

Triac Operation

Figure 21-54 shows the diagrammatic junction arrangement of a triac, often called a triode transistor. Both main terminals make contact with an N-type emitter as well as a P-type emitter. The N-type emitter at terminal 2 is located directly opposite the P-type emitter at terminal 1, and the P-type emitter at terminal 2 is located opposite the N-type emitter at terminal 1. This suggests that the triac consists of two four-layer devices (SCRs) in parallel, but oriented in opposite directions. The triac can therefore block or conduct current between the two main terminals in either direction. The gate terminal, which makes contact with both P-type and N-type materials, allows the application of either positive or negative trigger currents.

The voltage-current characteristic of the triac is shown in Fig. 21-55. Observe that the triac has the same OFF-state and ON-state regions as the SCR, but now for either polarity of voltage applied across the main terminals. When main terminal 2 is positive with respect to main terminal 1, the triac is *positively biased* and operates in the first quadrant of its characteristic. When main terminal 2 is negative with respect to main terminal 1, the triac is *negatively biased* and operates in the third quadrant of the characteristic. Like an SCR, the triac switches from the OFF state to the ON state at the breakover voltage, $V_{(BO)}$. The breakover point can be reached at lower main terminal voltages by applying a positive or a negative trigger pulse to the gate.

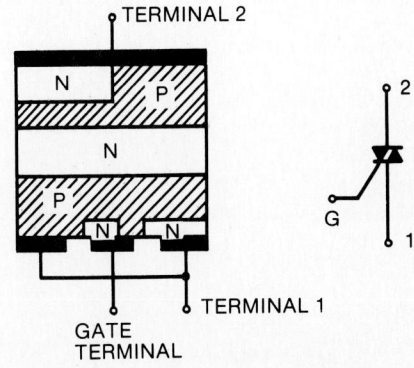

Fig. 21-54: Junction arrangement and schematic symbol of the triac.

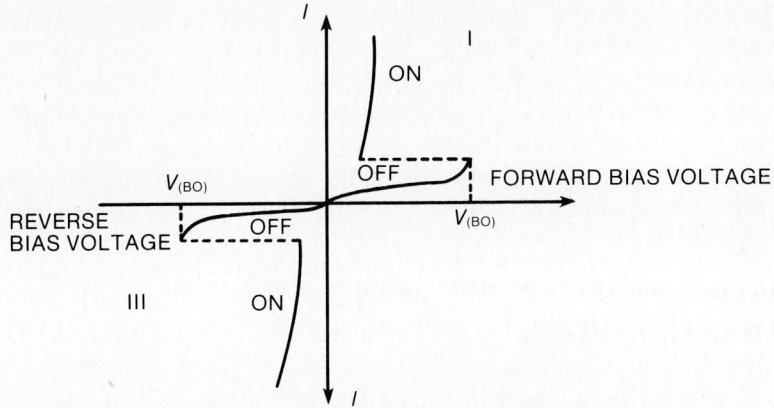

Fig. 21-55: Voltage-current characteristic of the triac.

Gate triggering occurs in any of the four operating modes summarized in Table 21-2. The gate-triggering requirements are different for each operating mode. The most sensitive trigger mode is usually mode I(+), where the triac is forward-biased and a positive gate trigger is applied. Mode I(+) is identical to conventional SCR operation.

The maximum trigger current rating published in the triac rating sheet is the largest value of gate current required to trigger the triac in any operating mode.

Table 21-2: Triac Triggering Modes

Main Terminal 2 to Main Terminal 1 Voltage	Gate to Main Terminal 1 Voltage	Operating Quadrant*
Positive	Positive	I(+)
Positive	Negative	I(−)
Negative	Positive	III(+)
Negative	Negative	III(−)

*A plus (+) or minus (−) sign in the quadrant designation indicates the polarity of the gate trigger value or current

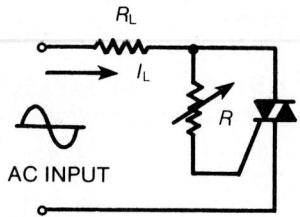

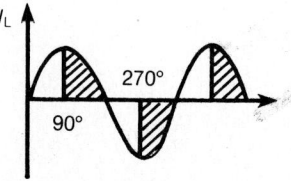

Fig. 21-56: Basic triac gate trigger circuit.

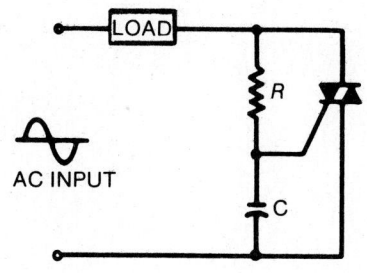

Fig. 21-57: RC network in fixed-phase control of triac conduction.

The method of triggering a triac is very similar to that for the SCR. Figure 21-56 shows a basic circuit where the variable resistance in the gate circuit controls the conduction angle of the triac. As the resistance is decreased, the gate current increases until the triac turns on at both the peak positive (90°) and peak negative (270°) points on the voltage waveform. The conduction angle can be increased by appropriate reduction of R. Nearly 360° can be obtained by moving the firing point back from 90° to zero on the positive half-cycle, and from 270° to 180° on the negative half-cycle.

Triacs are used extensively in full-wave power control applications to supply either fixed or variable power to the load. Fixed load power is delivered when the triac is used as a static off-on switch in the fixed-phase firing mode, which supplies any desired portion of the input voltage to the load. A simple *fixed-phase control* circuit is shown in Fig. 21-57. In this circuit, the triac is fired at fixed points on the positive and negative half-cycles of the AC input waveform, as determined by the time constant of the RC network in the gate circuit. Commutation of the triac, taking place at the zero crossing of each half-cycle, can pose problems with inductive loads, where the transition from the ON state to the OFF state can produce large voltage transients which tend to drive the triac into premature conduction.

Diac Operation

The diac is a two-terminal, three-layer PNP device, which can be made to conduct in either direction. Figure 21-58 shows the basic junction arrangement, the schematic symbol, and the voltage-current characteristic. The diac is also known as a *bilateral* or *bidirectional diode*.

Voltage of either polarity between the diac terminals initially causes a small leakage current across the PN junction which is reverse-biased. When the applied voltage exceeds the breakdown voltage of that reverse-biased junction, the junction enters the avalanche breakdown region and the diac current suddenly increases sharply. In this ON-state condition, the voltage across the diac decreases with increasing current so that the diac exhibits a negative resistance characteristic.

512

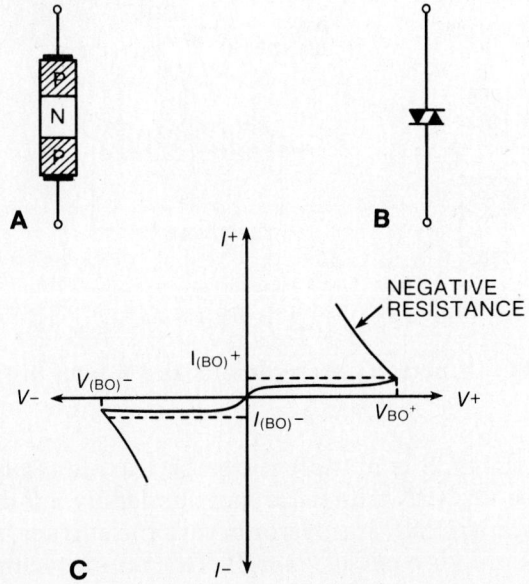

Fig. 21-58: (A) Junction diagram, (B) schematic symbol, and (C) voltage-current characteristic of a diac.

Unlike the bipolar junction transistor, the two PN junctions of the diac have equal doping concentrations, which produce symmetrical switching characteristics for both voltage polarities.

The diac is primarily used as a *trigger device* (it is sometimes called *AC trigger* diode) in triac power control circuits. A basic circuit for diac-triac phase control is shown in Fig. 21-59. The diac switches on when the capacitor voltage reaches the diac breakover voltage, $V_{(BO)}$, and the diac then supplies gate current to the triac. The conduction angle of the triac is controlled by the time constant of the RC network in the diac circuit.

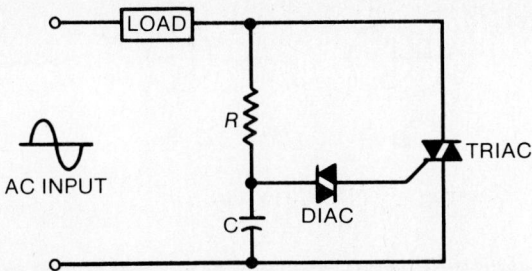

Fig. 21-59: General circuit of a diac-triac phase control.

Silicon-Controlled Switch (SCS)

The silicon-controlled switch (often called a tetrode thyristor) is a unilateral four-layer silicon device, with electrodes connected to the four terminals, as shown in the basic junction diagram of Fig. 21-60A. The four electrodes are known as anode (4), anode gate (3), cathode gate (2), and cathode (1).

513

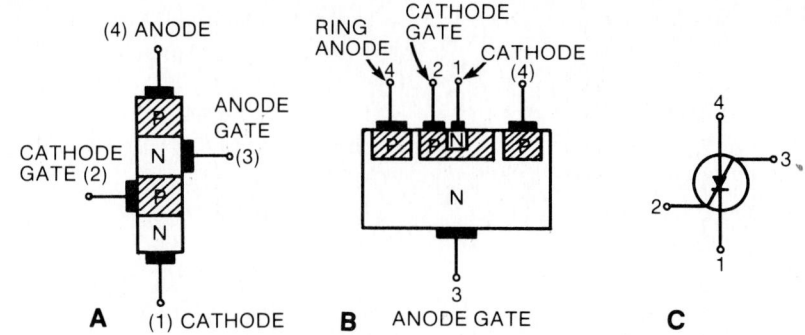

Fig. 21-60: Junction arrangement and schematic symbol of the SCS.

When the SCS is manufactured by the planar technique, it resembles an NPN transistor surrounded by a P-diffused ring which forms a PNP transistor across the surface, as shown in the sectional view of Fig. 21-60B. The planar technique makes all four terminals readily accessible, and the device enjoys the inherent parameter stability of all planar silicon structures. Figure 21-60C shows the schematic symbol of an SCS.

The basic junction diagram of Fig. 21-60A suggests the equivalent circuit of Fig. 21-61A which shows an NPN transistor with a diode in the collector circuit. Normal transistor action requires that the collector (3) is positive with respect to the emitter (1). To allow SCS current, the diode must be forward-biased, and the anode terminal (4) must, therefore, be positive with respect to the collector terminal (3). This reasoning leads to the SCS bias conditions and the recognition of three possible operating modes, which are summarized in Fig. 21-61.

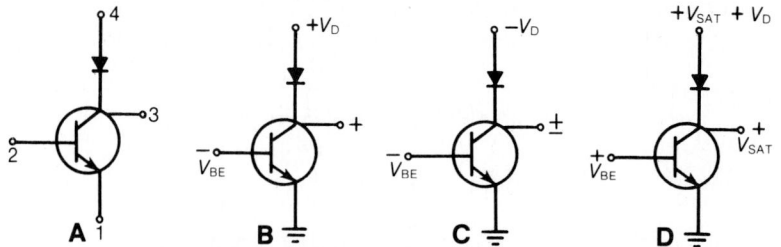

Fig. 21-61: Biasing a PNPN device.

In the *forward-blocking mode* of Fig. 21-61B, the anode (4) is positive, but the reverse-biased base-emitter junction (2) keeps the transistor cut off. (In lieu of negative V_{BE}, the base can be left disconnected or short-circuited to the emitter.) The maximum blocking voltage, defined as the maximum permissible positive anode voltage, is determined by the collector-to-base breakdown voltage.

In the reverse-bias mode of Fig. 21-61C, the anode is negative with respect to the emitter, the diode is reverse-biased, and there is no anode current. The maximum reverse voltage, defined as the maximum permissible negative anode voltage, is determined by the diode breakdown voltage plus the emitter breakdown voltage of the transistor.

In the *forward-conduction mode* of Fig. 21-61D, the anode is positive and the SCS is made to conduct by the positive bias on the base. The anode voltage and collector voltage differ only by the forward voltage drop across the conducting diode.

A different view of the SCS is obtained when considering the two-transistor equivalent circuit of Fig. 21-62B, which is derived from the basic junction diagram of Fig. 21-62A. This equivalent circuit considers the SCS as a *complementary pair with regenerative feedback,* which switches the device on when the product of the transistor gains is greater than unity. Base current I_B into the NPN transistor is multiplied by the NPN current gain β_1 and becomes base current $\beta_1 I_B$ for the PNP transistor. After multiplication by the PNP current gain β_2, it becomes the reinforcing base current $\beta_1 \beta_2 I_B$ for the NPN transistor. If this reinforcing base current exceeds the initial base current ($\beta_1 \beta_2 \geq 1$), regeneration takes place and both transistors are driven into saturation. This condition is known as the ON state for the PNPN device.

The product of the two transistor gains is the critical factor that determines whether the SCS switches on. Since the β of a transistor is a function of collector current I_C, collector voltage V_{CE}, base bias V_{BE}, and temperature T, it should be realized that the ON or OFF state of the SCS can be affected by changes in any one of these parameters.

For example, the typical forward characteristic of Fig. 21-63 shows that a reduction in anode current switches the SCS off. This is due to the fact that at low anode currents the transistor betas decrease, so their product becomes less than 1 and regeneration ceases.

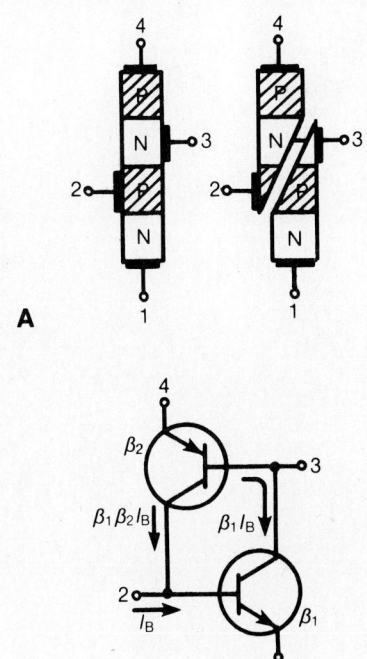

Fig. 21-62: Development of the two-transistor SCS equivalent circuit: (A) basic junction diagram and (B) two-transistor equivalent circuit.

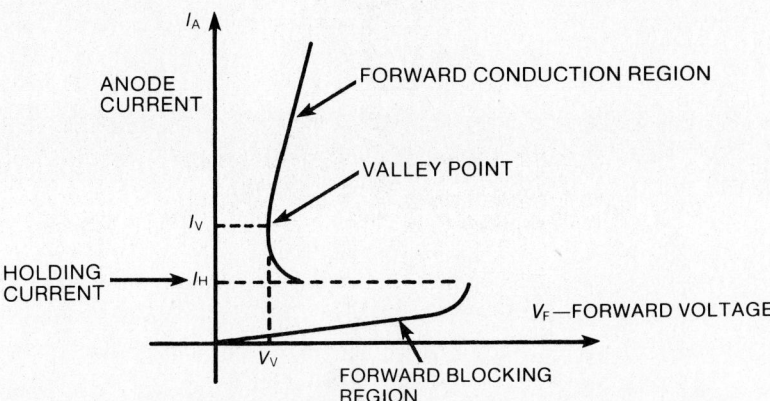

Fig. 21-63: Typical forward characteristic of an SCS, indicating that the device switches off when the anode current drops below the holding current I_H.

The SCS can be switched on accidentally if the anode voltage is applied suddenly or if the SCS is subjected to high-frequency transients. This phenomenon is known as the *rate effect*. The rate effect is caused by internal capacitance between electrodes (2) and (3), called interbase capacitance. The interbase capacitance presents a low-impedance path at high frequencies and can result in substantial base current which may trigger the SCS into conduction. If the circuit is such that the rate effect

515

causes device triggering, it can be eliminated by proper circuit design.

The equivalent circuit shown in Fig. 21-62 points out that the SCS can be used as a conventional transistor, simply by ignoring the anode connection. However, because of its inherent regenerative properties and high triggering sensitivity, the SCS is mainly used as a switching device which can be turned on or off at will.

The SCS turns on readily, simply by satisfying the stated voltage requirements for conduction. Figure 21-64 summarizes some typical circuits which can be used to trigger the SCS on.

In Fig. 21-64A, the cathode gate is left open and the voltage at the anode gate is set by the R_1 - R_2 voltage divider to a triggering threshold of $[R_1/(R_1 + R_2)]V$ volts. When the anode voltage rises above this threshold the SCS will trigger on.

Figure 21-64B is a variation of Fig. 21-64A using conventional transistor bias stabilization. When the anode voltage exceeds the stabilized collector voltage, the transistor saturates.

In Fig. 21-64C the Zener diode in the cathode circuits sets the threshold voltage. A positive input to the cathode gate, exceeding the Zener voltage, triggers the SCS on.

Figure 21-64D shows that the SCS can be triggered by applying a negative waveform of sufficient amplitude to the cathode terminal.

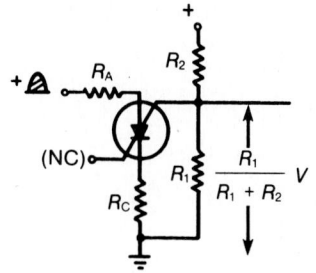

A

SCS TRIGGERING BY POSITIVE ANODE SIGNAL. THRESHOLD VOLTAGE IS SET BY THE R_1 - R_2 DIVIDER.

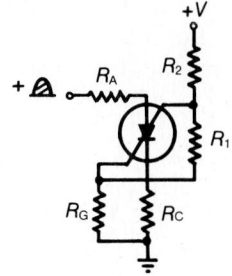

B

SCS TRIGGERING BY POSITIVE ANODE SIGNAL. BIAS STABILIZATION AND THRESHOLD VOLTAGE ARE SUPPLIED BY R_2, R_1, AND R_G.

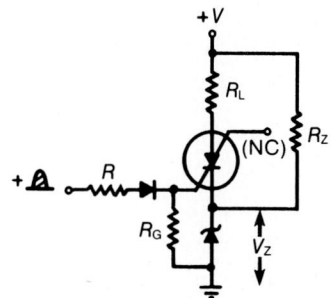

C

SCS TRIGGERING BY POSITIVE SIGNAL TO THE CATHODE GATE. THE ZENER DIODE PROVIDES THE THRESHOLD VOLTAGE.

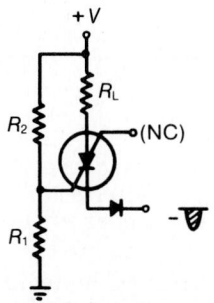

D

SCS TRIGGERING BY A NEGATIVE SIGNAL TO THE CATHODE. THRESHOLD VOLTAGE TO THE CATHODE GATE IS SUPPLIED BY R_1 AND R_2.

Fig. 21-64: Various methods of switching the SCS to the ON state.

Silicon-controlled switches turn on readily, but they are much more difficult to switch off again. Figure 21-65 summarizes a number of turn-off methods. The simplest turn-off is achieved by operating from AC, as in Fig. 21-65A. The PNPN device turns off when the anode becomes negative. When operating from a DC supply, the anode can be opened by a switch, as in Fig. 21-65B. Removal of the anode current stops transistor action and turns the SCS off. Resistor R_{RE} suppresses the rate effect when the switch is closed again.

The gates can also be used to turn the device off. In Fig. 21-65C, a negative cathode gate input is used, while in Fig. 21-65D a positive anode gate pulse is applied. In both instances, the gate pulse applies reverse bias to the PN junction and stops transistor action so the saturated transistor turns off as the current carriers recombine. A positive anode gate pulse in Fig. 21-65C, where the load is in the anode circuit, represents a poor method of turning the SCS off since it does not reverse bias the anode junction. Similarly, a negative cathode gate pulse in Fig. 21-65D, where the load is in the cathode circuit, does not reverse bias the cathode junction and poor turn-off results can be expected.

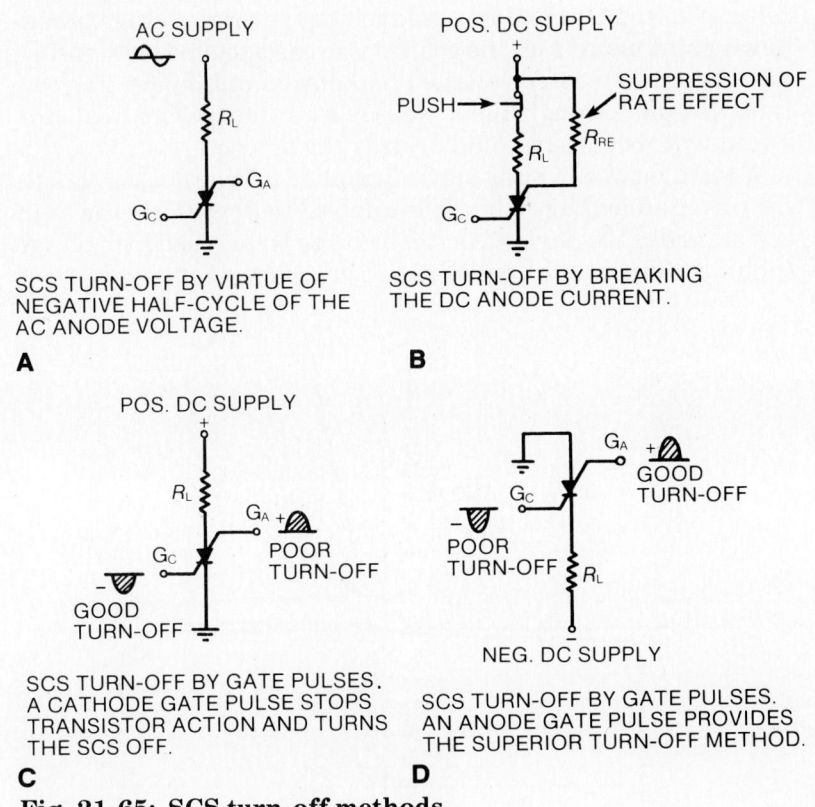

SCS TURN-OFF BY VIRTUE OF NEGATIVE HALF-CYCLE OF THE AC ANODE VOLTAGE.

A

SCS TURN-OFF BY BREAKING THE DC ANODE CURRENT.

B

SCS TURN-OFF BY GATE PULSES. A CATHODE GATE PULSE STOPS TRANSISTOR ACTION AND TURNS THE SCS OFF.

C

SCS TURN-OFF BY GATE PULSES. AN ANODE GATE PULSE PROVIDES THE SUPERIOR TURN-OFF METHOD.

D

Fig. 21-65: SCS turn-off methods.

Four-Layer Diodes

A semiconductor junction diode, as was mentioned in Chapter 19, is composed of two regions of semiconductor materials separated by a PN junction. A standard BJT consists of three

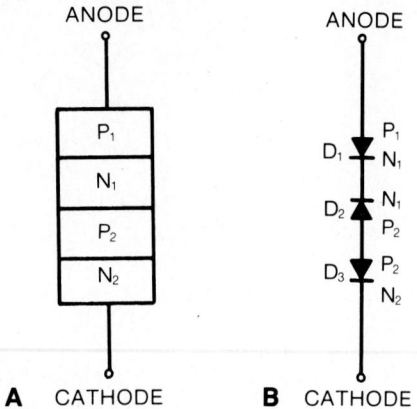

Fig. 21-66: (A) Graphic representation of the Shockley diode. (B) Equivalent circuit of the Shockley diode.

regions separated by two PN junctions. A *four-layer PNPN semiconductor device* with three PN junctions can be formed by adding another region to the conventional transistor.

The most simple type of PNPN semiconductor device is called a *Shockley diode.* Figure 21-66 illustrates the structural diagram and equivalent circuit for this device. It is called a diode because it has only two terminals.

If a voltage were applied, positive to the anode and negative to the cathode, diodes D_1 and D_3 would be forward-biased and would, therefore, offer very little resistance. On the other hand, D_2 would be reverse-biased and would offer a very high resistance. The entire resistance of the Shockley diode would be very high since all three diodes are connected in series.

If a voltage were applied to the Shockley diode in the reverse direction (negative to the anode and positive to the cathode), diodes D_1 and D_3 would be reverse-biased and would offer a very high value of resistance. Diode D_2 would be forward-biased, offering a very low value of resistance. Once again, the Shockley diode exhibits a very high total resistance.

Figure 21-67 shows a characteristic curve for a typical Shockley diode. The graph shows that a Shockley diode should never be operated with a reverse voltage high enough to reach the reverse voltage breakdown point. If the reverse voltage breakdown point (point 1 on the graph) is exceeded, diodes D_1 and D_3 would go into reverse voltage breakdown, and the reverse current flowing through them would rise rapidly. The heat produced by this current could destroy the device.

When a forward voltage is first applied to the Shockley diode, only a small leakage current is allowed to flow due to the high resistance of the device. As this voltage is increased, a voltage (point 2 of the graph), called the *forward-breakover voltage,* is

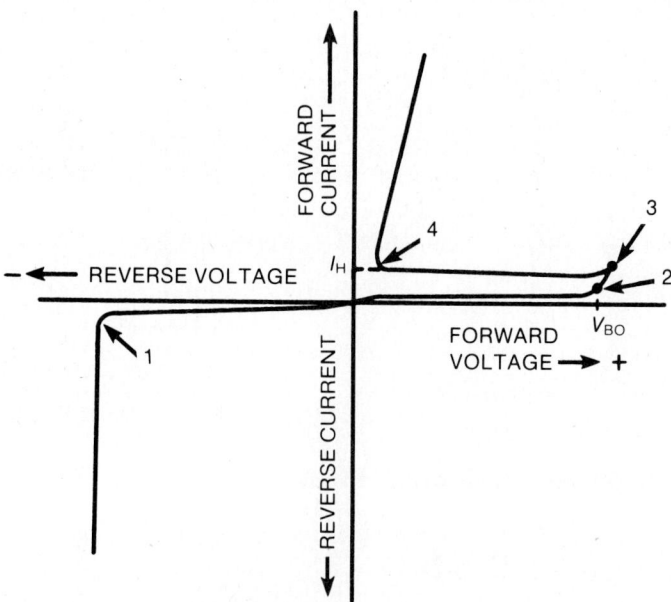

Fig. 21-67: Typical characteristic curve for a Shockley diode.

518

reached. At this point, diode D_2, which had been reverse-biased, goes into reverse voltage breakdown, resulting in reduced resistance and a rise in forward current corresponding to point 3 of the graph.

Here, the resistance of the diode drops swiftly, reducing the forward voltage across it to a very low value (point 4 on the graph). From this point on, the Shockley diode behaves as a conventional forward-biased diode with its forward current being determined primarily by the voltage applied to the diode and the resistance of its external load.

Thus, the Shockley diode functions as a switch. It acts like an open switch (high resistance) when its forward voltage is less than the forward breakover value and like a closed switch (low resistance) at voltages above this value. By reducing the voltage so that the current flowing through the device drops below its *holding current* (I_H) *value*, the switch can again be opened. That is, when the current flowing through the diode is below the holding current value, diode D_2 comes out of its reverse-breakdown state, and its high value of resistance is restored.

The schematic symbol for the Shockley diode is illustrated in Fig. 21-68.

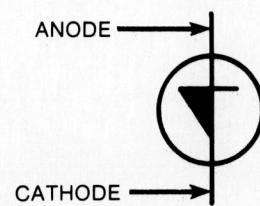

Fig. 21-68: Schematic symbol of a Shockley diode.

UNIJUNCTION TRANSISTORS

The unijunction transistor (UJT) is a three terminal semiconductor device. Sometimes referred to as a double base diode, its characteristics are quite different from those of the conventional two junction transistor. Some of its features are: (1) a stable triggering voltage, (2) a very low value of firing current, (3) a negative resistance characteristic, and (4) a high pulse current capability. They find use in oscillators, timing circuits, voltage and current sensing circuits, SCR trigger circuits, and bistable circuits.

Figure 21-69 is a block diagram of the unijunction transistor showing the leads, types of materials used, and the names of the elements. If terminals 2 and 3 were connected together, the resulting device would have the characteristics of a conventional junction diode.

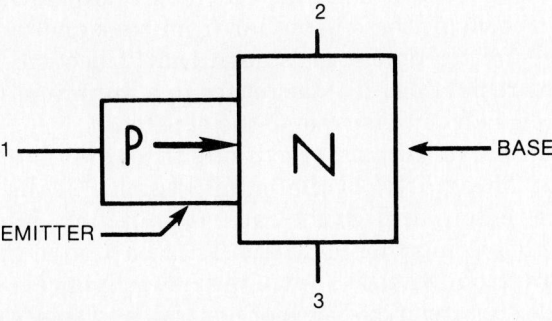

Fig. 21-69: Block diagram of UJT.

Figure 21-70A represents the schematic symbol for the unijunction transistor. A simplified equivalent circuit is shown in Fig. 21-70B. R_{B1} and R_{B2} in Fig. 21-70B represent the resistances of the N-type silicon bar between the ohmic contacts, called base one (B_1) and base two (B_2), at opposite ends of the base. A single rectifying contact called the emitter (E) is made between base one and base two. The diode in Fig. 21-70B represents the unijunction emitter diode. In normal circuit operation base one is grounded and a positive base voltage (V_{BB}) is applied at base two. For example, if the bias voltage applied at base two is 10V positive with respect to base one (R_{B1} is equal to R_{B2}), the voltage drop across each resistor is equal to 5V. There is no emitter current flowing, and the silicon bar acts as if it were a simple voltage divider. The emitter, being halfway along the bar, sees 5V (with respect to B_1) opposite it. So long as the emitter voltage (V_E) remains 5V or less, very little emitter current flows, as the emitter will be either in a neutral state or reverse-biased. When the emitter voltage (V_E) rises above 5V, the junction becomes forward-biased, emitter current increases very rapidly, and the resistance (R_{B1}) decreases very rapidly. Current flow at this time is from base one to the emitter.

Referring to Fig. 21-70B, R_{B1} is shown as a variable resistor and R_{B2} is fixed. Exact values are not important, but the realization that the combination of R_{B1} and R_{B2} represents the total resistance between base one and base two contacts will aid in the understanding of how the UJT functions.

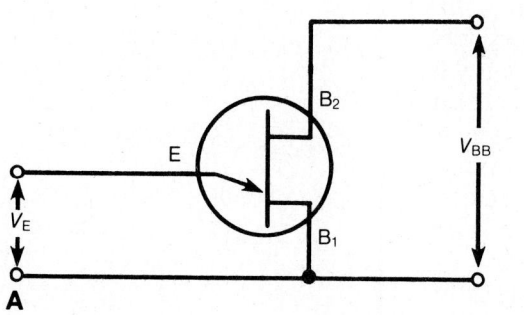 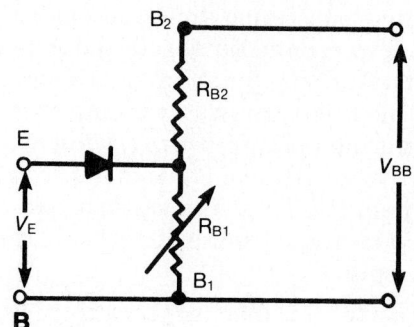

Fig. 21-70: UJT schematic symbol and simplified equivalent circuit.

When the emitter-base one contacts are in the reverse-biased condition the resistance R_{B1} is at its maximum value, and current flows through the silicon bar from base one to base two. When the emitter potential is such that it becomes forward-biased, the resistance, R_{B1}, decreases to a minimum value and current flow is from base one to the emitter.

The unijunction transistor exhibits three distinct regions of operation. These are: (1) the cutoff region, (2) the negative resistance region, and (3) the saturation region. Figure 21-71 is a curve depicting these regions. That portion of the emitter current graph shown to the left of the zero reference is where the emitter is either reverse-biased or neutral, and this is the cutoff

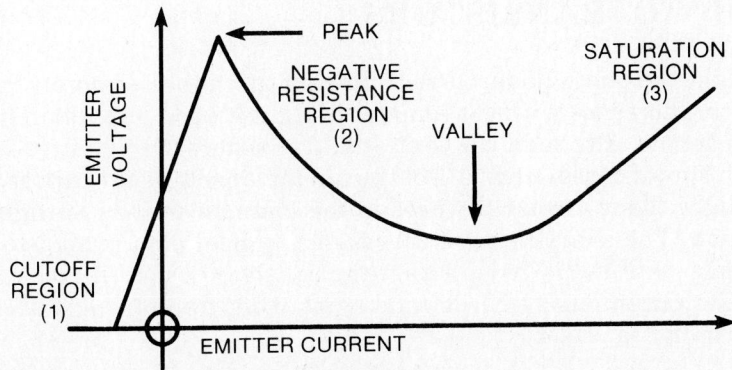

Fig. 21-71: UJT regions of operation.

region, which ends at the point marked PEAK. At the peak point the current between base one and the emitter increases rapidly because a forward bias condition now exists. That part of the resistance of the silicon bar, R_{B1}, in Fig. 21-70B, between base one and the emitter is now minimal, and emitter current is increasing while emitter voltage is decreasing. This, in effect, is the negative resistance region shown between the peak and valley points of Fig. 21-71. Beyond the valley point the resistance of the device behaves as a positive resistance; current again increases slowly with voltage. This area of positive resistance is the saturation region of the unijunction transistor.

A typical circuit application of the unijunction transistor is a sawtooth generator. Figure 21-72 shows the schematic diagram of such a circuit with the resultant sawtooth waveform output. Initially, C_1 (Fig. 21-72) is not charged, and the emitter potential is zero. When power is applied, C_1 will start to charge slowly since it is connected to the battery through R_1. This charging will continue until the potential on the emitter (the top of C_1) becomes positive with respect to base one and forward biases the emitter-base one junction. As soon as this point is reached, current from base one to the emitter rises rapidly and C_1 will discharge through base one and the low resistance of the PN junction. When C_1 is discharged the cycle repeats. The voltage across C_1, then, is a sawtooth waveform possessing a slow rise time (t_1 to t_2, t_3 to t_4, t_5 to t_6) and a rapid descent (t_2 to t_3, t_4 to t_5, t_6 to t_7). When the emitter-base one junction is forward-biased, current through the silicon bar also rises. As can be seen, this is a very simple method of obtaining a sawtooth waveform.

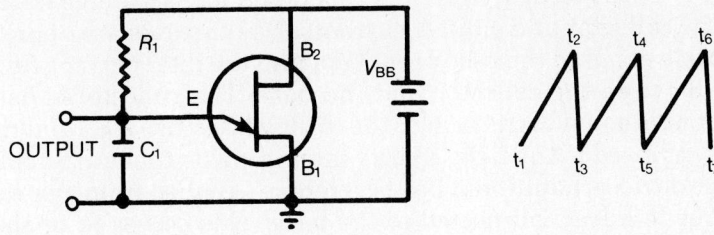

Fig. 21-72: UJT sawtooth generator.

PHOTOTRANSISTORS

Like the photodiode, current flowing through the phototransistor increases with the amount of light hitting a junction in the device. Although conventional transistors are mounted in light-proof cases, elements of the phototransistors are exposed to light. They are usually used in the common-emitter configuration. The characteristic curves of a typical phototransistor appear in Fig. 21-73. Note that the collector current increases with light intensity. The circuit symbol of a phototransistor is given in Fig. 21-74.

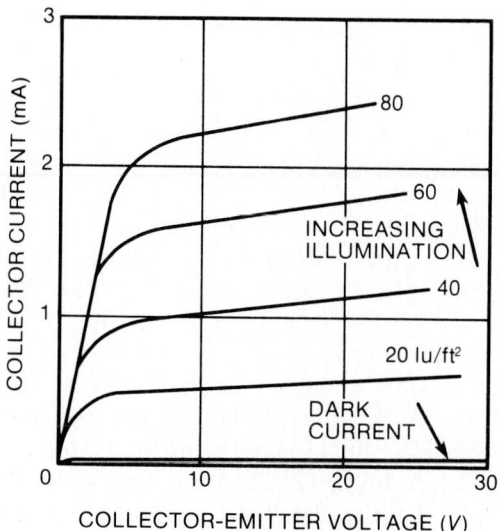

Fig. 21-73: **Typical characteristics of a phototransistor.**

Fig. 21-74: **Circuit symbol of a phototransistor.**

As is the case with the conventional transistor, a DC power supply is connected between the collector and emitter. Unlike the practice employed with other transistors, the base of the phototransistor is usually not connected to a source of current. If the transistor is an NPN device, the collector is invariably made positive with respect to the emitter. The collector-base junction is exposed to light. When there is no light impinging on this junction, ordinary leakage current, I_{CBO}, flows through it. Collector and emitter currents are βI_{CBO} or I_{CEO}, the same leakage current that flows through the junctions of ordinary transistors connected in common-emitter arrangements. As energy in the form of light hits this junction, I_{CBO} increases, as do the collector and emitter currents. The currents are approximately equal to the product of beta and the total current flowing between the collector and the base. The collector-to-base current is equal to the sum of the leakage current, I_{CBO}, and the current owed to the light energy impinging on the junction. In some devices, additional base current is supplied from a power source to a lead connected to the base. This serves to further increase the collector current by a factor equal to the product of β and the current flowing into the base.

522

When used in conjunction with a source of light, the photo-transistor can serve as an element of a light-sensitive smoke detector. Here, smoke obstructs the beam of light between the light source and phototransistor, altering the collector current. This change in collector current can be used to trigger an alarm device.

Phototransistors are frequently mounted in a single package with an LED to form an opto-isolator. There is no electrical contact between the two devices. Because of the isolation of the LED from the phototransistor, the opto-isolator can function as a relay. In this application, an applied current turns the LED on in one circuit. Light from the LED initiates an increase in the amount of current flowing through the phototransistor connected in a second circuit. Current in one circuit is thus due to the presence or absence of current (and light) in a second circuit. This is the description of a typical relay.

Besides being used as a relay switch, the opto-isolator can be used to control quantities of current flowing through the photo-transistor. At low collector voltage, the impedance of the tran-sistor is a function of the collector current flowing through it. Wired properly in the circuit, the opto-isolator can be used to optically couple audio signals from one circuit to another. Us-ing this very same variable resistance characteristic, the device can also be used as a remote optically coupled level control.

Factors such as rise time, delay time, storage time, and fall time (time for the phototransistor's collector current to drop from 90% to 10% of its maximum level) were described in the discussion of switching diodes. These factors also describe the time it takes for the phototransistor collector current to react to switching commands applied to the light-emitting device. The order of magnitude of these factors is in μs. Reaction time of the opto-isolator is equal to about 1/1,000 of the reaction time of a mechanical relay.

INTERPRETING DATA SHEETS

The *data sheet* (also called *specification sheet*) that accom-panies a solid-state device is fully as important to designers as the device itself. With proper interpretation, a designer obtains from this sheet nearly all the basic design information for that particular series of devices. In exacting work such as extremely high frequency, digital circuits, and very close tolerances, tests may still need to be performed under simulated circuit condi-tions. Most manufacturers of semiconductors publish data sheets of their products in either data books or catalogs. This readily available information enables a person to be thorough-ly familiar with a product's characteristics before making a purchase.

There are two important groups of data on the sheet: maxi-mum ratings and electrical characteristics. Notice that both of these groups specify a temperature of 25°C. This is room temperature; other temperature conditions will alter the performance of the device. It is important to remember that, unless otherwise specified, all ratings and characteristics are understood to be the performance standards at 25°C.

The *maximum ratings* (Figs. 21-75 and 77) section includes voltages, current, power, and temperature. Some data sheets specify absolute maximum ratings. These given values must be considered as absolute maximums, whether or not it is so specified. Exceeding one or more of these maximum ratings not only causes inferior performance, but also seriously jeopardizes the transistor.

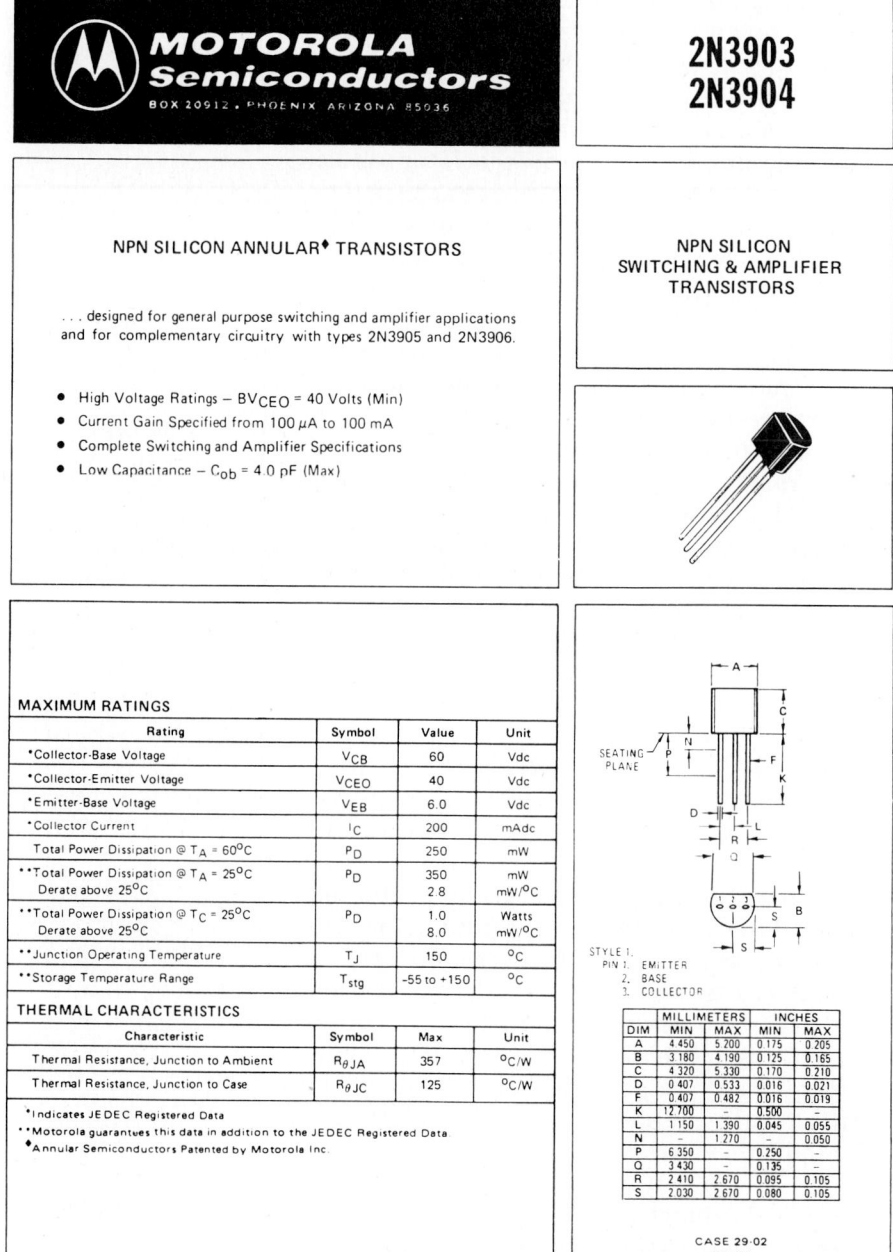

MOTOROLA Semiconductors
BOX 20912 . PHOENIX ARIZONA 85036

2N3903
2N3904

NPN SILICON ANNULAR* TRANSISTORS

. . . designed for general purpose switching and amplifier applications and for complementary circuitry with types 2N3905 and 2N3906.

- High Voltage Ratings – BV$_{CEO}$ = 40 Volts (Min)
- Current Gain Specified from 100 µA to 100 mA
- Complete Switching and Amplifier Specifications
- Low Capacitance – C$_{ob}$ = 4.0 pF (Max)

NPN SILICON SWITCHING & AMPLIFIER TRANSISTORS

MAXIMUM RATINGS

Rating	Symbol	Value	Unit
*Collector-Base Voltage	V$_{CB}$	60	Vdc
*Collector-Emitter Voltage	V$_{CEO}$	40	Vdc
*Emitter-Base Voltage	V$_{EB}$	6.0	Vdc
*Collector Current	I$_C$	200	mAdc
Total Power Dissipation @ T$_A$ = 60°C	P$_D$	250	mW
**Total Power Dissipation @ T$_A$ = 25°C Derate above 25°C	P$_D$	350 2.8	mW mW/°C
**Total Power Dissipation @ T$_C$ = 25°C Derate above 25°C	P$_D$	1.0 8.0	Watts mW/°C
**Junction Operating Temperature	T$_J$	150	°C
**Storage Temperature Range	T$_{stg}$	–55 to +150	°C

THERMAL CHARACTERISTICS

Characteristic	Symbol	Max	Unit
Thermal Resistance, Junction to Ambient	R$_{\theta JA}$	357	°C/W
Thermal Resistance, Junction to Case	R$_{\theta JC}$	125	°C/W

*Indicates JEDEC Registered Data
**Motorola guarantees this data in addition to the JEDEC Registered Data
*Annular Semiconductors Patented by Motorola Inc.

SEATING PLANE

STYLE 1.
PIN 1. EMITTER
2. BASE
3. COLLECTOR

DIM	MILLIMETERS		INCHES	
	MIN	MAX	MIN	MAX
A	4.450	5.200	0.175	0.205
B	3.180	4.190	0.125	0.165
C	4.320	5.330	0.170	0.210
D	0.407	0.533	0.016	0.021
F	0.407	0.482	0.016	0.019
K	12.700	–	0.500	–
L	1.150	1.390	0.045	0.055
N		1.270		0.050
P	6.350	–	0.250	–
Q	3.430	–	0.135	
R	2.410	2.670	0.095	0.105
S	2.030	2.670	0.080	0.105

CASE 29-02
TO-92

© MOTOROLA INC., 1973   DS 5127 R2

Fig. 21-75: Typical maximum rating data sheet. (Data sheet furnished by courtesy of Motorola Semiconductor Products, Inc.)

The *electrical characteristics* (Figs. 21-76 and 78) are subdivided into small-signal characteristics, high-frequency characteristics, direct-current characteristics, and switching characteristics. Some of these headings are generally omitted, depending upon the intended application of the device. For example, when a device is not recommended for switching, there would be little point in publishing switching characteristics.

*ELECTRICAL CHARACTERISTICS (T_A = 25°C unless otherwise noted)

Characteristic		Fig. No.	Symbol	Min	Max	Unit
OFF CHARACTERISTICS						
Collector-Base Breakdown Voltage ($I_C = 10 \mu Adc$, $I_E = 0$)			BV_{CBO}	60	-	Vdc
Collector-Emitter Breakdown Voltage (1) ($I_C = 1.0$ mAdc, $I_B = 0$)			BV_{CEO}	40	-	Vdc
Emitter-Base Breakdown Voltage ($I_E = 10 \mu Adc$, $I_C = 0$)			BV_{EBO}	6.0	-	Vdc
Collector Cutoff Current ($V_{CE} = 30$ Vdc, $V_{EB(off)} = 3.0$ Vdc)			I_{CEX}	-	50	nAdc
Base Cutoff Current ($V_{CE} = 30$ Vdc, $V_{EB(off)} = 3.0$ Vdc)			I_{BL}	-	50	nAdc
ON CHARACTERISTICS						
DC Current Gain (1) ($I_C = 0.1$ mAdc, $V_{CE} = 1.0$ Vdc)	2N3903 2N3904	15	h_{FE}	20 40	- -	
($I_C = 1.0$ mAdc, $V_{CE} = 1.0$ Vdc)	2N3903 2N3904			35 70	- -	
($I_C = 10$ mAdc, $V_{CE} = 1.0$ Vdc)	2N3903 2N3904			50 100	150 300	
($I_C = 50$ mAdc, $V_{CE} = 1.0$ Vdc)	2N3903 2N3904			30 60	- -	
($I_C = 100$ mAdc, $V_{CE} = 1.0$ Vdc)	2N3903 2N3904			15 30	- -	
Collector-Emitter Saturation Voltage (1) ($I_C = 10$ mAdc, $I_B = 1.0$ mAdc) ($I_C = 50$ mAdc, $I_B = 5.0$ mAdc)		16, 17	$V_{CE(sat)}$	- -	0.2 0.3	Vdc
Base-Emitter Saturation Voltage (1) ($I_C = 10$ mAdc, $I_B = 1.0$ mAdc) ($I_C = 50$ mAdc, $I_B = 5.0$ mAdc)		17	$V_{BE(sat)}$	0.65 -	0.85 0.95	Vdc
SMALL-SIGNAL CHARACTERISTICS						
Current-Gain—Bandwidth Product ($I_C = 10$ mAdc, $V_{CE} = 20$ Vdc, f = 100 MHz)	2N3903 2N3904		f_T	250 300	- -	MHz
Output Capacitance ($V_{CB} = 5.0$ Vdc, $I_E = 0$, f = 100 kHz)		3	C_{ob}	-	4.0	pF
Input Capacitance ($V_{BE} = 0.5$ Vdc, $I_C = 0$, f = 100 kHz)		3	C_{ib}	-	8.0	pF
Input Impedance ($I_C = 1.0$ mAdc, $V_{CE} = 10$ Vdc, f = 1.0 kHz)	2N3903 2N3904	13	h_{ie}	0.5 1.0	8.0 10	k ohms
Voltage Feedback Ratio ($I_C = 1.0$ mAdc, $V_{CE} = 10$ Vdc, f = 1.0 kHz)	2N3903 2N3904	14	h_{re}	0.1 0.5	5.0 8.0	X 10-4
Small-Signal Current Gain ($I_C = 1.0$ mAdc, $V_{CE} = 10$ Vdc, f = 1.0 kHz)	2N3903 2N3904	11	h_{fe}	50 100	200 400	
Output Admittance ($I_C = 1.0$ mAdc, $V_{CE} = 10$ Vdc, f = 1.0 kHz)		12	h_{oe}	1.0	40	μmhos
Noise Figure ($I_C = 100 \mu Adc$, $V_{CE} = 5.0$ Vdc, $R_S = 1.0$ k ohms, f = 10 Hz to 15.7 kHz)	2N3903 2N3904	9, 10	NF	- -	6.0 5.0	dB
SWITCHING CHARACTERISTICS						
Delay Time ($V_{CC} = 3.0$ Vdc, $V_{BE(off)} = 0.5$ Vdc,		1, 5	t_d	-	35	ns
Rise Time $I_C = 10$ mAdc, $I_{B1} = 1.0$ mAdc)		1, 5, 6	t_r	-	35	ns
Storage Time ($V_{CC} = 3.0$ Vdc, $I_C = 10$ mAdc,	2N3903 2N3904	2, 7	t_s	- -	175 200	ns
Fall Time $I_{B1} = I_{B2} = 1.0$ mAdc)		2, 8	t_f	-	50	ns

(1) Pulse Test: Pulse Width = 300 μs, Duty Cycle = 2.0%.
*Indicates JEDEC Registered Data

FIGURE 1 – DELAY AND RISE TIME EQUIVALENT TEST CIRCUIT

300 ns DUTY CYCLE = 2%   +10.9 V   +3.0 V   275   10 K   -0.5 V   < 1.0 ns C_s < 4.0 pF *

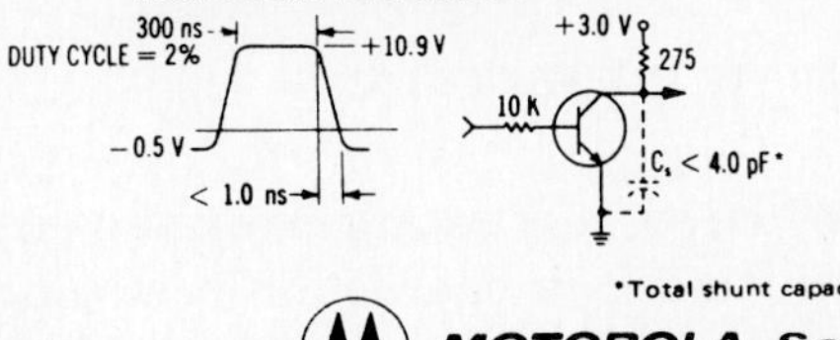

FIGURE 2 – STORAGE AND FALL TIME EQUIVALENT TEST CIRCUIT

$10 < t_1 < 500 \mu s$ t_1 +10.9 V   +3.0 V   275   DUTY CYCLE = 2%   0   10 K   -9.1 V   < 1.0 ns   1N916 C_s < 4.0 pF *

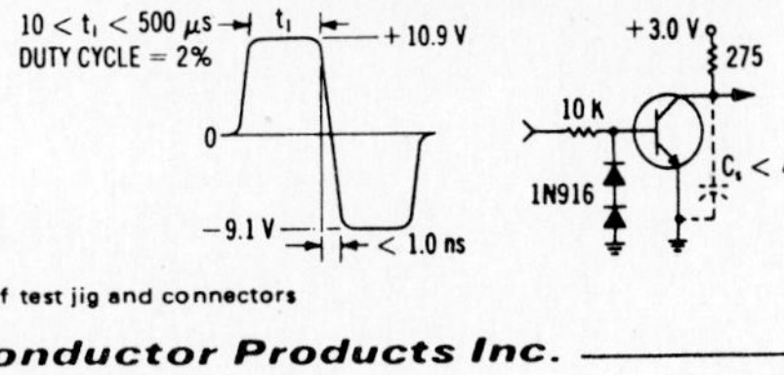

*Total shunt capacitance of test jig and connectors

Ⓜ **MOTOROLA** *Semiconductor Products Inc.*

Fig. 21-76: Typical electrical characteristics sheet. (Data sheet furnished by courtesy of Motorola Semiconductor Products, Inc.)

MOTOROLA
Semiconductors
BOX 20912 • PHOENIX, ARIZONA 85036

2N3905
2N3906

PNP SILICON
SWITCHING & AMPLIFIER
TRANSISTORS

PNP SILICON ANNULAR♦ TRANSISTORS

.... designed for general purpose switching and amplifier applications and for complementary circuitry with types 2N3903 and 2N3904.

- High Voltage Ratings — BV_{CEO} = 40 Volts (Min)
- Current Gain Specified from 100 μA to 100 mA
- Complete Switching and Amplifier Specifications
- Low Capacitance — C_{ob} = 4.5 pF (Max)

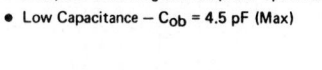

*MAXIMUM RATINGS

Rating	Symbol	Value	Unit
Collector-Base Voltage	V_{CB}	40	Vdc
Collector-Emitter Voltage	V_{CEO}	40	Vdc
Emitter-Base Voltage	V_{EB}	5.0	Vdc
Collector Current	I_C	200	mAdc
Total Power Dissipation @ T_A = 60°C	P_D	250	mW
Total Power Dissipation @ T_A = 25°C Derate above 25°C	P_D	350 2.8	mW mW/°C
Total Power Dissipation @ T_C = 25°C Derate above 25°C	P_D	1.0 8.0	Watt mW/°C
Junction Operating Temperature	T_J	+150	°C
Storage Temperature Range	T_{stg}	−55 to +150	°C

THERMAL CHARACTERISTICS

Characteristic	Symbol	Max	Unit
Thermal Resistance, Junction to Ambient	$R_{\theta JA}$	357	°C/W
Thermal Resistance, Junction to Case	$R_{\theta JC}$	125	°C/W

STYLE 1:
PIN 1. EMITTER
2. BASE
3. COLLECTOR

DIM	MILLIMETERS		INCHES	
	MIN	MAX	MIN	MAX
A	4.450	5.200	0.175	0.205
B	3.180	4.190	0.125	0.165
C	4.320	5.330	0.170	0.210
D	0.407	0.533	0.016	0.021
F	0.407	0.482	0.016	0.019
K	12.700	–	0.500	–
L	1.150	1.390	0.045	0.055
N	–	1.270	–	0.050
P	6.350	–	0.250	–
Q	3.430	–	0.135	–
R	2.410	2.670	0.095	0.105
S	2.030	2.670	0.080	0.105

CASE 29-02
(TO-92)

*Indicates JEDEC Registered Data.
♦Annular semiconductors patented by Motorola Inc.

©MOTOROLA INC., 1973   DS 5128 R2

Fig. 21-77: Typical maximum rating data sheet. (Data sheet furnished by courtesy of Motorola Semiconductor Products, Inc.)

TRANSISTOR TESTING

The first step in troubleshooting transistor circuits is to visually inspect printed circuit boards for loose connections, short circuits due to poor soldering, and open circuits due to cracked foil. Careful visual inspection may save hours of troubleshooting.

Whenever a particular transistor is suspected to be faulty, it may be tested in the circuit if there are no low resistance shunts

*ELECTRICAL CHARACTERISTICS (T_A = 25°C unless otherwise noted.)

Characteristic		Fig. No.	Symbol	Min	Max	Unit
OFF CHARACTERISTICS						
Collector-Base Breakdown Voltage (I_C = 10 μAdc, I_E = 0)			BV_{CBO}	40	—	Vdc
Collector-Emitter Breakdown Voltage (1) (I_C = 1.0 mAdc, I_B = 0)			BV_{CEO}	40	—	Vdc
Emitter-Base Breakdown Voltage (I_E = 10 μAdc, I_C = 0)			BV_{EBO}	5.0	—	Vdc
Collector Cutoff Current (V_{CE} = 30 Vdc, $V_{BE(off)}$ = 3.0 Vdc)			I_{CEX}	—	50	nAdc
Base Cutoff Current (V_{CE} = 30 Vdc, $V_{BE(off)}$ = 3.0 Vdc)			I_{BL}	—	50	nAdc
ON CHARACTERISTICS (1)						
DC Current Gain		15	h_{FE}			
(I_C = 0.1 mAdc, V_{CE} = 1.0 Vdc)	2N3905			30	—	
	2N3906			60	—	
(I_C = 1.0 mAdc, V_{CE} = 1.0 Vdc)	2N3905			40	—	
	2N3906			80	—	
(I_C = 10 mAdc, V_{CE} = 1.0 Vdc)	2N3905			50	150	
	2N3906			100	300	
(I_C = 50 mAdc, V_{CE} = 1.0 Vdc)	2N3905			30	—	
	2N3906			60	—	
(I_C = 100 mAdc, V_{CE} = 1.0 Vdc)	2N3905			15	—	
	2N3906			30	—	
Collector-Emitter Saturation Voltage		16, 17	$V_{CE(sat)}$			Vdc
(I_C = 10 mAdc, I_B = 1.0 mAdc)				—	0.25	
(I_C = 50 mAdc, I_B = 5.0 mAdc)				—	0.4	
Base-Emitter Saturation Voltage		17	$V_{BE(sat)}$			Vdc
(I_C = 10 mAdc, I_B = 1.0 mAdc)				0.65	0.85	
(I_C = 50 mAdc, I_B = 5.0 mAdc)				—	0.95	
SMALL-SIGNAL CHARACTERISTICS						
Current-Gain — Bandwidth Product			f_T			MHz
(I_C = 10 mAdc, V_{CE} = 20 Vdc, f = 100 MHz)	2N3905			200	—	
	2N3906			250	—	
Output Capacitance (V_{CB} = 5.0 Vdc, I_E = 0, f = 100 kHz)		3	C_{ob}	—	4.5	pF
Input Capacitance (V_{BE} = 0.5 Vdc, I_C = 0, f = 100 kHz)		3	C_{ib}	—	1.0	pF
Input Impedance		13	h_{ie}			k ohms
(I_C = 1.0 mAdc, V_{CE} = 10 Vdc, f = 1.0 kHz)	2N3905			0.5	8.0	
	2N3906			2.0	12	
Voltage Feedback Ratio		14	h_{re}			X 10^{-4}
(I_C = 1.0 mAdc, V_{CE} = 10 Vdc, f = 1.0 kHz)	2N3905			0.1	5.0	
	2N3906			1.0	10	
Small-Signal Current Gain		11	h_{fe}			—
(I_C = 1.0 mAdc, V_{CE} = 10 Vdc, f = 1.0 kHz)	2N3905			50	200	
	2N3906			100	400	
Output Admittance		12	h_{oe}			μmhos
(I_C = 1.0 mAdc, V_{CE} = 10 Vdc, f = 1.0 kHz)	2N3905			1.0	40	
	2N3906			3.0	60	
Noise Figure (I_C = 100 μAdc, V_{CE} = 5.0 Vdc, R_S = 1.0 k ohm,	2N3905	9, 10	NF	—	5.0	dB
f = 10 Hz to 15.7 kHz)	2N3906			—	4.0	
SWITCHING CHARACTERISTICS						
Delay Time (V_{CC} = 3.0 Vdc, $V_{BE(off)}$ = 0.5 Vdc		1, 5	t_d	—	35	ns
Rise Time I_C = 10 mAdc, I_{B1} = 1.0 mAdc)		1, 5, 6	t_r	—	35	ns
Storage Time	2N3905	2, 7	t_s	—	200	ns
(V_{CC} = 3.0 Vdc, I_C = 10 mAdc,	2N3906			—	225	
Fall Time I_{B1} = I_{B2} = 1.0 mAdc)	2N3905	2, 8	t_f	—	60	ns
	2N3906			—	75	

*Indicates JEDEC Registered Data. (1) Pulse Width = 300 μs, Duty Cycle = 2.0 %.

FIGURE 1 — DELAY AND RISE TIME EQUIVALENT TEST CIRCUIT

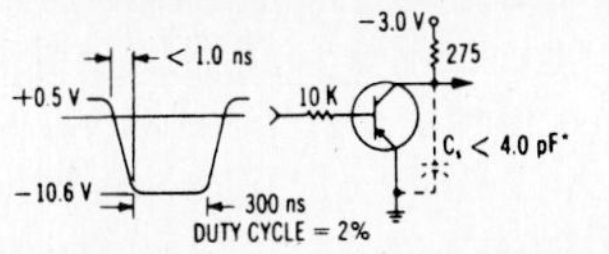

FIGURE 2 — STORAGE AND FALL TIME EQUIVALENT TEST CIRCUIT

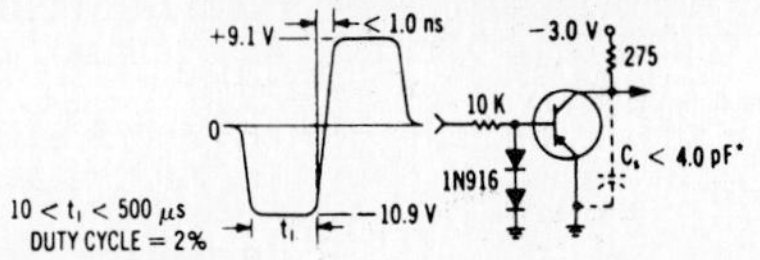

*Total shunt capacitance of test jig and connectors

Fig. 21-78: Typical electrical characteristics sheet. (Data sheet furnished by courtesy of Motorola Semiconductor Products, Inc.)

(such as coils, forward-biased diodes, low value resistors, etc.) across any of its leads. A low resistance shunt across any two of the transistor leads can easily cause erroneous indications on the transistor test set.

Erroneous indications can be eliminated by removing the suspected transistor from the circuit. Some manufacturers have simplified the removal and replacement of transistors by using transistor sockets. Whenever a circuit board without transistor sockets is encountered, transistors will have to be carefully unsoldered in order to test them out of circuits.

527

Any meter used for voltage measurements in transistor circuits should have a sensitivity of greater than $20,000\,\Omega$ per volt on all ranges. Meters with a lower sensitivity may draw excessive current from the circuit under test when using the lower voltage ranges. Meters with an input impedance of $100,000\,\Omega$ per volt or more are preferred. Precautions to be observed when taking measurements in transistor circuits are:

1. *Before* making any resistance measurements, make sure that all power to the circuit under test has been disconnected and that all capacitors have been discharged.

2. Do not use an ohmmeter which passes more than 1mA through the circuit under test.

3. Ensure that all test equipment is isolated from the power line either by the equipment's own power supply transformer or by an external isolation transformer.

4. Always connect a ground lead between *common* for the circuit under test and *common* on the test equipment.

5. Do not short circuit any portion of a transistor circuit. Short circuiting individual components or groups of components may allow excessive current to flow, thus damaging components.

6. Do not remove or replace any transistors in an energized circuit.

Transistor Testing Using an Ohmmeter

While an ohmmeter cannot test all the characteristics of transistors, especially transistors used for high frequencies or switching, it is capable of making many simple transistor tests. Because transistor testers might not always be available and the ohmmeter is usually available, some hints on how to test transistors using an ohmmeter follow.

Since damage to transistors could occur when using voltages above approximately 6V, care must be taken to avoid using resistance scales where the internal voltage of the ohmmeter is greater than 6V. This higher potential is usually found on the higher resistance scales. Excess current might also cause damage to the transistor under test. Since the internal current limiting resistance generally increases as the resistance range is increased, the low range of the resistance scale should also be avoided. Basically, if we stay away from the highest resistance range (possible excessive voltage) and the lowest resistance range (possible excessive current), the ohmmeter should present no problems in transistor testing. Generally speaking, the R × 10 and R × 100 scales may be considered safe.

The polarity of the battery, as well as the voltage value, must be known when using the ohmmeter for transistor testing. Although, in some cases, the ground or common lead (black) is negative and the hot lead (red) is positive, this battery setup is not always the case.

It is possible to reverse the polarity of the ohmmeter leads by changing the function switch position. This means that the black or common lead is now the positive battery and the red lead is the negative battery. **Caution:** Although the jacks of the meter may show a negative sign under the common and a positive sign under the other jack, these do not indicate the internal polarity of the battery connected to the jack. In addition, do not use any ohmmeter for transistor testing that draws more than 1mA.

SUMMARY

The transistor is a device that may function as an amplifier or an electronic switch.

The bipolar junction transistor is made up of either an N- or P-type material sandwiched between two pieces of the opposite type, thus forming two PN junctions. The center material is called the base, and the outer materials are called emitter and collector. A transistor conducts only when the emitter-base junction is forward-biased and the collector-base junction is reverse-biased.

When a BJT is used as the principal controlling element in an amplifier circuit, it can be connected in basically three different configurations: common-emitter, common-base, and common-collector.

The field effect transistor is a unipolar, voltage-sensitive device that uses an electrostatic field to control conduction. There are two general types of FETs: junction (JFET) and insulated gate (IGFET, or MOSFET). The FET is used in three basic circuit configurations: common-gate, common-source, and common-drain.

Thyristors are bistable solid-state devices that have two or more junctions and that can be switched between the ON/OFF conducting states. They are heavy-current devices used extensively in applications involving power rectification, control, and switching.

The silicon-controlled rectifier (SCR), a four-layer device with three electrodes, is a unidirectional reverse-blocking thyristor. It can be switched from the OFF state to the ON state by a positive trigger pulse applied to the gate.

The triac, also a four-layer device with three electrodes, is a bidirectional device which can block voltages of either polarity and conduct current in either direction. It can be switched from the OFF state to the ON state in either direction.

The silicon-controlled switch (SCS) is a unilateral four-layer silicon device with electrodes connected to the four terminals. This device has three base operating modes: forward-blocking mode, reverse-bias mode, and forward-conduction mode.

The Shockley diode is a four-layer device that functions as a switch. It acts like an open switch when its forward voltage is less than the forward breakover value and like a closed switch at voltages above this value.

The unijunction transistor (UJT), sometimes referred to as a double-base diode, features a stable triggering voltage, a very low value of firing current, a negative resistance characteristic, and a high-pulse current capability.

The phototransistor is usually a two-lead device with the collector-base junction exposed to light. The amount of current flowing through the device increases with the amount of light hitting the junction.

All the maximum voltages and electrical characteristics for a given solid-state device may be found on its data sheet (or specification sheet).

CHECK YOURSELF (Answers at back of book)

Questions

Fill in the blanks with the appropriate word or words.

1. By reducing the voltage across the Shockley diode so that the current flowing through the device drops below its _____ , the switch can again be opened.
2. The first step in designing any FET bias circuit is to determine the desired levels of _____ and _____ under quiescent conditions.
3. The _____ MOSFET takes its name from the fact that the conduction channel between source and drain is nonexistent until the proper bias is applied.
4. The _____ of a transistor is equal to I_C/I_B.
5. In the PNP transistor, the arrow on the emitter points toward the _____ .
6. The SCS can be switched on accidentally if subjected to high-frequency transients. This action is known as the _____ .
7. The diac is primarily used as a _____ in triac power control circuits.
8. Exceeding one or more of the _____ not only causes inferior performance, but also seriously jeopardizes the transistor.
9. The negatively-biased triac operates in the _____ of its characteristic.
10. Any meter used for voltage measurements in transistor circuits should have a sensitivity of at least _____ .
11. The static characteristic curves show only the _____ and _____ conditions of a transistor.
12. If the two bases of a UJT were connected together, the resulting device would have the characteristics of a _____ .
13. A typical application of the UJT is a _____ .
14. Phototransistors are frequently mounted in a single package with an LED to form an _____ .
15. _____ is equal to the ratio of I_C to I_E.
16. One of the transistor's leads must be common to both the _____ and the _____ of the circuit.
17. In a JFET, the size of the _____ defines the width of the conducting channel.
18. The input impedance of a JFET is much greater than the input impedance of a _____ .
19. The FET _____ configuration provides high power gain and makes a good buffer amplifier.
20. The FET _____ configuration provides good voltage amplification and reasonable gain in both current and power.

CHAPTER 22

Amplifier Circuits

As stated in the previous chapter, one of the major functions of transistors is acting as amplifier circuits. In other words, a transistor can increase the voltage, current, or power level of a signal applied to it.

AMPLIFIER-CIRCUIT CLASSIFICATION

There are several ways to classify amplifier circuits. The most common method, however, is to use the *operating point* of the transistors. That is, the particular circuit is classified by the amount of current flowing in the amplifier transistors under no-signal conditions. The following is a brief summary of the four basic operating-point classifications of class A, class B, class AB, and class C amplifiers.

In all four classifications, the base-collector (or gate-drain in the case of an FET) junction is always reverse-biased at the operating point, as well as under all signal conditions. Thus, no base-collector current flows (with the possible exception of reverse leakage current, called I_{CBO} on most transistor data sheets). The base-emitter (or gate-source) junction is biased such that base-emitter current flows under certain conditions and possibly under all conditions. When base-emitter current flows, emitter-collector current also flows.

Class A Amplifiers

Any amplifier that is biased so that its output current flows during the entire cycle of the AC input voltage is said to be operating as a class A amplifier. The output current of a class A amplifier used to amplify an AC sinusoidal signal is shown in Fig. 22-1. The level of current increases and decreases around its no-signal value but never drops to zero. The output voltage will vary in the same way. Because the class A amplifier is biased to operate at a point where any change in input current will result in a proportional amplified change in output current, the circuit operates in a linear manner. A minimum amount of distortion is produced by this type of amplifier.

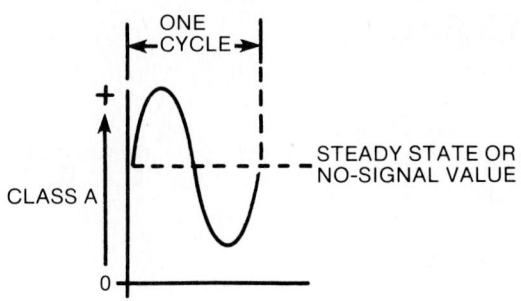

Fig. 22-1: Typical class A amplifier characteristic curves.

Class A operation is achieved by DC biasing the transistor midway between its upper current limit (saturation) and its lower current limit (cutoff). The input signal will cause the forward bias to increase and decrease around this DC (no-signal) value. Thus, the input current will vary; this, in turn, will cause the output current to vary. The input signal will be amplified without distortion as long as the amplitude of this signal is not high enough to overdrive the amplifier. If the amplitude of the input signal is too high, distortion will result, as shown in Fig. 22-2. Notice that both the positive and negative peaks are clipped off because the saturation and cutoff points are reached.

Class B Amplifiers

When an amplifier is biased so that its output current will flow for only one-half of the AC cycle (Fig. 22-3), it is said to be operating as a class B amplifier. Class B operation can be obtained by biasing the transistor at cutoff. When biased in this way, the output current can only flow during the input alternation, which causes the forward bias and input current to increase. Since the transistor's emitter junction is effectively reverse-biased during the other alternation, the circuit will not respond on that half-cycle. Thus the class B amplifier amplifies only one-half of the input AC signal.

Since the class B amplifier conducts only 50% of the time, it can deliver a much higher output power than the class A amplifier.

Class AB Amplifiers

The *class AB amplifier* is a compromise between a class A and class B amplifier. Like the class B amplifier, it must be used in a push-pull configuration (see Chapter 24). Although the class AB is not quite as power efficient as a class B amplifier, distortion is minimized.

An amplifier biased so that its output current flows for less than one full cycle of the AC input, but for more than one-half of the AC cycle, is said to be operating as a class AB amplifier. Class AB amplifier operation can be achieved by DC biasing

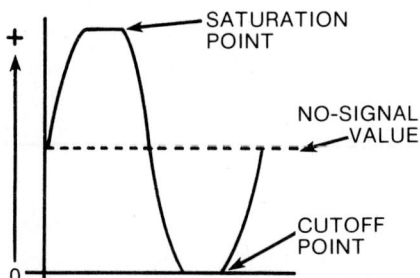

Fig. 22-2: Example of class A distortion.

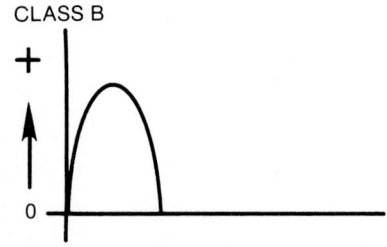

Fig. 22-3: Typical class B amplifier characteristic curve.

the transistor closer to cutoff than saturation. The output current waveform produced by a class AB amplifier is shown in Fig. 22-4. Notice that the current waveform drops to zero for a period of time which is less than the time needed to produce one-half of the AC cycle.

Class C Amplifiers

An amplifier biased so that its output current will flow for less than one-half of the AC input cycle (Fig. 22-5) is said to be operating as a *class C amplifier*. Class C operation is achieved by biasing the transistor beyond the cutoff point. Since the voltage during each alternation must rise high enough to overcome the reverse bias, the output current can only flow for a portion of each alternation.

Because current flows for only a short period of time during each half-cycle, very little power is drawn from the supply; therefore, the efficiency of the class C amplifier is high. Class C amplifiers cannot, however, be used to reproduce audio signals because of their inherent distortion. They are usually used in radio frequency circuitry with a tuned LC circuit as a load. Each time the amplitude of the input signal is large enough to turn on the transistor, energy is supplied to the LC circuit. An entire cycle appears across the resonant circuit due to its flywheeling characteristics (see Chapter 25).

GENERAL AMPLIFIER BIASING

All transistors (and many other solid-state devices) require, as mentioned in Chapter 21, a form of bias. The purpose of the bias circuit is to establish collector-base-emitter voltage and current relationships at the *operating point* of the circuit. (The operating point is also known as the *quiescent point,* or *Q-point.*) Since transistors rarely operate at this Q-point, the basic bias circuits are used as a reference or starting point for design. The actual circuit configuration and (especially) the bias circuit values must be selected on the basis of dynamic circuit conditions (desired output voltage swing, expected input signal level, etc.).

In a typical transistor, the collector-base circuit must be reverse-biased. That is, current should not flow between collector and base. Any collector-base current that does flow is the result of leakage or breakdown. Breakdown must be avoided by proper design. Leakage is an undesirable (but almost always present) condition that must be reckoned with in design.

The emitter-base circuit of a typical transistor must be forward-biased. That is, current should flow between base and emitter during normal operation. Some current flows all the time in certain circuits (class A amplifiers and most oscillators). In other circuits (class C amplifiers, switches, etc.) current flows only in the presence of an operating signal or trigger. Either way, the emitter-base circuit must be biased so that current can flow.

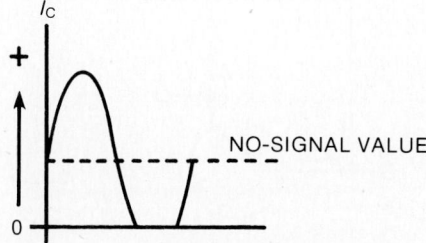

Fig. 22-4: Typical class AB amplifier characteristic curve.

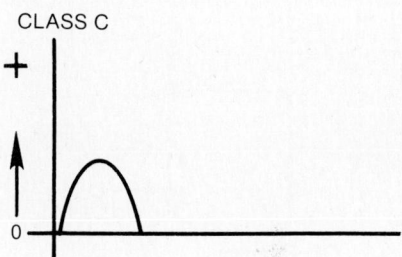

Fig. 22-5: Typical class C amplifier characteristic curve.

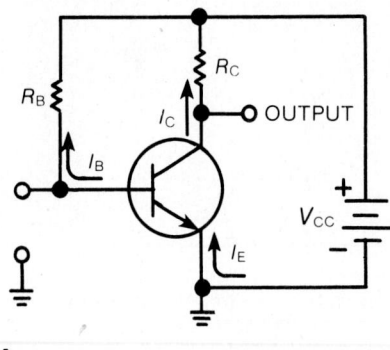

A

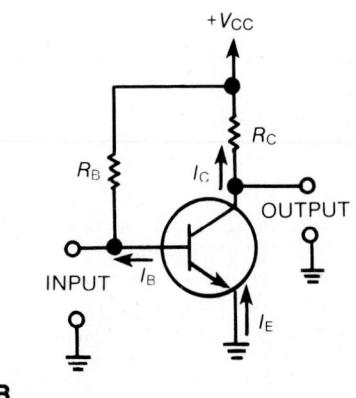

B

Fig. 22-6: Simple base-biased common-emitter amplifier circuit.

The desired bias is accomplished by applying voltages to the corresponding transistor elements, usually through resistances.

Biasing Arrangements

Although there are several bias networks, each with its own advantages and disadvantages, one major factor must be considered for any network. The basic network must maintain the desired base current in the presence of temperature (and frequency, in some cases) changes. This is often referred to as *bias stability*. Several of the more common types of bias follow.

Fixed Bias (Base Injection Bias). An extremely simple method of biasing a common-emitter transistor circuit is illustrated in Fig. 22-6. Notice that only one voltage source (V_{CC}) is employed to provide the forward and reverse-bias voltages needed for transistor conduction. Although an NPN transistor is used in this circuit, a PNP could be used if the polarity of V_{CC} was reversed.

The two resistors R_B and R_C are used to properly distribute the voltage. Resistor R_C is simply a collector load resistor that is in series with the collector. The collector current I_C flows through this resistor and produces a voltage across it. The sum of the voltage across R_C (V_{RC}) and the collector-to-emitter voltage of the transistor (V_{CE}) must always equal V_{CC}. Expressed mathematically,

$$V_{RC} + V_{CE} = V_{CC}$$

Therefore, the transistor's collector-to-emitter voltage is equal to V_{CC} minus the voltage developed across R_C. That is,

$$V_{CE} = V_{CC} - V_{RC}$$

The transistor's collector-to-emitter voltage has the proper polarity to reverse-bias the transistor's collector junction as required.

Resistor R_{B1}, which is in series with the base lead, controls the amount of base current (I_B) flowing out of the base. This base current flows through resistor R_B and develops a voltage (V_{RB}) across it. Most of source voltage V_{CC} is dropped across R_B and the difference appears across the base-to-emitter junction of the transistor to provide the required forward-bias voltage. Expressed as an equation,

$$V_{RB} = V_{CC} - V_{BE}$$

The circuit is often called a *base-biased* circuit because the base current (I_B) in the circuit is controlled by resistor R_B and voltage source V_{CC}. The value of R_B is usually chosen so that the value of base current (I_B) will be high enough to cause a substantial value of collector current (I_C) to flow through R_C. The I_B value is normally in the microampere range, while the value of I_C is likely to be in the milliampere range.

The circuit is also called a *fixed-biased* circuit because the value of base current is fixed. That is, it is determined by:

$$I_B = \frac{V_{CC} - V_{BE}}{R_B}$$

The input signal voltage either aids or opposes the existing forward-bias voltage across the emitter junction. This causes I_C to vary which, in turn, causes the voltage across R_C to vary. The transistor's collector-to-emitter voltage likewise varies and produces the output terminal and ground.

Figure 22-6B represents the same circuit shown in Fig. 22-6A; however, it is shown as it would appear on a typical schematic diagram. Instead of showing voltage source V_{CC}, only the upper positive terminal is shown and is identified as $+V_{CC}$. The negative side of V_{CC} is represented by the ground symbol. The base-biased circuit is extremely unstable and is, therefore, seldom used in electronic equipment. However, it serves as a logical introduction to the more complex and more practical circuits that follow. The base-biased circuit is said to be unstable because it is not capable of compensating for changes in its steady state (no-signal) bias currents.

For example, changes in temperature can cause the internal resistances of the transistor to vary and this can cause the bias currents (I_B and I_C) to vary. This, in turn, can cause the transistor's operating point to shift and it can also reduce the gain of the transistor. This entire process is known as *thermal instability* and is inherent in any base-biased circuit.

Fixed Bias with Emitter-Resistor. In Fig. 22-7A a small value of emitter resistance (typically one-tenth the value of R_B or less) is added to counteract instability. If the collector current increases due to an increase in junction temperature, the voltage across R_E will also increase, thereby decreasing the base-emitter voltage. This decrease in the forward bias of the base-emitter circuit causes a decrease in both the base and emitter currents, thereby counteracting the increase in collector current. This method of biasing stabilizes the circuit by reducing the effects of temperature drift and also compensates for differences in transistor parameters.

In Fig. 22-7B, C_E is added to minimize AC degeneration. The capacitor is made large in value so that its reactance is approximately one-tenth the value of R_E at the lowest operating frequency. Note that this capacitor provides a low impedance path for the AC signal without changing the DC bias conditions.

Collector Feedback Bias. The circuits shown in Fig. 22-8 use DC feedback. The purpose of returning R_B to the collector instead of returning it to the power supply is to stabilize the bias under variations of current flow due to changes in transistors or temperature. If the current flow should increase through a transistor, the voltage at the collector would decrease and thus decrease the potential at the base. This would decrease the forward bias on the transistor and reduce the current flow. On the other hand, should the current flow through the transistor decrease, the collector voltage would increase, increasing the potential on the base and increasing the forward bias which will allow more current to flow. This configuration stabilizes the DC bias level on the base. This degenerative voltage feed-

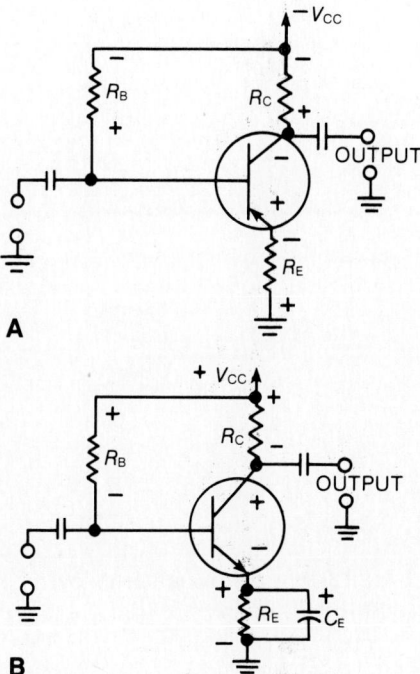

Fig. 22-7: (A) Fixed bias with emitter-resistor; (B) C_E added to minimize AC degeneration.

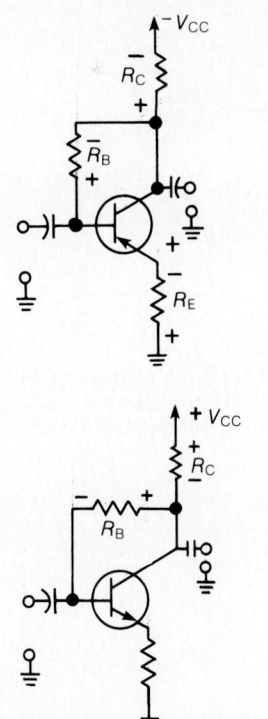

Fig. 22-8: Collector feedback bias.

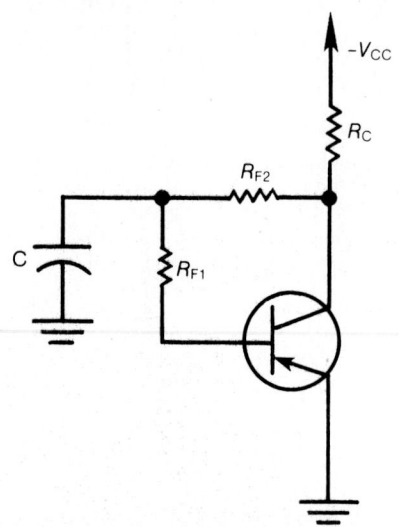

Fig. 22-9: Compensating for degenerative feedback.

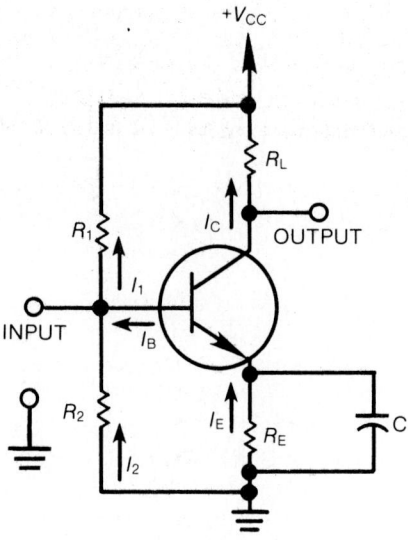

Fig. 22-10: A common-emitter circuit which uses a voltage-divider and emitter feedback.

back does, however, decrease the gain of the transistor because a portion of the signal is also fed back. Compensation for this may be made by eliminating the AC component from the collector potential. This is done by breaking up R_B into R_{F1} and R_{F2} with a decoupling capacitor between them (Fig. 22-9).

Voltage-Divider Bias. Figure 22-10 illustrates a voltage-divider biased circuit that can be used when temperature stability better than that of any of the previously described biasing arrangements is required. Note that in this circuit resistor R_B has been replaced by two resistors (R_1 and R_2). Resistors R_1 and R_2 are in series and are connected across the voltage source V_{CC}; therefore, the voltage dropped across R_1 plus the voltage dropped across R_2 must be equal to V_{CC}. In other words, these two resistors serve as a voltage-divider, effectively dividing voltage V_{CC} into two voltages. Expressed as an equation,

$$V_{CC} = V_{R_1} + V_{R_2}$$

The voltage across R_2 is substantially lower than source voltage V_{CC}. Also, the upper end of R_2 is connected to the base of the transistor; therefore, the voltage at the base with respect to ground is equal to the voltage across R_2. This voltage remains essentially constant. Furthermore, the voltage at the upper end of R_2 (the base voltage) is positive with respect to ground since I_2 flows up through the resistor.

Currents I_1 and I_2 are almost identical, but there is a minute difference. I_1 is just slightly higher than I_2 due to the fact that I_B also flows through R_1. However, I_B is extremely small and can, therefore, be ignored. The values of R_1 and R_2 are usually selected so that I_1 and I_2 will be much higher than I_B. As a result, any changes in I_B will not upset the voltage-divider and cause the voltage across R_2 to vary by a significant amount. The function of the voltage-divider is simply to establish a steady voltage from base to ground.

The I_C and I_E transistor currents flow in their normal manner. Since I_E flows up through R_E, the voltage developed across R_E causes the emitter of the transistor to be positive with respect to ground. However, the base of the transistor is also positive with respect to ground; therefore, the voltage across the emitter-base junction of the transistor is equal to the difference between these two positive voltages:

$$V_{BE} = V_B - V_E$$

In order to have the proper forward-bias voltage across the emitter junction of the NPN transistor, the base is just slightly more positive than the emitter. The voltage dropped across the emitter junction tends to remain essentially constant even when the current flowing through the junction varies over a wide range. When the temperature increases, the values of I_C and I_E also tend to increase. The increasing value of I_E causes the voltage across R_E to increase, which causes the emitter of the transistor to become more positive with respect to ground.

This action reduces the forward-bias voltage across the emitter junction, which, in turn, causes a decrease in I_B. The lower value of base current reduces the values of collector and emitter currents toward their normal values. If the temperature decreases, the exact opposite action will take place: I_B will increase to counteract the decreasing values of I_C and I_E.

AMPLIFIER CONFIGURATIONS

The three basic circuit amplifier configurations—common-emitter, common-base, and common-collector—as well as various FET circuits were discussed in Chapter 21. In this chapter, some of the more important considerations that go into the design of amplifier circuits will be explored.

Common-Emitter Amplifiers

Understanding the behavior of a diode that is forward-biased with DC and has an AC signal riding on that DC level will aid in the understanding of the transistor amplifier shown in Fig. 22-11. The voltage versus current curve for such a diode (the base-emitter junction of a silicon transistor) is illustrated in Fig. 22-12. The diode is forward-biased at 3mA DC and has an applied voltage of 0.7V DC. However, the AC signal riding on the DC level causes the current to swing from 4mA to 2mA at the peaks which run from 0.71V to 0.69V. Therefore, the resistance of the diode junction to the AC signal is

$$r_j = \frac{\Delta V}{\Delta I} = \frac{(0.71 - 0.69)V}{(4 - 2)mA} = 10\Omega$$

If the diode were biased at a higher DC current, the steeper slope of the curve would produce a lower ratio of $\Delta V/\Delta I$ or, in other words, a lower AC junction resistance r_j. Likewise, a lower DC bias current produces a higher junction resistance.

$$I_B = \frac{V_{CC} - V_{BE}}{R_B}$$

$$I_C = \beta I_B$$

$$V_C = V_{CC} - I_C R_C$$

$$r_j \approx \frac{30mV}{I_E} \approx \frac{30mV}{I_C}$$

$$A_v = \frac{-R_{c(line)}}{R_{e(line)}} = \frac{-R_c}{r_j}$$

$$Z_{in} = \beta r_j // R_B$$

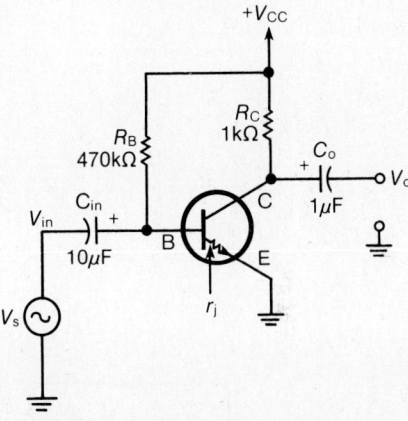

Fig. 22-11: Simple common-emitter amplifier.

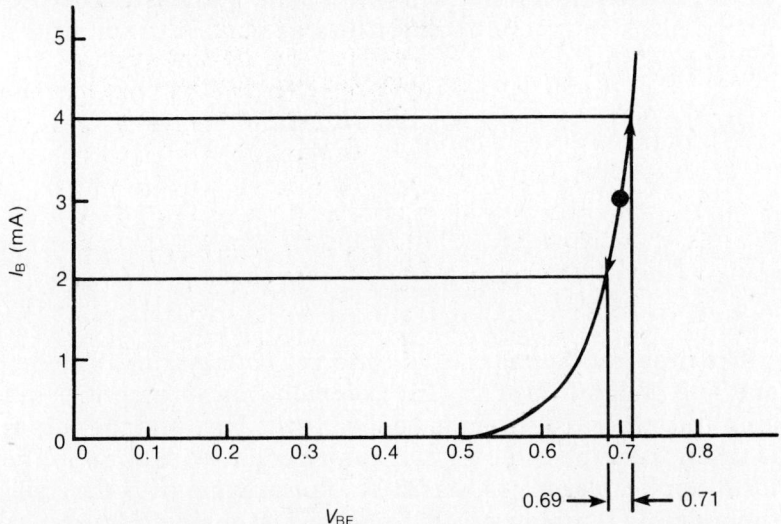

Fig. 22-12: Characteristic curve of base-emitter junction.

The AC junction resistance, often called *dynamic resistance,* can also be approximated, without the use of a characteristic curve, by using the following formula:

$$r_j = \frac{25\text{mV}}{I_E}$$

Static vs. Dynamic Resistance. Dynamic resistance is so called because of the active nature of the AC signal swings; conversely, the *static* or *DC resistance* at the bias point in Fig. 22-12 is simply

$$R_D = \frac{V}{I} = \frac{0.7\text{V}}{3\text{mA}} = 233\Omega$$

The fact that a device can possess a dynamic (AC) resistance that is so different from its static (DC) resistance is often difficult to comprehend. With nonlinear devices, however, it is common and it will soon seem quite natural to analyze circuits twice—first for the DC bias levels and then for the AC signals.

Amplifier Voltage Gain

Referring to Fig. 22-11, note that the AC signal (V_in) is connected directly to the base of the transistor, since C_in is chosen to have a reactance that is nearly equal to zero at the signal frequency. Thus, V_in appears across the base-emitter junction resistance r_j.

Since capacitor C_o offers essentially zero reactance and the voltage source V_{CC} is an AC ground, the output signal appears across the collector resistance (R_c). The DC emitter current I_E is equal to the DC collector current I_c (disregarding the negligible base current). The same is true for the AC current swings ($I_e \approx I_c$). Thus, R_c and r_j are somewhat like a voltage-divider with V_in across r_j, V_o across R_c, and the same current $I_e \approx I_c$ flowing through them (Fig. 22-13). If the resistance to AC from the collector to ground is referred to as $R_{c(\text{line})}$, and the resistance to AC from emitter to ground is referred to as $R_{e(\text{line})}$, the amplifier's voltage gain is

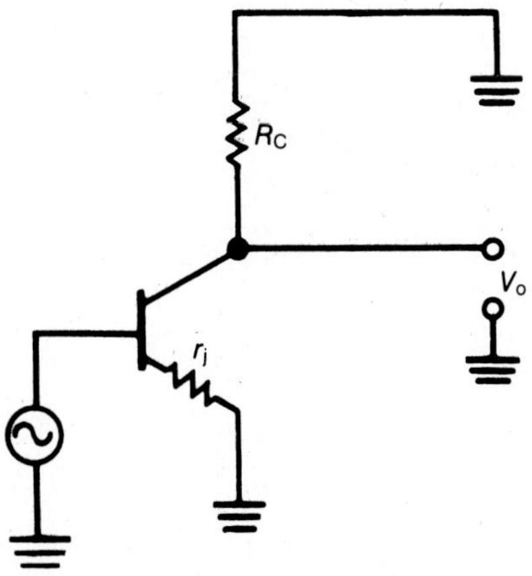

$$A_v = \frac{V_o}{V_\text{in}} = \frac{-I_c R_{c(\text{line})}}{-I_e R_{e(\text{line})}}$$

$$A_v = \frac{-R_{c(\text{line})}}{R_{e(\text{line})}}$$

$$A_v = \frac{-R_c}{r_j}$$

The first equation is the definition of *voltage gain.* The second is a general form for any *common-emitter* amplifier (an amplifier where the input is applied to the base and the output is taken from the collector). The third equation is the specific form, applicable only to Fig. 22-11. The minus sign in the equations simply represents a 180° *phase shift* or *signal inversion.*

Fig. 22-13: The basic transistor amplifier appears to AC as a voltage-divider with the input voltage across r_j and the output voltage across R_C.

538

That is, the positive half-cycle of the input signal produces the negative half-cycle of the output waveform, and the negative half-cycle of the input signal produces the positive half-cycle of the output waveform.

Amplifier Input Impedance

Input impedance (Z_{in}) is the term used to describe the ratio of the input voltage to the input current. Thus,

$$Z_{in} = \frac{V_{in}}{I_{in}}$$

The input impedance is the amount of opposition the AC signal (V_{in}) "sees" when it looks into the amplifier of Fig. 22-11. This opposition could be purely resistive, but it is common practice to call it Z_{in} rather than r_{in}.

It is important to know the input resistance of a circuit because low values of Z_{in} can cause loading of the source voltage. For example if r_s is 10kΩ and Z_{in} is 10kΩ resistive, half of V_s will be dropped across r_s and half will appear as V_{in} across the amplifier Z_{in}. However, if Z_{in} drops to 1kΩ, only one-eleventh of V_s will appear at V_{in}, the other ten-elevenths will be dropped across r_s.

The following facts must be known in order to determine Z_{in} for the circuit of Fig. 22-11:

- Capacitor C_{in} is chosen to have essentially zero reactance of the frequency of the signal V_s.
- Because the supply V_{CC} terminal is effectively at ground for AC, R_B is connected from the input to ground.
- The transistor's collector-base junction is reverse-biased and, therefore, presents an extremely high impedance (essentially an open circuit) to the AC signal at the base.
- The base-emitter junction of the transistor is a forward-biased diode with a dynamic resistance approximated by $25mV/I_E$.
- The emitter current is larger than the base current by a factor approximately equal to β. This is true for the AC signal current as well as for the DC bias current.

Base-Input Resistance r_b. It should be clear after examining Fig. 22-11 that there are two parallel paths for AC current from V_{in} to ground. The first is simply through R_B to V_{CC}, which is AC ground; the second is from the base of the transistor to the emitter through dynamic resistance r_j. However, because the base current is less than the base-emitter junction current I_j by an approximate factor of β, the dynamic resistance from the base to ground (r_b) is greater than r_j by a factor of β.

$$\Delta V_{be} = \Delta I_j r_j$$

$$\Delta I_b = \frac{\Delta I_j}{\beta}$$

$$r_b = \frac{\Delta V_{be}}{\Delta I_b} = \beta r_j$$

The input impedance of the amplifier is the combination of the bias resistor R_B and the base input resistance r_b:

$$Z_{in} = R_B \; // \; r_b$$
$$= R_B \; // \; \beta_{rj}$$
$$\approx R_B \; // \; \frac{(25\text{mV})(\beta)}{I_E}$$

Note that the double bars are a shorthand representation of the parallel resistance formula. Thus $R_1 // R_2$ is calculated as

$$\frac{1}{1/R_1 + 1/R_2}$$

Example: Determine the required value of R_B for $V_C = 5\text{V}$, and calculate the gain and input impedance for the circuit of Fig. 22-14.

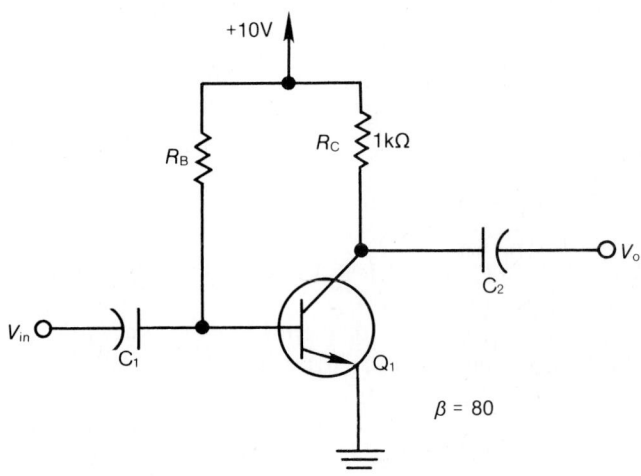

Fig. 22-14: Simple common-emitter amplifier.

Solution:

$$V_{RC} = V_{CC} - V_C = 10 - 5 = 5\text{V}$$
$$I_C = \frac{V_{RC}}{R_c} = \frac{5\text{V}}{1\text{k}\Omega} = 5\text{mA}$$
$$I_B = \frac{I_C}{\beta} = \frac{5\text{mA}}{80} = 0.0625\text{mA}$$
$$V_{RB} = V_{CC} - V_{BE} = 10 - 0.7 = 9.3\text{V}$$
$$R_B = \frac{V_{RB}}{I_B} = \frac{9.3\text{V}}{0.0625\text{mA}} = 148.8\text{k}\Omega$$
$$r_j = \frac{25\text{mV}}{I_C} = \frac{0.025}{0.005} = 5\Omega$$
$$r_b = \beta_{rj} = (80)\,(5) = 400\Omega$$
$$Z_{in} = r_b // R_B = 400\Omega // 148.8\text{k}\Omega = 399\Omega$$
$$A_v = \frac{-R_c}{r_j} = \frac{-1,000}{5} = -200$$

540

Amplifier Output Impedance

The output of an amplifier is generally expected to deliver current to some kind of load. This load, whether it be a loudspeaker or the input impedance to the next amplifier or even a servomotor, is usually represented as a resistance. Adding a load resistance to an amplifier causes its output voltage to be loaded down, lowering the voltage gain of the amplifier.

By knowing the *output impedance* (Z_o) of an amplifier, we can tell how much the output voltage and gain will drop when a given load is connected. For example, an amplifier having an output impedance of $1\mathrm{k}\Omega$ and a voltage gain of 100 will have a voltage gain of 50 if a load resistance of $1\mathrm{k}\Omega$ is connected.

The output impedance of an amplifier is the opposition seen by the load looking back into the output terminals of the amplifier. Z_o does not include the load resistor, R_L. For example, the output impedance of the simple circuit of Fig. 22-14 is approximately equal to R_C. The only other path from V_o to ground is through the collector of the transistor, which is a very high dynamic resistance.

However, in some circuits, where R_C is very high or where the transistor is a power transistor, the dynamic collector resistance (r_c) may be low enough to become important. In circuits such as these,

$$Z_o = R_C // r_c$$

The dynamic collector resistance for any transistor can be determined by using the slope of the characteristic-curve lines as shown in Fig. 22-15. Note that as I_C becomes higher, r_c becomes lower.

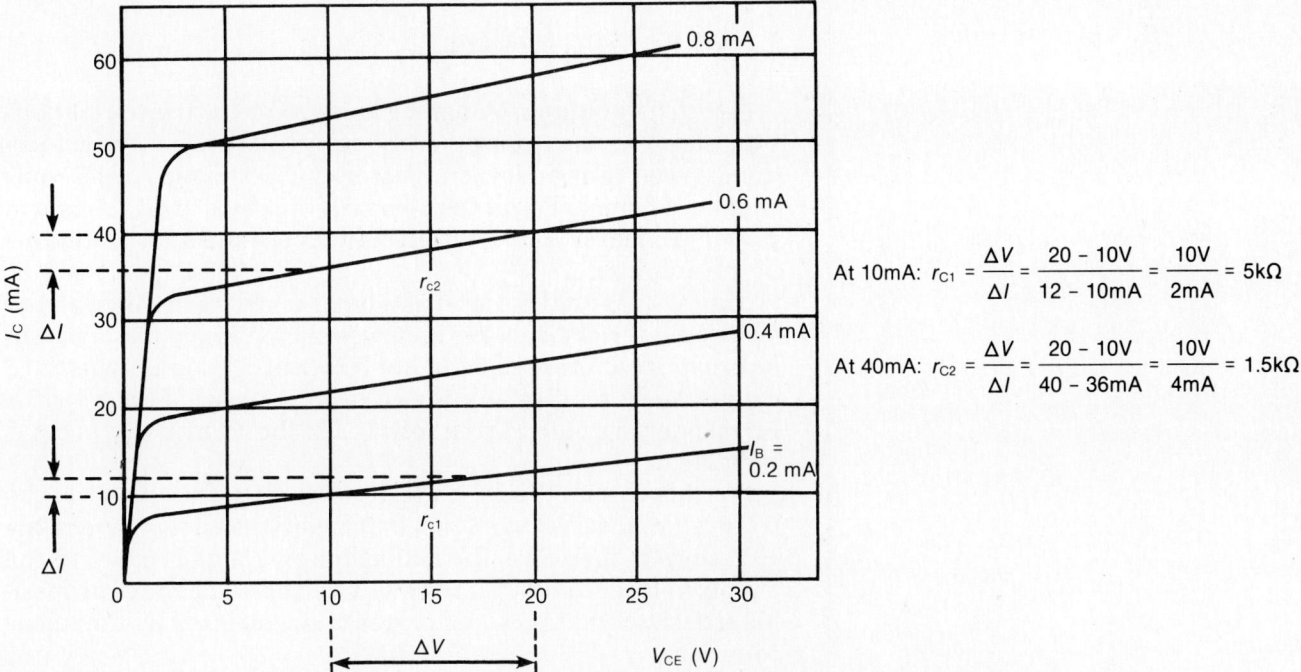

At 10mA: $r_{C1} = \dfrac{\Delta V}{\Delta I} = \dfrac{20 - 10\mathrm{V}}{12 - 10\mathrm{mA}} = \dfrac{10\mathrm{V}}{2\mathrm{mA}} = 5\mathrm{k}\Omega$

At 40mA: $r_{C2} = \dfrac{\Delta V}{\Delta I} = \dfrac{20 - 10\mathrm{V}}{40 - 36\mathrm{mA}} = \dfrac{10\mathrm{V}}{4\mathrm{mA}} = 1.5\mathrm{k}\Omega$

Fig. 22-15: A steeper slope of the characteristic curve line indicates a lower dynamic collector resistance.

541

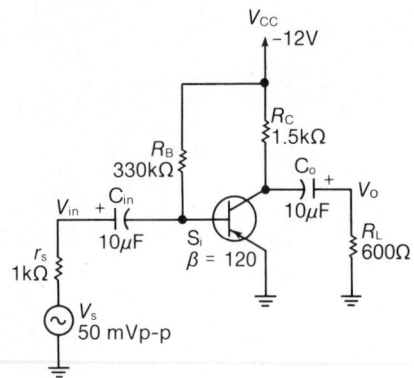

Fig. 22-16: R_L appears in parallel with R_C for AC, lowering the amplifier gain.

Amplifier With Load Resistance. A simple common-emitter amplifier with a load resistor is illustrated in Fig. 22-16. The collector signal current flows through R_C and R_L, which are in parallel for AC (capacitor C_o connects the bottom of R_C to the top of R_L, and the power supply output capacitor connects the top of R_C to ground). Thus, the voltage gain formula becomes

$$A_v = \frac{-R_{c(line)}}{R_{c(line)}} = \frac{R_C // R_L}{r_j}$$

Example: Determine V_C, Z_{in}, Z_o, and V_o for the circuit of Fig. 22-16.

$$I_B = \frac{V_{CC} - V_{BE}}{R_B} = \frac{12 - 0.6}{330k\Omega} = 34.5\mu A$$

$$I_C = \beta I_B = (120)\ (34.5\mu A) = 4.14mA$$

$$V_{RC} = I_C R_C = (4.14mA)\ (1.5k\Omega) = 6.21V$$

$$V_C = V_{CC} - V_{RC} = 12 - 6.21 = 5.79V$$

$$r_j = \frac{25mV}{4.14mA} = 6.04\Omega$$

$$Z_{in} = R_B // r_b = R_B // \beta r_j$$
$$= (300k\Omega) // (120)\ (6.04\Omega) = 723\Omega$$

$$V_{in} = V_s \frac{Z_{in}}{r_s + Z_{in}} = (50mV) \frac{723}{1,000 + 723} = 21mV_{p-p}$$

$$Z_o = R_C = 1.5k\Omega$$

$$A_v = \frac{R_C // R_L}{r_j} = \frac{1,500 // 600}{6.04} = 70.96$$

$$V_o = V_{in} A_v = (21mV)\ (70.96) = 1.49V_{p-p}$$

Load Line Analysis

A transistor amplifier can be analyzed by using the transistor's characteristic curves. Although graphical analysis is seldomly used in practice (transistor characteristic curves vary radically from one unit to the next), this type of analysis will provide a clearer visualization of the operation and limitations of the amplifier.

Figure 22-14 shall be used as a basis for this graphical analysis. Using the characteristic curves shown in Fig. 22-17, it can be proven that the transistor has a beta of 80. It is known that β equals I_C/I_B; by dividing the collector current of 8mA by the corresponding base current of 100μA, the correct value of β is indicated.

For this circuit, the DC collector current path consists of the transistor in series with the 1kΩ collector-resistor across the 10V supply; therefore, the transistor's voltage is equal to the supply voltage minus the resistor's voltage. The range of possible transistor voltages and currents is expressed by the linear equation:

$$I_C = \frac{V_{RC}}{R_C} = \frac{V_{CC} - V_C}{R_C}$$

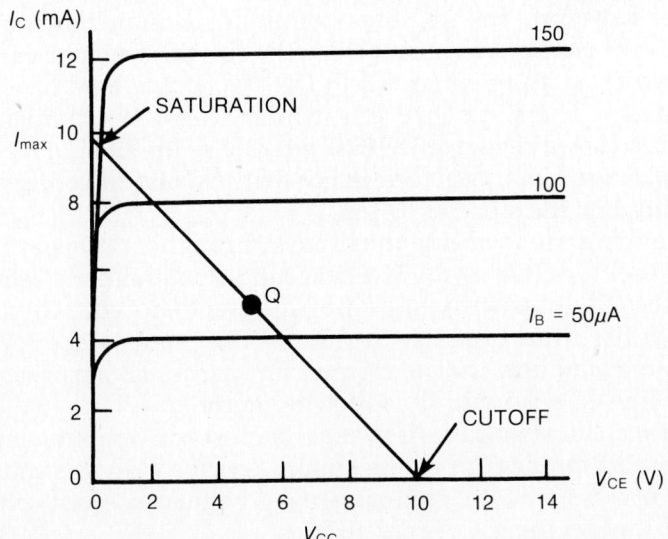

Fig. 22-17: Transistor characteristic curves with DC load line and Q-point.

All possible combinations of voltage and current for the transistor are indicated by a straight line on the V vs. I graph. Any other voltage and current combinations are impossible to obtain as long as the circuit has V_{CC} = 10V and R_C = 1kΩ.

Static Load Line. The DC or *static load line* described above can be readily positioned by its end points, which represent the extreme conduction conditions of the transistor (i.e., fully on which is saturation or fully off which is cutoff).

Since no voltage drop will be developed across the collector-resistor when zero collector current flows, the transistor's collector voltage will equal the supply voltage when I_C is zero. Thus, V_{CC} equals the cutoff point, as shown in Fig. 22-17.

When saturation occurs, only a few tenths of a volt appear from the collector to the emitter of the transistor. Virtually the entire supply voltage appears across R_C, so the collector current equals:

$$I_{C(sat)} = \frac{V_{CC}}{R_C}$$

$I_{C(sat)}$ determines the second end point of the static load line (at this time, collector voltage is very nearly zero). All possible circuit conditions lie on the straight line between saturation and cutoff.

Changes in transistor base current cause movement up and down the load line. As higher input voltages cause larger base currents, the collector output voltages will decrease. Hence, the amplifier gives an output signal which is inverted (180° phase shift) from the input signal.

Dynamic Load Line. Load line analysis can provide a better understanding of the effects of adding a load resistance to an amplifier. If a 1kΩ load is added to the circuit in Fig. 22-14, the resistance seen by an AC signal looking out of the collector is now $R_C // R_L$, or 500Ω, instead of 1,000Ω. This lower dynamic resistance is represented by a load line with a more vertical

543

slope. However, the DC bias conditions remain unchanged since the load resistor is coupled to the circuit through a capacitor, which is an open circuit to DC. Thus, two load lines are necessary. The static load line, which is used to establish the Q-point (the quiescent or DC bias point), has the slope of R_C. The *dynamic load line,* used to determine the AC output voltage and current, has the slope of $R_C//R_L$.

The dynamic load line must pass through the Q-point and have the slope of $R_C//R_L$. The first condition is satisfied easily. The second is met with the help of another line, called the slope line. Other than to assist in finding the proper slope for the dynamic load line, this additional line serves no purpose. The slope line is laid out in the following manner:

1. Select a test voltage (it is usually most convenient to make it one-half to one-fifth of the supply voltage V_{CC}) and mark a point at this point and zero current on the curves. A test voltage of 2V was chosen for Fig. 22-18.

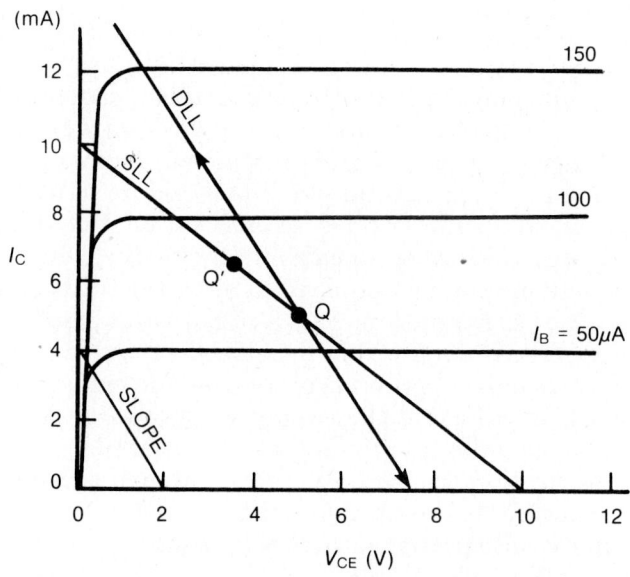

Fig. 22-18: **Transistor curves with DC (static) load line, slope line, and AC (dynamic) load line.**

2. Determine the parallel combination of R_C and R_L. In this example, this combination is $1k\Omega//1k\Omega$, or 500Ω.

3. Determine the amount of current the test voltage would push through this parallel resistance.

$$I = \frac{V}{R} = \frac{2}{500} = 4mA$$

Mark this current on the curves at zero voltage.

4. The line between the zero voltage and zero current points has the slope of $R_C//R_L$.

Using a triangle and straightedge, a line with the same slope can be drawn through the Q-point. The operating point is now shifted up and down along this dynamic load line when an AC signal is applied. Notice that the lower total resistance caused by adding R_L causes the dynamic load line to be steeper than the static load line. This results in less output voltage swing and lower voltage gain. Lower values of R_C or R_L will result in even steeper load lines and even lower voltage gain.

By observing the static load line, it becomes apparent that the collector should be biased midway between its saturation and cutoff voltage when no load resistance is used. This Q-point will allow maximum output swing up and down the load line. For example, if the transistor were biased at $V_C = 8V$, the voltage swing would be limited to 2V above the 8V Q-point, since 10V is the maximum possible peak output voltage. Thus, the total output voltage is limited to $4V_{p-p}$ (2V above and below the Q-point). Any attempt to produce a larger output waveform would result in distortion (the clipping off of the waveform peaks). Distortion would also result if the bias point were set too low. Therefore, the maximum swing up and down possible without distortion occurs with the bias point set in the center of the load line.

However, the optimum bias point for a circuit changes when a load resistance is added. Although the Q-point in Fig. 22-18 is midway between saturation and cutoff, a close examination will show that the voltage swing along the dynamic load line meets the zero current clipping point well before the zero voltage point is reached by an equal swing in the opposite direction.

A larger peak-to-peak swing will be possible, in Fig. 22-18, without waveform clipping, if a new dynamic load line of the same slope is drawn through a new bias point Q'. The optimum Q-point is best found experimentally by lowering the amount of circuit base resistance to raise the collector current while increasing the input signal level to observe waveform clipping. The output waveform will begin to show clipping at the top and bottom simultaneously as the input signal is increased, when the optimum Q-point is finally reached.

Common-Emitter Amplifier with Voltage-Divider Bias

The fundamental circuit of the preceding sections is not used in commercial applications because it allows the transistor's collector current to vary directly with the transistor beta. Beta may vary from 40 to 200 (a factor of 5) for typical small signal transistors; therefore, it would be virtually impossible to reliably position the Q-point near the center of the load line. A certain transistor may have such a high beta that it is biased into saturation, while another may have such a low beta that it is biased into cutoff. In either case, large output signal swings will be clipped off. Also, as beta changes, the emitter-current changes proportionally and the junction resistance (r_j) changes inversely. Thus the gain and input impedance are also affected by varying beta.

A circuit which avoids all of these problems is illustrated in Fig. 22-19. Note that this circuit utilizes voltage-divider bias. This type of bias tends to keep I_C constant, regardless of beta variations.

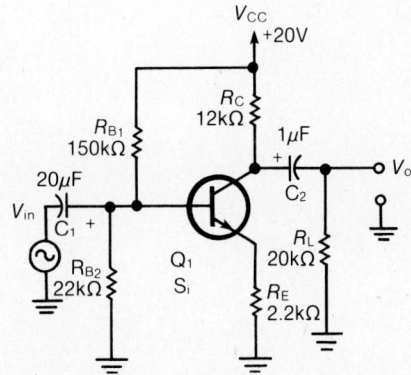

Fig. 22-19: Voltage-divider bias in a common-emitter amplifier provides bias stability, moderate voltage gain, and high input impedance.

545

As described earlier, R_{B1} and R_{B2} form a voltage-divider which keeps V_B constant.

$$V_B = V_{CC} \frac{R_{B2}}{R_{B1} + R_{B2}}$$

With the voltage of the transistor base fixed, V_E is fixed at a value of 0.7V lower than V_B for a silicon transistor and at a value of 0.3V lower than V_B for a germanium transistor.

$$I_E = \frac{V_E}{R_E}$$

Since collector current is nearly identical to emitter current, I_C and the voltage across R_C are fixed.

$$V_{RC} = I_C R_C \approx I_E R_C$$

The supply voltage minus the voltage drop across R_C is equal to the collector voltage.

$$V_C = V_{CC} - V_{RC}$$

Bias Stability

The proper operation of the circuit of Fig. 22-19 depends on two factors: (1) negligible changes in the base-to-emitter voltage and (2) negligible base current.

Changes in V_{BE} cause changes in bias current I_E because even with a fixed voltage at the base (V_B), V_E changes when V_{BE} changes. V_{BE} is typically 0.7V for a silicon transistor, but it may easily vary by ± 0.1V with temperature changes or from one transistor to the next. To swamp out the changes in the base-to-emitter voltage V_{BE}, V_{RE} must be much larger than the ± 0.1V changes in V_{BE}. Making $V_{RE} = 2.0$V would limit bias-current variations due to changes in V_{BE} to about $\pm 0.1/2.0$ (or $\pm 5\%$). This is the recommended value of minimum V_{RE} for general purpose low-power amplifiers.

Changes in I_B will occur in the circuit of Fig. 22-19 in order to keep I_E at a constant value as beta changes. R_{B1}-R_{B2} (the voltage-divider) must be capable of maintaining a fixed voltage regardless of these variations in transistor base current; therefore, I_B must be much lower than the main-line current through R_{B1} and R_{B2}.

Assume that I_E equals 10mA and $\beta = 100$. I_B must then be 0.1mA. If β drops to 50 (this might happen as a result of transistor replacement), then I_B would have to increase to 0.2mA. To insure that this current would not load the voltage-divider, the main-line current would have to be designed to be many times greater than 0.2mA (about 2mA, if β were 50).

An easy-to-apply rule of thumb is to make R_{B2} not larger than $10R_E$. If β_{min} is 40, this will limit the range of possible emitter-bias currents to a maximum value as given by the previously stated formulas for V_B, V_E, and I_E (at very high beta) and a minimum that is 20% less than this value. When it is required to limit the loading to 10%, the rule of thumb becomes $R_{B2} \leq 5R_E$.

Example: Determine the collector voltage for the circuit of Fig. 22-19 using the component values given and assuming a very high beta. Estimate the actual V_C if β drops to a minimum of 40.

Solution:

$$V_B \approx V_{CC} \, \frac{R_{B2}}{R_{B1} + R_{B2}} = (20) \, \frac{22}{150 + 22} = 2.56V$$

$$V_E = V_B - V_{BE} = 2.56 - 0.6 = 1.96V$$

$$I_E = \frac{V_E}{R_E} = \frac{1.96V}{2.2k\Omega} = 0.89mA$$

$$V_{RC} = I_C R_C \approx I_E R_C = (0.89mA)(12k\Omega) = 10.7V$$

$$V_C = V_{CC} - V_{RC} = 20 - 10.7 = 9.3V$$

If β were as low as 40, we would expect I_E and, consequently, V_{RC} to drop by about 20% since $R_{B1} = 10R_E$.

$$V_{RC\min\beta} \approx 80\% \times 10.7V = 8.6V$$

$$V_{C\min\beta} \approx V_{CC} - V_{RC} = 20 - 8.6 = 11.4V$$

AC Behavior of the Stabilized Amplifier

The AC behavior of the amplifier of Fig. 22-19 can be predicted by using the same methods employed for the fundamental circuit of Fig. 22-11.

AC Voltage Gain. A_v is the ratio of the resistance through which the AC collector current flows divided by the resistance through which the AC emitter current flows. V_{out} is taken across R_C and R_L, which appear in parallel with the AC signal. V_{in} is applied across the emitter junction resistance (r_j) of the transistor in series with R_E. Thus, for Fig. 22-19,

$$A_v = \frac{-R_{c(line)}}{R_{e(line)}} = \frac{-R_C // R_L}{r_j + R_E}$$

Input Impedance. Z_{in} is the parallel combination of three resistance paths to ground: R_{B1} (the source is AC ground), R_{B2}, and the base input resistance (r_b) of the transistor. As it was before, r_b is the product of the emitter-line resistance multiplied by beta, because I_b is less than I_e by a factor of beta. Thus, for Fig. 22-19,

$$Z_{in} = R_{B1} // R_{B2} // r_b$$

where

$$r_b = \beta(r_j + R_E)$$

and

$$r_j \approx \frac{25mV}{I_E}$$

Output Impedance. Z_o is simply R_C in parallel with the collector resistance (r_c) of the transistor, although r_c is usually high enough to be disregarded with R_C.

$$Z_o = R_C // r_c$$

Example: Referring to Fig. 22-19, calculate A_v, Z_{in}, and Z_o. Use $I_C = 0.73mA$ from the previous example and $\beta = 40$.

Solution:

$$r_j \approx \frac{25mV}{I_E} = \frac{25mV}{0.73mA} = 34.25\Omega$$

$$A_v = \frac{-R_C // R_L}{r_j + R_E} = \frac{-12 // 20}{0.03425 + 2.2} = -3.36$$

$$Z_{in} = R_{B1} // R_{B2} // \beta(r_j + R_E)$$
$$= 150 // 22 // 40(0.03425 + 2.2) = 15.8k\Omega$$

$$Z_o = R_C = 12k\Omega$$

Note that the voltage gain of this circuit is much lower than the voltage gain of the unstabilized circuit and that the input impedance is quite a bit higher.

Emitter Bypassing

Bias stability and high AC signal gain can be achieved together in a common-emitter amplifier by connecting a bypass capacitor across R_E as shown in Fig. 22-20. C_E appears as a short circuit to the AC signal, leaving r_j alone in the emitter line. Therefore, the voltage gain and input impedance for the circuit of Fig. 22-20 are expressed mathematically as

$$A_v = \frac{R_C // R_L}{r_j}$$
$$Z_{in} = R_{B1} // R_{B2} // r_b$$

where

$$r_b = \beta r_j$$

Observe that these formulas are the same as those that apply to the unstabilized circuit of Fig. 22-11, with the single addition of R_{B2}. The voltage gain is high, and the input impedance is relatively low.

Amplifier Linearity. *Amplifier linearity* (also called *fidelity*) is a desirable characteristic wherein the output-signal waveshape is an exact reproduction of the input-signal waveshape. Linearity is generally specified in terms of *percent harmonic distortion*. In the circuit of Fig. 22-20, the output voltage (V_o) is directly dependent on the circuit voltage gain (A_v), which varies inversely with the junction resistance (r_j), which varies inversely with the emitter current (i_E). If the amplifier handles a large signal, i_E may change considerably from the bias point to the peaks of the signal swing. Thus r_j and A_v will be affected by the signal. The net effect will be to increase A_v on the positive input-signal swing. This corresponds to the negative output-signal swing which will be overemphasized in the output-signal waveform as shown in Fig. 22-21.

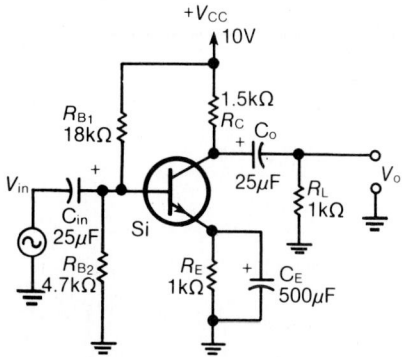

Fig. 22-20: Bypassing the emitter-resistor in a stabilized common-emitter amplifier provides maximum voltage gain but lower Z_{in}.

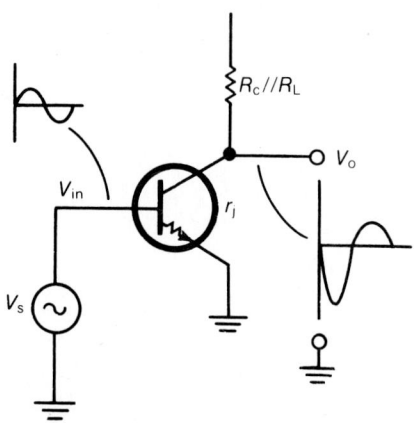

Fig. 22-21: When the AC collector-current swings are nearly as large as the DC collector-bias current, r_j changes with signal swings and the V_o is distorted.

548

Nonlinearity is not a problem unless the AC signal current becomes an appreciable fraction of the DC bias current. This can be recognized when $V_{o(AC)}$ becomes an appreciable fraction of V_{RC}.

Partial Bypassing. To increase the input impedance and improve the large-signal linearity of the amplifier of Fig. 22-20, it is common to connect an unbypassed resistor into the emitter leg. This unbypassed resistor (R_{E1}) is called a *swamping resistor*. The DC bias current sees $R_{E1} + R_{E2}$, and the voltage drop across them is made to be 2V or more for bias stability. The AC signal sees $r_j + R_{E1}$, and these determine the voltage gain and input impedance. For the circuit of Fig. 22-22,

$$A_v = \frac{R_C//R_L}{r_j + R_{E1}}$$

$$Z_{in} = R_{B1}//R_{B2}//r_b$$

$$r_b = \beta(r_j + R_{E1})$$

Example: Calculate V_C, Z_{in}, Z_o, and A_v for the circuit of Fig. 22-23.

Solution:

$$\frac{V_B}{6.8k\Omega} = \frac{-9V}{15k\Omega + 6.8k\Omega}$$

$$V_B = 2.8V$$

$$V_E = V_B - V_{BE} = 2.8 - 0.3 = 2.5V$$

$$\frac{V_{RC}}{R_C} = \frac{V_{RE}}{R_E}$$

$$\frac{V_{RC}}{3.9k\Omega} = \frac{2.5V}{2.7k\Omega + 0.1k\Omega}$$

$$V_{RC} = 3.5V$$

$$V_C = V_{CC} - V_{RC} = 9 - 3.5 = 5.5V$$

$$I_C = \frac{V_{RC}}{R_C} = \frac{3.5V}{3.9k\Omega} = 0.89mA$$

$$r_j = \frac{25mV}{I_E} = \frac{0.025}{0.00089} = 28\Omega$$

$$r_b = \beta(r_j + R_{E1}) = 35(28 + 100) = 4.5k\Omega$$

$$Z_{in} = R_{B1}//R_{B2}//r_b = 15k\Omega//6.8k\Omega//4.5k\Omega = 2.3k\Omega$$

$$Z_o = 3.9k\Omega$$

$$R_{c(line)} = R_C//R_L = 3.9k\Omega//5k\Omega = 2.19k\Omega$$

$$A_v = \frac{R_{c(line)}}{r_j + R_{E1}} = \frac{2.19k\Omega}{28 + 100} = 17.1$$

Emitter-Follower Amplifier

The characteristically low input impedance of a bipolar transistor frequently presents problems of matching a high-impedance signal source to an amplifier input. For example, if

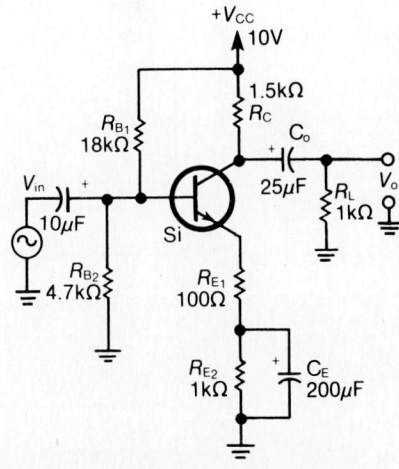

Fig. 22-22: Bypassing only a part of the total R_E permits a compromise between high A_v and high Z_{in}.

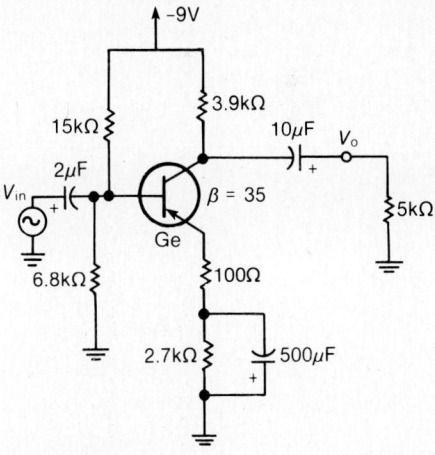

Fig. 22-23.

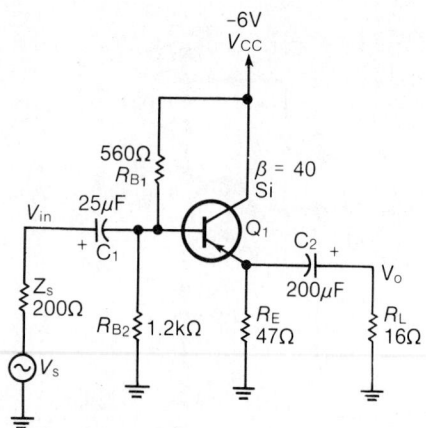

Fig. 22-24: Emitter-follower amplifier.

the circuit shown in Fig. 22-23 were to be driven with a typical high-impedance microphone having an impedance of 50kΩ and an output level of 10mV, only approximately 0.4V would appear across the amplifier's 2.3kΩ input impedance, while the other 9.6mV would be lost across the internal impedance of the microphone. A step-down transformer could be used to provide a proper impedance match, but transformers are heavy and expensive and have limited frequency response.

Figure 22-24 illustrates a circuit that solves the problem with none of the above disadvantages. Note that the input is taken from the base and the output is taken from the emitter, leaving the collector as the common element between them; this is a *common-collector* amplifier. However, because the emitter voltage tends to follow the base-signal waveform (the only difference between them being the 0.7V or 0.3V base-emitter junction drop), this circuit is more commonly called an *emitter-follower*. In an emitter-follower amplifier the emitter (output) current is greater than the base (input) current by an approximate factor of beta. This allows a high impedance source to deliver a relatively low input current to the base of the amplifier and to deliver a much larger current to the load, which may have a relatively low impedance.

Voltage Gain. In contrast to the common-emitter amplifier, the voltage gain of an emitter-follower is slightly less than unity, and there is no phase inversion of the input signal waveform. If the opposition seen looking out of the emitter, $R_{e(line)}$, is so small that the emitter junction resistance (r_j) drops an appreciable portion of the voltage, the voltage gain (A_v) will drop appreciably below unity.

$$A_v = \frac{R_E // R_L}{r_j + R_E // R_L}$$

Input Impedance. Since the base current is smaller than the emitter current by an approximate factor of beta, the resistance seen looking into the base is larger than the emitter line resistance by the same factor:

$$r_b = \beta(r_j + R_{e(line)})$$

$R_{e(line)}$ is usually the parallel combination of R_E and R_L, but to determine the total impedance, the shunting effect of R_{B1} and R_{B2} must also be accounted for:

$$Z_{in} = R_{B1} // R_{B2} // r_b$$

where

$$r_b = \beta(r_j + R_E // R_L)$$

Output Impedance. The output impedance of an emitter-follower primarily depends on the beta of the transistor and the impedance of the signal source. R_B, R_E, and r_j also contribute to the output impedance, but to a lesser extent. Therefore, the emitter-follower is often thought of as an impedance transforming device whose output impedance is approximately Z_s/β. The formula for the total output impedance of an emitter-follower is

550

$$Z_o = R_E // (r_j + \frac{R_{B1} // R_{B2} // Z_s}{\beta})$$

Example: Calculate V_E, A_v, Z_{in}, and Z_o for the circuit of Fig. 22-24.

Solution:

$$V_B = V_{CC} \frac{R_{B2}}{R_{B1} + R_{B2}} = (6) \frac{1,200}{560 + 1,200} = 4.09V$$

$$V_E = V_B - V_{BE} = 4.09 - 0.7 = 3.39V$$

$$I_E = \frac{V_E}{R_E} = \frac{3.39}{47} = 72.13mA$$

$$r_j = \frac{0.025}{0.07213} = 0.35\Omega$$

$$A_v = \frac{R_E // R_L}{r_j + R_E // R_L} = \frac{47 // 16}{0.35 + 47 // 16} = 0.97$$

$$Z_{in} = R_{B1} // R_{B2} // r_b$$

$$r_b = \beta(r_j + R_E // R_L) = 40(0.35 + 47 // 16) = 491.30\Omega$$

$$Z_{in} = 560 // 1,200 // 491.30 = 214.84\Omega$$

$$Z_o = R_E // (r_j + \frac{R_{B1} // R_{B2} // Z_s}{\beta})$$

$$= 47 // (0.35 + \frac{560 // 1,200 // 200}{40}) = 3.37\Omega$$

The Common-Source FET Amplifier

The circuit of Fig. 22-25 shows a basic FET (field effect transistor) audio amplifier. This circuit utilizes an N-channel FET as the amplifying device. It should be recalled (Chapter 21) that the FET is a high gain device which is quite similar in characteristics to the electron tube pentode. The circuit shown is a class A audio amplifier utilizing the common-source configuration. Self-bias is developed by the source-bias circuit consisting of R_s and C_2. C_1 is the input coupling capacitor, R_G is the

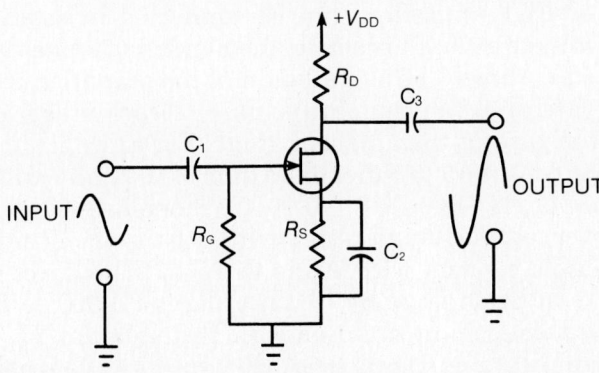

Fig. 22-25: FET audio amplifier.

input gate resistance used to develop the input signal, C_3 is the output coupling capacitor, and R_D is the drain load resistor. In Fig. 22-25, as the gate becomes less negative with respect to the source, source-drain current will increase. As the gate becomes more negative with respect to the source, source-drain current will decrease.

During the positive alternation of the input signal, the gate becomes less negative than its quiescent voltage, the depletion region decreases, and drain current increases. The increase in drain current is proportional to the change in junction bias. As drain current increases, the voltage drop across R_D increases, causing a corresponding decrease in drain voltage. Output voltage is consequently decreasing or passing through its negative half-cycle. As shown in Fig. 22-25, the common-source configuration provides a 180° phase inversion.

During the negative alternation of the input signal, the gate voltage becomes more negative than its quiescent value and the depletion region increases, causing drain current to decrease. The decreased drain current will drop less voltage across R_D, resulting in an increased drain voltage. Therefore, the output voltage is going through its positive alternation.

As mentioned previously, the FET is a high gain device, with gain values of 400 to 1,000 being common for practical amplifier circuits. FETs have a high input impedance which immediately makes them acceptable for use with high impedance input devices. Their output impedances are rated as medium to high and, therefore, are applicable for driving high-impedance earphones or high power amplifiers. FET circuits, as a rule, are less complex than standard transistor circuits as they require fewer circuit components. They have low noise ratings for the audio frequency range and produce minimal leakage currents.

MULTISTAGE AMPLIFIERS

When stable voltage gains greater than about 20 are required and it is not practical to bypass the emitter (or source) resistor of a single stage, two or more transistor amplifier stages can be used in *cascade* (where the output of one transistor is fed to the input of a second transistor). In theory, any number of two-junction or FET amplifiers can be connected in cascade to increase voltage gain. In practice, the number of stages is usually limited to three. The overall gain of the amplifier is equal (approximately) to the cumulative gain of each stage, multiplied by the gain of the adjacent stage. For example, if each stage of a three-stage amplifier has a gain of 10, the overall gain is (approximately) 1,000 ($10 \times 10 \times 10$). Since it is possible to design a very stable single stage with a gain of 10 and adequately stable circuits with gains of 15 to 20, a three-stage amplifier can provide gains in the range of 1,000 to 8,000. Generally, this is more than enough voltage gain for most practical applications. Using the 8,000 figure, a $1 \mu V$ signal (say from a low-voltage transducer or delicate electronic device) can be raised to the 8mV range while maintaining stability in the presence of temperature and power-supply variations.

The operation of coupling networks must be considered in any amplifier circuit. Even a single-stage audio amplifier must be coupled to the input and output devices. If the circuit is multistage, there must be *interstage coupling*. Amplifier circuits are often classified according to coupling design. For example, the four basic coupling techniques are RC (capacitor), inductive, direct, and transformer. In Chapters 13, 15, 16, and 17, the basic principles of coupling networks were discussed. In this chapter, the techniques of how these principles apply to multistage amplifier circuits are covered.

Bias Network Frequency Response. In each of the two preceding single-stage audio amplifiers, self-bias was utilized to develop class A operating bias. The amount of bias developed is determined by the quiescent (no-signal) current flow through the amplifying device. The bypass capacitor is in the circuit to prevent fluctuations in the bias voltage which would otherwise occur as the current varied. These unwanted fluctuations, if present, would cause degeneration and a loss of amplifier gain.

At the higher audio frequencies, it is relatively easy for a capacitor to bypass the signal variations to ground. However, at the lower frequencies, if the capacitor is improperly chosen, bias voltage variations will be present and degeneration will occur.

In order to prevent unwanted bias variations, degeneration, and a loss of gain at the lower audio frequencies, the capacitor must bypass even the variations of the lowest audio frequency to be encountered in the circuit. A rule of thumb is: The capacitive reactance of the capacitor must be one-tenth or less than the self-bias resistance value at the lowest audio frequency (20Hz).

RC Coupling

By far, the most common method of coupling two or more stages of audio amplification is RC coupling, or *capacitor coupling* as it is sometimes called. The RC network (shown within the dashed line in Fig. 22-26) used between the stages consists of a load resistor (R_1) for the first stage, a DC blocking capacitor (C_1), and a DC return resistor (R_2) for the input element of the second stage.

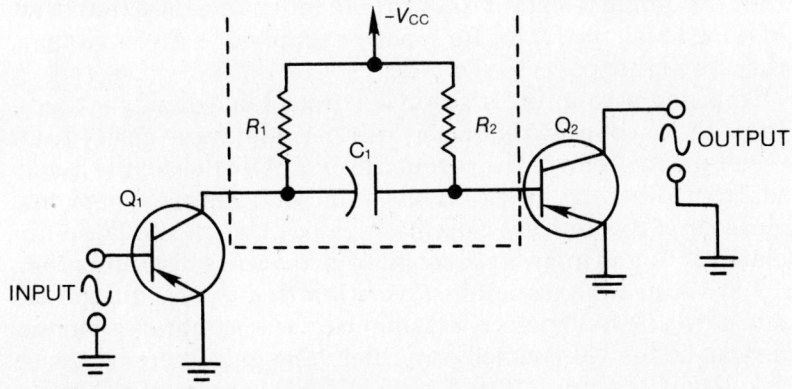

Fig. 22-26: An RC-coupling network.

Because of the dissipation of DC power in the load resistors, the efficiency (ratio of AC power out to DC power delivered to the stage) of the RC coupled amplifier is low.

The DC blocking capacitor prevents the DC voltage component of the output of the first stage from appearing on the input terminal of the second stage. To prevent a large signal voltage drop across the DC blocking capacitor, the reactance of the capacitor must be small compared to the input resistance of the following stage with which it is in series. Since the reactance must be low, the value of capacitance must be high. However, because of the low voltages used in transistor circuits, the physical size of the capacitor may be kept small. Physically larger capacitors are required in electron tube circuits due to the higher voltages. The resistance of the DC return resistor is usually much larger than the input resistance of the second stage. The upper limit of the value of this resistor is dictated by the DC bias considerations in the case of the transistor circuit and by the shunting capacitance in the case of the electron tube circuit.

The frequency response of the RC coupled audio amplifier is limited by the same factors that limited frequency response in other RC coupling circuits. In other words, the very low frequencies are attenuated by the coupling capacitor whose reactance increases with low frequencies. The high-frequency response of the amplifier is limited by the shunting effect of the output capacitance of the first stage and the input capacitance of the second stage. Circuit techniques (such as peaking) that can be used to extend the low frequency and high frequency response of RC coupled transistors and electron tube amplifiers are covered elsewhere in this text.

RC coupling is used extensively in audio amplifiers because of good frequency response, economy of circuit parts, and the small physical size which can be achieved with this method of coupling.

Inductive (Impedance) Coupling

With inductive or impedance coupling (Fig. 22-27), the load resistors R_{C1} and R_{C2} are replaced by inductors L_1 and L_2. The advantage of impedance coupling over resistance coupling is that the ohmic resistance of the load inductor is less than that of a load resistor. Thus, for a power supply of a given voltage, there is higher collector voltage.

Impedance coupling also suffers from some disadvantages. Impedance coupling is larger, heavier, and more costly than resistance coupling. To prevent the magnetic field of the inductor from affecting the signal, the inductor turns are wound on a closed iron core and are usually shielded extensively. The main disadvantage of impedance coupling is frequency discrimination.

With impedance coupling at very low frequencies, the gain is low owing to the capacitive reactance of the coupling capacitor, just as in the RC-coupled amplifier. The gain increases with frequency, leveling off at the middle frequencies of the audio range. (However, the frequency spread of this level portion is

554

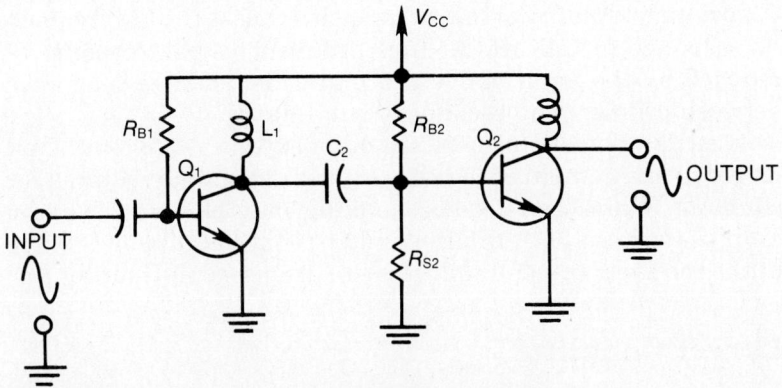

Fig. 22-27: A circuit with inductive coupling.

not as great as for the RC amplifier.) With impedance coupling at very high frequencies, the gain drops because of the increased reactance. Impedance coupling is rarely, if ever, used at frequencies well above the audio range.

Transformer Coupling

Interstage coupling of audio amplifiers by means of transformers is shown in Fig. 22-28. The primary winding of transformer T_1 (including the AC reflected load from the secondary winding) is the load impedance of the first stage. The secondary winding of transformer T_1 introduces the AC signal to the input of the second amplifier and also acts as the input DC return path.

Because there is no load resistance to dissipate power, the power efficiency of the transformer coupled amplifier approaches the theoretical maximum of 50%. For this reason, the transformer coupled amplifier is used extensively in portable equipment where battery power is used. Transformers facilitate the matching of the load to the output of the amplifier and the source to the amplifier to bring about maximum available power gain in a given stage.

The frequency response of a transformer coupled stage is not as good as that of the RC coupled stage. The shunt reactance of

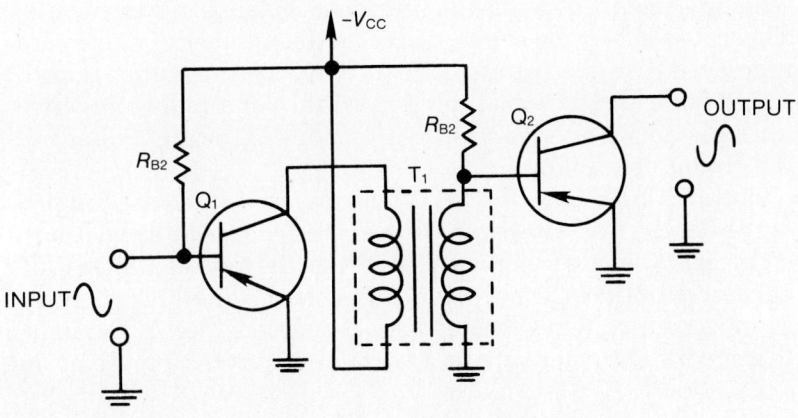

Fig. 22-28: A circuit with transformer coupling.

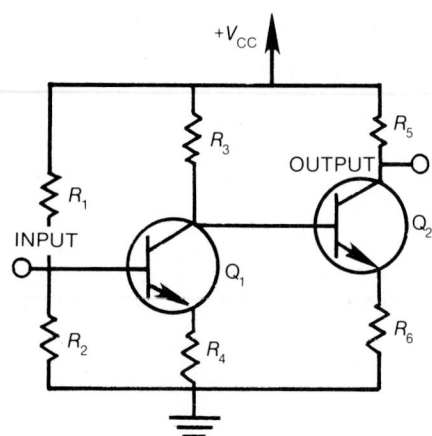

Fig. 22-29: A circuit with direct coupling.

the primary winding at low frequencies causes the low frequency response to fall off. At high frequencies the response is reduced by the input capacitance and the leakage reactance between primary and secondary windings.

In addition, transformers are more expensive, heavier, and larger in size compared to resistors and capacitors required for coupling. While transformer coupling may be used for audio amplifiers, its use is normally limited to the output power stage of the receiver. Power amplifiers are discussed in Chapter 24.

Direct Coupling

Figure 22-29 shows direct coupling between two stages of an audio amplifier. Since the base of Q_2 obtains its voltage from the collector of Q_1, a separate voltage-divider is not required to provide the base voltage for Q_2. This means that fewer components are required to construct the two common-emitter stages when they are coupled in this manner. However, the design of a direct-coupled circuit can be quite time consuming since there is much interaction between the bias voltages and current in the circuit. For example, R_3 must serve as a collector load resistor for Q_1 and as a base resistor for Q_2. This resistor must limit the collector current through Q_1, to the proper value but it must also control the base voltage of Q_2 so that the base current through Q_2 will have the proper value. Normally, resistor R_3 must have a high value so that both of these conditions can be met. The typical circuit values used in this circuit may, therefore, be considerably different than the values used in the RC-coupled amplifier previously described.

The major advantage of direct coupling is that reactive coupling elements (capacitors and inductors) are not required. Transformer coupling quite often gives a poor high-frequency response and, likewise, capacitors tend to give a poorer low-frequency response than desirable. Direct-coupled audio amplifiers provide a very flat frequency response for the range of frequencies that they are designed to cover. The major problem of direct coupling is the larger than normal requirements for power supplies. Each succeeding stage requires a higher source voltage than the preceding one. Usually direct coupling is a very inefficient method due to the high power losses in the load resistors, coupled with the intricate biasing requirements. These losses become even greater as the number of stages are increased. It is also often difficult to match the output impedance of one stage to the input impedance of a following stage utilizing only a single resistor. This adds to the inefficiency of the circuit operation.

Both MOSFETs and JFETs can be used in direct-coupled amplifier circuits. However, MOSFETs are especially well suited to direct-coupled applications. Since the gate of a MOSFET acts essentially as a capacitor rather than a diode junction, no coupling capacitor is needed between stages. For AC signals, this means that there are no low-frequency cutoff problems, in theory. In practical circuits, the input capacitance can form an RC filter with the source resistance and result in some low-frequency attenuation.

Figure 22-30 is the working schematic of an all-MOSFET three-stage amplifier. Note that all three MOSFETs are of the same type, and all three drain resistors (R_{D1}, R_{D2}, and R_{D3}) are the same value. This arrangement simplifies design. At first glance, it may appear that all three stages are operating at zero bias. However, when drain current flows, there is some drop across the corresponding drain resistor, producing a voltage at the drain of the stage and an identical voltage at the gate of the next stage. The gate of the first stage is at essentially the same voltage as the drain of the last stage because of feedback resistor (R_F). There is no current drain through R_F, with the possible exception of reverse gate current (which can be ignored for practical purposes).

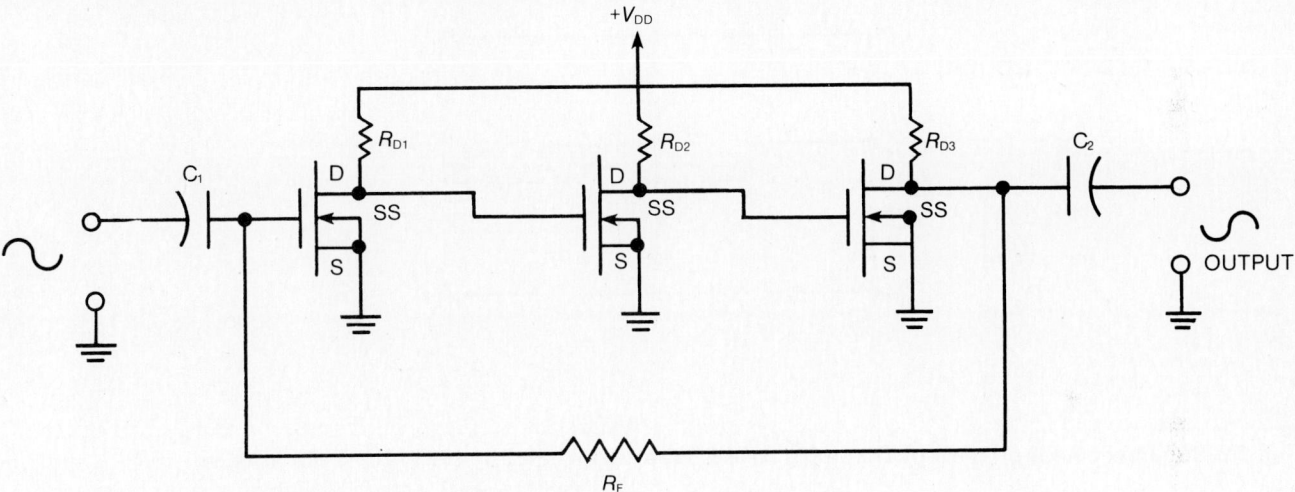

Fig. 22-30: An all-MOSFET three-stage amplifier.

The circuit of Fig. 22-30 requires one coupling capacitor (C_1) at the input. This is necessary to isolate the input gate from any DC voltage that may appear at the input. This makes the circuit of Fig. 22-30 unsuitable for use as a DC amplifier. The circuit can be converted to a DC amplifier when the coupling capacitor is replaced by a series resistor (R_{in}), as shown in Fig. 22-31. The considerations concerning operating point are the same for both circuits. However, the series resistance must be terminated at a DC level equivalent to the operating point. For example, if the operating point is –7V, point A must be at –7V. If point A is at some other DC level, the operating point is shifted.

In certain applications, FET stages and two-junction transistor stages can be combined (Fig. 22-32). The classic example is where a single FET stage is used at the input, followed by two, two-junction circuits. Such an arrangement takes advantage of both the FET and two-junction transistor characteristics.

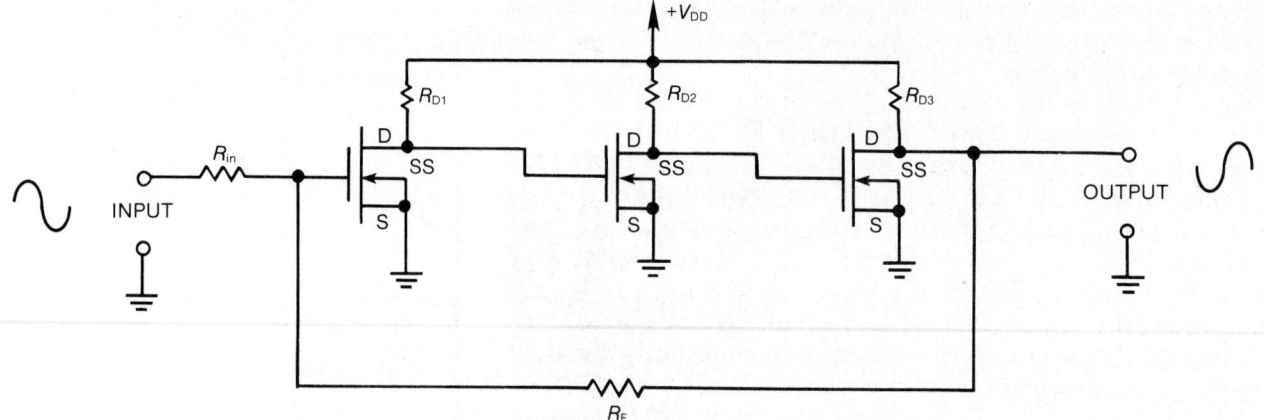

Fig. 22-31: All-MOSFET three-stage direct-coupled amplifier.

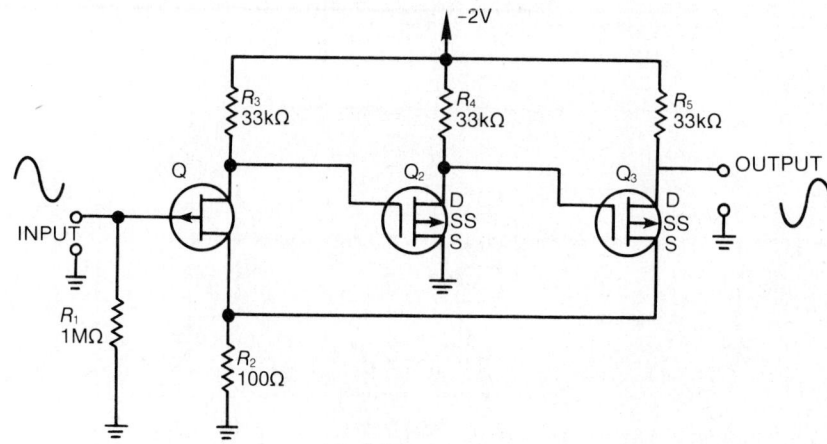

Fig. 22-32: Direct-coupled amplifier with JFET input, followed by two MOSFET stages.

FETs are voltage-operated devices. This permits large voltage swings with low currents, making it possible to use high-resistance values (resulting in high impedances) at the input and between stages. In turn, these high-resistance values permit the use of low-value coupling capacitors and eliminate the need for bulky, expensive electrolytic capacitors. If operated at the *zero-temperature coefficient* point (0TC point) with the gate-source biased at a specific voltage and held constant at the voltage, the drain current will not vary with changes in temperature. The 0TC point is different for each type of FET.

Two-junction transistors are essentially current-operated devices, permitting large currents at about the same voltage levels as the FET. Thus, with equal supply voltage and signal current swings, the two-junction transistor can supply much more current gain (and power gain) than the FET. Since currents are high, the impedances (input, interstage, and output) must be low in two-junction transistor circuits. This requires large-value coupling capacitors if low frequencies are involved. The low impedances also place a considerable load on devices feeding the amplifier, particularly if the devices are of high impedance. On the other hand, a low-output impedance is often a desirable characteristic for an amplifier.

558

When an FET is used as the input stage, the amplifier input impedance is high. This places a small load on the signal source and allows use of a low-value input coupling capacitor (if required). If the FET is operated at the 0TC point, the amplifier input is temperature-stable. (Generally, the input stage is the most critical in regards to temperature stability.) When two-junction transistors are used as the output stages, the output impedance is low and current gain (as well as power gain) is high.

Combination amplifiers can be direct-coupled or RC coupled, depending upon requirements. The direct-coupled design offers the best low-frequency response, permits DC amplification, and is generally simpler (uses less parts). The capacitor-coupled combination amplifier permits a more stable design and eliminates the voltage regulation problem. That is, a DC amplifier cannot tell the difference between signal-level changes and power-supply changes, so the power supply must be very stable for a DC amplifier.

Special Multistage Amplifier Circuits

In addition to the direct-coupled amplifiers just described, there are several other multistage amplifiers used to provide voltage amplification at audio frequencies. Included among these are the Darlington pair, the differential amplifier, and the complementary amplifier.

Darlington Pairs. Figure 22-33 illustrates how two transistors can be connected together so they function as a single unit. When this method of connecting transistors is used, the transistors are said to be arranged as a *Darlington pair*.

A Darlington pair is formed by connecting the emitter and collector leads of one transistor (Q_1) to the base and collector leads of a second transistor (Q_2) as shown. This permits the emitter current of Q_1 to serve as the base current for Q_2. If the base current of Q_1 is varied, the emitter current of Q_1 and the base current of Q_2 will also vary, causing the collector current of Q_2 to vary. Thus, Q_1 controls the conduction of Q_2. When con-

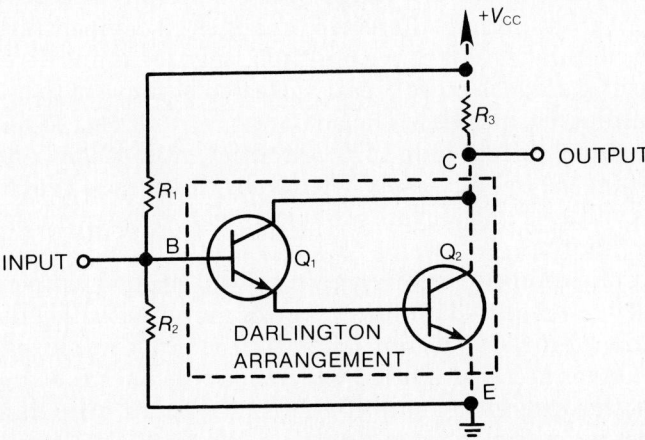

Fig. 22-33: A DC amplifier which uses a Darlington pair.

559

nected in this manner, Q_1 and Q_2 effectively function as a single transistor which has base (B), emitter (E), and collector (C) leads as shown in Fig. 22-33. The beta of the first transistor is effectively multiplied by the beta of the second so that the overall current gain is approximately equal to the product of the two beta values.

A Darlington pair can be employed in place of a single transistor in various applications. In Fig. 22-33, the Darlington arrangement is used in a simple DC amplifier circuit that can provide a very high voltage gain. Resistors R_1 and R_2 provide voltage-divider bias, and R_3 serves as the collector load resistor. No emitter-resistor is used in this circuit, since it would reduce the overall voltage gain of the circuit considerably and take away the primary advantage that the Darlington circuit has to offer. Although it is true that the emitter-resistor could be by-passed with a capacitor, a bypass capacitor would serve no useful function as far as DC signals are concerned. Bypass capacitors are useful only in preventing degeneration of AC signals.

In practice, the amplifier of Fig. 22-33 may provide slightly less gain than expected. This is because it is usually necessary to adjust the bias voltages so the input base current of Q_1 is very low. Low base current is necessary to keep the emitter current of Q_1 to a relatively low value, since this emitter current also serves as the base current for Q_2. If the input base current is allowed to increase by a significant amount, transistor Q_2 can be quickly driven into saturation. Q_1 must operate with a base current that is lower than normal. This means that the overall current gain of the Darlington pair will be reduced which, in turn, reduces the voltage gain.

Differential Amplifiers. All of the direct-coupled multistage amplifiers described up to this point have one major disadvantage: They exhibit a high degree of thermal instability. In all multistage amplifiers, the bias voltages in the second stage are determined in part by the bias voltages in the first stage. Variations in temperature can cause the bias voltages and currents in the first stage to vary, and this, in turn, can cause the DC output voltage of the first stage to change or drift from its steady-state (no-signal) value. The second stage will amplify this change in voltage in the same way that it would amplify a DC signal; therefore, changes in temperature can have a cumulative effect in a multiple-stage DC amplifier. Each succeeding stage is more greatly affected because of the amplified voltage variations.

In a three- or four-stage DC amplifier, it is possible for the final stage to be so adversely affected by the cumulative change in bias voltages and currents that it could be driven into saturation or cutoff.

In applications where high gain as well as good temperature stability are required, another type of amplifier called the *differential* (or *difference*) amplifier is commonly employed. Unlike a conventional amplifier, which has only one input and one output, the differential amplifier has two separate inputs and it can provide either one or two outputs. Figure 22-34 illustrates a

560

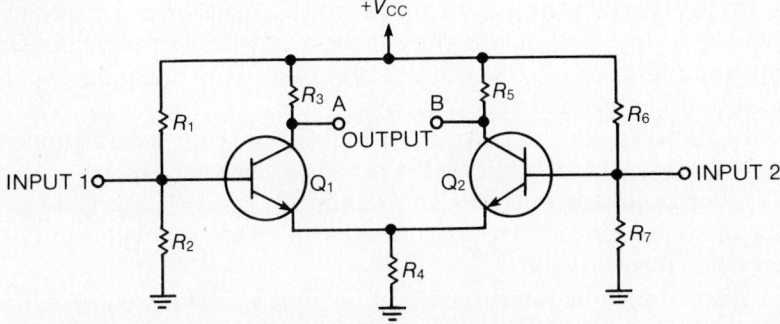

Fig. 22-34: A basic differential amplifier.

basic differential amplifier. Note that the circuit contains two transistors, Q_1 and Q_2, which share the same emitter-resistor (R_4), although each transistor has its own collector load resistor (R_3 and R_5). Resistors R_1 and R_2 provide voltage-divider bias for Q_1, and R_6 and R_7 provide voltage-divider bias for Q_2. One input signal can be applied to Q_1, while a second input signal can be applied to Q_2.

If transistors Q_1 and Q_2 have identical electrical characteristics and the bias voltages applied to each transistor are equal, the collector currents of the two transistors must also be equal. Thus, the voltage between the collector of Q_1 (point A) and ground must be equal to the voltage between the collector of Q_2 (point B) and ground. Under these conditions, the potential difference between points A and B must be zero. These are the conditions that should exist in an ideal differential amplifier when no input signal is applied to either transistor. At this time, the circuit is capable of providing an output between points A and B which is equal to zero, or it can provide two separate outputs (between points A or B and ground) which remain at some steady-state or no-signal value.

The differential amplifier can be used as a conventional amplifier if a signal voltage is applied to only one input. For example, if a signal is applied to input 1, transistor Q_1 amplifies the signal and produces an output between its collector (point A) and ground, as would be the case in an ordinary amplifier stage. However, the emitter-resistor (R_4) is not bypassed and a small signal voltage will be developed across it. This small signal voltage is applied to the emitter of Q_2, causing this transistor to function as if it were connected in a common-base amplifier configuration. Transistor Q_2 amplifies the signal at its emitter and produces an output signal which appears at point B. This output signal is 180° out of phase with the output signal produced at point A.

However, the differential amplifier is rarely used as a conventional amplifier to provide one or even two outputs with respect to ground. Instead, the output signal voltage is generally obtained between point A and point B. Since the voltage at point A is 180° out of phase with that at point B, a substantial voltage is developed between these two points. Furthermore, the signal voltage can be applied to either input and the output voltage can still be acquired between points A and B.

It is only when the output signal voltage is obtained between points A and B that a differential amplifier can provide its optimum degree of temperature stability. This high degree of stability is obtained because:

1. Q_1 and Q_2 are mounted close together so that both transistors are equally affected by the temperature variations.

2. The collector currents of Q_1 and Q_2 tend to increase or decrease by the same amount as temperature increases or decreases, respectively.

Since the collector currents of Q_1 and Q_2 are varying with temperature, the voltages between point A and ground and point B and ground will also tend to vary with temperatures by the same amount. This causes the potential difference between points A and B to remain essentially constant. The fluctuations in voltage between these two points occur only as a result of signal amplifications.

The differential amplifier is more stable and more versatile than any of the circuits that were previously described. In some applications, both inputs are used simultaneously to compare the amplitude of two signals, and an amplified output voltage is produced between points A and B that is proportional to the difference between the two input signals. As the difference between the two input signals becomes greater, the output signal voltage becomes greater. If the two input voltages are equal, the output voltage is equal to zero. This is why the circuit is referred to as a difference or differential amplifier.

Complementary Amplifiers. The working schematic of a two-stage, direct-coupled complementary amplifier is shown in Fig. 22-35. In this configuration, the output of an NPN is fed into a PNP transistor to increase stabilization. The increased stabilization for the complementary direct-coupled amplifier is due to the fact that a change in collector current of Q_1 (because of temperature, power supply variation, etc.) is opposed by an equal change in current of Q_2, but in the opposite direction. If additional stages are required, the complementary system is continued; that is, NPN and PNP transistors are used alternately.

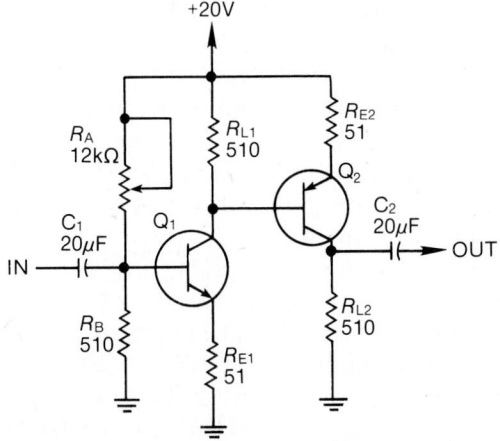

Fig. 22-35: Basic direct-coupled complementary amplifier circuit.

The overall gain for this circuit configuration is about 70% of the combined gains of each stage. Each stage has a gain of approximately 10, as set by the 10:1 ratio of collector and emitter resistances. The combined gain theoretically is 100, but practically about 70; thus, an input of 100mV is increased to about 7V at the output. This circuit is highly stable and has a very wide band-frequency response (generally from direct current [0Hz] on up to whatever maximum is set by the transistor high-frequency limitations).

Negative Feedback

High-gain audio amplifiers are frequently subject to regenerative (positive) feedback when several stages of amplification are used. In addition to the required decoupling, it may be necessary to provide some form of degenerative (negative) feedback. Negative feedback can then be used to cancel the effects of the regenerative feedback and provide increased amplifier stability and frequency response. This is because at the ends of the frequency response curve of an amplifier the gain decreases. This decrease causes a decrease in negative feedback and tends to increase the effective gain, thus extending the frequency response.

The most common method of achieving negative feedback is to leave the emitter/source/cathode resistor unbypassed. An example circuit using this method, in addition to decoupling, is shown in Fig. 22-36. As illustrated, the emitter-resistors are left unbypassed. This causes degeneration (negative feedback) by limiting the gain of the various amplifier stages. In addition to the prevention of regenerative feedback being aided, the fidelity of the amplifiers is improved.

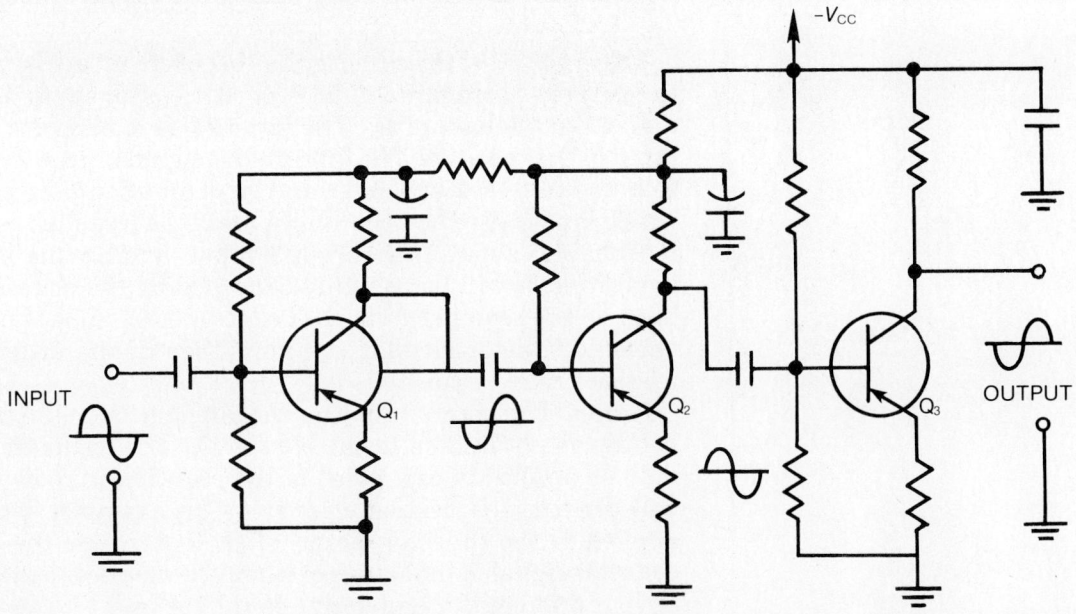

Fig. 22-36: A circuit using an unbypassed emitter/source/cathode resistor to achieve negative feedback.

High-gain audio amplifier chains are always subject to signal *distortion*. That is, large gains may often result in peak clipping, where one or more amplifiers are inadvertently allowed to be driven either to cutoff or a near saturation condition. By leaving the emitter-resistors unbypassed (no emitter-capacitors), the gain of the amplifiers is reduced and the chances of a loss of fidelity are reduced. The purpose of the audio section of any receiver is to reproduce the audio signal without any appreciable distortion. Clipping must, therefore, be eliminated in the circuitry, and this requires some means of reducing the overall amplifier gain. The method illustrated in Fig. 22-36 adequately meets these requirements.

An alternate method of obtaining negative feedback is illustrated in Fig. 22-37. In this method of obtaining negative feedback, a portion of the output signal is fed back to the input circuitry in such a way as to degenerate the input signal. In this way, the gain of the various amplifiers remains the same, but it is the signal which is attenuated.

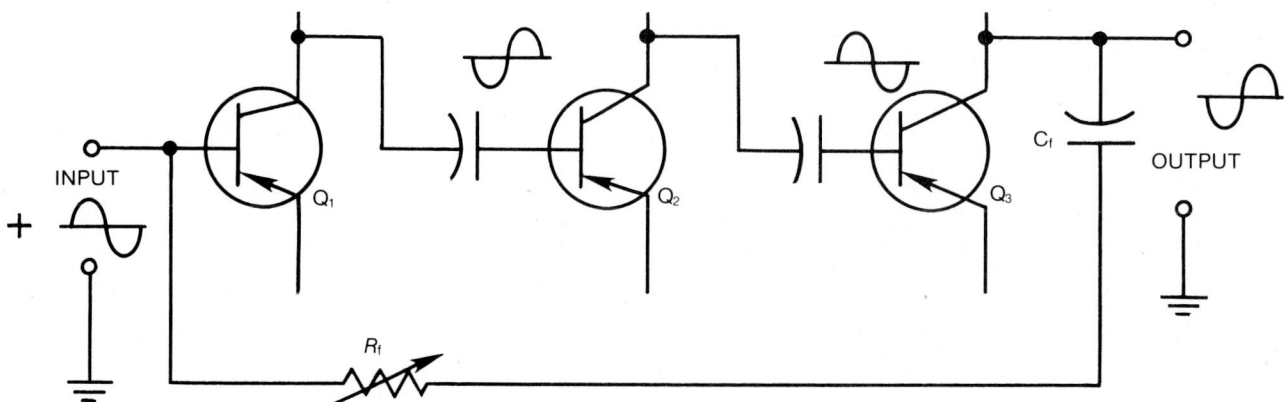

Fig. 22-37: **A circuit using a portion of the output fed back to the input to achieve negative feedback.**

The majority of the circuitry has been left out of Fig. 22-37 for the purpose of simplification. A positive input signal is shown applied to the base of Q_1. The input to Q_2 is negative and the input to Q_3 is positive. The final output signal is negative-going. C_f is the feedback capacitor and R_f is an attenuating resistor which is used to reduce the amplitude of the feedback voltage to reasonable values. As the original input signal on the base of Q_1 is positive, it will be partially cancelled by the negative feedback voltage coming from the collector of Q_3. R_f is shown as a variable resistor so that the amplitude of the degenerative feedback may be controlled.

It should be pointed out that the application of the feedback voltage is not limited to the base of Q_1. The feedback voltage may be applied to any point in the preceding circuit where it will produce the desired effects. As an example, it could be applied in the collector circuit of Q_2 to decrease the positive collector signal of that stage or it could be applied to the emitter of Q_2 if that emitter had been left unbypassed.

Impedance Matching in Audio Amplifiers

While the input impedance for transistors may be high in some configurations, transistor circuits are nevertheless more troublesome in this respect. For this reason, a more detailed discussion will now be given on the manner in which the effects of the input impedance determine the choice of transistor configurations.

The most desirable method of matching source impedance to input impedance is by transformer coupling; however, as mentioned previously, this is not always practical. When the preamplifier must be fed from a low-resistance source ($20\,\Omega$ to $1,500\,\Omega$), without the benefit of transformer coupling, either the common-base or the common-emitter configuration may be used. The common-base configuration has an input impedance which is normally between $30\,\Omega$ and $150\,\Omega$; the common-emitter configuration has an input impedance which is normally between $500\,\Omega$ and $1,500\,\Omega$.

If the signal source has a high internal impedance, a high input impedance can then be obtained by using one of the three following circuit arrangements.

It would appear that the easiest configuration to use would be the common-collector. The input resistance of the common-collector configuration is high because of the large negative voltage feedback in the base-emitter circuit. As the input voltage rises, the opposing voltage developed across the load resistance (Fig. 22-38) substantially reduces the net voltage across the base-emitter junction. By this action, the current drawn from the signal source remains low. From Ohm's Law it is known that a low current drawn by a relatively high voltage represents a high impedance. If a load impedance of $500\,\Omega$ is used, the input resistance of a typical common-collector configuration will be over $30,000\,\Omega$. The disadvantage of the common-collector configuration, however, is that small variations in the current drawn by the following stage cause large changes in the input impedance value.

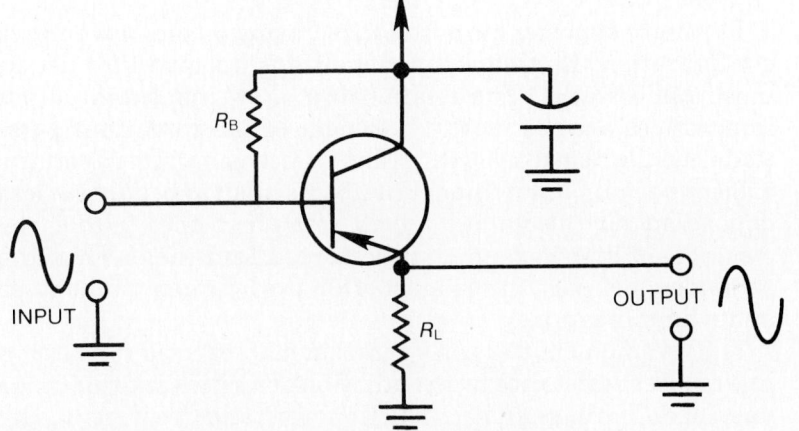

Fig. 22-38: A common-collector configuration.

The variation of input impedance, as a function of load impedance, for the common-emitter, common-base, and common-collector configurations is shown in Fig. 22-39. Notice that the common-collector configuration with an R_L of 500Ω has an input impedance of approximately 15,000Ω. Within the operating range of 1,000Ω to 100,000Ω of load resistance, the input impedance of the common-collector circuit will increase, r_i of the common-emitter circuit will decrease with an R_L increase, and r_i of the common-base circuit will increase with an R_L increase.

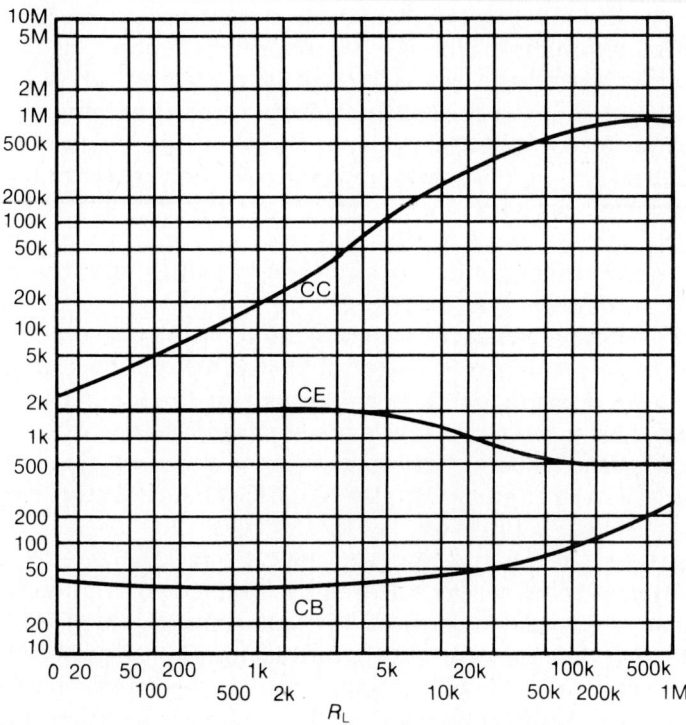

Fig. 22-39: Variations of input impedance, r_i, with R_L for each configuration.

The highest power gain is realized using a common-emitter configuration. Its frequency response is adequate for use in most audio stages. The relationship of its input and output impedances are the most convenient of the three configurations, and when used with transformer coupling it permits easier impedance matching. Finally, its input impedance is less dependent on the value of load resistance than either the common-emitter or common-collector configurations. Thus, the common-emitter circuit is most often used for general purpose audio amplification.

The common-emitter configuration may be used to match a high source resistance by the addition of a series resistor in the base lead. The base-emitter junction resistance (represented by r_i in Fig. 22-40) for a typical common-emitter configuration is approximately 1,000Ω, if a load resistance of 30,000Ω is used. The input resistance, r_i, may be increased by reducing the load resistance (R_L). For instance, decreasing the load resistance to

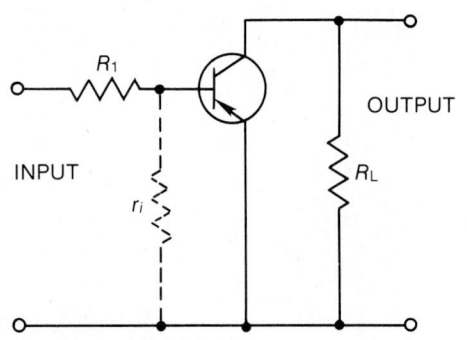

Fig. 22-40: Simplified schematic of common-emitter preamplifier with series resistor.

approximately $10,000\,\Omega$ will increase r_i to approximately $1,500\,\Omega$ as seen in the curve of Fig. 22-39.

Assume a source impedance of approximately $20,000\,\Omega$. In order to use a common-emitter configuration having a $10,000\,\Omega$ R_L, it would be necessary to insert a series resistance (R_1 in Fig. 22-40) of approximately $18,500\,\Omega$ in the base lead. Thus, the input resistance (r_i) of the amplifier will appear equal to the source. Notice that before the series resistor was added, the input resistance (r_i) varies considerably with change in R_L. When R_1 is added to the circuit, r_i will still vary with a change in R_L; however, r_i is now only a small portion of the total input resistance so that r_i remains essentially constant with load resistance changes. The fact that the total input resistance remains relatively constant, even with large variations in transistor parameters or current drain by the following stage, is the main advantage of this circuit arrangement. The disadvantage of this circuit, in addition to a small loss in current gain, is the large resistance in the base lead. This resistance leads to poor bias stability if the bias voltage is fed to the transistor through this resistor.

Another method of increasing the input resistance of a common-emitter configuration is shown in Fig. 22-41. This type of circuit is the previously mentioned degenerated common-emitter configuration. If an unbypassed resistor (R_E) is inserted in the emitter lead, the signal voltage developed across this resistor opposes the input signal voltage.

As in the case of the common-collector configuration, this negative-feedback voltage or degenerative voltage causes an increase in the input resistance. With a bypassed resistor in the emitter lead, the input resistance of the common-emitter configuration would be $2,000\,\Omega$ if a load resistor of $500\,\Omega$ were used (Fig. 22-39). With an unbypassed resistor (R_E) of $500\,\Omega$, the input resistance (r_i) will appear as approximately $20,000\,\Omega$. The input resistance may be made to appear as any desired value (within practical limits) by the proper choice of R_L and R_E. Like the common-emitter circuit with the series resistor, the total input resistance of the degenerated common-emitter circuit will remain relatively constant with a varying load. However, the advantage of the degenerated common-emitter configuration is that the unbypassed resistor (R_E) also acts as an emitter stabilizer and aids in stabilizing the transistor bias.

Figure 22-42 shows the normal frequency response for an audio amplifier chain (dotted lines). With proper feedback circuits this response curve can be improved to that shown in the solid line curve of Fig. 22-42.

In each of the feedback networks considered previously, it was considered that each feedback network was both proportional to signal amplitude and nonfrequency selective. It is, however, possible to develop a feedback network that will provide very little feedback at the low and high audio frequencies (allowing gain to remain high), while more feedback is developed for the middle audio frequencies (thus reducing mid-band amplifier gain). While such a system will reduce the overall gain of the system, it, nevertheless, provides a more uniform frequency response over the entire range of desired amplification.

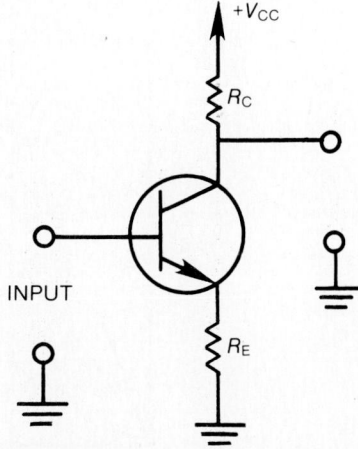

Fig. 22-41: Degenerated common-emitter configuration.

567

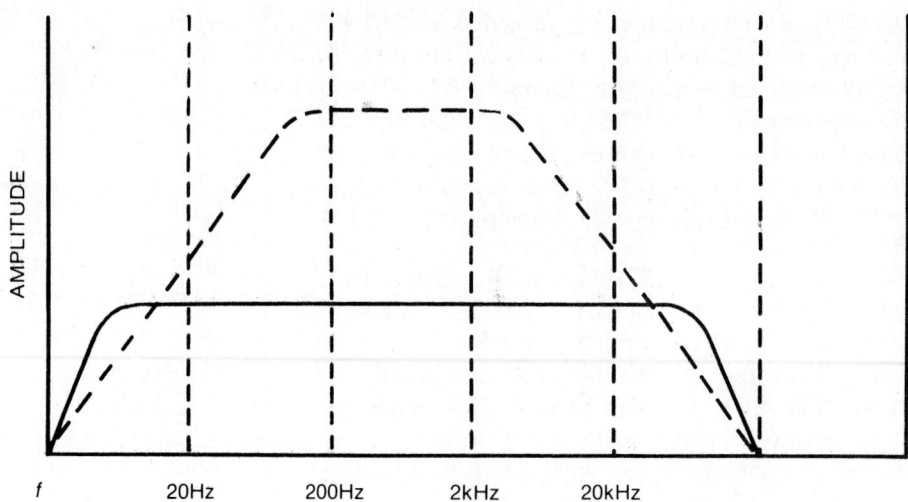

Fig. 22-42: Audio amplifier response curves.

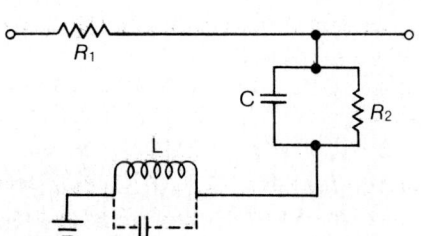

Fig. 22-43: Frequency sensitive feedback.

A simplified system for developing frequency sensitive feedback voltages is shown in Fig. 22-43. At low frequencies, R_1 and R_2 act as a voltage-divider which selects the initial amount of feedback. The inductance (L) is chosen so that it is essentially a short at these low frequencies. Capacitor (C) is chosen so as to offer a very high reactance at the lower audio frequencies. As the frequencies go into the mid-band range, the gain of the amplifier chain increases. This will increase the voltage applied to the feedback network. However, the capacitor at these frequencies begins to partially shunt the signal around R_2. This has the effect of holding the output at roughly the same amplitude rather than allowing it to rise, as the input signal does. As the frequency continues to increase, the shunting action of the capacitor increases. However, the reactance of the inductor now begins to become appreciable. The sum of the remaining voltage across R_2 plus the drop across L manages to keep the output amplitude roughly the same as before. In effect, the increasing reactance of L counteracts the increasing shunting effect of the capacitor.

Finally, at high frequencies (normally above the regular audio range), the interwinding capacitance shorts (effectively) the inductance. A capacitor may be added in parallel with the inductor to insure that this happens.

By using a circuit similar to the one shown in Fig. 22-43 it is, therefore, possible to obtain a much flatter response curve and a much wider bandpass. While the gain of the normal amplifier chain is much higher, note the bandwidths as shown in Fig. 22-42. As can be seen, the bandwidth is much wider when a frequency sensitive feedback network is utilized in conjunction with the amplifiers.

Biasing by Succeeding Stages

Bias considerations increase as an increasing number of stages are involved. It is possible to use a succeeding stage to change the bias of a previous one as shown in Fig. 22-44. In Q_1,

568

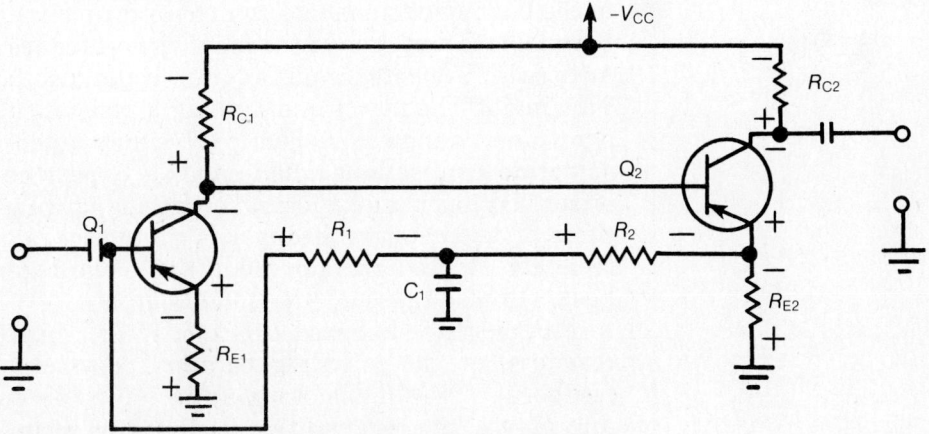

Fig. 22-44: Directly-coupled amplifiers with bias stabilization.

it may be seen that the base is negative in respect to the emitter which permits current flow from the collector source to the emitter terminal. In Q_2 the base, due to voltage-divider action, is negative with respect to the emitter; this forward bias allows current flow from the collector source through R_{C2} to the emitter source terminal and some current flow from the collector terminal of Q_1 to the base of Q_2 and to the emitter terminal. Lastly, a path for current flow exists from the collector of Q_2 through the transistor and from the top of R_{E2} through R_2 and R_1 to the base of Q_1. The purpose of C_1 is to isolate Q_1 from Q_2 by preventing the amplified AC signal variations from reaching the base of Q_1 while variations in the DC voltage level are fed back to the base of Q_1. The potential difference from the top of R_{E2} to ground is the source for base-emitter bias of Q_1. For instance, suppose the value of R_{E1} decreases, allowing more current flow through Q_1. This would cause a larger voltage drop across R_{C1}, the collector load, and a decrease in potential difference from collector to ground. Since the collector of Q_1 and base of Q_2 are directly coupled, this decrease in potential makes the base of Q_2 less negative in respect to its emitter and decreases forward bias. This will cause less current flow through Q_2 and will make the potential at the top of R_{E2} less negative. Since the base of Q_1 is tied to this point, this base is made less negative with respect to the emitter of Q_1, which decreases the current flow through Q_1, minimizing the effects caused by a small change in ohmic value of R_{E1}.

Failure Analysis

If there is no output, check for a loss of source voltage, open load resistors, or open biasing components which would cause the amplifying device to cut off. The transistor or tube could also be faulty. A check of the circuits' DC voltages will normally indicate the exact trouble. If the DC voltages are normal, check for faulty soldering connections of the coupling capacitors.

If the output is reduced, the amplifying device is probably defective. This is especially true if the reduction has taken place over a period of time. A second cause could be a change in value of the biasing components, which, in turn, changes the operational characteristics of the amplifier. A reduced load impedance (which lowers amplifier gain) could also be responsible. Turning off the power and making a resistance check should show any change in resistance that has taken place. Faulty soldering connections could cause a bypass capacitor to be effectively open and thereby introduce unwanted degenerations. If degenerative feedback is being used, a quick check of feedback attenuators is in order. Excessive degenerative feedback will also cause a reduction in output.

If the output is distorted, check for a loss of bias (open biasing resistors) or excessive regenerative feedback. Regenerative feedback can result in motorboating. Faulty decoupling capacitors may allow regenerative feedback to occur. If controlled feedback is being used, open feedback components may be responsible. Without the required amount of degenerative feedback, distortion will occur. If all other components check out good, the amplifying device itself may be faulty. If the cause of the distortion cannot easily be found, it is a good idea to start at the output and work backwards through the circuitry with an oscilloscope until the stage in which the trouble originates is located.

Types of Distortion

When the output signal waveform from an amplifier circuit is not an exact duplication of the input signal waveform, some type of distortion has been introduced into the circuit. The absence of distortion is necessary for high-fidelity sound reproduction.

Several of the different types of distortion (amplitude, overload, harmonic, intermodulation, frequency, phase, and crossover) will be described in the following paragraphs.

Amplitude Distortion. *Amplitude distortion* is produced when an amplifier is operated in the nonlinear region of the transfer characteristic of the amplifier. Low supply voltage, a weak amplifier, incorrect bias, and excessive signal are all possible causes for amplitude distortion. Amplitude distortion is a problem that arises more frequently in power amplifiers because they handle larger signals. In extreme cases of amplitude distortion, rough, hoarse, or garbled sound may result.

Overload Distortion. *Overload distortion* becomes apparent when the input signal is too great. Figure 22-45 illustrates the transfer characteristic curve of a silicon NPN transistor operating under overload conditions. The bias V_{BE} is 0.6V for class A operation. Linear operation on the curve includes only ± 0.1V (approximately) around the bias.

Below 0.5V, the collector current (I_C) approaches zero for cutoff. Above 0.7V, the collector current approaches saturation. The input signal is excessive enough to drive V_{BE} to saturation and cutoff, resulting in the flattening of the top and bottom of

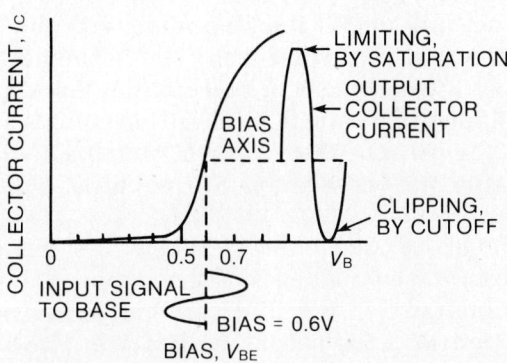

Fig. 22-45: Nonlinear amplitude distortion in a collector output signal produced by an excessive input signal causing overload.

the output waveform. There are three main factors affecting overload distortion. These are:

1. Signal changes near the saturation region are reduced or eliminated in the output.

2. Signal changes near the cutoff region are reduced or eliminated in the output.

3. The DC level of the output signal shifts.

The DC level should not change with or without signal in a class A amplifier. However, in the distorted output signal of Fig. 22-45, the bias axis is not at the center of the waveform. Determining the difference between the bias axis and the center axis of the waveform yields the shift in the DC level. This shift in the DC level in a class A amplifier is a measure of the amount of amplitude distortion.

Harmonic Distortion. When the input signal to a class A power amplifier is composed of a single frequency, the output signal may contain the *fundamental* and several *harmonics* of the fundamental frequency. Figure 22-46 shows that an approximate square wave results when a sine wave plus all the odd harmonics up to the fifteenth are added together. It is

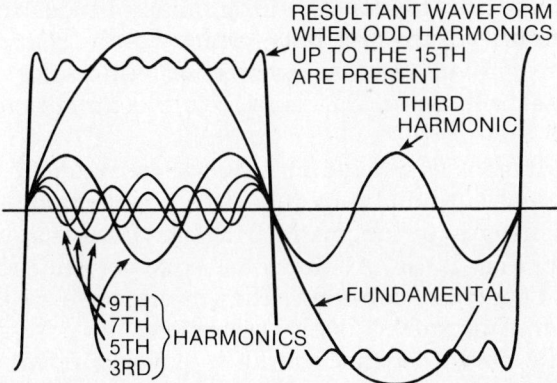

Fig. 22-46: A square wave consists of the fundamental plus all the odd harmonics out to an infinite frequency. The sum of the odd harmonics up to 15 is shown together with the appropriate odd harmonics up to the 9th.

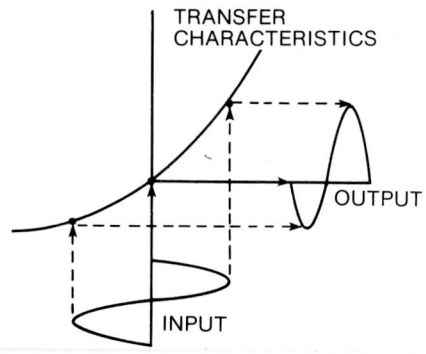

TRANSFER CHARACTERISTICS

OUTPUT

INPUT

Fig. 22-47: Harmonic distortion.

standard procedure to call the sine wave with the lowest frequency the fundamental waveform. The harmonics are those sine waves with frequencies of twice, three times, four times, etc., the fundamental frequency. These harmonics are then referred to as the second, third, fourth, etc., harmonics. These harmonics alter the wave shape of the output signal; this is *harmonic distortion.*

The nonlinear transfer characteristics of a transistor will generate harmonic components of the signal. This nonlinear function is illustrated in Fig. 22-47. The input is a perfect sine wave, one frequency; the output is distorted. The output contains the fundamental frequency and several harmonics of the fundamental. The amplitudes and phases of the harmonics have fixed values in respect to the fundamental. Only the amplitudes are considered here. The number of harmonics in the output is determined by the curvature of the transfer characteristics curve. The curve in Fig. 22-47 is rather severe, and the output contains a large number of harmonics. The greater the curve, the larger the number of harmonics and the greater the distortion.

Intermodulation Distortion. An audio signal, except for a single pure tone, is not a pure sine wave. The standard audio signal is a complex wave containing many frequencies and constant frequency changes. This complex signal tends to cause *intermodulation distortion* in an audio amplifier. This occurs when the harmonics introduced in the amplifier combine with each other or with the original frequencies to produce new frequencies that are not harmonics of the fundamental frequency. Intermodulation distortion is the reason for the rough, garbled sound of amplitude distortion, because that distortion is not harmonically related to the signal.

Frequency Distortion. *Frequency distortion* occurs when variations in frequency cause variations in amplifier gain. Music and speech have many different frequency components which correspond to the fundamental and harmonic frequencies produced by the sound source. Each of these different frequencies probably has its own different amplitude. However, in order to maintain the original character of the sound, the amplifier must provide the same gain for all frequencies so that the frequencies will have the same relative amplitudes in the output.

Generally, however, the amplifier does not respond as well at very low and high audio frequencies. As an example, refer to the frequency response curve for an RC-coupled audio amplifier shown in Fig. 22-48. At the low-frequency end, the series-coupling capacitors cause insufficient response, or *roll-off,* in the output. This roll-off is also present at the high-frequency end of the curve because of the bypassing effect of shunt capacitance.

Frequency distortion at low audio frequencies results in weak bass in the sound. Insufficient treble is caused by frequency distortion at the high audio frequencies. In the middle range of frequencies, the RC-coupled amplifier has flat response and

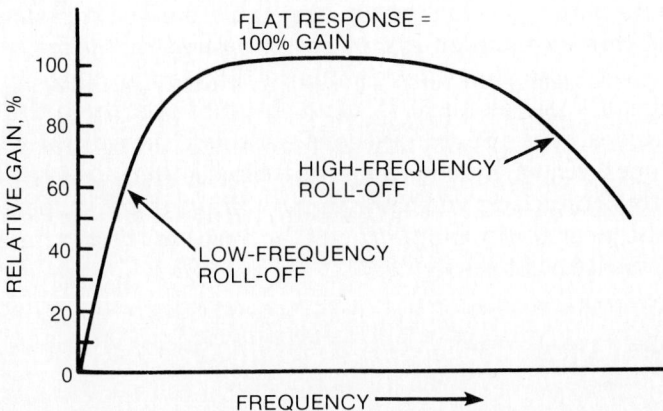

Fig. 22-48: Frequency distortion means the gain is not the same for all audio frequencies. The response curve shown is for an RC-coupled amplifier.

produces uniform gain. It is in this middle range of frequencies that the amplifiers normally operate.

Phase Distortion. *Phase distortion* is present when the relative phases of the harmonic components of a signal are not maintained. Such distortion is caused by the presence of reactive and resistive components in the circuit. In cathode-ray oscilloscope amplifiers, television video amplifiers, and radar circuits, phase distortion is highly undesirable. It is often said that phase distortion is unimportant in audio amplifiers because the ear is insensitive to moderate changes in phase. While it is true that the ear is insensitive in this respect, it is not true that demands on the audio amplifier can be relaxed. The effect of phase shift is of great importance to the speaker diaphragm from the transient point of view. The quality of a sound depends, among other things, upon the attack and decay times. To obtain similar attack and decay times in the reproduced sound, phase distortion should be reduced to a minimum.

Crossover Distortion. The waveform shown in Fig. 22-49 represents the severe distortion introduced in a typical class B, push-pull amplifier (described in Chapter 24). This type of distortion is termed *crossover distortion*, because it occurs during

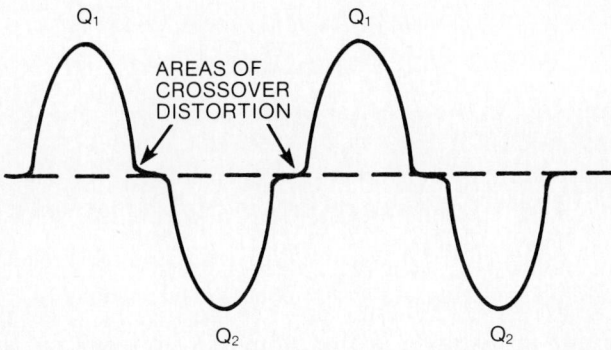

Fig. 22-49: Crossover distortion in a class B, push-pull amplifier.

573

the time that operation is crossing over from one transistor to the other in the push-pull amplifier. Crossover distortion occurs at the very low signal levels due to the turn-on base-emitter voltage of a transistor (0.7V) and the nonlinearity in the low-signal area. It can be reduced considerably if the transistors are each operated with a small forward bias so that they both are slightly conducting when the input voltage is zero. Operation over the most nonlinear portion of the base-emitter PN junction characteristic is thus avoided.

SUMMARY

The most common way of classifying amplifier circuits is to use the operating point of the transistors. The four basic operating-point classifications are known as Class A, Class B, Class AB, and Class C.

A form of bias is required to establish the voltage and current relationships at the operating point (Q-point) of the circuit. There are several types of bias networks, each with its own advantages and disadvantages. Among these are fixed-biased networks, collector feedback arrangements, and voltage-divider networks. The basic network must maintain the desired base current in the presence of temperature (and sometimes frequency) variations. This is referred to as bias stability.

The AC junction resistance (r_j) of the base-emitter diode of a transistor in an amplifier circuit can be determined using a characteristic curve and the following formula:

$$r_j = \frac{\Delta V}{\Delta I}$$

The AC junction resistance (sometimes called dynamic resistance) can also be approximated without the use of a characteristic curve by using the following formula:

$$r_j = \frac{25\text{mV}}{I_E}$$

The voltage gain of a common-emitter amplifier can be determined using the formula:

$$A_V = \frac{-R_{c(\text{line})}}{R_{e(\text{line})}}$$

The minus sign in the equation represents a 180° phase shift or signal inversion.

Input impedance (Z_{in}) is the term used to describe the ratio of the input voltage to the input current. Stated mathematically,

$$Z_{in} = \frac{V_{in}}{I_{in}}$$

The input impedance is the amount of opposition the AC signal sees when it looks into the amplifier.

The dynamic resistance from base to ground (r_b) is greater than r_j by a factor of β. That is,

574

$$r_b = \frac{\Delta V_{be}}{\Delta I_b} = \beta r_j$$

In a common-emitter this base input resistance must be considered when calculating input impedance:

$$Z_{in} = R_B // r_b$$

The output impedance of an amplifier is the opposition seen by the load looking back into the output terminals of the amplifier. In a common-emitter amplifier, $Z_o = R_C // r_c$ where r_c is the dynamic collector resistance.

A graphical analysis of amplifier circuits can be made using both static and dynamic load lines.

Amplifier linearity is a desirable characteristic wherein the output waveform is an exact reproduction of the input signal. Linearity is generally specified in terms of percent harmonic distortion.

An unbypassed resistor, called a swamping resistor, can be added into the emitter leg to increase the input impedance and improve the large-signal linearity.

The voltage gain of a common-collector (emitter-follower) amplifier is always slightly less than one.

$$A_v = \frac{R_E // R_L}{r_j + R_E // R_L}$$

There is no phase inversion of the input signal waveform.

The following formulas can be used to determine the input and output impedances of an emitter-follower amplifier.

$$Z_{in} = R_{B1} // R_{B2} // r_b$$

where

$$r_b = \beta (r_j + R_E // R_L)$$

$$Z_o = R_E // \left(\frac{r_j + R_{B1} // R_{B2} // Z_s}{\beta} \right)$$

$$\approx \frac{Z_s}{\beta}$$

Field effect transistors can also be used in amplifier circuit configurations. FET circuits, as a rule, are less complex than standard transistor circuits because they require fewer circuit components. The FET is a high-gain, high-input impedance, medium-to-high output impedance device that has low noise ratings for the audio frequency range and produces minimal leakage current. The FET is most commonly used in a common-source configuration.

Two or more amplifier stages can be used in cascade (where the output of one transistor is fed to the input of the next transistor) when stable voltage gains greater than about 20 are required and it is not practical to bypass the emitter (or source) resistor of a single stage.

The operation of coupling networks must be considered in any amplifier circuit (single-stage or multistage). A single-stage audio amplifier must be coupled to the input and output

575

devices. In a multistage circuit, there must be interstage coupling. There are four major coupling methods: RC coupling, inductive (impedance) coupling, transformer coupling, and direct coupling.

A Darlington pair is formed by connecting the emitter and collector leads of one transistor to the base and collector leads of a second transistor. In this arrangement, the two transistors effectively function as a single transistor, having a current gain that is approximately equal to the product of the beta values of each transistor. Darlington pairs are generally packaged in a single case having only one base lead, one emitter lead, and one collector lead.

Either a differential amplifier or a complementary amplifier can be employed in applications where high gain and good temperature stability are required.

Negative (degenerative) feedback can be used to cancel the effects of positive (regenerative) feedback that often occurs in high-gain audio amplifiers. Negative feedback also provides increased amplifier stability and frequency response.

High-gain audio amplifiers are always subject to signal distortion. Among the many different types of distortion are amplitude distortion, overload distortion, harmonic distortion, intermodulation distortion, frequency distortion, phase distortion, and crossover distortion.

CHECK YOURSELF (Answers at back of book)

Questions

Fill in the blanks with the appropriate word or words.

1. Any amplifier that is biased so that its output current flows during the entire cycle of the AC input voltage is said to be operating as a _____ amplifier.
2. Class B operation can be obtained by biasing the transistor at _____ .
3. Input impedance (Z_{in}) is the term used to describe the ratio of the _____ to the _____ .
4. The maximum swing up and down possible without distortion occurs with the bias point set at the _____ of the load line.
5. A_v is the ratio of the resistance through which the AC _____ current flows divided by the resistance through which the AC _____ current flows.
6. The output impedance of an emitter-follower primarily depends on the _____ of the transistor and the _____ of the signal source.
7. The overall gain of a multistage amplifier is approximately equal to the _____ gain of each stage multiplied by the gain of the adjacent stage.

8. The most common way of coupling two or more audio amplifier stages is _____ coupling, or _____ coupling as it is sometimes called.
9. Direct-coupled audio amplifiers provide a very _____ frequency response for the range of frequencies that they are designed to cover.
10. When the output signal waveform from an amplifier circuit is not an exact duplication of the input signal waveform, some type of _____ has been introduced into the circuit.

Determine whether the following statements are true or false.

11. Class AB amplifier operation can be achieved by DC biasing the transistor closer to saturation than to cutoff.
12. The efficiency of a class C amplifier is high.
13. Adding a load resistance to an amplifier causes its output voltage to be loaded down, lowering the voltage gain of the amplifier.
14. Voltage-divider bias tends to keep I_C constant, regardless of beta variations.
15. FETs are current-operated devices that permit large voltage swings with high current, making it possible to use low-resistance values (resulting in low impedances) at the input and between stages.

Problems

1. Calculate the required value of R_B, the gain, and the input impedance for the circuit of Fig. 22-50 if $V_C = 7V$.
2. Determine the values of V_C, Z_{in}, Z_o, and A_V for the circuit of Fig. 22-51.
3. Calculate V_E, A_V, Z_{in}, and Z_o for the circuit of Fig. 22-52.

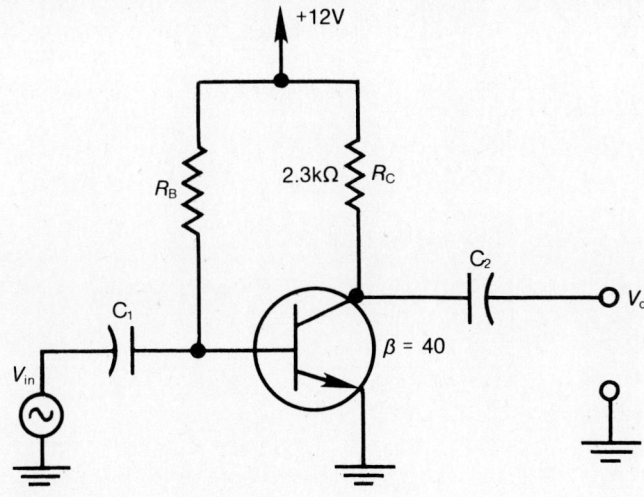

Fig. 22-50.

577

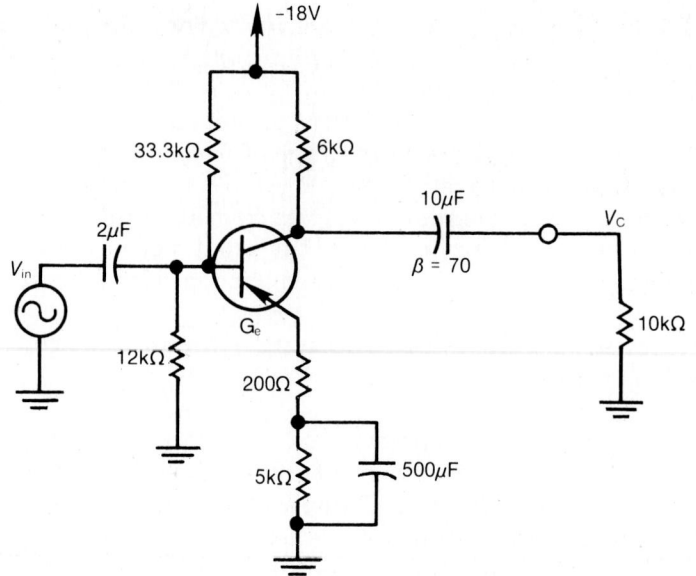

Fig. 22-51.

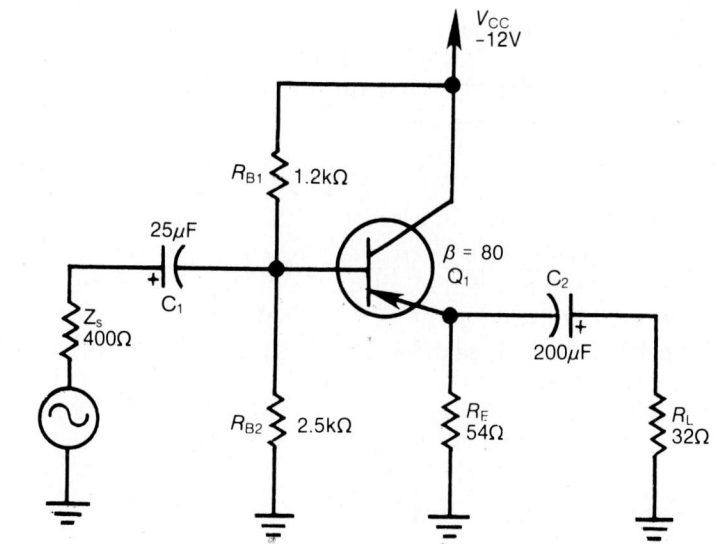

Fig. 22-52.

578

CHAPTER 23

Transducers

Sound is a vibration or mechanical disturbance in an elastic medium such as a gas, liquid, or solid. This vibration, which is considered a physical change in the medium, is propagated by the medium as a sound wave. A physical change in the medium is considered a change in its density, pressure, or displacement. These sound waves occur because the medium (a gas, liquid, or solid) possesses a property called elasticity. This means that the medium resists deformation in the same way that a spring resists extension or compression. If one part of the medium is moved or disturbed, this motion is transmitted to adjacent parts of the medium as a wave motion at some characteristic speed.

The area of science that is concerned with the production, propagation, reception, and control of sound waves is called *acoustics*. A wide range of phenomena is included in the science of acoustics, far wider than implied by the everyday meaning of the word "sound," which is limited to what can be heard. A sound wave may or may not be heard; it may be inaudible because its frequency lies outside the range to which our ears can respond or because it is being propagated in a medium, such as a solid, which is inaccessible to the ear. Therefore, the familiar process known as "hearing" is only one part of the subject of acoustics; it is a special case of the reception of a sound wave.

THE NATURE OF SOUND WAVES

To help understand how a basic sound wave is formed, consider a long cylinder or tube filled with air that has a movable piston at one end of the cylinder (Fig. 23-1A). If the piston is quickly moved into the cylinder and then stopped, the air directly in front of the piston is momentarily *compressed*. The compressed region expands by pushing the air immediately to its right, and the region of compression originally formed at the piston moves through the cylinder at a constant speed as shown in Fig. 23-1B. The wave produced in this manner is called a *pulse*.

Suppose that the piston in the cylinder just discussed is made to vibrate or move back and forth in the same way that a weight

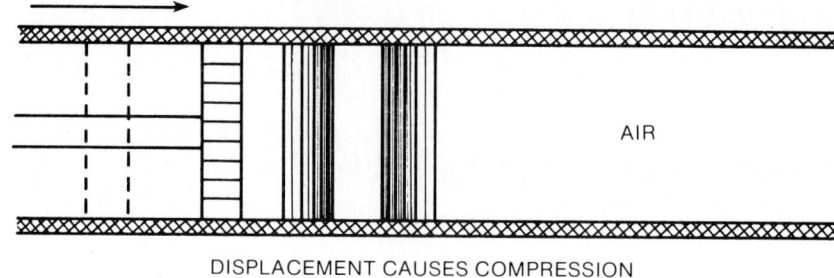

DISPLACEMENT CAUSES COMPRESSION

A

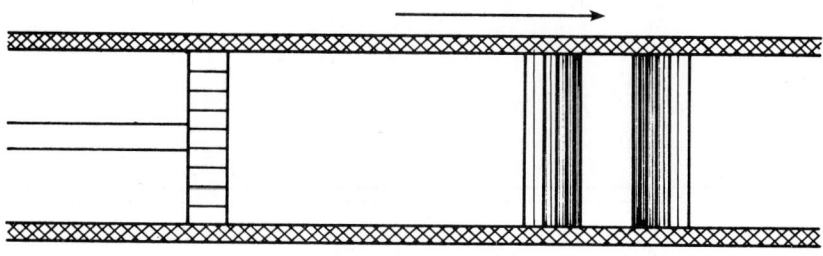

COMPRESSION MOVES

B

Fig. 23-1: Basic mechanism of a sound wave.

on a spring would vibrate; then when the piston moves to the right in the cylinder it would set up a *compression* as before. When the piston is moved to the left, the opposite of a compression will occur; that is, the pressure and density will be reduced below the equilibrium value of normal air pressure. This is called a *rarefaction*. A wave of alternate compressions and rarefactions would then be propagated down the cylinder as shown in Fig. 23-2 as sound waves.

The sound waves that are most familiar to the ear are mechanical vibrations occurring in air at a rate to which the ear is sensitive. In other words, the physical effect interpreted by the ear as sound consists of a pressure wave in air. It is a succession of variations in pressure above and below the normal pressure of air, which at sea level is 14.7 pounds per square inch (psi).

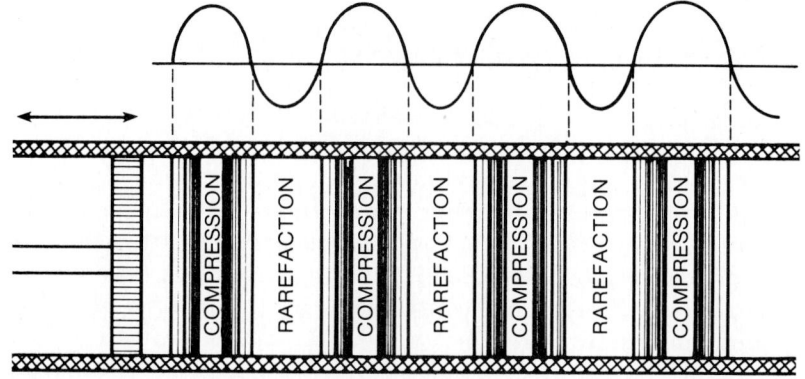

Fig. 23-2: Production of a sound wave. Compression is the point of maximum air pressure; minimum pressure is a rarefaction.

580

Frequency and Wavelength of Sound Waves

The rate of vibration of sound is expressed in hertz per second just as are electrical frequencies. The sounds of human speech lie between 100Hz and 10,000Hz; whereas, the human ear can hear sound in the frequency spectrum from 20Hz to 20,000Hz, although individual ears differ greatly. Of course, musical instruments and other mechanical devices extend the range in both directions.

As can be seen in Fig. 23-3, the range of the pipe organ encompasses the spectrum of human hearing, while the piano keyboard has a range of *fundamental* notes from 16Hz to 4,186Hz. Note the stress on "fundamental." When you strike the key to sound an A-note, for example, the string will vibrate at a *fundamental* frequency of 440Hz, which is the characteris-

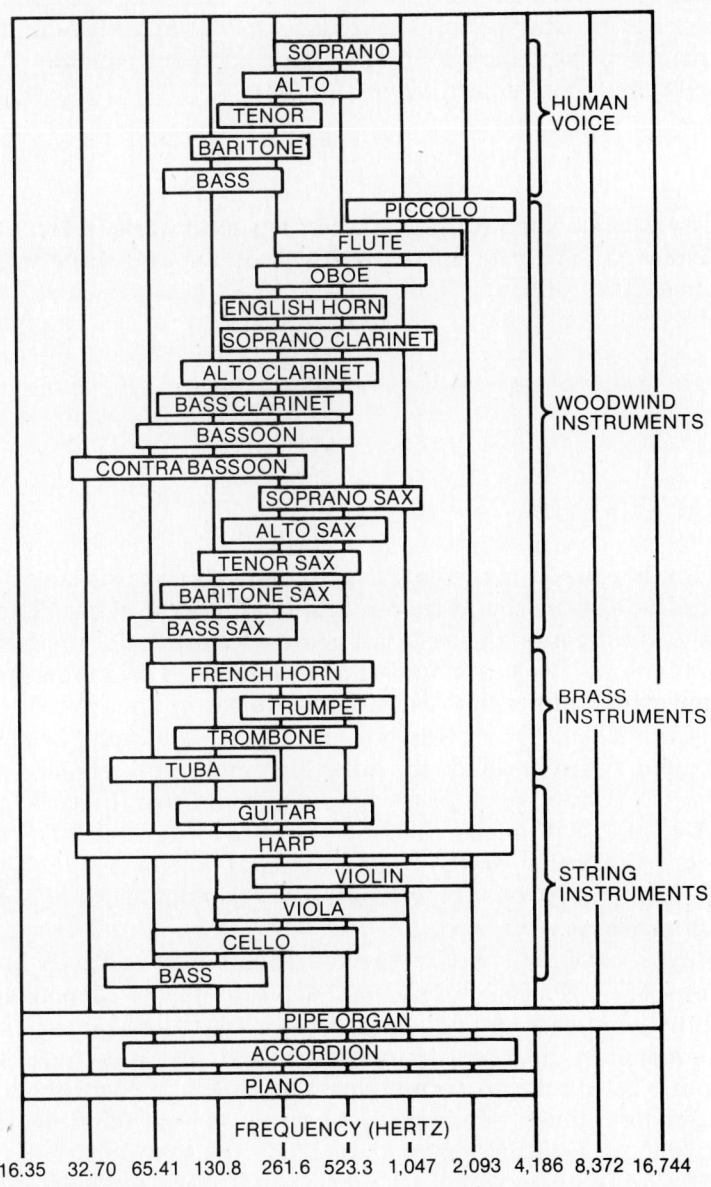

Fig. 23-3: Frequency range of musical instruments.

tic pitch to which all instruments of an orchestra are tuned. In addition to the fundamental, however, the string will also be excited to *sympathetic* vibrations or *overtones* at *multiples* of the fundamental; that is, an overtone of 880Hz (called the *second harmonic*), one of 1,320Hz (the *third harmonic*), one of 1,760Hz (the *fourth harmonic*), and so on. The fourth harmonic of the 4,186Hz piano note thus reaches up to 16,744Hz—the upper limit of human hearing. Although the intensity (loudness) of these harmonics is much less than that of the fundamentals and varies from instrument to instrument, their presence determines the characteristic quality and timbre of different instruments.

The speed or *velocity* (*v*) of sound waves is approximately 1,130 feet per second (ft/s) (344.43 meters per second [m/s]) in dry air at a temperature of 68°F (20°C). In solids and liquids the velocity is much greater. Sound cannot exist in a vacuum. As described in Chapter 11, the *wavelength* of any wave motion depends on the velocity of propagation and the frequency of the variations. For a sound wave, this is

$$\lambda = \frac{1,130\text{ft/s}}{f}$$

where λ is wavelength in feet and f is sound wave in Hz.

Example: What is the wavelength of a sound wave with a frequency of 900Hz?
Solution:

$$\lambda = \frac{1,130\text{ft/s}}{f} = \frac{1,130}{900} = 1.26'$$

LOUDNESS

How loud a sound seems to the ear depends on the intensity or amplitude of the sound waves. It should be kept in mind, however, that the human ear is rather slow to respond to increased stimulation. Thus, a sound that is increased in intensity to *twice* its original value (twice the power) is *not* heard twice as loud, but it seems to have hardly changed in loudness. To make it *sound* twice as loud, the intensity must be increased *ten times*. Moreover, to make the sound appear *three* times as loud as before, the intensity must be increased a *hundredfold* over the original value, and to make it appear *four* times as loud, the sound intensity must be increased a *thousand times* over the original value.

A relationship between two quantities, where one goes up by jumps of 1, 10, 100, 1,000, etc., while the other (dependent) quantity increases by 1, 2, 3, 4, and so on, is called *logarithmic*. The response of the ear to increased stimulation is thus logarithmic in nature; to obtain equal increases in loudness of a sound, its intensity must go up by ratios of ten each time. It is convenient, for this reason, to measure the sound intensity in *logarithmic units* which go up in equal steps for tenfold increases in intensity. Such a unit is the *decibel,* which is used to express ratios of intensity and increases in loudness. But remember that the decibel is *not* a unit of measurement, only a unit of difference in power levels.

The decibel is based on the manner in which the human ear recognizes sounds of different intensities. The ear responds to changes in sound intensity in a fashion that is logarithmic. Hence, the ear is much more responsive to changes in low intensity sound levels than it is to sound levels of higher intensity.

Since the decibel is based on the logarithmic response of the ear, the comparison of power changes in the various signals encountered in electronics and electricity is also based on logarithms. Hence, the word *decibel* indicates the change in volume level, power level, or intensity that occurs when the *average* ear is barely able to recognize a difference in a gradually changing amplitude of sound.

The unit *bel* was named after Alexander Graham Bell, the noted American scientist and inventor. The unit value, however, is not employed. Instead, one-tenth of a bel, the decibel (dB) is the common term.

Mathematically, the decibel for power has a relationship as expressed by the equation

$$dB = 10 \log \frac{P_1}{P_2}$$

In this formula the larger power is divided by the smaller to obtain the ratio of the two powers, and the logarithm of this ratio is multiplied by 10. Usage of this formula would indicate some basic shortcut methods for obtaining the decibel rating with respect to the relationship of two powers. A doubling of power, for instance, represents three decibels:

$$10 \log \frac{2}{1} = 10 \times 0.301 = 3.01 \ (\text{or } 3) \text{dB}$$

Thus, if the power of an electrical apparatus is increased 10 times, the difference in decibels would be 10.

$$10 \log 10/1 = 10 \times 1.0 = 10 \text{dB}$$

(*Note:* Values of logarithmic units are given in Appendix R.)

Audio Signals and Sound Waves

Figure 23-4 compares an audio signal with that of a sound wave. The audio signal is a variation in current (I) and voltage (V), while sound is a variation in air pressure. The audio signal travels through a conductor at a speed of 186,000mi/s, or 3×10^{10}cm/s, while a sound wave travels at only 1,130ft/s. The frequency of the audio signal is the same as that of the sound waves, including harmonic components.

Figure 23-5 shows how audio signals and sound waves interact. The human voice generates sound pressure waves by a vibrating vocal cord. These sound waves are composed of air

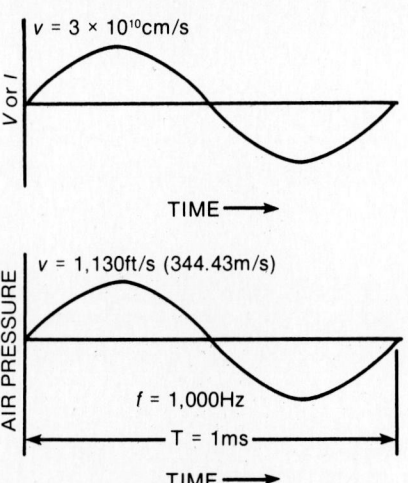

Fig. 23-4: Comparison of electrical audio signal with sound waves. Both are at a frequency of 100Hz.

583

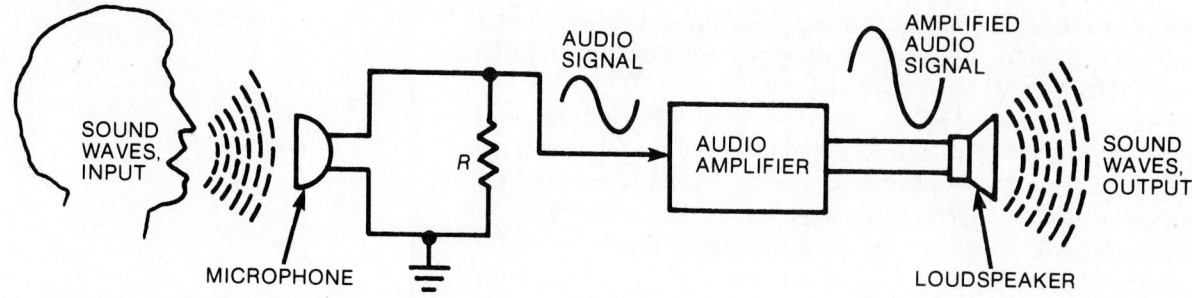

Fig. 23-5: Voice sound waves are converted by the microphone to an audio signal for amplification and then converted back to sound waves by a loudspeaker.

compressions and rarefactions that can be used to deform or displace a microphone element. The element, in turn, will convert the deformations into electrical energy that can be further amplified. After enough amplification, the audio signal is coupled to the loudspeaker. The speaker's vibrating cone converts the audio variations back into sound waves. The sound waves generated by the loudspeaker are also composed of air compressions and rarefactions that vibrate the ear drums, thereby allowing the listener to hear the sound.

The device that converts sound into audio signals and audio signals back into sound is called a *transducer*. The common sound-related transducers are:

• *Microphones*—devices for converting sound into electrical energy.

• *Phono Pickups*—devices for converting mechanical undulations in a record groove into electrical energy.

• *Loudspeakers*—devices that convert electrical energy into sound.

• *Earphones*—devices that convert electrical energy into sound; similar to a loudspeaker but used as an earpiece.

• *Magnetic Tape Recorders*—devices used to convert electrical energy into magnetic energy and to impress the magnetic pattern onto plastic tape, coated on one side with iron oxide.

MICROPHONES

A microphone is an electromechanical transducer that changes or converts sound energy (mechanical vibrations) into electrical energy (audio signals). When one speaks into a microphone, the sound pressure waves strike a transducer element in the microphone and cause it to move or vibrate in accordance with the instantaneous pressure delivered to it. The transducer element is attached to a device that causes current to flow in proportion to the instantaneous pressure applied to the element. Many devices can perform this function, each having characteristics that make its use advantageous under a given set of circumstances.

In many types of microphones, the output in electrical energy is considerably less than the input in acoustical energy. Some, however, are more efficient than others and some have a better frequency response than others. Microphones can be rated according to their (1) *frequency response*, (2) *sensitivity*, and (3) *impedance*.

Frequency Response. For good audio quality, it is important that the electrical signals from a microphone closely follow or repeat the magnitude and frequency of the sound waves that cause them so that no new frequencies are introduced. The addition of new frequencies will cause audio distortion when the original sound waves are reproduced. The *frequency range* over which the microphone is capable of responding must not be wider than the desired overall response limits of the system with which it is used. The microphone response must be uniform, or flat, within its frequency range and free from sharp peaks or dips in its response curve. To attain this condition, some form of damping may have to be used. To meet high-fidelity requirements, a good quality microphone must have a frequency response that is flat within ± 1dB from 20Hz to 20kHz. A typical response curve for a microphone is given in Fig. 23-6.

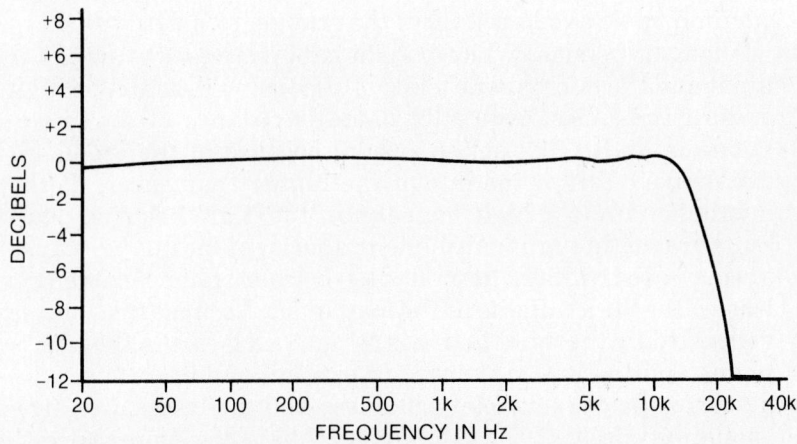

Fig. 23-6: Typical microphone response curve.

Sensitivity. The *sensitivity*, or efficiency, of a microphone relates the output voltage to the input sound pressure. It is specified at a given frequency between 200Hz and 1kHz in terms of the reference sensitivity of 1V output for a pressure of 1 Pascal (Pa) or 1 volt per microbar (1V/μB).

It is important to have the sensitivity of a microphone as high as possible. High sensitivity means a high output level for a given input sound level. High microphone output levels require less gain in the amplifier stages that follow in a circuit and provide optimum performance in overcoming thermal noise, amplifier hum, and noise pickup. Table 23-1 gives sensitivity and frequency response values for some typical microphones.

Impedance. The *impedance* of a microphone is usually measured between its terminals at some arbitrary frequency in

**Table 23-1: Sensitivity and Frequency Response
of Typical Microphones**

Type of Microphone	Sensitivity (dB at 1V/Pa)	Frequency Response, kHz (approximate)
Electret-condenser	-38	0.005 to 25
Air condenser	-26	0.005 to 20
Ceramic	-40	0.005 to 12

the audio range—for example, 1kHz. The actual impedance of a microphone is important because it relates to the load impedance into which the microphone is designed to operate. If the load has a high impedance, the microphone should have a high impedance and vice versa. Of course, impedance matching devices may be used between the microphone and its load (such as transformer, emitter-follower circuits, etc.).

Microphones can be further classified by their impedance values. Low impedance types include the carbon, dynamic, and ribbon microphones. High impedance types include the crystal, ceramic, and condenser microphones. Crystal microphones have impedances of several thousand ohms; whereas, dynamic microphones have impedances that range from 20Ω to 600Ω.

A long transmission line or audio cable between the amplifier input and the microphone tends to seriously alternate the high audio frequencies, especially if the impedance of the microphone is high. This action results because of the increased capacitive effect of the line at the higher frequencies. If the microphone has a high impedance, the high frequency currents drawn through the inherent capacity of the line will cause an increased voltage drop across the microphone; therefore, less voltage is available at the load input. Because the voltage generated by the microphone is minute, all losses in the microphone and the line must be kept to a minimum. At lower frequencies the capacitive effect is less and the losses are correspondingly less. If the microphone has a low impedance, a lower voltage drop will occur across the microphone and more voltage will be available at the load input.

Remember that microphone cables must be well shielded to prevent any possible interference pickup. Any unwanted signal introduced through the microphone cable can find its way to the amplifier output. To overcome the cable problem, the *wireless* microphone is used. This method of transmission employs a small radio transmitter in the microphone assembly to build up the signal to the amplifier.

Carbon Microphones

The carbon microphone is the most common type of microphone. It operates on the principle that a change in sound pressure on a diaphragm that is coupled to a small volume of carbon granules will cause a corresponding change in the electrical resistance of the granules.

The *single button* carbon microphone (Fig. 23-7) consists of a diaphragm mounted against carbon granules that are contained in a small cup or button. In order to produce an output current, this microphone is connected in a series circuit containing a battery and microphone transformer. The pressure of the sound wave on the diaphragm causes the carbon granules to be compressed and decompressed, thereby varying the resistance of the carbon granules. The resistance of the carbon granules will vary inversely with the sound pressure exerted upon them. The current flow from the battery through this resistance will vary inversely with the resistance; in other words, as the resistance increases, the current decreases and vice versa. These DC current variations in the primary of the transformer will develop an AC signal across the secondary of the transformer. The AC signal has essentially the same waveform as that of the sound waves striking the diaphragm. The current through the carbon microphone may be as great as 100mA and as low as 5mA (as measured on the ammeter in the circuit), and the resistance may vary from about 50Ω to 90Ω. The audio frequency signal developed across the secondary will, of course, depend on the turns ratio of the transformer and also upon the rate of change in the primary current.

Characteristically the carbon microphone is noted for its low impedance, limited frequency response (approximately 250Hz to 2,700 Hz), and a rated button current that varies from 5mA to 100mA. Table 23-2 outlines advantages and disadvantages of the typical single button carbon microphone.

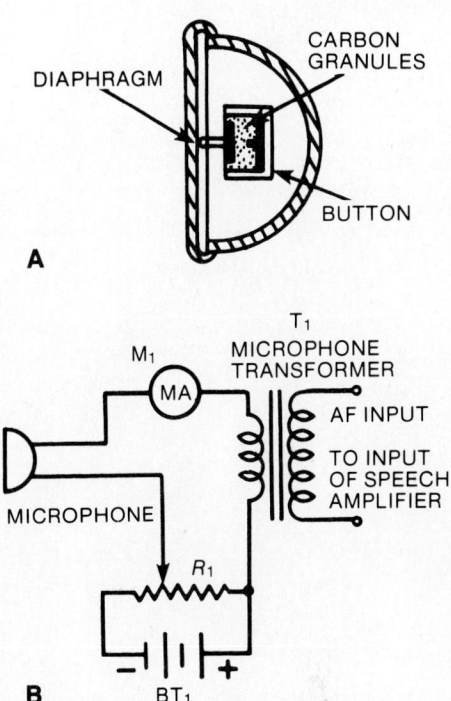

Fig. 23-7: A single button carbon microphone: (A) structural diagram and (B) schematic diagram.

Table 23-2: Single Button Carbon Microphone Advantages and Disadvantages

Advantages	Disadvantages
1. High sensitivity—on the order of –50dB (acts as an amplifier)	1. Carbon granules may stick and become packed (hitting the microphone will loosen the granules).
2. Lightweight and portable	2. Carbon hiss generated by DC bias current passing through carbon granules.
3. Low cost	3. Limited frequency response restricts its usage to voice communications such as telephone or radio.
4. Rugged construction	4. Requires a DC source for operation.

The *double button* carbon microphone, shown in Fig. 23-8, tends to overcome some of the shortcomings of the single button microphone. Here, one button is positioned on each side of the diaphragm so that an increase in pressure and resistance on one side is accompanied simultaneously by a decrease in pressure and resistance on the other side. Each button is in

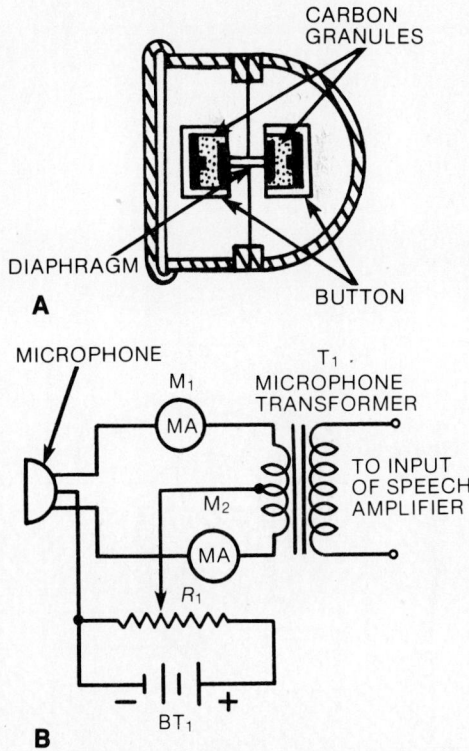

Fig. 23-8: A double button carbon microphone: (A) structural diagram and (B) schematic diagram.

587

series with the battery and one-half of the transformer primary. The decreasing current in one-half of the primary and the increasing current in the other half produce an AC signal in the secondary that is proportional to the sum of the primary signal components. This action can be related to that of a push-pull amplifier.

The double button carbon microphone has the advantage of a somewhat improved frequency response over a single button microphone. In addition, carbon hiss is balanced out to some extent by the push-pull feature and the microphone operation provides a no core saturation condition for the transformer.

Dynamic Microphones

The *dynamic*, or moving-coil, microphone (Fig. 23-9) consists of a coil of wire attached to a diaphragm and is so constructed that the coil is suspended and free to move in a magnetic field.

The dynamic microphone has good frequency response over varying output impedances and a high sensitivity second only to the carbon microphone, requires no external voltage source, and is lightweight and small in size. The impedance of this microphone is low, usually on the order of $600\,\Omega$. The dynamic microphone is also very directional for high frequency sounds because of the sensitive diaphragm used in the microphone.

The *dual impedance ("Z") dynamic* microphone (Fig. 23-10) allows the user to select a low or high impedance output for the application at hand. Closing the switch in the primary of the microphone transformer allows the use of the low impedance output of the microphone. This selection usually provides better audio quality; the lower the impedance, the greater the output current and the less noise on the output. Opening the switch on the transformer primary permits use of the high impedance output of the microphone. The impedance output selected will generally depend on the load impedance (input) of the audio amplifier to be used.

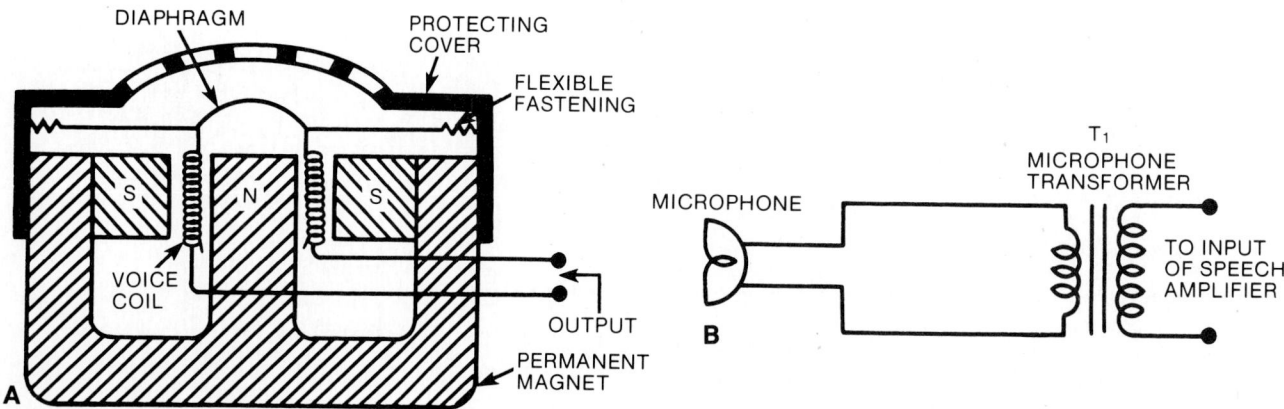

Fig. 23-9: A dynamic microphone: (A) structural diagram and (B) schematic diagram.

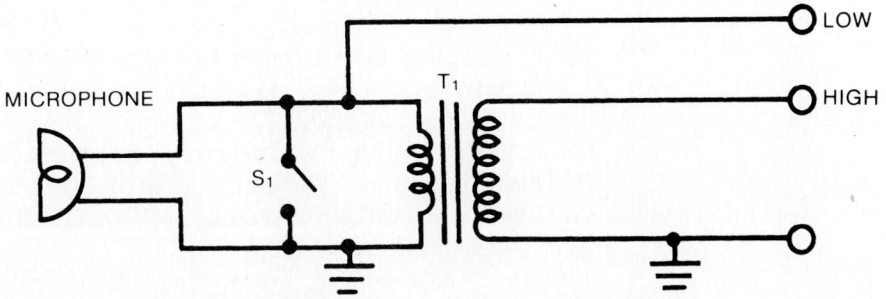

Fig. 23-10: Typical circuit for a dual impedance dynamic microphone.

Ribbon Microphones

The *ribbon,* or velocity, microphone (Fig. 23-11) uses a corrugated aluminum ribbon mounted or suspended in a magnetic field. Sound waves striking the ribbon cause it to vibrate. The vibrating ribbon cuts the lines of magnetic flux; this induces a voltage in the ribbon. The induced voltage is proportional to the air pressure variations striking the ribbon. This microphone is characterized by its low impedance and its very high directivity.

The ribbon microphone exhibits very good frequency response, has better sensitivity than other low impedance microphones, and requires no external voltage source. Because there is only one conducting ribbon, the microphone provides a low output power. In addition, the ribbon is very delicate and must be protected from air drafts and air blasting.

Crystal Microphones

The *crystal* microphone (Fig. 23-12) utilizes a property of crystals—such as quartz and Rochelle salt—known as the *piezoelectric effect* (see Chapter 2). The bending of the crystal resulting from the pressure of the sound wave produces a voltage across the faces of the crystal.

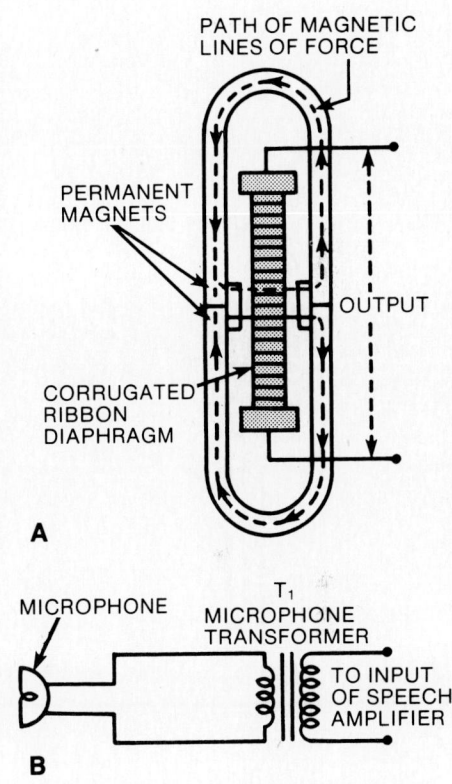

Fig. 23-11: A typical ribbon microphone: (A) structural diagram and (B) schematic diagram.

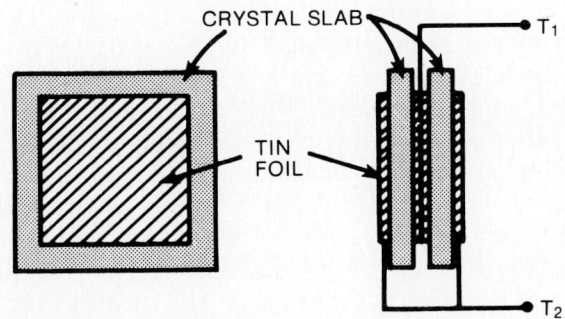

Fig. 23-12: Basic construction of a crystal microphone cell.

The crystal microphone consists of a diaphragm that is directly attached or cemented to one surface of the crystal. A metal plate, or electrode, is attached to the other surface of the crystal. When sound waves strike the diaphragm, the vibrations produce a varying pressure on the surface of the crystal; therefore, a voltage is induced across the electrode. The induced voltage has essentially the same waveform as that of the sound wave striking the diaphragm.

Rochelle salt is most commonly used in crystal microphones because of its relatively high voltage output; however, a large percentage of crystal microphones use some form of the *bimorph cell*. In this type of cell, two crystals are cemented together and used in lieu of the single crystal. The crystals are cut and oriented so that their output voltages are additive.

This type of microphone, whether it be Rochelle salt or bimorph cell, has high impedance (approximately 50kΩ to 100kΩ), is light in weight and somewhat rugged, requires no external voltage source, is nondirectional, has good frequency response with the emphasis on the high frequency end, and has a low cost. The crystal microphone is, however, sensitive to high temperature, humidity, and rough handling; therefore, it is restricted in use to certain environmental conditions. The crystals may also be damaged or destroyed by the battery voltage of an ohmmeter when voltage or resistance measurements are made. The crystal microphone finds widespread use in broadcast work where its relatively high output is an advantage.

Ceramic Microphones

Synthetic ceramic materials can also be manufactured to possess piezoelectric characteristics by mixing a solution of barium and calcium titanate. The structure and principle of operation of this type of microphone is the same as for that of the crystal unit.

The ceramic microphone provides a better immunity to high temperature and humidity, has a good frequency response, and is more durable than the crystal microphone. However, the ceramic unit has a somewhat lower output than the natural Rochelle salt or crystal units.

Condenser Microphones

The *condenser,* or capacitor, microphone consists of a diaphragm, an insulating spacer, and a back plate. The diaphragm and back plate form a pair of capacitor plates with the insulating spacer as the dielectric. Sound waves striking the diaphragm vary the thickness of the dielectric and, therefore, the overall capacitance of the cell. This change in capacitance causes a varying current (charging and discharging) to flow through the load resistor at the audio rate. Figure 23-13 illustrates the basic construction of a condenser microphone.

The condenser microphone has good frequency response and a very high output impedance; however, this type microphone

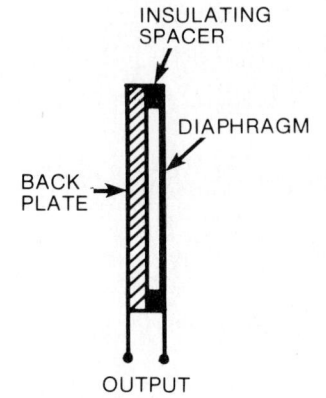

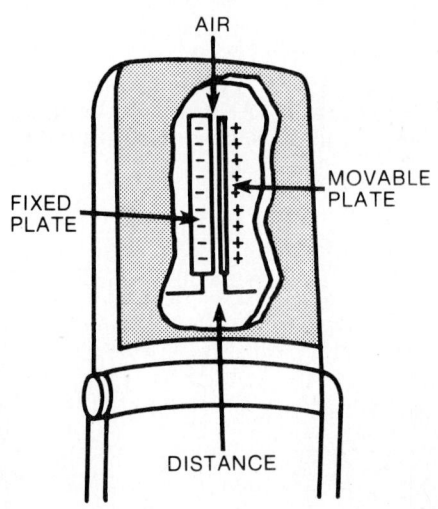

Fig. 23-13: Basic construction of a condenser microphone.

590

has a very low output (sensitivity), usually requiring a *pream-plifier stage,* and needs a DC voltage source to charge the capacitance.

Figure 23-14 shows a simple JFET preamplifier circuit for the condenser microphone. Usually the preamplifier circuit is self-contained in the microphone case.

The change in microphone capacitance varies the current flow through the load resistor which, in turn, charges and discharges the capacitor. The bias on the JFET is provided by the resistor-capacitor network (R_1 through R_4, C_2) between the source and ground. In this circuit, however, no microphone gain control is provided on the input; the gain of the stage is fixed. Microphone gain can be controlled by adding a potentiometer in the output or in a following stage.

A similar unit, called the *electret microphone,* consists of an insulator that contains a permanent electrical charge. This charge is produced by heating a material, such as carnauba wax, and placing it in a strong electric field during cooling. This type of microphone is better than the basic condenser microphone because the need for an external DC voltage source to charge the capacitance is eliminated.

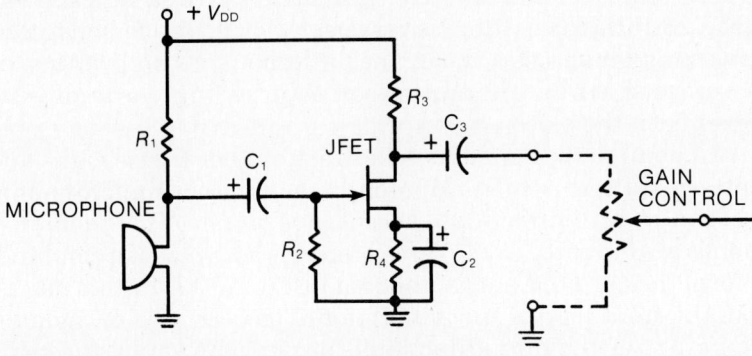

Fig. 23-14: JFET preamplifier circuit.

Microphone Pickup Patterns

No truly universal type of microphone has ever been devised. There are various situations in which a general-purpose microphone is not suitable and a special-purpose microphone is required. The most widely used types are the dynamic (moving-coil) microphone and the ribbon microphone. The ribbon microphone has acoustic characteristics that make it very useful in certain recording situations. Another design type in use is the condenser or electret microphone. The crystal or ceramic types of microphones are not classified as high fidelity transducers and find use in other communications areas. One of the important characteristics of any microphone is its acceptance pattern which indicates the microphone's directional characteristics.

The most widely used directional pattern is the *cardioid* characteristic as shown in Fig. 23-15. The cardioid somewhat

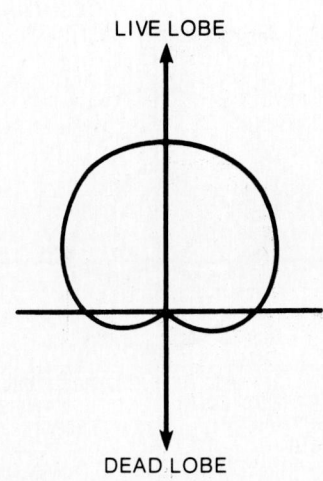

Fig. 23-15: Basic cardioid pattern.

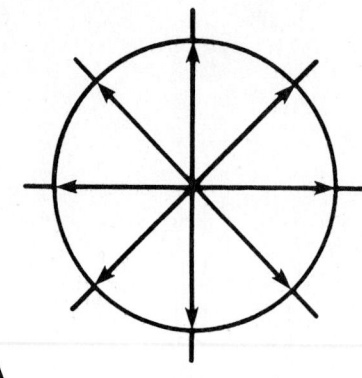

A

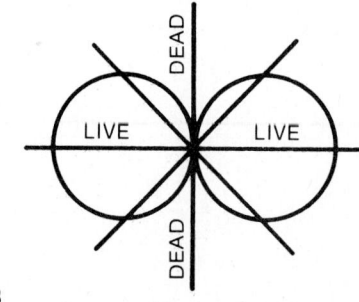

B

Fig. 23-16: Basic microphone acceptance patterns: (A) omnidirectional and (B) bidirectional.

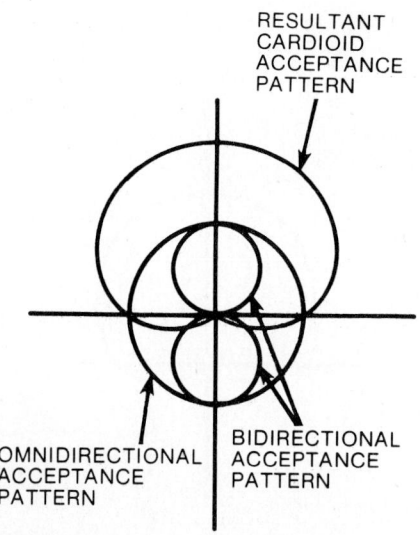

RESULTANT
CARDIOID
ACCEPTANCE
PATTERN

OMNIDIRECTIONAL
ACCEPTANCE
PATTERN

BIDIRECTIONAL
ACCEPTANCE
PATTERN

Fig. 23-17: Development of a cardioid directional pattern.

resembles a heart-shaped unidirectional pattern with its useful acceptance pattern encompassing approximately 180°. The pattern can be considered as having a live lobe and a dead lobe. The cardioid type microphone normally exhibits a so-called *proximity effect;* in other words, its frequency characteristics change when the sound wave is very close to the microphone; the bass response of the microphone increases. Some cardioid type microphones are provided with a switch to reduce the low frequency output in close-up applications. A cardioid pattern can be used in various applications. For example, a cardioid type microphone can be oriented with its live lobe toward a specific sound source and its dead lobe toward an ambient noise source, thereby enhancing the reception of the sound source.

An *omnidirectional acceptance pattern* is shown in Fig. 23-16A and is useful when there are various sound sources to be recorded from different directions, such as a round-table discussion. In this situation, an omnidirectional microphone may be placed in the center of the table to allow reception of all the speakers' voices. At a more formal banquet, a cardioid microphone could be placed before each speaker and their outputs connected to an audio mixer. This technique allows the volume of each speaker to be adjusted for optimum levels. If the levels are equalized, a more satisfactory recording will result. Observe the *bidirectional* or *figure eight* acceptance pattern shown in Fig. 23-16B. This pattern is very useful when one person interviews another across a table. The pattern allows both voices to be recorded while alternating or suppressing the ambient noises from the sides.

A ribbon microphone has a bidirectional or figure eight acceptance pattern while a dynamic (moving coil) microphone exhibits an omnidirectional acceptance pattern. If both microphones are mounted in the same housing, both outputs combine to form the cardioid pattern shown in Fig. 23-17. In this situation, the front lobe of the bidirectional pattern aids or adds to the output of the omnidirectional pattern; however, the rear lobe of the bidirectional pattern opposes the output of the omnidirectional pattern. The result of this combination is that the rearward sound that would normally be picked up by the bidirectional microphone is canceled and a cardioid pattern is produced.

Some bidirectional microphones have narrower lobes than others. These designs are classified into orders of acceptance patterns of lobe widths. The lobe widths can vary from full circular, to elliptical, to narrow. For a ribbon microphone, the lobe width is a function of acoustical design. It also follows that there are orders of acceptance patterns or lobe widths for omnidirectional microphones. Various microphone pickup patterns can be generated by using or combining the two microphones with different or varying lobe widths. The resultant patterns from the various combinations are shown in Fig. 23-18. The acceptance patterns include *omnidirectional, cardioid, supercardioid, hypercardioid,* and *bidirectional.* The supercardioid and hypercardioid produce patterns that discriminate much more effectively against ambient noise from the rear of the microphone.

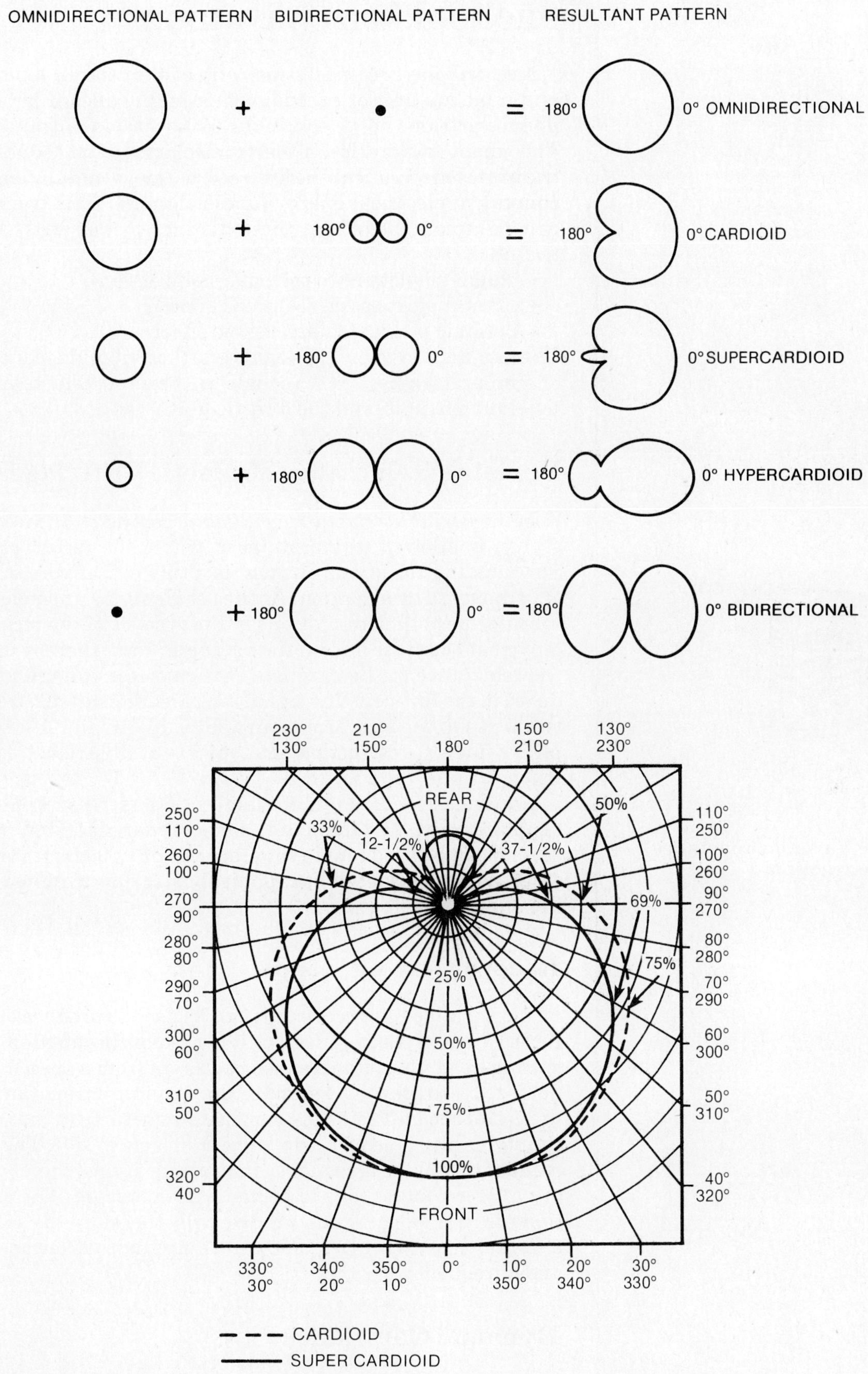

Fig. 23-18: Various microphone pickup patterns.

PHONOGRAPH CARTRIDGES

A record player consists basically of a turntable, a tone arm, and a pickup unit or cartridge. Our main concern here is the phono cartridge as we continue the discussion of transducers. The phono pickup is an electromechanical transducer that translates mechanical energy (record-groove undulations) into equivalent electrical energy (audio signals). This translation can be accomplished by three distinct pickup units or cartridges:

- Magnetic pickups (electromagnetic action).
- Crystal pickups (piezoelectric effect).
- Ceramic pickups (piezoelectric effect).

The magnetic type pickups can be further subdivided into units of similar designs. These include the variable reluctance unit, the dynamic unit, and the dynetic unit.

Variable Reluctance Magnetic Cartridges

In the *variable reluctance magnetic* cartridge, the needle, or stylus, is allowed to vibrate as it traces the record grooves. Surrounding the stylus element is a tiny coil of wire which is encompassed by a magnet. As the needle stylus and pole pieces vibrate due to the undulations in the grooves of the record, the movement cuts the lines of the magnetic force, thereby varying the reluctance of the magnet. This causes a voltage to be induced in the tiny coil. The signal is carried from the cartridge by two wires in the tone arm to an applicable preamplifier or amplifier stage. Construction of a typical variable reluctance cartridge is shown in Fig. 23-19.

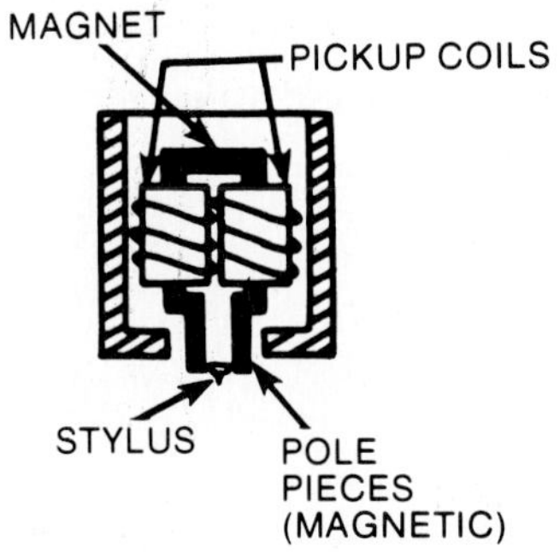

Materials used for the manufacture of the stylus include steel, sapphire, ruby, and diamond. A stylus manufactured from diamond can be used for playing records for 1,000 hours or more; whereas, a stylus made of sapphire or steel has a limited useful life of about 40 hours.

Fig. 23-19: Variable reluctance pickup construction.

Record Cutting Heads

The variable reluctance unit can be used to produce or "cut" records. When the magnetic unit is used in conjunction with a cutting head, the audio frequency current from a speech amplifier is fed into the coil terminals. The varying current flowing in the coil of the unit sets up a varying magnetic field that causes the stylus and pole pieces to move from side to side. The movement of the stylus also moves the cutting stylus, thereby causing a lateral-cut groove to be made in the record. The undulations in the sides of the groove will, therefore, be made in accordance with the frequency and amplitude variations of the audio-frequency current.

Dynamic Cartridges

The *dynamic* cartridge, a magnetic pickup, uses a coil which is mechanically coupled to a magnet by an upper and lower bearing. The coil of wire is on the order of one-fourth the thickness of human hair. The coil is made to vibrate in the magnetic field via the stylus which follows the grooves of the record. This vibration induces a voltage in the suspended coil. This type of cartridge exhibits good frequency response with low distortion, especially at the low frequencies. Its output, however, is lower than that of the basic variable reluctance unit. A stereo turntable utilizing a dynamic cartridge is manufactured with two magnets and two coils. A diagram of the dynamic cartridge is shown in Fig. 23-20.

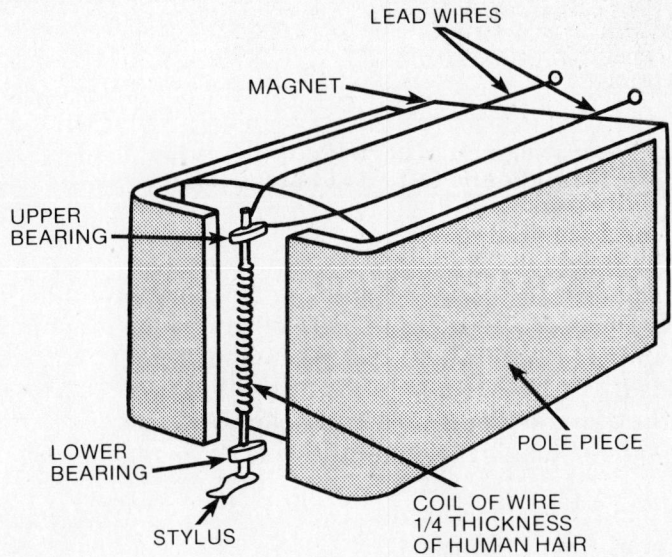

Fig. 23-20: Basic construction of a dynamic cartridge.

Dynetic Cartridges

The *dynetic* cartridge, also known as a magnetic pickup unit, uses the armature as a permanent magnet that is made to vibrate in a stationary coil. The vibrating armature induces an audio frequency current in the coil which is fed to a preamplifier or amplifier. A diagram of the dynetic cartridge is shown in Fig. 23-21.

This type of cartridge has an excellent frequency response over the entire audio range and a very low tracking force. Tracking force is the pressure applied to keep the stylus riding smoothly in the record grooves.

Magnetic Pickups for Stereo

A dynetic cartridge or dynamic cartridge can be used for playing back stereo records. A typical dynetic stereo cartridge (Fig. 23-22) consists of a stylus, a movable yoke, and two coils. As the stylus follows the record grooves it moves the yoke, thereby electromagnetically inducing the audio frequency signals into the left and right coils in the cartridge. The yoke is a

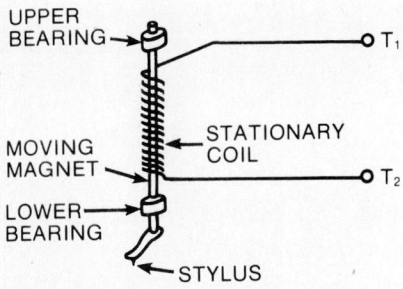

Fig. 23-21: Basic construction of a dynetic cartridge.

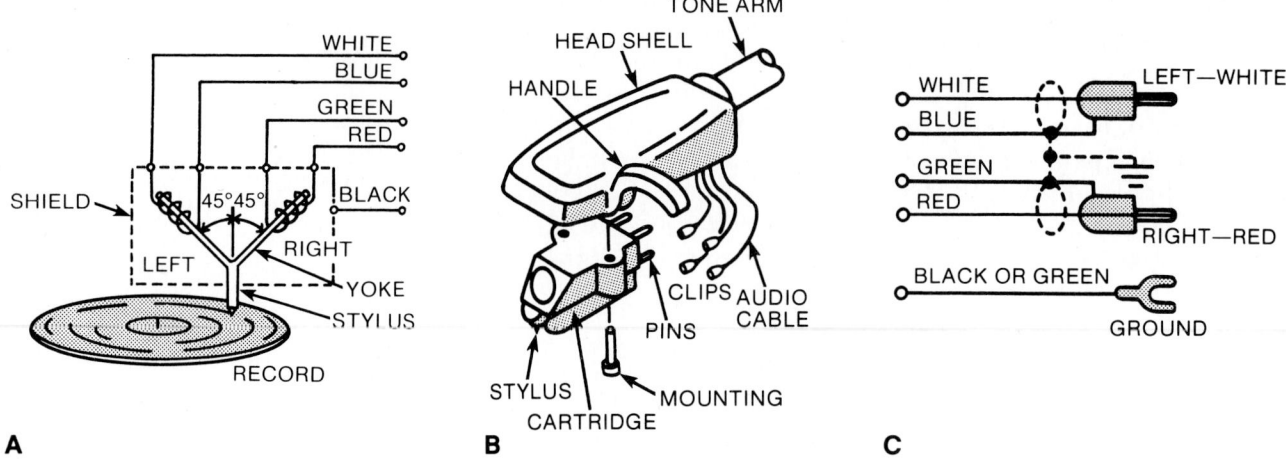

Fig. 23-22: Stereo magnetic pickup cartridge: (A) construction of cartridge with output leads, (B) cartridge mounted in head shell on tone arm, and (C) right and left phono plugs for connecting to the amplifier.

movable magnet core in the dynetic cartridge. The left and right coils are mounted 90° apart or 45° apart from the yoke center. Two sets of wires from each coil are fed to the applicable preamplifier or amplifier stage. A common or shield wire is also provided.

In the dynamic type of cartridge for stereo playback units, the magnetic flux is fixed and the coils are movable.

Crystal Cartridges

The *crystal* cartridge operates on the principle that certain crystals will produce a voltage when pressure is exerted upon their surface. This cartridge consists of a stylus arm assembly connected or mounted to a mechanical lever. Between the mechanical lever a bimorph crystal element is mounted. A diagram of the crystal cartridge is shown in Fig. 23-23.

The vibration of the stylus arm in the record grooves distorts the crystal element which, in turn, generates a voltage across the faces of the crystal. This type of cartridge has a high fidelity output, is lightweight (thereby reducing background noise and minimizing wear), and generally uses Rochelle salt for the

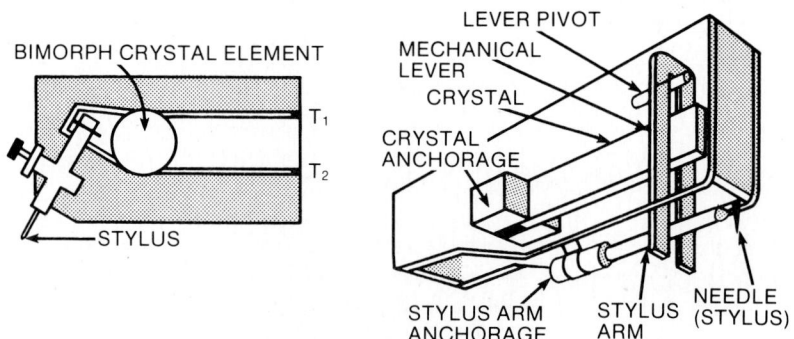

Fig. 23-23: Construction of a crystal cartridge.

596

crystal element. One disadvantage of this cartridge is that the crystal element may be damaged by extremes of temperature and moisture.

Ceramic Cartridges

The *ceramic* cartridge can be made to possess piezoelectric characteristics by mixing a solution of barium and calcium titanate. The basic structure and the principle of operation are the same as those discussed for the crystal unit. Figure 23-24 provides a diagram of a ceramic stereo cartridge.

The ceramic cartridge provides better immunity to heat and humidity over that of a crystal unit, is less expensive than that of magnetic types, and has a higher output over that of magnetic types.

The turnover lever shown in Fig. 23-24 permits the selection of a stylus. A three-unit stylus is provided on the cartridge for the older type 78rpm records. A fine groove stylus, 0.7 units or less, is also provided for use with modern type 33-1/3 or 45rpm records.

Phonograph Records

In the early days, phonograph records were manufactured as shellac records and were designed for 78rpm turntables. The grooves or units in these records were rather wide. Shellac records had several disadvantages. They were heavy, bulky, and very inflexible; cracked easily; had limited playing time; and were very susceptible to scratches that eventually produced surface noise.

Modern type records are made of vinyl and are designed for 45 and 33-1/3rpm turntables. This type of record uses fine groove cuts—approximately 250 grooves per inch. The newer vinyl records are noted for their longer playing time (LP records); they have less surface noise and are also very flexible and durable.

A typical stylus point-record groove interface is shown in Fig. 23-25. The stylus point is more rounded than pointed and rides in the record groove following its undulations. Stereo type records utilize a 45°-45° groove system; whereby, right and left audio signals are cut on opposite walls in the record groove.

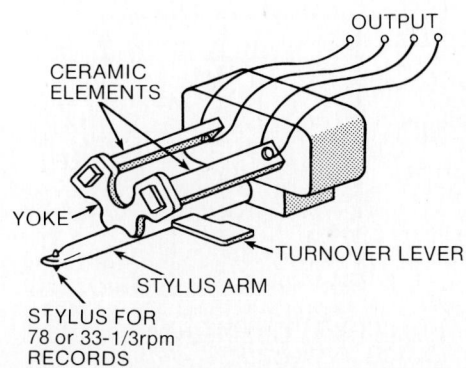

Fig. 23-24: Construction of a ceramic stereo cartridge.

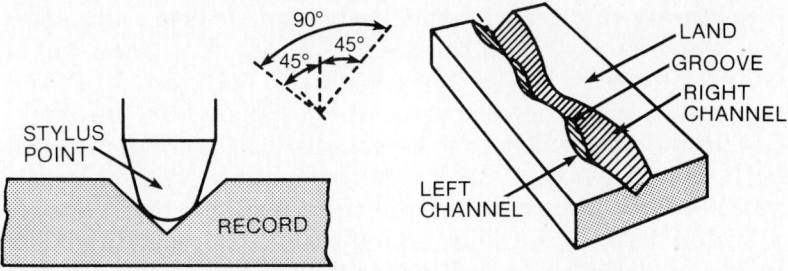

Fig. 23-25: Typical stylus record groove interface.

597

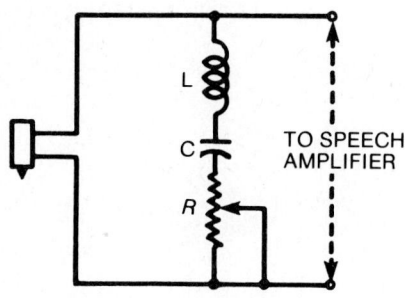

Fig. 23-26: Typical scratch filter circuit.

Scratch Filters

Surface scratch or needle scratch is produced by irregularities on the surface of the record during manufacture or by extended playing time. These noises have a frequency of about 5kHz and can be filtered out by the use of an appropriate circuit.

A typical scratch filter circuit is shown in Fig. 23-26. In this circuit a series resonant circuit, consisting of an inductor (L), capacitor (C), and a variable resistor (R), is used to filter out the interfering noise. The series circuit is approximately tuned to 5kHz, the variable resistor (R) allowing finer tuning to resonance. One main disadvantage of this circuit is that any audio signals in proximity of 5kHz will also be filtered out depending on the band width of the resonant circuit. This could cause a serious loss in frequency response of the filter not properly designed.

Antiskating Adjustment

The term "skating" refers to the tendency of the tone arm on a turntable to move inward and apply more force on the inside edge of the record grooves. If the tone arm or stylus should become clear of the record groove, the tone arm will skate across the top of the record toward the spindle. This skating action can severely scratch or damage the record grooves, thereby sacrificing frequency response and fidelity. Antiskating controls on a turntable apply a slight outward pressure on the tone arm to balance or counteract the skating effect. Usually more expensive turntables incorporate these antiskating controls. Tone arm sensitivity can be adjusted by moving weight to the end of the arm for the proper balance.

LOUDSPEAKERS

The loudspeaker is a transducer at the end of a chain of components comprising an audio system. The function of the loudspeaker is the reverse of the microphone and pickup cartridge which are transducers at the beginning of the chain. The loudspeaker converts electrical energy (audio signals) into mechanical energy (air pressure variations or sound). The most widely used unit is the dynamic loudspeaker.

The dynamic loudspeaker uses a permanent magnet or an electromagnet to generate the fixed magnetic field and a voice coil which is wound on the apex of a cone. The voice coil is connected electrically to the output of an audio amplifier and mechanically to a speaker cone. Audio signals from the amplifier through the coil change the magnetic field around the coil with respect to the magnet. The resultant vibrations of the voice coil are transmitted to the speaker cone which, in turn, vibrates. These vibrations, acting on air, produce sound. Dynamic loudspeakers can be classified as permanent magnet (PM) type or electrodynamic type.

Permanent Magnet (PM) Dynamic Loudspeakers

The *PM dynamic* loudspeaker (Fig. 23-27) uses an alnico permanent magnet to produce a strong, fixed magnetic field. The magnetic flux is concentrated in the air gap between the pole pieces and the paper cone assembly. The voice coil is mounted in the air gap. When audio signals flow in the coil, a magnetic field is set up about the coil proportional to the current flowing through the coil. This magnetic field interacts with the magnetic field from the permanent magnet, thereby causing the coil to vibrate in accordance with the audio signals. A speaker diaphragm attached to the voice coil is moved in accordance with the audio signals. The moving diaphragm vibrates a paper cone which, in turn, produces sound. The stiffer the diaphragm, the higher the frequency response of the loudspeaker; the looser the diaphragm, the more the movement of air, therefore, the lower the frequency response. Some characteristics of the PM-type dynamic loudspeaker are as follows:

- Light weight, low cost.
- Sizes range from about 2″ to 16″.
- Power output ranges from about 1W to over 100W.
- Impedance of the voice coil ranges from about 3Ω to 15Ω.
- No field coil or separate power supply is required.

The PM dynamic loudspeaker can also be used as a microphone or it can be used in an intercom system where the same speaker is used as a microphone and a loudspeaker. A typical intercom circuit using a simple microphone-loudspeaker is shown in Fig. 23-28. Utilizing the given switches, two-way communications can be provided using a PM speaker/microphone unit at each station. The variable potentiometer (*R*) serves as a volume control.

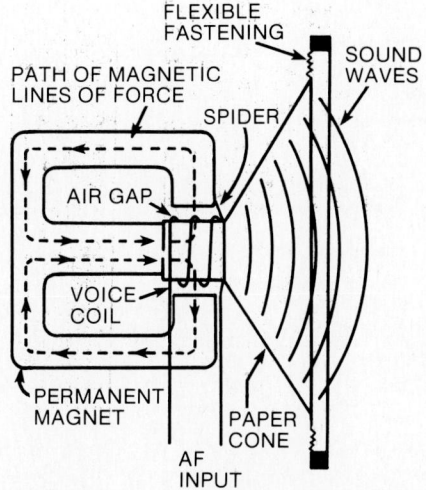

Fig. 23-27: **Permanent magnet dynamic loudspeaker.**

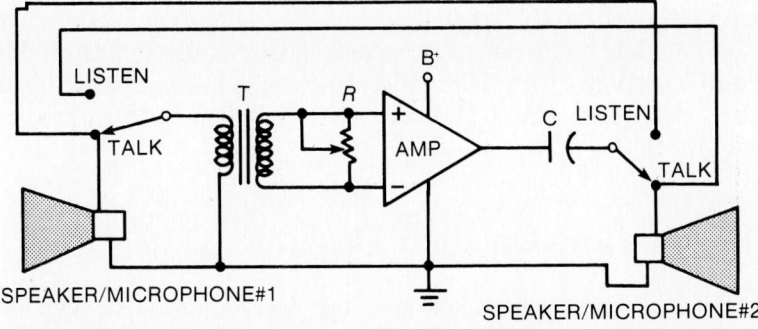

Fig. 23-28: **Typical two-way intercom circuit.**

The spider element shown in Fig. 23-27 is a flexible metal or a composition metal material that is used to connect the speaker cone to the frame and centers the cone on the core of the magnet. The flexible fastening in front of the loudspeaker protects the paper cone from damage.

Figure 23-29 illustrates an audio push-pull amplifier stage connected to a PM loudspeaker.

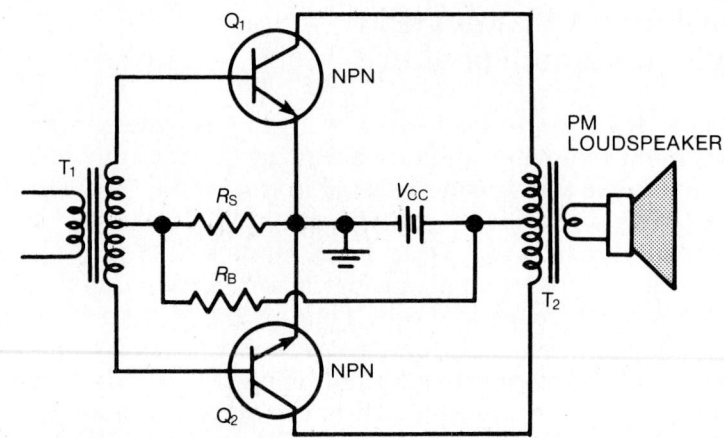

Fig. 23-29: Audio push-pull amplifier circuit suitable for PM loudspeaker.

Electrodynamic Loudspeakers

An electromagnet may be used instead of a permanent magnet to form an *electrodynamic* loudspeaker (Fig. 23-30). In this form, however, sufficient DC voltage must be available to energize the magnetic field. The operation of this type loudspeaker is otherwise much the same as that of the PM-type. Some features that are characterized by the electrodynamic loudspeaker are as follows:

- A field coil is used to produce a fixed magnetic field.
- The field coil must be energized by a DC source.
- Size ranges from about 3″ to 18″.
- Power output ranges from about 2W to 50W.
- Impedance of the voice coil ranges from about 2Ω to 10Ω.
- In the average radio receiver, the field coil was used as a filter choke in the power supply. The DC resistance of the coil was approximately $1,000\Omega$.
- The field coil was never used as a filter choke in automobile radio receivers. The field coil voltage was taken directly from the car battery. The DC resistance of the coil in this case was about 5Ω.

Various Loudspeaker Designs

Early magnetic type loudspeakers were of the horn type and the cone type (Fig. 23-31). The horn type utilized a coil wound on a permanent magnet which moved or vibrated a metal diaphragm. The diaphragm was directly connected to a horn-type speaker. The horn was a flared tube which increased the radiating efficiency of the vibrating diaphragm. The cone-type loudspeaker used a rigid wire connected to the diaphragm and a large paper cone. As the diaphragm vibrated, the rigid wire moved the paper cone to reproduce the sound.

The function of a horn can be thought of as a transformer or an impedance matching device. It takes the high amplitude air vibrations which are distributed over a small area at the throat

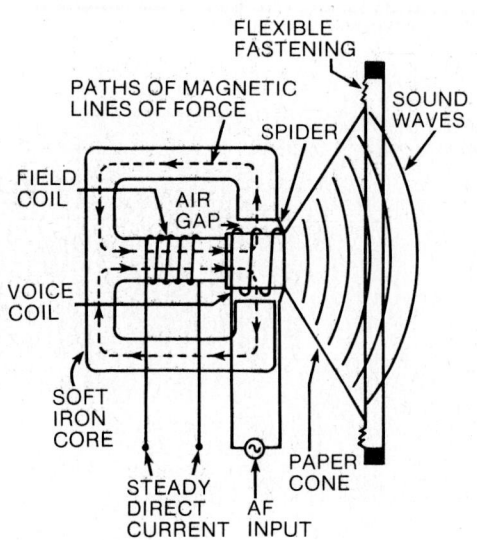

Fig. 23-30: Electrodynamic loudspeaker.

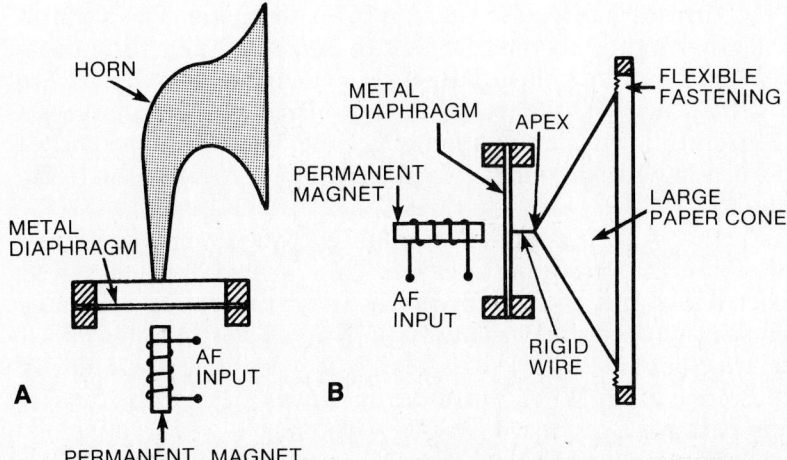

Fig. 23-31: Early magnetic type loudspeakers: (A) horn type and (B) cone type.

of the horn and converts them to air vibrations at the open end of the horn which are distributed over a larger area. Therefore, when a horn is coupled to a small vibrator or diaphragm, the result is that the horn concentrates the sound emitted from the diaphragm and also emits a larger amount of acoustic energy. The horn increases the radiation efficiency of a small source by effectively changing it to a large source. Figure 23-32 illustrates the basic structure of a horn-type loudspeaker and comparative horn shapes.

The properties of a horn are characterized by its flare, its length, and the size of the open end. The most common flare-types are *conical, parabolic, exponential,* and *hypex* (Fig. 23-32B). For proper operation, the size of the open end must be big enough to cut down the reflection back from the mouthpiece; therefore, the smaller the mouth of the horn, the less efficiently the bass notes are radiated. The rate of flare on the horn must also be gradual if it is to function properly.

Loudspeaker Frequency Response

Loudspeakers have been designed and constructed to accommodate certain frequency responses so that many subtypes of speakers have been produced. These subtypes can be combined and mounted into enclosures to provide the optimum in frequency response average. They include such speaker types as *woofers, tweeters,* and *midrange* devices.

A woofer is a low frequency, or bass tone, loudspeaker whose design enables it to reproduce the lower part of the frequency range, while limiting its ability in the higher audio range. Woofers may operate up to several thousand hertz, for example, up to about 3kHz. To reproduce the entire audio range of interest, a high frequency loudspeaker must also be used. This type of speaker is known as a tweeter, which can usually reproduce audio frequencies from about 3kHz up to about 20kHz.

Generally, woofers are characterized by large, heavy diaphragms and large voice coils that overhang the magnetic (air)

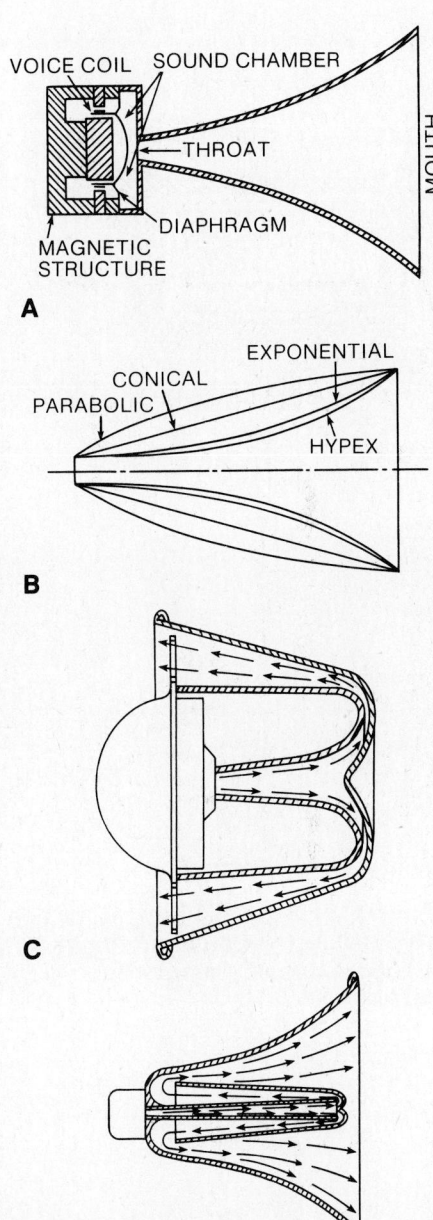

Fig. 23-32: Horn-type loudspeakers: (A) Basic structure of horn and driver unit. (B) Comparative shapes of different types of horns with the same throat and mouth diameters. (C) Single-fold horn reproducer for speech reproduction. (D) Two-fold horn reproducer for general reproduction of speech and music.

601

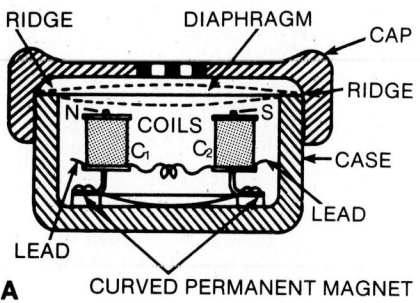

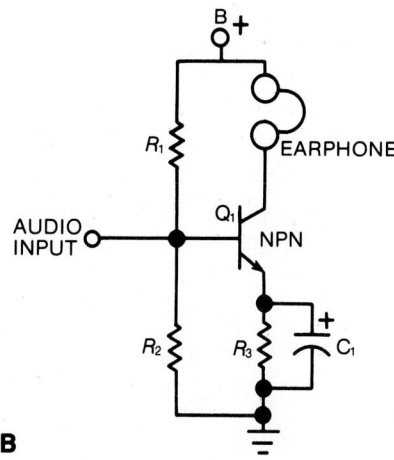

Fig. 23-33: (A) Construction and (B) operation of a magnetic earphone.

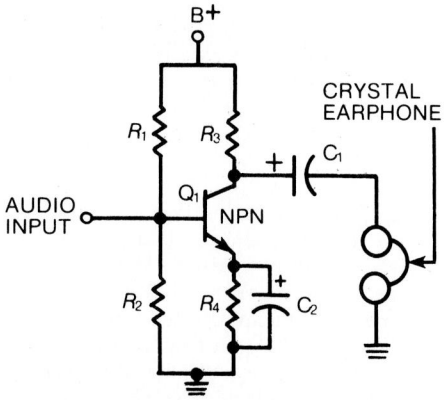

Fig. 23-34: Crystal earphone connected to an audio amplifier.

gap. The average woofer is 12″ to 15″ in diameter. Tweeters, on the other hand, are smaller in size (usually resembling cone-type woofers) with their diameters measuring from about 2″ to 6″. Another type of tweeter is the horn tweeter. This type uses a thin, flat diaphragm instead of a cone. Usually some sort of horn is mounted over the diaphragm to distribute the audio tones.

Further developments in improving frequency response have produced the midrange speaker. This device, physically and electrically, lies between a woofer and a tweeter and handles those frequencies between the two. When a woofer and a tweeter are combined into one enclosure, it is known as a two-way speaker system. When the midrange speaker is added, the system becomes a three-way speaker system.

Loudspeaker enclosures provide or serve two main functions. They prevent rear radiation of the speaker cones from cancelling the forward radiation, and the acoustics of the enclosure can contribute more uniform frequency response over the audio frequency range. The enclosure can also provide optimum sound dispersion and maximum operating efficiency.

EARPHONES

An earphone is a transducer similar to a loudspeaker. The earphone, however, is usually held to human ears by means of a flexible band over the head, thereby affording private listening to selected programming. Like the loudspeaker, the earphone converts electrical energy to mechanical or sound energy.

Two types of earphones are available: the *magnetic earphone* and the *crystal earphone*. The basic magnetic earphone (Fig. 23-33) consists of a curved permanent magnet, two coils (per earpiece), and a diaphragm. When no signal is present the permanent magnet exerts a steady pull on the diaphragm. Audio signals flowing through the coils produce a magnetic field that interacts (adds to or subtracts from) with the permanent magnetic field. This will cause the metallic diaphragm to vibrate in accordance with the audio frequency variations.

As far as circuits are concerned, signals applied to the base of an NPN transistor will cause the diaphragms in the earphone to vibrate or follow the signal variations. In this example, the earphone is connected in series with the collector of the transistor.

The audio signals can also be applied to a crystal in an earphone, thereby causing the crystal to vibrate in accordance with the audio signal variations. This type of operation is based on the piezoelectric effect. The crystal is mechanically connected to the diaphragm in the earphone. As the crystal vibrates, so does the diaphragm, therefore producing the required sounds. Figure 23-34 is a diagram of a crystal earphone connected to an audio amplifier. In this case, the earphone is connected across (or in parallel to) the transistor stage. Capacitor C_1 allows the audio signal variations to be passed to the earphone coils.

HEADPHONES

Headphones are very similar to earphones in construction and operation. Modern types of headphones employ miniature loudspeakers mounted directly in the earpieces of the headphone. The structure and principle of operation are basically the same as those discussed for a standard size PM dynamic loudspeaker.

Stereo headphones feed the left and right channel audio information independently to the left and right earpieces to produce the true stereo effect.

TAPE RECORDERS

Magnetic recording operation is one of transforming electrical impulses into magnetic patterns during recording and converting those patterns back into electrical signals for playback.

A typical tape recorder relies on mechanical, electrical, and electronic components to accomplish the recording and playback functions. During recording, a microphone converts sound energy into small pulses of electrical energy. This energy is fed to a recording amplifier whose output is applied to a recording head. The head is actually an electromagnet with a very narrow air gap between its two pole pieces. The recording head is usually made of many laminations of metal that are perfectly aligned with respect to the magnetic tape. The recording head receives a fixed amount of bias to prepare it for the recording. When the amplified signal from the microphone is fed to the recording head, an alternating magnetic field is produced in the air gap that corresponds to the original variations in the microphone.

In summary, magnetic tape recording requires the application of an audio signal (typically 30mV) to a coil in the tape head (Fig. 23-35). The coil, in turn, magnetizes the very small particles of the magnetic coating on the surface of the tape as it passes through the narrow air gap of the electromagnet in the recording head. When the tape passes the air gap of the recording head, the magnetic variations at the gap arrange the metallic particles on the tape so that the tape now carries a magnetic pattern of the original line sound. Figure 23-36 illustrates the effect produced on the magnetic particles when an audio signal is impressed on the magnetic tape.

During playback, the recording bias voltage is removed from the recording head and a playback amplifier is switched into the circuit. As the magnetized tape passes the recording head, the tape sets up a magnetic pattern or shield across the air gap. The variations in this pattern are amplified and fed to a loudspeaker to reproduce the original program material.

Magnetic Tape

The magnetic tape used to record or play back program material is composed of very small ferromagnetic particles (an iron-oxide compound) on one side of the tape. The tape itself is a

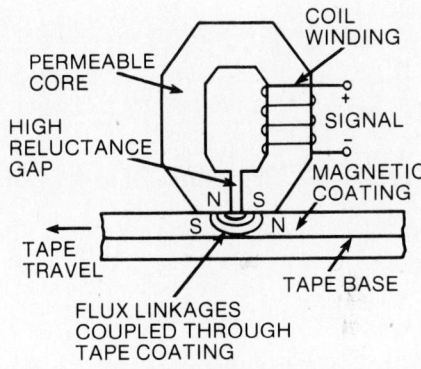

Fig. 23-35: Construction of a recording or playback head.

603

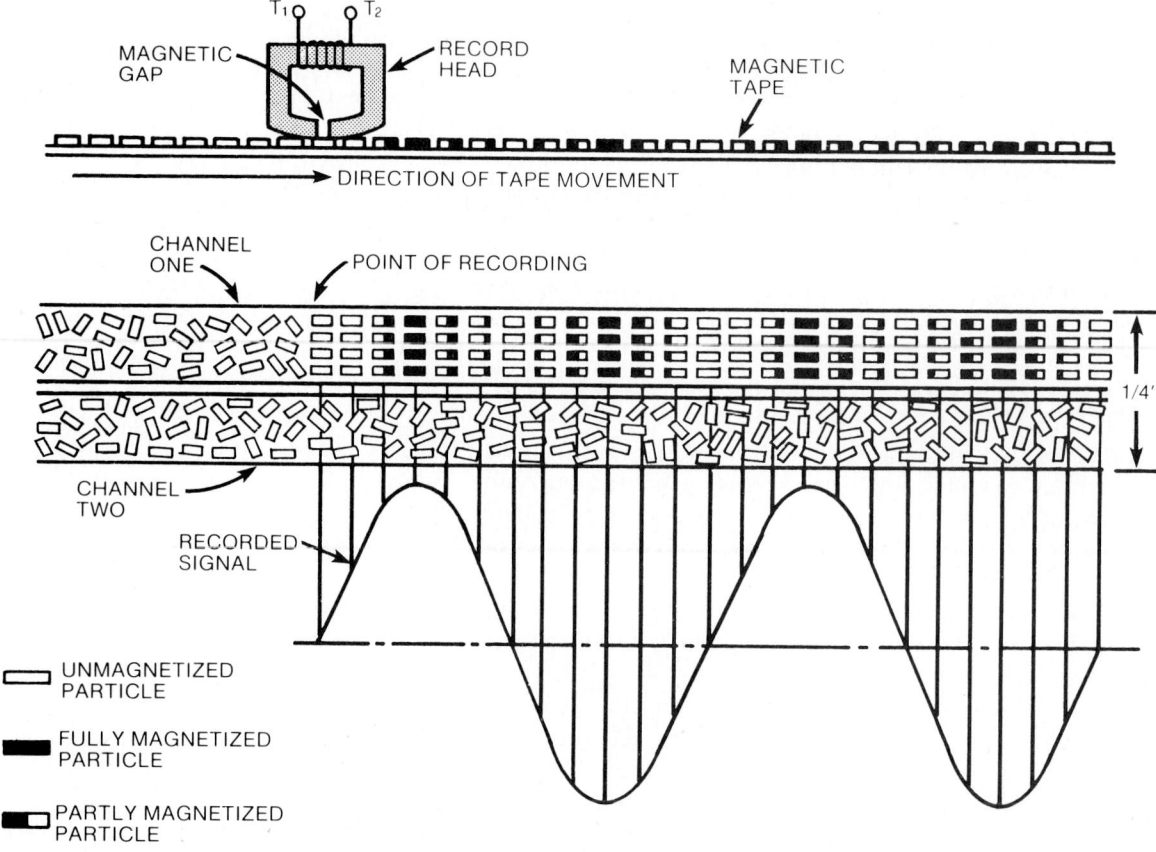

Fig. 23-36: Principles of magnetic tape recording.

sturdy plastic ribbon made of acetate or polyester (mylar). Most modern magnetic tapes use a mylar base because it is strong, durable, and does not stretch.

The typical magnetic tape in use today is approximately 0.001″ thick. The tape width for reel-to-reel and eight-track systems is generally 1/4″ (6.35mm). Cassette systems use a 1/8″ (3.92mm) tape width. The magnetic coating on a typical tape is ferric or iron oxide—chemically the same as iron rust. Coatings of chromium dioxide (CrO_2) and cobalt are used for improved frequency response and a higher signal-to-noise ratio. This type of coating usually requires a higher AC bias current because of the higher coercivity of this compound compared to iron oxide.

The use of magnetic tape provides better stereo separation, reduces surface noise and wear on the tape, and allows the same equipment to be utilized for both the recording and playback functions.

AC Bias for Recording

The audio signal from the amplifier is not directly applied to the tape recording head; instead, the audio signal is superimposed or combined with a high frequency AC bias and then fed

604

to the recording head. As a result the audio signal rides on the peaks of the AC bias. The AC bias is used to minimize recording distortion. It can also be used to erase signal variations from the tape and produce a clean tape for future recording. During the erase process, only the AC bias is applied to the recording head—no audio signals. Typical bias values range from 3mA to 6mA peak at a frequency from 80kHz to 100kHz. Figure 23-37 illustrates the signal-bias combination. The output of a power supply and a master bias oscillation combine with the audio input signal. The resultant signal-bias variations are then applied to the recording head. During the playback process, the AC bias is removed from the recording head.

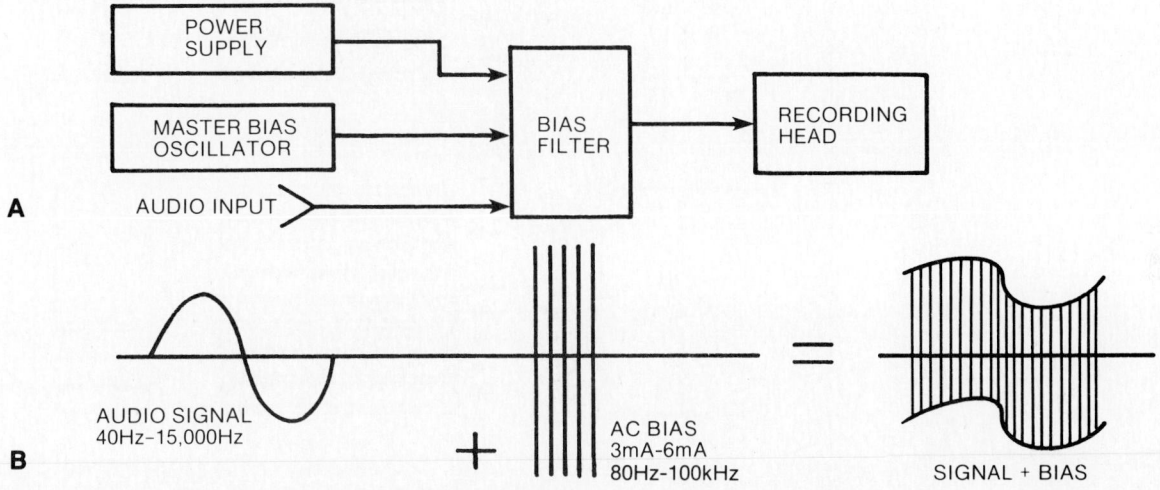

Fig. 23-37: Generation of AC bias for recording: (A) block diagram and (B) signal representation.

Tape Formats

Recording tape is standardized at 1/4″ width (Fig. 23-38). When the entire width is used, the process is called full-track recording. When one-half the width is used, it is called half-track or dual-track recording. When one-fourth the width is used, it is known as quarter-track or four-track recording. For most professional applications, full-track recordings are used to provide the ultimate in fidelity and facilitate editing and splicing. Most recorders for horn usage utilize half-track or four-track recording. To accommodate the various recording widths, the recording head must be designed to have the proper air gap for the recording width to be used (Fig. 23-38).

Reel-to-Reel Tape Systems

This type of system generally uses 1/4″ (6.35mm) tape on 7″ (17.78cm) reels. A selection of operating speeds are also provided: 7-1/2″, 3-3/4″, or 1-7/8″ per second. As a general rule, the higher speeds mean higher fidelity or frequency response, especially in the upper frequency regions. However, the higher

605

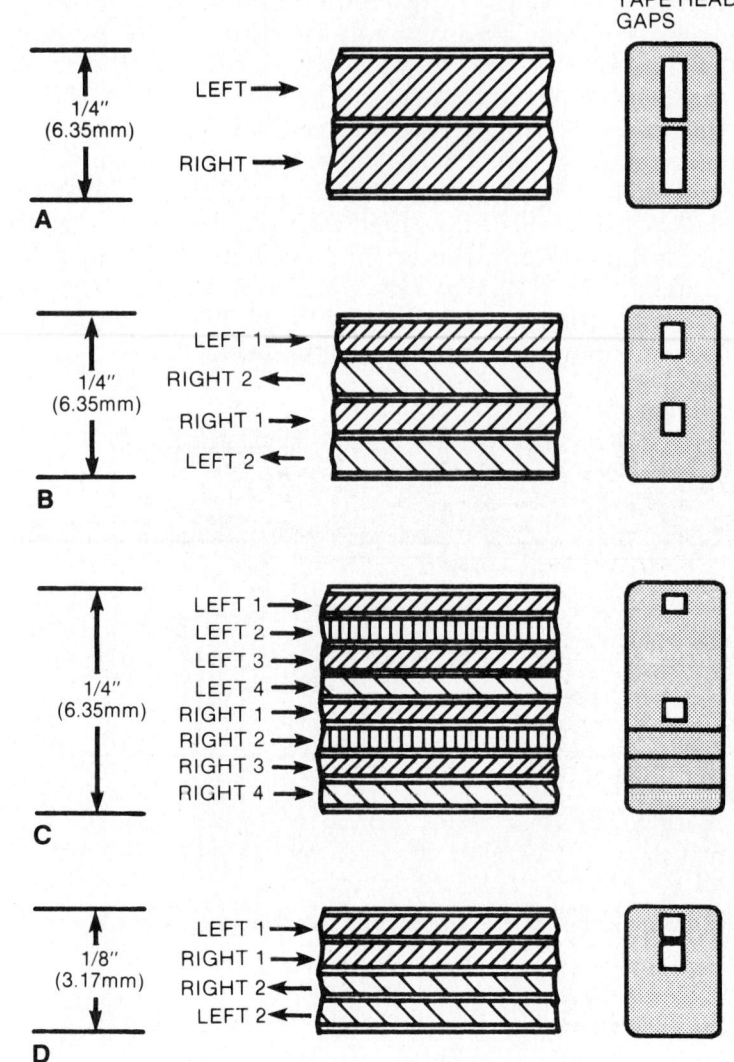

Fig. 23-38: Magnetic tape formats: (A) one-half track reel-to-reel, (B) one-fourth track reel-to-reel, (C) eight track cartridge, and (D) cassette.

tape speeds reduce the playing time. Professional recorders use the two highest speeds; home recorders run at the middle speed; and office-type machines, used largely for dictation purposes or notekeeping, operate adequately at the lower speed.

Cassette Tape Systems

The cassette system uses 1/8″ (3.17mm) tape that is pre-threaded on two reels and enclosed in a single package (Fig. 23-39). The cassette can be likened to a miniature reel-to-reel system. Typical cassette operating speed is 1-7/8″ per second (47.63mm/sec). Cassettes are available in models c-30, c-60, c-90, and c-120 that relate to 15, 30, 45, and 60 minutes playing

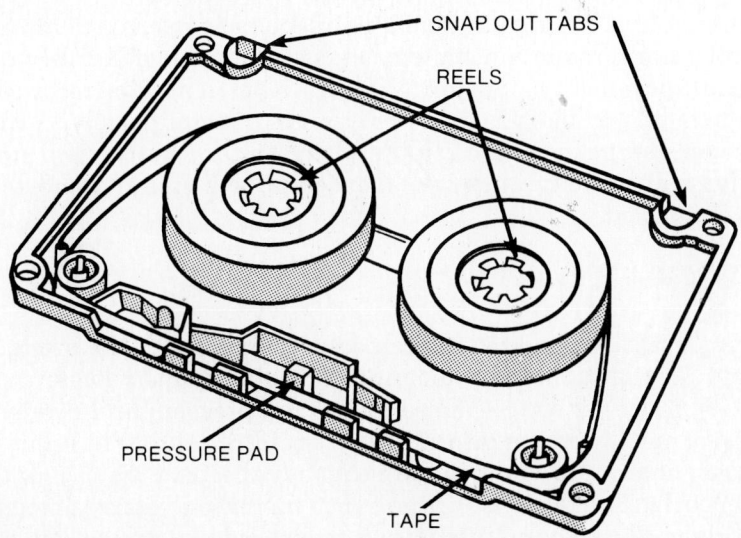

SNAP OUT TABS

REELS

PRESSURE PAD

TAPE

Fig. 23-39: Construction of a tape cassette.

time per side, respectively. The cassette provides good quality sound reproduction, is compact in size, and does not require tape threading.

Eight-Track Tape Systems

The eight-track tape system utilizes 1/4" (6.35mm) tape enclosed in a cartridge that has only one reel with an endless loop. The operating speed of this type of cartridge is 3-3/4" per second and playing time is generally 40 to 90 minutes for all tracks in succession without turning over the tape.

The eight-track system, because of its endless loop, offers continuous programming without the need for tape turnover. In addition, any one of the four pairs of tracks on the tape can be selected for program listening.

Dolby Noise Reduction System

Magnetic tape generates a noise or hiss during use because of the particles on the tape. This noise or hiss lies in a certain band of audio frequencies and can be recorded along with the program of interest. At the time of recording, this band of frequencies (noise plus audio) is increased or amplified by 10dB or so. The tape noise is usually at a constant level, while the audio signals are at various low and high levels. Amplification of this band of frequencies, therefore, increases the level of the audio signals which tend to block out the tape noise or hiss. During the playback function, these same frequencies are decreased in amplitude (by 10dB or so) by the same amount to provide a resultant output with uniform frequency response. The decrease in amplitude of these frequencies also decreases the tape

607

noise or hiss during playback. The scheme is referred to as the Dolby noise reduction system. This system can effectively reduce tape noise by more than 50%. However, if a Dolby recorded tape is played back in a system that is not equipped with the proper compensating circuitry, the resultant output will not have a uniform frequency response and will exhibit distorted audio output.

SUMMARY

Acoustics is the area of science concerned with the production, propagation, reception, and control of sound waves.

Sound waves are produced by the compression and rarefaction of molecules in a medium (gas, liquid, or solid). The human ear is sensitive to sound in the frequency spectrum from 20 Hz to 20,000 Hz. The response of the ear to increased volume is logarithmic. Thus, sound intensity is measured in decibels, a logarithmic unit of measure.

A microphone is an electromechanical transducer that converts sound energy into electric energy. It is rated according to its frequency response, sensitivity, and impedance. The most common type of microphone is the carbon microphone. Other types are the dynamic, ribbon, crystal, ceramic, and condenser microphones.

The phonograph cartridge is an electromechanical transducer that converts mechanical energy (record-groove undulations) into electrical energy. There are three types: magnetic, crystal, and ceramic pickups. The variable reluctance unit, the dynamic unit, and the dynetic unit are the three subdivisions of magnetic pickups.

The loudspeaker is a transducer that converts electrical energy into mechanical energy (air pressure variations or sounds). The dynamic loudspeaker is the most widely used unit. Woofers, tweeters, and midrange speakers are designed to accommodate certain frequency responses.

Tape recorders transform electrical impulses into magnetic patterns during recording and convert those patterns back into electrical signals for playback. The reel-to-reel, cassette, and eight-track are the three types of tape systems.

CHECK YOURSELF (Answers at back of book)

Questions

Fill in the blanks with the appropriate word or words.

1. The human ear can hear sound in the frequency spectrum from _____ Hz to _____ Hz.
2. The _____ indicates the change in volume level that occurs when the average ear is barely able to recognize a difference in a gradually changing amplitude of sound.
3. In a tape recorder, the audio signal is superimposed with a high frequency _____ and fed to the recording head.

4. The _____ cartridge uses the armature as a permanent magnet that is made to vibrate in a stationary coil.
5. The _____ microphone utilizes a moving coil in a stationary magnetic field.
6. Loudspeaker _____ prevent rear radiation of the speaker cones from canceling the forward radiation.
7. Two types of earphones are available: the _____ earphone and the _____ earphone.
8. The _____ system can effectively reduce tape noise by more than 50%.
9. The _____ , or efficiency, of a microphone relates the output voltage to the input sound pressure.
10. A device that converts sound into audio signals or audio signals back into sound is called a _____ .
11. The _____ microphone is the most common type of microphone.
12. The _____ microphone provides a low output power. Also, it is very delicate and must be protected from air drafts.
13. _____ controls on a turntable apply a slight outward pressure on the tone arm to balance the tone arm's tendency to move inward.
14. The _____ loudspeaker uses an alnico permanent magnet to produce a strong, fixed magnetic field.
15. _____ and cobalt coatings on tape provide improved frequency response and a higher signal-to-noise ratio.
16. Modern records are made of _____ and are designed for _____ and _____ rpm turntables.
17. Materials used for the manufacture of the stylus include _____ , _____ , _____ , and _____ .
18. The _____ and _____ microphones produce pickup patterns that discriminate more effectively against noise from the rear of the microphone.
19. The phono cartridge translates _____ energy into _____ energy.
20. When a woofer and tweeter are combined into one enclosure, it is known as a _____ speaker system.
21. _____ , _____ , and _____ are all magnetic type pickups.
22. _____ vibrations are overtones produced at multiples of the fundamental frequency.
23. _____ is the reduction of pressure below the equilibrium value of normal air pressure.
24. The area of science concerned with the production, propagation, reception, and control of sound waves is called _____ .
25. The normal pressure of air (at sea level) is _____ pounds per square inch.
26. The speed or velocity of sound waves is approximately _____ in dry air at 20°C.
   A. 1,130m/sec
   B. 344.43m/sec
   C. 3,444.3km/sec

D. 1,130km/sec
E. none of the above
27. Which of the following is not a low-impedance type microphone?
A. carbon
B. condenser
C. ribbon
D. dynamic
E. all of the above
28. The _____ cartridge has an excellent frequency response over the entire audio range and a very low tracking force.
A. dynetic
B. cardioid
C. dynamic
D. dynatron
E. none of the above
29. The properties of a horn-type speaker are characterized by _____ .
A. its flare
B. its length
C. the size of the open end
D. all of the above
E. none of the above
30. To meet high-fidelity requirements, a good quality microphone must have a frequency response that is flat within _____ .
A. ±1dB from 20Hz to 20kHz
B. ±1dB from 60Hz to 120Hz
C. ±25dB from 20kHz to 200kHz
D. ±10dB from 20Hz to 120Hz

CHAPTER 24

Power Amplifiers

In general, the final stage of a series of audio amplifiers is called the *power stage*. The power stage differs from the preceding stages in that it is usually designed to obtain *maximum power output* rather than maximum voltage gain.

The purpose of a power amplifier is to efficiently deliver distortion-free power to the load. In the case of the single-ended power amplifier, the load will be a speaker which is transformer coupled to the amplifier. Through the impedance matching properties of the transformer, a reasonably good efficiency is obtained. The circuit must, if possible, deliver the electrical power to the speaker in a distortion-free manner. Some distortion of the audio waveforms always takes place, but in a properly designed circuit this distortion will be held to a minimum. The input to the power amplifier is high amplitude voltage as developed by the preceding voltage amplifiers. The voltage output of the power amplifier is lower in amplitude but very high in power level. This output is sent to the speaker in order to develop audio sound power.

SINGLE-ENDED AUDIO POWER AMPLIFIER

Figure 24-1 shows a single-ended audio power amplifier. C_1 and R_1 form the input coupling network, R_3 works in conjunction with R_2 and C_2 to establish class A forward bias, Q_1 is the amplifying device, and T_1 serves as both the collector load for Q_1 and the coupling device to supply power to the speaker. Q_1 is a power transistor and is designed so as to have higher power handling abilities rather than producing large voltage gains. The transistor is operated class A so as to introduce as little distortion of the audio waveforms as possible.

As the positive half-cycle of the input signal is applied, the forward bias on the transistor is increased and collector current through the primary of T_1 expands the magnetic field of the primary and induces a voltage into the secondary. During the negative half-cycle of the input, the forward bias is reduced, collector current reduces, and the magnetic field about the primary of T_1 begins to collapse. The collapsing field induces the opposite polarity voltage (as compared to the previous half-

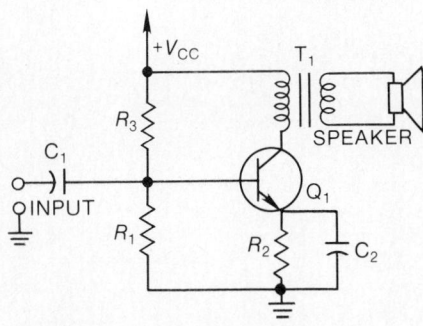

Fig. 24-1: Single-ended audio power amplifier.

611

cycle) into the secondary of T_1. Thus, AC audio signal power is supplied to the speaker, and the electrical audio signals are reproduced in the form of sound power.

PHASE SPLITTERS AND INVERTERS

An alternative to the single-ended power amplifier is the push-pull amplifier. In the cases where the power amplifier is a push-pull circuit, it is necessary to use a *phase splitter* or *phase inverter* as a driving stage. Driver stage is the term used to describe the amplifier which is used to supply the driving, or input, signal to the final power amplifier. The purpose of phase splitters and inverters, as drivers, is to supply two equal amplitude output signals, which differ in phase by 180° with respect to each other. These signals must be developed from a single source.

Through the use of a center-tapped secondary transformer, a single-ended audio stage can produce the signal requirements for push-pull operation. Figure 24-2 shows an audio driver stage with a transformer phase splitter.

The audio signal is coupled from the volume control (R_1) to the input of the driver stage by C_1. R_2 and R_3 provide fixed bias in conjunction with emitter network $R_4 C_2$. T_1 is used for signal coupling, phase splitting, and impedance matching. (It should be noted that while a maximum power transfer is not desired at this point in the circuit, T_1 must be designed to reflect the required load impedance to Q_1 and V_1.)

As the input signal goes through its positive alternation, the forward bias on Q_1 will be decreased with a resultant decrease in transistor conduction. Collector current from V_{CC} through T_1 primary decreases, and the collector potential (top of T_1 primary) becomes more negative. Through transformer action, the bottom of the secondary goes negative (with respect to the secondary center tap), while the top of the secondary goes positive.

When the input signal goes through its negative alternation, the forward bias of Q_1 is increased. Transistor conduction increases, and the top of T_1 primary goes in a positive direction (less negative). Through transformer action the bottom of T_1 secondary produces a positive output; whereas, the top of the secondary produces a negative output. It can be seen that the secondary signals are of equal amplitude, but 180° out of phase, satisfying the input signal requirements of a push-pull power amplifier.

Although the transformer phase splitter provides a simple means of developing the required signals, economy, size, and weight may dictate the use of other means of developing the necessary signals.

Figure 24-3 shows a split-load phase inverter which develops two signals, 180° out of phase, without the use of a transformer. The output current of transistor Q_1 flows through both the collector load resistor (R_3) and the emitter load resistor R_2. Resistors R_2 and R_3 are equal in value. Resistor R_1 establishes the base bias voltage. C_1 and R_4 are the standard input coupling

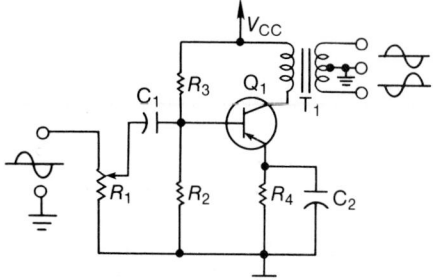

Fig. 24-2: Driver stage with transformer phase splitter.

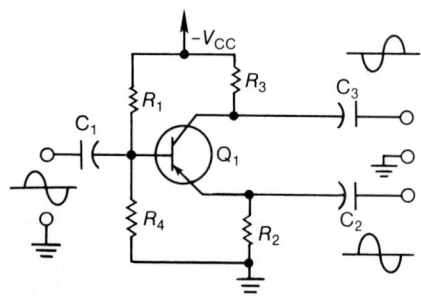

Fig. 24-3: Split-load phase inverter.

components. C_2 and C_3 are the output coupling capacitors for the two signals which are to be developed.

When the input signal aids the forward bias (base becomes more negative), the output current increases. The increased output current causes the collector side of resistor R_3 to become more positive with respect to ground and the emitter side of resistor R_2 to become more negative with respect to ground. When the input signal opposes the forward bias, the output current decreases and causes voltage polarities across resistors R_3 and R_2 opposite to those just indicated. This action produces two output signals that are reversed 180° with respect to each other.

In this circuit, equal voltages are obtained by making resistor R_2 equal in value to resistor R_3. However, an unbalanced impedance results because the collector output impedance of transistor Q_1 is higher than its emitter output impedance. This disadvantage is overcome by the addition of a series resistor between C_2 and the top of R_2. The values of R_2 and the series resistor are chosen so that the output impedance of the collector and emitter are balanced. This eliminates distortion at strong signal currents. The signal voltage loss across the series resistor is compensated for by making R_2 higher in value than R_3.

Notice that emitter-resistor R_2 is unbypassed in order to develop one of the output signals. Because of the large negative feedback voltage developed across R_2 (degeneration), a large signal input is required to drive the one-stage phase inverter. This disadvantage can be overcome by using two-stage phase inverters. In addition, a two-stage phase inverter provides more power output than a one-stage phase inverter. This advantage is important if the driver stage must feed a large amount of power to a high-level push-pull power output stage.

It should be noted that the single-stage phase inverter is also often referred to as a phase splitter, while the two-stage circuitry (covered next) is called only a phase inverter. Finally, it should be noted that due to the degeneration which occurs in the emitter circuit, a voltage gain of less than unity is all that is available from the one-stage phase inverter.

TWO-STAGE PHASE INVERTER

The two-stage inverter, also called the *paraphase amplifier*, provides, as the previous circuit, two signals of equal amplitude which are 180° out of phase with each other. This circuit is shown in Fig. 24-4. The circuit consists of two identical common-emitter configurations.

R_1 is utilized to develop forward bias for Q_1, C_1 and R_2 aid in the selection of the operating point, Q_1 is the first stage amplifying device, and R_3 is its collector load resistor. C_2 is an interstage coupling capacitor, R_4 is an attenuator resistor that reduces signal amplitude to Q_2, R_5 establishes forward bias for Q_2, C_3R_7 is the self-bias network for Q_2, R_6 is the collector load for Q_2, and C_4 and C_5 are the coupling capacitors for the two output signals.

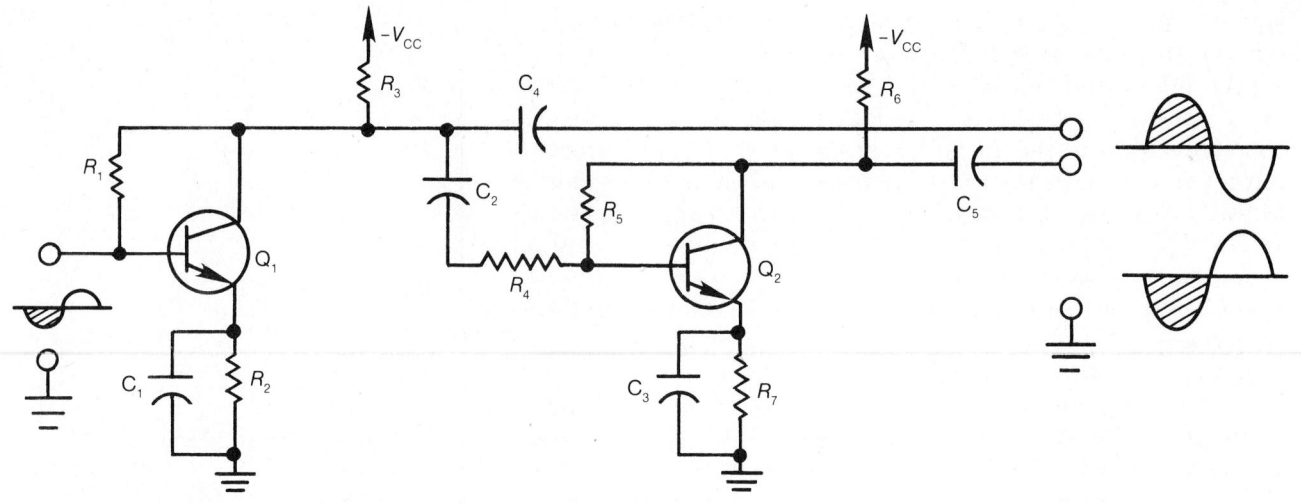

Fig. 24-4: Two-stage transistor paraphase amplifier.

Assume that each transistor configuration produces a voltage gain of 20 and that R_4 produces a 20 to 1 voltage reduction. Assume that a negative half-cycle is applied to the input at a 0.1V amplitude. Because of the 180° phase reversal in the common-emitter configuration, the signal at the collector of Q_1 will be positive-going. The signal will have been amplified and will appear in the collector circuit as a 2V signal. This 2V positive-going signal is coupled through C_4 as one of the required outputs. It is also coupled through C_2 to R_4 and Q_2. The action of R_4 will be such that it will drop 1.9V and allow only a positive-going 0.1V signal to be felt on the base of Q_2. Q_2 will amplify the signal and a negative-going 2V signal will appear on its collector. This signal is coupled out through C_5 as the second of the two required signals.

Since two identical common-emitter configurations are used, the source impedances are equal for the two input circuits of the push-pull output stage. In addition, the amount of power that can be delivered by the two-stage paraphase amplifier is much greater than that of the split-load phase inverter. Also, rather than having a voltage gain of less than unity, the paraphase amplifier is capable of producing voltage amplification at the same time that the phase-splitting action is taking place. In the case where the gains of Q_1 and Q_2 are not quite equal, R_4 may be made adjustable so that it can provide additional, or less, attenuation as required.

Failure Analysis

The following problems occasionally arise in a power amplifier:

No Output. In the single-stage phase splitter, a no-output condition would often be the result of an open collector resistor, an open emitter-resistor, or a defective amplifying device. Open coupling capacitors (poorly soldered leads) in either the input or output would also cause a no-signal condition. It should be

614

noted that as there are two outputs from the phase splitter, the loss of only one output would immediately indicate the problem area (collector or emitter).

In the two-stage phase inverter, the lack of one of the two outputs would immediately indicate a trouble in the second stage. If a no-signal condition existed in the first stage, there would be no input (and consequently no output) to the second stage. Once it is determined which stage is at fault, the possible troubles are the same as for the standard voltage amplifier.

Reduced Output. In the single-stage phase splitter, a reduced output could be caused by an amplifying device which has weakened with age. A change in value for the input resistance could possibly reduce the level of input signal. Finally, the following stage could load the circuitry so heavily that its output could be reduced. This, however, would be indicative of a problem in the following stages rather than in the phase splitter. Also, poorly soldered connections on the coupling capacitors could cause attenuation which would reduce the output signal.

In the two-stage phase inverter, as before, if both outputs are reduced, the problem is either in the original input or in the first stage. If a problem exists in the second stage, only its output will be reduced in amplitude. The possible troubles which could reduce the output amplitude of the two-stage phase inverter are the same as those for a standard voltage amplifier.

Distorted Output. In the single-stage phase splitter, improper bias, low supply voltages, or a defective amplifying device can cause a distorted output. If the bias is either too high or too low, peak clipping is likely to result. Low supply voltage can also result in peak clipping. Finally, too large an input signal will cause peak clipping by instantaneously causing cutoff of the amplifying device. An oscilloscope can be used to trace the signal through the circuitry and to locate the origin of the distortion.

The two-stage phase inverter, as previously covered, is two cascaded amplifiers. As such, they are subject to the same forms of distortions as other voltage amplifiers. If only one stage shows distortion, it will be a trouble in the second stage. If a series distortion is present in both stages, the trouble is to be located either in the first stage or the preceding circuitry. It should be noted that about 1% or 2% higher distortion is expected from the second stage. Distortion beyond this amount, however, indicates a malfunction which should be corrected.

CLASS A, PUSH-PULL POWER AMPLIFIERS

The push-pull amplifier's purpose is to develop the necessary driving power to operate the loudspeaker. It produces a higher distortion-free output power than does the single-ended amplifier. It is a two-stage device consisting of a transformer coupled to the speaker.

The input to the push-pull amplifier comes from a one- or two-stage phase inverter. It requires two input signals which are equal in amplitude and 180° out of phase.

Figure 24-5 illustrates the schematic diagram of a typical class A, push-pull power amplifier. It consists essentially of two transistors connected back to back. Both transistors are biased for class A operation. In this circuit, a transformer phase splitter is used to develop two signals that are equal in amplitude and 180° out of phase with each other. These signals are applied as the two inputs for the push-pull amplifier. The transformer also provides the path used by R_1 to develop class A bias for both Q_1 and Q_2.

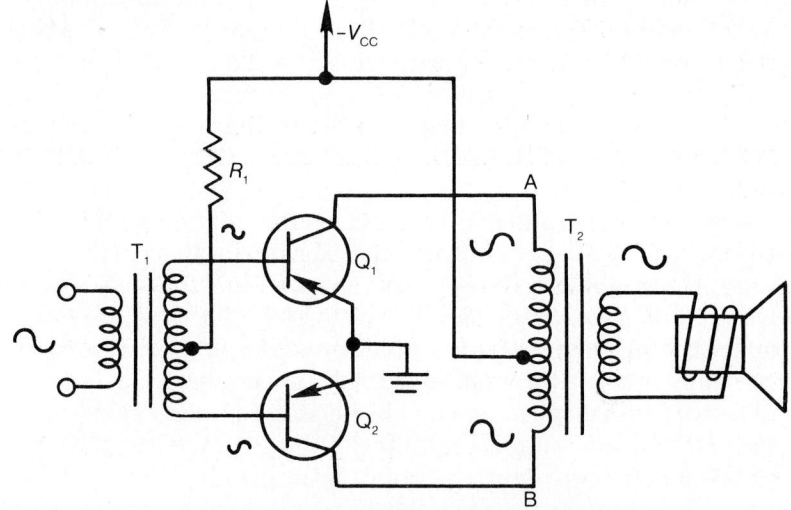

Fig. 24-5: Transistor class A, push-pull amplifier.

The output of the amplifier is transformer coupled to the speaker. One-half of the primary of T_2 represents the collector load impedance for Q_1, while the second half represents the collector load for Q_2. T_2 also provides impedance matching between the relatively high output impedance of the transistors and the low impedance of the speaker voice coil.

On the positive alternation of the input signal, the potential on the base of Q_1 will increase in a positive direction, while the potential on the base of Q_2 will increase in a negative direction. Since both transistors are of the PNP type, the voltage applied to their respective base elements will cause the conduction of Q_1 to decrease and the conduction of Q_2 to increase.

Currents flow from the center tap of T_2 toward points A and B. Under quiescent operating conditions these currents are equal and the magnitude and polarity of the voltage drops they produce are such that there is no difference of potential between points A and B. The center tap of T_2 is fixed at the maximum negative value in the circuit by virtue of being connected to V_{CC}. Thus, the potentials at points A and B are by some amount less negative than the center tap. As stated, the positive alternation of the input signal causes the conduction of Q_1 to decrease. In other words, point A becomes more negative because it is approaching the value of the potential at the center tap. The col-

616

lector voltage waveform of Q_1, which is actually a graph of the potential at point A, is seen to be increasing in a negative direction at this time.

The positive alternation of the input signal also causes the conduction of Q_2 to increase. This causes the potential difference between point B and the center tap to increase. Thus, the potential at point B is increasing in a positive direction (as shown by the collector voltage waveform of Q_2). Since point A is becoming more negative, there is a potential difference developed across the entire primary winding. By transformer action this potential difference is coupled to the secondary of T_2 and appears as the waveform output.

On the negative alternation of the input signal, the reverse of the above actions occurs, as the collector voltage waveforms of Q_1 and Q_2 show, and as the polarity of the voltage developed across the primary of T_2 is reversed.

The power output from this class A, push-pull circuit is more than twice that obtainable from a single-ended class A power amplifier. An added advantage of this circuit is that, due to the push-pull action of the output transformer, all even order harmonics are eliminated in the output if the transistor circuits are balanced. Balancing will be covered later. Since the distortion is caused mainly by second order harmonics, elimination of these harmonics will result in a relatively distortion-free output signal.

Failure Analysis

The following problems occasionally arise in a class A, push-pull power amplifier:

No Output. The no-output condition is seldom encountered in the push-pull amplifier. The most common cause of trouble in any amplifier circuit is the failure of the amplifying device. It is highly unlikely that both amplifying devices of the push-pull amplifier should fail at the same time. Therefore, a no-output condition will indicate the failure of a component common to both amplifying devices. The emitter-resistor could be open, the power supply may have failed, or there may be no input signal. The latter trouble would indicate that the trouble is located in one of the preceding amplifiers and not in the push-pull stage itself.

Reduced Output. If one of the two amplifying devices fails or if components which would disable one-half of the push-pull stage fail, a reduced output will result. An oscilloscope should be used to determine which of the two amplifiers is not operating. The individual device is then subject to the same failure analysis as a single-stage amplifier. If both devices are found to be operating satisfactorily, check the input signal. A preceding stage may be responsible for the problem. A reduced output could also be caused by a change in value of the balancing resistors. This would allow a mismatch in amplifier operation and possibly cause degeneration through the presence of a degenerative signal across the emitter.

617

Distorted Output. A distorted output could be caused through the improper biasing of one of the amplifiers or improper balancing or changes in component values in one of the amplifier circuits. An oscilloscope should be used to determine the source of the distortion and the trouble should be analyzed in the same manner as for a single-stage amplifier.

CLASS B, PUSH-PULL POWER AMPLIFIERS

For class B, push-pull power amplifiers, the common-emitter configuration is the most used for transistor circuits. However, on some occasions either the common-base or the common-collector configurations may prove advantageous. In many respects all three configurations are similar. They all deliver about the same power output, have the same maximum efficiency, and require about the same load impedance. They differ mainly, as might be expected, in their input impedance and power gain. The common-collector has a high input impedance, approximately equal to the peak collector voltage swing divided by the peak base current. The common-collector stage generally has a greater gain than the common-base stage, although less than the common-emitter. Another feature of the common-collector stage is its low output impedance. It is, therefore, attractive in cases where very low output impedance is desirable.

Balancing Networks. As mentioned previously, matched amplifying devices are used in push-pull amplifier circuits to insure equal values of current and voltage in the output. No matter how well matched the two devices may be, there will be some differences in their characteristics; therefore, unequal outputs will cause excessive distortion. The center tap on the transformer is critical to the output. If the tap is not physically located at the exact center of the winding, there will be unequal outputs from the amplifiers. A means must, therefore, be developed to insure that the outputs of the amplifying devices are the same. This is provided by the use of balancing networks.

When the DC components of the collector currents through the devices are not equal, an emitter balancing network is required. Such a circuit is shown in Fig. 24-6. As these circuits are used to balance the quiescent DC currents through the matched devices, they are sometimes called static balancing networks. By adjusting the potentiometer arm of R_3, the value of emitter self-bias to each device may be changed. In this way, the transistor currents may be adjusted so that they are exactly equal to each other. This adjustment is usually made while observing accurate ammeter measurements of the currents through the devices.

A second form of balancing is shown in Fig. 24-7. The network is called base balancing. The purpose of this balancing network is to insure that the outputs of both amplifying devices are equal. This network is necessary because regardless of how well two amplifier devices are matched, they will not possess exactly the same parameters.

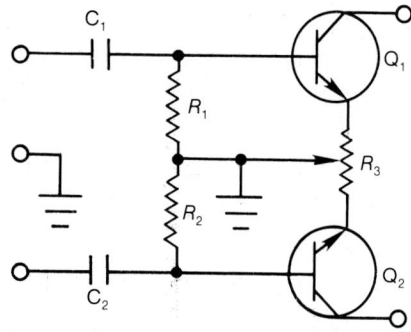

Fig. 24-6: Balancing network.

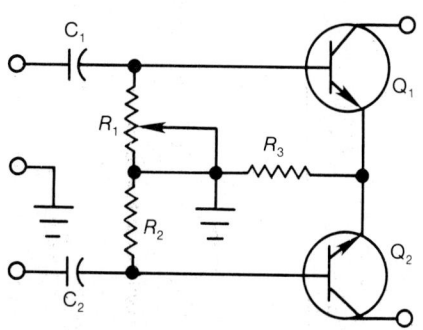

Fig. 24-7: Base balancing.

618

By changing the position of the wiper arm of potentiometer R_1, the amplitude of the input signals to the devices may be varied to cause equal outputs. For example, if Q_1 has a slightly greater gain than Q_2, the wiper arm should be moved toward the top of the potentiometer to decrease the magnitude of the input to Q_1. This adjustment is made by observing the voltage waveform across the emitter-resistor, R_3, on an oscilloscope. When the circuits are balanced, the current increase through one device will equal the current decrease through the other device and vice versa. Thus, the current through the emitter-resistor is constant, developing only a DC voltage equal to that at quiescence. Therefore, the circuits are balanced when a minimum signal appears across resistor R_3.

Comparison of Power Amplifiers. Some of the advantages of using the push-pull amplifiers instead of the single-ended amplifier have been mentioned. There are no adverse effects due to core saturation in the push-pull amplifier because of the balanced direct currents flowing in the primary of the output transformer. It was also shown that the push-pull amplifier eliminated even-order harmonic distortion by the process of cancellation. One other advantage of the push-pull amplifier is that there is no hum in the output caused by the power supply ripple. In a single-ended power amplifier driving a loudspeaker, the power supply ripple will cause the plate current to increase and decrease at the ripple frequency, causing a hum to be heard in the output speaker. This will not occur in a push-pull stage, since an increase in power supply voltage will cause an increased plate current through both amplifying devices. Equal and opposite fields will result about the transformer primary, and the effects will cancel. The same will occur when the power supply decreases its voltage. Providing the push-pull stage is balanced, no voltage will be induced in the secondary due to the power supply ripple.

Another advantage of a push-pull stage is that a decoupling network is not necessary, providing the circuit is properly balanced. The reason this network is not necessary in a push-pull stage is because the input signals are of opposite polarity, causing the current in one of the amplifying devices to increase at the same time that the current in the other device is decreasing. Thus, the current through the power supply is constant and a varying voltage is not developed across it. For this same reason, an emitter bypass capacitor is not required when using a common-emitter-resistor.

Figure 24-8 shows a simplified circuit of a class B, push-pull amplifier. The emitter-base junction is zero biased. In this circuit each transistor conducts on alternate half-cycles of the input signal. The output signal is combined in the secondary of the output transformer. Maximum efficiency is obtained even during an idling (no input signal) period, because neither transistor conducts during this period.

An indication of the output current waveform for a given signal current input can be obtained by considering the dynamic transfer characteristics for the amplifier. It is assumed that the two transistors have identical dynamic transfer characteris-

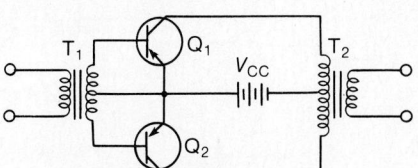

Fig. 24-8: Transistor class B, push-pull amplifier.

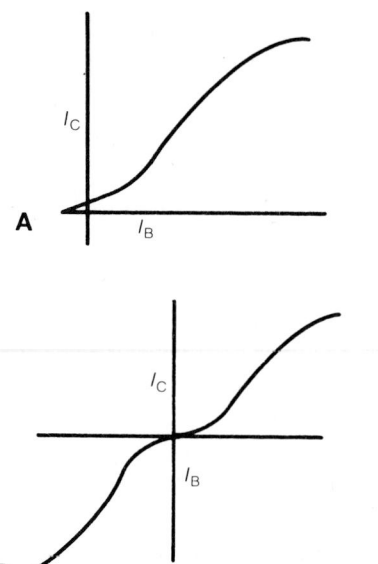

Fig. 24-9: Dynamic transfer characteristic curves.

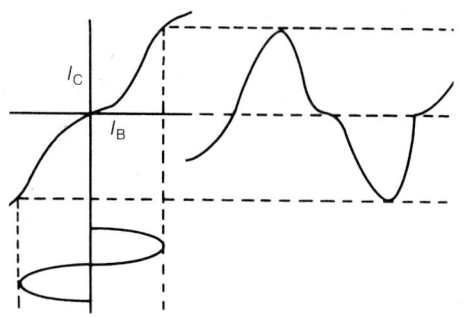

Fig. 24-10: Dynamic transfer curves of class B, push-pull amplifier.

tics. This characteristic, for one of the transistors, is shown in Fig. 24-9A. The variation in output (collector) current is plotted against input (base) current under load conditions. Since two transistors are used, the overall dynamic transfer characteristics for the push-pull amplifier are obtained by placing two of the curves (Fig. 24-9B) back to back.

Note that the zero line of each is lined up vertically to reflect the zero bias current. In Fig. 24-10, points on the input base current (a sine wave) are projected onto the dynamic transfer characteristic curve. The corresponding points are determined and projected as indicated to form the output collector current waveform. Note that severe distortion occurs at the crossover points; that is, the distortion occurs where the signal passes through zero value. This type of distortion becomes more severe with low input currents. Crossover distortion can be eliminated by using a small forward bias on both transistors of the push-pull amplifier.

A class B push-pull amplifier with a small forward bias applied to the base-emitter junction is shown in Fig. 24-11. A voltage-divider is formed by resistors R_1 and R_2. The voltage across R_2 supplies the base-emitter bias for both transistors. This small forward bias eliminates crossover distortion.

In the voltage-divider (Fig. 24-11A) current flow from the battery is in the direction of the arrow. This current established the indicated polarity across resistor R_2 to furnish the required small forward bias. Note that no bypass capacitor is across resistor R_2. If a bypass capacitor were used (Fig. 24-11B), the capacitor would charge (solid-line arrow) through the base-emitter junction of the conducting transistor (during the presence of a signal) and discharge (dashed-line arrow) through resistor R_2. The discharge current through resistor R_2 would develop a DC voltage with the polarity indicated. This is a reverse-bias polarity that could drive the amplifier into class C operation with the resultant distortion even more severe than crossover distortion.

A study of the dynamic transfer characteristic curve of the amplifier demonstrates the elimination of crossover distortion. In Fig. 24-12A, the dynamic transfer characteristic curve of each transistor is placed back to back for zero base current bias

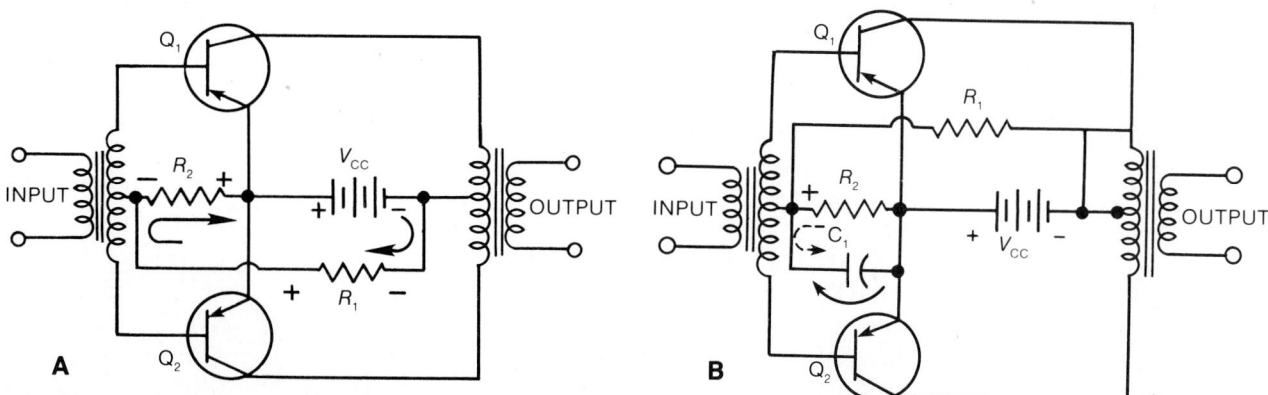

Fig. 24-11: Class B, push-pull amplifier with small bias voltage applied.

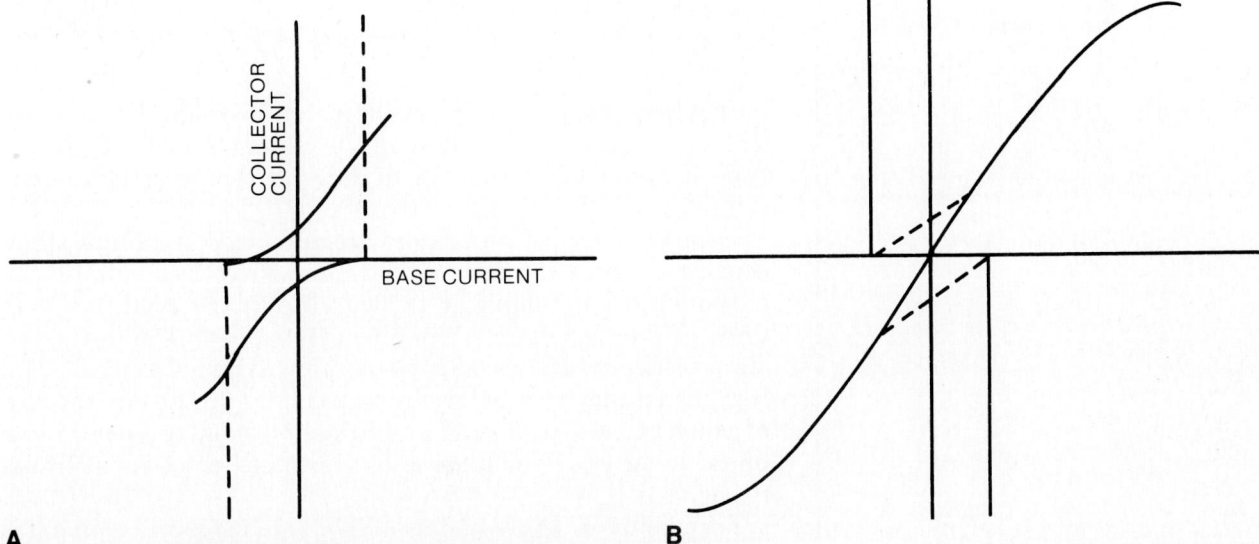

A
B

Fig. 24-12: Dynamic transfer curves, low bias, class B.

conditions. The two curves are back to back and not combined. The dashed lines indicate the base current values when forward bias is applied to obtain the overall dynamic characteristic curve of the amplifier. With forward bias applied, the separate curve of each transistor must be placed back to back and aligned at the base bias current line (dashed line). The zero base current lines (solid lines) are offset (Fig. 24-12B).

In Fig. 24-13, points on the input base current (a sine wave) are projected onto the dynamic transfer characteristic curve. The corresponding points are determined and projected as indicated to form the output collector current waveform. Compare this output current waveform with that shown in Fig. 24-10. Note that the crossover distortion does not occur when a small forward bias is applied.

CLASS AB, PUSH-PULL POWER AMPLIFIER

As mentioned, the class B, push-pull amplifier is easily converted to class AB. The class AB, push-pull amplifier has a small forward bias on both transistors which causes each transistor to conduct slightly more than 50% of the time. This bias causes a slight current through both transistors during the crossover phase. The crossover distortion is effectively eliminated by keeping the transistors on the linear portion of their transfer characteristics curve.

One method of biasing a push-pull power amplifier for class AB operation is illustrated in Fig. 24-14. The biasing network for both transistors is composed of R_1 and R_2. The voltage drop across R_2 is a very low value, normally about 0.75V. This voltage across R_2 is the bias voltage. Emitter-resistors, R_3 and R_4, provide degenerative feedback. The values of R_3 and R_4 are equal, and neither value should exceed $1\,\Omega$. About $0.47\,\Omega$ is usually sufficient.

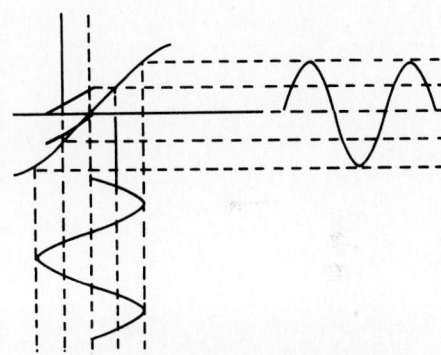

Fig. 24-13: Dynamic transfer characteristic curves of class B, push-pull amplifier with small forward bias, showing input and output waveforms.

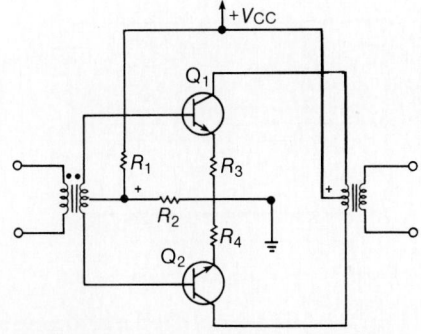

Fig. 24-14: Class AB, push-pull amplifier.

621

COMPLEMENTARY
SYMMETRY AMPLIFIERS

Junction transistors are available as PNP and NPN types. The direction of current flow in the terminal leads of the one type of transistor is opposite to that in the corresponding terminal lead of the other type transistor.

If the two types of transistors are connected in a single stage (Fig. 24-15), the DC current path (indicated by the arrows) in the output circuit is completed through the collector-emitter junctions of the transistors. When connected in this manner, the circuit is referred to as a complementary symmetry circuit. The complementary symmetry circuit provides all the advantages of conventional push-pull amplifiers without the need for a phase-inverter driver stage or for a center-tapped input transformer.

Transistor Q_1 (Fig. 24-15) is a PNP transistor and transistor Q_2 is an NPN type. A negative-going input signal forward biases transistor Q_1 and causes it to conduct. As one transistor conducts, the other is nonconducting, because the signal that forward biases one transistor reverse biases the other transistor.

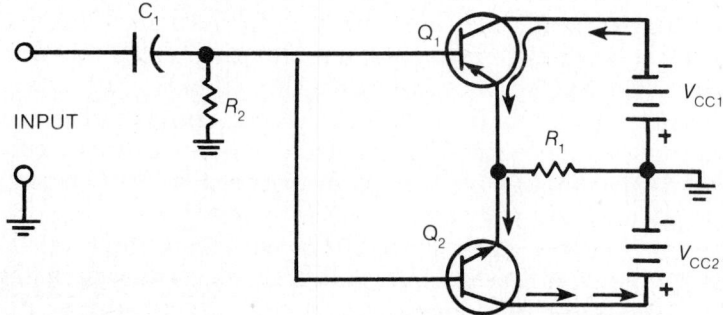

Fig. 24-15: Zero bias complementary symmetry circuit.

The resultant action in the output circuit can be understood by considering the circuit of Fig. 24-16. This is a simplified version of the output circuit. The internal emitter-collector circuit of transistor Q_1 is represented by a variable resistor R_1 and that of transistor Q_2 by variable resistor R_2. With no input signal and class B operation (zero emitter-base bias), the arms of the variable resistors can be considered to be in the OFF positions. No current flows through the transistors or through load resistor R_L. As the incoming signal goes positive, transistor Q_2 conducts and transistor Q_1 remains nonconducting. Variable resistor R_1 remains in the OFF position. The variable arm of resistor R_2 moves toward point 3 and current passes through the series circuit consisting of battery V_{CC2}, resistor R_L, and variable resistor R_2. The amount of current flow depends upon the amplitude of the incoming signal, the variable arm moving toward point 3 for increasing forward bias and toward point 4 for decreasing forward bias. The current flows in the direction of the dashed arrow, producing a voltage with the indicated polarity.

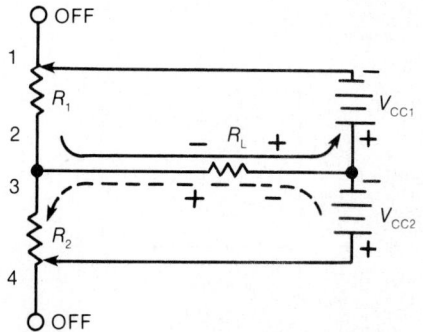

Fig. 24-16: Simplified version of output circuit of a complementary symmetry circuit.

622

When the input signal goes negative, transistor Q_1 conducts and transistor Q_2 becomes nonconducting. The same action is repeated with variable resistor R_1. Current flows through battery V_{CC1}, variable resistor R_1, and load resistor R_L with the polarity indicated.

For class A operation of a complementary symmetry circuit (such as Fig. 24-15), a voltage-divider network is used to apply forward bias to the two transistors so that collector current is not cut off at any time. In the simplified circuit (Fig. 24-16), the variable resistors will not be in the OFF position at any time. Bias current in the output circuit flows out of the negative terminal of battery V_{CC2} into the positive terminal of battery V_{CC1}, through variable resistor R_1, variable resistor R_2, and into the positive terminal of battery V_{CC2}. No resultant current flows through resistor R_L.

Under the above conditions, the output circuit can be considered a balanced bridge, the arms of the bridge consisting of resistors R_1 and R_2 and batteries V_{CC1} and V_{CC2}. When the input signal goes positive, transistor Q_2 conducts more and transistor Q_1 conducts less. In the simplified circuit, the variable arm of resistor R_1 moves toward point 1 and that of resistor R_2 moves toward point 3.

This action results in an unbalanced bridge and resultant current flows through resistor R_L in the direction of the dashed-line arrow, producing a voltage with the indicated polarity. When the input signal goes negative, transistor Q_1 conducts more and transistor Q_2 conducts less. In the simplified circuit, the variable arm of resistor R_1 moves toward point 2 and that of resistor R_2 moves toward point 4. Again the bridge is unbalanced and current flows through resistor R_L in the direction of the solid-line arrow, producing a voltage with the indicated polarity.

In either class B or class A operation, no resultant DC current flows through the load resistor or the primary of the output transformer, whichever is used as the load for the amplifier. Thus the voice coil of a loudspeaker may be connected directly in place of resistor R_L. The voice coil will not be offset by DC current flow through it, and thus distortion will not occur.

HEAT SINK

Because amplifiers, especially the power type, handle such large amounts of power, some components can get quite hot. These components should be mounted in such a way as to allow ventilation and heat dissipation. Sometimes all that is necessary is to space the components far enough apart from each other or the circuit board to allow air to pass freely around them. In other cases, a component may need to be mounted on a heat sink. This is a metal device which provides a larger area from which the heat can be radiated. A power transistor is capable of delivering more power safely when adequate heat sinking is provided (Fig. 24-17).

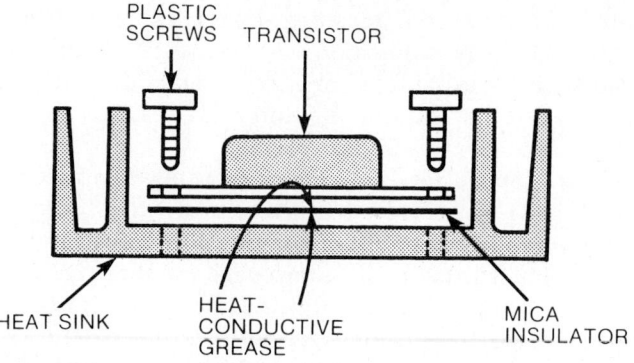

HEAT SINK

HEAT-CONDUCTIVE GREASE

MICA INSULATOR

Fig. 24-17: Typical mounting of a transistor to a heat sink with insulation of the transistor case.

When energy is dissipated in a transistor, its temperature rises until the rate of thermal energy flow from the transistor to the environment equals the rate of electrical energy input to the transistor. It is possible to visualize power (P) flowing through thermal resistance (θ) producing a temperature drop (ΔT). This is analogous to current (I) flowing through resistance (R) producing a voltage drop (V). The thermal "Ohm's Law" is stated as:

$$P = \frac{\Delta T}{\theta}$$

As shown in Fig. 24-18, thermal resistance (θ) consists of several series components: transistor junction to case (θ_{JC}), case to heat sink (θ_{CS}), and heat sink to ambient air (θ_{SA}). The first of these three is part of the specifications for any power transistor. The second (θ_{CS}) is zero unless a heat sink insulator is used, in which case it is about 0.35°C/W for a 1 square inch insulator area. The third (θ_{SA}) is specified by the heat sink manufacturer or read from the graph in Fig. 24-19.

Example: A TO–3-case transistor has a specified θ_{JC} of 1.2°C/W and $T_{J(max)}$ of 180°. It is to be used with a heat sink insulator having θ_{CS} of 0.3°C/W. Power dissipation is to be 30W. Ambient temperature may reach 75°C. Find the required heat sink area.

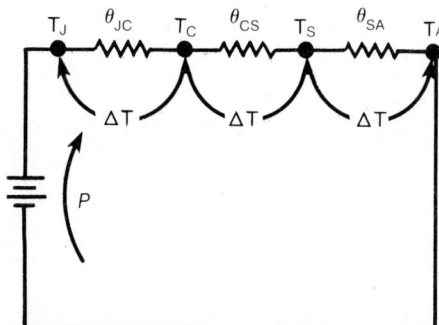

Fig. 24-18: Thermal power (watts) flow through the thermal resistance (θ) producing a temperature differential (ΔT). The subscripts refer to the junction, case, sink, and ambient. The use of an electric-circuit schematic to represent the process is common.

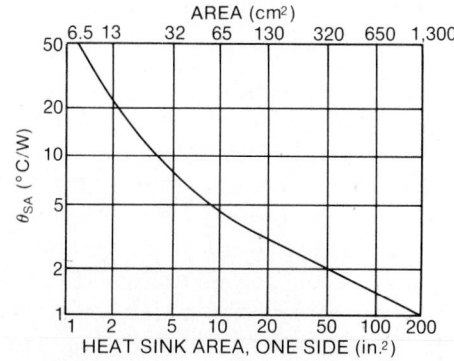

Fig. 24-19: Thermal resistance versus total surface area for a square sheet painted black and mounted vertically in free air.

624

Solution:

$$\theta_T = \frac{\Delta T}{P} = \frac{180 - 75}{30} = 2.5°\text{C/W}$$

$$\theta_{SA} = \theta_T - \theta_{JC} - \theta_{CS}$$

$$= 3.5 - 1.2 - 0.3 = 2.0°\text{C/W}$$

Figure 24-19 indicates that the plate area must be 50 square inches to achieve this value.

NOISE IN AUDIO AMPLIFIERS

In electronic circuit design, any undesirable signal is classified as noise. Some of the common sources of electronic noise are electrical storms, AC power supplies, arcing in electrical devices, and induction from nearby circuits. In various stages of design, it is important to plan filters to greatly reduce this external noise.

In audio circuits, the noise generated in the actual circuit components is of primary concern. When constructing a chain of several audio amplifiers, it is absolutely essential to minimize the noise in the first stage. Any noise generated in the first stage will be amplified through all subsequent stages.

Signal-to-Noise Ratio

The *signal-to-noise ratio* is a comparison of the signal strength of an intelligence signal to the strength of the noise that is carried with the signal. In audio circuits, any noise at all is objectionable; only the highest possible signal-to-noise ratio is acceptable.

Signal-to-noise ratios may be expressed as power ratios, voltage ratios, or decibels. When a ratio is expressed in decibels, an addition of each 3dB doubles the power, and each 6dB doubles the voltage. The decibel power ratio is defined as 10 times the ratio of two powers. The decibel voltage ratio is defined as 20 times the ratio of two voltages. The equations for both voltage and power signal-to-noise ratios are

$$\text{dB} = 10 \log \frac{P_{so}}{P_{no}} = 20 \log \frac{V_{so}}{V_{no}}$$

where P = power, V = voltage, s = signal, n = noise, and o = output.

The signal-to-noise ratio in terms of either voltage or power may be expressed as a simple ratio such as 2:1, 4:1, 10:1, and so on. In this case, the first number represents signal strength and the second number represents the relative strength of the noise. If the ratio is in terms of power, a ratio of 4:1 indicates the power of the signal is four times the power of the noise. This is equivalent to a decibel rating of 5dB. The same rating (4:1) for voltage means a voltage four times as strong as the noise and is equated to a decibel rating of 12dB and a power rating of nearly 16:1.

625

Table 24-1 compares several examples of the three ratio expressions.

Table 24-1: Comparison of Signal-to-Noise Ratio Expressions		
Power Ratio	Voltage Ratio	Decibel Ratio
1:1	1:1	0
2:1	1.5:1	3
4:1	2:1	6
8:1	3:1	9
16:1	4:1	12
100:1	10:1	20
10,000:1	100:1	40

Measuring Ratios

The easiest way to measure a signal-to-noise ratio involves the use of a signal generator and an oscilloscope. The signal generator is coupled to the input of an amplifier, and the oscilloscope is set to measure the voltage amplitude. This measurement will reveal the output signal amplitude. The connections are illustrated in Fig. 24-20. This measurement provides one of the two readings for determining the voltage signal-to-noise ratio. Record the peak-to-peak voltage amplitude reading.

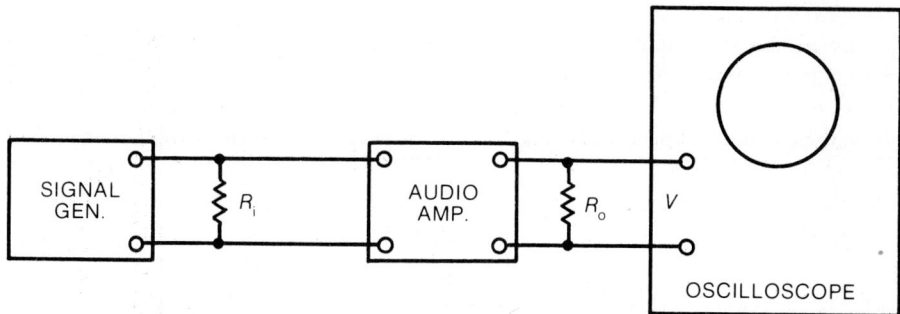

Fig. 24-20: Measuring the signal voltage amplitude.

The next step is to determine the noise level under normal conditions with no signal input. The oscilloscope connections remain the same, but the signal generator is removed from the input. Instead of an input signal, a resistance equal to the output resistance of the signal generator is connected in parallel to the input of the amplifier. This arrangement is illustrated in Fig. 24-16.

The arrangement illustrated here may be referred to as a shorted input. But it is readily apparent that an actual short across the input would alter the bias and shift the operating point.

The peak-to-peak noise amplitude measurement is taken using the arrangement illustrated in Fig. 24-21. This is the second reading for determining the signal-to-noise voltage ratio. The

626

next step is calculation. The ratio is signal amplitude divided by noise amplitude:

$$\text{voltage ratio} = \frac{\text{signal amplitude}}{\text{noise amplitude}}$$

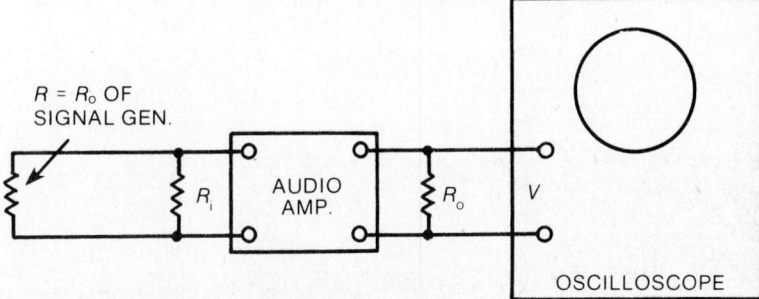

Fig. 24-21: **Measuring the noise level.**

SUMMARY

The final stage of a series of audio amplifiers is called the power stage. The power amplifier is designed to deliver maximum power output with a very low level of distortion. These characteristics can be obtained by a variety of basic circuits.

In a single-ended audio power amplifier, the high power output is coupled to the speaker through a transformer in the collector branch of the power transistor. This transistor is biased class A in the collector-emitter configuration.

A phase splitter or phase inverter acts as a driving stage for another type of power amplifier known as a push-pull. The purpose of phase splitters and inverters is to supply two equal amplitude output signals, 180° out of phase with respect to each other.

The push-pull amplifier produces a higher output power than does the single-ended amplifier. The push-pull consists of two transistors connected back to back. These transistors may be biased class A, B, or AB. Balancing networks are used in conjunction with push-pull amplifiers to insure the output of the two transistors is the same. Some advantages of the push-pull power amplifier are the elimination of core saturation, even order harmonic distortion, and the need for a decoupling network. It is also capable of greater efficiency when biased class B or AB.

The complementary symmetry amplifier provides all the advantages of a push-pull circuit without the need for a phase inverter driver stage.

A power transistor is capable of delivering more power safely when adequate heat sinking is provided.

Any undesirable signal in an electronic circuit is classified as noise. The signal-to-noise ratio is a comparison of the true signal strength to the strength of the noise that is carried with the signal. Signal-to-noise ratios may be expressed as power ratios, voltage ratios, or decibels.

627

Questions

Fill in the blanks with the appropriate word or words.

1. The two-stage phase inverter is also called a _____ .
   A. phase splitter
   B. push-pull amplifier
   C. bistable amplifier
   D. paraphase amplifier

2. A _____ provides all the advantages of conventional push-pull amplifiers without the need for a phase inverter.
   A. complementary symmetry amplifier
   B. single-ended power amplifier
   C. class AB, push-pull amplifier
   D. paraphase amplifier

3. The decibel power ratio is defined as _____ the ratio of two powers.
   A. twice
   B. 10 times
   C. 20 times
   D. 30 times

4. The _____ condition is seldom encountered in a push-pull amplifier.
   A. reduced output
   B. distorted output
   C. no output

5. Poorly soldered connections on a coupling capacitor could cause a _____ .
   A. no output condition
   B. reduced output condition
   C. distorted output condition
   D. A and B

6. The paraphase amplifier circuit consists of two identical _____ .
   A. common-base configurations
   B. common-collector configurations
   C. common-emitter configurations
   D. voltage-divider networks

7. Through the use of a _____ , a single-ended audio stage can produce the signal requirements for push-pull operation.
   A. complementary pair of transistors
   B. center-tapped transformer
   C. coupling capacitor
   D. B and C

8. The decibel voltage ratio is defined as _____ the ratio of two voltages.
   A. 10 times
   B. 20 times

C. 30 times

D. none of the above

9. _____ consists of several series components: transistor junction to case (θ_{JC}), case to heat sinks (θ_{CS}), and heat sink to ambient air (θ_{SA}).

A. The temperature coefficient

B. The cooling coefficient

C. The temperature transfer coefficient

D. Thermal resistance

10. In a single-stage phase splitter, improper bias, low supply voltages, or a defective amplifying device may cause _____ .

A. a reduced output

B. a no-output condition

C. a distorted output

D. none of the above

11. The decibel rating for a 2:1 power ratio is _____ .

A. 0dB

B. 3dB

C. 4dB

D. 6dB

12. The decibel rating for a 2:1 voltage ratio is _____ .

A. 0dB

B. 3dB

C. 4dB

D. 6dB

13. Signal-to-noise ratios may be expressed as _____ .

A. power ratios

B. voltage ratios

C. decibels

D. all of the above

14. When adequate heat sinking is provided, a power transistor delivers _____ .

A. less power

B. more power

C. the same amount of power it would deliver without heat sinking

D. more power until it heats up

15. The transistor in a single-ended audio power amplifier is operated _____ so as to introduce as little distortion of the audio waveform as possible.

A. class AB

B. class B

C. class C

D. class A

16. A class AB, push-pull amplifier has a small forward bias on both transistors which causes each transistor to conduct slightly more than _____ of the time.

A. 95%

B. 63.2%

C. 50%

D. 28%

17. _____ are used to insure that the outputs of the amplifying devices in a class B, push-pull amplifier are the same.
   A. Heat sinks
   B. Balancing networks
   C. Inductive networks
   D. RL circuits
18. A class A, push-pull amplifier requires two input signals which are equal in amplitude and _____ out of phase.
   A. 45°
   B. 180°
   C. 270°
   D. 90°

Problems

1. A 2N4347 power transistor has a specified θ of 1.5°C/W and $T_{j(max)}$ of 200°C. The maximum power dissipation of the device is 10W and the ambient temperature is not expected to exceed 50°C. Determine the required thermal resistance of the heat sink. (No heat sink insulator is used.)
2. What is the signal-to-noise ratio in decibels of a signal of 80V with 2V of noise?
3. What is the dB value for a power ratio of 400?
4. What is the dB value for a voltage ratio of 400?

CHAPTER 25

Oscillators

As mentioned in Chapter 12, AC can be generated mechanically and electronically. Where the frequencies are low, up to several thousand Hertz per second, the ordinary rotating-armature type of generator can be used. But where the frequencies are high, ranging up to millions or even billions of Hertz per second, such a mechanical generator cannot be used. The machine would simply fly apart by centrifugal force in its attempt to attain such a high state of speed to generate the required frequencies. However, as will be shown, transistors and applicable circuit components can be made to generate these high currents or RF frequencies. These electronic generators are known as *oscillators*.

An oscillator is a device that is capable of converting direct current to alternating current at a frequency determined by the values of the components in the circuit. The frequency-determining section of an oscillator may consist of an inductance-capacitance (LC) network, resistance-capacitance (RC) network, or a crystal. The LC network and the crystal are generally used with transistor circuits to generate radio or RF frequencies; on the other hand, oscillators using RC networks are used for low frequency generation, usually at audio frequencies. The waveform output of an oscillator may be either sinusoidal or nonsinusoidal. As mentioned earlier, the typical sine wave output is an example of a sinusoidal waveform. Sawtooth or square waves would, however, be considered nonsinusoidal. Oscillators that have a sinusoidal output generally employ LC networks or crystals, though such waveforms may also be generated by oscillators using RC networks. In this chapter, both sinusoidal and nonsinusoidal oscillator circuits will be considered.

BASIC OSCILLATOR OPERATION

The essential parts of an oscillator circuit (Fig. 25-1) are:
• an oscillatory or tank circuit where the oscillations originate and are maintained.
• an amplifier transistor stage where a portion of the current flowing in the tank circuit is amplified.
• a feedback circuit whereby the output of the amplifier is fed back to the tank circuit to maintain its oscillations.

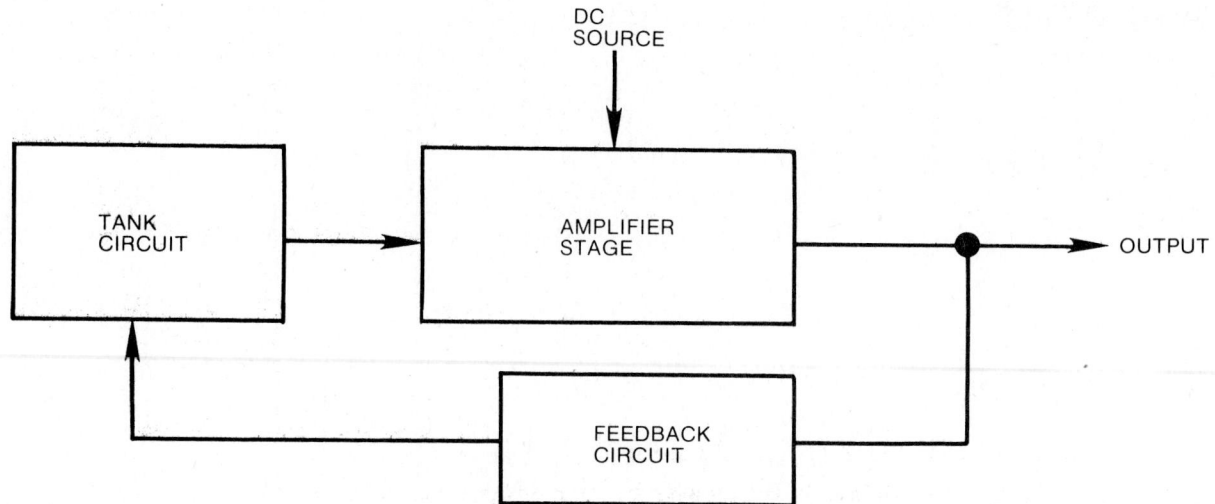

Fig. 25-1: Basic block diagram of an oscillator.

- a DC power source for the tank circuit and transistor amplifier stage.

The mere fact that a circuit is capable of oscillating does not make it a useful oscillator. The circuit must be able to generate sustained oscillations in a desired and controllable manner. A transistor oscillator is essentially an amplifier that is designed to supply its own input through a feedback system.

Two conditions must be met by this feedback network if sustained oscillations are to be produced. First, the voltage feedback from the output circuit must be in phase with the excitation voltage at the input circuit of the amplifier; in other words, the feedback must aid the action taking place in the tank circuit. This is called *positive* or *regenerative* feedback. Generally, amplifier circuits use a *negative* feedback system to inhibit unwanted oscillations in their outputs. Second, the amount of energy fed back to the tank or input circuit must be sufficient to compensate for circuit losses.

Feedback may be accomplished by LC or RC coupling between the output and input circuits. Various circuits have been developed to produce feedback with the proper phase and amplitude. Each of these circuits has certain characteristics that make its use advantageous under given circumstances.

Desired Oscillator Characteristics

Virtually every piece of electronic equipment that utilizes an oscillator necessitates two main requirements of the oscillator—*amplitude stability* and *frequency stability*.

Amplitude stability refers to the ability of the oscillator to maintain a constant amplitude output waveform over a specified frequency range and temperature range. Frequency stability refers to the ability of the oscillator to maintain the *desired* operating frequency over a specified temperature range.

The oscillator tank circuit consists of a closed loop containing an inductor and a capacitor connected in parallel. When a current is started flowing in this loop, the current will flow back and forth (oscillate) between the inductor and the capacitor until its energy is dissipated by the circuit resistance.

Examine the simple LC oscillatory circuit illustrated in Fig. 25-2. An inductor and a capacitor are connected in parallel. A battery is also connected across the tank with a series switch to apply or remove DC power from the tank. When the switch is closed for a short duration, the capacitor charges as shown in Fig. 25-2A. The closing and immediate opening of the switch in the circuit can be likened to shock excitation of the tank circuit. The inductance of the coil will prevent any appreciable current from flowing in the coil for the instant the switch is closed. As the switch is opened, the capacitor discharges through the coil as shown in Fig. 25-2B. The discharge current through the coil will set up an outward expanding magnetic field about the coil. When the capacitor is discharged (Fig. 25-2C), the magnetic

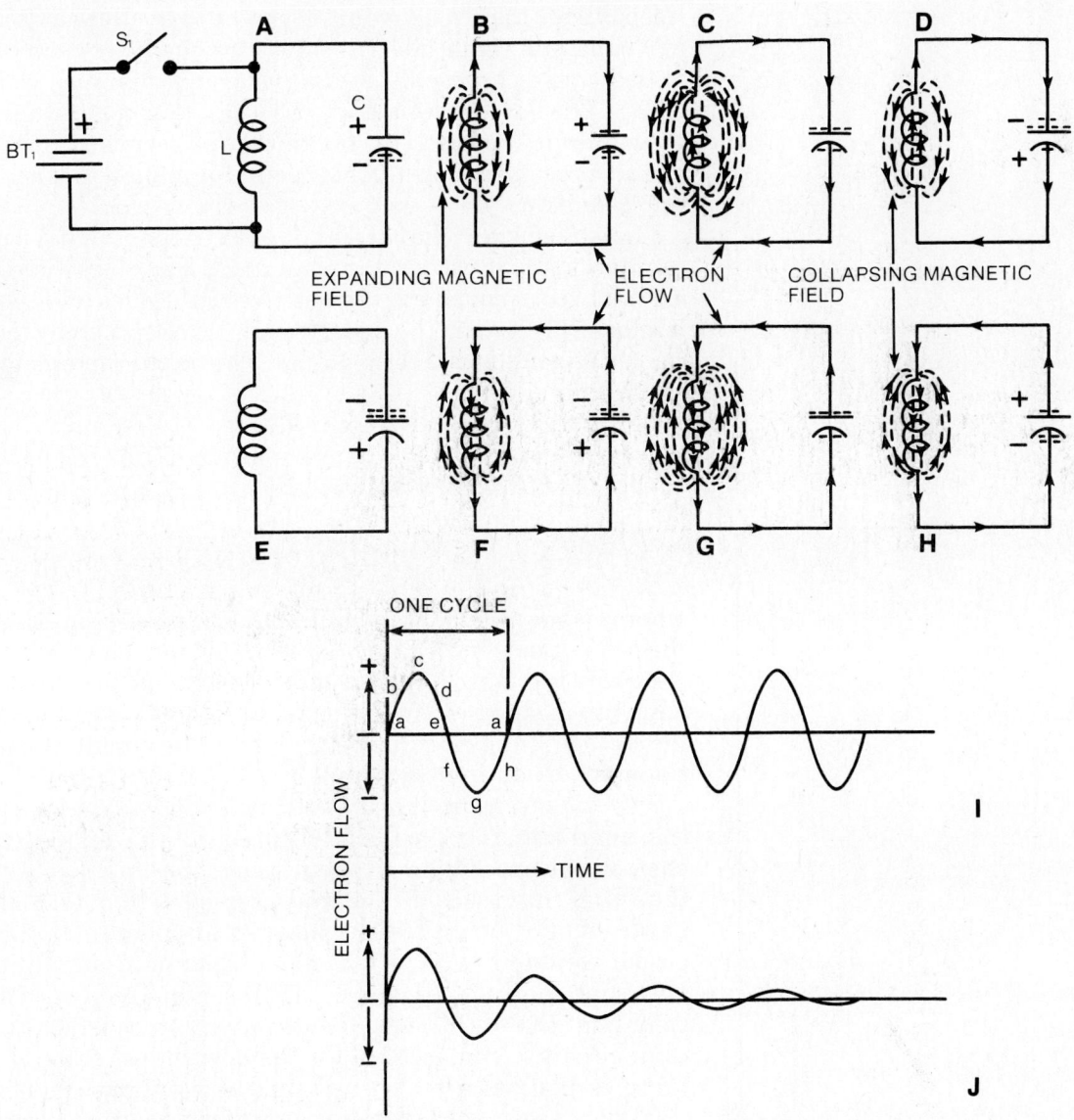

Fig. 25-2: Flywheel effect of a tank circuit.

field around the coil will begin to collapse inward, inducing a counter emf in the coil. The counter emf will then cause current to flow and charge the capacitor in the opposite direction (Fig. 25-2D). When the magnetic field completely collapses, the capacitor will be fully charged as shown in Fig. 25-2E. The capacitor, now charged in the opposite direction or polarity, can begin to discharge through the coil, setting up an expanding magnetic field about the coil (Fig. 25-2F). When the capacitor is fully discharged (Fig. 25-2G), the magnetic field around the coil collapses (Fig. 25-2H), charging the capacitor to its original polarity (Fig. 25-2A).

The action described here produces 1Hz or cycle of AC as shown in Fig. 25-2I. It should be noted that DC energy is transferred from an electrostatic charge on the capacitor to a magnetic field about the coil. This process is called the *flywheel effect* and best describes the circuit oscillation. If this energy transfer or oscillation continued without any tank losses (resistance), the circuit would continue to oscillate indefinitely at a constant amplitude. In reality, the successive tank circuit losses cause the oscillatory waveform to dampen out (Fig. 25-2J). The less the resistance or losses, the longer it takes the waveform to dampen out. It can now be seen why a feedback circuit is required for a basic oscillator stage—to sustain the oscillations.

The waveform of the current flowing in the circuit during the oscillations is illustrated in Fig. 25-2I. The clockwise flow of electrons corresponds to the positive half-cycles and the counterclockwise flow to the negative half-cycles, thereby generating a sinusoidal waveform output. This is characteristic of LC oscillators.

The frequency of these oscillations is determined by the values of L and C. A fair approximation of this frequency can be obtained from the formula:

$$f = \frac{1}{2\pi \sqrt{L \times C}}$$

where f is the frequency in Hz, L is the inductance in H, and C is the capacitance in F. The value of π (pi), of course, is 3.14.

As already stated, resistance in the circuit produces losses which soon stop the oscillations. But if there was some method for replacing the electric energy lost in the circuit, the oscillations would continue indefinitely.

Such replacement of lost electrical energy is the function of the amplifier and feedback sections of the oscillator. The energy developed in the output section of the amplifier is greater than that impressed on the input section. (Of course, the DC power supply furnishes the additional energy.) If the tank circuit were to be connected to the input of an amplifier, the amplified energy would appear in the output section. If a sufficient quantity of this output energy were to be fed back to the tank circuit to compensate for the energy lost through resistance, oscillations in that circuit could be maintained indefinitely.

If, as is usual, the transistor amplifier is used in a common-emitter configuration, there is a 180° phase difference between the input and output of the amplifier. To aid the oscillations in the tank circuit, the energy fed back must be in phase with these oscillations; that is, the feedback must be positive or regenerative. Thus the feedback section must not only feed back energy from the output of the amplifier to the tank circuit, but it must also reverse the phase of this feedback.

The amount of positive feedback must be carefully controlled. If the feedback is too small, the oscillations may die down. If it is too large, the oscillations may grow to a point where the transistor may be damaged.

The oscillator may be *triggered* or *self-excited*. In a triggered oscillator, a pulse from some external source starts oscillations in the tank circuit at full normal strength the moment the pulse is received. In the self-excited oscillator, any small disturbance in the oscillator circuit may start weak oscillations in the tank circuit. As a result of amplification and feedback, these weak oscillations are built up to normal strength very quickly.

Fundamental Oscillator Circuit

One of the simplest oscillators is the tickler-coil oscillator whose circuit is illustrated in Fig. 25-3. When circuit operating voltage V_{CC} is applied, transistor Q_1 conducts, causing collector current to flow through tickler-coil L_C. The current flow produces an expanding magnetic field around L_C and causes voltage to be induced into coil L_B. The induced voltage in L_B is such that it is positive with respect to the base of Q_1 and negative to the emitter of Q_1. The positive voltage on the base increases collector current and causes a further expansion of the magnetic field around L_C. The increased magnetic field increases the voltage induced into L_B. This cycle of operation continues until transistor Q_1 is driven to saturation. With Q_1 in saturation, capacitor C_B starts to discharge. The discharging of C_B decreases the positive voltage on the base of Q_1, thereby decreasing collector current and causing the magnetic field around L_C to collapse and induce a voltage into L_B that is negative to the base of Q_1. The negative voltage to the base

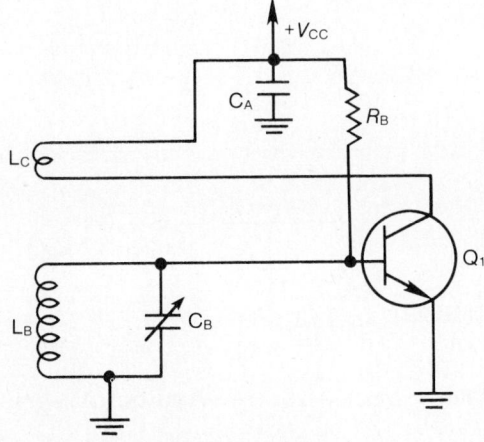

Fig. 25-3: Basic tickler-coil oscillator.

causes the collector current to decrease even more. This cycle of operation continues until transistor Q_1 is driven toward cutoff, at which time the oscillator cycle begins again.

Capacitor C_B is variable and can be adjusted to the desired operating frequency of the basic oscillator. Capacitor C_A acts as a filter to enable the AC components of the collector current to flow around the power source, V_{CC}, rather than through it.

SINUSOIDAL LC OSCILLATOR CIRCUITS

The sinusoidal LC oscillators are the most used of all oscillator circuits in electronics. There are several different types.

Armstrong Oscillator

An Armstrong oscillator circuit utilizes the tickler-coil principle for its operation. The circuit also allows the DC power source to be connected by one of two methods as shown in Fig. 25-4. One is the series-fed method shown in Fig. 25-4A. Here the V_{CC} is in series with the feedback circuit (tickler-coil L_1). Where it is necessary or desirable to keep the V_{CC} source out of the feedback circuit, the shunt-fed or parallel-fed method is used (Fig. 25-4B). The RF choke, RFC, keeps the AC components out

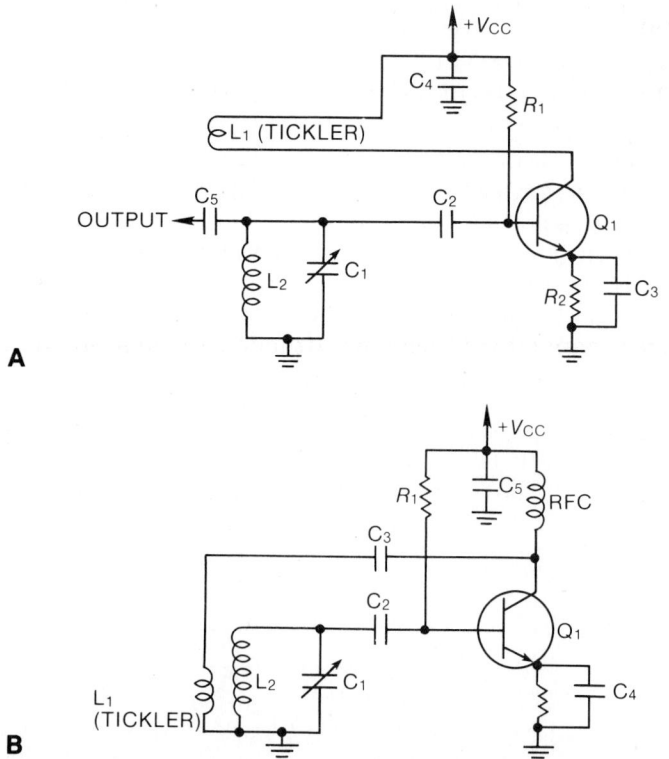

Fig. 25-4: Armstrong oscillator circuits: (A) series-fed and (B) shunt-fed.

636

of the V_{CC} source and, thereby, divorces the V_{CC} source from the feedback circuit.

Series-Fed Circuit. As mentioned, the diagram of a series-fed Armstrong oscillator is shown in Fig. 25-4A. Its operation is similar to that discussed for a fundamental tickler-coil oscillator. The oscillator voltage which is applied to the base of transistor Q_1 is developed by the L_2C_1 tank. The amplifier is suitably biased for this operation by the R_2C_4 and R_1C_2 networks. If the oscillator is to be self-starting, forward biasing of Q_1 is required. The output current appears in the collector circuit and flows through tickler-coil L_1. Coil L_1 is, therefore, used to feed back voltage by transformer action to the tank circuit to overcome losses and keep the oscillations going.

The amount of feedback can be controlled by varying the number of turns of coil L_1 and by varying the coupling between L_1 and L_2. The feedback voltage must be in phase with the voltage of the base circuit; otherwise, sustained oscillations will not occur. Since the base voltage is 180° out of phase with the collector voltage, the feedback must be accompanied by a 180° phase shift. This phase shift is accomplished by the direction of turns on tickler-coil L_1. If the feedback is negative (of the wrong phase), it can be changed to the correct positive feedback by interchanging the connections to the tickler-coil.

The frequency of this oscillator can be varied by changing the values of L_2C_1. Note that capacitor C_1 in Fig. 25-4A is variable. In some instances C_1 will be fixed and L_2 will have a movable iron core that can be moved in or out to vary the inductance and, hence, the frequency of the oscillator.

Only a small fraction of the power in the tank circuit is fed to the amplifier for feedback purposes. The largest portion is retained in the tank circuit to be applied to some external load. The amplitude of the feedback voltage can be adjusted by varying the degree of coupling between coils L_1 and L_2 or by varying the turns ratio between the coils.

There are several methods for obtaining the output from this oscillator. In Fig. 25-4A a coupling capacitor (C_5) is connected to the hot or ungrounded side of the tank coil L_2. If a lesser output is desired, capacitor C_5 can be attached to taps on the coil winding instead of the top. The other method that can be utilized requires the use of a coil inductively coupled to the tank coil.

Shunt-Fed Circuit. The principle of operations of the Armstrong shunt-fed oscillator illustrated in Fig. 25-4B is basically identical to that given for the series-fed circuit. The only difference is the isolation of the V_{CC} source from the feedback circuit by RFC_1 and C_3. Here, too, the output is capacitively coupled to the external load.

In summary, the following characteristics can be applied to series-fed or shunt-fed Armstrong oscillators:

- Inductive feedback is employed (tickler-coil).
- Energy is fed back in phase.
- Phase of the feedback voltage is determined by the direction of the turns of the tickler-coil.

- Amplitude of the feedback voltage is determined by the degree of coupling between the tickler-coil and the tank coil and by their respective turns ratio.
- Oscillator frequency is determined by the LC product of the tank circuit.
- Forward-biasing of the amplifier is required if the oscillator is to be self-starting.
- The base tank circuit may be tapped for impedance matching purposes.
- A capacitively coupled output technique is usually employed.

Hartley Oscillator

The noticeable feature of a Hartley oscillator is the tuned LC network with a tapped coil for inductive feedback instead of a separate tickler-coil. This type of oscillation is quite popular. Instead of using two separate coils inductively coupled, this oscillator employs a single-tapped coil. The bottom portion (from the top down) is the feedback coil; the entire coil is the base tank coil and, together with capacitor C_1, determines the frequency of the oscillator. Typical series-fed and shunt-fed Hartley oscillator circuits are shown in Fig. 25-5. The circuit shown in Fig. 25-5A is a Hartley series-fed oscillator because the feedback component is connected in series with the V_{CC} power source. The oscillator frequency is generated in the tank circuit $(L_1 C_1)$. Capacitor C_1 is variable so that the oscillator can be tuned to the desired frequency. Capacitor C_4 is used for

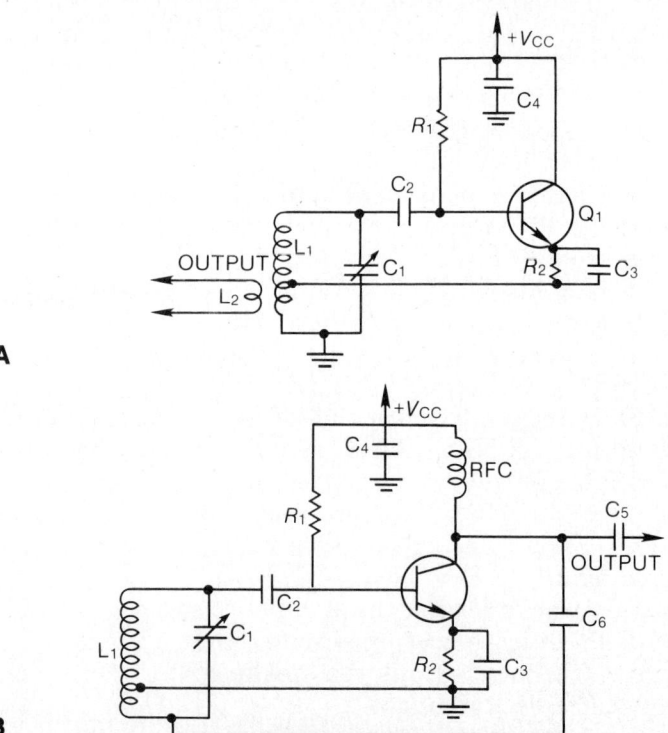

Fig. 25-5: Hartley oscillator circuits: (A) series-fed and (B) shunt-fed.

638

decoupling. Resistor R_2 and capacitor C_3 form a bias stabiliza-tion network and are connected in series with the emitter of transistor Q_1. Degeneration (or negative feedback), which would otherwise occur, is minimized by this network.

For the oscillations to begin, a small forward bias must be applied to the base-emitter junction. This bias establishes a flow of current through the transistor when the circuit is first energized and insures the circuit will be self-starting. Voltage dropping resistor R_1 provides the path for this bias current. Capacitor C_2 prevents the tank coil from shorting the base-emitter junction of the transistor.

When the circuit is turned on (V_{CC} source applied), collector current flows from the negative side of V_{CC} through the lower portion of tank coil L_1 to the emitter of Q_1 via the top on L_1. The rising value of the collector current induces a voltage in the upper half of L_1 that aids the flow of base-emitter current. The accompanying increase in base-emitter current further in-creases the collector current. This action continues until the transistor reaches saturation. At saturation there is no longer a change of current and the coupling between the taps on L_1 falls to zero. This action reduces the magnitude of the base-emitter bias current which reduces the magnitude of the collector cur-rent. The collapse of collector current induces a voltage in the upper half of L_1 of opposite polarity to that originally induced by the rise of collector current. This voltage opposes the base-emitter bias current and reduces the collector current to zero (transistor driven to cutoff). When the collector current is zero, the induced voltage in the upper half of L_1 becomes zero and the base-emitter bias current again increases. The increase in col-lector current now initiates the second cycle of operation. Dur-ing each half-cycle the tank capacitor charges and discharges with an interchange of energy occurring between L_1 and C_1. This action controls the oscillator frequency. In this circuit the tapped coil L_1 is in series with the collector circuit of the transis-tor. A shunt-fed circuit is shown in Fig. 25-5B. The operation is the same. Note that the DC path for the emitter-collector cur-rent is not through coil L_1. The AC signal path, however, is through capacitor C_6 and coil L_1. At the collector C_6 junction, the two current components are separated and required to take parallel paths—DC through RFC_1 to the V_{CC} source and AC through C_6 to coil L_1. Both oscillator circuits receive their feed-back energy by means of magnetic coupling.

The reason why the tap on L_1 provides positive feedback is illustrated in Fig. 25-6. Just L_1 alone is shown with the coil in two parts L_A and L_B so you can see the polarities of induced voltage. Consider electron flow into point A. This current flows through the turns of L_A between points A and G to the V^+ supply. The turns of L_B are not in that path. However, all the turns are on one coil; therefore, L_B is transformer-coupled to L_A.

In the AC variations, assume that I is increasing. Then, the self-induced voltage V_{AG} is negative at point A to oppose the increase of I. Furthermore, this induced voltage is negative for the entire coil. All the turns are wound in one direction, and they are in the same magnetic field.

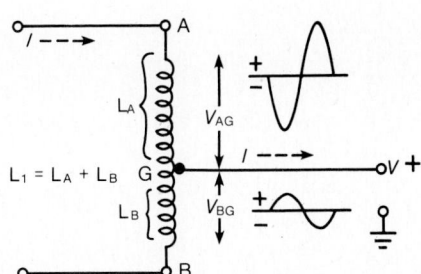

Fig. 25-6: Phase inversion with a tapped coil.

639

Point A is the negative end of the induced voltage compared with any turns farther down on the coil. The highest voltage is across all the turns from A to B. If we look at the induced voltage from B, the point is positive compared with any turns farther up on the coil. At any point tapped between A and B, therefore, the two points have opposite polarities with respect to the tap.

With AC variations in I, AC voltages are induced. Still, V_{AG} and V_{BG} are out of phase by 180°. When one is maximum negative, the other is maximum positive. Since the tap at G is grounded, V_{AG} and V_{BG} are AC voltages in opposite polarity with respect to chassis ground.

In the Hartley circuit of Fig. 25-5, V_{BG} provides the positive feedback required for input to the base. In general, the tap provides feedback voltage of about one-third the total voltage across the coil.

The amplification by Q_1 has one phase inversion, and the feedback from the tapped coil has another phase inversion. As a result, the positive feedback results in regeneration for the oscillator circuit.

In summary, the following characteristics can be applied to series-fed or shunt-fed Hartley oscillators:

• Inductive feedback is utilized (via the tapped coil).

• The tank coil is divided into two sections: a base section and a collector section.

• Energy is fed back in phase and is determined by the direction of the turns on the tapped coil.

• Amplitude of the feedback is determined by the ratio of the base-coil turns to the collector-coil turns.

Colpitts Oscillator

The Colpitts oscillator uses an LC circuit to establish its operating frequency. Capacitive rather than inductive tuning is usually employed, and feedback is obtained through a capacitive voltage-divider circuit. The same result may be achieved by using fixed capacitors and a variable inductor for tuning the circuit. The frequency stability of the Colpitts oscillator is good and considered better than that of the Hartley at lower and medium frequencies.

A transistorized Colpitts oscillator is illustrated in Fig. 25-7. Note that the tank circuit is composed of inductor L_1 and variable capacitors C_1 and C_2. The voltage developed across C_2 is the input signal to the transistor and the amount of feedback depends on the ratio between C_1 and C_2.

The bias voltage is determined by R_1, R_2, and C_4. DC isolation of the tank circuit is provided by C_3, the base coupling capacitor. A collector load for Q_1 is provided by choke RFC_1. Capacitor C_5, as well as providing isolation, acts as the coupling device for the feedback voltage to the tank circuit. Note that it would be impossible for a Colpitts oscillator to be series-fed. As capacitive coupling is used in the feedback circuit, emitter current does not flow in any portion of the tank circuit; therefore, all Colpitts oscillators will be shunt-fed.

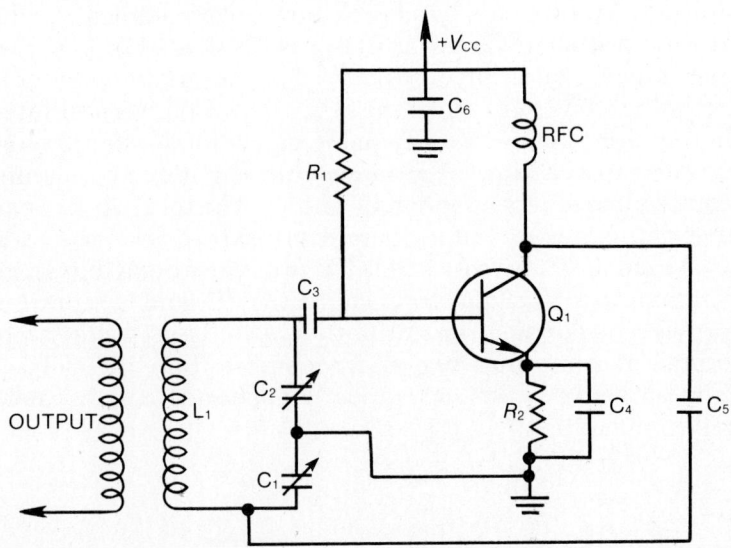

Fig. 25-7: Shunt-fed Colpitts oscillator.

Capacitors C_1 and C_2 form an AC voltage-divider across the output from the amplifier and are equivalent to a tapped coil for the oscillator signal. The collector current is divided across C_1 and C_2 with their polarities changing during the positive and negative cycles. The voltage developed across C_2 is the feedback voltage fed to the base and emitter of Q_1. As the base goes positive, the collector goes negative; therefore, the feedback is regenerative. During the next half-cycle of oscillation, the action is reversed and again the feedback is regenerative.

The Colpitts oscillator is used for frequencies from 100kHz to 300MHz; however, the oscillator has poor frequency stability because of variations in junction capacitance.

The following features are representative of a Colpitts oscillator circuit:

• Capacitive feedback is used (capacitive C_2).

• Amplitude of the feedback is determined by the ratio of the two tank circuit capacitances.

• Capacitance is divided into two sections: the base section (C_1) and the collector section (C_2).

• Frequency of oscillation is determined by the tank circuit values. The two variable capacitors can be gauged so that as the capacitance of one is increased, the capacitance of the other is decreased by the same amount. This will keep the overall frequency of the oscillation constant.

• The output of the oscillator is inductively coupled to the external load.

• The transistor configuration has very poor frequency stability because of the variations in junction capacitances.

Clapp Oscillator

The Clapp oscillator is a variation of the Colpitts oscillator. The principal difference between the two versions is that the

Colpitts is parallel-tuned and the Clapp is series-tuned. A schematic diagram of a Clapp oscillator is illustrated in Fig. 25-8. Other than the addition of capacitor C_S, the circuit operates in exactly the same manner as the equivalent Colpitts oscillator.

In this circuit, capacitors C_1 and C_2 can be made several times larger than capacitor C_S in series with the inductor. Tuning becomes almost dependent on C_1 and C_2. The total tank capacitance of the tank circuit is, therefore, about equal to the value of C_S which tends to reduce the effects of variations in transistor parameters. The two larger capacitors (C_1 and C_2) continue to provide a feedback transformation ratio. This circuit can be operated at somewhat higher frequencies than the Colpitts oscillator because the net series capacitance can be made somewhat smaller.

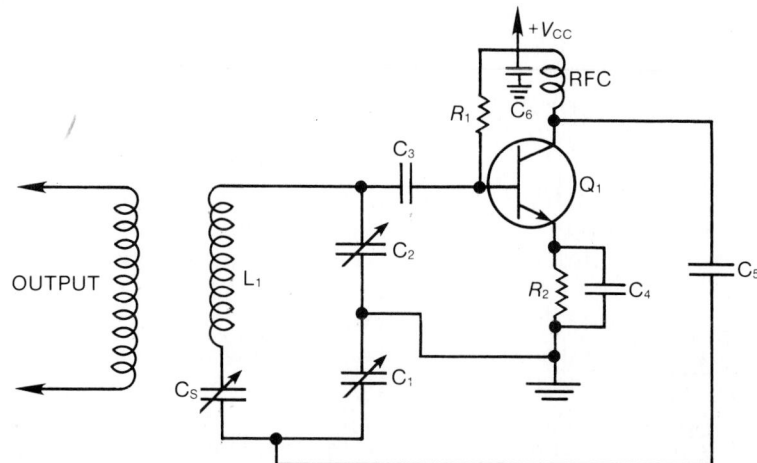

Fig. 25-8: Clapp oscillator schematic diagram.

Tuned-Input Tuned-Output Oscillator

The tuned-input tuned-output oscillator is used to produce a sine wave output of constant amplitude and fairly constant frequency within the RF range. This type of oscillator is generally used as a variable frequency oscillator (VFO) in radio/communications equipment. The tuned-input tuned-output oscillator contains two parallel tuned LC tank circuits, one in the base circuit and the other in the collector circuit of the transistor, and uses the internal base-collector capacitance of the transistor to accomplish feedback. A series-fed and a shunt-fed circuit are illustrated in Fig. 25-9.

When a tuned LC tank circuit is connected to the base circuit and to the collector circuit of a transistor, sustained oscillations will occur when the two circuits are resonated at the same frequency. The amplitude of the oscillations is determined by the amount of feedback from collector to base and the tuning of the two tank circuits. To provide proper operation, both tank circuits are tuned to a slightly higher frequency than their resonant frequency. At the operating frequency, therefore, both tank circuits appear slightly inductive. The two circuits are

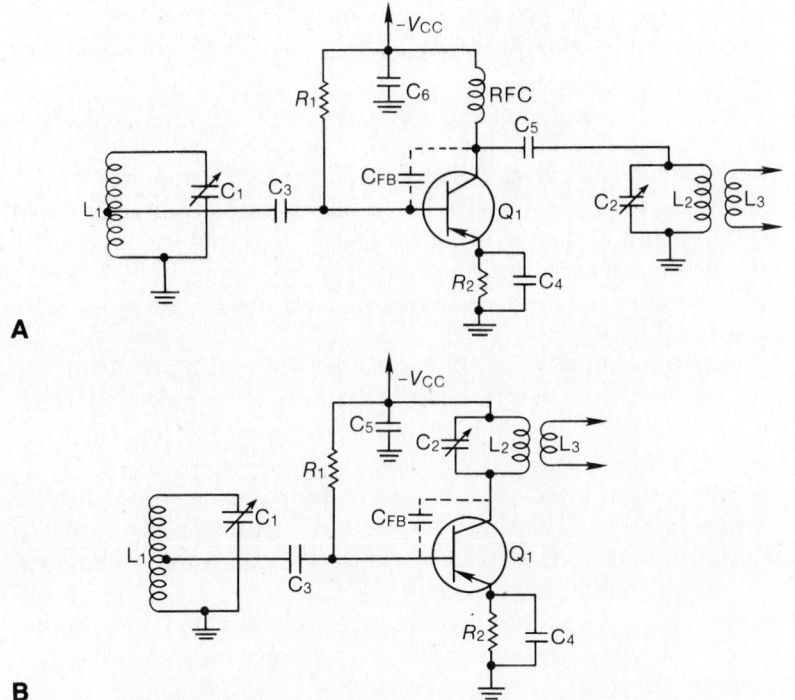

Fig. 25-9: Tuned-input tuned-output oscillator circuits: (A) shunt-fed and (B) series-fed.

substantially the same frequency, but are usually detuned from each other by a small amount to make the oscillations more stable. When they are identically tuned to the same frequency, the oscillation will tend to be unstable.

To sustain oscillation it is necessary only to supply the losses in the base circuit which are relatively small; consequently, only a small capacitance is needed to supply sufficient feedback. The base-collector interelement capacitance of the transistor is adequate for this purpose in most cases.

The base tank circuit of the shunt-fed oscillator (Fig. 25-9A) consists of L_1C_1 while the collector tank consists of L_2C_2, with no mutual coupling between L_1 and L_2. Feedback occurs through the base-collector capacitance, C_{FB} (shown in dotted lines). The feedback amplitude can be controlled by detuning the base tank with C_1; therefore, differences in interelement capacitances can be accommodated by merely adjusting C_1. Capacitor C_5 serves as the collector coupling and blocking device while RFC_1 isolates the power source from RF variations.

The collector-base current in effect creates an in-phase feedback source through C_{FB}. This forms a valid input signal for the transistor and causes the sustained oscillations to be amplified. The tank with the higher Q has a greater effect on the frequency. Usually, C_2 in the collector circuit is adjusted to set the oscillator frequency, while C_1 varies the amount of base excitations. The oscillator output is coupled by L_3 to the next stage.

A series-fed tuned-input tuned-output oscillator is easily designed by placing the collector tank circuit in series with the power source (V_{CC}) and then bypassing it to ground with C_5 (Fig. 25-9B). The series-fed oscillator eliminates the problems of

RFC oscillations, but the effects of DC in the collector tank circuit (insulation requirements) usually pose bigger disadvantages.

The following characteristics are typical of tuned-input tuned-output oscillator circuits:

- Capacitive feedback is employed through capacitor C_{FB}.

- Capacitor C_{FB} can be the interelement capacitance between the base and collector of a transistor or it can be externally connected and sometimes adjustable.

- Feedback is usually controlled by adjusting the input tank circuit.

- The frequency of oscillation is determined by the tank circuit with the highest Q (normally the output tank circuit). If the two tank circuits are physically placed close together, they will tend to drift together.

- The output tank circuit must be tuned to a slightly higher frequency than the input tank circuit in order to maintain oscillations. If both tanks are adjusted to the same frequency, the circuit looks resistive and the frequency tends to be unstable. If the two tank circuits are not adjusted to approximately the same frequency, the AC voltages may be sufficiently out of phase to prevent oscillations.

Negative Resistance Oscillator

The negative resistance oscillator is frequently called a *tunnel oscillator* because it uses a tunnel diode as the main circuit device. The tunnel diode, as described in Chapter 20, is a small two-lead device having a single PN junction which is formed from very heavily doped semiconductor materials. It differs from other PN junction diodes in that the doping levels are from 100 to several thousand times higher in the tunnel diode. The higher impurity levels in both the N- and P-type materials result in an extremely thin barrier region at the junction. The effects which occur at this junction produce the unusual current-voltage characteristics and high-frequency capabilities of the tunnel diode.

Figure 25-10 compares the current voltage characteristic curve of the tunnel diode with that of a conventional PN diode. The broken line shows the curve for the conventional diode and the solid line indicates the characteristic curve of the tunnel diode.

In the standard PN diode, the forward current does not begin to flow freely until the forward voltage reaches a value on the order of half a volt. This corresponds to point Y in Fig. 25-10. This forward voltage is sometimes referred to as the *offset* voltage. In the reverse direction, the conventional diode has a high resistance to current flow until the *breakdown* region is reached. The breakdown region is shown at point X.

The tunnel diode is much more conductive near zero voltage. Appreciable current flows when a small bias is applied to either the forward or the reverse direction. Because the active region of the tunnel diode is at a much lower voltage than the standard semiconductor devices, it is an extremely low power device.

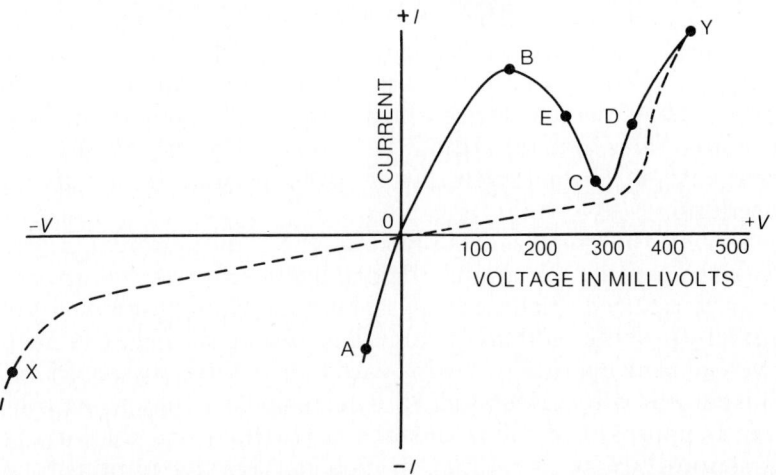

Fig. 25-10: Characteristic of a tunnel diode vs. standard PN junction.

The basic tunnel diode circuit is illustrated in Fig. 25-11. Resistors R_1 and R_2 are used to adjust the diode's operating point to the negative resistance point of its characteristic curve (point E, Fig. 25-10). This point or portion of the curve is relatively linear and operation over this linear region is required if the ideal output waveform is to be achieved.

When power is first applied there is a brief surge of current through C_1, through the junction capacitance of the tunnel diode, and through the resistor network to the power source ($+V$). This surge supplies enough power to C_1 to begin the operation of the resonant tank. The voltage-divider action of R_1 and R_2 set the DC bias level for the tunnel diode along the linear portion of the negative-resistance slope. The resistance values of R_1 and R_2 are usually on a 5:1 ratio to insure proper bias adjustments for the diode (approximately 0.15V for a germanium tunnel diode).

The voltage applied across the tunnel diode is determined not only by the resistor network, but also by the varying voltage in the tank circuit. The bias developed across the resistors would by themselves operate the diode at point E (Fig. 25-10) of its characteristic curve. Voltage variations (positive and negative half-cycles) of the tank circuit would then shift the operation of the diode back and forth along the negative-resistance portion of the curve. Recall that there is some resistance in the resonant circuit. This positive resistance of the tank circuit absorbs power and would eventually damp the tank circuit oscillations. However, if the negative resistance of the tunnel diode is equal to or greater than the resistance of the tank circuit, the diode will act as a switch which will resupply the lost energy to the tank circuit, and oscillation will be sustained.

As the forward bias on the tunnel diode is increased, the current reaches a sharp maximum as shown at point B in Fig. 25-10. This point is referred to as peak voltage and peak current point. The curve then drops to a deep minimum at point C. These are called *valley point voltage* and *valley point current*, respectively. The curve now increases exponentially with the applied voltage and finally coincides with the characteristic

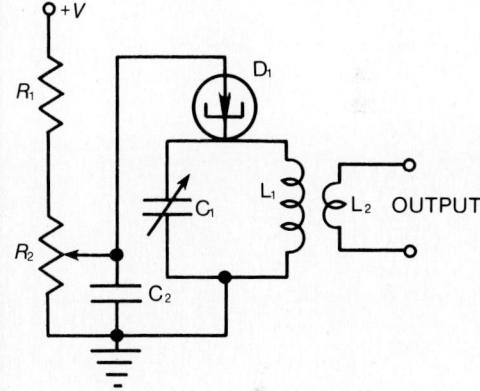

Fig. 25-11: Basic tunnel diode oscillator.

645

curve of a conventional rectifier. This is shown at point Y on the characteristic curve. The drop in current with an increasingly positive voltage (area B to C) gives the tunnel diode the property of negative resistance in this region. This negative resistance enables the tunnel diode to convert DC power supply current into AC circuit current and, thus, permits its use as an oscillator.

During the positive half-cycle of the tank's operation, the forward bias on the tunnel diode is being reduced. As forward bias is reduced, diode current increases. This resupplies the tank by providing additional current. During the negative half-cycle of tank operation, the forward bias is being increased but the current will correspondingly decrease as valley point voltage is approached. Little current is provided and the tank is essentially on its own. Thus, on each positive alternation of the tank, the diode conducts harder, energy is resupplied to the tank, and if this energy is sufficient, oscillations will be sustained.

In this circuit configuration, the tank circuit controls the frequency of operation and the tunnel diode essentially functions as a device to supply the energy required to maintain oscillations. Extremely high frequency operation can be obtained by utilizing the tunnel diode. The narrowness of the potential barrier at the junction allows a very fast reaction time for the device. The tunnel diode may also be tapped down the tank circuit instead of being across the coil as shown in Fig. 25-11.

Voltage-Controlled Oscillator

The voltage-controlled oscillator (VCO) circuit is used for electronic tuning of the oscillator frequency. The method requires the use of a semiconductor capacitive diode, also called a *varicap* or *varactor* (see Chapter 21). Its capacitance varies with the amount of reverse voltage. When there is a capacitive diode across L in the tuned circuit of the oscillator, its frequency is varied by controlling the DC voltage across the diode.

Varactor Diodes. A PN junction with reverse voltage is actually a capacitor. The P and N electrodes are the two conductor plates on the sides of the depletion zone at the junction. This zone serves as an insulator because it has no free charges. Capacitor C may be 80pF or more with a reverse voltage of 6V as typical values. The reverse voltage can be negative at the anode of the diode or positive at the cathode.

Most importantly, C changes with the amount of reverse voltage. The reason is that more reverse voltage widens the depletion area. The effect is equivalent to more insulation distance between the plates, which reduces the capacitance.

VCO Circuit. In Fig. 25-12, transistor Q_1 is a Hartley oscillator. The tuned circuit consists of the tapped coil L_1 across the capacitive diodes D_1 and D_2. Both cathodes have a positive DC control voltage for the reverse voltage to vary C_v. This capacitance controls the oscillator frequency. Two diodes are used in series. The purpose is to balance out the effect of oscillator voltage on the diodes.

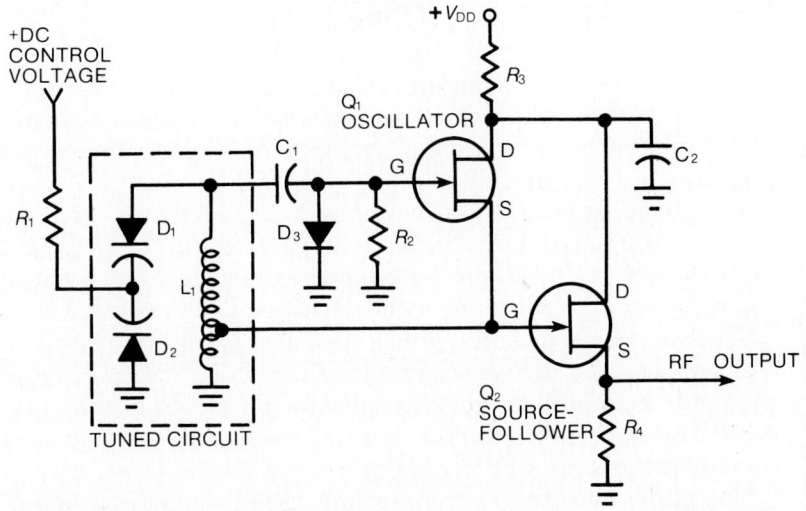

Fig. 25-12: Voltage-controlled oscillator (VCO) circuit.

The oscillator output from the source electrode of Q_1 is directly coupled to the gate of Q_2. Its output is from the source in the source-follower circuit, which corresponds to an emitter-follower. The purpose is to use Q_2 as a buffer stage, which isolates the oscillator output of Q_1 from the load connected to Q_2. The advantage is better frequency stability. Both Q_1 and Q_2 are of the N-channel JFET type.

The functions of all the components of Fig. 25-12 can be summarized as follows:

Q_1	Hartley oscillator transistor.
D_1 and D_2	Capacitive diodes to control the oscillator frequency.
L_1	Oscillator coil.
C_1	Coupling capacitor.
R_2	DC return for the gate of Q_1.
D_3	Rectifies signal for signal bias at the gate of Q_1.
C_2	RF bypass for the drain electrodes.
R_3	Develops the AC signal. Provides the drain voltage for Q_1 and Q_2.
V_{DD}	DC supply voltage for drain of Q_1 and Q_2.
Q_2	Source-follower transistor.
R_4	Output load resistance for the source electrode of Q_2.

The VCO has many applications because of its ability to control the oscillator frequency with a DC voltage. As an example, the push-button tuning on television receivers switches in the DC control voltage to set the frequency for each channel. This application is called *electronic tuning*, or varactor tuning, although the DC control voltage is adjusted manually. However, the DC voltage can also be adjusted automatically by an electronic control circuit.

647

Phase-Locked Loop Circuit

Modern electronic equipment differs from that made just a couple of years ago in the use of frequency synthesis to control the operating frequency. One type in general use today is the phase-locked loop circuit.

The purpose of the phase-locked loop (PLL) circuit is to make a variable-frequency oscillator lock in at the frequency and phase angle of a standard frequency used as the reference. Then the oscillator has the same accuracy as the standard.

The method is to use a phase detector to compare the two frequencies. Any difference of phase results in an error signal that indicates how much the oscillator differs from the standard. The error signal is a DC control voltage that can be used to correct the oscillator frequency.

The phase detector, or comparator, uses two diodes in a balanced rectifier circuit. It has two input signals. The amount of rectified DC output depends on the difference in phase between the two input frequencies. Essentially, the circuit is very similar to the FM detector used for FM signals. However, the phase detector is designed to compare the phase angles within one cycle.

A voltage controlled oscillator is perfect for use with a PLL circuit because the varactor uses a DC control voltage to determine the oscillator frequency and the PLL circuit produces a DC control voltage. A typical arrangement is shown in Fig. 25-13. The four main parts of the circuit are:

1. *Phase detector or comparator.* The two inputs are signals from the VCO with a frequency f_o and a separate oscillator with the standard frequency f_s used as the reference. The output is an error signal that indicates whether f_o is the same as f_s or has a different phase. The detector can have opposite polarities of DC output voltage for lagging and leading phase angles between the two input signals.

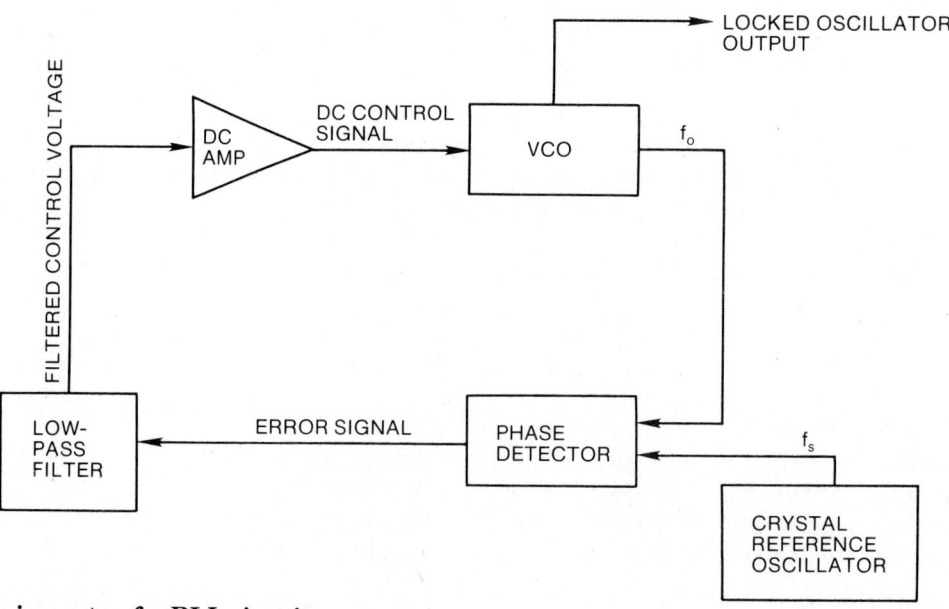

Fig. 25-13: Basic parts of a PLL circuit.

2. *Low-pass filter.* This RC circuit removes the AC signal variations of the two oscillators from the rectified DC output of the phase detector. Input to the filter is the DC error signal with AC ripple, but the output is a filtered DC control voltage for the DC amplifier.

3. *DC amplifier.* The DC amplifier increases the amount of DC control voltage for better control. The amplifier output provides the desired DC level for the control voltage in the polarity needed for the varactor in the VCO.

4. *Voltage-controlled oscillator.* The VCO uses a varactor to set the oscillator frequency, as in the circuit of Fig. 25-12. Input from the DC amplifier keeps the VCO locked into the frequency and phase of the standard oscillator.

The PLL circuit has three operating modes for the VCO. They are free-running, capture, and locked-in, or tracking. If f_o is too far from f_s, the PLL circuit cannot lock in the oscillator. Without lock-in, the VCO is a *free-running oscillator.* Also, the free-running frequency will be produced without any DC control voltage if the PLL circuit is disconnected. When the VCO frequency is within the range of the PLL circuit, however, it produces DC control voltage that reduces the difference between the two frequencies. Once the control voltage starts to change the VCO frequency, the oscillator is in the *capture state.* When f_o is exactly the same as f_s, the VCO is in the lock-in state. The PLL will hold the VCO locked in as long as the DC control voltage is applied.

There are many applications for the PLL circuit. In general, when the circuit is used to control the frequency of an oscillator, it is called *automatic frequency control* (AFC). That function in the RF tuner of television receivers is also called *automatic fine tuning* (AFT). In the horizontal synchronizing circuits, the application is *horizontal* AFC.

One of the most important uses is with a crystal-controlled oscillator as the standard source for a reference. Then the PLL circuit makes an oscillator without a crystal have the same frequency stability as the crystal oscillator.

LC Oscillator Characteristics

The primary function of an oscillator is to generate stable, self-sustained oscillations at a near constant amplitude. To that end certain circuit characteristics must be considered to minimize dynamic instability. Dynamic instability is defined as the change in output frequency of an oscillator due to the varying operating characteristics caused by varying operating potentials. Some of the characteristics that affect an oscillator's operation are presented here.

Inductive Output Coupling. Inductive coupling is accomplished when the inductance of a tank circuit is used as the primary of a transformer. The output is then taken from the secondary winding. In some cases the secondary winding is shunted by a capacitor to form a tank circuit which is resonant to the frequency of the oscillator. This is then known as a tuned secondary.

649

In either case, the output amplitude is determined by the degree of coupling between the tuned primary and the secondary winding. If the coupling is tight (primary and secondary close together), the output amplitude is high, and a considerable amount of power is transferred from the oscillator tank to the secondary. This has the disadvantage of reflecting a high I^2R loss back to the tank circuit, effectively loading it and lowering the tank Q. If the distance between the tank inductance and the secondary is increased, less power is transferred from the oscillator tank, and the output amplitude decreases. This is known as loose coupling. Loose coupling is usually used because the reflected losses are low and tank Q is not appreciably affected. This results in a greater degree of amplitude and frequency stability. When measuring oscillator frequency with a frequency meter, use loose coupling so that the oscillator tank circuit's stability and frequency will not be affected.

Capacitive Output Coupling. Capacitive coupling is achieved by selecting coupling capacitors which present a high capacitive reactance at the resonant frequency. In this manner the signal may be routed from one location to another, without excessively loading the tank circuit and lowering the Q. Slight adjustments may be necessary as capacitive coupling will often have slight effects on the frequency of operation.

Amplitude Stability. Amplitude stability may be affected by changes in bias, transistor gain, supply voltage, and reflected impedance. An increase in bias would cause a decrease in feedback amplitude in most circuits. The decreased feedback would lower the amplitude of oscillations. A possible decrease in bias would produce the opposite effects.

A decrease in the gain of the amplifier device would reduce the gain of the circuit and produce the same effect as increased bias. A change in supply voltage would change the operating point of the circuit and, consequently, change the amplitude of oscillations.

If the impedance of the load decreases, it will reflect an increased power drain back to the oscillator tank circuit in most cases. Without a change in feedback amplitude this loss would not be compensated for and the amplitude of the oscillations would decrease. Amplitude stability can be improved by utilizing base leak bias in transistor circuits. This improvement is due to the bias level being dependent on the amplitude of the tank signal. Most oscillators are operated class C for best efficiency.

Frequency Stability. Frequency stability can be affected by changes in temperature, tank Q, reflected losses, supply voltage, and vibration. Changes in temperature will cause the junction capacitances to vary in the transistor. They will also cause slight variations in the values of the tank's L and C. These changes will shift the output frequency. Since these changes occur slowly, the resultant shift in frequency is called *drift*. Frequency drift can be minimized by adequate ventilation, circuit components which can easily dissipate heat, or, in some cases, a temperature control system.

Variations in inductance values are minimized by adequate ventilation, limiting the DC through the coil, and by using a coil wound with large wire. Variations in capacitance values can be minimized by using temperature compensating capacitors and by using a high C-to-L ratio in the tuning circuit.

Mechanical vibration can cause the same variations in component values as temperature changes did. In this case, however, the resultant frequency shift is more rapid in nature. The oscillator chassis is often shock mounted to overcome this problem. Variations in supply voltage will change the operating point of the circuit and cause circuit instability. This can be overcome by use of a regulated power supply.

The load impedance reflected into the tank circuit has a great effect upon the frequency stability of the oscillator. The reflected impedance may contain both resistive and reactive components. The resistive component will lower the Q of the tank and the reactive component will alter the resonant frequency. When the Q of the tank is very high and the reflected impedance is small, the effect on the resonant frequency and tank Q is negligible. The lower the effective Q of the tank, the greater will be the effect of a change in load impedance on the resonant frequency. High Q tanks and loose coupling reduce the frequency instability caused by these factors.

Tuning and Tracking. In most radio receivers, the local oscillator frequency will always differ from the desired station frequency by an amount equal to the intermediate frequency. The local oscillator develops the intermediate frequency. It is usually tuned above the station frequency, although, in some cases, it may be tuned below. When a receiver, having an IF of 455kHz, is tuned to a station at the low end of the broadcast band (540kHz for example), the local oscillator frequency will be: 540kHz + 455kHz = 995kHz. When the same receiver is tuned to a station at the high end of the band (1,600kHz for example), the local oscillator will be: 1,600kHz + 455kHz = 2,055kHz.

Thus, the frequency range of a local oscillator in a superhet receiver, with a 455kHz IF, will vary from approximately 995kHz to 2,055kHz. The tuning component, usually the capacitor, of the tank circuit is mechanically ganged with the tuning component of the input circuit. This allows the local oscillator frequency to be changed as the station frequency is changed, thereby maintaining the LO frequency, at all times, separated from the station frequency by an amount equal to the IF.

Since the LO frequency follows the selected frequency as tuning occurs, the LO is said to *track* the capacitors in the RF stage; there will be slight electrical differences (due to manufacturing tolerances) in the oscillator tuning capacitors. For this reason, a trimmer capacitor is usually connected in parallel with the oscillator tuning capacitor. Adjustment of the trimmer at the high end of the band insures better tracking. Tracking adjustments at the low end of the band are usually accomplished either through the use of padder capacitors, slug tuned inductors, or slotted rotor end plates on the tuning capacitor.

Usually, Hartley type oscillators are tuned by means of a variable tank capacitor. This lends itself easily to the ganged tuning mentioned earlier, since the input frequency selector circuit is usually tuned by means of a variable capacitor. The rotors of these variable capacitors can be mechanically connected to a common shaft. It should also be recalled that the varactor is sometimes used to accomplish oscillator tuning.

Tracking is more difficult in a Colpitts oscillator when the tank capacitors are made variable. The two capacitors provide impedance matching between the output and input circuits. The ratio of impedances must be maintained throughout the range of frequencies or an impedance mismatch will occur. This results in a loss of feedback power. This difficulty encountered in varying Colpitts oscillator frequencies may be reduced by making the tank coil variable or by making the split capacitors of the tank circuit fixed and inserting a variable tuning capacitor in parallel with the tank. Operation of the Colpitts in this manner will allow the tank frequency to be varied, but will maintain the impedance ratio between the two fixed capacitors.

CRYSTALS AND CRYSTAL OSCILLATORS

In order to obtain a higher degree of frequency stability a *crystal* oscillator is often used. A properly cut crystal possesses the characteristics of a resonant circuit and may, therefore, be used in place of a tuned circuit as the frequency controlling element. Before discussing the use and operation of a crystal in a circuit, it will prove advantageous to discuss some of the properties and preparation of various types of crystals.

The control of frequency by means of crystals is based upon the *piezoelectric effect*. When certain crystals are compressed or stretched in specific directions, electric charges appear on the surface of the crystal. Conversely, when such crystals are placed between two metallic surfaces across which a difference of potential exists, the crystals expand or contract. This interrelation between the electrical and mechanical properties of a crystal is termed the piezoelectric effect. If a slice of crystal is stretched along its length, so that its width contracts, opposing electrical charges appear across its faces; thus, a difference of potential is generated. If the slice of crystal is now compressed along its length, so that its width expands, the charges across its faces will reverse polarity. Thus if alternately stretched and compressed, the slice of crystal becomes a source of alternating voltage.

Also, if a small alternating voltage is applied across the faces of a crystal, it will vibrate mechanically. The amplitude of these vibrations will be very vigorous when the frequency of the AC voltage is equal to the natural mechanical frequency of the crystal. If all mechanical losses are overcome, the vibrations at this natural frequency will sustain themselves and generate electrical oscillations of a constant frequency. Accordingly, a crystal can be substituted for the tuned tank circuit in transistor and tube circuits.

Practically all crystals exhibit the piezoelectric effect, but only a few are suitable as the equivalent of tuned circuits for frequency control purposes. Among these are quartz, Rochelle salt, and tourmaline. Of these, Rochelle salt is the most active piezoelectric substance; that is, it generates the greatest amount of voltage for a given mechanical strain. However, the operation of a Rochelle salt crystal is affected to a large extent by heat, aging, mechanical shock, and moisture. Although these factors render this substance unsuitable for use in frequency control applications, it does find use in some microphones and photograph pickups (see Chapter 22).

Tourmaline is almost as good as quartz over a considerable frequency range and is somewhat better than quartz in the 3MHz to 90MHz range, but it has the disadvantage of being a semiprecious stone, and its cost is, therefore, prohibitive.

Quartz, although less active than Rochelle salt, is used universally for oscillator frequency control because it is inexpensive, mechanically rugged, and expands very little with heat. Crystals used in oscillator circuits are cut from natural or artificially grown quartz crystals which have the general form of a hexagonal prism as shown in Fig. 25-14A; however, they are rarely as symmetrical as shown. Assuming a symmetrical crystal, the cross section is hexagonal, as in Figs. 25-14B and C.

The axis joining the points at each end, or apex of the crystal, is known as the optical or Z axis. Stresses along this axis produce no piezoelectric effect. The X axis passes through the hexagonal axis and cross-sectional area at right angles to the Z axis and these are known as the *electrical axes*. These are the directions of the greatest piezoelectric activity.

The Y axes, which are perpendicular to the faces of the crystal, as well as the Z axis, are called the *mechanical axes*. A mechanical stress in the direction of any Y axis produces an electrostatic stress, or charge, in the direction of that X axis which is perpendicular to the Y axis involved. The polarity of the charge depends on whether the mechanical strain is a compression or a tension. Conversely, an electrostatic stress, or voltage applied in the direction of any electrical axis, produces a mechanical strain (either an expansion or a contraction) along that mechanical axis which is at right angles to the electrical axis.

For example, if a crystal is compressed along the Y axis, a voltage will appear on the faces of the crystal along the X axis. If a voltage is applied along the X axis of a crystal, it will expand or contract in the direction of the Y axis. This interconnection between mechanical and electric properties is exhibited by practically all sections cut from a piezoelectric crystal.

A cut crystal will vibrate at a frequency determined mainly by its thickness. The following formula can be used to determine the fundamental frequency of a cut crystal:

$$f = \frac{K}{T}$$

where: f = frequency in kHz

T = thickness of the cut in inches

K = constant, which is 112.6 for an X cut, 77.0 for a Y cut, and 66.2 for an AT cut.

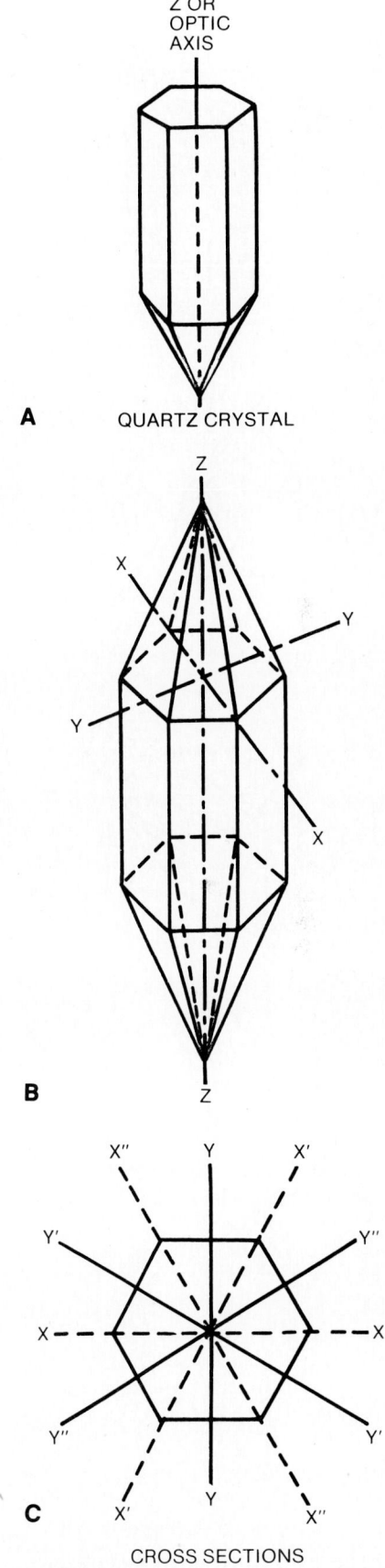

A QUARTZ CRYSTAL

B

C CROSS SECTIONS

Fig. 25-14: **Quartz crystal axes.**

653

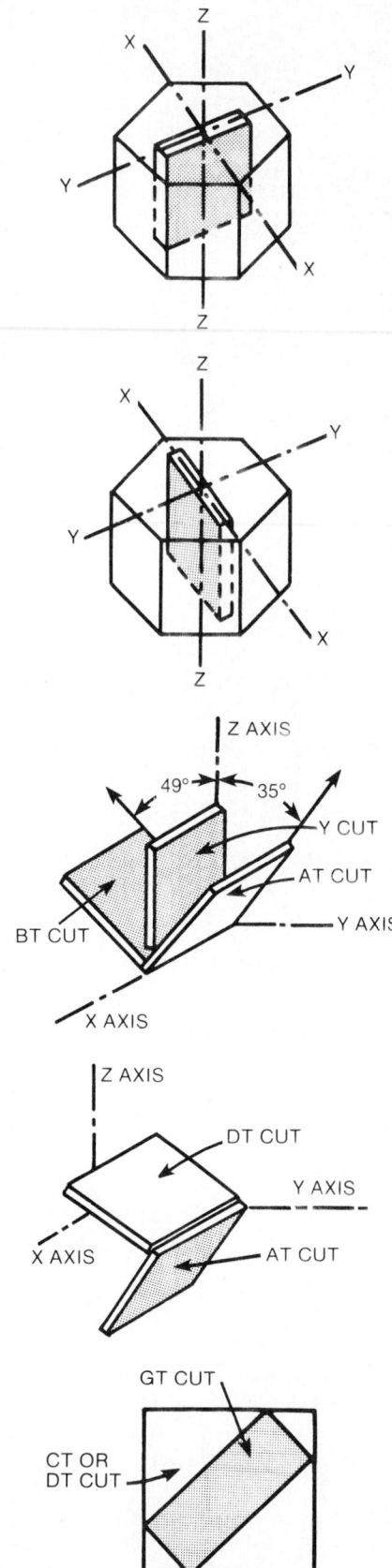

Fig. 25-15: Methods of cutting a quartz crystal.

Crystal wafers can be cut from the mother crystal in a variety of directions along the axes. They are known as *cuts* and are identified by designations such as X, Y, AT, etc. (Fig. 25-15). Each has certain advantages, but, in general, one or more of the following properties are desirable: ease of oscillation at the intended frequency, a single operating frequency, and minimum frequency variation due to temperature changes.

Both the X and Y cuts have unfavorable temperature characteristics. Better characteristics can be obtained by cutting wafers at different angles of rotation about X axes. The Y cut serves as the zero-degree reference, since it is lined up with both the X and Z axes; that is, it lies in a plane formed by the X and Z axes. By rotating this slice from its starting point, different cuts may be formed. Many other cuts exist, but they will not be discussed here.

Crystals used in oscillator circuits must be cut and ground to accurate dimensions. For instance, a typical quartz which is resonant at 1 MHz would measure 2.54cm by 2.54cm by 0.286cm. Crystals may be cut in various shapes, with crystals in the lower frequency range being square or rectangular and some of the crystals in the higher frequency range being disc shaped, similar to a coin. For precision test work, crystals may be cut in the form of a flat ring. The type of cut also determines how active the crystal will be. Also, for any given crystal cut, the thinner the crystal, the higher the resonant frequency. The more care that is taken in cutting and grinding the crystal, the more accurate will be the final result.

Crystals which are not ground to a uniform thickness may have two or more resonant frequencies. Usually one of these resonant frequencies will be more pronounced than the others, with the less prominent ones being termed *spurious* frequencies.

The resonant frequency of quartz crystals is practically unaffected by changes in the load. Like most other materials, however, quartz expands slightly with an increase in temperature. This affects the resonant frequency of the crystal.

Temperature Coefficient and Calculations

The temperature coefficient of a crystal refers to the number of Hertz the frequency of the crystal will vary per MHz of its operating frequency per degree Centigrade of change. The temperature coefficients vary widely from one crystal to another because of the type of cuts employed. A positive temperature coefficient is assigned to those cuts which produce an increase in frequency with an increase in temperature (i.e., a Y cut crystal). A negative coefficient is assigned to X cut crystals which decrease their natural resonant frequency when the temperature increases. On the other hand, the AT cut (also BT, CT, DT, and GT cuts) has a zero temperature coefficient over a wide range of temperature variations.

The temperature coefficient also depends on the surrounding temperature at which it is measured. Heating of the crystal can be caused by external conditions such as the high temperature of power supply circuits. Heating can also be caused by exces-

654

sive RF currents flowing through the crystal. The slow shift of the resonant frequency resulting from crystal heating is called *frequency drift*. This is avoided by use of crystals with nearly zero temperature coefficient and also by maintaining the crystal at a constant temperature.

Temperature coefficient can be calculated by using mathematical methods. The crystal frequency will drift with temperature (the manufacturer of the crystal will provide information regarding the amount of drift) and this is usually specified as the change for temperature in parts per million (PPM) per degree C.

Example: A 600kHz, X cut crystal is calibrated at 50°C and has a temperature coefficient of -20PPM/°C. At what frequency will the crystal oscillate when the temperature is 60°C?

Solution: The -20PPM/°C is interpreted as -20Hz/°C/MHz, since 600kHz is 0.6MHz. The change in temperature is 10°. The change in frequency is then:

$$-20 \times 10 \times 0.6 = -120 \text{Hz}$$

Therefore, at 60°C the 600kHz crystal oscillates at 599.88kHz (600,000Hz - 120Hz = 599,880Hz).

If a similar Y cut crystal were used, the 120Hz would have been added instead of subtracted from the fundamental frequency (600,000Hz + 120Hz = 600,120Hz or 600.12kHz).

Various Crystal Characteristics

Crystal oscillator circuits exhibit certain characteristics that must be observed or noted when used in equipment designs. Some of these characteristics are noted as follows:

• Never apply DC to the crystal element because it will fracture and destroy the crystal.

• The temperature of the crystal must be maintained constant to avoid frequency drift.

• The crystal should be kept clean in order to operate properly. The faces of the crystal should not be touched with the fingers. Doing so would allow human oil secretions to deposit on the faces and, therefore, affect the overall operating frequency. The crystal faces may be cleaned with soap and water or carbon tetrachloride.

• The frequency capability of the crystal is limited by its physical thickness. The X cut crystal can be produced having higher resonant frequencies than any other crystal cuts.

• At a slightly higher frequency than the natural resonant frequency of the crystal, the LCR circuit (Fig. 25-16) appears inductive and the net reactance will equal the reactance of capacitance C_{MH}. Capacitance C_{MH} represents the crystal metallic holding plates and aids in forming a parallel resonant circuit. This basic circuit can be considered an equivalent electrical circuit of a crystal.

• The equivalent inductance of the crystal causes the inductive reactance to be much greater than the equivalent resistance. The crystal, therefore, forms a very high Q resonant circuit providing excellent frequency stability.

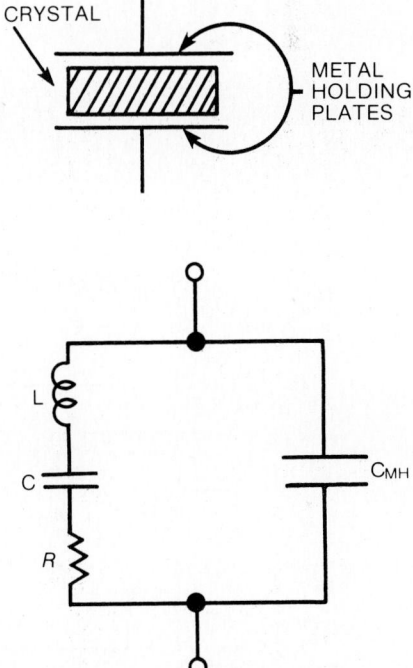

Fig. 25-16: Equivalent electrical circuit of a crystal.

655

Hartley Crystal-Controlled Oscillator

The Hartley crystal-controlled oscillator is very similar to the LC-type Hartley oscillator. In the crystal-controlled version (Fig. 25-17), the tank circuit is tuned to the crystal frequency. The crystal, Y_1, acting as a series resonant circuit, controls any frequency drift of the tank circuit. The crystal is also used to supply the necessary feedback to sustain oscillations. The circuit shown here is a typical shunt-fed oscillator.

Colpitts Crystal Oscillator

An example of a Colpitts crystal oscillator is shown in Fig. 25-18 and can be readily identified by the capacitive voltage divider C_1/C_2. The crystal Y_1 takes the place of the inductor in the LC tank. The split capacitance (C_1, C_2) is for feedback pur-

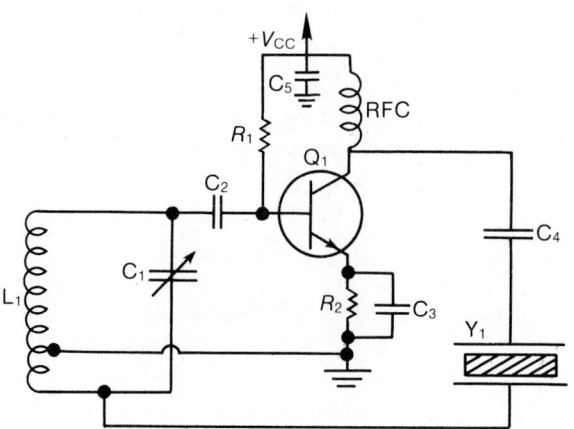

Fig. 25-17: Hartley crystal-controlled oscillator.

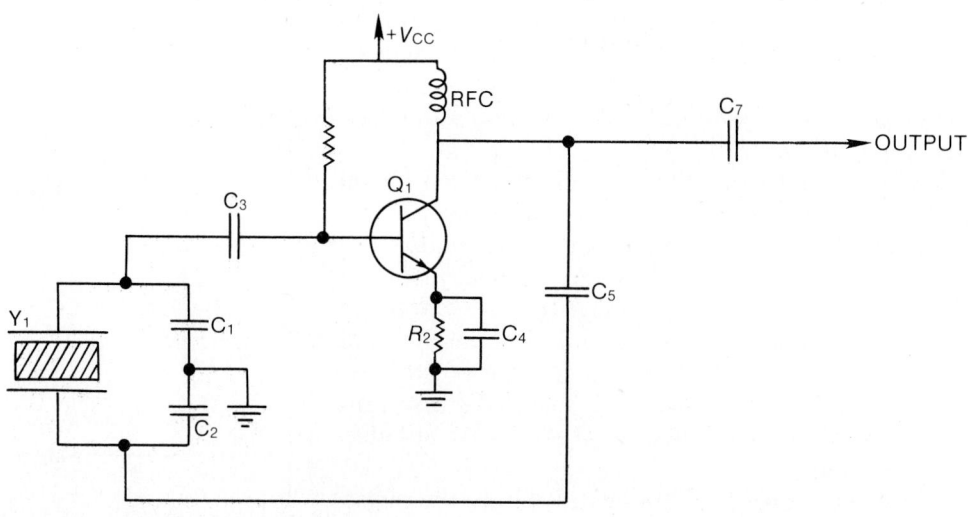

Fig. 25-18: Colpitts crystal oscillator.

poses. Other than the crystal, the oscillator circuit functions exactly the same as its LC tank circuit counterpart.

Pierce Crystal Oscillator

Circuits where the crystal element is connected directly from the collector to the base are called Pierce oscillators. Unlike most other crystal-controlled oscillators, no tuned tank circuit is required. In the Pierce oscillator circuit shown in Fig. 25-19, crystal Y_1 is the feedback path from collector to base. Oscillation occurs at the series-resonant frequency of the crystal. Resistor R_1 must be chosen to make base bias voltage high enough so that the net voltage across the crystal is low; otherwise, the crystal may crack. Collector current is fed through choke RF_1. The RF voltage developed across the choke, which is not resonant at the oscillator frequency, is fed out through capacitor C_1.

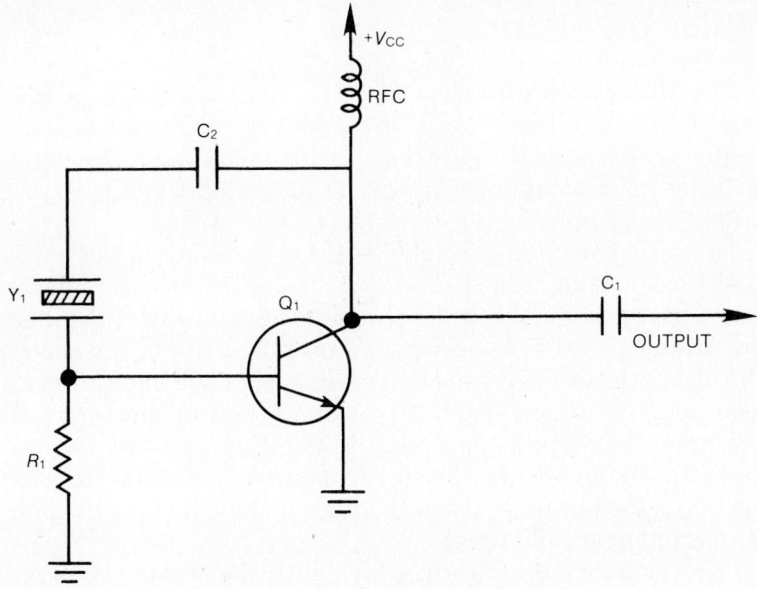

Fig. 25-19: Typical Pierce crystal oscillator.

Tuned-Input Tuned-Output Crystal Oscillator

The circuit in Fig. 25-20 is very similar to the tuned-input tuned-output oscillator in Fig. 25-9, except that a crystal replaces the input tank circuit. Feedback is provided through C_3, the internal capacitance between collector and base.

In the base circuit the R_1/R_2 voltage-divider network supplies forward voltage from $+V_{CC}$. The capacitor C_1 is used to keep the DC base voltage off the crystal. However, C_1 may be omitted because the crystal holder itself can exhibit capacitance. The RF choke RFC_1 provides a high impedance load for the crystal output to the base.

In the emitter circuit, R_E and C_E are used for bias stabilization. In the collector, the LC tuned circuit provides oscillator output inductively coupled by L_2 to the next stage.

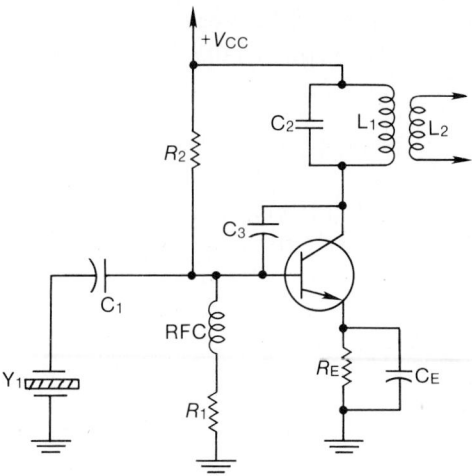

Fig. 25-20: Tuned-input tuned-output crystal oscillator.

Miller Oscillator

The Miller oscillator circuit of Fig. 25-21 is one of the most commonly found in VHF-FM circuits. The crystal is found connected between the gate (base, grid) and ground. The output is tuned to a frequency approximately equal to the crystal frequency by parallel resonant tank circuit L_1/C_1.

The Miller oscillator circuit will not oscillate if the output tank circuit is mistuned. The stability of the circuit is dependent also upon the tuning of the tank. The tuning will change the drain current of the oscillator, and the slope of the current change as the tank is tuned will be different on different sides of resonance. If we approach the resonant point from below, for example, the slope is less than if we approach the resonant frequency from above. The correct point is approximately 50% to 75% up the shallow side of the tuning slope. This is the point of maximum stability.

The variable capacitor in series with the crystal is used to vary the actual operating frequency of the oscillator. One of the factors that determines the oscillating frequency is the capacitance seen by the crystal when it looks into the circuit. By placing a series, or shunt, capacitance in the crystal circuit, the frequency can be pulled a small amount. This feature is used to net the transmitter onto the correct frequency. This same technique could be used in any of the oscillator circuits, but is shown only in one for simplicity's sake.

Overtone Oscillator

An *overtone* is a frequency that is almost a harmonic of the crystal's fundamental frequency. The overtone is always approximately an odd integer multiple of the fundamental. But if the frequency is an exact multiple, it is a harmonic. Some crystals can be made to oscillate in the overtone mode, which means that they will produce a signal at the overtone frequency. Over-

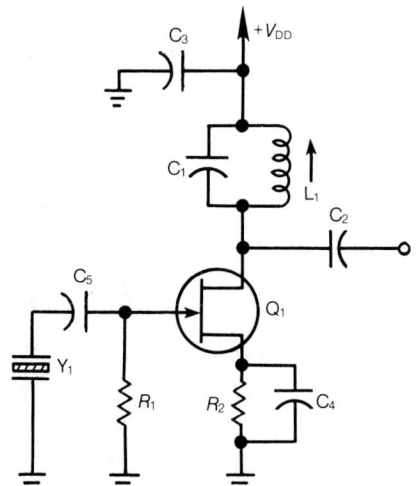

Fig. 25-21: Miller oscillator circuit.

tones of 5, 7, and 9 are commonly available, and this type of crystal is usually used at frequencies above 20MHz or 25MHz. Fundamental mode crystals become too thin at these frequencies, and this makes them too fragile for common use.

Overtone oscillators are built with LC tank circuits that are tuned to the overtone frequency, not the fundamental frequency. Note that the frequency marked on the case of an overtone crystal is calibrated at the overtone, not at the fundamental frequency.

A third overtone series resonant crystal is used in the oscillator whose circuit is shown in Fig. 25-22. The feedback path from the collector to the emitter to the transistor is through C_1 and the crystal. Since the base is grounded at the signal frequency by C_5, the transistor is connected in the common-base configuration. The output signal is taken from a low impedance point on collector tank coil L_1.

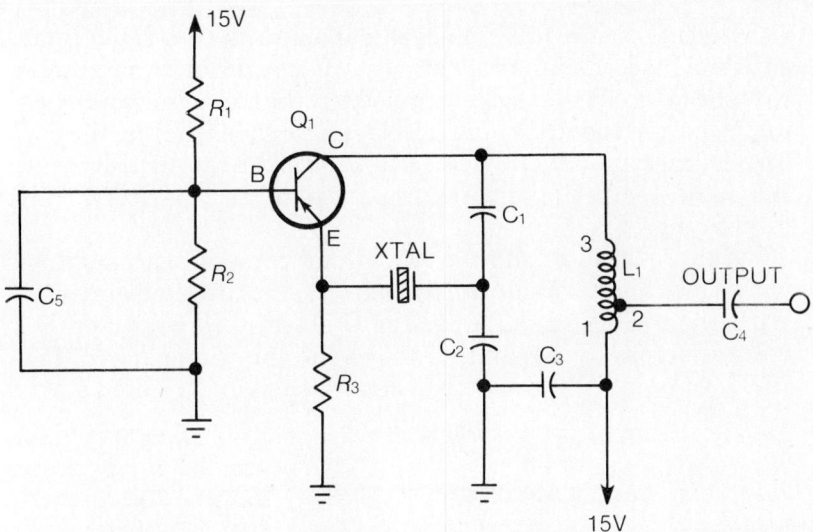

Fig. 25-22: Transistor overtone crystal-controlled oscillator circuit.

NONSINUSOIDAL OSCILLATORS

The outputs of the oscillators discussed in the previous portions of this chapter have sinusoidal outputs which are characteristic of oscillations in an LC tank or crystal circuit. There are other types of oscillators which can generate AC outputs with nonsinusoidal waveforms, such as sawtooth, square wave, etc. Such oscillators usually depend for their operation on the charge and discharge of a capacitor in series with a resistor.

Before taking a look at various types of nonsinusoidal oscillation, it is wise to review the performance of a series RC circuit that was fully discussed in Chapter 15. Remember that if a source of DC is connected in series with a resistor and capacitor (Fig. 25-23), the moment the circuit is completed, current starts to flow at its maximum rate, charging up the capacitor. When the capacitor is fully charged, current ceases to flow.

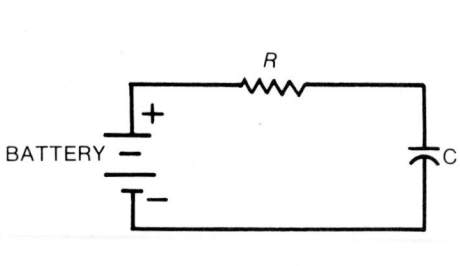

Fig. 25-23: RC series circuit.

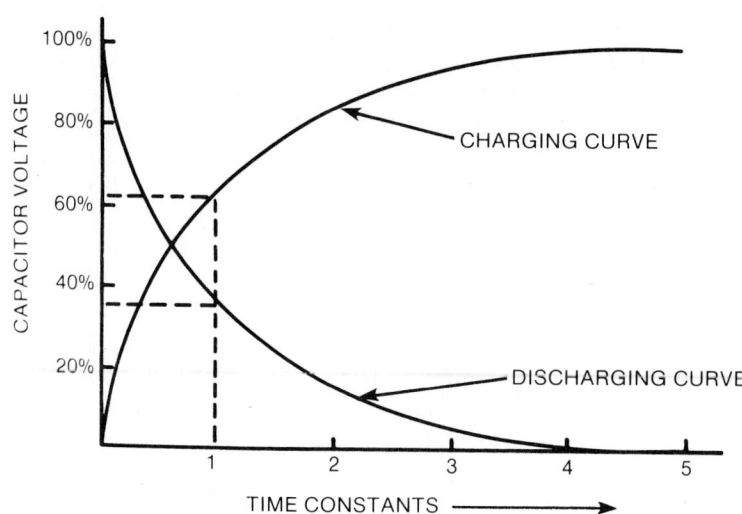

Conversely, if a fully charged capacitor is connected to an external load, the capacitor starts to discharge at its maximum rate the moment the load is applied. As the capacitor continues to discharge, the force of its electrostatic field (that is, the emf force of the capacitor) is reduced and so is the rate of discharge. When the electromotive force drops to zero, the capacitor is fully discharged.

Figure 25-24 shows how an RC circuit can produce an output with a sawtooth waveform. Switch (S_W) is connected across C. With the switch open, capacitor C starts to charge and the voltage across the capacitor follows the charging curve of Fig. 25-24. At point A, the switch is closed, shorting C which starts

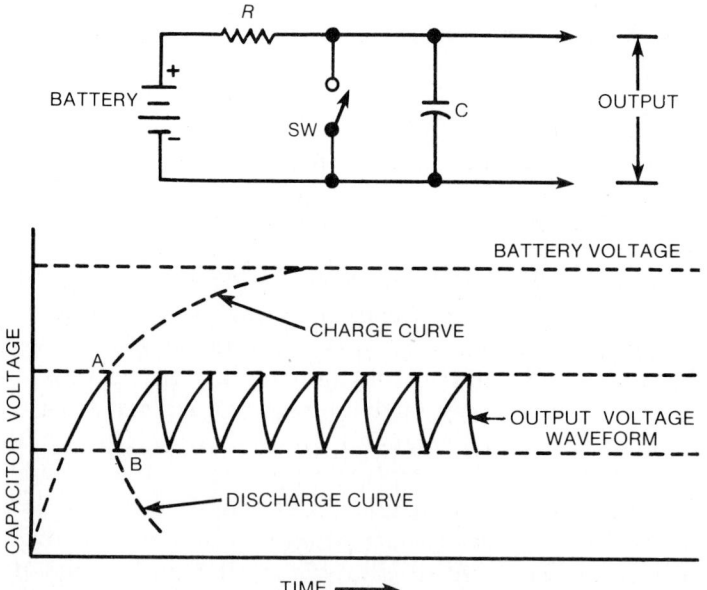

Fig. 25-24: Action of a simple relaxation oscillator.

to discharge through the switch. The voltage across C falls to point B, as indicated by the discharge curve. Thus the waveform of the voltage across C during the charge-discharge cycle resembles a sawtooth, rising on charge and falling on discharge. This is the voltage which is passed on to the load.

At point B, the switch is opened and a new cycle starts. Thus, by periodically opening and closing the switch, a periodic sawtooth voltage is applied to the load and the frequency of this voltage depends upon the frequency with which the switch is opened and closed.

In effect, the RC circuit illustrated in Fig. 25-24 is a sort of oscillator whereby the direct current of the source is converted to an alternating current output. Oscillators which depend for their operation upon the alternate charging and discharging of a capacitor are called *relaxation oscillators*.

Multivibrators

The multivibrator (MV) is a type of relaxation oscillator used extensively to produce various nonsinusoidal waveforms such as square, rectangular, and sawtooth. The multivibrator is basically a two-stage amplifier, or oscillator, circuit that operates in two *modes* or *states* controlled by the circuit conditions. Each amplifier stage supplies feedback to the other in such a manner that will drive the active device of one stage to saturation and the other to cutoff. This is followed by a new set of actions that causes the opposite effects, that is, the saturated stage is driven to cutoff, and the cutoff stage is driven to saturation. The successful operation of the multivibrator is based on the premise that no two devices have exactly identical characteristics. Multivibrators are classed as (1) *astable*, (2) *bistable*, and (3) *monostable*.

The astable multivibrator, also called a *free-running* multivibrator, alternates automatically between the two states and continues the alternations at a rate determined by the values of the circuit components. This type of multivibrator is self-starting and sustains oscillation under normal operation; however, in some applications synchronizing, or triggering, pulses may be used in conjunction with the multivibrator.

The bistable multivibrator, or *flip-flop* multivibrator, has two states of operation but requires the application of an external triggering pulse to change the operation from either one state to the other. Depending upon the circuit design, the circuit remains in the existing state (1) as long as the triggering pulse is applied or (2) until a second triggering pulse is applied.

The monostable multivibrator, or *one-shot* multivibrator, has (1) a normal or quiescent state, and (2) a transient state induced by an external triggering pulse. Following triggering, the circuit automatically returns to its initial quiescent state after a period of time determined by the values of the circuit components.

The Astable Multivibrator

The basic astable multivibrator circuit shown in Fig. 25-25 is a relaxation oscillator for generating square-wave pulses. Two amplifier stages are used. The output of Q_1 drives the input of Q_2, and the output of Q_2 is fed back to Q_1. Since each common-emitter amplifier inverts the polarity of the signal, the two stages provide positive feedback, which results in oscillations. The multivibrator oscillator has the practical feature that it does not need a transformer for feedback.

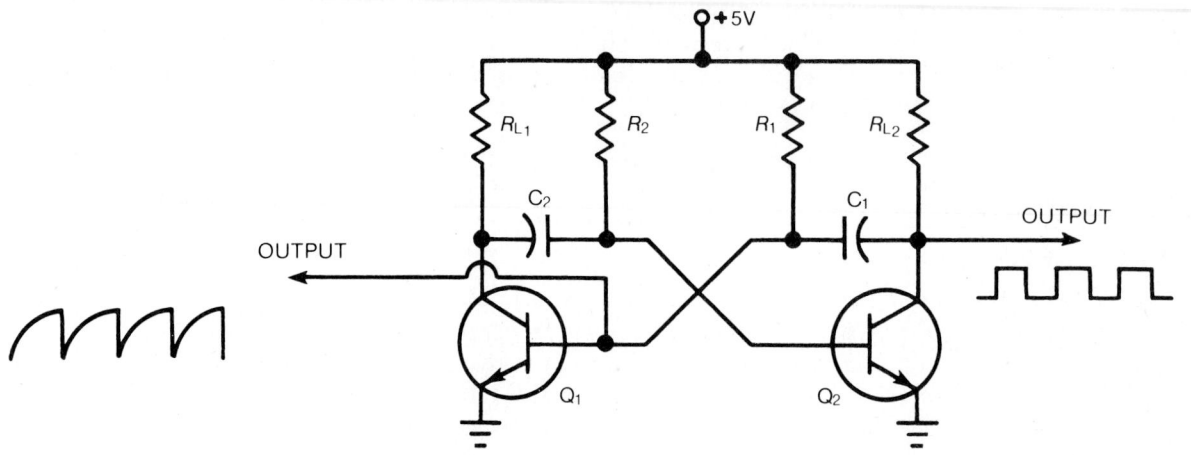

Fig. 25-25: Typical astable multivibrator circuit.

The oscillations are in the ON-OFF conditions for each amplifier. When one stage conducts, it cuts off the other. When the OFF stage starts to conduct, that action cuts off the stage that was on. The circuit is considered a relaxation oscillator because one stage is always resting while it is cut off. Any circuit that alternates between conduction and cutoff at a regular rate can be considered a relaxation oscillator.

The rate at which Q_1 and Q_2 are cut off determines the oscillator frequency. One cycle includes the time for both stages. As an example, when each stage is cut off for 0.5ms, the total cutoff period T is $2 \times 0.5\text{ms} = 1\text{ms}$. The frequency is then 1,000Hz. That would be the frequency of the square-wave signal output.

The multivibrator is a very popular circuit with many applications. It can be used as a simple, economical audio oscillator for a tone generator. In digital logic circuits, it is the clock generator that provides timing pulses. Also, it is used as a sawtooth generator with an RC waveshaping circuit. In this application, the output is taken from Q_1. A useful feature is that it can easily be synchronized with external trigger pulses to make it lock in at the frequency of the synchronizing signal.

Collector-Coupled Multivibrator. In the collector-coupled multivibrator circuit (Fig. 25-26) the collector of Q_1 is coupled through C_2 to the base of Q_2. Also, the collector of Q_2 feeds back to the base of Q_1 through C_1. The two stages are *cross-coupled*. Each stage has a collector load R_{L1} or R_{L2}. Forward bias for the base is provided by R_3 and R_4 from the DC supply of 5V. In the

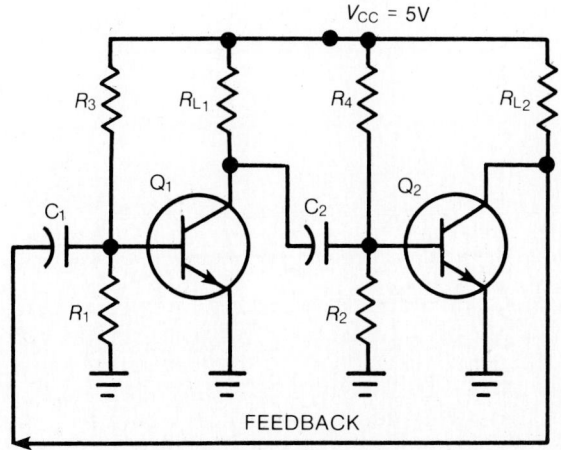
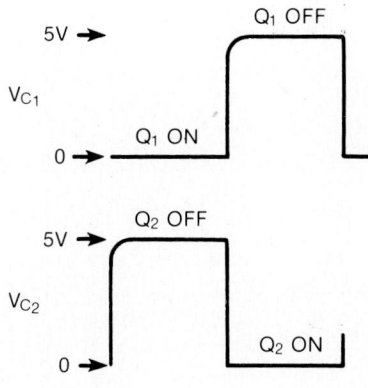

Fig. 25-26: Collector-coupled multivibrator circuit.

bias network R_1 is in the discharge path of C_1, and R_2 is in the discharge path of C_2.

Initially, one stage conducts slightly more than the other even when the amplifier circuits are symmetrical. Any slight difference is amplified by both stages because of the positive feedback. It should be noted that more I_C means less V_C, which provides negative going voltage drive to the next stage.

Assume Q_1 is conducting more than Q_2 at the start. The feedback makes Q_1 conduct more current, and there is less current in Q_2. Very quickly, then, Q_1 conducts maximum current to cut off Q_2. How long Q_2 remains off depends on the R_2C_2 time constant in the coupling for its base input circuit.

Then Q_2 starts to conduct. This change is amplified to produce maximum current in Q_2, which cuts off Q_1. As a result, the stages alternate between conduction and cutoff.

The collector voltage V_C provides a square-wave output voltage as shown in Fig. 25-26. The variations are between 5V and 0V. The 5V for V_C equals the DC supply voltage when the stage is cut off, without any I_C. The zero value of V_C results when the maximum I_C produces a 5V drop across R_L.

Remember that a drop in $+V_C$ is a negative change. When V_C goes from 5V to 0V, the base circuit of the next stage has a drive of –5V.

How long the base signal is at –5V depends on the time for the coupling capacitor to discharge with a lower value of V_C. As C_1 or C_2 discharges, the voltage across R_1 or R_2 decreases from –5V toward zero. When the negative drive at the base drops enough, the stage can start conducting. The stage that conducts can then cut off the other.

When the cutoff time is the same for each stage, the multivibrator is symmetrical and a square-wave output is produced. An unsymmetrical multivibrator provides unequal pulse widths. The output voltage can be taken from the collector of either Q_1 or Q_2 with inverted polarities.

Emitter-Coupled Multivibrator. A typical emitter-coupled multivibrator circuit is illustrated in Fig. 25-27. The collector of Q_1 drives the base of Q_2 through the coupling circuit with C_C

663

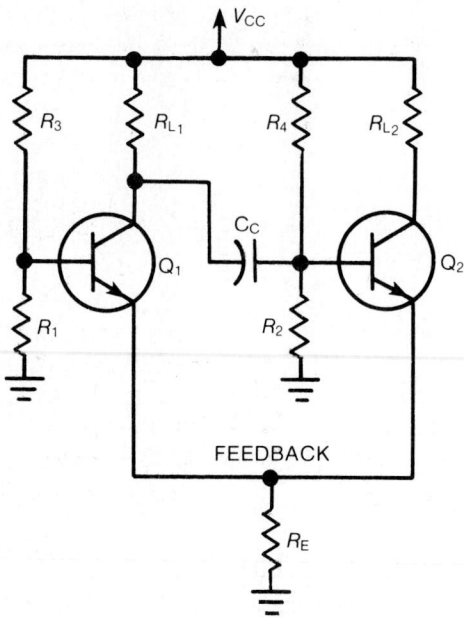

Fig. 25-27: Emitter-coupled multivibrator circuit.

and R_2. However, Q_2 feeds back to Q_1 only through the common R_E.

Q_2 is cut off by a negative base signal when Q_1 conducts. In the opposite case, Q_1 is cut off by a positive emitter voltage across R_E when Q_2 conducts. At that time, Q_2 cannot cut itself off because R_2 has a positive base drive from Q_1.

The emitter-coupled circuit is automatically unsymmetrical with Q_2 off for more time than Q_1. The reason is that Q_2 has an RC coupling circuit for negative base drive but Q_1 has not.

The astable multivibrators discussed here are considered free-running which means that they automatically oscillate without the need for any external signal or trigger to change states. The free-running multivibrator is astable because it is not stable when either stage is cut off. The circuit naturally changes states at the frequency of oscillations.

The Bistable Multivibrator

The bistable multivibrator or flip-flop needs a trigger pulse to change states. This type of multivibrator has two stable states. When an input pulse activates on the OFF stage, the conditions reverse and stay that way. Another trigger pulse is needed to put the multivibrator back to its original conditions. The name flip-flop describes this idea of switching states one way and then the opposite way.

The basic circuit for the bistable multivibrator is shown in Fig. 25-28. Note its resemblance to that of the astable multivibrator previously discussed. There are some differences; however, the emitters of both transistors are connected to ground and both base resistors (R_1, R_2) are connected to a source of positive voltage ($+V_{BB}$). The voltage from this positive source

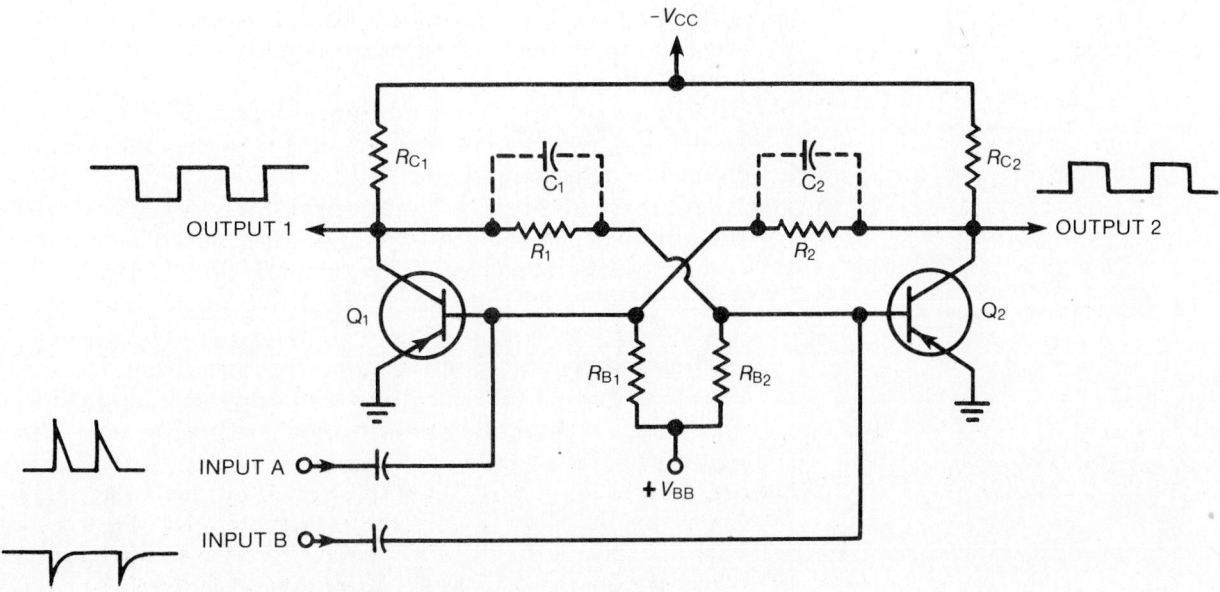

Fig. 25-28: Basic circuit of the bistable multivibrator.

is considerably smaller than the voltage from the negative source ($-V_{CC}$) and, by itself, is not large enough to reverse-bias the emitter-base junctions of the transistors. Also, coupling between the transistors is through resistors R_1 and R_2.

Assume, as power is applied, Q_1 conducts more heavily than Q_2. A chain of events, similar to those for the astable multivibrator, is started and ends in the state where Q_1 is conducting at saturation and Q_2 is cut off. In the astable multivibrator this is an unstable state and the multivibrator immediately starts changing to its other state where Q_1 is cut off and Q_2 is at saturation. In the bistable multivibrator, however, the base voltage $+V_{BB}$ holds Q_2 at cutoff. Thus the state is stable with Q_1 conducting and Q_2 cut off.

If a positive pulse, with sufficient amplitude to overcome the negative voltage at the base of Q_1, is applied to input A, the forward-bias on the emitter-base junction of Q_1 is changed to a reverse-bias and Q_1 is cut off momentarily. At this instant, Q_2 starts to conduct since the positive voltage applied to its base is removed and $+V_{BB}$ alone is insufficient to hold it at cutoff. As Q_2 becomes conductive, a chain of events is started that ends with the multivibrator in its other state where Q_1 is cut off and Q_2 is conducting at saturation. This state, too, is stable since $+V_{BB}$ now holds Q_1 at cutoff. (Note that capacitors C_1 and C_2 are not essential for the action of the multivibrator. Collector voltage of Q_1 is transmitted to the base of Q_2 through R_1 and, similarly, collector voltage of Q_2 is transmitted to the base of Q_1 through R_2. The capacitors usually are included, however, to speed up the response, especially if very short trigger pulses are employed. Otherwise, the internal capacitance of the transistors would have to charge through R_1 and R_2. Since this requires a certain amount of time, a short pulse might not be able to trigger the circuit.)

665

Another method for flipping the multivibrator from its first state to its second is to apply a negative pulse of sufficient amplitude to the base of Q_2 (which is in cutoff) through input B. This pulse overcomes the holding voltage applied to the base of Q_2 by $+V_{BB}$, Q_2 starts to conduct, and soon the multivibrator is flipped to its second state.

This method is better than the first since, by the first method, the amplitude of the positive pulse must be sufficient to overcome the large negative voltage on the base of Q_1. By the second method the amplitude of the negative pulse must be large enough only to overcome the relatively small voltage of $+V_{BB}$.

Thus the bistable multivibrator is flipped from one state to the other every time a positive trigger pulse is applied to the base of the conducting transistor or, preferably, every time a negative pulse is applied to the base of the transistor that is cut off.

The Monostable Multivibrator

The monostable or one-shot multivibrator circuit has one stable state with only one stage conducting. An input pulse or trigger is needed to trigger the OFF state into conduction. Then the multivibrator goes through one cycle of changes and back to its original conditions. This type of circuit has reverse bias at the base to hold off conduction until a trigger pulse is received.

The monostable multivibrator, whose basic circuit is illustrated in Fig. 25-29, is another type of multivibrator which is widely used. In this multivibrator a single narrow input trigger pulse produces a single rectangular output pulse whose waveshape, amplitude, and pulse width depend upon the values of the circuit components rather than upon the trigger pulse.

This multivibrator has two states, one stable and the other unstable. Normally, it is in its stable state with Q_2 conducting at saturation and Q_1 cut off. When the trigger pulse is received,

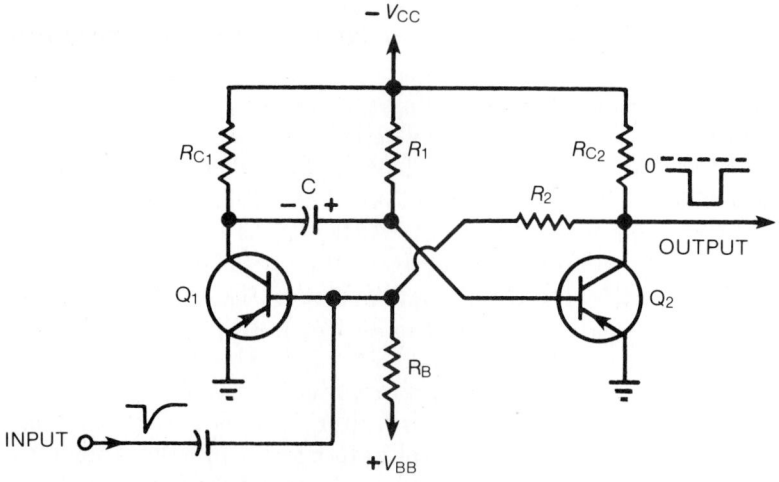

Fig. 25-29: Monostable multivibrator.

the multivibrator switches to its opposite unstable state where Q_2 is cut off and Q_1 is conducting. After a period of time, depending upon the circuit components, the multivibrator automatically switches back to its original stable state with Q_2 conducting and Q_1 cut off again.

When the multivibrator is in its original stable state, battery V_{CC} provides reverse-bias for the collector-base junctions of Q_1 and Q_2 and forward-bias for the emitter-base junction of Q_2. Transistor Q_2 thus is conducting at saturation and its collector voltage is practically at zero. Battery V_{BB} supplies reverse-bias to the base-emitter junction of Q_1, keeping that transistor at cutoff. As a result, the collector voltage of Q_1 is at the $-V_{CC}$ value. This voltage charges capacitor C, with the indicated polarity, to the V_{CC} value.

A negative trigger pulse, received at the input, overcomes the reverse-bias of the emitter-base junction of Q_1, which starts to conduct. As it does so, its collector voltage starts to fall (swing positive). This positive-going voltage, fed to the base of Q_2, decreases its forward-bias, causing its collector current to start decreasing and increasing its collector voltage (more negative). The negative-going collector voltage of Q_2 is fed to the base of Q_1 through resistor R_2. With its emitter-base forward-bias increased, Q_1 conducts more heavily. This action is cumulative and ends with Q_1 conducting at saturation and Q_2 cut off. Since C had been charged to the V_{CC} voltage, Q_2 is held at cutoff by the positive potential the capacitor places on its base.

This unstable state is maintained as C starts to discharge through R_1 to V_{CC} and to ground through the low resistance furnished by Q_1 at saturation. As C discharges, the positive potential at the base of Q_2 is reduced. When C has discharged sufficiently, the positive base potential of Q_2 has fallen low enough to bring Q_2 out of cutoff. As Q_2 starts to conduct, its collector voltage falls (swings positive). This positive voltage, fed to the base of Q_1 through R_2, reduces the collector current of Q_1 and increases its collector voltage (more negative). Again, the action is cumulative and ends with Q_1 cut off and Q_2 conducting heavily; that is, the multivibrator is back to its original stable state. The next negative trigger pulse causes the entire cycle to repeat itself.

Output is taken from the collector of Q_2. In the original stable state of the multivibrator, Q_2 is conducting heavily and its collector voltage is practically zero. When the negative trigger pulse is received, Q_2 is cut off and its collector voltage rises to $-V_{CC}$. While C is discharging, Q_2 is held at cutoff and its collector voltage is maintained at $-V_{CC}$. When C is discharged sufficiently, the multivibrator switches back to its original stable state and Q_2 conducts heavily once more. Its collector voltage thus falls back toward zero.

You can see, then, that the output from the multivibrator is a negative-going rectangular pulse whose width is determined by the time constant of C and R_1 during the discharge period. Because this multivibrator produces one output pulse for every input trigger pulse it receives, it often is called a *one-shot multivibrator*.

The Unijunction Relaxation Oscillator

The unijunction transistor (see Chapter 21) relaxation oscillator is the basic circuit for many applications such as sawtooth generators, pulse generators, timing circuits, and triggering circuits.

Figure 25-30 shows the basic configuration of a UJT relaxation oscillator. When initially switch S is closed, capacitor C_T charges through resistor R_T, and emitter voltage V_E rises exponentially toward the supply voltage V_{BB}. When V_E reaches the peak voltage V_P (V_P is approximately 60% of V_{BB}) of the UJT, the emitter diode becomes forward-biased and the UJT triggers on, providing a low-resistance path between E and B_1. The capacitor then discharges through the E-B_1 path and V_E decreases rapidly. When V_E can no longer sustain the bias required for UJT conduction [$V_{E(min)}$ is approximately 0.5 $V_{E(sat)}$], the E-B_1 discharge path is broken and the charging cycle repeats.

The frequency of oscillation is determined by the magnitudes of timing resistor R_T and timing capacitor C_T, as well as by the peak-point voltage of the UJT.

The various waveforms in the circuit are included in Fig. 25-30. Observe that the emitter voltage resembles a sawtooth waveform which shows a convex curvature due to the non-linear capacitor charging current. Base one terminal provides a positive pulse output; base two terminal provides a negative pulse output.

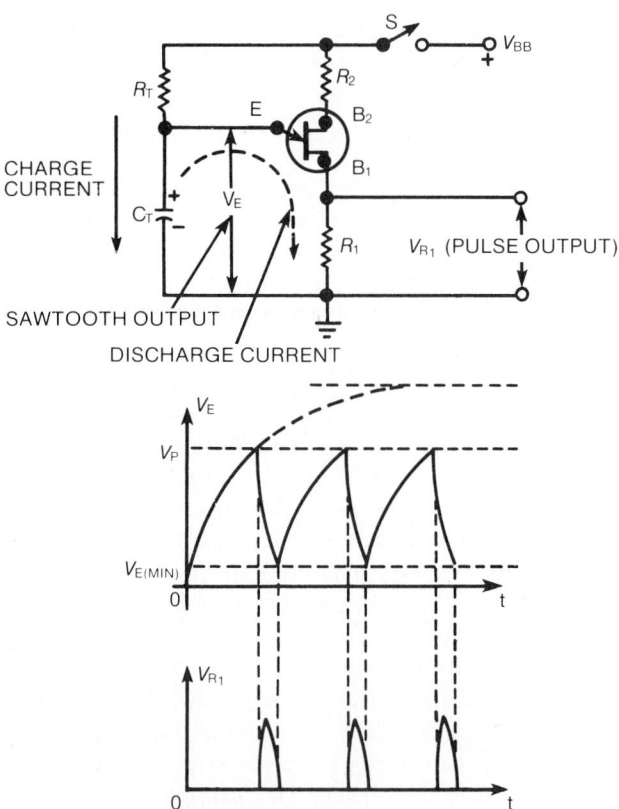

Fig. 25-30: UJT relaxation oscillator with typical waveforms.

SUMMARY

An oscillator is an electronic circuit that is used to convert the DC power of the circuit to an AC output signal. The oscillator can be divided into two basic groups, depending on the shape of the AC signal output: sinusoidal (sine wave output) or nonsinusoidal (square wave, sawtooth, or rectangular output). The essential parts of an oscillator circuit are: (1) an oscillatory tank circuit, (2) an amplifier, (3) a feedback circuit to sustain oscillations, and (4) a DC power source.

The AC output signal from the fundamental oscillator is produced by the flywheel effect exhibited by a typical LC tank. This AC signal will continue as long as the proper feedback voltage is applied to sustain oscillations and, therefore, overcome resistive tank losses.

Typical circuits that utilize the LC tank (or tickler-coil) for their operation include the following oscillators: Armstrong, Hartley, Colpitts, Clapp, the tuned-input tuned-output, the tunnel diode, the VCO, and the PLL. The Armstrong and Hartley oscillator circuits employ inductive feedback. The Colpitts and Clapp oscillator circuits employ capacitive feedback.

The tuned-input tuned-output oscillator circuit utilizes capacitive feedback and two tank circuits. One tank circuit is used by the transistor base (input) and the other tank circuit is used by the collector (output). The frequency of oscillation is determined by the tank circuit with the highest Q-factor, normally the output tank circuit.

The negative resistance oscillator, or tunnel diode oscillator, has an LC circuit across the diode which is normally biased along its negative resistance region. The tunnel diode is not widely used today and has been surpassed in performance by other modern devices.

Another modern sinusoidal oscillator is the voltage-controlled oscillator, or VCO. This circuit uses electronic tuning to control the oscillator frequency. A varactor, whose capacitance varies with the amount of reverse voltage, is placed across the tank coil in the tuned circuit. Controlling the DC voltage across the varactor varies its capacitance and, therefore, the oscillator frequency. The VCO is also used with phase-locked loops (PLL) to synthesize and generate highly stable oscillator frequencies.

The operation of a crystal oscillator depends on the piezoelectric effect. If pressure is applied to the electrical axis of a crystal, a voltage will be generated. Conversely, if a voltage is applied to the electrical axis, the crystal will expand and contract.

A properly cut crystal possesses the characteristics of a resonant circuit and can be used in place of a tuned input circuit of an oscillator. Typical circuits that use a crystal in lieu of a tuned-input tank circuit include the Hartley, Colpitts, Pierce, tuned-input tuned-output, and overtone oscillators. For the most part, these circuits operate the same as their LC tank circuit counterparts.

Nonsinusoidal oscillators usually depend on a series RC circuit for their operation and are referred to as relaxation oscillators. The multivibrator is a type of relaxation oscillator and can be classified as astable, bistable, and monostable. The astable multivibrator is free-running and automatically alternates between two states. The bistable multivibrator, or flip-flop, has two states of operation, but requires an external trigger pulse to change states. The monostable multivibrator, or single shot, has a normal or quiescent state and a transient stage caused by an external trigger pulse.

These multivibrator circuits alternate between a conduction state and a cutoff state at a predetermined rate and output a square wave.

The unijunction transistor (UJT) relaxation oscillator has many applications, such as for a sawtooth generator, pulse generator, timing circuit, or triggering circuit.

CHECK YOURSELF (Answers at back of book)

Questions

Fill in the blank with the appropriate word or words.

1. The feedback voltage for an oscillator circuit must be _____ .
   A. positive and out-of-phase
   B. negative and in-phase
   C. positive and in-phase
   D. negative and out-of-phase
2. The noticeable feature of a Hartley oscillator is the _____ .
   A. capacitive voltage-divider across the tuning coil
   B. RC tuning network in the base circuit
   C. DC error voltage to control output frequency
   D. tuned LC network with a tapped coil for feedback
3. A capacitive voltage-divider network across the tank coil for feedback best describes the _____ oscillator circuit.
   A. Colpitts
   B. Clapp
   C. VCO
   D. Armstrong
4. Variations in inductance values for oscillator circuits are minimized by _____ .
   A. using temperature compensated capacitors
   B. using a coil wound with very small gauge wire
   C. limiting the DC current through the coil
   D. operating the oscillator as a class A circuit
5. The electrical axis of a crystal is parallel to any flat side and is called the _____ .
   A. AT-axis
   B. X-axis
   C. Z-axis
   D. Y-axis

6. Which type of oscillator listed below generates a nonsinusoidal output waveform?
   A. relaxation oscillator
   B. VCO
   C. PLL
   D. Pierce oscillator
7. A free-running multivibrator is considered _____ .
   A. monostable
   B. bistable
   C. single-shot
   D. astable
8. Circuits where the crystal element is connected directly from the collector to the base of a transistor are called _____ .
   A. Hartley crystal-controlled oscillators
   B. Pierce oscillators
   C. negative resistance oscillators
   D. Colpitts crystal-controlled oscillators
9. What circuit makes a variable frequency oscillator lock in at the frequency and phase angle of a standard frequency used as a reference?
   A. VCO
   B. tunnel diode oscillator
   C. PLL
   D. shunt-fed Colpitts
10. Which of the following *is not* considered an essential part of an oscillator circuit?
    A. feedback circuit
    B. tank circuit
    C. DC power source
    D. a Z-cut crystal
11. To start a circuit oscillating we need to apply power from the collector to the base of a transistor. This is called _____ .
12. A Colpitts circuit is an oscillator circuit that always uses _____ feedback.
13. Any circuit that alternates at a constant rate between conduction and cutoff can be considered a _____ oscillator.
14. A free-running multivibrator is _____ because it is not stable when either stage is cut off.
15. The operating frequency of an astable multivibrator is determined mainly by the _____ of the coupling capacitors and resistors.
16. The _____ oscillator belongs to the inductive feedback class of oscillator circuits.
17. One major advantage of using a crystal-controlled oscillator is _____ stability.
18. The essential parts of an oscillator circuit are _____ , _____ , _____ , and _____ .
19. The number of Hertz the frequency of a crystal will vary per MHz of its operating frequency per degree centigrade of change is referred to as _____ .

20. The _____ axis of a crystal is parallel to any flat side of the crystal and is called the _____ axis.

Problems

1. Transistors Q_1 and Q_2 are used in an astable multivibrator circuit. During operation, each stage is cut off for 0.25ms. What is the frequency of the square wave output?
2. An X-cut crystal has been produced with a resultant thickness of 0.025″. At what frequency will the crystal vibrate? (*Note:* K = 112.6 for an X-cut.)
3. A 500kHz, X-cut crystal has a temperature coefficient of –10Hz/°C/MHz. The crystal was calibrated at a temperature of 50°C. At what frequency will the crystal oscillate when the temperature rises to 70°C?
4. If a similar Y-cut crystal were used in Problem No. 3, at what frequency will the crystal oscillate?
5. A Y-cut crystal is to be produced so that it will oscillate at a frequency of 2kHz. What is the required thickness of the crystal for this frequency? (*Note:* K = 77 for a Y-cut.)

Glossary of Common Electrical and Electronic Terms

Absorption Wavemeter. A device for measuring wavelength or frequency. Usually consists of a tunable resonant circuit and an indicator to show when maximum energy is being absorbed from circuit being tested.

Accelerate. To speed up or make go faster.

Accelerating Electrode. One or more internal elements or an electron tube used to increase the velocity of an electron stream.

AC-DC. Refers to an electrical device which operates on either alternating or direct current.

AC Generator. A generator that is a source of alternating current.

Acorn Tube. An acorn-shaped vacuum tube designed for ultra-high-frequency circuits. The tube has short electron transit time and low interelectrode capacitance due to close spacing and small size electrodes.

Acoustic. Pertaining to the generation, transmission, and effects of sound.

Activate. To make active; to put into motion or work.

Active Elements. Those components in a circuit which have gain or which direct current flow—diodes and transistors.

Adjustable Resistor. A resistor whose value can be changed mechanically. Also called *adjustable voltage divider*.

Admittance. The measure of the ease with which current flows in a circuit: the reciprocal of resistance. Admittance is measured in siemens and designated by Y.

Affix. To attach or connect one device or material to another.

Air Core. Descriptive term for coils or transformers with air cores.

Align. To adjust the tuned circuits of a receiver or transmitter for maximum signal response.

Aligning Tool. Small screwdriver or special tool, generally of noninductive material, for aligning circuits.

Alligator Clip. A long-nosed metal clip with meshing jaws, used to make temporary connections.

All-Wave Antenna. An antenna designed to receive or radiate over a wide range of radio frequencies.

All-Wave Receiver. A set capable of broadcast and shortwave reception. Usually tunes from about 500kc to 30MHz. Also called *general-coverage*.

Alternating Current. Electric current in which the electrons move first in one direction and then in the other.

Alternation. One-half of a complete cycle.

Ambient Temperature. The temperature of the air (or other coolant) surrounding electric components or equipment.

Ammeter. An instrument for measuring the amount of electron flow in amperes.

Ampacity. Current-carrying capacity expressed in amperes.

Ampere. The basic unit of electric current.

Ampere-Hour. Unit used to show energy storage capacity of cell or battery.

Ampere-Turn. The magnetizing force produced by a current of one ampere flowing through a coil of one turn.

Ampere-Turn per Meter. Base unit of magnetic field strength.

Amplification. The process of increasing the strength (current, power, or voltage) of a signal.

Amplification Factor. The ratio of a small change in plate voltage to a small change in grid voltage, with all other electrode voltages constant, required to produce the same small change in plate current.

Amplifier. A device used to increase the signal voltage, current, or power, generally composed of a vacuum tube or semiconductor device and associated circuit called a stage. It may contain several stages in order to obtain a desired gain.

Amplitude. The maximum, instantaneous value of an alternating voltage or current, measured in either the positive or negative direction.

Amplitude Distortion. The changing of a waveshape so that it is no longer proportional to its original form. Also called *harmonic distortion*.

Amplitude Modulation. The modulating of a carrier-frequency current by varying its amplitude above and below normal value at an audio rate, in accordance with the voice or music being transmitted.

Analyzer. A test instrument for checking electronic equipment, parts, or circuits.

Angle of Lag or Lead. Phase angle by which voltages or currents precede or follow one another. These relations are often indicated by plotting the sinusoidal curves along an axis of electric degrees. They may also be pictured by phasors.

Angle of Radiation. The angle between the center of a radiated radio beam and the earth's surface.

Anode. Positive electrode or terminal; the plate of a vacuum tube. In some cases the most positive electrode.

Antenna. A device used to radiate or absorb RF energy. Also called *aerial*.

Antenna Coil. The RF coil (or transformer) in a radio receiver or transmitter to which the antenna is connected.

Antenna Coupler. A device used to connect a receiver or transmitter to an antenna or antenna transmission line.

Antenna Current. The current flowing in the antenna and associated circuits.

Antenna Transmission Line. A system of conductors connecting an antenna to a receiver or transmitter, often through an antenna coupler.

A-Power Supply. Power source for the filaments of electron tubes. Also called *A-Battery*.

Apparent Power. The product of volts times amperes when they are in phase. (Not found in most AC circuits. Must be multiplied by a power factor for real power.)

Arc. A flash caused by an electric current ionizing a gas or vapor.

Armature. The rotating part of an electric motor or generator. Also the moving part of a relay or vibrator.

Array. An arrangement of antenna elements, usually dipoles, which results in desirable directional characteristics.

Atmospheric Interference. Crackling and hissing noises reproduced by a receiver as a result of electric disturbances in the atmosphere. Also called *static*.

Atom. The smallest particle of an element that can exist alone.

Atomize. To make a liquid into a fine spray.

Attenuation. The reduction in the strength of a signal.

Attenuator. A network of resistors used to reduce the voltage, current, or power delivered to a load.

Audible. Capable of being heard by the human ear.

Audio. Pertaining to voltages or currents in the audible frequency range.

Audio Amplifier. A device to strengthen audio-frequency signals.

Audio Frequency. A frequency which can be detected as a sound by the human ear. The range of audio frequencies extends approximately from 20 to 20,000 cycles per second.

Audio-Frequency Oscillator. A device which generates audio-frequency signals.

Audio Transformer. An iron-core transformer used in audio-frequency circuits.

Autodyne Circuit. A circuit in which the same elements and vacuum tube are used as an oscillator and as a detector. The output has a frequency equal to the difference between the frequencies of the received signal and the oscillator signal.

Autodyne Reception. Radio reception in which the incoming signal beats with an oscillating detector to produce an audible beat frequency; employed in regenerative receivers for the reception of CW (continuous wave) code signals.

Automatic Frequency Control. A circuit which keeps a receiver or transmitter accurately tuned to a predetermined frequency. Abbreviated AFC.

Automatic Gain Control. A method of automatically regulating the gain of a receiver so that the output tends to remain constant though the incoming signal may vary in strength. Also called *automatic volume control*.

Autotransformer. A transformer in which the primary and secondary coils are connected together in one winding.

Average Power. The power dissipated by the load. Also called real power or true power; in resistive loads it is equal to the product of effective current squared (I^2) and the resistance of the load. Unit of measure is the watt (W).

Axial Leads. Leads that are connected on the axis (center line) of a component.

Balanced Circuit. A circuit with equal voltage and current flow across and through all its branches.

Balance to Ground. A state in certain circuits (e.g., cathode ray tubes) where the voltages (such as on deflection plates) are equal above and below ground potential.

Ballast. An inductance coil or transformer which is used with a fluorescent tube.

Ballast Resistor. A special type of resistor used to compensate for fluctuations in AC power line voltage. The resistance of a ballast resistor increases as the current through it increases, thus maintaining the current essentially constant in spite of line voltage fluctuations.

Ballast Tube. A ballast resistor mounted inside an evacuated envelope.

Banana Jack. A receptacle that fits a banana plug.

Banana Plug. A banana-shaped plug. Elongated springs provide compression contact.

Band. In radio, frequencies which are within two definite limits. For example, the standard broadcast band extends from 550kc to 1,600kc.

Band of Frequencies. The frequencies existing between two definite limits.

Band-Pass Coupling. A type of coupling between stages that provides relatively linear energy transfer over a wide band of frequencies.

Band-Pass Filter. A circuit designed to pass, with nearly equal response, all currents having frequencies within a definite band and to reduce substantially the amplitudes of currents of all frequencies outside that band.

Bandspread. Any method, mechanical or electronic, of effectively widening the tuning scale of a receiver between radio stations.

Band Switch. Used to change one or more circuits of a multiband receiver or transmitter. Also called *band selector*.

Bandwidth. The space, expressed in hertz, between the lowest and highest frequencies to which an inductance-capacitive circuit will respond with at least 70.7% of its maximum response.

Base Insulator. A large insulator used at the base of some radio towers.

Bathtub Capacitor. A capacitor enclosed in a metal can with rounded corners like a bathtub.

Battery. Two or more primary or secondary cells connected together electrically. The term does not apply to a single cell.

Beam. (1) A colloquialism for an antenna array. (2) A stream of electrons flowing from the cathode to the plate in beam power tubes. (3) A stream of electrons flowing from the cathode to the screen of a television or cathode ray tube.

Beam Antenna. Antenna array that receives or transmits radio-frequency energy more sharply in one direction than others.

Beat Frequency. A frequency resulting from the heterodyning of two different frequencies. It is numerically equal to the difference between or the sum of these two frequencies. Also called *beat note*.

Beat-Frequency Oscillator. A device from which an audible signal is obtained by combining and rectifying two higher inaudible frequencies.

Beat Reception. Radio reception by combining a received external signal with an internal one generated in the receiver; the difference frequency is then amplified and detected. Also called *heterodyne reception*.

Bias. The average DC difference of potential between the cathode and control grid of a vacuum tube.

Biasing Resistor. A resistor used to provide the voltage drop for a required bias.

Bias Modulation. A means of amplitude modulation in which the modulating voltage is superimposed on the bias voltage of an RF stage. Control grid, suppressor grid, and cathode modulation are types of bias modulation.

Bilateral. Conducts current in both directions; conduction of current in one direction is termed unilateral.

Bimetal. Two dissimilar metals, with different rates of expansion, bonded together. When heated, the assembly will buckle or bend. Mostly used in temperature-actuated switches to make and break contact points.

Bimetal Strip. Temperature-regulating or indicating device which works on the principle that two dissimilar metals with unequal expansion rates, welded together, will bend as temperatures change. Also called *bimetallic blade*.

Binary. A system of numerical representation which uses only two symbols, 0 and 1.

Binding Post. A fixed terminal to which wires may be attached.

Bleeder. A resistance connected in parallel with a power supply output to protect equipment from excessive voltages if the load is removed or substantially reduced, to improve the voltage regulation and to drain the charge remaining in the filter capacitors when the unit is turned off.

Bleeder Current. A current drawn continuously from a power supply to improve its voltage regulation.

Bleeder Resistor. Connected across a power supply, it serves to discharge or bleed off the charge on the filter capacitors once the supply is turned off.

Block Diagram. Simplified outline of an electronic system with circuits or parts shown as functional boxes.

Blocking. The application of high negative grid bias to a vacuum tube, reducing tube current to zero.

Blocking Capacitor. A capacitor used to block the flow of direct current while permitting the flow of alternating current.

Body Capacitance. The capacitance existing between the human body and a piece of electronic equipment.

Bonding Jumper. A reliable conductor used to ensure the required electrical conductivity between metal parts required to be electrically connected.

Branch. A conducting path between two nodes in a circuit.

Branch Circuit. That portion of a wiring system extending beyond the final overcurrent device protecting the circuit. Also, a branch circuit that supplies a number of outlets for lighting and appliances.

Breakdown Voltage. The voltage at which an insulator or dielectric ruptures or at which ionization and conduction take place in a gas or vapor.

Breaker Points. Metal contacts that open and close a circuit at timed intervals.

Breaker Strip. A strip of wood or plastic used to cover the joint between the outside case and the inside liner of a refrigerator.

Bridge. A connection between two parallel circuits.

Bridge Circuit. Any one of a variety of electric circuit networks, one branch of which, the "bridge" proper, connects two points of equal potential and hence carries no current when the circuit is properly adjusted or balanced.

Bridge Rectifier. A full-wave rectifier with four elements connected in series as in a bridge circuit. Alternating voltage is applied to one pair of opposite junctions, and direct voltage is obtained from the other pair of junctions.

675

Broadband. Ability of a circuit or antenna to be effective over a relatively wide frequency range.

Broadband Amplifier. An amplifier that maintains flat response over a relatively wide range of frequencies.

Broadband Antenna. A transmitting or receiving antenna that is uniformly efficient over a relatively wide frequency band.

Broadband RF Stage. An amplifier stage that provides approximately uniform amplification over a wide band of frequencies.

Broadcast Band. The band of frequencies between 550kc and 1,600kc in which are assigned all standard AM broadcast stations operating in the United States.

Broadcasting. A general term applying to radio transmission of material for general public listening.

Brush. The conducting material, usually a block of carbon, bearing against the commutator or slip rings through which the current flows in or out.

B Supply. The plate voltage source for vacuum tubes. Also called *B-Battery*.

Buffer. Any part or circuit used to reduce undesirable interaction between two or more circuits.

Buffer Amplifier. An amplifier used to isolate the output of an oscillator from the effects produced by changes in voltage or loading in following circuits.

Buncher. The electrode of a velocity-modulated tube which alters the velocity of electrons in the constant current beam, causing the electrons to become bunched in a drift space beyond the bunch electrode.

Butterfly Capacitor. A variable capacitor whose plates roughly approximate the shape of a butterfly.

Bypass. Passage at one side of, or around, a regular passage.

Bypass Capacitor. A capacitor used to provide an alternating current path of comparatively low impedance around a circuit element.

Cable. Generally a heavy wire, solid or stranded, bare or insulated. Insulated small wires grouped in a common insulating sleeve are called multiconductor cables.

Calibrate. To adjust to a known standard or norm.

Calorie. The amount of heat required to raise the temperature of 1g of water 1°C.

Capacitance. The factor in an electric circuit that opposes a change in voltage. The ability of a circuit to store an electrostatic field. Distributed capacitance in a circuit is the result of adjacent loops on coils and parallel leads.

Capacitance Bridge. A variant of the Wheatstone bridge used to make exact comparisons of capacitances.

Capacitive Coupling. A method of transferring energy from one circuit to another by means of a capacitor that is common to both circuits.

Capacitive Reactance. The opposition to alternating current flow due to capacitance in the circuit.

Capacitor. Two electrodes or sets of electrodes in the form of plates separated from each other by an insulating material called the *dielectric*.

Capacitor-Input Filter. A type of power-supply filter in which a capacitor precedes an inductor or a resistor across the output of the rectifier.

Capacitor-Start Motor. A motor which has a capacitor in the starting circuit.

Carbon Resistor. A resistor made of carbon particles and a ceramic binder molded into a cylindrical shape, with axial leads.

Carrier. The radio frequency component of a transmitted wave upon which an audio signal or other form of intelligence can be impressed.

Carrier Suppression. Radio transmission in which the energy of the carrier wave is greatly reduced.

Cascade. Actually, in series; as in amplifier stages where the output of one stage is connected to the input of the next.

Catcher. The electrode of a velocity-modulated tube which receives energy from the bunched electrons.

Cathode. The electrode in a vacuum tube which is the source of electron emission. Also called *negative electrode*.

Cathode Bias. The method of biasing a tube by placing the biasing resistor in the common cathode return circuit, making the cathode more positive, rather than the grid more negative, with respect to ground.

Cathode Follower. A vacuum tube circuit in which the input signal is applied between the control grid and ground and the output is taken from the cathode and ground. A cathode follower has a high input impedance and a low output impedance.

Cathode Modulation. Amplitude modulation by varying the cathode bias of an RF amplifier in accordance with the modulating intelligence.

Cathode Ray Oscilloscope. A test instrument using a cathode ray tube, providing a visible graphic presentation of electrical impulses.

Cathode Ray Tube. A type of funnel-shaped tube in which a beam of electrons generated at the apex of the tube impinges on a fluorescent screen on the inner face of the tube, thereby causing a spot of light on the face. Voltages applied to vertical and horizontal pairs of deflection plates control the position of the beam and hence the spot on the face of the tube. Used in oscilloscopes and television receivers.

Cathode Ray Tube Screen. The fluorescent material (phosphor) that covers the inside surface of the face end of a cathode ray tube.

Cathode Ray Tuning Indicator. A very small cathode ray tube used in receivers to indicate when a station is properly tuned.

Cell. Chemical system that produces DC voltage.

Center-Fed Antenna. A type of transmitting or receiving antenna having a transmission line connected to its center.

Center Frequency. The assigned frequency of an FM station; frequency shifts take place in step with the audio signal.

Center Tap. A wire tap made to the center of a coil.

Centigrade Scale. The temperature scale used in the metric system. The freezing point of water is 0°C; the boiling point is 100°C.

Centimeter. A metric unit of linear measurement which equals 0.3937".

Ceramic Capacitor. A capacitor with a ceramic dielectric.

Channel. (1) A band of frequencies including the assigned carrier frequency, within which transmission is confined in order to prevent interference with stations on adjacent channels. (2) An electrical path over which signals travel; thus, an amplifier may have several input channels, such as microphone, tuner, or phonograph.

Characteristic Impedance. The ratio of the voltage to the current at every point along a transmission line on which there are no standing waves. Also called *surge impedance*.

Charge. Electrical property of electrons and protons.

Charge Carrier. A carrier of electrical charge within the crystal of a solid-state device, such as an electron or a hole.

Chassis. The frame or baseplate to which the electrons' components are attached.

Chip. A single substrate on which all the active and passive elements of an electronic circuit have been fabricated utilizing the semiconductor technologies of diffusion, passivation, masking, photoresist, and epitaxial growth. A chip is not ready for use until it is packaged and provided with terminals for connection to the outside world. Also called a *die*.

Choke. An inductance used in a circuit to present a high impedance to frequencies without appreciably limiting the flow of direct current. Also called a *choke coil* or *impedance coil*. A groove or other discontinuity in a waveguide so shaped and dimensioned that it impedes the passage of guided waves within a limited frequency range.

Circuit. The complete path of an electric current.

Circuit Breaker. A device designed to open and close a circuit by nonautomatic means and to open the circuit automatically on a predetermined overload of current, without injury to itself when properly applied within its rating.

Circuit, Parallel. Arrangement of electrical devices in which all positive terminals are joined to one conductor and all negative terminals to the other conductor.

Circuit, Series. An electric path (circuit) in which electricity to operate a second lamp or device must pass through the first, and so on; current flow travels through all devices connected together.

Circular Mil. An area equal to that of a circle with a diameter of 0.001". It is used for measuring the cross section of wires.

Clamping Circuit. A circuit which maintains either amplitude extreme of a waveform at a certain level of potential.

Class A Operation. Operation of a vacuum tube so that plate current flows throughout the entire operating cycle and distortion is kept to a minimum.

Class AB Operation. Operation of a vacuum tube with grid bias so that the operating point is approximately halfway between Class A and Class B.

Class B Operation. Operation of a vacuum tube with bias at or near cutoff so that plate current flows during approximately one-half of the cycle.

Class C Operation. Operation of a vacuum tube with bias considerably beyond cutoff so that plate current flows for less than one-half cycle.

Clipping. Distortion in amplifiers produced by flattening the positive and/or negative peaks of the signal due to tube saturation during positive grid swing or due to driving the grid below cutoff. Also, distortion in the AF component of a modulated wave when modulation amplitude exceeds 100%.

Clock. A pulse generator which controls the timing of switching circuits and memory states and equals the speed at which the major portion of the computer operates.

Closed Circuit. A circuit that is complete and allows current to flow.

Coaxial Cable. A transmission line consisting of two conductors concentric with and insulated from each other.

Coefficient of Coupling. A numerical indication of the degree of coupling existing between two circuits, expressed in terms of either a decimal or a percentage.

Coil. A wire conductor wound on a form.

Cold Cathode. A cathode that does not depend upon heat for electron emission.

Color Coding. An industry-supported system of indicating the values of components that are too small to take printing.

Colpitts Oscillator. A popular type of oscillator employing capacitive feedback to sustain oscillation.

Commutator. The copper segments on the armature of a motor or generator. The commutator is cylindrical in shape and is used to pass power into or from the brushes. It is a switching device.

Components. One (or more) of the parts of a whole.

Compound. Matter composed of two or more elements.

Compound Circuit. Circuits of resistances connected in series and in parallel.

Conductance. The ability of a material to conduct or carry an electric current. It is the reciprocal of the resistance of the material and is expressed in siemens.

Conductive Path. A carrier of electricity that is usually made of metal, such as copper.

Conductivity. The ease with which a substance transmits electricity.

Conductor. A material which offers little opposition to the continuous flow of electric current.

677

Conductor, Bare. A conductor having no covering or insulation whatsoever.

Conductor, Covered. A conductor having one or more layers of nonconducting materials that are not recognized as insulation under the Code.

Conductor, Insulated. A conductor covered with material recognized as insulation.

Connectives. Devices which connect or join two or more things together.

Connector, Pressure (solderless). A connector that establishes the connection between two or more conductors or between one or more conductors and a terminal by means of mechanical pressure and without the use of solder.

Contactor. A magnetic relay switch with contact points which usually open and close high-voltage circuits. It is actuated by a low-voltage magnetic coil.

Contact Resistance. The resistance, measured in ohms, between the contacts on a switch or relay.

Contacts. The points in a switch where circuits are made and broken.

Contaminant. A substance (dirt, moisture, etc.) foreign to the refrigerant or the refrigerant oil in a refrigeration system.

Continuity. Continuous path for current.

Continuous Load. A load in which the maximum current is expected to continue for three hours.

Continuous Waves. Radio waves which maintain a constant amplitude and a constant frequency.

Control. An automatic or manual device used to stop, start, and/or regulate the flow of electricity.

Control Device. Used to modify the movement of electricity in the electric current. The switch and the fuse are examples of control devices.

Control Grid. The electrode of a vacuum tube other than a diode upon which the signal voltage is impressed in order to control the plate current.

Conventional Current Direction. An arbitrary choice for current direction in an electric circuit in which current is assumed to move from the positive terminal of the source, through the electric circuit, and back to the negative terminal of the source.

Conversion Transconductance. A characteristic associated with the mixer function of vacuum tubes and used in the same manner as transconductance is used. It is the ratio of the IF current in the primary of the first IF transformer to the RF signal voltage producing it.

Converter. The circuit in a superheterodyne radio receiver which changes incoming signals to the intermediate frequency; the converter section includes the oscillator and the first detector. Also, a device changing electric energy from one form to another, such as AC to DC, etc.

Converter Tube. A multielement vacuum tube used both as a mixer and as an oscillator in a superheterodyne receiver. It creates a local oscillator frequency and combines it with an incoming signal to produce an intermediate frequency.

Core. A magnetic material that affords an easy path for magnetic flux lines in a coil.

Core Loss. Refers to the conversion of electrical energy to heat energy in the windings of magnetic devices.

Core Losses. Losses in a motor, generator, or a transformer caused by eddy currents and hysteresis.

Coulomb. The standard electrical unit of quantity measure. It is 6.3 billion billion electrons.

Counter. (1) A device capable of changing states in a specified sequence upon receiving appropriate input signals; (2) a circuit which provides an output pulse or other indication after receiving a specified number of input pulses.

Counter emf. Counter electromotive force; an emf induced in a coil or armature that opposes the applied voltage. Also called *back-electromotive force.*

Counterflow. Flow in an opposite direction.

Counting Circuit. A circuit which receives uniform pulses representing units to be counted and produces a voltage in proportion to their frequency.

Coupled Impedance. The effect produced in the primary winding of a transformer by the current flowing in the secondary winding. Same as *reflected impedance.*

Coupling. The means by which signals are transferred from one electronic circuit to another. Coupling can be direct through a conductor, electrostatic through a capacitor, or inductive through a transformer.

Coupling Element. The means by which energy is transferred from one circuit to another; the common impedance necessary for coupling.

C Power Supply. Source of bias voltage for a vacuum tube. Also called *C-battery.*

Critical Coupling. The degree of coupling which provides the maximum transfer of energy between two resonant circuits at the resonant frequency.

Cross-Modulation. Interference which modulates a signal undesirably, usually from some unwanted source.

Crystal. (1) A natural substance, such as quartz or tourmaline, which is capable of producing a voltage stress when under pressure or producing pressure when under an applied voltage. Under stress it has the property of responding only to a given frequency when cut to a given thickness. (2) A nonlinear element such as galena or silicon in which case the piezoelectric characteristic is not exhibited.

Crystal-Controlled Converter. A type of radio-frequency converter employing a piezoelectric crystal to establish the frequency of its oscillator.

Crystal-Controlled Oscillator. A type of radio-frequency oscillator employing a piezoelectric crystal to establish its operating frequency. Oscillators of this type provide high stability.

Crystal Mixer. A device which employs the nonlinear characteristic of a crystal (nonpiezoelectric type) and a point contact to mix two frequencies.

Crystal Oscillator. An oscillator circuit in which a piezoelectric crystal is used to control the frequency and to reduce frequency instability to a minimum.

Current. The flow of electrons through a conductor, measured in amperes.

Current Carrier. Charged particle (electron or ion).

Current Feed. The feeding current to an antenna at the point of maximum current amplitude.

Current Limited. The maximum current that a source may provide in a load. Some commercial power supplies have circuitry that automatically limits the current to a constant amount.

Current Limiter. A protective device similar to a fuse, usually used in high-amperage circuits.

Current Limiting Resistor. A resistor used in a circuit as a protective device for tubes and filter elements against overload from voltage surges.

Current Transformer. A transformer used to extend the range of an AC ammeter.

Cut-Off. The minimum value of negative grid bias which prevents the flow of plate current in a vacuum tube.

Cut-Off Frequency. The frequency at which the intrinsic gain of a device falls to some predetermined value. For example, the frequency at which the gain of a bipolar transistor in a common-emitter amplifier configuration is equal to unity.

Cut-Off Limiting. Limiting the maximum output voltage of a vacuum-tube circuit by driving the grid beyond cutoff.

Cut-Off Voltage. Negative grid bias beyond which plate current ceases to flow.

Cycle. One complete positive and one complete negative alternation of a current or voltage waveshape.

Damped Waves. Waves which decrease exponentially in amplitude.

DC Generator. A rotating device that converts mechanical energy into unidirectional electric energy.

DC Plate Resistance. In ohms, the DC plate voltage divided by the DC plate current of a vacuum tube.

DC Resistance. In ohms, the opposition to current flow offered by a circuit or component to DC current flow.

Deceleration. A slowing down in speed or motion.

Decibel. A term expressing a ratio between two amplitudes or energies. The dB unit between two amplitudes is computed as twenty times the log of the ratio and between two energies as ten times the log of the ratio. Practically, a decibel is approximately the smallest change in sound intensity that the human ear can detect.

Decoupling Circuit. A network of one or more resistors and capacitors that separate and bypass unwanted signals.

Decoupling Network. A network of capacitors, chokes, or resistors, placed in leads which are common to two or more circuits to prevent unwanted interstage coupling.

De-energize. To disconnect a component or circuit from the power source.

Deflecting Electrodes. Pair of cathode ray tube electrodes to which the electron beam moves in a horizontal or vertical direction, depending upon the applied potentials.

Deflecting Plate. The part of a certain type of electron tube which provides an electrical field to produce deflection of an electron beam.

Deflection. The deviation of a meter indicator from 0.

Deflection Sensitivity. The quotient of the displacement of the electron beam at the place of impact by the change in the deflecting field. It is usually expressed in millimeters per volt applied between the deflection electrodes or in millimeters per gauss of the deflecting magnetic field.

Degeneration. The process whereby a part of the output signal of an amplifying device is returned to its input circuit in such a manner that it tends to cancel part of the input.

Degrees Celsius. Unit commonly used for temperature measurement in electronic technology. Degrees Celsius (0°) is the same size as the base unit of temperature, the kelvin.

Deionization Potential. The potential at which ionization of the gas within a gas-filled tube ceases and conduction stops.

Delta Connection. The connection of three-phrase alternators and transformers in which the start end of one winding is connected to the end of the second.

Demagnetization. The removal of magnetism from a substance.

Demand Factor. In any system or part of a system, the ratio of the maximum demand of the system, or part of the system, to the total connected load of the system, or part of the system, under consideration.

Density. Closeness of texture or consistency.

Detection. The process of separating the modulation component from the received signal. Also called *demodulation*.

Detector. Commonly, the stage or circuit in a radio set that demodulates the RF signal. Generally, a device or circuit that changes the form of a received signal into a more useable form for a specific purpose.

Determine. To set limits, bounds; to decide upon; to be a deciding factor; to give a definite aim to; to direct.

Device. An invention; a mechanical contraption, tool, or method.

Diaphragm. A thin, flexible sheet which vibrates when struck by or when producing sound waves, as in a microphone, earphone, or loudspeaker.

Dielectric. An insulator; a term that refers to the insulating material between the plates of a capacitor.

Dielectric Constant. The ratio of the capacitance of a capacitor with a dielectric between the electrodes to the capacitance with air between the electrodes.

Dielectric Loss. Energy loss in the dielectric of a capacitor. It shows up as heat.

Dielectric Strength. The ability of an insulator to withstand a potential difference.

Differentiating Circuit. A circuit which produces an output voltage substantially in proportion to the rate of change of the input voltage.

Diffraction. The bending of waves (light, sound, or radio) around the edges of obstacles.

Digital Circuit. A circuit which operates like a switch, that is, it is either "on" or "off."

Diode. A two-element electron tube or device which will allow substantially more electron flow in one direction in a circuit than in the other direction.

Diode Detector. A detector circuit employing a diode tube.

Dip (Dual In-line Package). A circuit package somewhat longer than it is wide, with the leads coming out of the two long sides.

Dipole Antenna. Two metallic elements, each approximately one-quarter wavelength long, which radiate radio frequency energy fed to them by the transmission line.

Direct Coupling. The use of a conductor to connect two amplifier stages together and provide a direct path for the signal currents.

Direct Current. An electric current that flows in one direction only.

Directional Antenna. Any antenna which picks up or radiates signals better in one direction than another.

Directly Heated Cathode. A filament cathode which carries its own heating current for electron emission, as distinguished from an indirectly heated cathode. Also called *filament*.

Director (Antenna). A parasitic antenna placed in front of a radiating element so that radio frequency radiation is aided in the forward direction.

Direct Resistance-Coupled Amplifier. An amplifier in which the plate of one stage is connected either directly or through a resistor to the control grid of the next stage.

Dissipation. Unused or lost energy.

Dissipation Factor. A number obtained by dividing resistance by reactance. The reciprocal of Q, or coulomb. Used to indicate relative energy loss in a capacitor.

Distortion. The production of an output waveform which is not a true reproduction of the input waveform. Distortion may consist of irregularities in amplitude, frequency, or phase.

Distributed Capacitance. The capacitance that exists between the turns in a coil or choke or between adjacent conductors or circuits, as distinguished from the capacitance which is concentrated in a capacitor.

Distributed Inductance. The inductance that exists along the entire length of a conductor, as distinguished from the inductance concentrated in a coil.

Distribution. A system for supplying electric or electronic power to various points.

Doorknob Tube. A doorknob-shaped vacuum tube designed for ultra-high-frequency circuits. This tube has short electron transit time and low interelectrode

capacitance, due to the close spacing and small size of electrodes.

Double Diode. Two diodes in the same envelope. Also called *duodiode*.

Double-Pole Switch. A switch which simultaneously opens or closes two separate circuits or both sides of the same circuit.

Doubler. In a transmitter, a circuit in which the output is tuned to twice the frequency of the input circuit. Also power supplies that double the voltage.

Doublet Antenna. An antenna composed of two elements usually strung in a straight line and connected at the middle to a single insulator, each element being some definite fraction or whole of the desired wavelength.

Double-Throw Switch. A switch which will connect one (or one group) of its poles with either of two (or two groups) of its other poles, but not simultaneously.

Double Triode. Two triodes in the same tube envelope. Also called *duotriode*.

Drain. A term used to indicate the current taken from a voltage source.

Driver Stage. The amplifier stage preceding the high-power audio-frequency or radio-frequency output stage.

Driver Tube. The tube used in a driver stage.

Drop. Voltage drop, due to current flow through an electronic component.

Dropping Resistor. A resistor used to decrease a given voltage to a lower value.

Dry-Cell Battery. An electrical device, used to provide DC electricity, having no liquids in the cells.

Dry Electrolytic Capacitor. An electrolytic capacitor using a paste instead of a liquid electrolyte.

Dual Capacitor. Two capacitors in a single housing.

Dummy Antenna. A resistor or other device which duplicates the characteristics of a transmitting antenna without radiating signals. For testing and adjusting transmitters.

Duplex Cable. A two-wire conductor with the wires insulated from each other and enclosed in a single insulated covering.

Duty, Continuous. A requirement of service that demands operation at a substantially constant load for an indefinitely long time.

Duty, Intermittent. A requirement of service that demands operation for alternate intervals of (1) load and no load, (2) load and rest, or (3) load, no load, and rest.

Duty, Periodic. A type of intermittent duty in which the load conditions regularly recur.

Duty, Short-Time. A requirement of service that demands operations at loads and for intervals of time, both of which may be subject to wide variation.

Dynamic Characteristics. The relation between the instantaneous plate voltage and plate current of a vacuum tube as the voltage applied to the grid is moved; thus, the characteristics of a vacuum tube during operation.

Eccles-Jordan Circuit (trigger circuit). A direct coupled multivibrator circuit possessing two conditions of stable equilibrium. Also called *flipflop circuit* or *half-shift register.*

Eddy Current. Induced circulating currents in a conducting material that are caused by a varying magnetic field.

Effective Resistance. The ratio between the true power absorbed in a circuit and the square of the effective current flowing in a circuit.

Effective Value. The equivalent heating value of an alternating current or voltage, as compared to a direct current or voltage. It is 0.707 times the peak value of a sine wave. It is also called the RMS value.

Efficiency. The ratio of output power to input power, generally expressed as a percentage.

E Layer. An ionized layer in the E region of the ionosphere.

Electric Angle. A method of indicating a particular instant of segment in an alternating-current cycle. One cycle is considered equal to 360°, hence one-half cycle is 180° and one-fourth cycle is 90°.

Electric Current. A flow of electric charge (electricity) along a conductor in a closed circuit. Current (as it is usually called) carries energy in conductors (wires) by means of free electrons. The energy carried by the current may do work by producing heat, light, mechanical force, or magnetic attraction or repulsion.

Electric Degree. One-360th of a cycle of an alternating current or voltage.

Electric Energy Source. Causes electricity to move through the electric circuit. In the automobile, the storage battery and the alternator are examples of electric energy sources.

Electric Field. A space in which an electric charge will experience a force exerted upon it.

Electricity. A general label used to describe either a stationary or moving electric charge.

Electrode. A terminal at which electricity passes from one medium into another.

Electrolysis. The breakdown of two dissimilar materials caused by an electric current passing between them. The current may either be impressed or self-induced.

Electrolyte. A solution of a substance which is capable of conducting electricity. An electrolyte may be in the form of either a liquid or a paste.

Electrolytic Capacitor. A capacitor employing a metallic plate and an electrolyte as the second plate separated by a dielectric which is produced by electrochemical action.

Electromagnet. A magnet made by passing current through a coil of wire wound on a soft iron core.

Electromagnetic Field. A space field in which electric and magnetic phasors at right angles to each other travel in a direction at right angles to both.

Electromagnetic Spectrum. Range of frequencies of electromagnetic waves.

Electromagnetic Waves. Radiation taking many different forms, but all having in common the characteristic of a velocity that is about 3×10^{10} cm/sec.

Electromagnetism. Magnetic effects produced by electric currents.

Electromotive Force. The force that produces an electric current in a circuit.

Electron. A negatively charged particle of matter; high-speed, negatively charged particle forming the outer shell of an atom; the smallest electric charge that can exist.

Electron Coupling. A process of coupling two circuits through a vacuum tube, principally multigrid tubes.

Electron Emission. The liberation of electrons from a body into space under the influence of heat, light, impact, chemical disintegration, or potential difference.

Electron Gun. Tube electrodes designed for the production of a narrow beam of electrons intended for use in fluorescent screen or microwave tubes.

Electronic Switch. A circuit which causes a start-and-stop action or a switching action by electronic means.

Electrostatic. Having to do with static electricity, that is, electricity at rest.

Electrostatic Charge. An electric charge stored in a capacitor or on the surface of an insulated object.

Electrostatic Field. The field of influence between two charged bodies.

Electrostatic Shield. A grounded metal screen, sheet, or conductor placed to prevent the action of any electric field through the shield.

Element. Matter composed entirely of one type of atom.

Emission. (1) The process of radiating radio waves into space by a transmitter. (2) The process of ejecting electrons from a heated material.

End Effect. (1) The effect of capacitance at the ends of an antenna. (2) The effect of inductance at the end of a coil.

Energizing Current. The primary current in an unloaded transformer.

Energy. The capacity to do work by overcoming resistance.

Engineering Notation. A form of the power of ten notation in which the decimal point is adjusted so that the power of ten is a multiple of three. Also, one of the formatting functions on many scientific calculators.

Envelope. The glass or metal housing of a vacuum tube.

Equivalent Circuit. A diagrammatic arrangement of coils, resistors, and capacitors, representing the effects of a more complicated circuit in order to permit easier analysis.

E Region. A region in the ionosphere, from about 55 to 85 miles above the surface of the earth, containing ionized layers capable of bending or reflecting radio waves.

Excitation. Application of a signal to the input of a vacuum tube or similar solid-state device. Or, application of voltage to the field coils of a motor, generator, loudspeaker, or other device that requires a magnetic field.

Exciter. The oscillator that generates the carrier frequency of a transmitter.

Factor of Safety. A designated load, usually above the normal operating load of a circuit, component, or device, that can be handled without damage or hazard.

Fade. To change gradually in signal amplitude.

Fahrenheit Scale. On a Fahrenheit thermometer, under standard atmospheric pressure, the boiling point of water is 212° and the freezing point is 32° above zero on its scale.

Farad. The unit of capacitance. In the practical system of units, the farad is too large for ordinary use; so, measurements are made in terms of microfarads and picofarads.

Federal Communications Commission (FCC). A board of commissioners appointed by the President, having the power to regulate all communications systems originating in the United States.

Feedback. A transfer of energy from the output circuit of a device back to its input.

Feedthrough Capacitor. A very efficient type of bypass capacitor, designed so that the inner foil is in series with the wire to be bypassed and the outer shell is coaxial to the wire.

Feedthrough Insulator. A type of insulator which permits feeding of wire or cable through walls, etc., with minimum current leakage.

Ferromagnetic. A highly magnetic material such as iron, nickel, steel, etc.

Fidelity. The degree of exactness with which a system or portion of a system reproduces an input signal.

Field. A space containing electric or magnetic lines of force.

Field Coil. The stationary electromagnet in a commutator-type motor.

Field Intensity. Electrical or magnetic strength of a field.

Field Magnets. Electromagnets in the field of a generator or motor.

Field Pattern. Usually expressed as a polar diagram, indicating the horizontal field strength of an antenna.

Field-Strength Meter. A measuring instrument used to determine the strength of radiated energy (field strength) from an electronic transmitter.

Field Winding. The coil used to provide the magnetizing force in motors and generators.

Filament. A wire which, when heated with an electric current, emits electrons.

Filament Emission. Evolution of electrons from a heated filament in a vacuum tube.

Filament Winding. A separate secondary winding on the power transformer of AC-operated apparatus used as a filament voltage source.

Filter. A combination of circuit elements designed to pass a definite range of frequencies, attenuating all others.

Filter Capacitor. A capacitor used in a power-pack filter system to provide a low-reactance path for alternating currents.

Filter Choke. An iron-core coil used in a power-pack filter system to pass direct current while offering high impedance to pulsating or alternating currents.

Firing Potential. The controlled potential at which conduction through a gas-filled tube begins.

First Detector. The stage in a superheterodyne receiver in which the beat-frequency signal is combined with the incoming radio-frequency signal to produce the intermediate-frequency signal. Also called a *mixer*.

Fixed Bias. A bias voltage of constant value, such as one obtained from a battery, power supply, or generator.

Fixed Capacitor. A capacitor which has no provision for varying its capacitance.

Fixed Resistor. A resistor which has no provision for varying its resistance.

Fluctuating DC. A DC that varies in amplitude but does not periodically drop to zero.

Fluorescence. The property of emitting light as the immediate result of electronic bombardment.

Flux. Lines of magnetic force.

Flux Density. The number of lines in the flux of an area of magnetic force.

Flux Field. All the electric or magnetic lines of force in a given region.

Fly-Back. The portion of the time base during which the spot is returning to the starting point. This is usually not seen on the screen of the cathode ray tube due to gating action or the rapidity with which it occurs.

Free Electrons. Electrons which are loosely held, consequently, tend to move at random among the atoms of the material.

Free Oscillations. Oscillatory currents which continue to flow in a tuned circuit after the impressed voltage has been removed have a frequency that is the resonant frequency of the tuned circuit.

F Region. That region of the ionosphere extending from about 90 to 250 miles above the earth's surface.

Frequency. The number of complete cycles per second existing in any form of wave motion, such as the number of cycles per second of an alternating current.

Frequency Conversion. The process of converting the frequency of a signal to some other frequency by combining it with another frequency.

Frequency Discriminator. A circuit that converts a frequency-modulated signal into an audio signal.

Frequency Distortion. Distortion which occurs as a result of failure to amplify or attenuate equally all frequencies present in a complex wave.

Frequency Doubler. A vacuum-tube stage having a resonant plate circuit tuned to twice the input frequency.

Frequency Drift. An undesired change in the frequency of an oscillator, transmitter, or receiver.

Frequency Modulation. A method of modulating a carrier frequency by causing the frequency to vary above and below a center frequency in accordance with the sound to be transmitted. The amount of deviation in frequency above and below the center frequency is proportional to the amplitude of the modulating intelligence. The number of complete deviations per second above and below the center frequency corresponds to the frequency of the modulating intelligence.

Frequency Multiplier. A frequency changer used to multiply a frequency by an integral value, such as a frequency doubler.

Frequency Response. A rating or graph which expresses the manner in which a circuit or device handles the different frequencies falling within its operating range.

Frequency Shift. A deliberate change in the frequency of a radio transmitter or receiver.

Frequency Stability. The ability of an oscillator to maintain its operation at a constant frequency.

Frequency Standard. An oscillator used for frequency calibration.

Front-to-Back Ratio. The ratio of the effectiveness of a directional antenna or microphone between the front and the rear.

Full-Load Voltage. The source voltage available when full-load is being drawn.

Full Scale. The maximum value of the measured quantity that an instrument is calibrated to indicate.

Full-Wave Rectifier Circuit. A circuit which utilizes both the positive and the negative alternations of an alternating current to produce a direct current.

Fundamental. The lowest frequency component of a complex vibration, tone, or electronic signal.

Fuse. A protective device inserted in series with a circuit. It contains a metal that will melt or break when current is increased beyond a specific value for a definite period of time.

Gain. The ratio of the output power, voltage, or current to the input power, voltage, or current, respectively.

Gain Control. A control that can change the overall gain of an amplifier. A volume control.

Galvanometer. An instrument used to measure small direct currents.

Gang Capacitor. Two or more variable capacitors mechanically mounted so that they can be simultaneously turned by a single shaft.

Gang Control. A number of similar pieces of apparatus that can be simultaneously adjusted or tuned by a single control or shaft.

Gap Arrester. A type of antenna lightning arrester employing one or more air gaps connected between the antenna and ground.

Gas Tube. A tube filled with gas at low pressure in order to obtain certain desirable characteristics.

Gate. A device that makes an electronic circuit operable for a short time. Also, a circuit having two or more inputs and one output, the output depending upon the combination of logic signals at the inputs.

Gating (Cathode Ray Tube). Applying a rectangular voltage to the grid or cathode of a cathode ray tube to sensitize it during the sweep time only. Also called *blanking*.

Generator. A machine that converts mechanical energy into electrical energy.

Gilbert. A unit of measurement of magnetomotive force.

Grid. An electrode having one or more openings for the passage of electrons or ions.

Grid Bias. The DC voltage applied to the control grid of a vacuum tube to make it negative with respect to the cathode.

Grid-Bias Cell. A small cell used in the grid circuit of a vacuum tube to provide C bias voltage.

Grid Circuit. The circuit between the grid and cathode of a vacuum tube. The input circuit of the tube.

Grid Current. Current which flows between the cathode and the grid whenever the grid becomes positive with respect to the cathode.

Grid Detection. Detection by rectification in the grid circuit of a detector.

Grid-Dip Meter. An oscillator having in its input circuit a sensitive current-indicating meter that dips when energy is drawn from the oscillator by a coupled resonant circuit.

Grid Driving Power. The wattage applied to the grid circuit of a tube.

Grid Leak. A high resistance connected across the grid capacitor or between the grid and the cathode to provide a DC path from grid to cathode and to limit the accumulation of charge on the grid.

Grid Limiting. Limiting the positive grid voltage (minimum output voltage) of vacuum-tube circuit by means of a large series grid resistor.

Grid Modulation. Modulation produced by introduction of the modulating intelligence into the grid circuit.

Grid-Plate Capacitance. The capacitance between the grid and the plate within a vacuum tube.

Grid Return. The lead or connection which provides a path for electrons from the grid circuit to the cathode.

Grid Swing. The total grid signal voltage variation from the positive to negative peaks.

Grid Voltage. The voltage between grid and cathode.

Ground. A metallic connection with the earth to establish ground potential. Also, a common return to a point of zero RF potential, such as the chassis of a receiver of a transmitter.

Ground Absorption. Transmitted radio power dissipated in the ground.

Grounded. Connected to earth or to some conducting body that serves in place of the earth.

Grounded Conductor. A system or circuit conductor that is intentionally grounded.

Grounded-Grid Amplifier. A circuit in which the input is applied to the cathode rather than to the grid of a triode tube.

Grounding Conductor. A conductor used to connect equipment or the grounded circuit of a wiring system to a grounding electrode.

Ground-Return Circuit. A circuit that is completed by utilizing the earth as a conductive path.

Ground Wave. A radio wave that is propagated near or at the surface of the earth.

Ground Wire. A conductor leading to an electrical connection with the ground.

Gunn Oscillator. Microwave oscillator based on the velocity dependence of charge carriers as a function of an electric field in some semiconducting materials such as gallium arsenide.

Half-Wave Antenna. An antenna whose length is approximately equal to one-half the wavelength to be transmitted or received.

Half-Wave Line. A transmission line having an electric length equal to one-half the wavelength of the signal to be transmitted or received.

Half-Wave Rectification. The process of rectifying an alternating current wherein only one-half of the input cycle is passed and the other half is blocked by the action of the rectifier, thus producing pulsating direct current.

Half-Wave Rectifier. A radio tube or other device which converts alternating current into pulsating direct current by allowing current to pass during one-half of each alternating current cycle.

Hard Tube. A high vacuum electron tube.

Harmonic. An integral multiple of a fundamental frequency. (The second harmonic is twice the frequency of the fundamental or first harmonic.)

Harmonic Content. The degree or number of harmonics in a complex frequency output.

Harmonic Generator. A vacuum tube or other generator which produces an alternating current having many harmonics.

Harmonic Suppression. The prevention of harmonic generation in an oscillator or in circuits that follow it.

Harness. Wires and cables so arranged and tied together that they may be connected or disconnected as a unit.

Hartley Oscillator. A vacuum-tube oscillator circuit identified by a tuned circuit employing a tapped winding connected between the grid and plate.

Heater. The tube element used to indirectly heat a cathode.

Heater Current. The current flowing through a heater serving an indirectly heated cathode.

Heater Voltage. The voltage between the terminals of a filament used for supplying heat to an indirectly heated cathode.

Heaviside Layer. A layer of ionized gas in the region between 50 and 400 miles above the surface of the earth which reflects radio waves back to earth under certain conditions.

Helmoltz Coil. A variometer having horizontal and vertical balanced coil windings, used to vary the angle of phase difference between any two similar waveforms of the same frequency.

Henry. The basic unit of inductance.

Hertz. A unit of frequency equal to one cycle per second. Also called *voltage frequency*.

Heterodyne. To beat or mix two signals of different frequencies.

Heterodyne Frequency. The beat frequency, which is the sum or difference frequency of two signals.

Heterodyne Reception. The process of receiving radio waves by combining a received radio-frequency voltage with a locally generated alternating voltage to produce a beat frequency that is more readily amplified.

High Frequency. A frequency in the band extending from 3MHz to 30MHz.

High-Mu Tube. A vacuum tube having a high amplification factor.

High-Pass Filter. A filter designed to pass currents at all frequencies above a desired frequency while attenuating frequencies below the desired frequency.

High Q. Having a high ratio of reactance to effective resistance. Factor determining coil efficiency.

Hole. The absence of a valence electron in a semiconductor crystal. Motion of a hole is equivalent to motion of a positive charge.

Horizontal Blanking. The pulse which cuts off the electron beam while it is returning from the right side to the left side of the screen of a cathode ray tube.

Horn Radiator. Any open-ended device for concentrating energy from a waveguide and directing this energy into space.

Horsepower. The English unit of power nearly equal to work done at the rate of 550ft-lb/sec and equal in the U.S. to 746W of electric power.

Hot Wire. Any ungrounded supply-circuit wire.

Hum. A low and constant audio frequency, usually either 60 or 120 cycles, in the output of an audio amplifier. Hum is frequently caused by a faulty filter capacitor in the power supply or by heater-cathode leakage in a tube.

Hybrid. A method of manufacturing integrated circuits by using a combination of the monolithic and thin-film methods.

Hysteresis. A lagging of the magnetic flux in a magnetic material behind the magnetizing force which is producing it.

IC (Integrated Circuit). A circuit in which many elements are fabricated and interconnected by a single process, as opposed to a "nonintegrated" circuit in which the transistors, diodes, resistors, etc., are fabricated separately and then assembled.

Image Frequency. An undesired signal capable of beating with the local oscillator signal of a superheterodyne receiver which produces a difference frequency within the bandwidth of the IF channel.

Impedance. The total opposition offered to the flow of an alternating current. It may consist of any combination of resistance, inductive reactance, and capacitive reactance.

Impedance Angle. Angle of the impedance phasor with respect to the resistance vector, representing voltage lag or lead with respect to current.

Impedance Coil. A choke coil. An inductor.

Impedance Coupling. The use of a tuned circuit or an impedance coil as the common coupling element between two circuits.

Impulse. Any force acting over a comparatively short period of time, such as a momentary rise in voltage.

Indirectly Heated Cathode. A cathode which is brought to the temperature necessary for electron emission by a separate heater element.

Induced. Produced as a result of the influence of an electric or magnetic field.

Induced Current. The current flow caused by an induced voltage.

Induced Voltage. The voltage induced in a conductor or coil as it moves through a magnetic field.

Inductance. The property of a circuit which tends to oppose a change in the existing current.

Inductance Bridge. An instrument similar to a Wheatstone bridge, used to measure an unknown inductance by comparing it with a known inductance.

Induction. The act or process of producing voltage by the relative motion of a magnetic field across a conductor.

Induction Motor. An AC motor with field windings energized by out-of-phase currents. A rotating field is established. The rotor has no electric connections but receives its energy by the transformer action of the field winding. The motor torque is provided by the induced rotor current and the rotating field.

Inductive Circuit. Circuit containing for the most part inductive reactance, rather than capacitive reactance or simply pure resistance.

Inductive Coupling. A form of coupling in which energy is transferred from a coil in one circuit to a coil in another circuit by induction.

Inductive Feedback. Feedback of energy from the plate circuit of a vacuum tube to the grid circuit through an inductance or by means or inductive coupling.

Inductive Kick. The high cemf (counter electromotive force) that is generated when an inductive circuit is opened.

Inductive Load. The load on a source of alternating current which will cause the current to lag the voltage.

Inductive Reactance. The opposition to the flow of alternating or pulsating current caused by the inductance of a circuit. It is measured in ohms.

Inductor. A circuit element designed so that its inductance is its most important electrical property; a coil. Also called *reactor*.

Inferred Zero Point. In theory, the temperature at which the material of the conductor has zero resistance.

Infinite. Extending indefinitely; having innumerable parts, capable of endless division within itself.

In-Phase. Applied to the condition that exists when two waves of the same frequency pass through their maximum and minimum values of like polarity at the same instant.

Input Transformer. A transformer used to transfer incoming energy to the input of a circuit or device.

Instantaneous Value. The magnitude at any particular instant when a value is continually varying with respect to time.

Insulator. A material that has very high resistivity.

Integrating Circuit. A circuit which produces an output voltage substantially in proportion to the frequency and amplitude of the input voltage.

Intensify. To increase the brilliance of an image on the screen of a cathode ray tube.

Intensify Modulation. The control of the brilliance of the trace on the screen of a cathode ray tube in conformity with the signal. Also called *brilliance modulation*.

Interelectrode Capacitance. The capacitance existing between the electrodes in a vacuum tube.

Interface States. Extra donors, acceptors, or traps that may occur at a boundary between a semiconductor and some other material.

Interference Filter. A device used between a source of interference and a radio, to attenuate or eliminate noise.

Interlocking Device. A mechanism interposed between control elements which forbids the movement of the responding of one control until the other has been properly set for safety or for other reasons.

Intermediate Frequency. In superheterodyne reception, a frequency resulting from the combination of the received frequency with the locally generated frequency.

Intermediate Frequency Amplifier. That section of a superheterodyne receiver which is designed to amplify signals at a predetermined frequency called the intermediate frequency of the receiver.

Intermediate Frequency Transformer. A transformer employed at the input and output of each intermediate frequency amplifier stage in a superheterodyne receiver.

Intermittent. Starting and stopping again at different times.

Intermittent Cycle. A cycle which repeats itself at different intervals.

Internal Resistance. Resistance contained within a power or energy source.

Interpolation. The process of finding the value between two known values.

Interstage Coupling. Coupling between vacuum-tube or transistor stages.

Interstage Transformer. A transformer used to provide coupling between two amplifier stages.

Inverse Feedback. See Negative Feedback.

Inversely. Inverted or reversed in position or relationship.

Inverse Peak Voltage. The highest instantaneous negative potential which the plate can acquire with respect to the cathode without danger of injuring the tube.

Inverter. The output is always in the opposite logic state of the input. Also called a *not circuit*.

Ion. An elementary particle of matter or a small group of such particles having a net positive or negative charge.

Ion Exchange. Ion exchange occurs when one ion or group of ions replaces another ion or group of ions in a solution.

Ion Implantation. Introduction into a semiconductor of selected impurities in controlled regions (via high-voltage ion bombardment) to achieve desired electronic properties.

Ionization. Process by which ions are produced in solids, liquids, or gases.

Ionization Current. Current flow existing between two oppositely charged electrodes in an ionized gas.

Ionization Potential. The lowest potential at which ionization takes place within a gas-filled tube.

Ionosphere. A region composed of highly ionized layers of atmosphere from 70 to 250 miles above the surface of the earth.

IR Drop. An electrical term indicating the loss in a circuit expressed in amperes × resistance ($I \times R$), or voltage drop.

Iron-Core Transformer. A transformer in which iron forms part or all of the magnetic circuit linking the windings.

Iron-Vane Meter. A type of meter which uses a moving iron vane and a stationary coil. Commonly used to measure alternating current.

Isolate. To set apart from others; to place alone; to separate.

Isolation Transformer. A transformer with independent primary and secondary windings. Transformers of this type (generally 1:1 ratio) are used to isolate AC-DC equipment from the AC power line.

Joule. A unit of energy or work. 1J of energy is liberated by 1A flowing for 1 second through a resistance of 1Ω.

Jumper. A length of wire or a conductor used to make temporary connections such as a bypass around a switching component.

Kilo. A prefix meaning 1,000.

Kilocycle. One thousand cycles; conversationally used to indicate 1,000 cycles per second.

Kirchhoff's Current Law. A fundamental law of electricity which states that the sum of all the currents flowing to a point in a circuit must be equal to the sum of all the currents flowing away from that point.

Kirchhoff's Voltage Law. The algebraic sum of the voltages around a closed circuit is equal to zero.

Klystron. A tube in which oscillations are generated by the bunching of electrons (that is, velocity modulation).This tube utilizes the transit time between two given electrodes to deliver pulsating energy to a cavity resonator in order to sustain oscillations within the cavity.

Knife Switch. A switch in which one or more flat metal blades, pivoted at one end, serve as the moving connectors, making contact with flat, gripping spring clips.

Lag. The amount one wave is behind another in time; expressed in electrical degrees.

Lagging Load. Inductive load; the current lags behind the voltage.

Laminated Core. A core built up from thin sheets of metal and used in transformers and relays.

L Antenna. An antenna consisting of one or more horizontal wires with vertical lead-in connected at one end.

Layer Winding. A coil-winding method in which adjacent turns are laid evenly side by side along the length of the coil.

LC Product. Inductance L in henries multiplied by capacitance C in farads.

Lead. The opposite of lag. Also, a wire or connection.

Lead-In. The conductor or conductors that connect the antenna proper to electronic equipment.

Leakage. The electrical loss due to poor insulation.

Leakage Current. An undesirable flow of current through or over the surface of an insulating material or insulator or the flow of direct current through a capacitor. Also, the alternating current that passes through a rectifier without being rectified.

Leakage Resistance. The resistance of the path over which leakage current flows, normally a high value.

Lecher Line. A section of open-wire transmission line used for measurements of standing waves.

Left-Hand Rule. A rule for determining direction of magnetic lines of force around a single wire. If the fingers of the left hand are placed around the wire in such a way that the thumb points in the direction of electron flow, the fingers will then be pointed in the direction of the magnetic field.

Lenz's Law. States that the cemf will always oppose the force which created it.

Limit Control. A control used to open or close electric circuits as temperature or pressure limits are reached.

Limiting. Removal by electronic means of one or both extremities of a waveform at a predetermined level. Also called *chopping* or *clipping*.

Line. An electric conductor, usually a wire or cable.

Linear. Having an output which varies in direct proportion to the input.

Linear Amplification. Amplification in which the waveform is reproduced accurately, but in magnified form.

Linear Circuit. A circuit whose output is an amplified version of its input or whose output is a predetermined variation of its input.

Linear Modulation. Modulation which is equally proportional to the amplitude of the sound wave at all audio frequencies.

Linear Resistance. Resistance that remains constant for variation in voltage, current, or temperature.

Line-Balance Converter. A device used at the end of a coaxial line to isolate the outer conductor from ground.

Line Cord. A two- or three-wire cord terminating in a two- or three- prong plug at one end and used to connect equipment to a power outlet.

Line Drop. The voltage drop between two points on a power line or transmission line.

Line Filter. A device inserted in the power line to block noise impulses which might otherwise enter the equipment from the power line.

Line of Force. A line in an electric or magnetic field that shows the direction of the force.

Line-Voltage Regulator. A device such as a ballast voltage regulator or special transformer that delivers an essentially constant voltage to the load, regardless of minor variations in the line voltage.

Link Coupling. Two or more coils of separate circuits coupled by a transmission line.

Litz Wire. A special wire used to reduce the skin effect.

Load. The impedance to which energy is being supplied.

Loaded. In reference to a source, the source is loaded when a load is connected across its terminals and the source is supplying power to the load.

Load Line. A straight line drawn across a series of plate current-plate voltage characteristic curves on a graph to show how plate current will change with grid voltage when a specified plate load resistance is used.

Local Oscillator. The oscillator used in a superheterodyne receiver, the output of which is mixed with the desired RF carrier to form the intermediate frequency.

Loctal Tube. An eight-prong vacuum tube having a lock-in type of base.

Logic. The arrangement of circuitry designed to accomplish certain objectives.

Long Waves. Wavelengths longer than the longest broadcast-band wavelength of 545m. Long waves correspond to frequencies between about 30kc and 550kc.

Loop Antenna. An antenna consisting of one or more complete turns of wire. Loop antennas are also commonly used in direction-finding equipment and portable radios.

Loopstick Antenna. A built-in receiving antenna widely used in broadcast receivers. Loopstick antennas consist of a coil wound on a powdered-iron core. In some types the inductance is adjusted by moving the core.

Loose Coupling. Less than critical coupling; coupling providing little transfer of energy.

Loss. A circuit using low-voltage current to operate relay coils, which in turn control switches for operating higher-voltage circuits.

Loudspeaker. A device for converting audio-frequency current into sound waves.

Low-Energy Power Circuit. A circuit that is not a remote-control or signal circuit but whose power supply is limited in accordance with the requirements of Class 2 remote-control circuits.

Low Frequency. A frequency in the band extending from 30kc to 300kc in the radio spectrum.

Low-Pass Filter. A filter designed to pass currents at all frequencies below a critical frequency, while substantially attenuating the amplitude of other frequencies.

L Pad. A dual volume control presenting a constant load impedance at all control settings.

Magnet. A metallic material which attracts iron and steel, and, if free to move, aligns itself north and south because of the influence of the earth's magnetic field.

Magnetic Circuit. The complete path of magnetic lines of force.

Magnetic Deflection. Method of deflecting electrons in a cathode ray tube by means of the magnetic field generally produced by coils placed outside the tube.

Magnetic Field. The space in which a magnetic force exists.

Magnetic Flux. The total number of lines of force issuing from a pole of a magnet.

Magnetic Flux Density. The number of magnetic lines of force per unit area.

Magnetic Focusing. A method of focusing an electron stream in a cathode ray tube through the action of magnetic lens.

Magnetic Lines of Force. Imaginary lines used to designate the directions in which magnetic forces are acting throughout the magnetic field associated with a permanent magnet, electromagnet, or current-carrying conductor.

Magnetic Poles. Regions of a magnet near which the field is concentrated, usually the two ends of a magnet. The north pole, the south pole.

Magnetic Shield. A soft-iron housing used to protect equipment or components from the effects of stray magnetic fields.

Magnetize. To convert a material into a magnet by causing the molecules to rearrange.

Magnetron. A vacuum-tube oscillator containing two electrodes in which the flow of electrons from cathode to anode is controlled by an externally applied magnetic field.

Majority Carrier. The mobile charge carrier (hole or electron) that predominates in a semiconductor material—for example, electrons in an N-type region.

Mask. Pattern or template used in photolithographic-like processes employed in making integrated circuits. The pattern describes regions of the circuit electrical leads, etc.

Master Oscillator. An oscillator of comparatively low power used to establish the carrier frequency of a transmitter.

Matched Impedance. The condition which exists when two coupled circuits are so adjusted that their impedances are equal.

Matching. Connecting two circuits or components with a coupling device so that the impedance of either circuit will be equal to the impedance existing between them.

Maxwell. A single line of magnetic flux.

Mean Carrier Frequency. The center or resting frequency of a frequency-modulation transmission transmitter.

Medium Frequency. The band from 300kHz to 3,000kHz.

Meg. A prefix meaning one million.

Megacycle. One million cycles. Used conversationally to mean 1,000,000 cycles per second.

Megger. A test instrument used to measure insulation resistance and other high resistances. It is a portable, hand-operated DC generator used as an ohmmeter.

Metallic Insulator. A shorted quarter-wave section of a transmission line which acts as an electrical insulator at a frequency corresponding to its quarter-wave length.

Meter. An instrument or device for measuring quantities of electricity.

Meter Loading. Changes in circuit current or voltage caused by putting a meter in the circuit.

Metric System. A decimal system of measures and weights, based on the meter and gram. The length of 1m is 39.37".

Micro. A prefix meaning one-millionth.

Micromicro. The prefix meaning one-millionth of one-millionth.

Microsecond. One-millionth of a second.

Microswitch. Trade name for a small switch in which a minute motion makes or breaks contact.

Milli. A prefix meaning one-thousandth.

Milliampere. One-thousandth of an ampere.

Mismatch. The conditions in which the impedance of a source does not match or equal the impedance of a connected load.

Mixer. A vacuum tube or crystal and suitable circuit used to combine the incoming and local-oscillator frequencies to produce an intermediate frequency. Also called *first detector* or *converter*.

Mixing. Combining two or more signals.

Modulate. To vary the amplitude, frequency, or phase of a radio-frequency carrier in accordance with speech, music, or other forms of signals.

Modulation. The process of impressing intelligence on an RF carrier in such a manner as to vary the amplitude of the resultant wave (amplitude modulation), the frequency of the resultant wave (frequency modulation), or the phase of the resultant wave (phase modulation).

Modulation Envelope. A curve drawn through the peaks of a graph showing the waveform of an amplitude-modulated signal.

Modulator. The circuit which provides the signal that varies the amplitude, frequency, or phase of the resultant wave in a transmitter.

Molecule. The smallest division of matter. If subdivided further, the matter loses its identity.

Monitor. A device used for checking radio or audio signals.

Motorboating. Feedback occurring at a low audio-frequency rate in an audio amplifier. Resembles sounds made by a motorboat.

Multielectrode Tube. A vacuum tube containing more than three electrodes associated with a single electron stream.

Multimeter. Electric instrument designed to measure two or more electrical quantities.

Multiplier. Precision resistor used to extend voltage range of a meter movement.

Multiunit Tube. A vacuum tube containing within one envelope two or more groups of electrodes, each associated with separate electron streams.

Multivibrator. A type of relaxation oscillator for the generation of nonsinusoidal waves in which the output of each of its two tubes is coupled to the input of the other to sustain oscillations.

Mutual Inductance. A circuit property existing when the relative position of two inductors causes the magnetic lines of force from one to link with the turns of the other.

Negative. A term used to describe a terminal from which electrons flow.

Negative Charge. The electric charge carried by a body which has an excess of electrons.

Negative Feedback. An arrangement by which a signal is fed back from the plate circuit to the grid circuit 180° out of phase with the grid signal, thus decreasing gain. Also called *inverse or degenerative feedback.*

Neon Bulb. A glass bulb containing two electrodes in neon gas at low pressure.

Network. Any electrical circuit containing two or more interconnected elements.

Neutral. Neither positive nor negative.

Neutralization. The process of nullifying the voltage fed back through the interelectrode capacitance of an amplifier tube by providing an equal voltage of opposite phase; generally necessary only with triode tubes.

Neutralizing Capacitor. A capacitor, usually variable, employed in neutralizing circuits.

Neutron. A particle having the weight of a proton but carrying no electric charge. It is located in the nucleus of an atom.

Node. A zero point; specifically, a current node is a point of zero current and a voltage node is a point of zero voltage.

Noise. Interference characterized by undesirable random disturbances caused by internal circuit defects or from some external source.

Noise Filter. A combination of one or more choke coils and capacitors used to block noise interference.

Noise Level. Volume of noise, usually expressed in decibels.

Noise Suppressor. A circuit used in receiver or amplifier to reduce noise.

Nonconductor. An insulating material.

Noninductive Capacitor. A capacitor in which the inductive effects at high frequencies are reduced to the minimum.

Noninductive Circuit. A circuit in which inductance is reduced to a minimum or negligible value.

Noninductive Load. A load having no inductance.

Noninductive Winding. A winding made so that one turn or section cancels the field of the next adjacent turn or section. For example, the wire may be doubled before winding. Used particularly with resistors to prevent them from exhibiting resonant effects.

Nonlinear. Having an output which does not vary in direct proportion to the input.

Nonlinear Detection. Square law detection. Detection based on the curvature of a tube characteristic. This results in distortion without complete rectification. Either of these effects results in demodulation.

Nucleus. The central part of an atom; it is mainly comprised of protons and neutrons. It is the part of the atom that has the most mass.

Null. Zero.

Null Indicator. Any device that indicates when current, voltage, or power is zero.

Ohm. The unit of electrical resistance.

Ohmmeter. An instrument for measuring resistance.

Ohm's Law. A fundamental law of electricity which expresses the relationship between voltage, current, and resistance in a DC circuit, or the relationship between voltage, current, and impedance in an AC circuit.

Ohms-per-Volt. A sensitivity rating for voltage-measuring instruments, obtained by dividing the resistance of the instrument in ohms at a particular range by the full-scale voltage value at that range. The higher the ohms-per-volt rating, the more sensitive is a meter.

Oil Capacitor. A capacitor with oil-impregnated paper as a dielectric.

Omnidirectional. In all directions, such as the radiation pattern of a vertical antenna.

Open Circuit. A circuit which does not provide a complete path for the flow of current.

Open-Circuit Voltage. The voltage across the output terminals of an electric circuit or device operating under normal conditions without the load attached.

Order of Magnitude. Ten times greater than, as in $100\,\Omega$ is an order of magnitude greater than $10\,\Omega$. One-tenth as great, as in $100\,\Omega$ is an order of magnitude less than $1,000\,\Omega$.

Oscillation. Fluctuation, instability; a single swing of a swinging object. The variation between maximum and minimum values, as in alternating current.

Oscillator. A circuit capable of converting direct current into alternating current of a frequency determined by the constants of the circuit. It generally uses a vacuum tube.

Oscillator Coil. The transformer or coil used in an oscillator circuit.

Oscillatory Circuit. A circuit in which oscillations can be generated or sustained.

Oscilloscope. An instrument for visually showing graphical representations of the waveforms encountered in electrical circuits. Also called *cathode ray oscilloscope.*

Output Impedance. The impedance as measured between the output terminals of an electronic device, generally at a definite frequency or at a predominant frequency. For maximum power transfer, the load impedance should match or be equal to this output impedance.

Output Indicator. A meter or other device connected to indicate variations in signal strength of the output circuits.

Output Stage. The final stage of an electronic device.

Output Transformer. The iron-core audio-frequency transformer used to match the output stage of an audio-frequency amplifier to its loudspeaker or other load.

Output Tube. An amplifier tube used in an output stage.

Overdriven Amplifier. An amplifier designed to distort the input signal waveform by a combination of cut-off limiting and saturation limiting.

Overload. A load greater than the load for which the system was intended.

Overmodulation. Amplitude modulation in excess of 100%.

Padder. In a superheterodyne receiver, the capacitor placed in series with the oscillator tuning circuit to control the receiver calibration at the low-frequency end of a tuning range. Also, any small capacitor inserted in series with a main capacitor, for alignment purposes.

Paper Capacitor. A capacitor in which the foil plates are separated by waxed paper.

Parallel Circuit. A circuit containing two or more parallel paths for electrons supplied by a common source.

Parallel Connection. A manner of connecting a group of devices in which one pole of each is connected to one common wire and the other pole of each is connected to a second common wire. Opposed to Series Connection.

Parallel Feed. Application of a DC voltage to the plate or grid of a tube in parallel with an AC circuit so that the DC and AC components flow in separate paths. Also called *shunt feed*.

Parallel-Resonant Circuit. A tuning circuit in which the applied voltage is connected across a parallel circuit formed by a capacitor and an inductor.

Paraphase Amplifier. An amplifier which converts a single input into a push-pull output.

Parasitic Oscillations. Unwanted self-sustaining oscillations at a frequency different from the operating frequency.

Parasitic Suppressor. A resistor in a vacuum-tube circuit to prevent unwanted oscillations.

Passivation. Treatment of a region of a device to prevent deterioration of electronic properties through chemical action or corrosion. Usually passivation protects against moisture or contaminants.

Peak. The highest or maximum value of a sine wave reached during a cycle.

Peaking Circuit. A type of circuit which converts an input to a peaked output waveform.

Peak Load. The maximum load consumed or produced in a given period of time.

Peak Plate Current. The maximum instantaneous plate current passing through a tube.

Peaks. Momentary high amplitude levels occurring in electronic equipment.

Peak Value. The maximum instantaneous value of a varying current, voltage, or power. It is equal to 1.414 times the effective value of a sine wave.

Peak Voltmeter. A voltmeter that reads peak value of a voltage.

Pentagrid Converter. A tube employed as an oscillator-mixer in a superheterodyne receiver.

Pentode. A five-electrode vacuum tube containing a cathode, control grid, screen grid, suppressor grid, and plate.

Period. Time required to complete one cycle.

Permanent Magnet. Hardened steel or special alloy which retains the magnetic effect after a magnetizing force has been removed.

Permeability. A measure of the ease with which magnetic lines of force can flow through a material as compared with the ease with which magnetic lines can flow through air.

Phase. When two sine waves pass through their zero positions at the same time, they are in phase.

Phase Angle. The angular difference between the two vectors of two sine waves.

Phase Difference. The time in electrical degrees by which one wave leads or lags another.

Phase Inversion. A phase difference of 180° between two similar waveshapes of the same frequency.

Phase Shift. The result of two waveforms being out of step with each other.

Phase-Splitting Circuit. A circuit which produces from the same input waveform, two output waveforms which differ in phase from each other.

Phasing. Interconnecting transformer, generator, or motor windings so that they have the correct time (phase) relationships between them.

Phasor. A line representing alternating current or voltage at some instant of time.

Phosphorescence. The property of emitting light for some time after excitation by electronic bombardment.

Photovoltaic. Producing voltage from light energy.

Pierce Oscillator. A crystal oscillator circuit featuring a crystal connected between the grid and plate of the oscillator tube.

Piezoelectric Effect. The effect of producing a voltage by placing a stress, either by compression, by expansion, or by twisting, on a crystal, and, conversely, the effect of producing a stress in a crystal by applying a voltage to it.

Plate. The principal electrode in a tube to which the electron stream is attracted.

Plate Bypass Capacitor. A capacitor connected to the plate circuit of a vacuum tube to bypass high-frequency currents.

Plate Circuit. The complete electrical circuit connecting the cathode and plate of a vacuum tube.

Plate Current. The current flowing in the plate circuit of a vacuum tube.

Plate Detection. The operation of a vacuum-tube detector at or near cutoff so that the input signal is rectified in the plate circuit.

Plate Dissipation. The power in watts consumed at the plate in the form of heat.

Plate Efficiency. The ratio of the AC power output from a tube to the average DC power supplied to the plate circuit.

Plate-Load Impedance. The impedance in the plate circuit across which the output signal voltage is developed by the alternating component of the plate current.

Plate Modulation. Amplitude modulation of a Class C radio frequency amplifier by varying the plate voltage in accordance with the modulating signal.

Plate Resistance. The internal resistance to the flow of alternating current between the cathode and plate of tube. It is equal to a small change in plate voltage divided by the corresponding change in plate current and is expressed in ohms. It is also called AC resistance, internal impedance, plate impedance, and dynamic plate impedance. The static plate resistance or resistance to the flow of direct current if a different value. It is denoted by R_b.

Plate Supply. The voltage source used in a vacuum-tube circuit to put the plate at a high positive potential with respect to the cathode.

Plate Voltage. The direct voltage between the plate and the cathode of a vacuum tube.

Polarity. Having a definite sign, such as positive or negative voltage.

Polarization. Accumulation of gas ions around the electrode of a cell.

Pole. The section of a magnet where the flux lines are concentrated; also where they enter and leave the magnet. An electrode of a battery.

Polycrystalline. Descriptive of a material that consists of many small crystallites rather than a single crystal.

Polyphase. A circuit that utilizes more than one phase of alternating current.

Positive Bias. The condition in which the control grid is positive with respect to the cathode of a vacuum tube.

Positive Charge. The electric charge carried by a body which has become deficient in electrons.

Positive Feedback. See Regeneration.

Positive Terminal. The terminal of a battery or other voltage source toward which electrons flow in the external circuit.

Potential. The amount of charge held by a body as compared to another point or body. Usually measured in volts.

Potential Difference. Voltage between two points in an electric circuit, resulting from current doing work in converting electric energy into heat or light or some other energy form.

Potentiometer. A variable divider; a resistor which has a variable contact arm so that any portion of the potential applied between its ends may be selected.

Power. The rate of doing work or the rate of expending energy. The unit of electrical power is the watt.

Power Amplification. The process of amplifying a signal to produce a gain in power, as distinguished from voltage amplification. The gain in the ratio of the alternating power output to the alternating power input of an amplifier.

Power Amplifier. An audio- or radio-frequency amplifier designed to deliver a relatively large amount of output energy. Also, the last stage of an amplifier as distinguished from previous stages usually classed as voltage amplifiers.

Power Factor. The ratio of the actual power of an alternating or pulsating current, as measured by a wattmeter, to the apparent power, as indicated by ammeter and voltmeter readings. The power factor of an inductor, capacitor, or insulator is an expression of the losses.

Power Gain. The ratio of two powers such as output to input of a vacuum tube or output to input of an audio-frequency amplifier.

Power Level. The amount of electric power passing through a given point in a circuit. Power level can be expressed in watts or in decibels.

Power Line. Two or more wires used for conducting power from one location to another.

Power Loss. Loss in a circuit due to resistance of conductors.

Power Output. The power in watts delivered by an amplifier to a load, such as a speaker.

Power Pack. The power-supply unit of a radio receiver, amplifier, transmitter, or other radio apparatus.

Power Transformer. An iron-core transformer having a primary winding usually connected to an AC power line and having a number of secondary windings that provide different voltage values.

Power Tube. A vacuum tube designed to handle a greater amount of power than the ordinary voltage-amplifying tube.

Preamplifier. An extra stage of amplification at the input of an amplifier.

Preselector. A tuned-radio-frequency amplifier or antenna tuning device inserted between the receiver and the antenna to increase the amplitude of the incoming signal.

Primary. The incoming side of a transformer. The side from which the voltage or current is being changed to a different value by the transformer.

Primary Cell. Cell that is not meant to be recharged.

Primary Circuit. The first, in electrical order, of two or more coupled circuits in which a change in current induces a voltage in the other or secondary circuits, such as the primary winding of a transformer.

Printed Circuit. A method by which circuit connections and many of the components are printed or etched on a plane surface with conductive or resistive media for building compact circuits.

Propagation. The travel of electromagnetic waves or sound waves through a medium.

Proton. A positively charged particle in the nucleus of an atom.

Pulsating Current. A unidirectional current which increases and decreases in magnitude.

Push-Pull Circuit. A push-pull circuit usually refers to an amplifier circuit using two vacuum tubes in such a fashion that when one vacuum tube is operating on a positive alternation, the other vacuum tube operates on a negative alternation.

691

Push-Pull Oscillator. A vacuum-tube oscillator containing two tubes or a double-section tube connected in a phase relation similar to that of a push-pull amplifier.

Push-Pull Transformer. An audio transformer designed for use in a push-pull amplifier circuit.

Quality. A number which indicates the ratio of reactance to resistance.

Quarter-Wave Antenna. An antenna electrically equal to one-fourth the wavelength of the transmitted or received signal.

Quartz Crystal. A thin slice of quartz that vibrates at a frequency determined by its thickness and its original position in the natural quartz. Used to maintain high-frequency stability in oscillators.

Quenching Frequency. A locally generated frequency of a super-regenerative detector stage which prevents oscillation during reception of strong signals.

Radial Leads. Leads that are connected on the side of a component.

Radiate. To send out energy, such as radio frequency waves, into space.

Radiation Pattern. A diagram indicating the intensity of the radiation field of a transmitting antenna as a function of plane or solid angles. In the case of a receiving antenna, it is a diagram showing the response of the antenna to a unit field intensity signal arriving from different directions.

Radiation Resistance. A fictitious resistance which would dissipate the same power that the antenna dissipates.

Radio Broadcasting. A one-way transmission of voice and music.

Radio Channel. A band of frequencies having sufficient width for radio communication and broadcasting purposes. The width of a channel depends on the type of transmission and the tolerance for the frequency of emission.

Radio Converter. A unit for adapting a receiver for use on frequency bands other than the ones for which it was designed originally.

Radio Frequency. Any frequency of electrical energy capable of propagation into space. Radio frequencies normally are much higher than sound wave frequencies.

Radio-Frequency Amplification. The amplification of a radio wave by a receiver before detection or by a transmitter before radiation.

Radio-Frequency Choke. An air-core or powdered iron-core coil used to impede the flow of radio frequency currents.

Radio-Frequency Transformer. A transformer for radio-frequency currents having either an air core or some form of pulverized-iron core.

Radio Spectrum. The entire range of useful radio waves as classified into seven bands by the Federal Communications Commission.

Ratio. The value obtained by dividing one number by another, indicating their relative proportions.

RC, RC Circuit. Designation for any resistor-capacitor circuit.

RC Coupling. Resistor-capacitor coupling between two circuits.

Reactance. The opposition offered to the flow of an alternating current by the inductance, capacitance, or both, in any circuit.

Reactive. Pertaining to either inductive or capacitive reactance.

Reactive Load. A load consisting of inductive reactance which causes the current to lag the applied voltage.

Reciprocal. The value obtained by dividing the number 1 by any quantity.

Reciprocating. Action in which the motion is back and forth in a straight line.

Recombination. In a semiconductor, the combining of holes and electrons. Recombination tends to reduce the minority carriers to their equilibrium number after injection has taken place.

Rectifier. A device used to change alternating current to unidirectional current.

Rectifying Contact. An electrical contact through which current flows easily in one direction (the "forward" direction), but with difficulty or not at all in the reverse direction.

Reflected Wave. The sky radio wave, reflected back to the earth from an ionosphere layer.

Reflection. The turning back of a radio wave caused by re-radiation from any conducting surface which is large in comparison to the wavelength of the radio wave.

Reflector. A metallic object placed behind a radiating antenna to prevent radio frequency radiation in an undesired direction and to reinforce radiation in a desired direction.

Refracted Wave. The wave that is bent as it travels into a second medium, as from the atmosphere into an ionized layer of the stratosphere.

Regeneration. The process of returning a part of the output signal of an amplifier to its input circuit in such a manner that it reinforces the grid excitation and thereby increases the total amplification.

Regeneration Control. A potentiometer or variable condenser which is used to control the amount of signal fed back from output to input in the regenerative detector stage.

Regulated Power Supply. A power supply containing a regulator device for maintaining constant voltage or constant current under changing load conditions.

Regulation. Holding constant some conditions, like voltage, current, power, or position.

Regulator. A device that accomplishes regulation within desired limits such as a current or voltage regulator.

Relative Permeability. Permeability of a material compared with permeability of air.

Relaxation Oscillator. A circuit for the generation of nonsinusoidal waves by gradually storing and quickly releasing energy either in the electric field of a capacitor or in the magnetic field of an inductor.

Relay. An electromechanical switching device that can be used as a remote control.

Reluctance. A measure of the opposition that a material offers to magnetic lines of force.

Remote-Control Circuit. Any electrical circuit that controls any other circuit through a relay or equivalent device.

Remote-Cutoff Tube. A variable mu tube. A tetrode or pentode in which the spacing of the control-grid wires is wider at the center than at the ends. Thus, the amplification of the tube does not vary in direct proportion to the bias, and some plate current flows regardless of the negative bias on the grid. Used in RF amplifiers.

Residual. A remainder; a remaining force, as in magnetism; leftover.

Residual Magnetism. Magnetic flux left in a temporary magnet after the magnetizing force has been removed.

Resistance. The opposition to the flow of current caused by the nature and physical dimensions of a conductor.

Resistance-Capacitance-Coupled Amplifier. A vacuum-tube amplifier, the various stages of which employ resistors for the plate load and in the grid circuit; coupling between them is by capacitors. Also used with transistors.

Resistance-Coupled Amplifier. An amplifier in which the various stages are coupled solely by resistances between output and input. A direct-coupled amplifier.

Resistance Drop. Voltage drop due to flow of current through a resistance. Also known as *IR* drop.

Resistance Wire. Wire made from an alloy having high resistivity.

Resistivity. The resistance between opposite parallel faces of a 1m cube of a material; also called *specific resistance*. Resistivity ($\Omega \times$ m) depends only on the properties of the material itself, whereas resistance (Ω) depends not only on the properties of the material, but on its length and cross-sectional area.

Resistor. A circuit element whose chief characteristic is resistance; used to oppose the flow of current.

Resonance. The condition existing in a circuit in which inductive and capacitive reactances cancel.

Resonance Curve. A graphical representation of the manner in which a resonant circuit responds to various frequencies at and near the resonant frequency.

Resonant Frequency. The frequency at which inductive and capacitive reactance values cancel out.

Resonate. To bring to resonance, as by tuning.

Response. Frequency range, or response, within specific limitations of speakers, amplifiers, etc.

Retentivity. The measure of the ability of material to hold its magnetism.

Return Wire. A common wire, a ground wire, or the negative wire in a DC circuit.

Rheostat. A variable resistor.

Rhombic Antenna. A directional antenna array consisting of four long conductors laid out like an equal-sided parallelogram (rhombus).

Ripple. The AC component present in the output of a DC generator, rectifier system, or power supply.

Ripple Current. The AC component of a pulsating unidirectional current.

Ripple Factor. Defined as the effective value of the alternating component of voltage (or current) divided by the direct or average values of the voltage (or current).

Ripple Filter. A low-pass filter designed to attenuate the AC components of a pulsating unidirectional current while passing the direct current from the rectifier or DC generator.

Ripple Frequency. The frequency of the ripple current.

Ripple Voltage. The fluctuations in the output voltage of a rectifier, filter, or generator.

Rise Time. For an instantaneous change in voltage applied, the time required for the steady-state current to rise from 10% to 90% of its maximum value.

Root-Mean-Square. Or effective value; the square root of the average (mean) values of the squares of the instantaneous values of one cycle of a periodic quantity.

Rotary Beam Antenna. A highly directional antenna that can be rotated by hand or by motor to any desired position. Provides maximum concentration of radiated energy or reception.

Safety Factor. The load, above the normal operating rating, to which a device can be subjected without failure.

Saturation. The condition existing in any circuit when an increase in the driving signal produces no further change in the resultant effect.

Saturation Limiting. Limiting the minimum output voltage of a vacuum-tube circuit by operating the tube in the region of plate-current saturation (not to be confused with emission saturation).

Saturation Point. The point beyond which an increase in either grid voltage, plate voltage, or both produces no increase in the existing plate current.

Schematic Diagram. A diagram of the electric circuit showing the interconnection of the electric circuit components using standardized symbols to represent the parts.

Screen. A metal partition or shield to isolate a device or apparatus from external magnetic or electric field.

Also, the coated surface on the inside of the large end of a cathode ray tube.

Screen Dissipation. The power dissipated in the form of heat on the screen grid as the result of bombardment by the electron stream.

Screen Grid. An electrode placed between the control grid and the plate of a vacuum tube to reduce interelectrode capacitance.

Screen-Grid Modulation. A type of amplitude modulation wherein the modulating voltage is superimposed on the DC screen-grid voltage of the RF amplifer.

Screen-Grid Voltage. The direct voltage applied between the screen grid and the cathode in a vacuum tube.

Secondary. The output coil of a transformer.

Secondary Cell. Cell that can be recharged.

Secondary Controls. Controls which must respond to complete a function but which do not originate it.

Secondary Emission. The emission of electrons knocked loose from the plate grid or fluorescent screen of a vacuum tube by the impact or bombardment of electrons arriving from the cathode.

Secondary Voltage. The voltage across the secondary winding of a transformer.

Second Detector. In a superheterodyne receiver, the stage that separates the intelligence signal from the intermediate-frequency carrier signal.

Selective. The characteristic of responding to a desired frequency to a greater degree than to other frequencies.

Selective Interference. Radio interference in a narrow band of frequencies.

Selective Reflection. Reflection of waves of only a certain group of frequencies.

Selectivity. The degree to which a receiver is capable of discriminating between signals of different carrier frequencies.

Self-Bias. The bias of a tube created by the voltage drop developed across a resistor through which its cathode current flows.

Self-Exited Oscillator. An oscillator depending on its resonant circuits for frequency determination.

Self-Induction. The production of a counterelectromotive force in a conductor when its own magnetic field collapses or expands with a change in current in the conductor.

Semiconductor. An element such as silicon or germanium that is intermediate in electrical conductivity between the metals and insulators.

Sensitivity. The degree of response of a circuit to signals of the frequency to which it is tuned.

Series Circuit. A circuit which contains only one possible path for electrons through the circuit. This path may pass through more than one component or terminal.

Series Connection. A manner of connecting a group of devices in which one pole of one device is connected to one pole of another, leaving one pole on each end of the series for connecting to a supply or to another circuit.

Series Feed. Application of the DC voltage to the plate or grid of a tube through the same impedance in which the alternating current flows.

Series Parallel. An arrangement of the circuit elements so that some are in series and some are in parallel with each other.

Series Resonance. The condition existing in a circuit when the source of voltage is in series with an inductor and capacitor whose reactances cancel each other at the applied frequency and thus reduce the impedance to minimum.

Series Resonant Circuit. A resonant circuit in which the capacitor and the inductor are in series with the applied voltage.

Series Tester. A device for the preliminary testing of appliances in which a lamp and/or other resistance(s) are interposed in one pole of the supply circuit.

Series-Type Commutator Motor. A motor whose field coils are connected in series with its commutator brushes.

Series-Wound. A motor or generator in which the armature is wired in series with the field winding.

Sharp Cutoff. Term applied to a tube or the grid of a tube in which the control grid spirals are uniformly spaced. The result is that as grid voltage is made negative, plate current decreases steadily to cutoff.

Shielded Line. A transmission line whose elements confine propagated radio waves inside a tubular conducting surface called the sheath. This prevents the line from radiating radio waves.

Shielded Pair. A two-wire transmission line surrounded by a metallic sheath.

Shielded Wire. Insulated wire covered with a metal shield, usually of tinned, braided copper wire.

Shielding. A metallic covering used to prevent magnetic or electrostatic coupling between adjacent circuits.

Short-Circuit. A low-impedance or zero-impedance path between two points.

Shorted Out. Made inactive by connecting a heavy wire or other conductor path around a device or circuit, usually for protective purposes.

Shortwaves. A general term usually applied to wavelengths whose frequency is higher than 1,600kc.

Shunt. Parallel. A parallel resistor placed in an ammeter to increase its range.

Sidebands. Two bands of frequencies on either side of the carrier frequency of an amplitude-modulated carrier, including components whose frequencies are the sum and difference of the carrier and the modulation frequencies.

Siemen. The unit of conductance.

Signal Generator. An oscillator that generates audio- or radio-frequency signals at any frequency needed for aligning or servicing electronic equipment.

Signal Strength. A measure of the power output of a radio transmitter at a particular location. Usually expressed in microvolts or millivolts per meter of effective height of the receiving antenna employed.

Signal-to-Noise Ratio. The ratio of the radio field intensity of a desired, received radio wave to the radio noise field intensity received with the signal.

Sine Wave. The curve traced by the projection on a uniform time scale of the end of a rotating arm or phasor. Also called *sinusoidal wave.*

Single-Crystal Material. A material which consists of a single crystal as distinct from most materials which are polycrystalline.

Single-Phase. Pertaining to a circuit or device that is energized by a single alternating voltage. One of the phases of a polyphase system.

Single-Phase Motor. A motor which operates on single-phase alternating current.

Single-Pole, Double-Throw Switch. An electric switch with one blade and two contact points.

Single-Sideband Transmission. An important variation of amplitude-modulation transmission, wherein either of the two sidebands and the carrier wave are eliminated or greatly reduced. The remaining sideband carries the same intelligence as in the suppressed sideband, so normal communications is carried on. Commonly called SSB.

Sinusoidal. Varying in proportion to the sine of an angle or time function. Ordinary alternating current is sinusoidal.

Skin Effect. The tendency of alternating currents to flow near the surface of a conductor thus being restricted to a small part of the total cross-sectional area. This effect increases the resistance and becomes more marked as the frequency rises.

Skip Zone. A region around a transmitter within which there is no reception from the transmitter.

Sky Wave. A radio wave that is reflected back to earth from the ionosphere. Sometimes called ionospheric wave.

Slice. A thin slab of semiconductor sawed from an ingot for the purpose of making semiconductor devices.

Smoothing Choke. An iron-core inductor employed as a filter to remove pulsations in the unidirectional output current of a rectifier.

Smoothing Filter. A filter composed of inductance and capacitance (or either alone) to remove AC components from the unidirectional output current of a rectifier or DC generator.

Snap-Back Diode. A kind of diode which, when switched from a forward to a reverse direction, passes current for a short time and then very rapidly turns off. The sudden change in current is very rich in harmonics. Used as frequency multiplier or pulse generator.

Soft Tube. A vacuum tube, the characteristics of which are adversely affected by the presence of gas in the tube; not to be confused with tubes designed to operate with gas inside them.

Solenoid. A multiturn coil of wire wound in a uniform layer or layers on a hollow cylindrical form.

Solid Conductor. A single wire. A conductor that is not divided into strands.

Solid-State Electronics. Designation used to describe devices and circuits fabricated from solid materials such as semiconductors, ferrites, or films as distinct from devices and circuits making use of electron tube technology.

Source Resistance. The resistance presented to the input terminals of a circuit by any source of electric energy. The source resistance of an ideal voltage source is $0\,\Omega$, whereas that of an ideal current source is an infinite resistance.

Space Charge. The cloud of electrons existing in the space between the cathode and plate in a vacuum tube, formed by the electrons emitted from the cathode in excess of those immediately attracted to the plate.

Space Current. The total current flowing between the cathode and all the other electrodes in a tube. This includes the plate current, grid current, screen-grid current, and any other electrode current which may be present.

Spaghetti. Cloth or plastic tubing sometimes used to provide insulation over bare wires.

Splice. The joining together of the ends of two wires to form a conductive path.

Split-Phase Motor. A single-phase induction motor. It develops its starting torque by a phase difference between the starting windings and the running windings of the field.

Spreader. A short, light insulator used to hold apart the wires of an open transmission line.

Spreading Resistance. The ohmic resistance of a small ohmic contact to a large volume of semiconductor material.

Square Wave. The waveform that shifts abruptly from one to the other of two definite values, giving a square or rectangular pattern when amplitude is plotted against time.

Squealing. A condition in which a high-pitched note is heard along with the desired signal.

Squelch Circuit. An AVC circuit that reduces or attenuates the noise otherwise heard in a radio receiver between signals by blocking some stage when the signal amplitude is below a value called the squelch level.

Stability. Freedom from undesired variation.

Stacked Array. An arrangement in which two or more identical antennas are mounted, one above the other, on a single mast. Noted for strong directional effect.

Stage. All the components in a circuit containing one or more vacuum tubes or transistors performing a single function.

Standing Wave. A distribution of current and voltage on a transmission line formed by two sets of waves traveling in opposite directions and characterized by the presence of a number of points of successive maxima and minima in the distribution curves.

Starting Winding. The winding in an electric motor used only during the brief period when the motor is starting.

Static. A fixed nonvarying condition; without motion.

Static Characteristics. The characteristics of an electronic tube with no output load and with DC potentials applied to the grid and plate.

Static Charge. An electric charge accumulated on an object.

Static Electricity. Atmospheric electricity. Electricity without motion as compared with electric current, which is electricity in motion.

Stator. The stationary field coils of a generator or a motor.

Step Down. To reduce from a higher to a lower voltage, as in a step-down transformer.

Step-Down Transformer. A transformer in which the secondary delivers a lower voltage than is applied to the primary.

Step Up. To increase to a high voltage by use of a transformer.

Step-Up Transformer. A transformer in which the secondary delivers a higher voltage than is applied to the primary.

Step Voltage. A voltage that changes from one constant voltage level to another constant voltage level instantaneously (without a passage of time).

Storage Battery. A unit consisting of two or more storage cells.

Straight-Line Capacitance. A variable capacitor characteristic obtained when the rotor plates are shaped so that capacitance varies directly in proportion to the angle of rotation.

Stray Capacitance. Capacitance existing between circuit wires or parts or between the metal chassis of electronic apparatus and the parts mounted on it.

Stray Field. Stray inductance. Leakage magnetic flux from an inductor.

Superheterodyne. A receiver in which the incoming signal is mixed with a locally generated signal to produce a predetermined intermediate frequency.

Suppressor Grid. An electrode used in a vacuum tube to minimize the harmful effects of secondary emission from the plate.

Suppressor Modulation. A type of amplitude modulation in which the modulating voltage is superimposed on the suppressor grid.

Surge. Sudden changes of current or voltage in a circuit.

Sweep Circuit. The part of a cathode ray oscilloscope which provides a time-reference base.

Swing. The variation in frequency or amplitude of an electrical quantity.

Swinging Choke. A choke with an effective inductance which varies with the amount of current passing through it. It is used in some power-supply filter circuits.

Switch. A mechanical device for completing, interrupting, or changing the connections in an electric circuit.

Synchronous. Happening at the same time; having the same period and phase.

Tachometer. An instrument for measuring revolutions per minute.

Tank Circuit. A parallel inductive-capacitive circuit.

Tap. A connection point or contact made in the body of a resistor or coil.

Tapped Field. Motor field coils to which several leads have been tapped at various points on the winding during manufacture for connecting to a selector switch to provide speed control by varying the resistance in the field.

Tapped Resistor. A wire-wound fixed resistor having one or more taps. Also called an adjustable resistor.

Temperature Coefficient. The change in resistance of a material per degree of change in temperature. As a general rule, conductors have a positive temperature coefficient, whereas semiconductors and insulators have a negative temperature coefficient.

Temperature-Compensating Capacitor. A capacitor whose capacitance varies with temperature.

Terminal. A point of an electronic connection.

Terminated Line. A transmission line terminated in the characteristic impedance of the line.

Termination. The load connected to the output end of a transmission line.

Tertiary Winding. A third winding on a transformer or magnetic amplifier that is used as a second control winding.

Tetrode. A four-electrode vacuum tube containing a cathode, control grid, screen grid, and plate.

Thermal Cutout. An overcurrent protective device containing a heater element in addition to, and affecting, a renewable fusible member that opens the circuit. It is not designed to interrupt short-circuit currents.

Thermionic Emission. Electron emission caused by heating an emitter.

Thermistor. A resistor that is used to compensate for temperature variations in a circuit.

Thermocouple. A junction of two dissimilar metals that produces a voltage when heated at one end (the thermal junction) and connected to a temperature meter at the other.

Thermocouple Ammeter. An ammeter which operates by means of a voltage produced by the heating effect of a current passed through the junction of two dissimilar metals. It is used for radio frequency measurements.

Three-Phase. Three voltages or currents displaced from each other by 120 electrical degrees.

Thyratron. A hot-cathode, gas-discharge tube in which one or more electrodes are used to control electrostatically the starting of a unidirectional flow of current.

Tight Coupling. Degree of coupling in which practically all of the magnetic lines of force produced by one coil link a second coil.

Time Constant. The time a capacitor requires to charge or discharge (through a resistor) 63.2% of the available voltage.

Time-Delay Relay. A relay in which the energizing or de-energizing of the coil precedes movement of the contact armature by a determinable interval.

Timer. A mechanism used to control the on and off times of an electronic current.

Tinned Wire. Copper wire that has been coated with a layer of tin or solder during manufacture to simplify soldering.

Tolerance. Amount of variation or change allowed from a standard accuracy; especially the difference between the allowable maximum and minimum sizes of some mechanical part.

Tone Control. A device provided in electronic sound equipment to alter the proposition of bass and treble frequency response.

T-Pad. A special type of potentiometer with equal input and output impedance. Also called *T-network*.

Trace. A visible line or lines appearing on the screen of a cathode ray tube in operation.

Transceiver. A combination transmitter-receiver in which a single set of tuning elements is used interchangeably for transmission and reception.

Transconductance. The ratio of the change in plate current to the change in grid voltage, producing this change in plate current while all other electrode voltages remain constant.

Transducer. A device converting input into a different type of output.

Transformer. A device composed of two or more coils, linked by magnetic lines of force, and used to transfer energy from one circuit to another.

Transient. The voltage or current which exists as the result of a change from one steady-state condition to another.

Transient Oscillation. A momentary oscillation occuring in a circuit during switching.

Transistor. A compact unit consisting of semiconducting material.

Transmission. Transfer of electric energy from one location to another through conductors or by radiation or induction fields.

Transmission Lines. Any conductor or system of conductors used to carry electrical energy from its source to a load.

Transmission Loss. A term used to donate a loss in power during the transmission of energy from one point to another.

Transmitter. A term applying to the equipment used for generating an RF carrier signal, modulating this carrier with intelligence and radiating the modulated RF carrier into space.

Transpose Polarity. To interchange the connections to a pair of poles.

Trap. Tuned circuit used to eliminate a given signal or to keep it out of a given circuit. A common trap is simply a tuned circuit which absorbs the energy of the signal to be eliminated.

Triggering. Starting an action in another circuit which then functions for a time under its own control.

Trimmer Capacitor. A small, adjustable capacitor, used in the tuning circuits of radio receivers and other radio apparatus.

Triode. A three-electrode vacuum tube, containing a cathode, control grid, and plate.

Tubular Capacitor. A paper or electrolytic capacitor having the form of a cylinder, with leads projecting axially from one or both ends.

Tuned Antenna. An antenna designed to provide resonance at the desired operating frequency by means of its own inductance and capacitance.

Tuned Circuit. A resonant circuit.

Tuned Filter. An arrangement of electronic components tuned either to attenuate or pass signals at its resonant frequency.

Tuned-Grid Tuned-Plate Oscillator. A vacuum-tube oscillator with tuned grid and plate circuits. Maximum oscillation depends on maximum feedback, which occurs when the grid and plate circuits are tuned to resonance.

Tuned Radio-Frequency Amplifier. An amplifier employing vacuum tubes or transistors and tuned circuits for the purpose of amplifying radio-frequency energy.

Tuned Radio-Frequency Stage. A stage of amplification which is tunable to the radio frequency of the signal being received.

Tuning. The process of adjusting a radio circuit so that it resonates at the desired frequency.

Tuning Capacitor. A variable capacitor.

Tuning Coil. A variable inductor.

Tuning Meter. A DC meter connected to show when the receiver is accurately tuned to a desired frequency.

Tunnel Diode. A diode that exhibits negative resistance due to tunneling through a thin depletion layer.

Tunneling. In quantum mechanics, a process that explains how charge carriers can penetrate insulating regions that are sufficiently thin.

Turns per Volt Ratio. The number of turns for each volt in a transformer winding.

Twin Line. A type of transmission line which has a solid insulating material, in which the two conductors are placed parallel to each other. Several impedance values are in common use ($75\,\Omega$, $150\,\Omega$, and $300\,\Omega$).

Twisted Pair. A cable composed of two insulated conductors twisted together either with or without a common covering.

Ultrahigh Frequency. A Federal Communications Commission designation for the frequency band from 300MHz to 3,000MHz.

Unbalanced Line. A transmission line in which the voltages on the two conductors are not equal with respect to ground; for example, a coaxial line.

Unidirectional. In one direction only.

Universal Motor. A series motor which operates on either alternating or direct current.

Unloaded. In reference to a source, the source is unloaded when no load is connected across the terminals and no power is supplied to the load.

Vacuum. An enclosed space from which practically all air has been removed.

Vacuum Capacitor. A type of capacitor having tubular elements which are housed in an evacuated glass envelope. Vacuum capacitors are characterized by extremely high breakdown voltage.

Vacuum Switch. A switch enclosed in an evacuated bulb.

Vacuum Tube. Specifically, an evacuated enclosure including two or more electrodes between which conduction through the vacuum may take place. A general term used for all electronic tubes.

Vacuum-Tube Voltmeter. A device which uses either the amplifier characteristic or the rectifier characteristic of a vacuum tube or both to measure either DC or AC voltages. Its input impedance is very high, and the current used to actuate the meter movement is not taken from the circuit being measured. It can be used to obtain accurate measurements in sensitive circuits.

Valence Electrons. Electrons in the atom that are responsible for bonding between chemical elements. In metals, such as electric conductors, the valence electrons are often free electrons.

Varactor Diode. A diode making use of the variation of capacitance that takes place as reverse bias is varied. Can be used as a frequency multiplier, as a tuning element in a tuned circuit, or as a low-noise parametric amplifier.

Variable-Mu Tube. A remote cutoff tube. A vacuum tube with a grid designed so that the amplification factor and the mutual conductance are variable.

Variable Transformer. A transformer whose output voltage can be varied continuously.

Variable-U Tube. A vacuum tube in which the control grid is irregularly spaced, so that the grid exercises a different amount of control on the electron stream at different points within its operating range.

Varicoupler. Two independent inductors arranged mechanically so that their mutual inductance (coupling) can be varied.

Variometer. A varicoupler having its two coils connected in series and so mounted that the movable coil may be rotated within the fixed coil, thus changing the total inductance of the unit.

Velocity Modulation. A method of modulation in which the input signal voltage is used to change the velocity of electrons in a constant current electron beam so that the electrons are grouped into bunches.

Vernier. An auxiliary scale of slightly smaller divisions than the main measuring scale, permitting measurements with greater precision than allowed by the main scale.

Vertical Deflecting Electrodes. The pair of electrodes that serves to move the electron beam up and down on the fluorescent screen of a cathode ray tube employing electrostatic deflection.

Vertically Polarized Wave. A wave whose direction of electric polarization is perpendicular to the earth.

Vertical Polarization. The condition in which radio waves are transmitted with their plane of electric polarization initially perpendicular to the surface of the earth.

Very High Frequencies. A band of frequencies in the radio spectrum extending from 30MHz to 300MHz. In television, Channels 2 to 13, or 54MHz to 216MHz.

Very Low Frequencies. A band of frequencies in the radio spectrum extending from 10kHz to 30kHz.

Vibrating-Reed Meter. A meter used to measure frequency, especially the low frequencies used in power systems.

Video Amplifier. A circuit capable of amplifying a very wide range of frequencies, including and exceeding the audio band of frequencies.

Volt. The unit of electric potential. The voltage of a circuit is the greatest root-mean-square (effective difference of potential) between any two conductors.

Voltage Amplification. The process of amplifying a signal to produce a gain in voltage. The voltage gain of an amplifier is the ratio of its alternating voltage output to its alternating-voltage input.

Voltage Divider. An impedance connected across a voltage source. The load is connected across a fraction of this impedance so that the load voltage is substantially in proportion to this fraction.

Voltage Doubler. A method of increasing the voltage by rectifying both halves of a cycle and causing the outputs of both halves to be additive.

Voltage Drop. The voltage measured across a resistance. It is equal to the product of the current in amperes times the resistance in ohms.

Voltage Feed. Excitation of a transmitting antenna by applying voltage at a voltage loop or antinode.

Voltage Gain. Voltage amplification.

Voltage Multiplier. A precision resistor used in series with a voltmeter to extend its measuring range.

Voltage Node. A point having zero voltage in a system of stationary waves.

Voltage Rating. The maximum sustained voltage that can safely be applied to or taken from a device without risking damage.

Voltage Regulation. A measure of the degree to which a power source maintains its output-voltage stability under varying load conditions.

Voltage-Regulator Tube. A gas-filled electron tube used to keep voltage essentially constant despite wide variations in line voltage or to maintain an essentially constant direct voltage in a circuit.

Voltage to Ground. In grounded circuits the voltage between the given conductor and that point or conductor of the circuit which is grounded; in ungrounded circuits, the greatest voltage between the given conductor and any other conductor of the circuit.

Voltmeter. An instrument for measuring voltage.

Voltmeter Sensitivity. Ohms-per-volt rating of voltmeter. Numerically equal to reciprocal of full-scale current of the meter movement used in voltmeter.

Volt-Ohmmeter. A test instrument having provisions for measuring voltage, resistance, and current. Abbreviated VOM.

Volume. The intensity or loudness of the sound produced by a headphone or loudspeaker.

Volume Control. A potentiometer used to vary the audio-frequency output of an audio amplifier.

Watt. The practical unit of electric power. In a DC circuit, equal to volts multiplied by amperes. In an AC circuit, true watts are equal to effective volts multiplied by effective amperes, then multiplied by the circuit power factor.

Wattage Rating. A rating expressing the maximum power which a device or component can safely absorb or handle.

Watt-Hour. Unit of energy (3,600 joules).

Wattmeter. A meter used to measure power in an electric circuit in watts.

Watt-Second. Unit of energy (1 joule).

Wave. Loosely, an electromagnetic impulse periodically changing in intensity and traveling through space. More specifically, the graphical representation of the intensity of the impulse over a period of time.

Wave Band. A band of assigned frequencies.

Waveform. The shape of the wave obtained when instantaneous values of an AC quantity are plotted against time in rectangular coordinates.

Waveguide. An electric conductor consisting of a metal tubing used for the conduction or directional transmission of microwaves.

Wavelength. The distance, usually expressed in meters, traveled by a wave during the time interval of one complete cycle. It is equal to the velocity divided by the frequency.

Wavemeter. A calibrated variable-frequency resonator used to determine wavelengths of radio waves.

Wave Propagation. The transmission of RF energy through space.

Wave Trap. A device sometimes connected to the aerial system of a radio receiver to reduce the strength of signals at a particular frequency.

Weak Coupling. Loose coupling in a transformer.

Wet Cell. A cell in which the electrolyte is in liquid form.

Wet Electrolyte Capacitor. A capacitor employing a liquid electrolyte dielectric.

Wheatstone Bridge. Circuit configuration used to measure electrical qualities such as resistance.

Wien-Bridge Circuit. A circuit in which the various values of capacitance and resistance are made to balance with each other at a certain frequency.

Wiring Harness. A bundle of individually insulated wires.

Working Voltage. The maximum voltage that can be continuously applied to a capacitor without the danger of arcing between the plates.

Wye Connection. Connecting the phases of a three-phase system at a common point so that the line and phase currents are equal.

Zero Adjuster. A device for bringing the pointer of an electric instrument or meter to zero when the electric quantity is zero.

Zero Beat. The condition of a receiver in which an internal oscillator is at the exact frequency of an external radio wave so that no beat tone is produced or heard when the two are mixed.

Zero Bias. A condition in which the control grid and cathode of a vacuum tube are at the same potential.

Zero Potential. An expression usually applied to the potential of the earth, as a convenient reference for comparison.

Glossary of Semiconductor Terms

Acceptor. An impurity that can make a P-type semiconductor by accepting valence electrons, thereby leaving "holes" in the valence band. The holes act as carriers of positive charges.

Acceptor Impurity. A substance with three electrons in the outer orbit of its atom which, when added to a semiconductor crystal, provides one hole in the lattice structure of the crystal.

Active Elements. Those components in a circuit which have gain or which direct current flow—diodes and transistors. Also called *active component*.

Actuator. A device which converts a voltage or current input into a mechanical output.

A/D Converter. Analog-to-digital converter. A circuit which converts an analog (continuous) voltage or current into an output digital code.

Adder. Switching circuits which generate sum and carry bits.

Alloyed Junction. A PN junction formed by recrystallization of a molten region of P-type material on an N-type substrate, or vice versa.

Alpha. The ratio of collector to emitter current in a transistor.

Alpha Cutoff. The frequency at which alpha drops to 0.707 of its low-frequency value.

Alpha Particle. A particle consisting of two protons and two neutrons.

Analog. Representation of one quantity by means of another quantity proportional to the first.

Analog Multiplexer. An array of switches with a common output connection for selecting one of a number of analog inputs. The output signal follows the selected input within a small error.

AND. A Boolean logic operator analogous to multiplication. Of two variables, both must be true for the output to be true.

AND Circuit (AND Gate). A coincidence circuit that functions as a gate so that when all the inputs are applied simultaneously, a prescribed output condition exists.

AND-OR Circuit (AND-OR Gate). A gating circuit that produces a prescribed output condition when several possible combined input signals are applied; it exhibits the characteristics of the AND gate and the OR gate.

Astable Multivibrator. A multivibrator that can function in either of two semistable states switching rapidly from one to the other. Also called *free running*.

Asynchronous. Operation of a switching network by a free-running signal which triggers successive instructions; the completion of one instruction triggers the next.

Avalanche Diode. A diode that conducts current only above a certain "breakdown" voltage, by virtue of high-field impact ionization (avalanche). Useful for voltage-limiting or microwave generation.

Avalanche Multiplication. A high-field effect in semiconductors which leads to an increase in current. High-energy charge carriers create additional carriers by impact ionization of valence electrons.

Back Diode. A tunnel diode with very low peak tunnel current, used as a low-voltage rectifier conductive in the reverse direction.

Barrier. In a semiconductor, the electric field between the acceptor ions and the donor ions at the junction.

Barrier Height. In a semiconductor, the difference in potential from one side of a barrier to the other.

Base (Junction Transistor). The center semiconductor region of a double junction (NPN or PNP) transistor. The base is comparable to the grid of an electron tube.

Base Spreading Resistance. In a transistor, the resistance of the base region caused by the resistance of the bulk material of the base region.

Baud. In general, a unit of signaling speed. In data processing, it is a group of tracks on a magnetic drum or on a side of a magnetic disk. In data communication, the band is the frequency spectrum between two defined limits.

Beam Lead (for integrated circuits). A deposited metal lead, usually of gold, which projects beyond the edge of the semiconductor chip. Used for both mechanical and electrical contact to the chip.

Beta. Ratio of transistor collector current to base current.

Bifilar. Two wires or filaments wound side by side.

Bifurcated Contact. A contact having two fingers for redundant contact.

BIGFET (Bipolar Insulated Gate Field-Effect Transistor). A simple integrated circuit combining a field-effect transistor and a bipolar transistor.

Bilateral. Conducts current in both directions; conduction of current in one direction is termed unilateral.

Binary. The number system consisting of only the digits 0 and 1.

Bipolar Mode. For a data converter, when the analog signal range includes both positive and negative values.

Bipolar Offset. The analog displacement of one-half of a full scale range in a data converter operated in the bipolar mode. The offset is generally derived from the converter reference circuit.

Bipolar Transistor. A transistor consisting of an emitter, base, and collector whose action depends on the injection of minority carriers into the base by the emitter and the collection of these minority carriers from the base by the collector. Sometimes called NPN or PNP transistor to emphasize its layered structure.

Bit. Binary digit equal to 0 or 1.

Boron. The P-type dopant commonly used for the isolation and base diffusion in standard bipolar integrated circuit processing.

Buffer. A noninverting member of the digital family which may be used to handle a large fan-out or to convert input and output levels. Normally a buffer is an emitter-follower type of circuit.

Bus. One or more conductors used for transmitting signals or power. These buses include a microprocessor interface bus, address bus, data bus, and control bus.

Byte. An eight-bit binary number.

CCD (Charge Coupled Device). A semiconductor device whose action depends on the storage of electric charge within a semiconductor by an insulated electrode on its surface, with the possibility of selectively moving the charge to another electrode by proper manipulation of voltages on the electrode.

CDI (Collector Diffusion Isolation). A process for fabricating integrated circuits whereby the collector diffusion also electrically isolates the transistors from one another.

Channel. A region of surface conduction opposite in type from that expected from the bulk doping. Channels are often introduced intentionally by charging external electrodes or unintentionally by surface ionic contamination.

Character. A member of a set of elements for which agreement on meaning has been reached and that is used for the organization, control, or representation of data. Characters may be letters, digits, punctuation marks, or other symbols.

Clamping Circuit. A circuit that maintains either or both amplitude extremities of a waveform at a certain level or potential.

Clear. To restore a memory or storage device to a "standard" state, usually the "zero" state. Also called *reset*.

Clock. A pulse generator which controls the timing of switching circuits and memory states and equals the speed at which the major portion of the computer operates.

Clock Rate. The frequency of the timing pulses of the clock circuit in an A/D converter.

CML (Current-Mode Logic). Operates in the unsaturated mode as distinguished from all the other forms which operate in the saturated mode.

CMOS (Complementary Metal-Oxide Semiconductor). A low-power logic family using field-effect transistors.

Collector. The end semiconductor material of a double junction (NPN or PNP) transistor that is normally reverse biased with respect to the base. The collector is comparable to the plate of an electron tube.

Common-Base (CB) Amplifier. A transistor amplifier in which the base element is common to the input and the output circuit. This configuration is comparable to the grounded-grid triode electron tube.

Common-Collector (CC) Amplifier. A transistor amplifier in which the collector element is common to the input and the output circuit. This configuration is comparable to the electron tube cathode follower.

Common-Emitter (CE) Amplifier. A transistor amplifier in which the emitter element is common to the input and the output circuit. This configuration is comparable to the conventional electron tube amplifier.

Complementary Binary Code. A binary code which is the logical complement of straight binary. All 1's become 0's and vice versa.

Complementary-Symmetry Circuit. An arrangement of PNP-type and NPN-type transistors that provides push-pull operation from one input signal.

Compound-Connected Transistor. A combination of two transistors to increase the current amplification factor at high emitter currents. This combination is generally employed in power amplifier circuits.

Constant Power Dissipation Line. A line (superimposed on the output static characteristic curves) representing the points of collector voltage and current, the products of which represent the maximum collector power rating of a particular transistor.

Contamination. A general term used to describe unwanted material that adversely affects the physical or electrical characteristics of a semiconductor wafer.

Conversion Rate. The number of repetitive A/D or D/A conversions per second for a full scale change to a specified resolution and linearity.

Conversion Time. The time required for an A/D converter to complete a single conversion to a specified resolution and linearity for a full scale analog input change.

Counter. A device capable of changing states in a specified sequence upon receiving appropriate input signals; a circuit which provides an output pulse or other indication after receiving a specified number of input pulses.

Counter, Binary. A flip-flop having a single input. Each time a pulse appears at the input, the flip-flop changes state. Also called *"T" flip-flop*.

Counter, Ring. A loop or circuit of interconnected flip-flops so arranged that only one is "on" at any given time and that, as input signals are received, the position of the "on" state moves in sequence from one flip-flop to another around the loop.

Counter Type A/D Converter. A feedback method of A/D conversion whereby a digital counter drives a D/A converter which generates an output ramp which is compared with the analog input. When the two are equal, a comparator stops the counter and output data is ready. Also called *servo type A/D converter*.

Cross-Over Distortion. Distortion that occurs at the points of operation in a push-pull amplifier where the input signals cross (go through) the zero reference points.

Cross Talk. In an analog multiplexer, the ratio of output voltage to input voltage with all channels connected in parallel and "off." It is generally expressed as an input to output attenuation ratio in dB.

Cryogen. Device which becomes a superconductor at extremely cold temperatures.

Current Stability Factor. In a transistor, the ratio of a change in emitter current to a change in reverse-bias current flow between the collector and the base.

D/A Converter. Digital-to-analog converter. A circuit which converts a digital code word into an output analog (continuous) voltage or current.

Darlington Pair. A circuit, often integrated, consisting of two bipolar transistors with the collectors connected together and the emitter of one connected to the base of the other. The combination behaves like a very high gain transistor.

Data. Any representation of information such as characters or analog quantities to which meaning is or might be assigned.

Data Acquisition System. A system consisting of analog multiplexers, sample-holds, A/D converters, and other circuits which process one or more analog signals and convert them into digital form for use by a computer.

Data Communications. The transmission and reception of data via an electrical transmission system.

Data Distribution System. A system which uses D/A converters and other circuits to convert the digital outputs of a computer into analog form for control of a process or system.

Data Processing. The systematic performance of operations upon data, for example handling, emerging, sorting, and computing.

DCTL (Direct-Coupled Transistor Logic). Logic is performed by transistors.

Decimal. A system of numerical representation which uses ten symbols 0,1,2,3,...9.

Deglitched DAC. A D/A converter which incorporates a deglitching circuit to virtually eliminate output spikes (or glitches). These DACs are commonly used in CRT display systems.

Delay. Undesirable delay effects are caused by rise time and fall time which reduce circuit speed, but intentional delay units may be used to prevent inputs from changing while clock pulses are present. The delay time is always less than the clock pulse interval.

Dependent Variable. In a transistor, one of four variable currents and voltages that is arbitrarily chosen and considered to vary in accordance with other currents and voltages. (An independent variable.)

Depletion Layer. The region in a semiconductor where essentially all charge carriers have been swept out by the electric field which exists there.

Depletion Region (or Layer). The region in a semiconductor containing the uncompensated acceptor and donor ions. Also called *space-charge region* or *barrier region*.

Depletion Zone. A portion of a semiconductor near the junction void of charge carriers.

Derating. Of the maximum performance rating of a device or electric system to ensure long-term operation. The reduction in performance rating introduces a margin of safety in the design.

Differentiating Circuit. A circuit that produces an output voltage proportional to the rate of change of the input voltage.

Diffusion. Spontaneous intermingling of substances, as, for example, the diffusion of an impurity into a semiconductor at high temperature to create a desired concentration of N or P charge carriers.

Digital. A system in which characters or codes are used to represent numbers or physical quantities in discrete steps.

Digital Circuit. A circuit which operates like a switch, that is, it is either "on" or "off."

Digitizer. A device which converts analog into digital data; an A/D converter.

Diode. A two-terminal device that allows current to flow in one direction but not in the other. A diode is present at the intersection of a P-type and an N-type region of a semiconductor.

Disconnecting Means. A device, a group of devices, or other means whereby the conductors of a circuit can be disconnected from their source of supply.

Discrete. Electronic circuits built of separate, finished components.

Documentation. Instructions, notes, and diagrams prepared to assist in understanding and use of an instrument or computer program.

Donor. An impurity that can make a semiconductor N type by donating extra "free" electrons to the conduction band. The free electrons are carriers of negative charge.

Donor Impurity. A substance with electrons in the outer orbit of its atom; added to a semiconductor crystal, it provides one free electron.

Dopant. An element that alters the conductivity of a semiconductor by contributing either a hole or an electron to the conducting process.

Doping. The introduction of an impurity into the crystal lattice of a semiconductor to modify its electronic properties—for example, adding boron to silicon to make the material more P type.

Double-Diffused Transistor. A transistor in which two PN junctions are formed by alternately diffusing P- and N-type impurities into an N-type semiconductor.

Double-Junction Photosensitive Semiconductor. Three layers of semiconductor material with an electrode connection to each end layer. Light energy is used to control current flow.

Double-Level Multiplexing. A method of channel expansion in analog multiplexers whereby the outputs of a group of multiplexers connect to the inputs of another multiplexer.

Drain. Along with the source and gate, one of the three regions of a unipolar or field-effect transistor.

DTL (Diode-Transistor Logic). Logic is performed by diodes. The transistor acts as an amplifier and the output is inverted.

Dynamic Transfer Characteristic Curve. In transistors, a curve that shows the variation of output current (dependent variable) with the variation of an input current under load conditions.

Dynatron. A negative resistance device; particularly, a tetrode operating on that portion of its i_p versus v_p characteristic where secondary emission exists to such an extent that an increase in plate voltage actually causes a decrease in plate current and, therefore, makes the circuit behave like a negative resistance.

ECL (Emitter-Coupled Logic). A high speed logic family. Also called *current-mode logic* (CML).

Emitter. Along the base and collector, one of the three regions of the bipolar type of transistor that emits electrons.

Emitter (Junction Transistor). The end semiconductor material of a double junction (PNP or NPN) transistor that is forward biased with respect to the base. The emitter is comparable to the cathode of an electron tube.

Energy Gap (or a semiconductor). The region, according to quantum mechanics, of forbidden electron energies. The gap lies between the energy band for valence electrons and the energy band for conduction electrons.

Enhancement-Mode FET. An insulated-gate field-effect transistor that is nonconductive at zero gate-source voltage and turns on with forward gate bias.

Epitaxial. A very significant thin-film type of deposition for making certain devices in microcircuits, involving a realignment of molecules.

Epitaxial Growth. A chemical reaction in which silicon is precipitated from a gaseous solution and grows upon the surface of a silicon wafer present in the gaseous solution.

Epitaxial Layer. A deposited layer of material having the same crystallographic characteristics as the substrate material.

Exclusive "or." The output is true if either of two variables is true, but not if both are true.

Extrinsic Conductivity. Electrical conductivity dependent on the impurities in the material. Also called *extrinsic semiconductor.*

Fall Time. A measure of the time required for a circuit to change its output from a high level (1) to a low level (0).

Fan-In. The number of inputs available on a gate.

Fan-Out. The number of gates that a given gate can drive. The term is applicable only within a given logic family.

Feedback Type A/D Converter. A class of analog-to-digital converters in which a D/A converter is enclosed in the feedback loop of a digital control circuit which changes the D/A output until it equals the analog input.

FET (Field-Effect Transistor). A transistor consisting of a source, gate, and drain, whose action depends on the flow of majority carriers past the gate from source to drain. The flow is controlled by the transverse electric field under the gate.

First-Order Hold. A type of sample-hold, used as a recovery filter, which utilizes the present and previous analog samples to predict the slope to the next sample. Also called *extrapolative hold.*

Flip-Flop. An electronic circuit having two stable states and having the ability to change from one state to the other upon the application of a signal in a specified manner.

Flip-Flop "D." D stands for delay. A flip-flop whose output is a function of the input which appeared one pulse earlier, that is, if a 1 appears at its input, the output a pulse later will be a 1.

Flip-Flop "J-K." A flip-flop having two inputs designated J and K. At the application of a clock pulse, a 1 on the J input will set the flip-flop to the 1 or "on" state; a 1 on the K input will reset it to the 0 or "off" state, and 1's simultaneously on both inputs will cause it to change state regardless of what state it had been in.

Flip-Flop "R-S." A flip-flop having two inputs designated R and S. At the application of a clock pulse, a 1 on the S input will set the flip-flop to the 1 or "on" state, and a 1 on the R input will reset it to the 0 or "off" state. It is assumed that 1's will never appear simultaneously at both inputs.

Flip-Flop "R-S-T." A flip-flop having three inputs, R, S, and T. The R and S inputs produce states as described for the R-S flip-flop above; the T causes the flip-flop to change states.

Flip-Flop "T." A flip-flop having only one input. A pulse appearing on the input will cause the flip-flop to change states.

Forward Bias. In a transistor, an external potential applied to a PN junction so that the depletion region is narrowed and relatively high current flows through the junction.

703

Forward Short-Circuit Current Amplification Factor. In a transistor, the ratio of incremental values of output to input current when the output circuit is AC short-circuited

Four-Layer Diode. A diode constructed of semiconductor materials resulting in three PN junction. Electrode connections are made to each end layer.

Frequency Folding. In the recovery of sampled data, the overlap of adjacent spectra caused by insufficient sampling rate. The overlapping results in distortion in the recovered signal which cannot be eliminated by filtering the recovered signal.

Gate. A circuit having two or more inputs and one output, the output depending upon the combination of logic signals at the inputs. There are four gates which are called: AND, OR, NAND, NOR.

Gate, AND. All inputs must have 1-state signals to produce a 1-state output.

Gate, NAND. All inputs must have 1-state signals to produce a 0-state output.

Gate, NOR. Any one input or more having a 1-state signal will yield a 0-state output.

Gate, OR. Any one input or more having a 1-state signal is sufficient to produce a 1-state output.

Gate-Turnoff Switch. A switching device similar to an SCR but able to be turned off from its gate terminal.

Gating Circuit. A circuit operating as a switch, making use of a short or open circuit to apply or eliminate a signal.

Getter. A process by which unwanted impurities are segregated out of a material such as silicon or silicon dioxide.

Growler Junction. PN junction made by controlling the type of impurity in a single crystal while it is being grown from a melt.

Gunn Diode. A diode that produces gigahertz oscillations when DC biased at the proper voltage.

HIC (Hybrid Integrated Circuit). Consists of an assembly of one or more semiconductor devices and a thin-film integrated circuit on a single substrate, usually of ceramic.

High-Level Multiplexing. An analog multiplexing circuit in which the analog signal is first amplified to a higher level (1V to 10V) and then multiplexed. This is the preferred method of multiplexing to prevent noise contamination of the analog signal.

Hold-Mode. The operating mode of a sample-hold circuit in which the sampling switch is open.

Hole. A mobile vacancy in the electronic valence structure of a semiconductor. The hole acts similarly to a positive electronic charge having a positive mass.

Hybrid. A method of manufacturing integrated circuits by using a combination of monolithic and thin-film methods.

Hybrid Parameter. The parameters of an equivalent circuit of a transistor which are the result of selecting the input current and the output voltage as independent variables.

IC (Integrated Circuit). A circuit in which many elements are fabricated and interconnected by a single process, as opposed to a "nonintegrated" circuit in which the transistors, diodes, resistors, etc., are fabricated separately and then assembled.

IGFET. A field-effect transistor whose gate is insulated from the semiconductor by a thin intervening layer of insulator, usually thermal oxide.

I²L (Integrated Injection Logic). A direct-coupled bipolar-transistor logic family presently used primarily in large-scale specialized integrated circuits.

IMPATT Diode (Impact Avalanche and Transit Time Diode). An avalanche diode used as a high-frequency oscillator or amplifier. Its negative resistance depends upon the transit time of charge carriers through the depletion layer.

Increment. A small change in value.

Independent Variable. In a transistor, one of several voltages and currents chosen arbitrarily and considered to vary independently.

Indirect Type A/D Converter. A class of analog-to-digital converters which converts the unknown input voltage into a time period and then measures this period.

Inhibition Gate. A gate circuit used as a switch and placed in parallel with the circuit it is controlling.

Injection. In a semiconductor, the introduction of excess minority carriers. Injection can take place at a conducting PN junction or in a region illuminated by light.

Integral Linearity Error. The maximum deviation of a data converter transfer function from the ideal straight line with offset and gain errors zeroed. It is generally expressed in LSB's or in percent of FSR.

Integrated Circuit. The electronic Industries Association defines the semiconductor integrated circuit as "the physical realization of a number of electrical elements inseparably associated on or within a continuous body of semiconductor material to perform the functions of a circuit."

Integrating A/D Converter. One of several types of A/D conversion techniques whereby the analog input is integrated with time. This includes dual slope, triple slope, and charge balancing type A/D converters.

Interelement Capacitance. The capacitance caused by the PN junctions between the regions of a transistor; measured between the external leads of the transistor.

Interface. A shared boundary. As an example, the interface can be the boundary that links two devices.

Intrinsic Conductivity. Electrical conductivity of an apparently pure material—i.e., without the presence of any significant impurities. Also called *intrinsic semiconductor*.

Inverter. The output is always in the opposite logic state of the input. Also called a *NOT circuit*.

Isolation Mask. The second mask in standard bipolar integrated circuit fabrication. Boron diffuses into silicon in regions etched during the isolation photoresist process and electrically separates or isolates regions of silicon.

JFET (Junction Field-Effect Transistor). An FET in which the gate electrode is formed by a PN junction.

Junction. The interface between two semiconductor regions of differing conductivity. Usually refers to a PN junction, at which the conductivity changes from P-type to N-type.

Junction Transistor. A bipolar transistor constructed from interacting PN junctions. The term is used to distinguish junction transistors from other types, such as field-effect and point-contact transistors.

LASER (Light Amplification by Stimulated Emission of Radiation). In the laser, excited electrons give up their excitation in step with the light that is passing by to add energy to the transmitted light. Some lasers can generate or amplify extremely pure colors, in very narrow beams often with very high intensity.

Lattice Structure. In a crystal, a stable arrangement of atoms and their electron-pair bonds.

LED (Light-Emitting Diode). A semiconductor device in which the energy of minority carriers in combining with holes is converted to light. Usually, but not necessarily, constructed as PN junction device.

Lifetime. Term used to describe the life of a minority charge carrier in a semiconductor crystal—for example, how long a free electron exists in a P-type material before it combines with a hole and is neutralized.

Linear Circuit. A circuit whose output is an amplified version of its input or whose output is a predetermined variation of its input.

LPE (Liquid Phase Epitaxy). The formation of an epitaxial layer by placing the substrate in contact with a molten liquid and allowing crystallization onto the substrate to take place.

LSA Diode (Limited Space-Charge Accumulation Diode). A microwave oscillator diode similar to the Gunn diode but attaining higher power or higher frequency by avoiding the formation of "domains" or regions of nonuniform distribution of charge.

Majority Carriers. The holes or free electrons in P-type or N-type semiconductors respectively.

Major Transition. In a data converter, the change from a code of 1000...000 to 0111...111 or vice-versa. This transition is the most difficult one to make from a linearity standpoint since the MSB weight must ideally be precisely one LSB larger than the sum of all other bit weights.

Mesa. A device structure fabricated by selective etching which leaves flat portions of the original surface ("mesas") projecting above the neighboring regions. The mesa technique is often used to limit the extent of the electronically active material to the area of the mesa.

Minority Carriers. The holes or excess electrons found in N-type or P-type semiconductors respectively.

Modem. A device that modulates and then demodulates signals transmitted over communication facilities.

Monobrid. A method of manufacturing integrated circuits by using more than one monolithic chip within the same package.

Monolith. A semiconductor chip containing a multiplicity of devices interconnected into an integrated circuit. The term is contrasted with hybrid circuitry.

Monolithic Integrated Circuit. An integrated circuit fabricated from a single chip of semiconductor material, usually silicon.

Monostable Multivibrator. A multivibrator having one stable and one semistable condition. A trigger is used to drive the unit into the semistable state where it remains for a predetermined time before returning to the stable condition.

MOSFET. A field-effect transistor containing a metal gate over thermal oxide over silicon. The MOSFET structure is one way to make an IGFET.

MSB (Most Significant Bit). The leftmost bit in a data converter code. It has the largest weight, equal to one-half of full scale range.

Multivibrator. A type of relaxation oscillator for the generation of nonsinusoidal waves in which the output of each of two stages is coupled to the input of the other to sustain oscillations.

Negative Resist. Photoresist that remains in areas that were not protected from exposure by the opaque regions of a mask while being removed in regions that were protected by the develop cycle. A negative image of the mask remains following the develop process. Waycoat and microneg are two common negative resists.

Negative True Logic. A logic system in which the more negative of two voltage levels is defined as a logical 1 (true) and the more positive level is defined as a logical 0 (false).

Neutralization. The prevention of the oscillation of an amplifier by canceling possible changes in the reactive component of the input circuit caused by positive feedback.

Noise Rejection. The amount of suppression of normal mode analog input noise of an A/D converter or other circuit, generally expressed in dB. Good noise rejection is a characteristic of integrating type A/D converters.

Nonmonotonic. A D/A converter transfer characteristic in which the output does not continuously increase with increasing input. At one or more points there may be a dip in the output function.

NOR Circuit. An OR gating circuit that provides pulse-phase inversion.

Normal Mode Rejection. The attenuation of a specific frequency or band of frequencies appearing directly across two electrical terminals. In A/D converters, normal-mode rejection is determined by an input filter or by integration of the input signal.

NOT. A Boolean logic operator indicating negation. A variable designated "not" will be the opposite of its "and" or "or" function. A switching function for only one variable.

NOT AND Circuit. An AND gating circuit that provides pulse-phase inversion.

NPN. A semiconductor structure consisting of a layer of P-type material sandwiched between layers of N-type material, as commonly used in the bipolar type of transistor.

NPN Transistor. A device consisting of a P-type section and two N-type sections of semiconductor material with the P-type in the center.

N-Type. Semiconductor material in which the majority carriers are electrons and are therefore negative.

N-Type Semiconductor. A semiconductor into which a donor impurity has been introduced. It contains free electrons.

Nybble. One-half byte; four binary digits.

Offset Binary Code. Natural binary code in which the code word 000....0000 is displaced by one-half analog full scale. The code represents analog values between –FS and +FS (full scale). The code word 1000....0000 then corresponds to analog zero.

Ohmic Contact. Contact to a semiconductor or other part of a device having low electrical resistance and not showing rectifying behavior.

Open Circuit Parameters. The parameters of an equivalent circuit of a transistor which are the result of selecting the input current and output current as independent variables.

OR. A Boolean operator analogous to addition (except that the two truths will only add up to one truth). Of two variables, only one need be true for the output to be true.

OR Circuit (OR Gate). A gate circuit that produces the desired output with only one of several possible input signals applied.

Oxide Masking. Use of an oxide on a semiconductor to create a pattern in which impurities are diffused or implanted.

OXIM (Oxide-Isolated Monolith). A method of making integrated circuits in which an oxide layer is introduced to insulate semiconductor regions from each other.

Parallel Operation. Pertaining to the manipulation of information within computer circuitry in which the digits of a word are transmitted simultaneously on separate lines. It is faster than serial operation, but requires more equipment.

Parallel Type D/A Converter. The most commonly used type of D/A converter in which upon application of an input code, all bits change simultaneously to produce a new output.

Parameter. A derived or measured value which conveniently expresses performance; used in calculations.

Parity. A binary digit added to a word to make the sum of all bits always 1 or always 0. Used to catch machine errors.

Passivation. A layer of a material put over an integrated circuit to stabilize the surface of its devices. Silicon dioxides or silicon nitride are often used for passivation.

Passive Elements. Those components in a circuit which have no gain characteristics, capacitors, resistors, or inductors.

Phosphorus. The N-type dopant commonly used for the sinker and emitter diffusions in standard bipolar integrated circuit technology.

Photodiode. An optically sensitive diode. Often the current is accurately proportional to the intensity of the incident light.

Photolithographic Process. Technique used in making integrated circuits which uses light and selective masking to develop a fine-scaled pattern of areas in a semiconductor; analogous to processes using photo negatives in offset printing.

Photoresist. The light-sensitive film spun onto wafers and "exposed" using high-intensity light through a mask. The "exposed" photoresist can be dissolved off of the wafer using developers leaving a pattern of photoresist which allows etching to take place in some areas while preventing it in others.

Photosensitive Semiconductor. A semiconductor material in which light energy controls current carrier movement.

PIN Diode. Diode constructed of an I-type (intrinsic) layer between P and N layers and used in high-speed or high-power microwave switching.

Planar Structure. A flat-surfaced-device structure fabricated by diffusion and oxide-masking, with the junctions terminating on a single plane. The structural planarity is often advantageous for photoresist processing.

PNIP. Semiconductor crystal structure consisting of layers that are P-type, N-type, Intrinsic, and P-type. Used for high-voltage, high-frequency bipolar transistors.

PN Junction. Within a crystal, an interface between a P region that conducts primarily by holes and an N region that conducts primarily by electrons.

PNP. Semiconductor crystal structure consisting of an N-type region sandwiched between two P-type regions, as commonly used in bipolar transistors.

PNPN. A semiconductor crystal structure in which the layers are successively P-type, N-type, P-type, and N-type. When ohmic contacts are made to the various layers, a silicon controlled rectifier (SCR), or thyristor, results.

PNP Resistor. A transistor with an emitter and collector of P-type silicon and a base of N-type silicon.

Positive Logic. The more positive voltage (or current level) represents the 1-state; the less positive level represents the 0-state.

Positive Resist. Photoresist that is removed in areas that were not protected from exposure by the opaque regions of a mask while remaining in regions that were protected by the develop cycle. A positive image of the mask remains following the develop process. AZ-1350 is a common positive resist.

Positive True Logic. A logic system in which the more positive of two voltage levels is defined as a logical 1 (true) and the more negative level is defined as a logical 0 (false).

Preamplifier. A low-level stage of amplification usually following a transducer.

Predeposition. The process step during which a controlled amount of a dopant is introduced into the crystal structure of a semiconductor.

Propagation Delay. A measure of the time required for a change in logic level to propagate through a chain of circuit elements.

Propagation Type A/D Converter. A type of A/D conversion method which employs one comparator per bit to achieve ultra-fast A/D conversion. The conversion propagates down the series of cascaded comparators.

P-Type. Semiconductor material in which the majority carriers are holes and are, therefore, positive.

Pulse Repetition Frequency. The number of nonsinusoidal cycles (square waves) that occur in 1 second.

Pulse Time. The length of time a pulse remains at its maximum value.

Quantilizer. A circuit which transforms a continuous analog signal into a set of discrete output states. Its transfer function is the familiar staircase function.

Quasi-Intrinsic. Descriptive of a semiconductor material whose conductivity is kept low, close to the intrinsic value, by doping with impurities which create carrier-traps lying near the center of the energy gap.

Quiescence. The operating condition that exists in a circuit when no input signal is applied to the circuit.

RCTL (Resistor-Capacitor Transistor Logic). Same as RTL except that capacitors are used to enhance switching speed.

Register. A device used to store a certain number of digits within the computer circuitry, often one word. Certain registers may also include provisions for shifting, circulating, or other operations.

Relative Accuracy. The worst case input to output error of a data converter, as a percent of full scale, referred to the converter reference. The error consists of offset, gain, and linearity components.

Resolution. The smallest change that can be distinguished by an A/D converter or produced by a D/A converter. Resolution may be stated in percent of full scale, but is commonly expressed as the number of bits, n, where the converter has 2^n possible states.

Reverse Bias. An external potential applied to a PN junction such as to widen the depletion region and prevent the movement of majority current carriers.

Reverse Open-Circuit Voltage Amplification Factor. In a transistor, the ratio of incremental values of input voltage to output voltage measured with the input AC open circuited.

Rise Time. The length of time during which the leading edge of a pulse increases from 10% to 90% of its maximum value.

RTL (Resistor-Transistor Logic). Logic is performed by resistors. The transistor produces an inverted output from any positive input.

Sample-Hold. A circuit which accurately acquires and stores an analog voltage on a capacitor for a specified period of time.

Sampler. An electronic switch which is turned "on" and "off" at a fast rate to produce a train of analog sample pulses.

Saturation (Leakage) Current. The current flow between the base and collector or between the emitter and collector measured with the emitter lead or the base lead, respectively, open.

SBC (Standard Buried Collector). A method of making integrated circuits in which diffused collector areas are "buried" by overlaying layers.

S/C (Silicon Integrated Circuit). An integrated circuit where all the elements such as transistors, diodes, resistors, and capacitors are successively fabricated in or on the silicon and interconnected.

Schottky Barrier. A potential barrier formed between a metal and a semiconductor. The term usually refers to a barrier which is high enough and thick enough to serve as a rectifier but which avoids the slowing-down effects that result from injection of charge in PN junction rectifiers.

SCR (Silicon Controlled Rectifier). A PNPN device useful for switching and power applications because it can have both high breakdown voltage and high current-carrying capability.

Sealed Junction. A PN junction sealed by covering it with an inert material which does not allow troublesome impurities to reach the junction and cause changes in its electrical characteristics.

Semiconductor. A conductor whose resistivity is between that of metals and insulators in which electrical charge carrier concentration increases with increasing temperature over a specific temperature change.

Serial Operation. Pertaining to the manipulation of information within computer circuitry, in which the digits of a word are transmitted one at a time along a single line. Though slower than parallel operation, its circuitry is considerably less complex.

Sheet Resistivity. A measurement with dimensions of ohms per square that tells the number of N-type or P-type donor atoms in a semiconductor.

Short-Circuit Parameters. The parameters of an equivalent circuit of a transistor which are the result of selecting the input and output voltages as independent variables.

Short Cycling. The termination of an A/D conversion process at a resolution less than the full resolution of the converter. This results in a shorter conversion time for reduced resolution in A/D converters with a short cycling capability.

Single-Junction Photosensitive Semiconductor. Two layers of semiconductor materials with an electrode connection to each material. Light energy controls current flow.

Single-Level Multiplexing. A method of channel expansion in analog multiplexers whereby several multiplexers are operated in parallel by connecting their outputs together. Each multiplexer is controlled by a digital enable input.

Single-Slope A/D Converter. A simple A/D converter technique in which a ramp voltage generated from a voltage reference and integrator is compared with the analog input voltage by a comparator. The time required for the ramp to equal the input is measured by a clock and counter to produce the digital output word.

Sinker. An n+ diffusion from the surface of a device to the buried layer. The sinker provides a low-resistance path from the collector contact to the buried layer. Phosphorus is the dopant commonly used for sinkers.

Skewing. Refers to the time delay or offset between any two signals.

Skewing Rate. Refers to rate at which output can be driven from limit to limit over the dynamic range.

Slew Rate. The maximum rate of change of the output of an operational amplifier or other circuit. Slew rate is limited by internal charging currents and capacitances and is generally expressed in volts per microsecond.

Solar Cell. Large-area diode in which a PN junction close to the surface of a semiconductor generates electrical energy from light falling on the surface.

Source. Along with the gate and drain, one of the three regions of the unipolar or field-effect transistor.

Spacistor. A semiconductor device consisting of one PN junction and four electrode connections characterized by a low transient time for carriers to flow from the input element to the output element.

Span. For an A/D or D/A converter, the full scale range or difference between maximum and minimum analog values.

Status Output. The logic output of an A/D converter which indicates whether the device is in the process of making a conversion or the conversion has been completed and output data is ready. Also called *busy output* or *end of conversing output*.

Storage Time. The time required to withdraw the minority carriers from both sides of a PN junction when the junction is switched from a forward to a reverse bias.

Substrate. The underlying material upon which a device, circuit, or epitaxial layer is fabricated.

Supergain Transistor. A transistor with a common emitter current gain of about a thousand or more, usually fabricated as a Darlington pair.

Surface States. Extra donors, acceptors, traps, usually undesired, which may occur on a semiconductor surface due to crystal imperfections or contamination and which may vary undesirably with time.

Susceptor. The flat slab of material (usually graphite) on which wafers are heated during high-temperature deposition processes like epitaxial growth or nitride deposition.

Swamping Resistor. In transistor circuits, a resistor placed in the emitter lead to mask (or minimize the effects of) variations in emitter-base junction resistance caused by variations in temperature.

Synchronous. Operation of a switching network by a clock pulse generator. It is slower and more critical than asynchronous timing but requires less and simpler circuitry.

Tetrode Transistor. A junction transistor with two electrode connections to the base (one to the emitter and one to the collector) to reduce the interelement capacitance.

Thermal Agitation. In a semiconductor, the random movement of holes and electrons within a crystal due to the thermal (heat) energy.

Thermal Oxide. In silicon semiconductor devices, an oxide fabricated by exposing the silicon to oxygen at high temperatures. The resulting interface is outstandingly free of ionic impurities and defects.

Thick-Film Circuit. A circuit in which resistors, capacitors, and conductors are printed or painted onto a substrate, often by silk-screen printing.

Thin-Film Circuit. A circuit consisting of patterns of tantalum or other materials laid down on a substrate of glass or ceramic, typically larger than silicon integrated circuits. Sometimes designated *FIC*.

Three-State Output. A type of A/D converter output used to connect to a data bus. The three output states are logic 1, logic 0, and "off." An enable control turns the output "on" or "off."

Throughput Rate. The maximum repetitive rate at which a data conversion system can operate to give specified output accuracy. It is determined by adding the various times required for multiplexer settling, sample-hold acquisition, A/D conversion, etc. and then taking the inverse of total time.

Thyristor. A PNPN device useful as a controlled rectifier that can conduct high currents by injection of a highly conducting hole-electron plasma. So called by analogy to a thyratron electron tube.

Track-and-Hold. A sample-hold circuit which can continuously follow the input signal in the sample-mode and then go into hold-mode upon command.

Tracking A/D Converter. A counter-type analog-to-digital converter which can continuously follow the analog input at some specified maximum rate and continuously update its digital output as the input signal changes. The circuit uses a D/A converter driven by an up-down counter.

Transfer Function. The input to output characteristic of a device such as a data converter expressed either mathematically or graphically.

Transistor. A semiconductor device capable of transferring a signal from one circuit to another and producing amplification.

TRAPATT (Trapped Plasma Avalanche Transity Time Diode). A diode used as microwave oscillator in a manner analogous to the IMPATT diode.

Trigger Pulse Steering. In transistors, the routing or directing of trigger signals (usually pulses) through diodes or transistors (called steering diodes or steering transistors) so that the trigger signals affect only one circuit of several associated circuits.

Triple-Slope A/D Converter. A variation on the dual slope type A/D converter in which the time period measured by the clock and counter is divided into a coarse (fast slope) measurement and a fine (slow slope) measurement.

TTL (Transistor-Transistor Logic). A modification of DTL which replaces the diode cluster with a multiple-emitter transistor.

Tuned-Base Oscillator. A transistor oscillator with the frequency-determined device (resonant circuit) located in the base circuit. It is comparable to the tuned grid electron tube oscillator.

Tuned-Collector Oscillator. A transistor oscillator with the frequency-determined device located in the collector circuit. It is comparable to the tuned plate electron tube oscillator.

Turnoff Time. The time that it takes a switching circuit (gate) to completely stop the flow of current in the circuit it is controlling.

Two-Stage Parallel A/D Converter. An ultra-fast A/D converter in which two parallel type A/D's are operated in cascade to give higher resolution. In the usual case a 4-bit parallel converter first makes a conversion; the resulting output code drives an ultra-fast 4-bit D/A, the output of which is subtracted from the analog input to form a residual. This residual then goes to a second 4-bit parallel A/D. The result is an 8-bit word converted in two steps.

Unijunction Transistor. A PN junction transistor with one electrode connection to one of the semiconductor materials and two connections to the other semiconductor material.

Unilateralization. The process by which an amplifier is prevented from going into oscillation by canceling the resistive and reactive component changes in the input circuit of an amplifier caused by positive feedback.

Unipolar Mode. In a data converter, when the analog range includes values of one polarity only.

Unipolar Transistor. A transistor such as an FET whose action depends on majority charge carriers only.

Video A/D Converter. An ultra-fast A/D converter capable of conversion rates of 5MHz and higher. Resolution is usually 8 bits but can vary depending on the application. Conversion rates of 20MHz and higher are common.

Voltage-to-Frequency (V/F) Converter. A device which converts an analog voltage into a train of digital pulses with frequency proportional to the input voltage.

VPE (Vapor Phase Epitaxy). The formation of an epitaxial layer of deposition from the vapor phase onto the substrate.

Wafer. A usually round, thin slice of a semiconductor material. Often used when referring to a wafer of silicon.

Wafer Sort. The step at which integrated circuits are tested to see whether or not they work. Probes contact the pads of the circuit and they are measured by putting in an electrical signal and seeing if the correct one comes out.

Weighted Current Source D/A Converter. A digital-to-analog converter design based on a series of binary weighted transistor current sources which can be turned "on" or "off" by digital inputs.

Word. A group of digits handled by a digital computer as a single unit. A word is usually 8 bits in a microprocessor.

Xylene. A solvent often used in semiconductor processing to remove unexposed photoresist.

Zener Diode. A voltage-limiting diode with high impedance at low voltages but low impedance above a "breakdown" voltage. Most voltage-limiting diodes break down by impact ionization ("avalanche") rather than by cold-field (the zener effect).

Zero Error. The error at analog zero for a data converter operating in the unipolar mode.

Zero-Order Hold. A name for a sample-hold circuit used as a data recovery filter. It is used to accurately reconstruct an analog signal from a train of analog samples.

Zone Refining. Technique of purifying a semiconductor or other substance by "sweeping" or passing a molten zone through the otherwise solid material.

APPENDIX B

Frequency Spectrum

Table B-1 shows the complete range of frequencies generally included in the so-called electronic spectrum, along with the possible applications. It must be remembered that frequency and wavelength are inversely proportional to each other. The higher the frequency is, the shorter the wavelength and vice versa.

AF and RF waves are generally considered in terms of frequency because the wavelength is so long. The exception is microwaves, which are often designated by wavelength because their frequencies are so high. Light waves, x-rays, and gamma rays are also generally considered by wavelength because their frequencies are so high. The units of wavelength are

Table B-1: Frequency Ranges and Applications

Frequency or Wavelength	Name	Applications
0Hz	Steady direct current or voltage	DC motors, solenoids, relays, DC supply voltage for tubes, transistors, and IC Chips
16–16,000Hz	Audio frequencies	60Hz power, AC motors, audio amplifiers, microphones, loudspeakers, phonographs, tape recorders, high-fidelity equipment, public address systems, and intercoms
16–30kHz	Ultrasonic frequencies or very low radio frequencies	Sound waves for ultrasonic cleaning, vibration testing, thickness gauging, flow detection, and sonar; electromagnetic waves for induction heating
30kHz–30,000MHz	Radio frequencies	Radio communications and broadcasting, including television, radio navigation, radio astronomy, satellites, and industrial, medical, scientific, and military radio
30,000–300,000MHz or 1–0.1cm	Extra-high frequencies	Experimental, weather radar, amateur, government
300,000–7,600Å	Infrared light rays	Heating, infrared photography
7,600–3,900Å	Visible light rays	Color, illumination, photography
3,900–320Å	Ultraviolet rays	Sterilizing, deodorizing, medical
320–0.1Å	X-rays	Thickness gauges, inspection, medical
0.1–0.006Å	Gamma rays	Radiation detection; more penetrating than hardest x-rays
Shortest of all elecromagnetic waves	Cosmic rays	Exist in outer space; can penetrate 70m of water or 1m of lead

the micrometer, equal to 10^{-6}m; the nanometer, equal to 10^{-9}m; and the angstrom, Å, equal to 10^{-10}m.

The four main categories of electromagnetic radiation and their frequencies can be summarized as follows:

1. Radio frequency waves from 30kHz to 300,000MHz.

2. Heat waves or infrared rays from 1×10^{13} to 2.5×10^{14}Hz. (Infrared means below the frequency of visible red light.)

3. Visible light frequencies from about 2.5×10^{14}Hz for red up to 8×10^{14}Hz for blue and violet.

4. Ionizing radiation such as ultraviolet rays, x-rays, gamma rays, and cosmic rays from about 8×10^{14}Hz for ultraviolet light to above 5×10^{20}Hz for cosmic rays. (Ultraviolet means above the frequencies of blue and violet visible light.)

Note that radio waves have the lowest frequencies and, therefore, the longest wavelengths of any form of electromagnetic radiation. The radio frequencies from 30kHz to 300,000MHz can be subdivided as shown in Table B-2.

Table B-2: Allocations from 30kHz to 300,000MHz

Band	Allocation	Remarks
30–535kHz	Includes maritime communications and navigation, aeronautical radio navigation	Low and medium radio frequencies
535–1,605kHz	Standard radio broadcast band	AM broadcasting
1,605kHz–30MHz	Includes amateur radio, CB radio, loran, government radio, international shortwave broadcast, fixed and mobile communications, radio navigation, industrial, scientific, and medical radio	Amateur bands 3.5–4.0MHz and 28–29.7 MHz; industrial, scientific, and medical band 26.95–27.54MHz; citizens radio band class D for voice is 25.965–27.045MHz
30–50MHz	Government and nongovernment, fixed and mobile	Includes police, fire, forestry, highway, and railroad services; VHF band starts at 30 MHz
50–54MHz	Amateur	6m band
54–72MHz	Television broadcast channels 2 and 4	Also fixed and mobile services
72–76MHz	Government and nongovernment services	Aeronautical marker beacon on 75MHz
76–88MHz	Television broadcast channels 5 and 6	Also fixed and mobile services
88–108MHz	FM broadcast	Also available for facsimile broadcast; 88–92MHz educational FM broadcast
108–122MHz	Aeronautical navigation	Localizers, radio range, and airport control
122–174MHz	Government and nongovernment, fixed and mobile, amateur broadcast	144–148MHz amateur band
174–216MHz	Television broadcast channels 7 to 13	Also fixed and mobile services
216–470MHz	Amateur, government and nongovernment, fixed and mobile, aeronautical navigation, citizens band radio	Radio altimeter, glide path, and meteorological equipment; citizens radio band 462.5–465MHz; civil aviation 225–400MHz; UHF band starts at 300MHz
470–890MHz	Television broadcasting	UHF television broadcast channels 14 to 83
890–3,000MHz	Aeronautical radio navigation, amateur broadcast, studio-transmitter relay	Radar bands 1,300–1,600MHz

Table B-2: Allocations from 30kHz to 300,000MHz (Continued)		
Band	**Allocation**	**Remarks**
3,000–30,000MHz	Government and nongovernment, amateur, radio navigation	Super-high frequencies (SHF); satellite communications
30,000–300,000MHz	Experimental, government, amateur	Extra-high frequencies (EHF)

Note that each band has a 10:1 range in frequencies. In addition, each band is 10 times higher than the preceding lower band.

The SHF and EHF bands are listed in gigahertz (1GHz = 1 × 10^9Hz, or 1,000MHz). Microwaves are generally considered to be in the range of approximately 0.3GHz and higher for wavelengths of 1m or less.

AM radio broadcasting at 535kHz to 1,605kHz is in the MF band. FM radio broadcasting at 88MHz to 108MHz is in the VHF band. Television broadcasting is in the VHF and UHF bands.

Television Channels

Each channel is 6MHz wide for the FM audio signal and AM picture carrier signal, including the multiplexed 3.58MHz color signal (Table B-3). Channels 2 to 6 and 7 to 13 are in the VHF band. Channels 14 to 83 are in the UHF band.

Television Channel Carrier Frequency Locations. Remember that the picture carrier is always located 1.25MHz above frequency limit. The sound carrier is always located .25MHz below upper frequency limit. The difference between picture and sound carriers is 4.5MHz.

Example:

Channel 2:
Lower Frequency Limit = 54MHz
Picture Carrier Frequency = 55.25MHz

Upper Frequency Limit = 60MHz
Sound Carrier Frequency = 59.75MHz

Citizens Band (CB) Radio

Table B-4 lists the carrier frequencies for channels 1 through 40 of class D service.

Short-Wave Radio

The bands listed in Table B-5 are used for international radio communications.

Table B-3: Television Channel Allocations

Channel Number	Frequency Band, MHz	Channel Number	Frequency Band, MHz
1*	—	42	638–644
2	54–60	43	644–650
3	60–66	44	650–656
4	66–72	45	656–662
5	76–82	46	662–668
6	82–88	47	668–674
7	174–180	48	674–680
8	180–186	49	680–686
9	186–192	50	686–692
10	192–198	51	692–698
11	198–204	52	698–704
12	204–210	53	704–710
13	210–216	54	710–716
14	470–476	55	716–722
15	476–482	56	722–728
16	482–488	57	728–734
17	488–494	58	734–740
18	494–500	59	740–746
19	500–506	60	746–752
20	506–512	61	752–758
21	512–518	62	758–764
22	518–524	63	764–770
23	524–530	64	770–776
24	530–536	65	776–782
25	536–542	66	782–788
26	542–548	67	788–794
27	548–554	68	794–800
28	554–560	69	800–806
29	560–566	70	806–812
30	566–572	71	812–818
31	572–578	72	818–824
32	578–584	73	824–830
33	584–590	74	830–836
34	590–596	75	836–842
35	596–602	76	842–848
36	602–608	77	848–854
37	608–614	78	854–860
38	614–620	79	860–866
39	620–626	80	866–872
40	626–632	81	872–878
41	632–638	82	878–884
		83	884–890

*The 44 to 50MHz band was television channel 1, but is now assigned to other services.

Amateur Radio Bands

The amateur radio bands, together with permissible types of emission for the transmitter, can be found in Table B-6.

Table B-4: Channels for Citizens Band (CB) Radio*

Channel	RF Carrier, MHz	Channel	RF Carrier, MHz
1	26.965	24	27.235
2	26.975	25	27.245
3	26.985	26	27.265
4	27.005	27	27.275
5	27.015	28	27.285
6	27.025	29	27.295
7	27.035	30	27.305
8	27.055	31	27.315
9†	27.065	32	27.325
10	27.075	33	27.335
11‡	27.085	34	27.345
12	27.105	35	27.355
13	27.115	36	27.365
14	27.125	37	27.375
15	27.135	38	27.385
16	27.155	39	27.395
17	27.165	40	27.405
18	27.175		
19‡	27.185		
20	27.205		
21	27.215		
22	27.225		
23	27.255		

*Channels 1 through 23 were the original CB channels. Channels 24 through 40 were added in 1976.
†Channel 9 is for emergency communications only.
‡Channels 11 and 19 are for auto traffic monitoring.

Table B-5: International Short-Wave Radio Allocations

Minor Bands, MHz	Major Bands, MHz
3.200–3.400	9.500–9.775
4.750–5.060	11.700–11.975
5.950–6.200	15.100–15.450
7.100–7.300	17.700–17.900
	21.450–21.750
	25.600–26.100

Radar Band Frequencies

Table B-7 lists the radar band codes and frequencies.

Satellites

Frequency allocations for satellites include the following bands, in gigahertz: 3.7 to 42, 5.9 to 6.4, 10.7 to 13.2, 17.7 to 21.2, 30 to 31, and 36 to 41. These bands are used for communications, meteorology, and navigation.

Table B-6: U.S. Amateur Radio Frequency Allocations

Frequency Band	Emissions	Frequency Band	Emissions	Frequency Band	Emissions
kHz		28.000–29.700	A1		A4, A5, F0, F1, F2, F3, F4, F5, P (2,300–2,450)
1,800–1,900	A1, A3	28.000–28.500	F1	3,300–3,500	A0, A1, A2, A3, A4, A5, F0, F1, F2, F3, F4, F5, P
1,900–2,000	A1, A3	28.500–29.700	A3, A4, A5, F3, F4, F5	5,650–5,925	A0, A1, A2, A3, A4, A5, F0, F1, F2, F3, F4, F5, P
3,500–4,000	A1	50.000–54.000	A1	**GHz**	
3,500–3,775	F1	50.100–54.000	A2, A3, A4, A5, F1, F2, F3, F4, F5	10.0–10.5	A0, A1, A2, A3, A4, A5, F0, F1
3,775–4,000	A3, A4, A5, F3, F4, F5	51.000–54.000	A0	24.0–24.25	A0, A1, A2, A3, A4, A5, P, F2, F3, F4, F5
4,383.8	A3J/A3A	144–148	A1	48–50, 71–76, 165–170, 240–250	A0, A1, A2, A3, A4, A5, F0, F1, F2, F3, F4, F5, P
7,000–7,300	A1	144.100–148.000	A0, A2, A3, A4, A5, F0, F1, F2, F3, F4, F5	Above 300	A0, A1, A2, A3, A4, A5, F0, F1, F2, F3, F4, F5, P
7,000–7,150	F1	220–225	A0, A1, A2, A3, A4, A5, F0, F1, F2, F3, F4, F5		
7,075–7,100	A3, F3	420–450	A0, A1, A2, A3, A4, A5, F0, F1, F2, F3, F4, F5		
7,150–7,300	A3, A4, A5, F3, F4, F5	1,215–1,300	A0, A1, A2, A3, A4, A5, F0, F1, F2, F3, F4, F5		
14,000–14,350	A1	2,300–2,450	A0, A1, A2, A3		
14,000–14,200	F1				
14,200–14,350	A3, A4, A5, F3, F4, F5				
MHz					
21.000–21.450	A1				
21.000–21.250	F1				
21.250–21.450	A3, A4, A5, F3, F4, F5				

Note: The types of emission referred to in the amateur rules are as follows:
Type A0—Steady, unmodulated pure carrier.
Type A1—Telegraphy on pure continuous waves.
Type A2—Amplitude tone-modulated telegraphy.
Type A3—AM telephony including single and double sideband, with full, reduced, or suppressed carrier.
Type A4—Facsimile.
Type A5—Television.
Type F0—Steady, unmodulated pure carrier.
Type F1—Carrier-shift telegraphy.
Type F2—Audio frequency-shift telegraphy.
Type F3—Frequency- or phase-modulated telephony.
Type F4—FM facsimile.
Type F5—FM television.
Type P—Pulse emissions.

Table B-7: Radar Band Codes and Frequencies

Band P/L			Band S			Band X			Band K			Band Q		
Sub	**Freq**	**λ**	**Sub**	**Freq**	**λ**	**Sub**	**Freq**	**λ**	**Sub**	**Freq**	**λ**	**Sub**	**Freq**	**λ**
(Band P)	0.225 / 0.390	133.3 / 76.9	E	1.55 / 1.65	19.3 / 18.2	A	5.20 / 5.50	5.77 / 5.45	P	10.90 / 12.25	2.75 / 2.45	A	36.00 / 38.00	0.834 / 0.790
Band L			F	1.65 / 1.85	18.2 / 16.2	Q	5.50 / 5.75	5.45 / 5.22	S	12.25 / 13.25	2.45 / 2.26	B	38.00 / 40.00	0.790 / 0.750
P	0.390 / 0.465	76.9 / 64.5	T	1.85 / 2.00	16.2 / 15.0	Y	5.75 / 6.20	5.22 / 4.84	E	13.25 / 14.25	2.26 / 2.10	C	40.00 / 42.00	0.750 / 0.715
C	0.465 / 0.510	64.5 / 58.8	C	2.00 / 2.40	15.0 / 12.5	D	6.20 / 6.25	4.84 / 4.80	C	14.25 / 15.35	2.10 / 1.95	D	42.00 / 44.00	0.715 / 0.682
L	0.510 / 0.725	58.8 / 41.4	Q	2.40 / 2.60	12.5 / 11.5	B	6.25 / 6.90	4.80 / 4.35	U	15.35 / 17.25	1.95 / 1.74	E	44.00 / 46.00	0.682 / 0.652

Table B-7: Radar Band Codes and Frequencies (Continued)

Band L			Band S			Band X			Band K			Band V		
	Freq	λ	Sub	Freq	λ	Sub	Freq	λ	Sub	Freq	λ	Sub	Freq	λ
Y	0.725	41.4	Y	2.60	11.5	R	6.90	4.35	T	17.25	1.74	A	46.00	0.652
	0.780	38.4		2.70	11.1		7.00	4.29		20.50	1.46		48.00	0.625
T	0.780	38.4	G	2.70	11.1	C	7.00	4.29	Q	20.50	1.46	B	48.00	0.625
	0.900	33.3		2.90	10.3		8.50	3.53		24.50	1.22		50.00	0.600
S	0.900	33.3	S	2.90	10.3	L	8.50	3.53	R	24.50	1.22	C	50.00	0.600
	0.950	31.6		3.10	9.68		9.00	3.33		26.50	1.13		52.00	0.577
X	0.950	31.6	A	3.10	9.68	S	9.00	3.33	M	26.50	1.13	D	52.00	0.577
	1.150	26.1		3.40	8.83		9.60	3.13		28.50	1.05		54.00	0.556
K	1.150	26.1	W	3.40	8.83	X	9.60	3.13	N	28.50	1.05	E	54.00	0.556
	1.350	22.2		3.70	8.11		10.00	3.00		30.70	0.977		56.00	0.536
F	1.350	22.2	H	3.70	8.11	F	10.00	3.00	L	30.70	0.977			
	1.450	20.7		3.90	7.69		10.25	2.93		33.00	0.909			
Z	1.450	20.7	Z	3.90	7.69	K	10.25	2.93	A	33.00	0.909			
	1.550	19.3		4.20	7.15		10.90	2.75		36.00	0.834			
			D	4.20	7.15									
				5.20	5.77									

APPENDIX C

Periodic Table of the Elements

Symbol	Name	Atomic Number	Atomic Weight
Ac	Actinium	89	1(227)
Ag	Silver	47	107.868
Al	Aluminum	13	26.982
Am	Americium	95	(243)
Ar	Argon	18	39.95
As	Arsenic	33	74.922
At	Astatine	85	(210)
Au	Gold	79	196.967
B	Boron	5	10.81
Ba	Barium	56	137.34
Be	Beryllium	4	9.012
Bi	Bismuth	83	208.980
Bk	Berkelium	97	(247)
Br	Bromine	35	79.904
C	Carbon	6	12.011
Ca	Calcium	20	40.08
Cd	Cadmium	48	112.40
Ce	Cerium	58	140.12
Cf	Californium	98	(249)
Cl	Chlorine	17	35.453
Cm	Curium	96	(247)
Co	Cobalt	27	58.933
Cr	Chromium	24	51.996
Cs	Cesium	55	132.905
Cu	Copper	29	63.546
Dy	Dysprosium	66	162.50
Er	Erbium	68	167.26
Es	Einsteinium	99	(254)
Eu	Europium	63	151.96
F	Fluorine	9	18.998
Fe	Iron	26	55.847
Fm	Fermium	100	(257)
Fr	Francium	87	(223)
Ga	Gallium	31	69.72
Gd	Gadolinium	64	157.25
Ge	Germanium	32	72.59
H	Hydrogen	1	1.008
He	Helium	2	4.003
Hf	Hafnium	72	178.49
Hg	Mercury	80	200.59

Symbol	Name	Atomic Number	Atomic Weight
Ho	Holmium	67	164.930
I	Iodine	53	126.904
In	Indium	49	114.82
Ir	Iridium	77	192.2
K	Potassium	19	39.102
Kr	Krypton	36	83.80
*Ku	Kurchatovium	104	(257)
La	Lanthanum	57	138.91
Li	Lithium	3	6.94
Lr	Lawrencium	103	(256)
Lu	Lutetium	71	174.97
Md	Mendelevium	101	(258)
Mg	Magnesium	12	24.305
Mn	Manganese	25	54.938
Mo	Molybdenum	42	95.94
N	Nitrogen	7	14.007
Na	Sodium	11	22.990
Nb	Niobium	41	92.906
Nd	Neodymium	60	144.24
Ne	Neon	10	20.18
Ni	Nickel	28	58.71
No	Nobelium	102	(255)
Np	Neptunium	93	(237)
O	Oxygen	8	15.999
Os	Osmium	76	190.2
P	Phosphorus	15	30.974
Pa	Protactinium	91	(231)
Pb	Lead	82	207.2
Pd	Palladium	46	106.4
Pm	Promethium	61	(147)
Po	Polonium	84	(210)
Pr	Praseodymium	59	140.907
Pt	Platinum	78	195.09
Pu	Plutonium	94	(242)
Ra	Radium	88	(226)
Rb	Rubidium	37	85.47
Re	Rhenium	75	186.2
Rh	Rhodium	45	102.905
Rn	Radon	86	(222)
Ru	Ruthenium	44	101.07
S	Sulfur	16	32.06
Sb	Antimony	51	121.75
Sc	Scandium	21	44.956
Se	Selenium	34	78.96
Si	Silicon	14	28.086
Sm	Samarium	62	150.35
Sn	Tin	50	118.69
Sr	Strontium	38	87.62
Ta	Tantalum	73	180.948
Tb	Terbium	65	158.924
Tc	Technetium	43	(99)
Te	Tellurium	52	127.60
Th	Thorium	90	232.038
Ti	Titanium	22	47.90
Tl	Thallium	81	204.37
Tm	Thulium	69	158.934

*Element proposed but not confirmed.

Symbol	Name	Atomic Number	Atomic Weight
U	Uranium	92	238.03
V	Vanadium	23	50.942
W	Tungsten	74	183.85
Xe	Xenon	54	131.30
Y	Yttrium	39	88.905
Yb	Ytterbium	70	173.04
Zn	Zinc	30	65.37
Zr	Zirconium	40	91.22

Periodic Table of the Elements

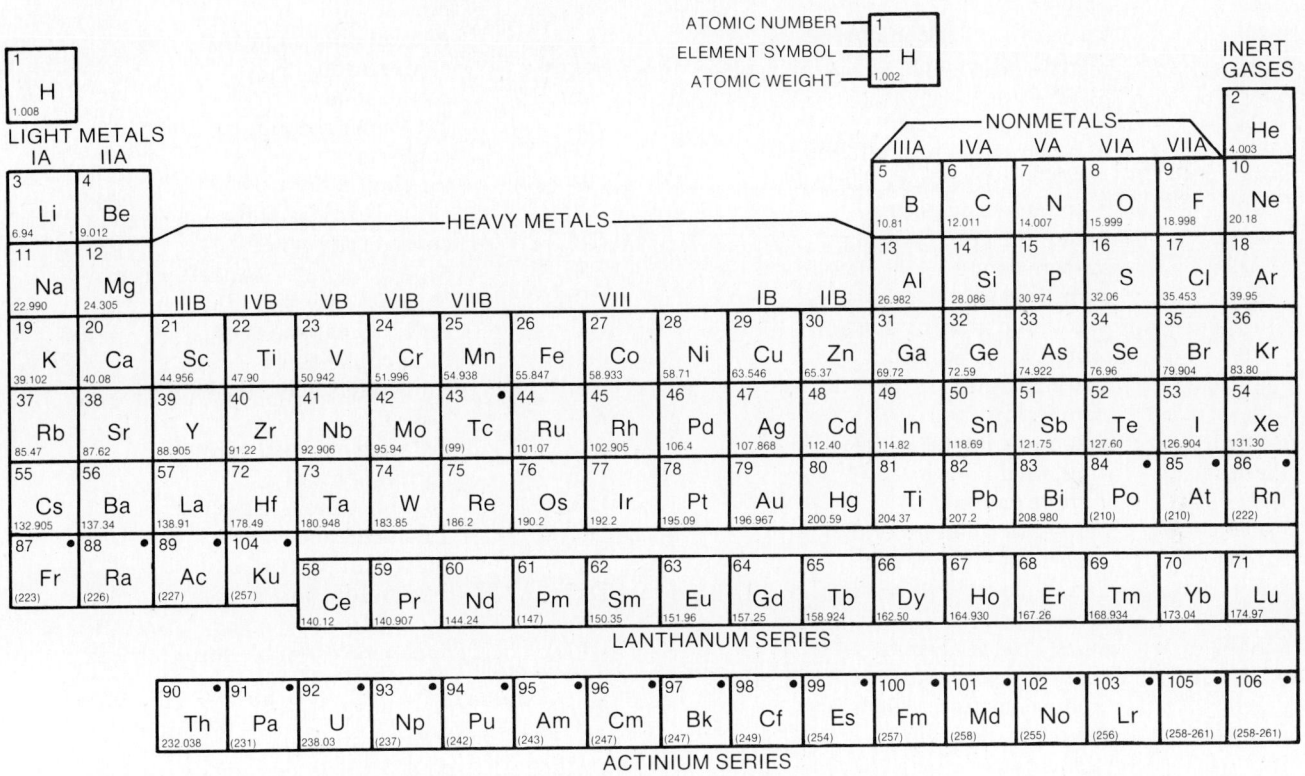

APPENDIX D

Common Abbreviations and Symbols

Common Abbreviations

A (amp)	ampere
A	ampere turn
a	turns ratio of a transformer
AC	alternating current
adj	adjustable
AF	audio frequency
A_f	band pass of a resonant circuit
afc	automatic frequency control
aft	automatic fine tuning
agc	automatic gain control
Ah	ampere-hour
A_i	current gain
alc	automatic limiting control
am	amplitude modulation
antilog	antilogarithm
A_p	power gain
apc	automatic phase control
auto	automatic
A_v	voltage gain
avc	automatic volume control
AWG	American Wire Gauge
B	flux density
BC	broadcast
BCD	binary coded decimal
bfo	beat frequency oscillator
BJT	bipolar junction transistor
Btu	British thermal unit
C	capacitance
c	centi (10^{-2})
c	common
CB	citizens band
CB	common-base configuration
CC	common-collector configuration
cc	current carrying
C_{cb}	interelement capacitance between base and collector
C_{cc}	interelement capacitance between collector and emitter
CE	common-emitter configuration
C_{eb}	interelement capacitance between base and emitter
cemf	instantaneous counter electromotive force
cm	centimeter

Common Abbreviations (Continued)

cmil (cir mil)	circular mil
C_N	neutralizing capacitor
comp	compressor
cont	continuous
cps	cycles per second
C(Q)	coulomb
CR, cr	junction diode
CRO	cathode ray oscilloscope
CRT	cathode ray tube
CSCR	capacitor start—capacitor run
CSIR	capacitor start—induction run
C_{sp}	reflected capacitance
C_T	total capacitance
cw	continuous wave transmission
d	deci (10^{-1})
d	distance between points
D-A	digital to analog
DBS	direct broadcast satellite
DC	direct current
di	change in current
DIP	dual in-line package
dk	deca (10)
DPDT	double-pole, double-throw switch
dq	change in charge
dsb	double sidebands
dt	change in time
DTL	diode-transistor logic
dv	change in voltage
DVM	digital voltmeter
E	difference in potential (DC or rms value)
e	difference in potential (instantaneous value)
EBS	Emergency Broadcast System
ECL	emitter-coupled oscillator
ECO	electron-coupled oscillator
emf	electromotive force
f	frequency
f_a	cutoff frequency
FET	field effect transistor
FF	flip-flop
fil	filament
fm	frequency modulation

f_r	frequency at resonance
fsk	frequency-shift keying
G	conductance
G	giga (10^9)
g_{bc}	base-collector conductance
g_{be}	base-emitter conductance
g_{ce}	collector-emittance conductance
g_{fe}	forward transfer conductance
g_{ie}	input conductance
g_{oe}	output conductance
g_{re}	reverse transfer conductance
H	henry
H	magnetic field intensity
H	magnetizing flux
h	hecto (10^2)
h	hybrid
herm	hermetic
hi	high
hp	horsepower
Hz	hertz
I	current
i	instantaneous current
I_B	DC base current
I_b	changing base current (AC)
IC	integrated circuit
I_c	capacitive current
I_C	DC collector current
i_c	changing collector current (AC)
i_c	instantaneous capacitive current
ID (i.d.)	inside diameter
I_e	total emitter current
I_{eff}	effective current
IF	intermediate frequency
I_L	inductive current
i_L	instantaneous inductive current
I_m, I_{max}	maximum current
I_{min}	minimum current
intlk	interlock
I_P	plate current
I_R	current through resistance
i_R	instantaneous current through resistance
I_S	secondary current
I_T	total current
K	coefficient of coupling
k	kilo (10^3)
kc	kilocycle
kHz	kilohertz
kV	kilovolt
kVA	kilovoltampere
kW	kilowatt
kWh	kilowatt-hour
L	inductance
lb/in²(psi)	pounds per square inch
L-C	inductance-capacitance
LCD	liquid crystal display
L-C-R	inductance-capacitance-resistance
LED	light-emitting diode

L_M	mutual inductance
LNA	low-noise amplifier
lo	low
ls	loudspeaker
L_T	total inductance
M	mega (10^6)
M	mutual conductance
M	mutual inductance
m	milli (10^{-3})
mA	milliampere
mag	magnetron
max	maximum
MCW	modulated continuous wave
mH	millihenry
MHz	megahertz
min	minimum
mm	millimeter
mmf	magnetomotive force
MMF, mmf	picofarad
MOS	metal-oxide semiconductor
MOSFET	metal-oxide semiconductor field effect transistor
mpx	multiplex
MSI	medium scale integrated circuit
mV	millivolt
mv	multivibrator
mW	milliwatt
Mx	maxwell
N	number of turns in an inductor
N	revolutions per minute
n	nano (10^{-9})
nA	nanoampere
NC	no connection
NC	normally closed
NEC	National Electrical Code
NEG, neg	negative
nF	nanofarad
nH	nanohenry
nm	nanometer
NO	normally open
norm	normal
N_p, N_1	primary turns
N_s, N_2	secondary turns
ns	nanosecond
nW	nanowatt
OD (o.d.)	outside diameter
o.l.	overload
OP AMP	operational amplifier
P	pico (1×10^{-12})
P	power
p	instantaneous power
PA	public address (system), power amplifier
pA	picoampere
PAM, pam	pulse-amplitude modulation
P_{ap}	apparent power
P_{av}	average power
pc	permanent capacitor

PCM, pcm	pulse-code modulation
PDM	pulse-duration modulation
pF	picofarad
PLL, pll	phase-locked loop
pm	phase modulation
PNP	transistor
POT, pot	potentiometer
P_p	primary power
PPM	pulse-position modulation
PRF	pulse-recurrence frequency, pulse repetition frequency
PRT	pulse-recurrence time
P_s	secondary power
ps	picosecond
PSC	permanent split capacitor
pw	pulse width
pW	picowatt
PWM, pwm	pulse-width modulation
Q	charge or quality
q	instantaneous charge
QAVC	quiet automatic volume control
R	potentiometer
R	resistance
RAM	random access memory
r_b	AC base resistance
rcvr	receiver
r_d	AC collector resistance
r_e	AC emitter resistance
rect	rectifier
ref	reference
RF	radio frequency
rf	radio frequencies
R_G	grid resistance
RI	repulsion-induction
R_L	load resistor
r_m	mutual resistance
r/min (rpm), N	revolutions per minute
rms	root mean square
R_N	neutralizing resistor
R_o	load resistance
r_{oe}	output resistance with input open
ROM	read-only memory
r_{re}	reverse transfer resistance with input open
r/s	revolutions per second
RSIR	resistance start—induction run
RT, Rt	thermistor
SCR	silicon controlled rectifier
S_I	current-stability factor
S/N	signal-to-noise
SPDT	single-pole, double-throw switch
SPI	split-phase induction switch
SPNC	single-pole, normally closed
SPNO	single-pole, normally open
SPST	single-pole, single-throw switch
sq cm	square centimeter
ssb	single sideband

STD	standard
S_v	voltage-stability factor
sw	switch
SWR	standing-wave ratio
SYNC	synchronous
T	tera (10^{12})
T	torque
T	transformer
t, TC	time constant
t	time (seconds)
TD	temperature differential
TE	transverse electric
temp	temperature
term. bd.	terminal board
t_f	fall time
thermo	thermostat
therm. prot.	thermo protector
THI	temperature humidity index
thm	therm
THz	teraherz
TM	transverse magnetic
t_p	pulse time
TR	transmit-receive
t_r	rise time
TRF	tuned radio frequency
t_s	storage time
TTL or T²L	transistor-transistor logic
TV	television
TVM	transistorized voltmeter
TWT	travelling wave tube
UHF	ultra high frequency
UJT	unijunction transistor
UL	Underwriters Laboratories Inc.
V	vacuum tube
V, v	volt
v	instantaneous voltage
VA	voltampere
V_{av}	voltage (average value)
V_{BE}	fixed base-emitter voltage
V_c	capacitive voltage
vc	instantaneous capacitive voltage
V_{CE}	fixed collector-emitter voltage
VCO	voltage controlled oscillator
VFO	variable frequency oscillator
V_g	source voltage
VHF	very high frequency
V_{in}	input voltage
V_{inst}	voltage (instantaneous value)
V_L	inductive voltage
v_L	instantaneous inductive voltage
V_m, V_{max}	maximum voltage
VOM	volt-ohm-milliammeter
V_{out}	output voltage
V_p	primary voltage
V_s	secondary voltage
V_s	source voltage
V_{SAT}	saturation voltage
$VSWR$	voltage standing wave ratio

Common Abbreviations (Continued)

V_T	total voltage
W	watt
X_C	capacitive reactance
xfmr	transformer
X_L	inductive reactance
Y	admittance
Z	impedance
Z_{in}	input impedance
Z_o	output impedance
Z_p	primary impedance
Z_s	secondary impedance
Z_T	total impedance
$I\phi$	phase current
$k\Omega$	kilohm
$M\Omega$	megohm
ϕ_M	mutual flux
ϕ_P	primary flux
ϕ_S	secondary flux
μA	microampere
μF (mfd)	microfarad
μH	microhenry
$\mu\mu F$	micromicrofarad
μV	microvolt
μW	microwatt
Ω	ohm
λ	wavelength
°	degrees
°C	degrees Centigrade
°F	degrees Fahrenheit

Mathematical Symbols

+	plus; addition; positive; "or" gate
-	minus; subtraction; negative
±	tolerance; plus or minus, positive or negative
$A \div B$, A/B, $\frac{A}{B}$, $A{:}B$	division; A divided by B
$A \times B$, $A \cdot B$, AB, $(A)(B)$	multiplication; A times B
( ), { }, [ ]	parentheses, braces, and brackets for grouping
$=$, $::$	equal to
$\cong$, $\approx$	approximately equal to
$\neq$	not equal to
$\equiv$	identical to
$<$	less than
$\not<$	not less than
$\leqq$	equal or less than
$>$	greater than
$\geqq$	equal or greater than
$\not>$	not greater than
$::$, $:$	proportional to (equals)
$:$	ratio
$\longrightarrow$	approaches as a limit
∞	infinity

Mathematical Symbols (Continued)

Δ	delta; increment of; change in
$\sqrt{\ }$, $\sqrt{\ }$	square root
$\sqrt[n]{\ }$	Nth root
X^n	exponent of X; Nth power of X
Log, Log_{10}	common logarithm
Ln, Log e	hyperbolic, natural or napierian logarithm
e, ϵ	epsilon; base of natural (hyperbolic) logarithm; 2.718
π	Pi; 3.1416
$\frac{1}{2\pi}$	0.159
$\angle$	angle
	right angle
$\perp$	perpendicular to
$\parallel$, $/\!/$	parallel to; parallel
$\lvert N \rvert$	absolute value of N
$\bar{N}$	average value of N
X^{-n}, $\frac{1}{X^n}$	reciprocal of Nth power of X
$N°$	N degrees
N'	N minutes; N feet
N''	N seconds; N inches
$\therefore$	therefore
$F(x)$	function of (X)
C, K	constant (when following an equal sign)
$\int$	integration (integral in calculus)
$_b\!\int^a$	integral between limits of a and b
α	varies directly as
!	factorial
Σ	summation
j	j operator, square root of minus one
ω	angular velocity $2\pi F$
lim	limit value of an expression
sin	sine
cos	cosine
tan, tg, tang	tangent
cot, ctg	cotangent
sec	secant
cosec	cosecant
versin	versed sine
covers	coversed sine

Abbreviations for Official Electronics Organizations*

ANSI	American National Standards Institute, 1430 Broadway, New York, NY 10018. Issues standards for graphic symbols in electricity and electronics.

Abbreviations for Official Electronics Organizations* (Continued)	
ARRL	American Radio Relay League, Newington, CT 06111. For amateur radio operators.
ASEE	American Society for Engineering Education, One Dupont Circle, Washington, DC 20036. Professional group for engineering and technology educators.
EIA	Electronic Industries Association, 2001 Eye Street, NW, Washington, DC 20015. The main industry group for standards on receivers, semiconductor devices, and tubes.
FCC	Federal Communication Commission, Washington, DC 20554. The United States government agency that issues rules and regulations for transmitters, including radio and television broadcasting and citizens band (CB) radio.
IEEE	Institute of Electrical and Electronics Engineers, 345 East 47 Street, New York, NY 10017. The professional group that provides standards for measurements, definitions, and graphic symbols diagrams in the electrical and electronics fields.
ISCET	International Society of Certified Electronics Technicians, 1715 Expo Lane, Indianapolis, IN 46224. Professional group for radio and television servicing.
JEDEC	Joint Electronic Devices Engineering Council. Professional group for standards on electron devices.
NAB	National Association of Broadcasters, 1771 N Street, NW, Washington, DC 20036. Professional group.
NEMA	National Electrical Manufacturers Association, 155 East 44 Street, New York, NY 10017. Trade group.
NESDA	National Electronic Service Dealers Association, 1715 Expo Lane, Indianapolis, IN 46224. Trade group.
NFPA	National Fire Protection Association, 470 Atlantic Avenue, Boston, MA 02210. Issues National Electrical Code (NEC).
WARC	World Allocations Resources Committee, Geneva, Switzerland. International government group for frequency allocations in the radio spectrum.

*The abbreviations are given here in alphabetical order. The organizations include professional groups, government regulating agencies, and industry associations that promote standards.

SEMICONDUCTOR LETTER SYMBOLS

Letter symbols used in solid-state circuits are those proposed as standard for industry or are special symbols not included as standard. Semiconductor symbols consist of a basic letter with subscripts, either alphabetical, numerical, or both, in accordance with the following rules:

1. A capital (uppercase) letter designates external circuit parameters and components; large-signal device parameters; and maximum (peak), average (DC), or root-mean-square values of current, voltage, and power (I, V, P, etc.)

2. Instantaneous values of current, voltage, and power, which vary with time, and small-signal values are represented by the lowercase (small) letter of the proper symbol (i, v, p, i_e, V_{eb}, etc.).

3. DC values, instantaneous total values, and large-signal values are indicated by capital subscripts (i_C, I_C, v_{EB}, V_{EB}, P_C, etc.).

4. Alternating component values are indicated by using lowercase subscripts; note the examples i_c, I_c, v_{eb}, V_{eb}, P_c, p_c.

5. When it is necessary to distinguish between maximum, average, or root-mean-square values, maximum or average values may be represented by addition of a subscript m or av; examples are i_{cm}, I_{CM}, I_{cav}, i_{CAV}.

6. For electrical quantities, the first subscript designates the electrode at which the measurement is made.

7. For device parameters, the first subscript designates the element of the four-pole matrix; examples are I or i for input, O or o for output, F or f for forward transfer, and R or r for reverse transfer.

8. The second subscript normally designates the reference electrode.

9. Supply voltages are indicated by repeating the associated device electrode subscript, in which case, the reference terminal is then designated by the third subscript; note the cases V_{EE}, V_{CC}, V_{EEB}, V_{CCB}.

10. In devices having more than one terminal of the same type (say two bases), the terminal subscripts are modified by adding a number following the subscript and placed on the same line, for example, V_{B1-B2}.

11. In multiple-unit devices the terminal subscripts are modified by a number preceding the electrode subscript; note the example V_{1B-2B}.

Semiconductor symbols change and new symbols are developed to cover new devices as the art changes; alphabetical lists of the complex symbols are presented here for easy reference. These lists are divided into several sections.

Signal and Rectifier Diodes

I_o	average rectifier forward current
I_r	average reverse current
I_{surge}	peak surge current
PRV	peak reverse voltage
V_F	average forward voltage drop
V_R	DC blocking voltage

Zener Diodes

I_F	forward current
I_R	reverse current
I_Z	zener current
I_{ZK}	zener current near breakdown knee
I_{ZM}	maximum DC zener current (limited by power dissipation)
I_{ZT}	zener test current
V_f	forward voltage
V_R	reverse test voltage
V_Z	nominal zener voltage
Z_Z	zener impedance
Z_{ZK}	zener impedance near breakdown knee
Z_{ZT}	zener impedance at zener test current

Thyristors and SCRs

I_f	Forward current, rms, value of forward anode current during the ON state.
$I_{FM(pulse)}$	Repetitive pulse current. Repetitive peak forward anode current after application of gate signal for specified pulse conditions.
$I_{FM(surge)}$	Peak forward surge current. The maximum forward current having a single forward cycle in a 60Hz single-phase resistive load system.
I_{FOM}	Peak forward blocking current, gate open. The maximum current through the thyristor when the device is in the OFF state for a specified anode-to-cathode voltage (anode positive) and junction temperature with the gate open.

Thyristors and SCRs (Continued)

I_{FXM}	Peak forward blocking current. Same as I_{FOM} except that the gate terminal is returned to the cathode through a stated impedance and/or bias voltage.
I_{GFM}	Peak forward gate current. The maximum instantaneous value of current which may flow between gate and cathode.
I_{GT}	Gate trigger current (continuous DC). The minimum DC gate current required to cause switching from the OFF state at a specified condition.
I_H	Holding current. That value of forward anode current below which the controlled rectifier switches from the conducting state to the forward blocking condition with the gate open, at stated conditions.
I_{HX}	Holding current (gate connected). The value of forward anode current below which the controlled rectifier switches from the conducting state to the forward blocking condition with the gate terminal returned to the cathode terminal through specified impedance and/or bias voltage.
I_{ROM}	Peak reverse blocking current. The maximum current through the thyristor when the device is in the reverse blocking state (anode negative) for a stated anode-to-cathode voltage and junction temperature with the gate open.
I_{RXM}	Peak reverse blocking current. Same as I_{ROM} except that the gate terminal is returned to the cathode through a stated impedance and/or bias voltage.
$P_{F(AV)}$	Average forward power. Average value of power dissipation between anode and cathode.
$P_{GF(AV)}$	Average forward gate power. The value of maximum allowable gate power dissipation averaged over a full cycle.
P_{GFM}	Peak gate power. The maximum instantaneous value of gate power dissipation permitted.
V_F	Forward "on" voltage. The voltage measured between anode and cathode during the ON condition for specified conditions of anode and temperature.
$V_{F(on)}$	Dynamic forward "on" voltage. The voltage measured between anode and cathode at a specified time after turn-on function has been initiated at stated conditions.

Thyristors and SCRs (Continued)	
V_{FOM}	Peak forward blocking voltage, gate open. The peak repetitive forward voltage which may be applied to the thyristor between anode and cathode (anode positive) with the gate open at stated conditions.
V_{FXM}	Peak forward blocking voltage. Same as V_{FOM} except that the gate terminal is returned to the cathode through a stated impedance and/or voltage.
V_{GFM}	Peak forward gate voltage. The maximum instantaneous voltage between the gate terminal and the cathode terminal resulting from the flow of forward gate current.
V_{GRM}	Peak reverse gate voltage. The maximum instantaneous voltage which may be applied between the gate terminal and the cathode terminal when the junction between the gate region and the adjacent cathode region is reverse biased.
V_{GT}	Gate trigger voltage (continuous DC). The DC voltage between the gate and the cathode required to produce the DC gate trigger current.
$V_{ROM(rep)}$	Peak reverse blocking voltage, gate open. The maximum allowable value of reverse voltage (repetitive or continuous DC) which can be applied between anode and cathode (anode negative) with the gate open for stated conditions.
V_{RXM}	Peak reverse blocking voltage. Same as V_{ROM} except that the gate terminal is returned to the cathode through a stated impedance and/or bias voltage.

Transistors	
A_G	Available gain.
A_P	Power gain.
A_I	Current gain.
B or b	Base electrode.
BV_{BCO}	DC base-to-collector breakdown voltage, base reverse-biased with respect to collector, emitter open.
BV_{BEO}	DC base-to-emitter breakdown voltage, base reverse-biased with respect to emitter, collector open.
BV_{CBO}	DC collector-to-base breakdown voltage, collector reverse-biased with respect to base, emitter open.
BV_{CEO}	DC collector-to-emitter breakdown voltage, collector reverse-biased with respect to emitter, base open.

Transistors (Continued)	
BV_{EBO}	DC emitter-to-base breakdown voltage, emitter reverse-biased with respect to base, collector open.
BV_{ECO}	DC emitter-to-collector breakdown voltage, emitter reverse-biased with respect to collector, base open.
C or c	Collector electrode.
C_c	Collector junction capacitance.
C_e	Emitter junction capacitance.
C_{ib}, C_{ic} C_{ie}	Input capacitance for common base, collector, and emitter, respectively.
C_{ob}, C_{oc} C_{oe}	Output terminal capacitance, AC input open, for common base, collector and emitter, respectively.
D	Distortion.
E or e	Emitter electrode.
$f_{\alpha b}, f_{\alpha c},$ $f_{\alpha e}$	Alpha cutoff frequency for common base, collector, and emitter, respectively.
f_{co}	Cutoff frequency.
f_{max}	Maximum frequency of oscillation.
GC (CB), GC (CC), GE (CE)	Grounded (or common) base, collector, and emitter, respectively.
h	Hybrid parameter.
$h_{fe}, h_{fb},$ h_{fc}	Small signal forward current transfer ratio, AC output shorted, common emitter, common base, common collector, respectively.
h_{ib}	Small-signal input impedance, AC output shorted, common base.
h_{ob}	Small-signal output admittance, AC input open, common base.
I	Direct current (DC).
I_B, I_C, I_E	DC current for base, collector, and emitter, respectively.
i_b	Base current (instantaneous).
i_c	Collector current (instantaneous).
I_{CBO}	DC collector current, collector reverse-biased with respect to base, emitter-to-base open.
I_{CES}	DC collector current, collector reverse-biased with respect to emitter, base shorted to emitter.
I_{EBO}	DC emitter current, emitter reverse-biased with respect to base, collector-to-base open.
NF	Noise figure.
P_D	Total average power dissipation of all electrodes of a semiconductor device.

Transistors (Continued)

P_{in}	Input power.
P_{out}	Output power.
P_T	Over-all power gain.
R_B	External base resistance.
r_b	Base spreading resistance.
$r'b$	Equivalent base resistance, high frequencies.
r_i	Input junction resistance.
T_j	Junction temperature.
T_{stg}	Storage temperature.
t_f	Fall time, from 90% to 10% of pulse (switching applications).
t_r	Rise time, from 10% to 90% pulse (switching applications).
t_s	Storage time (switching applications).
V_{BB}	Base supply voltage.
V_{BE}	Base-to-emitter DC voltage.
V_C	Collector voltage (with respect to ground or common point).
V_{CB}	Collector-to-base DC voltage.
V_{CC}	Collector supply voltage.
V_{CE}	Collector-to-emitter DC voltage.
$V_{CE}(sat)$	Collector-to-emitter saturation voltage.
V_{ce}	Collector-to-emitter voltage (rms).
v_{ce}	Collector-to-emitter voltage (instantaneous).
V_{CEO}	DC collector-to-emitter voltage with collector junction reverse-biased, zero base current.
V_{CER}	Similar to V_{CEO}, except with a resistor (of value R) between base and emitter.
V_{CES}	Similar to V_{CEO}, except with base shorted to emitter.
V_{CEV}	DC collector-to-emitter voltage, used when only voltage bias is used.
V_{CEX}	DC collector-to-emitter voltage, base-emitter back biased.
V_{EB}	Emitter-to-base DC voltage.
V_{EBO}	Emitter-to-base voltage (static).
V_{EE}	Entire supply voltage.
V_{pt}	Punch-through voltage.

Semiconductor, General

BV	breakdown voltage
T_A	ambient temperature
T_{ep}	operating temperature

Unijunction Transistors

I_E	Emitter current.
$I_{B2(mod)}$	Interbase modulation current. B2 current modulation due to firing. Measured at a specified interbase voltage, emitter, and temperature.
I_{EO}	Emitter reverse current. Measured between emitter and base-two at a specified voltage, and base-one open-circuited.
I_p	Peak point emitter current. The maximum emitter current that can flow without allowing the UJT to go into the negative resistance region.
I_V	Valley point emitter. The current flowing in the emitter when the device is biased to the valley point.
r_{BB}	Interbase resistance. Resistance between base-two and base-one measured at a specified interbase voltage.
V_{B2B1}	Voltage between base-two and base-one. Positive at base-two.
V_p	Peak point emitter voltage. The maximum voltage seen at the emitter before the UJT goes into the negative resistance region.
V_D	Forward voltage drop of the emitter junction.
V_{EB1}	Emitter to base-one voltage.
$V_{EB1(SAT)}$	Emitter saturation voltage. Forward voltage drop from emitter to base-one at a specified emitter current (larger than I_V) and specified interbase voltage.
V_{OB1}	Base-one peak pulse voltage. The peak voltage measured across a resistor in series with base-one when the UJT is operated as a relaxation oscillator in a specified circuit.
V_v	Valley point emitter voltage. The voltage at which the valley point occurs with a specified V_{B2B1}.
α_{rBB}	Interbase resistance temperature coefficient. Variation of resistance between B_2 and B_1 over the specified temperature range and measured at the specific interbase voltage and temperature with emitter open circuited.

Field Effect Transistors

I_D	Drain current.
I_{DGO}	Maximum leakage from drain to gate with source open.
I_{DSS}	Drain current with gate connected to source.

Field Effect Transistors (Continued)

I_G	Gate current.
I_{GSS}	Maximum gate current (leakage) with drain connected to source.
$V_{(BR)DGO}$	Drain to gate, source open.
$V_{(BR)DGS}$	Drain to gate, source connected to drain.
$V_{(BR)DS}$	Drain to source, gate connection not specified.
$V_{(BR)DSX}$	Drain to source, gate biased to cutoff or beyond.
$V_{(BR)GS}$	Gate to source, drain connection not specified.
$V_{(BR)GSS}$	Gate to source, drain connected to source.
$V_{(BR)GD}$	Gate to drain, source connection not specified.
$V_{(BR)GDS}$	Gate to drain, source connected to drain.
V_D	DC drain voltage.
V_G	DC gate voltage.
$V_{G1S(OFF)}$	Gate 1 source cutoff voltage (with gate 2 connected to source).
$V_{G2S(OFF)}$	Gate 2 source cutoff voltage (with gate 1 connected to source).
$V_{GS(OFF)}$	Cutoff.

Transistor Letter Symbols for Current

A_i	Current gain.
H_{Fb}	Common base forward current ratio (large signal, short circuit).
h_{FB}	Forward current transfer ratio for common base (static value).
h_{fb}	Forward current transfer ratio for common base (small signal, short circuit).
H_{FC}	Common collector forward current ratio (large signal, short circuit).
h_{FC}	Forward current transfer ratio for common collector (static value).
h_{fc}	Forward current transfer ratio for common collector (small signal, short circuit).
H_{Fe}	Common emitter forward current ratio (large signal, short circuit).
h_{FE}	Forward current transfer ratio for common emitter (static value).
h_{fe}	Forward current transfer ratio for common emitter (small signal, short circuit).

Transistor Letter Symbols for Current (Continued)

H_{21}	Current amplification with output short circuited.
I_B	DC base current.
I_b	Changing base current (AC—rms).
i_b	Base current (instantaneous—AC).
I_C	DC collector current.
I_c	Collector current (rms).
i_c	Changing collector current (AC).
I_{CBO}	Collector cutoff DC current (emitter open).
I_{CEO}	Collector cutoff DC current (base open).
I_{CER}	Collector cutoff DC current (specified resistance between base and emitter).
I_{CES}	Collector cutoff DC current (base short circuited to emitter).
I_{CEV}	Collector cutoff DC current (specified voltage between base and emitter).
I_{CEX}	Collector current (DC*—between base and emitter). *specified circuit
I_D	Drain current (DC).
$I_{D(off)}$	Drain cutoff current.
I_{DSS}	Drain current-zero gate voltage.
I_E	Emitter DC current.
I_e	DC emitter current (total).
i_e	Emitter current (instantaneous).
I_{EBO}	Emitter cutoff DC current (collector open).
$I_{EC(OFS)}$	Emitter-collector offset current.
I_{ECS}	Emitter cutoff DC current (base short circuited to collector).
I_F	Forward current (DC).
i_F	Forward current (instantaneous).
I_f	Forward current (alternating component).
$I_{F(AV)}$	Forward current (DC with alternating component).
I_{FM}	Forward current (peak total value).
$I_{F(OV)}$	Forward current (overload).
I_{FSM}	Forward current (peak surge).
I_G	Gate current (DC).
I_{GF}	Gate current forward.
I_{GR}	Gate current (reverse).
I_O	Average output rectified current.

Transistor Letter Symbols for Current (Continued)

I_P	Peak point current (double-base transistors).
I_R	Reverse DC current.
I_r	Reverse current (alternating components—rms).
i_r	Reverse current (instantaneous).
$i_{R(REC)}$	Reverse recovery current.
$I_{R(rms)}$	Reverse current (total—rms).
I_{SDS}	Source current (zero gate voltage).
I_V	Valley point current (double base transistors).
I_Z	Regulator current (DC reference).
I_{ZK}	Regulator current-reference current (DC near breakdown knee).
I_{ZM}	Regulator current-reference current (DC max rated).
S_I	Current stability factor.

Transistor Letter Symbols for Voltages

A_v	Voltage gain.
BV_{CBO}	Breakdown voltage between collector-base (emitter open).
BV_{CEO}	Breakdown voltage between collector-emitter (base open).
BV_{EBO}	Breakdown voltage between emitter-base (collector open).
BV_R	Breakdown reverse voltage.
H_{12}	Voltage feedback ratio with input open circuited.
h_{rb}	Reverse voltage transfer ratio for common base (small signal, open circuit).
h_{rc}	Reverse voltage transfer ratio for common collector (small signal, open circuit).
h_{re}	Reverse voltage transfer ratio for common emitter (small signal, open circuit).
S_V	Voltage stability factor.
V_{BB}	Base supply DC voltage.
V_{BC}	Base-collector DC voltage.
V_{bc}	Base-collector voltage (rms).
V_{bc}	Changing voltage (AC) between base-collector.
v_{bc}	Base-collector voltage (instantaneous).
V_{BE}	Fixed base-emitter DC voltage.
V_{be}	Base-emitter voltage (rms).

Transistor Letter Symbols for Voltages (Continued)

V_{cb}	Collector to base voltage (rms).
v_{cb}	Collector to base voltage (instantaneous).
V_{CC}	Collector DC supply voltage.
V_{CE}	Fixed collector-emitter DC voltage.
$V_{CE(sat)}$	Collector to emitter saturation voltage.
V_{ce}	Changing voltage (AC) between collector-emitter.
v_{ce}	Collector to emitter voltage (instantaneous).
V_{EB}	Emitter to base DC voltage.
V_{eb}	Emitter to base voltage (rms).
v_{eb}	Emitter to base voltage (instantaneous).
V_{EC}	Emitter to collector DC voltage.
V_{ec}	Emitter to collector voltage (rms).
v_{ec}	Emitter to collector voltage (instantaneous).
V_{EE}	Emitter supply DC voltage.
V_F	Forward DC voltage.
v_f	Forward voltage (instantaneous).
V_g, V_s	Source voltage.
V_{in}	Input voltage.
V_{out}	Output voltage.
V_R	Reverse DC voltage.
v_r	Reverse voltage (instantaneous).
V_{RT}	Reach voltage.
V_{sat}	Saturation voltage.

Transistor Resistances Letter Symbols

h_{IB}	Common base input resistance (static value).
h_{ib}	Input resistance of common base (output short circuited).
h_{IC}	Common collector input resistance (static value).
h_{ic}	Input resistance of common collector (output short circuited).
h_{IE}	Common emitter input resistance (static value).
h_{ie}	Input resistance of common emitter (output short circuited).
r_b	AC base resistance.
$r_{CE(sat)}$	Collector-emitter saturation resistance.
r_c	AC collector resistance.

Transistor Resistances
Letter Symbols (Continued)

r_e	AC emitter resistance.
r_{fe}	Forward transfer resistance with input open.
r_{ie}	Input resistance with output open.
R_L	Load resistance.
r_m	Mutual resistance.
r_{oe}	Output resistance with input open.
r_{re}	Reverse transfer resistance with input open.

Transistor Amplification Letter Symbols

G_{PB}	Average power gain of common base (large signal).
G_{pb}	Average power gain of common base (small signal).
G_{PC}	Average power gain of common collector (large signal).
G_{pc}	Average power gain of common collector (small signal).
G_{PE}	Average power gain of common emitter (large signal).
G_{pe}	Average power gain of common emitter (small signal).
$\propto fb$	Forward short circuit current amplification factor (common base).
$\propto fc$	Forward short circuit current amplification factor (common collector).
$\propto fe$	Forward short circuit current amplification factor (common emitter).
$\mu \, rb$	Reverse open circuited voltage amplification factor (common base).
$\mu \, rc$	Reverse common circuited voltage amplification factor (common collector).
$\mu \, re$	Reverse open circuited voltage amplification factor (common emitter).

Transistor Conductance Letter Symbols

g_{bc}	Base-collector conductance.
g_{be}	Base-emitter conductance.
g_{ce}	Collector-emitter conductance.
g_{fe}	Forward transfer conductance.
g_{ie}	Input conductance.
g_{oe}	Output conductance.

Transistor Conductance
Letter Symbols (Continued)

g_{re}	Reverse transfer conductance.
h_{OB}	Common base output conductance (open circuit, static value)
h_{ob}	Output conductance of common base with input open.
h_{OC}	Common collector output conductance (open circuit, static value).
h_{oc}	Output conductance of common collector with input open.
h_{OE}	Common emitter output conductance (open circuit, static value).
h_{oe}	Output conductance of common emitter with input open.

Transistor Capacitance Letter Symbols

C_{cb}	Interelement capacitance (base-collector).
C_{ce}	Interelement capacitance (collector-emitter).
C_{eb}	Interelement capacitance (base-emitter).
C_{ib}	Common base input capacitance.
C_{ibo}	Common base input capacitance (open circuit).
C_{ibs}	Common base input capacitance (short circuit).
C_{ic}	Common collector input capacitance.
C_{ie}	Common emitter input capacitance.
C_{ieo}	Common emitter input capacitance (open circuit).
C_{ies}	Common emitter input capacitance (short circuit).
C_{iss}	Gate source capacitance (drain shorted to source).
C_{ob}	Common base output capacitance.
C_{obo}	Common base output capacitance (open circuit).
C_{obs}	Common base output capacitance (short circuit).
C_{oc}	Common collector output capacitance.
C_{oe}	Common emitter output capacitance.
C_{oeo}	Common emitter output capacitance (open circuit).
C_{oes}	Common emitter output capacitance (short circuit).
C_{oss}	Drain source capacitance (gate shorted to source).

Transistor Capacitance Letter Symbols (Continued)

C_{rbs}	Common base reverse transfer capacitance (short circuit).
C_{rcs}	Common collector reverse transfer capacitance.
C_{res}	Common emitter reverse transfer capacitance (short circuit).
C_{sp}	Reflected capacitance.

Transistor Power Letter Symbols

P_{BE}	Base-emitter total power input (DC, average).
P_{BE}	Base-emitter total power input (instantaneous).
P_{CB}	Collector-base total power input (DC, average).
p_{CB}	Collector-base total power input (instantaneous).
P_{CE}	Collector-emitter total power input (DC, average).
p_{CE}	Collector-emitter total power input (instantaneous).
P_{EB}	Emitter-base total power input (DC, average).
P_{ES}	Emitter-base total power input (instantaneous).
P_{IB}	Common base input power (large signal).
P_{ib}	Common base input power (small signal).
P_{IC}	Common collector input power (large signal).
P_{ic}	Common collector input power (small signal).
P_{IE}	Common emitter input power (large signal).
P_{ie}	Common emitter input power (small signal).
P_{OB}	Common base output power (large signal).
P_{ob}	Common base output power (small signal).
P_{OC}	Common collector output power (large signal).
P_{oc}	Common collector output power (small signal).
P_{OE}	Common emitter output power (large signal).
P_{oe}	Common emitter output power (small signal).

Miscellaneous Transistor Letter Symbols

B, b	Base electrode.
B_Z	Large signal breakdown impedance.
b_z	Small signal breakdown impedance.
CB	Common base configuration.
C, c	Collector electrode.
CC	Common collector configuration.
CE	Common emitter configuration.
E, e	Emitter electrode.
$f\propto$	Cutoff frequency.
G_{mb}	Large signal transconductance (common base).
gM^b	Common base static transconductance.
g_{mb}	Small signal transconductance (common base).
gM^c	Common collector static transconductance.
G_{mc}	Large signal transconductance (common collector).
g_{mc}	Small signal transconductance (common collector).
gM^e	Common emitter static transconductance.
G_{me}	Large signal transconductance (common emitter).
g_{me}	Small signal transconductance (common emitter).
h	Hybrid.
h_{ie}	Common emitter input impedance (small signal, short circuited).
H_{11}	Input impedance with output short circuited.
H_{22}	Output admittance with input open circuited.
L_c	Conversion loss.
NF	Noise figure.
NPN	Transistor with one P-type and two N-type semiconductors.
N-type	Semiconductor with donor impurity.
PN	Combination of N-type and P-type semiconductor.
PNP	Transistor with one N-type and two P-type semiconductors.
P_T	Total power input (DC, average).
p_T	Total power input (instantaneous).
P-type	Semiconductor with acceptor impurity.
T	Temperature.

Miscellaneous Transistor Letter Symbols (Continued)

T_A	Ambient temperature.
T_C	Case temperature.
t_d	Time delay.
t_f	Fall time.
t_{fr}	Forward recovery time.
T_j or T_J	Junction temperature.
T_{opr}	Operating temperature.
t_p	Pulse time.
t_r	Rise time.
t_{rr}	Reverse recovery time.
t_s	Storage time.
T_{stg}	Storage temperature.
t_w	Pulse average time.

Greek Alphabet and Common Prefixes

Name	Capital	Lower-case	Designates
Alpha	A	α	Angles
Beta	B	β	Angles, flux density
Gamma	Γ	γ	Conductivity
Delta	Δ	δ	Variation of a quantity, increment changes
Epsilon	E	ϵ	Base of natural logarithms (2.71828)
Zeta	Z	ζ	Impedance, coefficients, coordinates
Eta	H	η	Hysteresis coefficient, efficiency, magnetizing force
Theta	Θ	θ	Phase angle
Iota	I	ι	
Kappa	K	κ	Dielectric constant, coupling coefficient, susceptibility.

Greek Alphabet and Common Prefixes (Continued)

Name	Capital	Lower-case	Designates
Lambda	Λ	λ	Wavelength
Mu	M	μ	Permeability, micro, amplification factor
Nu	N	ν	Reluctivity
Xi	Ξ	ξ	
Omicron	O	o	
Pi	Π	π	3.1416
Rho	P	ρ	Resistivity
Sigma	Σ	σ	
Tau	T	τ	Time constant, time-phase displacement
Upsilon	Υ	υ	
Phi	Φ	ϕ	Angles, magnetic flux
Chi	X	χ	
Psi	Ψ	ψ	Dielectric flux, phase difference
Omega	Ω	ω	Ohms (capital), angular velocity ($2\pi f$)

Semiconductor Greek Letters

α	Alpha; current amplification factor.
α_{fb}	Forward short-circuit current amplification factor for the CB configuration.
α_{fc}	Forward short-circuit current amplification factor for the CC configuration.
α_{fe}	Forward short-circuit current amplification factor for the CE configuration.
∞	Infinity.
μA	Microampere.
μ_{rb}	Reverse open-circuited voltage amplification factor for CB configuration.
μ_{rc}	Reverse open-circuited voltage amplification factor for CC configuration.
μ_{re}	Reverse open-circuited voltage amplification factor for CE configuration.

APPENDIX E

Electronic Color Coding

Tables E-1 through E-5 are color code systems which are generally used in the electronic industry.

RESISTOR COLOR CODING

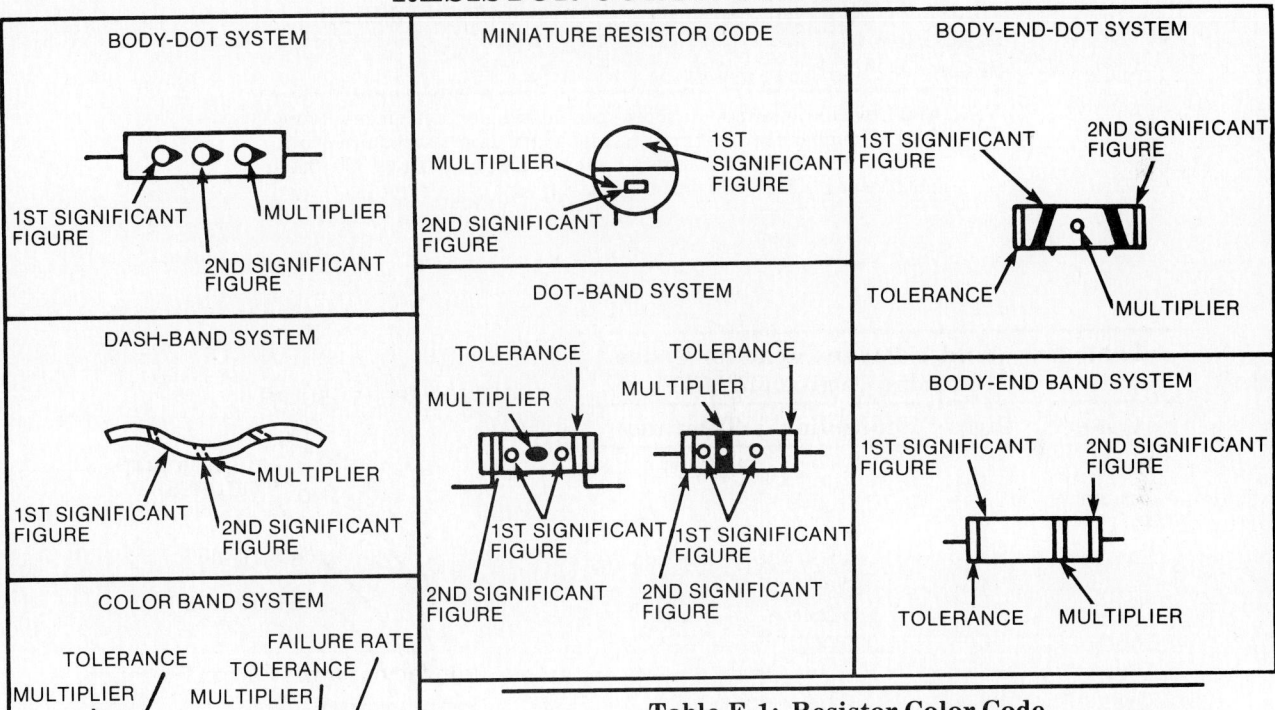

Color	Digit	Multiplier	Carbon ± Tolerance	Film-type Tolerance
Black	0	1	20%	0
Brown	1	10	1%	1%
Red	2	100	2%	2%
Orange	3	1,000	3%	
Yellow	4	10,000	GMV	
Green	5	100,000	5% (alt.)	0.5%
Blue	6	1,000,000	6%	0.25%
Violet	7	10,000,000	12.5%	0.1%
Gray	8	0.01 (alt.)	30%	0.05%
White	9	0.1 (alt.)	10% (alt.)	
Silver		0.01 (pref.)	10% (pref.)	10%
Gold		0.1 (pref.)	5% (pref.)	5%
No color			20%	

Table E-1: Resistor Color Code

RESISTORS WITH BLACK BODY COLOR ARE COMPOSITION, NONINSULATED. RESISTORS WITH COLORED BODIES ARE COMPOSITION INSULATED. WIREWOUND RESISTORS HAVE THE 1ST DIGIT COLOR BAND DOUBLE WIDTH.

CAPACITOR COLOR CODING

Table E-2: Molded Mica Capacitor Codes
(Capacitance given in MMF)

Color	Digit	Multiplier	Tolerance	Class or Characteristic
Black	0	1	20%	A
Brown	1	10	1%	B
Red	2	100	2%	C
Orange	3	1,000	3%	D
Yellow	4	10,000	—	E
Green	5		5% (ELA)	F(JAN)
Blue	6			G(JAN)
Violet	7			
Gray	8			I(ELA)
White	9			J(ELA)
Gold		0.1	5% (JAN)	
Silver		0.01	10%	

Note: Class or characteristic denotes specifications of design involving Q factors, temperature coefficients, and production test requirements. All axial lead mica capacitors have a voltage rating of 300, 500, or 1,000V for 4.0MMF whichever is greater.

Table E-3: Molded Paper Capacitor Codes
(Capacitance given in MMF)

Color	Digit	Multiplier	Tolerance
Black	0	1	20%
Brown	1	10	
Red	2	100	
Orange	3	1,000	
Yellow	4	10,000	
Green	5	100,000	5%
Blue	6	1,000,000	
Violet	7		
Gray	8		
White	9		10%
Gold			5%
Silver			10%
No color			20%

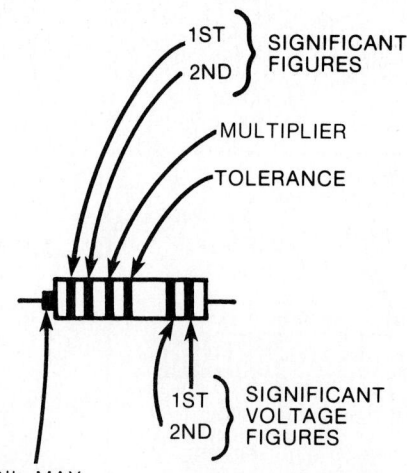

MOLDED PAPER TUBULAR

1ST } SIGNIFICANT
2ND } FIGURES

MULTIPLIER

TOLERANCE

1ST } SIGNIFICANT
2ND } VOLTAGE FIGURES

INDICATES OUTER FOIL. MAY BE ON EITHER END. MAY ALSO BE INDICATED BY OTHER METHODS SUCH AS TYPOGRAPH-ICAL MARKING OR BLACK STRIP.

ADD TWO ZEROS TO SIGNIFICANT VOLTAGE FIGURES. ONE BAND INDICATES VOLTAGE RATINGS UNDER 1,000V.

734

Table E-4: Tolerances		
	Tolerance	
JAN letter	**10 MMF or less**	**Over 10 MMF**
C	± 0.25 MMF	
D	± 0.6 MMF	
F	± 1.0 MMF	± 1%
G	± 2.0 MMF	± 2%
J		± 5%
K		± 10%
M		± 20%

MOLDED-INSULATED AXIAL
LEAD CERAMICS

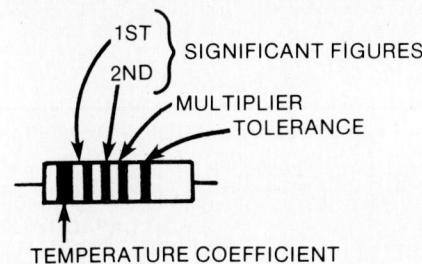

TYPOGRAPHICALLY MARKED
CERAMICS

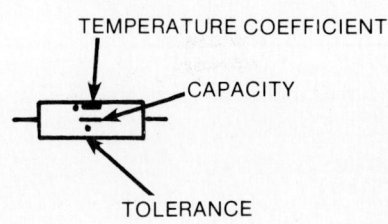

EXTENDED RANGE T.C.
TUBULAR CERAMICS

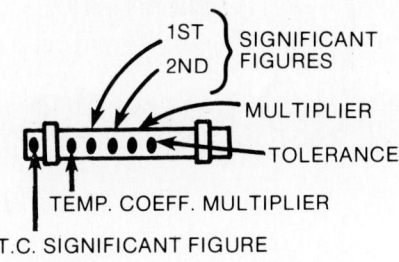

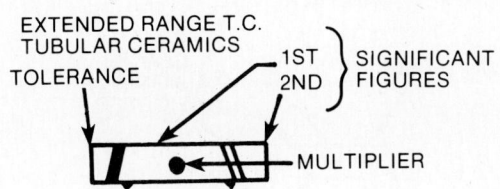

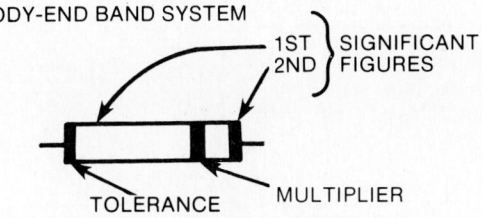

Table E-5: Ceramic Capacitor Codes (Capacity given in MMF)

Color	Digit	Multiplier	Tolerance 10 MMF or Less	Tolerance Over 10 MMF	Temperature Coefficient PPM/°C	Extended Range Temp Coeff. Significant Figure	Extended Range Temp Coeff. Multiplier
Black	0	1	± 2.0 MMF	± 20%	0 (NP0)	0.9	–1
Brown	1	10	± 0.1 MMF	± 1%	–33 (N033)		–10
Red	2	100		± 2%	–75 (N075)	1.0	–100
Orange	3	1,000		± 2.5%	–150 (N150)	1.5	–1,000
Yellow	4	10,000			–220 (N220)	2.2	–10,000
Green	5		± 0.5 MMF	± 5%	–330 (N330)	3.3	+1
Blue	6				–470 (N470)	4.7	+10
Violet	7				–750 (N750)	7.5	+100
Gray	8	0.01	± 0.25 MMF		–30 (P030)		+1,000
White	9	0.1	± 1.0 MMF	± 10%	General purpose bypass and coupling		+100,000
Silver							
Gold					+100 (P100) (Jan)		

Note: Ceramic capacitor voltage ratings are standard 500V for some manufacturers and 1,000V for other manufacturers, unless otherwise specified.

DISC CERAMICS
(5-DOT SYSTEM)

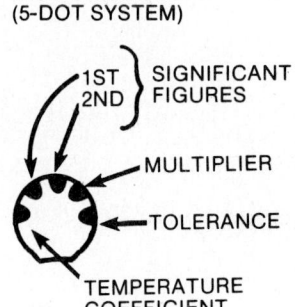

1ST 2ND SIGNIFICANT FIGURES
MULTIPLIER
TOLERANCE
TEMPERATURE COEFFICIENT

DISC-CERAMICS
(3-DOT SYSTEM)

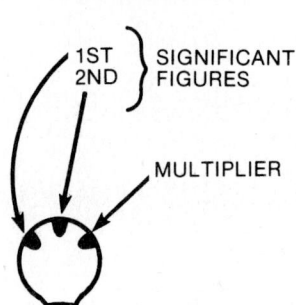

1ST 2ND SIGNIFICANT FIGURES
MULTIPLIER

HIGH CAPACITANCE TUBULAR CERAMICS INSULATED OR NONINSULATED

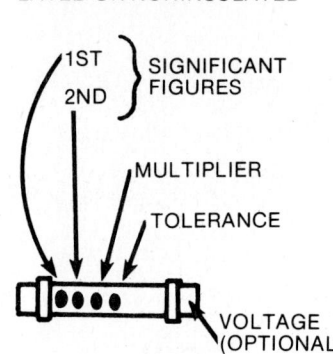

1ST 2ND SIGNIFICANT FIGURES
MULTIPLIER
TOLERANCE
VOLTAGE (OPTIONAL)

TEMPERATURE COMPENSATING TUBULAR CERAMICS

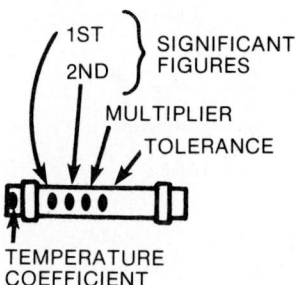

1ST 2ND SIGNIFICANT FIGURES
MULTIPLIER
TOLERANCE
TEMPERATURE COEFFICIENT

CURRENT STANDARD JAN AND ELA CODE

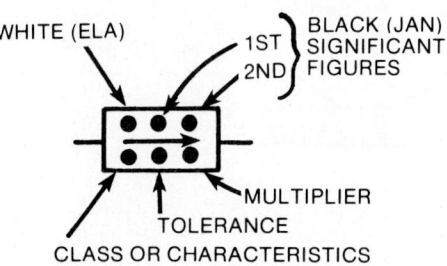

WHITE (ELA)
BLACK (JAN)
1ST 2ND SIGNIFICANT FIGURES
MULTIPLIER
TOLERANCE
CLASS OR CHARACTERISTICS

736

BUTTON SILVER MICA

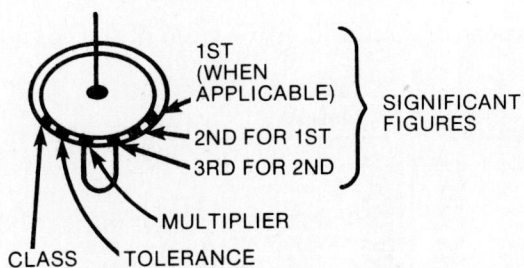

1ST (WHEN APPLICABLE)
2ND FOR 1ST
3RD FOR 2ND
} SIGNIFICANT FIGURES
MULTIPLIER
CLASS   TOLERANCE

MOLDED FLAT PAPER CAPACITORS (COMMERICAL CODE)

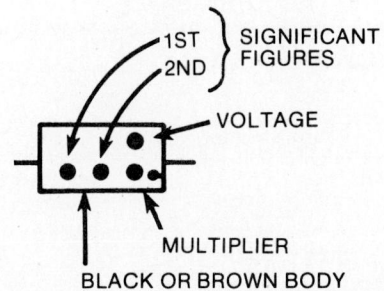

1ST
2ND
} SIGNIFICANT FIGURES
VOLTAGE
MULTIPLIER
BLACK OR BROWN BODY

MOLDED FLAT PAPER CAPACITORS (JAN CODE)

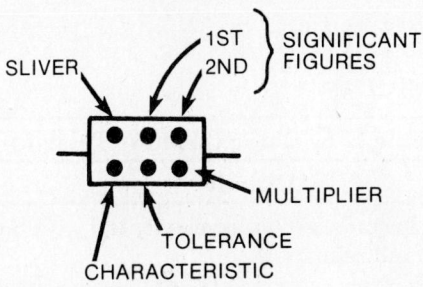

SLIVER
1ST
2ND
} SIGNIFICANT FIGURES
MULTIPLIER
TOLERANCE
CHARACTERISTIC

MOLDED CERAMICS USING STANDARD RESISTOR COLOR-CODE

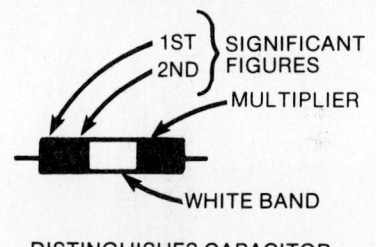

1ST
2ND
} SIGNIFICANT FIGURES
MULTIPLIER
WHITE BAND
DISTINGUISHES CAPACITOR FROM RESISTOR

BUTTON CERAMICS

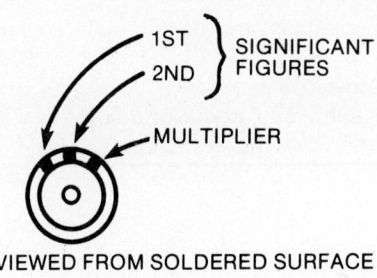

1ST
2ND
} SIGNIFICANT FIGURES
MULTIPLIER
VIEWED FROM SOLDERED SURFACE

STAND-OFF CERAMICS

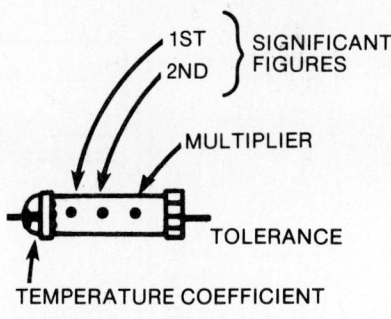

1ST
2ND
} SIGNIFICANT FIGURES
MULTIPLIER
TOLERANCE
TEMPERATURE COEFFICIENT

FEED-THRU CERAMICS

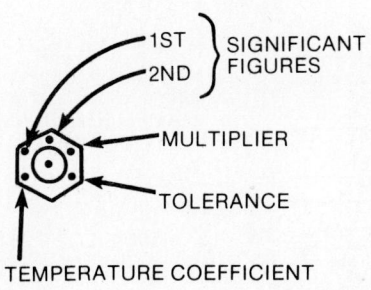

1ST
2ND
} SIGNIFICANT FIGURES
MULTIPLIER
TOLERANCE
TEMPERATURE COEFFICIENT

737

TRANSFORMER COLOR CODING

Standard colors used in chassis wiring for the purpose of circuit identification of the equipment are shown in Table E-6.

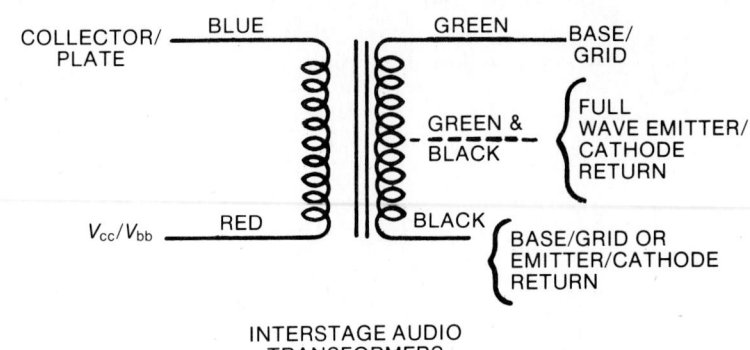

INTERSTAGE AUDIO
TRANSFORMERS

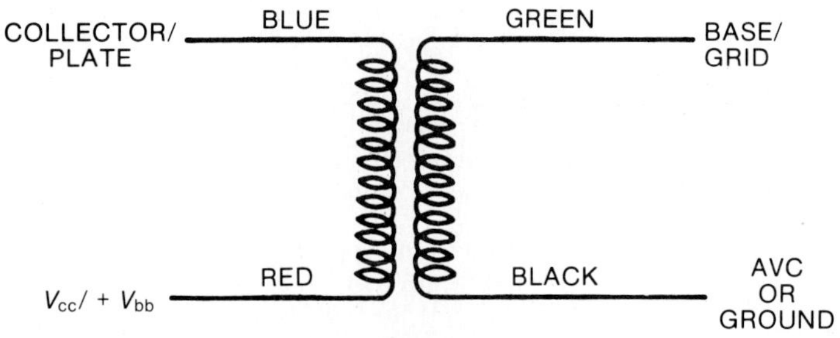

Table E-6: Color Code for Transformers

Circuit	Color
Grounds, grounded elements, and returns	Black
Heaters or filaments, off ground	Brown
Power supply $+V_{cc}/+V_{bb}$	Red
Screen grids	Orange
Emitters/cathodes	Yellow
Bases/control grids	Green
Collectors/plates	Blue
Power supply, minus	Violet (Purple)
AC power lines	Gray
Miscellaneous, above or below ground returns, AVC, etc.	White

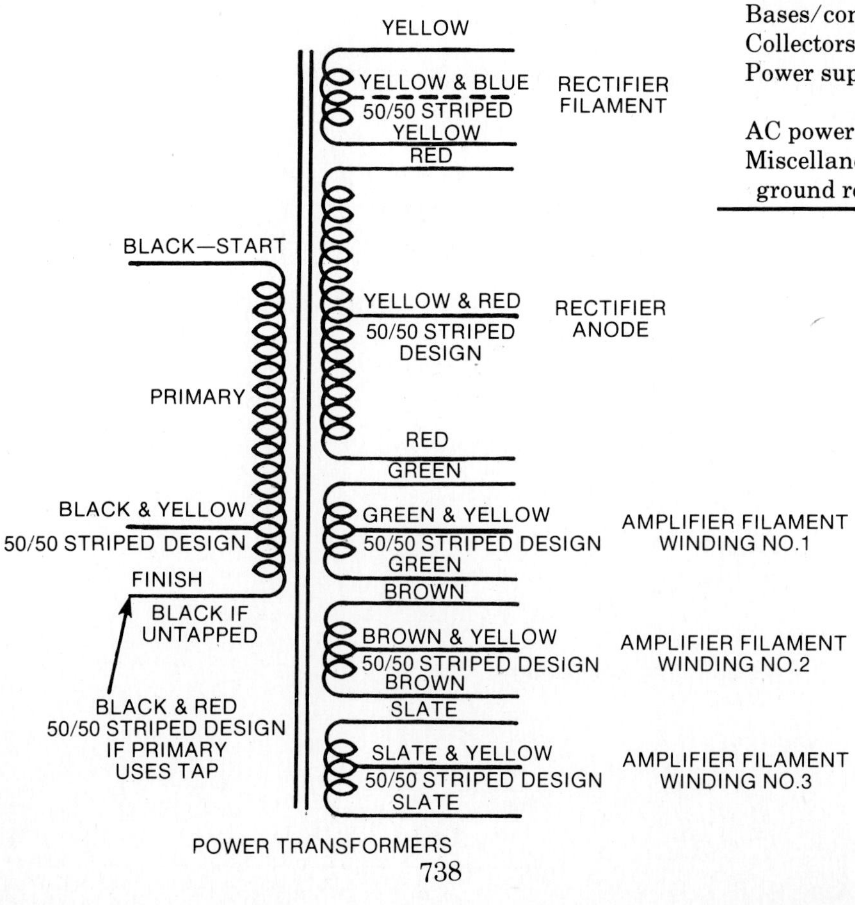

IF TRANSFORMERS

POWER TRANSFORMERS

738

WIRE COLOR CODING

For fast and easy servicing, many manufacturers use a component-terminal-marking and wiring-color-coding system. These terminal markings appear in the wiring diagram as well as the wiring color-coding symbols. Table E-7 lists the color coding most generally used.

Table E-7: Commercial Radio-Television Chassis Wiring Color Code			
	Digit	**Color**	**Circuitry Function**
Grounds	0	Black (solid)	Grounded components and returns
		Black-brown	Identified grounds i.e., filament
		Black-red	Identified grounds i.e., B—
		Black-yellow	Identified grounds i.e., cathode/emitter
		Black-green	Identified grounds i.e., grid/base
Heater/Filaments	1	Brown (solid)	Above or below ground
		Brown-red	Identified purpose i.e., rectifier
		Brown-yellow	Identified purpose
		Brown-green	Identified purpose
		Brown-white	Identified purpose
Power Supplies	2	Red (solid)	Main source, B+
		Red-black	Intermediate source potential
		Red-yellow	Identified source potential
		Red-green	Intermediate source potential
		Red-blue	Intermediate source potential
		Red-white	Identified source
		Red-blue-yellow	Intermediate source potential
Screen Grids	3	Orange (solid)	Positive potential
Cathodes/Emitter	4	Yellow (solid)	Above and below ground
		Yellow-red	Identified circuit i.e., output
		Yellow-green	Identified circuit i.e., oscillator
Control Grids/Bases	5	Green (solid)	Bias potentials
		Green-red	Identified element
		Green-yellow	Identified element
		Green-white	Identified element
Plate/Collector	6	Blue (solid)	Plate or collector potentials
		Blue-red	Identified element
		Blue-yellow	Identified element
Miscellaneous	7	Violet (solid)	Biases, returns

Table E-7: Commercial Radio-Television Chassis
Wiring Color Code (Continued)

	Digit	Color	Circuitry Function
AC Power Source	8	Gray (solid)	AC power potentials
Bias Supply Source	9	White (solid)	Main source
		White-black	Alternate/offground connection
		White-brown	Intermediate AVC bias
		White-red	Below ground, maximum value
		White-orange	Intermediate fixed value
		White-yellow	Intermediate fixed value
		White-green	Preferred AVC bias
		White-blue	Internal antenna or connection to

APPENDIX F

Electronics Symbols

AMPLIFIER

general

with two inputs

with two outputs

with adjustable gain

with associated power supply

with associated attenuator

with external feedback path

Amplifier Letter Combinations (amplifier-use identification in symbol if required)

BDG Bridging
BST Booster
CMP Compression
DC   Direct Current
EXP Expansion
LIM Limiting
MON Monitoring
PGM Program
PRE Preliminary
PWR Power
TRQ Torque

ANTENNA

general

dipole

loop

counterpoise

ARRESTER, LIGHTNING

general

carbon block

electrolytic or aluminum cell

horn gap

protective gap

sphere gap

valve or film element

multigap

ATTENUATOR, FIXED

(same symbol as variable attenuator, without vari-ability)

ATTENUATOR, VARIABLE

balanced

unbalanced

AUDIBLE SIGNALING DEVICE

bell, electrical; ringer, telephone

buzzer

horn, electrical; loud-speaker; siren; under-water sound hydrophone, projector, or transducer

Horn, Letter Combinations (if required)

*HN   Horn, electrical
*HW   Howler
*LS   Loudspeaker
*SN   Siren
‡EM   Electromagnetic with moving coil
‡EMN Electromagnetic with moving coil and neutralizing winding
‡MG   Magnetic armature
‡PM   Permanent magnet with moving coil

identification replaces (*) asterisk and (‡) dagger

sounder, telegraph

BATTERY

generalized DC source; one cell

multicell

CAPACITOR

general

polarized

adjustable or variable

continuously adjustable or variable differential

phase-shifter

split-stator

feed-through

CELL, PHOTOSENSITIVE (Semiconductor)

asymmetrical photocon-ductive transducer

symmetrical photocon-ductive transducer

photovoltaic transducer; solar cell

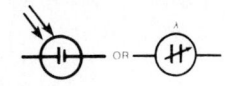

CIRCUIT BREAKER

general

with magnetic overload

drawout type

CIRCUIT ELEMENT

general

Circuit Element Letter Combinations (replaces asterisk)

EG	Equalizer
FAX	Facsimile set
FL	Filter
FL-BE	Filter, band elimination
FL-BP	Filter, band pass
FL-HP	Filter, high pass
FL-LP	Filter, low pass
PS	Power supply
RG	Recording unit
RU	Reproducing unit
DIAL	Telephone dial
TEL	Telephone station
TPR	Teleprinter
TTY	Teletypewriter

Additional Letter Combinations (symbols preferred)

AR	Amplifier
AT	Attenuator
C	Capacitor
CB	Circuit breaker
HS	Handset
I	Indicating or switch board lamp
L	Inductor
J	Jack
LS	Loudspeaker
MIC	Microphone
OSC	Oscillator
PAD	Pad
P	Plug
HT	Receiver, headset
K	Relay
R	Resistor
S	Switch or key switch
T	Transformer
WR	Wall receptacle

CLUTCH; BRAKE

disengaged when operating means is de-energized

engaged when operating means is de-energized

COIL, RELAY, AND OPERATING

semicircular dot indicates inner end of wiring

CONNECTOR

assembly, movable or stationary portion; jack, plug, or receptacle

jack or receptacle

plug

separable connectors

two-conductor switchboard jack

two-conductor switchboard plug

jacks normalled through one way

jacks normalled through both ways

2-conductor nonpolarized, female contacts

2-conductor polarized, male contacts

waveguide flange

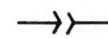

plain, rectangular

choke, rectangular

engaged 4-conductor; the plug has 1 male and 3 female contacts, individual contact designations shown

coaxial, outside conductor shown carried through

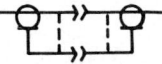

coaxial, center conductor shown carried through; outside conductor not carried through

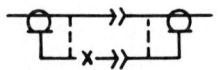

mated choke flanges in rectangular waveguide

COUNTER, ELECTROMAGNETIC; MESSAGE REGISTER

general

with a make contact

COUPLER, DIRECTIONAL

(common coaxial/waveguide usage)

(common coaxial/waveguide usage)

E-plane aperture-coupling, 30dB transmission loss

COUPLING

by loop from coaxial to circular waveguide, DC grounds connected

CRYSTAL, PIEZO-ELECTRIC

DELAY LINE

general

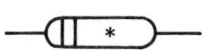

tapped delay

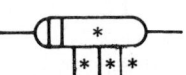

bifilar slow-wave structure (commonly used in traveling-wave tubes)

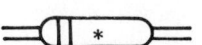

(length of delay indication replaces asterisk)

DETECTOR, PRIMARY; MEASURING TRANSDUCER

DISCONTINUITY
(common coaxial/ waveguide usage)

equivalent series element, general

capacitive reactance

inductive reactance

inductance-capacitance circuit, infinite reactance at resonance

inductance-capacitance circuit, zero reactance at resonance

resistance

equivalent shunt element, general

capacitive susceptance

conductance

inductive susceptance

inductance-capacitance circuit, infinite susceptance at resonance

inductance-capacitance circuit, zero susceptance at resonance

ELECTRON TUBE

triode

pentode, envelope connected to base terminal

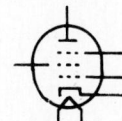

twin triode, equipotential cathode

typical wiring figure to show tube symbols placed in any convenient position

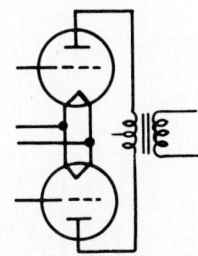

rectifier; voltage regulator (see LAMP, GLOW)

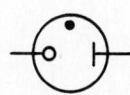

phototube, single and multiplier

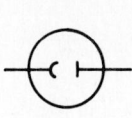 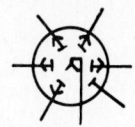

cathode ray tube, electrostatic and magnetic deflection

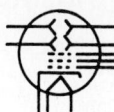 

mercury-pool tube, ignitor and control grid (see RECTIFIER)

resonant magnetron, coaxial output and permanent magnet

reflex klystron, integral cavity, aperture coupled

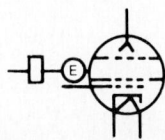

transmit-receive (TR) tube gas filled, tunable integral cavity, aperture coupled, with starter

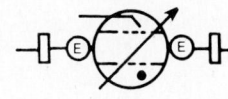

traveling-wave tube (typical)

forward-wave traveling-wave-tube amplifier shown with four grids, having slow-wave structure with attenuation, magnetic focusing by external permanent magnet, rf input and rf output coupling each E-plane aperture to external rectangular waveguide

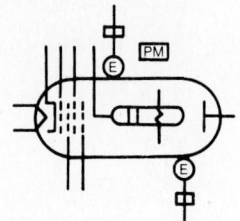

FERRITE DEVICES

field polarization rotator

field polarization amplitude modulator

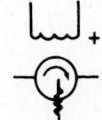

FUSE

high-voltage primary cutout, dry

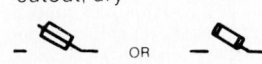

high-voltage primary cutout, oil

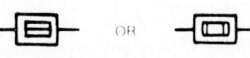

GOVERNOR (Contact-making)

contacts shown here as closed

HALL GENERATOR

HANDSET

general

operator's set with push-to talk switch

HYBRID

general

junction (common coaxial/waveguide usage)

circular

(E, H, or HE transverse field indicators replace asterisk)

rectangular waveguide and coaxial coupling

INDUCTOR

general

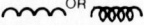

iron core

tapped

adjustable, continuously adjustable

KEY, TELEGRAPH

LAMP

ballast lamp; ballast tube

lamp, fluorescent, 2 and 4 terminal

lamp, glow; neon lamp

AC

DC

lamp, incandescent

indicating lamp; switch-board lamp

LINES AND CONNECTIONS

integral conductor

harness wire

crossover

crossover

shield

permanent connection

terminal

ground (earth)

ground (chassis)

plug connector

grounded service cord (3 prong)

service cord (2 prong)

mechanical connection

terminal board (number and arrangement as required)

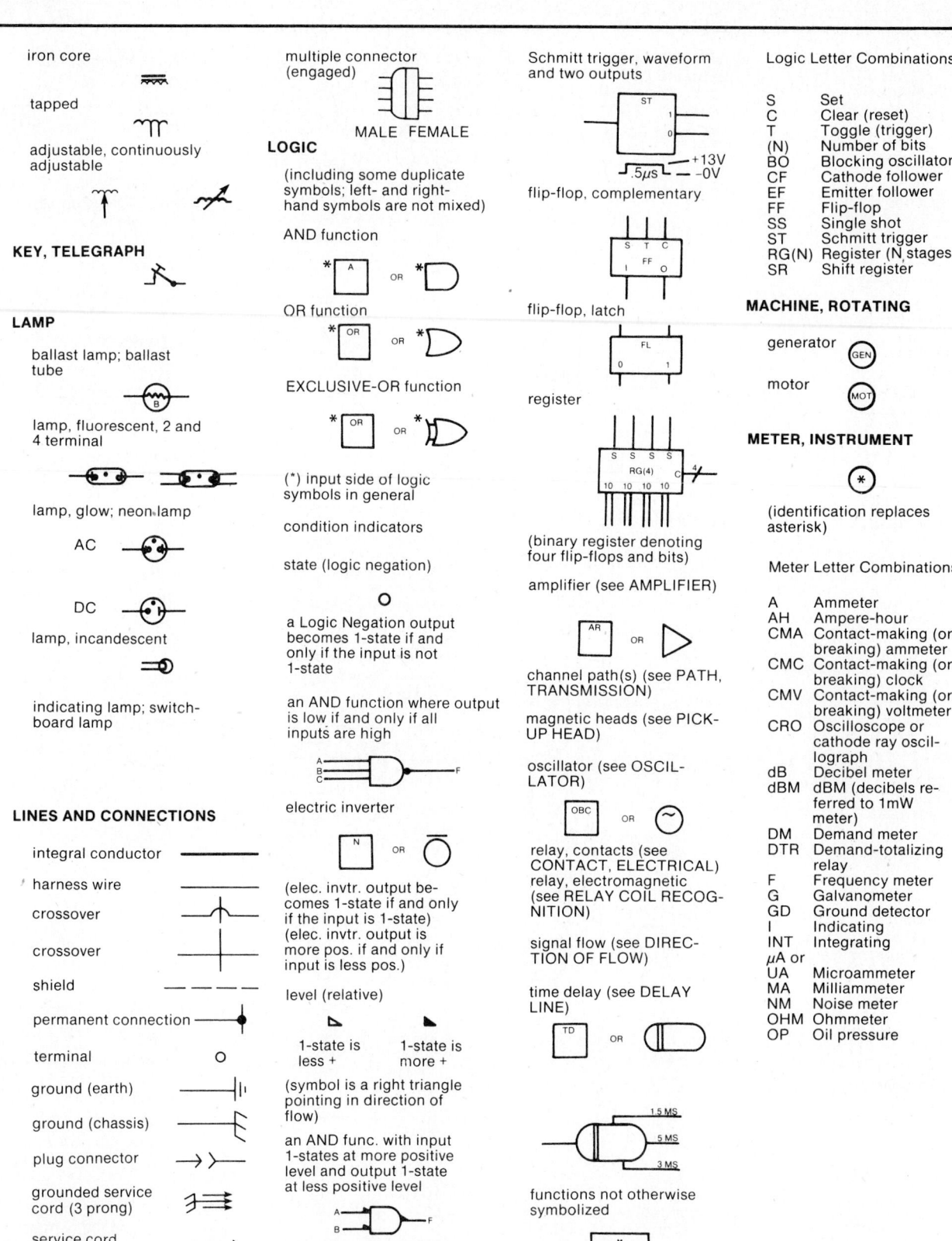

multiple connector (engaged)

MALE   FEMALE

LOGIC

(including some duplicate symbols; left- and right-hand symbols are not mixed)

AND function

* A   OR   *

OR function

* OR   OR   *

EXCLUSIVE-OR function

* OR   OR   *

(*) input side of logic symbols in general

condition indicators

state (logic negation)

o

a Logic Negation output becomes 1-state if and only if the input is not 1-state

an AND function where output is low if and only if all inputs are high

A B C — F

electric inverter

N   OR

(elec. invtr. output becomes 1-state if and only if the input is 1-state)
(elec. invtr. output is more pos. if and only if input is less pos.)

level (relative)

1-state is less +   1-state is more +

(symbol is a right triangle pointing in direction of flow)

an AND func. with input 1-states at more positive level and output 1-state at less positive level

A B — F

single shot (one output)

SS * *

waveform data replaces inside/outside asterisk

Schmitt trigger, waveform and two outputs

ST 1 0   .5μs   +13V −0V

flip-flop, complementary

S T C   FF   1 O

flip-flop, latch

FL   0 1

register

S S S S   RG(4)   10 10 10 10   C

(binary register denoting four flip-flops and bits)

amplifier (see AMPLIFIER)

AR   OR

channel path(s) (see PATH, TRANSMISSION)

magnetic heads (see PICK-UP HEAD)

oscillator (see OSCIL-LATOR)

OBC   OR

relay, contacts (see CONTACT, ELECTRICAL)
relay, electromagnetic (see RELAY COIL RECOG-NITION)

signal flow (see DIREC-TION OF FLOW)

time delay (see DELAY LINE)

TD   OR

1.5 MS   5 MS   3 MS

functions not otherwise symbolized

*

(identification replaces asterisk)

Logic Letter Combinations

S	Set
C	Clear (reset)
T	Toggle (trigger)
(N)	Number of bits
BO	Blocking oscillator
CF	Cathode follower
EF	Emitter follower
FF	Flip-flop
SS	Single shot
ST	Schmitt trigger
RG(N)	Register (N stages)
SR	Shift register

MACHINE, ROTATING

generator   GEN

motor   MOT

METER, INSTRUMENT

*

(identification replaces asterisk)

Meter Letter Combinations

A	Ammeter
AH	Ampere-hour
CMA	Contact-making (or breaking) ammeter
CMC	Contact-making (or breaking) clock
CMV	Contact-making (or breaking) voltmeter
CRO	Oscilloscope or cathode ray oscil-lograph
dB	Decibel meter
dBM	dBM (decibels referred to 1mW meter)
DM	Demand meter
DTR	Demand-totalizing relay
F	Frequency meter
G	Galvanometer
GD	Ground detector
I	Indicating
INT	Integrating
μA or UA	Microammeter
MA	Milliammeter
NM	Noise meter
OHM	Ohmmeter
OP	Oil pressure

MODE TRANSDUCER

(common coaxial/waveguide usage)

transducer from rectangular waveguide to coaxial with mode suppression, DC grounds connected

MOTION, MECHANICAL

rotation applied to a resistor

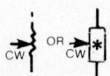

(identification replaces asterisk)

NUCLEAR-RADIATION DETECTOR, gas filled; IONIZATION CHAMBER; PROPORTIONAL COUNTER TUBE; GEIGER-MULLER COUNTER TUBE (50)

MOTORS

timer motor

single-speed

two-speed

three-speed

Note: Internal motor circuitry may be shown if required.

compressor motor

single-speed motor

two-speed motor

three-speed motor

multispeed motor (show internal circuitry as req'd.)

PATH, TRANSMISSION

cable; 2-conductor, shield grounded and 5-conductor shielded

PICKUP HEAD

general

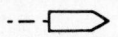

writing; recording

reading; playback

erasing

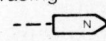

writing, reading, and erasing

stereo

RECTIFIER

semiconductor diode; metallic rectifier; electrolytic rectifier; asymmetrical varistor

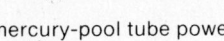

mercury-pool tube power rectifier

fullwave bridge-type

RESISTOR

general

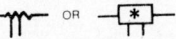

tapped

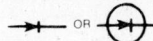

heating

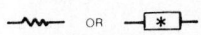

symmetrical varistor resistor, voltage sensitive (silicon carbide, etc.)

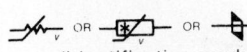

(identification marks replace asterisk)

with adjustable contact

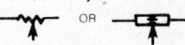

adjustable or continuously adjustable (variable)

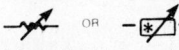

(identification replaces asterisk)

RESONATOR, TUNED CAVITY

(common coaxial/waveguide usage)

resonator with mode suppression coupled by an E-plane aperture to a guided transmission path and by a loop to a coaxial path

tunable resonator with DC ground connected to an electron device and adjustably coupled by an E-plane aperture to a rectangular waveguide

ROTARY JOINT, RF (COUPLER)

general; with rectangular waveguide

(transmission path recognition symbol replaces asterisk)

coaxial type in rectangular waveguide

circular waveguide type in rectangular waveguide

SEMICONDUCTOR DEVICE
(Two Terminal, diode)

semiconductor diode; rectifier

capacitive diode (also Varicap, Varactor, reactance diode, parametric diode)

 OR

breakdown diode, unidirectional (also backward diode, avalanche diode, voltage regulator diode, Zener diode, voltage reference diode)

 OR

breakdown diode, bidirectional and backward diode (also bipolar voltage limiter)

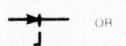 OR

tunnel diode (also Esaki diode)

 OR

temperature-dependent diode

 OR

photodiode (also solar cell)

 OR

semiconductor diode, PNPN switch (also Shockley diode, four-layer diode and SCR)

  OR

(Multiterminal, transistor, etc.)

PNP transistor

NPN transistor

unijunction transistor, N-type base

unijunction transistor,
P-type base

field-effect transistor,
N-type base

field-effect transistor,
P-type base

semiconductor triode,
PNPN-type switch

semiconductor triode,
NPNP-type switch

NPN transistor, trans-
verse-biased based

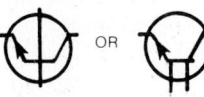

PNIP transistor, ohmic
connection to the intrin-
sic region

NPIN transistor, ohmic
connection to the in-
trinsic region

PNIN transistor, ohmic
connection to the in-
trinsic region

NPIP transistor, ohmic
connection to the intrin-
sic region

SQUIB

explosive

igniter

sensing link; fusible
link operated

SWITCH

push button, circuit closing (make)

push button, circuit op-
ening (break)

nonlocking; momentary
circuit closing (make)

nonlocking; momentary
circuit opening (break)

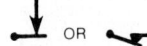

transfer

locking, circuit closing
(make)

locking, circuit opening
(break)

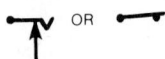

transfer, 3-position

wafer

(example shown: 3-pole
3-circuit with 2 non-
shorting and 1 shorting
moving contacts)

safety interlock, circuit
opening and closing

2-pole field-discharge
knife, with terminals and
discharge resistor

(identification replaces
asterisk)

SYNCHRO

Synchro Letter Combina-
tions
CDX Control-differential
         transmitter
CT   Control transformer
CX   Control transmitter
TDR Torque-differential
         receiver
TDX Torque-differential
         transmitter
TR   Torque receiver
TX   Torque transmitter
RS   Resolver
B   Outer winding rotat-
         able in bearings

THERMAL ELEMENT

actuating device

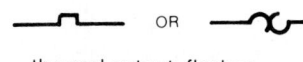

thermal cutout; flasher

thermal relay

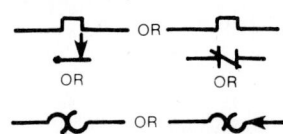

thermostat (operates on
rising temperature), con-
tact)

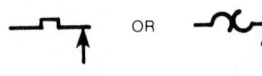

thermostat, make contact

thermostat, integral
heater and transfer
contacts

THERMISTOR; THERMAL
RESISTOR

with integral heater

THERMOCOUPLE

temperature-measuring

current-measuring, inte-
gral heater connected

HEATER

current-measuring, inte-
gral heater insulated

HEATER

temperature-measuring,
semiconductor

current-measuring, semi-
conductor

TRANSFORMER

general

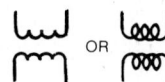

magnetic-core

one winding with adjust-
able inductance

separately adjustable
inductance

adjustable mutual induc-
tor, constant-current

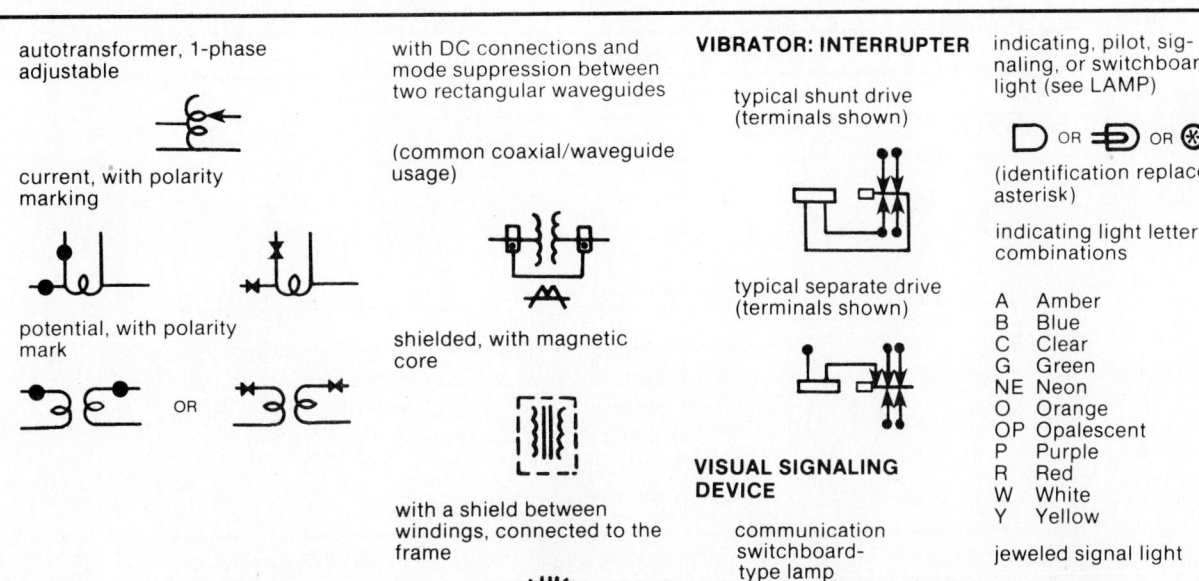

autotransformer, 1-phase adjustable

current, with polarity marking

potential, with polarity mark

OR

with DC connections and mode suppression between two rectangular waveguides

(common coaxial/waveguide usage)

shielded, with magnetic core

with a shield between windings, connected to the frame

VIBRATOR: INTERRUPTER

typical shunt drive (terminals shown)

typical separate drive (terminals shown)

VISUAL SIGNALING DEVICE

communication switchboard-type lamp

indicating, pilot, signaling, or switchboard light (see LAMP)

OR OR

(identification replaces asterisk)

indicating light letter combinations

A   Amber
B   Blue
C   Clear
G   Green
NE Neon
O   Orange
OP Opalescent
P   Purple
R   Red
W   White
Y   Yellow

jeweled signal light

DIGITAL LOGIC SYMBOLS

Name	Symbol	Notes	Name	Symbol	Notes
AND gate		$AB = Y$	J-K flip-flop		CP is clock pulse
OR gate		$A + B = Y$	Clock pulse		Falling edge
Inverter		$A = \overline{A}$	Trigger		Rising edge
NAND gate		$\overline{AB} = Y$	Operational amplifier, op amp		For digital-analog circuits
NOR gate		$\overline{A + B} = Y$			
Single shot	SS				
Schmitt trigger	ST				
Amplifier					

Bipolar Junction Transistors*

Type	Symbol	Electrodes	Notes
NPN		C = Collector B = Base E = Emitter; electron current into base	Needs $+V_c$ reverse voltage for collector; forward bias V_{BE} of 0.6V for silicon or 0.2V for germanium
PNP		C = Collector B = Base E = Emitter; electron current out from base	Needs $-V_c$ reverse voltage for collector; forward bias V_{BE} of 0.2V for germanium or 0.6V for silicon

*Type numbers for bipolar junction transistors start with 2N, as in 2N1482, for two semiconductor junctions in the EIA system. Japanese transistors are marked 2SA for small-signal PNP, 2SB for power PNP, 2SC for small-signal NPN, and 2SD for power NPN types.

Field-Effect Transistors* (FET)

Type	Symbol	Electrodes	Notes
JFET		D = Drain G = Gate S = Source	Junction-type FET; arrow pointing to channel indicates P-gate and N-channel; reverse bias at junction
IGFET or MOSFET N-channel		D = Drain G = Gate S = Source	Insulated-gate; depletion or depletion-enhancement type; SS substrate connected internally to source
IGFET or MOSFET P-channel		D = Drain G = Gate S = Source	Arrow pointing away from channel indicates P-channel
IGFET or MOSFET, N-channel, enhancement		D = Drain G = Gate S = Source	Broken lines for channel show enhancement type
Dual-gate IGFET or MOSFET, N-channel		D = Drain G_2 = Gate 2 G_1 = Gate 1 S = Source	Either or both gates control amount of drain current I_D

*Classified by EIA as type A, B, or C. For an N-channel: depletion type A takes negative gate bias for a middle value of I_D; depletion-enhancement type B can operate with zero gate bias; enhancement type C needs positive gate bias.

748

Semiconductor Diodes*

Type	Symbol	Applications
Rectifier	○—▸|—○	Conducts forward current with positive anode voltage
Zener	○—▸|⌐—○	Voltage regulator; takes reverse voltage
Varactor	○—▸|ǁ—○	Capacitive diode; C varies with amount of reverse voltage; used for electronic tuning
Tunnel	○—▸|⌐—○	Microwave diode; Esaki diode; same symbol for Schottky diode and Gunn diode
Photoconductive	○—⊗—○	Takes reverse voltage; high dark-resistance decreases with light input
Photovoltaic	○—⊕▸|—○	No voltage applied; light input generates voltage output; solar cell
Photoemissive	○—⊕▸|—○	Light emitting diode (LED) produces red, green, or yellow light output; takes forward voltage

*Type numbers for diodes start with 1N, for one semiconductor junction. An example is 1N2864.

Types of Thyristors

Name	Symbol	Electrodes	Notes
Silicon-controlled rectifier (SCR)		C = Cathode G = Gate A = Anode	Unidirectional with positive anode voltage; G can turn SCR on but not off
Triac		T1 = Terminal 1 G = Gate T2 = Terminal 2	Bidirectional with gate; G can turn triac on for either polarity, but not off
Diac		T1 = Terminal 1 T2 = Terminal 2	Bidirectional trigger without gate
Unijunction transistor (UJT)		B2 = Base 2 E = Emitter B1 = Base 1	Base has standoff voltage from DC supply; V_E applied to start conduction

Protective Devices for Semiconductors

Type	Symbol	Notes
Thermistor		Resistance decreases with temperature; used as a series component
Voltage-dependent resistor (VDR)		Resistance decreases with V continuously; used as a series component
Varistor		Resistance decreases abruptly at specific voltage; used as a shunt component to short-circuit voltage transients

DIODE POLARITY IDENTIFICATION

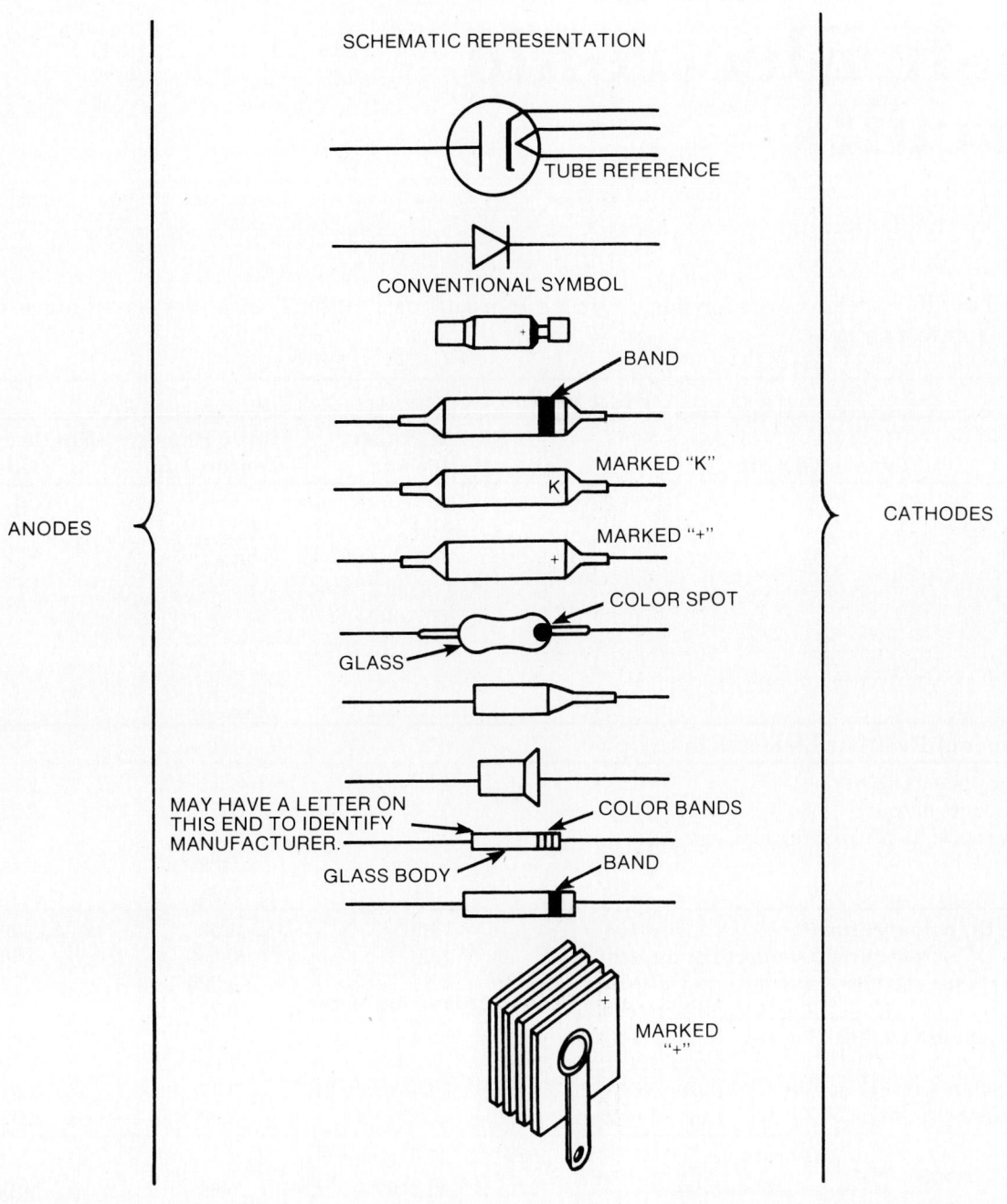

SCHEMATIC REPRESENTATION

TUBE REFERENCE

CONVENTIONAL SYMBOL

ANODES

CATHODES

BAND

MARKED "K"

MARKED "+"

COLOR SPOT

GLASS

MAY HAVE A LETTER ON
THIS END TO IDENTIFY
MANUFACTURER.

COLOR BANDS

GLASS BODY

BAND

MARKED
"+"

751

APPENDIX G

Basic Electronic Circuits

Included in this appendix are such basic electronic circuits as: rectifiers, amplifiers, oscillators, filters, and transistor test circuits.

Basic Rectifier Circuits			
Type of Circuit	**Single Phase Half-Wave**	**Single Phase Center-Tap**	**Single Phase Bridge**
Primary Secondary One cycle wave of rectifier output voltage (no overlap)			
Number of Rectifier Elements in Circuit	**1**	**2**	**4**
rms DC volts output	1.57	1.11	1.11
Peak DC volts output	3.14	1.57	1.57
Peak reverse volts per rectifier element	3.14	3.14	1.57
	1.41	2.82	1.41
	1.41	1.41	1.41
Average DC output current	1.00	1.00	1.00
Average DC output current per rectifier element	1.00	0.500	0.500
rms current per rectifier element Resistive load	1.57	0.785	0.785
Inductive load		0.707	0.707
Peak current per rectifier element Resistive load	3.14	1.57	1.57
Inductive load		1.00	1.00
Ratio: Peak to average current Resistive load	3.14	3.14	3.14
per element Inductive load		2.00	2.00
% ripple $\dfrac{\text{rms of ripple}}{(\text{average output voltage})}$	121%	48%	48%
	Resistive Load		**Inductive**
Transformer secondary rms volts per leg	2.22	1.11 (To center-tap)	1.11 (Total)
Transformer secondary rms volts line-to-line	2.22	2.22	1.11
Secondary line current	1.57	0.707	1.00
Transformer secondary volt-amperes per leg	3.49	1.57	1.11
Transformer primary rms amperes per leg	1.57	1.00	1.00
Transformer primary volt-amperes per leg	3.49	1.11	1.11
Average of primary and secondary volt-amperes	3.49	1.34	1.11
Primary line current	1.57	1.00	1.00
Line power factor		0.900	0.900

VACUUM TUBE AMPLIFIER CIRCUITS

Resistance-capacitance coupled has simple, economical construction and wide frequency range.

$$\frac{\text{Stage}}{\text{Gain}} = \frac{V_{\text{out}}}{V_{\text{in}}} = \mu \times \frac{R_{\text{L}}}{R_{\text{L}} + P_{\text{P}}}$$

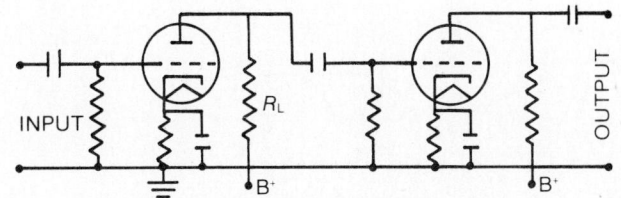

Transformer coupled has no coupling capacitor and uses full +B voltage on plates. It receives step-up from transformer and restricts frequency range.

$$\frac{\text{Stage}}{\text{Gain}} = \frac{V_{\text{out}}}{V_{\text{in}}} = \frac{N_2}{N_1} + \frac{X_2}{X_{\text{L}} + R_{\text{P}}}$$

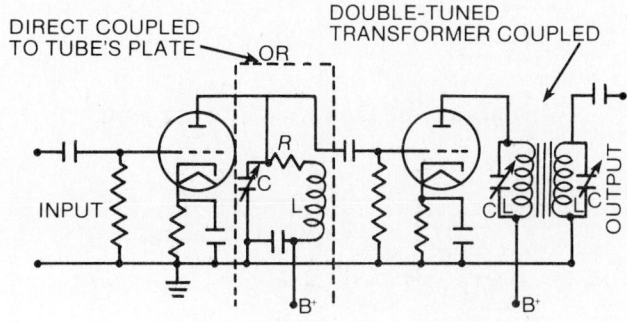

Impedance coupled has single winding plus coupling capacitor. Restricts frequency range; full +B voltage on plates.

$$\frac{\text{Stage}}{\text{Gain}} = \frac{V_{\text{out}}}{V_{\text{in}}} = \mu \times \frac{X_{\text{L}}}{X_{\text{L}} + R_{\text{P}}}$$

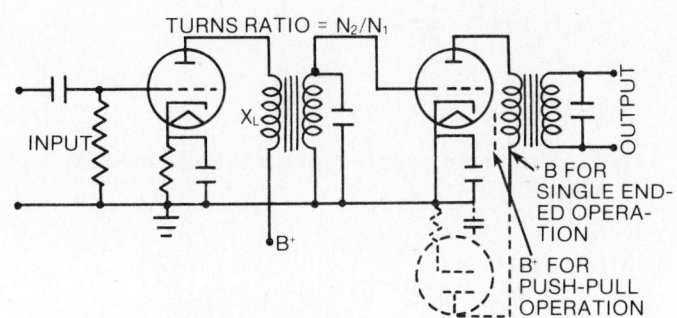

Tuned circuit coupled has high gain and good selectivity and tunes out grid and plate capacities.

$$\frac{\text{Stage}}{\text{Gain}} = \frac{V_{\text{out}}}{V_{\text{in}}} = G_{\text{M}} \times \frac{L}{CR}$$

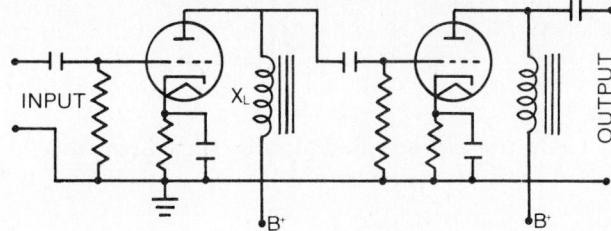

Direct coupled has no coupling capacitor nor grid and cathode resistors. It uses large, expensive voltage divider and bypassing capacitors. State gain same as resistance coupled stage, with wide frequency range.

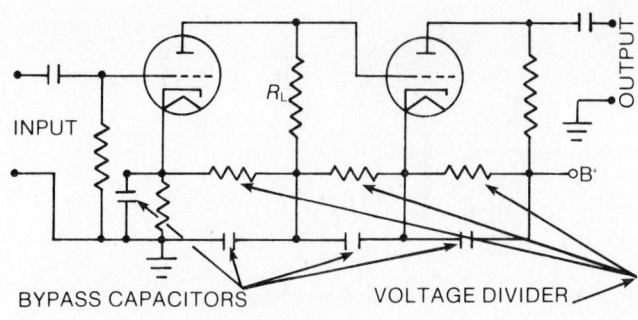

753

TRANSISTOR AMPLIFIER CIRCUITS

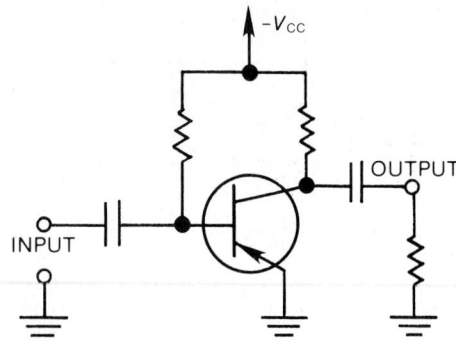

Resistance-capacitance (*RC*) coupled.

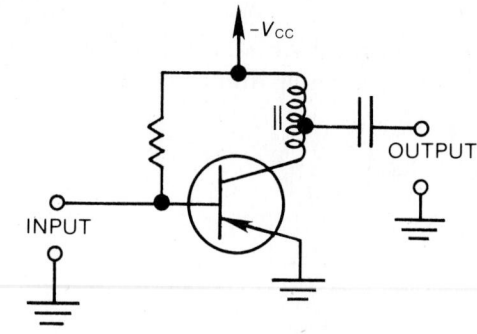

Impedance-coupled (video).

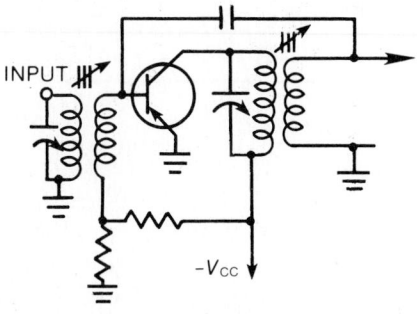

Transformer-coupled (neutralized video IF stage).

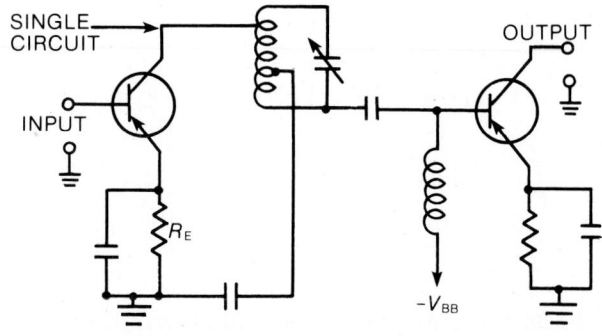

Impedance-coupled (tuned circuit).

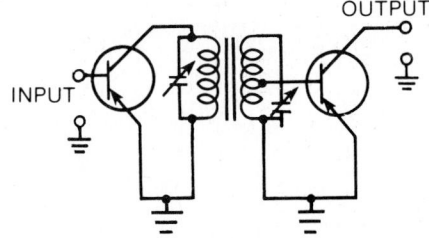

Transformer-coupled (double tuned circuit).

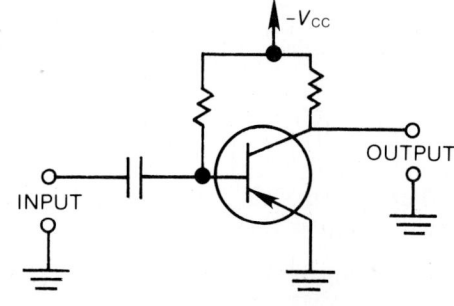

Direct-coupled.

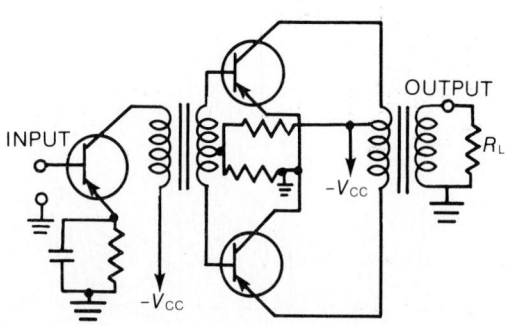

Push-pull, class B.

VACUUM TUBE OSCILLATOR CIRCUITS

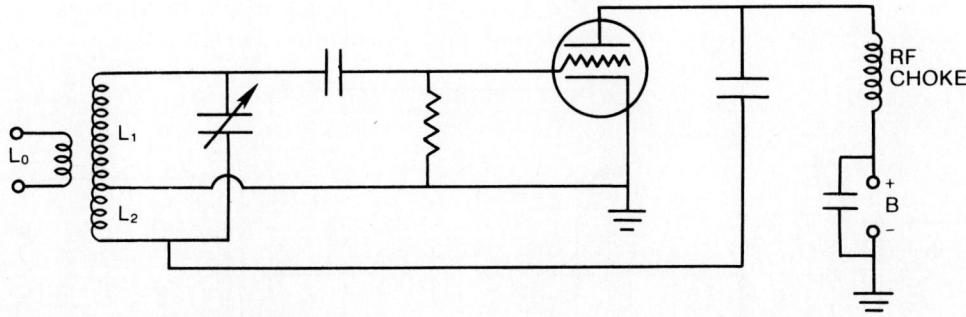

Hartley oscillator has inductive feedback due to coupling between L_1 and L_2.

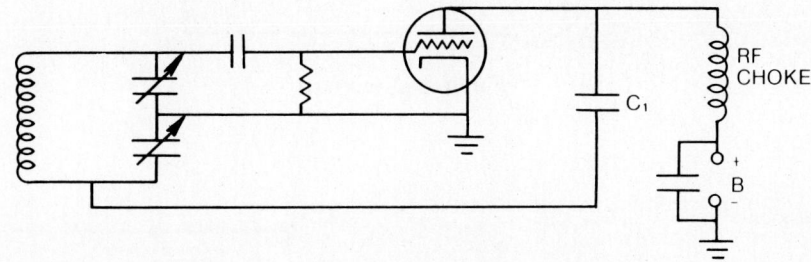

Colpitts oscillator has capacitive feedback due to voltage across C_1.

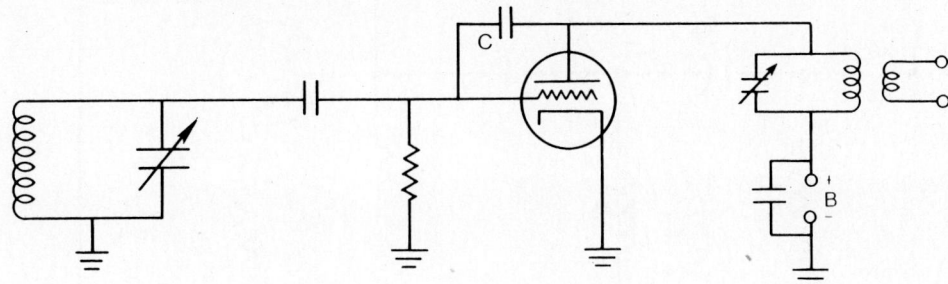

Tuned grid-tuned plate oscillator derives feedback to grid from plate through grid-plate capacity C.

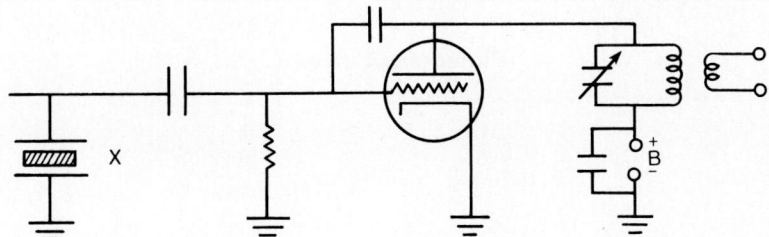

Crystal oscillator utilizes resonant properties of crystal X, derives feedback from resonant circuit LC through C.

TRANSISTOR SINE WAVE OSCILLATOR CIRCUITS

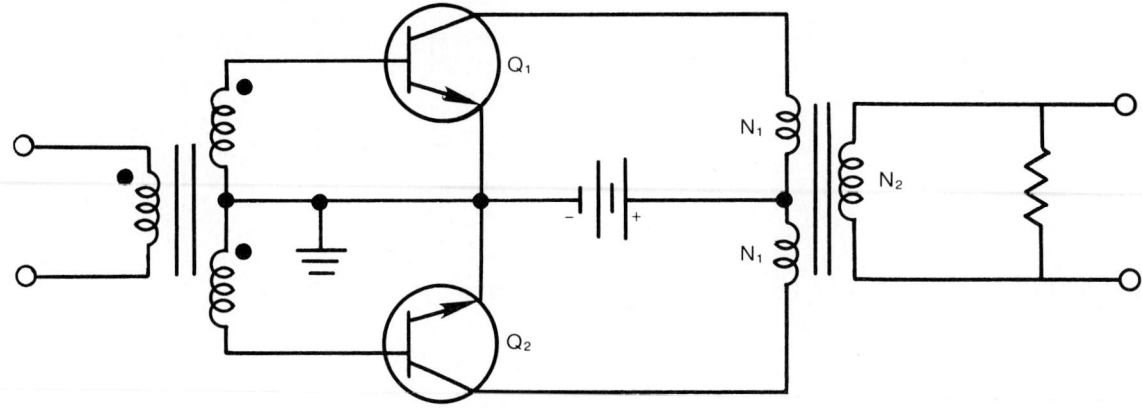

Push-pull circuit

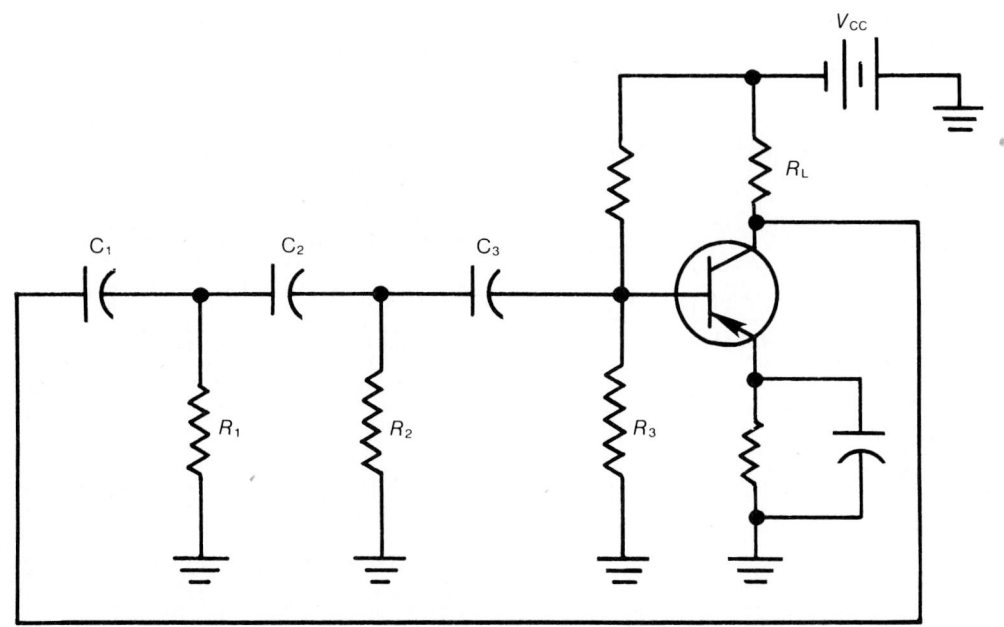

Phase-shift circuit.

ELEMENTARY FILTER CIRCUITS

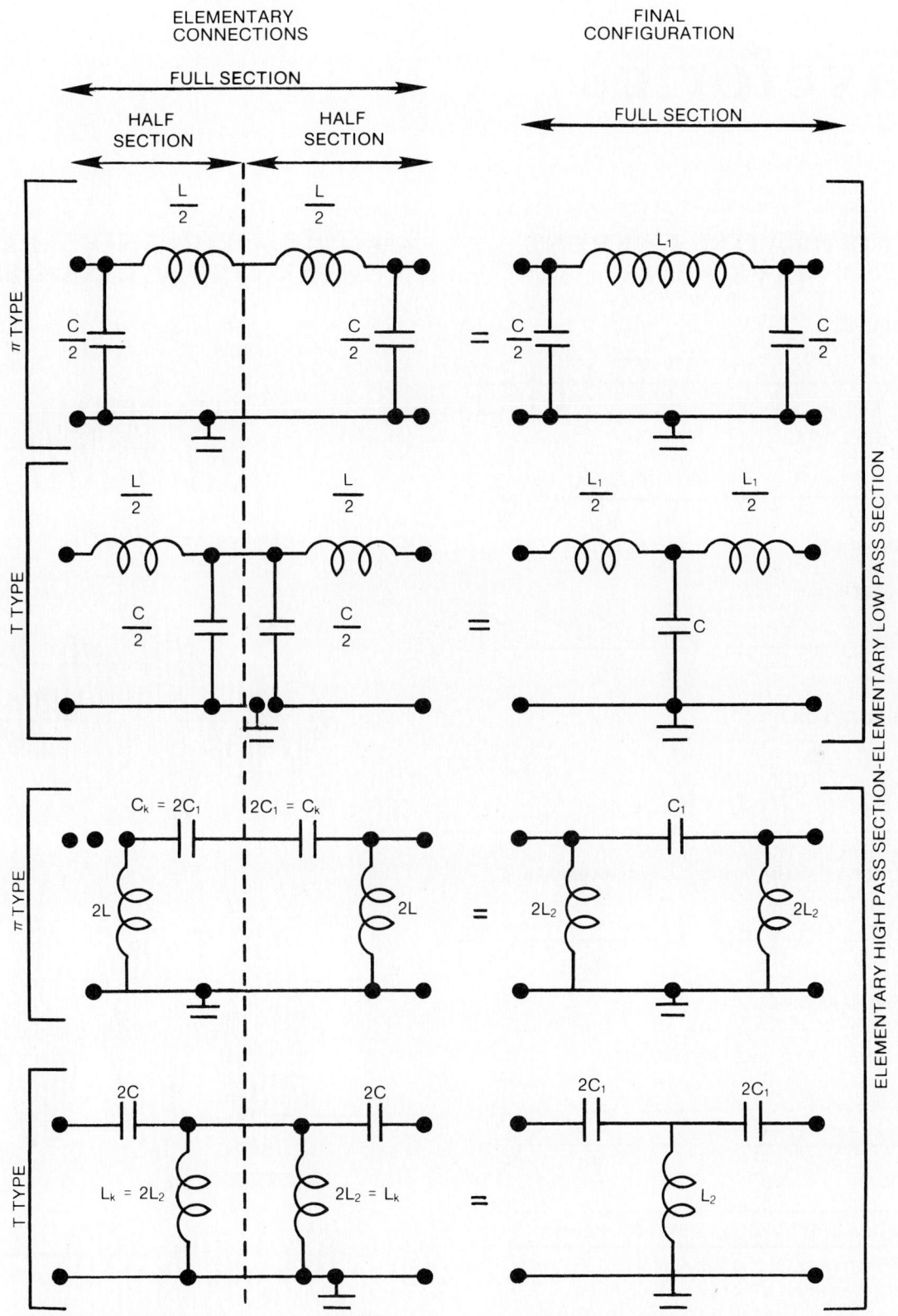

ELEMENTARY
CONNECTIONS

FINAL
CONFIGURATION

FULL SECTION

HALF SECTION

HALF SECTION

FULL SECTION

π TYPE

T TYPE

π TYPE

T TYPE

ELEMENTARY HIGH PASS SECTION—ELEMENTARY LOW PASS SECTION

APPENDIX H

Waveforms

ALTERNATING CURRENT WAVEFORMS

SYMMETRICAL

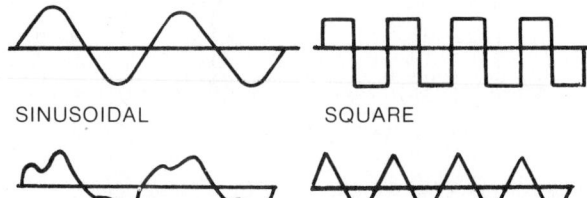

SINUSOIDAL   SQUARE

NONSINUSOIDAL   TRIANGULAR

NONSYMMETRICAL

SAW-TOOTH

SUPERIMPOSED ON DC

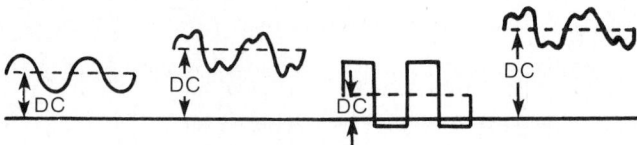

COMPOUNDED OF FUNDAMENTAL AND HARMONICS

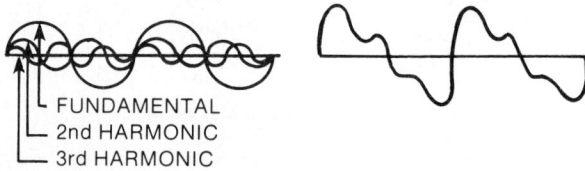

FUNDAMENTAL
2nd HARMONIC
3rd HARMONIC

COMPOUNDED OF UNLIKE FREQUENCIES

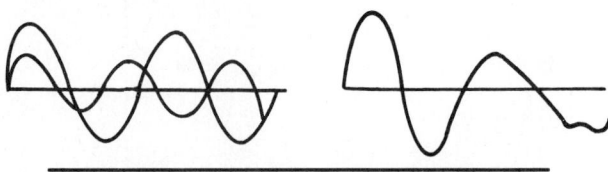

Sinusoidal Voltages and Currents
Effective value = 0.707 × peak value
Average value = 0.637 × peak value
Peak value = 1.414 × effective value
Effective value = 1.11 × average value
Peak value = 1.57 × average value
Average value = 0.9 × effective value

MODULATION AND CARRIER WAVE RELATIONSHIPS

UNMODULATED RF OR IF
CARRIER WAVEFORM

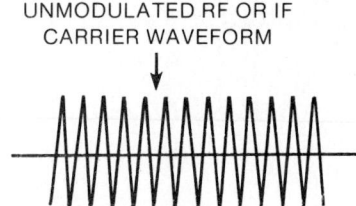

MODULATED CARRIER WAVEFORMS

SPEECH OR MUSIC
MODULATED
(ABOUT 40%)

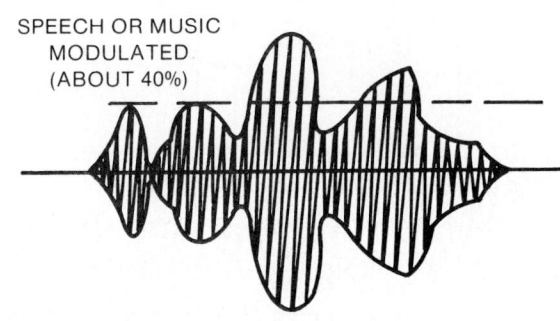

100% SINE WAVE
MODULATED

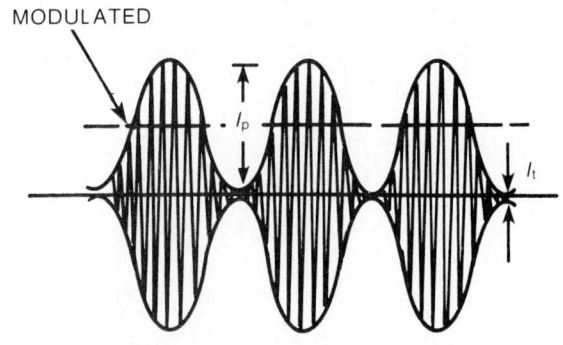

50% SINE WAVE
MODULATED

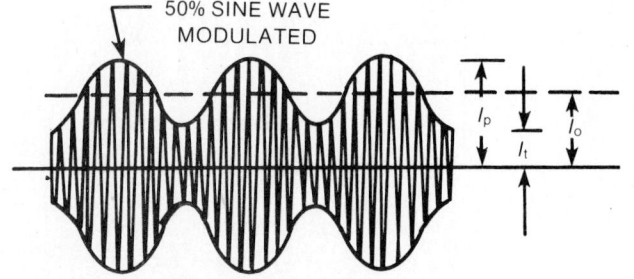

PERCENT MODULATION—50%

$$M = \frac{I_p - I_t}{2I_o} \times 100$$

$$= \frac{3 - 1}{4} \times 100$$

$$= 50\%$$

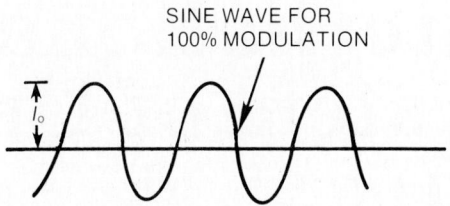

SINE WAVE FOR
100% MODULATION

**MODULATING AUDIO
FREQUENCY WAVEFORMS**

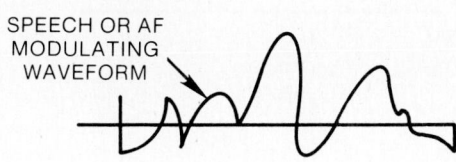

SPEECH OR AF
MODULATING
WAVEFORM

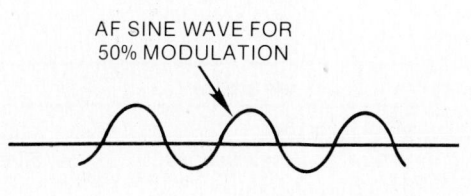

AF SINE WAVE FOR
50% MODULATION

Summary of Sine Wave, FM, and Pulse Modulation				
Type of Modulation	**Multiplex System**	**Waveform**	**System Requirements**	**Display**
AM Amplitude	Frequency division, analog		AM Detector, amplitude reference	Voltmeter, scope
FM Frequency	Frequency division, analog		FM Detector, frequency reference	Freq. meter, scope
PM Phase	Frequency division, analog		Phase detector, phase reference	Voltmeter
PAM Pulse amplitude	Time division, analog		AM Peak detector, amplitude ref., sync. pulses	Voltmeter, scope
PDM Pulse duration	Time division, analog		PD Detector, sync. pulses	Voltmeter, scope
PPM Pulse position	Time division, analog		Pulse interval detector, sync. pulses	Voltmeter, scope
PCM Pulse code (RZ) (Return to zero)	Time division, digital		Digital decoder, sync. pulses	Digital readout
PCM Pulse code (NZR) (Non-return to zero)	Time division, digital		Digital decoder, sync. pulses	Digital readout

759

APPENDIX I

International System of Units (SI)

Quantity	Unit	Symbol	
Acceleration	meter per second squared	m/s^2	
Angular acceleration	radian per second squared	rad/s^2	
Angular velocity	radian per second	rad/s	
Area	square meter	m^2	
Density	kilogram per cubic meter	kg/m^3	
Dynamic viscosity	newton-second per square meter	N-s/m^2	
Electric capacitance	farad	F	(A · s/V)
Electric charge	coulomb	C	(A · s)
Electric field strength	volt per meter	V/m	
Electric resistance	ohm	Ω	(V/A)
Flux of light	lumen	lm	(cd · sr)
Force	newton	N	(kg · m/s^2)
Frequency	hertz	Hz	(s^{-1})
Illumination	lux	lx	(lm/m^2)
Inductance	henry	H	(V · s/A)
Kinematic viscosity	square meter per second	m^2/s	
Luminance	candela per square meter	cd/m^2	
Magnetic field strength	ampere per meter	A/m	
Magnetic flux	weber	Wb	(V · s)
Magnetic flux density	tesla	T	(Wb/m^2)
Magnetomotive force	ampere	A	
Power	watt	W	(J/s)
Pressure	newton per square meter	N/m^2	
Velocity	meter per second	m/s	
Voltage, potential difference, electromotive force	volt	V	(W/A)
Volume	cubic meter	m^3	
Work, energy, quantity of heat	joule	J	(N · m)

Table of Electrical and Electronic Conversion Factors

Multiply	By	To Obtain
Abamperes	10	amperes
abamperes	3×10^{10}	statamperes
abamperes per sq cm	64.52	amperes per sq inch
abampere-turns	10	ampere-turns
abampere-turns	12.57	gilberts
abampere-turns per cm	25.40	ampere-turns per inch
abcoulombs	10	coulombs
abcoulombs	3×10^{10}	statcoulombs
abcoulombs per sq cm	64.52	coulombs per sq inch
abfarads	10^9	farads
abfarads	10^{15}	microfarads
abfarads	9×10^{20}	statfarads
abhenries	10^{-9}	henries
abhenries	10^{-6}	millihenries
abhenries	$1/9 \times 10^{-20}$	stathenries
abmhos per cm cube	1.662×10^2	mhos per mil foot
abmhos per cm cube	10^3	megmhos per cm cube
abohms	10^{-15}	megohms
abohms	10^{-3}	microhms
abohms	10^{-9}	ohms
abohms	$1/9 \times 10^{-20}$	statohms
abohms per cm cube	10^{-3}	microhms per cm cube
abohms per cm cube	6.015×10^{-3}	ohms per mil foot
abvolts	$1/3 \times 10^{-10}$	statvolts
abvolts	10^{-8}	volts
amperes	1/10	abamperes
amperes	3×10^9	statamperes
amperes per sq cm	6.452	amperes per sq inch
amperes per sq inch	0.01550	abamperes per sq cm
amperes per sq inch	0.1550	amperes per sq cm
amperes per sq inch	4.650×10^8	statamperes per sq cm
ampere-turns	1/10	abampere-turns
ampere-turns	1.257	gilberts
ampere-turns per cm	2.540	ampere-turns per inch
ampere-turns per inch	0.03937	abampere-turns per cm
ampere-turns per inch	0.3937	ampere-turns per cm
ampere-turns per inch	0.4950	gilberts per cm
British thermal units	144 sq in. × 1 in.	kilogram-calories
British thermal units	777.5	foot-pounds
British thermal units	3.927×10^{-4}	horsepower-hours
British thermal units	1054	joules

Multiply	By	To Obtain
British thermal units	107.5	kilogram-meters
British thermal units	2.928×10^{-4}	kilowatt-hours
Centigrams	0.01	grams
centiliters	0.01	liters
centimeters	0.3937	inches
centimeters	0.01	meters
centimeters	393.7	mils
centimeters	10	millimeters
circular mils	5.067×10^{-6}	square centimeters
circular mils	7.854×10^{-7}	square inches
circular mils	0.7854	square mils
coulombs	1/10	abcoulombs
coulombs	3×9^9	statcoulombs
coulombs per sq inch	0.01550	abcoulombs per sq cm
coulombs per sq inch	0.1550	coulombs per sq cm
coulombs per sq inch	4.650×10^8	statcouls per sq cm
cubic centimeters	3.531×10^{-5}	cubic feet
cubic centimeters	6.102×10^{-2}	cubic inches
cubic centimeters	10^{-6}	cubic meters
cubic centimeters	1.308×10^{-6}	cubic yards
cubic centimeters	2.642×10^{-4}	gallons
cubic centimeters	10^{-3}	liters
cubic centimeters	2.113×10^{-8}	pints (liq)
cubic centimeters	1.057×10^{-8}	quarts (liq)
cubic feet	2.832×10^4	cubic cm
cubic feet	1728	cubic inches
cubic feet	0.02832	cubic meters
cubic feet	0.03704	cubic yards
cubic feet	7.481	gallons
cubic feet	28.32	liters
cubic feet	59.84	pints (liq)
cubic feet	29.92	quarts (liq)
cubic inches	16.39	cubic centimeters
cubic inches	5.787×10^{-4}	cubic feet
cubic inches	1.639×10^{-5}	cubic meters
cubic inches	2.143×10^{-5}	cubic yards
cubic inches	4.329×10^{-3}	gallons
cubic inches	1.639×10^{-2}	liters
cubic inches	0.03463	pints (liq)
cubic inches	0.01732	quarts (liq)
cubic meters	10^6	cubic centimeters
cubic meters	35.31	cubic feet
cubic meters	61,023	cubic inches
cubic meters	1.308	cubic yards
cubic meters	264.2	gallons
cubic meters	10^9	liters
cubic meters	2113	pints (liq)
cubic meters	1057	quarts (liq)
cubic yards	7.646×10^5	cubic centimeters
cubic yards	27	cubic feet
cubic yards	46,656	cubic inches
cubic yards	0.7646	cubic meters
cubic yards	202.0	gallons
cubic yards	764.6	liters
cubic yards	1616	pints (liq)
cubic yards	807.9	quarts (liq)

Multiply	By	To Obtain
Decigrams	0.1	grams
decimeters	0.1	meters
degrees (angle)	60	minutes
degrees (angle)	0.01745	radians
degrees (angle)	3600	seconds
degrees per second	0.01745	radians per second
degrees per second	0.1667	revolutions per min
degrees per second	0.002778	revolutions per sec
dekagrams	10	grams
dekaliters	10	liters
dekameters	10	meters
drams	1.772	grams
drams	0.0625	ounces
dynes	1.020×10^{-3}	grams
dynes	2.248×10^{-6}	pounds
Farads	10^{-9}	abfarads
farads	10^{6}	microfarads
farads	9×10^{-11}	statfarads
feet	30.48	centimeters
feet	12	inches
feet	0.3048	meters
feet	.36	varas
feet	1/3	yards
Gallons	3785	cubic centimeters
gallons	0.1337	cubic feet
gallons	231	cubic inches
gallons	3.784×10^{-3}	cubic meters
gallons	4.951×10^{-3}	cubic yards
gallons	3.785	liters
gallons	8	pints (liq)
gallons	4	quarts (liq)
gallons per minute	2.228×10^{-3}	cubic feet per second
gallons per minute	0.06308	liters per second
gausses	6.452	lines per square inch
gilberts	0.07958	abampere-turns
gilberts	0.7958	ampere-turns
gilberts per centimeter	2.021	ampere-turns per inch
grains (troy)	1	grains (av)
grains (troy)	0.06480	grams
grams	980.7	dynes
grams	15.43	grains (troy)
grams	10^{-3}	kilograms
grams	10^{3}	milligrams
grams	0.03527	ounces
grams	0.03215	ounces (troy)
grams	2.205×10^{-3}	pounds
Hectograms	100	grams
hectoliters	100	liters
hectometers	100	meters
hectowatts	100	watts
henries	10^{9}	abhenries
henries	10^{3}	millihenries
henries	$1/9 \times 10^{-11}$	stathenries
horsepower	42.44	Btu per min
horsepower	33,000	foot-pounds per min

Multiply	By	To Obtain
horsepower	550	foot-pounds per sec
horsepower	1.014	horsepower (metric)
horsepower	10.70	kg.-calories per min
horsepower	0.7457	kilowatts
horsepower	745.7	watts
Inches	2.540	centimeters
inches	10^3	mils
inches	.03	varas
Joules	9.486×10^{-4}	British thermal units
joules	10^7	ergs
joules	0.7376	foot-pounds
joules	2.390×10^{-4}	kilogram-calories
joules	0.1020	kilogram-meters
joules	2.778×10^{-4}	watt-hours
Kilograms	980,665	dynes
kilograms	10^2	grams
kilograms	70.93	poundals
kilograms	2.2046	pounds
kilolines	10^9	maxwells
kiloliters	10^3	liters
kilometers	10^5	centimeters
kilometers	3281	feet
kilometers	10^3	meters
kilometers	0.6214	miles
kilometers	1093.6	yards
kilowatts	56.92	Btu per min
kilowatts	4.425×10^4	foot-pounds per min
kilowatts	737.6	foot-pounds per sec
kilowatts	1.341	horsepower
kilowatts	14.34	kg.-calories per min
kilowatts	10^3	watts
kilowatt-hours	3415	British thermal units
kilowatt-hours	2.655×10^6	foot-pounds
kilowatt-hours	1.341	horsepower-hours
kilowatt-hours	3.6×10^6	joules
kilowatt-hours	860.5	kilogram-calories
kilowatt-hours	3.671×10^5	kilogram-meters
knots	6080	feet
knots	1.853	kilometers
knots	1.152	miles
knots	2027	yards
Lines per square cm	1	gausses
lines per square inch	0.1550	gausses
liters	10^3	cubic centimeters
liters	0.03531	cubic feet
liters	61.02	cubic inches
liters	10^3	cubic meters
liters	1.308×10^{-3}	cubic yards
liters	0.2642	gallons
liters	2.113	pints (liq)
liters	1.057	quarts (liq)
$\log^{10} N$	2.303	$\log_e N$ or In N
$\log_e N$ or In N	0.4343	$\log^{10} N$
lumens per sq ft	1	foot-candles

Multiply	By	To Obtain
Maxwells	10^{-3}	kilolines
megalines	10^6	maxwells
megmhos per cm cube	10^{-3}	abmhos per cm cube
megmhos per cm cube	2.540	megmhos per in cube
megmhos per cm cube	0.1662	mhos per mil foot
megmhos per inch cube	0.3937	megmhos per cm cube
megohms	10^6	ohms
meters	100	centimeters
meters	3.2808	feet
meters	39.37	inches
meters	10^{-3}	kilometers
meters	10^3	millimeters
meters	1.0936	yards
mhos per mil foot	6.015×10^{-3}	abmhos per cm cube
mhos per mil foot	6.015	megmhos per cm cube
mhos per mil foot	15.28	megmhos per inch cube
microfarads	10^{-15}	abfarads
microfarads	10^{-6}	farads
microfarads	9×10^5	staffarads
micrograms	10^{-6}	grams
microliters	10^{-6}	liters
microhms	10^3	abohms
microhms	10^{-12}	megohms
microhms	10^{-6}	ohms
microhms	$1/9 \times 10^{-17}$	statohms
microhms per cm cube	10^3	abohms per cm cube
microhms per cm cube	0.3937	microhms per inch cube
microhms per cm cube	6.015	ohms per mil foot
microhms per inch cube	2.540	microhms per cm cube
microns	10^{-6}	meters
miles	1.609×10^5	centimeters
miles	5280	feet
miles	1.6093	kilometers
miles	1760	yards
milligrams	10^{-2}	grams
millihenries	10^6	aghenries
millihenries	10^{-3}	henries
millihenries	$1/9 \times 10^{-14}$	stathenries
millimeters	0.1	centimeters
millimeters	0.03937	inches
millimeters	39.37	mils
mils	0.002540	centimeters
mils	10^{-3}	inches
minutes (angle)	2.909×10^{-4}	radians
minutes (angle)	60	seconds (angle)
Ohms	10^9	abohms
ohms	10^{-6}	megohms
ohms	10^6	microhms
ohms	$1/9 \times 10^{-11}$	statohms
ohms per mil foot	166.2	abohms per cm cube
ohms per mil foot	0.1662	microhms per cm cube
ohms per mil foot	0.06524	microhms per inch cube
ounces	8	drams
ounces	437.5	grains
ounces	28.35	grams
ounces	0.0625	pounds

Multiply	By	To Obtain
ounces (fluid)	1.805	cubic inches
ounces (fluid)	0.02957	liters
ounces (troy)	480	grains (troy)
ounces (troy)	31.10	grams
ounces (troy)	0.08333	pounds (troy)
Pints (dry)	33.60	cubic inches
pints (liquid)	28.87	cubic inches
pounds	444,823	dynes
pounds	7000	grains
pounds	453.6	grams
pounds	16	ounces
pounds (troy)	0.8229	pounds (av)
Quadrants (angle)	90	degrees
quadrants (angle)	5400	minutes
quadrants (angle)	1.571	radians
quarts (dry)	67.20	cubic inches
quarts (liq)	57.75	cubic inches
Radians	57.30	degrees
radians	3438	minutes
radians	0.637	quadrants
radians per second	57.30	degrees per second
radians per second	0.1592	revolutions per second
radians per second	9.549	revolutions per min
radians per sec per sec	573.0	revs per min per min
radians per sec per sec	9.549	revs per min per sec
radians per sec per sec	0.1592	revs per sec per sec
revolutions	360	degrees
revolutions	4	quadrants
revolutions	6.283	radians
rods	16.5	feet
Seconds (angle)	4.848×10^{-6}	radians
spheres (solid angle)	12.57	steradians
spherical right angles	0.25	hemispheres
spherical right angles	0.125	spheres
square centimeters	1.973×10^{5}	circular mils
square centimeters	1.076×10^{-3}	square feet
square centimeters	0.1550	square inches
square centimeters	10^{-6}	square meters
square centimeters	100	square millimeters
square feet	2.296×10^{-5}	acres
square feet	929.0	square centimeters
square feet	144	square inches
square feet	0.09290	square meters
square feet	3.587×10^{-3}	square miles
square feet	1/9	square yards
square inches	1.273×10^{6}	circular mils
square inches	6.452	square centimeters
square inches	6.944×10^{-3}	square feet
square inches	10^{6}	square mils
square inches	645.2	square millimeters
square kilometers	247.1	acres
square kilometers	10.76×10^{6}	square feet
square kilometers	10^{6}	square meters
square kilometers	0.3861	square miles
square kilometers	1.196×10^{6}	square yards
square meters	2.471×10^{-4}	acres
square meters	10.764	square feet

Multiply	By	To Obtain
square meters	3.861×10^{-7}	square miles
square meters	1.196	square yards
square miles	640	acres
square miles	27.88×10^6	square feet
square miles	2.590	square kilometers
square miles	3.098×16^6	square yards
square millimeters	1.973×10^3	circular mils
square millimeters	0.01	square centimeters
square millimeters	1.550×10^{-3}	square inches
square mils	1.273	circular mils
square mils	6.452×10^{-6}	square centimeters
square mils	10^{-6}	square inches
statamperes	$1/3 \times 10^{-10}$	abamperes
statamperes	$1/3 \times 10^{-9}$	amperes
statcoulombs	$1/3 \times 10^{-10}$	abcoulombs
statcoulombs	$1/3 \times 10^{-9}$	coulombs
statfarads	$1/9 \times 10^{-20}$	abfarads
statfarads	$1/9 \times 10^{-11}$	farads
statfarads	$1/9 \times 10^{-5}$	microfarads
stathenries	9×20^{20}	abhenries
stathenries	9×10^{11}	henries
stathenries	9×10^{14}	millihenries
statohms	9×20^{20}	abohms
statohms	9×10^5	megohms
statohms	9×10^{17}	microhms
statohms	9×10^{11}	ohms
statvolts	3×10^{10}	abvolts
statvolts	300	volts
steradians	0.1592	hemispheres
steradians	0.07958	spheres
steradians	0.6366	spherical right angles
Temp (°C) + 17.8	1	abs temp (°C)
temp (°F) + 460	1.8	temp (°Fahr)
temp (°F) – 32	1	abs temp (°F)
temp (°C) + 273	5/9	temp (°Cent)
Volts	10^8	abvolts
volts	1/300	statvolts
volts per inch	3.937×10^7	abvolts per cm
volts per inch	1.312×10^{-3}	statvolts per cm
Watts	0.05692	Btu per min
watts	10^7	ergs per second
watts	44.26	foot-pounds per min
watts	0.7376	foot-pounds per sec
watts	1.341×10^{-3}	horsepower
watts	0.01434	kg-calories per min
watts	10^{-3}	kilowatts
watt-hours	3.415	British thermal units
watt-hours	2655	foot-pounds
watt-hours	1.341×10^{-3}	horsepower-hours
watt-hours	0.8605	kilogram-calories
watt-hours	367.1	kilogram-meters
watt-hours	10^{-3}	kilowatt-hours
webers	10^8	maxwells
Yards	91.44	centimeters
yards	3	feet
yards	36	inches
yards	0.9144	meters

APPENDIX K

Temperature Conversions

Find the number to be converted in the center (boldface) column. If converting Fahrenheit degrees, read the Celsius equivalent in the column headed "°C." If converting Celsius degrees, read the Fahrenheit equivalent in the column headed "°F."

°C		°F	°C		°F	°C		°F	°C		°F
-273	**-459**		-40	**-40**	-40	24.4	**76**	168.8	199	**390**	734
-268	**-450**		-34	**-30**	-22	25.6	**78**	172.4	204	**400**	752
-262	**-440**		-29	**-20**	-4	26.7	**80**	176.0	210	**410**	770
-257	**-430**		-23	**-10**	14	27.8	**82**	179.6	216	**420**	788
-251	**-420**		-17.8	**0**	32	28.9	**84**	183.2	221	**430**	806
-246	**-410**		-16.7	**2**	35.6	30.0	**86**	186.8	227	**440**	824
-240	**-400**		-15.6	**4**	39.2	31.1	**88**	190.4	232	**450**	842
-234	**-390**		-14.4	**6**	42.8	32.2	**90**	194.0	238	**460**	860
-229	**-380**		-13.3	**8**	46.4	33.3	**92**	197.6	243	**470**	878
-223	**-370**		-12.2	**10**	50.0	34.4	**94**	201.2	249	**480**	896
-218	**-360**		-11.1	**12**	53.6	35.6	**96**	204.8	254	**490**	914
-212	**-350**		-10.0	**14**	57.2	36.7	**98**	208.4	260	**500**	932
-207	**-340**		-8.9	**16**	60.8	37.8	**100**	212.0	266	**510**	950
-201	**-330**		-7.8	**18**	64.4	43	**110**	230	271	**520**	968
-196	**-320**		-6.7	**20**	68.0	49	**120**	248	277	**530**	986
-190	**-310**		-5.6	**22**	71.6	54	**130**	266	282	**540**	1004
-184	**-300**		-4.4	**24**	75.2	60	**140**	284	288	**550**	1022
-179	**-290**		-3.3	**26**	78.8	66	**150**	302	293	**560**	1040
-173	**-280**		-2.2	**28**	82.4	71	**160**	320	299	**570**	1058
-168	**-270**	-454	-1.1	**30**	86.0	77	**170**	338	304	**580**	1076
-162	**-260**	-436	0.0	**32**	89.6	82	**180**	356	310	**590**	1094
-157	**-250**	-418	1.1	**34**	93.2	88	**190**	374	316	**600**	1112
-151	**-240**	-400	2.2	**36**	96.8	93	**200**	392	321	**610**	1130
-146	**-230**	-382	3.3	**38**	100.4	99	**210**	410	327	**620**	1148
-140	**-220**	-364	4.4	**40**	104.0	100	**212**	414	332	**630**	1166
-134	**-210**	-346	5.6	**42**	107.6	104	**220**	428	338	**640**	1184
-129	**-200**	-328	6.7	**44**	111.2	110	**230**	446	343	**650**	1202
-123	**-190**	-310	7.8	**46**	114.8	116	**240**	464	349	**660**	1220
-118	**-180**	-292	8.9	**48**	118.4	121	**250**	482	354	**670**	1238
-112	**-170**	-274	10.0	**50**	122.0	127	**260**	500	360	**680**	1256
-107	**-160**	-256	11.1	**52**	125.6	132	**270**	518	366	**690**	1274
-101	**-150**	-238	12.2	**54**	129.2	138	**280**	536	371	**700**	1292
-96	**-140**	-220	13.3	**56**	132.8	143	**290**	554	377	**710**	1310
-90	**-130**	-202	14.4	**58**	136.4	149	**300**	572	382	**720**	1328
-84	**-120**	-184	15.6	**60**	140.0	154	**310**	590	388	**730**	1346
-79	**-110**	-166	16.7	**62**	143.6	160	**320**	608	393	**740**	1364
-73	**-100**	-148	17.8	**64**	147.2	166	**330**	626	399	**750**	1382
-68	**-90**	-130	18.9	**66**	150.8	171	**340**	644	404	**760**	1400
-62	**-80**	-112	20.0	**68**	154.4	177	**350**	662	410	**770**	1418
-57	**-70**	-94	21.1	**70**	158.0	182	**360**	680	416	**780**	1436
-51	**-60**	-76	22.2	**72**	161.6	188	**370**	698	421	**790**	1454
-46	**-50**	-58	23.3	**74**	165.2	193	**380**	716	427	**800**	1472

°C		°F	°C		°F	°C		°F	°C		°F
432	**810**	1490	738	**1360**	2480	1043	**1910**	3470	1349	**2460**	4460
438	**820**	1508	743	**1370**	2498	1049	**1920**	3488	1354	**2470**	4478
443	**830**	1526	749	**1380**	2516	1054	**1930**	3506	1360	**2480**	4496
449	**840**	1544	754	**1390**	2534	1060	**1940**	3524	1366	**2490**	4514
454	**850**	1562	760	**1400**	2552	1066	**1950**	3542	1371	**2500**	4532
460	**860**	1580	766	**1410**	2570	1071	**1960**	3560	1377	**2510**	4550
466	**870**	1598	771	**1420**	2588	1077	**1970**	3578	1382	**2520**	4568
471	**880**	1616	777	**1430**	2606	1082	**1980**	3596	1388	**2530**	4586
477	**890**	1634	782	**1440**	2624	1088	**1990**	3614	1393	**2540**	4604
482	**900**	1652	788	**1450**	2642	1093	**2000**	3632	1399	**2550**	4622
488	**910**	1670	793	**1460**	2660	1099	**2010**	3650	1404	**2560**	4640
493	**920**	1688	799	**1470**	2678	1104	**2020**	3668	1410	**2570**	4658
499	**930**	1706	804	**1480**	2696	1110	**2030**	3686	1416	**2580**	4676
504	**940**	1724	810	**1490**	2714	1116	**2040**	3704	1421	**2590**	4696
510	**950**	1742	816	**1500**	2732	1121	**2050**	3722	1427	**2600**	4712
516	**960**	1760	821	**1510**	2750	1127	**2060**	3740	1432	**2610**	4730
521	**970**	1778	827	**1520**	2768	1132	**2070**	3758	1438	**2620**	4748
527	**980**	1796	832	**1530**	2786	1138	**2080**	3776	1443	**2630**	4766
532	**990**	1814	838	**1540**	2804	1143	**2090**	3794	1449	**2640**	4784
538	**1000**	1832	843	**1550**	2822	1149	**2100**	3812	1454	**2650**	4802
543	**1010**	1850	849	**1560**	2840	1154	**2110**	3830	1460	**2660**	4820
549	**1020**	1868	854	**1570**	2858	1160	**2120**	3848	1466	**2670**	4838
554	**1030**	1886	860	**1580**	2876	1166	**2130**	3866	1471	**2680**	4856
560	**1040**	1904	866	**1590**	2894	1171	**2140**	3884	1477	**2690**	4874
566	**1050**	1922	871	**1600**	2912	1177	**2150**	3902	1482	**2700**	4892
571	**1060**	1940	877	**1610**	2930	1182	**2160**	3920	1488	**2710**	4910
577	**1070**	1958	882	**1620**	2948	1188	**2170**	3938	1493	**2720**	4928
582	**1080**	1976	888	**1630**	2966	1193	**2180**	3956	1499	**2730**	4946
588	**1090**	1994	893	**1640**	2984	1199	**2190**	3974	1504	**2740**	4964
593	**1100**	2012	899	**1650**	3002	1204	**2200**	3992	1510	**2750**	4982
599	**1110**	2030	904	**1660**	3020	1210	**2210**	4010	1516	**2760**	5000
604	**1120**	2048	910	**1670**	3038	1216	**2220**	4028	1521	**2770**	5018
610	**1130**	2066	916	**1680**	3056	1221	**2230**	4046	1527	**2780**	5036
616	**1140**	2084	921	**1690**	3074	1227	**2240**	4064	1532	**2790**	5054
621	**1150**	2102	927	**1700**	3092	1232	**2250**	4082	1538	**2800**	5072
627	**1160**	2120	932	**1710**	3110	1238	**2260**	4100	1543	**2810**	5090
632	**1170**	2138	938	**1720**	3128	1243	**2270**	4118	1549	**2820**	5108
638	**1180**	2156	943	**1730**	3146	1249	**2280**	4136	1554	**2830**	5126
643	**1190**	2174	949	**1740**	3164	1254	**2290**	4154	1560	**2840**	5144
649	**1200**	2192	954	**1750**	3182	1260	**2300**	4172	1566	**2850**	5162
654	**1210**	2210	960	**1760**	3200	1266	**2310**	4190	1571	**2860**	5180
660	**1220**	2228	966	**1770**	3218	1271	**2320**	4208	1577	**2870**	5198
666	**1230**	2246	971	**1780**	3236	1277	**2330**	4226	1582	**2880**	5216
671	**1240**	2264	977	**1790**	3254	1282	**2340**	4244	1588	**2890**	5234
677	**1250**	2282	982	**1800**	3272	1288	**2350**	4262	1593	**2900**	5252
682	**1260**	2300	988	**1810**	3290	1293	**2360**	4280	1599	**2910**	5270
688	**1270**	2318	993	**1820**	3308	1299	**2370**	4298	1604	**2920**	5288
693	**1280**	2336	999	**1830**	3326	1304	**2380**	4316	1610	**2930**	5306
699	**1290**	2354	1004	**1840**	3344	1310	**2390**	4334	1616	**2940**	5324
704	**1300**	2372	1010	**1850**	3362	1316	**2400**	4352	1621	**2950**	5342
710	**1310**	2390	1016	**1860**	3380	1321	**2410**	4370	1627	**2960**	5360
716	**1320**	2408	1021	**1870**	3398	1327	**2420**	4388	1632	**2970**	5378
721	**1330**	2426	1027	**1880**	3416	1332	**2430**	4406	1638	**2980**	5396
727	**1340**	2444	1032	**1890**	3434	1338	**2440**	4424	1643	**2990**	5414
732	**1350**	2462	1038	**1900**	3452	1343	**2450**	4442	1649	**3000**	5432

CONVERSION FORMULAS

To change degrees Fahrenheit to degrees Celsius:

$$°C = \frac{5}{9}\ (°F - 32)$$

To change degrees Celsius to degrees Fahrenheit:

$$°F = (\frac{9}{5} \times °C) + 32$$

To change degrees Celsius to degrees Kelvin:

$$°K = °C + 273°$$

To change degrees Kelvin to degrees Celsius:

$$°C = °K - 273°$$

APPENDIX L

American Wire Standards

Gauge	Dia. (mm)	Copper Wire Resistance (Ω/km)	Dia. (mil)	Copper Wire Resistance (Ω/1,000')
0000	11.68	0.160	460	0.049
000	10.40	0.203	409.6	0.062
00	9.266	0.255	364.8	0.078
0	8.252	0.316	324.9	0.098
1	7.348	0.406	289.3	0.124
2	6.543	0.511	257.6	0.156
3	5.827	0.645	229.4	0.197
4	5.189	0.813	204.3	0.248
5	4.620	1.026	181.9	0.313
6	4.115	1.29	162	0.395
7	3.665	1.63	144.3	0.498
8	3.264	2.06	128.5	0.628
9	2.906	2.59	114.4	0.792
10	2.588	3.27	101.9	0.999
11	2.30	4.10	90.7	1.26
12	2.05	5.20	80.8	1.59
13	1.83	6.55	72	2
14	1.63	8.26	64.1	2.52
15	1.45	10.4	57.1	3.18
16	1.29	13.1	50.8	4.02
17	1.15	16.6	45.3	5.06
18	1.02	21.0	40.3	6.39
19	0.912	26.3	35.9	8.05
20	0.813	33.2	32	10.1
21	0.723	41.9	28.5	12.8
22	0.644	52.8	25.3	16.1
23	0.573	66.7	22.6	20.3
24	0.511	83.9	20.1	25.7
25	0.455	106	17.9	32.4
26	0.405	134	15.9	41
27	0.361	168	14.2	51.4
28	0.321	213	12.6	64.9
29	0.286	267	11.3	81.4
30	0.255	337	10	103
31	0.227	425	8.9	130
32	0.202	537	8	164
33	0.180	676	7.1	206
34	0.160	855	6.3	261
35	0.143	1,071	5.6	329
36	0.127	1,360	5	415
37	0.113	1,715	4.5	523
38	0.101	2,147	4	655
39	0.090	2,704	3.5	832
40	0.080	3,422	3.1	1,044

Standard Annealed Solid Copper Wire

Gauge Number	Diameter (mils)	Cross Section		Ohms per 1,000'		Pounds per 1,000'
		Circular Mils	Square Inches	25°C (=77°F)	65°C (=149°F)	
0000	460.0	212,000.0	0.166	0.0500	0.0577	641.0
000	410.0	168,000.0	.132	.0630	.0727	508.0
00	365.0	133,000.0	.105	.0795	.0917	403.0
0	325.0	106,000.0	.0829	.100	.116	319.0
1	289.0	83,700.0	.0657	.126	.146	253.0
2	258.0	66,400.0	.0521	.159	.184	201.0
3	229.0	52,600.0	.0413	.201	.232	159.0
4	204.0	41,700.0	.0328	.253	.292	126.0
5	182.0	33,100.0	.0260	.319	.369	100.0
6	162.0	26,300.0	.0206	.403	.465	79.5
7	144.0	20,800.0	.0164	.508	.586	63.0
8	128.0	16,500.0	.0130	.641	.739	50.0
9	114.0	13,100.0	.0103	.808	.932	39.6
10	102.0	10,400.0	.00815	1.02	1.18	31.4
11	91.0	8,230.0	.00647	1.28	1.48	24.9
12	81.0	6,530.0	.00513	1.62	1.87	19.8
13	72.0	5,180.0	.00407	2.04	2.36	15.7
14	64.0	4,110.0	.00323	2.58	2.97	12.4
15	57.0	3,260.0	.00256	3.25	3.75	9.86
16	51.0	2,580.0	.00203	4.09	4.73	7.82
17	45.0	2,050.0	.00161	5.16	5.96	6.20
18	40.0	1,620.0	.00128	6.51	7.51	4.92
19	36.0	1,290.0	.00101	8.21	9.48	3.90
20	32.0	1,020.0	.000802	10.4	11.9	3.09
21	28.5	810.0	.000636	13.1	15.1	2.45
22	25.3	642.0	.000505	16.5	19.0	1.94
23	22.6	509.0	.000400	20.8	24.0	1.54
24	20.1	404.0	.000317	26.2	30.2	1.22
25	17.9	320.0	.000252	33.0	38.1	0.970
26	15.9	254.0	.000200	41.6	48.0	0.769
27	14.2	202.0	.000158	52.5	60.6	0.610
28	12.6	160.0	.000126	66.2	76.4	0.484
29	11.3	127.0	.0000995	83.4	96.3	0.384
30	10.0	101.0	.0000789	105.0	121.0	0.304
31	8.9	79.7	.0000626	133.0	153.0	0.241
32	8.0	63.2	.0000496	167.0	193.0	0.191
33	7.1	50.1	.0000394	211.0	243.0	0.152
34	6.3	39.8	.0000312	266.0	307.0	0.120
35	5.6	31.5	.0000248	335.0	387.0	0.0954
36	5.0	25.0	.0000196	423.0	488.0	0.0757
37	4.5	19.8	.0000156	533.0	616.0	0.0600
38	4.0	15.7	.0000123	673.0	776.0	0.0476
39	3.5	12.5	.0000098	848.0	979.0	0.0377
40	3.1	9.9	.0000078	1,070.0	1,230.0	0.0299

APPENDIX M

Decimal Equivalents of Common Fractions

8ths	16ths	32nds	64ths	Decimal Equivalent in.	Decimal Equivalent mm
			1	.016	0.397
		1	2	.031	0.794
			3	.047	1.191
	1	2	4	.063	1.587
			5	.078	1.984
		3	6	.094	2.381
			7	.109	2.778
1	2	4	8	.125	3.175
			9	.141	3.572
		5	10	.156	3.969
			11	.172	4.366
	3	6	12	.188	4.762
			13	.203	5.159
		7	14	.219	5.556
			15	.234	5.953
2	4	8	16	.25	6.350
			17	.266	6.747
		9	18	.281	7.144
			19	.297	7.541
	5	10	20	.313	7.937
			21	.328	8.334
		11	22	.344	8.731
			23	.359	9.128
3	6	12	24	.375	9.525
			25	.391	9.922
		13	26	.406	10.319
			27	.422	10.715
	7	14	28	.438	11.112
			29	.453	11.509
		15	30	.469	11.906
			31	.484	12.303
4	8	16	32	.50	12.700
			33	.516	13.097
		17	34	.531	13.494
			35	.547	13.890
	9	18	36	.563	14.287
			37	.578	14.684
		19	38	.594	15.081
			39	.609	15.478
5	10	20	40	.625	15.875

8ths	16ths	32nds	64ths	Decimal Equivalent in.	Decimal Equivalent mm
			42	.641	16.272
		21	43	.656	16.669
			43	.672	17.065
	11	22	44	.688	17.462
			45	.703	17.859
		23	46	.719	18.256
			47	.734	18.653
6	12	24	48	.75	19.050
			49	.766	19.447
		25	50	.781	19.844
			51	.797	20.240
	13	26	52	.813	20.637
			53	.828	21.034
		27	54	.844	21.431
			55	.859	21.828
7	14	28	56	.875	22.225
			57	.891	22.622
		29	58	.906	23.019
			59	.922	23.415
	15	30	60	.938	23.812
			61	.953	24.209
		31	62	.969	24.606
			63	.984	25.003

APPENDIX N

Component Circuit Formulas

Ohm's Law for DC Circuits

$$I = \frac{V}{R} = \frac{P}{V} = \sqrt{\frac{P}{R}}$$

$$R = \frac{V}{I} = \frac{P}{I^2} = \frac{V^2}{P}$$

$$V = IR = \frac{P}{I} = \sqrt{PR}$$

$$P = VI = \frac{V^2}{R} = I^2R$$

Resistors in Series

$$R_T = R_1 + R_2 + ...$$

Resistors in Parallel
  Two resistors

$$R_T = \frac{R_1R_2}{R_1+R_2}$$

  More than two

$$\frac{1}{R_T} = \frac{1}{R_1} + \frac{1}{R_2} + \frac{1}{R_3} + ...$$

RL Circuit Time Constant

$$\frac{L \text{ (in henrys)}}{R \text{ (in ohms)}} = t \text{ (in seconds), or}$$

$$\frac{L \text{ (in microhenrys)}}{R \text{ (in ohms)}} = t \text{ (in microseconds)}$$

RC Circuit Time Constant

R (ohms) × C (farads) = t (seconds)

R (megohms) × C (microfarads) = t (seconds)

R (ohms) × C (microfarads) = t (microseconds)

R (megohms) × C (micromicrofarads) = t (microseconds)

Capacitors in Series
  Two capacitors

$$C_T = \frac{C_1 C_2}{C_1 + C_2}$$

More than two

$$\frac{1}{C_T} = \frac{1}{C_1} + \frac{1}{C_2} + \frac{1}{C_3} + \dots$$

Capacitors in Parallel: $C_T = C_1 + C_2 + \dots$

Capacitive Reactance: $X_C = \dfrac{1}{2\pi f C}$

Impedance in an *RC* Circuit (Series)

$$Z = \sqrt{R^2 + (X_C)^2}$$

Inductors in Series

$L_T = L_1 + L_2 + \dots$ (No coupling between coils)

Inductors in Parallel
  Two inductors

$$L_T = \frac{L_1 L_2}{L_1 + L_2} \text{ (No coupling between coils)}$$

More than two

$$\frac{1}{L_T} = \frac{1}{L_1} + \frac{1}{L_2} + \frac{1}{L_3} + \dots \text{ (No coupling between coils)}$$

Inductive Reactance

$X_L = 2\pi f L$

Q of a Coil

$$Q = \frac{X_L}{R}$$

Impedance of an *RC* Circuit (Series)

$Z = \sqrt{R^2 + X_C{}^2}$

Impedance of an *RL* Circuit (Series)

$Z = \sqrt{R^2 + (X_L)^2}$

Impedance with *R*, C, and L in Series

$Z = \sqrt{R^2 + (X_L - X_C)^2}$

Parallel Circuit Impedance

$$Z = \frac{Z_1 Z_2}{Z_1 + Z_2}$$

Sine-Wave Voltage Relationships
  Average value

$$V_{\text{ave}} = \frac{2}{\pi} \times V_{\text{max}} = 0.637 V_{\text{max}}$$

Effective or rms value

$$V_{\text{eff}} = \frac{V_{\text{max}}}{\sqrt{2}} = \frac{V_{\text{max}}}{1.414} = 0.707\,V_{\text{max}}$$
$$= 1.11\,V_{\text{ave}}$$

Maximum value

$$V_{\text{max}} = \sqrt{2}\,(V_{\text{eff}}) = 1.414\,V_{\text{eff}}$$
$$= 1.57\,V_{\text{ave}}$$

Power in AC Circuit

Apparent power: $P = VI$

True power: $P = VI \cos \theta = VI \times PF$

Power Factor

$$PF = \frac{P}{VI} = \cos \theta$$

$$\cos \theta = \frac{\text{true power}}{\text{apparent power}}$$

RL Circuit Time Constant

$$\frac{\text{L (in henrys)}}{R \text{ (in ohms)}} = t \text{ (in seconds), or}$$

$$\frac{\text{L (in microhenrys)}}{R \text{ (in ohms)}} = t \text{ (in microseconds)}$$

RC Circuit Time Constant

R (ohms) × C (farads) = t (seconds)

R (megohms) × C (microfarads) = t (seconds)

R (ohms) × C (microfarads) = t (microseconds)

R (megohms) × C (micromicrofarads) = t (microseconds)

Transformers
Voltage relationship

$$\frac{V_p}{V_s} = \frac{N_p}{N_s} \text{ or } V_s = V_p \times \frac{N_s}{N_p}$$

Current relationship

$$\frac{I_p}{I_s} = \frac{N_s}{N_p}$$

Induced voltage

$$V_{\text{eff}} = 4.44 \times BAf\text{N} \times 10^{-8}$$

Turns ratio

$$\frac{N_p}{N_s} = \sqrt{\frac{Z_p}{Z_s}}$$

Secondary current

$$I_s = I_p \times \frac{N_p}{N_s}$$

Secondary voltage

$$V_s = V_p \times \frac{N_s}{N_p}$$

Comparison of Units in Electric and Magnetic Circuits		
	Electric Circuit	**Magnetic Circuit**
Force	Volt, V, or emf	Gilberts, F, or mmf
Flow	Ampere, I	Flux, ϕ, in maxwells
Opposition	Ohms, R	Reluctance, R
Law	Ohm's Law, $I = \dfrac{V}{R}$	Rowland's law, $\phi = \dfrac{F}{R}$
Intensity of force	Volts per cm of length	$H = \dfrac{1.2571N}{L}$, gilberts per centimeter of length
Density	Current density—for example, amperes per cm^2	Flux density—for example, lines per cm^2, or gausses

RESONANCE IN AC CIRCUITS

General

Resonance occurs when the effects of inductance and capacitance exactly oppose each other. This occurs at a single particular frequency: $f = 1/(2\pi\sqrt{LC})$

Series Resonance

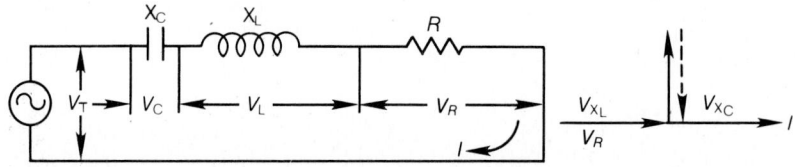

Series resonance occurs when:

$$V_c = V_L \quad IX_C = IX_L \quad \text{or} \quad V_T = V_R = IR$$

AC voltages (and sometimes reactances) are added phasorally. Voltage across inductor (or inductive reactance) opposes voltage across capacitor (or capacitive reactance).

Series resonance causes voltage step-up. Voltage step-up is the ratio of the voltage across either of the reactive elements to the voltage across the source. Its value is

$$\frac{V_{X_C}}{V_T} = \frac{V_{X_L}}{V_T} = Q = \frac{X_C}{R} = \frac{X_L}{R}$$

Impedance = $Z = R$

778

Parallel Resonance

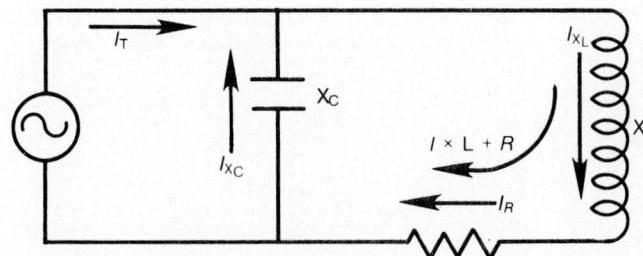 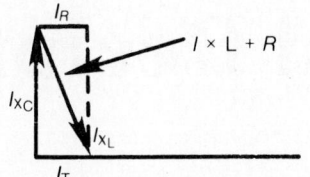

Parallel resonance occurs when:

$$IX_C = IX_L \quad I_T = I_R$$

Branch currents are added phasorally. Current through inductor opposes capacitor current.

Parallel resonance causes current step-up. Current step-up is the ratio of the current through either of the reactive branches to the total line current. Its value is:

$$\frac{I_{XC}}{I_T} = \frac{I_{XL}}{I_T} = Q = \frac{X_C}{R} = \frac{X_L}{R}$$

Impedance = $Z = \dfrac{L}{CR}$

Resistive

Impedance = $Z_R = R$

Current is in phase with voltage.
Therefore, $\theta = 0$ and $\cos\theta = 1$
        Power factor = 1
        Power = $VI\cos\theta = VI$

Inductive

Impedance = $X_L = 2\pi fL$

Current lags voltage by 90°.
Therefore, $\theta = 90°$ and $\cos\theta = 0$
        Power factor = 0
        Power = $VI\cos\theta = 0$

Capacitive

Impedance = $X_C = \dfrac{1}{2\pi fC}$

Current leads voltage by 90°.
Therefore, $\theta = 90°$ and $\cos\theta = 0$
        Power factor = 0
        Power = $VI\cos\theta = 0$

Resistive and Inductive (Capacitive)

Impedance = $Z = \sqrt{R^2 + X^2}$

Current lags (leads) voltage by less than 90°
Therefore, θ is variable = $\tan^{-1} X/R$
        Power factor = $\cos\theta$
        Power = $VI\cos\theta$

779

Powers, Roots, and Reciprocals

Number	Square	Cube	Square Root	Cube Root	Reciprocal
1	1	1	1.000000	1.000000	1.0000000
2	4	8	1.414214	1.259912	.5000000
3	9	27	1.732051	1.442250	.3333333
4	16	64	2.000000	1.587401	.2500000
5	25	125	2.236068	1.709976	.2000000
6	36	216	2.449490	1.817112	.1666667
7	49	343	2.645751	1.912931	.1428571
8	64	512	2.828427	2.000000	.1250000
9	81	729	3.000000	2.080084	.1111111
10	100	1000	3.162278	2.154435	.1000000
11	121	1331	3.316625	2.223980	.0909091
12	144	1728	3.464102	2.289429	.0833333
13	169	2197	3.605551	2.351335	.0769231
14	196	2744	3.741657	2.410142	.0714286
15	225	3375	3.872983	2.466212	.0666667
16	256	4096	4.000000	2.519842	.0625000
17	289	4913	4.123106	2.571282	.0588235
18	324	5832	4.242641	2.620741	.0555556
19	361	6859	4.358899	2.668402	.0526316
20	400	8000	4.472136	2.714418	.0500000
21	441	9261	4.582576	2.758924	.0476190
22	484	10,648	4.690416	2.802039	.0454545
23	529	12,167	4.795832	2.843867	.0434783
24	576	13,824	4.898980	2.884499	.0416667
25	625	15,625	5.000000	2.924018	.0400000
26	676	17,576	5.099020	2.962496	.0384615
27	729	19,683	5.196152	3.000000	.0370370
28	784	21,952	5.291503	3.036589	.0357143
29	841	24,389	5.385165	3.072317	.0344828
30	900	27,000	5.477226	3.107233	.0333333
31	961	29,791	5.567764	3.141381	.0322581
32	1024	32,768	5.656854	3.174802	.0312500
33	1089	35,937	5.744563	3.207534	.0303030
34	1156	39,304	5.830952	3.239612	.0294118
35	1225	42,875	5.916080	3.271066	.0285714

Number	Square	Cube	Square Root	Cube Root	Reciprocal
36	1296	46,656	6.000000	3.301927	.0277778
37	1369	50,653	6.082763	3.332222	.0270270
38	1444	54,872	6.164414	3.361975	.0263158
39	1521	59,319	6.244998	3.391211	.0256410
40	1600	64,000	6.324555	3.419952	.0250000
41	1681	68,921	6.403124	3.448217	.0243902
42	1764	74,088	6.480741	3.476027	.0238095
43	1849	79,507	6.557439	3.503398	.0232558
44	1936	85,184	6.633250	3.530348	.0227273
45	2025	91,125	6.708204	3.556893	.0222222
46	2116	97,336	6.782330	3.583048	.0217391
47	2209	103,823	6.855655	3.608826	.0212766
48	2304	110,592	6.928203	3.634241	.0208333
49	2401	117,649	7.000000	3.659306	.0204082
50	2500	125,000	7.071068	3.684013	.0200000
51	2601	132,651	7.141428	3.708430	.0196078
52	2704	140,608	7.211103	3.732511	.0192308
53	2809	148,877	7.280110	3.756286	.0188679
54	2916	157,464	7.348469	3.779763	.0185185
55	3025	166,375	7.416199	3.802953	.0181818
56	3136	175,616	7.483315	3.825862	.0178571
57	3249	185,193	7.549834	3.848501	.0175439
58	3364	195,112	7.615773	3.870877	.0172414
59	3481	205,379	7.681146	3.892997	.0169492
60	3600	216,000	7.745967	3.914868	.0166667
61	3721	226,981	7.810250	3.936497	.0163934
62	3844	238,328	7.874008	3.957892	.0161290
63	3969	250,047	7.937254	3.979057	.0158730
64	4096	262,144	8.000000	4.000000	.0156250
65	4225	274,625	8.062258	4.020726	.0153846
66	4356	287,496	8.124038	4.041240	.0151515
67	4489	300,763	8.185353	4.061548	.0149254
68	4624	314,432	8.246211	4.081655	.0147059
69	4761	328,509	8.306624	4.101566	.0144928
70	4900	343,000	8.366600	4.121285	.0142857
71	5041	357,911	8.426150	4.140818	.0140845
72	5184	373,248	8.485281	4.160168	.0138889
73	5329	389,017	8.544004	4.179339	.0136986
74	5476	405,224	8.602325	4.198336	.0135135
75	5625	421,875	8.660254	4.217163	.0133333
76	5776	438,976	8.717798	4.235824	.0131579
77	5929	456,533	8.774964	4.254321	.0129870
78	6084	474,552	8.831761	4.272659	.0128205
79	6241	493,039	8.888194	4.290840	.0126582
80	6400	512,000	8.944272	4.308870	.0125000
81	6561	531,441	9.000000	4.326749	.0123457
82	6724	551,368	9.055385	4.344482	.0121951
83	6889	571,787	9.110434	4.362071	.0120482
84	7056	592,704	9.165151	4.379519	.0110048
85	7225	614,125	9.219545	4.396830	.0117647

Number	Square	Cube	Square Root	Cube Root	Reciprocal
86	7396	636,056	9.273619	4.414005	.0116279
87	7569	658,503	9.327379	4.431048	.0114943
88	7744	681,472	9.380832	4.447960	.0113636
89	7921	704,969	9.433981	4.464745	.0112360
90	8100	729,000	9.486833	4.481405	.0111111
91	8281	753,571	9.539392	4.497941	.0109890
92	8464	778,688	9.591663	4.514357	.0108696
93	8649	804,357	9.643651	4.530655	.0107527
94	8836	830,584	9.695360	4.546836	.0106383
95	9025	857,375	9.746794	4.562903	.0105263
96	9216	884,736	9.797959	4.578857	.0104167
97	9409	912,673	9.848858	4.594701	.0103093
98	9604	941,192	9.899495	4.610436	.0102041
99	9801	970,299	9.949874	4.626065	.0101010
100	10,000	1,000,000	10.000000	4.641589	.0100000

APPENDIX P

Laws of Exponents

The International Symbols Committee has adopted prefixes for denoting decimal multiples of units. The National Bureau of Standards has followed the recommendations of this committee and has adopted the following list of prefixes:

Multiplication Factor	Prefix	Symbol
1 000 000 000 000 000 000 = 10^{18}	exa	E
1 000 000 000 000 000 = 10^{15}	peta	P
1 000 000 000 000 = 10^{12}	tera	T
1 000 000 000 = 10^{9}	giga	G
1 000 000 = 10^{6}	mega	M
1 000 = 10^{3}	kilo	k
100 = 10^{2}	hecto	h
10 = 10	deka	da
0.1 = 10^{-1}	deci	d
0.01 = 10^{-2}	centi	c
0.001 = 10^{-3}	milli	m
0.000 001 = 10^{-6}	micro	μ
0.000 000 001 = 10^{-9}	nano	n
0.000 000 000 001 = 10^{-12}	pico	p
0.000 000 000 000 001 = 10^{-15}	femto	f
0.000 000 000 000 000 001 = 10^{-18}	atto	a

To multiply like (with same base) exponential quantities, add the exponents. In the language of algebra the rule is $a^m \times a^n = a^{m+n}$.

$$10^4 \times 10^2 = 10^{4+2} = 10^6$$
$$0.003 \times 8.252 = 3 \times 10^{-3} \times 8.252 \times 10^2$$
$$= 24.756 \times 10^{-1} = 2.4756$$

To divide exponential quantities, subtract the exponents. In the language of algebra the rule is

$$\frac{a^m}{a^n} = a^{m-n} \text{ or}$$
$$10^8 \div 10^2 = 10^6$$
$$3{,}000 \div 0.015 = (3 \times 10^3) \div (1.5 \times 10^{-2})$$
$$= 2 \times 10^5 = 200{,}000$$

To raise an exponential quantity to a power, multiply the exponents. In the language of algebra $(x^m)^n = x^{mn}$.

$$(10^3)^4 = 10^{3\times4} = 10^{12}$$
$$2{,}500^2 = (2.5 \times 10^3)^2 = 6.25 \times 10^6 = 6{,}250{,}000$$

Any number (except zero) raised to the zero power is one. In the language of algebra $x^0 = 1$.

$$x^3 \div x^3 = 1$$
$$10^4 \div 10^4 = 1$$

Any base with a negative exponent is equal to 1 divided by the base with an equal positive exponent. In the language of algebra $x^{-a} = \dfrac{1}{x^a}$.

$$10^{-2} = \frac{1}{10^2} = \frac{1}{100}$$

$$5a^{-3} = \frac{5}{a^3}$$

$$(6a)^{-1} = \frac{1}{6a}$$

To raise a product to a power, raise each factor of the product to that power.

$$(2 \times 10)^2 = 2^2 \times 10^2$$
$$3{,}000^3 = (3 \times 10^3)^3 = 27 \times 10^9$$

To find the nth root of an exponential quantity, divide the exponent by the index of the root. Thus, the nth root of $a^m = a^{m/n}$.

$$\sqrt{x^6} = x^{6/2} = x^3$$
$$\sqrt[3]{64 \times 10^3} = 4 \times 10 = 40$$

APPENDIX Q

Trigonometric Functions and Their Use

Trigonometry, as the term implies, is a method of measuring triangles by numerical computation. With the development of alternating-current generators, it became impossible to analyze or understand the behavior of even the simplest circuit without considering angles and their computation.

The law of trigonometric function states that the ratio of any pair of sides of a right triangle is a function of the acute angle $\propto$. The six possible ratios are called the trigonometric functions of the acute angle $\propto$.

In Fig. Q-1, $\propto$ represents any acute angle and BC is a line drawn from point B perpendicular to C to form a right triangle ABC. The small letters a, b, and c represent the lengths of the sides respectively opposite the vertices A, B, and C. Referring to this illustration, we can now list the names and abbreviations given to the six ratios or trigonometric functions (Table Q-1).

Table Q-1: Trigonometric Functions

Ratio	Name	Abbreviation
a/b	tangent of $\propto$	$\tan \propto$
b/a	cotangent of $\propto$	$\cot \propto$
a/c	sine of $\propto$	$\sin \propto$
b/c	cosine of $\propto$	$\cos \propto$
c/b	secant of $\propto$	$\sec \propto$
c/a	cosecant of $\propto$	$\csc \propto$

Only three of the six trigonometric functions are used for solving AC problems in this book—namely, the sine, cosine, and tangent. In the right triangle of Fig. Q-1, a is the leg opposite to $\propto$, b is the leg adjacent to $\propto$, and c is the hypotenuse; then,

$$\sin \propto = \frac{a}{c} = \frac{\text{opposite}}{\text{hypotenuse}}$$

$$\cos \propto = \frac{b}{c} = \frac{\text{adjacent}}{\text{hypotenuse}}$$

$$\tan \propto = \frac{a}{b} = \frac{\text{opposite}}{\text{adjacent}}$$

Example: ∝ and β are the acute angles in the right triangle of Fig. Q-1; then,

sin ∝ = 3/5 = 0.6	sin β = 4/5 = 0.8
cos ∝ = 4/5 = 0.8	cos β = 3/5 = 0.6
tan ∝ = 3/4 = 0.75	tan β = 4/3 = 1.33

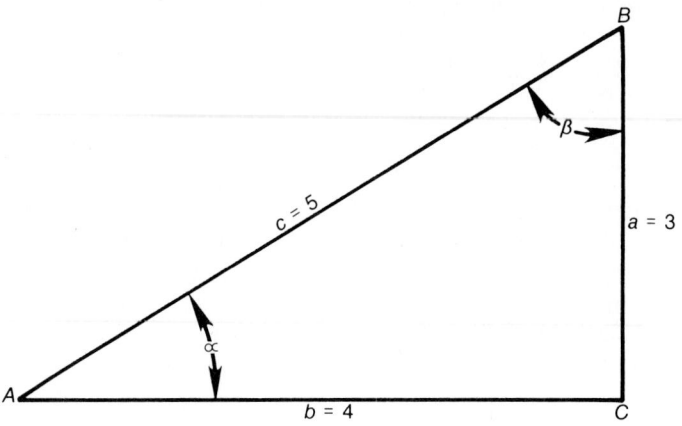

Fig. Q-1: Right triangle functions.

Solutions by Pythagorean Theorem

The Pythagorean theorem states that in a right triangle, the square of the hypotenuse equals the sum of the squares of the other two sides.

$$c^2 = a^2 + b^2$$

Example 1: In Fig. Q-2, $c = 6$ and $b = 5$. Find a.

Solution: $c^2 = a^2 + b^2$

$$6^2 = a^2 + 5^2$$
$$a^2 = 36 - 25 = 11$$
$$a = \sqrt{11} = 3.317$$

Example 2: Find the trigonometric functions of the acute angle ∝ in Fig. Q-2.

Solution:

$$\sin \propto = \frac{\text{opposite}}{\text{hypotenuse}} = \frac{\sqrt{11}}{6}$$

$$\cos \propto = \frac{\text{adjacent}}{\text{hypotenuse}} = \frac{5}{6}$$

$$\tan \propto = \frac{\text{opposite}}{\text{adjacent}} = \frac{\sqrt{11}}{5}$$

Example 3: If sin ∝ = 3/4, find the cosine and tangent of ∝.
Solution: Since sin ∝ = 3/4 = a/c, construct a right triangle with $a = 3$ and $c = 4$ (Fig. Q-3).

786

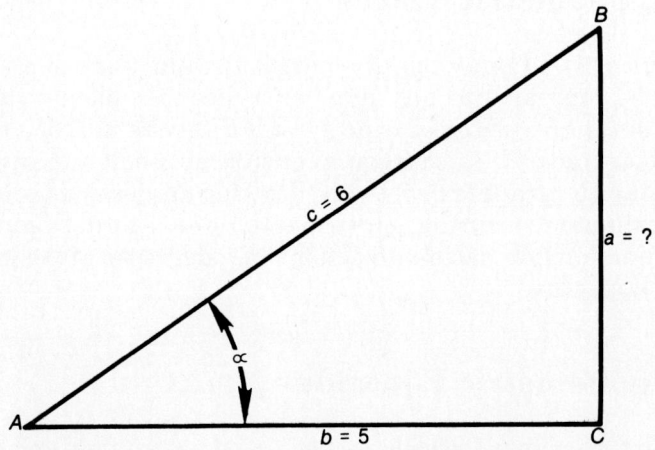

Fig. Q-2: Trigonometric functions of acute angle.

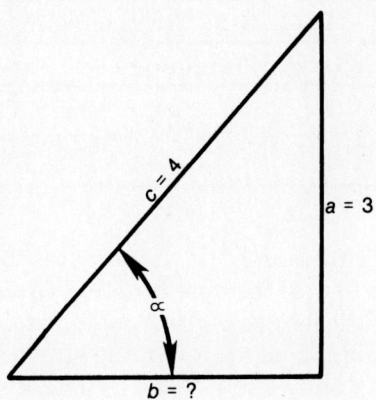

Fig. Q-3: Right triangle.

From the Pythagorean theorem,

$$c^2 = a^2 + b^2$$
$$b^2 = c^2 - a^2$$
$$b^2 = 16 - 9 = 7$$
$$b = \sqrt{7} = 2.64 \text{ (approx.)}$$

$$\cos \alpha = \frac{\text{adjacent}}{\text{hypotenuse}} = \frac{b}{c} = \frac{\sqrt{7}}{4}$$

$$\tan \alpha = \frac{\text{opposite}}{\text{adjacent}} = \frac{a}{b} = \frac{3}{\sqrt{7}} = \frac{3\sqrt{7}}{7}$$

Trigonometric Tables

By use of advanced mathematics, the functions of an angle can be computed to any number of decimal places. In such computations, the error is no greater than one-half of a unit in the last place. This accuracy is considerably better than can be obtained by graphical methods (drawing angles and their sides to scale and measuring lengths). A table of natural trigonometric functions for decimal fractions of a degree is given later in this appendix.

Trigonometric Equations

Partly because Greek letters are used as symbols and partly because new terms are introduced, students sometimes feel that trigonometry is a difficult subject. This is not true. Trigonometric functions are merely ratios, the quotient obtained by dividing the length of one side by the length of another side of a right triangle. A direct comparison shows that the function equations are as simple as those expressing Ohm's Law (Table Q-2).

Table Q-2: Comparison	
Ohm's Law	**Trigonometric Function**
$I = V/R$	$\sin \theta = a/c$
$V = I \times R$	$a = c \times \sin \theta$
$R = V/I$	$c = a/\sin \theta$

The Ohm's Law equation $I = V/R$ indicates that, if V and R are known, I can be found; the sine function equation, $\sin \theta = a/c$, indicates that if the length of the hypotenuse, c, and the length of the opposite side, a, are known, the sine can be found. Similarly $a = c \times \sin \theta$ shows that if the hypotenuse, c, and the sine of θ are known, the opposite side, a, can be determined; the equation $c = a/\sin \theta$ shows that if the opposite side, a, and the sine of θ are known, the hypotenuse, c, can be determined. In other words, these equations for trigonometric functions are as simple and as readily solved as the familiar Ohm's Law equations.

Example 1: In triangle ABC find the unknown side by use of the Pythagorean theorem and obtain all the functions of $\propto$ when $b = 5$ and $a = 12$.

Solution:
$$c^2 = a^2 + b^2$$
$$c^2 = 144 + 25$$
$$c^2 = 169$$
$$c = 13$$

Since the three sides are now known, the functions may be read directly: $\sin \propto = a/c = 12/32$; $\cos \propto = b/c = 5/13$; $\tan \propto = a/b = 12/5$.

Example 2: Cos ∝ = 15/17; construct the angle and find its other functions.

Solution:

$$\cos \propto = b/c = 15/17$$
$$a^2 = c^2 - b^2$$
$$a^2 = 17^2 - 15^2 = 64$$
$$a = \sqrt{64} = 8$$

Then, sin ∝ = a/c = 8/17; cos ∝ = b/c = 15/17; tan ∝ = a/b = 8/15. The angle ∝ can be found from any one of its functions in the trigonometric table given later in this appendix.

$$\sin \propto = 8/17 = 0.4705 \text{ (approx.)}$$
$$\sin^{-1} 0.4705 = 28°$$

Note. $\sin^{-1} 0.4705 = 28°$ may also be written as arcsin 0.4705 = 28°. Both are read: The angle whose sine is 0.4705 is 28°. Consider the statement: The tangent of 37.2° is 0.7590. Stated inversely, this is written,

$$37.2° = \tan^{-1} 0.7590 \text{ or}$$
$$37.2° = \arctan 0.7590$$

Both are read: 37.2° is the angle whose tangent is 0.7590.

Trigonometry and AC Voltages

Figure Q-4 is a graph of the amplitude variations of a certain type of AC voltage (or current) with time. Time is plotted along the x-axis and amplitude variations are indicated by the varying height of the curve relative to the x-axis. Actually, as is explained in this section, Fig. Q-4 is essentially a graph of the equation

$$a = c \sin \theta$$

when c is held constant and θ is varied. Voltages (or currents) which vary, as shown in Fig. Q-4 are called sinusoidal voltages (or currents). The generation of sinusoidal voltages (or currents) is discussed in Chapter 11. These voltages and currents form a basis for all AC theory; therefore, an understanding of the trigonometry involved is of considerable importance.

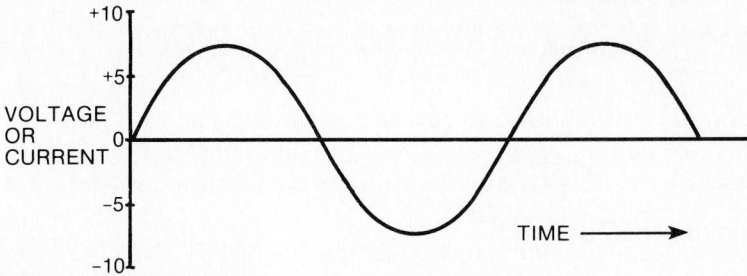

Fig. Q-4: Sine wave.

789

The Radian

In discussion of AC theory, the radian is the commonly used unit of angular measure. In Fig. Q-5, point A is the center of the circle, AC and AB are radii, and the length of the arc CB equals the length of the radius. By definition angle CAB is one radian. Figure Q-6 shows that there are approximately 6.28 radians in a circle. The Greek letter π represents a constant equal to 3.14; therefore, $2\pi = 6.28$. Thus, one revolution of side AB generates an angle of 2π radians.

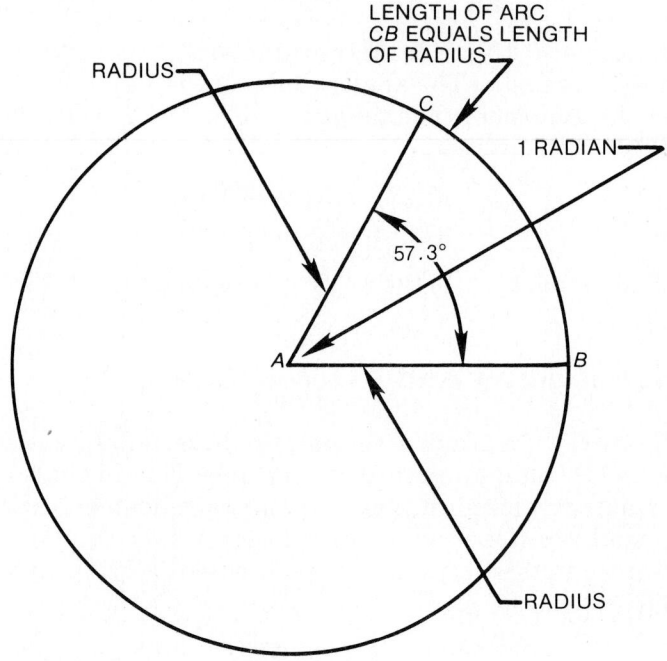

Fig. Q-5: Relation of radius, radian, and arc of a circle.

Conversion Formulas. As explained, if AB makes a complete revolution, it generates an angle of 360°, or 2π radians; therefore,

$$2\pi \text{ radians} = 360°$$

and, $\quad 1 \text{ radian} = \dfrac{360°}{2\pi} = \dfrac{180°}{\pi} = 57.3° \text{ (approx.)}$

This formula is used to convert radians to degrees and vice versa. To convert degrees to radians, multiply the number of degrees by $\pi/180$; to convert radians to degrees, multiply the number of radians by $\dfrac{180}{\pi}$.

790

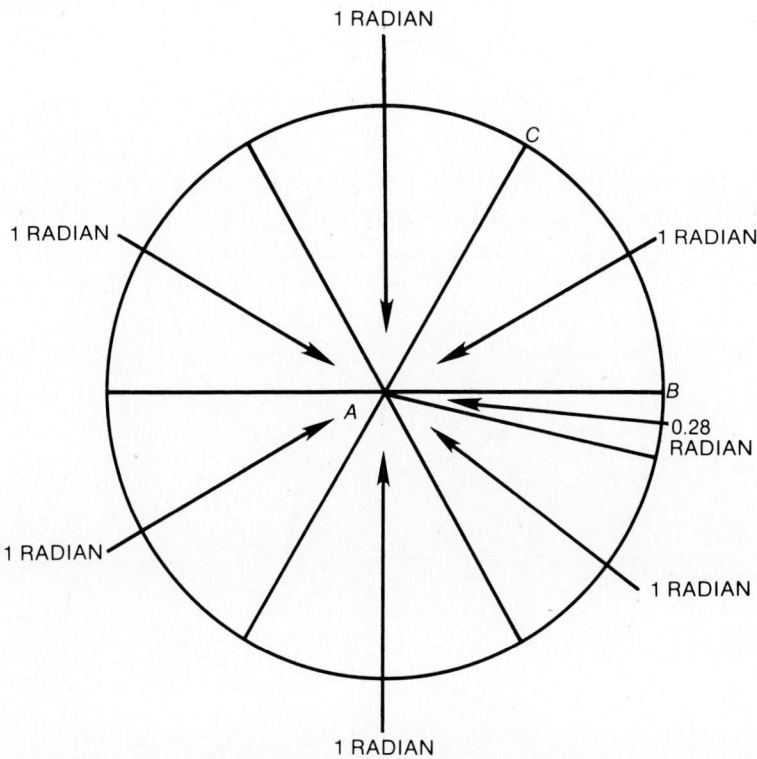

Fig. Q-6: Number of radians in a circle.

Trigonometric Functions

Let θ be any varying angle (Fig. Q-7A). This angle θ varies from one value to another by rotation of its terminal side OP. Thus, if side OP makes three complete revolutions counterclockwise from its initial position OX, it generates an angle of 1,080°, or $3 \times 2\pi$ radians. The positions of the terminal side OP repeat themselves at intervals of 360°, or 2π radians; consequently, the trigonometric functions also repeat themselves.

Example: Consider Fig. Q-7A. The side OP has rotated counterclockwise from its initial position OX to generate angle θ of 45°. The sine of 45° = a/c. Now consider that OP is rotated an additional 360° (Fig. Q-7B) to generate an angle of $(\theta + 360°)$, or

$$\propto = (\theta + 360°) = 45° + 360° = 405°.$$

We see that OP' coincides with OP; that is, the 45° angle θ and the 405° angle $\propto$ are coterminal. Because of this, a and c have the same numerical values for both angles, thus:

$$\sin \propto = \sin (405° - 360°) = \sin 45° = a/c.$$

This shows that the trigonometric functions repeat and we say that any function of θ is periodic, with 360°, or 2π radians, as the period.

791

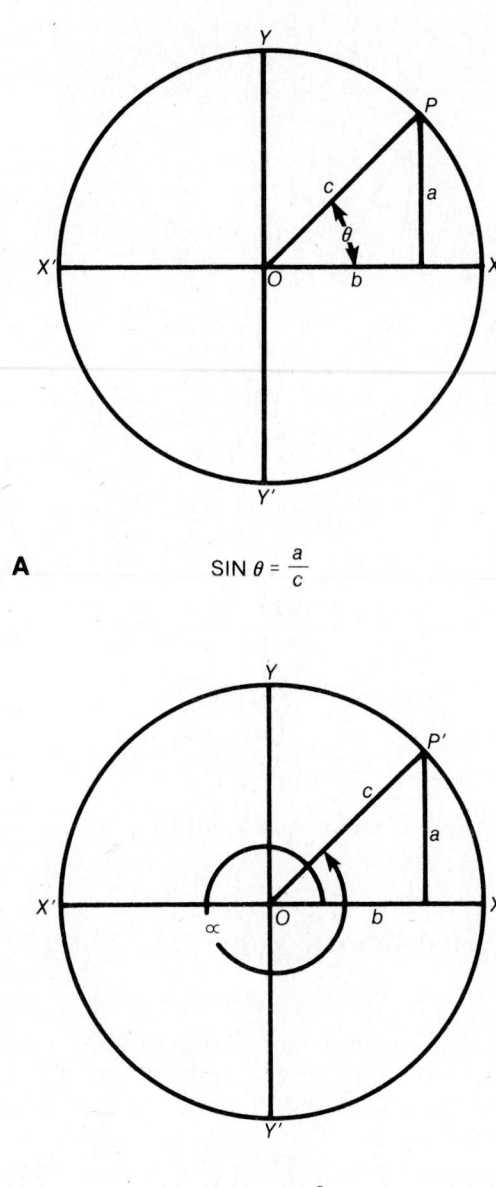

$$\text{SIN } \theta = \frac{a}{c}$$

A

$$\text{SIN } \propto = \frac{a}{c}$$

B

Fig. Q-7: Periodic trigonometric functions: (A) 45° and (B) 405°.

Graph of Sine Curve, y = sin x

From Fig. Q-8, we read $\sin \theta = y/r$ and, if r is a unit length, $\sin \theta = y/1 = y$. Suppose that we now use the symbol x to represent any angle θ, so that the equation becomes $y = \sin x$. The graph of this equation is plotted just as the graphs of algebraic equations are plotted as follows: Values are assigned to x and the corresponding values of y are calculated, points are plotted from the coordinates obtained, and a smooth curve is drawn through the points.

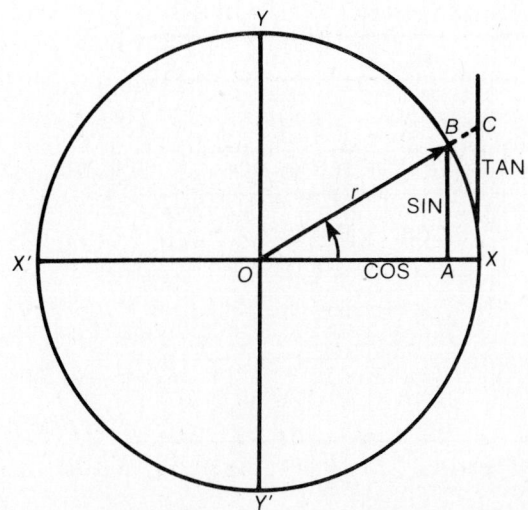

Fig. Q-8: Sine θ, cosine θ, and tangent θ are all in the first quadrant.

Example: In AC problems, radians are usually used as the unit of angular measure. The sine curve is plotted with the angle x and its function y having the same scale, one unit on the y-axis being equal to the length representing one radian on the x-axis. Thus, in setting up the tabulation, the first column lists angles in degrees; the second gives x in radians; the third gives the number of unit lengths representing x; the fourth gives the number of unit lengths representing y; and the fifth column lists the coordinates of a number of points on the curve (see Table Q-3). Assuming that a unit length represents one radian along the x-axis of Fig. Q-9, the calculation for changing 30° to radians is as follows:

$$360° = 2\pi \text{ radians}$$

$$30 \times \frac{\pi}{180} = \frac{30\pi}{180} = \frac{\pi}{10} \text{ radians}$$

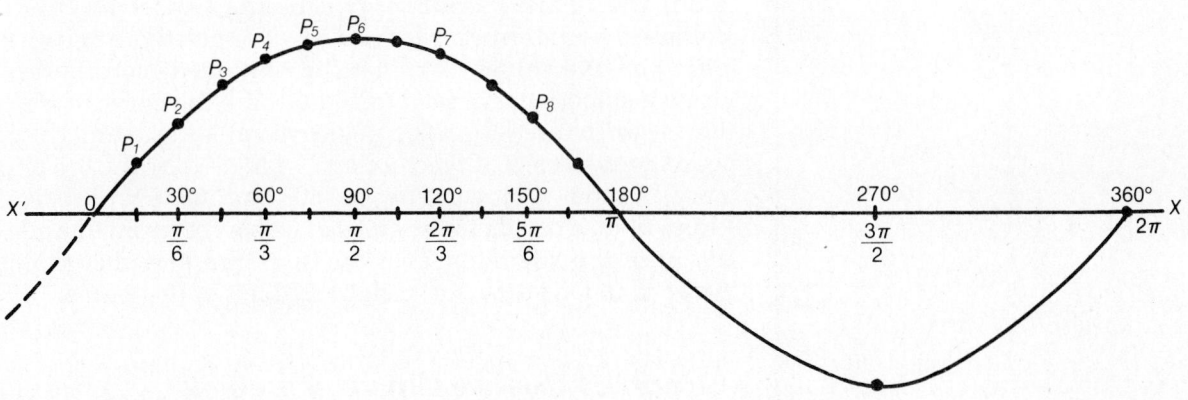

Fig. Q-9: Graph of sine curve.

793

Table Q-3: Table of Sine Curves

1	2	3	4	5
X Degrees	X Radians	X Unit Measure	Y (Sin X)	Point
0	0	0	0	$P_0 = (0, 0)$
* 15	$\dfrac{\pi}{12}$	0.261	0.258	$P_1 = (0.261, 0.258)$
30	$\dfrac{\pi}{6}$	0.52	0.50	$P_2 = (0.52, 0.50)$
* 45	$\dfrac{\pi}{4}$	0.785	0.707	$P_3 = (0.785, 0.707)$
60	$\dfrac{\pi}{3}$	1.05	0.87	$P_4 = (1.05, 0.87)$
* 75	$\dfrac{5\pi}{12}$	1.30	0.965	$P_5 = (1.30, 0.965)$
90	$\dfrac{\pi}{2}$	1.57	1.00	$P_6 = (1.57, 1.00)$
120	$\dfrac{2\pi}{3}$	2.09	0.87	$P_7 = (2.09, 0.87)$
150	$\dfrac{5\pi}{6}$	2.62	0.50	$P_8 = (2.62, 0.50)$
180	π	3.14	0.00	$P_9 = (3.14, 0)$

*Not Indicated on Graph.

This is entered in column 2. To obtain the number of unit lengths for column 3, complete the division indicated by $\pi/6$, getting

$$\frac{3.14}{6} = 0.52 \text{ unit length.}$$

From Table Q-3, we find that sin 30° = 0.5. This means that y is equal to one-half the unit length chosen to represent one radian along the x-axis. The values in columns 3 and 4 are entered in column 5 as coordinates for point P_2. For positive angles assign only positive values to x, plot the computed coordinates, and draw a smooth curve to obtain the solid line curve of Fig. Q-9. For negative angles assign negative values to x, and the curve extends to the left of the y-axis as indicated by the broken line curve. Values of sine x between 180° and 360° are negative, but equal in magnitude to the sines of angles between 0° and 180°. Because the trigonometric functions are periodic, the curve extends indefinitely to the right and left of the y-axis.

Graph of Cosine Curve, y = cos x

The cosine curve of Fig. Q-10 is obtained by plotting the graph of the equation $y = \cos x$. The graphs of $y = \sin x$ and

$y = \cos x$ have the same shape, but differ in their positions with relation to the axes. The cosine of angle x is equal to the sine of an angle which is 90° greater. Thus, if the graph of $y = \sin x$ is shifted 90° to the left, we obtain the graph of $y = \cos x$.

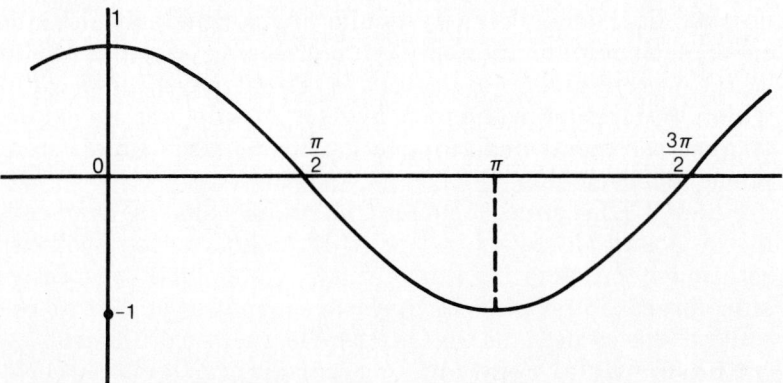

Fig. Q-10: Graph of $y = \cos x$.

Graph of Tangent Curve, y = tan x

There is no finite tangent for 90°; tan 90° = ∞. This causes breaks in the curve, and the graph of $y = \tan x$ (Fig. Q-11) consists of an infinite number of separated branches. The broken lines, called asymptotes, do not touch the curve because there is no tangent (the tangent is infinite) and, hence, no point on the graph corresponding to $x = \pi/2$ and $x = 3\pi/2$.

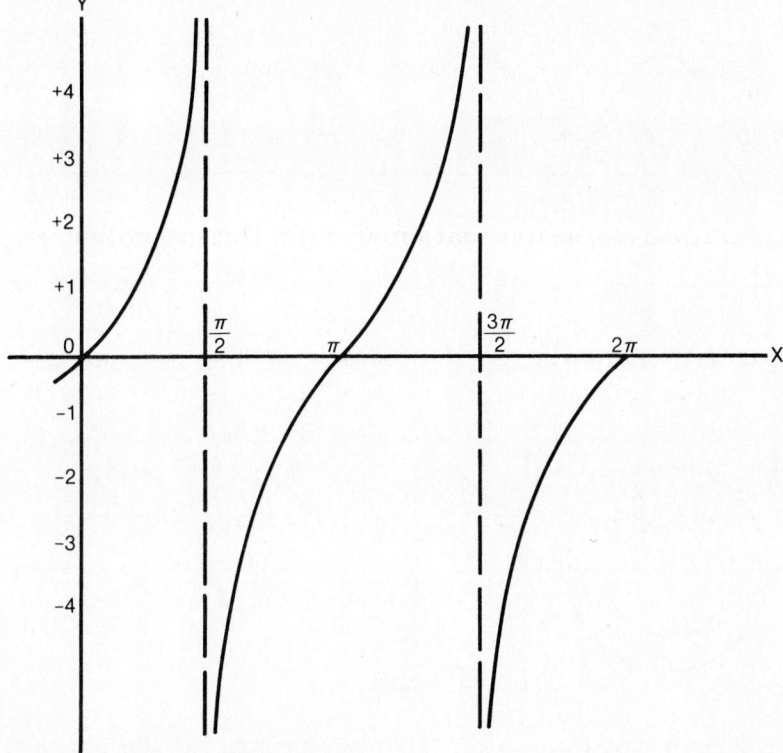

Fig. Q-11: Graph of $y = \tan x$.

Phase

When a series of events repeats itself with regularity, we speak of a particular event as a phase of the series. Thus, we speak of the moon as being in its first, second, third, or fourth quarter. Each quarter is a particular phase. Similarly, in a sine curve, each point on the curve is a phase in a cycle; the point at which y is maximum and positive is one phase, the point at which y is maximum and negative is another phase, and so on. It must be remembered that every point represents a different phase in a cycle (Fig. Q-12).

Phase Difference. Assume that radius phasors r_1 and r are in the positions shown in Fig. Q-13. Start counterclockwise rotation at the same angular velocity, ω, and plot the respective sine curves A and B. Note that corresponding phases do not occur at the same time; that is, there is a phase difference.

Phase Angle. Points on A occur earlier than the corresponding points on B; consequently, A is said to lead B. When $t = 0$, r starts to generate the curve B, $y = r \sin \omega t$. At the same time, r_1 is ahead of r by an angle θ. Consequently, r_1 generates curve A, $y_1 = r_1 \sin (\omega t + \theta)$. This moves the y_1 curve along the time axis by an angle θ. The angle θ is the angular difference between the curves and it is called the phase angle. Since it is ahead of y, y_1 leads y; therefore, in the equation $y_1 = r_1 \sin (\omega t + \theta)$, the angle θ is called the angle of lead. If the positions of r_1 and r are interchanged, r_1 will be behind r. The phase angle θ will be the same, but y_1 will now lag y. Consequently, the curve generated by r_1 is $y_1 = r_1 \sin (\omega t - \theta)$ and θ is the angle of lag.

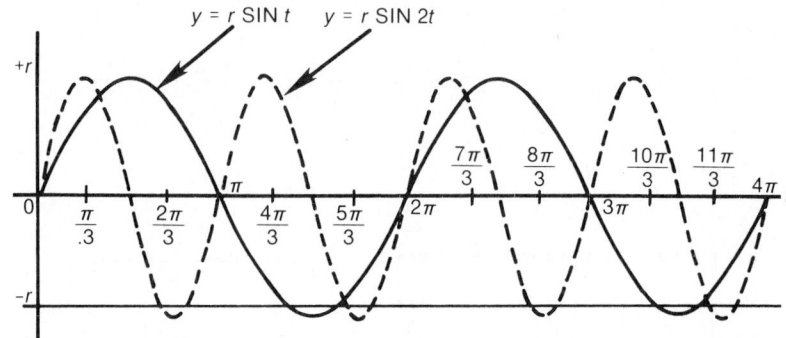

Fig. Q-12: Graph of $y = r \sin t$ and $y = r \sin 2\,t$.

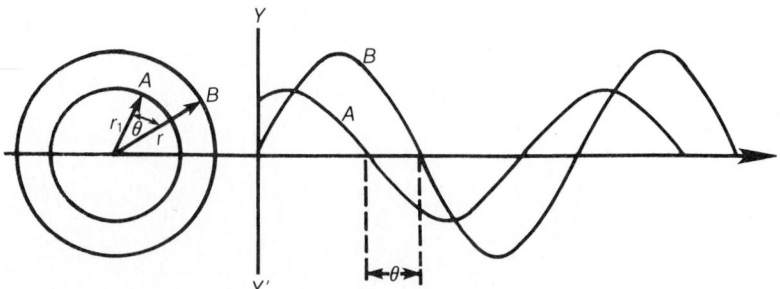

Fig. Q-13: Rotation of two radii, showing phase difference.

796

TABLE OF NATURAL-TRIGONOMETRIC FUNCTIONS

′	Sin	Tan	Cot	Cos	′
0	.00000	.00000	———	1.0000	60
1	.00029	.00029	3437.7	1.0000	59
2	.00058	.00058	1718.9	1.0000	58
3	.00087	.00087	1145.9	1.0000	57
4	.00116	.00116	859.44	1.0000	56
5	.00145	.00145	687.55	1.0000	55
6	.00175	.00175	572.96	1.0000	54
7	.00204	.00204	491.11	1.0000	53
8	.00233	.00233	429.72	1.0000	52
9	.00262	.00262	381.97	1.0000	51
10	.00291	.00291	343.77	1.0000	50
11	.00320	.00320	312.52	.99999	49
12	.00349	.00349	286.48	.99999	48
13	.00378	.00378	264.44	.99999	47
14	.00407	.00407	245.55	.99999	46
15	.00436	.00436	229.18	.99999	45
16	.00465	.00465	214.86	.99999	44
17	.00495	.00495	202.22	.99999	43
18	.00524	.00524	190.98	.99999	42
19	.00553	.00553	180.93	.99998	41
20	.00582	.00582	171.89	.99998	40
21	.00611	.00611	163.70	.99998	39
22	.00640	.00640	156.26	.99998	38
23	.00669	.00669	149.47	.99998	37
24	.00698	.00698	143.24	.99998	36
25	.00727	.00727	137.51	.99997	35
26	.00756	.00756	132.22	.99997	34
27	.00785	.00785	127.32	.99997	33
28	.00814	.00815	122.77	.99997	32
29	.00844	.00844	118.54	.99996	31
30	.00873	.00873	114.59	.99996	30
31	.00902	.00902	110.89	.99996	29
32	.00931	.00931	107.43	.99996	28
33	.00960	.00960	104.17	.99995	27
34	.00989	.00989	101.11	.99995	26
35	.01018	.01018	98.218	.99995	25
36	.01047	.01047	95.489	.99995	24
37	.01076	.01076	92.908	.99994	23
38	.01105	.01105	90.463	.99994	22
39	.01134	.01135	88.144	.99994	21
40	.01164	.01164	85.940	.99993	20
41	.01193	.01193	83.844	.99993	19
42	.01222	.01222	81.847	.99993	18
43	.01251	.01251	79.943	.99992	17
44	.01280	.01280	78.126	.99992	16
45	.01309	.01309	76.390	.99991	15
46	.01338	.01338	74.729	.99991	14
47	.01367	.01367	73.139	.99991	13
48	.01396	.01396	71.615	.99990	12
49	.01425	.01425	70.153	.99990	11
50	.01454	.01455	68.750	.99989	10
51	.01483	.01484	67.402	.99989	9
52	.01513	.01513	66.105	.99989	8
53	.01542	.01542	64.858	.99988	7
54	.01571	.01571	63.657	.99988	6
55	.01600	.01600	62.499	.99987	5
56	.01629	.01629	61.383	.99987	4
57	.01658	.01658	60.306	.99986	3
58	.01687	.01687	59.266	.99986	2
59	.01716	.01716	58.261	.99985	1
60	.01745	.01746	57.290	.99985	0
′	Cos	Cot	Tan	Sin	′

′	Sin	Tan	Cot	Cos	′
0	.01745	.01746	57.290	.99985	60
1	.01774	.01775	56.351	.99984	59
2	.01803	.01804	55.442	.99984	58
3	.01832	.01833	54.561	.99983	57
4	.01862	.01862	53.709	.99983	56
5	.01891	.01891	52.882	.99982	55
6	.01920	.01920	52.081	.99982	54
7	.01949	.01949	51.303	.99981	53
8	.01978	.01978	50.549	.99980	52
9	.02007	.02007	49.816	.99980	51
10	.02036	.02036	49.104	.99979	50
11	.02065	.02066	48.412	.99979	49
12	.02094	.02095	47.740	.99978	48
13	.02123	.02124	47.085	.99977	47
14	.02152	.02153	46.449	.99977	46
15	.02181	.02182	45.829	.99976	45
16	.02211	.02211	45.226	.99976	44
17	.02240	.02240	44.639	.99975	43
18	.02269	.02269	44.066	.99974	42
19	.02298	.02298	43.508	.99974	41
20	.02327	.02328	42.964	.99973	40
21	.02356	.02357	42.433	.99972	39
22	.02385	.02386	41.916	.99972	38
23	.02414	.02415	41.411	.99971	37
24	.02443	.02444	40.917	.99970	36
25	.02472	.02473	40.436	.99969	35
26	.02501	.02502	39.965	.99969	34
27	.02530	.02531	39.506	.99968	33
28	.02560	.02560	39.057	.99967	32
29	.02589	.02589	38.618	.99966	31
30	.02618	.02619	38.188	.99966	30
31	.02647	.02648	37.769	.99965	29
32	.02676	.02677	37.358	.99964	28
33	.02705	.02706	36.956	.99963	27
34	.02734	.02735	36.563	.99963	26
35	.02763	.02764	36.178	.99962	25
36	.02792	.02793	35.801	.99961	24
37	.02821	.02822	35.431	.99960	23
38	.02850	.02851	35.070	.99959	22
39	.02879	.02881	34.715	.99959	21
40	.02908	.02910	34.368	.99958	20
41	.02938	.02939	34.027	.99957	19
42	.02967	.02968	33.694	.99956	18
43	.02996	.02997	33.366	.99955	17
44	.03025	.03026	33.045	.99954	16
45	.03054	.03055	32.730	.99953	15
46	.03083	.03084	32.421	.99952	14
47	.03112	.03114	32.118	.99952	13
48	.03141	.03143	31.821	.99951	12
49	.03170	.03172	31.528	.99950	11
50	.03199	.03201	31.242	.99949	10
51	.03228	.03230	30.960	.99948	9
52	.03257	.03259	30.683	.99947	8
53	.03286	.03288	30.412	.99946	7
54	.03316	.03317	30.145	.99945	6
55	.03345	.03346	29.882	.99944	5
56	.03374	.03376	29.624	.99943	4
57	.03403	.03405	29.371	.99942	3
58	.03432	.03434	29.122	.99941	2
59	.03461	.03463	28.877	.99940	1
60	.03490	.03492	28.636	.99939	0
′	Cos	Cot	Tan	Sin	′

′	Sin	Tan	Cot	Cos	′
0	.03490	.03492	28.636	.99939	60
1	.03519	.03521	28.399	.99938	59
2	.03548	.03550	28.166	.99937	58
3	.03577	.03579	27.937	.99936	57
4	.03606	.03609	27.712	.99935	56
5	.03635	.03638	27.490	.99934	55
6	.03664	.03667	27.271	.99933	54
7	.03693	.03696	27.057	.99932	53
8	.03723	.03725	26.845	.99931	52
9	.03752	.03754	26.637	.99930	51
10	.03781	.03783	26.432	.99929	50
11	.03810	.03812	26.230	.99927	49
12	.03839	.03842	26.031	.99926	48
13	.03868	.03871	25.835	.99925	47
14	.03897	.03900	25.642	.99924	46
15	.03926	.03929	25.452	.99923	45
16	.03955	.03958	25.264	.99922	44
17	.03984	.03987	25.080	.99921	43
18	.04013	.04016	24.898	.99919	42
19	.04042	.04046	24.719	.99918	41
20	.04071	.04075	24.542	.99917	40
21	.04100	.04104	24.368	.99916	39
22	.04129	.04133	24.196	.99915	38
23	.04159	.04162	24.026	.99913	37
24	.04188	.04191	23.859	.99912	36
25	.04217	.04220	23.695	.99911	35
26	.04246	.04250	23.532	.99910	34
27	.04275	.04279	23.372	.99909	33
28	.04304	.04308	23.214	.99907	32
29	.04333	.04337	23.058	.99906	31
30	.04362	.04366	22.904	.99905	30
31	.04391	.04395	22.752	.99904	29
32	.04420	.04424	22.602	.99902	28
33	.04449	.04454	22.454	.99901	27
34	.04478	.04483	22.308	.99900	26
35	.04507	.04512	22.164	.99898	25
36	.04536	.04541	22.022	.99897	24
37	.04565	.04570	21.881	.99896	23
38	.04594	.04599	21.743	.99894	22
39	.04623	.04628	21.606	.99893	21
40	.04653	.04658	21.470	.99892	20
41	.04682	.04687	21.337	.99890	19
42	.04711	.04716	21.205	.99889	18
43	.04740	.04745	21.075	.99888	17
44	.04769	.04774	20.946	.99886	16
45	.04798	.04803	20.819	.99885	15
46	.04827	.04833	20.693	.99883	14
47	.04856	.04862	20.569	.99882	13
48	.04885	.04891	20.446	.99881	12
49	.04914	.04920	20.325	.99879	11
50	.04943	.04949	20.206	.99878	10
51	.04972	.04978	20.087	.99876	9
52	.05001	.05007	19.970	.99875	8
53	.05030	.05037	19.855	.99873	7
54	.05059	.05066	19.740	.99872	6
55	.05088	.05095	19.627	.99870	5
56	.05117	.05124	19.516	.99869	4
57	.05146	.05153	19.405	.99867	3
58	.05175	.05182	19.296	.99866	2
59	.05205	.05212	19.188	.99864	1
60	.05234	.05241	19.081	.99863	0
′	Cos	Cot	Tan	Sin	′

'	Sin	Tan	Cot	Cos	'
0	.05234	.05241	19.081	.99863	**60**
1	.05263	.05270	18.976	.99861	59
2	.05292	.05299	18.871	.99860	58
3	.05321	.05328	18.768	.99858	57
4	.05350	.05357	18.666	.99857	56
5	.05379	.05387	18.564	.99855	**55**
6	.05408	.05416	18.464	.99854	54
7	.05437	.05445	18.366	.99852	53
8	.05466	.05474	18.268	.99851	52
9	.05495	.05503	18.171	.99849	51
10	.05524	.05533	18.075	.99847	**50**
11	.05553	.05562	17.980	.99846	49
12	.05582	.05591	17.886	.99844	48
13	.05611	.05620	17.793	.99842	47
14	.05640	.05649	17.702	.99841	46
15	.05669	.05678	17.611	.99839	**45**
16	.05698	.05708	17.521	.99838	44
17	.05727	.05737	17.431	.99836	43
18	.05756	.05766	17.343	.99834	42
19	.05785	.05795	17.256	.99833	41
20	.05814	.05824	17.169	.99831	**40**
21	.05844	.05854	17.084	.99829	39
22	.05873	.05883	16.999	.99827	38
23	.05902	.05912	16.915	.99826	37
24	.05931	.05941	16.832	.99824	36
25	.05960	.05970	16.750	.99822	**35**
26	.05989	.05999	16.668	.99821	34
27	.06018	.06029	16.587	.99819	33
28	.06047	.06058	16.507	.99817	32
29	.06076	.06087	16.428	.99815	31
30	.06105	.06116	16.350	.99813	**30**
31	.06134	.06145	16.272	.99812	29
32	.06163	.06175	16.195	.99810	28
33	.06192	.06204	16.119	.99808	27
34	.06221	.06233	16.043	.99806	26
35	.06250	.06262	15.969	.99804	**25**
36	.06279	.06291	15.895	.99803	24
37	.06308	.06321	15.821	.99801	23
38	.06337	.06350	15.748	.99799	22
39	.06366	.06379	15.676	.99797	21
40	.06395	.06408	15.605	.99795	**20**
41	.06424	.06438	15.534	.99793	19
42	.06453	.06467	15.464	.99792	18
43	.06482	.06496	15.394	.99790	17
44	.06511	.06525	15.325	.99788	16
45	.06540	.06554	15.257	.99786	**15**
46	.06569	.06584	15.189	.99784	14
47	.06598	.06613	15.122	.99782	13
48	.06627	.06642	15.056	.99780	12
49	.06656	.06671	14.990	.99778	11
50	.06685	.06700	14.924	.99776	**10**
51	.06714	.06730	14.860	.99774	9
52	.06743	.06759	14.795	.99772	8
53	.06773	.06788	14.732	.99770	7
54	.06802	.06817	14.669	.99768	6
55	.06831	.06847	14.606	.99766	**5**
56	.06860	.06876	14.544	.99764	4
57	.06889	.06905	14.482	.99762	3
58	.06918	.06934	14.421	.99760	2
59	.06947	.06963	14.361	.99758	1
60	.06976	.06993	14.301	.99756	**0**
'	Cos	Cot	Tan	Sin	'

'	Sin	Tan	Cot	Cos	'
0	.06976	.06993	14.301	.99756	**60**
1	.07005	.07022	14.241	.99754	59
2	.07034	.07051	14.182	.99752	58
3	.07063	.07080	14.124	.99750	57
4	.07092	.07110	14.065	.99748	56
5	.07121	.07139	14.008	.99746	**55**
6	.07150	.07168	13.951	.99744	54
7	.07179	.07197	13.894	.99742	53
8	.07208	.07227	13.838	.99740	52
9	.07237	.07256	13.782	.99738	51
10	.07266	.07285	13.727	.99736	**50**
11	.07295	.07314	13.672	.99734	49
12	.07324	.07344	13.617	.99731	48
13	.07353	.07373	13.563	.99729	47
14	.07382	.07402	13.510	.99727	46
15	.07411	.07431	13.457	.99725	**45**
16	.07440	.07461	13.404	.99723	44
17	.07469	.07490	13.352	.99721	43
18	.07498	.07519	13.300	.99719	42
19	.07527	.07548	13.248	.99716	41
20	.07556	.07578	13.197	.99714	**40**
21	.07585	.07607	13.146	.99712	39
22	.07614	.07636	13.096	.99710	38
23	.07643	.07665	13.046	.99708	37
24	.07672	.07695	12.996	.99705	36
25	.07701	.07724	12.947	.99703	**35**
26	.07730	.07753	12.898	.99701	34
27	.07759	.07782	12.850	.99699	33
28	.07788	.07812	12.801	.99696	32
29	.07817	.07841	12.754	.99694	31
30	.07846	.07870	12.706	.99692	**30**
31	.07875	.07899	12.659	.99689	29
32	.07904	.07929	12.612	.99687	28
33	.07933	.07958	12.566	.99685	27
34	.07962	.07987	12.520	.99683	26
35	.07991	.08017	12.474	.99680	**25**
36	.08020	.08046	12.429	.99678	24
37	.08049	.08075	12.384	.99676	23
38	.08078	.08104	12.339	.99673	22
39	.08107	.08134	12.295	.99671	21
40	.08136	.08163	12.251	.99668	**20**
41	.08165	.08192	12.207	.99666	19
42	.08194	.08221	12.163	.99664	18
43	.08223	.08251	12.120	.99661	17
44	.08252	.08280	12.077	.99659	16
45	.08281	.08309	12.035	.99657	**15**
46	.08310	.08339	11.992	.99654	14
47	.08339	.08368	11.950	.99652	13
48	.08368	.08397	11.909	.99649	12
49	.08397	.08427	11.867	.99647	11
50	.08426	.08456	11.826	.99644	**10**
51	.08455	.08485	11.785	.99642	9
52	.08484	.08514	11.745	.99639	8
53	.08513	.08544	11.705	.99637	7
54	.08542	.08573	11.664	.99635	6
55	.08571	.08602	11.625	.99632	**5**
56	.08600	.08632	11.585	.99630	4
57	.08629	.08661	11.546	.99627	3
58	.08658	.08690	11.507	.99625	2
59	.08687	.08720	11.468	.99622	1
60	.08716	.08749	11.430	.99619	**0**
'	Cos	Cot	Tan	Sin	'

'	Sin	Tan	Cot	Cos	'
0	.08716	.08749	11.430	.99619	**60**
1	.08745	.08778	11.392	.99617	59
2	.08774	.08807	11.354	.99614	58
3	.08803	.08837	11.316	.99612	57
4	.08831	.08866	11.279	.99609	56
5	.08860	.08895	11.242	.99607	**55**
6	.08889	.08925	11.205	.99604	54
7	.08918	.08954	11.168	.99602	53
8	.08947	.08983	11.132	.99599	52
9	.08976	.09013	11.095	.99596	51
10	.09005	.09042	11.059	.99594	**50**
11	.09034	.09071	11.024	.99591	49
12	.09063	.09101	10.988	.99588	48
13	.09092	.09130	10.953	.99586	47
14	.09121	.09159	10.918	.99583	46
15	.09150	.09189	10.883	.99580	**45**
16	.09179	.09218	10.848	.99578	44
17	.09208	.09247	10.814	.99575	43
18	.09237	.09277	10.780	.99572	42
19	.09266	.09306	10.746	.99570	41
20	.09295	.09335	10.712	.99567	**40**
21	.09324	.09365	10.678	.99564	39
22	.09353	.09394	10.645	.99562	38
23	.09382	.09423	10.612	.99558	37
24	.09411	.09453	10.579	.99556	36
25	.09440	.09482	10.546	.99553	**35**
26	.09469	.09511	10.514	.99551	34
27	.09498	.09541	10.481	.99548	33
28	.09527	.09570	10.449	.99545	32
29	.09556	.09600	10.417	.99542	31
30	.09585	.09629	10.385	.99540	**30**
31	.09614	.09658	10.354	.99537	29
32	.09642	.09688	10.322	.99534	28
33	.09671	.09717	10.291	.99531	27
34	.09700	.09746	10.260	.99528	26
35	.09729	.09776	10.229	.99526	**25**
36	.09758	.09805	10.199	.99523	24
37	.09787	.09834	10.168	.99520	23
38	.09816	.09864	10.138	.99517	22
39	.09845	.09893	10.108	.99514	21
40	.09874	.09923	10.078	.99511	**20**
41	.09903	.09952	10.048	.99508	19
42	.09932	.09981	10.019	.99506	18
43	.09961	.10011	9.9893	.99503	17
44	.09990	.10040	9.9601	.99500	16
45	.10019	.10069	9.9310	.99497	**15**
46	.10048	.10099	9.9021	.99494	14
47	.10077	.10128	9.8734	.99491	13
48	.10106	.10158	9.8448	.99488	12
49	.10135	.10187	9.8164	.99485	11
50	.10164	.10216	9.7882	.99482	**10**
51	.10192	.10246	9.7601	.99479	9
52	.10221	.10275	9.7322	.99476	8
53	.10250	.10305	9.7044	.99473	7
54	.10279	.10334	9.6768	.99470	6
55	.10308	.10363	9.6493	.99467	**5**
56	.10337	.10393	9.6220	.99464	4
57	.10366	.10422	9.5949	.99461	3
58	.10395	.10452	9.5679	.99458	2
59	.10424	.10481	9.5411	.99455	1
60	.10453	.10510	9.5144	.99452	**0**
'	Cos	Cot	Tan	Sin	'

798

′	Sin	Tan	Cot	Cos	′
0	.10453	.10510	9.5144	.99452	60
1	.10482	.10540	9.4878	.99449	59
2	.10511	.10569	9.4614	.99446	58
3	.10540	.10599	9.4352	.99443	57
4	.10569	.10628	9.4090	.99440	56
5	.10597	.10657	9.3831	.99437	55
6	.10626	.106S7	9.3572	.99434	54
7	.10655	.10716	9.3315	.99431	53
8	.10684	.10746	9.3060	.99428	52
9	.10713	.10775	9.2806	.99424	51
10	.10742	.10805	9.2553	.99421	50
11	.10771	.10S34	9.2302	.99418	49
12	.10800	.10863	9.2052	.99415	48
13	.10829	.10893	9.1803	.99412	47
14	.10858	.10922	9.1555	.99409	46
15	.10887	.10952	9.1309	.99406	45
16	.10916	.10981	9.1065	.99402	44
17	.10945	.11011	9.0821	.99399	43
18	.10973	.11040	9.0579	.99396	42
19	.11002	.11070	9.0338	.99393	41
20	.11031	.11099	9.0098	.99390	40
21	.11060	.11128	8.9860	.99386	39
22	.11089	.11158	8.9623	.99383	38
23	.11118	.11187	8.9387	.99380	37
24	.11147	.11217	8.9152	.99377	36
25	.11176	.11246	8.8919	.99374	35
26	.11205	.11276	8.8686	.99370	34
27	.11234	.11305	8.8455	.99367	33
28	.11263	.11335	8.8225	.99364	32
29	.11291	.11364	8.7996	.99360	31
30	.11320	.11394	8.7769	.99357	30
31	.11349	.11423	8.7542	.99354	29
32	.11378	.11452	8.7317	.99351	28
33	.11407	.11482	8.7093	.99347	27
34	.11436	.11511	8.6870	.99344	26
35	.11465	.11541	8.6648	.99341	25
36	.11494	.11570	8.6427	.99337	24
37	.11523	.11600	8.6208	.99334	23
38	.11552	.11629	8.5989	.99331	22
39	.11580	.11659	8.5772	.99327	21
40	.11609	.11688	8.5555	.99324	20
41	.11638	.11718	8.5340	.99320	19
42	.11667	.11747	8.5126	.99317	18
43	.11696	.11777	8.4913	.99314	17
44	.11725	.11806	8.4701	.99310	16
45	.11754	.11836	8.4490	.99307	15
46	.11783	.11865	8.4280	.99303	14
47	.11812	.11895	8.4071	.99300	13
48	.11840	.11924	8.3863	.99297	12
49	.11869	.11954	8.3656	.99293	11
50	.11898	.11983	8.3450	.99290	10
51	.11927	.12013	8.3245	.99286	9
52	.11956	.12042	8.3041	.99283	8
53	.11985	.12072	8.2838	.99279	7
54	.12014	.12101	8.2636	.99276	6
55	.12043	.12131	8.2434	.99272	5
56	.12071	.12160	8.2234	.99269	4
57	.12100	.12190	8.2035	.99265	3
58	.12129	.12219	8.1837	.99262	2
59	.12158	.12249	8.1640	.99258	1
60	.12187	.12278	8.1443	.99255	0
′	Cos	Cot	Tan	Sin	′

′	Sin	Tan	Cot	Cos	′
0	.12187	.12278	8.1443	.99255	60
1	.12216	.12308	8.1248	.99251	59
2	.12245	.12338	8.1054	.99248	58
3	.12274	.12367	8.0860	.99244	57
4	.12302	.12397	8.0667	.99240	56
5	.12331	.12426	8.0476	.99237	55
6	.12360	.12456	8.0285	.99233	54
7	.12389	.12485	8.0095	.99230	53
8	.12418	.12515	7.9906	.99226	52
9	.12447	.12544	7.9718	.99222	51
10	.12476	.12574	7.9530	.99219	50
11	.12504	.12603	7.9344	.99215	49
12	.12533	.12633	7.9158	.99211	48
13	.12562	.12662	7.8973	.99208	47
14	.12591	.12692	7.8789	.99204	46
15	.12620	.12722	7.8606	.99200	45
16	.12649	.12751	7.8424	.99197	44
17	.12678	.12781	7.8243	.99193	43
18	.12706	.12810	7.8062	.99189	42
19	.12735	.12840	7.7882	.99186	41
20	.12764	.12869	7.7704	.99182	40
21	.12793	.12899	7.7525	.99178	39
22	.12822	.12929	7.7348	.99175	38
23	.12851	.12958	7.7171	.99171	37
24	.12880	.12988	7.6996	.99137	36
25	.12908	.13017	7.6821	.99163	35
26	.12937	.13047	7.6647	.99160	34
27	.12966	.13076	7.6473	.99156	33
28	.12995	.13106	7.6301	.99152	32
29	.13024	.13136	7.6129	.99148	31
30	.13053	.13165	7.5958	.99144	30
31	.13081	.13195	7.5787	.99141	29
32	.13110	.13224	7.5618	.99137	28
33	.13139	.13254	7.5449	.99133	27
34	.13168	.13284	7.5281	.99129	26
35	.13197	.13313	7.5113	.99125	25
36	.13226	.13343	7.4947	.99122	24
37	.13254	.13372	7.4781	.99118	23
38	.13283	.13402	7.4615	.99114	22
39	.13312	.13432	7.4451	.99110	21
40	.13341	.13461	7.4287	.99106	20
41	.13370	.13491	7.4124	.99102	19
42	.13399	.13521	7.3962	.99098	18
43	.13427	.13550	7.3800	.99094	17
44	.13456	.13580	7.3639	.99091	16
45	.13485	.13609	7.3479	.99087	15
46	.13514	.13639	7.3319	.99083	14
47	.13543	.13669	7.3160	.99079	13
48	.13572	.13698	7.3002	.99075	12
49	.13600	.13728	7.2844	.99071	11
50	.13629	.13758	7.2687	.99067	10
51	.13658	.13787	7.2531	.99063	9
52	.13687	.13817	7.2375	.99059	8
53	.13716	.13846	7.2220	.99055	7
54	.13744	.13876	7.2066	.99051	6
55	.13773	.13906	7.1912	.99047	5
56	.13802	.13935	7.1759	.99043	4
57	.13831	.13965	7.1607	.99039	3
58	.13860	.13995	7.1455	.99035	2
59	.13889	.14024	7.1304	.99031	1
60	.13917	.14054	7.1154	.99027	0
′	Cos	Cot	Tan	Sin	′

′	Sin	Tan	Cot	Cos	′
0	.13917	.14054	7.1154	.99027	60
1	.13946	.14084	7.1004	.99023	59
2	.13975	.14113	7.0855	.99019	58
3	.14004	.14143	7.0706	.99015	57
4	.14033	.14173	7.0558	.99011	56
5	.14061	.14202	7.0410	.99006	55
6	.14090	.14232	7.0264	.99002	54
7	.14119	.14262	7.0117	.98998	53
8	.14148	.14291	6.9972	.98994	52
9	.14177	.14321	6.9827	.98990	51
10	.14205	.14351	6.9682	.98986	50
11	.14234	.14381	6.9538	.98982	49
12	.14263	.14410	6.9395	.98978	48
13	.14292	.14440	6.9252	.98973	47
14	.14320	.14470	6.9110	.98969	46
15	.14349	.14499	6.8969	.98965	45
16	.14378	.14529	6.8828	.98961	44
17	.14407	.14559	6.8687	.98957	43
18	14436	.14588	6.8548	.98953	42
19	.14464	.14618	6.8408	.98948	41
20	.14493	.14648	6.8269	.98944	40
21	.14522	.14678	6.8131	.98940	39
22	.14551	.14707	6.7994	.98936	38
23	.14580	.14737	6.7856	.98931	37
24	.14608	.14767	6.7720	.98927	36
25	.14637	.14796	6.7584	.98923	35
26	.14666	.14826	6.7448	.98919	34
27	.14695	.14856	6.7313	.98914	33
28	.14723	.14886	6.7179	.98910	32
29	.14752	.14915	6.7045	.98906	31
30	.14781	.14945	6.6912	.98902	30
31	.14810	.14975	6.6779	.98897	29
32	.14838	.15005	6.6646	.98893	28
33	.14867	.15034	6.6514	.98889	27
34	.14896	.15064	6.6383	.98884	26
35	.14925	.15094	6.6252	.98880	25
36	.14954	.15124	6.6122	.98876	24
37	.14982	.15153	6.5992	.98871	23
38	.15011	.15183	6.5863	.98867	22
39	.15040	.15213	6.5734	.98863	21
40	.15069	.15243	6.5606	.98858	20
41	.15097	.15272	6.5478	.98854	19
42	.15126	.15302	6.5350	.98849	18
43	.15155	.15332	6.5223	.98845	17
44	.15184	.15362	6.5097	.98841	16
45	.15212	.15391	6.4971	.98836	15
46	.15241	.15421	6.4846	.98832	14
47	.15270	.15451	6.4721	.98827	13
48	.15299	.15481	6.4596	.98823	12
49	.15327	.15511	6.4472	.98818	11
50	.15356	.15540	6.4348	.98814	10
51	.15385	.15570	6.4225	.98809	9
52	.15414	.15600	6.4103	.98805	8
53	.15442	.15630	6.3980	.98800	7
54	.15471	.15660	6.3859	.98796	6
55	.15500	.15689	6.3737	.98791	5
56	.15529	.15719	6.3617	.98787	4
57	.15557	.15749	6.3496	.98782	3
58	.15586	.15779	6.3376	.98778	2
59	.15615	.15809	6.3257	.98773	1
60	.15643	.15838	6.3138	.98769	0
′	Cos	Cot	Tan	Sin	′

′	Sin	Tan	Cot	Cos	′
0	.15643	.15838	6.3138	.98769	60
1	.15672	.15868	6.3019	.98764	59
2	.15701	.15898	6.2901	.98760	58
3	.15730	.15928	6.2783	.98755	57
4	.15758	.15958	6.2666	.98751	56
5	.15787	.15988	6.2549	.98746	55
6	.15816	.16017	6.2432	.98741	54
7	.15845	.16047	6.2316	.98737	53
8	.15873	.16077	6.2200	.98732	52
9	.15902	.16107	6.2085	.98728	51
10	.15931	.16137	6.1970	.98723	50
11	.15959	.16167	6.1856	.98718	49
12	.15988	.16196	6.1742	.98714	48
13	.16017	.16226	6.1628	.98709	47
14	.16046	.16256	6.1515	.98704	46
15	.16074	.16286	6.1402	.98700	45
16	.16103	.16316	6.1290	.98695	44
17	.16132	.16346	6.1178	.98690	43
18	.16160	.16376	6.1066	.98686	42
19	.16189	.16405	6.0955	.98681	41
20	.16218	.16435	6.0844	.98676	40
21	.16246	.16465	6.0734	.98671	39
22	.16275	.16495	6.0624	.98667	38
23	.16304	.16525	6.0514	.98662	37
24	.16333	.16555	6.0405	.98657	36
25	.16361	.16585	6.0296	.98652	35
26	.16390	.16615	6.0188	.98648	34
27	.16419	.16645	6.0080	.98643	33
28	.16447	.16674	5.9972	.98638	32
29	.16476	.16704	5.9865	.98633	31
30	.16505	.16734	5.9758	.98629	30
31	.16533	.16764	5.9651	.98624	29
32	.16562	.16794	5.9545	.98619	28
33	.16591	.16824	5.9439	.98614	27
34	.16620	.16854	5.9333	.98609	26
35	.16648	.16884	5.9228	.98604	25
36	.16677	.16914	5.9124	.98600	24
37	.16706	.16944	5.9019	.98595	23
38	.16734	.16974	5.8915	.98590	22
39	.16763	.17004	5.8811	.98585	21
40	.16792	.17033	5.8708	.98580	20
41	.16820	.17063	5.8605	.98575	19
42	.16849	.17093	5.8502	.98570	18
43	.16878	.17123	5.8400	.98565	17
44	.16906	.17153	5.8298	.98561	16
45	.16935	.17183	5.8197	.98556	15
46	.16964	.17213	5.8095	.98551	14
47	.16992	.17243	5.7994	.98546	13
48	.17021	.17273	5.7894	.98541	12
49	.17050	.17303	5.7794	.98536	11
50	.17078	.17333	5.7694	.98531	10
51	.17107	.17363	5.7594	.98526	9
52	.17136	.17393	5.7495	.98521	8
53	.17164	.17423	5.7396	.98516	7
54	.17193	.17453	5.7297	.98511	6
55	.17222	.17483	5.7199	.98506	5
56	.17250	.17513	5.7101	.98501	4
57	.17279	.17543	5.7004	.98496	3
58	.17308	.17573	5.6906	.98491	2
59	.17336	.17603	5.6809	.98486	1
60	.17365	.17633	5.6713	.98481	0
′	Cos	Cot	Tan	Sin	′

′	Sin	Tan	Cot	Cos	′
0	.17365	.17633	5.6713	.98481	60
1	.17393	.17663	5.6617	.98476	59
2	.17422	.17693	5.6521	.98471	58
3	.17451	.17723	5.6425	.98466	57
4	.17479	.17753	5.6329	.98461	56
5	.17508	.17783	5.6234	.98455	55
6	.17537	.17813	5.6140	.98450	54
7	.17565	.17843	5.6045	.98445	53
8	.17594	.17873	5.5951	.98440	52
9	.17623	.17903	5.5857	.98435	51
10	.17651	.17933	5.5764	.98430	50
11	.17680	.17963	5.5671	.98425	49
12	.17708	.17993	5.5578	.98420	48
13	.17737	.18023	5.5485	.98414	47
14	.17766	.18053	5.5393	.98409	46
15	.17794	.18083	5.5301	.98404	45
16	.17823	.18113	5.5209	.98399	44
17	.17852	.18143	5.5118	.98394	43
18	.17880	.18173	5.5026	.98389	42
19	.17909	.18203	5.4936	.98383	41
20	.17937	.18233	5.4845	.98378	40
21	.17966	.18263	5.4755	.98373	39
22	.17995	.18293	5.4665	.98368	38
23	.18023	.18323	5.4575	.98362	37
24	.18052	.18353	5.4486	.98357	36
25	.18081	.18384	5.4397	.98352	35
26	.18109	.18414	5.4308	.98347	34
27	.18138	.18444	5.4219	.98341	33
28	.18166	.18474	5.4131	.98336	32
29	.18195	.18504	5.4043	.98331	31
30	.18224	.18534	5.3955	.98325	30
31	.18252	.18564	5.3868	.98320	29
32	.18281	.18594	5.3781	.98315	28
33	.18309	.18624	5.3694	.98310	27
34	.18338	.18654	5.3607	.98304	26
35	.18367	.18684	5.3521	.98299	25
36	.18395	.18714	5.3435	.98294	24
37	.18424	.18745	5.3349	.98288	23
38	.18452	.18775	5.3263	.98283	22
39	.18481	.18805	5.3178	.98277	21
40	.18509	.18835	5.3093	.98272	20
41	.18538	.18865	5.3008	.98267	19
42	.18567	.18895	5.2924	.98261	18
43	.18595	.18925	5.2839	.98256	17
44	.18624	.18955	5.2755	.98250	16
45	.18652	.18986	5.2672	.98245	15
46	.18681	.19016	5.2588	.98240	14
47	.18710	.19046	5.2505	.98234	13
48	.18738	.19076	5.2422	.98229	12
49	.18767	.19106	5.2339	.98223	11
50	.18795	.19136	5.2257	.98218	10
51	.18824	.19166	5.2174	.98212	9
52	.18852	.19197	5.2092	.98207	8
53	.18881	.19227	5.2011	.98201	7
54	.18910	.19257	5.1929	.98196	6
55	.18938	.19287	5.1848	.98190	5
56	.18967	.19317	5.1767	.98185	4
57	.18995	.19347	5.1686	.98179	3
58	.19024	.19378	5.1606	.98174	2
59	.19052	.19408	5.1526	.98168	1
60	.19081	.19438	5.1446	.98163	0
′	Cos	Cot	Tan	Sin	′

′	Sin	Tan	Cot	Cos	′
0	.19081	.19438	5.1446	.98163	60
1	.19109	.19468	5.1366	.98157	59
2	.19138	.19498	5.1286	.98152	58
3	.19167	.19529	5.1207	.98146	57
4	.19195	.19559	5.1128	.98140	56
5	.19224	.19589	5.1049	.98135	55
6	.19252	.19619	5.0970	.98129	54
7	.19281	.19649	5.0892	.98124	53
8	.19309	.19680	5.0814	.98118	52
9	.19338	.19710	5.0736	.98112	51
10	.19366	.19740	5.0658	.98107	50
11	.19395	.19770	5.0581	.98101	49
12	.19423	.19801	5.0504	.98096	48
13	.19452	.19831	5.0427	.98090	47
14	.19481	.19861	5.0350	.98084	46
15	.19509	.19891	5.0273	.98079	45
16	.19538	.19921	5.0197	.98073	44
17	.19566	.19952	5.0121	.98067	43
18	.19595	.19982	5.0045	.98061	42
19	.19623	.20012	4.9969	.98056	41
20	.19652	.20042	4.9894	.98050	40
21	.19680	.20073	4.9819	.98044	39
22	.19709	.20103	4.9744	.98039	38
23	.19737	.20133	4.9669	.98033	37
24	.19766	.20164	4.9594	.98027	36
25	.19794	.20194	4.9520	.98021	35
26	.19823	.20224	4.9446	.98016	34
27	.19851	.20254	4.9372	.98010	33
28	.19880	.20285	4.9298	.98004	32
29	.19908	.20315	4.9225	.97998	31
30	.19937	.20345	4.9152	.97992	30
31	.19965	.20376	4.9078	.97987	29
32	.19994	.20406	4.9006	.97981	28
33	.20022	.20436	4.8933	.97975	27
34	.20051	.20466	4.8860	.97969	26
35	.20079	.20497	4.8788	.97963	25
36	.20108	.20527	4.8716	.97958	24
37	.20136	.20557	4.8644	.97952	23
38	.20165	.20588	4.8573	.97946	22
39	.20193	.20618	4.8501	.97940	21
40	.20222	.20648	4.8430	.97934	20
41	.20250	.20679	4.8359	.97928	19
42	.20279	.20709	4.8288	.97922	18
43	.20307	.20739	4.8218	.97916	17
44	.20336	.20770	4.8147	.97910	16
45	.20364	.20800	4.8077	.97905	15
46	.20393	.20830	4.8007	.97899	14
47	.20421	.20861	4.7937	.97893	13
48	.20450	.20891	4.7867	.97887	12
49	.20478	.20921	4.7798	.97881	11
50	.20507	.20952	4.7729	.97875	10
51	.20535	.20982	4.7659	.97869	9
52	.20563	.21013	4.7591	.97863	8
53	.20592	.21043	4.7522	.97857	7
54	.20620	.21073	4.7453	.97851	6
55	.20649	.21104	4.7385	.97845	5
56	.20677	.21134	4.7317	.97839	4
57	.20706	.21164	4.7249	.97833	3
58	.20734	.21195	4.7181	.97827	2
59	.20763	.21225	4.7114	.97821	1
60	.20791	.21256	4.7046	.97815	0
′	Cos	Cot	Tan	Sin	′

′	Sin	Tan	Cot	Cos	′
0	.20791	.21256	4.7046	.97815	60
1	.20820	.21286	4.6979	.97809	59
2	.20848	.21316	4.6912	.97803	58
3	.20877	.21347	4.6845	.97797	57
4	.20905	.21377	4.6779	.97791	56
5	.20933	.21408	4.6712	.97784	55
6	.20962	.21438	4.6646	.97778	54
7	.20990	.21469	4.6580	.97772	53
8	.21019	.21499	4.6514	.97766	52
9	.21047	.21529	4.6448	.97760	51
10	.21076	.21560	4.6382	.97754	50
11	.21104	.21590	4.6317	.97748	49
12	.21132	.21621	4.6252	.97742	48
13	.21161	.21651	4.6187	.97735	47
14	.21189	.21682	4.6122	.97729	46
15	.21218	.21712	4.6057	.97723	45
16	.21246	.21743	4.5993	.97717	44
17	.21275	.21773	4.5928	.97711	43
18	.21303	.21804	4.5864	.97705	42
19	.21331	.21834	4.5800	.97698	41
20	.21360	.21864	4.5736	.97692	40
21	.21388	.21895	4.5673	.97686	39
22	.21417	.21925	4.5609	.97680	38
23	.21445	.21956	4.5546	.97673	37
24	.21474	.21986	4.5483	.97667	36
25	.21502	.22017	4.5420	.97661	35
26	.21530	.22047	4.5357	.97655	34
27	.21559	.22078	4.5294	.97648	33
28	.21587	.22108	4.5232	.97642	32
29	.21616	.22139	4.5169	.97636	31
30	.21644	.22169	4.5107	.97630	30
31	.21672	.22200	4.5045	.97623	29
32	.21701	.22231	4.4983	.97617	28
33	.21729	.22261	4.4922	.97611	27
34	.21758	.22292	4.4860	.97604	26
35	.21786	.22322	4.4799	.97598	25
36	.21814	.22353	4.4737	.97592	24
37	.21843	.22383	4.4676	.97585	23
38	.21871	.22414	4.4615	.97579	22
39	.21899	.22444	4.4555	.97573	21
40	.21928	.22475	4.4494	.97566	20
41	.21956	.22505	4.4434	.97560	19
42	.21985	.22536	4.4373	.97553	18
43	.22013	.22567	4.4313	.97547	17
44	.22041	.22597	4.4253	.97541	16
45	.22070	.22628	4.4194	.97534	15
46	.22098	.22658	4.4134	.97528	14
47	.22126	.22689	4.4075	.97521	13
48	.22155	.22719	4.4015	.97515	12
49	.22183	.22750	4.3956	.97508	11
50	.22212	.22781	4.3897	.97502	10
51	.22240	.22811	4.3838	.97496	9
52	.22268	.22842	4.3779	.97489	8
53	.22297	.22872	4.3721	.97483	7
54	.22325	.22903	4.3662	.97476	6
55	.22353	.22934	4.3604	.97470	5
56	.22382	.22964	4.3546	.97463	4
57	.22410	.22995	4.3488	.97457	3
58	.22438	.23026	4.3430	.97450	2
59	.22467	.23056	4.3372	.97444	1
60	.22495	.23087	4.3315	.97437	0
′	Cos	Cot	Tan	Sin	′

′	Sin	Tan	Cot	Cos	′
0	.22495	.23087	4.3315	.97437	60
1	.22523	.23117	4.3257	.97430	59
2	.22552	.23148	4.3200	.97424	58
3	.22580	.23179	4.3143	.97417	57
4	.22608	.23209	4.3086	.97411	56
5	.22637	.23240	4.3029	.97404	55
6	.22665	.23271	4.2972	.97398	54
7	.22693	.23301	4.2916	.97391	53
8	.22722	.23332	4.2859	.97384	52
9	.22750	.23363	4.2803	.97378	51
10	.22778	.23393	4.2747	.97371	50
11	.22807	.23424	4.2691	.97365	49
12	.22835	.23455	4.2635	.97358	48
13	.22863	.23485	4.2580	.97351	47
14	.22892	.23516	4.2524	.97345	46
15	.22920	.23547	4.2468	.97338	45
16	.22948	.23578	4.2413	.97331	44
17	.22977	.23608	4.2358	.97325	43
18	.23005	.23639	4.2303	.97318	42
19	.23033	.23670	4.2248	.97311	41
20	.23062	.23700	4.2193	.97304	40
21	.23090	.23731	4.2139	.97298	39
22	.23118	.23762	4.2084	.97291	38
23	.23146	.23793	4.2030	.97284	37
24	.23175	.23823	4.1976	.97278	36
25	.23203	.23854	4.1922	.97271	35
26	.23231	.23885	4.1868	.97264	34
27	.23260	.23916	4.1814	.97257	33
28	.23288	.23946	4.1760	.97251	32
29	.23316	.23977	4.1706	.97244	31
30	.23345	.24008	4.1653	.97237	30
31	.23373	.24039	4.1600	.97230	29
32	.23401	.24069	4.1547	.97223	28
33	.23429	.24100	4.1493	.97217	27
34	.23458	.24131	4.1441	.97210	26
35	.23486	.24162	4.1388	.97203	25
36	.23514	.24193	4.1335	.97196	24
37	.23542	.24223	4.1282	.97189	23
38	.23571	.24254	4.1230	.97182	22
39	.23599	.24285	4.1178	.97176	21
40	.23627	.24316	4.1126	.97169	20
41	.23656	.24347	4.1074	.97162	19
42	.23684	.24377	4.1022	.97155	18
43	.23712	.24408	4.0970	.97148	17
44	.23740	.24439	4.0918	.97141	16
45	.23769	.24470	4.0867	.97134	15
46	.23797	.24501	4.0815	.97127	14
47	.23825	.24532	4.0764	.97120	13
48	.23853	.24562	4.0713	.97113	12
49	.23882	.24593	4.0662	.97106	11
50	.23910	.24624	4.0611	.97100	10
51	.23938	.24655	4.0560	.97093	9
52	.23966	.24686	4.0509	.97086	8
53	.23995	.24717	4.0459	.97079	7
54	.24023	.24747	4.0408	.97072	6
55	.24051	.24778	4.0358	.97065	5
56	.24079	.24809	4.0308	.97058	4
57	.24108	.24840	4.0257	.97051	3
58	.24136	.24871	4.0207	.97044	2
59	.24164	.24902	4.0158	.97037	1
60	.24192	.24933	4.0108	.97030	0
′	Cos	Cot	Tan	Sin	′

′	Sin	Tan	Cot	Cos	′
0	.24192	.24933	4.0108	.97030	60
1	.24220	.24964	4.0058	.97023	59
2	.24249	.24995	4.0009	.97015	58
3	.24277	.25026	3.9959	.97008	57
4	.24305	.25056	3.9910	.97001	56
5	.24333	.25087	3.9861	.96994	55
6	.24362	.25118	3.9812	.96987	54
7	.24390	.25149	3.9763	.96980	53
8	.24418	.25180	3.9714	.96973	52
9	.24446	.25211	3.9665	.96966	51
10	.24474	.25242	3.9617	.96959	50
11	.24503	.25273	3.9568	.96952	49
12	.24531	.25304	3.9520	.96945	48
13	.24559	.25335	3.9471	.96937	47
14	.24587	.25366	3.9423	.96930	46
15	.24615	.25397	3.9375	.96923	45
16	.24644	.25428	3.9327	.96916	44
17	.24672	.25459	3.9279	.96909	43
18	.24700	.25490	3.9232	.96902	42
19	.24728	.25521	3.9184	.96894	41
20	.24756	.25552	3.9136	.96887	40
21	.24784	.25583	3.9089	.96880	39
22	.24813	.25614	3.9042	.96873	38
23	.24841	.25645	3.8995	.96866	37
24	.24869	.25676	3.8947	.96858	36
25	.24897	.25707	3.8900	.96851	35
26	.24925	.25738	3.8854	.96844	34
27	.24954	.25769	3.8807	.96837	33
28	.24982	.25800	3.8760	.96829	32
29	.25010	.25831	3.8714	.96822	31
30	.25038	.25862	3.8667	.96815	30
31	.25066	.25893	3.8621	.96807	29
32	.25094	.25924	3.8575	.96800	28
33	.25122	.25955	3.8528	.96793	27
34	.25151	.25986	3.8482	.96786	26
35	.25179	.26017	3.8436	.96778	25
36	.25207	.26048	3.8391	.96771	24
37	.25235	.26079	3.8345	.96764	23
38	.25263	.26110	3.8299	.96756	22
39	.25291	.26141	3.8254	.96749	21
40	.25320	.26172	3.8208	.96742	20
41	.25348	.26203	3.8163	.96734	19
42	.25376	.26235	3.8118	.96727	18
43	.25404	.26266	3.8073	.96719	17
44	.25432	.26297	3.8028	.96712	16
45	.25460	.26328	3.7983	.96705	15
46	.25488	.26359	3.7938	.96697	14
47	.25516	.26390	3.7893	.96690	13
48	.25545	.26421	3.7848	.96682	12
49	.25573	.26452	3.7804	.96675	11
50	.25601	.26483	3.7760	.96667	10
51	.25629	.26515	3.7715	.96660	9
52	.25657	.26546	3.7671	.96653	8
53	.25685	.26577	3.7627	.96645	7
54	.25713	.26608	3.7583	.96638	6
55	.25741	.26639	3.7539	.96630	5
56	.25769	.26670	3.7495	.96623	4
57	.25798	.26701	3.7451	.96615	3
58	.25826	.26733	3.7408	.96608	2
59	.25854	.26764	3.7364	.96600	1
60	.25882	.26795	3.7321	.96593	0
′	Cos	Cot	Tan	Sin	′

′	Sin.	Tan	Cot	Cos	′
0	.25882	.26795	3.7321	.96593	60
1	.25910	.26826	3.7277	.96585	59
2	.25938	.26857	3.7234	.96578	58
3	.25966	.26888	3.7191	.96570	57
4	.25994	.26920	3.7148	.96562	56
5	.26022	.26951	3.7105	.96555	55
6	.26050	.26982	3.7062	.96547	54
7	.26079	.27013	3.7019	.96540	53
8	.26107	.27044	3.6976	.96532	52
9	.26135	.27076	3.6933	.96524	51
10	.26163	.27107	3.6891	.96517	50
11	.26191	.27138	3.6848	.96509	49
12	.26219	.27169	3.6806	.96502	48
13	.26247	.27201	3.6764	.96494	47
14	.26275	.27232	3.6722	.96486	46
15	.26303	.27263	3.6680	.96479	45
16	.26331	.27294	3.6638	.96471	44
17	.26359	.27326	3.6596	.96463	43
18	.26387	.27357	3.6554	.96456	42
19	.26415	.27388	3.6512	.96448	41
20	.26443	.27419	3.6470	.96440	40
21	.26471	.27451	3.6429	.96433	39
22	.26500	.27482	3.6387	.96425	38
23	.26528	.27513	3.6346	.96417	37
24	.26556	.27545	3.6305	.96410	36
25	.26584	.27576	3.6264	.96402	35
26	.26612	.27607	3.6222	.96394	34
27	.26640	.27638	3.6181	.96386	33
28	.26668	.27670	3.6140	.96379	32
29	.26696	.27701	3.6100	.96371	31
30	.26724	.27732	3.6059	.96363	30
31	.26752	.27764	3.6018	.96355	29
32	.26780	.27795	3.5978	.96347	28
33	.26808	.27826	3.5937	.96340	27
34	.26836	.27858	3.5897	.96332	26
35	.26864	.27889	3.5856	.96324	25
36	.26892	.27921	3.5816	.96316	24
37	.26920	.27952	3.5776	.96308	23
38	.26948	.27983	3.5736	.96301	22
39	.26976	.28015	3.5696	.96293	21
40	.27004	.28046	3.5656	.96285	20
41	.27032	.28077	3.5616	.96277	19
42	.27060	.28109	3.5576	.96269	18
43	.27088	.28140	3.5536	.96261	17
44	.27116	.28172	3.5497	.96253	16
45	.27144	.28203	3.5457	.96246	15
46	.27172	.28234	3.5418	.96238	14
47	.27200	.28266	3.5379	.96230	13
48	.27228	.28297	3.5339	.96222	12
49	.27256	.28329	3.5300	.96214	11
50	.27284	.28360	3.5261	.96206	10
51	.27312	.28391	3.5222	.96198	9
52	.27340	.28423	3.5183	.96190	8
53	.27368	.28454	3.5144	.96182	7
54	.27396	.28486	3.5105	.96174	6
55	.27424	.28517	3.5067	.96166	5
56	.27452	.28549	3.5028	.96158	4
57	.27480	.28580	3.4989	.96150	3
58	.27508	.28612	3.4951	.96142	2
59	.27536	.28643	3.4912	.96134	1
60	.27564	.28675	3.4874	.96126	0
′	Cos	Cot	Tan	Sin	′

′	Sin	Tan	Cot	Cos	′
0	.27564	.28675	3.4874	.96126	60
1	.27592	.28706	3.4836	.96118	59
2	.27620	.28738	3.4798	.96110	58
3	.27648	.28769	3.4760	.96102	57
4	.27676	.28801	3.4722	.96094	56
5	.27704	.28832	3.4684	.96086	55
6	.27731	.28864	3.4646	.96078	54
7	.27759	.28895	3.4608	.96070	53
8	.27787	.28927	3.4570	.96062	52
9	.27815	.28958	3.4533	.96054	51
10	.27843	.28990	3.4495	.96046	50
11	.27871	.29021	3.4458	.96037	49
12	.27899	.29053	3.4420	.96029	48
13	.27927	.29084	3.4383	.96021	47
14	.27955	.29116	3.4346	.96013	46
15	.27983	.29147	3.4308	.96005	45
16	.28011	.29179	3.4271	.95997	44
17	.28039	.29210	3.4234	.95989	43
18	.28067	.29242	3.4197	.95981	42
19	.28095	.29274	3.4160	.95972	41
20	.28123	.29305	3.4124	.95964	40
21	.28150	.29337	2.4087	.95956	39
22	.28178	.29368	3.4050	.95948	38
23	.28206	.29400	3.4014	.95940	37
24	.28234	.29432	3.3977	.95931	36
25	.28262	.29463	3.3941	.95923	35
26	.28290	.29495	3.3904	.95915	34
27	.28318	.29526	3.3868	.95907	33
28	.28346	.29558	3.3832	.95898	32
29	.28374	.29590	3.3796	.95890	31
30	.28402	.29621	3.3759	.95882	30
31	.28429	.29653	3.3723	.95874	29
32	.28457	.29685	3.3687	.95865	28
33	.28485	.29716	3.3652	.95857	27
34	.28513	.29748	3.3616	.95849	26
35	.28541	.29780	3.3580	.95841	25
36	.28569	.29811	3.3544	.95832	24
37	.28597	.29843	3.3509	.95824	23
38	.28625	.29875	3.3473	.95816	22
39	.28652	.29906	3.3438	.95807	21
40	.28680	.29938	3.3402	.95799	20
41	.28708	.29970	3.3367	.95791	19
42	.28736	.30001	3.3332	.95782	18
43	.28764	.30033	3.3297	.95774	17
44	.28792	.30065	3.3261	.95766	16
45	.28820	.30097	3.3226	.95757	15
46	.28847	.30128	3.3191	.95749	14
47	.28875	.30160	3.3156	.95740	13
48	.28903	.30192	3.3122	.95732	12
49	.28931	.30224	3.3087	.95724	11
50	.28959	.30255	3.3052	.95715	10
51	.28987	.30287	3.3017	.95707	9
52	.29015	.30319	3.2983	.95698	8
53	.29042	.30351	3.2948	.95690	7
54	.29070	.30382	3.2914	.95681	6
55	.29098	.30414	3.2879	.95673	5
56	.29126	.30446	3.2845	.95664	4
57	.29154	.30478	3.2811	.95656	3
58	.29182	.30509	3.2777	.95647	2
59	.29209	.30541	3.2743	.95639	1
60	.29237	.30573	3.2709	.95630	0
′	Cos	Cct	Tan	Sin	′

′	Sin	Tan	Cot	Cos	′
0	.29237	.30573	3.2709	.95630	60
1	.29265	.30605	3.2675	.95622	59
2	.29293	.30637	3.2641	.95613	58
3	.29321	.30669	3.2607	.95605	57
4	.29348	.30700	3.2573	.95596	56
5	.29376	.30732	3.2539	.95588	55
6	.29404	.30764	3.2506	.95579	54
7	.29432	.30796	3.2472	.95571	53
8	.29460	.30828	3.2438	.95562	52
9	.29487	.30860	3.2405	.95554	51
10	.29515	.30891	3.2371	.95545	50
11	.29543	.30923	3.2338	.95536	49
12	.29571	.30955	3.2305	.95528	48
13	.29599	.30987	3.2272	.95519	47
14	.29626	.31019	3.2238	.95511	46
15	.29654	.31051	3.2205	.95502	45
16	.29682	.31083	3.2172	.95493	44
17	.29710	.31115	3.2139	.95485	43
18	.29737	.31147	3.2106	.95476	42
19	.29765	.31178	3.2073	.95467	41
20	.29793	.31210	3.2041	.95459	40
21	.29821	.31242	3.2003	.95450	39
22	.29849	.31274	3.1975	.95441	38
23	.29876	.31306	3.1943	.95433	37
24	.29904	.31338	3.1910	.95424	36
25	.29932	.31370	3.1878	.95415	35
26	.29960	.31402	3.1845	.95407	34
27	.29987	.31434	3.1813	.95398	33
28	.30015	.31466	3.1780	.95389	32
29	.30043	.31498	3.1748	.95380	31
30	.30071	.31530	3.1716	.95372	30
31	.30098	.31562	3.1684	.95363	29
32	.30126	.31594	3.1652	.95354	28
33	.30154	.31626	3.1620	.95345	27
34	.30182	.31658	3.1588	.95337	26
35	.30209	.31690	3.1556	.95328	25
36	.30237	.31722	3.1524	.95319	24
37	.30265	.31754	3.1492	.95310	23
38	.30292	.31786	3.1460	.95301	22
39	.30320	.31818	3.1429	.95293	21
40	.30348	.31850	3.1397	.95284	20
41	.30376	.31882	3.1366	.95275	19
42	.30403	.31914	3.1334	.95266	18
43	.30431	.31946	3.1303	.95257	17
44	.30459	.31978	3.1271	.95248	16
45	.30486	.32010	3.1240	.95240	15
46	.30514	.32042	3.1209	.95231	14
47	.30542	.32074	3.1178	.95222	13
48	.30570	.32106	3.1146	.95213	12
49	.30597	.32139	3.1115	.95204	11
50	.30625	.32171	3.1084	.95195	10
51	.30653	.32203	3.1053	.95186	9
52	.30680	.32235	3.1022	.95177	8
53	.30708	.32267	3.0991	.95168	7
54	.30736	.32299	3.0961	.95159	6
55	.30763	.32331	3.0930	.95150	5
56	.30791	.32363	3.0899	.95142	4
57	.30819	.32396	3.0868	.95133	3
58	.30846	.32428	3.0838	.95124	2
59	.30874	.32460	3.0807	.95115	1
60	.30902	.32492	3.0777	.95106	0
′	Cos	Cot	Tan	Sin	′

′	Sin	Tan	Cot	Cos	′
0	.30902	.32492	3.0777	.95106	60
1	.30929	.32524	3.0746	.95097	59
2	.30957	.32556	3.0716	.95088	58
3	.30985	.32588	3.0686	.95079	57
4	.31012	.32621	3.0655	.95070	56
5	.31040	.32653	3.0625	.95061	55
6	.31068	.32685	3.0595	.95052	54
7	.31095	.32717	3.0565	.95043	53
8	.31123	.32749	3.0535	.95033	52
9	.31151	.32782	3.0505	.95024	51
10	.31178	.32814	3.0475	.95015	50
11	.31206	.32846	3.0445	.95006	49
12	.31233	.32878	3.0415	.94997	48
13	.31261	.32911	3.0385	.94988	47
14	.31289	.32943	3.0356	.94979	46
15	.31316	.32975	3.0326	.94970	45
16	.31344	.33007	3.0296	.94961	44
17	.31372	.33040	3.0267	.94952	43
18	.31399	.33072	3.0237	.94943	42
19	.31427	.33104	3.0208	.94933	41
20	.31454	.33136	3.0178	.94924	40
21	.31482	.33169	3.0149	.94915	29
22	.31510	.33201	3.0120	.94906	38
23	.31537	.33233	3.0090	.94897	37
24	.31565	.33266	3.0061	.94888	36
25	.31593	.33298	3.0032	.94878	35
26	.31620	.33330	3.0003	.94869	34
27	.31648	.33363	2.9974	.94860	33
28	.31675	.33395	2.9945	.94851	32
29	.31703	.33427	2.9916	.94842	31
30	.31730	.33460	2.9887	.94832	30
31	.31758	.33492	2.9858	.94823	29
32	.31786	.33524	2.9829	.94814	28
33	.31813	.33557	2.9800	.94805	27
34	.31841	.33589	2.9772	.94795	26
35	.31868	.33621	2.9743	.94786	25
36	.31896	.33654	2.9714	.94777	24
37	.31923	.33686	2.9686	.94768	23
38	.31951	.33718	2.9657	.94758	22
39	.31979	.33751	2.9629	.94749	21
40	.32006	.33783	2.9600	.94740	20
41	.32034	.33816	2.9572	.94730	19
42	.32061	.33848	2.9544	.94721	18
43	.32089	.33881	2.9515	.94712	17
44	.32116	.33913	2.9487	94702	16
45	.32144	.33945	2.9459	.94693	15
46	.32171	.33978	2.9431	.94684	14
47	.32199	.34010	2.9403	.94674	13
48	.32227	.34043	2.9375	.94665	12
49	.32254	.34075	2.9347	.94656	11
50	.32282	.34108	2.9319	.94646	10
51	.32309	.34140	2.9291	.94637	9
52	.32337	.34173	2.9263	.94627	8
53	.32364	.34205	2.9235	.94618	7
54	.32392	.34238	2.9208	.94609	6
55	.32419	.34270	2.9180	.94599	5
56	.32447	.34303	2.9152	.94590	4
57	.32474	.34335	2.9125	.94580	3
58	.32502	.34368	2.9097	.94571	2
59	.32529	.34400	2.9070	.94561	1
60	.32557	.34433	2.9042	.94552	0
′	Cos	Cot	Tan	Sin	′

′	Sin	Tan	Cot	Cos	′
0	.32557	.34433	2.9042	.94552	60
1	.32584	.34465	2.9015	.94542	59
2	.32612	.34498	2.8987	.94533	58
3	.32639	.34530	2.8960	.94523	57
4	.32667	.34563	2.8933	.94514	56
5	.32694	.34596	2.8905	.94504	55
6	.32722	.34628	2.8878	.94495	54
7	.32749	.34661	2.8851	.94485	53
8	.32777	.34693	2.8824	.94476	52
9	.32804	.34726	2.8797	.94466	51
10	.32832	.34758	2.8770	.94457	50
11	.32859	.34791	2.8743	.94447	49
12	.32887	.34824	2.8716	.94438	48
13	.32914	.34856	2.8689	.94428	47
14	.32942	.34889	2.8662	.94418	46
15	.32969	.34922	2.8636	.94409	45
16	.32997	.34954	2.8609	.94399	44
17	.33024	.34987	2.8582	.94390	43
18	.33051	.35020	2.8556	.94380	42
19	.33079	.35052	2.8529	.94370	41
20	.33106	.35085	2.8502	.94361	40
21	.33134	.35118	2.8476	.94351	39
22	.33161	.35150	2.8449	.94342	38
23	.33189	.35183	2.8423	.94332	37
24	.33216	.35216	2.8397	.94322	36
25	.33244	.35248	2.8370	.94313	35
26	.33271	.35281	2.8344	.94303	34
27	.33298	.35314	2.8318	.94293	33
28	.33326	.35346	2.8291	.94284	32
29	.33353	.35379	2.8265	.94274	31
30	.33381	.35412	2.8239	.94264	30
31	.33408	.35445	2.8213	.94254	29
32	.33436	.35477	2.8187	.94245	28
33	.33463	.35510	2.8161	.94235	27
34	.33490	.35543	2.8135	.94225	26
35	.33518	.35576	2.8109	.94215	25
36	.33545	.35608	2.8083	.94206	24
37	.33573	.35641	2.8057	.94196	23
38	.33600	.35674	2.8032	.94186	22
39	.33627	.35707	2.8006	.94176	21
40	.33655	.35740	2.7980	.94167	20
41	.33682	.35772	2.7955	.94157	19
42	.33710	.35805	2.7929	.94147	18
43	.33737	.35838	2.7903	.94137	17
44	.33764	.35871	2.7878	.94127	16
45	.33792	.35904	2.7852	.94118	15
46	.33819	.35937	2.7827	.94108	14
47	.33846	.35969	2.7801	.94098	13
48	.33874	.36002	2.7776	.94078	12
49	.33901	.36035	2.7751	.94078	11
50	.33929	.36068	2.7725	.94068	10
51	.33956	.36101	2.7700	.94058	9
52	.33983	.36134	2.7675	.94049	8
53	.34011	.36167	2.7650	.94039	7
54	.34038	.36199	2.7625	.94029	6
55	.34065	.36232	2.7600	.94019	5
56	.34093	.36265	2.7575	.94009	4
57	.34120	.36298	2.7550	.93999	3
58	.34147	.36331	2.7525	.93989	2
59	.34175	.36364	2.7500	.93979	1
60	.34202	.36397	2.7475	.93969	0
′	Cos	Cot	Tan	Sin	′

′	Sin	Tan	Cot	Cos	′
0	.34202	.36397	2.7475	.93969	60
1	.34229	.36430	2.7450	.93959	59
2	.34257	.36463	2.7425	.93949	58
3	.34284	.36496	2.7400	.93939	57
4	.34311	.36529	2.7376	.93929	56
5	.34339	.36562	2.7351	.93919	55
6	.34366	.36595	2.7326	.93909	54
7	.34393	.36628	2.7302	.93899	53
8	.34421	.36661	2.7277	.93889	52
9	.34448	.36694	2.7253	.93879	51
10	.34475	.36727	2.7228	.93869	50
11	.34503	.36760	2.7204	.93859	49
12	.34530	.36793	2.7179	.93849	48
13	.34557	.36826	2.7155	.93839	47
14	.34584	.36859	2.7130	.93829	46
15	.34612	.36892	2.7106	.93819	45
16	.34639	.36925	2.7082	.93809	44
17	.34666	.36958	2.7058	.93799	43
18	.34694	.36991	2.7034	.93789	42
19	.34721	.37024	2.7009	.93779	41
20	.34748	.37057	2.6985	.93769	40
21	.34775	.37090	2.6961	.93759	39
22	.34803	.37123	2.6937	.93748	38
23	.34830	.37157	2.6913	.93738	37
24	.34857	.37190	2.6889	.93728	36
25	.34884	.37223	2.6865	.93718	35
26	.34912	.37256	2.6841	.93708	34
27	.34939	.37289	2.6818	.93698	33
28	.34966	.37322	2.6794	.93688	32
29	.34993	.37355	2.6770	.93677	31
30	.35021	.37388	2.6746	.93667	30
31	.35048	.37422	2.6723	.93657	29
32	.35075	.37455	2.6699	.93647	28
33	.35102	.37488	2.6675	.93637	27
34	.35130	.37521	2.6652	.93626	26
35	.35157	.37554	2.6628	.93616	25
36	.35184	.37588	2.6605	.93606	24
37	.35211	.37621	2.6581	.93596	23
38	.35239	.37654	2.6558	.93585	22
39	.35266	.37687	2.6534	.93575	21
40	.35293	.37720	2.6511	.93565	20
41	.35320	.37754	2.6488	.93555	19
42	.35347	.37787	2.6464	.93544	18
43	.35375	.37820	2.6441	.93534	17
44	.35402	.37853	2.6418	.93524	16
45	.35429	.37887	2.6395	.93514	15
46	.35456	.37920	2.6371	.93503	14
47	.35484	.37953	2.6348	.93493	13
48	.35511	.37986	2.6325	.93483	12
49	.35538	.38020	2.6302	.93472	11
50	.35565	.38053	2.6279	.93462	10
51	.35592	.38086	2.6256	.93452	9
52	.35619	.38120	2.6233	.93441	8
53	.35647	.38153	2.6210	.93431	7
54	.35674	.38186	2.6187	.93420	6
55	.35701	.38220	2.6165	.93410	5
56	.35728	.38253	2.6142	.93400	4
57	.35755	.38286	2.6119	.93389	3
58	.35782	.38320	2.6096	.93379	2
59	.35810	.38353	2.6074	.93368	1
60	.35837	.38386	2.6051	.93358	0
′	Cos	Cot	Tan	Sin	′

′	Sin	Tan	Cot	Cos	′
0	.35837	.38386	2.6051	.93358	60
1	.35864	.38420	2.6028	.93348	59
2	.35891	.38453	2.6006	.93337	58
3	.35918	.38487	2.5983	.93327	57
4	.35945	.38520	2.5961	.93316	56
5	.35973	.38553	2.5938	.93306	55
6	.36000	.38587	2.5916	.93295	54
7	.36027	.38620	2.5893	.93285	53
8	.36054	.38654	2.5871	.93274	52
9	.36081	.38687	2.5848	.93264	51
10	.36108	.38721	2.5826	.93253	50
11	.36135	.38754	2.5804	.93243	49
12	.36162	.38787	2.5782	.93232	48
13	.36190	.38821	2.5759	.93222	47
14	.36217	.38854	2.5737	.93211	46
15	.36244	.38888	2.5715	.93201	45
16	.36271	.38921	2.5693	.93190	44
17	.36298	.38955	2.5671	.93180	43
18	.36325	.38988	2.5649	.93169	42
19	.36352	.39022	2.5627	.93159	41
20	.36379	.39055	2.5605	.93148	40
21	.36406	.39089	2.5583	.93137	39
22	.36434	.39122	2.5561	.93127	38
23	.36461	.39156	2.5539	.93116	37
24	.36488	.39190	2.5517	.93106	36
25	.36515	.39223	2.5495	.93095	35
26	.36542	.39257	2.5473	.93084	34
27	.36569	.39290	2.5452	.93074	33
28	.36596	.39324	2.5430	.93063	32
29	.36623	.39357	2.5408	.93052	31
30	.36650	.39391	2.5386	.93042	30
31	.36677	.39425	2.5365	.93031	29
32	.36704	.39458	2.5343	.93020	28
33	.36731	.39492	2.5322	.93010	27
34	.36758	.39526	2.5300	.92999	26
35	.36785	.39559	2.5279	.92988	25
36	.36812	.39593	2.5257	.92978	24
37	.36839	.39626	2.5236	.92967	23
38	.36867	.39660	2.5214	.92956	22
39	.36894	.39694	2.5193	.92945	21
40	.36921	.39727	2.5172	.92935	20
41	.36948	.39761	2.5150	.92924	19
42	.36975	.39795	2.5129	.92913	18
43	.37002	.39829	2.5108	.92902	17
44	.37029	.39862	2.5086	.92892	16
45	.37056	.39896	2.5065	.92881	15
46	.37083	.39930	2.5044	.92870	14
47	.37110	.39963	2.5023	.92859	13
48	.37137	.39997	2.5002	.92849	12
49	.37164	.40031	2.4981	.92838	11
50	.37191	.40065	2.4960	.92827	10
51	.37218	.40098	2.4939	.92816	9
52	.37245	.40132	2.4918	.92805	8
53	.37272	.40166	2.4897	.92794	7
54	.37299	.40200	2.4876	.92784	6
55	.37326	.40234	2.4855	.92773	5
56	.37353	.40267	2.4834	.92762	4
57	.37380	.40301	2.4813	.92751	3
58	.37407	.40335	2.4792	.92740	2
59	.37434	.40369	2.4772	.92729	1
60	.37461	.40403	2.4751	.92718	0
′	Cos	Cot	Tan	Sin	′

′	Sin	Tan	Cot	Cos	′
0	.37461	.40403	2.4751	.92718	60
1	.37488	.40436	2.4730	.92707	59
2	.37515	.40470	2.4709	.92697	58
3	.37542	.40504	2.4689	.92686	57
4	.37569	.40538	2.4668	.92675	56
5	.37595	.40572	2.4648	.92664	55
6	.37622	.40606	2.4627	.92653	54
7	.37649	.40640	2.4606	.92642	53
8	.37676	.40674	2.4586	.92631	52
9	.37703	.40707	2.4566	.92620	51
10	.37730	.40741	2.4545	.92609	50
11	.37757	.40775	2.4525	.92598	49
12	.37784	.40809	2.4504	.92587	48
13	.37811	.40843	2.4484	.92576	47
14	.37838	.40877	2.4464	.92565	46
15	.37865	.40911	2.4443	.92554	45
16	.37892	.40945	2.4423	.92543	44
17	.37919	.40979	2.4403	.92532	43
18	.37946	.41013	2.4383	.92521	42
19	.37973	.41047	2.4362	.92510	41
20	.37999	.41081	2.4342	.92499	40
21	.38026	.41115	2.4322	.92488	39
22	.38053	.41149	2.4302	.92477	38
23	.38080	.41183	2.4282	.92466	37
24	.38107	.41217	2.4262	.92455	36
25	.38134	.41251	2.4242	.92444	35
26	.38161	.41285	2.4222	.92432	34
27	.38188	.41319	2.4202	.92421	33
28	.38215	.41353	2.4182	.92410	32
29	.38241	.41387	2.4162	.92399	31
30	.38268	.41421	2.4142	.92388	30
31	.38295	.41455	2.4122	.92377	29
32	.38322	.41490	2.4102	.92366	28
33	.38349	.41524	2.4083	.92355	27
34	.38376	.41558	2.4063	.92343	26
35	.38403	.41592	2.4043	.92332	25
36	.38430	.41626	2.4023	.92321	24
37	.38456	.41660	2.4004	.92310	23
38	.38483	.41694	2.3984	.92299	22
39	.38510	.41728	2.3964	.92287	21
40	.38537	.41763	2.3945	.92276	20
41	.38564	.41797	2.3925	.92265	19
42	.38591	.41831	2.3906	.92254	18
43	.38617	.41865	2.3886	.92243	17
44	.38644	.41899	2.3867	.92231	16
45	.38671	.41933	2.3847	.92220	15
46	.38698	.41968	2.3828	.92209	14
47	.38725	.42002	2.3808	.92198	13
48	.38752	.42036	2.3789	.92186	12
49	.38778	.42070	2.3770	.92175	11
50	.38805	.42105	2.3750	.92164	10
51	.38832	.42139	2.3731	.92152	9
52	.38859	.42173	2.3712	.92141	8
53	.38886	.42207	2.3693	.92130	7
54	.38912	.42242	2.3673	.92119	6
55	.38939	.42276	2.3654	.92107	5
56	.38966	.42310	2.3635	.92096	4
57	.38993	.42345	2.3616	.92085	3
58	.39020	.42379	2.3597	.92073	2
59	.39046	.42413	2.3578	.92062	1
60	.39073	.42447	2.3559	.92050	0
′	Cos	Cot	Tan	Sin	′

′	Sin	Tan	Cot	Cos	′
0	.39073	.42447	2.3559	.92050	60
1	.39100	.42482	2.3539	.92039	59
2	.39127	.42516	2.3520	.92028	58
3	.39153	.42551	2.3501	.92016	57
4	.39180	.42585	2.3483	.92005	56
5	.39207	.42619	2.3464	.91994	55
6	.39234	.42654	2.3445	.91982	54
7	.39260	.42688	2.3426	.91971	53
8	.39287	.42722	2.3407	.91959	52
9	.39314	.42757	2.3388	.91948	51
10	.39341	.42791	2.3369	.91936	50
11	.39367	.42826	2.3351	.91925	49
12	.39394	.42860	2.3332	.91914	48
13	.39421	.42894	2.3313	.91902	47
14	.39448	.42929	2.3294	.91891	46
15	.39474	.42963	2.3276	.91879	45
16	.39501	.42998	2.3257	.91868	44
17	.39528	.43032	2.3238	.91856	43
18	.39555	.43067	2.3220	.91845	42
19	.39581	.43101	2.3201	.91833	41
20	.39608	.43136	2.3183	.91822	40
21	.39635	.43170	2.3164	.91810	39
22	.39661	.43205	2.3146	.91799	38
23	.39688	.43239	2.3127	.91787	37
24	.39715	.43274	2.3109	.91775	36
25	.39741	.43308	2.3090	.91764	35
26	.39768	.43343	2.3072	.91752	34
27	.39795	.43378	2.3053	.91741	33
28	.39822	.43412	2.3035	.91729	32
29	.39848	.43447	2.3017	.91718	31
30	.39875	.43481	2.2998	.91706	30
31	.39902	.43516	2.2980	.91694	29
32	.39928	.43550	2.2962	.91683	28
33	.39955	.43585	2.2944	.91671	27
34	.39982	.43620	2.2925	.91660	26
35	.40008	.43654	2.2907	.91648	25
36	.40035	.43689	2.2889	.91636	24
37	.40062	.43724	2.2871	.91625	23
38	.40088	.43758	2.2853	.91613	22
39	.40115	.43793	2.2835	.91601	21
40	.40141	.43828	2.2817	.91590	20
41	.40168	.43862	2.2799	.91578	19
42	.40195	.43897	2.2781	.91566	18
43	.40221	.43932	2.2763	.91555	17
44	.40248	.43966	2.2745	.91543	16
45	.40275	.44001	2.2727	.91531	15
46	.40301	.44036	2.2709	.91519	14
47	.40328	.44071	2.2691	.91508	13
48	.40355	.44105	2.2673	.91496	12
49	.40381	.44140	2.2655	.91484	11
50	.40408	.44175	2.2637	.91472	10
51	.40434	.44210	2.2620	.91461	9
52	.40461	.44244	2.2602	.91449	8
53	.40488	.44279	2.2584	.91437	7
54	.40514	.44314	2.2566	.91425	6
55	.40541	.44349	2.2549	.91414	5
56	.40567	.44384	2.2531	.91402	4
57	.40594	.44418	2.2513	.91390	3
58	.40621	.44453	2.2496	.91378	2
59	.40647	.44488	2.2478	.91366	1
60	.40674	.44523	2.2460	.91355	0
′	Cos	Cot	Tan	Sin	′

′	Sin	Tan	Cot	Cos	′
0	.40674	.44523	2.2460	.91355	60
1	.40700	.44558	2.2443	.91343	59
2	.40727	.44593	2.2425	.91331	58
3	.40753	.44627	2.2408	.91319	57
4	.40780	.44662	2.2390	.91307	56
5	.40806	.44697	2.2373	.91295	55
6	.40833	.44732	2.2355	.91283	54
7	.40860	.44767	2.2338	.91272	53
8	.40886	.44802	2.2320	.91260	52
9	.40913	.44837	2.2303	.91248	51
10	.40939	.44872	2.2286	.91236	50
11	.40966	.44907	2.2268	.91224	49
12	.40992	.44942	2.2251	.91212	48
13	.41019	.44977	2.2234	.91200	47
14	.41045	.45012	2.2216	.91188	46
15	.41072	.45047	2.2199	.91176	45
16	.41098	.45082	2.2182	.91164	44
17	.41125	.45117	2.2165	.91152	43
18	.41151	.45152	2.2148	.91140	42
19	.41178	.45187	2.2130	.91128	41
20	.41204	.45222	2.2113	.91116	40
21	.41231	.45257	2.2096	.91104	39
22	.41257	.45292	2.2079	.91092	38
23	.41284	.45327	2.2062	.91080	37
24	.41310	.45362	2.2045	.91068	36
25	.41337	.45397	2.2028	.91056	35
26	.41363	.45432	2.2011	.91044	34
27	.41390	.45467	2.1994	.91032	33
28	.41416	.45502	2.1977	.91020	32
29	.41443	.45538	2.1960	.91008	31
30	.41469	.45573	2.1943	.90996	30
31	.41496	.45608	2.1926	.90984	29
32	.41522	.45643	2.1909	.90972	28
33	.41549	.45678	2.1892	.90960	27
34	.41575	.45713	2.1876	.90948	26
35	.41602	.45748	2.1859	.90936	25
36	.41628	.45784	2.1842	.90924	24
37	.41655	.45819	2.1825	.90911	23
38	.41681	.45854	2.1808	.90899	22
39	.41707	.45889	2.1792	.90887	21
40	.41734	.45924	2.1775	.90875	20
41	.41760	.45960	2.1758	.90863	19
42	.41787	.45995	2.1742	.90851	18
43	.41813	.46030	2.1725	.90839	17
44	.41840	.46065	2.1708	.90826	16
45	.41866	.46101	2.1692	.90814	15
46	.41892	.46136	2.1675	.90802	14
47	.41919	.46171	2.1659	.90790	13
48	.41945	.46206	2.1642	.90778	12
49	.41972	.46242	2.1625	.90766	11
50	.41998	.46277	2.1609	.90753	10
51	.42024	.46312	2.1592	.90741	9
52	.42051	.46348	2.1576	.90729	8
53	.42077	.46383	2.1560	.90717	7
54	.42104	.46418	2.1543	.90704	6
55	.42130	.46454	2.1527	.90692	5
56	.42156	.46489	2.1510	.90680	4
57	.42183	.46525	2.1494	.90668	3
58	.42209	.46560	2.1478	.90655	2
59	.42235	.46595	2.1461	.90643	1
60	.42262	.46631	2.1445	.90631	0
′	Cos	Cot	Tan	Sin	′

′	Sin	Tan	Cot	Cos	′
0	.42262	.46631	2.1445	.90631	60
1	.42288	.46666	2.1429	.90618	59
2	.42315	.46702	2.1413	.90606	58
3	.42341	.46737	2.1396	.90594	57
4	.42367	.46772	2.1380	.90582	56
5	.42394	.46808	2.1364	.90569	55
6	.42420	.46843	2.1348	.90557	54
7	.42446	.46879	2.1332	.90545	53
8	.42473	.46914	2.1315	.90532	52
9	.42499	.46950	2.1299	.90520	51
10	.42525	.46985	2.1283	.90507	50
11	.42552	.47021	2.1267	.90495	49
12	.42578	.47056	2.1251	.90483	48
13	.42604	.47092	2.1235	.90470	47
14	.42631	.47128	2.1219	.90458	46
15	.42657	.47163	2.1203	.90446	45
16	.42683	.47199	2.1187	.90433	44
17	.42709	.47234	2.1171	.90421	43
18	.42736	.47270	2.1155	.90408	42
19	.42762	.47305	2.1139	.90396	41
20	.42788	.47341	2.1123	.90383	40
21	.42815	.47377	2.1107	.90371	39
22	.42841	.47412	2.1092	.90358	38
23	.42867	.47448	2.1076	.90346	37
24	.42894	.47483	2.1060	.90334	36
25	.42920	.47519	2.1044	.90321	35
26	.42946	.47555	2.1028	.90309	34
27	.42972	.47590	2.1013	.90296	33
28	.42999	.47626	2.0997	.90284	32
29	.43025	.47662	2.0981	.90271	31
30	.43051	.47698	2.0965	.90259	30
31	.43077	.47733	2.0950	.90246	29
32	.43104	.47769	2.0934	.90233	28
33	.43130	.47805	2.0918	.90221	27
34	.43156	.47840	2.0903	.90208	26
35	.43182	.47876	2.0887	.90196	25
36	.43209	.47912	2.0872	.90183	24
37	.43235	.47948	2.0856	.90171	23
38	.43261	.47984	2.0840	.90158	22
39	.43287	.48019	2.0825	.90146	21
40	.43313	.48055	2.0809	.90133	20
41	.43340	.48091	2.0794	.90120	19
42	.43366	.48127	2.0778	.90108	18
43	.43392	.48163	2.0763	.90095	17
44	.43418	.48198	2.0748	.90082	16
45	.43445	.48234	2.0732	.90070	15
46	.43471	.48270	2.0717	.90057	14
47	.43497	.48306	2.0701	.90045	13
48	.43523	.48342	2.0686	.90032	12
49	.43549	.48378	2.0671	.90019	11
50	.43575	.48414	2.0655	.90007	10
51	.43602	.48450	2.0640	.89994	9
52	.43628	.48486	2.0625	.89981	8
53	.43654	.48521	2.0609	.89968	7
54	.43680	.48557	2.0594	.89956	6
55	.43706	.48593	2.0579	.89943	5
56	.43733	.48629	2.0564	.89930	4
57	.43759	.48665	2.0549	.89918	3
58	.43785	.48701	2.0533	.89905	2
59	.43811	.48737	2.0518	.89892	1
60	.43837	.48773	2.0503	.89879	0
′	Cos	Cot	Tan	Sin	′

′	Sin	Tan	Cot	Cos	′
0	.43837	.48773	2.0503	.89879	60
1	.43863	.48809	2.0488	.89867	59
2	.43889	.48845	2.0473	.89854	58
3	.43916	.48881	2.0458	.89841	57
4	.43942	.48917	2.0443	.89828	56
5	.43968	.48953	2.0428	.89816	55
6	.43994	.48989	2.0413	.89803	54
7	.44020	.49026	2.0398	.89790	53
8	.44046	.49062	2.0383	.89777	52
9	.44072	.49098	2.0368	.89764	51
10	.44098	.49134	2.0353	.89752	50
11	.44124	.49170	2.0338	.89739	49
12	.44151	.49206	2.0323	.89726	48
13	.44177	.49242	2.0308	.89713	47
14	.44203	.49278	2.0293	.89700	46
15	.44229	.49315	2.0278	.89687	45
16	.44255	.49351	2.0263	.89674	44
17	.44281	.49387	2.0248	.89662	43
18	.44307	.49423	2.0233	.89649	42
19	.44333	.49459	2.0219	.89636	41
20	.44359	.49495	2.0204	.89623	40
21	.44385	.49532	2.0189	.89610	39
22	.44411	.49568	2.0174	.89597	38
23	.44437	.49604	2.0160	.89584	37
24	.44464	.49640	2.0145	.89571	36
25	.44490	.49677	2.0130	.89558	35
26	.44516	.49713	2.0115	.89545	34
27	.44542	.49749	2.0101	.89532	33
28	.44568	.49786	2.0086	.89519	32
29	.44594	.49822	2.0072	.89506	31
30	.44620	.49858	2.0057	.89493	30
31	.44646	.49894	2.0042	.89480	29
32	.44672	.49931	2.0028	.89467	28
33	.44698	.49967	2.0013	.89454	27
34	.44724	.50004	1.9999	.89441	26
35	.44750	.50040	1.9984	.89428	25
36	.44776	.50076	1.9970	.89415	24
37	.44802	.50113	1.9955	.89402	23
38	.44828	.50149	1.9941	.89389	22
39	.44854	.50185	1.9926	.89376	21
40	.44880	.50222	1.9912	.89363	20
41	.44906	.50258	1.9897	.89350	19
42	.44932	.50295	1.9883	.89337	18
43	.44958	.50331	1.9868	.89324	17
44	.44984	.50368	1.9854	.89311	16
45	.45010	.50404	1.9840	.89298	15
46	.45036	.50441	1.9825	.89285	14
47	.45062	.50477	1.9811	.89272	13
48	.45088	.50514	1.9797	.89259	12
49	.45114	.50550	1.9782	.89245	11
50	.45140	.50587	1.9768	.89232	10
51	.45166	.50623	1.9754	.89219	9
52	.45192	.50660	1.9740	.89206	8
53	.45218	.50696	1.9725	.89193	7
54	.45243	.50733	1.9711	.89180	6
55	.45269	.50769	1.9697	.89167	5
56	.45295	.50806	1.9683	.89153	4
57	.45321	.50843	1.9669	.89140	3
58	.45347	.50879	1.9654	.89127	2
59	.45373	.50916	1.9640	.89114	1
60	.45399	.50953	1.9626	.89101	0
′	Cos	Cot	Tan	Sin	′

′	Sin	Tan	Cot	Cos	′
0	.45399	.50953	1.9626	.89101	60
1	.45425	.50989	1.9612	.89087	59
2	.45451	.51026	1.9598	.89074	58
3	.45477	.51063	1.9584	.89061	57
4	.45503	.51099	1.9570	.89048	56
5	.45529	.51136	1.9556	.89035	55
6	.45554	.51173	1.9542	.89021	54
7	.45580	.51209	1.9528	.89008	53
8	.45606	.51246	1.9514	.88995	52
9	.45632	.51283	1.9500	.88981	51
10	.45658	.51319	1.9486	.88968	50
11	.45684	.51356	1.9472	.88955	49
12	.45710	.51393	1.9458	.88942	48
13	.45736	.51430	1.9444	.88928	47
14	.45762	.51467	1.9430	.88915	46
15	.45787	.51503	1.9416	.88902	45
16	.45813	.51540	1.9402	.88888	44
17	.45839	.51577	1.9388	.88875	43
18	.45865	.51614	1.9375	.88862	42
19	.45891	.51651	1.9361	.88848	41
20	.45917	.51688	1.9347	.88835	40
21	.45942	.51724	1.9333	.88822	39
22	.45968	.51761	1.9319	.88808	38
23	.45994	.51798	1.9306	.88795	37
24	.46020	.51835	1.9292	.88782	36
25	.46046	.51872	1.9278	.88768	35
26	.46072	.51909	1.9265	.88755	34
27	.46097	.51946	1.9251	.88741	33
28	.46123	.51983	1.9237	.88728	32
29	.46149	.52020	1.9223	.88715	31
30	.46175	.52057	1.9210	.88701	30
31	.46201	.52094	1.9196	.88688	29
32	.46226	.52131	1.9183	.88674	28
33	.46252	.52168	1.9169	.88661	27
34	.46278	.52205	1.9155	.88647	26
35	.46304	.52242	1.9142	.88634	25
36	.46330	.52279	1.9128	.88620	24
37	.46355	.52316	1.9115	.88607	23
38	.46381	.52353	1.9101	.88593	22
39	.46407	.52390	1.9088	.88580	21
40	.46433	.52427	1.9074	.88566	20
41	.46458	.52464	1.9061	.88553	19
42	.46484	.52501	1.9047	.88539	18
43	.46510	.52538	1.9034	.88526	17
44	.46536	.52575	1.9020	.88512	16
45	.46561	.52613	1.9007	.88499	15
46	.46587	.52650	1.8993	.88485	14
47	.46613	.52687	1.8980	.88472	13
48	.46639	.52724	1.8967	.88458	12
49	.46664	.52761	1.8953	.88445	11
50	.46690	.52798	1.8940	.88431	10
51	.46716	.52836	1.8927	.88417	9
52	.46742	.52873	1.8913	.88404	8
53	.46767	.52910	1.8900	.88390	7
54	.46793	.52947	1.8887	.88377	6
55	.46819	.52985	1.8873	.88363	5
56	.46844	.53022	1.8860	.88349	4
57	.46870	.53059	1.8847	.88336	3
58	.46896	.53096	1.8834	.88322	2
59	.46921	.53134	1.8820	.88308	1
60	.46947	.53171	1.8807	.88295	0
′	Cos	Cot	Tan	Sin	′

′	Sin	Tan	Cot	Cos	′
0	.46947	.53171	1.8807	.88295	60
1	.46973	.53208	1.8794	.88281	59
2	.46999	.53246	1.8781	.88267	58
3	.47024	.53283	1.8768	.88254	57
4	.47050	.53320	1.8755	.88240	56
5	.47076	.53358	1.8741	.88226	55
6	.47101	.53395	1.8728	.88213	54
7	.47127	.53432	1.8715	.88199	53
8	.47153	.53470	1.8702	.88185	52
9	.47178	.53507	1.8689	.88172	51
10	.47204	.53545	1.8676	.88158	50
11	.47229	.53582	1.8663	.88144	49
12	.47255	.53620	1.8650	.88130	48
13	.47281	.53657	1.8637	.88117	47
14	.47306	.53694	1.8624	.88103	46
15	.47332	.53732	1.8611	.88089	45
16	.47358	.53769	1.8598	.88075	44
17	.47383	.53807	1.8585	.88062	43
18	.47409	.53844	1.8572	.88048	42
19	.47434	.53882	1.8559	.88034	41
20	.47460	.53920	1.8546	.88020	40
21	.47486	.53957	1.8533	.88006	39
22	.47511	.53995	1.8520	.87993	38
23	.47537	.54032	1.8507	.87979	37
24	.47562	.54070	1.8495	.87965	36
25	.47588	.54107	1.8482	.87951	35
26	.47614	.54145	1.8469	.87937	34
27	.47639	.54183	1.8456	.87923	33
28	.47665	.54220	1.8443	.87909	32
29	.47690	.54258	1.8430	.87896	31
30	.47716	.54296	1.8418	.87882	30
31	.47741	.54333	1.8405	.87868	29
32	.47767	.54371	1.8392	.87854	28
33	.47793	.54409	1.8379	.87840	27
34	.47818	.54446	1.8367	.87826	26
35	.47844	.54484	1.8354	.87812	25
36	.47869	.54522	1.8341	.87798	24
37	.47895	.54560	1.8329	.87784	23
38	.47920	.54597	1.8316	.87770	22
39	.47946	.54635	1.8303	.87756	21
40	.47971	.54673	1.8291	.87743	20
41	.47997	.54711	1.8278	.87729	19
42	.48022	.54748	1.8265	.87715	18
43	.48048	.54786	1.8253	.87701	17
44	.48073	.54824	1.8240	.87687	16
45	.48099	.54862	1.8228	.87673	15
46	.48124	.54900	1.8215	.87659	14
47	.48150	.54938	1.8202	.87645	13
48	.48175	.54975	1.8190	.87631	12
49	.48201	.55013	1.8177	.87617	11
50	.48226	.55051	1.8165	.87603	10
51	.48252	.55089	1.8152	.87589	9
52	.48277	.55127	1.8140	.87575	8
53	.48303	.55165	1.8127	.87561	7
54	.48328	.55203	1.8115	.87546	6
55	.48354	.55241	1.8103	.87532	5
56	.48379	.55279	1.8090	.87518	4
57	.48405	.55317	1.8078	.87504	3
58	.48430	.55355	1.8065	.87490	2
59	.48456	.55393	1.8053	.87476	1
60	.48481	.55431	1.8040	.87462	0
′	Cos	Cot	Tan	Sin	′

′	Sin	Tan	Cot	Cos	′
0	.48481	.55431	1.8040	.87462	60
1	.48506	.55469	1.8028	.87448	59
2	.48532	.55507	1.8016	.87434	58
3	.48557	.55545	1.8003	.87420	57
4	.48583	.55583	1.7991	.87406	56
5	.48608	.55621	1.7979	.87391	55
6	.48634	.55659	1.7966	.87377	54
7	.48659	.55697	1.7954	.87363	53
8	.48684	.55736	1.7942	.87349	52
9	.48710	.55774	1.7930	.87335	51
10	.48735	.55812	1.7917	.87321	50
11	.48761	.55850	1.7905	.87306	49
12	.48786	.55888	1.7893	.87292	48
13	.48811	.55926	1.7881	.87278	47
14	.48837	.55964	1.7868	.87264	46
15	.48862	.56003	1.7856	.87250	45
16	.48888	.56041	1.7844	.87235	44
17	.48913	.56079	1.7832	.87221	43
18	.48938	.56117	1.7820	.87207	42
19	.48964	.56156	1.7808	.87193	41
20	.48989	.56194	1.7796	.87178	40
21	.49014	.56232	1.7783	.87164	39
22	.49040	.56270	1.7771	.87150	38
23	.49065	.56309	1.7759	.87136	37
24	.49090	.56347	1.7747	.87121	36
25	.49116	.56385	1.7735	.87107	35
26	.49141	.56424	1.7723	.87093	34
27	.49166	.56462	1.7711	.87079	33
28	.49192	.56501	1.7699	.87064	32
29	.49217	.56539	1.7687	.87050	31
30	.49242	.56577	1.7675	.87036	30
31	.49268	.56616	1.7663	.87021	29
32	.49293	.56654	1.7651	.87007	28
33	.49318	.56693	1.7639	.86993	27
34	.49344	.56731	1.7627	.86978	26
35	.49369	.56769	1.7615	.86964	25
36	.49394	.56808	1.7603	.86949	24
37	.49419	.56846	1.7591	.86935	23
38	.49445	.56885	1.7579	.86921	22
39	.49470	.56923	1.7567	.86906	21
40	.49495	.56962	1.7556	.86892	20
41	.49521	.57000	1.7544	.86878	19
42	.49546	.57039	1.7532	.86863	18
43	.49571	.57078	1.7520	.86849	17
44	.49596	.57116	1.7508	.86834	16
45	.49622	.57155	1.7496	.86820	15
46	.49647	.57193	1.7485	.86805	14
47	.49672	.57232	1.7473	.86791	13
48	.49697	.57271	1.7461	.86777	12
49	.49723	.57309	1.7449	.86762	11
50	.49748	.57348	1.7437	.86748	10
51	.49773	.57386	1.7426	.86733	9
52	.49798	.57425	1.7414	.86719	8
53	.49824	.57464	1.7402	.86704	7
54	.49849	.57503	1.7391	.86690	6
55	.49874	.57541	1.7379	.86675	5
56	.49899	.57580	1.7367	.86661	4
57	.49924	.57619	1.7355	.86646	3
58	.49950	.57657	1.7344	.86632	2
59	.49975	.57696	1.7332	.86617	1
60	.50000	.57735	1.7321	.86603	0
′	Cos	Cot	Tan	Sin	′

30° (210°)   (329°) **149°**

'	Sin	Tan	Cot	Cos	'
0	.50000	.57735	1.7321	.86603	60
1	.50025	.57774	1.7309	.86588	59
2	.50050	.57813	1.7297	.86573	58
3	.50076	.57851	1.7286	.86559	57
4	.50101	.57890	1.7274	.86544	56
5	.50126	.57929	1.7262	.86530	55
6	.50151	.57968	1.7251	.86515	54
7	.50176	.58007	1.7239	.86501	53
8	.50201	.58046	1.7228	.86486	52
9	.50227	.58085	1.7216	.86471	51
10	.50252	.58124	1.7205	.86457	50
11	.50277	.58162	1.7193	.86442	49
12	.50302	.58201	1.7182	.86427	48
13	.50327	.58240	1.7170	.86413	47
14	.50352	.58279	1.7159	.86398	46
15	.50377	.58318	1.7147	.86384	45
16	.50403	.58357	1.7136	.86369	44
17	.50428	.58396	1.7124	.86354	43
18	.50453	.58435	1.7113	.86340	42
19	.50478	.58474	1.7102	.86325	41
20	.50503	.58513	1.7090	.86310	40
21	.50528	.58552	1.7079	.86295	39
22	.50553	.58591	1.7067	.86281	38
23	.50578	.58631	1.7056	.86266	37
24	.50603	.58670	1.7045	.86251	36
25	.50628	.58709	1.7033	.86237	35
26	.50654	.58748	1.7022	.86222	34
27	.50679	.58787	1.7011	.86207	33
28	.50704	.58826	1.6999	.86192	32
29	.50729	.58865	1.6988	.86178	31
30	.50754	.58905	1.6977	.86163	30
31	.50779	.58944	1.6965	.86148	29
32	.50804	.58983	1.6954	.86133	28
33	.50829	.59022	1.6943	.86119	27
34	.50854	.59061	1.6932	.86104	26
35	.50879	.59101	1.6920	.86089	25
36	.50904	.59140	1.6909	.86074	24
37	.50929	.59179	1.6898	.86059	23
38	.50954	.59218	1.6887	.86045	22
39	.50979	.59258	1.6875	.86030	21
40	.51004	.59297	1.6864	.86015	20
41	.51029	.59336	1.6853	.86000	19
42	.51054	.59376	1.6842	.85985	18
43	.51079	.59415	1.6831	.85970	17
44	.51104	.59454	1.6820	.85956	16
45	.51129	.59494	1.6808	.85941	15
46	.51154	.59533	1.6797	.85926	14
47	.51179	.59573	1.6786	.85911	13
48	.51204	.59612	1.6775	.85896	12
49	.51229	.59651	1.6764	.85881	11
50	.51254	.59691	1.6753	.85866	10
51	.51279	.59730	1.6742	.85851	9
52	.51304	.59770	1.6731	.85836	8
53	.51329	.59809	1.6720	.85821	7
54	.51354	.59849	1.6709	.85806	6
55	.51379	.59888	1.6698	.85792	5
56	.51404	.59928	1.6687	.85777	4
57	.51429	.59967	1.6676	.85762	3
58	.51454	.60007	1.6665	.85747	2
59	.51479	.60046	1.6654	.85732	1
60	.51504	.60086	1.6643	.85717	0
'	Cos	Cot	Tan	Sin	'

120° (300°)   (239°) **59°**

31° (211°)   (328°) **148°**

'	Sin	Tan	Cot	Cos	'
0	.51504	.60086	1.6643	.85717	60
1	.51529	.60126	1.6632	.85702	59
2	.51554	.60165	1.6621	.85687	58
3	.51579	.60205	1.6610	.85672	57
4	.51604	.60245	1.6599	.85657	56
5	.51628	.60284	1.6588	.85642	55
6	.51653	.60324	1.6577	.85627	54
7	.51678	.60364	1.6566	.85612	53
8	.51703	.60403	1.6555	.85597	52
9	.51728	.60443	1.6545	.85582	51
10	.51753	.60483	1.6534	.85567	50
11	.51778	.60522	1.6523	.85551	49
12	.51803	.60562	1.6512	.85536	48
13	.51828	.60602	1.6501	.85521	47
14	.51852	.60642	1.6490	.85506	46
15	.51877	.60681	1.6479	.85491	45
16	.51902	.60721	1.6469	.85476	44
17	.51927	.60761	1.6458	.85461	43
18	.51952	.60801	1.6447	.85446	42
19	.51977	.60841	1.6436	.85431	41
20	.52002	.60881	1.6426	.85416	40
21	.52026	.60921	1.6415	.85401	39
22	.52051	.60960	1.6404	.85385	38
23	.52076	.61000	1.6393	.85370	37
24	.52101	.61040	1.6383	.85355	36
25	.52126	.61080	1.6372	.85340	35
26	.52151	.61120	1.6361	.85325	34
27	.52175	.61160	1.6351	.85310	33
28	.52200	.61200	1.6340	.85294	32
29	.52225	.61240	1.6329	.85279	31
30	.52250	.61280	1.6319	.85264	30
31	.52275	.61320	1.6308	.85249	29
32	.52299	.61360	1.6297	.85234	28
33	.52324	.61400	1.6287	.85218	27
34	.52349	.61440	1.6276	.85203	26
35	.52374	.61480	1.6265	.85188	25
36	.52399	.61520	1.6255	.85173	24
37	.52423	.61561	1.6244	.85157	23
38	.52448	.61601	1.6234	.85142	22
39	.52473	.61641	1.6223	.85127	21
40	.52498	.61681	1.6212	.85112	20
41	.52522	.61721	1.6202	.85096	19
42	.52547	.61761	1.6191	.85081	18
43	.52572	.61801	1.6181	.85066	17
44	.52597	.61842	1.6170	.85051	16
45	.52621	.61882	1.6160	.85035	15
46	.52646	.61922	1.6149	.85020	14
47	.52671	.61962	1.6139	.85005	13
48	.52696	.62003	1.6128	.84989	12
49	.52720	.62043	1.6118	.84974	11
50	.52745	.62083	1.6107	.84959	10
51	.52770	.62124	1.6097	.84943	9
52	.52794	.62164	1.6087	.84928	8
53	.52819	.62204	1.6076	.84913	7
54	.52844	.62245	1.6066	.84897	6
55	.52869	.62285	1.6055	.84882	5
56	.52893	.62325	1.6045	.84866	4
57	.52918	.62366	1.6034	.84851	3
58	.52943	.62406	1.6024	.84836	2
59	.52967	.62446	1.6014	.84820	1
60	.52992	.62487	1.6003	.84805	0
'	Cos	Cot	Tan	Sin	'

121° (301°)   (238°) **58°**

32° (212°)   (327°) **147°**

'	Sin	Tan	Cot	Cos	'
0	.52992	.62487	1.6003	.84805	60
1	.53017	.62527	1.5993	.84789	59
2	.53041	.62568	1.5983	.84774	58
3	.53066	.62608	1.5972	.84759	57
4	.53091	.62649	1.5962	.84743	56
5	.53115	.62689	1.5952	.84728	55
6	.53140	.62730	1.5941	.84712	54
7	.53164	.62770	1.5931	.84697	53
8	.53189	.62811	1.5921	.84681	52
9	.53214	.62852	1.5911	.84666	51
10	.53238	.62892	1.5900	.84650	50
11	.53263	.62933	1.5890	.84635	49
12	.53288	.62973	1.5880	.84619	48
13	.53312	.63014	1.5869	.84604	47
14	.53337	.63055	1.5859	.84588	46
15	.53361	.63095	1.5849	.84573	45
16	.53386	.63136	1.5839	.84557	44
17	.53411	.63177	1.5829	.84542	43
18	.53435	.63217	1.5818	.84526	42
19	.53460	.63258	1.5808	.84511	41
20	.53484	.63299	1.5798	.84495	40
21	.53509	.63340	1.5788	.84480	39
22	.53534	.63380	1.5778	.84464	38
23	.53558	.63421	1.5768	.84448	37
24	.53583	.63462	1.5757	.84433	36
25	.53607	.63503	1.5747	.84417	35
26	.53632	.63544	1.5737	.84402	34
27	.53656	.63584	1.5727	.84386	33
28	.53681	.63625	1.5717	.84370	32
29	.53705	.63666	1.5707	.84355	31
30	.53730	.63707	1.5697	.84339	30
31	.53754	.63748	1.5687	.84324	29
32	.53779	.63789	1.5677	.84308	28
33	.53804	.63830	1.5667	.84292	27
34	.53828	.63871	1.5657	.84277	26
35	.53853	.63912	1.5647	.84261	25
36	.53877	.63953	1.5637	.84245	24
37	.53902	.63994	1.5627	.84230	23
38	.53926	.64035	1.5617	.84214	22
39	.53951	.64076	1.5607	.84198	21
40	.53975	.64117	1.5597	.84182	20
41	.54000	.64158	1.5587	.84167	19
42	.54024	.64199	1.5577	.84151	18
43	.54049	.64240	1.5567	.84135	17
44	.54073	.64281	1.5557	.84120	16
45	.54097	.64322	1.5547	.84104	15
46	.54122	.64363	1.5537	.84088	14
47	.54146	.64404	1.5527	.84072	13
48	.54171	.64446	1.5517	.84057	12
49	.54195	.64487	1.5507	.84041	11
50	.54220	.64528	1.5497	.84025	10
51	.54244	.64569	1.5487	.84009	9
52	.54269	.64610	1.5477	.83994	8
53	.54293	.64652	1.5468	.83978	7
54	.54317	.64693	1.5458	.83962	6
55	.54342	.64734	1.5448	.83946	5
56	.54366	.64775	1.5438	.83930	4
57	.54391	.64817	1.5428	.83915	3
58	.54415	.64858	1.5418	.83899	2
59	.54440	.64899	1.5408	.83883	1
60	.54464	.64941	1.5399	.83867	0
'	Cos	Cot	Tan	Sin	'

122° (302°)   (237°) **57°**

'	Sin	Tan	Cot	Cos	'
0	.54464	.64941	1.5399	.83867	60
1	.54488	.64982	1.5389	.83851	59
2	.54513	.65024	1.5379	.83835	58
3	.54537	.65065	1.5369	.83819	57
4	.54561	.65106	1.5359	.83804	56
5	.54586	.65148	1.5350	.83788	55
6	.54610	.65189	1.5340	.83772	54
7	.54635	.65231	1.5330	.83756	53
8	.54659	.65272	1.5320	.83740	52
9	.54683	.65314	1.5311	.83724	51
10	.54708	.65355	1.5301	.83708	50
11	.54732	.65397	1.5291	.83692	49
12	.54756	.65438	1.5282	.83676	48
13	.54781	.65480	1.5272	.83660	47
14	.54805	.65521	1.5262	.83645	46
15	.54829	.65563	1.5253	.83629	45
16	.54854	.65604	1.5243	.83613	44
17	.54878	.65646	1.5233	.83597	43
18	.54902	.65688	1.5224	.83581	42
19	.54927	.65729	1.5214	.83565	41
20	.54951	.65771	1.5204	.83549	40
21	.54975	.65813	1.5195	.83533	39
22	.54999	.65854	1.5185	.83517	38
23	.55024	.65896	1.5175	.83501	37
24	.55048	.65938	1.5166	.83485	36
25	.55072	.65980	1.5156	.83469	35
26	.55097	.66021	1.5147	.83453	34
27	.55121	.66063	1.5137	.83437	33
28	.55145	.66105	1.5127	.83421	32
29	.55169	.66147	1.5118	.83405	31
30	.55194	.66189	1.5108	.83389	30
31	.55218	.66230	1.5099	.83373	29
32	.55242	.66272	1.5089	.83356	28
33	.55266	.66314	1.5080	.83340	27
34	.55291	.66356	1.5070	.83324	26
35	.55315	.66398	1.5061	.83308	25
36	.55339	.66440	1.5051	.83292	24
37	.55363	.66482	1.5042	.83276	23
38	.55388	.66524	1.5032	.83260	22
39	.55412	.66566	1.5023	.83244	21
40	.55436	.66608	1.5013	.83228	20
41	.55460	.66650	1.5004	.83212	19
42	.55484	.66692	1.4994	.83195	18
43	.55509	.66734	1.4985	.83179	17
44	.55533	.66776	1.4975	.83163	16
45	.55557	.66818	1.4966	.83147	15
46	.55581	.66860	1.4957	.83131	14
47	.55605	.66902	1.4947	.83115	13
48	.55630	.66944	1.4938	.83098	12
49	.55654	.66986	1.4928	.83082	11
50	.55678	.67028	1.4919	.83066	10
51	.55702	.67071	1.4910	.83050	9
52	.55726	.67113	1.4900	.83034	8
53	.55750	.67155	1.4891	.83017	7
54	.55775	.67197	1.4882	.83001	6
55	.55799	.67239	1.4872	.82985	5
56	.55823	.67282	1.4863	.82969	4
57	.55847	.67324	1.4854	.82953	3
58	.55871	.67366	1.4844	.82936	2
59	.55895	.67409	1.4835	.82920	1
60	.55919	.67451	1.4826	.82904	0
'	Cos	Cot	Tan	Sin	'

'	Sin	Tan	Cot	Cos	'
0	.55919	.67451	1.4826	.82904	60
1	.55943	.67493	1.4816	.82887	59
2	.55968	.67536	1.4807	.82871	58
3	.55992	.67578	1.4798	.82855	57
4	.56016	.67620	1.4788	.82839	56
5	.56040	.67663	1.4779	.82822	55
6	.56064	.67705	1.4770	.82806	54
7	.56088	.67748	1.4761	.82790	53
8	.56112	.67790	1.4751	.82773	52
9	.56136	.67832	1.4742	.82757	51
10	.56160	.67875	1.4733	.82741	50
11	.56184	.67917	1.4724	.82724	49
12	.56208	.67960	1.4715	.82708	48
13	.56232	.68002	1.4705	.82692	47
14	.56256	.68045	1.4696	.82675	46
15	.56280	.68088	1.4687	.82659	45
16	.56305	.68130	1.4678	.82643	44
17	.56329	.68173	1.4669	.82626	43
18	.56353	.68215	1.4659	.82610	42
19	.56377	.68258	1.4650	.82593	41
20	.56401	.68301	1.4641	.82577	40
21	.56425	.68343	1.4632	.82561	39
22	.56449	.68386	1.4623	.82544	38
23	.56473	.68429	1.4614	.82528	37
24	.56497	.68471	1.4605	.82511	36
25	.56521	.68514	1.4596	.82495	35
26	.56545	.68557	1.4586	.82478	34
27	.56569	.68600	1.4577	.82462	33
28	.56593	.68642	1.4568	.82446	32
29	.56617	.68685	1.4559	.82429	31
30	.56641	.68728	1.4550	.82413	30
31	.56665	.68771	1.4541	.82396	29
32	.56689	.68814	1.4532	.82380	28
33	.56713	.68857	1.4523	.82363	27
34	.56736	.68900	1.4514	.82347	26
35	.56760	.68942	1.4505	.82330	25
36	.56784	.68985	1.4496	.82314	24
37	.56808	.69028	1.4487	.82297	23
38	.56832	.69071	1.4478	.82281	22
39	.56856	.69114	1.4469	.82264	21
40	.56880	.69157	1.4460	.82248	20
41	.56904	.69200	1.4451	.82231	19
42	.56928	.69243	1.4442	.82214	18
43	.56952	.69286	1.4433	.82198	17
44	.56976	.69329	1.4424	.82181	16
45	.57000	.69372	1.4415	.82165	15
46	.57024	.69416	1.4406	.82148	14
47	.57047	.69459	1.4397	.82132	13
48	.57071	.69502	1.4388	.82115	12
49	.57095	.69545	1.4379	.82098	11
50	.57119	.69588	1.4370	.82082	10
51	.57143	.69631	1.4361	.82065	9
52	.57167	.69675	1.4352	.82048	8
53	.57191	.69718	1.4344	.82032	7
54	.57215	.69761	1.4335	.82015	6
55	.57238	.69804	1.4326	.81999	5
56	.57262	.69847	1.4317	.81982	4
57	.57286	.69891	1.4308	.81965	3
58	.57310	.69934	1.4299	.81949	2
59	.57334	.69977	1.4290	.81932	1
60	.57358	.70021	1.4281	.81915	0
'	Cos	Cot	Tan	Sin	'

'	Sin	Tan	Cot	Cos	'
0	.57358	.70021	1.4281	.81915	60
1	.57381	.70064	1.4273	.81899	59
2	.57405	.70107	1.4264	.81882	58
3	.57429	.70151	1.4255	.81865	57
4	.57453	.70194	1.4246	.81848	56
5	.57477	.70238	1.4237	.81832	55
6	.57501	.70281	1.4229	.81815	54
7	.57524	.70325	1.4220	.81798	53
8	.57548	.70368	1.4211	.81782	52
9	.57572	.70412	1.4202	.81765	51
10	.57596	.70455	1.4193	.81748	50
11	.57619	.70499	1.4185	.81731	49
12	.57643	.70542	1.4176	.81714	48
13	.57667	.70586	1.4167	.81698	47
14	.57691	.70629	1.4158	.81681	46
15	.57715	.70673	1.4150	.81664	45
16	.57738	.70717	1.4141	.81647	44
17	.57762	.70760	1.4132	.81631	43
18	.57786	.70804	1.4124	.81614	42
19	.57810	.70848	1.4115	.81597	41
20	.57833	.70891	1.4106	.81580	40
21	.57857	.70935	1.4097	.81563	39
22	.57881	.70979	1.4089	.81546	38
23	.57904	.71023	1.4080	.81530	37
24	.57928	.71066	1.4071	.81513	36
25	.57952	.71110	1.4063	.81496	35
26	.57976	.71154	1.4054	.81479	34
27	.57999	.71198	1.4045	.81462	33
28	.58023	.71242	1.4037	.81445	32
29	.58047	.71285	1.4028	.81428	31
30	.58070	.71329	1.4019	.81412	30
31	.58094	.71373	1.4011	.81395	29
32	.58118	.71417	1.4002	.81378	28
33	.58141	.71461	1.3994	.81361	27
34	.58165	.71505	1.3985	.81344	26
35	.58189	.71549	1.3976	.81327	25
36	.58212	.71593	1.3968	.81310	24
37	.58236	.71637	1.3959	.81293	23
38	.58260	.71681	1.3951	.81276	22
39	.58283	.71725	1.3942	.81259	21
40	.58307	.71769	1.3934	.81242	20
41	.58330	.71813	1.3925	.81225	19
42	.58354	.71857	1.3916	.81208	18
43	.58378	.71901	1.3908	.81191	17
44	.58401	.71946	1.3899	.81174	16
45	.58425	.71990	1.3891	.81157	15
46	.58449	.72034	1.3882	.81140	14
47	.58472	.72078	1.3874	.81123	13
48	.58496	.72122	1.3865	.81106	12
49	.58519	.72167	1.3857	.81089	11
50	.58543	.72211	1.3848	.81072	10
51	.58567	.72255	1.3840	.81055	9
52	.58590	.72299	1.3831	.81038	8
53	.58614	.72344	1.3823	.81021	7
54	.58637	.72388	1.3814	.81004	6
55	.58661	.72432	1.3806	.80987	5
56	.58684	.72477	1.3798	.80970	4
57	.58708	.72521	1.3789	.80953	3
58	.58731	.72565	1.3781	.80936	2
59	.58755	.72610	1.3772	.80919	1
60	.58779	.72654	1.3764	.80902	0
'	Cos	Cot	Tan	Sin	'

′	Sin	Tan	Cot	Cos	′
0	.58779	.72654	1.3764	.80902	**60**
1	.58802	.72699	1.3755	.80885	59
2	.58826	.72743	1.3747	.80867	58
3	.58849	.72788	1.3739	.80850	57
4	.58873	.72832	1.3730	.80833	56
5	.58896	.72877	1.3722	.80816	**55**
6	.58920	.72921	1.3713	.80799	54
7	.58943	.72966	1.3705	.80782	53
8	.58967	.73010	1.3697	.80765	52
9	.58990	.73055	1.3688	.80748	51
10	.59014	.73100	1.3680	.80730	**50**
11	.59037	.73144	1.3672	.80713	49
12	.59061	.73189	1.3663	.80696	48
13	.59084	.73234	1.3655	.80679	47
14	.59108	.73278	1.3647	.80662	46
15	.59131	.73323	1.3638	.80644	**45**
16	.59154	.73368	1.3630	.80627	44
17	.59178	.73413	1.3622	.80610	43
18	.59201	.73457	1.3613	.80593	42
19	.59225	.73502	1.3605	.80576	41
20	.59248	.73547	1.3597	.80558	**40**
21	.59272	.73592	1.3588	.80541	39
22	.59295	.73637	1.3580	.80524	38
23	.59318	.73681	1.3572	.80507	37
24	.59342	.73726	1.3564	.80489	36
25	.59365	.73771	1.3555	.80472	**35**
26	.59389	.73816	1.3547	.80455	34
27	.59412	.73861	1.3539	.80438	33
28	.59436	.73906	1.3531	.80420	32
29	.59459	.73951	1.3522	.80403	31
30	.59482	.73996	1.3514	.80386	**30**
31	.59506	.74041	1.3506	.80368	29
32	.59529	.74086	1.3498	.80351	28
33	.59552	.74131	1.3490	.80334	27
34	.59576	.74176	1.3481	.80316	26
35	.59599	.74221	1.3473	.80299	**25**
36	.59622	.74267	1.3465	.80282	24
37	.59646	.74312	1.3457	.80264	23
38	.59669	.74357	1.3449	.80247	22
39	.59693	.74402	1.3440	.80230	21
40	.59716	.74447	1.3432	.80212	**20**
41	.59739	.74492	1.3424	.80195	19
42	.59763	.74538	1.3416	.80178	18
43	.59786	.74583	1.3408	.80160	17
44	.59809	.74628	1.3400	.80143	16
45	.59832	.74674	1.3392	.80125	**15**
46	.59856	.74719	1.3384	.80108	14
47	.59879	.74764	1.3375	.80091	13
48	.59902	.74810	1.3367	.80073	12
49	.59926	.74855	1.3359	.80056	11
50	.59949	.74900	1.3351	.80038	**10**
51	.59972	.74946	1.3343	.80021	9
52	.59995	.74991	1.3335	.80003	8
53	.60019	.75037	1.3327	.79986	7
54	.60042	.75082	1.3319	.79968	6
55	.60065	.75128	1.3311	.79951	**5**
56	.60089	.75173	1.3303	.79934	4
57	.60112	.75219	1.3295	.79916	3
58	.60135	.75264	1.3287	.79899	2
59	.60158	.75310	1.3278	.79881	1
60	.60182	.75355	1.3270	.79864	**0**
′	Cos	Cot	Tan	Sin	′

′	Sin	Tan	Cot	Cos	′
0	.60182	.75355	1.3270	.79864	**60**
1	.60205	.75401	1.3262	.79846	59
2	.60228	.75447	1.3254	.79829	58
3	.60251	.75492	1.3246	.79811	57
4	.60274	.75538	1.3238	.79793	56
5	.60298	.75584	1.3230	.79776	**55**
6	.60321	.75629	1.3222	.79758	54
7	.60344	.75675	1.3214	.79741	53
8	.60367	.75721	1.3206	.79723	52
9	.60390	.75767	1.3198	.79706	51
10	.60414	.75812	1.3190	.79688	**50**
11	.60437	.75858	1.3182	.79671	49
12	.60460	.75904	1.3175	.79653	48
13	.60483	.75950	1.3167	.79635	47
14	.60506	.75996	1.3159	.79618	46
15	.60529	.76042	1.3151	.79600	**45**
16	.60553	.76088	1.3143	.79583	44
17	.60576	.76134	1.3135	.79565	43
18	.60599	.76180	1.3127	.79547	42
19	.60622	.76226	1.3119	.79530	41
20	.60645	.76272	1.3111	.79512	**40**
21	.60668	.76318	1.3103	.79494	39
22	.60691	.76364	1.3095	.79477	38
23	.60714	.76410	1.3087	.79459	37
24	.60738	.76456	1.3079	.79441	36
25	.60761	.76502	1.3072	.79424	**35**
26	.60784	.76548	1.3064	.79406	34
27	.60807	.76594	1.3056	.79388	33
28	.60830	.76640	1.3048	.79371	32
29	.60853	.76686	1.3040	.79353	31
30	.60876	.76733	1.3032	.79335	**30**
31	.60899	.76779	1.3024	.79318	29
32	.60922	.76825	1.3017	.79300	28
33	.60945	.76871	1.3009	.79282	27
34	.60968	.76918	1.3001	.79264	26
35	.60991	.76964	1.2993	.79247	**25**
36	.61015	.77010	1.2985	.79229	24
37	.61038	.77057	1.2977	.79211	23
38	.61061	.77103	1.2970	.79193	22
39	.61084	.77149	1.2962	.79176	21
40	.61107	.77196	1.2954	.79158	**20**
41	.61130	.77242	1.2946	.79140	19
42	.61153	.77289	1.2938	.79122	18
43	.61176	.77335	1.2931	.79105	17
44	.61199	.77382	1.2923	.79087	16
45	.61222	.77428	1.2915	.79069	**15**
46	.61245	.77475	1.2907	.79051	14
47	.61268	.77521	1.2900	.79033	13
48	.61291	.77568	1.2892	.79016	12
49	.61314	.77615	1.2884	.78998	11
50	.61337	.77661	1.2876	.78980	**10**
51	.61360	.77708	1.2869	.78962	9
52	.61383	.77754	1.2861	.78944	8
53	.61406	.77801	1.2853	.78926	7
54	.61429	.77848	1.2846	.78908	6
55	.61451	.77895	1.2838	.78891	**5**
56	.61474	.77941	1.2830	.78873	4
57	.61497	.77988	1.2822	.78855	3
58	.61520	.78035	1.2815	.78837	2
59	.61543	.78082	1.2807	.78819	1
60	.61566	.78129	1.2799	.78801	**0**
′	Cos	Cot	Tan	Sin	′

′	Sin	Tan	Cot	Cos	′
0	.61566	.78129	1.2799	.78801	**60**
1	.61589	.78175	1.2792	.78783	59
2	.61612	.78222	1.2784	.78765	58
3	.61635	.78269	1.2776	.78747	57
4	.61658	.78316	1.2769	.78729	56
5	.61681	.78363	1.2761	.78711	**55**
6	.61704	.78410	1.2753	.78694	54
7	.61726	.78457	1.2746	.78676	53
8	.61749	.78504	1.2738	.78658	52
9	.61772	.78551	1.2731	.78640	51
10	.61795	.78598	1.2723	.78622	**50**
11	.61818	.78645	1.2715	.78604	49
12	.61841	.78692	1.2708	.78586	48
13	.61864	.78739	1.2700	.78568	47
14	.61887	.78786	1.2693	.78550	46
15	.61909	.78834	1.2685	.78532	**45**
16	.61932	.78881	1.2677	.78514	44
17	.61955	.78928	1.2670	.78496	43
18	.61978	.78975	1.2662	.78478	42
19	.62001	.79022	1.2655	.78460	41
20	.62024	.79070	1.2647	.78442	**40**
21	.62046	.79117	1.2640	.78424	39
22	.62069	.79164	1.2632	.78405	38
23	.62092	.79212	1.2624	.78387	37
24	.62115	.79259	1.2617	.78369	36
25	.62138	.79306	1.2609	.78351	**35**
26	.62160	.79354	1.2602	.78333	34
27	.62183	.79401	1.2594	.78315	33
28	.62206	.79449	1.2587	.78297	32
29	.62229	.79496	1.2579	.78279	31
30	.62251	.79544	1.2572	.78261	**30**
31	.62274	.79591	1.2564	.78243	29
32	.62297	.79639	1.2557	.78225	28
33	.62320	.79686	1.2549	.78206	27
34	.62342	.79734	1.2542	.78188	26
35	.62365	.79781	1.2534	.78170	**25**
36	.62388	.79829	1.2527	.78152	24
37	.62411	.79877	1.2519	.78134	23
38	.62433	.79924	1.2512	.78116	22
39	.62456	.79972	1.2504	.78098	21
40	.62479	.80020	1.2497	.78079	**20**
41	.62502	.80067	1.2489	.78061	19
42	.62524	.80115	1.2482	.78043	18
43	.62547	.80163	1.2475	.78025	17
44	.62570	.80211	1.2467	.78007	16
45	.62592	.80258	1.2460	.77988	**15**
46	.62615	.80306	1.2452	.77970	14
47	.62638	.80354	1.2445	.77952	13
48	.62660	.80402	1.2437	.77934	12
49	.62683	.80450	1.2430	.77916	11
50	.62706	.80498	1.2423	.77897	**10**
51	.62728	.80546	1.2415	.77879	9
52	.62751	.80594	1.2408	.77861	8
53	.62774	.80642	1.2401	.77843	7
54	.62796	.80690	1.2393	.77824	6
55	.62819	.80738	1.2386	.77806	**5**
56	.62842	.80786	1.2378	.77788	4
57	.62864	.80834	1.2371	.77769	3
58	.62887	.80882	1.2364	.77751	2
59	.62909	.80930	1.2356	.77733	1
60	.62932	.80978	1.2349	.77715	**0**
′	Cos	Cot	Tan	Sin	′

′	Sin	Tan	Cot	Cos	′
0	.62932	.80978	1.2349	.77715	60
1	.62955	.81027	1.2342	.77696	59
2	.62977	.81075	1.2334	.77678	58
3	.63000	.81123	1.2327	.77660	57
4	.63022	.81171	1.2320	.77641	56
5	.63045	.81220	1.2312	.77623	55
6	.63068	.81268	1.2305	.77605	54
7	.63090	.81316	1.2298	.77586	53
8	.63113	.81364	1.2290	.77568	52
9	.63135	.81413	1.2283	.77550	51
10	.63158	.81461	1.2276	.77531	50
11	.63180	.81510	1.2268	.77513	49
12	.63203	.81558	1.2261	.77494	48
13	.63225	.81606	1.2254	.77476	47
14	.63248	.81655	1.2247	.77458	46
15	.63271	.81703	1.2239	.77439	45
16	.63293	.81752	1.2232	.77421	44
17	.63316	.81800	1.2225	.77402	43
18	.63338	.81849	1.2218	.77384	42
19	.63361	.81898	1.2210	.77366	41
20	.63383	.81946	1.2203	.77347	40
21	.63406	.81995	1.2196	.77329	39
22	.63428	.82044	1.2189	.77310	38
23	.63451	.82092	1.2181	.77292	37
24	.63473	.82141	1.2174	.77273	36
25	.63496	.82190	1.2167	.77255	35
26	.63518	.82238	1.2160	.77236	34
27	.63540	.82287	1.2153	.77218	33
28	.63563	.82336	1.2145	.77199	32
29	.63585	.82385	1.2138	.77181	31
30	.63608	.82434	1.2131	.77162	30
31	.63630	.82483	1.2124	.77144	29
32	.63653	.82531	1.2117	.77125	28
33	.63675	.82580	1.2109	.77107	27
34	.63698	.82629	1.2102	.77088	26
35	.63720	.82678	1.2095	.77070	25
36	.63742	.82727	1.2088	.77051	24
37	.63765	.82776	1.2081	.77033	23
38	.63787	.82825	1.2074	.77014	22
39	.63810	.82874	1.2066	.76996	21
40	.63832	.82923	1.2059	.76977	20
41	.63854	.82972	1.2052	.76959	19
42	.63877	.83022	1.2045	.76940	18
43	.63899	.83071	1.2038	.76921	17
44	.63922	.83120	1.2031	.76903	16
45	.63944	.83169	1.2024	.76884	15
46	.63966	.83218	1.2017	.76866	14
47	.63989	.83268	1.2009	.76847	13
48	.64011	.83317	1.2002	.76828	12
49	.64033	.83366	1.1995	.76810	11
50	.64056	.83415	1.1988	.76791	10
51	.64078	.83465	1.1981	.76772	9
52	.64100	.83514	1.1974	.76754	8
53	.64123	.83564	1.1967	.76735	7
54	.64145	.83613	1.1960	.76717	6
55	.64167	.83662	1.1953	.76698	5
56	.64190	.83712	1.1946	.76679	4
57	.64212	.83761	1.1939	.76661	3
58	.64234	.83811	1.1932	.76642	2
59	.64256	.83860	1.1925	.76623	1
60	.64279	.83910	1.1918	.76604	0
′	Cos	Cot	Tan	Sin	′

′	Sin	Tan	Cot	Cos	′
0	.64279	.83910	1.1918	.76604	60
1	.64301	.83960	1.1910	.76586	59
2	.64323	.84009	1.1903	.76567	58
3	.64346	.84059	1.1896	.76548	57
4	.64368	.84108	1.1889	.76530	56
5	.64390	.84158	1.1882	.76511	55
6	.64412	.84208	1.1875	.76492	54
7	.64435	.84258	1.1868	.76473	53
8	.64457	.84307	1.1861	.76455	52
9	.64479	.84357	1.1854	.76436	51
10	.64501	.84407	1.1847	.76417	50
11	.64524	.84457	1.1840	.76398	49
12	.64546	.84506	1.1833	.76380	48
13	.64568	.84556	1.1826	.76361	47
14	.64590	.84606	1.1819	.76342	46
15	.64612	.84656	1.1812	.76323	45
16	.64635	.84706	1.1806	.76304	44
17	.64657	.84756	1.1799	.76286	43
18	.64679	.84806	1.1792	.76267	42
19	.64701	.84856	1.1785	.76248	41
20	.64723	.84906	1.1778	.76229	40
21	.64746	.84956	1.1771	.76210	39
22	.64768	.85006	1.1764	.76192	38
23	.64790	.85057	1.1757	.76173	37
24	.64812	.85107	1.1750	.76154	36
25	.64834	.85157	1.1743	.76135	35
26	.64856	.85207	1.1736	.76116	34
27	.64878	.85257	1.1729	.76097	33
28	.64901	.85308	1.1722	.76078	32
29	.64923	.85358	1.1715	.76059	31
30	.64945	.85408	1.1708	.76041	30
31	.64967	.85458	1.1702	.76022	29
32	.64989	.85509	1.1695	.76003	28
33	.65011	.85559	1.1688	.75984	27
34	.65033	.85609	1.1681	.75965	26
35	.65055	.85660	1.1674	.75946	25
36	.65077	.85710	1.1667	.75927	24
37	.65100	.85761	1.1660	.75908	23
38	.65122	.85811	1.1653	.75889	22
39	.65144	.85862	1.1647	.75870	21
40	.65166	.85912	1.1640	.75851	20
41	.65188	.85963	1.1633	.75832	19
42	.65210	.86014	1.1626	.75813	18
43	.65232	.86064	1.1619	.75794	17
44	.65254	.86115	1.1612	.75775	16
45	.65276	.86166	1.1606	.75756	15
46	.65298	.86216	1.1599	.75738	14
47	.65320	.86267	1.1592	.75719	13
48	.65342	.86318	1.1585	.75700	12
49	.65364	.86368	1.1578	.75680	11
50	.65386	.86419	1.1571	.75661	10
51	.65408	.86470	1.1565	.75642	9
52	.65430	.86521	1.1558	.75623	8
53	.65452	.86572	1.1551	.75604	7
54	.65474	.86623	1.1544	.75585	6
55	.65496	.86674	1.1538	.75566	5
56	.65518	.86725	1.1531	.75547	4
57	.65540	.86776	1.1524	.75528	3
58	.65562	.86827	1.1517	.75509	2
59	.65584	.86878	1.1510	.75490	1
60	.65606	.86929	1.1504	.75471	0
′	Cos	Cot	Tan	Sin	′

′	Sin	Tan	Cot	Cos	′
0	.65606	.86929	1.1504	.75471	60
1	.65628	.86980	1.1497	.75452	59
2	.65650	.87031	1.1490	.75433	58
3	.65672	.87082	1.1483	.75414	57
4	.65694	.87133	1.1477	.75395	56
5	.65716	.87184	1.1470	.75375	55
6	.65738	.87236	1.1463	.75356	54
7	.65759	.87287	1.1456	.75337	53
8	.65781	.87338	1.1450	.75318	52
9	.65803	.87389	1.1443	.75299	51
10	.65825	.87441	1.1436	.75280	50
11	.65847	.87492	1.1430	.75261	49
12	.65869	.87543	1.1423	.75241	48
13	.65891	.87595	1.1416	.75222	47
14	.65913	.87646	1.1410	.75203	46
15	.65935	.87698	1.1403	.75184	45
16	.65956	.87749	1.1396	.75165	44
17	.65978	.87801	1.1389	.75146	43
18	.66000	.87852	1.1383	.75126	42
19	.66022	.87904	1.1376	.75107	41
20	.66044	.87955	1.1369	.75088	40
21	.66066	.88007	1.1363	.75069	39
22	.66088	.88059	1.1356	.75050	38
23	.66109	.88110	1.1349	.75030	37
24	.66131	.88162	1.1343	.75011	36
25	.66153	.88214	1.1336	.74992	35
26	.66175	.88265	1.1329	.74973	34
27	.66197	.88317	1.1323	.74953	33
28	.66218	.88369	1.1316	.74934	32
29	.66240	.88421	1.1310	.74915	31
30	.66262	.88473	1.1303	.74896	30
31	.66284	.88524	1.1296	.74876	29
32	.66306	.88576	1.1290	.74857	28
33	.66327	.88628	1.1283	.74838	27
34	.66349	.88680	1.1276	.74818	26
35	.66371	.88732	1.1270	.74799	25
36	.66393	.88784	1.1263	.74780	24
37	.66414	.88836	1.1257	.74760	23
38	.66436	.88888	1.1250	.74741	22
39	.66458	.88940	1.1243	.74722	21
40	.66480	.88992	1.1237	.74703	20
41	.66501	.89045	1.1230	.74683	19
42	.66523	.89097	1.1224	.74664	18
43	.66545	.89149	1.1217	.74644	17
44	.66566	.89201	1.1211	.74625	16
45	.66588	.89253	1.1204	.74606	15
46	.66610	.89306	1.1197	.74586	14
47	.66632	.89358	1.1191	.74567	13
48	.66653	.89410	1.1184	.74548	12
49	.66675	.89463	1.1178	.74528	11
50	.66697	.89515	1.1171	.74509	10
51	.66718	.89567	1.1165	.74489	9
52	.66740	.89620	1.1158	.74470	8
53	.66762	.89672	1.1152	.74451	7
54	.66783	.89725	1.1145	.74431	6
55	.66805	.89777	1.1139	.74412	5
56	.66827	.89830	1.1132	.74392	4
57	.66848	.89883	1.1126	.74373	3
58	.66870	.89935	1.1119	.74353	2
59	.66891	.89988	1.1113	.74334	1
60	.66913	.90040	1.1106	.74314	0
′	Cos	Cot	Tan	Sin	′

′	Sin	Tan	Cot	Cos	′
0	.66913	.90040	1.1106	.74314	60
1	.66935	.90093	1.1100	.74295	59
2	.66956	.90146	1.1093	.74276	58
3	.66978	.90199	1.1087	.74256	57
4	.66999	.90251	1.1080	.74237	56
5	.67021	.90304	1.1074	.74217	55
6	.67043	.90357	1.1067	.74198	54
7	.67064	.90410	1.1061	.74178	53
8	.67086	.90463	1.1054	.74159	52
9	.67107	.90516	1.1048	.74139	51
10	.67129	.90569	1.1041	.74120	50
11	.67151	.90621	1.1035	.74100	49
12	.67172	.90674	1.1028	.74080	48
13	.67194	.90727	1.1022	.74061	47
14	.67215	.90781	1.1016	.74041	46
15	.67237	.90834	1.1009	.74022	45
16	.67258	.90887	1.1003	.74002	44
17	.67280	.90940	1.0996	.73983	43
18	.67301	.90993	1.0990	.73963	42
19	.67323	.91046	1.0983	.73944	41
20	.67344	.91099	1.0977	.73924	40
21	.67366	.91153	1.0971	.73904	39
22	.67387	.91206	1.0964	.73885	38
23	.67409	.91259	1.0958	.73865	37
24	.67430	.91313	1.0951	.73846	36
25	.67452	.91366	1.0945	.73826	35
26	.67473	.91419	1.0939	.73806	34
27	.67495	.91473	1.0932	.73787	33
28	.67516	.91526	1.0926	.73767	32
29	.67538	.91580	1.0919	.73747	31
30	.67559	.91633	1.0913	.73728	30
31	.67580	.91687	1.0907	.73708	29
32	.67602	.91740	1.0900	.73688	28
33	.67623	.91794	1.0894	.73669	27
34	.67645	.91847	1.0888	.73649	26
35	.67666	.91901	1.0881	.73629	25
36	.67688	.91955	1.0875	.73610	24
37	.67709	.92008	1.0869	.73590	23
38	.67730	.92062	1.0862	.73570	22
39	.67752	.92116	1.0856	.73551	21
40	.67773	.92170	1.0850	.73531	20
41	.67795	.92224	1.0843	.73511	19
42	.67816	.92277	1.0837	.73491	18
43	.67837	.92331	1.0831	.73472	17
44	.67859	.92385	1.0824	.73452	16
45	.67880	.92439	1.0818	.73432	15
46	.67901	.92493	1.0812	.73413	14
47	.67923	.92547	1.0805	.73393	13
48	.67944	.92601	1.0799	.73373	12
49	.67965	.92655	1.0793	.73353	11
50	.67987	.92709	1.0786	.73333	10
51	.68008	.92763	1.0780	.73314	9
52	.68029	.92817	1.0774	.73294	8
53	.68051	.92872	1.0768	.73274	7
54	.68072	.92926	1.0761	.73254	6
55	.68093	.92980	1.0755	.73234	5
56	.68115	.93034	1.0749	.73215	4
57	.68136	.93088	1.0742	.73195	3
58	.68157	.93143	1.0736	.73175	2
59	.68179	.93197	1.0730	.73155	1
60	.68200	.93252	1.0724	.73135	0
′	Cos	Cot	Tan	Sin	′

′	Sin	Tan	Cot	Cos	′
0	.68200	.93252	1.0724	.73135	60
1	.68221	.93306	1.0717	.73116	59
2	.68242	.93360	1.0711	.73096	58
3	.68264	.93415	1.0705	.73076	57
4	.68285	.93469	1.0699	.73056	56
5	.68306	.93524	1.0692	.73036	55
6	.68327	.93578	1.0686	.73016	54
7	.68349	.93633	1.0680	.72996	53
8	.68370	.93688	1.0674	.72976	52
9	.68391	.93742	1.0668	.72957	51
10	.68412	.93797	1.0661	.72937	50
11	.68434	.93852	1.0655	.72917	49
12	.68455	.93906	1.0649	.72897	48
13	.68476	.93961	1.0643	.72877	47
14	.68497	.94016	1.0637	.72857	46
15	.68518	.94071	1.0630	.72837	45
16	.68539	.94125	1.0624	.72817	44
17	.68561	.94180	1.0618	.72797	43
18	.68582	.94235	1.0612	.72777	42
19	.68603	.94290	1.0606	.72757	41
20	.68624	.94345	1.0599	.72737	40
21	.68645	.94400	1.0593	.72717	39
22	.68666	.94455	1.0587	.72697	38
23	.68688	.94510	1.0581	.72677	37
24	.68709	.94565	1.0575	.72657	36
25	.68730	.94620	1.0569	.72637	35
26	.68751	.94676	1.0562	.72617	34
27	.68772	.94731	1.0556	.72597	33
28	.68793	.94786	1.0550	.72577	32
29	.68814	.94841	1.0544	.72557	31
30	.68835	.94896	1.0538	.72537	30
31	.68857	.94952	1.0532	.72517	29
32	.68878	.95007	1.0526	.72497	28
33	.68899	.95062	1.0519	.72477	27
34	.68920	.95118	1.0513	.72457	26
35	.68941	.95173	1.0507	.72437	25
36	.68962	.95229	1.0501	.72417	24
37	.68983	.95284	1.0495	.72397	23
38	.69004	.95340	1.0489	.72377	22
39	.69025	.95395	1.0483	.72357	21
40	.69046	.95451	1.0477	.72337	20
41	.69067	.95506	1.0470	.72317	19
42	.69088	.95562	1.0464	.72297	18
43	.69109	.95618	1.0458	.72277	17
44	.69130	.95673	1.0452	.72257	16
45	.69151	.95729	1.0446	.72236	15
46	.69172	.95785	1.0440	.72216	14
47	.69193	.95841	1.0434	.72196	13
48	.69214	.95897	1.0428	.72176	12
49	.69235	.95952	1.0422	.72156	11
50	.69256	.96008	1.0416	.72136	10
51	.69277	.96064	1.0410	.72116	9
52	.69298	.96120	1.0404	.72095	8
53	.69319	.96176	1.0398	.72075	7
54	.69340	.96232	1.0392	.72055	6
55	.69361	.96288	1.0385	.72035	5
56	.69382	.96344	1.0379	.72015	4
57	.69403	.96400	1.0373	.71995	3
58	.69424	.96457	1.0367	.71974	2
59	.69445	.96513	1.0361	.71954	1
60	.69466	.96569	1.0355	.71934	0
′	Cos	Cot	Tan	Sin	′

′	Sin	Tan	Cot	Cos	′
0	.69466	.96569	1.0355	.71934	60
1	.69487	.96625	1.0349	.71914	59
2	.69508	.96681	1.0343	.71894	58
3	.69529	.96738	1.0337	.71873	57
4	.69549	.96794	1.0331	.71853	56
5	.69570	.96850	1.0325	.71833	55
6	.69591	.96907	1.0319	.71813	54
7	.69612	.96963	1.0313	.71792	53
8	.69633	.97020	1.0307	.71772	52
9	.69654	.97076	1.0301	.71752	51
10	.69675	.97133	1.0295	.71732	50
11	.69696	.97189	1.0289	.71711	49
12	.69717	.97246	1.0283	.71691	48
13	.69737	.97302	1.0277	.71671	47
14	.69758	.97359	1.0271	.71650	46
15	.69779	.97416	1.0265	.71630	45
16	.69800	.97472	1.0259	.71610	44
17	.69821	.97529	1.0253	.71590	43
18	.69842	.97586	1.0247	.71569	42
19	.69862	.97643	1.0241	.71549	41
20	.69883	.97700	1.0235	.71529	40
21	.69904	.97756	1.0230	.71508	39
22	.69925	.97813	1.0224	.71488	38
23	.69946	.97870	1.0218	.71468	37
24	.69966	.97927	1.0212	.71447	36
25	.69987	.97984	1.0206	.71427	35
26	.70008	.98041	1.0200	.71407	34
27	.70029	.98098	1.0194	.71386	33
28	.70049	.98155	1.0188	.71366	32
29	.70070	.98213	1.0182	.71345	31
30	.70091	.98270	1.0176	.71325	30
31	.70112	.98327	1.0170	.71305	29
32	.70132	.98384	1.0164	.71284	28
33	.70153	.98441	1.0158	.71264	27
34	.70174	.98499	1.0152	.71243	26
35	.70195	.98556	1.0147	.71223	25
36	.70215	.98613	1.0141	.71203	24
37	.70236	.98671	1.0135	.71182	23
38	.70257	.98728	1.0129	.71162	22
39	.70277	.98786	1.0123	.71141	21
40	.70298	.98843	1.0117	.71121	20
41	.70319	.98901	1.0111	.71100	19
42	.70339	.98958	1.0105	.71080	18
43	.70360	.99016	1.0099	.71059	17
44	.70381	.99073	1.0094	.71039	16
45	.70401	.99131	1.0088	.71019	15
46	.70422	.99189	1.0082	.70998	14
47	.70443	.99247	1.0076	.70978	13
48	.70463	.99304	1.0070	.70957	12
49	.70484	.99362	1.0064	.70937	11
50	.70505	.99420	1.0058	.70916	10
51	.70525	.99478	1.0052	.70896	9
52	.70546	.99536	1.0047	.70875	8
53	.70567	.99594	1.0041	.70855	7
54	.70587	.99652	1.0035	.70834	6
55	.70608	.99710	1.0029	.70813	5
56	.70628	.99768	1.0023	.70793	4
57	.70649	.99826	1.0017	.70772	3
58	.70670	.99884	1.0012	.70752	2
59	.70690	.99942	1.0006	.70731	1
60	.70711	1.0000	1.0000	.70711	0
′	Cos	Cot	Tan	Sin	′

APPENDIX R

Logarithmic Units

The Decibel

Two powers P_1 and P_2 differ by D decibels where

$$D = 10 \log_{10}\left(\frac{P_1}{P_2}\right)$$

A positive number of decibels indicates that P_1 is greater than P_2; a negative number indicates that P_1 is less than P_2. For example, 50W is greater than 10W by $D = 10 \log_{10}(50/10) = 7$ decibels (dB); whereas, 10W is less than 50W by $D = 10 \log_{10}(10/50) = -7$dB. The decibel is one-tenth of a larger unit, the bel, defined as $\log_{10}(P_1/P_2)$; in practice, the bel is inconveniently large and the decibel is the preferred unit.

Although the decibel is based on power ratios, it can also be used to express voltage and current ratios. Let the power P_1 be developed in a resistor R_1 across which the voltage is V_1 and the current through which is I_1; let the corresponding quantities for P_2 be R_2, V_2, and I_2. Then,

$$D = 10 \log_{10}\left(\frac{V_1{}^2/R_1}{V_2{}^2/R_2}\right)$$

$$= 20 \log_{10}\left(\frac{V_1}{V_2}\right) + 10 \log_{10}\left(\frac{R_2}{R_1}\right)$$

By similar reasoning, it is easily shown that D can be expressed as

$$D = 20 \log_{10}\left(\frac{I_1}{I_2}\right) + 10 \log_{10}\left(\frac{R_1}{R_2}\right)$$

Therefore, it can be seen that a knowledge of the resistance values is needed, in addition to voltage or current, in order to determine the decibel value. Because of the widespread use of the decibel, its meaning has been extended to cover voltage and current ratios. By definition, two voltages, V_1 and V_2, differ by D_v decibels where

$$D_v = 20 \log_{10}\left(\frac{V_1}{V_2}\right)$$

Similarly, two currents, I_1 and I_2, differ by D_i decibels where

$$D_i = 20 \log_{10}\left(\frac{I_1}{I_2}\right)$$

As already shown, D_i and D_v can only be related to D when the ratio R_1/R_2 is known.

It is meaningless to state absolute values of voltage, current, or power, in decibels. The decibel represents a ratio, and therefore a reference level must also be stated if absolute values are required. For example, the statement "a power of 20dB" is meaningless, but "20dB relative to 1W" means 100W. Very often the reference level is understood implicitly; a common example of this being where noise levels are quoted in decibels, the reference level being the threshold of hearing. Another common example in telecommunications engineering is the selectivity curve of a tuned circuit, where the resonance value is used as reference.

Sometimes the reference level is indicated in abbreviated form, the best example of this being the dBm, meaning "decibels relative to 1mW." Other examples are dBμV and dBV, these being decibels relative to $1\mu V$ and 1V, respectively.

Simplified Decibel Calculations. As we have seen decibels express nothing more than a power ratio. Calculations are based on the familiar powers-of-ten method. In other words, for integral powers of ten the dB value corresponds to 10 times the exponent. For this conversion we take the following steps:

1. List power gain ratio, say   100 to 1
2. Rewrite exponentially 10^2 to 1
3. Multiply exponent by 10dB 2×10dB
4. Rewrite in final form   20dB

An easier way is to count the number of zeros or places to the right of the first digit, write this number down, and multiply it by 10. This procedure is summarized in the ratio conversion table below.

Ratio	Decibels
1:1	0
10:1	10
100:1	20
1,000:1	30
10,000:1	40
100,000:1	50
1,000,000:1	60
10,000,000:1	70
100,000,000:1	80

Extending this procedure, our system must be able to express intervening values in 1dB steps. For this no charts, books, tables, or slide rule are necessary. Only three ratios are used: 1.25, 1.6, and 2.0. These correspond respectively to 1, 2, and 3dB.

813

Next, to use these less-than-10dB figures, we must remember that multiplying numbers is the same as adding exponents or the same as adding decibels. Thus 10 (10dB) times 10,000 gives an answer equal to $10^{1+4} = 10^{+5} = 100,000$, or 50dB. Now using these smaller ratios to multiply 1.25 (1dB) by 1.6 (2dB), we see that the product (2) is equal to 3dB. Likewise 2 (3dB) times 2.5 (4dB) = 7dB. These decibel values between 1 and 10 plus their corresponding ratios are given in the table below.

Ratio	Decibels
1.25:1	1
1.6 :1	2
2.0 :1	3
2.5 :1	4
3.2 :1	5
4.0 :1	6
5.0 :1	7
6.4 :1	8
8.0 :1	9
10.0 :1	10

For applications that require greater accuracy—that is, for tenth-decibel values—we start with the first of the basic numbers: 1.25, and derive a table in .1dB steps, by setting 1.25 at the top of the list and subtracting from each succeeding value the amount in hundredths, as indicated by the following sequence.

$$1.0dB = 1.25 - \quad 0 = 1.25:1$$
$$0.9dB = 1.25 - .02 = 1.23:1$$
$$0.8dB = 1.23 - .03 = 1.20:1$$
$$0.7dB = 1.20 - .03 = 1.17:1$$
$$0.6dB = 1.17 - .02 = 1.15:1$$
$$0.5dB = 1.15 - .03 = 1.12:1$$

A full list of tenths of dB's are as follows:

Ratio	Decibels
1.25:1	1.0
1.23:1	0.9
1.20:1	0.8
1.17:1	0.7
1.15:1	0.6
1.12:1	0.5
1.09:1	0.4
1.07:1	0.3
1.04:1	0.2
1.01:1	0.1

Example 1: A 400W amplifier with 1W input has a power ratio of 400:1.

1. Breakdown: $400 = 100 \times 4$
                                       $= 10^2 \times 4$
2. Converting integral powers: $100 = 20dB$
3. Converting digital factors: $4 = 6dB$
4. Adding factor decibels: $= 26dB$
                                      Total Gain

Example 2: Find the power ratio that corresponds to 28.4 dB.

Factor and convert:	20dB =	100 to 1 ratio
Convert digits:	8dB =	6.4 to 1 ratio
Convert decimals:	.4dB =	1.09 to 1 ratio
	(add)	(multiply)
	28.4dB =	100 × 6.4
		× 1.09 =
		697.6 to 1 ratio

The Neper

The neper is a logarithmic unit originally introduced to express the attenuation of current along a transmission line, using natural (or naperian, hence the name "neper") logarithms rather than common logarithms. Two currents, I_1 and I_2, differ by N nepers where

$$N = \ln \frac{I_1}{I_2} = \log_e \frac{I_1}{I_2}$$

The relationship between decibels and Nepers is easily established.

$$D_i = 20 \log_{10} \frac{I_1}{I_2}$$

and on changing the logarithmic base

$$D_i = (20 \log_{10} e) \ \ln \frac{I_1}{I_2}$$

$$= 8.686N$$

Example: Express (a) 3 nepers in decibels; (b) 3 decibels in nepers

(a) $D = 8.686 \times 3$
    $= 26.06$dB

(b) $N = \dfrac{3}{8.686}$
    $= 0.345$N

Logarithmics Scales

By graduating graph axes in lengths proportional to the logarithms of numbers rather than to the numbers themselves, a much wider range of values can be accommodated on a given scale. Scales based on common logarithms are usually employed. Figure R-1 illustrates graph paper that has one axis graduated logarithmically and the other, linearly. This is known as *semilog* graph paper, and because the logarithmic scale pattern repeats itself four times, it is referred to as *four-cycle*. The logarithmic scale of Fig. R-1 can accommodate values ranging over four orders of magnitude (e.g., 1 to 10^4, 10 to 10^5, 0.01 to 10^2, etc.). It will be apparent that a zero origin cannot be shown on a logarithmic scale, and care must be taken when interpreting graphs plotted on logarithmic scales. Also, the slope of the graph must be carefully interpreted, as it involves the logarithms of numbers. Figure R-1 shows the decibel equations (D, D_i, and D_v) plotted against a ratio scale, which is logarithmic. The slope of the voltage or current ratio line should be 20, and this is obtained from

$$\text{slope} = \frac{D_3 - D_2}{\log_{10}X_2 - \log_{10}X_1} \left(\text{not } \frac{D_3 - D_2}{X_2 - X_1}\right)$$

$$= \frac{9.5 - 6}{0.4771 - 0.3010}$$

$$= 20$$

Similarly, the slope of the power ratio line is

$$\text{slope} = \frac{D_2 - D_1}{\log_{10}X_3 - \log_{10}X_1} \left(\text{not } \frac{D_2 - D_1}{X_3 - X_1}\right)$$

$$= 10$$

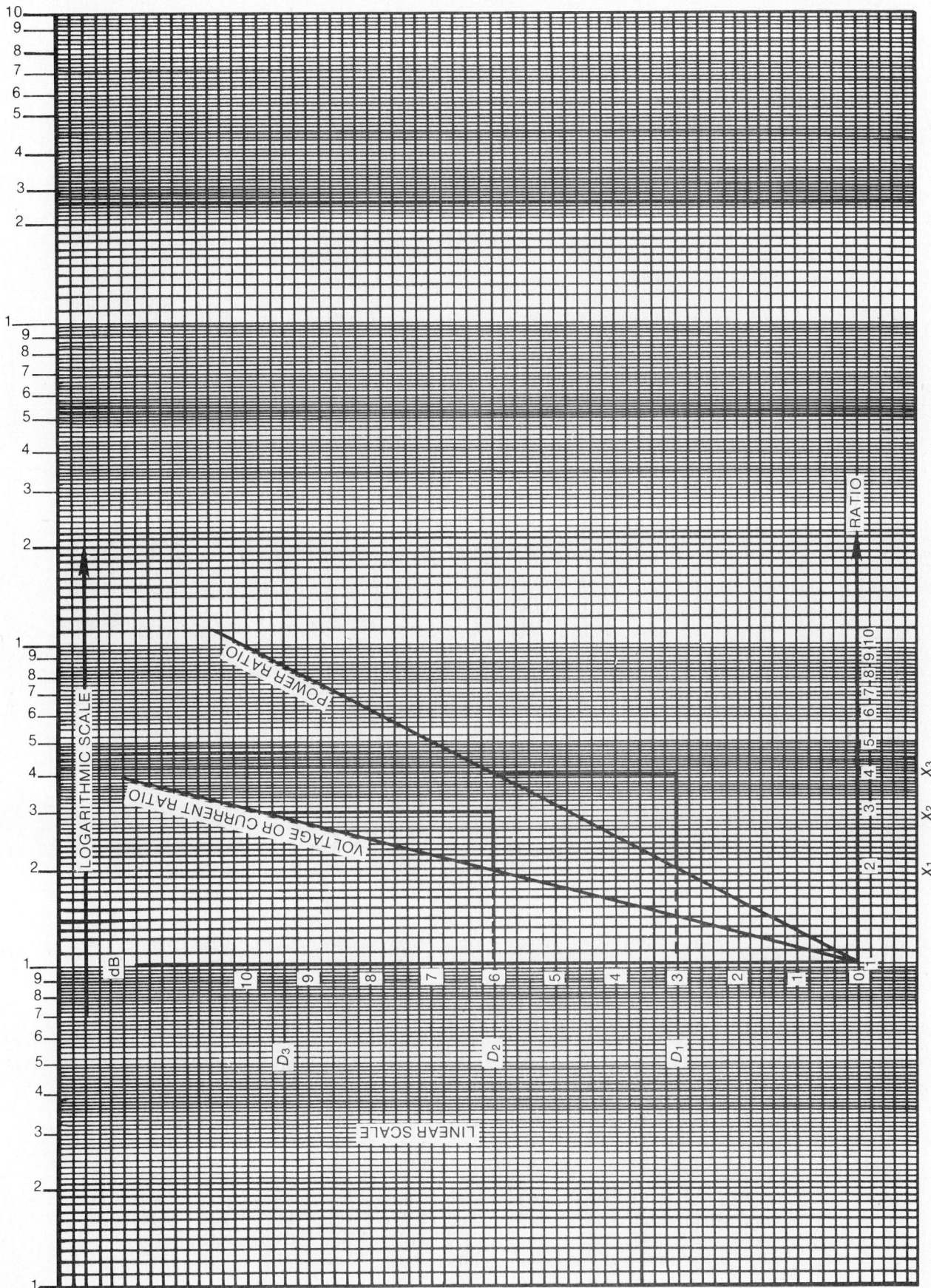

Fig. R-1: Conversion chart—ratios to decibels—using semilog graph.

Table of Logarithms

N	0	1	2	3	4	5	6	7	8	9
0	—	0000	3010	4771	6021	6990	7782	8451	9031	9542
1	0000	0414	0792	1139	1461	1761	2041	2304	2553	2788
2	3010	3222	3424	3617	3802	3979	4150	4314	4472	4624
3	4771	4914	5051	5185	5315	5441	5563	5682	5798	5911
4	6021	6128	6232	6335	6435	6532	6628	6721	6812	6902
5	6990	7076	7160	7243	7324	7404	7482	7559	7634	7709
6	7782	7853	7924	7993	8062	8129	8195	8261	8325	8388
7	8451	8513	8573	8633	8692	8751	8808	8865	8921	8976
8	9031	9085	9138	9191	9243	9294	9345	9395	9445	9494
9	9542	9590	9638	9685	9731	9777	9823	9868	9912	9956
10	0000	0043	0086	0128	0170	0212	0253	0294	0334	0374
11	0414	0453	0492	0531	0569	0607	0645	0682	0719	0755
12	0792	0828	0864	0899	0934	0969	1004	1038	1072	1106
13	1139	1173	1206	1239	1271	1303	1335	1367	1399	1430
14	1461	1492	1523	1553	1584	1614	1644	1673	1703	1732
15	1761	1790	1818	1847	1875	1903	1931	1959	1987	2014
16	2041	2068	2095	2122	2148	2175	2201	2227	2253	2279
17	2304	2330	2355	2380	2405	2430	2455	2480	2504	2529
18	2553	2577	2601	2625	2648	2672	2695	2718	2742	2765
19	2788	2810	2833	2856	2878	2900	2923	2945	2967	2989
20	3010	3032	3054	3075	3096	3118	3139	3160	3181	3201
21	3222	3243	3263	3284	3304	3324	3345	3365	3385	3404
22	3424	3444	3464	3483	3502	3522	3541	3560	3579	3598
23	3617	3636	3655	3674	3692	3711	3729	3747	3766	3784
24	3802	3820	3838	3856	3874	3892	3909	3927	3945	3962
25	3979	3997	4014	4031	4048	4065	4082	4099	4116	4133
26	4150	4166	4183	4200	4216	4232	4249	4265	4281	4298
27	4314	4330	4346	4362	4378	4393	4409	4425	4440	4456
28	4472	4487	4502	4518	4533	4548	4564	4579	4594	4609
29	4624	4639	4654	4669	4683	4698	4713	4728	4742	4757
30	4771	4786	4800	4814	4829	4843	4857	4871	4886	4900
31	4914	4928	4942	4955	4969	4983	4997	5011	5024	5038
32	5051	5065	5079	5092	5105	5119	5132	5145	5159	5172
33	5185	5198	5211	5224	5237	5250	5263	5276	5289	5302
34	5315	5328	5340	5353	5366	5378	5391	5403	5416	5428
35	5441	5453	5465	5478	5490	5502	5514	5527	5539	5551
36	5563	5575	5587	5599	5611	5623	5635	5647	5658	5670
37	5682	5694	5705	5717	5729	5740	5752	5763	5775	5786
38	5798	5809	5821	5832	5843	5855	5866	5877	5888	5899
39	5911	5922	5933	5944	5955	5966	5977	5988	5999	6010
40	6021	6031	6042	6053	6064	6075	6085	6096	6107	6117
41	6128	6138	6149	6160	6170	6180	6191	6201	6212	6222
42	6232	6243	6253	6263	6274	6284	6294	6304	6314	6325
43	6335	6345	6355	6365	6375	6385	6395	6405	6415	6425
44	6435	6444	6454	6464	6474	6484	6493	6503	6513	6522
45	6532	6542	6551	6561	6571	6580	6590	6599	6609	6618
46	6628	6637	6646	6656	6665	6675	6684	6693	6702	6712
47	6721	6730	6739	6749	6758	6767	6776	6785	6794	6803
48	6812	6821	6830	6839	6848	6857	6866	6875	6884	6893
49	6902	6911	6920	6928	6937	6946	6955	6964	6972	6981
50	6990	6998	7007	7016	7024	7033	7042	7050	7059	7067
N	0	1	2	3	4	5	6	7	8	9

Table of Logarithms

N	0	1	2	3	4	5	6	7	8	9
51	7076	7084	7093	7101	7110	7118	7126	7135	7143	7152
52	7160	7168	7177	7185	7193	7202	7210	7218	7226	7235
53	7243	7251	7259	7267	7275	7284	7292	7300	7308	7316
54	7324	7332	7340	7348	7356	7364	7372	7380	7388	7396
55	7404	7412	7419	7427	7435	7443	7451	7459	7466	7474
56	7482	7490	7497	7505	7513	7520	7528	7536	7543	7551
57	7559	7566	7574	7582	7589	7597	7604	7612	7619	7627
58	7634	7642	7649	7657	7664	7672	7679	7686	7694	7701
59	7709	7716	7723	7731	7738	7745	7752	7760	7767	7774
60	7782	7789	7796	7803	7810	7818	7825	7832	7839	7846
61	7853	7860	7868	7875	7882	7889	7896	7903	7910	7917
62	7924	7931	7938	7945	7952	7959	7966	7973	7980	7987
63	7993	8000	8007	8014	8021	8028	8035	8041	8048	8055
64	8062	8069	8075	8082	8089	8096	8102	8109	8116	8122
65	8129	8136	8142	8149	8156	8162	8169	8176	8182	8189
66	8195	8202	8209	8215	8222	8228	8235	8241	8248	8254
67	8261	8267	8274	8280	8287	8293	8299	8306	8312	8319
68	8325	8331	8338	8344	8351	8357	8363	8370	8376	8382
69	8388	8395	8401	8407	8414	8420	8426	8432	8439	8445
70	8451	8457	8463	8470	8476	8482	8488	8494	8500	8506
71	8513	8519	8525	8531	8537	8543	8549	8555	8561	8567
72	8573	8579	8585	8591	8597	8603	8609	8615	8621	8627
73	8633	8639	8645	8651	8657	8663	8669	8675	8681	8686
74	8692	8698	8704	8710	8716	8722	8727	8733	8739	8745
75	8751	8756	8762	8768	8774	8779	8785	8791	8797	8802
76	8808	8814	8820	8825	8831	8837	8842	8848	8854	8859
77	8865	8871	8876	8882	8887	8893	8899	8904	8910	8915
78	8921	8927	8932	8938	8943	8949	8954	8960	8965	8971
79	8976	8982	8987	8993	8998	9004	9009	9015	9020	9025
80	9031	9036	9042	9047	9053	9058	9063	9069	9074	9079
81	9085	9090	9096	9101	9106	9112	9117	9122	9128	9133
82	9138	9143	9149	9154	9159	9165	9170	9175	9180	9186
83	9191	9196	9201	9206	9212	9217	9222	9227	9232	9238
84	9243	9248	9253	9258	9263	9269	9274	9279	9284	9289
85	9294	9299	9304	9309	9315	9320	9325	9330	9335	9340
86	9345	9350	9355	9360	9365	9370	9375	9380	9385	9390
87	9395	9400	9405	9410	9415	9420	9425	9430	9435	9440
88	9445	9450	9455	9460	9465	9469	9474	9479	9484	9489
89	9494	9499	9504	9509	9513	9518	9523	9528	9533	9538
90	9542	9547	9552	9557	9562	9566	9571	9576	9581	9586
91	9590	9595	9600	9605	9609	9614	9619	9624	9628	9633
92	9638	9643	9647	9652	9657	9661	9666	9671	9675	9680
93	9685	9689	9694	9699	9703	9708	9713	9717	9722	9727
94	9731	9736	9741	9745	9750	9754	9759	9763	9768	9773
95	9777	9782	9786	9791	9795	9800	9805	9809	9814	9818
96	9823	9827	9832	9836	9841	9845	9850	9854	9859	9863
97	9868	9872	9877	9881	9886	9890	9894	9899	9903	9908
98	9912	9917	9921	9926	9930	9934	9939	9943	9948	9952
99	9956	9961	9965	9969	9974	9978	9983	9987	9991	9996
100	0000	0004	0009	0013	0017	0022	0026	0030	0035	0039
N	0	1	2	3	4	5	6	7	8	9

APPENDIX S

Nomographs and Charts

Nomographs, or collinear diagrams as they are sometimes called, are great time-savers. Various electronic engineering reference books contain hundreds of charts of this general type for solving all kinds of practical mathematical problems. Usually straight or dotted lines are drawn to show how to use the charts.

Figure S-1 shows an Ohm's Law Nomograph. The unknown is found at the intersecting straightedge between the two given quantities. For example: 20V across 5Ω dissipates 80W, or using left-hand ohm, milliampere and watt scales: 20V across 5,000Ω (4mA) dissipates 80mW.

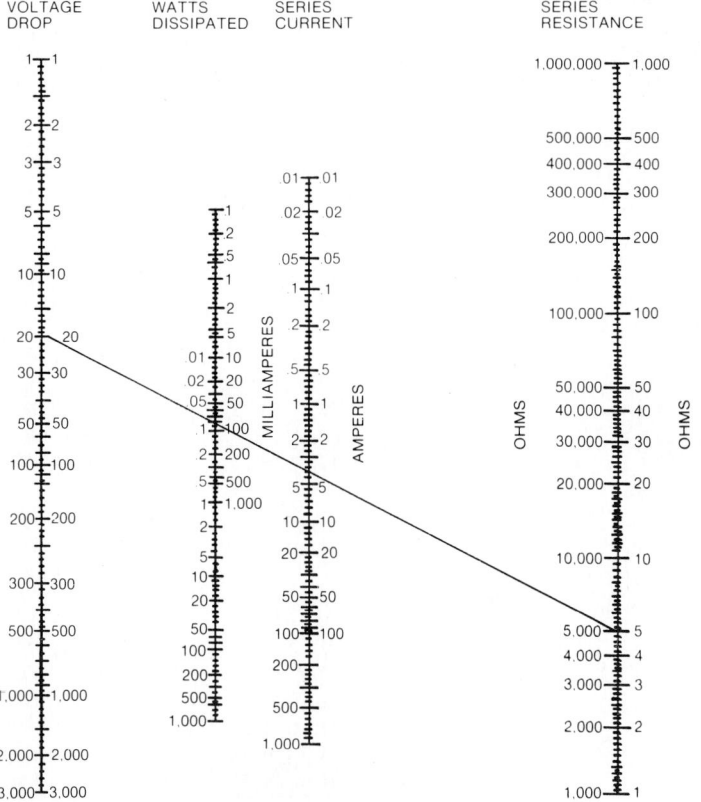

Fig. S-1: Ohm's Law Nomograph.

Two-Resistor Nomograph. Solving for the equivalent value of two parallel resistors:

1. Place straightedge (or draw a line) so that it intersects the two given values on the angularly diverging scales (R_1 and R_2).
2. Find the desired equivalent resistance (Q_T) at intersection on the vertical scale.

Example (Fig. S-2):

1. A line through $R_1 = 6\Omega$ and $R_2 = 8\Omega$
2. intersects at $R_T = 3.42\Omega$.

Range extension can be affected multiplying the value of all scales by 10, 100, 1,000, or as required.

For three resistances in parallel, first find equivalent resistance of two of them and then solve for this one in parallel with the third.

Figure S-3 is a nomograph which can be used for solving resonance problems. The accuracy of this graph is about 95% for most problems. It can be used for solving three types of resonance problems.

It can be used to determine the resonant frequency of a series or parallel resonant circuit when the inductance and capacitance values are known. For example, if a 100mH coil is connected in series with a $0.022\mu F$ capacitor, what is the resonant frequency? A straight line is drawn from $0.022\mu F$ in the left column to 100mH in the right column. The resonant frequency is read from the center column. Figure S-3 shows a resonant frequency of about 3.4kHz. The exact value should be 3.383kHz. So as you can see, the graph is fairly accurate.

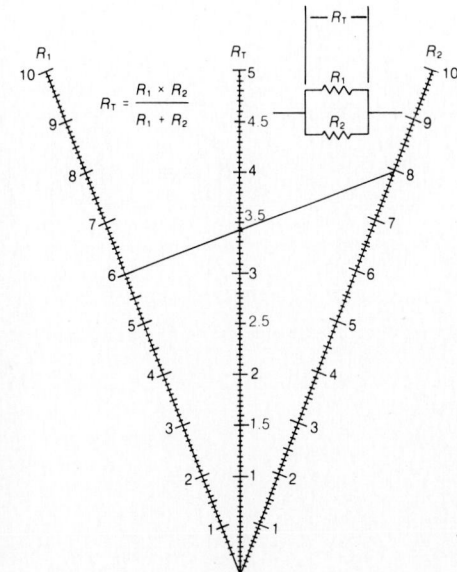

Fig. S-2: **Resistances in parallel.**

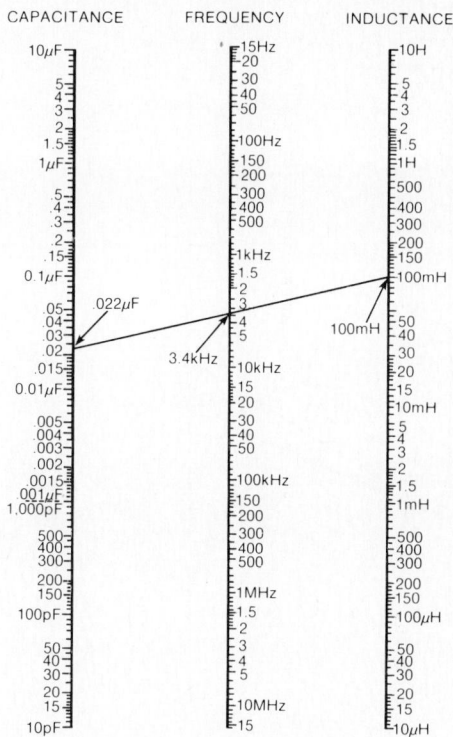

Fig. S-3: **Resonance nomograph.**

The graph can also be used to determine the inductance when the capacitance and resonant frequency are known. For example, what value of inductance will be resonant at 3.4kHz when connected in parallel with a 0.022µF capacitor? A straight line is drawn from 0.022µF on the left column through 3.4kHz in the center column. The inductance is read from the point at which this line crosses the right column.

Finally, the graph can be used to find the capacitance when the resonant frequency and the inductance are known. A straight line is drawn from the two known quantities. The capacitance is read from the point at which this line crosses the left column.

As stated earlier, nomographs are used for many electronic math solutions. It would be impossible to include all that can be drawn. However, Figs. S-4 and S-5 are two filter circuit design nomographs that are typical of many that are available.

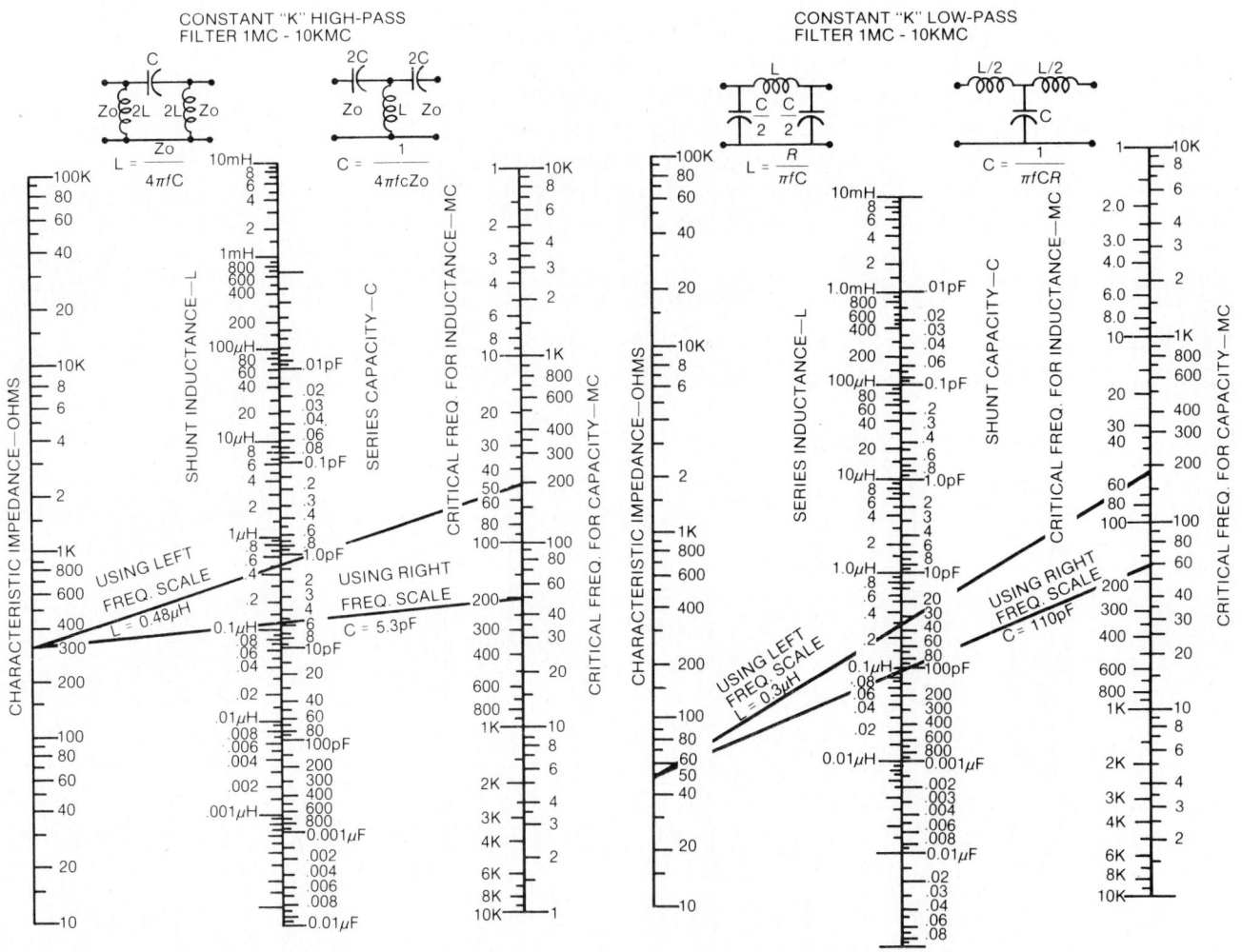

Fig. S-4: Filter circuit design nomograph.   **Fig. S-5: Filter circuit nomograph.**

Charts

There are also charts available that will reduce time spent on electronic math problems. For example, the accompanying reactance charts (Figs. S-6 and S-7) provide a simple direct method for obtaining, at any frequency: (1) the reactance (in ohms) of any size inductance, (2) the reactance (in ohms) of any size capacitor, (3) the resonant frequency determined by any capacitor-inductance circuit combination.

Figure S-6 is used for rough calculations to obtain the approximate range of values needed, and Fig. S-7 is used as a magnified replica (7:1) of a single square area of Fig. S-6 in order to obtain accurate values surrounding the point of interest. Both charts have the following four sets of scale lines superimposed on one another.

One set of horizontal lines	Inductance scale	Extending right to left with number of ohms on left termination.
One set of vertical lines	Frequency scale	Extending from top to bottom with numbers on bottom end of lines
One set of slant lines	Frequency scale	Extending upward to right from lower left to upper right. Numbered on outside border
One set of slant lines	Capacitance scale	Extending downward to right from upper left to lower right. Numbered on inside border

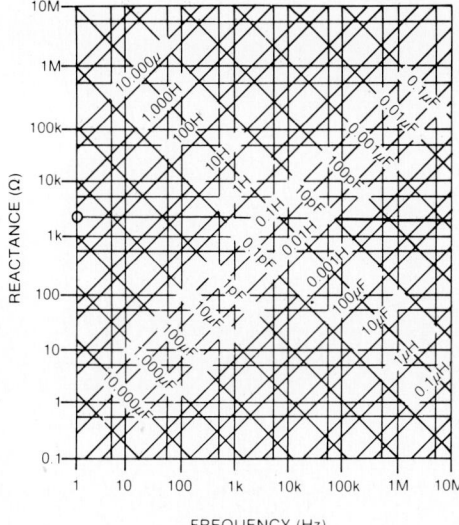

Fig. S-6: Use for approximation of entry on Fig. S-7.

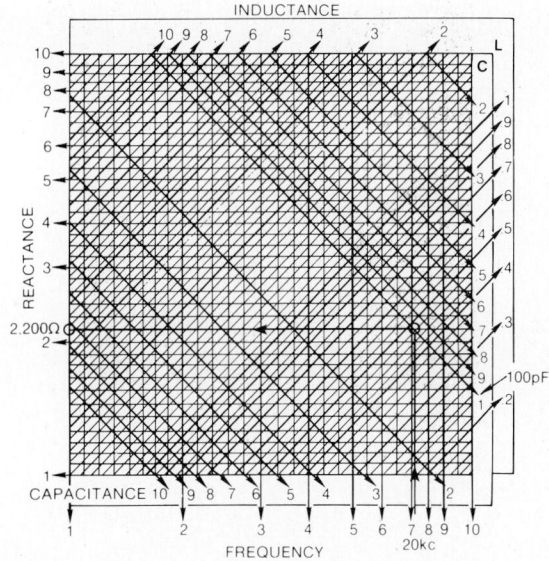

Fig. S-7: Use for obtaining exact value.

Correct use of the chart arrives at the desired electrical quantity by determining the intersection of two scale lines, each representing a given electrical quantity (inductance, capacitance, frequency, or reactance) and projecting this point along another line to the final value scaled on an edge of the chart.

After locating an intersection on the rough scale chart, Fig. S-6, notice in which part or general area of a heavy-line-bounded-square the intersection lies; then go to the same general area of Fig. S-7 (fine scale) since it is an expanded

Example 1: Find reactance in ohms (see point + on rough scale) for $100\mu F$ at 720kHz.

1. Locate intersection of 100pF capacitance line and 720kHz frequency line.

2. Draw horizontal line from intersection over to the left-hand reactance scale where it intersects the reactance scale at $2,200\Omega$—the answer.

3. On fine scale (Fig. S-7) follow same procedure.

Example 2: Find reactance of $500\mu H$ at 720kHz.

1. Locate intersection of $100\mu F$ capacitance line and 720kHz frequency line.

2. From this intersection draw horizontal line as above to $2,200\Omega$ point.

To determine the resonant frequency of a 100pF capacitor and a $500\mu H$ coil, select these two slant lines and locate their intersection at 720kHz.

824

Answers to Check Yourself Exercises

CHAPTER 1

1. gas, liquid, solid
2. proton
3. crystals
4. Inert
5. ionization potential
6. shells
7. protons
8. Compounds, synthesis
9. Semiconductors
10. negative, positive
11. gaseous
12. four
13. Conductors
14. molecule
15. gain, lose
16. protons, neutrons
17. photons, phonons
18. oxygen

CHAPTER 2

Questions

1. False
2. True
3. False
4. False
5. True
6. True
7. True
8. False
9. True
10. True
11. True
12. True
13. False
14. False
15. True

CHAPTER 2 (Continued)

Problems

1. $F = (9 \times 10^9) \dfrac{(2.5 \times 10^{-8})(2.5 \times 10^{-8})}{(0.1)^2}$

$F = (9 \times 10^9) \dfrac{6.25 \times 10^{-16}}{1 \times 10^{-2}}$

$F = (9 \times 10^9) \, 6.25 \times 10^{-14}$

$F = 5.625 \times 10^{-4}$ Newtons

2. $\dfrac{400 \text{ coulombs}}{25 \text{ seconds}} = 16\text{A}$

3. $\dfrac{1}{4\Omega} = 0.25S$

4. $\dfrac{25 \times 10^{18} \text{ electrons}}{6.25 \times 10^{18} \, (1 \text{ coulomb})} = 4 \text{ coulombs}$

5. 5Ω

CHAPTER 3

Questions

1. watts
2. kilo
3. 10^{-3}
4. joule
5. directly, indirectly

Problems

1. $I = V/R$ $P = V^2/R$
   $I = 6/5\text{k}\Omega$ $P = 6^2/5\text{k}\Omega$
   $I = 1.2\text{mA}$ $P = 7.2\text{mW}$
2. $V = IR$ $P = I^2R$
   $V = 25\text{mA} \cdot 400\Omega$ $P = (25\text{mA})^2 \cdot 400\Omega$
   $V = 10\text{V}$ $P = 250\text{mW}$
3. $R = V/I$ $P = VI$
   $R = 20\text{V}/125\text{mA}$ $P = 20\text{V} \cdot 125\text{mA}$
   $R = 160\Omega$ $P = 2.5\text{W}$
4. $I = P/V$ $R = V^2/P$
   $I = 50\text{mW}/10\text{V}$ $R = 10\text{V}^2/50\text{mW}$
   $I = 5\text{mA}$ $R = 2\text{k}\Omega$
5. $V = P/I$ $R = P/I^2$
   $V = 3\mu\text{W}/1\mu\text{W}$ $R = 3\mu\text{W}/(1\mu\text{W})^2$
   $V = 3\text{V}$ $R = 3\text{M}\Omega$
6. $V = PR$ $I = P/R$
   $V = (60\text{mW})120\Omega$ $I = 60\text{mW}/120\Omega$
   $V = 2.68\text{V}$ $I = 22.36\text{mA}$
7. 1.56×10^4; 2.3×10^{-7}; 1×10^6; 3×10^{-2}; 1.56×10^3; 2×10^{-3}.

CHAPTER 3 (Continued)

8. 3k; 50m; 13M; 10n; 400m; 27k

9. $= \dfrac{(1.4 \times 3.9 \times 5.1)\,(10^2 \times 10^3 \times 10^1)}{4.2 \times 10^5}$

$= \dfrac{27.846 \times 10^6}{4.2 \times 10^5}$

$= (27.846 \div 4.2)\,(10^6 \div 10^5)$

$= 6.63 \times 10^1$

$= 66.3$

10. $V = IR;\ P = I^2R$

11. $V = PR;\ I = P/R$

CHAPTER 4

1. True
2. False
3. True
4. False
5. False
6. False
7. True
8. True
9. True
10. True
11. False
12. True
13. False
14. False
15. False
16. True
17. False
18. True
19. True
20. True
21. A
22. B
23. B
24. C
25. B

CHAPTER 5

Questions

1. True
2. False
3. True
4. True
5. False

CHAPTER 5 (Continued)

6. False
7. False
8. False

Problems

1. A. The total resistance required by the circuit (to limit total current to 75mA) is given by Ohm's Law as:

$$R = \frac{V}{I}$$

$$= \frac{100V \text{ (full-scale voltage)}}{75mA}$$

$$= 1.3333M\Omega$$

The meter coil already contains 900Ω, so the multiplier resistance is therefore 1.3333MΩ – 900Ω = 1.332433MΩ.

   B. Using the same procedure, the multiplier resistance is seen to be:

$$R = \frac{V}{I}$$

$$= \frac{30V}{75mA}$$

$$= 400k\Omega$$

so, 400kΩ – 900Ω = 399.1kΩ

   C.

$$R = \frac{V}{I}$$

$$= \frac{5V}{75mA}$$

$$= 66.666k\Omega$$

66.666kΩ – 900Ω = 65.766kΩ

2. Sensitivity $= \dfrac{1}{\text{full-scale current}}$

$$= \frac{1}{30mA}$$

$$= 33.333k\Omega$$

3. A 200mA meter with a specified accuracy of ±3% may have an error of ±3% at full-scale or 6mA anywhere on its scale. So, when the meter indicates 130mA, the minimum and maximum values of current will be:

$$130mA + 6mA = 136mA \text{ (max)}$$

$$130mA - 6mA = 124mA \text{ (min)}$$

CHAPTER 6

Questions

1. current
2. $V = IR$ (Ohm's Law)
   $R = V/I = 100\text{V}/20\text{A} = 5\Omega$
3. $V = IR$
   $V = (3\text{mA})(50\text{k}\Omega)$
   $V = 150\text{V}$
4. $R_\text{T} = R_1 + R_2 + R_3$
   $R_\text{T} = 5\Omega + 20\Omega + 50\Omega = 75\Omega$
5. $V_\text{T} = V_{R_1} + V_{R_2}$
   $V_\text{T} = 20\text{V} + 20\text{V} = 40\text{V}$
6. In a series circuit, $I_\text{T} = I_{R_1} = I_{R_2} = I_{R_3}$, so $I_{R_1} = 5\text{A}$. Using Ohm's Law, $R_1 = V_{R_1}/I_{R_1} = 10\Omega$. Thus, there are three 10Ω resistors in series. $R_\text{T} = R_1 + R_2 + R_3 = 30\Omega$.
7. $V_\text{T} = V_{R_1} + V_{R_2} + V_{R_3}$
   $200\text{V} = 50\text{V} + 78\text{V} + V_{R_3}$
   $V_{R_3} = 200\text{V} - 50\text{V} - 78\text{V}$
   $V_{R_3} = 72\text{V}$
8. $R_\text{T} = V_\text{T}/I_\text{T} = 800/2 = 400\Omega$
   $R_\text{T} = R_1 + R_2 + R_3$
   $400\Omega = 50\Omega + 60\Omega + R_3$
   $R_3 = 400\Omega - 50\Omega - 60\Omega = 290\Omega$
9. 0V ($V = IR$. If $I = 0$, then $V = 0$.)
10. add, resistances, add, voltage drops, same

Problems

1. $R_\text{T} = R_1 + R_2 = 103\Omega$
   $I_\text{T} = V_\text{T}/R_\text{T} = 19.417\text{mA}$
   $V_{R_1} = I_{R_1}R_1 = 1.087\text{V}$
2. $R_\text{T} = 10\text{k}\Omega + 500\Omega = 10.5\text{k}\Omega$
   $I_\text{T} = V_\text{T}/R_\text{T} = 1.429\text{mA}$
   $V_{R_1} = IR_1 = 14.286\text{V}$
   $V_{R_2} = IR_2 = 0.714\text{V}$
   $P_\text{T} = I_\text{T}V_\text{T} = 21.429\text{mW}$
3. $R_\text{T} = 470\Omega + 330\Omega + 220\Omega = 1{,}020\Omega$
   $I_\text{T} = 51\text{V}/1{,}020\Omega = 50\text{mA}$
   $V_{R_1} = IR_1 = 23.5\text{V}$
   $P_{R_1} = IV_{R_1} = 1.175\text{W}$
   $V_{R_2} = IR_2 = 16.5\text{V}$
   $P_{R_2} = IV_{R_2} = 0.825\text{W}$
   $V_{R_3} = IR_3 = 11.0\text{V}$
   $P_{R_3} + IV_{R_3} = 0.550\text{W}$
   $P_\text{T} = P_{R_1} + P_{R_2} + P_{R_3} = 1.175\text{W} + 0.825\text{W} + 0.550\text{W} = 2.55\text{W}$
4. $R_3 = V_{R_3}/I_\text{T} = 2\text{V}/0.1\text{A} = 20\Omega$
   $R_\text{T} = 30\Omega + 10\Omega + 20\Omega = 60\Omega$
   $V = IR = 6\text{V}$

CHAPTER 6 (Continued)

5. $R_T = 500\Omega \times 5 = 2,500\Omega$
$I_T = 50V/2,500\Omega = 20mA$
$V_{R_1} = V_{R_2} = V_{R_3} = V_{R_4} = V_{R_5} = IR_1 = 10V$
$P_{R_1} = P_{R_2} = P_{R_3} = P_{R_4} = P_{R_5} = IV_{R_1} = 200mW$

6. $R_T = R_1 + R_2 + R_3,$ etc. $= 1.2k\Omega + 1.5M\Omega + 250k\Omega + 10\Omega + 500.4k\Omega = 2.25161M\Omega$

7. $R_T = 200\Omega + 600\Omega = 800\Omega$
$I_T = 12V/800\Omega = 15mA$
The voltage across the 600Ω resistor is
$15mA \times 600\Omega = 9V$ $(V = IR)$.

8. $R_T = 500\Omega + 450\Omega + 50\Omega = 1,000\Omega$
The current flowing in the circuit (I_T) equals $V_T/R_T = 50mA$.
$V_{R_1} = I_T R_1 = 25V$
$V_{R_2} = I_T R_2 = 22.5V$
$V_{R_3} = I_T R_3 = 2.5V$
$P_{R_1} = I_T V_{R_1} = 1.25W$
$P_{R_2} = I_T V_{R_2} = 1.125W$
$P_{R_3} = I_T V_{R_3} = 0.125W$

CHAPTER 7

Questions

1. open circuit
2. short
3. voltage
4. current, current
5. equal to zero

Problems

1. $R_{eq} = \dfrac{R}{N}$

$R_{eq} = \dfrac{1k\Omega}{2}$

$R_{eq} = 500\Omega$

2. $I_T = \dfrac{V_T}{R_T}$

$I_T = \dfrac{10V}{500\Omega}$

$I_T = 0.02A$ or $20mA$

3. $I_T + I_1 + I_2 + I_3 = 0$
$(100mA) + (-20mA) + (-60mA) + I_3 = 0$
$(100mA) + (-80mA) + I_3 = 0$
$I_3 = -(100mA) - (-80mA)$
$I_3 = -(100mA) + 80mA$
$I_3 = -20mA$

CHAPTER 7 (Continued)

4. $R_{eq} = \dfrac{1}{\dfrac{1}{R_1} + \dfrac{1}{R_2} + \dfrac{1}{R_3}}$

$R_{eq} = \dfrac{1}{\dfrac{1}{800\Omega} + \dfrac{1}{10k\Omega} + \dfrac{1}{1k\Omega}}$

$R_{eq} = \dfrac{1}{0.00125\Omega + 0.0001\Omega + 0.001\Omega}$

$R_{eq} = \dfrac{1}{0.00235\Omega}$

$R_{eq} = 425.5\Omega$

5. $I_1 = \dfrac{V_T}{R_1}$

$I_1 = \dfrac{5V}{800\Omega}$

$I_1 = 6.25mA$

$I_2 = \dfrac{V_T}{R_2}$

$I_2 = \dfrac{5V}{10k\Omega}$

$I_2 = 500\mu A$

$I_3 = \dfrac{V_T}{R_3}$

$I_3 = \dfrac{5V}{1k\Omega}$

$I_3 = 5mA$

$I_T = I_1 + I_2 + I_3$

$I_T = 6.25mA + 500\mu A + 5mA$

$I_T = 11.75mA$

6. $P_{R_1} = V_T I_1$

$P_{R_1} = 5V \times 6.25mA$

$P_{R_1} = 31.25mW$

$P_{R_2} = V_T I_2$

$P_{R_2} = 5V \times 500\mu A$

$P_{R_2} = 2.5mW$

$P_{R_3} = V_T I_3$

$P_{R_3} = 5V \times 5mA$

$P_{R_3} = 25mW$

$P_T = P_{R_1} + P_{R_2} + P_{R_3}$

$P_T = 31.25mW + 2.5mW + 25mW$

$P_T = 58.75mW$

7. $R_{eq} = \dfrac{R_1 R_2}{R_1 + R_2}$

$R_{eq} = \dfrac{20k\Omega \times 30k\Omega}{20k\Omega + 30k\Omega}$

$$R_{eq} = \frac{600k\Omega}{50k\Omega}$$

$$R_{eq} = 12k\Omega$$

8. $G_1 = \dfrac{1}{R_1}$

$G_1 = \dfrac{1}{700\Omega}$

$G_1 = 1.429mS$

$G_2 = \dfrac{1}{R_2}$

$G_2 = \dfrac{1}{900\Omega}$

$G_2 = 1.111mS$

$G_3 = \dfrac{1}{R_3}$

$G_3 = \dfrac{1}{850\Omega}$

$G_3 = 1.176mS$
$G_T = G_1 + G_2 + G_3$
$G_T = 1.429mS + 1.111mS + 1.176mS$
$G_T = 3.716mS$

$R_T = \dfrac{1}{G_T}$

$R_T = \dfrac{1}{3.716mS}$

$R_T = 269\Omega$

9. $I_1 = \dfrac{V_T}{R_1}$

$I_1 = \dfrac{5V}{700\Omega}$

$I_1 = 7.143mA$

$I_2 = \dfrac{V_T}{R_2}$

$I_2 = \dfrac{5V}{900\Omega}$

$I_2 = 5.556mA$

$I_3 = \dfrac{V_T}{R_3}$

$I_3 = \dfrac{5V}{850\Omega}$

$I_3 = 5.882mA$
$I_T = I_1 + I_2 + I_3$

CHAPTER 7 (Continued)

$I_T = 7.143\text{mA} + 5.556\text{mA} + 5.882\text{mA}$
$I_T = 18.581\text{mA}$
10. $P_T = V_T I_T$
$P_T = 5\text{V} \times 18.581\text{mA}$
$P_T = 92.905\text{mW}$

CHAPTER 8

Questions

1. C
2. A
3. C
4. A
5. B
6. A
7. B
8. B
9. B
10. D
11. $66.666\text{k}\Omega$
12. 4V
13. 8.8mA, 8.8mA
14. 1,100W
15. 68.5Ω

Problems

1. $R_T = 55.33\Omega$
2. $R_T = 10\text{k}\Omega$
$I_T = 3\text{mA}$
$P_T = 90\text{mW}$
$I_2 = 2\text{mA}$
$I_3 = 1\text{mA}$
$V_{R1} = 18\text{V}$
$V_{R2} = 12\text{V}$
$V_{R3} = 12\text{V}$
$P_1 = 54\text{mW}$
$P_2 = 24\text{mW}$
$P_3 = 12\text{mW}$
3. $R_x = 900\Omega$
4. $R_2 = 377.77\Omega$
5. $R_{AB} = 12\Omega$
6. $R_1 = 5\Omega$
$I_{R4} = 1.25\text{A}$
7. $R_B = 500\Omega$
$R_2 = 125\Omega$
$R_1 = 111\Omega$

CHAPTER 8 (Continued)

$P_{R_B} = 50\text{mW}$
$P_{R_2} = 50\text{mW}$
$P_{R_1} = 225\text{mW}$
8. $I_1 = 2.5\text{mA}$
$I_T = 2.5\text{mA}$
$V_a = 62.5\text{V}$
$R_T = 25\text{k}\Omega$
9. $R_T = 10\Omega$
$I_T = 10\text{A}$
$P_T = 1{,}000\text{W}$
10. $R_T = 1\text{k}\Omega$
$I_T = 10\text{mA}$
$V_{R_1} = 6.75\text{V}$
$V_{R_4} = 1.458\text{V}$
$V_{R_6} = 1.789\text{V}$
$P_{R_1} = 67.5\text{mW}$
$P_{R_5} = 3.34\text{mW}$
$P_{R_2} = 4.25\text{mW}$
11. $R_T = 95\Omega$
$I_T = 4\text{A}$
$I_{R_2} = 1\text{A}$
12. $I_T = 6\text{mA}$
13. $R_T = 10\text{k}\Omega$
14. $R_T = 34\Omega$
15. $R_T = 560\Omega$

CHAPTER 9

1. magnet
2. induction
3. Artificial
4. high, low
5. north pole, south pole
6. Weber
7. domain
8. magnetized, unmagnetized
9. poles
10. magnetic line of force
11. magnetic field
12. flux
13. Flux density
14. repel, attract
15. decreases
16. air gap
17. magnetism
18. retentivity
19. magnetic screen or shield
20. penetrate
21. shape
22. Paramagnetic

CHAPTER 9 (Continued)

23. Diamagnetic
24. Ferromagnetic
25. Ferrites
26. stronger
27. horseshoe
28. within

CHAPTER 10

Questions

1. Magnetism
2. amount of flux, number of turns, time rate of cutting
3. flux
4. ampere-turn
5. four, two-pole
6. number of turns, amount of current flow, ratio of coil's length to its width, type of core
7. Hysteresis
8. Degaussing
9. magnetomotive force, reluctance
10. Permeance
11. saturation
12. permeability
13. permeability, retentivity
14. left-hand rule for a conductor
15. Flux density
16. webers
17. teslas
18. Permeability
19. Hard steel
20. pulsating DC

Problems

1. $\text{mmf} = NI$
   $\text{mmf} = 1{,}000 \times 500\text{mA}$
   $\text{mmf} = 500 \text{ ampere-turns}$

2. $\mathcal{R} = \dfrac{\text{mmf}}{\Phi}$

   $\mathcal{R} = \dfrac{(300)(6)}{2 \times 10^{-3}\text{Wb}}$

   $\mathcal{R} = \dfrac{1{,}800 \text{ ampere-turns}}{2 \times 10^{-3}\text{Wb}}$

   $\mathcal{R} = 9 \times 10^{5}\text{A/Wb}$

CHAPTER 10 (Continued)

3. mmf = NI
   mmf = (200)(200mA)
   mmf = 40 ampere-turns

   $$H = \frac{mmf}{1}$$

   $$H = \frac{40A}{10 \times 10^{-2}m}$$

   H = 400A/m

4. $$V = \frac{Nd\Phi}{dt}$$

   $$V = \frac{75 \times 471\mu Wb}{0.1s}$$

   $V = 0.35V$

5. $$H = \frac{mmf}{1}$$

   $$H = \frac{760}{5 \times 10^{-2}m}$$

   H = 15.2kA/m

CHAPTER 11

Questions

1. 16
2. alternation
3. 0.637
4. 1.57
5. 0.707
6. 0.707
7. V_{rms}
8. Square waves, triangular waves (sawtooth also)
9. square wave
10. lag
11. armature
12. counterclockwise
13. A
14. D
15. A
16. C
17. C
18. A
19. B
20. C
21. B
22. A
23. C
24. 100
25. 100

CHAPTER 11 (Continued)

Problems

1. $t = \dfrac{1}{f}$

   A. $t = \dfrac{1}{1,000} = 1\text{ms}$

   B. $t = 1/60 = 16.67\text{ms}$
   C. $t = 1/2\text{M} = .5\mu\text{S}$
   D. $t = 1/15\text{k} = 66.67\mu\text{S}$

2. $V_{\text{avg}} = V_{\text{peak}} \times 0.637$
   $V_{\text{avg}} = 120 \times 0.637$
   $V_{\text{avg}} = 76.44\text{V}$

3. $v = V_{\text{m}} \sin \theta$
   A. $v = 500 \sin 63°$
   $\quad v = 445.5\text{V}$
   B. $v = 500 \sin 90°$
   $\quad v = 500\text{V}$
   C. $v = 500 \sin 178°$
   $\quad v = 17.45\text{V}$
   D. $v = 500 \sin 240°$
   $\quad v = -433\text{V}$

4. $V_{\text{avg}} = V_{\text{rms}} \times 0.899$
   $V_{\text{avg}} = 269.7\text{V}$
   $V_{\text{peak}} = V_{\text{rms}} \times 1.414$
   $V_{\text{peak}} = 424.2\text{V}$

5. $V_{\text{rms}} = V_{\text{peak}} \times 0.707$
   $V_{\text{rms}} = 141.4\text{V}$
   $V_{\text{avg}} = 127.4\text{V}$

6. $f = 1/t$
   A. $f = 1/5 = .2\text{Hz}$
   B. $f = 1/.0002 = 5\text{kHz}$
   C. $f = 1/1\mu\text{S} = 1\text{MHz}$
   D. $f = 1/.05 = 20\text{Hz}$

CHAPTER 12

Questions

1. mutual
2. inductance
3. secondary, primary
4. Dot rotation
5. oppose
6. toroidal
7. overcoupling
8. Hysteresis
9. reflected
10. AC, DC
11. Newtons, force
12. number of turns

CHAPTER 12 (Continued)

13. length
14. Air core
15. RF chokes
16. henry

Problems

1. $L_T = L_1 + L_2 + L_3$
   $L_T = 30m + 40m + 50m$
   $L_T = 120m$

2. $L_T = \dfrac{1}{\dfrac{1}{L_1} + \dfrac{1}{L_2} + \dfrac{1}{L_3}}$

   $L_T = \dfrac{1}{\dfrac{1}{30m} + \dfrac{1}{40m} + \dfrac{1}{50m}}$

   $L_T = \dfrac{1}{33.33 + 25 + 20}$

   $L_T = \dfrac{1}{78.33}$

   $L_T = 12.77mH$

3. $cemf = L\,\dfrac{di}{dt}$

   $= 10\,\dfrac{10 - 7}{90m}$

   $= 10 \times 33.3$
   $cemf = 333.3V$

4. $L = \dfrac{v_L}{\dfrac{di}{dt}}$

   $L = \dfrac{50mV}{\dfrac{200mA}{1}}$

   $L = .25H$

5. $a = \dfrac{N_p}{N_s}$

   $a = 17{:}1$

6. $K = \dfrac{L_m}{\sqrt{L_1 \times L_2}}$

   $K = \dfrac{8.18}{\sqrt{24 \times 9}}$

   $K = \dfrac{8.18}{14.6969}$

   $K = 0.557$

CHAPTER 12 (Continued)

7. $a = \sqrt{\dfrac{Z_p}{Z_s}}$

   $a = \sqrt{\dfrac{4,200}{8}}$

   $a = \sqrt{525}$

   $a = 22.9{:}1$

CHAPTER 13

Questions

1. magnetic field
2. inductor
3. 5
4. reactance
5. frequency, inductance
6. inductance, resistance
7. low, resistor
8. cutoff
9. Inductive susceptance
10. resistance, inductive reactance
11. decrease
12. resistance of the wire
13. applied voltage, circuit current, cosine
14. high
15. A
16. D
17. B

Problems

1. $X_L = 2\pi f L$

   $f = \dfrac{X_L}{2\pi L}$

   $f = \dfrac{2,000}{2\pi(0.5)}$

   $f = 636.6 \text{Hz}$

2. $Z = \sqrt{(R)^2 + (X_L)^2}$
   $Z = \sqrt{(1,500)^2 + (1,000)^2}$
   $Z = \sqrt{3.25 \times 10^6}$
   $Z = 1,803\Omega$

3. $\cos\theta = R/Z$
   $\theta = \cos^{-1} R/Z$

   $\theta = \cos^{-1} \dfrac{1,500}{1,803}$

   $\theta = 33.7°$

4. $5 \times \dfrac{L}{R}$ = full charge

$5 \times \dfrac{500m}{1K}$ = full charge

$2.5ms$ = full charge

5. $\text{Cos}\,\theta = \dfrac{\text{true power}}{\text{apparent power}}$

$\text{Cos}\,30° = \dfrac{\text{true power}}{5}$

$\text{Cos}\,30° \times 5$ = true power
$4.33W$ = true power

6. $X_L = 2\pi f L$
$X_L = 2\pi(4.5M)(250m)$
$X_L = 7.0686M\Omega$

7. $X_L = 2\pi(10K)\,(50m)$
$X_L = 3{,}141.6$
$Z = \sqrt{(500)^2 + (3{,}141.6)^2}$
$Z = 3{,}181\Omega$

8. $\text{Cos}\,\theta = R/Z$
   $\theta = \text{Cos}^{-1} R/Z$

   $\theta = \text{Cos}^{-1}\dfrac{500}{3{,}181}$

   $\theta = 80.96°$

9. $L_T = 30mH + 50mH + 80mH$
   $L_T = 160mH$

   $\tau = \dfrac{160mH}{10K}$

   $\tau = 16\mu s$

10. $X_L = 2\pi f L$
    $X_L = 2\pi(100K)(160m)$
    $X_L = 100.531k\Omega$
    $Z = \sqrt{(10K)^2 + (100.531K)^2}$
    $Z = 101.027k\Omega$

11. $\text{Cos}\,\theta = \dfrac{\text{true power}}{\text{apparent power}}$

    $\text{Cos}\,\theta \times$ apparent power = true power
    $\text{Cos}\,70° \times 20$ = true power
    $6.84W$ = true power

12. $X_L = 2\pi f L$
    $X_L = 2\pi(20K)(98m)$
    $X_L = 12.315\Omega$

    $X_L = 2\pi f L$
    $X_L = 2\pi(5K)(98m)$
    $X_L = 3.079\Omega$

CHAPTER 14

Questions

1. True
2. False
3. True
4. False
5. False
6. False
7. True
8. D
9. B
10. C

Problems

1. $Q = CV$

$$V = \frac{Q}{C} = \frac{1.6 \times 10^{-2}}{8 \times 10^{-6}} = 0.2 \times 10^4 = 2,000\text{V}$$

2. $C = 0.2249\left(\dfrac{KA}{d}\right)$

$\quad\ = 0.2249\left(\dfrac{(1)(9)}{0.001}\right)$

$\quad\ = 0.2249(9,000)$

$\quad\ = 2,024\text{pF} = 0.002024\mu\text{F}$

3. In series,

$$\frac{1}{C_T} = \frac{1}{C_1} + \frac{1}{C_2} + \dots \frac{1}{C_n}$$

$$\frac{1}{24\mu\text{F}} = \frac{1}{40\mu\text{F}} + \frac{1}{\text{X}\mu\text{F}}$$

$$\frac{1}{\text{X}\mu\text{F}} = \frac{1}{24\mu\text{F}} - \frac{1}{40\mu\text{F}}$$

$$\frac{1}{\text{X}\mu\text{F}} = 416,666.66 - 25,000$$

$$\frac{1}{\text{X}\mu\text{F}} = 16,666.66$$

$$\text{X}\mu\text{F} = \frac{1}{16,666.66} = 60\mu\text{F}$$

4. $C = 0.2249\,\dfrac{KA}{d}$

$\quad C = 0.2249\,\dfrac{(7)(300)}{0.25}$

$\quad C = (0.2249)(8,400)$

$\quad C = 1,889.16\text{pF}$

CHAPTER 14 (Continued)

5. $Q = CV$
   $Q = (12\mu F)(500V)$
   $Q = 0.006$ coulombs

6. $C = 0.2249 \dfrac{KA}{d}$

   $C = 0.2249 \dfrac{(2.6)(64.56)}{1.25 \times 10^{-3}}$

   $C = (0.2249)(134,284)$
   $C = 30,200.65pF = 0.003\mu F$

CHAPTER 15

Questions

1. C
2. D
3. E
4. A
5. E
6. resistor
7. capacitor
8. decreases
9. 63.2, 1
10. block, pass
11. leads, 90
12. resistance, capacitance
13. True
14. True
15. False
16. True
17. True

Problems

1. $X_c = \dfrac{1}{2\pi fC}$

   $C = \dfrac{1}{2\pi fX_c}$

   $C = \dfrac{1}{6.28 \times 400 \times 50}$

   $C = 7.96\mu F$

2. A. $X_c = \dfrac{1}{2\pi fC}$

   $= \dfrac{1}{6.28 \times 60 \times 0.002 \times 10^{-6}}$

   $= 1.327M\Omega$

B. $X_c = \dfrac{1}{2\pi fC}$

$\quad = \dfrac{1}{6.28 \times 60 \times 10 \times 10^{-6}}$

$\quad = 265.4\Omega$

C. $X_c = \dfrac{1}{2\pi fC}$

$\quad = \dfrac{1}{6.28 \times 60 \times 270 \times 10^{-12}}$

$\quad = 9.829\text{M}\Omega$

D. $X_c = \dfrac{1}{2\pi fC}$

$\quad = \dfrac{1}{6.28 \times 60 \times 3{,}000 \times 10^{-9}}$

$\quad = 884.6\Omega$

3. $1TC = RC$

$\quad = 750\text{k}\Omega \times 500\text{pF}$

$\quad = 0.375\text{ms}$

It takes 5TC to fully charge a capacitor, so:

$$(5)(0.375\text{ms}) = 1.875\text{ms}$$

4. A. After 1TC the voltage will be 63.2% of its full value (160V):

$$160\text{V} \times 63.2\% = 101.12\text{V}$$

After the second time constant, the capacitor will have charged an additional 63.2% of the remaining voltage (58.88V):

$$58.88 \times 63.2\% = 37.212\text{V}$$

Thus, after 2TC the capacitor has charged to 101.12 + 37.212 = 138.33V.

B. $1TC = RC$

$\quad = 250\text{k}\Omega \times 0.1\mu\text{F}$

$\quad = 0.025\text{s}$

$2TC = (0.025)(2)$

$\quad = 0.050$

$\quad = 50\text{ms}$

5. $1TC = RC$

$\quad 4\text{ms} = 2{,}000 \times C$

$\quad \dfrac{4\text{ms}}{2{,}000} = C$

$\quad 2\mu\text{F} = C$

6. $X_c = \dfrac{1}{2\pi f C}$

$f = \dfrac{1}{2\pi X_c C}$

$f = \dfrac{1}{6.28 \times 80 \times 0.002 \times 10^{-6}}$

$f = 995.22 \text{kHz}$

7. $X_c = \dfrac{1}{2\pi f C}$

$C = \dfrac{1}{2\pi f X_c}$

$C = \dfrac{1}{6.28 \times 20 \times 47 \times 10^3}$

$C = 169.39 \text{nF}$

8. A. $X_c = \dfrac{1}{2\pi f C}$

$= \dfrac{1}{6.28 \times 60 \times 0.01 \times 10^{-6}}$

$= 265.392 \text{k}\Omega$

By Ohm's Law,

$I = \dfrac{V_s}{X_C}$

$= \dfrac{120\text{V}}{265.392 \text{k}\Omega}$

$= 452.16 \mu\text{A}$

B. 0A, because a capacitor blocks DC.

9. $X_c = \dfrac{1}{2\pi f C}$

$= \dfrac{1}{6.28 \times 400 \times 0.1 \times 10^{-6}}$

$= 3.98 \text{k}\Omega$

$V = IX_c$

$= (5\text{mA})(3.98 \text{k}\Omega)$

$= 19.9 \text{V}$

10. A. $X_{c1} = \dfrac{1}{2\pi f C_1}$

$= \dfrac{1}{6.28 \times 60 \times 0.1 \times 10^{-6}}$

$= 26.539 \text{k}\Omega$

$X_{c2} = \dfrac{1}{2\pi f C_2}$

$$= \frac{1}{6.28 \times 60 \times 0.15 \times 10^{-6}}$$

$$= 17.693 \text{k}\Omega$$

$$X_{c3} = \frac{1}{2\pi f C_3}$$

$$= \frac{1}{6.28 \times 60 \times 0.2 \times 10^{-6}}$$

$$= 13.270 \text{k}\Omega$$

$$\frac{1}{X_{CT}} = \frac{1}{X_{c1}} + \frac{1}{X_{c2}} + \frac{1}{X_{c3}} = \frac{1}{26.539} + \frac{1}{17.693} + \frac{1}{13.270} = 5.897 \text{k}\Omega$$

B. $I_1 = \dfrac{V}{X_{c1}} = \dfrac{48}{26.539} = 1.81 \text{mA}$

$I_2 = \dfrac{V}{X_{c2}} = \dfrac{48}{17.693} = 2.71 \text{mA}$

$I_3 = \dfrac{V}{X_{c3}} = \dfrac{48}{13.270} = 3.61 \text{mA}$

C. $I_T = I_1 + I_2 + I_3 = 8.13 \text{mA}$

D. $C_T = \dfrac{1}{2\pi f X_{cT}}$

$C = \dfrac{1}{6.28 \times 60 \times 5.897 \times 10^3}$

$C_T = 0.45 \mu \text{F}$

CHAPTER 16

Questions

1. C
2. H
3. A
4. E
5. F
6. B
7. I
8. D
9. G
10. J
11. False
12. False
13. True
14. False
15. True
16. False
17. True
18. False

CHAPTER 16 (Continued)

Problems

1. $Z_1 = 0 - j100 = 100\Omega \angle -90°$
   $Z_2 = 100\Omega - j1,000\Omega = 1,005\Omega \angle -84°$
   $Z_3 = 1,000\Omega + j500\Omega = 1,118\Omega \angle 26°$

   $I_1 = \dfrac{1}{Z_1} = 10\text{mA} \angle 90° = 0 + j0.01$

   $I_2 = \dfrac{1}{Z_2} = 995\mu\text{A} \angle 84° = 104\mu + j990\mu$

   $I_3 = \dfrac{1}{Z_3} = 894\mu\text{A} \angle -26° = 803\mu - j392\mu$

   $I_T = I_1 + I_2 + I_3 = 907\mu + 10.598\text{m} \Rightarrow 10.64\text{m} \angle 85°$

   $Z = \dfrac{V_T}{I_T} = \dfrac{1}{10.64\text{m} \angle 85°}$

   $Z = 94\Omega \angle -85°$

2. $\begin{array}{l} 60 + j0 \\ 0 + j80 \\ 80 + j0 \\ 0 + j80 \\ 80 + j0 \\ 0 - j80 \\ \underline{60 + 0} \end{array}$

   $280 + j80 = 291.2\Omega \angle 15.95°$
   $I = V/Z = 34.34\text{mA} \angle -15.95°$

CHAPTER 17

Questions

1. widen
2. 0.707, 45°
3. capacitive
4. maximum
5. balanced-type filter
6. loose coupling, critical coupling, tight coupling
7. wave traps
8. minimum
9. Q
10. inductive
11. shunt resistance
12. parallel-tuned circuits
13. bandpass
14. capacitive

CHAPTER 17 (Continued)

Problems

1. $f_r = \dfrac{1}{2\pi \sqrt{LC}}$

$f_r = \dfrac{1}{2\pi \sqrt{(20m)(2\mu)}}$

$f_r = \dfrac{1}{2\pi(.0002)}$

$f_r = \dfrac{1}{0.001256}$

$f_r = 796Hz$

2. $f_r = \dfrac{1}{2\pi \sqrt{LC}}$

$= \dfrac{1}{2\pi \sqrt{(200y)(50p)}}$

$= \dfrac{1}{6.29 \times 10^7}$

$= 1.59MHz$

$Q = \dfrac{X_L}{R}$

$Q = \dfrac{2\pi fL}{R}$

$Q = \dfrac{2\pi \times 1.59M \times 200\mu}{10}$

$Q = \dfrac{1,998}{10}$

$Q \approx 200$

3. $BW = \dfrac{f_r}{Q}$

$= \dfrac{1.59M}{200}$

$= 7.95kHz$

4. $f_{decrease} = \dfrac{1}{2Q} \times f_r$

$= \dfrac{1}{2(10)} \times 15M$

$= .05 \times 15M$

$= 750kHz$

$f_{new} = f_r - f_{decrease}$
$= 15M - 750kHz$
$= 14.25M$

CHAPTER 17 (Continued)

5. $f_r = \dfrac{1}{2\pi \sqrt{LC}}$

$= \dfrac{1}{2\pi \sqrt{(10y)(10p)}}$

$= \dfrac{1}{6.28 \times 10^{-8}}$

$= 15.9\text{MHz}$

CHAPTER 18

1. anode, cathode, envelope
2. voltage regulation
3. Interelectrode capacitance
4. the speed of the electrons in the beam, the number of electrons that strike the screen at a given point
5. getter
6. B
7. C
8. D
9. D
10. True
11. True
12. False
13. True
14. False
15. False

CHAPTER 19

Questions

1. False
2. True
3. True
4. False
5. True
6. False
7. True
8. True
9. True
10. True
11. False
12. False
13. True
14. False
15. False
16. True
17. False

18. False
19. False
20. True
21. silicon, germanium, four
22. reverse, voltage
23. forward, 0.7, 0.3, forward
24. hole, recombination
25. doping, free electrons
26. transition capacitance
27. negative resistance
28. varactors
29. recombine, light
30. zener, regulation

Problems

1. Since the diode behaves as a closed switch (a short circuit),

$$I = \frac{V}{R} = \frac{30V}{100\Omega} = 0.3A$$

2. A. $V_{o(max)} = 30V - 0.7V = 29.3V$

$$I_{max} = \frac{V_{o(max)}}{R} = \frac{29.3V}{100\Omega} = 0.293A$$

B. Convert $50\,V_{rms}$ to a peak value (for maximum voltage)

$$50 \times 1.414 = 70.7V$$
$$V_{o(max)} = 70.7V - 0.7V = 70V$$
$$I_{max} = \frac{V_{o(max)}}{R} = \frac{70V}{220\Omega} = 0.318A$$

3. $V_T = V_R + V_{diode}$
   $10V = V_R + 1.8V$
   $V_R = 10V - 1.8V$
   $\quad = 8.2V$

$$I_R = I_{diode} = \frac{V_R}{R} = \frac{8.2V}{820\Omega} = 10mA$$

CHAPTER 20

Questions

1. center-tapped transformer, matched
2. pi
3. Zero
4. SCR
5. load circuits, network resistances
6. isolation

CHAPTER 20 (Continued)

7. bridge
8. ripple, average, output
9. sampling circuit, error detector, reference voltage, error amplifier, series regulator
10. charges, capacitors
11. average forward
12. surge
13. Line transients
14. stacked
15. silicone grease
16. 80
17. short circuit

Problems

1. $\% \text{ Reg} = \dfrac{V_{\text{no load}} - V_{\text{full load}}}{V_{\text{full load}}} \times 100$

   $\% \text{ Reg} = \dfrac{50 - 45}{45} \times 100$

   $\% \text{ Reg} = \dfrac{5}{45} \times 100$

   $\% \text{ Reg} = 11.1111\%$

2. $R_s = \dfrac{V_{\text{in(min)}} - V_z}{1.1 I_l(\text{max})}$

   $R_s = \dfrac{105\text{V} - 100\text{V}}{1.1(200\text{mA})}$

   $R_s = \dfrac{100}{.22}$

   $R_s = 454.5\Omega$

3. $R_a = \dfrac{V_a}{I_a}$

   $R_a = \dfrac{90\text{V}}{15\text{mA}}$

   $R_a = 6\text{k}\Omega$

   $R_b = \dfrac{V_b}{I_b}$

   $R_b = \dfrac{60\text{V}}{25\text{mA}}$

   $R_b = 2.4\text{k}\Omega$

   $R_c = \dfrac{V_c}{I_c}$

   $R_c = \dfrac{30\text{V}}{30\text{mA}}$

CHAPTER 20 (Continued)

$$R_c = 1k\Omega$$

$$R_d = \frac{V_d}{I_d}$$

$$R_d = \frac{90V}{60mA}$$

$$R_d = 1.5k\Omega$$

4. $\%\ Ripple = \dfrac{rms\ of\ ripple}{V_{out}} \times 100$

V_{rms} of $2V = 1.414V$

$\%\ Ripple = \dfrac{1.414}{40} \times 100$

$\%\ Ripple = 3.535$

CHAPTER 21

1. holding current value
2. V_{DS}, I_D
3. channel-enhancement
4. DC beta
5. base
6. rate effect
7. trigger device
8. maximum ratings
9. third quadrant
10. $20,000\Omega$ per volt
11. direct current, direct voltage
12. conventional junction diode
13. sawtooth generator
14. opto-isolator
15. DC alpha
16. input, output
17. depletion regions
18. BJT
19. common-drain
20. common-source

CHAPTER 22

Questions

1. class A
2. cutoff
3. input voltage, input current
4. center
5. collector, emitter

CHAPTER 22 (Continued)

6. beta, impedance
7. cumulative
8. RC, capacitor
9. flat
10. distortion
11. False
12. True
13. True
14. True
15. False

Problems

1. $V_{RC} = V_{CC} - V_C = 12 - 7 = 5V$

$$I_C = \frac{V_{RC}}{R_C} = \frac{5V}{2.3k\Omega} = 2.174mA$$

$$I_B = \frac{I_C}{\beta} = \frac{2.174mA}{40} = 54.35\mu A$$

$$V_{RB} = V_{CC} - V_{BE} = 12 - 0.7 = 11.3V$$

$$R_B = \frac{V_{RB}}{I_B} = \frac{11.3V}{54.35\mu A} = 207.9k\Omega$$

$$r_j = \frac{25mV}{I_C} = \frac{0.025}{0.002174} = 11.5\Omega$$

$$r_b = \beta r_j = (40)(11.5) = 460\Omega$$

$$Z_{in} = R_B//r_b = 207.9k\Omega//460\Omega = 458.98\Omega$$

$$A_v = \frac{-R_C}{r_j} = \frac{-2.3k\Omega}{11.5} = -200$$

2. $$\frac{V_B}{12k\Omega} = \frac{-18V}{33.3k\Omega + 12k\Omega}$$

$$V_B = 4.77V$$

$$V_E = V_B - V_{BE} = 4.77 - 0.3 = 4.47V$$

$$\frac{V_{RC}}{R_C} = \frac{V_{RE}}{R_E}$$

$$\frac{V_{RC}}{6k\Omega} = \frac{4.47}{5k\Omega + 200\Omega}$$

$$V_{RC} = 5.16V$$

$$V_C = V_{BC} - V_{RC} = 18 - 5.16 = 12.84V$$

$$I_C = \frac{V_{RC}}{R_C} = \frac{5.16V}{6k\Omega} = 0.86mA$$

$$r_j = \frac{25mV}{I_E} = \frac{0.025}{0.00086} = 29\Omega$$

$$r_b = \beta(r_j + R_{E1}) = 70(29 + 200) = 16.035k\Omega$$

$$Z_{in} = R_{B1}//R_{B2}//r_b = 33.3k\Omega//12k\Omega//16.035k\Omega = 5.69k\Omega$$

$$Z_o = 6k\Omega$$

$$R_{c(line)} = R_C//R_L = 6k\Omega//10k\Omega = 3.75k\Omega$$

$$A_v = \frac{R_{c(line)}}{r_j + R_{E1}} = \frac{3.75k\Omega}{29 + 200} = 16.38$$

3. $V_B = V_{CC} \dfrac{R_{B2}}{R_{B1} + R_{B2}} = (12) \dfrac{2.5k\Omega}{1.2k\Omega + 2.5k\Omega} = 8.12V$

$$V_E = V_B - V_{BE} = 8.12V - 0.7 = 7.42V$$

$$I_E = \frac{V_E}{R_E} = \frac{7.42}{54} = 137.4mA$$

$$r_j = \frac{25mV}{I_E} = \frac{0.025}{0.1374} = 0.18\Omega$$

$$A_v = \frac{R_E//R_L}{r_j + R_E//R_L} = \frac{54//32}{0.18 + 54//32} = 0.99$$

$$Z_{in} = R_{B1}//R_{B2}//r_b$$

$$r_b = \beta(r_j + R_E//R_L) = 80(0.18 + 20.09) = 1.622k\Omega$$

$$Z_{in} = 1.2k\Omega//2.5k\Omega//1.622k\Omega = 540.66\Omega$$

$$Z_o = R_E//\left(r_j + \frac{R_{B1}//R_{B2}//Z_s}{\beta}\right)$$

$$= 54//\left(0.18 + \frac{1.2k\Omega//2.5k\Omega//400\Omega}{80}\right) = 3.53\Omega$$

CHAPTER 23

Questions

1. 20, 20k
2. decibel
3. AC bias
4. dynetic
5. dynamic
6. enclosures
7. magnetic, crystal
8. Dolby noise reduction
9. sensitivity
10. transducer
11. carbon
12. ribbon
13. Antiskating
14. PM dynamic
15. Chromium dioxide
16. vinyl, 45, 33-1/3
17. steel, sapphire, ruby, diamond
18. supercardioid, hypercardioid
19. mechanical, electrical
20. two-way
21. Variable reluctance, dynamic, dynetic
22. Sympathetic

CHAPTER 23 (Continued)

23. Rarefaction
24. acoustics
25. 14.7
26. B
27. B
28. A
29. D
30. A

CHAPTER 24

Questions

1. D
2. A
3. B
4. C
5. D
6. C
7. B
8. B
9. D
10. C
11. B
12. D
13. D
14. C
15. D
16. C
17. B
18. B

Problems

1. $\theta_T = \dfrac{\Delta T}{P} = \dfrac{200°C - 50°C}{10W} = 15°C/W$

   $\theta_{SA} = \theta - \theta_{JC} - \theta_{CS}$

   $= 15° - 1.5° - 0°$

   $= 13.5°C/W$

2. $dB = 20 \log \dfrac{80V}{20V}$

   $dB = 20 \log 40MV$

   $dB = 32$

3. $dB = 10 \log 400$

   $dB = 26$

4. $dB = 20 \log 400$

   $dB = 52$

CHAPTER 25

Questions

1. C
2. D
3. A
4. C
5. B
6. A
7. D
8. B
9. C
10. D
11. feedback
12. capacitive
13. relaxation
14. astable
15. RC time constant
16. Hartley (or Armstrong)
17. frequency
18. tank circuit, amplifier transistor or tube, feedback circuit, DC power source
19. temperature coefficient
20. X, electrical

Problems

1. The rates at which Q_1 and Q_2 are cut off determine the oscillator frequency. One cycle includes the time for both stages; therefore, the total cutoff period is:

$$T = 2 \times 0.25ms = 0.5ms$$

Therefore, square wave frequency is:

$$f = \frac{1}{T}$$

$$f = \frac{1}{0.5ms}$$

$$f = 2,000Hz \text{ or } 2kHz$$

2. $f = \dfrac{K}{T}$ where f = frequency in kHz

   T = thickness of the cut in inches

   K = a constant

$$f = \frac{112.6}{0.025''}$$

$$f = 4,504kHz \text{ or } 4.504MHz$$

3. An X-cut crystal has a negative temperature coefficient (TC); therefore, the frequency of the crystal goes down with an increase in temperature. The change in temperature in

CHAPTER 25 (Continued)

this problem is 20° (from 50°C to 70°C = 20° change). The crystal frequency of 500kHz is also the same as 0.5MHz.
   The frequency of oscillation is, therefore,

f = TC × temp. change × fundamental freq. of crystal

f = –10Hz/°C/MHz × 20° × 0.5MHz

f = –100Hz

Therefore, at 70°C the 500kHz crystal will oscillate at 499,900Hz or 499.9kHz.

$$500,000Hz - 100Hz = 499,900Hz$$

4. A Y-cut crystal has a positive temperature coefficient (TC) which produces an increase in frequency with an increase in temperature; therefore, at 70°C the 500kHz Y-cut crystal will oscillate at 500.1kHz.
   The 100Hz calculated in Problem No. 3 must be added to the fundamental crystal frequency because of the positive TC.

f = TC × temp. change × fundamental freq. of crystal

f = +10 × 20 × 0.5MHz

f = +100Hz

∴ 500,000Hz + 100Hz = 500,100Hz or 500.1kHz

5.
$$f = \frac{K}{T}$$

$$2kHz = \frac{77}{T}$$

$$2,000T = 77$$

$$T = 0.0385'' \text{ thick}$$

Index